AF871794

Lecture Notes in Mechanical Engineering

Series Editors

Fakher Chaari, National School of Engineers, University of Sfax, Sfax, Tunisia

Francesco Gherardini, Dipartimento di Ingegneria "Enzo Ferrari", Università di Modena e Reggio Emilia, Modena, Italy

Vitalii Ivanov, Department of Manufacturing Engineering, Machines and Tools, Sumy State University, Sumy, Ukraine

Mohamed Haddar, National School of Engineers of Sfax (ENIS), Sfax, Tunisia

Editorial Board

Francisco Cavas-Martínez, Departamento de Estructuras, Construcción y Expresión Gráfica Universidad Politécnica de Cartagena, Cartagena, Spain

Francesca di Mare, Institute of Energy Technology, Ruhr-Universität Bochum, Bochum, Germany

Young W. Kwon, Department of Manufacturing Engineering and Aerospace Engineering, Graduate School of Engineering and Applied Science, Monterey, USA

Tullio A. M. Tolio, Department of Mechanical Engineering, Politecnico di Milano, Milano, Italy

Justyna Trojanowska, Poznan University of Technology, Poznan, Poland

Robert Schmitt, RWTH Aachen University, Aachen, Germany

Jinyang Xu, School of Mechanical Engineering, Shanghai Jiao Tong University, Shanghai, China

Lecture Notes in Mechanical Engineering (LNME) publishes the latest developments in Mechanical Engineering—quickly, informally and with high quality. Original research or contributions reported in proceedings and post-proceedings represents the core of LNME. Volumes published in LNME embrace all aspects, subfields and new challenges of mechanical engineering.

To submit a proposal or request further information, please contact the Springer Editor of your location:

Europe, USA, Africa: Leontina Di Cecco at Leontina.dicecco@springer.com

China: Ella Zhang at ella.zhang@cn.springernature.com

India, Rest of Asia, Australia, New Zealand: Swati Meherishi at swati.meherishi@springer.com

Topics in the series include:

- Engineering Design
- Machinery and Machine Elements
- Mechanical Structures and Stress Analysis
- Automotive Engineering
- Engine Technology
- Aerospace Technology and Astronautics
- Nanotechnology and Microengineering
- Control, Robotics, Mechatronics
- MEMS
- Theoretical and Applied Mechanics
- Dynamical Systems, Control
- Fluid Mechanics
- Engineering Thermodynamics, Heat and Mass Transfer
- Manufacturing Engineering and Smart Manufacturing
- Precision Engineering, Instrumentation, Measurement
- Materials Engineering
- Tribology and Surface Technology

Indexed by SCOPUS, EI Compendex, and INSPEC.

All books published in the series are evaluated by Web of Science for the Conference Proceedings Citation Index (CPCI).

To submit a proposal for a monograph, please check our Springer Tracts in Mechanical Engineering at https://link.springer.com/bookseries/11693.

Zhenyu Zhang • Paolo Mercorelli
Editors

Mechanical Design and Simulation: Insights and Innovations

Proceedings of the 5th International Conference on Mechanical Design and Simulation

Volume I

Springer

Editors
Zhenyu Zhang
School of Mechanical Engineering
Dalian University of Technology
Dalian, Liaoning, China

Paolo Mercorelli
Institute for Production Technology and Systems
Leuphana University of Lueneburg
Lüneburg, Niedersachsen, Germany

ISSN 2195-4356 ISSN 2195-4364 (electronic)
Lecture Notes in Mechanical Engineering
ISBN 978-981-95-2404-4 ISBN 978-981-95-2405-1 (eBook)
https://doi.org/10.1007/978-981-95-2405-1

© The Editor(s) (if applicable) and The Author(s) 2026

This book is an open access publication.

Open Access This book is licensed under the terms of the Creative Commons Attribution-NonCommercial-NoDerivatives 4.0 International License (http://creativecommons.org/licenses/by-nc-nd/4.0/), which permits any noncommercial use, sharing, distribution and reproduction in any medium or format, as long as you give appropriate credit to the original author(s) and the source, provide a link to the Creative Commons license and indicate if you modified the licensed material. You do not have permission under this license to share adapted material derived from this book or parts of it.
The images or other third party material in this book are included in the book's Creative Commons license, unless indicated otherwise in a credit line to the material. If material is not included in the book's Creative Commons license and your intended use is not permitted by statutory regulation or exceeds the permitted use, you will need to obtain permission directly from the copyright holder.
This work is subject to copyright. All commercial rights are reserved by the author(s), whether the whole or part of the material is concerned, specifically the rights of translation, reprinting, reuse of illustrations, recitation, broadcasting, reproduction on microfilms or in any other physical way, and transmission or information storage and retrieval, electronic adaptation, computer software, or by similar or dissimilar methodology now known or hereafter developed. Regarding these commercial rights a non-exclusive license has been granted to the publisher.
The use of general descriptive names, registered names, trademarks, service marks, etc. in this publication does not imply, even in the absence of a specific statement, that such names are exempt from the relevant protective laws and regulations and therefore free for general use.
The publisher, the authors and the editors are safe to assume that the advice and information in this book are believed to be true and accurate at the date of publication. Neither the publisher nor the authors or the editors give a warranty, expressed or implied, with respect to the material contained herein or for any errors or omissions that may have been made. The publisher remains neutral with regard to jurisdictional claims in published maps and institutional affiliations.

This Springer imprint is published by the registered company Springer Nature Singapore Pte Ltd.
The registered company address is: 152 Beach Road, #21-01/04 Gateway East, Singapore 189721, Singapore

If disposing of this product, please recycle the paper.

Preface

It is a great pleasure to present the preface for the 2025 5th International Conference on Mechanical Design and Simulation (MDS 2025) in Dalian, China, from February 28 to March 2, 2025. This prestigious event convened leading experts, scholars, and industry professionals from around the globe to explore cutting-edge advancements in mechanical design, mechanical engineering, and modeling and simulation technologies.

MDS 2025 aimed to foster interdisciplinary collaboration and intellectual exchange, providing a dynamic platform for sharing innovative research, expanding professional networks, and addressing critical challenges in the field. By bridging academia and industry, the conference sought to accelerate the translation of theoretical breakthroughs into practical applications, ultimately driving progress in both educational and industrial domains.

The conference featured keynote presentations by distinguished scholars, including Prof. Zhenyu Zhang (Dalian University of Technology), renowned for his pioneering work in ultra-precision machining and additive manufacturing; Prof. Wenbin Hou (Dalian University of Technology), a leading authority in automotive digitalization and lightweight structural optimization; Prof. Caixu Yue (Harbin University of Science and Technology), an expert in machine vision and intelligent manufacturing; and Prof. Yajun Liu (South China University of Technology), whose contributions to mechanical signal analysis and smart manufacturing systems have set new industry benchmarks.

The technical program encompassed a rich agenda of thematic sessions, oral presentations, and interactive discussions, covering diverse topics such as green chemical-mechanical polishing, intelligent tool condition monitoring, fiber-metal hybrid materials, and advanced simulations for industrial systems. Highlights include research on high-speed food processing equipment, deep-sea mining resistance analysis, and innovations in aerospace structural design. Attendees also benefited from networking opportunities during tea breaks, group photos, and dedicated discussion periods.

The excellent papers were selected from a large quantity of submissions for publication after rigorous peer review processes. The accepted papers cover topics including mechanical design and engineering, modeling and simulation technologies, advanced materials and applications, intelligent manufacturing and industrial applications, aerospace and heavy machinery, sustainable and green technologies, cross-disciplinary innovations, etc.

As we gather in Dalian—a hub of technological innovation—we extend our heartfelt gratitude to all participants. May MDS 2025 have inspired fruitful collaborations, sparked transformative ideas, and paved the way for future advancements in mechanical design and simulation.

MDS 2025 Conference Committee

Committee Member

Conference General Chair

Prof. Paolo Mercorelli, Leuphana University of Lueneburg, Lueneburg, Germany
Prof. Zhenyu Zhang, Dalian University of Technology, China

Technical Program Committee Chairs

Prof. Wenbin Hou, Dalian University of Technology, China
Prof. Xiaoguang Guo, Dalian University of Technology, China
Prof. Xiaohong Lu, Dalian University of Technology, China
Prof. Jianwei Ma, Dalian University of Technology, China

Publication Chairs

Prof. Caixu Yue, Harbin University of Science and Technology, China
Prof. Ir. Dr. YUSRI BIN YUSOF, Universiti Tun Hussien Onn Malaysia, Malaysia

Contents

Simulation Methods and Applications

Simulation and Optimization of Temperature Field for Arc Additive Manufacturing Based on Cloud Computing

Zhibo Yang

Abstract With the development of cloud computing technology, it shows unique advantages in high-performance computing applications. In this study, the cloud computing platform AWS was used to simulate and optimize the temperature field during arc additive manufacturing (AM) of 7055 aluminum alloy. The mathematical model of the temperature field is established by defining the thermal physical properties of the material and applying the Fourier heat conduction law and is solved by the finite element method and finite difference method. In the study, the temperature control strategy was optimized by adjusting parameters such as scanning speed, laser power, and cooling system, and the temperature gradient and maximum temperature in the manufacturing process were significantly improved. The results show that the model can effectively predict and control temperature change, and improve the manufacturing precision and material properties. This study demonstrates the application potential of cloud computing in complex industrial simulation and provides the theoretical basis and technical support for efficient arc additive manufacturing.

Keywords Arc additive manufacturing · Temperature field simulation · Cloud computing · 7055 aluminum alloy

1 Introduction

Arc additive manufacturing (AM) technology has attracted much attention in the modern manufacturing industry because of its high efficiency and complex structure manufacturing capability. However, precise control of the temperature field is the key to ensuring product quality during the AM process. With the development of cloud computing technology, its powerful data processing capabilities and resource allocation flexibility provide new possibilities for efficient simulation [1]. In this

Z. Yang (✉)
Shenyang Aerospace University, Shenyang, Liaoning, China

© The Author(s) 2026

Z. Zhang, P. Mercorelli (eds.), *Mechanical Design and Simulation: Insights and Innovations*, Lecture Notes in Mechanical Engineering,
https://doi.org/10.1007/978-981-95-2405-1_1

study, the AWS cloud platform was used to simulate and optimize the AM temperature field of 7055 aluminum alloy, and the influence of temperature control on manufacturing accuracy and material properties was explored. Through this research, we aim to demonstrate how cloud computing can help improve the efficiency and product quality of AM technology, and provide a theoretical basis and practical guidance for the advancement of related technologies.

2 Model Construction

2.1 Cloud Computing Platform Selection and Configuration

Choosing the right cloud computing platform is very important when constructing the temperature field simulation model of arc additive manufacturing. By comparing AWS, Azure, and Google Cloud, the study found that AWS offers a wide range of location-based services and mature security measures, while Google Cloud has advantages in data analytics and machine learning integration, and Azure stands out for its performance in enterprise-class services and hybrid cloud solutions. Based on the consideration of simulation requirements [2], AWS was selected in this study because it has the most balanced performance in terms of processing speed and data security. Specific configuration includes selecting compute instances suitable for HPC, and Amazon EC2 C5 instances, configuring sufficient storage space, utilizing scalable storage provided by Amazon S3, and setting up private virtual networks to ensure the security and stability of data transmission [3]. This configuration ensures an efficient and stable simulation environment so that dynamic changes in the temperature field can be accurately simulated.

2.2 Construction of Simulation Model

In constructing the temperature field simulation model of arc additive manufacturing, the physical and thermal-physical properties of 7055 aluminum alloy were first defined, including thermal conductivity of 130 W/m·K, specific heat capacity of 960 J/kg·K, melting point of 635 °C, and thermal diffusion coefficient of 25 mm^2/s [4]. Based on Fourier's law of heat conduction, the mathematical model is established by using the heat conduction equation:

$$\frac{\partial T}{\partial t} = \alpha \nabla^2 T + Q \tag{1}$$

where T is the temperature, t is the time, α is the thermal diffusion coefficient, and Q is the heat source intensity per unit volume. The boundary condition settings include a fixed temperature boundary, with the bottom and side boundaries of the model set to 25 °C, and an adiabatic boundary at the top of the model where the heat flow derivative is zero. The calculation formula is:

$$T(X,0) = T_0, \frac{\partial T}{\partial n} = 0 \tag{2}$$

where T_0 is the initial temperature and $\partial T/\partial n$ is the temperature gradient along the boundary normal. The heat source area simulates arc action, and the heat source intensity is set at 500 W/cm^3 [5]. The calculation formula is:

$$Q = Q_0 \cdot \exp\left\{-\frac{(X - X_0)^2 + (Y - Y_0)^2}{2\sigma^2}\right\} \tag{3}$$

where Q_0 is the intensity of the heat source, (X_0, Y_0) is the central location of the heat source, and σ represents the width of the heat source distribution. For the numerical solution methods, finite element method (FEM) and finite difference method (FDM) were selected, and the Crank-Nicolson method was combined for time integration to ensure the accuracy and efficiency of the simulation. The model is implemented on the AWS cloud platform and uses its high-performance computing resources to carry out spatial grid division and time step setting to accurately simulate the dynamic changes of the temperature field [6]. This comprehensive method takes into account both calculation accuracy and execution efficiency and provides a scientific basis for temperature control in arc additive manufacturing.

3 Simulation Process

3.1 Experiment Parameters Setup

In the simulation experiment of the temperature field of arc additive manufacturing, the key of the experiment design is to precisely set the parameters to simulate the actual manufacturing environment. In Table 1 below, first, the heating temperature is set to 705 °C, slightly higher than the melting point of 7055 aluminum alloy, to ensure that the material can be fully melted. The cooling rate is set at 1 °C/S to

Table 1 Parameter setting

Parameter	Set value
Heating temperature	705 °C
Cooling rate	1 °C/s
Layer thickness	0.5 mm
Scanning speed	10 mm/s

simulate the actual cooling environment and help analyze the microstructure and properties of the material after cooling. The layer thickness is set to 0.5 mm, based on the output accuracy of the print head and the level of detail of the required component. In addition, the scanning speed of 10 mm/s is designed to ensure that the material receives sufficient heat while stable deposition, thereby optimizing the mechanical properties and surface quality of the material. Through the precise control of these parameters, the simulation environment can be close to the real manufacturing conditions, to effectively verify and optimize the simulation model of the temperature field on the cloud platform [7].

3.2 Experiment Implementation and Data Processing

In the temperature field simulation experiment of arc additive manufacturing, the automatic data collection and processing process on the cloud platform ensures the validity and accuracy of the data. Figure 1 below, by connecting to AWS cloud services, we use APIs to capture key parameters such as temperature data, cooling rate, and layer thickness changes during simulation in real time and automatically store these data in the Amazon RDS database [8]. In the data processing stage, Amazon Athena is used for data query and statistical analysis, including data cleaning, normalization processing, and feature extraction [9]. The processed data was visualized by using the Amazon QuickSight tool, creating dynamic dashboards with temperature profiles and time-temperature curves to visualize the simulation results. In addition, based on the analysis results, the simulation parameters are adjusted through AWS Lambda automation scripts, such as modifying heat source intensity or scanning speed, to optimize the simulation model. This automatic data collection and processing process not only improves the experimental efficiency but also enhances the science and practicability of simulation [10].

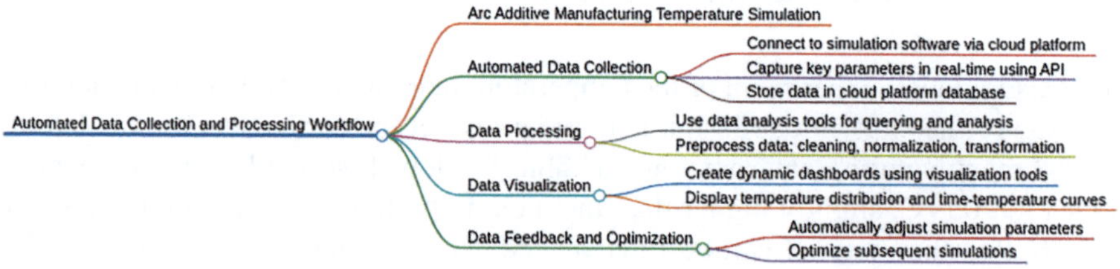

Fig. 1 Data collection and processing workflow

4 Experimental Results

4.1 Display of Simulation Results

In the temperature field simulation experiment of arc additive manufacturing, we analyzed the temperature distribution and variation in detail through the data collected at nine monitoring locations. In Fig. 2, at the beginning of the simulation, the temperature at all locations was room temperature 25 °C, showing the consistency of the initial state. As the time progresses to 60 s, the temperature at Positions 2 and 5 rises to 300 °C, and these regions are the hottest due to the high heat source. At 120 s, the temperature at Position 5 rises to 450 °C, making it the hottest zone. By 180 s, the peak temperature at Position 5 reached 700 °C, indicating a high-temperature state under continuous heating. During the 240 and 300-second cooling phases, the temperature at each position was gradually reduced, from 700 °C to 400 °C at Position 5, demonstrating an effective cooling process. Through these time-temperature curves (Fig. 3), we observe how the temperature at each monitoring point changes over time, with Location 5 showing the fastest rate of temperature rise and a slower rate of cooling, possibly due to heat source proximity or heat accumulation effects.

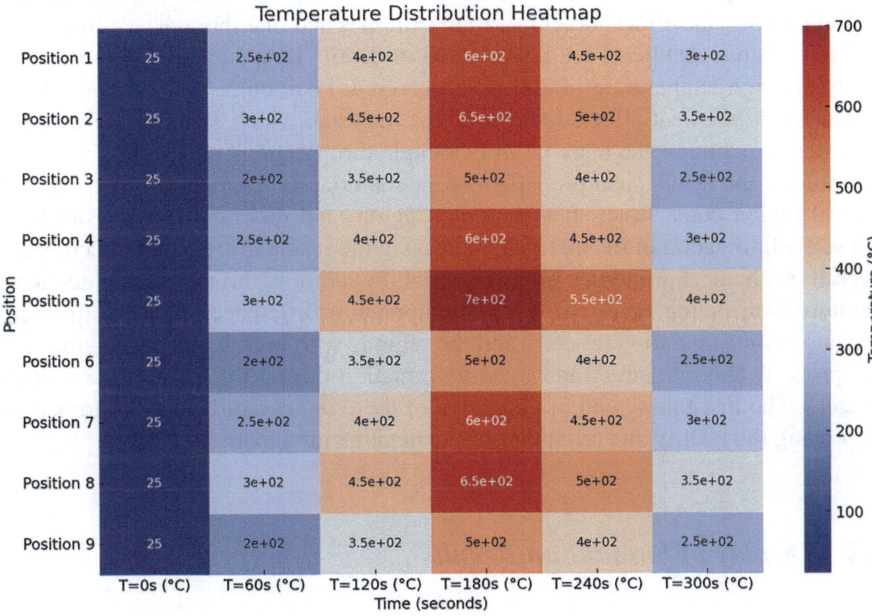

Fig. 2 Temperature distribution heat map

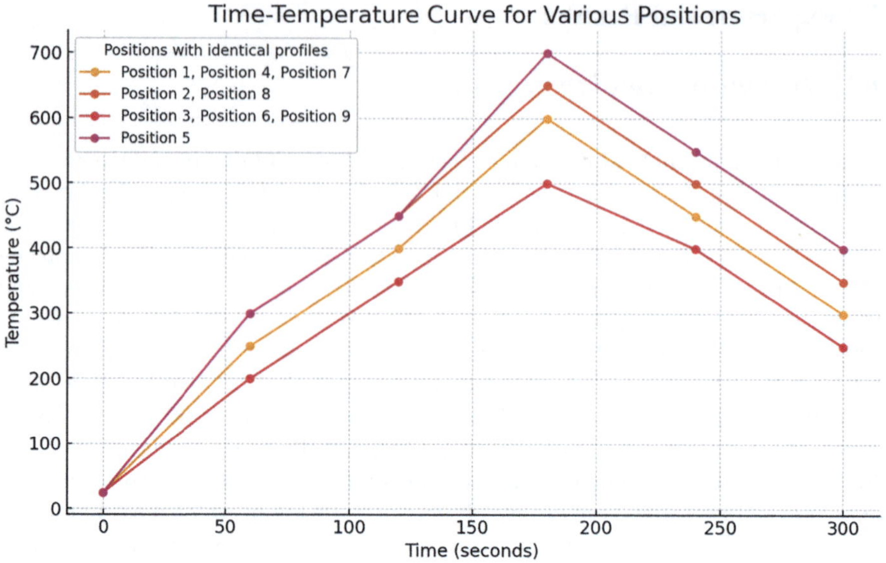

Fig. 3 Time-temperature curve

In the analysis of the simulation results of the temperature field of arc additive manufacturing, the effects of temperature extremes and distributions, cooling rates, and temperature gradients are investigated in detail. Temperature extremes, especially the high temperature of 700 °C observed at Position 5, lead to a significant expansion of the heat-affected region of the material, which significantly changes the microstructure of the material, for example through grain coarsening, which in turn affects the mechanical properties of the material. In terms of cooling rate, the central region cools faster than the edge region, and this rapid cooling leads to uneven solidification of the material, increasing the residual stress and affecting the overall mechanical properties of the material. In addition, through the plotted temperature distribution map, we see the temperature differences between different regions in the simulation process, and the regions with high-temperature gradients are prone to thermal stress and local deformation during the material deposition process. The fine-tuning and optimization of these temperature controls are key to improving the quality and reliability of the manufactured parts.

4.2 *Display of Simulation Results*

Aiming at the problem of temperature control in the arc additive manufacturing process, we have developed a series of specific optimization strategies according to the detailed simulation results. We adjust the scanning speed and power, increase the scanning speed from the original 10 mm/s to 15 mm/s, and reduce the laser

Table 2 Model optimization comparison

Parameter/Result	Before optimization	After optimization	Change
Maximum temperature (°C)	700	650	−50 °C
Cooling rate (°C/s)	1.00 (central area)	1.25 (central area)	+0.25 °C/s
Temperature gradient (°C/mm)	50	30	−20 °C/mm
Inter-layer temperature difference (°C)	40	20	−20 °C

power from the original 200 W to 180 W. This adjustment is designed to reduce the heat input per unit area, control the maximum temperature in the high-temperature area, and avoid overheating of the material. Improved cooling system design: Additional cooling channels were installed in key high-temperature areas and a more efficient cooling medium (from water-based coolant to argon) was adopted. These improvements can improve the cooling efficiency of these areas, accelerating the cooling rate to 1.5 °C/s. Interlayer wait time optimization: Increasing the wait time from 30 to 45 s after each layer is printed allows the material more time to cool before each layer is applied, reducing the temperature gradient between the layers and thus reducing heat accumulation (Table 2).

5 Conclusion

In this study, cloud computing technology is successfully applied to simulate and optimize the temperature field in arc additive manufacturing. Through detailed model building and simulation processes, we not only improve the accuracy of temperature control but also significantly improve thermal management in the manufacturing process through optimization strategies. The techniques used in the study include adjusting the scanning speed and power, improving the cooling system, and optimizing the interlayer wait time, all of which effectively reduce the maximum temperature and temperature gradient and reduce the thermal stress of the material. The results show that these optimization measures have a significant impact on improving the quality and performance of the final product. Therefore, cloud computing shows great potential in improving the efficiency and product quality of arc additive manufacturing, providing a valuable reference and practical basis for the development of related technologies in the future.

References

1. Wang L, Zhang Z, Xu D et al (2019) Numerical simulation and mechanism study of grain refinement during double pulsed wire arc additive manufacturing. Trans China Weld Inst 40(4):137–140
2. Wang H, Huang J, Chen S et al (2023) Analysis of the theory and temperature field of additive manufacturing with powder core wire based on Cu-Al-Fe alloy. Trans China Weld Inst 44(4):111–119

3. Wu X, Luo W, Wang Y et al (2021) Simulation on friction stir welding 7055 aluminum alloy based on CEL model. Trans China Weld Inst 42(7):44–50, 59
4. Wu X, Lin H, Wang Y et al (2023) The microstructure evolution in the nugget zone of Q&P980 steel joints under the condition of friction stir welding thermal-mechanical couple effects. Trans China Weld Inst 44(6):120–128
5. Wang JS, Ji LS (2023) Effect of scanning speed on temperature field of laser cutting 6061 aluminum alloy sheet. Manufact Technol Machine Tool 6:39–43
6. Gao Q, Li WB, Guo GY (2022) Simulation and experimental study on temperature field in micro-milling of single crystal nickel-based Superalloy. Machine Tool & Hydraulics 50(2):65–69
7. Wang L, Xie LY, Zhang D et al (2008) Dynamic simulation of temperature field during friction stir welding. J Northeastern Univ (Nat Sci) 29(7):1–025
8. Zhu J, Wang H, Shi D et al (2022) Element loss behavior and compensation of additive manufacturing memory alloy. Trans China Weld Inst 43(9):50–55
9. Liang H, Li P, Shen X et al (2023) Finite element analysis of the effect of ultrasonic impact on the stress of aluminum alloy arc additive manufacturing. Trans China Weld Inst 44(10):79–85, 119
10. Wang F, Song Y, Zhang J (2010) Detailed effect of the spot-welding process on spot quality of high-strength aluminum alloy used in astronavigation. Trans China Welding Inst 12:37–40, 44

Open Access This chapter is licensed under the terms of the Creative Commons Attribution-NonCommercial-NoDerivatives 4.0 International License (http://creativecommons.org/licenses/by-nc-nd/4.0/), which permits any noncommercial use, sharing, distribution and reproduction in any medium or format, as long as you give appropriate credit to the original author(s) and the source, provide a link to the Creative Commons license and indicate if you modified the licensed material. You do not have permission under this license to share adapted material derived from this chapter or parts of it.

The images or other third party material in this chapter are included in the chapter's Creative Commons license, unless indicated otherwise in a credit line to the material. If material is not included in the chapter's Creative Commons license and your intended use is not permitted by statutory regulation or exceeds the permitted use, you will need to obtain permission directly from the copyright holder.

Analysis, Improvement, and Verification of Stress in Aircraft Harness Bending

Yin Xie, Hujun Wang, Danyang Wang, Lina Xia, Junfeng Dai, Yi Wang, Su Wang, Fengyun Tan, and Hongfei Xiao

Abstract This study proposes innovative wire harness bending manufacturing processes and fatigue test platforms to address the issue of wire breakage caused by bending stress during aircraft wire harness assembly. Through modeling analysis, the impact of wire harness bending length changes on internal wire core stress is clarified, and optimized manufacturing processes are designed to reduce the risk of wire breakage effectively. Simultaneously, the developed fatigue test platform can simulate different bending conditions, verifying the reliability and effectiveness of the new processes. This technology addresses manufacturing challenges under assembly space constraints and bending requirements, improving the flexibility and assembly quality of wire harnesses by reserving process allowance and optimizing bending molds. Experimental results show that the improved wire harnesses exhibit higher fatigue resistance during simulated docking tests, significantly reducing damage and providing a strong guarantee for the safety and reliability of aircraft wire harnesses. This research not only provides new solutions for the field of aircraft wire harness manufacturing and assembly but also has significant implications for improving the overall performance and safety of aircraft.

Keywords Aircraft wiring harness · Bending length model · Bending manufacturing process · Wiring harness fatigue test platform

Y. Xie (✉) · H. Wang · D. Wang · L. Xia · J. Dai · Y. Wang · S. Wang · F. Tan
Aviation Industry Corporation of China, Chengdu Aircraft Industry (Group) Co., Ltd., Chengdu, Sichuan, China

H. Xiao
Unit 93147, People's Liberation Army, Chengdu, Sichuan, China

© The Author(s) 2026

Z. Zhang, P. Mercorelli (eds.), *Mechanical Design and Simulation: Insights and Innovations*, Lecture Notes in Mechanical Engineering,
https://doi.org/10.1007/978-981-95-2405-1_2

1 Introduction

The manufacturing and assembly techniques of aircraft cables stand out as a cardinal phase in aircraft construction, encapsulated with a holistic consideration of structural, electrical, and assembly compatibility attributes. These techniques must satisfy not only functional demands but also performance requisites. Unlike rigid structural elements, cables, on account of their flexibility, pose a considerable challenge to design engineers in controlling their shape and location in the design of wire harness paths [1]. During the product assembly process, handling the assembly of cables alongside rigid structures often involves a crisscross method. The practical assembly skills of the assemblers play a crucial role in determining the assembly quality. While complex to capture, these skills typically resist formulation by mathematical Eq. (2). The count of cables also tends to rise in correlation with the complexity of products, stirring up an exponential increase in cable assembly sequence, with a diversified selection for cable assembly paths.

The majority of the planners of aircraft cable assembly technology accomplish the task after rigid structural items are produced, undergoing trial mounting on the actual physical model. Subject to repeated mounts, the process seeks to determine relevant information on flexible cables, hence making it strenuous to accomplish a unified process of cable assembly. As a result, it hardly suffices to maintain the quality, reliability, consistency, and coordination in cable assembly, making precise assembly an utmost priority [3]. Control over bending in cables represents a critical limiting factor in the assembly phase, requiring discrimination based on cable type and outer diameter, among other physical aspects. However, the stringent space constraints often shadow the need for conditions of assembly. Additionally, the conventional method of equal-length flat production cannot support the mounting needs in these confined spots, inevitably leading to frequent open circuits in individual conductors within the cables [4, 5].

Cable harness assembly, an indispensable quality control phase in aircraft assembly, involves the strict channeling of various cable harnesses to distinct compartments post-installation. Following their anchorage through clamps post-installation, cable harnesses incur various degrees of bending under the impact of factors such as material attributes, slack in length, and installation methodology [6]. Under different bending states, the harness experiences differing levels of tension. The landscape changes further within the cable core in different areas within the harness [7, 8]. If a harnessed cable is subjected to continually increased stress post-bending, it can compromise both the longevity of the harness and the electrical connectors' reliability of contact [9].

Currently, the installation of wiring harnesses on aircraft often primarily draws on empirical knowledge to specify reference ranges for technical requirements such as length and bend degree, owing to the intricacy of onboard installation conditions. However, the internal stress conditions of the wiring harness vary in response to different technical requirements. No definitive data exists to support this. In this paper, we have taken the initiative to model the force situation encountered by the bending harness and have also undertaken a theoretical analysis of the relationship between the bending radius and the occurring stress. Based on this groundwork, we have presented a crafting process for cables with inadequate radii in the design stage, also providing the requisite precautions for harness assembly derived from curved manufacturing processes. Furthermore, a fatigue testing platform for the cable harness has been designed, which can simulate a wide gamut of bending conditions for cable harnesses in an off-aircraft setting. This testing platform substantiates that our approach in bending manufacturing substantially mitigates wire breaks due to over-stress in cable harnesses, thereby improving the quality of cable harness assembly.

2 Cable Harness Bend Modeling

2.1 The Cable Harness Bends the Theoretical Length Model

The cable harness bend theoretical length model, employing parameters such as the lengths of cable segments and bending angles, serves to compute the alteration in length experienced by the cable corresponding to any pin location on the terminal face during the course of cable deformation. A graphical representation of this model is shown in Fig. 1.

Upon the induction of curvature in a segment of the cable harness, the parameter detailing the bending radius specific to the midpoint of said cable segment is defined as:

$$R = \frac{L * 360^\circ}{2\pi * \varphi} = \frac{L * 180^\circ}{\pi * \varphi} \tag{1}$$

When the cable segment bends, a triangle is formed by the bending radius at the center point R, the bending radius r at any pinhole, and the distance d from the

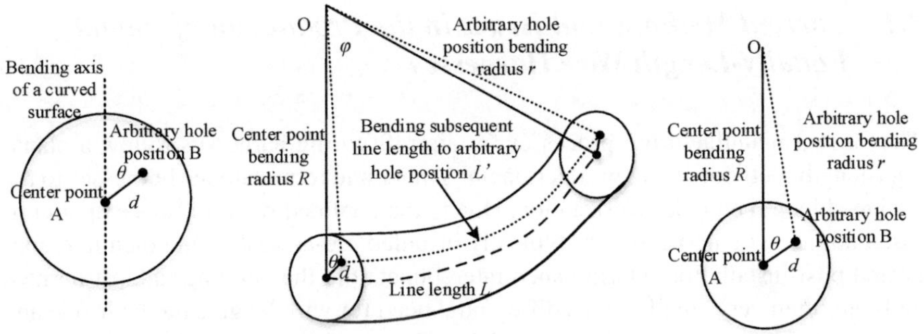

Fig. 1 The cable harness bends the theoretical length model

pinhole to the center point on the terminal face of the electrical connector, with an inner angle θ. According to the Cosine Law, the following relationship can be deduced:

$$r = \sqrt{R^2 + d^2 - 2Rdcos\theta} \tag{2}$$

Therefore, the theoretical length $L^{`}$ of the cable at any pin position after bending can be calculated accordingly.

$$L^{`} = 2\pi r * \frac{\varphi}{360°} \tag{3}$$

Consequently, the length variation of the core wire at any pinhole during the bending process can be formulated as $\Delta L = |L - L^{`}|$, substituting the corresponding parameters we have:

$$\Delta L = \left| L - \frac{2\pi\varphi}{360°}\sqrt{\left[\frac{L^2}{\pi^2} * \left(\frac{180°}{\varphi}\right)^2 + d^2 - \frac{360°}{\varphi} * \frac{L * d * \cos\theta}{\pi}\right]} \right| \tag{4}$$

When $0 \leq \theta \leq 90°$, the core wire is situated on the inner side of the bending channel, resulting in wa ire length decrease; when $90° \leq \theta \leq 180°$, the core wire is positioned on the outer side of the bending channel, causing the wire length to increase. Here, 'd' represents the distance from any pinhole to the center of the terminal face; 'L' represents the testing cable length; 'φ' represents the cable bending angle; 'θ' represents the angle between the line connecting a given pinhole on the terminal face to the center point and the plane in which the cable segment bends.

3 Methods and Effects of Wire Harness Bending in Manufacturing

3.1 Current Methods and Issues in the Production of Planar Equally-Length Wire Harnesses

The current manufacturing processes for aircraft wiring harnesses follow a planar equi-length method, where production units mark and terminate the harnesses to the designed length on a flat workstation. Due to the confined working space in aircraft installations, wire harnesses are typically installed pre-assembly and then interconnected post-installation. During such interconnection, the bending radii of harness ends are often very small, with a differential need for wire lengths on the inside and outside of the curve. The planar equal-length manufacturing method cannot adequately meet the assembly requirements. Moreover, due to insufficient flexibility at

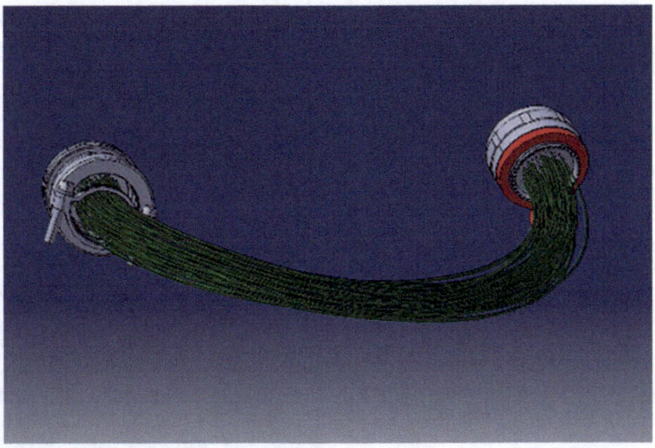

Fig. 2 The required lengths of the inner and outer wires are not consistent when the wiring harness branches bend

the branch ends, external wires can become overstretched and lead to wire breakage upon bending.

A model of theoretical bending length for wiring harnesses provides quantifiable predictions for the lengths of wire after curving, which can be utilized to refine existing manufacturing procedures for harness production. Owing to spatial limitations in certain aircraft compartments, it is imperative to bend wiring harness branches for connection post-production. This necessitates the wires from the outer sides of the connectors to be longer than those from the inner sides; thus, we simulate and model the internal wires of the interconnection wire harnesses, as shown in Fig. 2. However, under the traditional planar equal-length manufacturing process, the lengths of wires situated on the inner and outer sides of the harness curvature are identical. As a result, some harnesses with short overall lengths and poor limberness result in the outer wires being stressed upon bending, which may lead to wire fracture and failure.

3.2 *Wire Harness Bending Manufacturing Process Method*

To address the aforementioned issues, we propose an innovative manufacturing technique for wire harness branch end bending. The characteristics of the wire harness branch end (as shown in Fig. 3) are extracted, which include three quantifiable features (as shown in Fig. 4): the diameter of the wire harness branch end bending, the total length of the wire harness branch end (including the connector) and the relative position of the main keyway of the connector with respect to the direction of wire harness bending. These characteristics are communicated to the wire harness manufacturing units. Subsequently, the units customize the bending manufacturing molds and transfer conformal fixtures based on these feature requirements. This allows for the bending of the wire harness during manufacturing and the conformal transportation of the harnessed wire, as shown in Fig. 5.

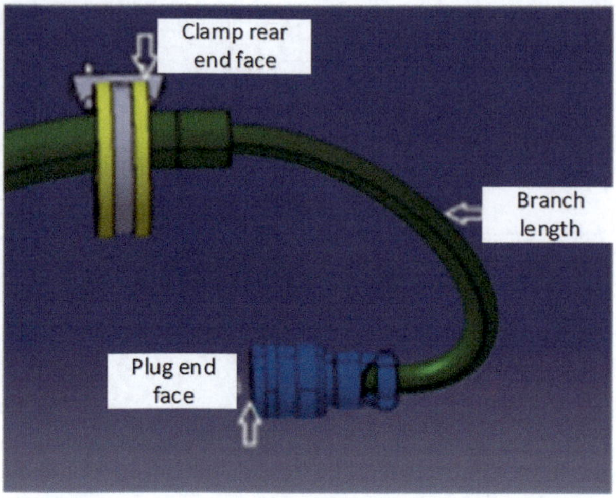

Fig. 3 Wiring harness branching model

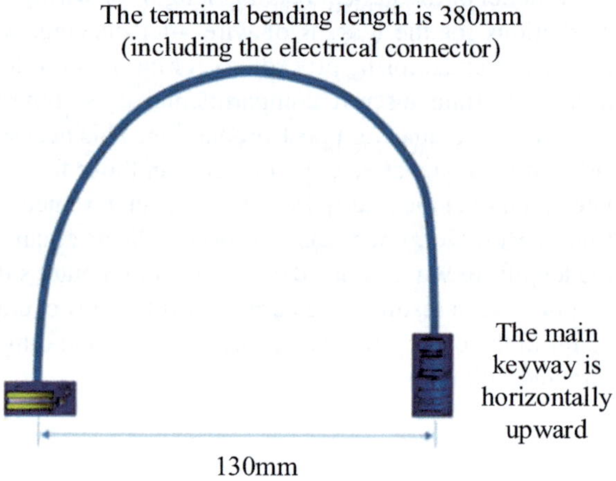

Fig. 4 Description of manufacturing characteristics for wire harness bending

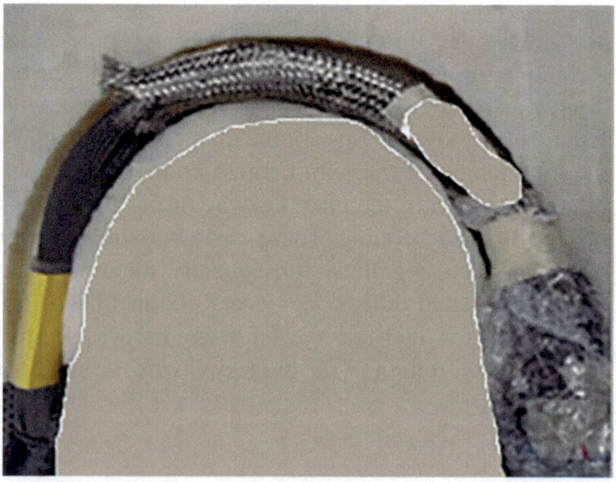

Fig. 5 Wire harness bending manufacturing mold

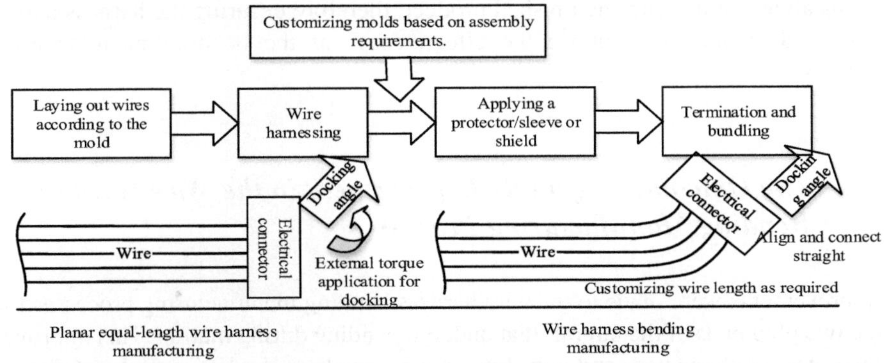

Fig. 6 The cable harness bends the theoretical length model

In accordance with specified requirements, the harness manufacturing entity marks out wires with a 40 mm technological allowance retained at the terminations. Upon entering the harness bending mold for wire arrangement, a maximum discrepancy of 39 mm in length is observed between inner and outer wires at the curvature, as calculated through three-dimensional simulation shown in Fig. 2. Adhering to the constraints of the bending mold, the outer wires are allocated a 40 mm technological allowance, while the inner wires are only allotted a mere 1 mm leeway for manufacturing variability. Wires at the midpoint are assigned a 20 mm technological allowance. Following a graduated system, corresponding allowances are systematically allotted to other wires in the harness. The process reaches closure with the termination and assembly of connectors, culminating in the creation of a harnessed branch that meets the requisite criteria for bent manufacturing. This manufacturing methodology is graphically elucidated in Fig. 6.

3.3 *Precautions for the Laying and Installation of Harnesses after Bending Manufacturing*

In the process of cable harness installation, the bay where the bending manufacturing branch resides should be selected as the starting bay to avoid deformation of the shaped branch harness during laying. If it cannot be guaranteed that this bay can serve as the initial start point, the cable harness branch's layout path should be specifically defined in the laying Assembly Operation (AO). This ensures that the bending status of the harness does not change due to overframe issues or force deformation during the process of laying. During cable harness installation, when adjustments are being made for bay harness installation, priority should be given to securing the clips at the ends of branches involving bending manufacturing connectors. Before fixing the clips, a simulation of the connector mating status should be undertaken. This guarantees that the harness's bending status is basically consistent with its status after mating with the finished product, therefore ensuring the harness deformation does not compromise the effectiveness of the bending manufacturing process.

3.4 The Outcome of Quality Improvement in the Wire Harness Bending Manufacturing Process

Improvements were made to the wire harness bending manufacturing processes for the two plug ends of the aircraft that undergo bending during mating. This improvement was implemented, and adjustments were made to the laying and installation requirements after the bending manufacturing process. Post implementation, there have been no re-occurrences of wire breakage at the ends of these two plugs, effectively solving the issue of wire disconnection. This suggests that improvements in the bending manufacturing process have been effective.

4 Wire Harness Bending Fatigue Test

4.1 Overall Design of the Wire Harness Fatigue Test Platform

To verify the fatigue resistance of the wire harness after bending manufacturing, we designed a wire harness fatigue test platform, as shown in Fig. 7. The structure of the test platform includes:

- A triaxial moving platform.
- A workbench.
- A control cabinet.

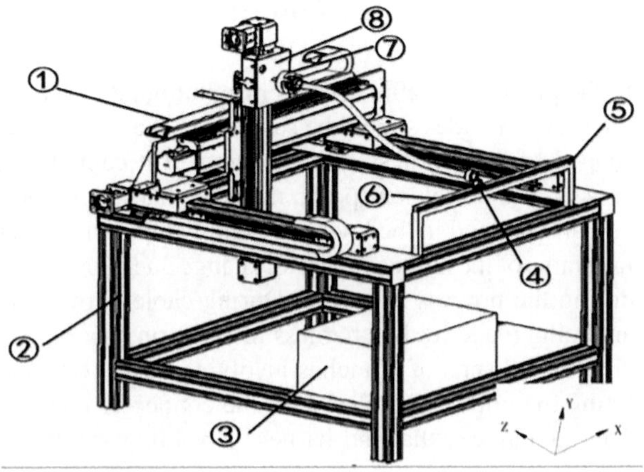

Fig. 7 Wire harness fatigue test platform

- Clamp slider for wire harness.
- Clamp support for wire harness.
- Slider track for wire harness clamps.
- A three-jaw chuck.
- A spherically hinged base.
- Harness clamp.
- Testing wire harness.

The wire harness fatigue test platform is designed to simulate the conditions that a wire harness could encounter over its service life. Its purpose is to repetitively bend the harness until cumulative strain leads to fatigue damage. This method, forcing the harness to repeatedly withstand such stress that it might only experience once or periodically during actual use, tests the product's limits. The aim of this testing is to assess the harness's durability under the stipulated conditions (which are often more severe than typical operational conditions). The objective of testing under more stringent conditions is to ensure the reliability and safety of the wire harness under both standard and extreme conditions, providing detailed insight into the harness's durability and robustness.

4.2 *The Design of Wire Harness Fatigue Testing*

In order to validate the performance of the wire harness post-bending manufacturing using the test platform, a wire harness fatigue comparative test was designed. Harness 1 is a 99-core planar isometric manufactured cable, with all wire cores having a length of 400 mm. Harness 2 is a 99-core bending manufactured wire harness with wire core lengths ranging from 380 mm to 420 mm. The outer side of the curved harness is 40 mm longer than the inner side. The test platform drives the bending motion of the harnesses, simulating the bending and mating process of the harness connectors. Harness 1 and Harness 2 undergo the same test conditions, as shown in Fig. 8.

4.3 *The Analysis of Wire Harness Fatigue Test Results*

The test harness was installed on the fatigue test platform. We simulated the installation and mating process of the harness in the aircraft under the settings on the simulation test platform. The parameters were configured as follows: bending angle of 180° downward, bending movement speed of 0.3 m/s, and an ambient temperature of 25 °C. We proceeded to simulate the mating process 10, 000 times and 50, 000 times, observing the test results after each simulation. Detailed results are shown in Tables 1 and 2.

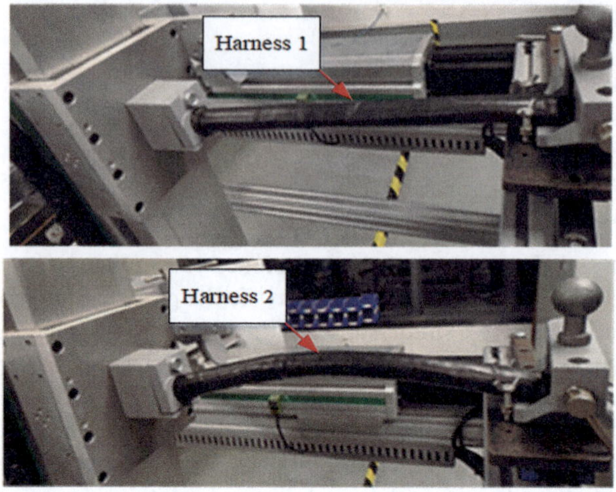

Fig. 8 Wire Harness fatigue comparative test

Table 1 Results of the bending test after 10, 000 cycles

	No. of bending cycles	Damage outcome
Harness 1	10, 000 times	Two conductors are broken at the upper edge
Harness 2	10, 000 times	0 conductors broken

Table 2 Results of the bending test after 50, 000 cycles

	No. of bending cycles	Damage outcome
Harness 1	50, 000 times	Five conductors are broken at the upper edge
Harness 2	50, 000 times	0 conductors broken

The planar equi-length manufactured wire harness exhibited conductor breaks −2 conductors after being subjected to 10, 000 simulated bending cycles, and a further increase to 5 conductors after 50, 000 bending cycles. In contrast, the bending manufactured wire harness did not record any conductor breakage through the entire simulation process (both at 10, 000 and 50, 000 bending cycles). This test result unequivocally suggests that the wire harness produced through the bending manufacturing method can effectively reduce fatigue-induced damage during the bending process when compared to those made with planar isometric manufacturing. As can be seen from the previous theoretical analysis, differences in wire core lengths directly affect the bending performance and internal stress distribution of the harness. In harnesses with equal-length wire cores, stress concentration occurs on the outer wire cores during bending, which can easily lead to wire breakage due to bending. However, in curved harnesses with unequal wire core lengths, the harness is manufactured using molds to achieve the desired curved shape. During bending, there is no stress concentration on the outer side of the harness, thus preventing wire breakage due to bending.

5 Conclusion

This research focuses on the wire breakage issue caused by bending stress during the assembly of aircraft wiring harnesses. Modeling analysis clarifies the impact of changes in harness bending length on internal wire core stress, revealing the differences in the stress conditions of the harness and its internal wire cores under different bending states. This provides a theoretical basis for optimizing the manufacturing process. Based on the results of harness bending stress analysis, an optimized harness bending manufacturing process is designed. By reserving process allowance and optimizing the bending manufacturing mold, the risk of wire breakage caused by excessive bending stress is effectively reduced, improving the flexibility and assembly quality of the harness. A fatigue test platform has been developed to simulate different bending conditions to verify the optimized harness manufacturing process. Experimental results show that the improved harness exhibits higher fatigue resistance and significantly reduces damage in simulated docking tests. Through practical application, the harness bending manufacturing process and fatigue test platform proposed in this research has been verified, solving the harness manufacturing challenges under assembly space limitations and bending requirements and providing strong support for the safety and reliability of aircraft wiring harnesses.

References

1. Wang LF (2017) An initial study on the design of bending radii for EWIS harnesses on civil aircraft. Technology and Market 24(12):52–53
2. Luo YT, Peng J, Gao JY, Luo RM, Chen J, Zhou B (2024) Research on multi-constrained harness layout algorithm for harness pre-assembly. J Graph 45(1):139–147
3. Zhang Y, Zhang HQ, Mao XP, Jia X (2017) YH cable extravehicular bending moment evaluation method. Electr Product Reliability Environ Testing 35(06):34–39
4. Gao YF (2019) Research and analysis on fatigue characteristics of multi-dimensional robot wire and cable. Soochow University
5. Zhang Y, Zhang B, Zhao P, Wang L (2023) Reliability design and verification of radio frequency coaxial cable assembly for spacecraft. In: 10th international forum on electrical engineering and automation (IFEEA), Nanjing, China, pp 454–460
6. Ding CF, Zhang S, Wu D (2020) Optimum design of cabling process for aviation control wire. Optical Fiber & Electric Cable and Their Applications 03(07):25–27
7. Li HJ, Wang LT, Wang W, Yang FJ (2016) The control of the bend radius of spacecraft low-frequency cables. Spacecraft Environmental Engineering 33(05):550–554

8. 1985 Requirements for interconnection cables, harnesses and wiring: NHB 5300.4(3G) [S]. [S.l.: s.n.]: 13–15
9. Gao S, Wu YB, Shao JM, Bai HM (2014) Design of helical cable for space docking mechanism. Aerosp Shanghai 31(03):56–63

Open Access This chapter is licensed under the terms of the Creative Commons Attribution-NonCommercial-NoDerivatives 4.0 International License (http://creativecommons.org/licenses/by-nc-nd/4.0/), which permits any noncommercial use, sharing, distribution and reproduction in any medium or format, as long as you give appropriate credit to the original author(s) and the source, provide a link to the Creative Commons license and indicate if you modified the licensed material. You do not have permission under this license to share adapted material derived from this chapter or parts of it.

The images or other third party material in this chapter are included in the chapter's Creative Commons license, unless indicated otherwise in a credit line to the material. If material is not included in the chapter's Creative Commons license and your intended use is not permitted by statutory regulation or exceeds the permitted use, you will need to obtain permission directly from the copyright holder.

Study on Vibration Isolation Performance Analysis and Optimization Design of Ship Floating Raft

Rongming Ni and Yuefang Wang

Abstract The problem of vibration and noise caused by the propulsion shaft system in modern ships has become increasingly prominent. Floating raft isolation system is an emerging technology used to reduce vibration and noise, which can effectively reduce the transmission of shaft system vibration and noise to the hull, improving the safety and comfort of ship navigation. Based on a small experimental platform for marine propulsion shaft system, this paper constructs a finite element model of the floating raft isolation system and conducts simulation analysis using ANSYS finite element software. The vibration level drop of the system is used as the evaluation index to analyse the isolation performance of the floating raft isolation system under vibration excitation. Finally, the middle raft structure in the floating raft is optimized by opening lightening holes on the rib plates of the raft structure to reduce its mass and achieve the goal of vibration reduction and noise reduction. The results show that the optimized raft structure has better isolation effect, which can effectively solve the vibration isolation problem of the propulsion shaft system and provide reference for the design of the floating raft isolation system.

Keywords Propulsion shafting · A floating raft · Vibration and noise reduction · Optimal design

1 Introduction

With the development of the shipbuilding industry and advancements in technology, the speed and power demands of ships have been constantly increasing. However, a major issue faced by modern ships is how to effectively reduce mechanical noise, particularly the vibrations and noise [1] generated by the propulsion system. To address this problem, the adoption of a floating raft isolation system is one of the

R. Ni (✉) · Y. Wang
Dalian University of Technology, School of Mechanics and Aeronautics, Dalian, China
e-mail: ayfwang@dlut.edu.cn

© The Author(s) 2026

Z. Zhang, P. Mercorelli (eds.), *Mechanical Design and Simulation: Insights and Innovations*, Lecture Notes in Mechanical Engineering,
https://doi.org/10.1007/978-981-95-2405-1_3

widely applied and effective approaches. Therefore, optimizing the structural design of the floating raft isolation system and enhancing its isolation performance have become important research directions in the field of ship isolation [2, 3].

The floating raft isolation technology has been widely applied in the field of ship vibration reduction and noise reduction, effectively reducing the transmission of mechanical vibrations in ships, playing an important role in improving ship stealth performance and combat effectiveness [4]. Scholars at home and abroad have conducted a large amount of research on floating raft isolation systems and have achieved some results. Among them, Zhang Heng [5] and others studied the mechanical properties and isolation performance of plate-frame floating rafts through finite element modeling, demonstrating its good performance. Guo Qi-xing [6] and others used vibration intensity and vibration level drop as indicators to compare the influence of raft frame structural parameters and shapes on isolation performance, and evaluated the isolation effect of the floating raft system. Liu Yang [7] and others improved the optimization process of the finite element method through topology optimization design, designing a floating raft isolation system with good mechanical performance. Zhang [8] et al. proposed a novel floating raft frame structure and designed floating rafts made of carbon fiber polymer materials and metal materials. They compared the frequencies and vibration characteristics of the two types of rafts through finite element modal analysis. The results showed that the raft frame made of carbon fiber polymer materials had better vibration reduction performance. Sun [9] studied the influence of the parameters of the intermediate raft frame and the elasticity of the base on the isolation performance of the floating raft. The results showed that appropriate intermediate raft frame stiffness and damping can improve the effectiveness of the floating raft isolation system. In this study, a finite element model of the propulsion shaft system and the floating raft isolation system will be established. Simulation analysis will be conducted using ANSYS Workbench, and the vibration level drop will be used to evaluate the isolation performance of the floating raft system, providing reference for practical engineering.

2 Establishment of Finite Element Model

2.1 Shafting Parameters and Modeling

To reduce computational load, a two-dimensional axisymmetric model was employed in this study for the analysis of the shaft system. The shaft system test rig consists of a propeller, stern shaft, stern bearing and pedestal, intermediate shaft, thrust shaft, thrust bearing and pedestal, sliding bearing and pedestal, and coupling, as shown in Fig. 1. The finite element model considers the stiffness of each bearing, as presented in Table 1. To handle the large size of the bearing pedestals, a solid modeling approach was utilized. This approach effectively reduces the computational load while maintaining accurate analysis of the shaft system.

Fig. 1 Finite element model of shafting test bed

Table 1 Individual bearing stiffness

	Stern bearing	Thrust bearing radial
K_{xx}/(N/m)	2×10^9	2×10^7
K_{yy}(N/m)	2×10^9	2×10^7
K_{xy}(N/m)	-2×10^{10}	1.49×10^5
Kyx(N/m)	-2×10^{10}	1.49×10^5

2.2 *Parameters and Modeling of Floating Raft Vibration Isolation Systems*

The floating raft isolation system consists of three main components: the central raft frame, vibration isolators, and vibration equipment. The raft frame is modeled as a solid structure and consists of an upper and lower board, with dimensions of 2 m × 0.6 m × 0.01 m. The central raft frame includes transverse and longitudinal rib plates, with a uniform thickness of 5 mm. In the transverse direction, the raft frame has 13 rib plates, with a distance of 0.0975 m between each rib plate and the ends of the raft frame, and a spacing of 0.145 m between adjacent rib plates. In the longitudinal direction, the raft frame has 5 rib plates, with a distance of 0.0575 m between each rib plate and the ends of the raft frame, and a spacing of 0.115 m between adjacent rib plates. The initial mass of the raft frame is 222.146 kg. The finite element model of the system is shown in Fig. 2, and its internal structure is depicted in Fig. 3.

This study utilized a total of 18 isolators as crucial components for connecting the shaft system and the floating raft. Among these, the upper isolators employed WHG-150 rubber isolators [10], with a lateral stiffness of 128 N/mm, longitudinal stiffness of 260 N/mm, vertical stiffness of 220 N/mm, and a damping ratio of 0.07. These upper isolators were positioned between three bearing bases and the bottom plate of the raft, with four isolators at each location. The lower isolators utilized JSD-530 isolators, which are composed of a combination of metal and rubber. They had a lateral stiffness of 750 N/mm, longitudinal stiffness of 1000 N/mm, vertical stiffness of 750 N/mm, and a damping ratio of 0.07. The isolator parameters are presented in Tables 2 and 3.

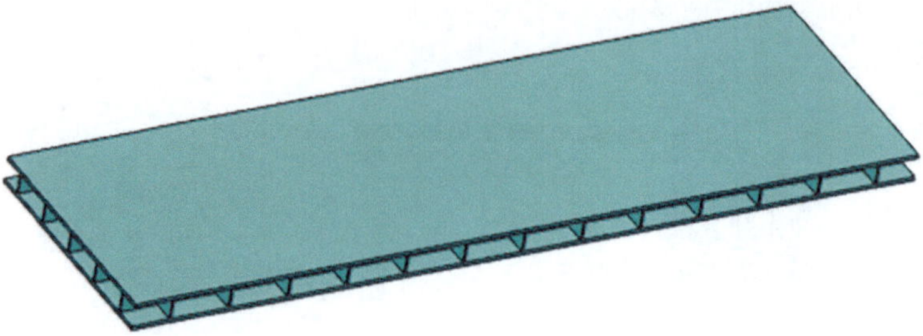

Fig. 2 Finite element model of raft

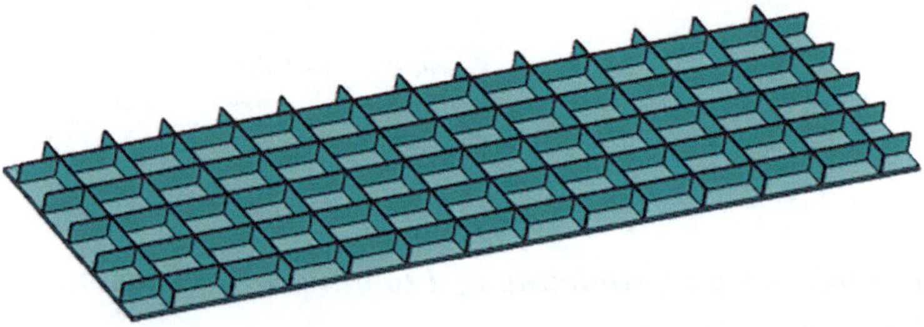

Fig. 3 Internal structure of raft

Table 2 Upper vibration isolator parameter table

	Horizontal	Longitudinal	Vertical
Stiffness coefficient $\boldsymbol{k}$ (N/mm)	128	260	220
Critical damping coefficient $\mathbf{C_0}$ (N·s/mm)	8	11.402	10.488
Damping ratioζ	0.07	0.07	0.07
Damping coefficient **C** (N·s/mm)	2.192	0.798	0.734

Table 3 Lower vibration isolator parameter table

	Horizontal	Longitudinal	Vertical
Stiffness coefficient $\boldsymbol{k}$ (N/mm)	750	1000	750
Critical damping coefficient $\mathbf{C_0}$ (N·s/mm)	31.309	36.152	31.309
Damping ratioζ	0.07	0.07	0.07
Damping coefficient **C** (N·s/mm)	2.192	2.531	2.192

Install the base to simulate the hull structure, and use SOLID185 unit to simulate the size of 2.3 m × 0.8 m × 0.03 m, and do fixed constraint treatment on the bottom surface of the base. The modeling parameters of floating raft vibration isolation system are shown in Table 4 below.

Table 4 Modeling parameter

	Floating raft rack	Base
Size	2 m × 0.6 m × 0.07 m	2.3 m × 0.8 m × 0.03 m
Young modulus	2×10^{11} Pa	2×10^{11} Pa
Poisson's ratio	0.3	0.3
Density	7850 kg/m^3	7850 kg/m^3

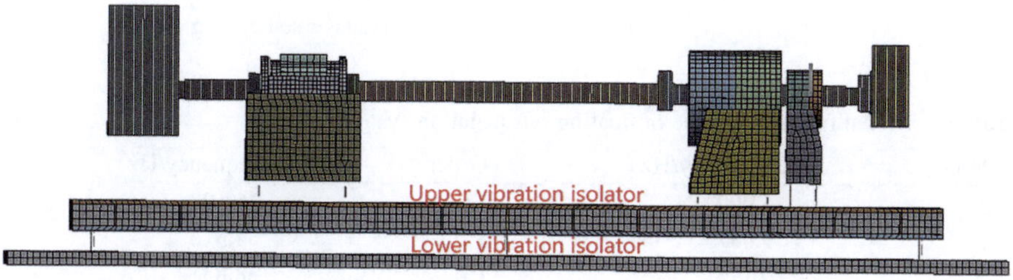

Fig. 4 Internal structure of raft

3 Fem Simulative Analysis

The floating raft isolation system is a common vibration isolation device on ships, mainly composed of the propulsion shaft system, upper isolator, lower isolator, and intermediate raft frame. In the finite element analysis, the isolator is simulated using the COMBIN14 element. The propulsion shaft system and the base are connected to the raft frame through the upper isolator, while the raft frame is installed on the base through the lower isolator. Fixed constraints are applied to the bottom surface of the base to ensure the stability of the model. The finite element simulation model of the floating raft isolation system established using ANSYS Workbench software is shown in Fig. 4.

3.1 Modal Analysis

Modal analysis is the most fundamental method in dynamic analysis and serves as the basis for other dynamic analyses. It has a wide range of applications and can assist in determining the natural frequencies and mode shapes of a structure, thereby enabling the avoidance of resonance in structural design. The natural frequency of a floating raft structure is crucial in the design of isolation systems as it reflects the inherent vibration characteristics of the raft structure. This forms the foundation for analyzing the dynamic properties of the isolation system for a floating raft.

Table 5 The first six modes of the raft

Order	Frequency/ Hz	Modal characteristics
1	122.51	First order longitudinal bending
2	203.47	Torsion
3	313.38	Second order longitudinal bending
4	402.77	Transverse bending and torsion compound
5	556.45	Third order longitudinal bending
6	595.42	Transverse bending

Table 6 Modal frequency value of floating raft isolation system

Order	Frequency/Hz	Order	Frequency/Hz
1	6.909	11	35.107
2	9.6873	12	34.316
3	10.196	13	38.831
4	11.411	14	82.476
5	14.029	15	89.084
6	15.475	16	90.541
7	16.517	17	94.516
8	26.309	18	120.88
9	27.687	19	124.77
10	32.275	20	176.37

When conducting modal analysis on the floating raft isolation system, the focus is on the influence of low-order modes on vibration characteristics [11]. Therefore, we only consider the results of the first 6 modes, excluding the first six rigid body modes, as shown in Table 5.

From Table 5, it can be seen that the natural frequency of the floating raft frame is much higher than the operating frequency of the shaft system, indicating that there will be no resonance between them. Additionally, the floating raft frame possesses a certain level of rigidity, indicating a reasonable design.

A modal analysis of the entire floating raft isolation system was conducted, and the results for the first 20 modes are presented in Table 6. It can be observed from Table 6 that the floating raft isolation system exhibits several closely spaced modal frequencies, such as the third and fourth natural frequencies, as well as the eighth and ninth natural frequencies. The modal shapes are depicted in Figs. 5, 6, 7 and 8.

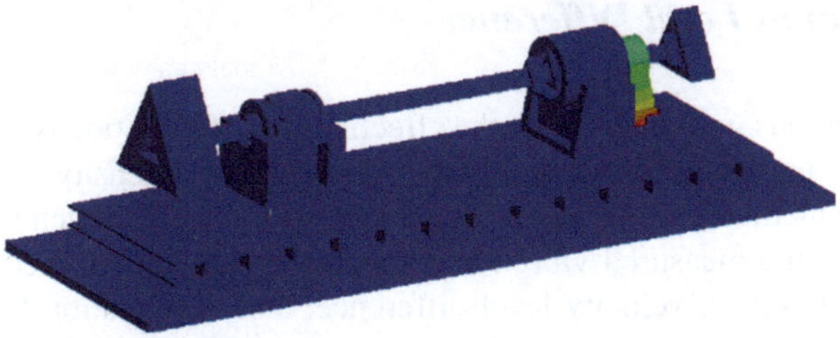

Fig. 5 The third MODE (10.196 Hz)

Fig. 6 The fourth MODE (11.411 Hz)

Fig. 7 The eighth MODE (26.309 Hz)

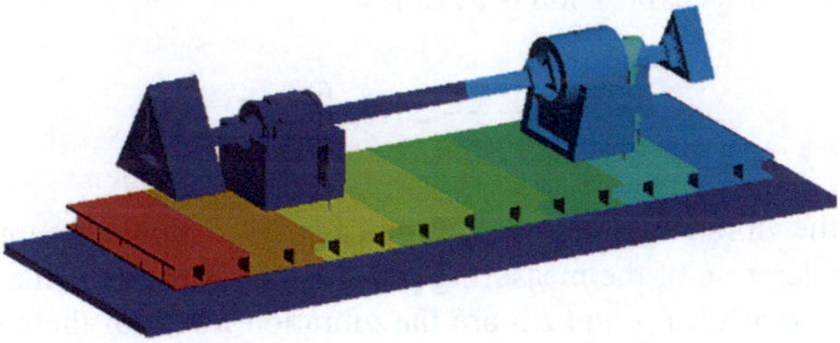

Fig. 8 The ninth MODE (27.687 Hz)

3.2 Vibration Level Difference

The main methods used to evaluate the effectiveness of vibration isolation currently include force transmissibility, insertion loss, vibration intensity, and power flow. Among these methods, vibration level difference is an easily implemented approach. Depending on the measured vibration response, it can be classified into displacement level difference, velocity level difference, and acceleration level difference. Due to the common use of acceleration sensors for measuring acceleration in practical engineering, acceleration level difference is more widely applied [12].

Vibration acceleration is expressed as vibration acceleration level by level, referred to as vibration level L, and its expression is as follows:

$$L = 20\log\frac{a_1}{a_0} \tag{1}$$

Where, a_1 is the acceleration of the measuring point; a_0 is the base acceleration, $a_0 = 10^{-6}$ mm/s^2.

In order to reduce the measurement error, n nodes of a certain layer are usually taken to form a response node group, and it is reasonable to analyze the average vibration level of this response node group. The average value of n measuring points is taken as the measurement result, and the calculation expression of the average vibration level $\overline{L}$ is as follows:

$$\overline{L} = 10\log\left(\frac{1}{n}\sum_{i=1}^{n}10^{\frac{L_i}{10}}\right) \tag{2}$$

Where, L_i is the vibration level of the i th measuring point.

The vibration level difference is obtained by taking the logarithm of the ratio between the effective value of the vibration response of the isolation device and the effective value of the corresponding base response, and then multiplying it by a constant of 20, and its expression is as follows:

$$L_D = 20\log\frac{a_1}{a_0} - 20\log\frac{a_2}{a_0} = L_{a_1} - L_{a_2} \tag{3}$$

Where, L_D is the vibration level drop between measured points, the unit is dB; a_1 and a_2 are the acceleration of the measuring point respectively; a_0 is the base acceleration, $a_0 = 10^{-6}$ mm/s2; L_{a1} and L_{a2} are the vibration levels of the measured points respectively, the unit is dB.

When the vibration level drop L_d of the total acceleration level is taken for calculation, there are:

$$L_d = 20\lg\frac{\sqrt{\left(a_{i1}^2 + a_{i2}^2 + \ldots + a_{in}^2\right)}}{\sqrt{\left(a_{j1}^2 + a_{j2}^2 + \ldots + a_{jn}^2\right)}} \tag{4}$$

3.3 Analysis of Isolation Characteristics of Floating Raft Isolation System

Using ANSYS software, a complete method is employed to analyze the harmonic response of a floating raft isolation system. The frequency range considered is from 0 to 400 Hz, with a solution interval of 200. A excitation force is applied at the center of the propeller hub, while the bottom of the base is fixed using a supported boundary condition. By outputting the calculation results, the vibration response of the floating raft isolation system under the excitation force can be obtained.

In this study, we used the vibration level difference as an indicator to evaluate the isolation effect. The vibration levels of each axis with respect to the base were calculated, and the vibration level curve and level difference curve of the bearing base were obtained using the calculation formula, as shown in Figs. 9, 10, 11 and 12. From the level difference curve, it can be observed that the isolation effects of the stern bearing and the sliding bearing are similar, while the isolation effect of the

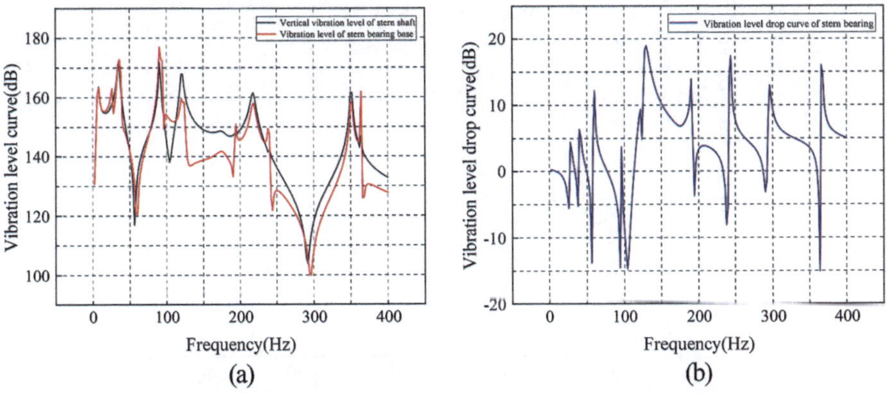

Fig. 9 (**a**) Vibration level curve of stern bearing; (**b**) Vibration level drop curve of stern bearing

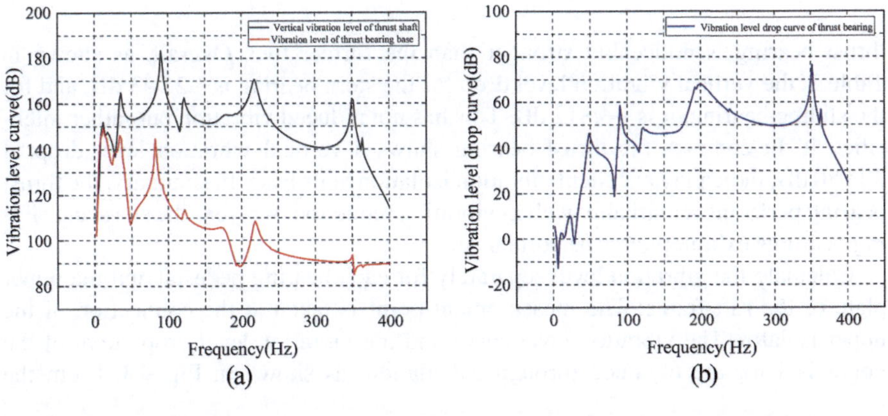

Fig. 10 (**a**)Vibration level curve of thrust bearing; (**b**) Vibration level drop curve of thrust bearing

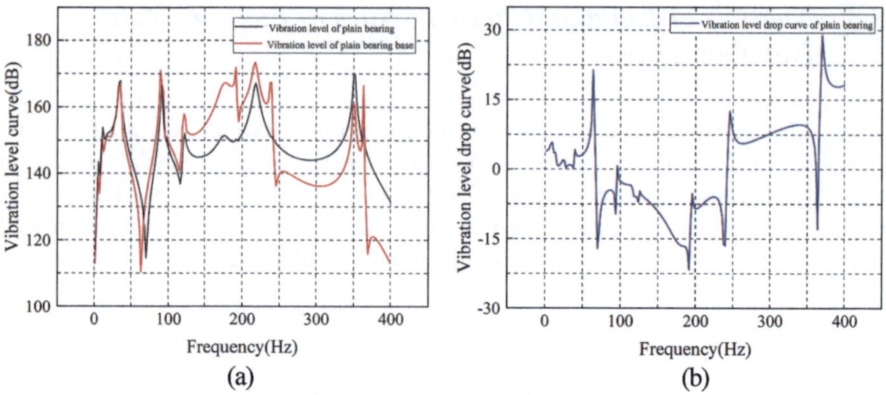

Fig. 11 (**a**) Vibration level curve of plain bearing; (**b**) Vibration level drop curve of plain bearing

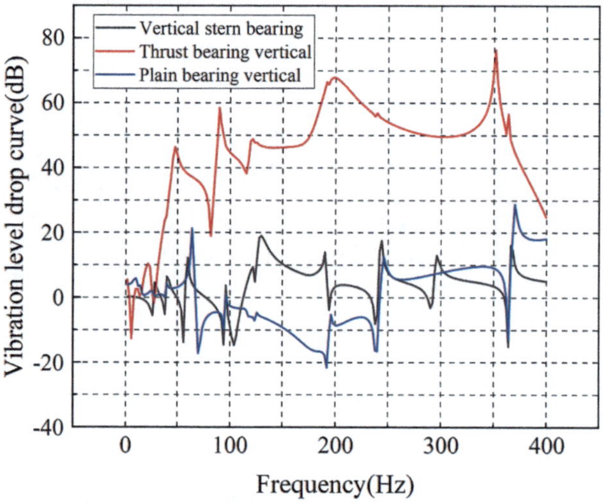

Fig. 12 Vibration level drop curve of bearing base

thrust bearing is noticeably stronger than the former two. Overall, as shown in Table 7, the vertical vibration level drop for the stern bearing is −2.043 dB, and for the sliding bearing, it is −4.812 dB. This has not reduced vibration but rather intensified it. In contrast, the thrust bearing shows a vertical vibration level drop of 30.080 dB, indicating excellent vibration isolation at its base. In summary, the thrust bearing performs remarkably well in vibration reduction, whereas the stern and sliding bearings exhibit poorer performance.

Calculate the vibration level separately for each bearing pedestal and the upper plate of the raft frame. The measurement point is taken at the connection of the upper isolator. The vibration level curve and the vibration level drop curve of the upper isolator are obtained through calculation, as shown in Fig. 13. From the

Table 7 The calculation result of vibration level drop

	Vertical vibration level drop total level/dB
Stern bearing	−2.043
Thrust bearing	30.080
Sliding bearing	−4.812
Upper vibration isolator	15.538
Lower vibration isolator	130.282

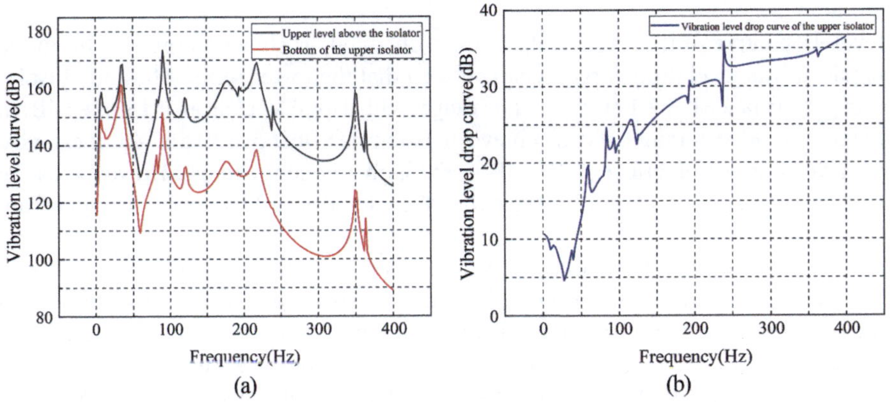

Fig. 13 (**a**) Vibration level curve of the upper isolator; (**b**) Vibration level drop curve of the upper isolator

vibration level drop curve, it can be seen that the upper isolator has good medium to high-frequency isolation performance, with a total level drop of 15.538 dB. It can achieve a vibration isolation effect of 5–25 dB in the low-frequency range and up to 35 dB in the high-frequency range, and the isolation effect shows a significant upward trend with increasing frequency. Significant peaks and valleys appear in the vibration level drop near the positions of 27 Hz, 38 Hz, 83 Hz, and 123 Hz. Comparing with the overall natural frequencies of the floating raft isolation system, the ninth, 13th, 14th, and 19th natural frequencies are 27.687 Hz, 38.381 Hz, 82.476 Hz, and 124.77 Hz, respectively. This may be due to the resonance between the excitation force and the isolation system.

When measuring the vibration level of the board under the raft frame and the base separately, the connection point at the lower isolator was selected as the measurement point. By using a calculation formula, the vibration level curve and the

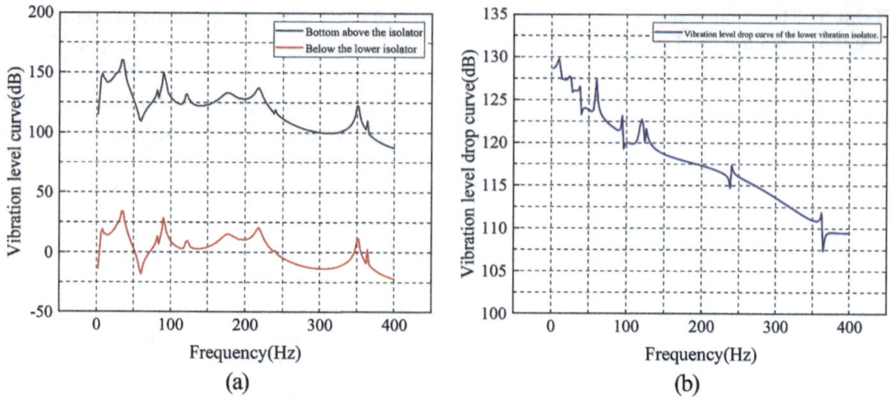

Fig. 14 (**a**) Vibration level curve of the lower isolator; (**b**) Vibration level drop curve of the lower vibration isolator

level drop curve of the lower isolator were obtained, as shown in Fig. 14. From the results of the level drop curve, it can be seen that the lower isolator has good isolation performance in the low-frequency range, with a total level drop of 130.282 dB. In the low-frequency range, it can achieve an isolation effect of 130 dB, and it can also achieve an isolation effect of 107 dB in the high-frequency range. However, as the frequency increases, it is observed that the isolation effect shows a significant decrease.

In conclusion, the raft isolation system has demonstrated good vibration isolation performance. By comparing the modal analysis results of the raft structure and the isolation system, it is observed that both exhibit similar vibration characteristics at 122.51 Hz in bending and 203.47 Hz in torsion. This indicates that the inherent modes of the raft structure are also present in the isolation system, highlighting the significant influence of the raft's vibration characteristics on the isolation effectiveness. Furthermore, it implies that modifying the raft structure can have a considerable impact on the isolation performance, thus optimizing the raft structure can further enhance the isolation capability.

4 Optimization of Raft Structure

In this section, the floating platform raft frame was optimized by introducing lightening holes in the transverse and longitudinal rib plates of the middle raft frame. After optimization, the weight of the floating platform raft frame was reduced by 7.4 kg, reaching 214.746 kg. The optimized finite element model is shown in Fig. 15.

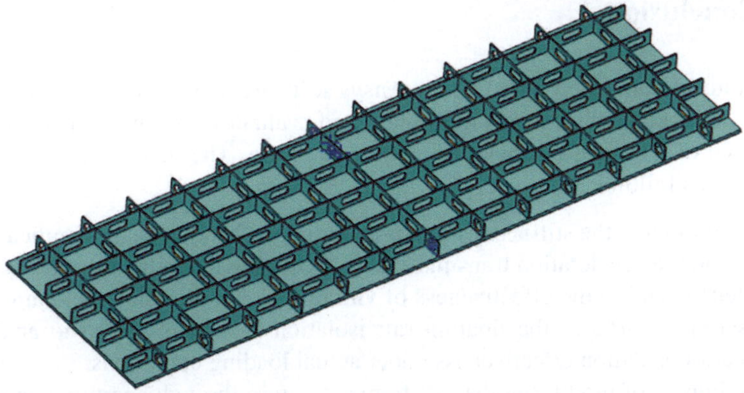

Fig. 15 Finite element model of floating raft raft frame after opening holes

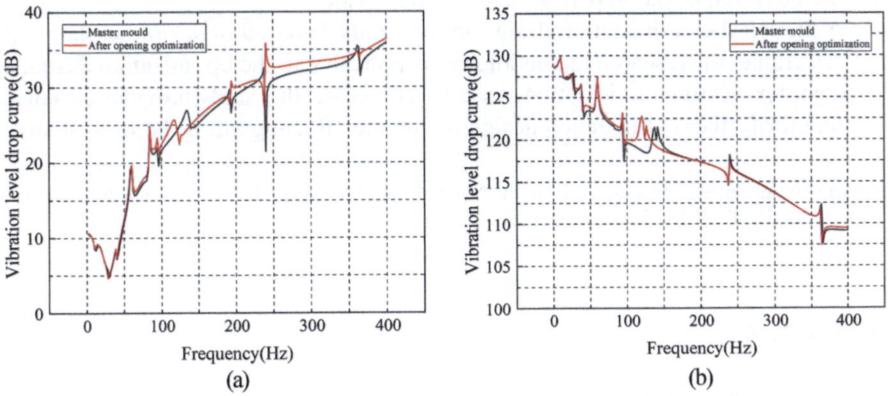

Fig. 16 (**a**) Vibration level drop curve of the upper isolator; (**b**) Vibration level drop curve of the lower vibration isolator

We investigated the vibration reduction performance of the floating raft system through finite element analysis. By introducing openings in the raft frame and conducting finite element analysis, we obtained the vibration level difference curves of the modified floating raft system's upper-side and lower-layer isolators, as shown in Fig. 16. These curves indicate that the modified floating raft system exhibits higher vibration level differences across the entire frequency range compared to the original model. This suggests that the vibration reduction performance of the floating raft system has been improved by modifying the rib plate structure of the raft frame. The introduction of openings has impacted the rigid and damping characteristics of the raft frame structure, thereby enhancing the isolation effectiveness of the raft body towards vibrational energy.

5 Conclusion

This study is conducted based on the Ansys software to establish a coupled system model of propulsion shaft-raft frame-hull and evaluates the vibration isolation performance of a typical flat-plate raft frame structure. The main conclusions of the study are as follows:

1. It is found that the stiffness of the intermediate raft frame has a significant influence on the acceleration transmissibility. Reducing the stiffness of the raft body is detrimental to the effectiveness of vibration isolation. The acceleration transmissibility curve of the floating raft isolation system is solved to analyze its vibration isolation effectiveness under actual loading conditions.
2. The impact of modifying the raft frame structure through aperture openings on the transmission of vibration energy by the damping device is analyzed. The results demonstrate that opening apertures in the rib plate can effectively improve the vibration isolation performance of the device.
3. Although the optimization design in this study is based on a simple plate-type raft frame, the research methodology is applicable to the optimization design of complex floating raft isolation systems. Therefore, this study provides a foundation for further optimization design of complex floating raft isolation systems.

Acknowledgments The authors thank the National Science and Technology Major Project (Grant 2019-IV-0019-0087) for financial support.

References

1. Huaicheng T, Yini Y, Ye L (2023) Research review on vibration reduction and isolation technology and calculation methods of marine machinery. Appl Math Mech 44(12):1413–1427. https://d.wanfangdata.com.cn/periodical/hjgcdxxb202104010
2. Chengshi X, Shaowei W, Guowen H (2022) Design and optimization of floating raft vibration isolation systems for large marine diesel generator sets. Noise Vib Control 42(03):25–29+35. https://lib.cqvip.com/Qikan/Article/Detail?id=7107484475&from=Qikan_Search_Index
3. Siwei M, Baocheng Z, Cuizhen M, Guanghua L (2024) Optimization and vibration reduction effect analysis of double bottom structures. Noise Vib Control 44(01):86–91. https://nvc.sjtu.edu.cn/CN/Y2024/V44/I1/86
4. Mingcheng X, Shaoyu X, Ruhang W (2024) Performance optimization and lightweight design of floating raft vibration isolation system based on RBF-PSO algorithm. Chin Ship Res 20:185. https://doi.org/10.19693/j.issn.1673-3185.03687
5. Heng Z, Xia-xia H, Shi-gang S (2013) Design and study of vibration isolation characteristics of frame floating raft, master's thesis, Zhejiang University of Technology. https://doi.org/10.7666/d.y2142894

6. Qi-xing G, Xiaotian W (2018) The design and study of the structure type of ship flat slab floating raft, master's thesis. Harbin Engineering University CNKI: CDMD 2(1018):048078
7. Yang L, Guo-ping C (2016) Optimization design of floating raft isolation system based on the finite element method, master's thesis, Nanjing University of Aeronautics and Astronautics. https://xueshu.baidu.com/usercenter/paper/show?paperid=1m0h0x40ag4e0ev09f7n00809k324387
8. Jinguang Z, Hairu Y, Guozhi C, Zeng Z (2018) Structure and modal analysis of carbon fiber reinforced polymer raft frame. J Low Freq Noise Vib Act Control 37(3):577–589. https://doi.org/10.1177/1461348417725960
9. Sun H (2016) Simplified performance indices and active control force of vibration isolation systems with elastic base. Shengxue Xuebao/Acta Acustica 41(2):227–235
10. Meng S, Changlin Q, Shiyang Z (2023) Research on vibration isolation performance of modular floating raft system of pump equipment. Ship Sci Technol 45(03):32–38. https://xueshu.baidu.com/usercenter/paper/show?paperid=1b0d0mm03d0b08808f760cy0yk343747&site=xueshu_se
11. Peng G, Qizheng Z, Deshi W (2023) Study on low frequency vibration isolation performance of local resonant floating raft isolation system. J Dyn Control 21(07):59–67. https://d.wanfangdata.com.cn/periodical/dlxykzxb202307009
12. Nan Z, Yu W, Lin C (2022) Study on vibration isolation performance of distributed acoustic black hole floating raft system. Vib Shock 41(13):75–80. https://doi.org/10.13465/j.cnki.jvs.2022.13.010

Open Access This chapter is licensed under the terms of the Creative Commons Attribution-NonCommercial-NoDerivatives 4.0 International License (http://creativecommons.org/licenses/by-nc-nd/4.0/), which permits any noncommercial use, sharing, distribution and reproduction in any medium or format, as long as you give appropriate credit to the original author(s) and the source, provide a link to the Creative Commons license and indicate if you modified the licensed material. You do not have permission under this license to share adapted material derived from this chapter or parts of it.

The images or other third party material in this chapter are included in the chapter's Creative Commons license, unless indicated otherwise in a credit line to the material. If material is not included in the chapter's Creative Commons license and your intended use is not permitted by statutory regulation or exceeds the permitted use, you will need to obtain permission directly from the copyright holder.

Research on the Field Evaluation Technology of Airborne Test Data in the Mode of Flight Test Mission Surge

Jiang Xie, Ruchang Huang, and Fenglei Shi

Abstract In order to meet the requirements of fine, reliable and efficient preparation at the flight test site, an airborne test data detection method was proposed. By obtaining the topology of the airborne test system, the analytic hierarchy process is used to quantitatively analyze the task support capacity of the test system, the airborne test data detection method based on the improved rule model is designed, and the remote control command state expectation-response matching algorithm is established to improve the digital level of the test work at the test site. Based on the airborne embedded platform, a set of test system data inspection and maintenance platforms is developed. The system has been installed and applied in multi-type testing machines, which greatly improves the on-site detection efficiency of airborne test data, realizes the decoupling of on-site test preparation of the testing machine from the test supervisor, allows the on-site supervisor to focus more on exception handling, reduces the impact of test failures on the test flight task, and reduces the on-site detection and evaluation time of airborne test data by more than 70%.

Keywords Task surge · Hierarchical analysis · Rule model · Remote control · Graded management

1 Introduction

The results of the detection and evaluation of airborne test data are an important determinant of on-site flight test decisions [1]. At present, scientific research flight test tasks have the following characteristics: The number of testing machines has increased significantly, requiring on-site test engineers to achieve high efficiency work quality; The rapid growth of test flight sorties requires reducing the working pressure at the test flight site and reducing human errors; The continuous

J. Xie (✉) · R. Huang · F. Shi
Chinese Flight Test Establishment, Xi'an, China

© The Author(s) 2026

Z. Zhang, P. Mercorelli (eds.), *Mechanical Design and Simulation: Insights and Innovations*, Lecture Notes in Mechanical Engineering,
https://doi.org/10.1007/978-981-95-2405-1_4

improvement of the test point model requires an increase in the correlation level between the parameter state and the task requirements. In the face of the new task situation, it is urgent to improve the digital level of the test site test work, and meet the fine, reliable and efficient work requirements of the test guarantee and the flight test task in the mode of task surge by standardized, refined and efficient means.

In view of the requirements of on-site flight test mission support under the new form, this paper proposes an airborne test data detection method, which upgrades the existing test site working mode from various aspects such as evaluation methods, detection methods, and control methods, in order to solve the bottleneck that restricts the on-site mission test support capability, so as to achieve the purpose of improving the on-site test support capability of the test flight mission under the new situation [2].

2 Schematic Design

2.1 Overall Scheme Design

The main function of the airborne test system status detection and maintenance platform is to detect the operating status of each subsystem of the test system, which mainly includes: the operating status of the collector, recorder and telemetry subsystem, and the GPS and sensor adjustment subsystem reflect the operating status through the parameters of the collector. Through the monitoring platform, the state of the collection, recording, telemetry, GPS, and sensing adjustment subsystem is completed, the rapid status report of the test system is realized, and the online configuration of telemetry frequency points can be realized, so as to realize the rapid maintenance of the status of the test system. This is shown in Fig. 1.

Through the establishment of the field detection platform of the airborne test system, the model algorithm is used to detect the airborne test data, obtain the

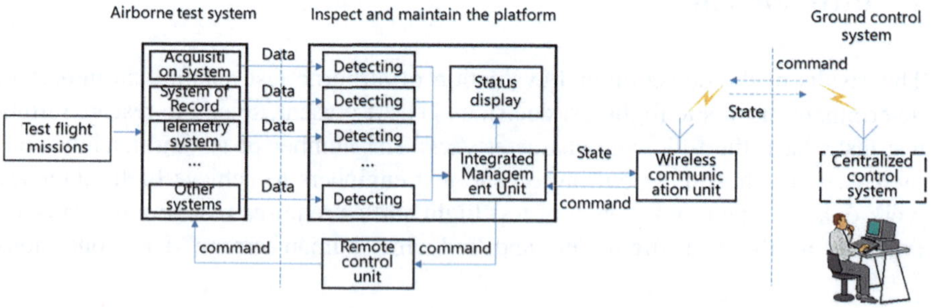

Fig. 1 Overall scheme diagram

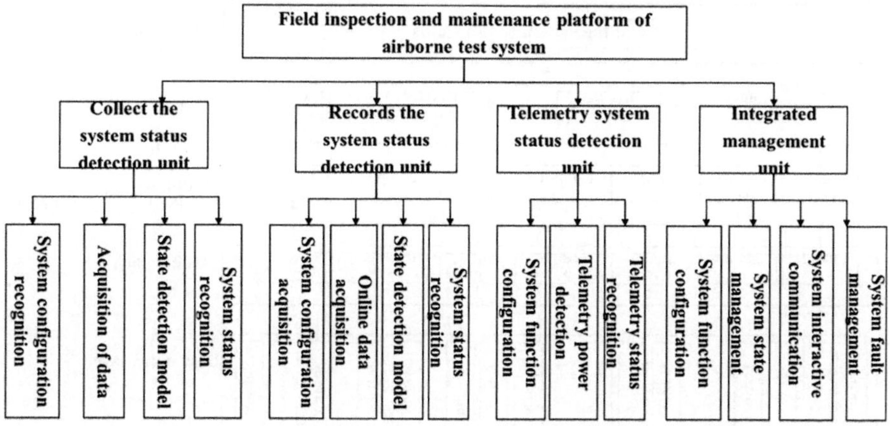

Fig. 2 System function composition

system status evaluation index and feedback, and support the fine, reliable and efficient work requirements of the test flight field test. The on-site detection platform of the airborne test system mainly consists of four parts: acquisition system inspection unit, recording system detection unit, telemetry system detection unit and system integrated management unit. The function composition of the system is shown in the Fig. 2:

2.2 *Health Status Evaluation Method of Test System Based on Hierarchical Analysis*

Typical airborne test systems mainly include a signal sensing layer, signal acquisition layer, data exchange layer, transmission application layer, etc., which is characterized by complex system cross-linking structure and clear data flow direction. Traditional function-based hierarchical division methods make it difficult to describe the membership relationship between different states, and cannot meet the requirements of efficient evaluation of system state. It can be seen from Fig. 3 that the airborne test system has a complex topology but a clear data flow direction. Based on the data flow direction of the airborne test system, a five-layer state evaluation index is established and the membership relationship between states at each level of the system is clear, which is conducive to state evaluation and analysis.

The accuracy of status evaluation is subject to the weight of indicators. As a subjective weighting method, the analytic hierarchy Process (AHP) determines its weight accuracy by the prior knowledge of experts [3]. In order to improve the accuracy of weight indicators, a combination of subjective and objective weighting methods is adopted to improve the accuracy of weights [4]. This is shown in Fig. 4.

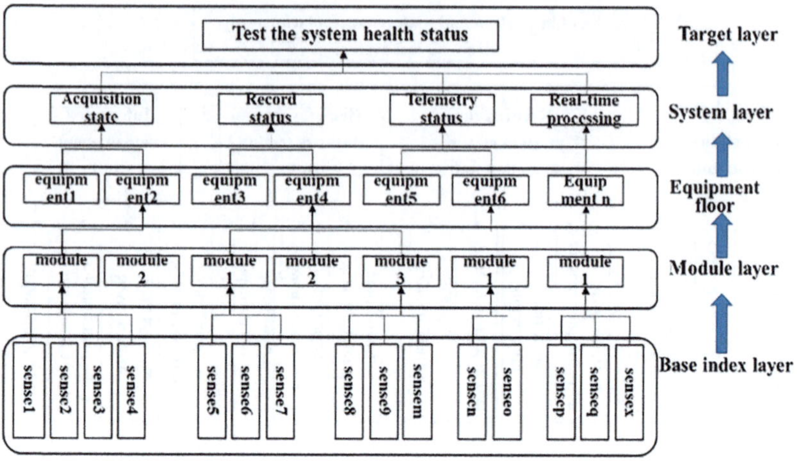

Fig. 3 Test system five-level state evaluation index

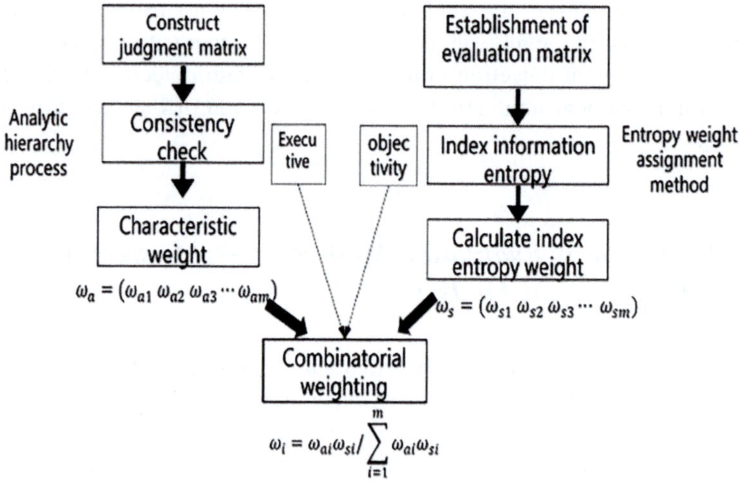

Fig. 4 Combinatorial weighting

ω_i is the combined weight, which is suitable for different evaluation tasks by adjusting the validity ratio. By synthesizing the weight set of four layers of status evaluation indicators, the influence of different systems, devices, modules and sensors on system health status can be obtained [0, 1]. According to whether each state is healthy or not, and its influence degree, the quantitative analysis of the health state of the test system is realized.

2.3 *Test Parameter Detection Method Based on Improved Rule Model*

In order to obtain good interpretability and detection classification performance at the same time, a test parameter detection method based on a rule representation learner is designed. By extracting the objective relation and law between the test parameters, the analytical model is established to realize the automatic detection of the test parameters. This is shown in Fig. 5.

Parameter Screening. Aiming at the problem that the identification of a single parameter state is easily affected by the state of the carrier, the correlation analysis algorithm is used to analyze the correlation degree among parameters, obtain the associated parameter group, and improve the accuracy and applicability of the model through multi-parameter fusion analysis. The parameter screening process is as follows.

According to the target parameter xt, the parameters are selected and the primary parameter group is established, A = {x_1, x_2, …, x_n};

The correlation index between each parameter and the target parameter xt is calculated, C = {c_1, c_2, …, c_n};

According to the calculation results, the correlation index threshold is selected and the parameter groups are screened, $c_i > c_{(t\text{-threshold})}$;

The parameter whose correlation indicator is greater than the threshold is selected and the associated parameter group for subsequent analysis is set up, B = {x_1, x_2, …, x_m};

Data Reconstruction. To solve the problem of inaccurate detection and identification of sub-class data caused by unbalanced distribution of parameter values, the parameter values are classified. By increasing the residual weight of sub-class data, the importance of different categories of data is balanced, and the accuracy of parameter identification is improved.

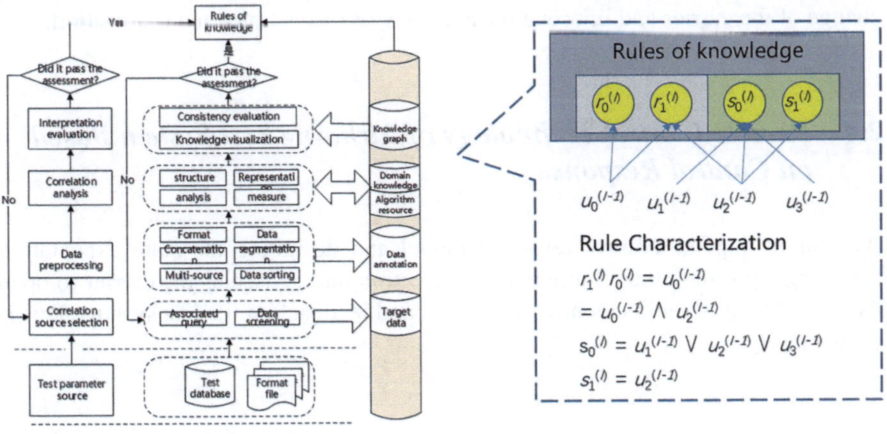

Fig. 5 Schematic diagram of typical rule model establishment

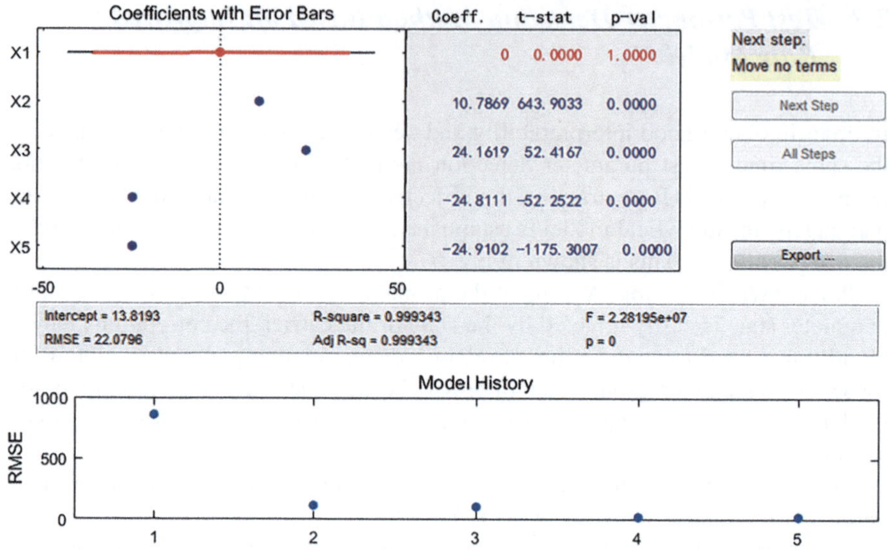

Fig. 6 Training process schematic based on improved rule model

The value distribution of target parameter xi is obtained, $x_i = \{x_{i1}, x_{i2}, \ldots, x_{il}\}$;

The x_i values are divided into k classes, $\{a_1, a_2, \ldots, a_k\}$;

The quantity of each type is obtained, $\{l_1, l_2, \ldots, l_k\}$;

The residual value is modified to increase the residual weight of muscle data, $\omega_i = \sqrt[t]{\frac{l - l_i}{l}}$.

Model Verification. Taking the outlet pressure of an engine compressor as the target parameter, the model algorithm is verified by selecting the associated parameter group and extracting the implicit association rule.

This is shown in Fig. 6. The results show that the model can adapt to the state change of the carrier and improve the accuracy of the test parameter detection.

2.4 Remote Control Technology of Airborne Test System Based on Control Response

By establishing the control response model and designing the state expectation matching algorithm, the remote control detection unit is used as the carrier to open the closed loop of remote control, detection and feedback of the test system. This is shown in Fig. 7.

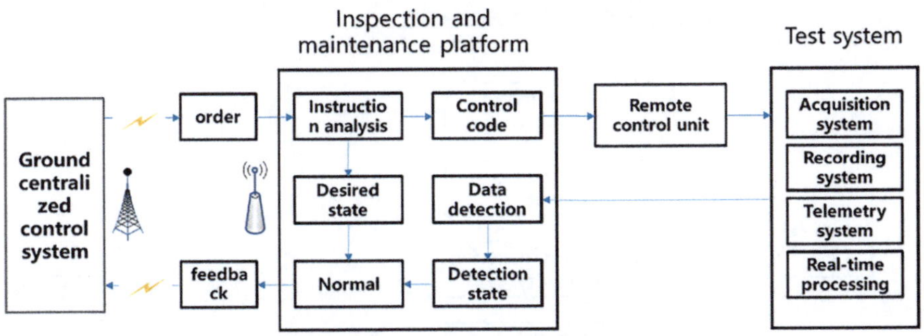

Fig. 7 Remote control loop diagram of airborne test system

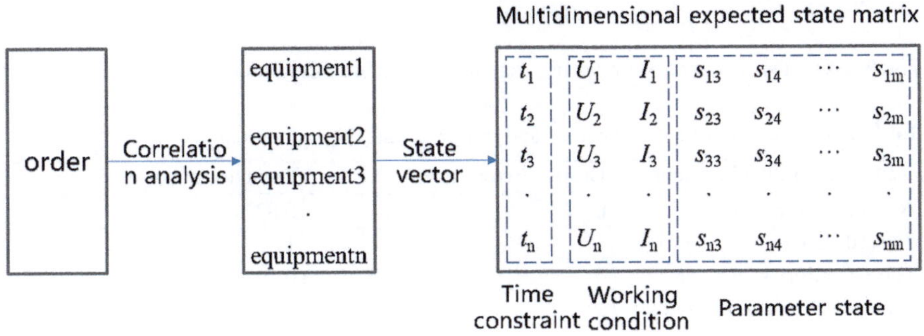

Fig. 8 Test the system instruction state matrix

State Expectation Matrix. Based on the topological analysis of the test system, the remote instruction associated device set is established, and the key state feature vector based on the time-constrained response is extracted, which mainly includes the time constraint condition, working state, parameter state, etc., to form an n×m multidimensional expected state matrix, This is shown in Figure 8, which is used to describe the expected response characteristics of the remote instruction.

Response Matching Evaluation. Through the data detection algorithm model, the response characteristic matrix of the associated device set is obtained, the "response-expectation" matching operator is defined, and the multidimensional evaluation result of instruction response is obtained, so as to judge whether the remote instruction meets the expectation [5]. This is shown in Fig. 9.

Taking a recording instruction as an example, the response-expected similarity level is calculated. The closer the similarity is to 1, the better the instruction execution result is, as shown in Table 1.

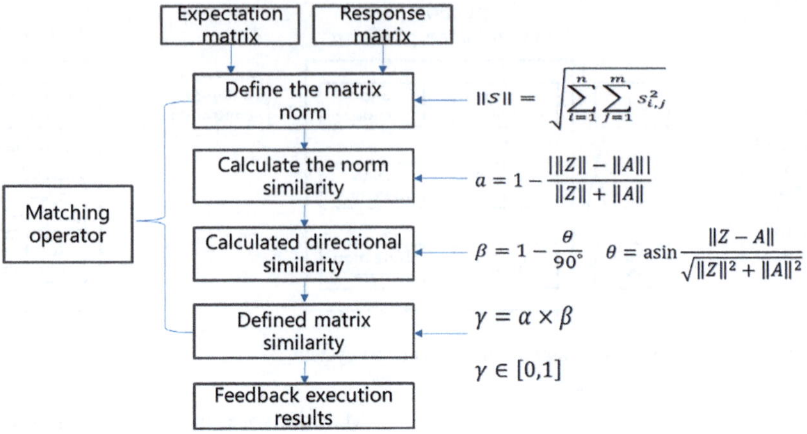

Fig. 9 Expected state matrix—response state matrix matching algorithm flow

Table 1 Remote recording of expected and response states of instructions

Remote recording instruction	Desired State	Detection status
Response time (%)	0.87	0.82
Record status	1	1
Fault conditions	0	0
Time synchronization status	1	1
Total recorded data rate (%)	0.21	0.22
Network data rate (%)	1	1.02
PCM data rate (%)	1	1.01
1553 data rate (%)	0.35	0.36

3 Experimental and Engineering Applications

On the basis of scheme design and key technology research, relying on embedded hardware systems, an airborne test data detection and evaluation application software is developed, and an experimental environment for verification is built. This is shown in Fig. 10.

Firstly, the applicability of the airborne testing system status online detection platform to different types of airborne testing systems was verified, including KAM series collectors, aNTS collection systems, XMA collection systems, UMA collection systems, recording systems, and telemetry systems. The results are shown in Table 2:

Remote control commands support operations such as power on control, recording control, and telemetry frequency point setting for testing equipment. The response expectation matching algorithm can accurately reflect the execution effect of remote commands. The system supports real-time detection, with a parameter detection capability of over 3000, a fault recognition accuracy rate of over 98%, and the ability to provide information on task impact and solutions, and support rapid troubleshooting, as shown in Fig. 11.

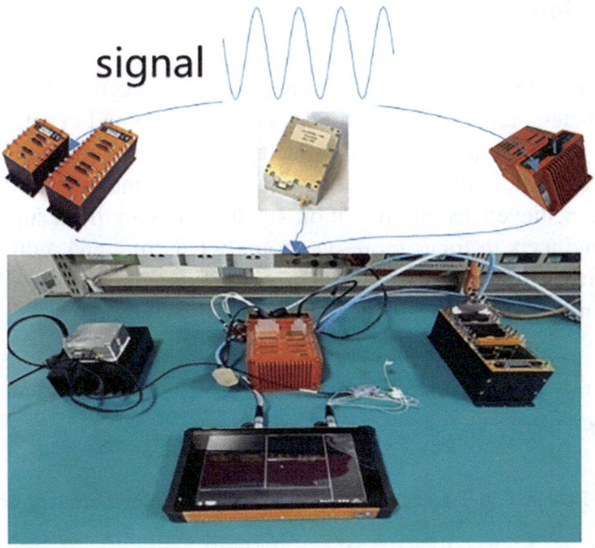

Fig. 10 Experimental verification and test environment construction

Table 2 Adaptive verification—collection system

Acquisition system	KAM	aNTS	XMA	UMA
PCM architecture	√	√	√	√
Network architecture	√	/	√	√

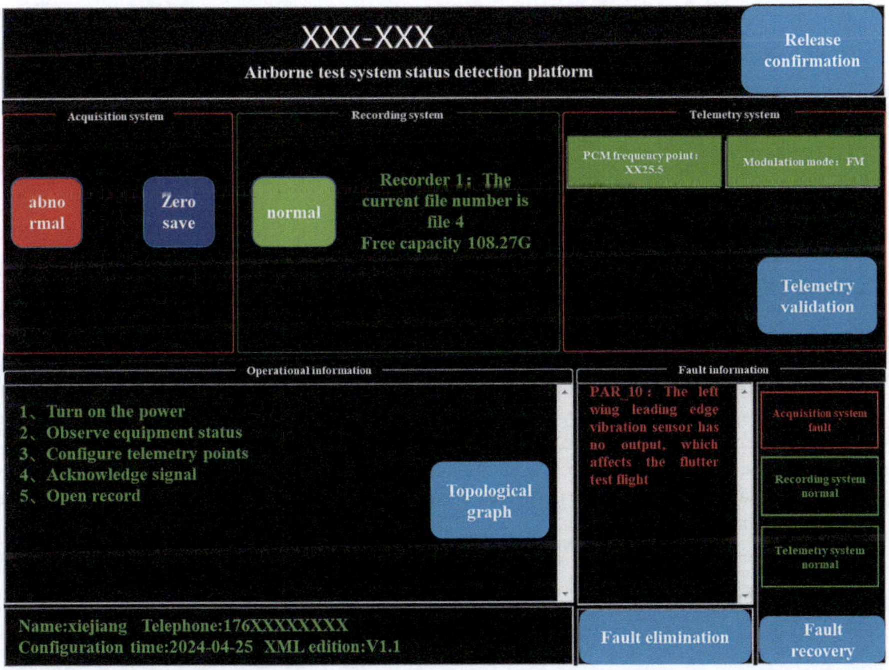

Fig. 11 Data detection result

4 Conclusion

The algorithm designed in this article can quickly and accurately detect and evaluate test parameters, and is suitable for on-site detection and evaluation of airborne test data for various types of testing machines. The system has been applied on multiple test aircraft, and the model-based testing parameter detection and evaluation mode has achieved decoupling of on-site testing work from supervisors, allowing on-site engineers to focus more on anomaly handling and reducing the impact of testing failures on flight test tasks. The on-site detection and evaluation time of airborne testing data has been reduced by more than 70%.

References

1. Li X (2023) Design of intelligent temperature controller for airborne test system. China Testing 05:151–157
2. Liu M (2018) Research on key technologies of intelligent airborne test system. China Testing 44(7):99–104
3. Zhao X (2024) Research on multi-transmission channel dynamic selection algorithm based on analytic hierarchy process. Yangtze River Inform Commun 04:86–88
4. Lin C (2021) Research on the application of a combinatorial weighting method. Electron Technol 06:106–107
5. Dong Y (2023) Failure mode and influence analysis method based on distance interval number operator. J Tongji Univ (Nat Sci Ed) 04:633–642

Open Access This chapter is licensed under the terms of the Creative Commons Attribution-NonCommercial-NoDerivatives 4.0 International License (http://creativecommons.org/licenses/by-nc-nd/4.0/), which permits any noncommercial use, sharing, distribution and reproduction in any medium or format, as long as you give appropriate credit to the original author(s) and the source, provide a link to the Creative Commons license and indicate if you modified the licensed material. You do not have permission under this license to share adapted material derived from this chapter or parts of it.

The images or other third party material in this chapter are included in the chapter's Creative Commons license, unless indicated otherwise in a credit line to the material. If material is not included in the chapter's Creative Commons license and your intended use is not permitted by statutory regulation or exceeds the permitted use, you will need to obtain permission directly from the copyright holder.

Aerodynamic Effects of Load Deformation on Civil Aircraft Forward Fuselage

Ling Ding, Xiaohui Zhao, and Yaru Tian

Abstract This paper investigates the aerodynamic effects of load deformation on the forward fuselage of civil aircraft. The study begins by analyzing the typical structural characteristics and load deformation distribution of the forward fuselage. Based on the finite element analysis results, the deformed shape of the forward fuselage under typical cruise conditions is simulated. The aerodynamic influence of local deformation is evaluated by using computational fluid dynamics (CFD) methods. The results show that the aerodynamic performance, including drag coefficient and pressure distribution, is sensitive to local shape changes, especially in the boarding gate area. Finally, two control strategies, passive control, and active control, are proposed to minimize the adverse aerodynamic effects caused by load deformation of the forward fuselage.

Keywords Aerodynamic effects · Local deformation · Aircraft

1 Introduction

At present, the research on the overall design and aerodynamic design of the forward fuselage of civil aircraft has been very extensive and in-depth [1, 2], but it is limited to the theoretical shape of the jig state. While the cruise state is affected by cabin pressurization and aerodynamic load, fuselage deformation is inevitable. This paper analyzes the simulation of the forward fuselage deformation and aerodynamic influence after cabin pressurization in a typical cruise state, providing design ideas for the forward fuselage shape control.

L. Ding (✉) · X. Zhao · Y. Tian
Shanghai Aircraft Design and Research Institute, Shanghai, China
e-mail: dingling@comac.cc

© The Author(s) 2026

Z. Zhang, P. Mercorelli (eds.), *Mechanical Design and Simulation: Insights and Innovations*, Lecture Notes in Mechanical Engineering,
https://doi.org/10.1007/978-981-95-2405-1_5

2 Geometric Characteristics of Civil Aircraft Forward Fuselage

The forward fuselage shape is the difficulty of the shape design of civil aircraft, especially the geometric design. To minimize the resistance generated, the forward fuselage shape generally tends to be cone-like, because it also needs to meet the installation space requirements of weather radar, front-end, cockpit equipment, and other systems, and the final appearance of the forward fuselage shape looks full. At present, the typical forward fuselage shape of civil aircraft can be divided into two types: (1) Traditional forward fuselage, such as B737 and A320. The shape of the lower edge of the windshield has a clear bend and a flat windshield; (2) A streamlined forward fuselage, such as B787 and C919, has a curved windshield without turning at the lower edge.

3 Typical Structure of Civil Aircraft Forward Fuselage

The main structure of the civil aircraft head adopts a typical semi-hard shell metal structure composed of a frame, girder, and skin, and is divided into a hatch cover, front lift, floor, wall girder, various types of frames, door frame reinforcement structure, transparent parts, radome, and other components according to different parts. There are many openings in the civil aircraft forward fuselage structure, including the windshield transparent part opening, the forward fuselage landing gear compartment, the boarding door, the service door, and other large openings, the EE cabin opening, the emergency door exit on the top of the cockpit, the wiper, the angle of attack sensor and other equipment small openings and maintenance covers, etc., which are locally strengthened.

4 Load Deformation on the Forward Fuselage

The internal and external pressure difference caused by the flight load and the pressurized cabin load leads to the deformation of the forward fuselage. Due to the curvature distribution characteristics of the forward fuselage and the structural design characteristics of the forward fuselage, the deformation varies from place to place. We take a research proposal as an example to analyze its macroscopic and microscopic deformation characteristics.

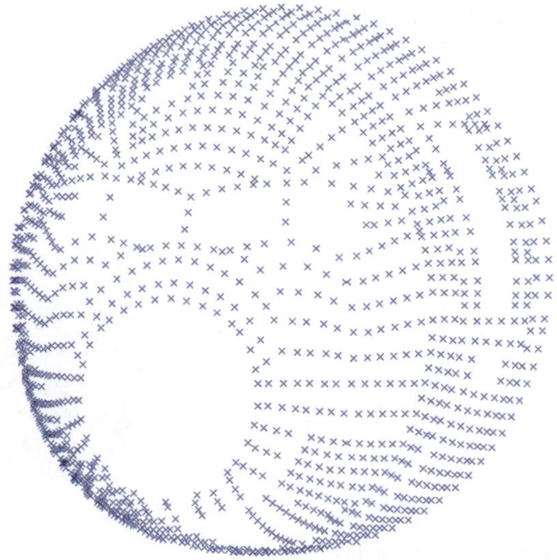

Fig. 1 Finite element mesh nodes

4.1 *Macroscopic Characteristic Analysis*

The finite element model mesh node is shown in Fig. 1. The node displacement after loading is shown in Fig. 2.

The macroscopic characteristics of the deformation distribution are as follows:

- Front service area overhead recessed.
- The boarding gate area is convex.
- Forward lifting is convex.
- In the windscreen front area, the main windscreen is recessed, and the side windscreen is convex.

4.2 *Microscopic Feature Analysis*

Based on the nodal displacement information of the finite element model, the shape of the machine head after loading is simulated (refer to Chap. 5 for the simulation method), and the surface quality is analyzed. Figure 3 is the result of Gaussian curvature analysis. There are obvious ripples (color-mottled area) in the local part, and the maximum amplitude is about 10 mm.

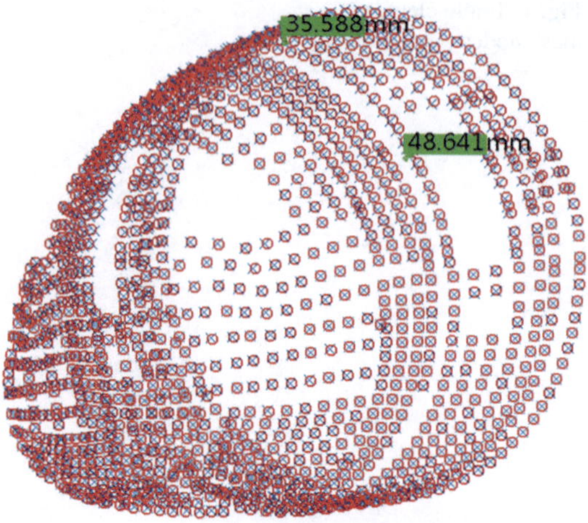

Fig. 2 Node before and after loading

Fig. 3 Surface quality analysis

5 Load Deformation on the Forward Fuselage

Based on the data from Chap. 3, we simulate the shape of the loaded forward fuselage. The shape of the deformed forward fuselage is reconstructed according to the finite element node loaded, which is close to the node while maintaining the original forming law of the theoretical shape, wherein the radome, the cockpit windshield, the boarding gate, the service gate, the front lifting cabin door and the like are missing in the finite element model, deforms with the adjacent nodes, and keeps tangent continuous with the straight segment of the fuselage.

The distance between the finite element node and the simulated shape loaded is shown in Fig. 4, all of which are less than 1 mm, indicating that the deformation simulation method is accurate and reliable.

Fig. 4 Surface quality analysis

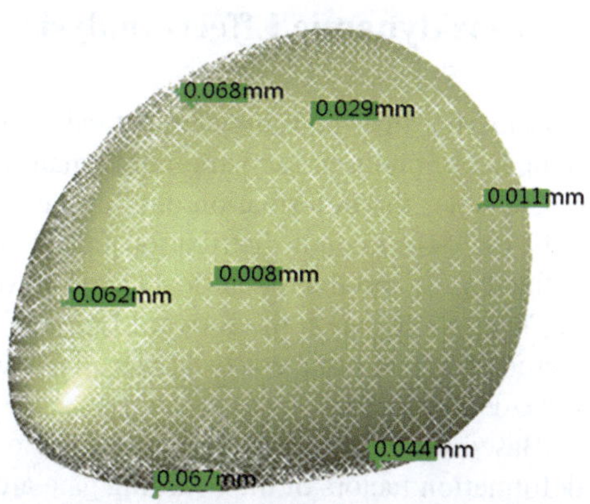

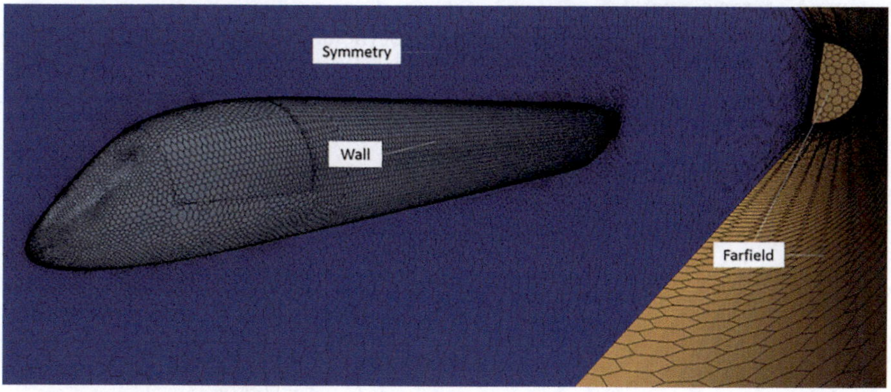

Fig. 5 Mesh and boundary condition settings

6 Aerodynamic Analysis Method

Computational Fluid Dynamics (CFD) has emerged as a powerful tool for aerodynamic analysis in the aerospace industry [3–5]. CFD simulations allow engineers to predict and analyze the flow behavior around complex geometries, such as the forward fuselage, with high accuracy and efficiency. Considering the efficiency of calculation and the symmetry of the fuselage, the half-mold fuselage configuration is adopted. A convection field domain generates a non-structured body mesh, and defines a semi-cylinder far-field domain with a flight direction of 1, 300 m and a radial direction of 30 m, wherein the Reynolds number Re is 4.65×10^7, the first layer mesh ensures $y^+ = 1$, the thickness of the boundary layer is 0.005 mm, the total thickness of the boundary layer is 86 mm, and there are 45 layers of grid in total. Considering the curvature distribution of the forward fuselage shape, the mesh is sparsely treated, and the local transition of the mesh at the boundary layer and the flow field are adjusted to ensure the convergence of the calculation. The mesh and boundary condition settings are shown in Fig. 5. Reynolds number average NS equation (RANS), high Reynolds number SA turbulence model, and steady implicit coupling solver were used to solve the flow field.

7 Aerodynamic Effect Analysis

Considering that the finite element calculation result corresponds to two times airtight load condition, and the aerodynamic effect analysis focuses on the typical cruise state, the deformation data in the finite element model is approximated by half, the deformation distribution factor contained in the finite element model is defined as 1, and the theoretical shape deformation distribution factor is defined as 0. Then, we focus on the forward fuselage shape with the deformation distribution factor of 0.5 and make a local sensitivity analysis for the boarding gate area with large deformation.

Based on the forward fuselage shape of the deformation factor 0.5, the design deformation factors of the boarding gate area are 0.3, 0.4, 0.6, and 0.7, respectively. Considering the large scale and large deformation here, the gentle transition around requires a large space.

In the typical cruise state, the resistance coefficient changes obviously in different degrees of deformation (see Table 1), and the pressure distribution pattern shows a worsening trend (see Figs. 6, 7, 8, 9 and 10), indicating that this area is more sensitive to shape changes.

Table 1 Drag coefficient

Factor	Resistance coefficient increment
0.3	−10.92%
0.4	0.57%
0.5	0
0.6	−10.34%
0.7	−11.90%

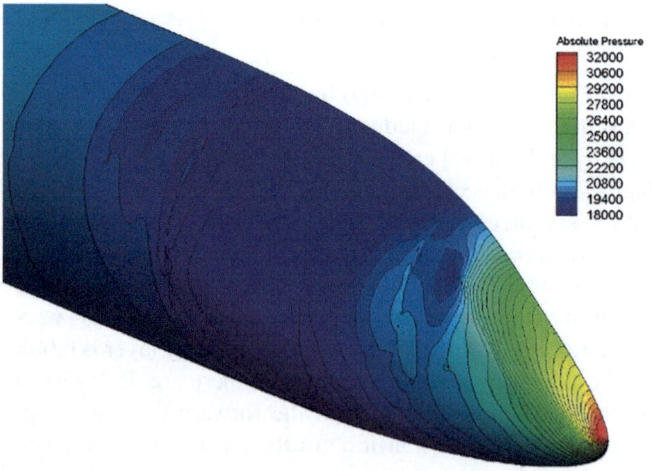

Fig. 6 Factor 0.3

Fig. 7 Factor 0.4

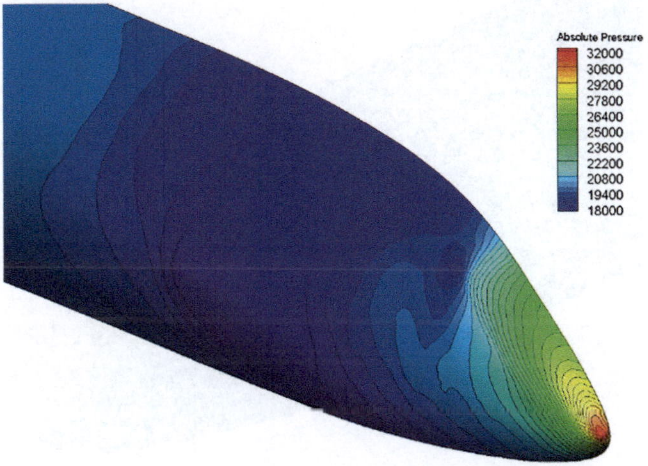

Fig. 8 Factor 0.5

Fig. 9 Factor 0.6

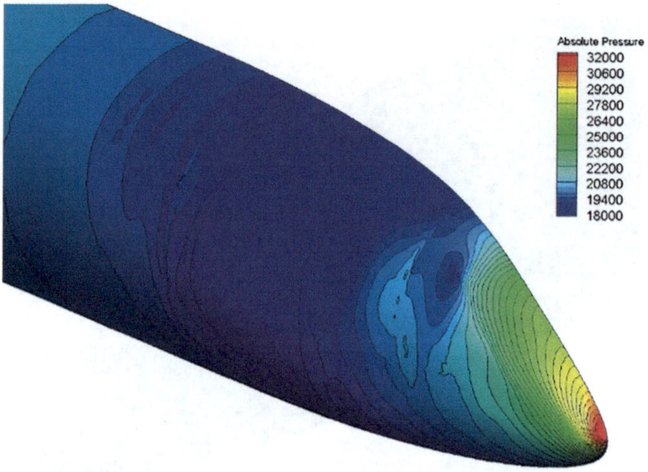

Fig. 10 Factor 0.7

8 Conclusion and Recommendation

The aerodynamic influence of the load deformation of the forward fuselage cannot be ignored. The forward fuselage shape emphasizes overall coordination, and the drag coefficient and pressure distribution pattern have significant changes after local deformation.

Two ideas can be considered to realize the load deformation control of the forward fuselage shape of civil aircraft. The first idea, passive control, is to design the structure based on the theoretical shape, obtain the distribution of the load-deformation, conduct different deformation simulations and comprehensive evaluation, formulate the load deformation index of the forward fuselage shape, and locally optimize the structure design according to the deformation index. The second idea, active control, is to redesign the structure based on the theoretical shape, obtain the load deformation distribution, reversely modify the theoretical shape according to the load deformation distribution, and update the structural design and system layout based on the modified shape, to hope that the deformation after load returns to the original theoretical shape.

References

1. Mattos B, Papa R, Carlos SL (2004) Considerations about forward fuselage aerodynamic design of a transport aircraft. In: 42nd AIAA aerospace sciences meeting and exhibit. Reno
2. Nicolosi F, Vecchia P, Ciliberti D, Cusati V (2016) Fuselage aerodynamic prediction methods. Aerosp Sci Technol 55:332–343
3. Fang Y, Yang H, Guan J (2023) Aerodynamic analysis of sideslip angle sensor layout design on civil aircraft. J Phys Conf Ser 2659:012005
4. Tahani M, Masdari M, Bargestan A (2021) Aerodynamic ground effects on DLRF6 vehicle. Aircraft Eng Aerosp Technol 93:755
5. Aprovitola A, Nuzzo PED, Pezzella G et al (2021) Aerodynamic analysis of a supersonic transport Aircraft at Landing Speed Conditions. Energies 14:6615

Open Access This chapter is licensed under the terms of the Creative Commons Attribution-NonCommercial-NoDerivatives 4.0 International License (http://creativecommons.org/licenses/by-nc-nd/4.0/), which permits any noncommercial use, sharing, distribution and reproduction in any medium or format, as long as you give appropriate credit to the original author(s) and the source, provide a link to the Creative Commons license and indicate if you modified the licensed material. You do not have permission under this license to share adapted material derived from this chapter or parts of it.

The images or other third party material in this chapter are included in the chapter's Creative Commons license, unless indicated otherwise in a credit line to the material. If material is not included in the chapter's Creative Commons license and your intended use is not permitted by statutory regulation or exceeds the permitted use, you will need to obtain permission directly from the copyright holder.

Normal Probability Distribution and Process Capability Index Analysis Determine the Reference Value for Continuity Requirements of Aircraft External Surface

Ling Ding

Abstract To ensure the aerodynamic performance of the aircraft, the continuity requirements of the external surface are determined to restrain the deviation between the manufacturing external surface and the design external surface of the aircraft. When formulating the continuity requirements of the external surface, we need to consider whether the manufacturing can be realized. This paper proposes the normal probability distribution analysis and process capability index analysis of the existing manufacturing detection data to provide the manufacturing reference value for the continuity requirements of the external surface. At present, this method has been verified and applied in practical work.

Keywords Aircraft external surface · Normal probability distribution · Process ability index · Reference value

1 Introduction

The aircraft's external surface has the design external surface and the manufacturing external surface. The designed external surface is the appearance generated in the design stage; the manufacturing external surface is the actual external surface of the aircraft after production, and there is an inevitable deviation from the design external surface due to manufacturing errors and other terms.[1] To ensure the aerodynamic performance of the aircraft, the deviation between the manufacturing external surface and the design external surface is controlled by the continuity requirements, and the allowable deviation between the manufacturing external surface and the design external surface is specified.

L. Ding (✉)
Shanghai Aircraft Design and Research Institute, Shanghai, China
e-mail: dingling@comac.cc

© The Author(s) 2026

Z. Zhang, P. Mercorelli (eds.), *Mechanical Design and Simulation: Insights and Innovations*, Lecture Notes in Mechanical Engineering,
https://doi.org/10.1007/978-981-95-2405-1_6

Determining the appropriate continuity requirements for the aircraft's external surface is a challenging task. It requires striking a balance between the aerodynamic performance requirements and the manufacturing feasibility. The traditional practice is to set continuity requirements with the main goal of aerodynamic smoothness, and the continuity requirements are often set too high, which may result in increased manufacturing costs or even make it impossible to achieve. On the other hand, if the continuity requirements are set too low, it may compromise the aircraft's performance and competitiveness in the market.

To address this challenge, it is crucial to analyze the existing aircraft manufacturing test data using statistical methods. By examining the actual manufacturing data, we can gain insights into the current manufacturing capabilities and levels. This analysis provides valuable support and reference for determining the appropriate continuity requirements for the external surface.

The purpose of this paper is to propose a method for determining the reference value for the continuity requirements of the aircraft's external surface using normal probability distribution and process capability index analysis. By analyzing the existing manufacturing detection data, we aim to establish a data-driven approach to setting realistic and achievable continuity requirements.

In the following sections, we will delve into the details of the proposed method. We will discuss the continuity requirements for the aircraft's external surface, explain the concepts of normal probability distribution and process capability index analysis, and outline the steps involved in determining the reference value. Additionally, we will present an application case to demonstrate the effectiveness of the proposed method in a practical scenario.

The findings of this study have significant implications for the aircraft manufacturing industry. By establishing a robust method for determining the continuity requirements of the external surface, we can ensure that the manufacturing process is optimized to meet the desired aerodynamic performance while maintaining cost-effectiveness and feasibility. This approach can lead to improved aircraft quality, reduced manufacturing errors, and enhanced competitiveness in the market.

2 Continuity Requirements for Aircraft External Surface

The external surface of the aircraft is composed of skin, outer surface equipment, access covers, and other structures. Continuity requirements for an aircraft's external surface refer to the allowed deviation between the manufacturing external surface and the designed external surface. The items include: steps (shown in Fig. 1), gap (shown in Fig. 2), etc.

The continuity requirements for external surface is the control requirement for aircraft external surface quality, which must be achieved through manufacturing. If the continuity requirements of the external surface are too much higher, the manufacturing cost will be too high or impossible; if the continuity requirements of the

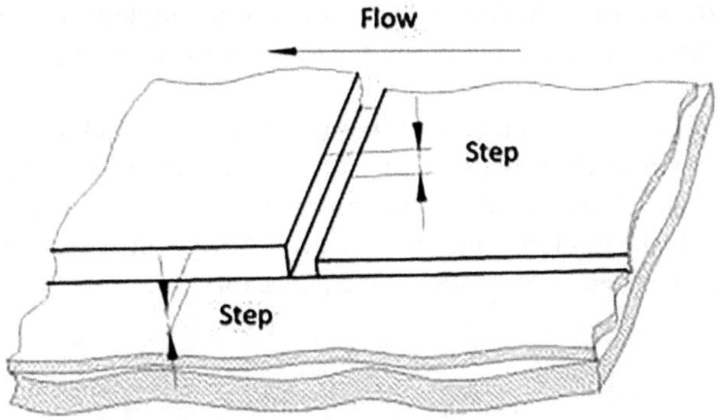

Fig. 1 Step

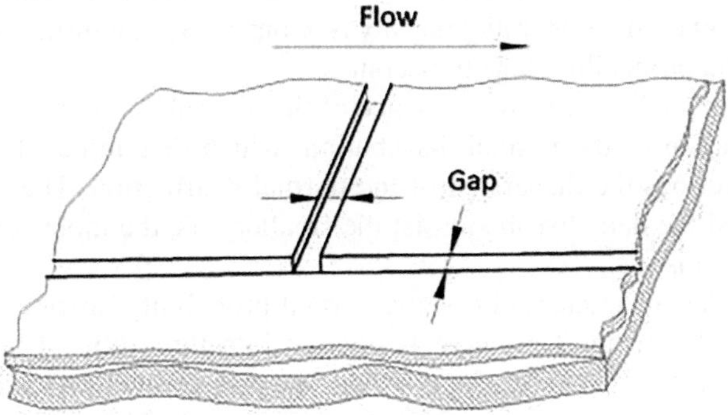

Fig. 2 Gap

external surface are much lower than the actual manufacturing level, the aircraft performance will be affected, and the competitiveness of the aircraft.

In the process of determining the continuity requirements for external surfaces, it is particularly important to analyze the manufacturing data of existing aircraft and understand the implementation level in the actual implementation. This requires recording and analyzing the relevant dimensional data in the manufacturing process, reflecting the actual manufacturing capacity and level, and providing support and reference for determining the continuity requirements of the external surface.

3 Normal Probability Distribution and Process Capability Index Analysis

The normal distribution has an extremely wide practical background, and the probability distribution of many random variables in production and scientific experiments can be described by the normal distribution.[2, 3] Aircraft manufacturing and assembly are the statistical control state process of the system specification. Large amounts of detection data (step etc.) also conform to the law of normal distribution.

It is a continuous probability distribution that is symmetric and bell-shaped (shown in Fig. 3), with the probability density function defined by the mean (μ) and variance (σ^2) of the distribution. The normal distribution is often denoted as $x \sim N(\mu, \sigma^2)$, where x is the random variable. The probability density function of the normal distribution is given by the following equation[4]:

$$p(x) = \frac{1}{\sigma\sqrt{2\pi}} e^{-(x-\mu)^2/2\sigma^2}, \quad -\infty < x < +\infty \tag{1}$$

where μ presents the mean of the normal distribution, which is the center of the curve and describes the central tendency of the distribution. The closer the value of x is to μ, the higher the probability density is. Conversely, the further away x is from μ, the lower the probability density becomes.

The parameter σ^2 presents the variance of the normal distribution, while σ is the standard deviation of the normal distribution, which determines the shape of the curve and describes the dispersion of the normal distribution. The larger σ is, the more dispersed the data distribution is; the smaller σ is, the more concentrated the data distribution is.

One of the key characteristics of the normal probability distribution is its symmetry around the mean. The curve is divided into three typical intervals: $[\mu-\sigma, \mu + \sigma]$, $[\mu-2\sigma, \mu + 2\sigma]$, and $[\mu-3\sigma, \mu + 3\sigma]$. The probability of a random variable x falling within these intervals is 68.26%, 95.44%, and 99.73%, respectively (shown in Fig. 4). This property makes the normal distribution a powerful tool for analyzing and interpreting data in various contexts.

The process capability index is another important concept that is closely related to the normal probability distribution. It is a measure of the relationship between the

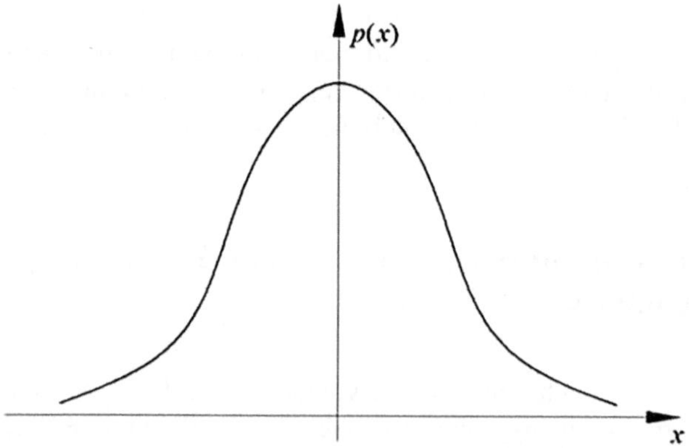

Fig. 3 A normally distributed probability density function curve

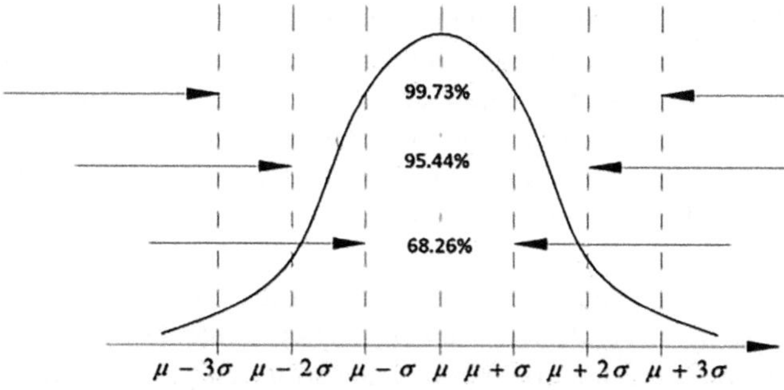

Fig. 4 Characteristics of the normal probability distribution

tolerance requirements and the manufacturing capability. The potential process capability index, denoted as C_p, is calculated by using Eq. (2).[5]

$$C_p = \frac{USL - LSL}{6\sigma} \tag{2}$$

where *USL* and *LSL* represent the upper and lower specification limits of the tolerance, respectively. The process capability index provides valuable insights into the manufacturing process and its ability to meet the specified requirements.

When $C_p < 1$, it indicates that the tolerance requirements exceed the manufacturing capability. In other words, the process is not capable of consistently producing parts within the specified tolerance limits. This situation requires immediate attention and corrective actions to improve the process and bring it within the acceptable range.

When $1 \leq C_p < 1.33$, it suggests that the tolerance requirements match the manufacturing capability. The process is considered marginally capable, and there is room for improvement to ensure a more robust and reliable manufacturing process.

When $1.33 \leq C_p < 1.67$, the process is deemed capable, and the manufacturing capability is sufficient to meet the tolerance requirements consistently. This range indicates a well-controlled and stable manufacturing process.

4 The Reference Value Step Determined by Normal Probability Distribution Analysis

Determining the reference value for the smooth requirements of the aircraft's external surface is a critical task that requires a systematic and data-driven approach. By analyzing the existing manufacturing detection data using normal probability distribution analysis, we can establish realistic and achievable smooth requirements that align with the manufacturing capabilities. The following steps outline the process of determining the reference value for smooth requirements:

Step 1: Random sampling of sufficient manufacturing test data

The first step in the analysis is to collect a representative sample of manufacturing test data. It is essential to ensure that the sample size is sufficient to capture the variability and characteristics of the manufacturing process. Typically, a sample size greater than 100 is recommended to obtain reliable results. The data should be randomly sampled from the existing manufacturing detection records to avoid any bias or selective sampling.

Step 2: The mean value and standard deviation of the sample data

Once the sample data is collected, the next step is to calculate the key statistical parameters, the mean value, and the standard deviation. These parameters provide a quantitative description of the manufacturing process and are essential for subsequent analysis.

Step 3: The normal distribution curve of the manufacturing detection data

Using the calculated mean value and standard deviation, the normal distribution curve of the manufacturing detection data can be plotted. This curve visually represents the probability distribution of the data and helps identify the reference values corresponding to different qualification rates.

Step 4: The reference values corresponding to different qualification rates

By analyzing the normal distribution curve, engineers can determine the reference values corresponding to different qualification rates. The qualification rate refers to the percentage of manufactured parts that meet the specified smooth requirements. Commonly used qualification rates include 68.26%, 95.44%, and 99.73%, which correspond to the intervals of $[\mu-\sigma, \mu + \sigma]$, $[\mu-2\sigma, \mu + 2\sigma]$, and $[\mu-3\sigma, \mu + 3\sigma]$, respectively. These reference values provide a range of options for setting smooth requirements based on the desired level of quality and manufacturing capability.

Step 5: The potential process capability index

To assess whether the initial smooth requirements match the manufacturing capability, the potential process capability index (C_p) is calculated. A C_p value greater than 1 indicates that the manufacturing process is capable of meeting the specified requirements consistently. If the C_p value is less than 1, it suggests that the initial smooth requirements may be too stringent and exceed the manufacturing capability.

Step 6: The initial smooth requirements based on the reference values

If the potential process capability index indicates that the initial smooth requirements are not achievable or exceed the manufacturing capability, adjustments can be made based on the reference values obtained from the normal probability distribution analysis. The reference values corresponding to different qualification rates provide a range of options for setting the smooth requirements. Engineers can select the appropriate reference value that balances the desired surface quality with the practical limitations of the manufacturing process.

By following these steps, engineers can determine the reference value for the smooth requirements of the aircraft's external surface using normal probability distribution analysis. This systematic approach ensures that the smooth requirements are realistic, achievable, and aligned with the manufacturing capabilities. It allows for the optimization of the manufacturing process, reducing the risk of nonconformance and improving the overall quality of the aircraft components.

5 Application Case

We take an aircraft to determine the different requirements of external surface equipment and skin as an example. The initial value of the step requirement is [−2.08, −0.8] mm.

The measured data of the existing aircraft were randomly sampled to obtain the data of 1, 284 samples shown in Table 1.

The mean value of the sample data was −1.8 mm with a standard deviation of 0.289 mm. Normal distribution plots of the sample data are shown in Fig. 5.

As can be determined from the figure, the step data intervals of different pass rates are shown in Table 2.

In addition, the step requires an initial value of [−2.08, −0.8] mm, and the calculated C_p is equal to 0.74. $C_p < 1$, beyond the manufacturing capacity. Correction of the step requirement is required. The tolerance requirement and the manufacturing capacity match $1 \leq C_p < 1.33$ (at least $C_p = 1$), when the reference interval of the step requirement [−2.67, −0.94] mm, and the qualified rate is expected to reach 99.73%. The step requirement of this place is adjusted from [−2.08, −0.8] mm to [−2.67, −0.94] mm.

The step requirements adjusted after the normal probability distribution and process capability index analysis of the existing manufacturing detection data are used to determine the steps between the equipment and skin of an aircraft. Step data of 100 points are recorded, as shown in Table 3.

According to Table 3, according to the requirements of [−2.67, −0.94] mm, the qualified rate of test data is 97%, and the difference from the expected qualified rate of 99.73% when the difference requirement is within 3%.

Moreover, the step requirement is adjusted from [−2.08, −0.8] mm to [−2.67, −0.94] mm, the lower difference is 0.59 mm, the upper difference is 0.14 mm, and the order difference is 0.59 mm, resulting in the influence of aerodynamic resistance increment is less than 0.00001 count, which can be ignored.

The application case demonstrates the effectiveness of using normal probability distribution and process capability index analysis to determine the reference value for the step requirement between the equipment and skin of an aircraft. By analyzing the existing manufacturing detection data, realistic and achievable step requirements can be established, ensuring a balance between the desired surface quality and manufacturing feasibility.

Table 1 Sample data

No.	Steps for surface-installed equipment and skin of an aircraft (unit: mm)											
1	−0.832	−1.88	−1.928	−1.856	−2.312	−1.84	−2.512	−1.936	−2.224	−2.056	−2.24	−1.768
2	−1.936	−2.424	−1.984	−1.4	−1.52	−1.928	−2.01	−1.584	−2. 12	−2.248	−1.736	−2.056
3	−1.608	−2.384	−2.48	−2.264	−2.36	−2.096	−2.176	−2.3	−2.248	−1.896	−2.176	−2.272
4	−2.184	−2.424	−2.048	−1.656	−1.968	−1.488	−0.952	−0.984	−1.104	−1.616	−1.768	−2.264
5	−2.16	−1.92	−2.776	−2.176	−2.264	−1.856	−2.24	−2.096	−1.928	−2.216	−1.584	−1.44
6	−2.064	−2.384	−1.56	−1.712	−1.736	−1.488	−2.24	−2. 112	−2.544	−1.968	−1.776	−1.672
7	−2.176	−1.56	−2.152	−1.912	−1.824	−1.584	−1.84	−1.808	−1.744	−1.84	−2.016	−2.104
……	……	……	……	……	……	……	……	……	……	……	……	……

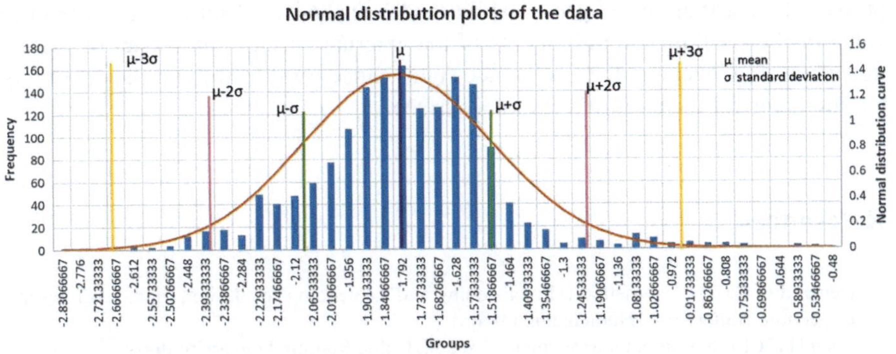

Fig. 5 Normal distribution map of sample data

Table 2 Reference step interval

Steps interval	Lower value (mm)	High value (mm)	Percent of pass
[μ−σ, μ + σ]	−2.6716	−0.9382	99.73%
[μ−2σ, μ + 2σ]	−2.3827	−1.2271	95.44%
[μ−3σ, μ + 3σ]	−2.0938	−1.5160	68.26%

Table 3 Validation data

NO.	Validation data							
1	−1.944	−2.008	−1.856	−1.872	−2.36	−1.976	−2. 144	−2. 112
2	−2.024	−1.968	−1.848	−1.952	−2.136	−1.584	−2.072	−1.336
3	−2.088	−0.888	−1.936	−2.008	−2.296	−1.888	−2.128	−1.752
4	−2.152	−1.96	−1.928	−1.48	−1.576	−1.608	−1.696	−1.544
5	−2.16	−1.816	−2.208	−1.184	−1.648	−1.584	−1.464	−1.52
6	−2.288	−1.8	−2.2	−1.992	−2.24	−2.256	−2.344	−2. 112
7	−2.224	−1.92	−1.944	−1.816	−2.072	−1.848	−2.04	−1.84
8	−1.424	−1.872	−2.36	−1.88	−1.848	−1.768	−1.896	−1.688
9	−2.176	−1.688	−2.104	−1.528	−2.056	−1.672	−2. 112	−1.928
10	−1.92	−0.584	−1.808	−1.448	−2.096	−0.672	−2. 112	−1.744
11	−2.328	−1.976	−2.184	−1.856	−1.976	−1.96	−2.072	−1.616
12	−2.184	−1.856	−2.136	−1.464	−2.024	−1.728	−2.208	−1.864
13	−2.01	−1.864	−2.104	−1.736	–	–	–	–

6 Conclusion

This paper introduced the methods and steps of the normal probability distribution and process capability index analysis of the existing manufacturing detection data and then applied them in the work practice. Taking the reference value of the step between the equipment and skin as an example, the method is demonstrated and the reference values are compared with the actual results. The method of normal probability distribution and processability index analysis to determine the reference value of the continuity requirements of the external surface can reflect the actual manufacturing level and have reference significance for determining the continuity requirements for the aircraft's external surface.

References

1. Henzold G (2021) Geometrical dimensioning and tolerancing for design, manufacturing and inspection. Butterworth-Heinemann, Oxford
2. Zhen H (2014) Six sigma management, 3rd edn. China Renmin University Press
3. Ramadan MA, Al Dhaheri MK, M. M et al (2023) Application of six sigma methodology to enhance the productivity and performance of a hotel in the UAE. TQM J 35:554
4. Shaou Gang Miaou, Jin Syan Chou (2012) Fundamentals of probability and statistics. Gao Li Books
5. Sediyama JAS, Alassane D, Silva RHT, Ribeiro JJ, I. (2023) Consistencies of the capability indices based on the normal probability distribution. Gestão Produção 30:e5722

Open Access This chapter is licensed under the terms of the Creative Commons Attribution-NonCommercial-NoDerivatives 4.0 International License (http://creativecommons.org/licenses/by-nc-nd/4.0/), which permits any noncommercial use, sharing, distribution and reproduction in any medium or format, as long as you give appropriate credit to the original author(s) and the source, provide a link to the Creative Commons license and indicate if you modified the licensed material. You do not have permission under this license to share adapted material derived from this chapter or parts of it.

The images or other third party material in this chapter are included in the chapter's Creative Commons license, unless indicated otherwise in a credit line to the material. If material is not included in the chapter's Creative Commons license and your intended use is not permitted by statutory regulation or exceeds the permitted use, you will need to obtain permission directly from the copyright holder.

Research on the Design and Analysis of Forced Air Cooling for Airborne Modular Integrated Rack

Jianzhao Lu, Wei Fang, Dong Wang, and Youguo Liu

Abstract The avionics equipment encounters changeable and harsh thermal boundary conditions in its task profile, and its thermal performance directly affects electrical function, performance and reliability. The design flow of structure and heat dissipation for electronic equipment was proposed based on the engineering experience. An airborne modular integrated rack was taken as the research object to carry out the related research on design and analysis of forced air cooling. The structural configuration, heat dissipation scheme and thermal control measures of the airborne modular integrated rack were discussed, and the heat dissipation analysis was carried out. The results of simulation calculation and physical prototype test show that the thermal control design is reasonable and feasible. There will be certain reference and guiding significance to the design of similar electronic equipment.

Keywords Integrated rack · Forced air cooling · Thermal design · Thermal analysis

J. Lu (✉) · W. Fang · D. Wang · Y. Liu
The 10th Research Institute of CETC, Chengdu, China

© The Author(s) 2026

Z. Zhang, P. Mercorelli (eds.), *Mechanical Design and Simulation: Insights and Innovations*, Lecture Notes in Mechanical Engineering,
https://doi.org/10.1007/978-981-95-2405-1_7

1 Introduction

With the booming development of avionics in recent years, especially the increasing degree of integration, avionics equipment shows a trend of high heat flux density. Avionics equipment faces variable and harsh thermal environments in its task profiles. Among the reasons for electronic equipment failures, 55% are due to poor heat dissipation [1–4]. When the operating temperature of the components exceed the allowable value, the components are likely to fail, which has a direct impact on the electrical function. Therefore, it is necessary to carry out the related research on thermal design and analysis for avionics equipment to ensure their excellent working performance and high reliability [5, 6].

Most aircraft platforms are not able to provide special thermal control conditions such as liquid cooling system and environmental control air system [7, 8]. The forced air cooling system design is required for most electronic equipment of large heat flux to ensure the requirements of the task are met.

The design flow of structure and heat dissipation for electronic equipment was proposed based on the engineering experience. According to the characteristics and performance requirements of a certain task, the thermal design and analysis of an airborne modular integrated rack were carried out. Based on the detailed introduction of the structural configuration, the thermal design scheme and the design work of structure and heat dissipation were carried out. The thermal simulation analysis of the equipment was carried out by using FloEFD software [9], and the prototype production was carried out based on the analysis results. The results of simulation calculation and prototype test showed that the thermal control design of the equipment is reasonable and feasible.

2 Design Flow of Structure and Heat Dissipation

The design of structure and heat dissipation should fully consider and evaluate the size, mass, thermal interface, environmental conditions, reliability, ergonomics and other multi-dimensional engineering factors of electronic equipment. Combined with engineering experience, the basic design flow of structure and heat dissipation for electronic equipment was proposed in this work, as shown in Fig. 1.

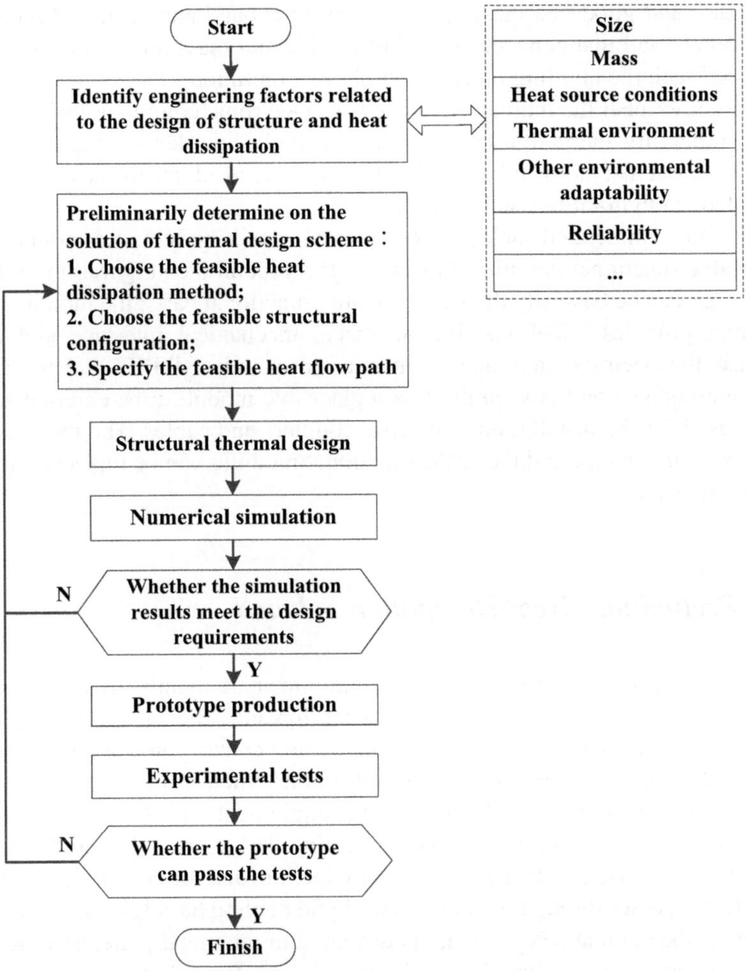

Fig. 1 Design flow of structure and heat dissipation for electronic equipment

3 Structural Thermal Design

3.1 Structural Configuration

The structural configuration of avionics equipment mainly includes the integrated configuration and building block assembly configuration. The integrated configuration takes the line replaceable module (LRM) as the basic unit, which has high integration and good maintainability, is easy to expand and refit, and reduces the development and maintenance costs, but its size and mass are slightly insufficient compared with the building block assembly configuration.

In order to meet the maintenance needs of electronic equipment, while improving the reliability and fault tolerance of increasingly complex and highly integrated electronic systems, more and more avionics equipment adopts the design concept of integrated configuration based on LRM.

A modular integrated rack [10] consists of a rack body and a number of line replaceable functional modules. In general, the structural configuration of the integrated rack can be basically divided into three functional parts: the module installation area provides LRM installation space, mechanical interface and thermal interface; the electrical interconnection area is used to install PCB and provide electrical interconnection between the line replaceable modules; the external interface area is used for the installation of electric couplers and cables. The integrated rack has good maintenance and the LRM functional modules can be replaced quickly in daily maintenance.

3.2 Preliminary Heat Dissipation Scheme

The heat dissipation methods of avionics equipment are mainly divided into natural heat dissipation, forced air cooling, liquid cooling, etc. The natural cooling method mainly dissipates heat through conduction, natural convection and radiation, which has the advantages of reliability, safety and simple structure, but the heat dissipation efficiency is low, and it is only suitable for equipment with low heat. The design scheme of forced air cooling is to fix the heating device or module to the radiators, the heat is transferred to the radiators by conduction, and the cool air takes away the heat when it passes through the radiators. Liquid cooling has higher heat dissipation efficiency, the coolant flow channel is designed in the metal plate, and the heat of the equipment is taken away by the liquid coolant. Forced air cooling and liquid cooling have the advantages of high heat dissipation efficiency, but they need fans or liquid cooling system, and the structures are more complicated and the cost is higher.

According to the task requirements, the modular integrated rack contains a total of 18 electrical functional modules in 17 types. The heating power of the integrated rack is 725 W. The maximum surface heat flux of a single module is 0.075 W/cm^2, the natural cooling method cannot meet the needs of use. Considering that the aircraft platform cannot provide heat dissipation conditions such as environmental control air or liquid cooling system, it is proposed to use the forced air cooling method of simple structure and low cost.

According to whether the cooling air directly passes through the module, the commonly used forced air cooling methods can divided into the following two forms, the main difference is that the heat transfer path is different, as shown in Fig. 2. In contrast, method A has lower thermal resistance than method B, relatively higher efficiency of heat transfer.

Table 1 is the summary of the heating power of modules.

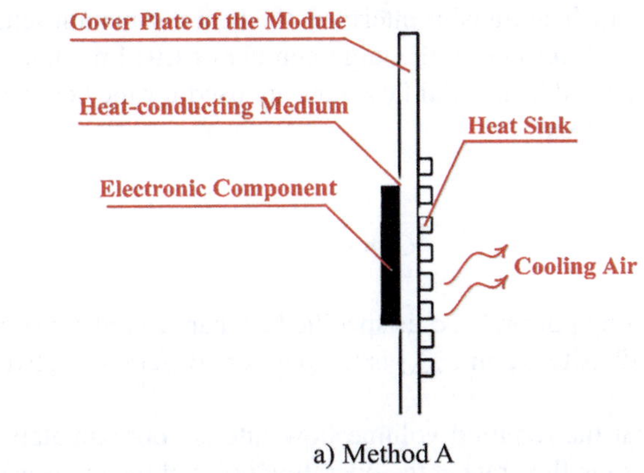

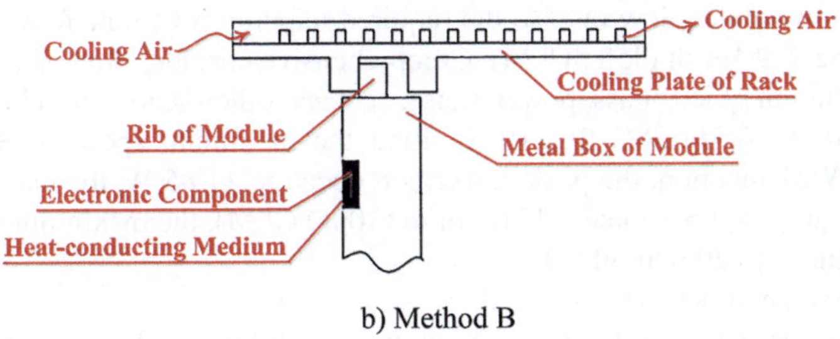

Fig. 2 Heat transfer path of forced air cooling methods. (**a**) Method A. (**b**) Method B

Table 1 Summary of the Heating power of modules

	Summary of Heating Power			
No.	Module name	Number of work modules	Heating power of a single module (W)	Subtotal of heating power (W)
1.	Module 1	1	25	25
2.	Module 2	1	20	20
3.	Module 3	1	40	40
4.	Module 4	1	40	40
5.	Module 5	1	35	35
6.	Module 6	1	65	65
7.	Module 7	1	65	65
8.	Module 8	1	36	36
9.	Module 9	1	30	30
10.	Module 10	2	32	64
11.	Module 11	1	55	55
12.	Module 12	1	70	70
13.	Module 13	1	30	30
14.	Module 14	1	20	20
15.	Module 15	1	25	25
16.	Module 16	1	70	70
17.	Module 17	1	35	35
Summary		18		725

In order to reduce the heat transfer interface, the heat dissipation scheme adopts the method A in Fig. 2. Considering the large number of LRM modules in the rack and the relatively dispersed heat, axial flow fans are used for the heat dissipation of this modular integrated rack.

$$q_m = \frac{Q}{C_p \bullet \rho \bullet \Delta_T} \tag{1}$$

Where, Q is total heating power, W; c_p is specific heat capacity at constant pressure, J/(kg • °C); ρ is air density, kg/m^3; $\triangle_T$ is temperature difference between air inlet and outlet, °C.

It is calculated that the required volume flow rate is approximately 519.4m^3/h (305.5CFM). The volume flow rate at the operating point of the fan is usually 2/3 of its maximum volume flow rate, so the required maximum volume flow rate is estimated to be 779.1m^3/h (458.3CFM). Comprehensive consideration of volume flow rate, installation space, mass, power, noise, cost and other factors, we choose 5 low-cost fan J85FZW511-38G-BC. By limiting the maximum speed to 8000 r/min through PWM function, the peak power is reduced to 11.85 W, the maximum volume flow rate is approximately 171.5 m^3/h (100.9 CFM), the maximum static pressure is reduced to 29.62mmH_2O.

Fan curve chart is shown in Fig. 3.

Through the speed limit, the maximum noise of 5 fans working at the same time is approximately 65.5 dB.

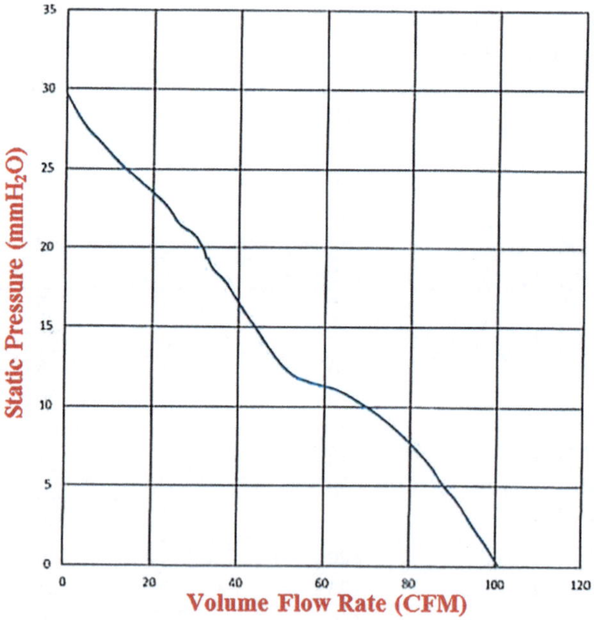

Fig. 3 Fan curve chart (limit the maximum speed to 8000 r/min)

Modular integrated rack is a very complex integrated equipment, it is difficult to get a reasonable thermal design scheme based on empirical formula. After determining the preliminary heat dissipation scheme according to the thermal design engineering experience, more detailed structural thermal design work should be carried out, and heat dissipation analysis and test verification should be carried out to obtain more reasonable heat dissipation design results.

3.3 Structure Design of Modular Integrated Rack

The modular integrated rack mainly consists of front panel, rear panel, upper side plate, lower side plate, left side plate, right side plate, supporting plate, frame, printed circuit board, fan assembly and vibration isolators, as shown in Fig. 4.

The module installation area consists of upper side plate, lower side plate, support board, left side plate and right side plate, which provides the installation space for the functional modules, and provides the mechanical and environmental interfaces for insertion, extraction, locking and heat transfer. The frame supports the installation of printed-circuit board, electric couplers and cables. The fan assembly adopts the quick disassembling method, which is designed with the guided ribs on both sides to guide and position during quick assembly and disassembly. Handles are designed on both sides of the rack for carrying equipment. Between the modules, there is a space of 3 mm–5 mm to form heat dissipation channels.

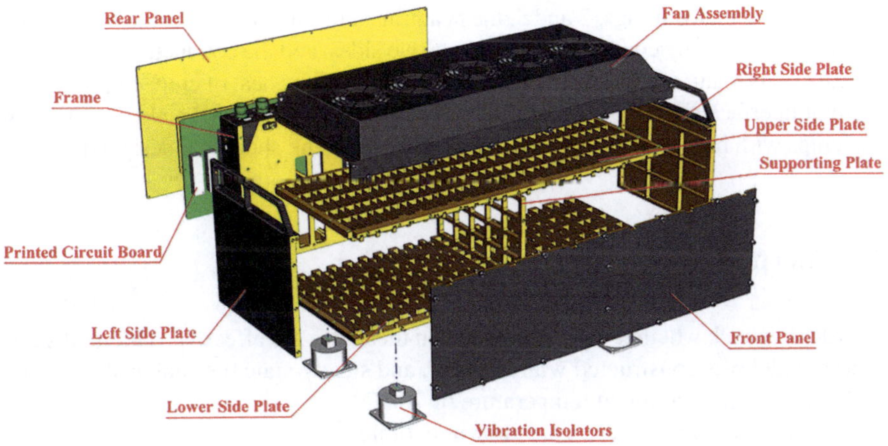

Fig. 4 Three dimensional diagram of the modular integrated rack

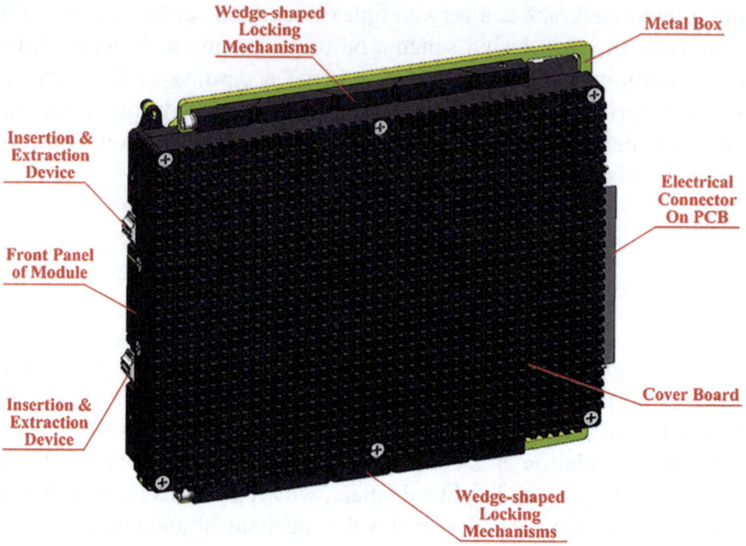

Fig. 5 LRM module structure

3.4 Structure Design of LRM Module

The structure of the LRM electrical functional modules is shown in Fig. 5, which consists of metal box, cover board, circuit board, insertion & extraction device, wedge-shaped locking mechanisms and electrical connector.

It can be seen from Figs. 2 and 5, the heat transfer of the module using method A is mainly carried out by convection through its two sides, and the conduction effect of the module rib is relatively weak. The main transfer path of the heat of chip: chip surface→ thermal interface material→ cover plate of the module and its heat sink→ cooling air. For chips with high heat flux, the corresponding heat sink should be designed.

4 Thermal Analysis

In order to check whether the thermal design measures are effective, a thermal simulation model was constructed with FloEFD, and steady-state thermal analysis of the rack was solved at ambient temperature of 70 °C.

The simulation parameters are shown in Table 2.

Table 2 Simulation parameters

No.	Simulation Parameters			
	Name	Thermal conductivity (W/m/°C)	Emissivity	Contact thermal resistance (°C■m²/W)
1	Printed circuit board	(30, 30, 0.3)	0.6	/
2	Metal box of module (5xxx series aluminum alloy)	117	0.3(internal surface)/0.7 (outside surface)	/
3	Interface between modules and rack ribs	/	/	6×10^{-4}
4	Interface between other metal structures	/	/	1×10^{-3}

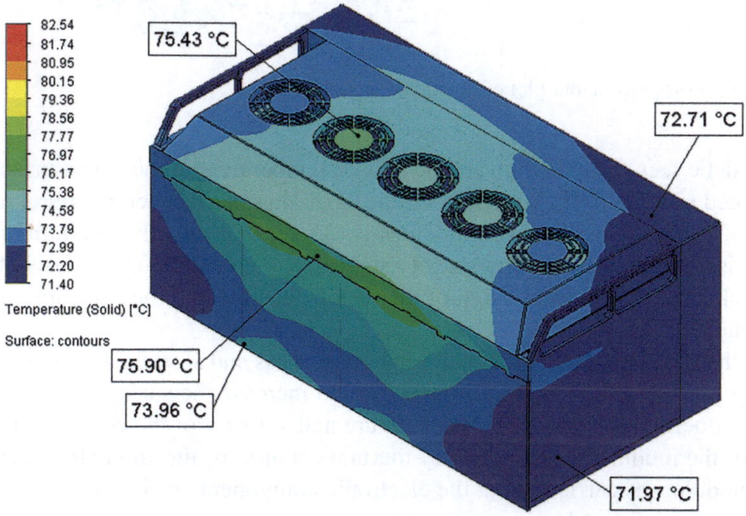

Fig. 6 Temperature contour plot of rack

The steady-state thermal simulation results of the modular integrated rack were obtained. The temperature contour plot is shown in Figs. 6 and 7. The calculation results of flow field are shown as in Fig. 8. The operating point of a fan is shown in Fig. 9.

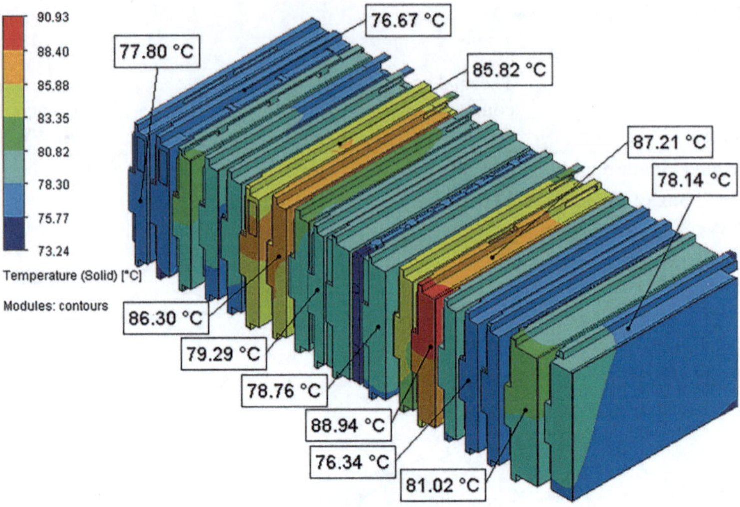

Fig. 7 Temperature contour plot of functional modules

It can be seen from the figures, the highest temperature of the outer surface of the integrated rack is 75.9 °C; module 12 has the highest surface temperature, reaching 88.94 °C; wind speed between modules is (3.6 ~ 10.6) m/s; the operating point of the fan is ideal; the volume flow rate of a single fan is about 0.0303 m^3/s(109.08 m^3/h). In conclusion, the airduct design of the rack is feasible and the selection of fans is reasonable.

On the premise of meet the requirements of mass and size, the heat sinks of high heating power module should be improved to increase the surface convective heat transfer coefficient. Taking the temperature and velocity of the cooling air on both sides of the module as the boundary thermal conditions, the simulation analysis of each module was carried out at the electronic component level.

The prototype production of the integrated rack was carried out, and the prototype passed the environmental test successfully, which verified the reasonability of the thermal design scheme.

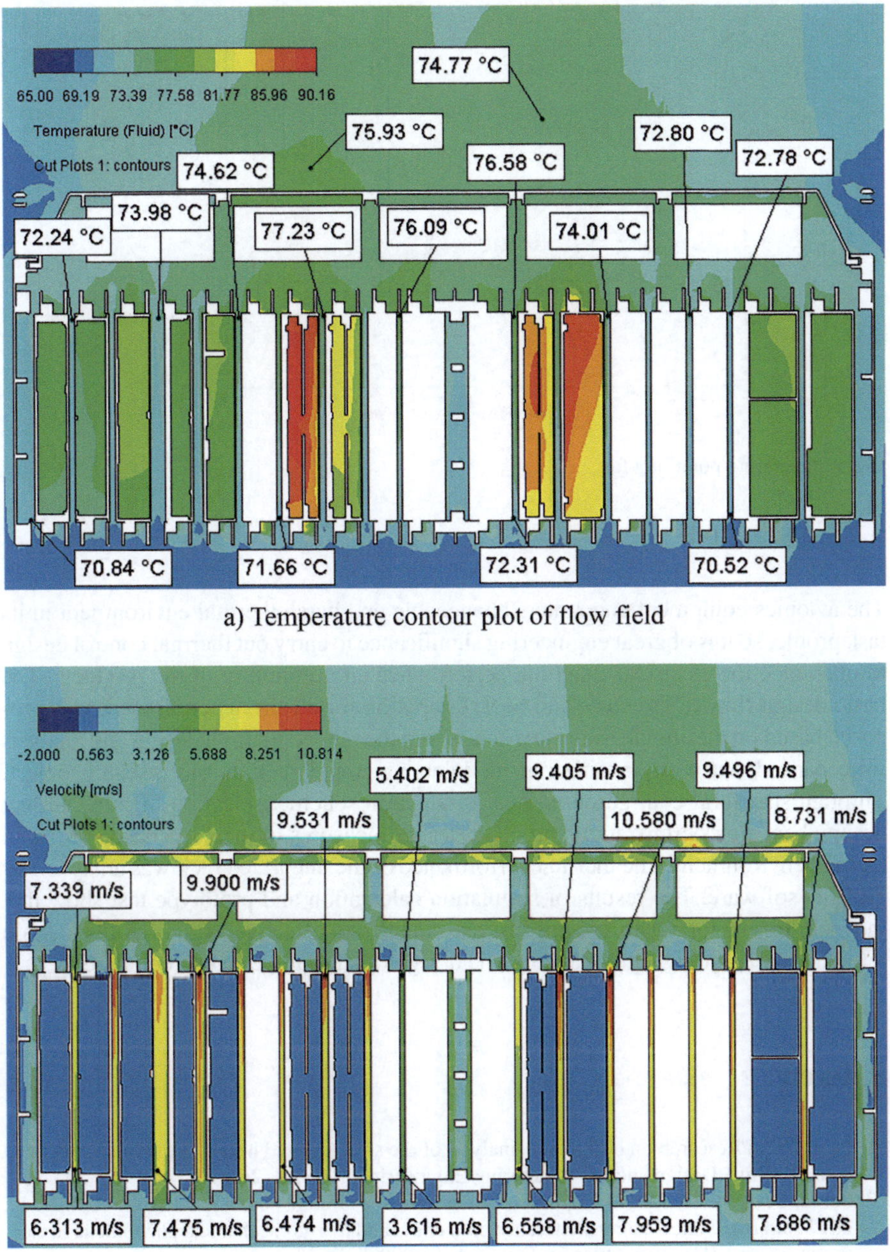

a) Temperature contour plot of flow field

b) Velocity contour plot of flow field

Fig. 8 Calculation results of flow field. (**a**) Temperature contour plot of flow field. (**b**) Velocity contour plot of flow field

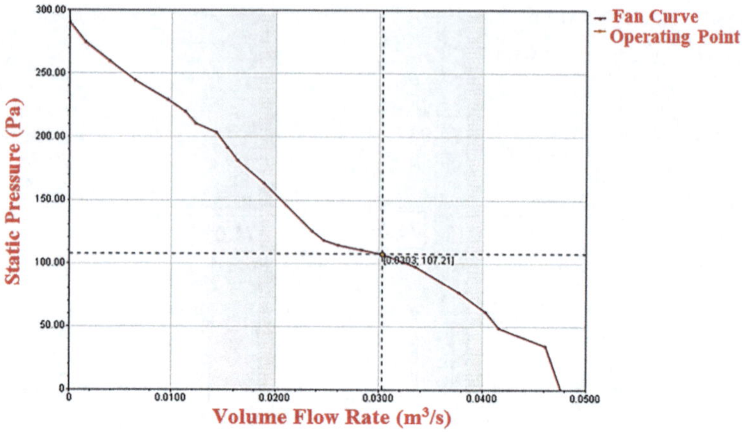

Fig. 9 Operating point of a fan

5 Conclusions

The avionics equipment encounters changeable and harsh thermal environment in its task profile, so it is of great engineering significance to carry out thermal control design to guarantee the electrical function, performance and reliability of the products. The basic design flow of structure and heat dissipation for electronic equipment was proposed based on the engineering experience, and the related thermal design and analysis were carried out with an airborne modular integrated rack as the research object. Through structural configuration, heat dissipation scheme, selection of fans, airduct design, heat sink optimization and other design measures to meet the needs of use in thermal environment. The thermal performance of the integrated rack was analyzed by FloEFD software. The results of simulation calculation and prototype test show that the measures of forced air cooling are reasonable and effective, which will be certain reference and guiding significance to the design of similar electronic equipment.

References

1. Lu J (2022) Research on design and analysis of the structure and heat of a modular integrated rack. Machine Design and Manufacturing Engineering 51(1):23–26
2. Pal D, Severson M (2017) Liquid cooled system for aircraft power electronics cooling. In: 2017 16th IEEE intersociety conference on thermal and thermomechanical phenomena in electronic systems (ITherm), Orlando, FL, USA, pp 800–805
3. Pedram M, Nazarian S (2006) Thermal modeling, analysis, and management in VLSI circuits: principles and methods. Proc IEEE 94(8):1487–1501
4. Liu X, Li B et al (2017) Structure Design of Forced air Cooling for airborne equipment of a data link. Shipbroard Electr Countermeasure 40:109–113
5. Cheng H (2022) Large-scale integrated modular rack air cooling design. In: 2022 Global reliability and prognostics and health management (PHM-Yantai), Yantai, China, pp 1–4

6. Yan N, Guan Z et al (2021) Design and verification of high heat dissipation structure of airborne air-cooled rack. Electro-mech Eng 37:28–34
7. Li M (2022) Thermal design and simulation optimization of an airborne equipment structure. In: 2022 Global Reliability and Prognostics and Health Management (PHM-Yantai), Yantai, China, pp 1–6
8. Cheng Huang (2020) Design of large integrated rack liquid cooling system. Environ Technol 38(4):27–30
9. Jia T, Zhang J, Wang L (2014) The thermal simulation of electromechanical platform system. In: 2014 IEEE conference and expo transportation electrification Asia-Pacific (ITEC Asia-Pacific), Beijing, China, pp 1–4
10. Lu J, Fang W, Hu J, Wang D (2022) Thermal design and transient analysis of liquid cooling integrated rack in spacecraft cabin. In: 2022 IEEE 5th International conference on electronics technology (ICET), Chengdu, China, pp 521–525

Open Access This chapter is licensed under the terms of the Creative Commons Attribution-NonCommercial-NoDerivatives 4.0 International License (http://creativecommons.org/licenses/by-nc-nd/4.0/), which permits any noncommercial use, sharing, distribution and reproduction in any medium or format, as long as you give appropriate credit to the original author(s) and the source, provide a link to the Creative Commons license and indicate if you modified the licensed material. You do not have permission under this license to share adapted material derived from this chapter or parts of it.

The images or other third party material in this chapter are included in the chapter's Creative Commons license, unless indicated otherwise in a credit line to the material. If material is not included in the chapter's Creative Commons license and your intended use is not permitted by statutory regulation or exceeds the permitted use, you will need to obtain permission directly from the copyright holder.

Investigation of the Impact of Connection Stiffness on Random Vibration Testing

Liuyuan Tao

Abstract This investigation aimed to examine the impact of connection stiffness on random vibration testing of heat sink products. The approach involved developing a finite element model of the test piece and fixture, and conducting simulated studies on random vibration testing under varying connection stiffness conditions, in conjunction with actual test results. The findings demonstrated that by adjusting the stiffness of the test piece and fixture models based on real test data, the calculated results closely aligned with experimental outcomes, effectively rectifying inaccuracies in finite element analysis. In summary, it was observed that different forms of connection led to diverse connection stiffness, which emerged as a primary factor influencing overall modal characteristics of both the test piece and fixture as well as significantly impacting simulation outcomes.

Keywords Connection stiffness · Random vibration

1 Introduction

Finite Element Simulation Analysis: Enhancing Accuracy in Vibration Testing. Vibration testing using finite element simulation has become integral to the product development process due to its ability to identify potential defects early on while providing a foundation for subsequent measurements and monitoring activities. By establishing comprehensive finite element models incorporating dynamic calculation methods such as modal analysis, frequency response analysis, and random vibration analysis for both fixtures and test pieces (including the vibration table model where applicable), numerical simulations enable cost savings across testing equipment operation costs, product production costs, and human resources while expediting the research cycle.

L. Tao (✉)
Yonghong Aviation Machinery Co., Ltd., Guiyang, Guizhou Province, China

© The Author(s) 2026

Z. Zhang, P. Mercorelli (eds.), *Mechanical Design and Simulation: Insights and Innovations*, Lecture Notes in Mechanical Engineering,
https://doi.org/10.1007/978-981-95-2405-1_8

Vibration test simulations have revealed discrepancies, leading to failures during actual tests. This highlights significant differences between simulated results and experimental outcomes. Our study identifies varying connection structures when designing test fixtures or selecting installation forms for impacting simulation results, a critical factor contributing to disparities between analytical predictions and experimental findings.

2 Foundational Theory

Random vibration analysis typically employs two methods [1]: time-domain analysis, which is based on statistical counting, and frequency-domain analysis, which relies on the power spectral density function. Frequency-domain analysis is commonly utilized [2], where the PSD function at critical points of the structure can be obtained through finite element modeling or testing. This information is then combined with the stress probability density function (PDF) to select appropriate cumulative damage criteria and failure criteria for estimating structural fatigue life.

The stress response PSD function can be calculated by inputting the load's PSD and the frequency response function (FRF) in practical engineering [3, 4].

$$\mathrm{Sout}(\omega) = \mathrm{H}(\omega) 2 \sin(\omega) \tag{1}$$

where Sout(ω) denotes the stress response PSD function; Sin(ω) denotes the known input PSD function.

The random vibration stress σ_{RMS} can be calculated by using the stress response PSD function as follows:

$$\sigma_{RMS} = \mathrm{Sout}(\omega) d\omega \tag{2}$$

where H(ω) is the inherent characteristic of the structure, which reflects the amplitude and phase changes of the steady-state response of the mechanism under a harmonic external excitation.

$$\mathrm{H}_{\omega} = \mathrm{k} - \mathrm{m}\omega^2 + jc\omega \tag{3}$$

where K represents the structural stiffness; M represents the structural mass; C represents the damping.

From the above formula, it is evident that, given a known input load, the stress response is influenced by three factors: mass, stiffness, and damping. Structural modeling software currently provides relatively accurate estimations of mass; thus, the distribution of mass in the finite element model has minimal impact on the accuracy of analysis results. Consequently, only 2.5% of structural damping is typically considered during the analysis [5]. When the stiffness of the model structure closely aligns with that of the test piece's actual situation, random vibration analysis results will closely approximate test results. Therefore, a finite element model is established for different connection forms to calculate their modal values and compare them with experimental results to investigate how simulation models' representation of stiffness influences random vibration when structures adopt different connections. Furthermore, this study aims to analyze how connection stiffness affects random vibration.

3 Typical Forms of Connection

The manufacturing processes for trial fixtures include overall machining, casting, welding, bolt connection, and adhesion. Overall, machining is known for its high precision, short cycle times, and excellent vibration transmission properties, making it suitable for the production of small trial fixtures. Casting can be used to fabricate intricate large fixtures with robust rigidity and damping characteristics. Welding is appropriate for fixtures with relatively simple structures and low cost. While bolted fixtures may have inferior dynamic characteristics, they are easy to install and disassemble. Adhesion is primarily utilized in the production of small fixtures but is generally not employed.

3.1 Monolithic Molded Structure

3.1.1 Structural and Finite Element Modeling

The monolithic molded structure is constructed from sheet metal and machining or additive manufacturing, which can be regarded as a monolithic material. As depicted in Fig. 1, a partition is formed through sheet metal forming and stamping, utilizing 1Cr18Ni9Ti stainless steel as the material. It possesses a modulus of elasticity of 199 GPa and a density of 7.9 g/mm^3. Utilizing ANSYS for diaphragm modeling and

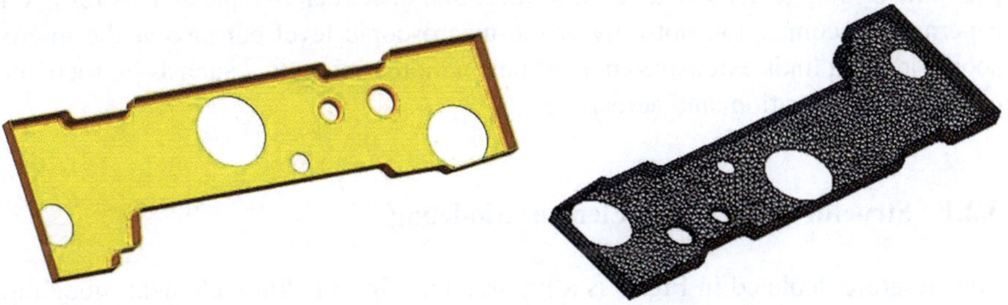

Fig. 1 Schematic of shear wall structure and its finite element model

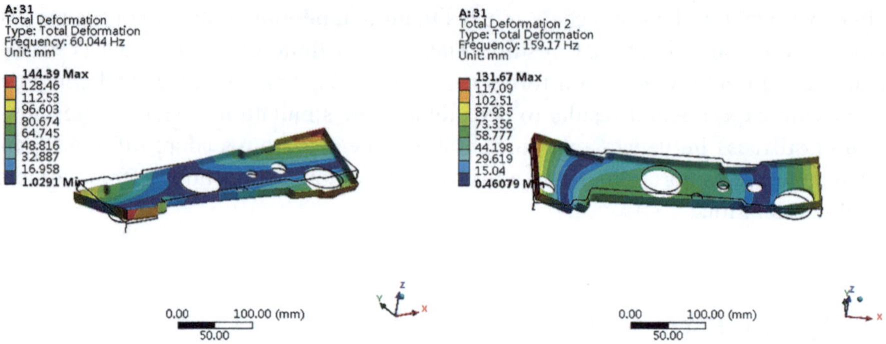

Fig. 2 The first two modalities of the shear wall structure

modal analysis, the first two natural frequencies in the free state were determined to be 60.0 Hz and 159.2 Hz, as illustrated in Fig. 2.

3.1.2 Empirical Comparative Analysis

The modal frequencies of the diaphragm were determined through modal testing using an Impact Hammer, as depicted in Figs. 3 and 4. The obtained test results indicate frequencies of 55.63 Hz, 107.50 Hz, and 163.13 Hz, which are slightly higher than the calculated values at 107.50 Hz. This discrepancy may be attributed to the rigid support system. A comparison between finite element analysis and experimental results reveals errors of 9.1% and 2.5%, respectively, for the diaphragm's behavior. Despite these discrepancies, the calculated results align with the requirements for practical engineering applications, demonstrating higher accuracy in predicting overall structural behavior when isolated from other connections.

3.2 *Welded Structural Assembly*

Welding, also referred to as joining engineering, involves the application of heat, pressure, or both to achieve an atomic-level bond between two parts. This results in a permanent connection not only at the macroscopic level but also at the microscopic level. It finds extensive applications in industrial sectors such as petrochemicals, power generation, and aerospace.

3.2.1 Structural and Finite Element Modeling

The structure depicted in Fig. 5 is a fixture, and Fig. 6 is finite element modeling. We predominantly utilize welding connections, which are fabricated from aluminum alloy with a modulus of elasticity of 71 GPa and a density of 2.7 g/mm^3. The

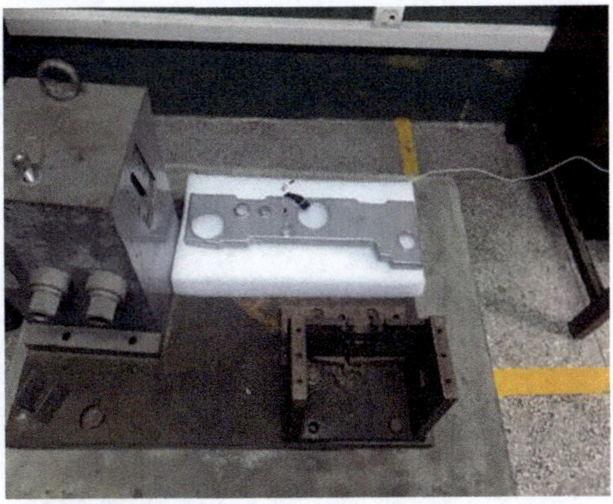

Fig. 3 Location of test measurement points

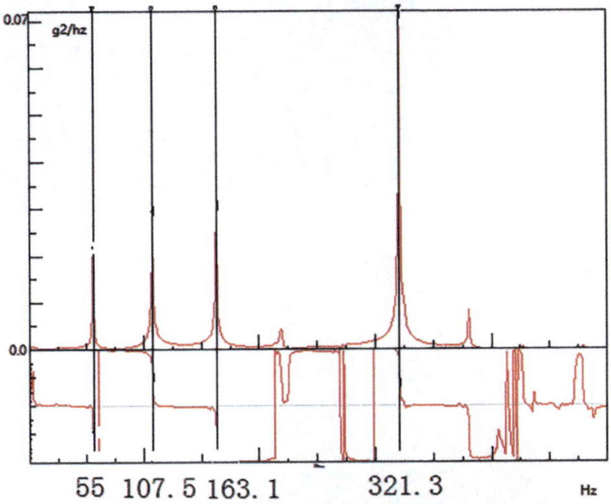

Fig. 4 Test findings

finite element model employs the contact (Bonded) feature provided by ANSYS to simulate the welding connection between the two components. The modal analysis results for the fixture, derived from the calculations, are presented in Figs. 7, 8, 9 and 10, showing the first four modal frequencies as 200.0 Hz, 350.8 Hz, 431.9 Hz, and 508.2 Hz.

3.2.2 Empirical Comparative Analysis

The data in Table 1 indicates the influence of welded joints on the inherent characteristics of the structure is investigated through experimental and analytical comparisons. Modal frequencies are determined by using a 0.5 g sinusoidal sweep method, with acceleration sensors strategically placed based on the structure's mode

Fig. 5 Location of test measurement points

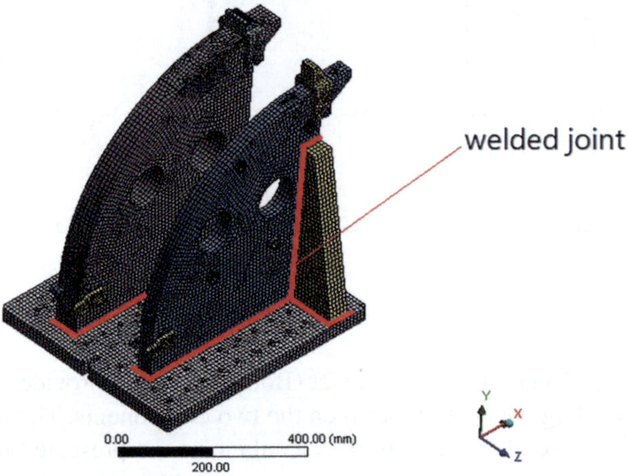

Fig. 6 Finite element modeling

shape diagram, and the test results are depicted in Figs. 11 and 12. The test results indicate that the welded connection permanently joins the two parts into a single entity, obviating the need for any special treatment when simulating the model.

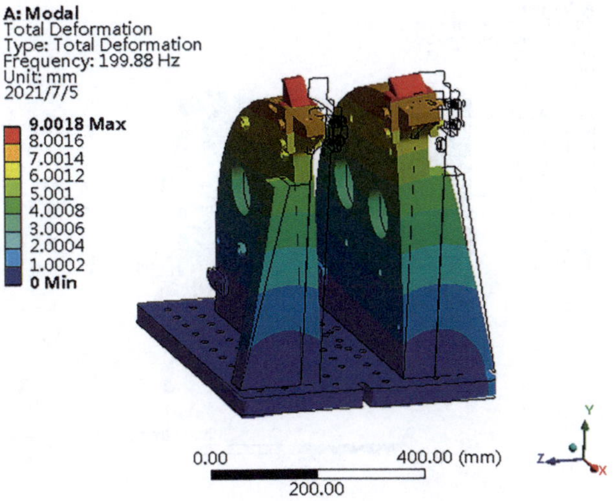

Fig. 7 First model

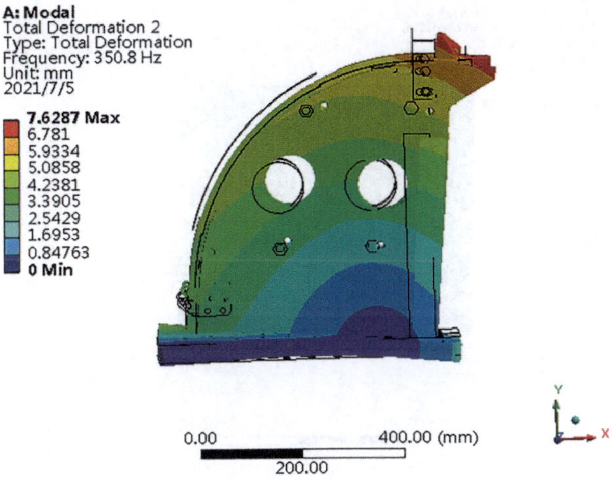

Fig. 8 Second model

3.3 *Bolted Connection Structure*

Bolt connections often exhibit nonlinear connection stiffness due to clearance fit, posing challenges for finite element models in simulation. This paper presents the finite element simulation and experimental comparison of the elliptical hole bolt connection used in the test fixture design process.

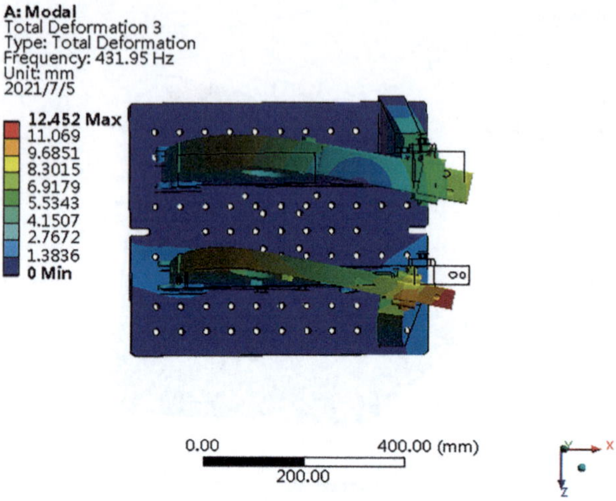

Fig. 9 Third model

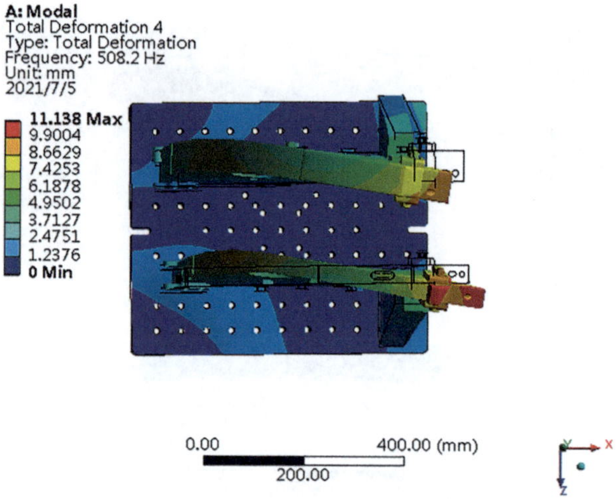

Fig. 10 Fourth model

Table 1 Computational and experimental error analysis of welded structural assembly

Modal order	Computed value (Hz)	Measured value (Hz)	Error (%) [a]
1	200.0	210.5	5.0
2	350.8	325.0	−7.9
3	431.9	447.6	3.5

[a] Error = (Computed value − Measured value)/Measured value × 100

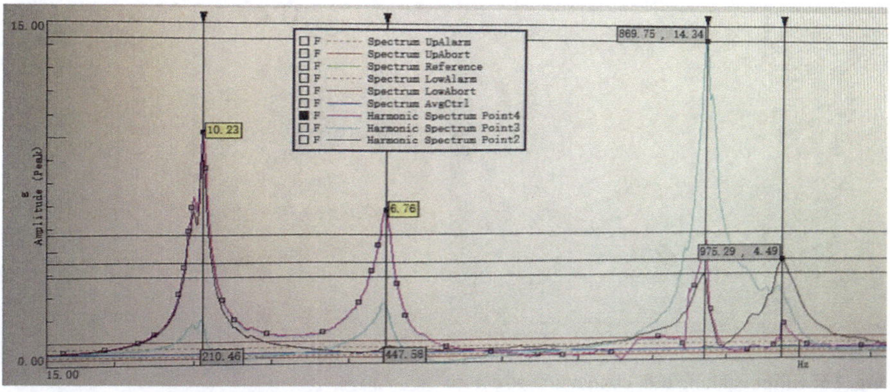

Fig. 11 Results of first and third frequency sweep experiment

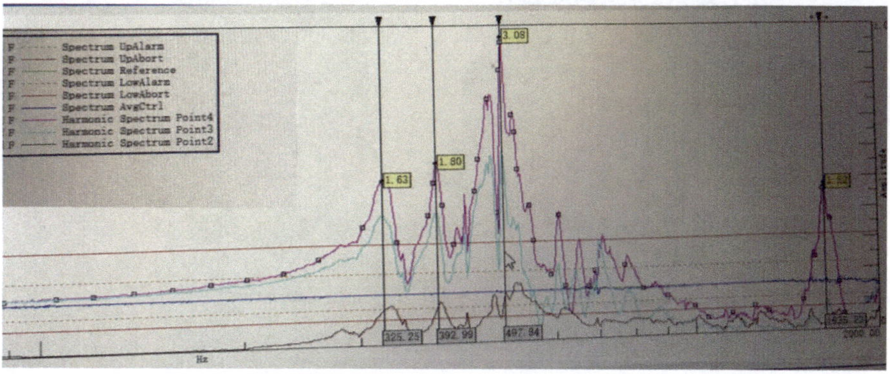

Fig. 12 Results of second and fourth frequency sweep experiment

3.3.1 Structural and Finite Element Modeling

The overall structure is depicted in Figs. 13 and 14, with both the product and the fixture constructed from aluminum alloy featuring a modulus of elasticity of 71 GPa and a density of 2.8 g/mm^3. To facilitate test piece installation, elliptical holes were incorporated into the fixture structure, providing ample clearance and adjustable positioning. A finite element model was developed for the structure, with bolts and L support tightly fitting in the X and Y directions. The model employs the same contact method as welded connections, establishing the Z direction through frictional force between parts. Model 1 utilizes the first modeling method, assuming a non-sliding connection (6 degrees of freedom), while Model 2 adopts the second method, assuming no connection in the Z direction (remaining 5 degrees of freedom). Modal analysis for each model was separately conducted, with results presented in Figs. 15, 16, 17, 18, 19 and 20.

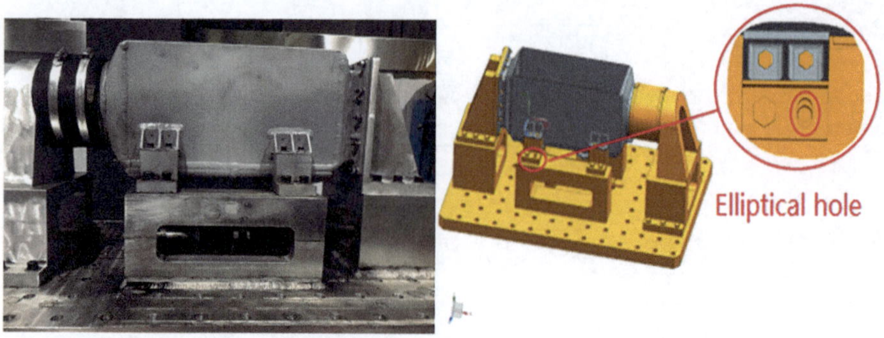

Fig. 13 Elliptical hole bolt connection structure

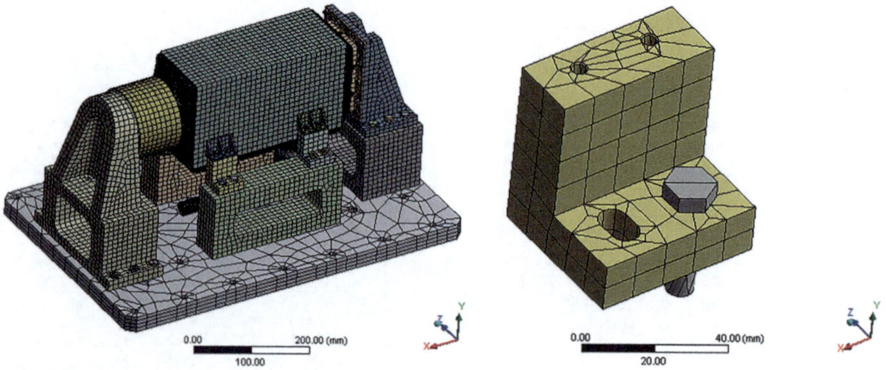

Fig. 14 Finite element modeling

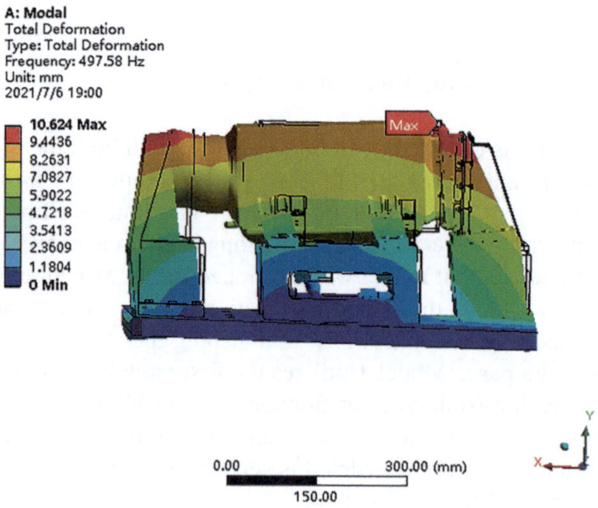

Fig. 15 First model of Mode 1

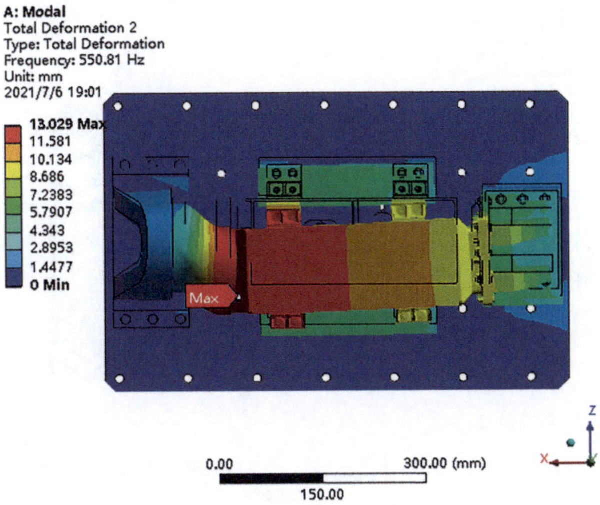

Fig. 16 Second model of Mode 1

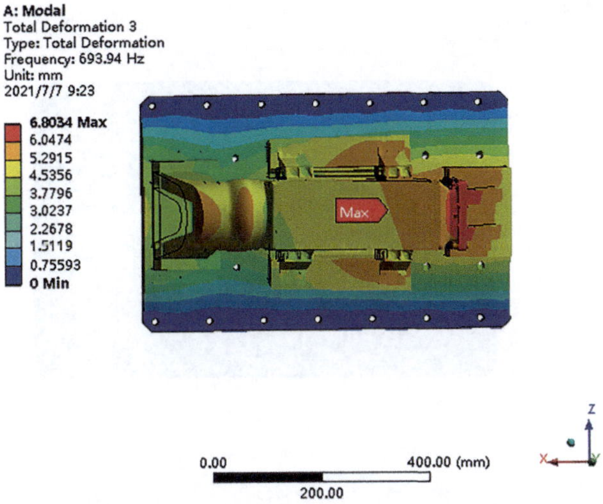

Fig. 17 Third model of Mode 1

The modal analysis results of the two-state model indicate that, when connected, the frequencies of the six degrees of freedom are 497.6 Hz, 550.8 Hz, 693.9 Hz, and 757.9 Hz. Upon releasing the Z-axis connection, the first mode remains almost unchanged at 489.7 Hz, while the second mode decreases to 359.9 Hz, the third mode becomes 585.8 Hz, and the fourth mode is measured at 696.0 Hz.

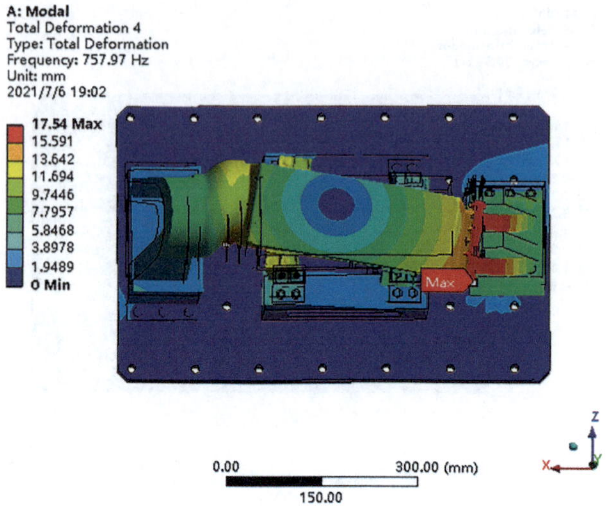

Fig. 18 Fourth model of Mode 1

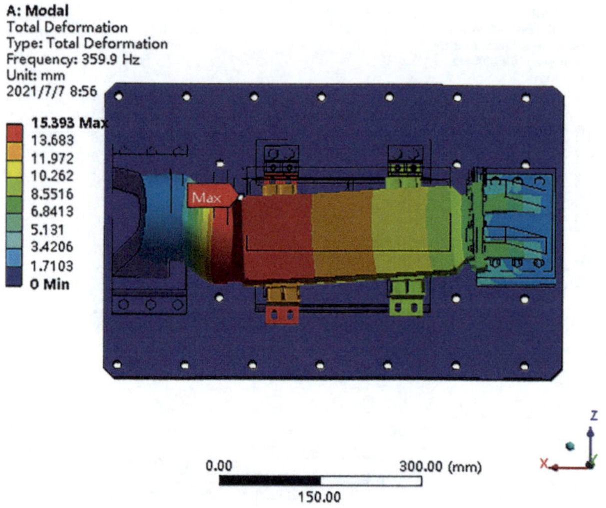

Fig. 19 Fifth model of Mode 1

3.3.2 Empirical Comparative Analysis

Through experimental and comparative analysis, the study investigates the impact of bolt gaps on the structural inherent characteristics and validates the effectiveness of two modeling methods. The structure's modal frequencies are examined by using a 0.5 g sinusoidal sweep method, with acceleration sensors strategically placed based on the structure's mode shape diagram, as depicted in Figs. 16 and 17.

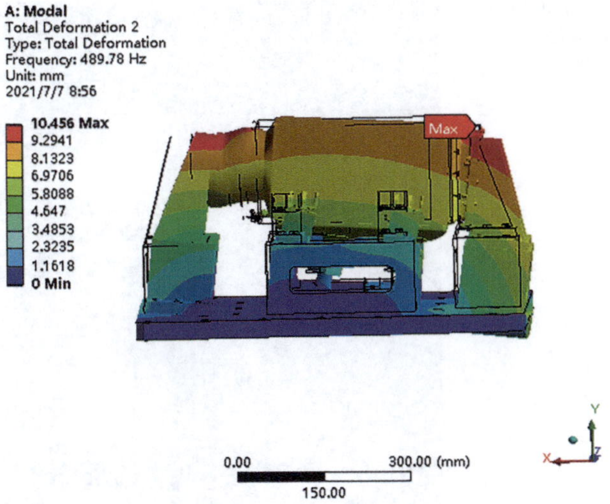

Fig. 20 Sixth model of Mode 1

Fig. 21 Validation of the sensor placements for the second and fourth models

The test results are presented in Figs. 21, 22, 23, 24, indicating modal frequencies of 345.2 Hz, 473.2 Hz, 543.6 Hz, and 789.3 Hz. A comparison between the test results and the analysis findings is detailed in Table 2, revealing a significantly larger error for Model 1 than for Model 2. This suggests that the connection stiffness provided by friction force in the gap bolt connection direction may be disregarded; hence, modifications to the model based on test results are necessary to achieve more accurate calculation outcomes [6].

Fig. 22 Validation of the sensor placements for the first and third models

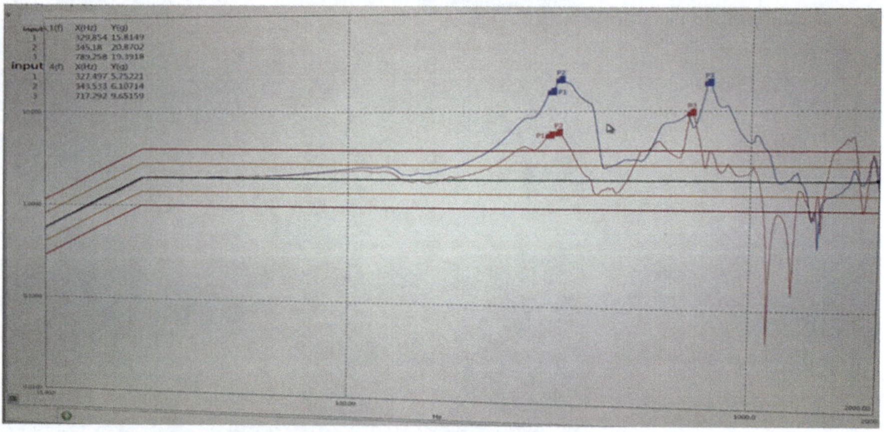

Fig. 23 Results of first and third frequency sweep experiment

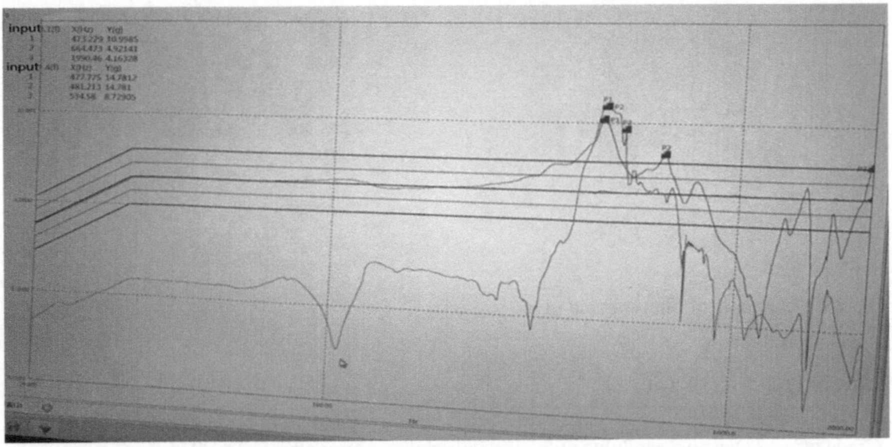

Fig. 24 Results of second and fourth frequency sweep experiment

Table 2 Computational and experimental error analysis of bolted connection structure

Modal order	Model 1 computed value (Hz)	Model 2 computed value (Hz)	Measured value (Hz)	Error1 (%) [a]	Error2 (%) [b]
1	497.6	359.9	345.2	−44.1%	−4.3%
2	550.8	489.7	473.2	−16.4%	−3.5%
3	693.9	585.8	543.6	−27.6%	−7.8%
4	757.9	696.0	789.3	4.0%	11.8%
5	497.6	359.9	345.2	−44.1%	−4.3%
6	550.8	489.7	473.2	−16.4%	−3.5%

[a] Error1 = (Model 1 computed value − Measured value)/Measured value × 100
[b] Error2 = (Model 2 computed value − Measured value)/Measured value × 100

4 An Investigation into the Impacts of Random Vibration Testing

In Sect. 3.3, the structure contains two pipe fittings that require pressurization during random vibration testing. As a result, two L-shaped caps (in Fig. 25) were installed and secured using clamps. The axial movement of the caps was constrained by the frictional force between the clamps and the caps.

The product is constructed from 6061 aluminum alloy, and its mechanical properties are detailed in Table 3 of the "China Materials Engineering Canon".

Fig. 25 Schematic of pipe cap and plug

Table 3 Properties of materials

Grade of material	Density g/cm^3	Modulus of elasticity (GPa)	Poisson's ratio	Yield strength (MPa)	Tensile strength (MPa)	Fatigue limit (MPa)
6061	2.7	72	0.33	262	310	97

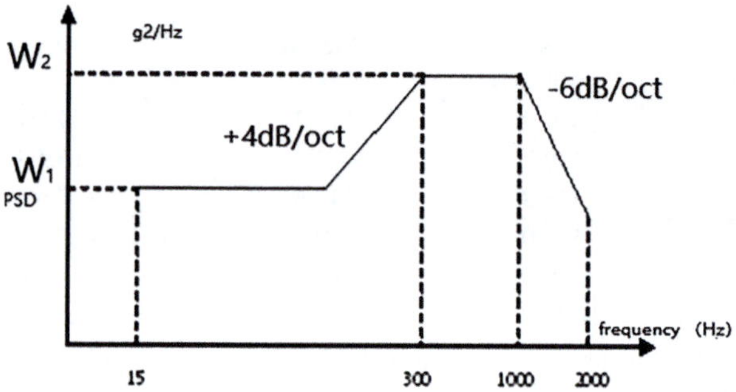

Fig. 26 Vibration test power spectrum

In accordance with the specified requirements, the product will undergo endurance vibration tests as per GJB150.16A-2009, Methods for Environmental Testing of Military Equipment, Part 16: Vibration Test. The product will be subjected to 30 hours of endurance testing in each axis (X, Y, and Z), and it must demonstrate no signs of damage or impairment that would affect its functionality. The input load for the test is Sin(ω), as depicted in Fig. 26.

4.1 The Initial Phase of Computational Analysis

It is assumed that full clamping of the L-shaped plug provides radial and axial restraint as well as free lateral rotation, as depicted in Fig. 27. Under these connection stiffness conditions and assuming a structural damping factor of 2.5%, the

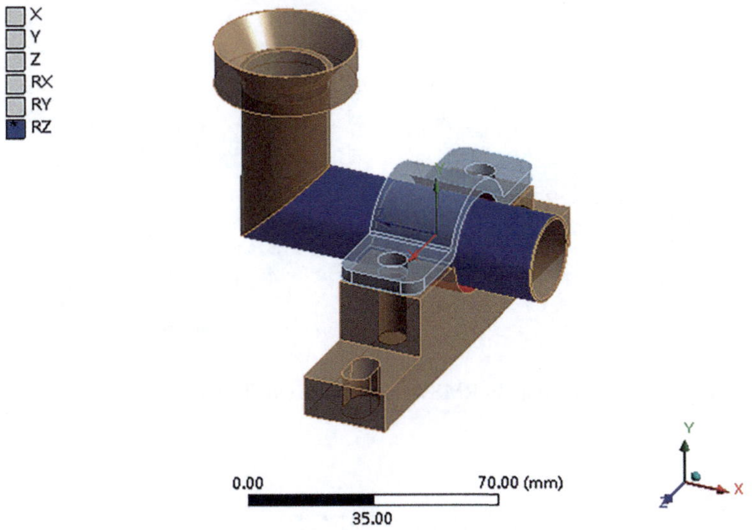

Fig. 27 L-shaped plug connection with a single degree of freedom, specifically azimuthal RZ freedom only

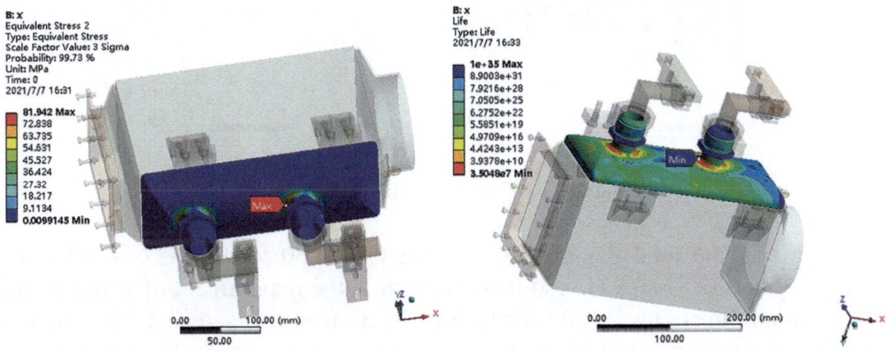

Fig. 28 X-axis random vibration 3σ RMS stress and life cloud char

results of random vibration analysis for the product are illustrated in Figs. 28, 29 and 30.

The operational lifespan of the nozzle tip in three directions is as follows: X-axis—9, 735.6 h (35, 048, 000 seconds), Y-axis—14, 525.0 h (522, 900 s). The total damage calculation for the three-axis 90 h test considering damage accumulation is as follows [7, 8]:

$$\mathrm{D} = \mathrm{D}_x + \mathrm{D}_y + D_z \tag{4}$$

where D denotes the total damage; D_x denotes the X-axis damage; D_y denotes the Y-axis damage; D_z denotes the Z-axis damage.

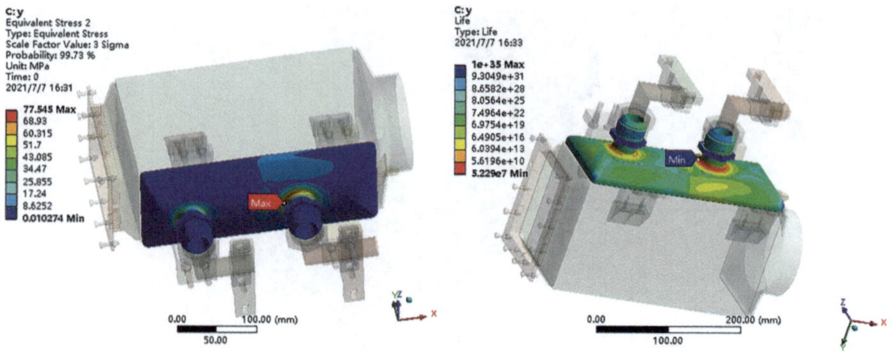

Fig. 29 Y-axis random vibration 3σ RMS stress and life cloud char

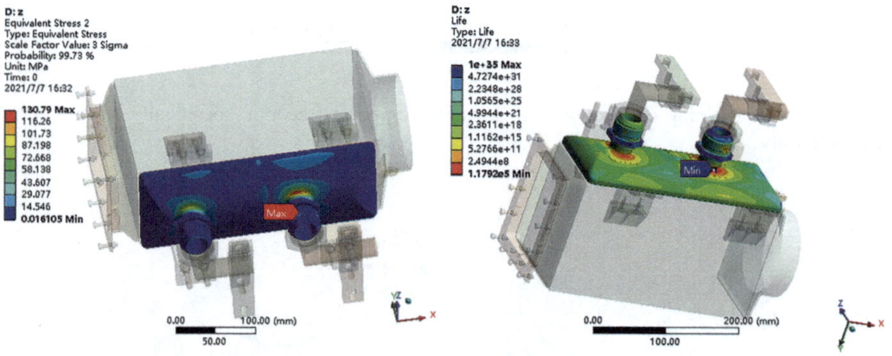

Fig. 30 Z-axis random vibration 3σ RMS stress and life cloud char

Given that the total damage D is calculated as 30 × 19, 735.6 + 30 × 11, 452.5 + 30 × 132.8 = 0.921, and it is less than 1, the requirement of withstanding testing in all directions for 30 h without being destroyed is satisfied. The analysis indicates that the structural fatigue life meets the specified criteria. However, despite only undergoing a two-hour test, the nozzle detached (in Fig. 31). Discrepancies between finite element calculations and actual test results reveal a significantly shorter actual lifespan compared to the calculated one.

4.2 *The Subsequent Computational Analysis*

When considering the axial connection stiffness of the L-shaped plug, it is prudent to release the Z-axis degree of freedom conservatively during estimation, as there may be a slightly lower stiffness in practice (which can be simulated by spring elements). The results of random vibration analysis for the product are illustrated in Figs. 32, 33, 34 and 35.

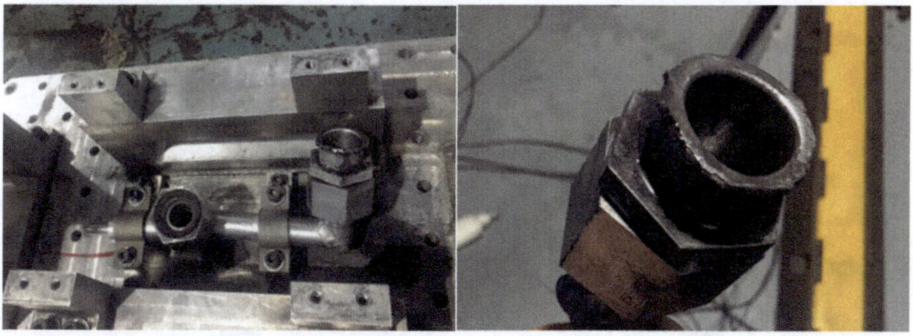

Fig. 31 Product test failure diagram

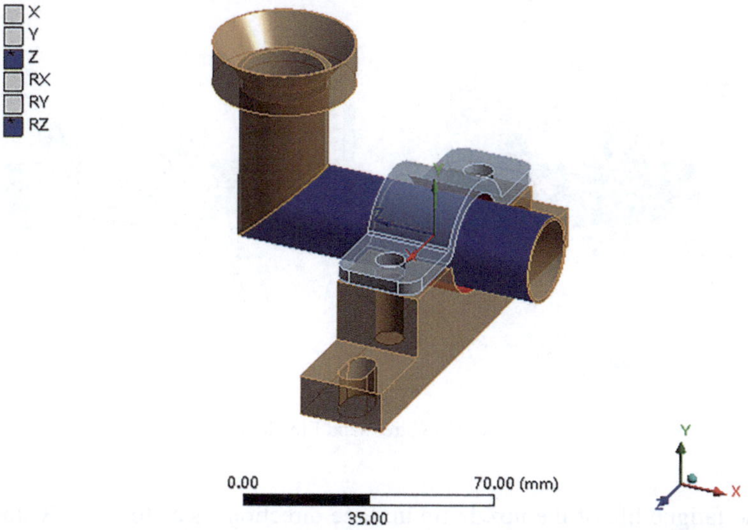

Fig. 32 L shaped plug connection with a single degree of freedom, specifically azimuthal RZ freedom only

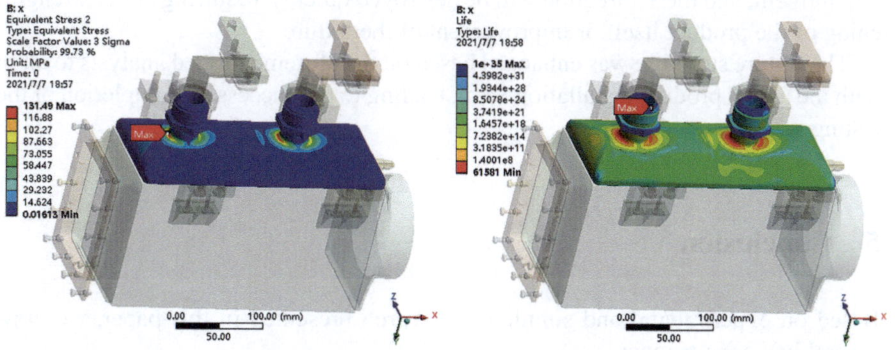

Fig. 33 X-axis random vibration 3σ RMS stress and life cloud char

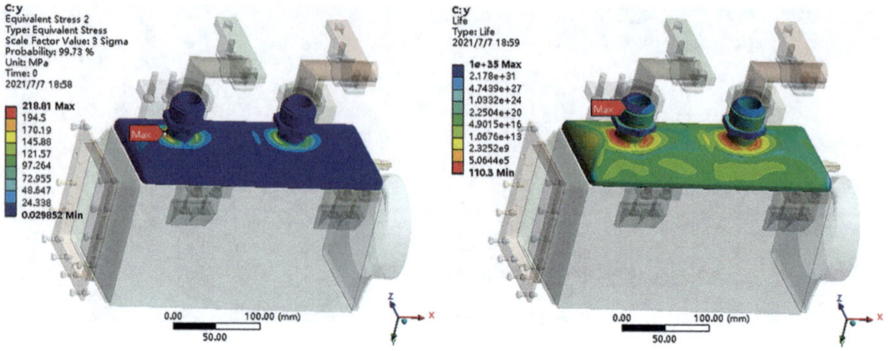

Fig. 34 Y-axis random vibration 3σ RMS stress and life cloud char

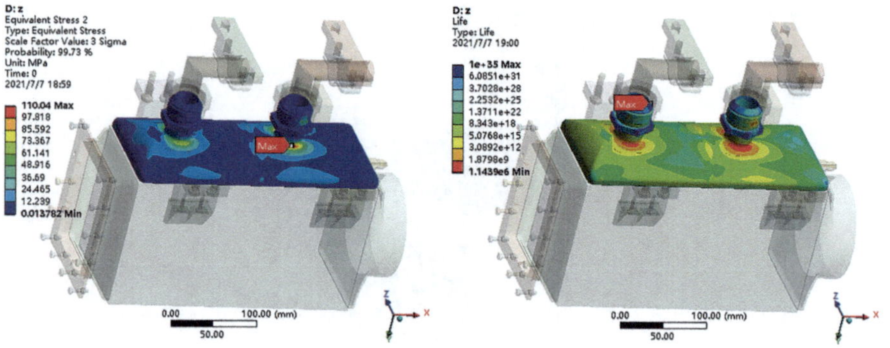

Fig. 35 Z-axis random vibration 3σ RMS stress and life cloud char

The fatigue life of the nozzle tip in three directions is as follows: X direction - 17.1 h (61, 581 s), Y direction - 0.03 h (110.3 s), and Z direction - 317.8 h (1, 143, 900 s). The calculation results show that the product only meets the 30-h durability test requirement in the Z direction, while the X and Y directions do not meet the requirement, and the Y direction will be destroyed quickly, requiring either strengthening of the product itself or improvement of the fixture.

The fixture structure was enhanced based on the aforementioned analysis to align with the actual product installation form, leading to the successful completion of the testing process.

5 Conclusion

Based on experimental and simulation research presented in this paper, we draw several key conclusions:

- The stiffness of connections between whole parts and welded components are equivalent, allowing for their treatment as an integrated unit within finite element modeling.
- Adjustments are necessary for local stiffness at gaps in bolt connections with substantial clearances.
- Connection stiffness significantly influences simulation outcomes and should therefore take precedence during vibration test design.

References

1. Huiqiang J (2023) LSTM-based dynamic response prediction method and vibration fatigue life analysis for aerospace structures. Nanchang University, Nanchang, Jiangxi
2. Yue S, Xiaofeng L, Wei S (2022) Vibration analysis of a hard-coating thin plate under random excitation using FEM and estimation of vibration reduction effect. J Vib Shock 41(4):63–69
3. Zuo P, Shi XJ, Ge RW et al (2022) Unified series solution for thermal vibration analysis of composite laminated joined conical-cylindrical shell with general boundary conditions. Thin-Walled Struct 178(109):525
4. Liu F, Wang J, Li X (2020) Research on random vibration fatigue life of water tank welds in motor trains using time-domain method. Mater Protect 53(8):63–67
5. Zhu Y (2003) Simulation of system-level product vibration testing. Strength Environ 11(4):59–62
6. Dong L (2022) Reduction and correction of finite element model for aero-engine rotor dynamics. Beijing University of Chemical Technology, Beijing
7. Aodi Y (2022) Fatigue life prediction and reliability analysis of aero-engine Main bearings. University of Electronic Science and Technology of China, Sichuan
8. Gao D (2020) Frequency domain method for multiaxial vibration fatigue life prediction of structures. Nanjing University of Aeronautics and Astronautics

Open Access This chapter is licensed under the terms of the Creative Commons Attribution-NonCommercial-NoDerivatives 4.0 International License (http://creativecommons.org/licenses/by-nc-nd/4.0/), which permits any noncommercial use, sharing, distribution and reproduction in any medium or format, as long as you give appropriate credit to the original author(s) and the source, provide a link to the Creative Commons license and indicate if you modified the licensed material. You do not have permission under this license to share adapted material derived from this chapter or parts of it.

The images or other third party material in this chapter are included in the chapter's Creative Commons license, unless indicated otherwise in a credit line to the material. If material is not included in the chapter's Creative Commons license and your intended use is not permitted by statutory regulation or exceeds the permitted use, you will need to obtain permission directly from the copyright holder.

Analysis of Influence of Volute Tongue Angle on Flow-Induced Vibration and Noise Characteristics of Centrifugal Pump

Yuqin Wang, Mengle Han, Sheng Zhou, Zhenhu Cao, and Jiale Shao

Abstract In order to reduce the adverse effects of flow-induced vibration and noise on the performance of centrifugal pumps, this study focuses on the impact of the volute tongue angle on the internal flow characteristics and the underlying mechanisms. By constructing a three-dimensional geometric model of a centrifugal pump and using computational fluid dynamics (CFD) software for meshing and simulation, this research analyzes how different volute tongue angles influence the internal flow field, pressure pulsation, and noise characteristics. It is found that the volute tongue angle significantly affects the natural frequency, pressure pulsation, and sound pressure level of the centrifugal pump. As the volute tongue angle increases, the natural frequency of the pump rises, leading to a shorter vibration period. This results in an improved internal pressure distribution and a decrease in the average pressure value. Furthermore, through acoustic simulation analysis using Actran software, it is observed that an increase in the volute tongue angle correlates with a reduction in the sound pressure level of the pump. In the experimental verification phase, a centrifugal pump vibration test bench is constructed. By measuring the Root Mean Square (RMS) values of the vibration intensity of centrifugal pumps with varying tongue angles, it is further confirmed that increasing the volute tongue angle helps to reduce the vibration of the pump body. This study demonstrates that optimizing the design of the volute tongue angle is an effective method to mitigate vibration and noise in centrifugal pumps, thereby enhancing the operational performance and stability of the pump.

Y. Wang (✉)
School of Mechanical Engineering, Chaohu University, Hefei, China

Process Industry Digital Service Engineering Research, Center of Anhui Province, Hefei, China

College of Engineering, Technological University of the Philippines, Manila, Philippines

M. Han · S. Zhou · Z. Cao · J. Shao
School of Mechanical Engineering, Chaohu University, Hefei, China

© The Author(s) 2026

Z. Zhang, P. Mercorelli (eds.), *Mechanical Design and Simulation: Insights and Innovations*, Lecture Notes in Mechanical Engineering,
https://doi.org/10.1007/978-981-95-2405-1_9

Keywords Centrifugal pump · Volute tongue angle · Vibration and noise characteristics · Modal analysis

1 Introduction

Centrifugal pumps, as a common type of fluid transmission equipment, are widely used in industries including agriculture and urban water supply [1]. However, in practical applications, centrifugal pumps often face issues of low work efficiency and short service life, caused by vibrations and noise. The noise generated by the operation of centrifugal pumps is primarily due to two factors. Firstly, it is due to the insufficient installation precision, which can lead to collisions between parts and subsequent vibration noise when the motor drives the pump. Secondly, it is due to the turbulence produced by the internal fluid movement during operation, which causes vibration noise upon colliding with the volute shell. The flow-induced vibration and noise in centrifugal pumps represent a forced vibration situation, primarily the result of forces generated by fluid movement within the pump's internal flow field. These forces are mainly the result of unstable phenomena such as turbulence pulsation, rotating stall, cavitation, and dynamic-static interference [2].

In recent years, scholars both domestically and abroad have conducted in-depth research on the hydraulic performance analysis of centrifugal pumps. A. He et al. [3] explored the noise issues induced by gas-liquid two-phase flow within centrifugal pumps, finding that an increase in gas would reduce flow stability and increase noise. R. Gangipamula et al. [4] conducted a comparative study on airborne noise and flow-induced noise in double-suction centrifugal pumps, proposing a method for predicting flow-induced noise. C. Dai et al. [5] studied the drag reduction and noise reduction effects and mechanisms of biomimetic blade centrifugal pumps, finding that biomimetic design can effectively improve pump performance. M. Kaya et al. [6] experimentally studied cavitation-induced noise and vibration in centrifugal pumps, finding that cavitation structures affect the noise and vibration of the pump. Q. Ma et al. [7] investigated the impact of shaft misalignment on the vibration and noise characteristics of ship centrifugal pumps, finding that the misalignment state of the shaft can be judged through vibration characteristics. Y. Yuan et al. [8] designed two new types of composite impellers to replace the original long blade impellers, and verified their effectiveness, significantly improving the vibration characteristics of high-speed centrifugal pumps.

This paper investigates the impact of the volute tongue angle on the flow-induced vibration and noise characteristics of centrifugal pumps. The study employs software simulation to analyze the effects of the volute tongue angle on these characteristics, aiming to offer innovative solutions for reducing the vibration and noise in centrifugal pumps.

2 Geometric Model and Modal Analysis

2.1 Structural Parameters

The IS 80–50-250 centrifugal pump is selected as the test pump, and its main structural parameters are shown in Table 1.

2.2 Grid Division

Based on the centrifugal pump structural parameters in Table 1, a three-dimensional model is established. The grid division is carried out using ICEM-CFD software, as shown in Fig. 1.

Table 1 Structural parameters

Parameters	Values	Parameters	Values
Flow rate $Q/\mathrm{m}^3 \cdot \mathrm{h}^{-1}$	50	Blade number Z	6
Head H/m	80	Impeller inlet placement angle $\beta_1/°$	30
Rotation speed $n/\mathrm{r} \cdot \mathrm{min}^{-1}$	2900	Impeller outlet placement angle $\beta_2/°$	20
Impeller inlet diameter D_j/mm	80	Blade wrap angle $\varphi/°$	110
Impeller outlet diameter D_2/mm	240	Blade thickness S/mm	4
Impeller inlet width b_1/mm	16	Volute Base circle diamete D_3/mm	250
Impeller outlet width b_2/mm	6	Volute inlet width b_3/mm	22

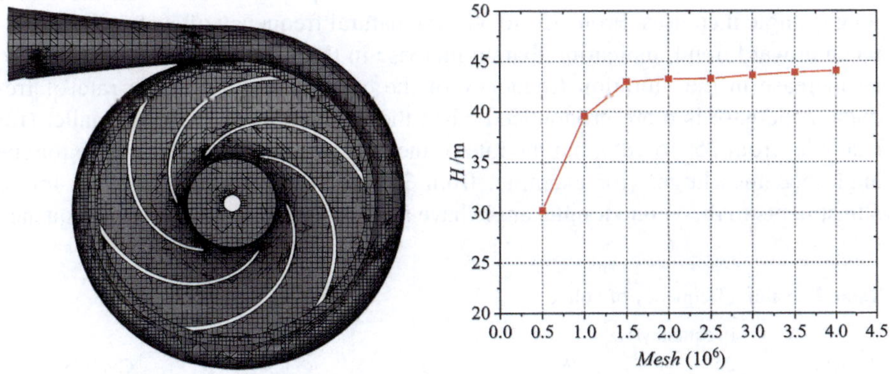

(a) Grid division (b) Grid independence

Fig. 1 Centrifugal pump structure model. (**a**) Grid division. (**b**) Grid independence

As shown in Fig. 1a, the impeller part is defined as the rotor, and the volute shell part is defined as the wall. The number of three-dimensional grids generated by the centrifugal pump is 1.513922E+06. To test the extent to which the number of grids affects the external performance characteristics of the pump, internal simulation experiments with different numbers of grids are conducted on the centrifugal pump, as shown in Fig. 1b.

The simulation results indicate that the pump head becomes stable when the grid count reaches 1.5E+06. Furthermore, as the grid count increases, the variation in this performance index remains below 1%. This suggests that the designed grid count satisfies the design requirements.

2.3 *Modal Analysis of Volute*

Modal analysis, the primary method for studying structural characteristics, is widely used in engineering vibration [9]. When analyzing the vibration characteristics of centrifugal pumps, the volute tongue angle is a critical design parameter because it directly affects the pump's natural frequency. This natural frequency, which is the frequency at which the structure vibrates freely, is crucial for predicting and controlling the pump's vibration behavior during operation. To assess the impact of various volute tongue angles on the flow-induced vibration characteristics, four centrifugal pump models with tongue angles of 25°, 28°, 31°, and 34° are designed, each exhibiting unique natural frequencies, damping ratios, and modal shapes. Determining the natural frequencies through modal analysis helps to understand how the volute tongue angle influences the pump's vibration characteristics. The calculated natural frequencies for these angles are detailed in Table 2.

According to the data in Table 2, the natural frequency increases as the volute tongue angle increases. From 25° to 34°, the natural frequency of each order exhibits an upward trend, indicating that an increase in the volute tongue angle leads to an increase in the vibration frequency of the centrifugal pump. The rate of frequency increase is more pronounced when the volute tongue angle is smaller (for example, from 25° to 28°), but the rate of increase diminishes as the volute tongue angle becomes larger (for example, from 31° to 34°). This suggests that minor adjustments to the volute tongue angle have a more significant impact on frequency

Table 2 Natural frequency of volute

Orders	Frequency/Hz				Cycle/S
	25°	*28°*	*31°*	*34°*	
1	350.79	361.94	372.22	380.86	0.55556
2	577.68	634.16	694.98	759.92	0.66667
3	1316.02	1357.73	1404.14	1445.73	0.77778
4	2940.03	2933.82	2949.32	2937.02	0.88889
5	3143.54	3226.13	3307.24	3357.43	1

during the early stages of design. Furthermore, as the volute tongue angle increases, the vibration period of the centrifugal pump is shortened, since the period is the reciprocal of the frequency, and an increase in frequency results in a decrease in the period. Consequently, a larger volute tongue angle indicates that the centrifugal pump requires less time to complete a vibration cycle. Therefore, proper adjustment of the volute tongue angle can significantly influence the vibration characteristics of the centrifugal pump.

3 Analysis of Vibration and Noise Characteristics

3.1 Selection of Volute Monitoring Points

To better analyze the characteristics of the centrifugal pump, eight monitoring points are placed in various parts of the volute, as depicted in Fig. 2. By analyzing the pressure pulsation characteristics at these points, a deeper understanding of the pump's vibration behavior can be achieved [10].

3.2 Analysis of Pressure Pulsation Characteristics

To ensure the accuracy of the numerical simulation results from the analysis of centrifugal pumps, the instantaneous values monitored at each point by the last set of data from unsteady flow are selected for pressure pulsation analysis. The average pressure values at each monitoring point are presented in Table 3.

According to the data in Table 3, the average pressure values at each monitoring point exhibit a similar trend across different volute tongue angles. As the volute

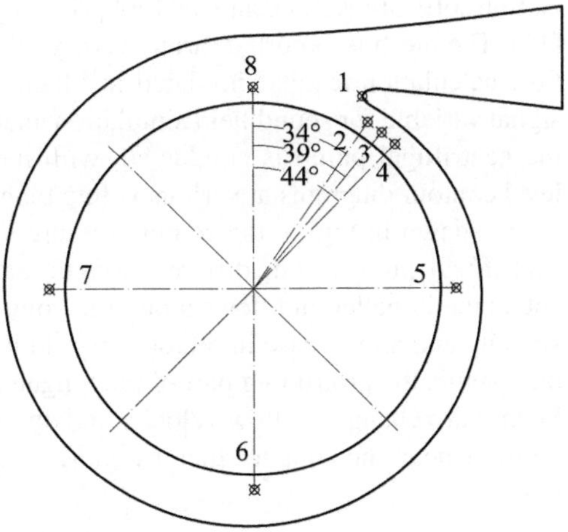

Fig. 2 Distribution of monitoring points

Table 3 Average pressure value (Pa)

Monitoring points	25°	28°	31°	34°
Point 1	69248.15	155952.54	97828.47	105208.24
Point 2	386377.99	357862.93	331582.99	305833.70
Point 3	446965.48	417088.66	389407.99	369252.35
Point 4	476046.59	448274.24	412277.98	398581.42
Point 5	572422.94	551597.29	520309.86	514216.98
Point 6	506000.22	498464.55	491055.42	491805.94
Point 7	390649.14	387574.79	367447.28	369886.06
Point 8	221323.48	198148.99	181194.49	159641.26

tongue angle increases, the average pressure values at most monitoring points decrease, indicating that the volute tongue angle significantly affects the internal pressure distribution within the pump. Among all monitoring points, Point 1 and Point 8 have relatively low average pressure values, while Point 2 and Point 3 have relatively high values. This is primarily because when the impeller periodically sweeps over the tongue, the outlet velocity of the blade directly interferes with the tongue, causing strong extrusion and disturbance of the fluid near the tongue. The outlet velocity of the blade drops sharply due to the tongue's obstruction, and this change in velocity is converted into a pressure change, leading to an increase in pressure near the tongue. Point 5 records the maximum value because when the water flow first enters the volute, the speed has not yet stabilized, resulting in the largest pressure pulsation value. Numerical analysis indicates that when the volute tongue angle is set at 34°, the average pressure pulsation value during centrifugal pump operation is minimized.

3.3 Analysis of Noise Characteristics

Actran software is a specialized tool primarily used for acoustic simulation analysis [11]. The mesh is exported through Ansys Workbench, and the density of unsteady flow calculation results simulated in Fluent software is taken as the time-domain signal variable for sound field simulation analysis. A flow-induced noise analysis of the centrifugal pump is conducted, with the simulated results of sound pressure level contour diagrams at various volute tongue angles illustrated in Fig. 3.

As shown in Fig. 3, the sound pressure level near the tongue is the highest for centrifugal pumps with different tongue angles. This is because when the fluid enters the impeller and flows along the volute channel, it causes a decrease in flow velocity and an increase in pressure within the channel. When the fluid passes near the tongue, the protruding part of the tongue changes the direction of the fluid flow. Such sharp changes in flow velocity and direction can induce vortex and reflux phenomena near the tongue, thereby generating dynamic-static interference effects

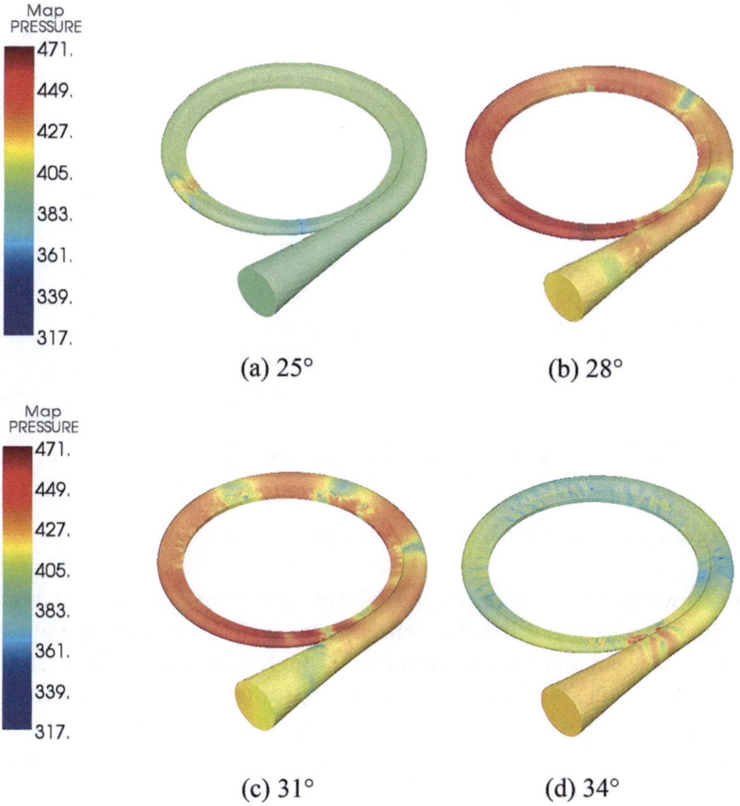

Fig. 3 Sound pressure level cloud map of centrifugal pump. (**a**) 25°. (**b**) 28°. (**c**) 31°. (**d**) 34°

between the impeller and the volute. Therefore, the sound pressure level near the tongue increases.

The frequency domain response curves of sound pressure levels for centrifugal pumps with different volute tongue angles are shown in Fig. 4.

As shown in Fig. 4, with the increase of the tongue angle from 25° to 34°, the sound pressure level shows a gradually decreasing trend at most frequency points. In the low-frequency region (48.333 Hz to 241.665 Hz), the centrifugal pump has a generally higher sound pressure level because fluid fluctuations are more likely to resonate with the pump's structure at low frequencies. As the tongue angle increases, the sound pressure level decreases significantly. In the medium and high-frequency region (above 241.665 Hz), the sound pressure level shows a decreasing trend because, as the frequency increases, the period of fluid fluctuation becomes shorter, reducing the resonance effect with the pump's structure. At these frequencies, the decreasing trend of the sound pressure level still exists, but compared with the low-frequency region, the amplitude is reduced.

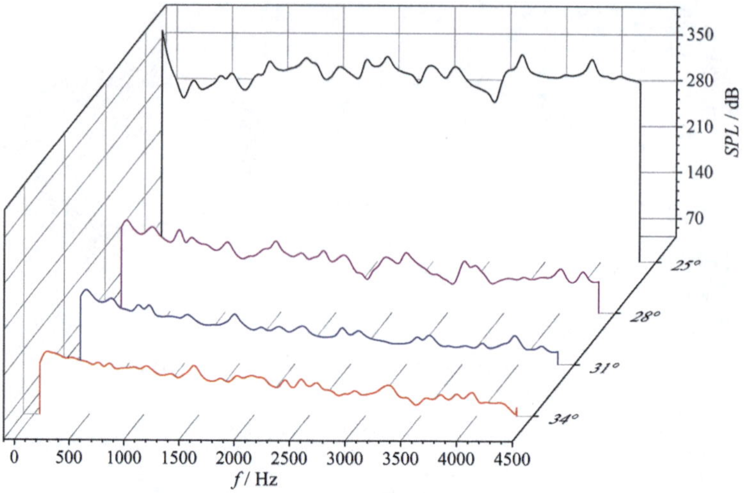

Fig. 4 Frequency domain response curve of sound pressure level

In the low-frequency band, the sound pressure level generally decreases with increasing frequency, and the difference between different tongue angles is very pronounced. The smaller the tongue angle, the higher the sound pressure level, indicating that the cochlear tongue angle significantly affects the sound pressure level in the low-frequency band. After entering the high-frequency band, the trend of the sound pressure level becomes more complex. Although the overall trend is still to decrease with increasing frequency, the differences between different tongue angles are reduced, and some frequency bands may even exhibit a crossover phenomenon. This is because, at high frequencies, fluid flow is more complex and influenced by multiple factors. Therefore, an appropriate increase in the tongue angle can effectively reduce pressure pulsations. By optimizing the design of the tongue angle in the volute, the noise level in the pump can be significantly reduced, thereby improving the pump's operating performance and stability.

4 Experimental Verification

To further examine the impact of different tongue angles on the internal vibration noise of centrifugal pumps, a centrifugal pump vibration test bench is set up, as shown in Fig. 5.

The RMS value of vibration intensity is a key indicator for evaluating the degree of vibration of centrifugal pumps. By detecting the RMS value of vibration intensity, it can be judged whether the operation of the pump is normal, and potential problems can be discovered in time for maintenance, thus avoiding possible equipment damage or production interruption. Acceleration sensors are used to measure

Fig. 5 Centrifugal pump vibration test bench

Table 4 RMS Values of vibration intensity (m/s²)

Flow rate	Measuring position	Pump body			Pump corner		
	Measuring direction	*X*	*Y*	*Z*	*X*	*Y*	*Z*
0.6*Q*	25°	5.325	3.932	2.896	4.656	2.897	1.872
	28°	4.116	2.645	1.643	4.026	1.849	1.326
	31°	3.598	1.862	1.246	2.883	1.256	0.657
	34°	1.863	0.887	0.635	0.559	0.329	0.369
1.0*Q*	25°	7.241	5.662	4.658	6.553	4.298	3.226
	28°	6.133	4.273	3.348	5.992	3.543	2.548
	31°	5.563	3.624	2.762	4.687	2.776	1.645
	34°	3.259	2.364	1.165	2.336	1.872	0.649
1.4*Q*	25°	8.861	7.236	6.262	8.223	5.925	4.883
	28°	7.534	5.956	5.653	7.349	5.332	3.956
	31°	7.065	4.993	4.392	6.562	4.462	3.265
	34°	4.668	3.882	2.598	3.954	3.658	2.113

the vibration intensity at the positions of the centrifugal pump body and pump corner. Through test experiments on centrifugal pumps with different tongue angles, the RMS values of vibration intensity under different working conditions are obtained as shown in Table 4.

As shown in Table 4, the vibration response of the pump body and the pump angle on the three coordinate axes of X, Y and Z increases with the increase of the amount of fluid flowing through the pump body. This phenomenon can be attributed to the corresponding increase of the flow velocity in the pump cavity due to the increase of the flow rate, which in turn stimulates a more significant hydrodynamic effect. When the fluid passes through the pump body, it will cause pressure fluctuation and velocity difference inside it, and these differential flows will exert a changing force on the pump body, which will lead to a significant increase in vibration response. Meanwhile, with the increase of the tongue angle, the RMS values of vibration intensity in the X, Y, and Z directions gradually decrease. At the pump body position, the vibration intensity in the X direction decreases from 7.241 m/s^2 at a tongue angle of 25° to 3.259 m/s^2 at a tongue angle of 34°, and similar decreasing trends are also observed in the Y and Z directions. This indicates that increasing the tongue angle is beneficial in reducing the internal vibration of the pump body. Similarly, the vibration intensity in the X, Y, and Z directions at the pump casing also decreases gradually with the increase of the tongue angle. At the pump casing position, the vibration intensity in the X direction decreases from 6.553 m/s^2 to 2.336 m/s^2, and the vibration intensity in the Y and Z directions also significantly decreases. This trend is consistent with the change in vibration of the pump body, indicating that the design of the tongue angle has a significant effect on the overall vibration characteristics of the pump.

The design of the tongue angle directly affects the flow path and velocity distribution of the fluid in the pump. A smaller tongue angle may cause a larger impact and reflux near the tongue, increasing the instability of the flow, which will then cause stronger vibrations and noise. As the tongue angle increases, the smoothness of the fluid flow is improved, reducing the impact and reflux, and thus reducing the vibration and noise levels. In addition, changes in the tongue angle may also affect the turbulence intensity inside the pump. A larger tongue angle may help suppress the development of turbulence, making the fluid flow smoother and reducing the generation of turbulence noise.

It can be seen that the sound pressure level and vibration intensity near the tongue of the centrifugal pump with different tongue angles both show a decreasing trend as the tongue angle increases. The internal mechanism of this trend mainly involves the improvement of flow stability, optimization of turbulence control, and reduction of structural response. By reasonably designing the tongue angle, the vibration and noise levels of the centrifugal pump can be effectively reduced, improving the operating efficiency and stability of the pump.

5 Conclusion

1. The volute tongue angle is a key design parameter affecting the internal flow characteristics and noise generation of centrifugal pumps. Through CFD simulation, it can be observed that the adjustment of the volute tongue angle can significantly alter the flow field distribution inside the pump, thereby affecting the pressure pulsation and noise level. The results of modal analysis indicate that an increase in the volute tongue angle leads to an increase in the natural frequency of the centrifugal pump, which helps to avoid the coincidence of the working frequency with the natural frequency of the pump, thereby reducing vibration caused by resonance.

2. The noise characteristic analysis reveals the influence of the tongue-baffle angle on the sound pressure level of a centrifugal pump. With an increase in the tongue-baffle angle, the sound pressure level of the centrifugal pump near the tongue baffle decreases, indicating that the acoustic performance of the pump can be optimized by adjusting the tongue-baffle angle.
3. In the experimental verification section, this paper sets up a centrifugal pump vibration test bench to conduct vibration intensity tests on centrifugal pumps with different tongue angles. The experimental results show that with the increase of flow rate, the vibration response of centrifugal pump is significantly enhanced. In addition, as the tongue angle increases, the RMS values of the vibration intensity in the X, Y, and Z directions of the pump body and casing all significantly decrease. This trend is consistent with the numerical simulation results, further verifying the impact of the volute tongue angle on the vibration and noise characteristics of centrifugal pumps.
4. The optimization strategy proposed in this study offers a new perspective for centrifugal pump design. By reasonably adjusting the volute tongue angle, not only can the vibration and noise of the pump be reduced, but also the operating efficiency and stability of the pump can be improved. This finding provides an important theoretical basis and experimental support for the optimization design of centrifugal pumps, which is of great significance for enhancing the operating efficiency and stability of centrifugal pumps and prolonging their service life. In the future, more design parameters affecting the vibration and noise characteristics of centrifugal pumps can be further explored to promote the continuous development of centrifugal pump technology. Especially, the multi-objective function of impeller and volute structural parameters is established, and the parameters are optimized by intelligent control algorithm in order to improve the vibration characteristics of centrifugal pump in an all-round way.

Acknowledgements This project is supported by Anhui Province Young Backbone Teachers Domestic Visiting and Training Funding Project (No. JNFX2024051); University Natural Science Research Project of Anhui Province (No. 2024AH051336, No. KJ2021A1026); and Anhui Province University Discipline (Professional) Top Talent Academic Funding Project (No. gxbjZD2021076).

References

1. Zhang C, Xu S, Yu P (2020) Numerical analysis of the effects of gas-phase properties on the internal characteristics and wear in a centrifugal pump. Aquac Eng 91:102126
2. Wang Y, Liu S, Cao Z (2024) The influence of blade curvature radius on the internal unsteady characteristics of a centrifugal pump. Energy Sci Eng 12(3):1148–1173
3. He A, Shao C (2021) Study on the induced noise of gas–liquid two-phase flow in a centrifugal pump. Appl Acoust 176(12):107892
4. Gangipamula R, Ranjan P, Patil RS (2023) Comparative studies on air borne noise and flow induced noise of a double suction centrifugal pump. Appl Acoust 202:109148

5. Dai C, Guo C, Ge Z, Liu H, Dong L (2021) Study on drag and noise reduction of bionic blade of centrifugal pump and mechanism. J Bionic Eng 18(2):428–440
6. Kaya M, Erol H (2022) Experimental investigation of cavitation-induced noise and vibration in a centrifugal pump. Noise Control Eng J 70(6):497–506
7. Ma Q, Wang K, Liu H, Cheng Z (2022) Influence of shaft combined misalignment on vibration and noise characteristics in a marine centrifugal pump. J Low Freq Noise Vib Active Control 41(4):1286–1306
8. Yuan Y, Yuan S, Tang L (2020) Investigation on the effect of complex impeller on vibration characteristics for a high-speed centrifugal pump. Proc Instit Mech Eng Part A: J Power Energy 234(5):611–624
9. Oza MN, Shah DS (2020) Theoretical and experimental modal analysis of centrifugal pump radial flow impeller. IOP Conf Ser Mater Sci Eng 992(1):012003
10. Wang Y, Shao J, Zhang H (2024) Analysis of the effect of cavitation on internal fluid excitation characteristics of centrifugal pumps. Phys Fluids 36(8):082104
11. Espinel E, Prada G, Rojas JP (2021) Cavitation study in centrifugal pumps through acoustic signal analysis. Int J Eng Appl (IREA) 9(5):258–267

Open Access This chapter is licensed under the terms of the Creative Commons Attribution-NonCommercial-NoDerivatives 4.0 International License (http://creativecommons.org/licenses/by-nc-nd/4.0/), which permits any noncommercial use, sharing, distribution and reproduction in any medium or format, as long as you give appropriate credit to the original author(s) and the source, provide a link to the Creative Commons license and indicate if you modified the licensed material. You do not have permission under this license to share adapted material derived from this chapter or parts of it.

The images or other third party material in this chapter are included in the chapter's Creative Commons license, unless indicated otherwise in a credit line to the material. If material is not included in the chapter's Creative Commons license and your intended use is not permitted by statutory regulation or exceeds the permitted use, you will need to obtain permission directly from the copyright holder.

Vibration Analysis and Noise Prediction of Gear Transmission System

Huiyang Sun, Lei Liu, Shihua Zhang, and Jun Zhao

Abstract As mechanical equipment develops towards high reliability, high speed and silence, the vibration and noise of gear transmission systems have attracted more and more attention. Aiming at the vibration and noise problems generated by gearboxes during operation, the dynamic model of the gear-shaft-bearing-gearbox coupling system is established based on the dynamics and structural acoustics of the gear transmission system. The dynamic force of the bearing is calculated through advanced whine analysis as the excitation force of the gear structure system. The harmonic response analysis of the gearbox is carried out using the modal superposition method in finite element analysis to solve the vibration response of the gearbox. The acoustic boundary element mesh model of the gearbox is established using the boundary element method. The results of the harmonic response analysis are used as boundary conditions to predict the vibration and noise characteristics of the gearbox. Finally, the gear transmission system test device is established through the FZG gear testing machine, the vibration signal of the gearbox is collected, and its main frequency components are analyzed and compared with the simulation results. The results show that this method can effectively predict the vibration and noise problems of the gear transmission system, and the accuracy of the theoretical simulation is verified, which provides a reliable basis for the noise control and structural optimization of the gear transmission system.

Keywords Gear transmission system · FEM · Vibration analysis · Structural acoustics · BEM · Noise prediction

H. Sun · L. Liu · S. Zhang · J. Zhao (✉)
School of Mechanical Engineering, Shandong University, Jinan, China

© The Author(s) 2026

Z. Zhang, P. Mercorelli (eds.), *Mechanical Design and Simulation: Insights and Innovations*, Lecture Notes in Mechanical Engineering,
https://doi.org/10.1007/978-981-95-2405-1_10

1 Introduction

The gear transmission system is an important power and motion transmission device in industrial production. It has the advantages of high transmission efficiency and constant transmission power. It has been widely used in aerospace, automobile, shipbuilding, power equipment, mining machinery, cutting machine tools, petrochemical and other industries. In view of the vibration and noise problems of gear transmission systems during operation, domestic and foreign scholars have conducted a series of studies. Optiz [1] studied the generation mechanism of gear system noise. Gao [2] established a dynamic model of the gear transmission system by the lumped mass method and obtained the vibration response of the gear transmission by numerical solution. Chang [3] established the unit motion equation based on the finite element method and formed a system dynamic model according to the connection relationship. Zhou [4, 5] obtained the excitation force of the gearbox through the transmission system dynamic model. But ignoring the separate modeling of the transmission shaft and bearing will lead to inaccurate vibration and noise prediction. He [6] established a rigid-flexible coupling multi-body dynamic model including all components of the gear transmission system and analyzed the radiation sound field of the gearbox. Park [7] reduces the gear system to a shaft-bearing-plate model and uses a circular plate in an infinite baffle to model radiated noise. Wang [8] used the FEM to establish a dynamic model that takes into account shaft flexibility. The above scholars provided a theoretical basis for vibration analysis and noise prediction, but did not systematically connect these theoretical methods. How to accurately establish an analysis model and improve calculation efficiency is the key to realizing the prediction of vibration and noise of gear transmission systems.

This paper uses the dynamic analysis software Romax to establish a model of the gear-shaft-bearing-gearbox coupling system to obtain the simple harmonic excitation force on the gearbox. The gearbox model is imported into the finite element analysis software Ansys Workbench, and the modal analysis is first performed to obtain the natural frequency and modal vibration shape of the gearbox. The exciting force is applied to the bearing seat of the gearbox, and the harmonic response analysis of the gearbox is performed using the modal superposition method to obtain the node displacement of the gearbox vibration. The response data solved by the vibration analysis is imported into acoustic software Simcenter 3D to establish an acoustic boundary element mesh, and a far-field acoustic field mesh is constructed around the gearbox. The surface sound pressure and far-field sound pressure of the gearbox are calculated using the boundary element method. Finally, a gear transmission system test device is established to collect the vibration signal of the gearbox and compare it with the simulation results to verify the feasibility of this prediction method, which provides a reliable basis for noise control and structural optimization of the gear transmission system.

2 Dynamics Simulation

2.1 Dynamic Model of Gear Transmission System

The main reason for the vibration and noise generated by the gear transmission system is the dynamic excitation of gear meshing [9]. Gear transmission system mainly consists of the gear pair, the transmission shaft, the bearing and the gearbox, etc. In the dynamic analysis software Romax, the gear pair, the transmission shaft and the bearing are modeled parametrically. The gearbox is a flexible part relative to the bearing, and the error caused by the force deformation cannot be ignored for the transmission error, which is already very small in magnitude. In order to be more in line with the actual situation, the gearbox needs to be pre-processed by finite element and imported into the dynamic analysis software Romax for correct positioning and matching. As shown in Fig. 1, the dynamic model of the gear-shaft-bearing-gearbox coupling system is constructed. Finally, the load conditions required for the gear transmission are defined.

2.2 Bearing Force Equation of Gear Transmission System

Based on the Hertz contact theory, a rolling bearing model was established and introduced into the gear transmission system to make the research more complete. A 6406 deep groove ball bearing was selected, and its inner ring was fixed to the transmission shaft that performs circular motion. It is assumed that the balls in the rolling bearing are evenly arranged between the inner and outer raceways at equal intervals, and the contact mode between each ball and the inner and outer raceways is pure rolling without slipping [10].

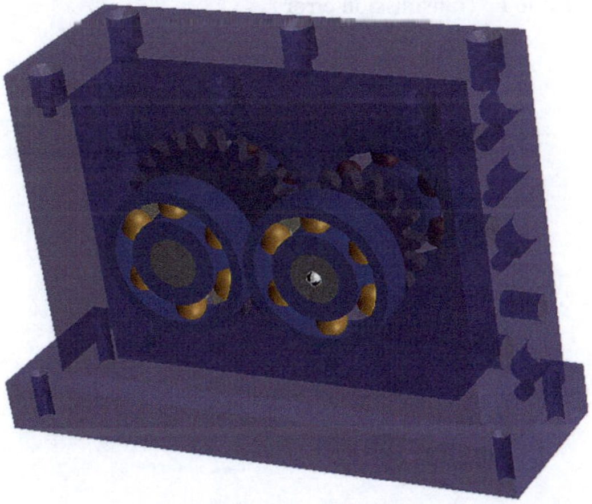

Fig. 1 Dynamic model of gear transmission system

When the jth ball contacts the inner and outer rings of the bearing, the pressure generated is F_j, and the nonlinear bearing force of the bearing is:

$$F_{bx} = \sum_{j=1}^{N_b} F_j \cos\theta_j \quad (1)$$

$$F_{by} = \sum_{j=1}^{N_b} F_j sin\theta_j \quad (2)$$

Where N_b is the number of rolling elements in the bearing; θ_j is the angle to which the jth rolling element of the bearing rotates.

2.3 Dynamic Response of Gear Transmission System

After defining the torque and speed, the dynamic response analysis of the gear transmission system is performed. The dynamic response of the bearing is one of the more critical components in the dynamic analysis. Gear transmission error is the main excitation source causing the squeal. Therefore, it is necessary to analyze the dynamic response of the bearing of the gear transmission system under excitation such as transmission error.

The transmission error is analyzed using FFT to extract the harmonics of each order, as shown in Table 1. Since the amplitudes of the second-order and higher harmonics are negligible, they can be disregarded. Therefore, this study focuses only on the first-order harmonic of the gear pair as the excitation source for NVH.

Taking the 6406 radial ball bearing at the input end as an example, in Romax, the frequency spectrum of the dynamic force of the bearing in the x, y, and z directions is obtained through the whine analysis calculation, as shown in Fig. 2.

Table 1 Transmission error

Order	Value/μm
First	14.49
Second	3.20
Third	1.95

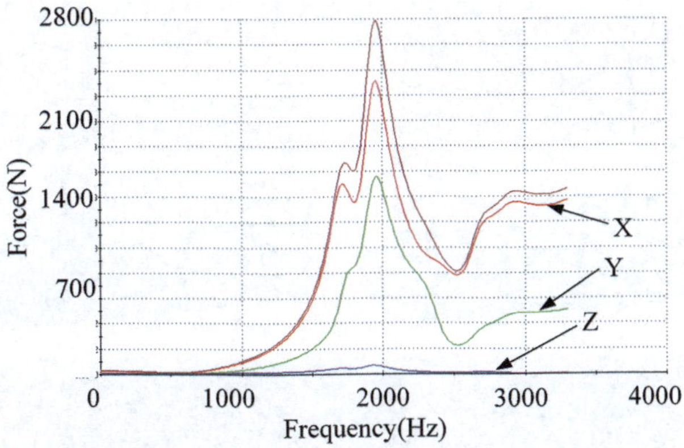

Fig. 2 Bearing dynamic force frequency response curve

Table 2 The first ten order of natural frequency

Order	Frequency/Hz	Order	Frequency/Hz
1	617.76	6	3283.4
2	1321.6	7	3399
3	1625.2	8	3619.2
4	1967.7	9	3839.8
5	2968.4	10	4313.7

3 Analysis of Vibration Characteristics

3.1 Modal Analysis

To solve for the vibration response of the gearbox, the first step is to perform a modal analysis. The gearbox model is constructed in SolidWorks and then imported into Ansys Workbench. The material of the gearbox is defined as structural steel with an elastic modulus of $E = 2 \times 10^{11}$ Pa, Poisson's ratio $\mu = 0.3$, and density $\rho = 7850$ kg/m^3. Fixed constraints are applied to the four bolt holes at the base of the gearbox, and the model is meshed using tetrahedral elements. There are many methods for extracting modes. This paper uses the Block Lanczos method to solve for the first ten natural frequencies of the gearbox, as shown in Table 2.

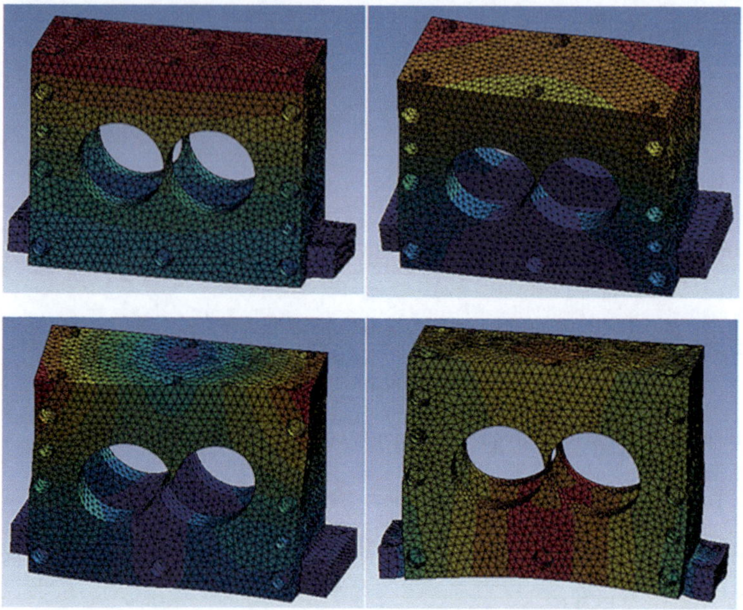

Fig. 3 First four orders of natural frequency

In Figs. 3, the first four mode shapes are shown. It can be seen that the natural frequencies of the gearbox range from 617.76 Hz to 4313.7 Hz, with the primary mode shapes involving bending and torsion. The areas with the greatest deformation are the upper end cover, bearing seats, and fixed supports. There is a noticeable overlap between the gearbox's natural frequencies and the operating frequencies of the gear transmission system. Simply increasing the stiffness of the housing to raise the natural frequencies and avoid the gearbox's operating frequencies is difficult to achieve. A comprehensive approach must be taken to address the issues of resonance and noise in the gearbox.

3.2 Harmonic Response Analysis

Before conducting acoustic radiation simulation, it is necessary to understand the response of the gear transmission system under dynamic excitation and the vibration law of the gear transmission system during operation. Therefore, it is necessary to perform a vibration response analysis on the gearbox. During the operation of the gear pair, although there is an instantaneous meshing impact, the overall vibration process of the gearbox is a periodic steady-state response. Therefore, the harmonic response analysis method is used to perform a vibration response analysis on the gearbox.

For the transmission system dynamics equation:

$$M\ddot{U}+C\dot{U}+KU=F(t) \tag{3}$$

If the node displacement and external load are in simple harmonic form, the displacement U and external load F can be expressed as follows:

$$U=U_{\max}e^{i\varphi}e^{i\omega t} \tag{4}$$

$$F=F_{\max}e^{i\psi}e^{i\omega t} \tag{5}$$

Where: M is the mass matrix of the system; C is the damping matrix of the system; K is the stiffness matrix of the system; F is the external load of the system; U_{max} and F_{max} are displacement amplitude and load amplitude; φ, ψ are the angular displacement phase and load phase; ω is the circular frequency;

This paper uses the modal superposition method to solve the dynamic differential equation. The principle is to decouple the dynamic equation through coordinate transformation, solve each independent equation separately, and finally superimpose the results of the solution. This method does not need to input the entire system matrix, which greatly improves the efficiency of the solution.

The bearing dynamic force is applied to the gearbox as a dynamic excitation. First, the model is simplified. At the center of the input and output ends, the equivalent mass point of the gear transmission system is established using the mass point mass21 unit. The bearing is simulated using the combin14 unit, and the bearing dynamic force is applied to the equivalent mass point, as shown in Fig. 4. The calculation frequency is set from 0 to 3300 Hz.

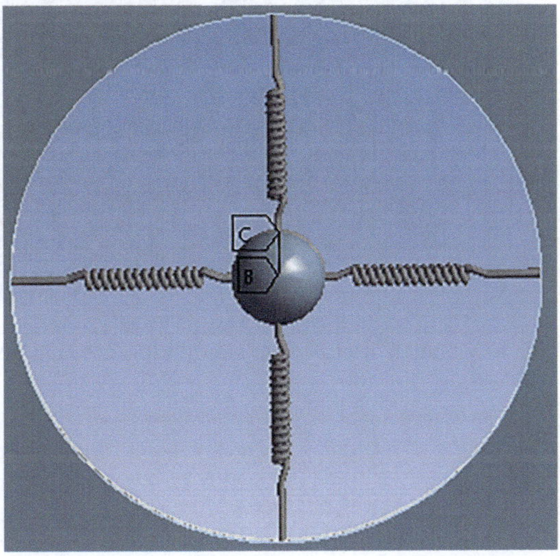

Fig. 4 Spring mass unit

The vibration displacement of the gearbox surface nodes is extracted, and the following trends are observed: the amplitude of the vibration displacement is larger near the bearing seat, the two narrow side plates on the left and right sides of the gearbox, and the upper cover, and as the frequency increases, the response amplitude at the input end is larger than that at other parts. The peak value of the vibration displacement appears at the meshing frequency and its multiples, and the larger the excitation value, the more intense the corresponding vibration. As shown in Figs. 5 and 6, the cloud diagrams of the vibration displacement at one and three times the meshing frequency.

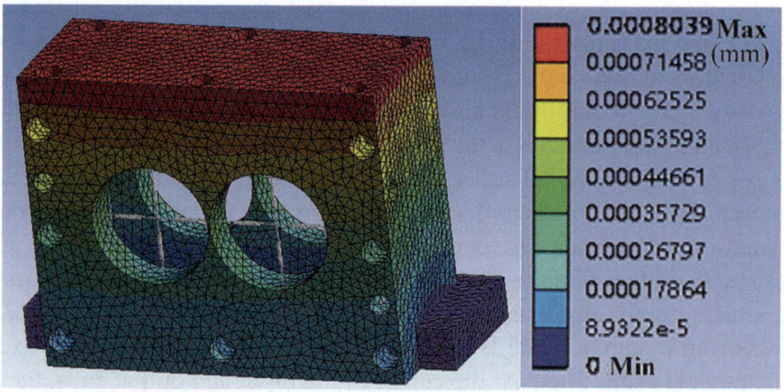

Fig. 5 Vibration displacement map at the meshing frequency

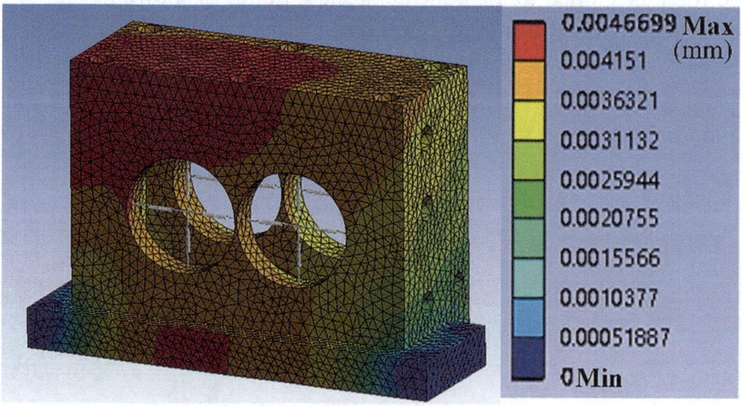

Fig. 6 Vibration displacement map at triple meshing frequency

4 Noise Prediction

4.1 Acoustic Characteristics Analysis

The gearbox is the vibrating body with the largest contact area with the outside world. The noise generated by its surface vibration accounts for 90%- 95% of the overall transmission energy of the gear transmission system. Therefore, it can be considered that the system radiation noise is mainly generated by the vibration of the gearbox. When calculating the acoustic radiation in infinite space, the finite element method has certain limitations. Therefore, the boundary element method can be used to solve the infinite space radiation problem of gearbox radiation noise [11].

The model file of the harmonic response analysis is imported into the acoustic analysis software Simcenter 3D. The acoustic boundary element grid is divided on the surface of the gearbox, as shown in Fig. 7. The displacement, torque and other results obtained from the harmonic response analysis are mapped to the acoustic grid. At the same time, the microphone grids of the far field and the surface are constructed.

The gear meshing frequency and its multiple frequency are the primary components that excite the gearbox and radiate noise. Therefore, these frequencies are analyzed in detail. Figures 8 show the surface sound pressure distribution of the gearbox at the single and triple meshing frequencies (around 600 Hz and 1800 Hz). The figures reveal that the bearing support, directly excited by the bearing's dynamic force, exhibits significant vibration and a relatively high surface sound pressure level. The gearbox's bottom support, which is constrained by fixed boundary conditions, also shows a relatively high sound pressure level. Comparing different

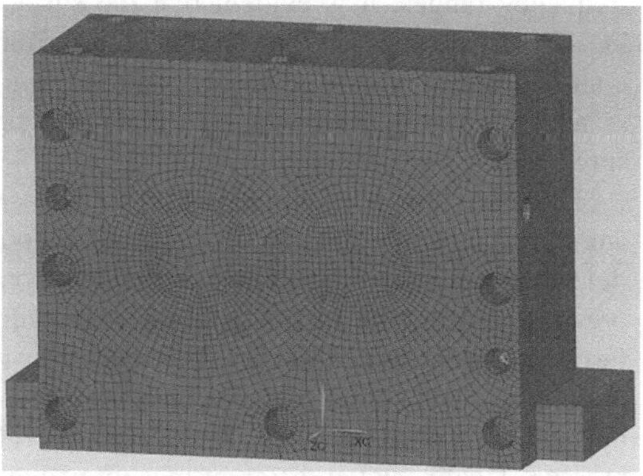

Fig. 7 Acoustic boundary element mesh

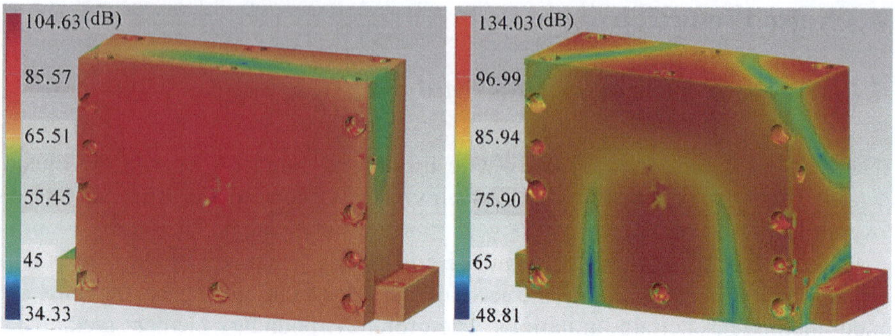

Fig. 8 Surface pressure level map

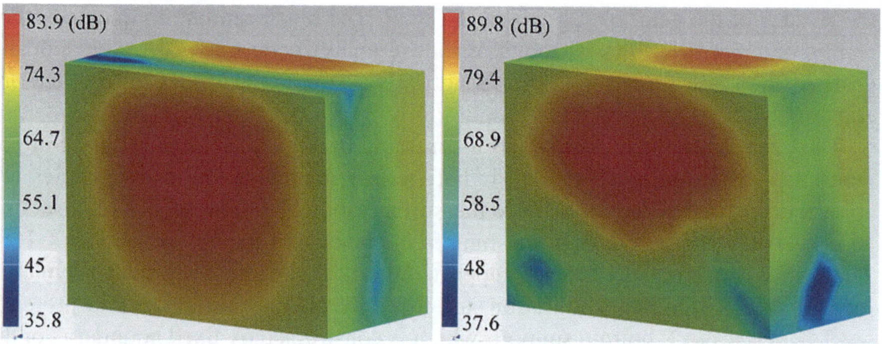

Fig. 9 Far-field pressure level map

frequency ranges, it is evident that the surface sound pressure peak at the third meshing frequency is higher. Due to high-frequency excitation, the gearbox exhibits a more complex vibration mode, resulting in a wider distribution of high-pressure areas, with the non-support side plate showing a particularly high sound pressure level.

The space in which sound pressure exists is referred to as the sound field. A rectangular grid of microphones is positioned around the gearbox to calculate the far-field sound pressure distribution at this location. Figures 9 show the far-field sound pressure distribution of the gearbox at the single and triple meshing frequencies (around 600 Hz and 1800 Hz) of the gear transmission system. The figures indicate that the far-field sound pressure level is higher near the input end, output end, and upper surface of the gearbox. There are noticeable differences in the far-field sound pressure level peaks at different meshing frequencies, with higher peaks observed under high-frequency excitation, which corresponds to the distribution of surface sound pressure and vibration signals. At lower frequencies, sound pressure level changes occur gradually, while at higher frequencies, these changes become more pronounced.

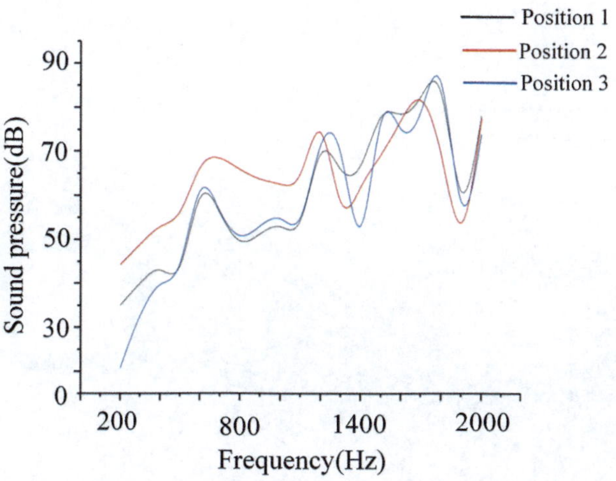

Fig. 10 Sound pressure level at different positions

Three observation points at different positions on the surface of the gearbox were selected to solve the sound pressure level frequency response curve, as shown in Fig. 10. It can be seen from the figure that the trend of the sound pressure level of the selected observation points changing with frequency is roughly similar, and a peak value appears near the meshing frequency and its multiples.

4.2 Experimental Verification

In this section, the FZG gear testing machine platform is used, as shown in Fig. 11, to carry out dynamic experiments and vibration signal measurement experiments, verify the accuracy of the results of the theoretical model simulation analysis above, and provide a basis for noise prediction for the gear transmission system.

This paper uses an IEPE three-axis acceleration sensor, as shown in Fig. 12. Two sensors are placed at the input end cover and the upper box cover respectively. When the sensor is working, based on the piezoelectric effect of the internal piezoelectric body, the stress and deformation are converted into voltage output. When the vibration test control frequency is much lower than the natural frequency of the sensor, the sensor output voltage is proportional to the force applied [12].

Since the experiment uses oil bath lubrication, the vibration amplitude at the gear mesh point cannot be directly measured. As an example, the voltage signal in the y-direction from the sensor placed on the upper cover was extracted, as shown in Fig. 13, and a time-domain plot of the voltage signal over time was obtained. To further analyze the frequency components of the vibration, the time-domain signal was transformed into the frequency domain using FFT, as illustrated in Fig. 14. The

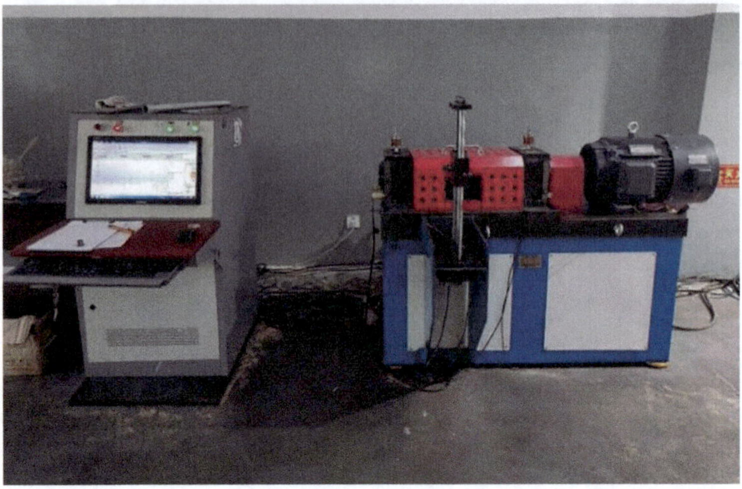

Fig. 11 Test equipment

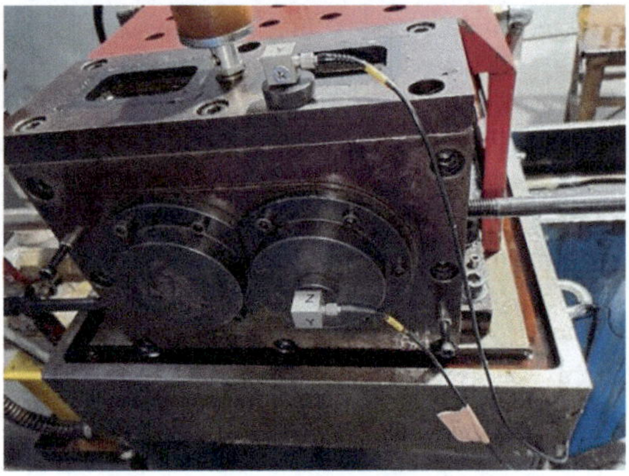

Fig. 12 Extraction of vibration signals

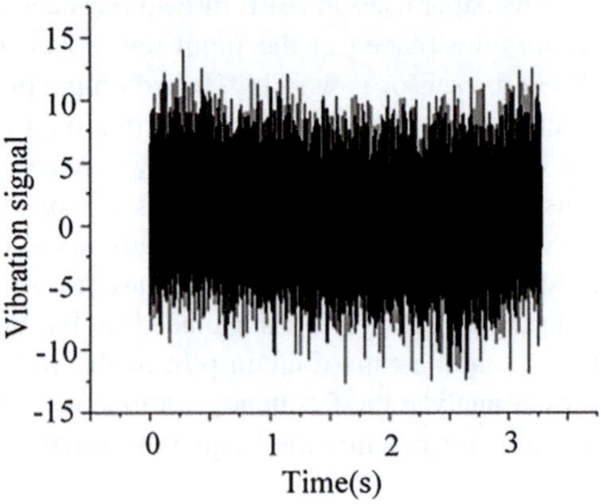

Fig. 13 Time domain map of vibration signal

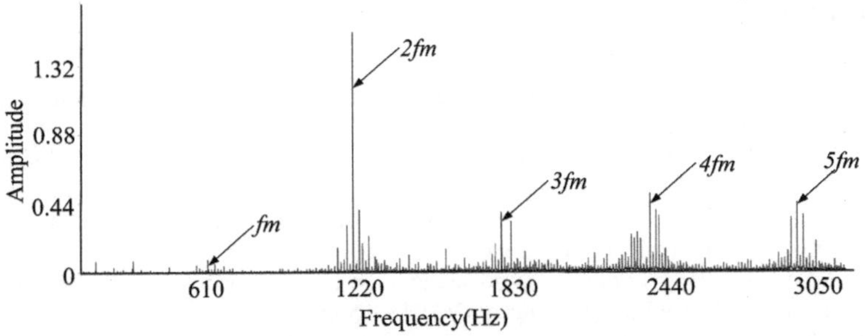

Fig. 14 Frequency domain map of vibration signal

analysis reveals that the primary frequency components with higher peak values in the experimental results are the meshing frequency and its multiple frequencies, which align with previous findings, verifying the validity of the analysis method. Additionally, rotational frequencies and their harmonics are also present in the frequency components. Due to experimental limitations, installation errors during gear assembly may differ from the theoretical values, leading to increased vibration in the gear transmission system and the appearance of other frequency components in the frequency domain map.

5 Conclusions

This paper establishes a model of the gear transmission system and conducts vibration analysis and noise prediction, which is summarized as follows:

1. The dynamic model of the gear-shaft-bearing-gearbox coupling system is established in Romax. The dynamic force of the bearing under dynamic excitation is calculated and used as the simple harmonic excitation force borne by the gearbox.
2. Through modal and harmonic response analysis in Ansys, the displacements of the gearbox surface nodes are obtained by FEM. In Simcenter 3D, the sound pressure cloud map of the gearbox and sound pressure level frequency response curve are calculated by BEM. It is known that the meshing frequency and its multiple frequencies are the main components that excite the gear transmission system and radiate noise outwards.
3. The vibration signal of the gearbox was collected using the FZG gear testing machine platform, and the frequency domain diagram of the vibration signal was obtained through FFT. The main frequency components of the vibration signal were analyzed. It can be seen that this method can provide a reliable basis for noise control and structural optimization of the gear transmission system.

References

1. Optiz H (1967) Dynamic behavior of super and helical gears. In: JSME Semi-international symposium papers, gears
2. Gao Z, Li S, Fu S, Polyakov R (2022) Analysis of spur cylindrical gear pair vibration characteristic considering error and randomness of tooth surface friction. J Mech Transm 46(2):8
3. Lehao C, Zhaoxia H, Geng L (2016) Dynamic modeling of parallel shaft gear transmissions using finite element method. J Vib Shock 35(20):47–53
4. Zhou J, Sun W (2014) Vibration and noise radiation character-istics of gear transmission system. J Low Freq Noise Vib Act Control 33(4):485–502
5. Zhou J, Sun W, Cao L (2019) Vibration and noise characteristics of a gear reducer under different operation conditions. J Low Freq Noise Vib Act Control 38(2):574–591
6. Zeyin He, Tianhong Luo, Xiangyang Xu, Lin TI (2016) An approach to calculate radiation noise ofgear system. J Vibroeng 18(1):649–660
7. Park CI (2015) Characteristics of noise generated by axial excitation of helical gears in shaftbearing-plate system. J Mech Sci Technol 29(4):1571–1579
8. Shengnan W, Xieeryazidan A, Xiangfeng Z, Jianxing Z (2020) Vibration and noise analysis for two-stage gear reducer considering shaft flexibility. J Xian Jiaotong Univ 54(09):117–127
9. Beinstingel A, Schabcrt S, Heider M et al (2023) Computational prediction of stmcture-bome noise in a two-stage double-helical gearbox using harmonic balance method. Mech Syst Signal Process 189:110112
10. Guo C, Meijiao Q (2019) Modeling and analysis of fit clearance between rolling bearing outer ring and housing. J Sound Vib 438(6):419–440
11. Meng WH (2023) A new periodic FM-BEM for solving the acoustic transmission problems in periodic media. Eng Anal Bound Elem 154:54–63
12. Wen-jing D (2024) Research on the impact of the process control of random vibration testing by piezoelectric accelerometer sensitivity. Environ Technol 2:6–11

Open Access This chapter is licensed under the terms of the Creative Commons Attribution-NonCommercial-NoDerivatives 4.0 International License (http://creativecommons.org/licenses/by-nc-nd/4.0/), which permits any noncommercial use, sharing, distribution and reproduction in any medium or format, as long as you give appropriate credit to the original author(s) and the source, provide a link to the Creative Commons license and indicate if you modified the licensed material. You do not have permission under this license to share adapted material derived from this chapter or parts of it.

The images or other third party material in this chapter are included in the chapter's Creative Commons license, unless indicated otherwise in a credit line to the material. If material is not included in the chapter's Creative Commons license and your intended use is not permitted by statutory regulation or exceeds the permitted use, you will need to obtain permission directly from the copyright holder.

Study on Stability of Aero-Engine Combustion Chamber

Zhixin Li

Abstract Through a comprehensive investigation into the stability of the combustion chamber in aircraft engines, this study aims to enhance its operational efficiency and safety under variable and complex conditions. By constructing and validating models for flame stability, acoustic coupling effects, and reaction kinetics, the research delves into the specific impacts of the interactions between the flow field, thermal field, and acoustic field on flame stability within the combustion chamber. The findings indicate that a finely optimized combustion chamber design and control parameters significantly improve flame stability, while also reducing the adverse effects of turbulence and acoustic waves on combustion efficiency. These insights provide valuable theoretical and practical guidance for the design and optimization of aircraft engine combustion chambers.

Keywords Aero-engine · Combustion chamber · Flame stability

1 Introduction

The combustion chamber of an aero-engine is the core component that produces thrust in flight, and its performance directly affects the thrust output, combustion efficiency, emission, and flight safety of the engine. Under complex working conditions such as high altitude, low temperature, and high pressure, the combustion chamber is easily disturbed by turbulence, sound waves, and airflow fluctuation, which leads to the decline of flame stability and thus affects the overall performance of the engine. The engine used is a Gao Han turbofan engine of CFM56 -5B, with a maximum thrust of 133 kN, a working temperature range of 300–1600 degrees Celsius, and a maximum fuel injection pressure of 8 MPa. As flight altitude and environmental conditions change, the fluid dynamics within the combustion chamber become increasingly complex, directly affecting combustion efficiency and

Z. Li (✉)
Taiyuan University of Technology, Taiyuan, China

© The Author(s) 2026

Z. Zhang, P. Mercorelli (eds.), *Mechanical Design and Simulation: Insights and Innovations*, Lecture Notes in Mechanical Engineering,
https://doi.org/10.1007/978-981-95-2405-1_11

thrust stability. To better understand the mechanisms behind these factors, flame stability models, acoustic coupling models, and reaction kinetics models were developed to analyze the interactions between the flow field, thermal field, and acoustic field in the combustion chamber from multiple perspectives. Through numerical simulations and experimental validation, the combustion chamber design and control parameters were optimized to ensure stable flame performance under various complex operating conditions. This also reduces the disruptive effects of turbulence and acoustic waves on the combustion process, further enhancing overall combustion efficiency and flight safety.

2 The Importance of Combustion Chamber in Aero-Engine

In the aero-engine, the combustion chamber is the place where fuel and air are mixed and burned at high temperatures. Its design and operation directly affect the thermal efficiency and thrust output of the engine [1]. The stability of the combustion chamber is very important to ensure the stable operation of the engine under complex working conditions such as high altitude, low temperature, and high pressure, which are directly related to combustion efficiency and pollutant emission levels. In this paper, the CFM56 - 5B Gao Han turbofan engine is used in the test, with a maximum thrust of 133 kN, a working temperature range from 300 °C to 1600 °C, and a maximum fuel injection pressure of 8 MPa. During the experiment, the combustion chamber realizes the uniform distribution of fuel through the high-pressure injection system, and accurately controls the air-fuel ratio (ranging from 0.8 to 1.5) by using the adjustable premixing system to simulate the operating environment of the combustion chamber under complex flight conditions. Key equipment such as the fuel injection system, air preheating device, and high-precision temperature and pressure sensors ensure the accurate operation of the combustion chamber and the accuracy of data in the experiment. Under such strict experimental conditions, the running stability of the combustion chamber plays a decisive role in the overall performance, reliability, and flight safety of the engine. The design, material selection, cooling technology, and fuel injection control of the combustion chamber are all key factors that determine its stability. Any deficiency in design and operation will lead to a decrease in combustion efficiency, incomplete combustion, flameout, or excessive vibration, and even lead to engine failure and even safety accidents in serious cases [2].

3 Aeroengine Combustion Chamber Stability Model Design

3.1 Basic Flame Stability Performance Model

In the design of aircraft engine combustion chambers, the flame stability model provides theoretical support for combustion chamber stability by accurately describing key parameters such as fuel-air mixture ratio, pressure, temperature, and flow velocity. By analyzing the behavior of these variables under extreme conditions like high altitude and low temperature, the model ensures the continuity and stability of the combustion process. Numerical simulations using Computational Fluid Dynamics (CFD) software are conducted, with selected turbulence models capturing the dynamic characteristics of complex flows within the combustion chamber. The reaction kinetics model reveals the chemical reaction mechanisms between fuel and oxidizer under high-temperature and high-pressure conditions, ensuring precise simulation of the combustion process.

The acoustic coupling effect model quantifies the interaction between the acoustic field and the combustion field, explaining the flame instability caused by acoustic waves during combustion. These models are chosen because they address the complex coupling between the flow field, thermal field, and acoustic field within the combustion chamber. The numerical simulation methods based on these models offer high flexibility, allowing accurate prediction of combustion chamber performance under different flight conditions, while providing reliable reference points for subsequent experimental validation. The experimental setup integrates high-precision sensors and complex combustion simulation equipment to ensure data accuracy and reproducibility, faithfully replicating actual operating conditions [3]. In the model setup, boundary conditions include a pre-mixed fuel-air ratio of 1.0 at the combustion chamber inlet, an inlet pressure of 8 MPa, an inlet temperature of 300 °C, adiabatic boundary conditions for the wall, and ambient pressure at the outlet, set to 1 atmosphere. The initial conditions are shown in Table 1.

Table 1 Initial condition settings of numerical simulation

Parameter	Value	Unit
Initial pressure	8	MPa
Initial temperature	300	°C
Oil-gas mixing ratio	One	Non-dimensional
Turbulence intensity	5%	Non-dimensional
Combustion chamber inlet velocity	60	m/s
Reaction rate constant	2.5 × 10^9	$m^3/(mol·s)$
Activation energy	150	kJ/mol
Stefan-Boltzmann constant	5.67 × 10^-8	$W/(m^2·K^4)$

A well-designed flame stability model integrates reaction rate and turbulence intensity to predict the flame behavior under different operating conditions, thus ensuring the efficient operation of the combustion chamber. The following are two core models.

1. Reaction rate equation:

$$r = k \cdot [Fuel]^{\alpha} \cdot [O_2]^{\beta} \cdot e^{-\frac{E_a}{RT}} \tag{1}$$

In the reaction rate equation, the reaction rate is represented by the reaction rate constant, which respectively represents the concentration of fuel and oxygen, the activation energy, the gas constant, the temperature, and the reaction order. r represents the reaction rate, k is the reaction rate constant, $[Fuel]$ and $[O_2]$ represent the concentrations of fuel and oxygen, respectively,E_a is the activation energy, R is the gas constant, T is the temperature, and β denotes the reaction order.

2. Flame Stability Equation: α

$$S = \frac{\dot{m} \cdot C_p \cdot T}{\sigma \cdot A \cdot \Delta T_{ad}} \tag{2}$$

Here, it represents the flame stability index, is the mass flow rate, is the constant pressure-specific heat, is the temperature, is the Stefan-Boltzmann constant, is the combustion area, is the adiabatic combustion temperature difference. Using the above model, the combustor designer can predict the flame behavior under different flight altitudes and environmental conditions. These equations not only emphasize the importance of precisely controlling the fuel-oxygen mixing ratio and supply rate, but also illustrate the important effects of temperature and pressure on combustion stability. S represents the flame stability index, $\dot{m}$ is the mass flow rate, C_p is the specific heat at constant pressure, T is the temperature, σ is the Stefan-Boltzmann constant, A is the combustion area, and ΔT_{ad} is the adiabatic combustion temperature difference. Using the above model, combustion chamber designers can predict flame behavior under varying flight altitudes and environmental conditions [4].

3.2 *Acoustic Coupling Effect Model*

The interaction between turbulent airflow and flame in a combustion chamber, especially under high temperature and high-pressure conditions, can easily lead to acoustic instability. This leads to the coupling of sound waves and combustion waves, forming self-excited oscillation called thermoacoustic instability. The acoustic coupling effect amplifies the pressure fluctuation in the combustion chamber, increases the instability of the flame, intensifies the structural vibration, shortens the life of the combustion chamber, and may even lead to failure [5]. The model quantifies the coupling between sound field and combustion field through an acoustic wave equation, and describes the complex interaction between pressure disturbance and heat release. The equation is as follows:

$$\frac{\partial^2 p(x,t)}{\partial t^2} - c^2\nabla^2 p(x,t) = \gamma\left(\dot{q}(x,t) = \rho\frac{\partial u(x,t)}{\partial x}\right) \tag{3}$$

Here, the pressure disturbance in the sound field is the sound speed, the specific heat ratio, the heat release rate of combustion, the air density, and the change of sound speed. The equation describes how the sound pressure changes due to the fluctuation of heat release during the propagation of sound waves, and how these changes are coupled with sound waves through heat release in the combustion chamber. In addition, the acoustic damping characteristics of the combustion chamber play an important role in controlling thermoacoustic instability. These damping characteristics are closely related to the structure, material, and flow field distribution of the combustion chamber [6]. In order to ensure acoustic stability under complex working conditions, an accurate damping model must be introduced to represent the energy dissipation under different conditions. Acoustic resistance can be described by the following equation: $p(x,t)$ representatives the pressure disturbance in the acoustic field, c is the speed of sound, γ is the specific heat ratio, $\dot{q}(x,t)$ is the rate of heat release from combustion, ρ is the air density, and $u(x,t)$ is the change in sound speed. This equation describes how sound pressure changes during sound wave propagation due to fluctuations in heat release and how these changes are coupled with acoustic waves through the heat release in the combustion chamber. Additionally, the acoustic damping characteristics of the combustion chamber play a crucial role in controlling thermoacoustic instability. These damping properties are closely related to the structure, materials, and flow field distribution within the combustion chamber [6]. To ensure acoustic stability under complex working conditions, a refined damping model must be introduced to characterize energy dissipation under different conditions. Acoustic damping can be described by the following equation:

$$D = \frac{1}{2}\rho_0 c_0 S \int \left(\frac{\partial p}{\partial t}\right)^2 dt \tag{4}$$

In this equation, the acoustic damping is represented, and the sum is the static density and the static speed, the cross-sectional area of the combustion chamber, and the pressure change rate. The equation reveals the process of energy dissipation in sound waves, especially how the attenuation of sound wave energy through damping can improve the stability of the combustion chamber. Energy dissipation caused by acoustic damping plays a key role in reducing pressure fluctuation and thermoacoustic instability, and ultimately contributes to the overall stability of the combustion chamber during operation. In this equation, D represents the acoustic damping, ρ_0 and c_0 are the static density and static sound speed, respectively, S is the cross-sectional area of the combustion chamber, and $\frac{\partial p}{\partial t}$ is the rate of pressure change. This equation reveals the process of energy dissipation in sound waves, particularly how the attenuation of sound wave energy through damping improves the stability of the combustion chamber. The energy dissipation caused by acoustic damping plays a critical role in reducing pressure fluctuations and mitigating thermoacoustic instability, ultimately contributing to the overall stability of the combustion chamber during operation.

3.3 Reaction Kinetics Model

The reaction kinetic model is the key to studying the stability of aero-engine combustion chambers because it reveals the chemical reaction rate and energy release process of fuel and oxidants under high temperatures and high pressures. This model directly affects the formation, maintenance, and efficiency of flame [7]. The model uses nonlinear equations to describe the relationship among reaction rate, temperature, pressure, and reactant concentration. The dependence of reaction rate on temperature accords with the Arrhenius formula, which ensures the stability analysis of the combustion chamber under complex working conditions. The equation is as follows:

$$k = A \cdot \exp\left(-\frac{E_a}{RT}\right)[Fuel]^m [O_X]^n \tag{5}$$

In this equation, the reaction rate is represented by the reaction rate constant, the activation energy, the gas constant, and the absolute temperature, which represent the concentration of fuel and oxidant and the reaction order, respectively. The equation can accurately predict the combustion reaction rate at different temperatures and pressures, thus determining the conditions required for flame maintenance and evaluating the stability of the combustion process [8]. In addition, the distribution of heat release in the combustion process directly affects the coupling between temperature field and flow field in the combustion chamber, and then has a far-reaching impact on flame stability. The heat release rate can be described by the following equation: k represents the reaction rate, A is the reaction rate constant, e is the activation energy, r is the gas constant, t is the absolute temperature, Fuel and O_x represent the concentrations of fuel and oxidizer, respectively, and m and n are the reaction orders. This equation allows for accurate prediction of the combustion reaction rate under different temperatures and pressures, thereby determining the conditions required for flame maintenance and assessing the stability of the combustion process [8]. Furthermore, the distribution of heat release during combustion directly affects the coupling between the temperature field and flow field within the combustion chamber, which in turn has a profound impact on flame stability. The heat release rate can be described by the following equation:

$$\dot{q} = \frac{\Delta H \cdot \rho \cdot u}{1 + \frac{\gamma - 1}{2} M^2} \tag{6}$$

In this equation, the exothermic rate is the enthalpy change of the reaction, the density, the flow rate, the specific heat ratio, and the Mach number. The equation is helpful to analyze the influence of heat release on the flow field in the combustion process, thus evaluating the thermodynamic stability of the combustion chamber. The establishment of a reaction dynamics model not only helps to deeply explore the micro-mechanism of the combustion process, but also provides a theoretical basis for optimizing the performance of the combustion chamber under various flight conditions. $\dot{q}$ represents the heat release rate, ΔH is the enthalpy change of the reaction, ρ is the density, u is the flow velocity, γ is the specific heat ratio, and M is the Mach number. This equation helps analyze the impact of heat release on the flow field during the combustion process, allowing for an assessment of the thermodynamic stability of the combustion chamber. The establishment of the reaction kinetics model not only enables a deep er exploration of the microscopic mechanisms of the combustion process but also provides a theoretical foundation for optimizing the performance of the combustion chamber under various flight conditions.

3.4 Numerical Simulation Method

Numerical simulation, through precise modeling of complex physical and chemical processes, accurately captures the dynamic evolution of multiple factors such as airflow, heat transfer, and chemical reactions within the combustion chamber [9]. Using CFD technology, the Navier-Stokes equations are discretely solved, combined with reaction kinetics and acoustic models to describe in detail the turbulent field, pressure distribution, and flame propagation path within the combustion chamber. This method can also couple turbulent reactions with heat release effects, analyzing flame stability under different operating conditions and flexibly adjusting parameters for multi-dimensional analysis, such as the synergistic effects of fuel flow rate, temperature, and chemical reaction rates, providing in-depth understanding and prediction of the combustion process.

4 Experimental Results and Analysis

4.1 Experimental Settings and Methods

The experiment was carried out using a highly accurate simulated combustion chamber, which accurately reproduced the high temperature and high-pressure conditions encountered in real-world scenes to ensure the accuracy of experimental data. The core components of the combustion chamber include THE fuel injection system, air preheating device, and high-precision temperature and pressure sensors, which can accurately control and monitor the fuel-air mixing ratio, combustion temperature, and airflow speed [10]. During the experiment, the fuel is evenly distributed in the combustion chamber through the high-pressure injection system, and the air-fuel ratio is finely controlled through the adjustable premixing system. Combined with the CFD numerical simulation model, the system realizes dynamic parameter adjustment and ensures the consistency between experimental and real flight conditions (see Table 2).

4.2 Experimental Results of Flame Stability

The results of flame stability experiments show significant differences in flame behavior under different conditions. Firstly, the flame stability under different injection pressures and air-fuel ratios was tested. Table 3 shows that the flame stability index increases significantly with the increase in injection pressure. At low pressure, the flame is more susceptible to turbulence and shows instability, while at high pressure, the flame is more concentrated and stable. The change of air-fuel ratio also has a profound influence on flame stability. When the air-fuel ratio is close to the ideal value of 1.0, the combustion efficiency is the highest, and the flame stability is the best. However, an air-fuel ratio that is too high or too low will lead to increased flame oscillation and decreased stability.

Table 3 shows the changes in the flame stability index under different injection pressures, indicating that high-pressure injection is helpful in improving flame stability and combustion efficiency. Based on these experimental results, the influence of temperature on flame stability is further analyzed, as shown in Table 4. With the increase in combustion temperature, the flame stability is also improved. However,

Table 2 Key parameters of the experimental device

Parameter	Equipment function	Adjustment range
Injection pressure	Uniform fuel distribution and flame formation	0–10 MPa
Temperature sensor	Real-time monitoring of combustion temperature	300–2000 degrees Celsius
Air-fuel ratio control	Control of fuel and air mixing ratio	0.8–1.5

Table 3 Influence of injection pressure on flame stability and combustion efficiency

Injection pressure (MPa)	Flame stability index	Combustion efficiency (%)
2	0.65	Seventy-eight
Five	0.83	Eighty-five
Eight	0.91	Eighty-nine

Table 4 Influence of combustion temperature on flame stability and combustion efficiency

Temperature (°C)	Flame stability index	Combustion efficiency (%)
1000	0.78	82
1500	0.85	88
2000	0.92	90

excessive temperature will lead to incomplete combustion and increase pollutant discharge.

4.3 Experimental Results of the Acoustic Coupling Effect

When exposed to sound waves with different frequencies and amplitudes, changes in flame stability are observed, especially at certain frequencies where resonance occurs between sound waves and heat release in the combustion chamber. This resonance intensifies the flame oscillation. Figure 1 shows the change in flame stability index at different sound wave frequencies. With the increase in frequency, the flame stability index decreases nonlinearly, and the most significant acoustic coupling effect occurs at 600 Hz. At this frequency, the oscillation amplitude of the flame increases significantly, and the combustion efficiency decreases accordingly.

Figure 1 shows the change in flame stability at different sound wave frequencies. The higher the frequency, the stronger the coupling between sound wave and heat release, which leads to an increase in flame volatility. In order to further analyze the influence of sound wave amplitude, Fig. 2 shows the influence of different vibration amplitudes on flame stability at a fixed frequency. It can be seen that with the increase in amplitude, the stability of the flame further deteriorates. At high vibration amplitude, the flame is easily disturbed by sound, which leads to a discontinuous combustion process and reduced efficiency.

4.4 Comparative Analysis of Numerical Simulation Results and Experiments

The comparative analysis of numerical simulation and experimental results shows that the predicted flame stability under different working conditions has high consistency, but there are some differences in some extreme working conditions.

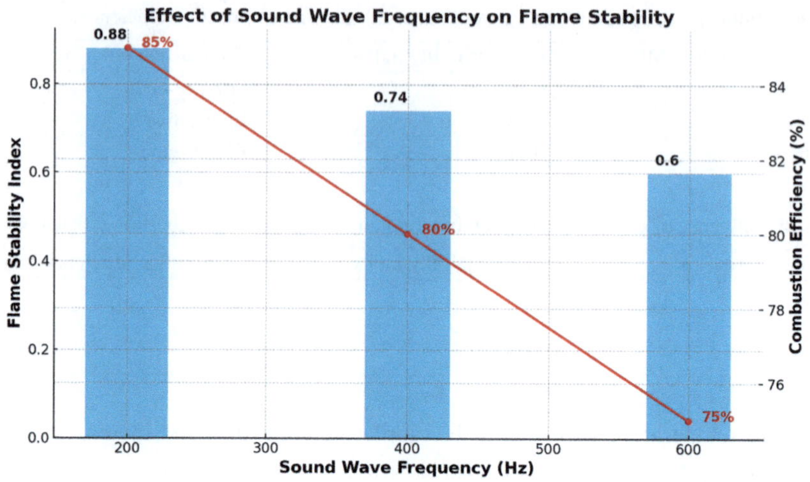

Fig. 1 Influence of different sound wave frequencies on flame stability

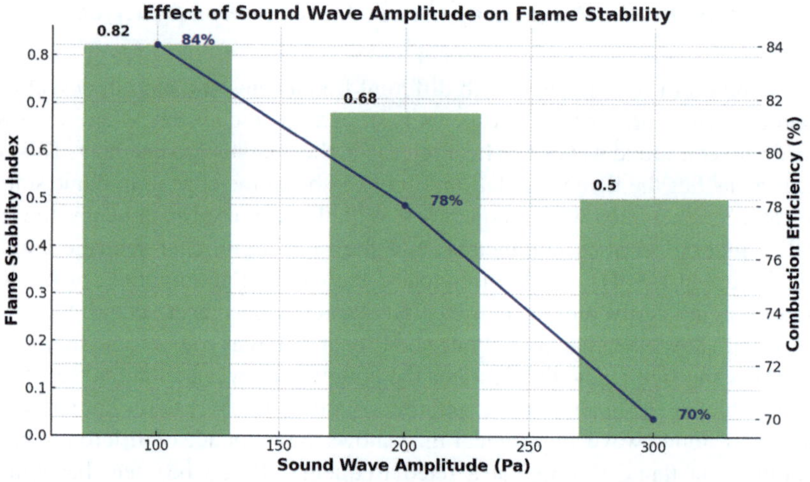

Fig. 2 Influence of different sound wave amplitudes on flame stability

Numerical simulation uses the CFD model to calculate the distribution of airflow, pressure, and temperature field in the combustion chamber, and accurately capture the dynamic behavior of turbulence and heat release. The difference between the numerical simulation results and the experimental data mainly occurs at low fuel ratio, and the flame stability in the simulation is slightly higher than the actual experimental results, indicating that the model has certain limitations in capturing the influence of turbulence at low fuel ratio (see Fig. 3).

In order to further verify the influence of the acoustic coupling effect, the simulation results and experimental results are compared and analyzed at acoustic frequency. It can be seen that numerical simulation can better capture the coupling

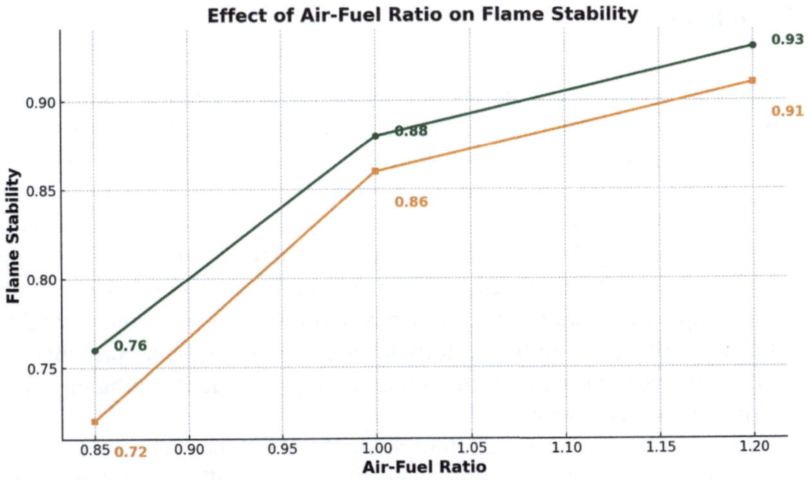

Fig. 3 Influence of air-fuel ratio on flame stability

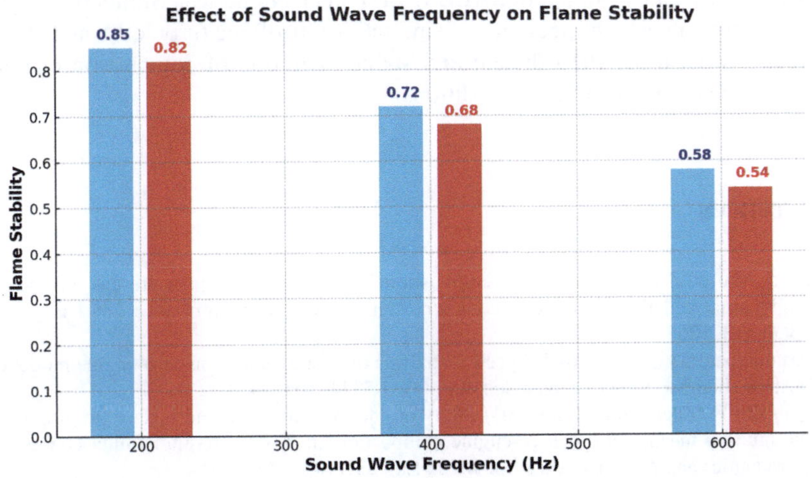

Fig. 4 Influence of sound wave frequency on flame stability

trend of sound wave and flame, but the simulation accuracy decreases at high frequency, as shown in Fig. 4.

Numerical simulation methods demonstrate high accuracy in predicting flame stability and acoustic coupling effects in aircraft engine combustion chambers, particularly under conventional operating conditions where experimental and simulation results show strong consistency, validating their effectiveness. However, under extreme conditions, discrepancies between the two reveal the limitations of existing models in handling complex turbulence and high-frequency acoustic coupling. Future improvements to these models are necessary to enhance their predictive capabilities across a broader range of operating conditions.

5 Conclusion

Under complex operating conditions such as high altitude and low temperature, the stability of the combustion chamber directly impacts the performance of aircraft engines and flight safety. By establishing flame stability, acoustic coupling, and reaction kinetics models, combined with numerical simulations and experimental analysis, this study reveals the critical role of the coupled flow, thermal, and acoustic fields in maintaining flame stability within the combustion chamber. These analyses provide essential theoretical support for optimizing combustion chamber design and improving combustion efficiency. However, under extreme conditions, accurately capturing the coupling of turbulence and acoustic waves presents certain technical challenges, particularly in simulating high-frequency turbulence and its interaction with acoustic waves.

To overcome these limitations, future research should focus on developing more precise turbulence-acoustic coupling algorithms and utilizing higher-resolution experimental equipment to conduct in-depth testing of high-frequency acoustic phenomena. Additionally, employing more advanced multi-scale simulation techniques, such as Large Eddy Simulation (LES) or Direct Numerical Simulation (DNS), could further enhance the accuracy of simulations involving turbulence and acoustic interactions, ensuring the efficient and stable operation of combustion chambers under a wide range of extreme conditions.

References

1. Bo J, Mai et al (2024) Hydrogen combustion in two-stage high-pressure gas turbine- thermodynamic and flow simulation compared with fossil fuel performance. Int J Engine Res 25(3):513–526
2. Arumapperuma G et al (2024) Spectral analysis of soot dynamics in aero-engine model combustion chamber. Proc Combust Soc 40(1–4):105344–105344
3. Qun Zhi W, Zhengyu F et al (2003) Study on the influence of flame surface development on the ignition mechanism of aero-engine combustion chamber . Beijing: China University of Aeronautics and Astronautics. J Soc Mech Eng 237(14):3255–3271
4. Shuang L, Lang L, Ye T et al (2023) Multi-objective optimization algorithm of aero-engine combustion chamber based on artificial neural network-multi-objective particle swarm optimization. J Soc Mech Eng J Aerospace Eng 237(11):2577–2593
5. Anon (2023) Inexact P-box sensitivity analysis of aero-engine combustor performance simulation model considering related variables. Energy 16(5):2362–2362
6. Hu X, Xiaoming W (2023) Analysis of exhaust pollution of aero-engine combustion chamber. Beijing: University of Aeronautics and Astronautics, 2001. Aircr Eng Aerosp Technol 95(4):569–579
7. B (2023) Large eddy simulation for evaluating intermittent soot formation in aero-engine model combustion chamber with high-speed measurement method. Proc Combustion Soc 39(4):4821–4829
8. Rao KVLN, Prasad BVSSS, Degaonkar GK et al (2022) Influence of intake vortex on flow pattern factor and pressure loss of aero-engine combustion chamber. J Soc Mech Eng, Ser C: J Mech Eng Sci 236(5):2631–2645

9. Guo J et al (2011) Numerical simulation of combustion chamber process of solid rocket engine under stable design conditions. Fluid Dynamics Sci 46(3):467–478
10. Mi (short for meter) Yu, doctor of medicine, Minghui L (2008) Influence of overload on low-frequency instability of solid rocket motor combustion chamber. Beijing: Science Press, 2001. J Vib Eng 10(3):272–276

Open Access This chapter is licensed under the terms of the Creative Commons Attribution-NonCommercial-NoDerivatives 4.0 International License (http://creativecommons.org/licenses/by-nc-nd/4.0/), which permits any noncommercial use, sharing, distribution and reproduction in any medium or format, as long as you give appropriate credit to the original author(s) and the source, provide a link to the Creative Commons license and indicate if you modified the licensed material. You do not have permission under this license to share adapted material derived from this chapter or parts of it.

The images or other third party material in this chapter are included in the chapter's Creative Commons license, unless indicated otherwise in a credit line to the material. If material is not included in the chapter's Creative Commons license and your intended use is not permitted by statutory regulation or exceeds the permitted use, you will need to obtain permission directly from the copyright holder.

Topology Optimization of Bit Feed Device Based on SIMP Method

Wei Shang, Sheng Chen, Jianjun Lv, Bing Feng, Ji'an Zhou, and Zhenyang Fu

Abstract This paper explores the use of the Solid Isotropic Material with Penalization (SIMP) method for topology optimization to enhance the performance of drill feed devices. The study focuses on improving the structural stability, reducing weight, and increasing material efficiency of the feed device, all of which are critical for drilling efficiency and processing quality in modern mechanical design. Software such as Abaqus was used for model creation, stress analysis, and optimization. After optimizing the feed device, its weight was reduced by 35.6% without compromising strength, demonstrating the effectiveness of topology optimization in promoting sustainability in mechanical design.

Keywords Topology optimization · SIMP method · Drill feed device · Structural stability · Weight reduction

1 Introduction

In modern mechanical design, the performance of drill feed devices directly affects drilling efficiency and processing quality. Topology optimization, as an innovative design approach, can significantly reduce structural weight and material consumption while meeting mechanical performance requirements. This paper explores how to enhance the overall performance of drill feed devices through topology optimization based on the Solid Isotropic Material with Penalization (SIMP) method. The research aims to provide an effective design strategy to promote the intelligence and efficiency of drilling equipment, contributing to the sustainable development of the related industries.

The feed body, as part of the feed device, provides a platform for the installation of components such as the power head, hydraulic cylinder, and gripper. Its structural

W. Shang · S. Chen (✉) · J. Lv · B. Feng · J. Zhou · Z. Fu
Department of Shaoxing Daming Electric Power Design Institute Co., Ltd, Shaoxing, China

© The Author(s) 2026

Z. Zhang, P. Mercorelli (eds.), *Mechanical Design and Simulation: Insights and Innovations*, Lecture Notes in Mechanical Engineering,
https://doi.org/10.1007/978-981-95-2405-1_12

stability directly impacts drilling efficiency and hole quality. To meet different practical requirements, structural optimization of the feed body has been performed.

Zhang Youzhen et al. [1] utilized ANSYS software to conduct modal and harmonic response analysis of the feed body of the ZDY4000L drilling rig, successfully optimizing the structure of this series of drilling rigs, providing a basis for subsequent design optimization. Wang Hejian [2] built a model of the feed body using Solid Edge software and analyzed the stress and strain using ANSYS. Based on the weak areas identified in the stress contour plots, structural optimization was performed, ultimately ensuring the strength and stiffness of the feed body. Zhang Rui [3] performed modal analysis using Abaqus and identified weak parts of the structure from the mode shape diagrams, laying a foundation for the optimized design of the feed body. Wang Tianlong [4] optimized the feed body according to the actual working environment, increasing the travel of the drilling rig, making the structure more compact, and adapting it to confined workspaces. Hao Yongjin [5] optimized the feed cylinder, increasing its travel from 0.95 m to 1.15 m while reducing the overall length of the machine from 2.73 m to 2.47 m. Lü Meng [6] implemented a combined design for the feed cylinder, effectively shortening the length of the feed body. In 1904, Michell [7] first proposed the theory of topology optimization, but it was only able to solve relatively simple problems and was not effectively applied to engineering. Hegerniner [8] extended Michell's theory to better address the optimization of nonlinear problems. Achtziger et al. [9] proposed the application of topology optimization to the solution of discrete systems.

Based on the above research achievements, topology optimization has gradually become an important method for improving the performance of mechanical structures. Through reasonable topology optimization design, the weight of the drill feed device can be significantly reduced while ensuring structural strength and stability under different working conditions. Therefore, integrating topology optimization with the practical requirements of complex mechanical systems to further enhance the overall performance of the device has become a key research direction today.

2 Topology Optimization Process for the Drill Feed Mechanism

As shown in Fig. 1, the topology optimization process is conducted using Abaqus software. This process includes creating the model, defining the design domain, setting topology optimization parameters, and performing the optimization calculations. The goal is to gradually enhance the structural performance of the drill feed mechanism, achieving both weight reduction and increased strength.

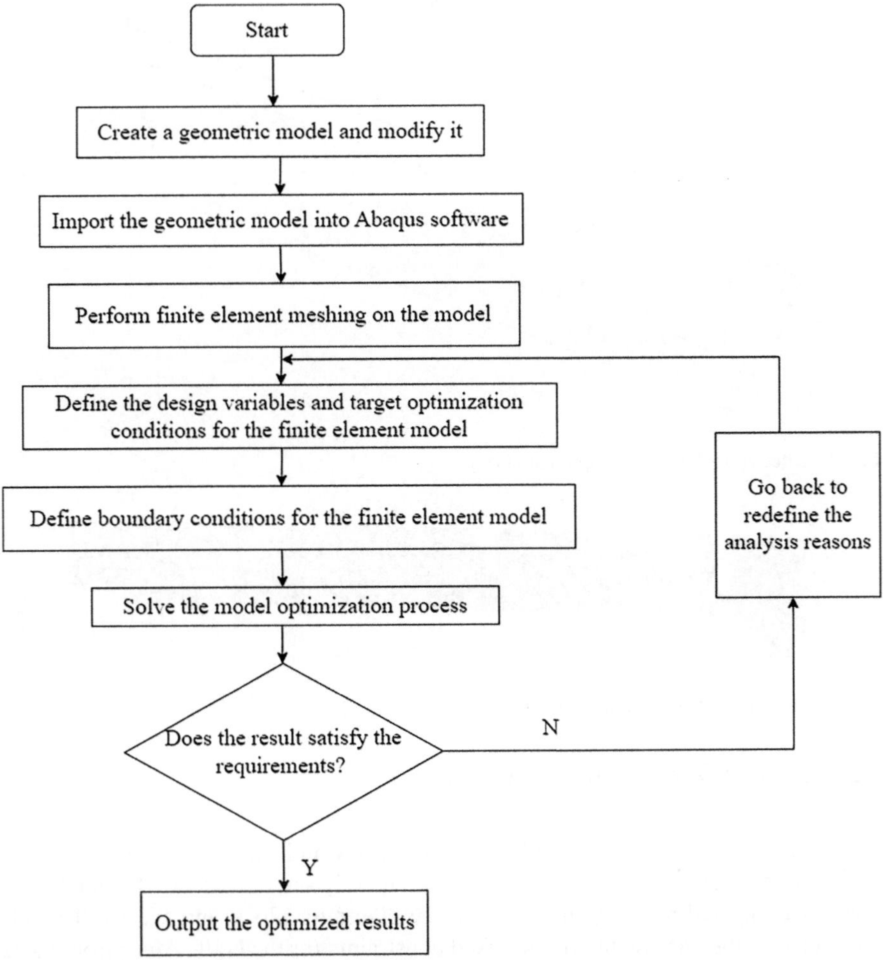

Fig. 1 Flowchart of topology optimization in abaqus

3 Model Creation

3.1 Preprocessing

Figure 2 shows the finite element model of the feed mechanism, which is defined using Q345 material properties. During the model creation process, the 3D model was imported into Abaqus in. step format and underwent appropriate meshing operations. The figure displays the filled model of the feed mechanism, which maintains consistency with the real mechanical structure. Details such as the transmission device and main components of the feed mechanism are clearly visible, providing a reliable foundation for subsequent topology optimization.

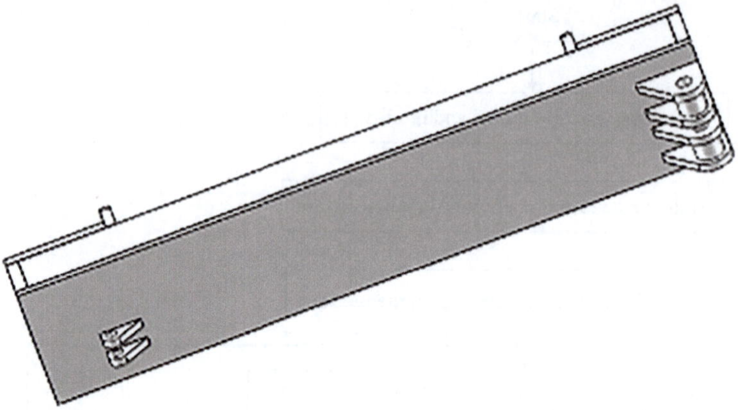

Fig. 2 Filled model of feed mechanism body

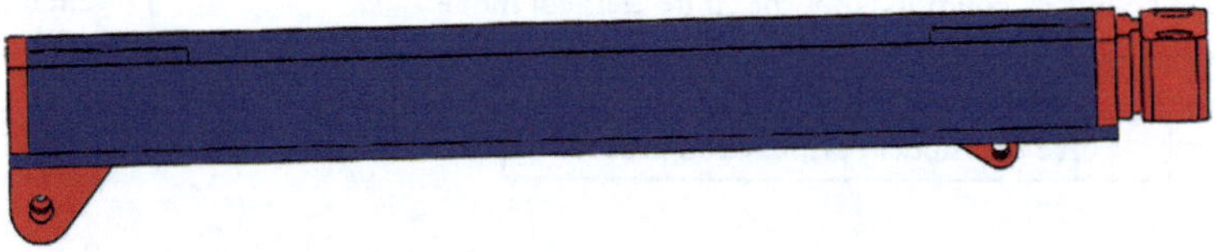

Fig. 3 Design and Non-design regions

3.2 Determination of Design Region

Based on the structure of the feed mechanism body and actual working conditions, the front and rear plates of the feed mechanism body serve to secure the double-rod double-acting hydraulic cylinder, while the clamping block and the fixed blocks connected to the base plate provide fixed constraints for the body. Apart from these, the other plates are designated as the design region. As shown in Fig. 3, the blue areas of the feed mechanism body represent the design region, while the red areas represent the non-design region.

3.3 Optimization Mathematical Model

Since the feed mechanism body is entirely welded from Q345 steel, with a consistent material density, the mass-constrained strain energy minimization problem can be considered equivalent to a volume-constrained strain energy minimization problem. Based on the three key elements of optimization and the predefined volume ratio $\lambda = 0.5$, the mathematical model for topology optimization can be expressed as:

$$\begin{cases} find\rho = (\rho_1, \ldots, \rho_n) \\ \min C(\rho) = \mathbf{U}^T\mathbf{K}\mathbf{U} = \sum_{s=1}^{n} (\rho_s)^p \mathbf{u}_s^T \mathbf{k}_s \mathbf{u}_s \\ subjectto\mathbf{K}(\rho)\mathbf{U} = \mathbf{F} \\ V(\rho) \le 0.5 \cdot V_0 \\ 0 < \rho_{\min} \le \rho_s \le 1 \end{cases} \quad (1)$$

4 Optimization Results Analysis

4.1 Topology Optimization Results of the Machine Body

The topology optimization of the feed machine body was initially set to terminate after 50 iterations. However, no optimal result was found when the optimization stopped. After adjusting the iteration count to 100, the optimization process ceased at 62 iterations, during which the strain energy decreased and stabilized, indicating that the optimal solution had been found. The optimization results are shown in Figs. 4, 5, and 6.

From the topology optimization results, it can be observed that the side plates of the machine body often exhibit triangular hole structures. Considering the stability of triangular structures and the issue of stress concentration, as well as the common use of rectangular structures in the optimization of lifting machinery and vehicles, two adjacent triangular structures in the topology result were approximated and merged into a single rectangular shape. Additionally, semicircular holes were added on both sides of the rectangle, connected to its edges, replacing the chamfered corners of the rectangle. A rounded rectangular structure was adopted for the bottom plate to avoid stress concentration. The 3D model is shown in Fig. 7.

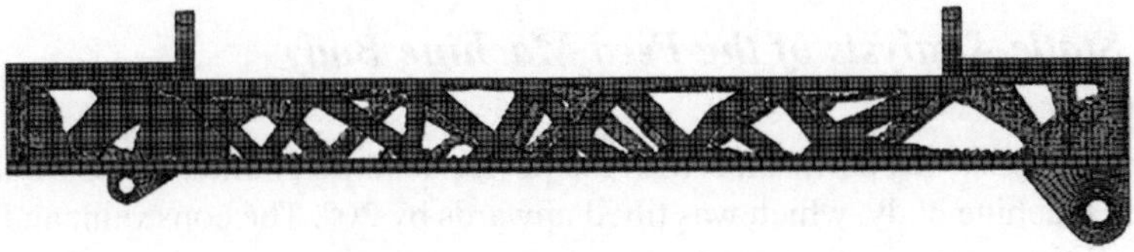

Fig. 4 Left side plate of the feed machine body

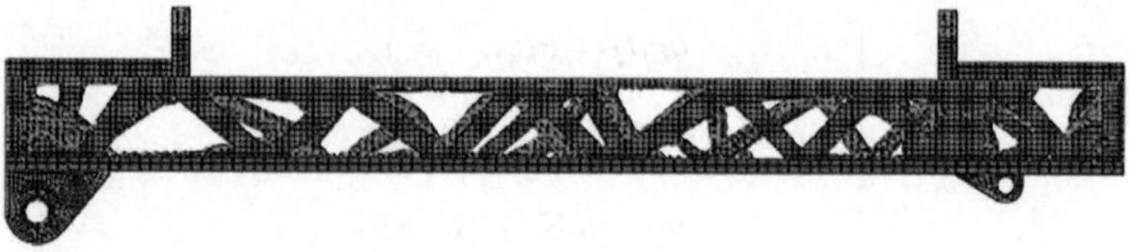

Fig. 5 Right side plate of the feed machine body

Fig. 6 Bottom side plate of the feed machine body

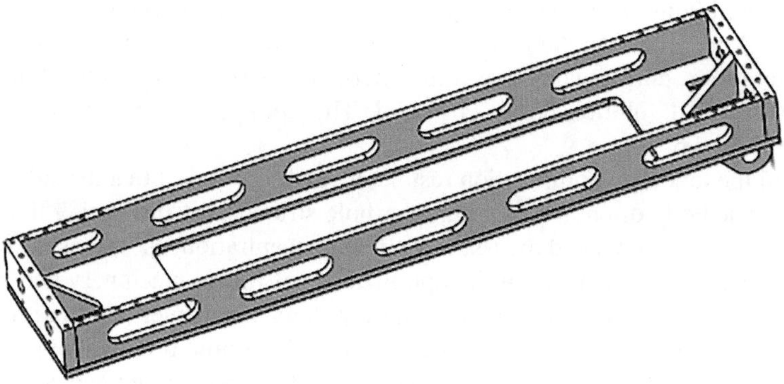

Fig. 7 Reconstructed model after topology optimization

4.2 *Static Analysis of the Feed Machine Body*

A static analysis was performed with the power head positioned at the front end of the feed machine body, which was tilted upwards by 20°. The constraint and boundary conditions remained consistent with the previous setup. Figure 8 shows the stress contour plot of the feed machine body, with a maximum stress of 129.7 MPa.

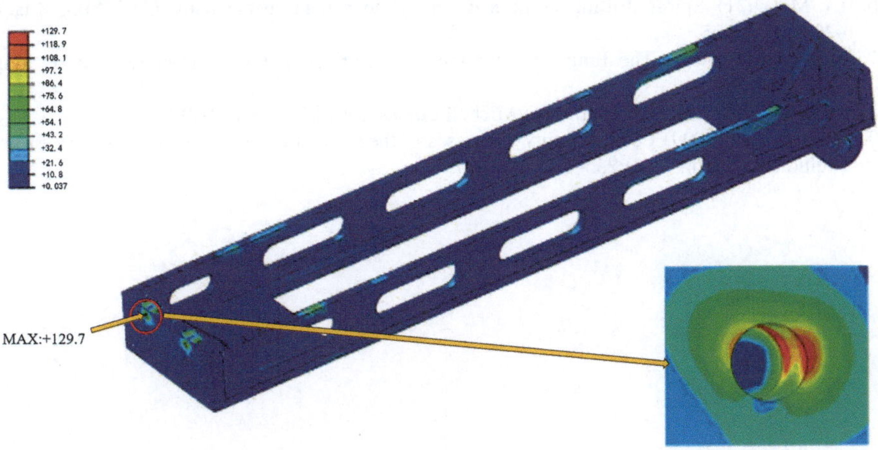

Fig. 8 Stress contour plot of the feed machine body after topology optimization

5 Conclusion

In this study, a volume-constrained mathematical model was established by integrating relevant theoretical knowledge of topology optimization. Abaqus software was used to perform topology optimization on the feed machine body after multi-objective optimization, and the model was reconstructed using SolidWorks. A static analysis of the reconstructed 3D model was then conducted using Abaqus, with results showing that the allowable stress of the material was satisfied. Through topology optimization, the feed machine body achieved a weight reduction of 55.65 kg, reducing the weight by 35.6%. Compared to the original model, the weight reduction was 70.88 kg, representing a decrease of 41.4%.

References

1. Zhang Y, Chang J, Zhang J (2008) Dynamic analysis study of the feed machine body of a fully hydraulic power head drilling rig. Coal Mine Machinery 2008(03):71–73
2. Wang H (2015) Design and simulation analysis of a multi-degree-of-freedom tunnel drilling rig frame. Coal Mine Machinery 36(07):109–111
3. Zhang R (2015) Dynamic analysis study of the feed machine body of a fully hydraulic drilling rig. Coal Technology 34(08):245–247
4. Wang T (2020) Analysis and design of the feed device for the drilling rig passing through faults in fully mechanized mining faces. Coal Mine Mach 41(09):101–103
5. Hao Y (2021) Optimization and application of the multi-variable crawler drilling rig. Inner Mongolia Petrochemical 47(03):23–26

6. Lv M (2021) Spiral drilling combined oil cylinder feed mechanism. Coal Mine Mach 42(11):114–115
7. Michell AG (1904) The limits of economy of materials in frame structures. Philos Mag 8(47):589–597
8. Hegemier GA, Prager W (1969) On Michell trusses. Int J Mech Sci 11(2):209–215
9. Achtziger W (1999) Local stability of trusses in the context of topology optimization. Struct Multidiscip Optim 17(4):235–246

Open Access This chapter is licensed under the terms of the Creative Commons Attribution-NonCommercial-NoDerivatives 4.0 International License (http://creativecommons.org/licenses/by-nc-nd/4.0/), which permits any noncommercial use, sharing, distribution and reproduction in any medium or format, as long as you give appropriate credit to the original author(s) and the source, provide a link to the Creative Commons license and indicate if you modified the licensed material. You do not have permission under this license to share adapted material derived from this chapter or parts of it.

The images or other third party material in this chapter are included in the chapter's Creative Commons license, unless indicated otherwise in a credit line to the material. If material is not included in the chapter's Creative Commons license and your intended use is not permitted by statutory regulation or exceeds the permitted use, you will need to obtain permission directly from the copyright holder.

Motion Performance Analysis of New Tracked Vehicle Based on Plateau Characteristic Agriculture

Yiwei Shi, Shichang Han, and Tang Xue

Abstract Aiming at the field of agricultural transportation with plateau characteristics, a bionic X-type tracked vehicle with an inerter is proposed. By comparing the theoretical calculation torque and simulation result torque of different road conditions and slopes, the feasibility and correctness of the tracked vehicle are verified. In the extreme case of climbing and crossing the ditch, the pitch angle is used as the evaluation standard to confirm that the tracked vehicle will not roll over under this road condition, which provides theoretical support for subsequent research.

Keywords Tracked vehicle · Inerter · Crossing ability · Pitch

1 Introduction

According to the survey, the cultivated land below the 0–25 degree slope of the plateau characteristic agriculture in Yunnan Province accounts for 81.36%, and the tracked vehicle has the advantages of low grounding specific pressure, good mobility, and strong obstacle crossing ability [1]. It has good trafficability and adaptability in the face of gravel, slopes, obstacles, and other roads, and is widely used in the field of agricultural transportation.

The emergence of the inerter [2] provides a new research direction for vehicle vibration reduction. Chen et al. [3] of Jiangsu University established a 1/4 model of the vehicle and optimized the suspension performance of the vehicle by improving the ISD (inerter-spring-damping) three-element structure. Yang et al. [4] studied a variety of vehicle suspension structures and found that the inertance coefficient provided by the inerter can effectively reduce the main frequency and offset the frequency of the suspension. Li et al. [5], based on the bionic vibration isolation theory,

Y. Shi · S. Han (✉) · T. Xue
Yunnan International Joint Laboratory of Intelligent Control and Application of Advanced Equipment, Kunming University of Science and Technology, Kunming, China

Kunming University of Science and Technology, Kunming, China

© The Author(s) 2026

Z. Zhang, P. Mercorelli (eds.), *Mechanical Design and Simulation: Insights and Innovations*, Lecture Notes in Mechanical Engineering,
https://doi.org/10.1007/978-981-95-2405-1_13

the research group designed a seat structure with an inerter and obtained the conclusion that the structure can effectively isolate low-frequency vibration isolation. The new bionic ISD structure applied to vehicle suspension is a new type of structure. Yang et al. [6] used displacement transmissibility and nod acceleration as evaluation criteria in the bionic suspension with inerter and verified that the bionic ISD structure of various pavements was superior to the SD structure, and the displacement transmissibility decreased with the increase of pavement complexity.

The new ISD structure is further applied to the plateau characteristic agriculture. In this paper, the pavement and ditch with plateau agricultural characteristics are taken as the research object, and a new bionic tracked vehicle with an inerter is studied to meet the trafficability and adaptability of the vehicle.

2 Structural Design of Tracked Vehicle

The design of the tracked vehicle is an important part of its driving stability, which is convenient for subsequent maintenance and design, and modularizes the chassis of the tracked vehicle. The tracked vehicle is divided into a suspension module, track module, and motor module. The overall structure of the tracked vehicle design is shown in Fig. 1.

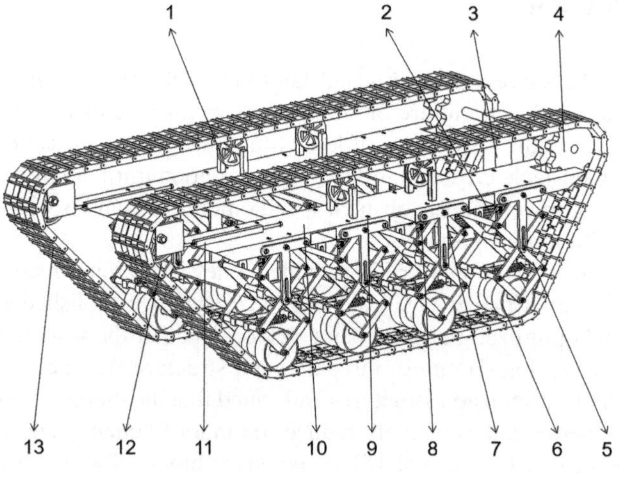

1-Branch chain wheel 2-Electrical machine 3-Gear box 4-Driving wheel 5-Bionic suspension 6-Loading wheel 7-T-steel 8-Inerter 9-Suspension mounting 10-Frame 11-Tensioning device 12-Tensioning wheel 13-Track

Fig. 1 Overall layout diagram of tracked vehicle. 1-Branch chain wheel 2-Electrical machine 3-Gear box 4-Driving wheel 5-Bionic suspension 6-Loading wheel 7-T-steel 8-Inerter 9-Suspension mounting 10-Frame 11-Tensioning device 12-Tensioning wheel 13-Track

2.1 Overall Layout of the Tracked Vehicle

2.2 Selection of Drive Module

The driving mode of the inverted ladder-tracked vehicle is generally divided into front drive, rear drive, and high drive [7]. The rear drive is adopted, and the contact end between the track and the ground is a tight edge, which will not produce large energy loss during operation. Electric drive is suitable for the driving mode of small tracked vehicles, which can instantaneously provide large torque to speed up. In summary, the vehicle uses a rear electric drive mode. The maximum torque generally appears when the maximum angle of the design climbing is designed. β is the climbing angle, r_x is the radius of the driving wheel, and the torque of the driving motor is calculated T_x:

$$T_x = \frac{r_x \left[ma + mg\left(f \cos\beta + \sin\beta \right) \right]}{i\eta} \tag{1}$$

The vehicle is designed as two drive motors, and the motor selection is calculated according to one half of the total weight of the vehicle [8], where i=9, f=0.2, a = 0.3m/s^2 the calculation results in the preset maximum slope driving torque is 17.9 N·m. According to the calculated torque, the DC rare earth permanent magnet high-performance servo brushless motor is selected. The model is 180 M-0250030C-E. The rated torque of the motor is 25 N·m, and the instantaneous torque of the motor is 55 N·m.

2.3 Track Module Design

Wheel-hole track [9] provides a reliable and effective way of power transmission, which is suitable for off-road conditions of complex terrain. When the vehicle is fully loaded, the bandwidth of the track and the length of the ground are set. According to the width of the track, it can be obtained by the empirical formula:

$$b = \left(0.9 \sim 1.1\right) \times 209 \times \sqrt[3]{m \times 10^{-3}} \tag{2}$$

The relationship ratio between the track grounding length d_1 of the tracked vehicle and the track gauge width B of the vehicle is as follows:

$$\varepsilon_b = d_1 / B \tag{3}$$

Table 1 Tracked vehicle design standards

Name of indicator	Parameter	Name of indicator	Parameter
Vehicle mass (kg)	250	Body size	1050 × 798 × 600
Load mass (kg)	100	Design speed	1.2
Climbing angle (°)	25	Energy	Electricity

According to the vehicle structure, the appropriate driving wheel pitch circle radius r_x=115 mm and the radius of the tension wheel r_t=104 are selected. The design index is shown in Table 1.

3 Tracked Vehicle Design Standards

Based on the motion state of the tracked vehicle in the 0–25° slope, the climbing and climbing ditches are calculated and analyzed.

3.1 Longitudinal Stability Analysis

The reaction force of the ground to the vehicle is F_m, the vertical distance between the center of mass of the vehicle and the direction of the ramp is h_c, the distance between the center of mass of the vehicle and the direction of the ramp and the rear end and the front end is c_1 and c_2 respectively, and the longitudinal stability angle of the vehicle isβ_1. The relationship is shown in Fig. 2.

According to the analysis of the force of the tracked vehicle during the uniform climbing process. The moment of the reaction force of the tracked vehicle to the forward and backward bearing wheel connection point A_a is obtained:

$$F_m n_{t1} + G h_c \sin\beta_1 = G c_1 \cos\beta_1 \tag{4}$$

The simplified condition for the tracked vehicle not to overturn is:

$$\beta_1 \le \arctan\frac{c_1}{h_c} \tag{5}$$

Similarly, the rollover conditions of the vehicle during the downhill process are:

$$\beta_1 \le \arctan\frac{c_2}{h_c} \tag{6}$$

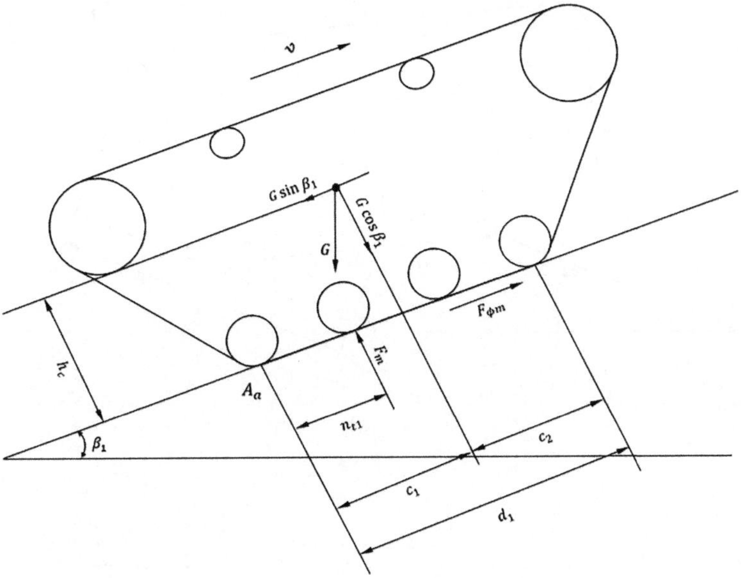

Fig. 2 Longitudinal stability analysis

No-load and full-load h_c are 358 mm and 350 mm respectively,c_1 are 604 mm and 578 mm respectively, and$c_2 = d_1 - c_1$. Considering the no-load and full-load conditions, the maximum angle of longitudinal stability of the vehicle is 51.2°.

3.2 *Analysis of Tracked Vehicle Climbing across Ditches*

The uphill crossing is divided into two stages. The first stage is when the center of mass falls into the ditch, and the tension wheel at the forward direction of the tracked vehicle contacts with the right wall, as shown in Fig. 3a. In the second stage, the vehicle has a driving wheel in contact with the left wall. In this case, the vehicle has an elevation angle greater than 25°, and the rear end of the vehicle is in the ditch, and the center of gravity has been separated from the ditch, as shown in Fig. 3b.

The maximum width of the slope gully is:

$$\begin{cases} d_g = \sqrt{(c_2 + h_h \tan\beta + L_{t1})^2 + h_g^{\ 2}} + r_t \\ d_g = \sqrt{(c_1 - h_h \tan\beta)^2 + r_w^{\ 2}} + r_w \end{cases} \quad (7)$$

The maximum width of the upper 25° slope is 466 mm.

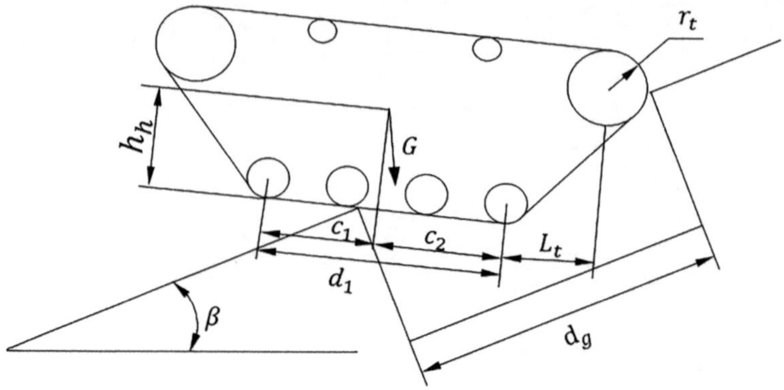

(a)

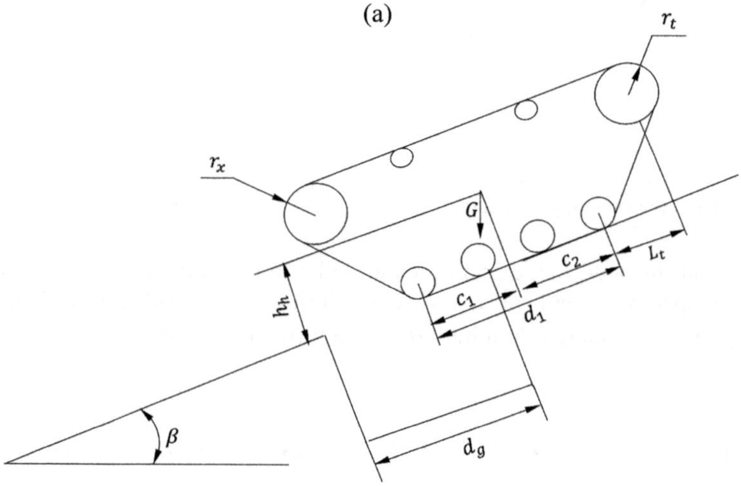

(b)

Fig. 3 Uphill ditch schematic diagram

3.3 Linear Dynamics Analysis of Tracked Vehicle

The pressure per unit area of the crawler is expressed by the Bekker formula:

$$p = z^n \left(\frac{k_c}{b} + k_\varphi \right) \tag{8}$$

p is the pressure per unit area of the track, z is the ground subsidence, k_c is the cohesive deformation modulus of the ground, k_φ is the friction deformation modulus of the ground. In practice, the tracked vehicle is not evenly distributed in the actual operation process, and its maximum pressure generally occurs at the position of the load-bearing wheel.

$$\begin{cases} \tan\alpha_h = t_p / D_c \\ \cos 2\alpha_h = 1 - 2\tan^2\alpha_h \end{cases} \tag{9}$$

α_h is the angle between the ground arc and the half of the track pitch. The sinkage is quite complicated in the actual situation. The front-end or back-end load-bearing wheel is consistent with the grounding arc angle of the crawler, and the force length of the grounding end is L_x, and then the sinkage of the front end of the crawler is z':

$$\begin{cases} L_x = 0.5D_c\left(\sin 2\alpha_h + \sin\alpha_h\right) \\ z' = \left[\dfrac{\dfrac{2F_{Ns}}{3bD_c}}{\left(\dfrac{k_c}{b} + k_\varphi\right)}\right]^{\frac{1}{n}} \end{cases} \tag{10}$$

When the tracked vehicle is running on the road, the work done by compacting the ground when passing through the land is E_p equal to the work done by the track consumption E_r, and then the driving resistance F_{Rc} is obtained:

$$F_{Rc} = \frac{1}{(n+1)\left(k_c + bk_\varphi\right)^{\frac{1}{n}}}\left(\frac{F_{Ns}}{L}\right)^{\frac{n+1}{n}} \tag{11}$$

Compaction resistance is usually simplified to rolling resistance F_{fm}:

$$F_{fm} = F_{Ns} f \tag{12}$$

When the tracked vehicle passes through the soft road, it will push the land under the ground all the time, and the bulldozing resistance is F_{Rb}:

$$F_{Rb} = b\left(czK_c + 0.5z^2\gamma K_r\right) \tag{13}$$

c is the cohesion coefficient, γ is the soil density, K_c and K_φ are calculated as follows:

$$\begin{cases} K_c = \left(N_c - \tan\varphi\right)\cos^2\varphi \\ K_\varphi = \left(\dfrac{2N_\gamma}{\tan\varphi} + 1\right)\cos^2\varphi \end{cases} \tag{14}$$

Table 2 Linear dynamic calculation results of soft ground

	Clayey soil		Sandy loam		Dry sand	
Parameter name	No-load	Full load	No-load	Full load	No-load	Full load
Sinkage (mm)	2.4	4.6	11.8	17.2	25.6	34.7
Torque (N•m)	28.2	35.4	31.3	43.8	37.1	51.7

φ is the internal friction angle. $N_c$and N_γ are the bearing capacity coefficient of Taisha. The friction resistance F_i is opposite to the direction of the vehicle's movement, f_i is the internal friction coefficient, and 0.08 is taken. The dynamic equation of the average speed straight line is as follows:

$$\begin{cases} F_{fr} = F_P + F_{fm} \\ F_j = F_{fm} + F_P + F_i \end{cases} \tag{15}$$

The commonly used sandy loam and dry sandy soil were selected. The results are shown in Table 2:

The following will verify the calculation results of the above subsidence, climbing angle, and the angle of crossing the ditch on the slope.

4 Simulation Analysis and Verification of the Motion Performance of the Tracked Vehicle

Recurdyn is based on a fully recursive algorithm and is suitable for large-scale calculations. It has a dedicated crawler module. The scheme uses Recurdyn multi-body dynamics for verification.

4.1 Tracked Vehicle in Different Road Straight Line Simulation Check

In the simulation, the soft soil foundation is set up, and the linear motion simulation analysis is carried out in the horizontal direction. The desert is dry sand, and the plateau environment is mostly humus soil. The simulation results are shown in Fig. 4.

According to the simulation results of Fig. 4a, the subsidence of the vehicle on different roads under no-load and full-load conditions is shown in Table 3.

Through Figs. 4b–d, the subsidence is gradually deepened, because the bulldozing resistance and the grounding area increase the torque, and the vehicle begins to accelerate to the specified speed within 1 ~ 3 s. The vehicle torque increases and fluctuates when the speed reaches the maximum value. In dry sand, the no-load and

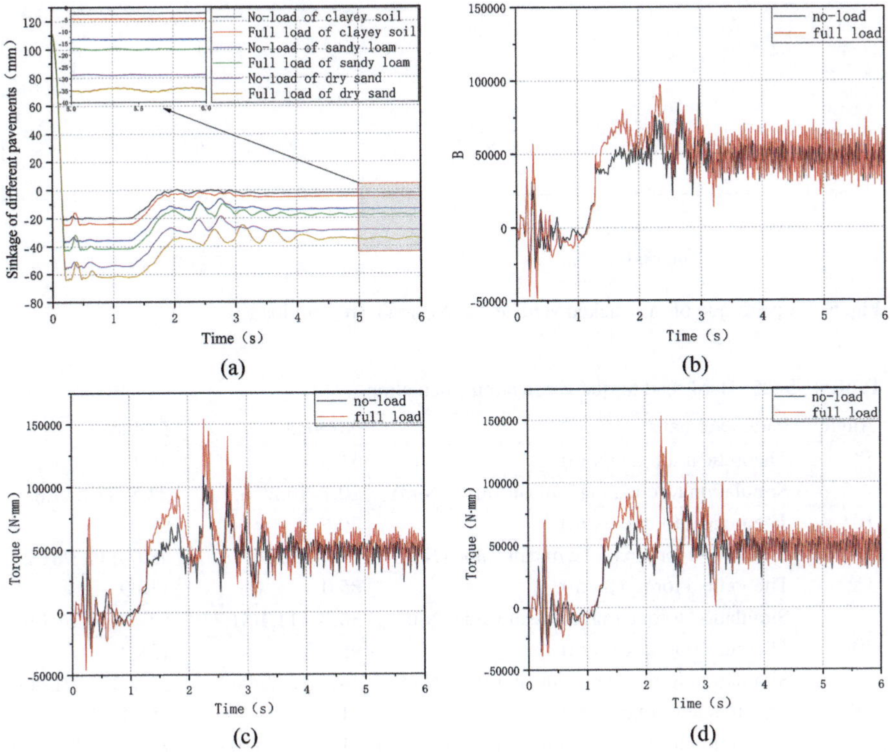

Fig. 4 Straight line soft ground simulation results

Table 3 Comparison table of subsidence of tracked vehicles

Parameter name	Clayey soil		Sandy loam		Dry sand	
	No-load	Full load	No-load	Full load	No-load	Full load
Theoretical sinkage (mm)	2.4	4.6	11.8	17.2	25.6	34.7
Practical sinkage (mm)	2.6	4.9	13.5	18.1	28.4	36.0
Percentage error	8.3%	6.5%	14.4%	5.2%	10.9%	3.7%

full-load values were stable at 23.4–58.1 N · m and 29.0–70.4 N · m after 3.5 s, respectively. In the sandy loam environment, the no-load and full-load are stable at 30.1 ~ 60.0 N · m and 32.8 ~ 67.8 N · m. In operation, there is a gap between the track and the wheel, which makes the torque of the vehicle fluctuate, and the torque fluctuation decreases with the increase of the subsidence. In the clayey soil environment, the subsidence is small, and the percentage difference will be enlarged. There are differences in the weight of the other two ground environments. The weight is not linearly related to the subsidence, but the subsidence increment will gradually decrease with the increase of weight, so the difference of the vehicle is greater than that of the full load.

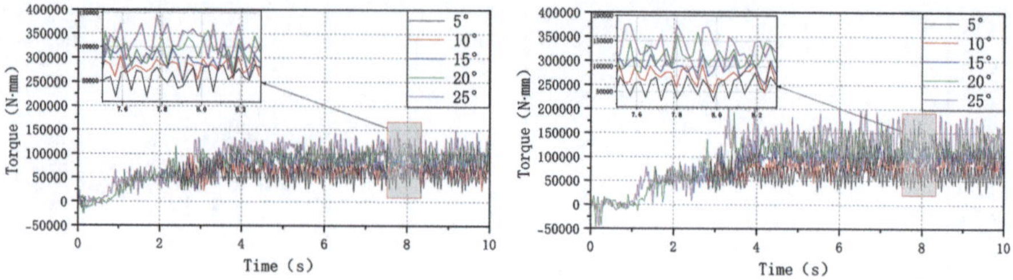

(a) No-load (b) Full load

Fig. 5 Slope torque of the tracked vehicle. (**a**) No-load. (**b**) Full load

Table 4 5–25° Hard road torque verification calculation

Angle	Parameter name	No-load	Full load
5°	Theoretical torque (N·m)	55.8	78.1
	Simulation torque range /mean value (N·m)	20.7–76.2/56.9	28.5–87.1/61.6
10°	Theoretical torque (N·m)	70.6	98.9
	Simulation torque range /mean value (N·m)	36.4–93.7/67.7	48.0–117.3/81.2
15°	Theoretical torque (N·m)	85.0	118.9
	Simulation torque range /mean value (N·m)	38.7–111.4/81.7	57.6–130.2/97.9
20°	Theoretical torque (N·m)	98.7	138.2
	Simulation torque range /mean value (N·m)	44.8–128.9/94.1	62.9–175.7/124.6
25°	Theoretical torque (N·m)	111.7	156.4
	Simulation torque range /mean value (N·m)	61.7–147.4/103.7	74.3–202.3/143.9

4.2 Tracked Vehicle Simulation on Slope Road

In the field of agricultural transportation, clear requirements are put forward for stability and obstacle-surmounting performance. In the software simulation, a slope with a slope of 25° is established, and the driving parameter is set to step (time, 1, 0, 3, 597.86D). The simulation results are shown in Fig. 5:

Figure 5 shows that the torque increases with the increase of the slope angle. Table 4 is the comparison table of theoretical calculation and simulation results:

In summary, the torque difference between no-load and full-load increases with the increase of slope angle. The calculated value of the 5°-25° tracked vehicle on a slope is in good agreement with the simulation range and mean value, so the simulation results are general. The maximum torque is 202.3 N·m, which is in line with the motor selection standard. Table 5 shows the mean percentage error of torque.

According to Table 5, the error of the tracked vehicle when it passes through the slope of 5°, 10°, and 15° is greater than that of other roads. Because the vehicle passes through the process from the horizontal to the slope, the broken surface of more than 20 degrees has a smooth buffer slope, so the impact is generated, which leads to a large percentage error in the upper slope. However, on the whole, the comparative analysis of the torque of the tracked vehicle structure is

Table 5 Percentage error of climbing torque of the tracked vehicle

Angle	Percentage error of torque	
	No-load	Full load
5°	2.0%	21.1%
10°	4.1%	17.8%
15°	4.0%	17.6%
20°	4.9%	9.8%
25°	7.7%	8.0%

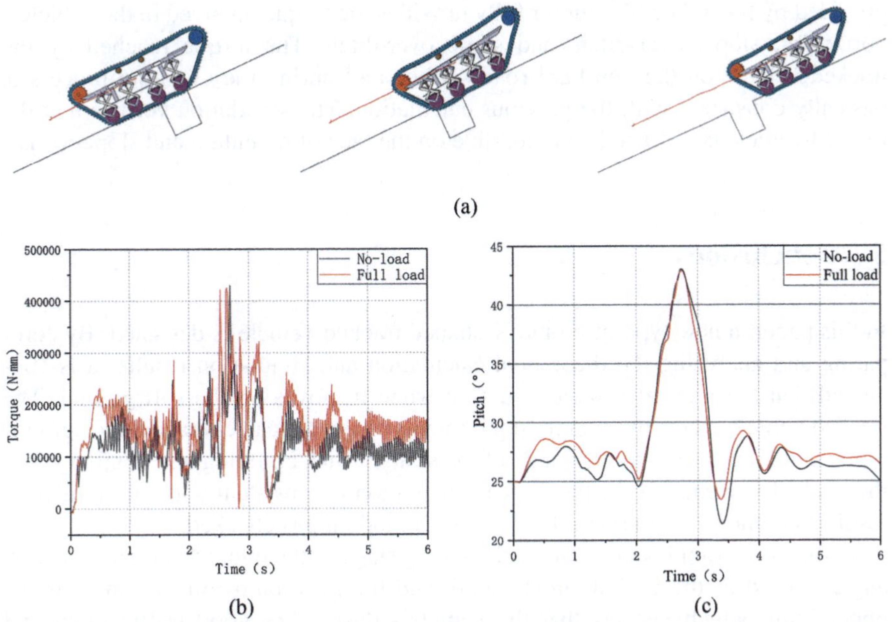

Fig. 6 Process and results of slope climbing and gully crossing

reliable, which verifies the unity of the simulation results and theoretical calculations of the vehicle on the slope.

4.3 *Simulation of a Tracked Vehicle Crossing a Ditch on a Slope*

The previous theoretical calculation shows that the vehicle can pass through the ditch with a width of 466 mm on the 25° slope. To verify the trafficability of the vehicle on the slope, a slope with a ditch is established in the software, and the driving function of the driving wheel is step (time, 0.1, 0,1, 597.86D). Its crossing ditch and simulation results are shown in Fig. 6.

As shown in Figs. 6b, c, the vehicle initially stops on the slope, and the speed rises from 0.1–1 s to 1.2 m/s. At this time, the torque increases rapidly, and the peak value of the fully-loaded vehicle reaches 228.9 N*m. In 1.6–3 s, the vehicle begins to cross the ditch, and the peak value reaches 412.5 N*m during the whole process of crossing the ditch and then decreases to gradually stable. When climbing over the ditch, the pitch angle of the vehicle is analyzed by the pitch angle, and the pitch angle of the vehicle is within the rollover angle.

In summary, the maximum torque appearing in the vehicle simulation is in line with the maximum value provided by the internal transmission ratio of the torque provided by the motor. The motor fully provides the torque involved in the vehicle's horizontal, slope, over-ditch, and slope over-ditch. The torque reached by the tracked vehicle on the non-hard road surface and during the climbing process is basically consistent with the previous calculation. The simulation verifies that the tracked vehicle is trafficable and feasible on the face of mountain and slope roads.

5 Conclusions

In this paper, a new type of bionic X-shaped tracked vehicle is designed. By comparing and analyzing the theoretical calculation and simulation results, it is concluded that the tracked vehicle will sink when it moves on the soft ground. The tracked vehicle meets the design requirements of the plateau characteristic agricultural vehicle under the motion state of the straight line, climbing and climbing over the ditch. The torque of the theoretical calculation is consistent with the simulation results, and the correctness of the theoretical modeling is checked.

In the most complex climbing and ditching stage of the tracked vehicle, the pitch angle is used as the evaluation standard, and the maximum rollover angle is the upper limit, which verifies that the vehicle's design has good trafficability and stability.

References

1. Weiwei W, Liqing C, Yang Y et al (2021) Research status and prospect of chassis technology of agricultural machinery. Agric Mach J 52(08):1–15
2. Smith MC, Wang FC (2004) Performance benefits in passive vehicle suspensions employing inerters. Veh Syst Dyn 42(4):235–257
3. Chen L, Yang X, Wang R et al (2014) Research on the performance of improved ISD three-element vehicle passive suspension. J Automot Eng 36(03):340–345
4. Yang X, Du Y, Liu Y et al (2018) Study on the influence of inertia coefficient on the frequency characteristics of vehicle ISD suspension system. J Vib Shock 37(07):240–246
5. Li X, Han S, Yang S (2024) Research on vibration isolation performance of bionic seat with inerter. Mech Sci Technol 44:1–10
6. Yang S, Han S, Li X (2024) Design and vibration reduction performance analysis of bionic X-shaped suspension with inerter. J Ordnance Equip Eng 45(06):260–266

7. Hao L (2023) Stability study of inverted trapezoidal tracked vehicle based on Recurdyn. Central North University
8. Zhitong W (2018) Research on key technologies of coal mine rescue tracked vehicle. Taiyuan University of Technology
9. Feng L (2016) Design and research on the chassis of rubber crawler bucket wheel reclaimer for salt. Tianjin University of Science and Technology

Open Access This chapter is licensed under the terms of the Creative Commons Attribution-NonCommercial-NoDerivatives 4.0 International License (http://creativecommons.org/licenses/by-nc-nd/4.0/), which permits any noncommercial use, sharing, distribution and reproduction in any medium or format, as long as you give appropriate credit to the original author(s) and the source, provide a link to the Creative Commons license and indicate if you modified the licensed material. You do not have permission under this license to share adapted material derived from this chapter or parts of it.

The images or other third party material in this chapter are included in the chapter's Creative Commons license, unless indicated otherwise in a credit line to the material. If material is not included in the chapter's Creative Commons license and your intended use is not permitted by statutory regulation or exceeds the permitted use, you will need to obtain permission directly from the copyright holder.

Hydrodynamic Performance Analysis of Large-Amplitude Undulating Fin Propulsion

Rui Zhu, Zhi-Dong Wang, and Guohuai Sun

Abstract Apteronotus albifrons possesses an elongated anal fin that enhances its cruising efficiency and maneuverability in complex environments. This study utilizes a three-dimensional numerical model in conjunction with dynamic mesh technology to analyze the hydrodynamic properties of a large-amplitude undulating fin under varying flow velocities and oscillation amplitudes. Based on the "differential first, then integral" approach, the thrust coefficient and propulsion efficiency of the undulating fin are evaluated using computational simulation methods. The results indicate that as the advance ratio increases, the thrust coefficient gradually decreases, becoming negative when the advance ratio reaches 1. At an advance ratio of 0.75, the average propulsion efficiency peaks at approximately 65.8%. Furthermore, larger oscillation amplitudes contribute to an increase in the thrust generated by the undulating fin, significantly enhancing its overall operational performance. These findings provide essential methodologies and theoretical support for the future optimization and design of undulating fins.

Keywords Large-amplitude undulating fin · Propulsive efficiency · Numerical simulation · Hydrodynamic performance

1 Introduction

Through long-term natural selection and evolution, fish have developed efficient propulsion systems and exceptional maneuverability in water. In recent years, biomimetic propulsion technology has emerged as a key focus in underwater vehicle research, attracting significant attention [1]. The remarkable swimming abilities of fish have fueled advancements in biomimetic research, which categorize fish propulsion into two primary methods: body and/or caudal fin (BCF) propulsion and

R. Zhu (✉) · Z.-D. Wang · G. Sun
School of Naval Architecture and Ocean Engineering, Jiangsu University of Science and Technology, Zhenjiang, Jiangsu, China

© The Author(s) 2026

Z. Zhang, P. Mercorelli (eds.), *Mechanical Design and Simulation: Insights and Innovations*, Lecture Notes in Mechanical Engineering,
https://doi.org/10.1007/978-981-95-2405-1_14

median and/or paired fin (MPF) propulsion. BCF propulsion, utilized by most fish, enables high-speed cruising and rapid acceleration, although it tends to be less efficient and generates more turbulence. In contrast, the MPF propulsion mode provides improved maneuverability and efficiency in complex underwater environments. Undulating fin propulsion is regarded as having significant potential for the design of new uncrewed underwater robotics.

Numerous scholars, both domestically and internationally, have extensively studied oscillating fin propulsion systems, focusing on their configurations, hydrodynamic properties, and propulsion mechanisms. In 2005, Epstein et al. [2] investigated the propulsion performance of the weakly electric knifefish (Apteronotus albifrons, Fig. 1) using computational fluid dynamics (CFD). They explored the relationships between thrust and parameters such as frequency, amplitude, and wavelength, revealing the impact of vortices on efficiency.

In 2007, Wang et al. [3] conducted both theoretical and experimental studies on long dorsal fin undulating propulsion, examining the oscillatory mechanism, propulsion efficiency, and flow field characteristics. The maximum amplitude of the fin rays reached 30°. Rahman et al. (2010) performed numerical simulations on dual undulating fins to investigate the mechanisms of thrust generation. Their findings indicated that thrust and propulsion efficiency are primarily influenced by the aspect ratio and fin angle.

Hu et al. [4] focused on analyzing the hydrodynamic performance of undulating fin propulsion systems. Through computational fluid dynamics (CFD) simulations and experiments, they investigated how parameters like undulation frequency, amplitude, and speed affect thrust and efficiency. In 2020, Liu et al. [5] designed an undulating fin robot that mimics aquatic swimming patterns. Computational fluid dynamics (CFD) simulations and tests demonstrated excellent maneuverability and efficiency; however, the robot exhibited a limitation in speed.

In 2023, Zhang et al. [6] studied the hydrodynamics of undulating fins using the knifefish as a biomimetic model. Through simulations and experiments, they analyzed thrust, efficiency, and fluid dynamics, providing insights for optimizing bio-inspired propulsion. The maximum amplitude of the fin rays was 30°.

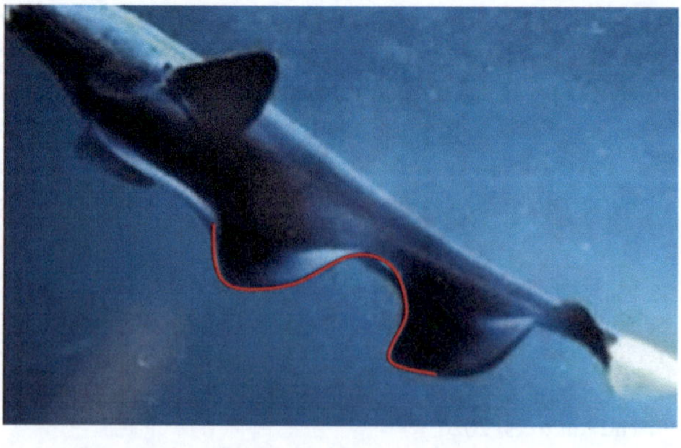

Fig. 1 Photo of Apteronotus albifrons

In the studies mentioned above, the amplitude of the undulating fins typically does not exceed 45°, and there is relatively limited research on large-amplitude torsional wave propulsion fins. Large-amplitude movements can significantly influence hydrodynamic characteristics, particularly thrust, propulsion efficiency, and flow field properties. Existing studies provide limited insight into how large-amplitude fin movements affect hydrodynamic performance, particularly in terms of thrust and efficiency, with minimal emphasis on the balance between amplitude and propulsion.

Consequently, this paper utilizes the knifefish (Apteronotus albifrons) as a model organism, designing undulating fins with a 65° amplitude and high fidelity. The study examines their hydrodynamic performance across a range of flow velocities and oscillation amplitudes. By employing a "differentiation followed by integration" method to calculate propulsion efficiency, the research enhances the accuracy of assessments, enabling a more precise evaluation of large amplitude undulating fins. This study provides valuable insights for optimizing biomimetic propulsion systems and lays the groundwork for developing more efficient and flexible underwater robots and unmanned underwater vehicles (UUVs).

The structure of this study is outlined as follows: Section 2 discusses motion control equations for large-amplitude undulating fins and outlines the numerical methods employed for the calculations, utilizing a "differentiation followed by integration" approach to enhance the accuracy of propulsion efficiency. In Sect. 3, the effects of various flow velocities and oscillation amplitudes on propulsion performance are examined, along with an exploration of changes in vorticity contours for the undulating fins. Finally, Sect. 4 summarizes the key findings and conclusions of the study.

2 Problem Description

2.1 The Equation of Motion for Undulating Fin

The dynamics of biomimetic fish fins are modeled using a fundamental frequency simple harmonic wave, with each fin ray oscillating like a simple harmonic pendulum about the X-axis. By controlling the variation in the oscillation angle along the X-axis, the overall movement pattern of the fin is regulated. Figure 2 illustrates both the three-dimensional view and the side view of the large-amplitude undulating fin.

This study examines the hydrodynamic performance of a large-amplitude undulating fin while excluding the effects of fluid-structure interaction. The computational model employed has a thickness of 0 mm, a length of 1380 mm, and a width of 150 mm. The parametric equation of the fin surface in Cartesian coordinates is expressed as:

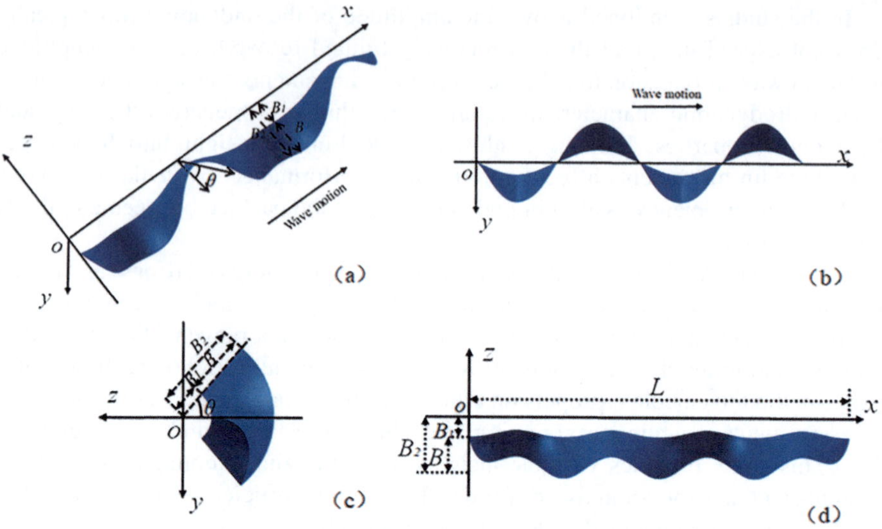

Fig. 2 Wave fin schematic diagram. (**a**) 3D view, (**b**) bottom view, (**c**) front view, (**d**) side view

$$\begin{cases} p_x(r,l,t) = l \\ p_y(r,l,t) = r \cdot \cos(\theta(l,t)) \\ p_z(r,l,t) = r \cdot \sin(\theta(l,t)) \\ 0 < r < B, 0 < l < L \end{cases} \tag{1}$$

In the equation, r represents the variation along the length of the fin ray, with a maximum value of B; *l* denotes the variation along the baseline of the fin surface, with a maximum value of L; *t* is the time variable. The angular variation of the fin ray at position *l* can be described by the following sine curve:

$$\theta(l,t) = \theta_{\max} \cdot \sin\left(\frac{2\pi}{T}t - \frac{2\pi}{\lambda}l + \phi_0\right) \tag{2}$$

where θ_{max} is used to describe the maximum swing angle of the fin, *f* denotes the oscillation frequency, T represents the period, λ indicates the wavelength, and Φ_0 is the initial displacement of the first fin ray, signifying the initial stage of the fin's undulating motion. The initial phases of all fin rays follow an arithmetic sequence.

2.2 Numerical Method

The undulating fin in this study is used in the propulsion system of an amphibious robot and features higher stiffness for land movement. With minimal fluid impact, the simulations focus solely on torsional wave propulsion, excluding fluid-structure interaction. Instead, the commercial computational fluid dynamics (CFD) solver Fluent, along with custom user-defined functions (UDFs), is utilized to ensure accurate modeling of fin dynamics and flow behavior. For a viscous three-dimensional incompressible fluid, this study employs the Reynolds-Averaged Navier-Stokes (RANS) equations for calculations [7]. The continuity and momentum equations are expressed as follows:

$$\rho\frac{\partial u_i}{\partial x_i} = 0 \tag{3}$$

$$\rho\frac{\partial u_i}{\partial t} + \rho\frac{\partial(u_i u_j)}{\partial x_j} = -\frac{\partial p}{\partial x_i} + \frac{\partial}{\partial x_j}\left(\mu\frac{\partial u_i}{\partial x_j} - \rho\overline{\mu_i^{'}\mu_j^{'}}\right) + S_i \tag{4}$$

In the equations, ρ represents fluid density, μ_i is the average velocity component, P is the average pressure, μ denotes the dynamic viscosity, $\rho\mu_i^{'}\mu_j^{'}$ is the Reynolds stress term that accounts for turbulence effects, and S_i is the volume force.

The Reynolds number calculated from L and U_0 is Re = 1.6×10^6, t, assuming turbulent flow over the fin surface. The *SST k-ω* [8] model is adopted for its ability to accurately capture shear stress and effectively control pressure gradients near the wall. For incompressible fluids, spatial discretization is performed using the finite volume method, with a second-order upwind scheme employed to enhance accuracy. A segregated solution approach is applied to the discrete equations, with the SIMPLE [9] algorithm implemented for simulations involving small time steps. Time discretization employs an implicit first-order scheme with a time step of 0.001 seconds over five wave periods.

2.3 Calculation Method for Propulsion Efficiency of Undulating Fin

The advance ratio, derived from propeller theory, represents the speed at which a propeller moves along its axis in relation to the water. In this study, it is defined as the propagation speed of the traveling wave on the undulating fin.

$$J = \frac{U}{f \cdot \lambda} \tag{5}$$

The non-dimensional thrust coefficient is defined as follows [4]:

$$CT = \frac{T}{\rho \cdot f^2 \cdot B^2 \cdot L^2} \tag{6}$$

where T represents the thrust generated during the propulsion of the long fin with torsional oscillation, f denotes the oscillation frequency of the fin, B represents the height of the fin, L represents the length of the fin, and ρ represents the density of water.

In the field of robotics, efficiency is precisely defined as the ratio of output power to input power. Several parameters are used to describe the hydrodynamic performance of an undulating fin: thrust (T_x), output power (P_x), input power (P), and propulsion efficiency (η). For a single propeller, the input power (P) is calculated as the product of motor torque (M) and angular velocity (Ω). Meanwhile, the effective output power (P_{out}) is determined by multiplying the measured thrust (T) by the flow velocity (U). The propulsion efficiency is defined as follows:

$$\eta = \frac{Pout}{P} = \frac{TU}{M\Omega} \tag{7}$$

In the equation, the thrust T corresponds to the undulating fin's thrust T_x and the flow velocity U matches the incoming flow velocity U_0 in numerical calculations.

Due to the periodic changes in angular velocity and torque over time at various positions across the entire undulating fin, the average torque ultimately equals zero. This study employs a followed-by-integration approach to calculate propulsion efficiency, dividing the undulating fin into several differential cells, as illustrated in Fig. 3.

The undulating fin is segmented into three sections, comprising a total of 30 differential cells. The head and tail of the oscillating fin are more sensitive to the influences of incoming and outgoing water flows, the width of the head and tail regions is set to $n_1 = 0.044$ m, while the width of the central portion is $n_2 = 0.05$ m. The final propulsion efficiency is calculated as follows [10]:

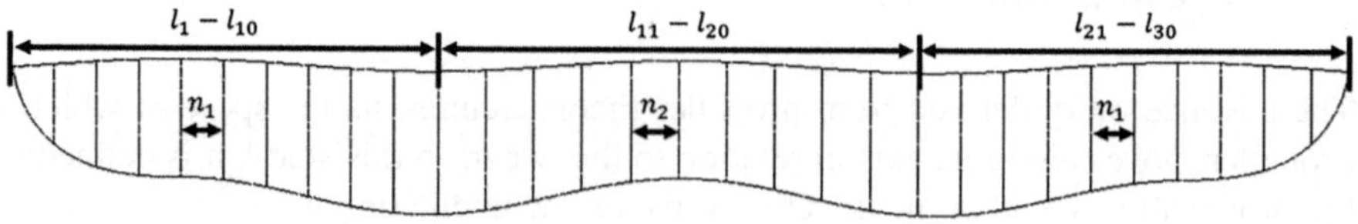

Fig. 3 Schematic diagram of differential unit division

$$\eta = \frac{Px(t)}{P(t)} = \frac{\int_0^L Tx(l,t)U0dl}{\int_0^L Mx(l,t)\Omega(l,t)dl} \tag{8}$$

2.4 *Computational Domain, Boundary Conditions, and Mesh Generation*

In the numerical simulation of large-amplitude undulating fins, two key challenges emerge. First, the propulsion mechanism of these fins involves complex fluid dynamics. The periodic and large-amplitude undulating motions cause the fin surface to undergo intricate deformations continuously. This necessitates constant adjustments to the shape and position of the computational grid near the fin. Therefore, dynamic mesh technology is required to update the grid at the moving boundaries and precisely capture the vortex structures caused by the motion of the fin. Second, precise motion equations must be established using User-Defined Functions (UDFs) to control the motion of the undulating fin. Addressing these two factors is critical for ensuring accurate simulations and optimizing the hydrodynamic performance of large-amplitude undulating fin.

Therefore, a hybrid mesh division is employed in the computational domain. Figure 4 illustrates this division. The dimensions of the outer computational domain are 3.5 L × L × L, while the inner domain measures 2 L × 0.5 L × 0.5 L, where L represents the fin length (1380 mm). The velocity inlet is located along the negative X-axis, and the pressure outlet is located behind the trailing edge, at a distance twice the length of the fin. The upper, lower, left, and right boundaries serve as symmetry planes.

Due to the large amplitude motion of the undulating fin, significant mesh deformations occur. To prevent the formation of negative-volume cells, the dynamic grid is updated using a diffusion smoothing technique and a local mesh reconstruction

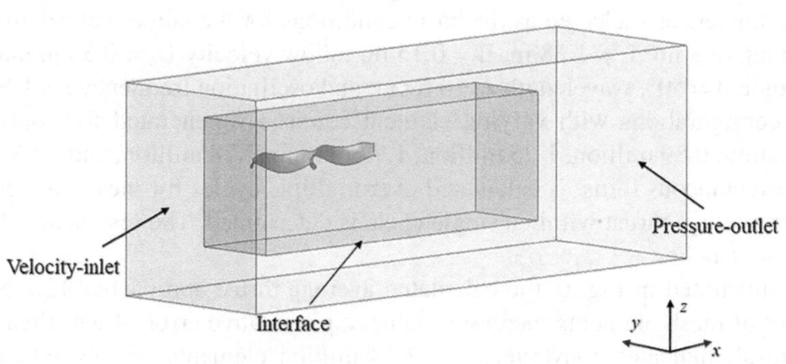

Fig. 4 Division of the computational domain

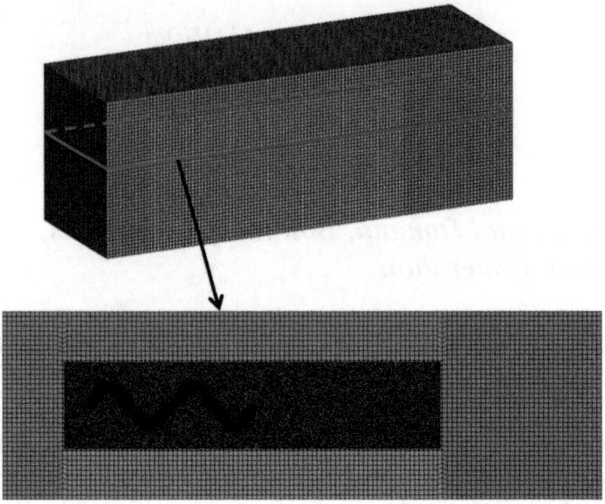

Fig. 5 Hydrodynamic computational mesh for large-amplitude oscillating flexible long-fin propulsion

method. Local mesh refinement is applied around the fin, while coarser mesh sizes are utilized in other fluid regions. Data is transferred between the outer and inner domains through an interface, with the mesh size gradually expanding from this interface. The fin surface mesh is generated separately, resulting in a total of 2.82 million mesh elements. The maximum skewness is below 0.69, indicating good orthogonality. Figure 5 illustrates the results of the mesh generation.

2.5 *Grid Independence Verification*

To guarantee that the quantity of mesh elements satisfies the accuracy requirements for numerical simulations, mesh convergence verification is essential. The following parameters are selected as the basic conditions for the numerical calculation of the undulating fin: L = 1.38 m, B = 0.15 m, inflow velocity $U_0 = 0.5$ kn, maximum amplitude $A = 50°$, wavelength $\lambda = 0.69$ m, and oscillation frequency $f = 1$ Hz. Five mesh configurations with varying element counts are generated for convergence verification: 0.89 million, 1.25 million, 1.77 million, 2.78 million, and 3.56 million. The instantaneous thrust is calculated over multiple cycles for mesh configuration, and the average thrust within a single cycle is determined. The results are shown in the Table 1 below.

As illustrated in Fig. 6, the calculated average thrust approaches 13.1 N as the number of mesh elements increases, achieving a relative error of less than 0.01%. This result indicates convergence at 2.78 million elements. Taking into account memory usage, computation time, and numerical accuracy, the mesh comprising

Table 1 Average thrust of an undulating fin over a single cycle at varying grid resolutions

Number of Grid ($\times 10^4$)	89 W	125 W	177 W	278 W	353 W
T_x (N)	12.9733	12.9978	13.1063	13.1074	13.1079

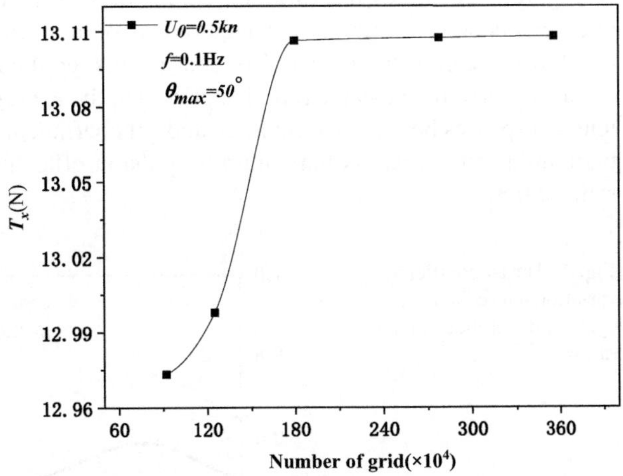

Fig. 6 Mesh convergence verification

2.78 million elements is selected for the numerical simulation of the hydrodynamic performance of the large amplitude undulating fin.

3 Hydrodynamic Calculation Results and Analysis

The control variable method is utilized for a hydrodynamic numerical study of the primary parameters of undulating fins. To effectively illustrate the general motion patterns, typical working conditions are established: an inflow velocity of 0.5 knots, a wavelength to fin length ratio of 1, a maximum oscillation angle of 50°, and an oscillation frequency of 1.00 Hz. Once the computed thrust, lateral force, lift, and input power achieve periodic convergence, the average values over multiple cycles are calculated as single-cycle averages and non-dimensionalized using dimensionless numbers.

3.1 *Influence of Inflow Velocity*

With a wavelength-to-fin-length ratio of $\lambda/L = 0.5$, an oscillation amplitude angle of $\theta_{max} = 50°$, and an oscillation frequency of $f = 1$ Hz, the advance ratio J is determined exclusively by the inflow velocity U_0.

Figure 7 illustrates that the thrust coefficient varies twice within a single cycle for different advance ratios, exhibiting a "W" shape. With an increase in the advance ratio, the thrust coefficient decreases, ultimately becoming negative at an advance ratio of 1.

Figure 8 illustrates the variation curves of average thrust, lateral force, and lift coefficients across different advance ratios. With an increase in the advance ratio, the thrust, lateral force, and lift coefficients gradually decrease and stabilize. Notably, when the advance ratio is below 0.8, the average thrust coefficient consistently surpasses both the lateral force and lift coefficients. Figure 9 indicates that the undulating fin attains its maximum propulsion efficiency of 65.8% at an advance ratio of 0.8.

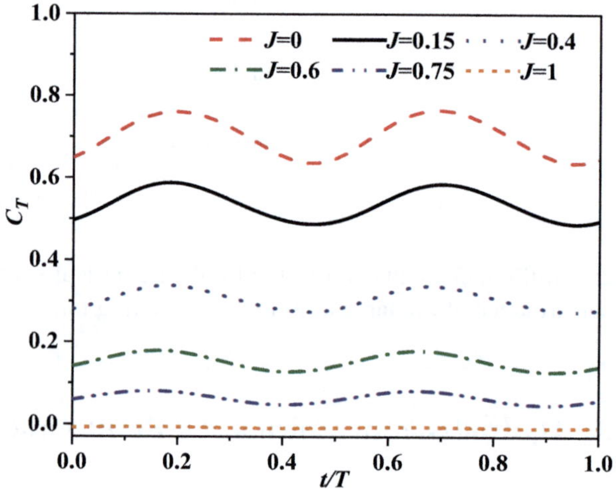

Fig. 7 Thrust coefficient variation curve over one cycle at different advance ratios

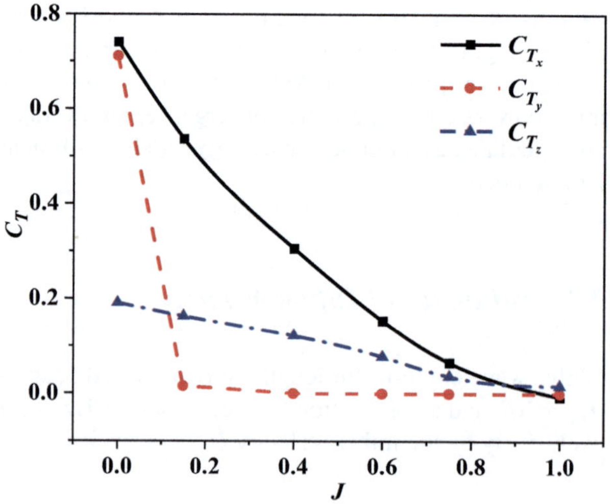

Fig. 8 Force coefficient curves at varying advance ratios

This paper employs the Q criterion to evaluate vortex intensity around the undulating fin. Figure 10 illustrates the vorticity curves of the fins at various advance ratios on the z = −0.14 m plane.

When the advance ratio is low, the vortex distribution is dense and robust. With an increase in the advance ratio, the formation of vortices at the fin's leading edge becomes increasingly challenging, while the dissipation rate at the trailing edge accelerates. This leads to a gradual decrease in both the number and strength of the vortices. These findings indicate that a higher incoming flow velocity directly influences the vortex structure and improves the hydrodynamic performance of the undulating fin.

Fig. 9 The variation curves of propulsive efficiency under different advance ratios

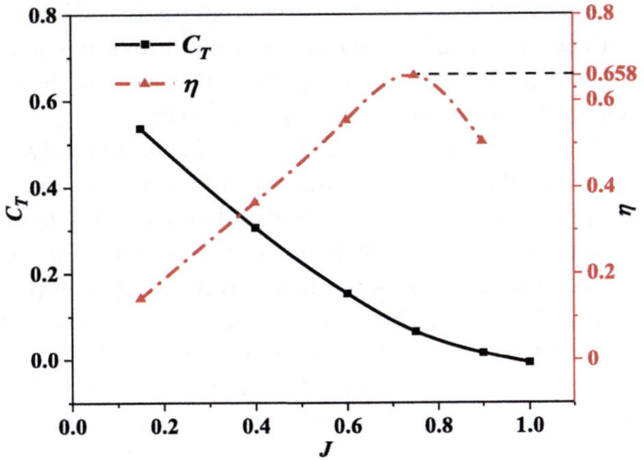

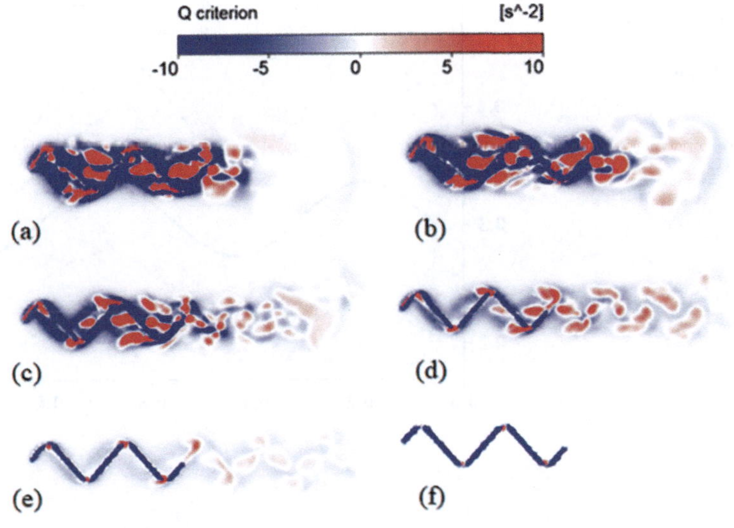

Fig. 10 Vorticity contours of the undulatory fin at different advance ratios: **(a)** J = 0, **(b)** J = 0.15, **(c)** J = 0.4, **(d)** J = 0.6, **(e)** J = 0.75, **(f)** J = 0.1

3.2 *Influence of Amplitude*

When the fin height is fixed, varying the oscillation angle affects the amplitude of the undulating fin. Numerical calculations produce a curve that illustrates the variation of the thrust coefficient over a single cycle at various oscillation angles, as shown in Fig. 11.

The variation of the thrust coefficient at different oscillation angles displays a periodic pattern akin to that observed at various advanced ratios. As the oscillation amplitude increases, the thrust coefficient steadily rises, and its fluctuations become more pronounced. Notably, when the maximum angle reaches 65°, the fluctuation period exhibits significant changes compared to smaller amplitudes.

Figure 12 illustrates the curves representing the average thrust, lateral force, and lift coefficients of the undulating fin at different oscillation amplitudes. It is clear that thrust increases significantly with larger oscillation amplitudes, while the lift and lateral force remain relatively constant.

Moreover, as illustrated in Fig. 13, the hydrodynamic characteristics demonstrate that the propulsion efficiency of the undulating fin remains relatively constant at approximately 36.5% as the oscillation amplitude increases. Simultaneously, the thrust coefficient significantly increases, allowing the undulating fin to maintain effective operational performance even at larger amplitudes.

As illustrated in Fig. 14, the vorticity contours of the undulating fin at varying amplitudes illustrate the formation of a typical reverse Kármán vortex street at the trailing edge, characterized by alternating vortices. The study reveals that as the amplitude increases, both the number and intensity of the vortices significantly

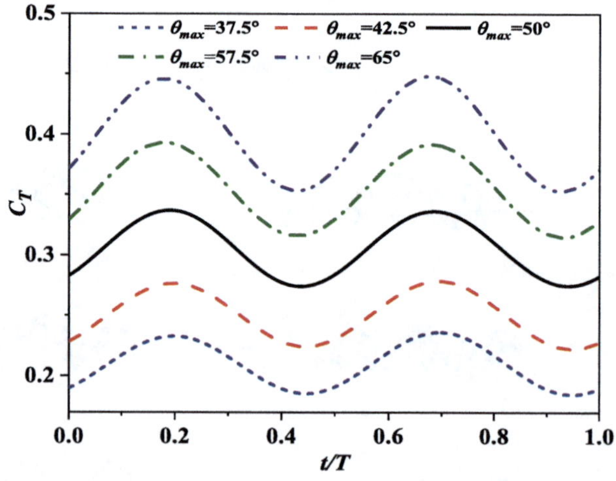

Fig. 11 Thrust coefficient variation within a single cycle at different oscillation angles

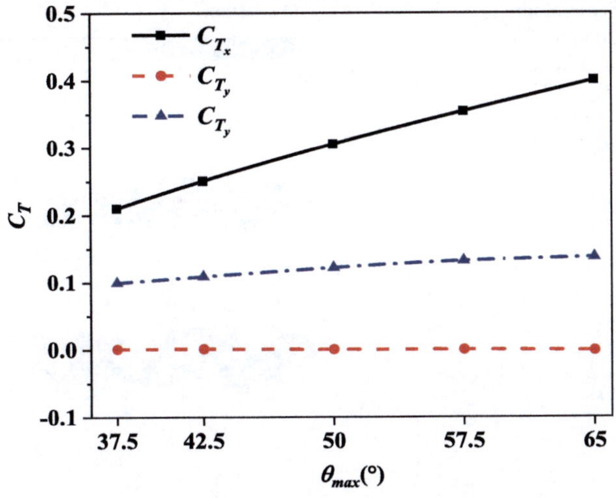

Fig. 12 Variation curves of force coefficients at different oscillation angles

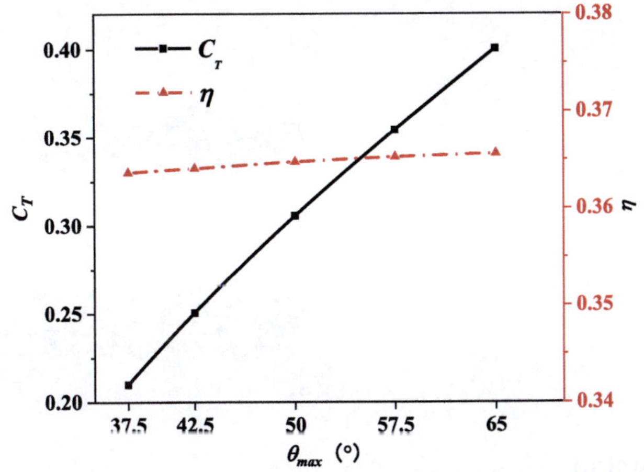

Fig. 13 Hydrodynamic characteristic curves at different angles of attack

increase. This suggests that larger amplitude generates stronger vortices, and the increase in oscillation amplitude exerts a positive and consistent influence on thrust, thereby confirming the previously mentioned variations in the thrust coefficient and enhancing the propulsion capability of the fins.

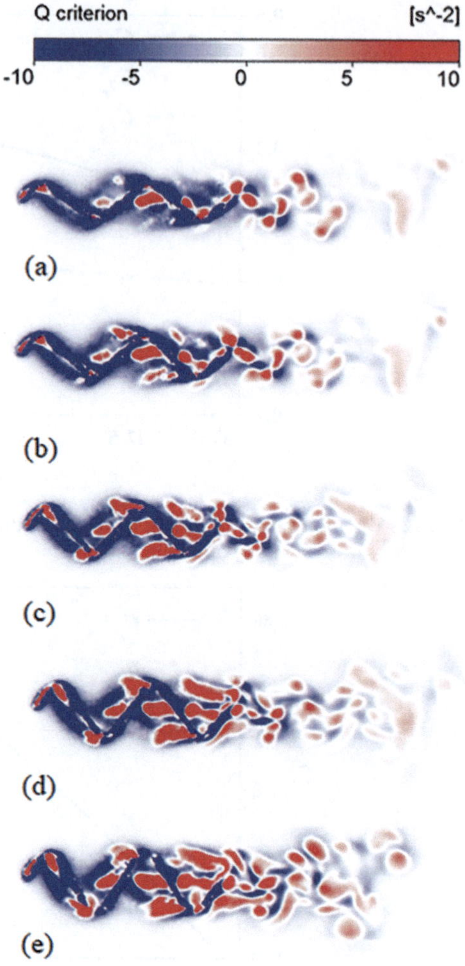

Fig. 14 Vorticity contour plots of the undulating fin under various amplitude conditions: (**a**) $\theta_{max} = 37.5°$, (**b**) $\theta_{max} = 42.5°$, (**c**) $\theta_{max} = 50°$, (**d**) $\theta_{max} = 57.5°$, (**e**) $\theta_{max} = 65°$

4 Conclusion

To enhance the understanding of the propulsion performance of three-dimensional, large-amplitude undulating fins, this study presents several significant contributions. Using the knifefish (Apteronotus albifrons) as a biomimetic model, we design a novel large-amplitude torsional wave propulsion fin. The research employs dynamic grid techniques, including diffusion smoothing and local grid reconstruction models, to effectively capture the complex movements of the fin and their effects on hydrodynamic characteristics. This study simulates large-amplitude torsional wave propulsion with a maximum oscillation amplitude of ±65°. Additionally, it elucidates the impact of large amplitudes on the thrust and propulsion efficiency of the undulating fin. The primary research findings are presented as follows:

1. This paper examines the thrust and propulsion efficiency of an undulating fin, examining the effects of flow velocity and oscillation amplitude on its performance. The results indicate that as the advance ratio (i.e. the flow velocity) increases, the thrust coefficient gradually decreases, while the average propulsion efficiency initially increases before subsequently decreasing. When $J = 0.75$ ($U_0 = 1$ kn), $\theta_{max} = 50°$, and $f = 1$ Hz, the maximum average efficiency η reaches 68.5% in the calculated working conditions.
2. As the amplitude of the undulating fin's oscillation increases, the average propulsion efficiency of the undulating fin remains stable at approximately 36.5%, demonstrating consistent performance. Simultaneously, the thrust coefficient significantly increases, allowing the undulating fin to maintain effective operational performance even at larger amplitudes.
3. A more detailed analysis is conducted on the vorticity curves of the undulating fin under varying advance ratios and amplitude conditions. As the amplitude increases, both the quantity and intensity of the vortices significantly increase. Consequently, this enhancement increases the propulsion efficiency of the undulating fin.

This research further investigates the intricate interactions between large-amplitude undulating fins and fluid flow. However, due to simplifications in the model, certain limitations exist, as factors such as fin thickness and material properties are not fully addressed. This oversight may result in suboptimal flexibility and efficiency within the propulsion system. Future research should incorporate fin thickness, material properties, and fluid-structure interactions, along with multi-condition testing and experimental validation, to optimize biomimetic design and enhance the performance and practicality of the propulsion system.

References

1. Zeng X, Xia M, Luo Z et al (2022) Design and control of an underwater robot based on hybrid propulsion of quadrotor and bionic undulating fin. J Mar Sci Eng 10(9):1327. https://doi.org/10.3390/jmse10091327
2. Epstein M, Colgate JE, MacIver MA (2005) A biologically inspired robotic ribbon fin. In: Proceedings of the 2005 IEEE/RSJ international conference on intelligent robots and systems (IROS), workshop on morphology, control, and passive dynamics. https://doi.org/10.1109/IROS.2006.281681
3. Guang-ming W (2007) Theoretic and experimental research on propulsion by bionic undulatory fin. National University of Defense Technology, Changsha
4. Chang W, Qiao H, Shijie L, Tangjia Z, Xindong S (2023) Hydrodynamic performance analysis of undulating fin propulsion. Phys Fluids 35(9):091906. https://doi.org/10.1063/5.0170156
5. Liu L, Li Y, Wang Y et al (2020) Design and preliminary evaluation of a biomimetic underwater robot with undulating fin propulsion. IOP Conf Ser Mater Sci Eng 790(1):012160. https://doi.org/10.1088/1757-899X/790/1/012160
6. Zhang M, Wang C, Su Y (2023) Hydrodynamic characteristics of an electric eel-like undulating fin. J Appl Fluid Mech 16(5):1030–1043

7. Liu F, Ji S (1996) Unsteady flow calculations with a multigrid Navier-stokes method. AIAA J 34(10):2047–2053. https://doi.org/10.2514/3.13351
8. Wilcox DC (1988) Reassessment of the scale-determining equation for advanced turbulence models. AIAA J 26(11):1299–1310. https://doi.org/10.2514/3.10041
9. Maertens AP, Triantafyllou MS, Yue DKP (2015) Efficiency of fish propulsion. Bioinspir Biomim 10(4):046013. https://doi.org/10.1088/1748-3190/10/4/046013
10. Sun G, Wang Z, Ling H et al (2024) Investigation on the propulsive efficiency of undulating fin propulsor. Ocean Eng 312:119113. https://doi.org/10.1016/j.oceaneng.2024.119113

Open Access This chapter is licensed under the terms of the Creative Commons Attribution-NonCommercial-NoDerivatives 4.0 International License (http://creativecommons.org/licenses/by-nc-nd/4.0/), which permits any noncommercial use, sharing, distribution and reproduction in any medium or format, as long as you give appropriate credit to the original author(s) and the source, provide a link to the Creative Commons license and indicate if you modified the licensed material. You do not have permission under this license to share adapted material derived from this chapter or parts of it.

The images or other third party material in this chapter are included in the chapter's Creative Commons license, unless indicated otherwise in a credit line to the material. If material is not included in the chapter's Creative Commons license and your intended use is not permitted by statutory regulation or exceeds the permitted use, you will need to obtain permission directly from the copyright holder.

Simulation Study on the Power of the Load Sensitive Proportional Variable Pump

Tianzhou Ma

Abstract Regarding the study of load sensitive proportional variable pump power, this article constructs a mathematical model of the variable pump and its control mechanism to obtain its key parameters. Using the Amesim platform to control the key parameters obtained from the mathematical model for simulation analysis, and finally comparing the output power simulation curves of the proportional variable pump system and the load sensitive variable pump system, it is found that the load sensitive variable pump has lower power when completing the same task.

Keywords Variable pump · Load sensitive · Simulation study

1 Introduction

The study of power characteristics of load sensitive proportional variable pumps is a hot topic in the field of hydraulic pump technology. It not only helps to deepen the understanding of the performance of pumps under different operating conditions but also provides a solid theoretical basis for the optimization design and performance improvement of load sensitive proportional variable pumps. Through this study, the working mechanism of the pump can be revealed. Its behavioral characteristics under different loads and flow demands can be grasped, providing strong support for the efficient operation of the pump in practical applications.

T. Ma (✉)
CCTEG Taiyuan Research Institute Co., Ltd., Taiyuan, China

Shanxi Tiandi Coal Machinery Equipment Co., Ltd., Inner Mongolia Branch, Ordos, China

© The Author(s) 2026

Z. Zhang, P. Mercorelli (eds.), *Mechanical Design and Simulation: Insights and Innovations*, Lecture Notes in Mechanical Engineering,
https://doi.org/10.1007/978-981-95-2405-1_15

In order to further study the characteristics of the variable displacement pump, this article decomposes it into three modules based on its functional structure: the inclined disc axial piston pump, the inclined disc angle adjustment cylinder, and the inclined disc angle adjustment proportional valve. These three modules each undertake different functions and together determine the overall performance of the pump. We have established mathematical models for these three modules to accurately describe their working processes and key parameters.

On the basis of constructing a mathematical model, this article successfully built a simulation model of a load sensitive proportional variable pump using the Amesim simulation platform. Through this simulation model, it is possible to simulate the operation of the pump under different working conditions and visually observe its core performance indicators, such as power output and flow rate changes. This simulation research method not only has the characteristics of high efficiency and accuracy but also can predict and verify the performance of the pump before practical application, thereby greatly reducing development costs and risks [1, 2].

2 Mathematical Model of Variable Pump

The variable displacement pump is roughly composed of three parts [3], namely the inclined disc axial piston pump 1, the inclined disc angle adjustment cylinder 2, and the inclined disc angle adjustment proportional valve 3, as shown in Fig. 1.

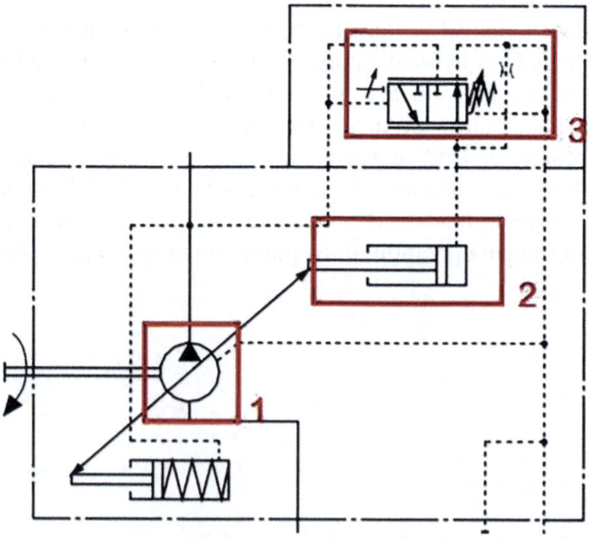

Fig. 1 Control principle of variable pump

2.1 The Mathematical Model of the Proportional Valve for Adjusting the Angle of the Inclined Plate

A proportional electromagnet converts the current signal input from a proportional control amplifier into force. The force acting on the valve core is proportional to the input current signal. The larger the input current is, the greater the force acts on the valve core. The magnitude of the force is expressed as:

$$F_1 = K_i I \tag{1}$$

In the formula:

F_1—Force exerted on the valve core (N).

K_i—Thrust gain of a proportional electromagnet (N/A).

I—Input current (A).

According to the force balance of the proportional valve, it can be expressed as:

$$F_2 = m_v \frac{d^2 x_v}{dt^2} + B_v \frac{dx_v}{dt} + K_v x_v \tag{2}$$

In the formula:

m_v—Mass of the valve core of the proportional valve (kg).

x_v—Valve core opening degree (m).

B_v—Damping coefficient generated by transient liquid flow (N · m/s).

K_v—Elastic coefficient of liquid flow force (N/m).

2.2 The Mathematical Model of the Inclined Plate Angle Adjustment Oil Cylinder

As shown in Fig. 2, the inclined plate angle adjustment cylinder is connected to the inclined plate angle adjustment proportional valve at both ends. Its force state is shown in the figure:

The expression for the flow rate at its inlet is:

$$q_1 = C_{ep} P_1 + C_{ip} P_L + q_t + \frac{dV_1}{dt} + \frac{V_1}{\beta t} \frac{dP_1}{dt} \tag{3}$$

The expression for the flow rate of its return oil chamber is:

$$q_2 = C_{ip} P_L - C_{ep} P_L + q_t - \frac{dV_2}{dt} - \frac{V_2}{\beta t} \frac{dP_2}{dt} \tag{4}$$

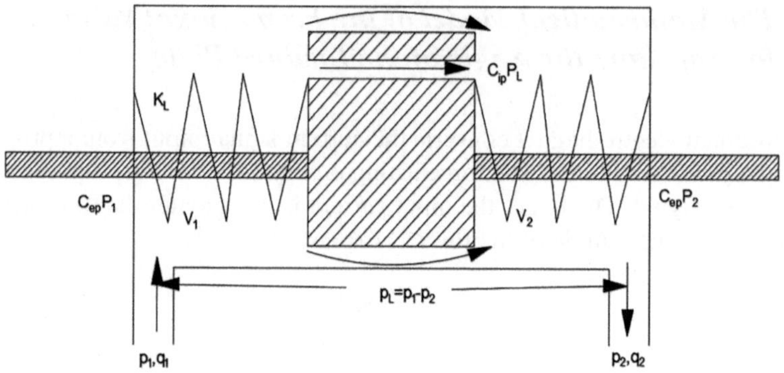

Fig. 2 Tilt plate angle adjustment cylinder

In the formula:

q_1—Flow rate at the inlet of the oil cylinder (L/min).
q_2—Flow rate of oil cylinder outlet (L/min).
C_{ep}—Leakage coefficient ($m^3 \cdot$ s-1/MPa).
C_{ip}—Internal leakage coefficient ($m^3 \cdot$ s-1/MPa).
q_t—Flow rate through the orifice (L/min).
P_1—Pressure at the inlet of the oil cylinder (MPa).
P_2—Pressure at the outlet of the oil cylinder (MPa).
t—Effective volume modulus of hydraulic oil.
V_1—Left volume of oil cylinder (m^3).
V_2—Right side volume of the oil cylinder (m^3).

By combining the flow equation and the force balance equation of the oil cylinder, the transfer function of the inclined plate angle adjustment oil cylinder can be obtained:

$$x_c = \frac{4\mu\beta_\tau A_c K_q K_i I - \left[4\beta_\tau\left(C_p - \mu K_c\right) + \mu V_c s\right]\left(m_v s^2 + B_v s + K_v\right)F_c}{\left[4\beta_\tau\left(C_p + \mu K_c\right) + \mu V_c s\right]\left[\left(m_a s^2 + B_p S + K_p\right) + 4\mu\beta_\tau A_c K_q s\right]\left(m_v s^2 + B_v s + K_v\right)} \tag{5}$$

2.3 Mathematical Model of Variable Pump

The relationship between the piston displacement of the oil cylinder and the inclination angle of the pump swash plate can be written as:

$$x_c = L_c \sin\alpha \tag{6}$$

In the formula:

x_c—Displacement of the oil cylinder piston (m).

L_c—The distance (m) from the point of force exerted by the oil cylinder on the inclined plate to the center of the inclined plate when the angle of the variable pump inclined plate is zero.

α—The inclination angle of the inclined plate of the pump (m).

In the relationship between the valve-controlled cylinder, there is a vibration link. The natural frequency of the vibration link is:

$$\omega_c = \sqrt{\frac{4\beta_\tau A_p^2}{V_P m_a}} \tag{7}$$

In the formula: Ap—the area per unit angle of the variable displacement pump plunger (m^2/rad).

When the angle $\frac{\pi}{2} + x$ between the distribution plate and the inclined plate shaft is, the force analysis of the variable displacement pump shows that the relationship between the force exerted by the oil cylinder on the inclined plate can be written as:

$$F_P = \frac{2RA_p P_s \sin x - \pi R^2 m_a \sin \alpha \omega_p^2 \alpha}{L_c} \tag{8}$$

In the formula:

R—When the inclination angle of the inclined plate is zero, it is the distance between the variable pump slide shoe and the main shaft on the projection of the inclined plate (m).

p—Speed of variable displacement pump (rad/s).

P_s—Pressure difference of variable displacement pump (kg).

Substituting Eqs. (6) and (8) into Eq. (5), the functional expression for the displacement of the variable pump inclined plate is:

$$\alpha = \frac{4\mu\beta_\tau A_c K_q K_i I - 4\beta_\tau \left(C_p + \mu K_c\right)\left(m_v s^2 + B_v s + K_v\right) 2RA_p P_s \sin x}{4\beta_\tau \left(m_v s^2 + B_v s + K_v\right)\left[\mu A_c^2 L_c^2 s - \pi R^2 m_a \omega_p^2 \left(C_p + \mu K_c\right)\right]} \tag{9}$$

3 Simulation Modeling

Using the hydraulic and HCD libraries of AMEsim software, models of various components of the variable displacement pump are established for simulation modeling. Based on component selection and design, simulation parameters are set, and then simulation analysis is conducted. The traditional quantitative pump load sensitive system has flow loss, but the system stability is better. The traditional load sensitive quantitative pump system can automatically adjust the output flow rate, but the system stability is poor. The system uses a proportional variable pump as the power source. When the input signal remains constant, the variable pump is equivalent to a quantitative pump. In order to highlight that the proportional variable pump load sensitive system is more stable and energy-saving than these two traditional load sensitive systems, this section conducts Amesim modeling on the proportional variable pump, and Amesim modeling on the load sensitive variable pump and the quantitative pump. Finally, a comparative analysis is conducted [4, 5].

3.1 Quantitative Pump Model

In order to compare the energy-saving performance with the load sensitive system of the proportional variable pump, an Amesim simulation model of the load sensitive quantitative pump was established, as shown in Fig. 3. This model is established through the fixed displacement pump model PU001 sub-model. The displacement of the quantitative pump is 71 mL/r, with a mechanical efficiency of 0.95 and a volumetric efficiency of 0.95.

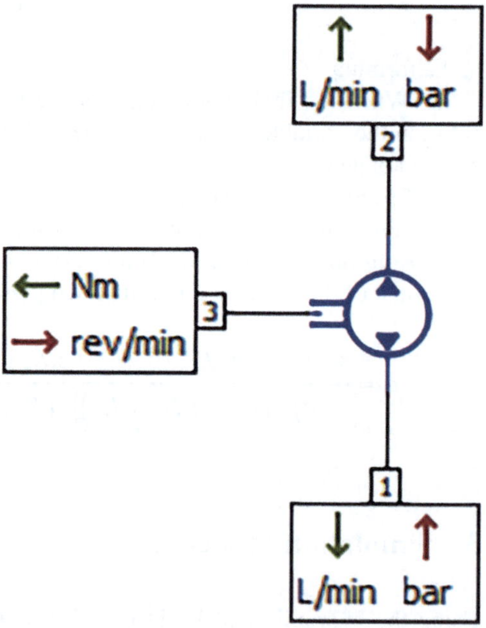

Fig. 3 Quantitative pump model

3.2 Load Sensitive Variable Pump Model

The load sensitive variable pump model mainly consists of a variable pump, a load sensitive valve, a constant pressure valve, and a variable cylinder. When the variable pump is in the regulating state, the load pressure is fed back to the spring chamber of the load sensitive valve, which is connected to the variable cylinder, thereby changing the displacement of the variable pump by controlling the piston displacement of the variable cylinder. When the system pressure is higher than the set pressure of the constant pressure valve spring chamber, it will push the piston of the constant pressure valve to move and close the connection channel between the load sensitive valve and the variable cylinder, thereby reducing the tilt angle of the variable pump to near zero by controlling the piston displacement of the variable cylinder. The Amesim simulation model of the load sensitive valve is shown in Fig. 4.

The parameters in the Amesim simulation model of the load sensitive valve are shown in Table 1.

The parameters in the Amesim simulation model of the load sensitive valve are shown in Table 2 below.

The Amesim simulation model of the variable cylinder is shown in Fig. 5.

The parameters in the Amesim simulation model of the variable cylinder are shown in Table 3.

The simulation model of the load sensitive variable pump is shown in Fig. 6, where 1 is the load sensitive valve, 2 is the constant pressure valve, and 3 is the variable cylinder.

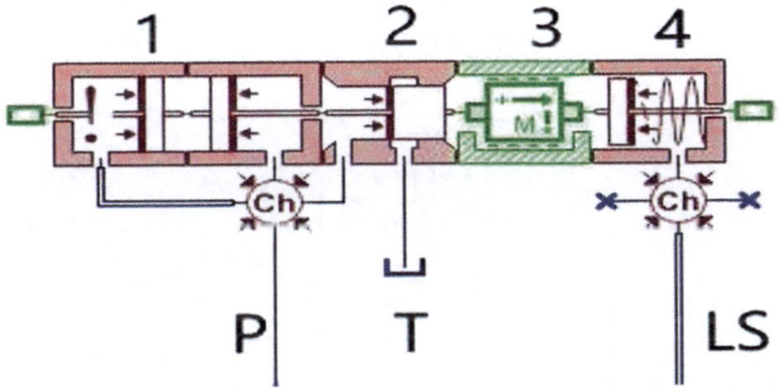

Fig. 4 Load sensitive valve model

Table 1 Simulation parameter settings for load sensitive valves

Number	Parameter	Setting value
1	Valve stem outer diameter	10 mm
	Valve stem bottom diameter	6 mm
2	Valve stem outer diameter	10 mm
	Valve stem bottom diameter	6 mm
3	Opening amount at zero displacement	−6 mm
	Viscous friction coefficient	0.5 N/(m/s)
4	The lower limit of displacement	0
	Displacement upper limit	15 mm
5	Spring set force	125 N

Table 2 Simulation parameter settings for constant pressure valve

Number	Parameter	Setting value
1	Valve stem outer diameter	10 mm
	Valve stem bottom diameter	6 mm
2	Valve stem outer diameter	10 mm
	Valve stem bottom diameter	6 mm
3	Opening amount at zero displacement	−8 mm
	Viscous friction coefficient	0.1 N/(m/s)
4	The lower limit of displacement	0
	Displacement upper limit	15 mm
5	Spring set force	1250 N

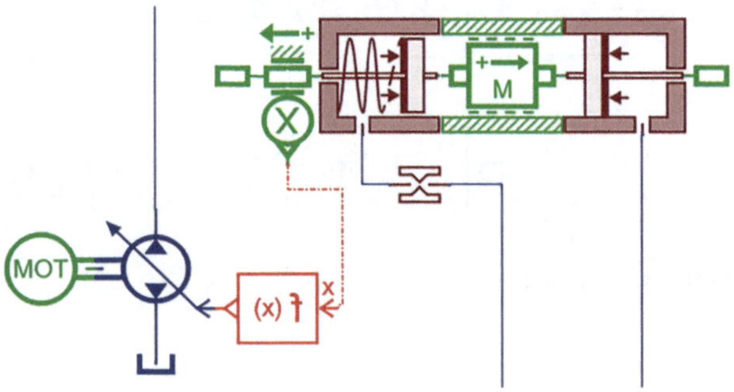

Fig. 5 Variable cylinder model

Table 3 Parameter settings for variable cylinder simulation

Number	Parameter	Setting value
1	The outer diameter of the valve stem	10 mm
	The bottom diameter of the valve stem	6 mm
2	Spring setting force	5 N
	Coefficient of viscous friction	0.1 N/(m/s)
	The upper limit of displacement	15 mm
3	The outer diameter of the valve stem	10 mm
	The bottom diameter of the valve stem	0 mm

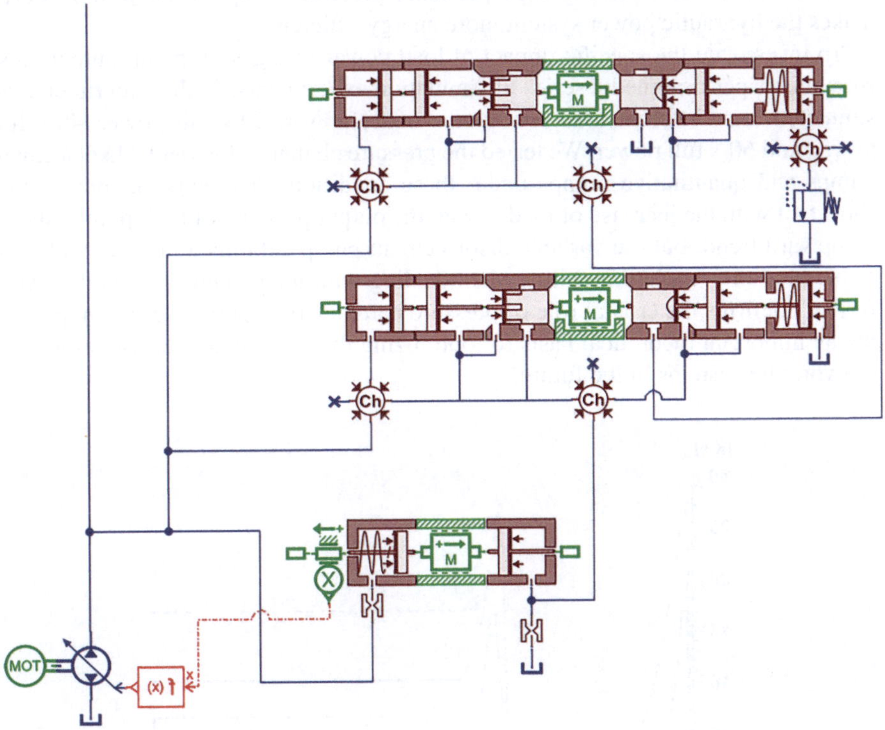

Fig. 6 Load sensitive variable pump model

4 Simulation Analysis

As shown in Fig. 7, the simulation time is 30 s, with 0 load for 0–10 s, no load for 10–20 s, and 6 kW load for 10–20 s. There is a comparison of output power simulation curves between the proportional variable pump system and load sensitive variable pump system when no signal is given to the hydraulic cylinder circuit.

From the Figs. 8 and 9, it can be seen that due to the pressure shock generated by the system at the beginning of the operation, the power curve will fluctuate at the beginning. After the system stabilizes, the output power of the proportional variable pump system under no load is 1.2 kW, and the output power of the quantitative pump load sensitive system is 2.5 kW. Starting from 10 s, the power curve of the system fluctuated again due to the sudden load. After the system stabilized, the output power of the proportional variable pump system was 10.4 kW during the sudden load, and the output power of the quantitative pump load sensitive system was 17 kW. Therefore, choosing a proportional variable pump as the power source makes the hydraulic power system more energy-efficient.

To investigate the specific impact of load power changes on pump output pressure, this paper designed a series of simulation experiments. In the experiment, we simulated four different load conditions: under power, 20-s full power, 40-s full power, and 60-s full power. We tested the pressure changes of variable displacement pumps and quantitative pumps under these conditions. The experimental results show that with the increase of load power, the output pressure of both pumps shows an upward trend, but the variable displacement pump exhibits a more flexible and stable pressure regulation ability when dealing with load changes. This discovery further confirms the crucial role of pressure parameters in power variation, providing an important theoretical basis for optimizing the stability and energy efficiency of hydraulic systems in the future.

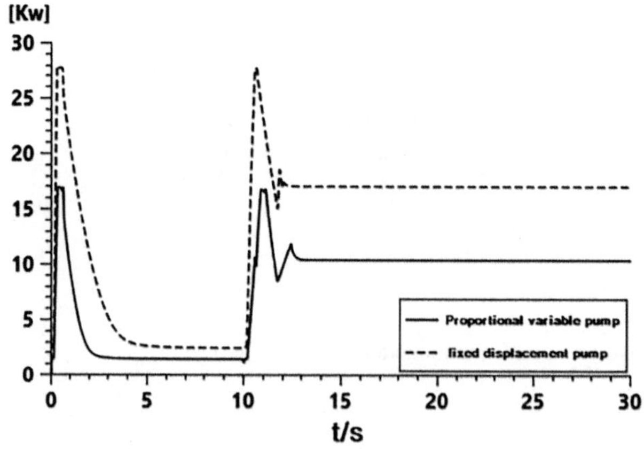

Fig. 7 Comparison of simulation results

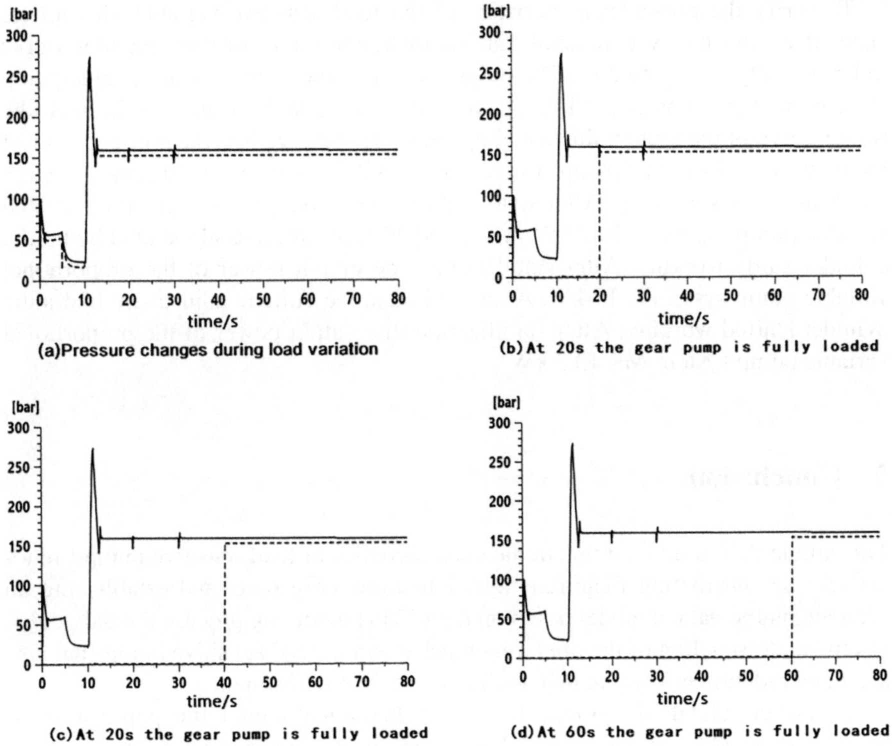

Fig. 8 Simulation of the effect of load on pump pressure

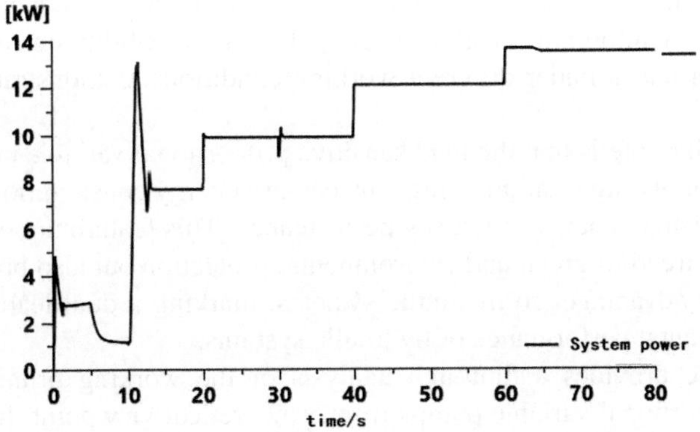

Fig. 9 Simulation of local power characteristics of load sensitive variables

To verify the power characteristics of the load sensitive variable, this article states that after the system stabilizes, the hydraulic motor of the generator works independently from 0 to 20 s. The output power of the proportional variable pump system under no load is 1.2 kW. At 10 s, due to the sudden increase in load, the power curve of the system fluctuated again. After stabilization, the output power of the proportional variable pump system was 7.6 kW. Starting from 20 s, the oil pump motor also starts working. After stabilization, the output power of the proportional variable pump system is 10.7 kW. At around 40 s, the attitude adjustment hydraulic cylinder starts working. After stabilization, the output power of the proportional variable pump system is 12 kW. At around 60 s, the attitude adjustment hydraulic cylinder started working. After stabilization, the output power of the proportional variable pump system was 13.3 kW.

5 Conclusion

This article delves into the functional characteristics of load sensitive pumps, innovatively deconstructing their functions into three core parts and establishing an accurate mathematical model based on them. This modeling process not only deepens the understanding of the operating mechanism of load sensitive pumps but also accurately identifies key parameters that affect their performance.

To further validate the practicality of the theoretical model, this paper successfully constructed a simulation model of a load sensitive proportional variable pump based on the Amesim simulation platform. Through simulation, it can be observed intuitively that the pump can flexibly and accurately adjust the output flow rate and pressure according to the actual needs of the system, effectively reducing power loss and significantly improving work efficiency. Its high flexibility ensures stable and reliable performance under different working conditions, demonstrating excellent adaptability.

Of particular note is that the load sensitive proportional variable pump achieves precise control of output, minimizing unnecessary energy consumption and demonstrating astonishing energy efficiency performance. This feature not only conforms to the current trend of green and environmental protection but also brings unprecedented power advantages to hydraulic systems, marking a dual leap in efficiency and environmental performance of hydraulic systems.

This article provides a thorough analysis of the working principle of load-sensitive proportional variable pumps from a theoretical viewpoint. It also demonstrates their significant role in enhancing hydraulic system efficiency, ensuring operational stability, and promoting sustainable development. By establishing and verifying simulation models, the article offers substantial theoretical and practical support for future innovations in hydraulic technology.

Acknowledgments This article expresses gratitude for the support of the key project 2023-2-TD-ZD010 "Research and Development of Mining Electro Hydraulic Servo Valve and Controller" funded by Tian Di Technology Innovation and Entrepreneurship Fund.

References

1. Zhang Y (2024) Non highway mining dump truck line controlled hydraulic steering system based on electro-hydraulic proportional and load sensitive technology. Hydraulic Pneumatic Sealing 44(07):118–123
2. Deng L, Dong H (2023) Integrated control Technology of Hydraulic Power Based on load sensitivity. Chinese Foreign Energy 28(S1):71–75
3. Zhentao Q (2023) Design and simulation of vehicle mounted hydraulic power system based on load sensitivity principle. Tianjin University of Technology
4. Wang R, Wei Y, Yang X (2024) Performance study of pressure refueling control valve based on AMESim. Mech Eng Automat 04:47–49
5. Fang M, Hong W, Tang B (2024) Dynamic characteristics analysis of hydraulic control valve for submersible pump based on AMESim. Mech Electr Eng Technol 53(06):166–169+200

Open Access This chapter is licensed under the terms of the Creative Commons Attribution-NonCommercial-NoDerivatives 4.0 International License (http://creativecommons.org/licenses/by-nc-nd/4.0/), which permits any noncommercial use, sharing, distribution and reproduction in any medium or format, as long as you give appropriate credit to the original author(s) and the source, provide a link to the Creative Commons license and indicate if you modified the licensed material. You do not have permission under this license to share adapted material derived from this chapter or parts of it.

The images or other third party material in this chapter are included in the chapter's Creative Commons license, unless indicated otherwise in a credit line to the material. If material is not included in the chapter's Creative Commons license and your intended use is not permitted by statutory regulation or exceeds the permitted use, you will need to obtain permission directly from the copyright holder.

CFD Calculation and Analysis of a 1.2NA Intake Duct

Hongxia Zhao, Qingsong Zhu, Minjie Liu, and Lin Zhou

Abstract In the development process of a 1.2NA engine air duct, to improve the overall power performance, we performed an optimization based on the original air duct, and the flow performance of the modified air duct was re-evaluated. This article applies AVL FIRE software to perform steady-state simulation analysis on the modified airway, obtaining the flow coefficient, roll ratio, velocity field, and pressure field distribution inside the cylinder under different fixed valve lifts. According to the AVL evaluation method, the average flow coefficient of the V1 intake duct of a certain 1.2NA engine is 0.3623, which is 4.8% lower than that of the V0 intake duct, and the flow coefficient is at a moderately low level. According to the FEV evaluation method, the average standard flow coefficient of the intake duct is 0.1008, which is 8.4% higher than that of the V0 duct, and the flow coefficient is at a moderate level. According to the AVL evaluation method, the average roll flow intensity of the V1 intake duct is 1.3953, which is 17.1% higher than that of the V0 intake duct, and the resulting roll flow intensity is at a moderate level. Airway tests are recommended to verify the calculation results.

Keywords CFD · Simulation analysis · Gasoline engine · Intake duct · Rolling ratio

1 Introduction

In the early development process of a 1.2NA engine air duct, the overall power performance was optimized based on the original air duct, and the flow performance of the modified air duct was re-evaluated. This article applies AVL FIRE software to conduct steady-state simulation analysis on the modified airway, obtaining the flow coefficient, roll ratio, velocity field, and pressure field distribution inside the

H. Zhao (✉) · Q. Zhu · M. Liu · L. Zhou
Automotive Engineering Institute, Beijing Polytechnic, Beijing, China
e-mail: 103052@bpi.edu.cn

© The Author(s) 2026

Z. Zhang, P. Mercorelli (eds.), *Mechanical Design and Simulation: Insights and Innovations*, Lecture Notes in Mechanical Engineering,
https://doi.org/10.1007/978-981-95-2405-1_16

cylinder under different fixed valve lifts, and comparing them with the original airway [1]. In the calculation of gasoline engine intake ports, the main parameters of concern are flow coefficient and tumble ratio. The flow coefficient directly determines the amount of charge in the cylinder, while the roll ratio has a significant impact on the formation, development, combustion diffusion speed, and stability of the mixture in the cylinder. Because increasing the flow coefficient will result in a decrease in the tumble ratio, and vice versa, it is often necessary to strike a compromise between the two [2–4].

This calculation is a steady-state calculation, and the inlet and outlet boundary conditions adopt a pressure difference boundary. The inlet condition is total pressure, and the outlet condition is static pressure. The multidimensional numerical simulation calculation of the working process inside the engine cylinder satisfies the equations of mass conservation, momentum conservation, and energy conservation, as well as the ideal gas state equation while selecting an appropriate turbulence model to close the equation system. The turbulence model used in this article is the k-ζ-f four equation model, which has a wide range of applications in the engineering field. For example, in the multidimensional numerical simulation of internal combustion engine cylinder flow, this model is used to describe the motion of turbulence in the cylinder [5]. By simulating the flow inside the cylinder at different injection times, the influence of injection timing on combustion and emissions can be analyzed, providing a reference for the improvement and optimization of the injection system. It is a high-precision and high-stability turbulence model with wide application prospects and important engineering value. The k-ζ-f four equation model is a method of assuming and modeling the second-order related moment terms in the Reynolds equation, thereby enclosing the Reynolds equation system. Compared to the classical k-εtwo equation model, this model has improved in terms of computational accuracy and stability [6]. The k-ζ-f four equation model consists of the following four equations:

- The transport equation of turbulent kinetic energy k describes the variation of turbulent kinetic energy.
- The transport equation of turbulent dissipation rate ε, together with the k equation, describes the energy dissipation process of turbulence.
- The transport equation of the ζ equation: ζ is a variable related to the turbulence scale, which describes the variation of ζ.
- The transport equation of the f equation: f is a variable related to the turbulent vortex structure, which further describes the detailed characteristics of turbulence.

These four equations are interrelated and together form the complete system of the k-ζ-f four equation model. The characteristics and advantages of these four equations mainly include the following three aspects. High precision: The k-ζ-f four equation model can more accurately describe the complex characteristics of turbulence by introducing more variables and equations. High stability: Compared to other turbulence models, the k-ζ-f four equation model exhibits higher stability in the calculation process. Wide applicability: This model is suitable for various types of turbulent flows, including complex flow situations such as rotational flow and shear flow [7, 8].

2 Computational Model

2.1 *Airway Test*

Figure 1 shows the layout of the airway steady flow test bench, which consists of a cylinder head, a simulated cylinder, blades, a blade anemometer, a pressure gauge, a pressure regulator box, a blower, and a bypass. The test bench device operates in a suction mode. During operation, the blower sucks in air, and the air in the atmosphere flows in through the intake duct. It is then discharged into the atmosphere through simulated cylinder liners, pressure stabilizing boxes, orifice flow meters, etc. The U-shaped pressure gauge measures airway resistance, the orifice flowmeter measures airflow, the rotating airflow drives the anemometer to rotate, the counter measures the number of revolutions, and the air volume is controlled by the regulating valve [9].

2.2 *Calculation Grid*

The computational model consists of an intake duct, intake valve, valve seat ring, combustion chamber top surface, and simulated cylinder liner. To ensure consistency between the computing environment and the experimental setup, we added a cubic stabilizing chamber at the inlet of the airway, simulating a cylinder liner length of 2.5 times the cylinder diameter. The lift point of the intake valve was calculated as 1 mm to 9 mm, and a calculation model was established every 1 mm.

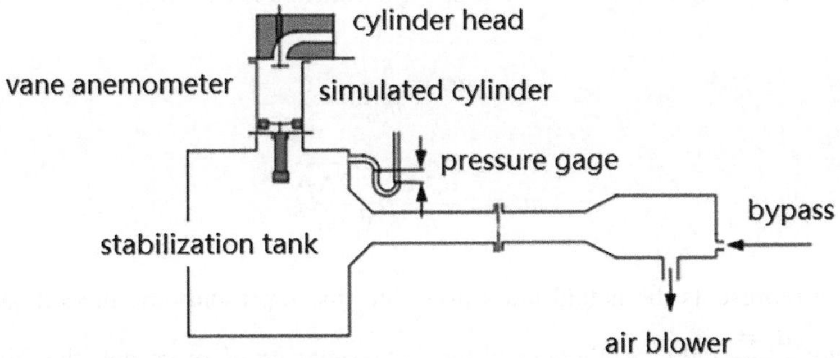

Fig. 1 Layout of airway test stand

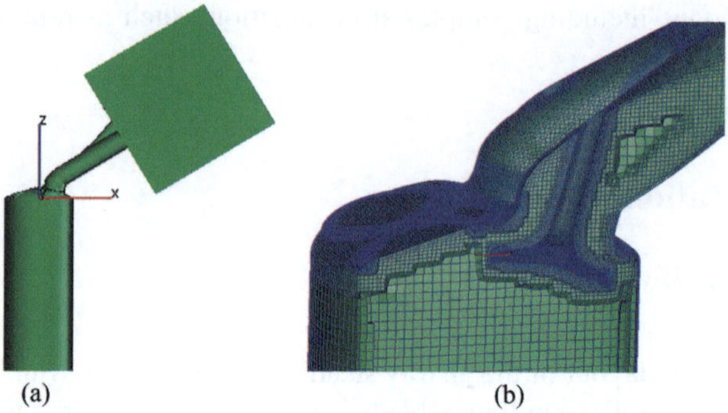

Fig. 2 Calculation grid

Using the preprocessing tool FAME Meshing of AVL-FIRE software for mesh generation, we set the basic mesh size to 2 mm, and some areas are further refined, including valve seats, intake valves, combustion chamber top surfaces, etc. The minimum mesh size is 0.125 mm. The total number of computational grids is approximately 1.2 million, as shown in Fig. 2. The top view of the original airway V0 is shown in Fig. 3, and the top view of the modified airway V1 is shown in Fig. 4.

2.3 Evaluation Method

This article applies FIRE software to establish a calculation model based on the AVL airway test bench. The flow coefficient and roll ratio of the intake duct are calculated according to the AVL method, and the flow coefficient and roll ratio are evaluated and analyzed using the AVL method.

Flow Coefficient. According to AVL's evaluation method, the flow coefficient refers to the ratio of actual mass flow to theoretical mass flow. The theoretical mass flow rate is calculated based on the flow velocity and pressure drop over the valve seat area, assuming no flow loss. The specific formula is as follows:

$$\mu\sigma = \frac{m}{m_{th}} \tag{1}$$

$$m_{th} = z\frac{d_v^{\,2}\pi}{4}\rho\sqrt{\frac{2\Delta p}{\rho}} \tag{2}$$

where m represents the actual mass flow rate; m_{th} represents the theoretical mass flow rate; $\frac{d_v^{\,2}\neq}{4}$ indicates the area of the valve seat ring; d_v represents the minimum diameter of the valve seat ring; ρ represents gas density; Δp represents pressure drop.

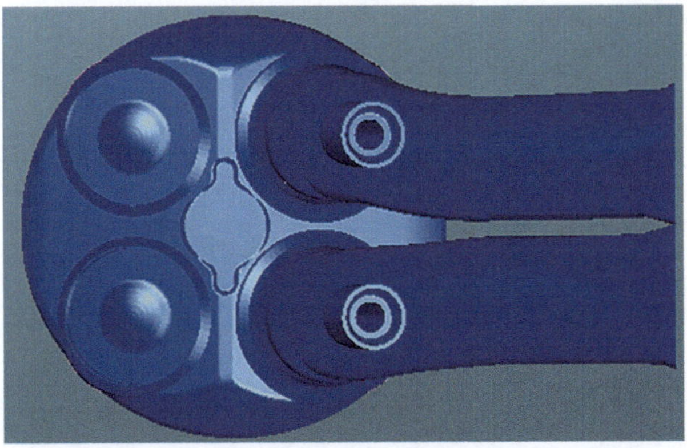

Fig. 3 Top view of a 1.2NA original airway V0

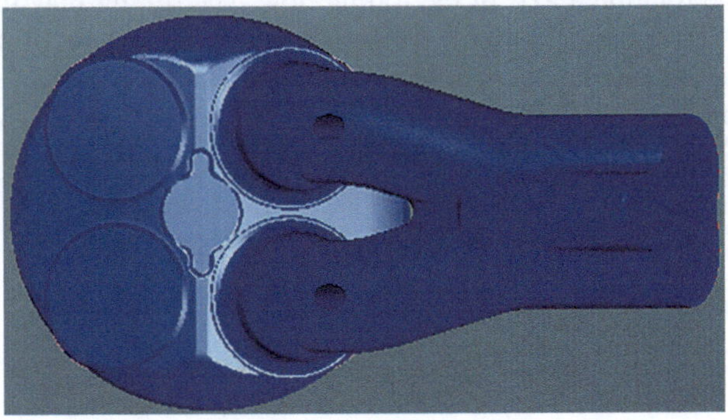

Fig. 4 Top view of a 1.2NA modified airway V1

The average flow coefficient is expressed as follows:

$$(\mu\sigma)_m = \frac{1}{\sqrt{\frac{1}{\pi}\int_0^{\pi}\left(\frac{c(\alpha)}{c_m}\right)^3 \frac{1}{(\mu\sigma)^2} d_\alpha}} \tag{3}$$

When the valve lift is small, the AVL standard pressure difference is set to 6500 Pa, and when the valve lift is large, the AVL standard pressure difference is set to 2500 Pa.

Rolling Ratio. The rolling flow is a macroscopic large-scale vortex formed during the engine intake process. The rotational axis of rolling flow is perpendicular to the cylinder axis, and the definition of rolling flow ratio is shown in Fig. 5. The rotational axis of rolling flow depends on the direction of the tangential airway.

In steady-state calculations, it is generally believed that the line perpendicular to the cylinder axis and the airway section on the surface 0.5 times the cylinder head

Fig. 5 Schematic diagram of rolling flow definition

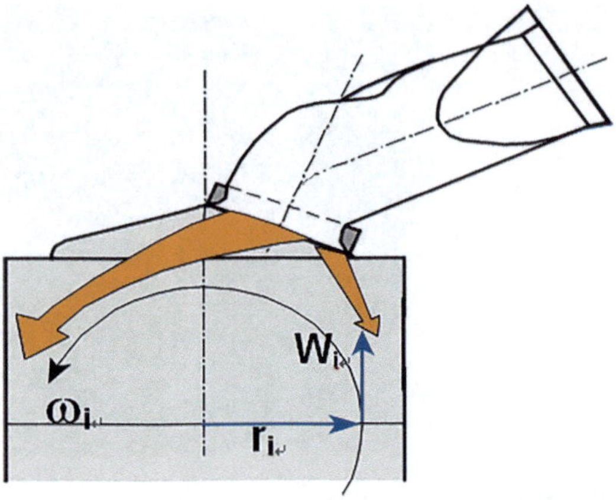

diameter is the rotational axis of the rolling flow. The rolling flow intensity calculated using FIRE software is based on the tangential component of the velocity on this plane (the velocity component parallel to the cylinder center axis). We assume that the average velocity of the piston is equal to the axial average velocity of the steady-state flow, and then, the calculation formula is as follows:

$$tumble\ ratio = \frac{n_D}{n} \tag{4}$$

$$c_m = \frac{sn}{30} = c_a = \frac{m}{\rho A_p} \tag{5}$$

$$n = \frac{30m}{\rho s A_p} \tag{6}$$

$$\frac{n_D}{n} = \frac{n_D \rho s A_p}{30m} \tag{7}$$

where c_m represents the average speed of the piston; s represents the engine stroke; n represents the engine speed; n_D represents the blade speed; c_a represents the average axial velocity of steady-state flow; m represents gas mass flow rate; ρ represents gas density; A_p represents the piston area.

Average tumble ratio:

$$\left(\frac{n_D}{n}\right)_m = \frac{1}{\pi}\int_0^{\pi} \frac{n_D}{n}\left(\frac{c(\alpha)}{c_m}\right)^2 d\alpha \tag{8}$$

In order to facilitate the evaluation of the engine flow coefficient between different cylinder diameters and valve seat diameters, the flow coefficient is standardized, and the standardized flow coefficient is represented as μσβ,

$$\beta = z\left(\frac{d_v}{D}\right)^2 \tag{9}$$

The simplified roll flow ratio is expressed as follows:

$$\left(\frac{n_D}{n}\right)_{red} = \frac{n_D}{n}\frac{D}{s} \tag{10}$$

where z represents the number of valves; D represents the cylinder diameter of the engine; s represents the engine stroke.

3 Boundary Conditions and Parameter Settings

3.1 *Airway Test*

This analysis is a steady-state calculation, assuming that the fluid is a compressible adiabatic flow process. Import and export use pressure difference boundary conditions. The inlet is the total pressure, and the outlet is the static pressure. To obtain accurate results in the experiment, we used different airway pressure differentials at different valve lifts, with a small valve lift pressure differential of 6500 Pa and a large valve lift airway pressure differential of 2500 Pa.

3.2 *Input Parameters*

The data that need to be input in the calculation settings file should be consistent with the experiment, as shown in Table 1.

Table 1 Input parameter table

Name	Value/Unit
Cylinder diameter	69.70/mm
Stroke	79.00/mm
Minimum diameter of valve seat	26.00/mm
Number of valves	2/each

3.3 Solver Settings

For spatial discretization settings, extrapolate is chosen as the boundary value calculation method and Least Sq. Fit with higher calculation accuracy is chosen as the derivative calculation method; The coupling of speed and pressure adopts the SIMPLE algorithm; The turbulence model adopts k-ζ-f; The momentum equation difference scheme adopts second-order accuracy with a softening factor of 0.5, the continuous equation difference scheme adopts center difference, and the others are upwind difference schemes; The wall treatment adopts a hybrid wall function, and the wall heat conduction adopts a standard wall function.

4 Result Analysis

4.1 Calculation Results 2D Data Analysis

Flow Coefficient and Roll Ratio of an Engine Intake Duct. According to the internal formula of FIRE, the flow coefficient and roll ratio under different valve lifts are obtained, as shown in Table 2. The average flow coefficient of the V1 intake duct of a certain 1.2NA engine is 0.3623, which is 4.8% lower than that of the V0 intake duct 0.3806; The average rolling flow intensity is 1.3953, which is 17.1% higher than the V0 airway 1.192.

According to the variation curves of flow coefficient and roll ratio, it can be seen that at low lift (1–6 mm), the flow coefficient of modified airway V1 is slightly lower than that of the original airway, while the roll ratio is comparable to that of the original airway; At high lift (6–9 mm), the flow coefficient is lower than that of the original airway, and the roll to roll ratio is significantly higher than that of the original airway. The modified airway V1 has good mixing ability for the mixture under atmospheric valve lift, which is shown in Figs. 6 and 7.

Table 2 Flow coefficient and tumble ratio for fixed valve lift

Intake Valve Lift hv (mm)	Flow Coefficient (μσ)	Rolling Ratio
1	0.1345	0.1001
2	0.2570	0.0654
3	0.3614	0.2306
4	0.4398	0.4683
5	0.4857	0.6850
6	0.5101	0.9304
7	0.5105	1.2975
8	0.5250	1.5038
9	0.5200	1.5484
Average value	0.3623	1.3953

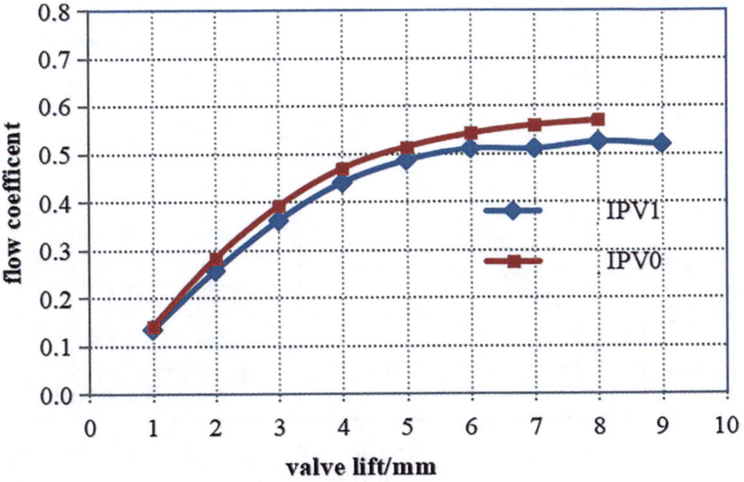

Fig. 6 1.2NA airway flow coefficient variation curve

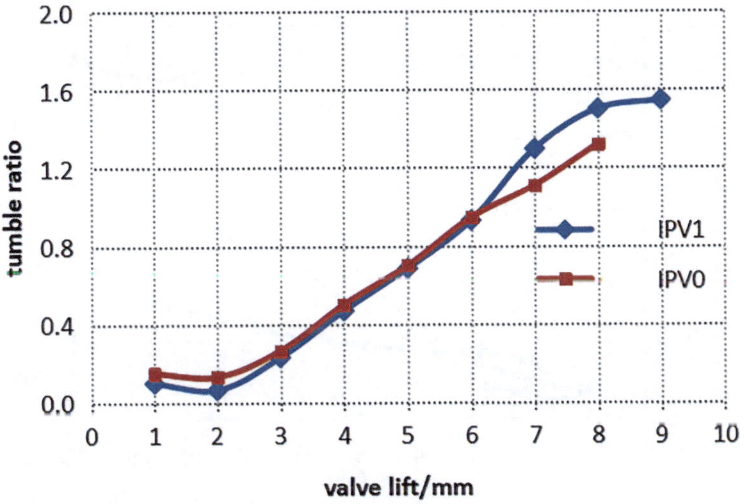

Fig. 7 1.2NA airway roll flow ratio variation curve

The flow coefficient and roll ratio curves of different models are shown in Figs. 8 and 9. It can be seen from the figures that the flow coefficient of a certain V1 airway is at a medium-low level, and the roll ratio is at a medium level.

Standardized Flow Coefficient and Simplified Tumble Ratio. According to the AVL evaluation method, the calculated flow coefficient is the ratio of the actual flow rate through the airway to the theoretical flow rate. It only represents the flow performance of the airway itself and cannot represent the flow performance of the airway after matching with the engine. Therefore, in the following analysis, the

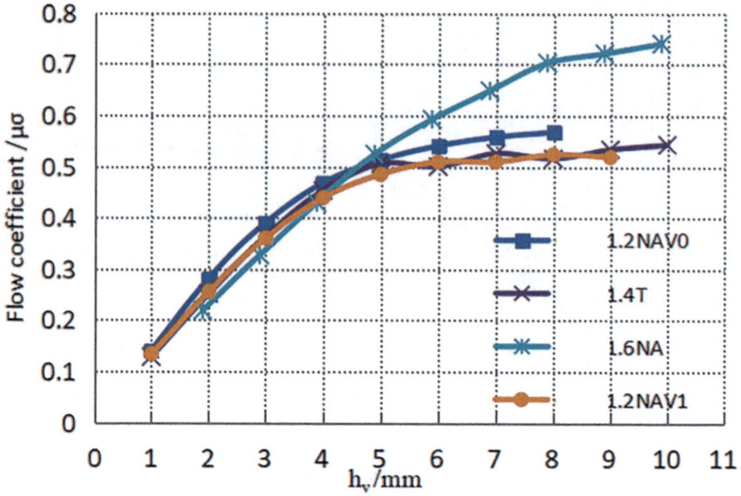

Fig. 8 Flow coefficient variation curves of different models

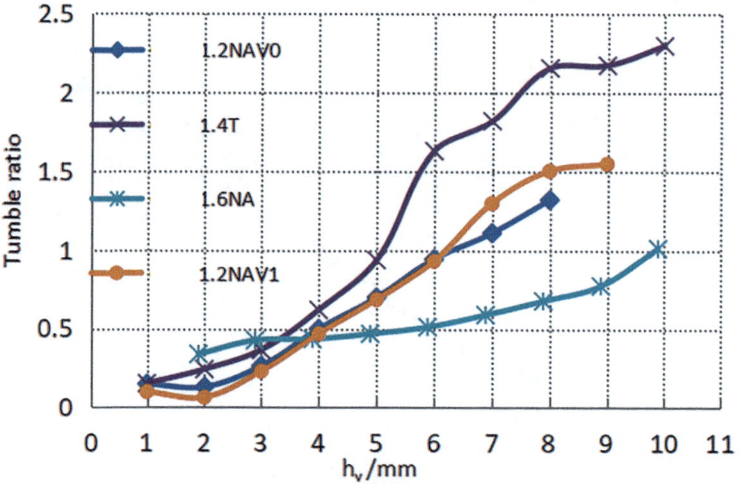

Fig. 9 Rolling ratio variation curves of different models

standardized flow coefficient and simplified roll ratio of parameters converted from the engine cylinder diameter and valve seat diameter are used.

Table 3 shows the standard flow coefficient and simplified roll ratio values corresponding to each lift point. After matching the engine cylinder diameter, the average standard flow coefficient of this intake passage is 0.1008, which is 8.4% higher than that of V0 intake passage 0.093; The average value of rolling flow intensity is 1.2310, which is 17.1% higher than that of V0 airway 1.052.

Table 3 Standardized flow coefficient and simplified roll flow ratio for fixed valve lift

Lift hv (mm)	Standardized Flow Coefficient (μσβ)	Reduced Tumble Ratio
1	0.0374	0.0883
2	0.0715	0.0577
3	0.1006	0.2034
4	0.1224	0.4132
5	0.1352	0.6043
6	0.1420	0.8209
7	0.1421	1.1448
8	0.1461	1.3268
9	0.1447	1.3661
Average value	0.1008	1.2310

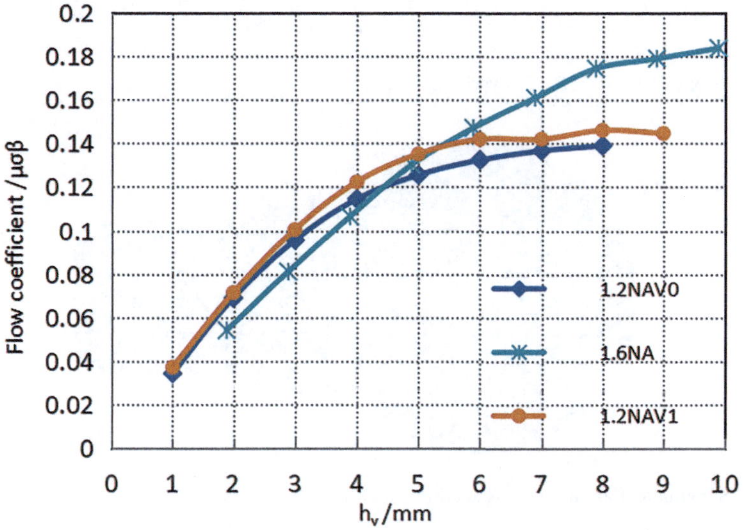

Fig. 10 Standardized flow coefficient variation curve

As shown in Figs. 10 and 11, the standardized flow coefficient and simplified roll ratio of different models vary with valve lift. It can be seen that the standardized flow coefficient of a certain 1.2NA airway V1 is higher than that of the original airway at full valve lift because the inlet valve seat diameter of the modified airway V1 is 26 mm, which is higher than the original 24.63 mm; Due to the unchanged cylinder diameter and stroke, the simplified roll ratio and roll ratio are the same. Under atmospheric valve lift (6–9 mm), the modified airway V1 is higher than the original airway.

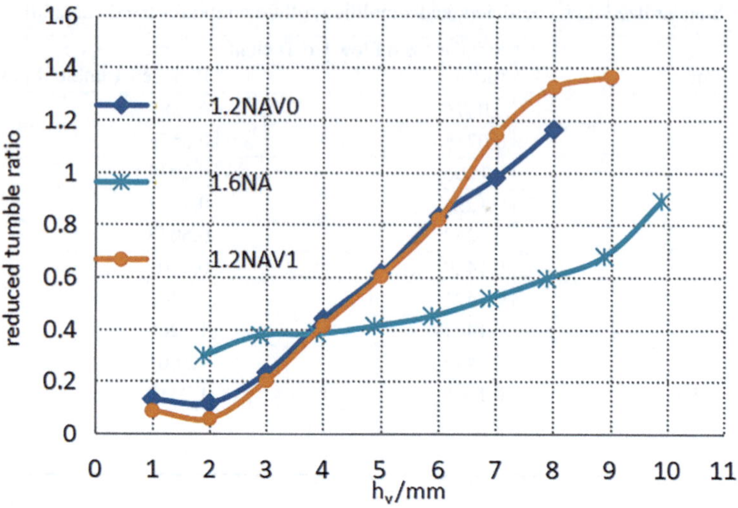

Fig. 11 Simplified roll flow ratio variation curve

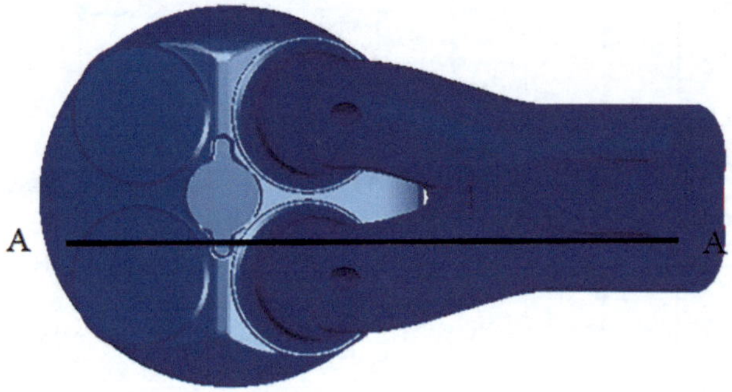

Fig. 12 Schematic diagram of A-A section

4.2 Slice Chart of Calculation Results

Figure 12 is a schematic diagram of the cross-sectional position of the calculation results in this article.

The velocity field, pressure field, and turbulent energy field in Section A-A when the valve lift is 4 mm and 8 mm, respectively are shown in Figs. 13, 14, 15, 16, 17 and 18.

From Figs. 13 and 14, it can be seen that at a valve lift of 4 mm, due to the throttling effect at the valve, the velocity at the intake valve throat is significantly higher than at a lift of 8 mm. From Fig. 14, it can be seen that the velocity of the air

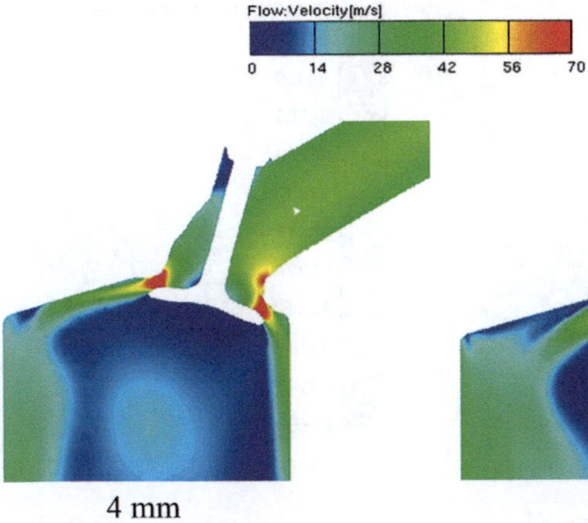

Fig. 13 Velocity field distribution_V0

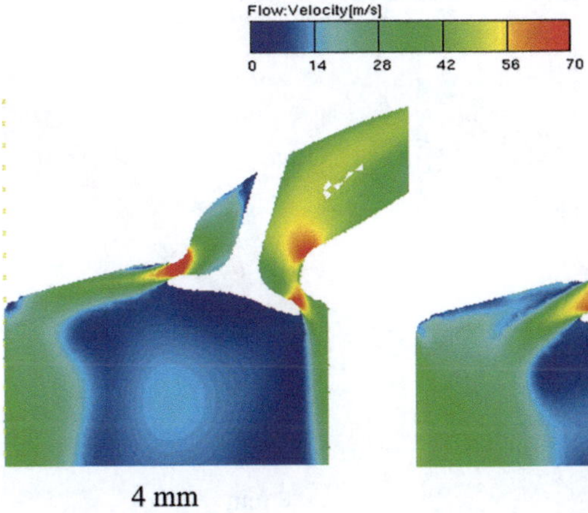

Fig. 14 Velocity field distribution_V1

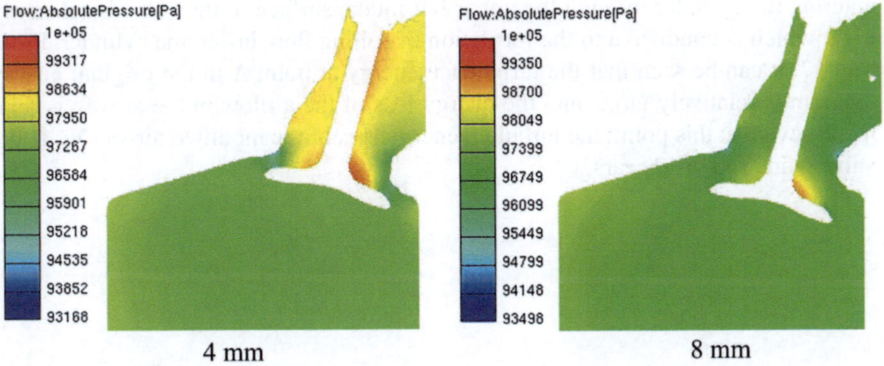

Fig. 15 Pressure field distribution_V0

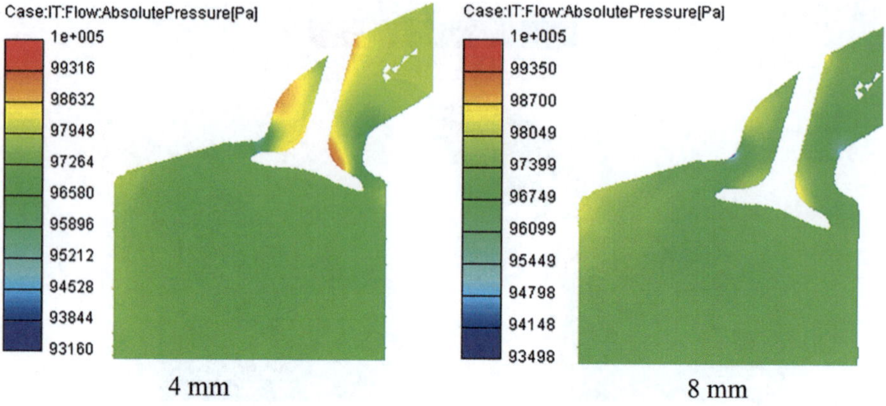

Fig. 16 Pressure field distribution_V1

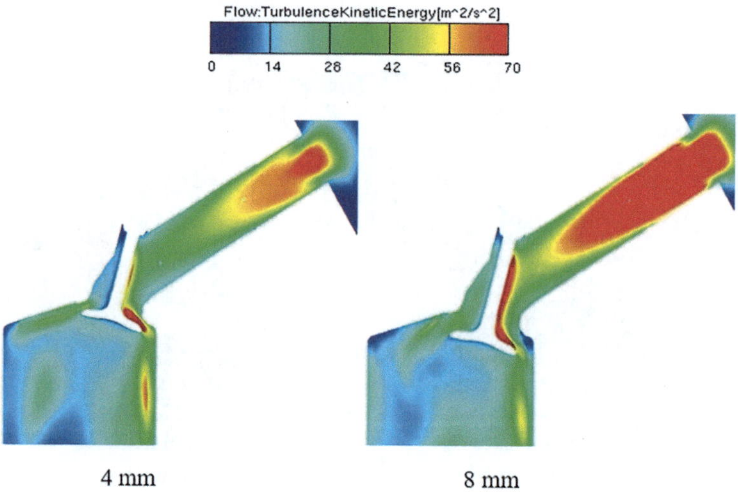

Fig. 17 Turbulent kinetic energy distribution_V0

entering the cylinder through the upper left intake surface of the valve is relatively high, which is conducive to the formation of rolling flow inside the cylinder. From Fig. 17, it can be seen that the turbulence energy at point A in the original airway diagram is relatively large, and the energy loss of the airflow in the airway is relatively severe at this point; the turbulent energy loss of the modified airway V1 at the valve guide rod disappears.

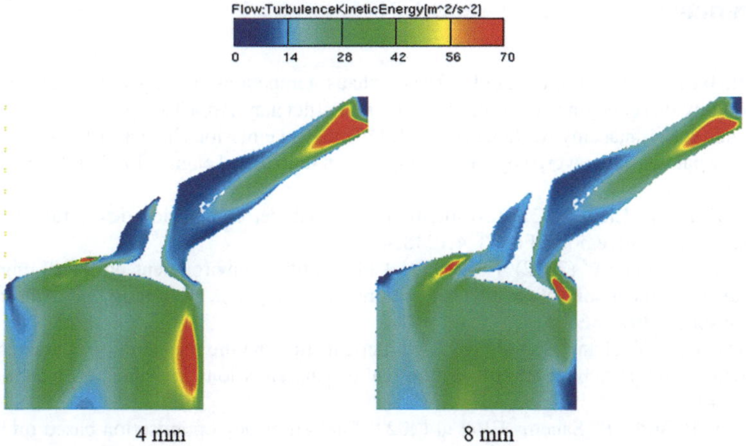

Fig. 18 Turbulent kinetic energy distribution_V1

5 Conclusion

By establishing a computational model for the intake duct and conducting CFD steady-state calculations, the conclusions are summarized as follows:

- According to the AVL evaluation method, the average flow coefficient of the V1 intake duct of a certain 1.2NA engine is 0.3623, which is 4.8% lower than that of the V0 intake duct 0.3806. The flow coefficient is at a moderately low level. It is recommended to conduct airway tests to verify the calculation results.
- According to the FEV evaluation method, the average standard flow coefficient of the intake duct is 0.1008, which is 8.4% higher than the V0 duct 0.093. The flow coefficient is at a moderate level.
- According to the AVL evaluation method, the average roll flow intensity of the V1 intake duct is 1.3953, which is 17.1% higher than that of the V0 intake duct 1.192. The rolling flow intensity formed is at a moderate level.

Airway tests are conducted to verify the calculation results.

Acknowledgments This work was financially supported by the R&D Program of Beijing Municipal Education Commission (KM202310858005).

References

1. Wu B, Jia Z, Li Z g et al (2021) Different exhaust temperature management technologies for heavy-duty diesel engines with regard to thermal efficiency. Appl Therm Eng 186:116495
2. Vellandi V, Krishnasamy A, Ramesh A (2021) Transient emission characteristics of a light duty commercial vehicle powered by a low compression ratio diesel engine. In: SAE technical paper 2021-01-1181
3. Meng Z, Liu Z, Liu J (2022) Investigation of in-cylinder combustion deterioration of diesel engines in plateau regions. Fuel 324:124824
4. Liu J, Li Y, Zhang C (2022) The effect of high altitude environment on diesel engine performance: comparison of engine operations in Hangzhou, Kunming and Lhasa cities. Chemosphere 309:136621
5. Xiao G, Wang Y, Zhang H et al (2022) Numerical study on the regeneration characteristics of catalytic diesel particulate filter based on real driving emissions in plateau environment. Fuel 321:124020
6. Piqueras P, Burke R, Sanchis EJ et al (2022) Fuel efficiency optimisation based on boosting control of the particulate filter active regeneration at high driving altitude. Fuel 319:123734
7. Szedlmayer M, Kweon CM. Effect of altitude conditions on combustion and performance of a multi-cylinder turbocharged direct-injection diesel engine. In: SAE technical paper 2016-01-0742
8. Broatch A, Bermudez V, Serrano JR et al (2019) Analysis of passenger car turbocharged diesel engines performance when tested at altitude and of the altitude simulator device used. J Eng Gas Turbines Power 141(8):1–9
9. Wang X, Ge Y, Yu L et al (2013) Effects of altitude on the thermal efficiency of a heavy-duty diesel engine. Energy 59:543–548

Open Access This chapter is licensed under the terms of the Creative Commons Attribution-NonCommercial-NoDerivatives 4.0 International License (http://creativecommons.org/licenses/by-nc-nd/4.0/), which permits any noncommercial use, sharing, distribution and reproduction in any medium or format, as long as you give appropriate credit to the original author(s) and the source, provide a link to the Creative Commons license and indicate if you modified the licensed material. You do not have permission under this license to share adapted material derived from this chapter or parts of it.

The images or other third party material in this chapter are included in the chapter's Creative Commons license, unless indicated otherwise in a credit line to the material. If material is not included in the chapter's Creative Commons license and your intended use is not permitted by statutory regulation or exceeds the permitted use, you will need to obtain permission directly from the copyright holder.

Research on Modeling and Effectiveness Evaluation of Space System Based on ABMS

Wenjie Sun and Hongwei Zhou

Abstract Aiming at the effectiveness evaluation requirements of the space information system, this paper analyzes the difficulties in the modeling and effectiveness evaluation of the space system. A composable and reusable model framework for the space information system and an agent behavior modeling method oriented towards multi-domain connectivity, information processing, and resource management and control are established. A modeling, simulation, and effectiveness evaluation method based on ABMS (Agent Based Modeling and Simulation) for the space cyber information system is proposed. Eventually, the effectiveness of this technology is verified through typical application cases, thereby providing a quantitative analysis tool for system design and evaluation in relevant fields.

Keywords Space information system · Model framework · ABMS · Resource management and control · Effectiveness evaluation

1 Introduction

The space-based information and communication system consists of various types of satellites such as remote sensing, transmission, communication, and navigation satellites, as well as constellations, ground stations, operation control centers, and inter-satellite and satellite-ground links. Remote sensing satellites include electronic, optical, and SAR types of satellites, while communication satellites include both geostationary and low Earth orbit communication satellites. These different types of satellites can form constellations or operate autonomously through inter-satellite links or independently through ground scheduling, each playing different roles in the system. With the scenario of detecting sea surface targets as a hypothetical background, this paper experimentally analyzes the impact of the resource architecture, operational strategies, and equipment capabilities of the space-based

W. Sun (✉) · H. Zhou
China Academy of Space Technology, Beijing, China

© The Author(s) 2026

Z. Zhang, P. Mercorelli (eds.), *Mechanical Design and Simulation: Insights and Innovations*, Lecture Notes in Mechanical Engineering,
https://doi.org/10.1007/978-981-95-2405-1_17

information and communication system on sea surface detection, identifies key influencing factors, and provides suggestions and basis for the construction of the space-based information and communication system [1]. This system plays an important role in multiple fields, especially in the detection, communication, and navigation of sea surface targets, where its complexity and multi-domain nature pose many challenges to its effectiveness evaluation [2].

Currently, the modeling methods for the space-based information and communication system mainly include system dynamics, network flow models, Petri nets, etc. These methods can describe the static and dynamic characteristics of the system to a certain extent, but they have limitations when dealing with complex scenarios involving multiple domains, tasks, and resources. System dynamics is mainly used to describe the macroscopic behavior of the system, but it is difficult to accurately simulate the microscopic behavior [3]; network flow models are mainly used to describe the flow of information in the network, but they struggle to handle complex interactions across multiple domains [4]; Petri net methods are mainly used to describe the concurrent behavior of the system, but they are not well-suited for handling large-scale complex systems [5]; qualitative evaluation methods mainly rely on expert experience and historical data, making it difficult to accurately describe the dynamic changes of the system [6]; quantitative evaluation methods can provide more accurate assessment results, but traditional simulation methods often fall short in terms of computational efficiency and accuracy when dealing with large-scale, high-complexity space-based information and communication systems [7]. Therefore, a new method is needed to overcome these limitations.

2 Composable Model Framework for Space-Based Information and Communication Systems

In the design process of the system model framework, it is necessary to start from the operational concepts and behaviors of the system, abstract operational entity models based on the performance indicators, effectiveness indicators, and interaction relationships of the system hierarchy at the physical domain, information domain, cognitive domain, and organizational domain. This ensures the generality of abstraction and modeling in these domains. This allows analysts to flexibly and agilely combine and adjust models from the system level for different system analysis problems, while also ensuring the simplicity and ease of use of system analysis [6–8].

The physical domain is the physical space where conflicts between hostile parties occur, containing all objective elements involved in combat processes. The essence of physical domain elements is their objective reality. The information domain is the information space where conflicts between hostile parties occur, containing all combat elements related to battlefield situation information, as well as the creation, management, and sharing of information. The cognitive domain is the thinking space of combat personnel, containing their perceptions, understandings, and shared situational awareness and decision-making based on this. The social domain is the space where human social behaviors are implemented, consisting of elements related to human social behavior.

Based on this, a hierarchical structure composed of agent models required for analysis of the space-based information and communication system is presented. Among them, operational users include combat forces executing tasks, which in turn command different combat units under various combat formations, and these combat units are composed of various combat platforms, personnel, and facilities. These combat entities constitute a multi-level inclusion relationship, communication relationship, command relationship, and maintenance relationship.

Combat entities facing the space-based information and communication system need to represent not only the composition of the physical domain but also the behaviors of the information domain and cognitive domain in the system. Therefore, it is necessary to determine a reasonable operational entity structure to support the model representation of combat entities. The components in the operational entity model mainly include physical domain components, information domain components, and cognitive domain components [9].

Physical Domain Components: Physical domain components include satellites, ground stations, operation control centers, etc. These components, most of which are equipment objects, provide flight, movement, detection, communication, and combat capabilities for the entity model, representing the inherent capabilities of the combat entity.

Information Domain Components: Information domain components need to represent the situational information perceived by the operational entity. This situational information includes threats from the enemy, friendly combat entities, as well as combat actions and tactical coordination information that need to be executed. For this purpose, the information domain components in the operational entity model mainly include local target lists, broadcast variables, and local command lists.

Cognitive Domain Components: Cognitive domain components are the core components that control the behavior of the agent model, reflecting the decision-making behavior of the operational entity model. Analysts can reflect the differences in combat behavior under different combat scenarios through the decision-making behavior model of the cognitive domain, as well as examine how information superiority in system operations is transformed into decision-making superiority and ultimately into combat superiority.

The model framework for the space-based information and communication system is shown in Fig. 1.

The components and data in the figure represent only the components that users need to define and use when developing simulation applications for the space-based system. In the model framework of the space-based information and communication system, combat forces include combat units that can be commanded. Combat units are derived from combat entity objects, so combat entities can represent combat units composed of multiple combat entities, as well as individual combat entities.

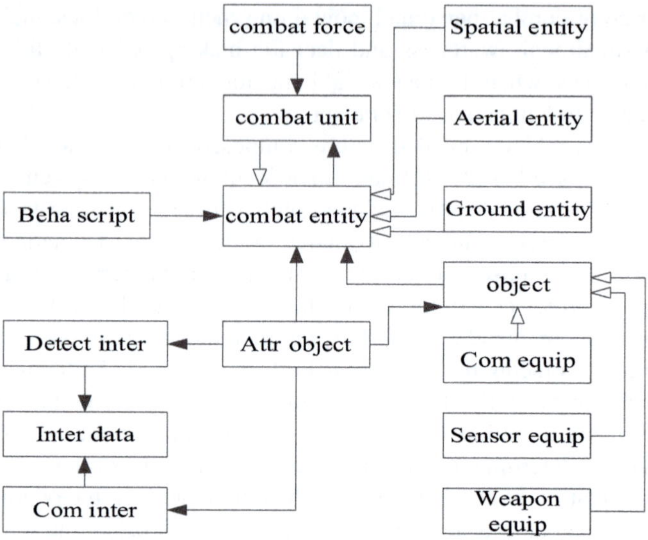

Fig. 1 Composable model framework for the space-based information and communication system

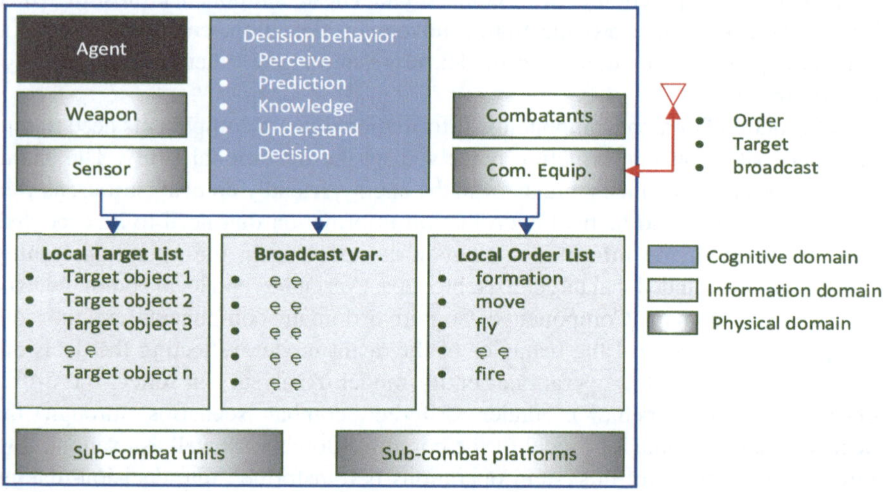

Fig. 2 Composition of the entity model for the space-based information and communication system

Based on the above framework design, the model structure for each entity of the space-based information and communication system is shown in the following Fig. 2:

The above abstract models all contain the physical domain, information domain, and cognitive domain.

3 Agent Behavior Modeling Method

The decision-making process of an agent's behavior reflects the complex process of transforming information superiority into decision-making superiority and ultimately into combat superiority under informationized conditions. This process includes a series of processes such as information processing, situational awareness, situation prediction, threat judgment, and decision-making actions. At the same time, this process is also the most variable part of system research, and how to reasonably and effectively support the representation of agent decision-making behavior is a core content of system effectiveness evaluation research. The modeling of agent decision-making behavior is a complex modeling process, and its decision results have the most significant impact on combat outcomes.

To describe decision-making behavior in simulations, several behavior representation methods have been formed, including reactive behavior representation methods, rule-based behavior representation methods, and activity-based behavior combination methods [9].

- **Reactive Behavior Representation Method:** Generally used for simple agent behavior modeling. Reactive decision rules allow agents to interact with other agents and respond to environmental changes, usually associating external stimuli with specific related reactions. Its advantage is that the rules are very simple, but the adaptability is usually very limited or cannot continue to evolve.
- **Rule-Based Behavior Representation Method:** Uses production rules to describe changes in combat entity decision-making behavior. The advantage is that knowledge is easy to represent, and the entire reasoning process only involves matching antecedents and performing consequents, but it cannot reflect the dynamic nature of combat processes over time, and flexibility is not strong, making development more difficult.
- **Activity-Based Behavior Combination Method:** Is a more flexible agent behavior representation method, which can represent the different activities of an agent at different times and under different conditions according to the order of activities executed in the simulation process.

This paper draws on the activity-based behavior combination method, strengthens the process description mode, and proposes a process-based behavior representation method. This method requires that the behavior of combat entities can be described in a serial manner according to processes or threads, and when delays occur in the simulation process, the execution of behaviors can be paused to ensure the causal correctness of the activity execution process in time among different agents. This approach is particularly suitable for the description and modeling of combat activities.

Supported by the process-based activity representation method, the decision-making behavior of each operational entity can be represented as a simulation process. Each entity's decision-making process consists of a series of activities, which can describe the complete process experienced by an operational entity from the start of the activity to the end of the simulation, including several events and activities that occur during this period, as well as the logic and timing relationships between these events and activities. The order of expression of these activities in the process represents the sequence of activity execution, and the execution conditions of activities can be controlled through flexible control statements. Processes are combinations of events and activities, so they can more comprehensively describe the state transition process in combat entity decision-making activities.

4 ABMS-Based Effectiveness Evaluation for Space-Based Information and Communication Systems

4.1 Features of the ABMS Method

Compared with traditional effectiveness simulation evaluation methods, the ABMS-based method can clearly represent the behavior modeling of specific individuals, while many traditional effectiveness simulations average or aggregate the entities in the system, and the corresponding computational models also attempt to simulate only the changes in the aggregated characteristics of these entities. The most important and difficult part of modeling is closely related to the interactions and adaptive behaviors of the entity Agent ensemble.

In terms of model structure, ABMS can represent the internal behavior of each individual. An Agent's behavior may depend on observations of other individuals but does not directly access these individuals' behavioral representations. This natural modularity follows the boundaries between individuals, and the encapsulation in model structure gives ABMS a clear advantage in system experimentation. In an ABMS, each entity has its own Agent or Agents model. An Agent makes decisions based on the information it has obtained or its "perception of the world," and its internal behavior does not need to be directly accessed by the rest of the system, so each Agent model in the system can only make decisions and take actions based on its maintained "perception of the world."

In terms of system representation, the combat process is determined by the discrete decisions of entities, and ABMS can most naturally represent this step-by-step decision-making execution process. The ABMS method is easier to depict the entities in the system. An Agent in an ABMS corresponds one-to-one with the individuals in the modeled system, making it easier for modelers to abstract Agent models and interaction relationships, characterizing the overall system behavior from local Agent behaviors. ABMS also makes it easier to depict the interaction space in the system, allowing for the definition of any topology form of Agent interaction. It is convenient to independently abstract the detection, communication, engagement relationships between entities, and their interactions with the environment.

The evaluation of space-based network information systems belongs to typical system experiments, and using the ABMS method can more effectively construct problem scenarios and carry out effectiveness evaluation experiments.

4.2 ABMS Evaluation Method for Space-Based Network Information Systems

Based on the idea of ABMS, designing and developing simulation models for the space-based information and communication system mainly includes five stages: problem description and requirement analysis, discovery and identification of agents, definition of agent interaction relationships and environment, agent behavior representation, and agent model definition. These processes constitute an iterative process of agent modeling and evaluation analysis.

1. Problem Description and Requirement Analysis: This stage mainly determines the problems that need to be solved by agent simulation, establishes the goals and requirements of the simulation research. In this stage, it is necessary to analyze the composition, boundaries, and environment of the system under study; clarify the quantitative indicators used for system analysis and experimentation.
2. Discovery and Identification of Agents: This stage mainly determines the entities and objects that need to be represented in the simulation according to the idea of agent simulation. ABMS cannot model all entities as agents, but needs to abstract and simplify the system according to the needs of the research problem.
3. Determination of Relationships and Interactions Between Agents: This stage needs to consider the relationships and interactions between agents and other agents as well as the environment based on the discovery and identification of agents. Similarly, these interactions and relationships need to be selected, abstracted, and aggregated according to the needs of problem research.
4. Agent Behavior Representation: This stage mainly uses the process-based behavior representation method to build relevant behavior decision-making models.
5. Agent Model Definition: In this stage, according to the determined agents, agent interaction relationships, and agent behavior representations, the attributes, states, parameters, and behaviors of agents and the environment can be represented through variables, parameters, and algorithms.

ABMS research generally undergoes multiple iterations in these stages. With each iteration, the model becomes more detailed, leading to repeated iterations of these stages. Since multiple iterations are required for each experiment scheme, and there are many uncertain random variables in the combat system, the number of simulations runs for each experimental scheme needs to be controlled within a certain range to ensure that the computational load is manageable. That is, under certain computational accuracy requirements, the number of simulations runs required to achieve the desired accuracy should be determined. This approach allows control over the length of the confidence interval, avoiding conclusions that are not applicable. For this purpose, the following method is proposed for sample size control.

The simulation evaluation for the space-based information and communication system is mainly based on a large number of simulation sample data, and the effectiveness evaluation analysis is carried out based on the following methods:

1. Important influencing factor analysis based on stepwise regression;
2. Importance analysis of influencing factors based on decision trees;
3. Trend analysis of key influencing factors.

The logical relationship between the three types of evaluation analysis methods is shown in Fig. 3.

1. Important influencing factor analysis based on stepwise regression mainly uses classical least squares method to fit the significant indicator factors in the simulation of space-based information and communication system effectiveness with survivability, establishing a fitting regression model that reflects the impact relationship between experimental design schemes and combat effects of the space-based system. This is used to analyze the statistical characteristics of simulation results to verify the rationality and effectiveness of the simulation process. On the other hand, it is used to identify the factors that have a significant impact on the capabilities of the space-based information and communication system, but it is not possible to determine the marginal benefit of a specific factor;

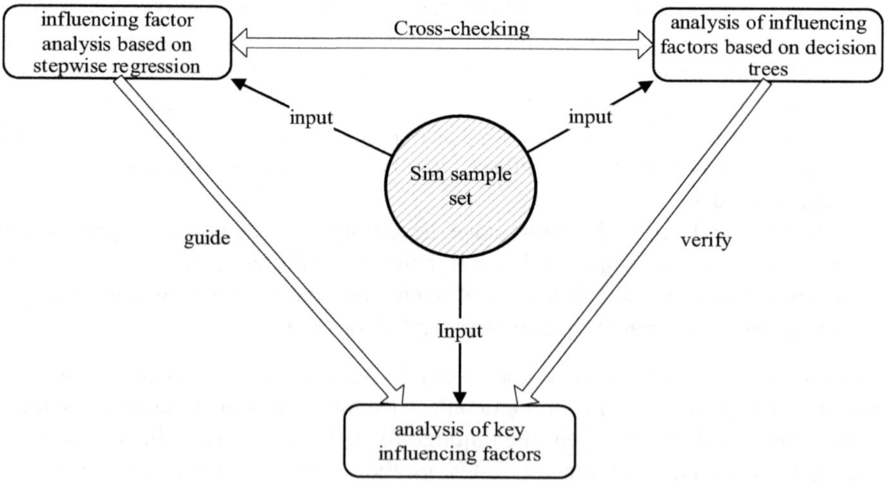

Fig. 3 Logical relationship between evaluation analysis methods

2. Importance analysis of influencing factors based on decision trees mainly uses classification and regression tree algorithms to further statistically analyze experimental samples, identifying the ranking of important influencing factors and determining the interval value relationship of the impact of different experimental factors on system capabilities. Compared to 1), the fitting regression model may have certain errors and limitations, and it is difficult to determine the key value of a certain influencing factor, while the analysis based on decision trees can provide factor rankings and specific values, but it cannot give the causal trend of the overall system capability;
3. Trend analysis of key influencing factors is to statistically analyze the change trends between various important influencing factors and assessment indicators such as system capabilities. Since the simulation sample set is the comprehensive impact of all influencing factors and random factors on combat effects under combat confrontation, the trend change between a single influencing factor and the assessment indicators will also contain the impact of other associated factors and will not show a simple linear increase or decrease trend, but the most important factors will show a certain trend.

Among the above three types of analyses, all original inputs are simulation experiment sample sets. Among them, the important influencing factor analysis based on stepwise regression and the importance analysis of influencing factors based on decision trees are mutually corroborative and complementary. The importance analysis of influencing factors based on decision trees is a further quantitative calculation of the important influencing factor analysis based on stepwise regression. The former identifies important factors that are generally a subset of the latter. The important influencing factor analysis based on stepwise regression can give the overall trend between system schemes and combat effects, while the importance analysis of influencing factors based on decision trees can give the key values of important factors; the three types of analyses are interlocked, mutually supportive and verified, ensuring that the experimental results and evaluation analysis are comprehensive, reasonable, and credible, providing scientific method support for subsequent conclusions and recommendations.

4.3 Model Validation

After the development of the relevant models, it is necessary to verify through model testing whether the established models meet the experimental design requirements of the space-based network information system. Model validation is a process that determines whether the computer implementation of the model accurately represents the model developer's conceptual expression and description of the actual system, and from the perspective of the expected application, determines the accuracy of the model in expressing the actual system.

Since the experiment of the space-based network information system belongs to a complex system experiment, it is not possible to obtain data through real experiments, and the model's resolution and abstraction level are relatively high, with not much real data for model confirmation. High-level abstract simulation models need to be supported by technical data from low-level simulation models for model validation. In this experiment, the main method is to use low-level model data to validate high-level models. The basic idea is to use hierarchical simulation experiments to gradually reduce the uncertain space; in different levels of simulation experiments, use models of different resolutions for different problems, reduce unnecessary computational overhead, and ensure the efficiency of simulation validation.

At the same time, for the testing of behavior models in the space-based network information system, it is possible to examine the correctness of the simulation model based on the process display of model operation tracking and visualization. If problems are found in the model testing process, it is necessary to adjust and modify the model's data definition, composition relationships, and behavior models according to the scenario and experimental design for correction.

5 Simulation Experiment and Analysis of the Information and Communication System

5.1 Experimental Design

The space-based information and communication system needs to describe a series of detection activities completed with the support of combat personnel and resources. These activities are mainly used for combat, including data collection, data transmission, data processing, etc. Based on the aforementioned combat business processes, relevant business models are constructed based on the aforementioned methods, taking a certain type of satellite combat business as the combat activity requirements, combining the space-based information and communication system combat process with system simulation, analyzing the impact of space-based information and communication system resources and combat activities on satellite combat activities, and the indicators should cover detection rate, mission completion rate, resource utilization rate, and other combat-related indicators.

Due to the many factors affecting the aforementioned evaluation indicators, it is possible to make choices about related factors based on analysis and evaluation needs, discard obviously unimportant experimental factors based on experience, narrow the experimental exploration space, and improve experimental calculation efficiency. According to the evaluation requirements of the space-based network information system in the task of sea surface target detection, eight factors such as electronic satellite detection accuracy, detailed survey task planning time, optical satellite maximum side swing angle, optical satellite projection radius, detailed survey planning strategy, launch task planning time, and launch preparation time are identified as experimental factors, and the impact of these factors is analyzed.

To evaluate the effectiveness of the space-based information and communication system in sea surface target detection tasks, this paper designed 8 experimental factors with different value ranges, adopting the Nearly Orthogonal Latin Hypercube (NOLHs) experimental design method. NOLHs absorbs the ideas of orthogonal Latin hypercube and uniform design, giving the method approximate orthogonality and excellent spatial filling properties. Hernandez et al. proposed some variants of this design method that are more effective and allow for the study of a larger number of design factors [10, 11]. The filling and uniformity of the experimental design scheme are shown in the following Fig. 4.

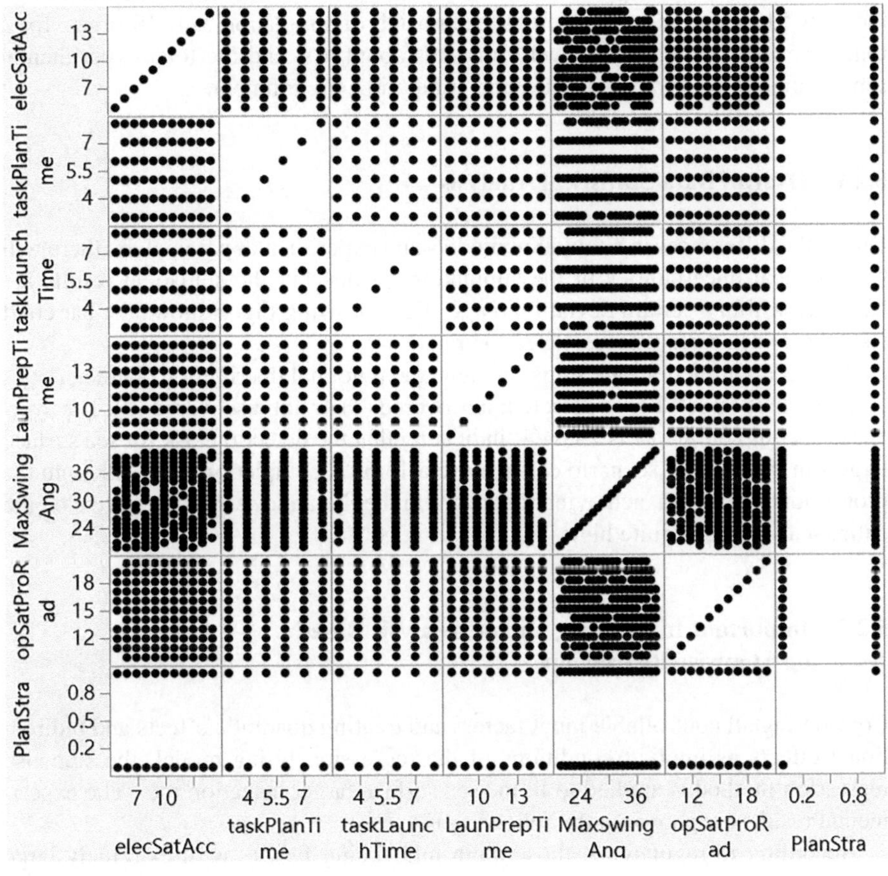

Fig. 4 Filling property of experimental design scheme

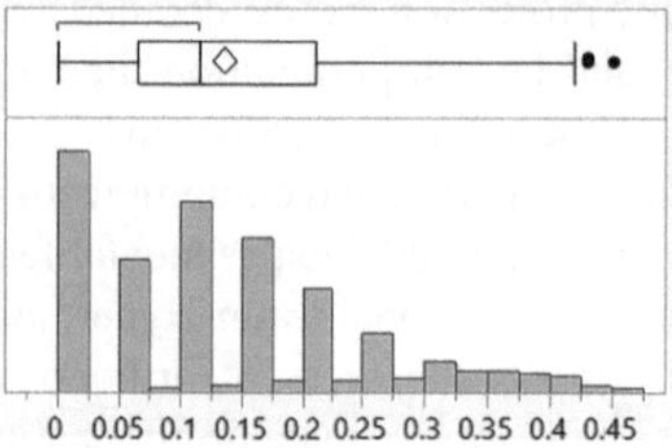

Quantiles

100.0%	maximum	0.45215
99.5%		0.431347856795
97.5%		0.37975
90.0%		0.3068
75.0%	quartile	0.2105875
50.0%	median	0.115425
25.0%	quartile	0.06525
10.0%		0
2.5%		0
0.5%		0
0.0%	minimum	0

Summary Statistics

Mean	0.136424
Std Dev	0.1092921
Std Err Mean	0.0048301
Upper 95% Mean	0.1459133
Lower 95% Mean	0.1269348
N	512

Fig. 5 Statistical chart of sea surface target detection rate

5.2 Simulation Result Analysis

Based on the above experimental design scheme, the final experimental design includes 512 experimental schemes, each batch of simulation runs 10 times, for a total of 51,200 simulations. Each run takes about 4 s, and the total experimental time is about 57 h in a general desktop computing environment.

5.2.1 Overall Data Statistical Analysis

Due to the differences in simulation models and experimental parameters, there will be statistical uncertainties in the simulation results, but the statistical results are basically within a certain deviation range. The following Fig. 5 shows the bar chart statistics of sea surface target detection rate.

The task completion rate statistics show the binomial distribution characteristics of sea surface targets being detected: discovered (1) or not discovered (0). The average task completion rate is 13.64%, indicating that the detection task for sea surface targets in this type of scenario can achieve a 13.64% completion rate. Although the proportion is not high, achieving this result in the dynamic search for targets on the entire sea surface is quite high.

5.2.2 Important Influencing Factor Analysis Based on Stepwise Regression

Considering all controllable input factors and treating quadratic effects and bidirectional effects as implicit conditions of the regression fitting model, the stepwise regression method is applied to fit the sea surface target detection rate. The experimental results are shown in the following Fig. 6.

According to regulations, those main influencing factors with extremely large associations or quadratic effects will be retained in the regression model, as shown in the following Fig. 7.

The figure above shows that considering the experimental samples, factors with a probability > |t| value <0.001 or smaller parameter estimates are considered to

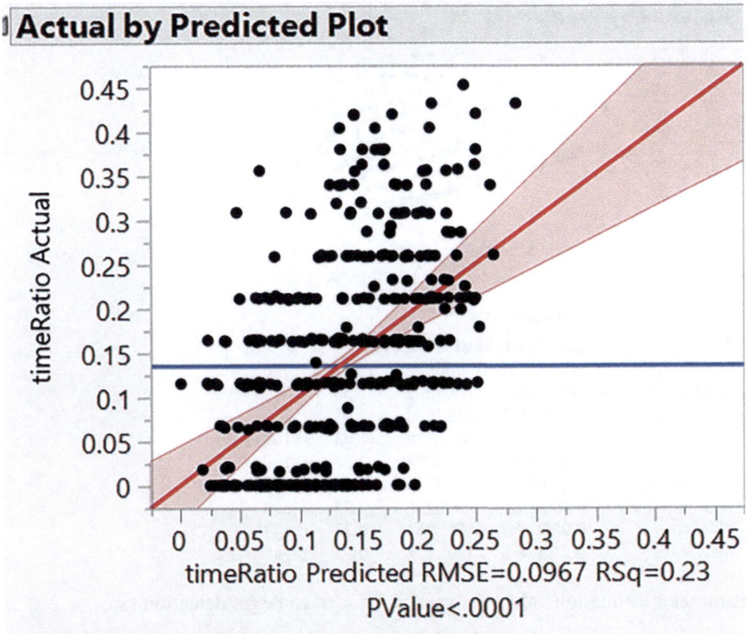

Summary of Fit

RSquare	0.229081
RSquare Adj	0.21682
Root Mean Square Error	0.096721
Mean of Response	0.136424
Observations (or Sum Wgts)	512

Fig. 6 Summary of fitting and variance analysis of sea surface target detection rate

have significant differences or a great association with influencing factors. It shows the basic order of optical satellite maximum side swing angle, launch preparation time, optical satellite projection radius, etc. This indicates that in this scenario, the optical satellite maximum side swing angle, task planning time, and optical satellite projection radius are the main factors affecting the detection rate of sea surface targets by the space-based network information system, and other factors such as launch preparation time and task planning strategy are general strategies. Accordingly, a multiple linear regression model is fitted between these experimental factors and the sea surface target detection rate as follows:

$$
\begin{aligned}
Y_{\text{timeRatio}} = & -0.09000675 + 0.00008669\,\bar{x}_{\text{elecSatAcc}} + 0.0048911\,\bar{x}_{\text{taskPlanTime}} \\
& -0.00850602\,\bar{x}_{\text{taskLaunchTime}} + 0.00143051\,\bar{x}_{\text{launPrepTime}} + \\
& 0.00992897\,\bar{x}_{\text{maxSwingAng}} - 0.00341101\,\bar{x}_{\text{opSatProRad}} - 0.0364856\,\bar{x}_{\text{trackTarget}}
\end{aligned}
$$

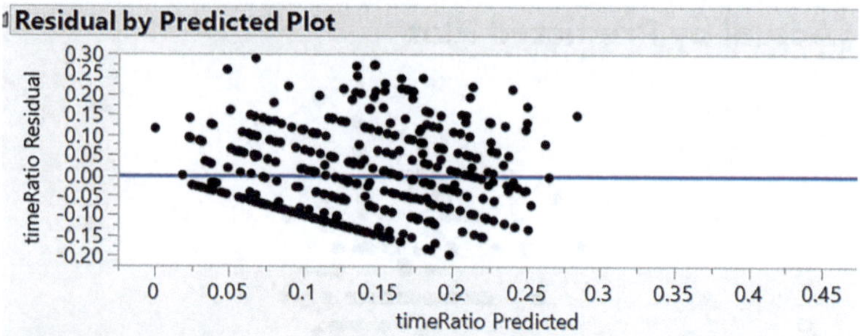

Parameter Estimates

Term	Estimate	Std Error	t Ratio	Prob>|t|
MaxSwingAng	0.009929	0.000926	10.72	<.0001*
taskLaunchTime	-0.008506	0.003209	-2.65	0.0083*
opSatProRad	-0.003411	0.001699	-2.01	0.0452*
Intercept	-0.090001	0.055115	-1.63	0.1031
taskPlanTime	0.0048911	0.003228	1.52	0.1304
TrackTarget	-0.036486	0.024116	-1.51	0.1309
LaunPrepTime	0.0014305	0.002242	0.64	0.5238
PlanStra	0.0050389	0.015361	0.33	0.7430
elecSatAcc	8.6686e-5	0.00137	0.06	0.9496

Fig. 7 Parameter estimation value analysis of sea surface target detection rate

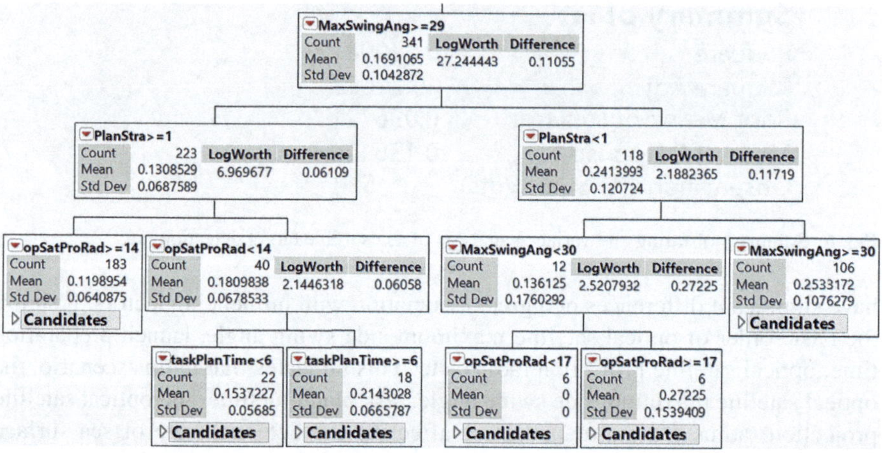

Fig. 8 Decision tree of sea

5.2.3 Importance Analysis of Influencing Factors Based on Decision Trees

Decision tree analysis is an important tool for determining influencing factors. The Fig. 8 below shows the decision tree using the average detection rate as the response variable and all controllable input factors. The decision tree has 9 branches with an R-squared value of 60%, which has a good significant impact. Analyzing the decision tree from top to bottom, the top elements have a greater impact on the corresponding variables. Since the higher the average detection rate, the better, the right

Table 1 Ranking of key influencing factors for sea surface target detection rate based on decision trees

Importance	Key factor	Analysis result
1	Detailed task planning strategy	When adopting an efficiency-first strategy for task planning, the average detection rate of sea surface targets is 24.14%
2	Maximum side swing angle of detailed survey satellite	When greater than or equal to 30, the average detection rate of sea surface targets is 25.33%

branches are the better branches, and the left branches are the worse branches. For example, the first branch on the far right, when using an efficiency-priority task detailed survey strategy, the average detection rate can reach 24.14%, and the far left branch can only reach 13.08%; then based on this branch, the second branch is the optical satellite maximum side swing angle. When the maximum side swing angle reaches more than 30 degrees, the average detection rate can reach 25.33%, otherwise, it can only reach 13.61%.

Many factors proven to be important in stepwise regression are also proven to be important in decision trees. The important influencing factors in stepwise regression analysis are the optical satellite maximum side swing angle, task planning time, and optical satellite projection radius, etc., and task planning strategy is a general influencing factor. The factors analyzed by the decision tree all include them, and the order is basically consistent.

The decision tree analysis conclusion in Table 1 quantifies the quantitative requirements for related influencing factors:

When the maximum side swing angle of the detailed survey satellite is greater than or equal to 30 and an efficiency-first strategy is adopted for detailed survey task planning, the average detection rate of sea surface targets is 25.33%.

5.2.4 Trend Analysis of the Impact of Maximum Side Swing Angle of Detailed Survey Satellite on Whether Sea Surface Targets Are Detected

The trend of the impact of the maximum side swing angle of the detailed survey satellite on whether sea surface targets are detected is shown in the Fig. 9 below. As the maximum side swing angle of the detailed survey satellite increases from 26 degrees to 31 degrees, the detection rate continues to rise and tends to level off after the maximum side swing angle exceeds 31 degrees. According to the aforementioned recursive partitioning decision tree, when the maximum side swing angle of the detailed survey satellite is greater than or equal to 30, the average detection rate is 25.33%. From the figure below, it can be seen that when the maximum side swing angle of the detailed survey satellite is greater than 30, the average detection rate does not increase, but it increases again from 38 degrees. Therefore, increasing the maximum side swing angle of the detailed survey satellite beyond 30 degrees has little marginal benefit.

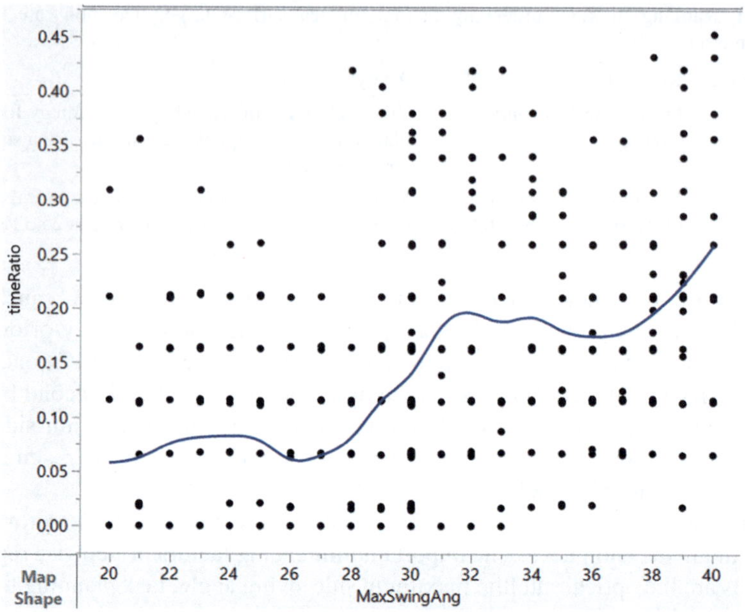

Fig. 9 Impact curve of the number of maximum side swing angle of detailed survey satellite on sea surface target detection

6 Conclusion

This paper aims at the effectiveness evaluation requirements of the space-based information and communication system, establishes a composable and reusable model framework for the space-based information and communication system oriented towards multi-domain connectivity, information processing, and resource management and control, and an agent behavior modeling method. A modeling, simulation, and effectiveness evaluation method based on ABMS is proposed. Experimental results of a certain space-based information and communication system for sea surface target detection applications show that the modeling method and framework can meet the needs of simulation evaluation of the space-based information and communication system. Using decision tree quantitative analysis methods to statistically analyze large-scale experimental results quantifies the contribution rate of different equipment elements of the space-based information and communication system to the detection effectiveness of sea surface targets and the impact trends of key elements. However, it is also found that when facing a much larger combat system, the description capability of agents needs to be improved, and related ABMS methods need further improvement, which will be the focus of future research directions.

References

1. Han C, Xiong W, Liu W, Yu X, Jian P (2022) Effectiveness evaluation method of space-based information systems based on SEM. J Syst Simul 34(8):1799–1810
2. Yu C, Peng S, Li M (2024) Modeling and effectiveness evaluation method of space-based information support system. Syst Eng Electron 46(10):3407–3415
3. Yu X, Xiong W (2024) Operational effectiveness evaluation of space-based information support equipment system based on Bayesian network. Ordnance Industry Automation 43(3):74–81
4. Wei X, Xiaolan Y, Yali L (2023) Operational effectiveness analysis method based on Bayesian network. Sci Technol Eng 23(17):7428–7435
5. Bintuan W, Jinfeng L, Puni D (2022) Research on the design factors and assessment for aircraft structures combat survivability. Adv Aeronaut Sci Eng 13(3):50–58
6. Bo Z, Depei K, Jianlu W (2023) Meta-synthesis method of SoS contribution rate for electronic information equipment. J Terahertz Sci Electr Inform Technol 21(3):325–331
7. Jamshidi M (2005) Theme of the IEEE SMC 2005[EB/OL], Waikoloa, Hawaii, USA. http://ieeesmc2005.unm.edu/
8. Jamshidi M (ed) (2008) System of systems Engineering—innovations for the 21st century. Wiley, New York
9. Wang C, Dong Y, Su Q, Zhang X (2022) Research on intelligent countermeasure behavior modeling methods in combat simulation. Command Control Simul 1:79–85
10. Hernandez R et al (2015) A review of variants and applications of the Latin hypercube sampling method. Reliab Eng Syst Saf 142:327–342
11. Cioppa TM (2000) A Latin hypercube design approach to parameter uncertainty and sensitivity analysis. In: Proceedings of the 2000 Winter simulation conference, pp 291–298

Open Access This chapter is licensed under the terms of the Creative Commons Attribution-NonCommercial-NoDerivatives 4.0 International License (http://creativecommons.org/licenses/by-nc-nd/4.0/), which permits any noncommercial use, sharing, distribution and reproduction in any medium or format, as long as you give appropriate credit to the original author(s) and the source, provide a link to the Creative Commons license and indicate if you modified the licensed material. You do not have permission under this license to share adapted material derived from this chapter or parts of it.

The images or other third party material in this chapter are included in the chapter's Creative Commons license, unless indicated otherwise in a credit line to the material. If material is not included in the chapter's Creative Commons license and your intended use is not permitted by statutory regulation or exceeds the permitted use, you will need to obtain permission directly from the copyright holder.

Climbing Motion Analysis of the Mobile Automatic Lifting and Lowering Dual-Arm Tower Assembly Device on Dirt Roads

Zhen Chen, Hao Chang, Bo Yu, Zhongpan Wang, Shida She, and Jian Sun

Abstract This paper systematically analyzes the motion characteristics of the mobile automatic lifting and landing double-arm clamping tower assembly device in soft terrain and slope environments through theoretical analysis and numerical research. First, a force model for the tracked chassis was established to analyze the nonlinear relationship between the depth of track front end sinking and ground pressure under soft soil conditions, and the influence of soil parameters on the sinking depth was explored. Next, the variation of pushback resistance under different slope angles was investigated. Finally, a model of the tower assembly device was established using RecurDyn, and slope climbing simulations were conducted to obtain the driving torque curve of the drive wheels. The results show that soil type and slope angle significantly affect track sinking depth and pushback resistance. Pushback resistance decreases with the increase of slope angle under sloped conditions. The research results provide a theoretical basis for the optimization design and enhancement of the device's adaptability to complex terrains.

Keywords Mobile tower assembly device · Tracked chassis · Soft soil · Bulldozing resistance · Dynamic analysis

Z. Chen (✉) · H. Chang · B. Yu · J. Sun
Liaoning Power Transmission and Transformation Engineering Co., Ltd, Shenyang, Liaoning, People's Republic of China

Emergency Rescue Center of State Grid Liaoning Electric Power Supply Co., Ltd, Shenyang, Liaoning, People's Republic of China

Z. Wang · S. She
State Grid Liaoning Electric Power Supply Co., Ltd, Shenyang, Liaoning, People's Republic of China

© The Author(s) 2026

Z. Zhang, P. Mercorelli (eds.), *Mechanical Design and Simulation: Insights and Innovations*, Lecture Notes in Mechanical Engineering,
https://doi.org/10.1007/978-981-95-2405-1_18

1 Introduction

The mobile automatic lifting and lowering dual-arm tower assembly device, as a new type of tower assembly equipment, offers flexibility and efficiency, making it suitable for constructing transmission towers in complex terrains. However, the motion stability and operational performance of the device in such environments still face numerous technical challenges [1]. Liu et al. [2] established a pressure distribution model for tracked vehicles on soft ground, Ding et al. [3] developed a prediction model for the sinking and sliding of tracked vehicles during turns, and Dong et al. [4] and Said Al-Milli et al. [5] created a comprehensive framework to evaluate the stability of tracked vehicles on soft soil.

This paper focuses on the mobile tower assembly device, using theoretical analysis and numerical simulation to primarily analyze the motion characteristics of the tracked chassis on soft, sloping surfaces, revealing the influence of soil parameters and slope angles on the device's motion performance. The research findings not only provide a theoretical basis for the optimization of tower assembly devices but also offer valuable reference for the development of similar equipment in complex terrains.

2 The Climbing Dynamics Analysis of the Tower Assembly Device

When the tracked chassis of the tower assembly device moves in a straight line on soft ground, the tracks will sink into the soil to some extent under the influence of gravity. The contact pressure per unit area of the tracked chassis has a nonlinear relationship with the sinking depth of the tracks [6], which is influenced by the soil's cohesive deformation modulus k_c and frictional deformation modulus k_ϕ,and expressed as:

$$p_e = \left(\frac{k_c}{b} + k_\phi \right) h_t^n \tag{1}$$

When the tower assembly device climbs uphill, the point of action of the pressure exerted by the tracked chassis on the slope surface changes, and the pressure varies with the slope angle [7]. The distribution of the ground pressure ratio on the outer track of the tracked chassis is shown in Fig. 1. In the figure: y_o represents the positive direction of the y-axis in the horizontal reference frame yDx; y_c represents the lateral offset of the supporting force from the center of the tower assembly device; P_{yNo} represents the ground pressure on the bottom of the outer track of the tower

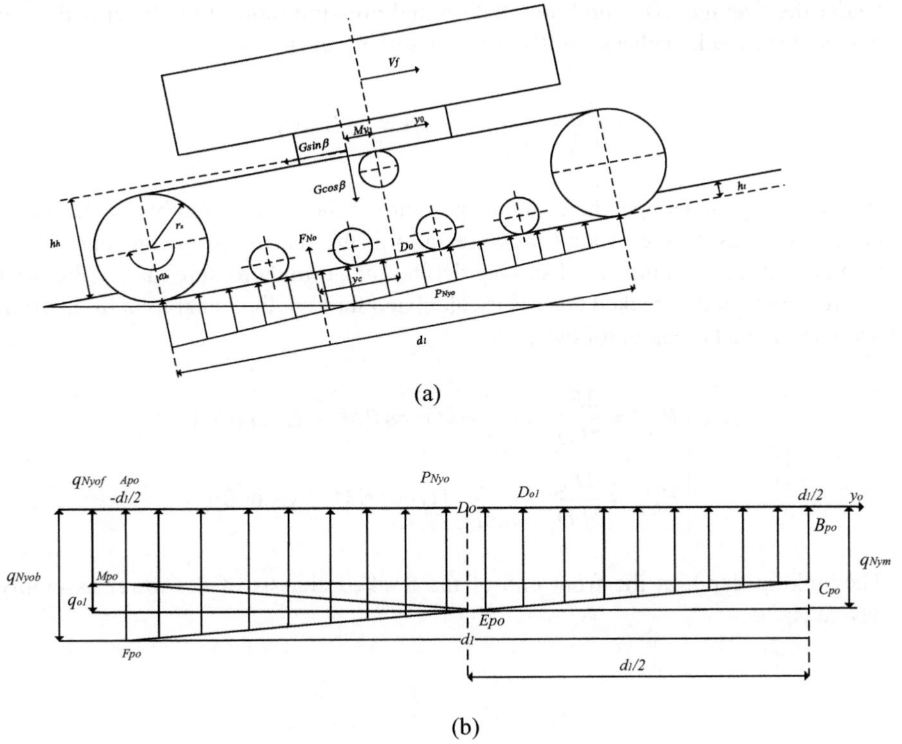

Fig. 1 Schematic Diagram of Ground Pressure Ratio of the Tracked Chassis of the Tower Assembly Device

assembly device; q_{Nyob}、q_{Nyom}、q_{Nyof} represent the normal loads borne by the respective parts of the outer track of the tower assembly device; F_{No} represents the total normal load borne by the outer track of the tower assembly device.

The average unit length of the outer track of the tower assembly device along the horizontal direction is denoted as the total normal load F_{No}, and the ratio of d_1 to the contact length. The forward pressure q_{Nyof} and the rear pressure q_{yNob} at the ends have an offset compared to the average pressure q_{o1}, which causes the normal load to change in direction. The normal load is expressed as:

$$\begin{cases} F_{yo} = F_{No} - \dfrac{6\left(G\cos\beta M_y + G\sin\beta h_h\right)}{d_1^2} y_o \\ F_{No} y_c = \dfrac{1}{6} q_{o1} d_1^2 \end{cases} \quad (2)$$

Under the average load condition, the ground pressure ratio of the inner and outer tracks of the tracked chassis of the tower assembly device is:

$$P_{Nys} = \frac{F_{Ns}}{bd_1} - \frac{6}{bd_1^3}\left(G\cos\beta M_y + G\sin\beta h_h\right)y_s \tag{3}$$

When driving on a slope, the center of pressure of the tower assembly device shifts backward, causing the ground pressure ratio at the rear to be greater than at the front, resulting in an increased sinking depth. The ground pressure ratio at the front and rear ends of the tracked chassis is calculated through the integration of the front and rear contact areas as follows:

$$\begin{cases} P_{Nyqs} = \dfrac{2F_{Ns}}{3bD_t} - \dfrac{2}{d_1bD_t}\left(G\cos\beta M_y + G\sin\beta h_h\right) \\ P_{Nyhs} = \dfrac{2F_{Ns}}{3bD_x} + \dfrac{2}{d_1bD_x}\left(G\cos\beta M_y + G\sin\beta h_h\right) \end{cases} \tag{4}$$

The sinking depth at the front end of the tracked chassis of the tower assembly device is:

$$h'_{tqs} = \left[P_{Nyqs} / \left(\frac{k_c}{b} + k_\phi\right)\right]^{\frac{1}{n}} \tag{5}$$

From Eqs. (4), and (5), it can be concluded that the push-pull resistance generated by the soil on the tracked chassis of the tower assembly device when driving on a sloped soft surface is [8, 9]:

$$F'_{ps} = b\left(0.67ch'_{tqs}k'_c + 0.5h'^2_{tqs}rk'_r\right) \tag{6}$$

When the tower assembly device experiences lateral mass offset, the load on the contact area between the tracked chassis and the ground is uniformly distributed. The traction force F_{js} of the inner and outer tracks of the tower assembly device is given by:

$$\begin{cases} F_p = b\left(0.67ch'_{tqs}k'_c + 0.5h'^2_{tqs}rk'_r\right) \\ k'_c = \left(N'_c - \tan\varphi'\right)\cos^2\varphi' \\ k'_r = \left(\dfrac{2N'_r}{\tan\varphi'} + 1\right)\cos^2\varphi' \\ \tan\varphi' = 2/3\tan\varphi \end{cases} \tag{7}$$

In the formula, $N_c^{'}$ and $N_r^{'}$ represents the bearing capacity coefficient for localized shear failure of the sandy foundation.

The driving resistance experienced by the tower assembly device $F_{ffs}^{'}$ is:

$$\begin{cases} F_{ffs}^{'} = F_{fm1s}^{'} + F_{ps}^{'} + F_{is}^{'} \\ F_{js}^{'} = F_{ffs}^{'} \\ F_{is}^{'} = f_i F_{Ns} \end{cases} \tag{8}$$

3 The Impact of Different Soils on Track Sinking Depth

Through the formula derivation, the sinking depth at the front end of the track is related to the soil parameters, as shown in Table 1. By substituting the parameters of the mobile automatic lifting and landing double-arm clamping tower assembly device and the soil parameters into the formula, the theoretical sinking depth at the front end of the track can be obtained [10]. The relationship between the sinking depth at the front end of the track and the slope angle is shown in Fig. 2.

From Fig. 2, it can be observed that as the slope angle increases, the sinking depth at the front end of the inner and outer tracks of the tower assembly device decreases. The sinking depth at the front end of the tracks in dry sandy soil is greater than that in sandy humus soil. On heavy clay surfaces, the sinking depth at the front end of the tracks is negligible, meaning the push-pull resistance experienced by the tower assembly device is zero. Additionally, the sinking depth of the right track is greater than that of the left track.

The relationship between the push resistance at the front end of the tracks and the slope angle is shown in Fig. 3. The trend of push resistance aligns with that of the sinking depth: as the slope angle increases, the sinking depth decreases, and the push resistance decreases. The push resistance of the left track is less than that of the right track.

Table 1 Soil parameters

Soil species	Humidness (%)	n	K_c (kN·m$^{-(n+1)}$)	K_φ (kN·m$^{-(n+2)}$)	c(kPa)	φ (°)	ρ (g/cm^3)
Dry sandy soil	0	1.1	0.95	1523.43	1.04	28	1.8
Sandy humus soil	11	0.9	52.53	1127.97	4.83	20	1.9
Heavy clay	25	0.13	2.7	1555.95	68.95	34	2.15

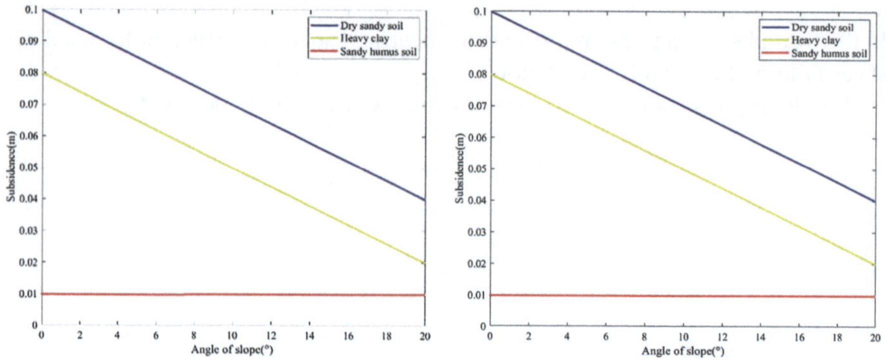

(a)The sinking depth at the front end of the left track

(b)The sinking depth at the front end of the right track

Fig. 2 Relationship between the sinkage of the front end of the track and the slope angle.(**a**) The sinking depth at the front end of the left track. (**b**) The sinking depth at the front end of the right track

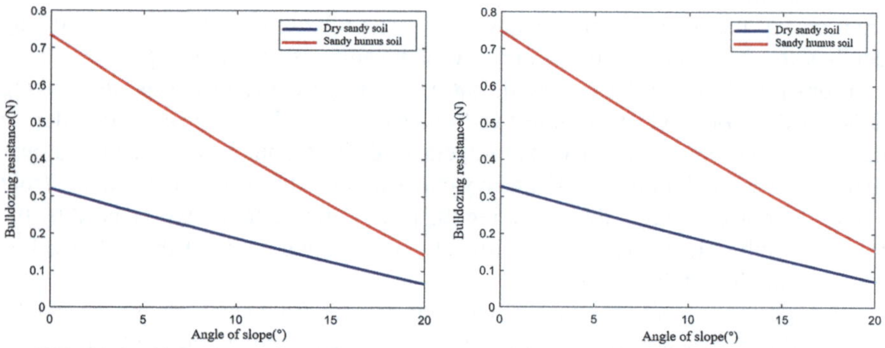

(a)The relationship between pushing resistance and slope Angle at the front end of left track

(b) The relationship between pushing resistance and slope Angle at the front end of right track

Fig. 3 Relationship between bulldozing resistance at the front end of track and slope angle. (**a**) The relationship between pushing resistance and slope Angle at the front end of left track. (**b**) The relationship between pushing resistance and slope Angle at the front end of right track

4 Tower Assembly Device Slope Climbing Simulation

This chapter will conduct a simulation of the mobile automatic lifting and landing double-arm clamping tower assembly device in RecurDyn. In line with the design requirements of the mobile automatic lifting and landing double-arm clamping

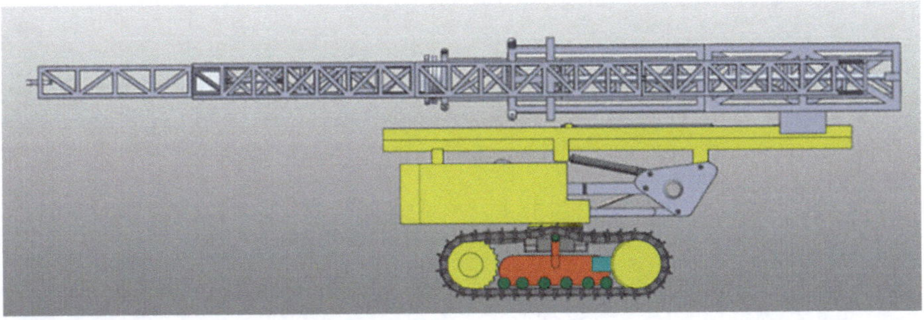

Fig. 4 The dynamic model of mobile automatic lifting and landing tower assembly device with floor-typed double horizontal arm holding pole

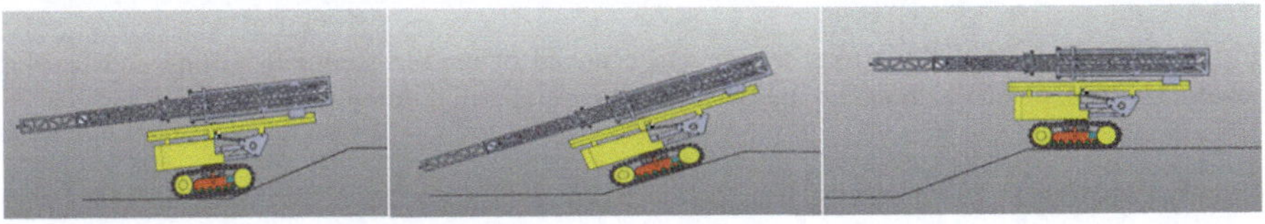

Fig. 5 Climbing process of mobile automatic lifting and landing tower assembly device with floor-typed double horizontal arm holding pole

tower assembly device, the tracked chassis model is established in RecurDyn using the High-Speed Track Module (HM). The model established is shown in Fig. 4.

According to the design requirements, the mobile automatic lifting and landing double-arm clamping tower assembly device should be able to climb a 20-degree slope. This section focuses on the obstacle-crossing simulation of the tower assembly device. First, a 4-meter-long 20-degree slope is created in the virtual environment, with the road surface condition set as hard terrain. The tower assembly device completes the slope climbing simulation at a speed of 0.5 m/s, with the driving wheel function defined as STEP(TIME,1,0,6,84D). The simulation process of climbing the slope is depicted in Fig. 5, the analysis results of the changes of torque and center of mass during the climbing process are shown in Fig. 6.

The analysis results indicate that the fluctuation trend of the driving torque is consistent with the motion process of the tracked chassis, and the mobile automatic lifting and landing double-arm clamping tower assembly device can meet the requirement of climbing a 20-degree slope.

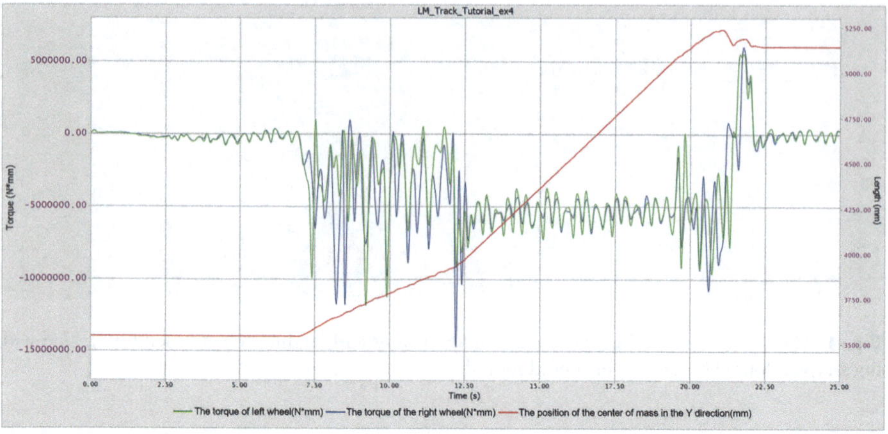

Fig. 6 The changes of slope road torque and centroid of mobile automatic lifting and landing tower assembly device with floor-typed double horizontal arm holding pole

5 Conclusions

This paper focuses on the motion characteristics of the mobile automatic lifting and landing dual-arm tower assembly device. Under soft slope road conditions, a detailed study was conducted on the force analysis of its tracked chassis, including the sink depth and pushing resistance. Through theoretical derivation, numerical analysis, and simulation analysis, the impact of soil parameters and slope angle on the device's motion performance was revealed, as well as the driving torque provided by the tracked chassis during the climbing process. The results show that variations in soil density and slope angle significantly affect the sink depth and pushing resistance at the front end of the track, with clear differences in sink depth and pushing resistance when the device moves on different soils.

This study provides a theoretical basis for the performance optimization and design improvement of the mobile tower assembly device under complex terrain conditions and offers a reference for the development and application of similar equipment in the future. Further research can combine experimental validation and multi-field coupling analysis to expand the applicability and reliability of the findings.

Acknowledgements This work was financially supported by the State Grid Liaoning Electric Power Supply Co. (2024YF-83).

References

1. Lee Won kyo F, Mun Sung Duk S, Shin Kooyong T (2019) Development of eco-friendly electric transmission towers in KEPCO. KEPCO J Electr Power Energy 5:135
2. Weiwei Liu F, Kai Cheng S (2020) An analytical model for predicting ground pressure under a rigid-flexible tracked vehicle on soft ground. Math Probl Eng 2020:6734121

3. Zhao Ding F, Yaoming Li S, Zhong Tang S (2020) Theoretical model for prediction of turning resistance of tracked vehicle on soft terrain. Math Probl Eng 2020:4247904
4. Al-Milli Said F, Lakmal D. Seneviratne S, Kaspar Althoefer T (2010) Track–terrain modelling and traversability prediction for tracked vehicles on soft terrain. J Terramech 47(3):151–160
5. Youkou Dong F, Dingtao Yan S, Xiaowei Feng T (2023) Stability of tracked vehicles on soft grounds under multidirectional loading conditions. Can Geotech J 60(11):1743–1755
6. Murakami H, Watanabe K, Kitano M (1992) a mathematical model for spatial motion of tracked vehicles on soft ground. J Terrramech 29(1):71–81
7. Edwin P, Shankar K, Kannan K (2018) Soft soil track interaction modeling in single rigid body tracked vehicle models. J Terrramech 77:1–14
8. Wong JY (2001) Theory of ground vehicles, 3th edn. Wiley, Hoboken
9. Wong JY (2008) Theory of ground vehicles, 4th edn. Wiley, Hoboken
10. Li Qiang F, Xintian Liu S (2023) a soil mechanical model for deformable grounds under the dynamic action of high-mobility tracked vehicles. Recent Patents Eng 17(2):92–104

Open Access This chapter is licensed under the terms of the Creative Commons Attribution-NonCommercial-NoDerivatives 4.0 International License (http://creativecommons.org/licenses/by-nc-nd/4.0/), which permits any noncommercial use, sharing, distribution and reproduction in any medium or format, as long as you give appropriate credit to the original author(s) and the source, provide a link to the Creative Commons license and indicate if you modified the licensed material. You do not have permission under this license to share adapted material derived from this chapter or parts of it.

The images or other third party material in this chapter are included in the chapter's Creative Commons license, unless indicated otherwise in a credit line to the material. If material is not included in the chapter's Creative Commons license and your intended use is not permitted by statutory regulation or exceeds the permitted use, you will need to obtain permission directly from the copyright holder.

Added Mass Solution Methods for Water Entry Problem

Fan Wang, Jun Jiao, Tengjiao Sang, Sheng-zhe Shi, and Tao Liu

Abstract The solution of added mass for water inflow can be summarized as the potential flow solution problem of the corresponding model. Using boundary element method and commercial software to solve UDS (User Defined Scalars) to solve the Laplace equation and obtain the potential flow solution of the corresponding problem. Firstly, the added mass of the plate water entry problem is calculated and compared with the theoretical solution to verify the reliability of the two methods. Furthermore, the added mass of wedges with different deadrise angles was calculated, and the results showed that the added mass values calculated by the two methods were close, while the added mass of wedges increased with the decrease of deadrise angles, gradually approaching the added mass value of flat plates entering the water.

Keywords Water entry · Boundary element method · UDS · Added mass

1 Introduction

The water entry problem is a complex hydrodynamics problem, which is often applied to the calculation of landing loads on navigation vehicles and slamming loads generated by ship motion in waves. Around the 1930s, Von Karman [1] and Wanger [2] were recognized as the first two scholars who carried out the research on slamming. Von Karman simplified the wedge water entry problem into the problem of solving added mass, and considered that the slamming force on the wedge during water entry was simplified into an inertial force, which was equal to the product of the added mass and the acceleration of the object. The momentum theorem was

F. Wang (✉) · J. Jiao · T. Sang · S.-z. Shi · T. Liu
Aviation Industry Corporation of China Special Vehicle Research Institute, Jingmen, Hubei, People's Republic of China

Key Aviation Scientific and Technological Laboratory of High-Speed Hydrodynamic, Jingmen, China

© The Author(s) 2026

Z. Zhang, P. Mercorelli (eds.), *Mechanical Design and Simulation: Insights and Innovations*, Lecture Notes in Mechanical Engineering,
https://doi.org/10.1007/978-981-95-2405-1_19

used to complete the solution of the water entry problem, and the added mass of the wedge was simplified into the calculation of the added mass of the plate with the same wetting width. In addition to Wagner, Zhao and Faltinsen [3] and Carcaterra [4] have also revised Von Karman's added mass formula, but the revised results are different. According to the theory of stripe, Wang and Fang [5] calculated the half width and added mass of the cross sections of the boat considering the rise of the liquid level during the water entry, and calculates impact force of the boat by the theory of momentum and kinetic energy respectively. Boundary element method, also known as boundary integral equation method, is a high-precision method developed after finite element method. It transforms the governing equation into boundary integral equation and discretely solves the boundary of the computational domain. Compared with other regional discrete solving methods for partial differential equations, the number of unknowns and computation amount are significantly reduced due to the reduction of solving dimension. At the same time, because it uses the basic solution of the differential equation as the kernel function of the boundary integral equation, it has the characteristics of combining analytic and numerical values, and usually has a high calculation precision. Dong et al. [6, 7] used the fully nonlinear boundary element method to solve the water entry problem of a two-dimensional wing. The User Defined Scalars (UDS) has a wide range of application scenarios in fluid dynamics simulation. With UDS, users can simulate complex fluid dynamics problems more flexibly. UDS allows users to define custom scalar transfer equations in Fluent. These equations can include the transient term, the flux term, the diffusivity term, and the source term. UDS allows users to simulate physical processes that are not included in Fluent's standard model. In this paper, the boundary element method and UDS method are used to solve the added mass problem. After comparison and verification with the theoretical solution of the added mass of the binary plate, the variation rule of the added mass of the wedges with different deadrise angles is studied. Zheng [8] proposed a method of calculating the added mass based on commercial software ANSYS fluent, which reduces the grid requirements and greatly improves the calculation speed while ensuring the calculation accuracy.

2 Theoretical Foundation

2.1 Test Model

The added mass problem of water entry can be solved by the corresponding potential flow problem. That is, the Laplace equation and the corresponding boundary conditions are satisfied. In order to simplify the calculation, only the two-dimensional problem is considered, and the plate enters the water at a unit velocity. The corresponding governing equation and boundary conditions are as follows [9]:

Governing equation:

$$\Delta\phi = 0 \tag{1}$$

Surface boundary conditions:

$$\frac{\partial\phi}{\partial n} = -n_3 \tag{2}$$

Infinite water bottom boundary conditions:

$$\nabla\phi = \vec{0} \tag{3}$$

Linear free surface boundary conditions:

$$\frac{\partial\phi}{\partial z} + \frac{\omega^2}{g}\phi = 0 \tag{4}$$

Where ϕ is the velocity potential, n_3 represents the vertical component of the normal direction of the surface, g is the gravitational acceleration, and ω is the circular frequency. For the water entry bang problem, it can be considered a high frequency response problem, that is, ω tends to infinity and the vertical derivative of the velocity potential is a finite definite value, so the linear free surface boundary condition holds only at $\phi = 0$ at the free liquid surface. After solving the above potential flow problem, the added mass M can be expressed as follows:

$$M = \rho \oint_s \phi \frac{\partial\phi}{\partial n} ds \tag{5}$$

The above equation represents the boundary integral on the surface of the computational domain. For the problem of water entry, since the free surface boundary condition is $\phi = 0$ and the boundary condition on the bottom is $\nabla \cdot \phi = 0$, and the velocity potential at infinity is also close to 0, the integral of the above equation can be directly integrated on the boundary of the surface.

2.2 Boundary Element Method

Boundary element method, also known as boundary integral equation method, is a high-precision method developed after finite element method. It transforms the governing equation into boundary integral equation and discretely solves the boundary of the computational domain. Compared with other regional discrete solving methods for partial differential equations, the number of unknowns and computation amount are significantly reduced due to the reduction of solving dimension. At the same time, because it uses the basic solution of the differential equation as the kernel function of the boundary integral equation, it has the characteristics of combining analytic and numerical values, and usually has a high calculation precision [10].

By using Green's second Formula (6), the calculated solution is transformed into the boundary integral equation. In the equation, ϕ is the velocity potential function to be solved, and $G(P, Q)$ becomes Green's function, which satisfies the basic solution of the governing equation: $\Delta G(P, Q) = \delta(P - Q)$, P and Q are the source point and the field point respectively. δ is the Dirac function. It can be simplified as Formula (7).

$$\int_{\Omega} \phi \nabla^2 G(P,Q) - G(P,Q) \nabla^2 \phi d\Omega = \int_{\partial\Omega} \phi \frac{\partial G(P,Q)}{\partial n} - G(P,Q) \frac{\partial \phi}{\partial n} ds \tag{6}$$

$$\phi(P) = \int_{\partial\Omega} \phi \frac{\partial G(P,Q)}{\partial n} - G(P,Q) \frac{\partial \phi}{\partial n} ds \tag{7}$$

There are two main methods to solve the above integral equations: one is the direct boundary element method, which constructs discrete equations according to Formula (7), and the solved unknown is the velocity potential; The other is called the indirect boundary element method, that is, the basic solution of the governing equation is multiplied by a constant distributed on the boundary of the calculation domain, the solution is called the source strength, and then the velocity potential of each point in the calculation domain is solved. Both methods have higher calculation accuracy. The indirect boundary element method is used in this paper.

Lnr/2π is the basic solution of two-dimensional Laplace equation, which only satisfies the bottom-bottom boundary conditions of infinite water depth, and the source points should be arranged on the boundary conditions of the free liquid surface and the boundary of the object surface. In this paper, a constant element is used, that is, the two-dimensional boundary is discretized into multiple line segments, and the point source strength on each line segment is a constant. Singularity exists when the spot point coincides with the source point, which needs special treatment. One way is to stagger the source point and the field point to avoid singularity. Another way is to treat singularity in a special way. This article takes the latter approach.

Singularity exists when calculating the contribution of the element itself to the velocity potential and normal derivative of the central field point of the element. As shown in Fig. 1, when calculating the partial derivative, the area around the field point is approximated to an infinitely small semicircle segment, where the point sources distributed in the straight segment have no contribution to the field point

Fig. 1 Singularity treatment of elements

velocity potential partial derivative, so it can be considered that the partial derivative calculation can be approximated to the contribution of the point sources distributed on the semicircle, that is $\frac{\pi \mathrm{r} S}{2\pi}\cdot\frac{\partial \ln r}{\partial r}=\frac{S}{2}$, where S is the intensity of the point sources distributed on the unit. For the calculation of the velocity potential, only when the field point coincides with the source point, there is singularity, and Cauchy principal value integral can be used.

2.3 UDS Method

General fluid business software solves the Reynolds transport equation in the form of Eq. (8). In the formula, ϕ is the transport volume and $\vec{u}$ is the transport velocity in the flow field. The time derivative term and flux term in the Reynolds diffusion equation at steady state and zero velocity flow field are 0, leaving only the diffusion term on the right of the equation. If the diffusion coefficient Γ is set at 1, the Reynolds transport equation directly degenerates into Laplace equation [11]. The Laplace equation can be solved by solving the UDS equation if certain boundary conditions are specified.

$$\frac{\partial \phi}{\partial t}+\vec{u}\cdot\nabla\phi=\nabla\cdot(\Gamma\nabla\phi) \tag{8}$$

The general boundary conditions can be divided into three categories: the first type of boundary conditions, namely Dirichlet boundary conditions, specify the value at the boundary, namely $\phi_f = \phi_{specfied}$; The second type of boundary condition, the Neuman boundary condition, specifies the normal derivative on the boundary, determining $\phi_{n\_f} = \phi_{n\_specfied}$. The third type of boundary condition, Robin boundary condition, gives the linear combination of the physical quantity and its normal derivative on the boundary, namely $\alpha\phi_f + \beta\phi_{n\_f} = \gamma$, where α, β, γ are the given known quantities. The first two types of boundary conditions are given in general commercial software, and the third type of boundary conditions can be indirectly transformed into the form of the first or second type of boundary conditions and solved iteratively.

3 Model Verification

The solution of the added mass in water of the plate is simplified into a two-dimensional model, as shown in Fig. 2. Its governing equation and boundary conditions are shown in Sect. 2, where the wetting length of the plate is 2C and the normal vertical component of the plate is 1, that is, the surface boundary condition is $\partial\phi/\partial n = -1$.

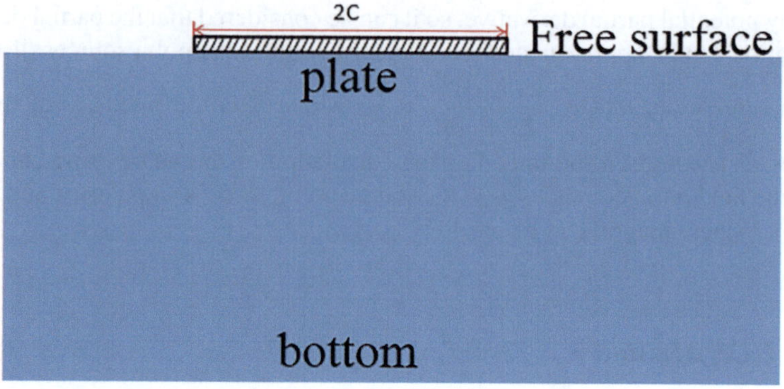

Fig. 2 Schematic diagram of two-dimensional plate model

To solve the above problem, complex variable Z = x + iz is introduced, and the analytical solution of complex potential can be expressed as:

$$\Phi = \phi + i\varphi = iZ - i\frac{Z}{\sqrt{Z^2 - C^2}} \tag{9}$$

ϕ represents the velocity potential and φ represents the flow function. According to the above formula, the velocity potential distribution on the plate can be obtained as follows:

$$\phi_w = -\sqrt{C^2 - x^2} \tag{10}$$

The vertical velocity at the free liquid surface is expressed as:

$$\frac{\partial \phi}{\partial z} = \frac{|x|}{\sqrt{x^2 - C^2}} - 1 \tag{11}$$

For the boundary element method, the discretization is shown in Fig. 3. Only the free surface and the surface of the object surface are discretized into corresponding line units, and point sources of equal intensity are distributed on each line unit. The length of the free liquid surface should ensure that it has no obvious influence on the calculation results. Ten times the wetting length was taken for the calculation. According to the above discretization and solution methods, the flow field information such as velocity potential distribution on the surface and normal velocity distribution on the free liquid surface can be obtained.

However, the UDS method needs to discretely process the computational domain of the two-dimensional plane problem, as shown in Fig. 4. Similarly, only a finite

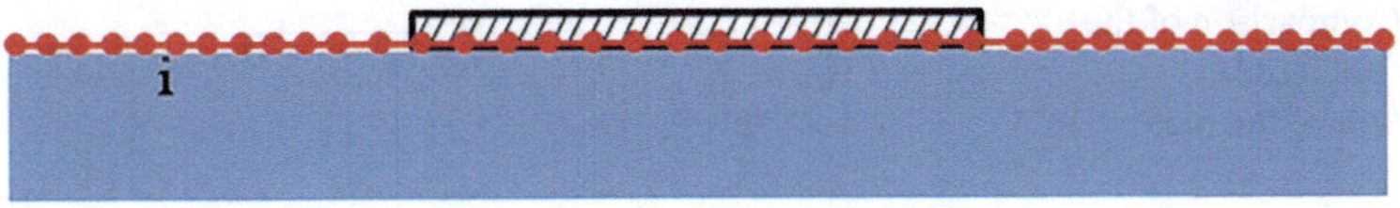

Fig. 3 Schematic diagram of boundary element dispersion

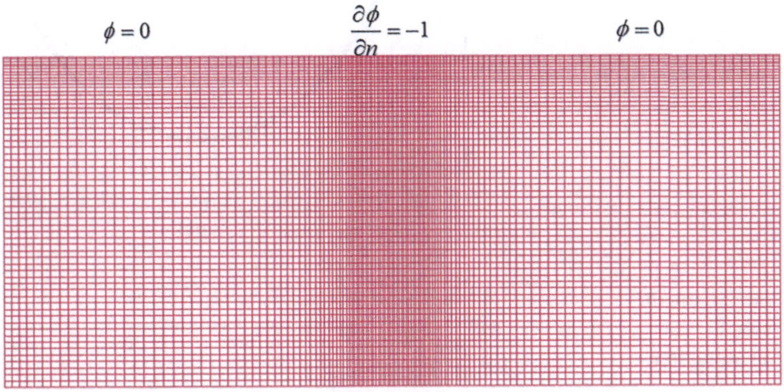

Fig. 4 Grid diagram of UDS

length free liquid surface is selected, and the boundary of two sides and bottom is set as a symmetric surface, and the normal derivative of the velocity potential on the boundary is 0. Theoretically, this is inconsistent with the real situation. However, considering that when the calculation domain is large enough, the water entry disturbance of the plate has relatively little influence on both sides and bottom, and the symmetric boundary conditions of both sides and bottom have relatively little influence on the calculation results, The boundary conditions of the free liquid surface and the plate wall are set according to the previous discussion.

During the calculation, the half-width of the plate was set to 0.5 m. The velocity potential distribution of the plate wall and the vertical velocity distribution of the free liquid surface obtained by the two methods and the corresponding theoretical solutions are shown in Figs. 5 and 6. The theoretical results show that the vertical velocity of the free surface is singular at both ends of the plate. The results show that the velocity potential distribution of the plate and the vertical velocity distribution of the free liquid surface obtained by the two methods are in good agreement with the theoretical solution, but the deviation is obvious at the end of the plate. Then the added mass of the plate is calculated, and the calculation results are shown in Table 1. The theoretical value of added mass is 391.91 kg, the calculated value of added mass is 396.9 kg by BEM and 391.52 kg by UDS. The added mass deviation of the two methods is 1.27% and 0.10% respectively. The results show that both methods are suitable for solving added mass problems without considering the water surface uplift effect, and the calculation accuracy is high.

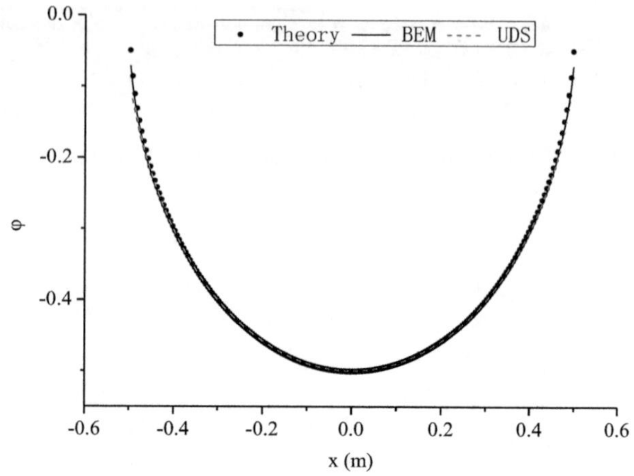

Fig. 5 Comparison of the velocity potential distribution of the plate

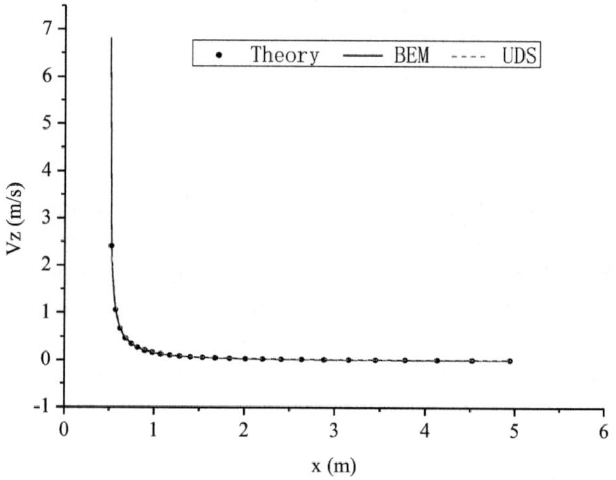

Fig. 6 Comparison of vertical velocity distribution at the free surface

Table 1 Comparison of solution results for added mass

Solution method	Added mass/kg		deviation
	Calculated value	Theoretical value	
BEM	396.90	391.91	1.27%
UDS	391.52	391.91	0.10%

4 Calculation of Added Mass of Wedges

Nine wedge shapes with different deadrise angles of 0, 5, 10, 15, 20, 30, 40, 50 and 60° were selected for calculation, and the wetted half-width was set at 0.5 m. The boundary element method and the UDS boundary element method are used

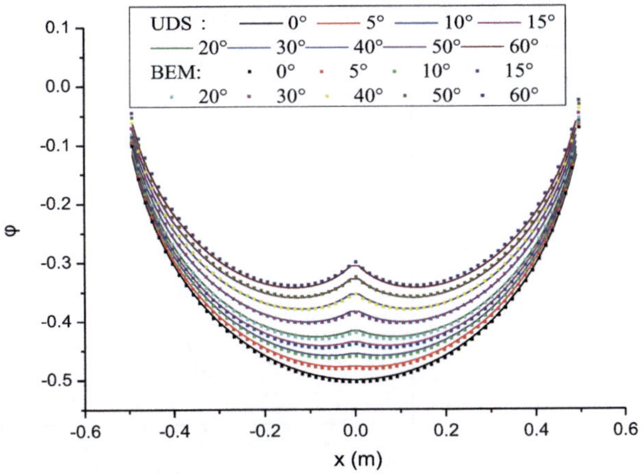

Fig. 7 Velocity potential distribution of wedge with different deadrise angles

respectively to solve the added mass, and the water surface uplift and air cushion effect are not considered in the calculation process.

The velocity potential distribution on the wetting surface of the wedge is shown in Fig. 7. The velocity potential distributions obtained by the two methods are close, and the differences are also concentrated in the two end points where the wedge body intersects with the free surface of still water. The velocity potential distribution of the wedge with different deadrise angles is mainly concentrated at the lower edge of the wedge, showing a local peak value, and the peak value distribution becomes more obvious with the increase of the deadrise angle.

Figure 8 shows the vertical velocity distributions on the free surface of the wedges with different deadrise angles. Theoretically, the velocity distributions are symmetrical. As shown in the figure, the calculation results of the two methods are quite different, which are also mainly concentrated on the two ends of the wedge. The UDS method fails to calculate the velocity distribution of the two ends of the wedge well, while the BEM method fails to capture the velocity of the large gradient at the two ends of the wedge. It is worth explaining that the UDS method mainly adopts the finite volume method, and the gradient calculation on the boundary of the calculation domain mainly adopts the interpolation method, which may be one of the reasons for the large deviation in the calculation of the vertical velocity of the free water surface in UDS. At the same time, the figure shows that the vertical velocity near the two ends of the wedge decreases with the increase of the angular velocity.

Then, the variation of the added mass of the 2D wedge model with different incline angles can be obtained by using Formula 5. As shown in Fig. 9, the added mass obtained by the two methods is relatively close, and the added mass of the wedge with the same wetted width also decreases with the increase of the deadrise angle.

Fig. 8 Vertical velocity distribution of free surface of wedge with different deadrise angles

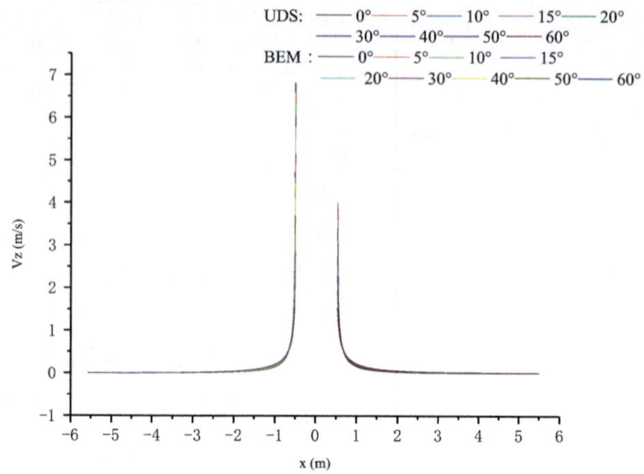

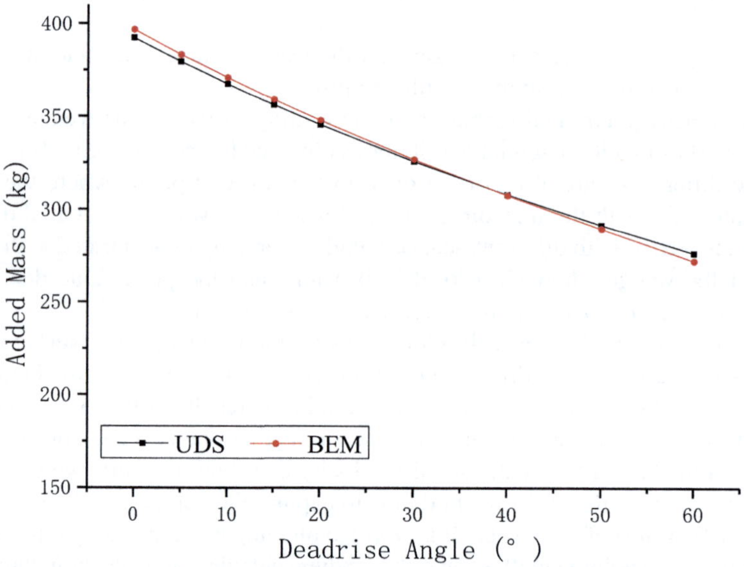

Fig. 9 The variation of the added mass of 2D wedge with the deadrise angle

5 Conclusion

The core advantage of the Boundary Element Method (BEM) lies in its direct handling of boundary conditions. It relates the boundary information to the internal field through the Green's function, which is particularly efficient for problems with strong analyticity, such as potential flow problems and acoustic problems. However, this also brings a key challenge: it relies on the exact Green's function corresponding to the equation, which means that not all types of equations, especially hyperbolic equations, are suitable for using BEM, because the Green's function of such equations is usually difficult to obtain.

The advantage of the User-Defined Scalar Equation Method (UDS) is its strong flexibility. Users can define equations according to their own understanding of the problem and specific requirements without relying on the Green's function. This is very useful when dealing with some special, non-standard physical phenomena or engineering problems. It can also conveniently combine the actual physical process to build a model and can reflect the physical essence of the problem more intuitively. However, this also leads to the fact that the user-defined equations may lack the support of universal theories, and users need to have profound professional knowledge and practical experience to ensure the rationality and correctness of the equations. Moreover, in terms of solution efficiency, it may not be as good as some mature general methods. Especially for complex large-scale problems, it may take more time to debug and optimize the user-defined equations.

In summary, both BEM and UDS have their respective applicable scenarios. Each method has its own unique advantages and limitations. Which one to choose depends on the nature of the specific problem and the needs of engineers.

In this paper, the added mass of two-dimensional wedge model is solved by using boundary element and UDS, and the reliability and accuracy of the calculation model are verified by comparing with the theoretical solution of two-dimensional plate added mass. The results show that both methods are suitable for solving added mass problems without considering the water surface uplift effect, and the calculation accuracy is high. Furthermore, the added mass of wedges with different deadrise angles was analyzed. The calculation results showed that under the same wetting half width, the added mass of wedges decreased with the increase of deadrise angle. When the inclination angle approached 0°, the added mass of wedges approached the value of the added mass of flat plates.

Because the calculation model refers to the neglected water surface uplift effect, the added mass solved will be smaller than the real situation. The next step is to use the instantaneous calculation model to solve the water entry process of the wedge according to the momentum theorem. At the same time, the position of the free surface is explicitly traced according to the vertical velocity of the free surface to consider the free surface uplift effect. At the same time, Bernoulli formula is used to calculate the distribution of wet surface pressure during the process of wedge entering water, and the calculation model is perfected. The reliability and accuracy of the method are further verified by wedge experiments.

References

1. Von Karman T (1929) The impact on sea-plane floats during landing. NACA Technical noteno 321
2. Wagner H (1932) Phenomena associated with impacts and sliding on liquid surfaces. J Appl Math Mech 12(4):193–215

3. Zhao R, Faltinsen O, Aarsnes J (1996) Water Entry of Arbitrary Two-dimensional Sections with and Without Flow Separation[C]. In: Proceedings ofthe21stSymposium on Naval Hydrodynamics, pp 408–423
4. Carcaterra A, Ciappi E, Iafrati A et al (2000) Shock spectral analysis of elastic systems impacting on the water Sur-face. J Sound Vib 229(3):579–605
5. Wang L, Fang X (2022) Research on the launch of free-fall lifeboat. Ship Eng 002:044
6. Dong L, Ding W, Wei Z et al (2023) Numerical study on the water entry of two-dimensional foils by BEM. Eng Anal Boundary Elem
7. Dong L, Wei Z, Yao B et al (2024) Numerical study on the water entry of two-dimensional freefall foils. J Huazhong Univ Sci Technol (Nat Sci Edn)
8. Zheng Z et al (2024) A method of calculating the added mass of underwater vehicle in direct flight by CFD. Comput Appl Softw 41(09):41–47
9. Ship wave load (2007) National Defense Industry Press
10. Yang D, Zhao Z. Boundary element theory and application. Beijing Institute of Technology Press, 2002

Open Access This chapter is licensed under the terms of the Creative Commons Attribution-NonCommercial-NoDerivatives 4.0 International License (http://creativecommons.org/licenses/by-nc-nd/4.0/), which permits any noncommercial use, sharing, distribution and reproduction in any medium or format, as long as you give appropriate credit to the original author(s) and the source, provide a link to the Creative Commons license and indicate if you modified the licensed material. You do not have permission under this license to share adapted material derived from this chapter or parts of it.

The images or other third party material in this chapter are included in the chapter's Creative Commons license, unless indicated otherwise in a credit line to the material. If material is not included in the chapter's Creative Commons license and your intended use is not permitted by statutory regulation or exceeds the permitted use, you will need to obtain permission directly from the copyright holder.

Validation of Human Finite Element Model for Non-lethal Blunt Ballistic Impact Damage Assessment

Zhaohui Zhang, Song Wang, Bin Chen, Ying Chen, and Jiakai Wu

Abstract To validate digital human thorax targets suitable for blunt ballistic impact damage assessment, based on the NATO STANDARD AEP-99 THORAX INJURY RISK ASSESSMENT OF NON-LETHAL PROJECTILES numerical validation method, The L5 digital model of a non-lethal kinetic energy projectile is constructed by finite element method and the digital target of Hybrid III 50th% Dummy is hit. The data is processed and verified by Bir biomechanical corridor and viscosity standard. The results show that Hybrid III 50th% Dummy digital target can be used for assessment of blunt ballistic impact damage of non-lethal kinetic energy projectile on human thorax, providing a reference for simulation setting of blunt impact study of non-lethal kinetic energy projectile on human thorax. It can lay a theoretical foundation for constructing and verifying the Chinese 50 percentile human thorax digital target for blunt ballistic impact damage assessment.

Keywords Thorax digital target · Biomechanical corridor · Viscosity criteria · Passive ballistic impact · Endpoint effect

---

Z. Zhang · B. Chen · Y. Chen · J. Wu
Equipment Management and Guarantee Institute, Engineering University of PAP, Xi'an, China

Graduate Regiment, Engineering University of PAP, Xi'an, China

S. Wang (✉)
Equipment Management and Guarantee Institute, Engineering University of PAP, Xi'an, China

© The Author(s) 2026

Z. Zhang, P. Mercorelli (eds.), *Mechanical Design and Simulation: Insights and Innovations*, Lecture Notes in Mechanical Engineering,
https://doi.org/10.1007/978-981-95-2405-1_20

1 Introduction

Non-fatal kinetic energy impact damage is blunt ballistic damage [1]. Impact to the thorax is a major consideration when assessing the power of blunt ballistic damage, since it is one of the most commonly impacted areas, and since the thorax includes vital human organs, a non-lethal kinetic energy projectile can cause serious injury or even death by impacting the area. Foreign researchers often use Finite Element models of Human body or bionic human thorax to simulate experimental working conditions. Nsiampa [2] constructed and verified the Surrogate Human Thorax For Impact Model (SHTIM), a finite element model for thorax damage assessment, and conducted a comparative study on impact damage assessment of six kinds of non-lethal kinetic energy projectile based on viscosity criteria. Bracq et al. [3] used a non-lethal kinetic energy projectile Flash-Ball finite element model to impact the thorax of HUByx finite element model to identify model-related numerical indicators for predicting blunt trauma to the lung and heart. Compared with foreign countries, non-lethal blunt ballistic impact damage assessment started late in China and is in the stage of tracking simulation. By constructing and verifying the 18.4 mm rubber bullet finite element model and impacting the improved thorax blunt impact damage assessment model MTHOTA-I, Wang Song [4] et al. gave suggestions on the safe use of 18.4 mm rubber bullets at different shooting distances by using the viscosity standard. LU Haitao [5] et al. built and verified a finite element model of human torso with relatively complete anatomical features based on the Chinese visual human body dataset, analyzed the damage to human tissues and organs caused by non-lethal kinetic energy projectile impacting the thorax, and put forward suggestions on its effectiveness and safety.

Although the domestic results maintain a good consistency with the foreign research results, the specific details of the existing studies are rarely disclosed, which makes it difficult to provide referable simulation samples of non-lethal blunt ballistic impact experimental conditions. In this paper, a non-lethal kinetic energy missile L5 finite element model and a blunt impact Hybrid III 50th% finite element model are constructed, and data processing is performed according to the NATO STANDARD AEP-99 THORAX INJURY RISK ASSESSMENT OF NON-LETHAL PROJECTILES (AEP-99). By comparing and analyzing the data of Bir biomechanical corridor and viscosity standard, it is verified that Hybrid III 50th% finite element model is suitable for the assessment of blunt ballistic impact damage of non-lethal kinetic energy projectile on human thorax. It can provide a reference for the simulation setting of the endpoint effect of different types of non-lethal kinetic energy projectiles on the human body, and then lay a theoretical foundation for the construction and verification of Chinese 50th% human thorax physical and digital targets for the assessment of blunt ballistic impact damage of non-lethal kinetic energy projectiles.

2 Hybrid III 50th% Finite Element Model

Hybrid III 50th% finite element model (Hereinafter referred to as dummy) is a digital model developed by Livermore Software Technology Corporation based on the anthropomorphic test device Hybrid III 50th% solid sitting dummy. It contains 279,203 nodes, 256 beam units, 238,052 shell units, 227,632 solid units, using the mm-ms-kg-kN unit system. Compared with the human thorax, the dummy model has a simplified thorax design, lacking organs and soft tissue fillings. Qi Wei et al. [6] verified the feasibility of using dummies for the assessment of human thorax damage caused by impact of most non-lethal kinetic energy weapons through blunt impact tests. Zhao Fadong and Chen Chaoming et al. [7, 8] carried out a study on verification and damage assessment of the finite element model of non-lethal kinetic energy projectile impact Hybrid III. Its appearance is shown in Fig. 1, and its main parameters are shown in Table 1.

According to the requirements of Bir impact on cadaver experiment conditions [9], the projectile was selected to impact the thorax of the dummy in a positive direction, and the impact position was the center of the third rib of the dummy directly to the skin surface node. The impact point is shown in the Fig. 2.

(a) Front view (b) Side view (c) Top view

Fig. 1 Hybrid III finite element model. (**a**) Front view (**b**) Side view (**c**) Top view

Table 1 Hybrid III finite element model parameters

Part	Material	Density (kg/m^3)	Young modulus (Pa)	PR
Jacket	Viscoelastic	794	/	/
Chestpad	Viscous foam	450	/	0.05
Sternum Assembly	Elastic	3200	7E+10	0.3
ClavicleLink Assembly	Elastic	2700	7E+10	0.3
RibsSteel	Elastic	7890	2.1E+11	0.3

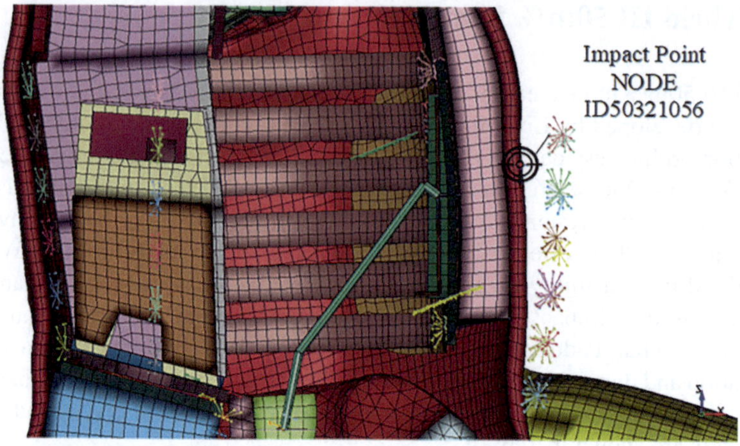

Fig. 2 Hybrid III finite element model impact position

(a) Projectile physical diagram (b) Projectile meshing model diagram

Fig. 3 L5 projectile. (**a**) Projectile physical diagram (**b**) Projectile meshing model diagram

3 Construction of L5 Projectile Finite Element Model

The L5 cylindrical projectile, made of incompressible polyvinyl chloride (PVC) material, comes in two types, long and short, with a diameter of 37 mm and a mass of 140 g and 30 g respectively. These projectiles, which are generally considered rigid projectiles, are used to define biomechanical corridors in Bir impact cadaver experiments [1]. The L5 projectiles are made of *MAT-001 ELASTIC [10], and its physical diagram and grid division diagram are shown in Fig. 3. Their physical and grid drawings are shown in Fig. 3.

4 Verification Method and Simulation Setting of Human Thorax Digital Model

AEP-99 [10, 11] specifies the verification method of human thorax model. Take the L5 long projectile to the vertical impact reference point of 20 m/s and 40 m/s, and the L5 short projectile to the vertical impact reference point of 60 m/s, and measure the displacement signal output as a function of time, and calculate VCmax according to the viscosity standard. The finite element model of human thorax suitable for non-fatal blunt ballistic impact damage assessment was successfully validated when it met the corridor interval defined by AEP-99.

Lagrange algorithm is used to characterize the precise change of impact process. At the same time, several materials of the dummy are bio-soft materials with low elastic modulus and large calculation step size, which is defined as *CONTACT_TIMESTEP TSSFAC and assigned a value of 0.67. Execute the *SET_PART_LIST command to participate in the entire impact simulation as a whole. For contact, use *CONTACT_AUTOMATIC_SURFACE_TO_SURFACE. To consider the volume cell calculations during simulation, execute the *CONTROL_SOLID command. Enter the termination command with the experimental data, and run the *CONTROL_TERMINATION command. Set ENDTIME to 4. Run *INTIAL_VELOCITY to add the initial speed. The simulation of impact test conditions is shown in Fig. 4.

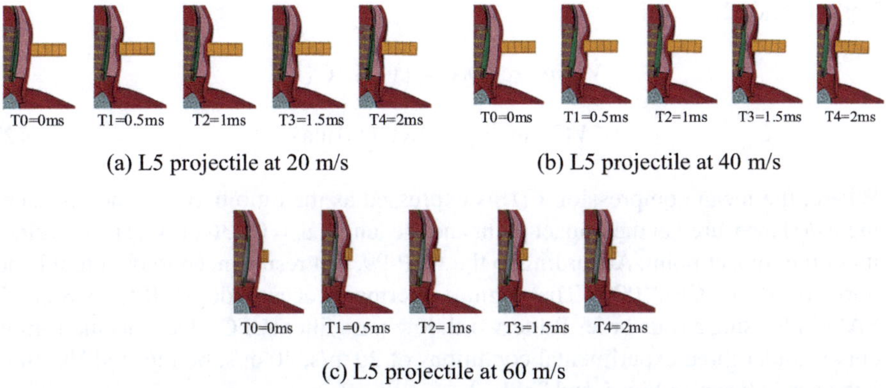

Fig. 4 Sectional impact sequence of L5 projectiles at different velocity. (**a**) L5 projectile at 20 m/s (**b**) L5 projectile at 40 m/s (**c**) L5 projectile at 60 m/s

The L5 projectile, as a rigid body, has no obvious deformation during the whole impact process, and the impact site of the dummy has two stages: compression (loading) and rebound (unloading). In the compression (loading) stage, the thorax is continuously compressed in the direction of projectile velocity by projectile impact. The L5 projectile with an initial velocity of 20 m/s reaches the maximum compression displacement peak after T4, and the L5 projectile with an initial velocity of 40 m/s reaches the maximum compression displacement peak between T1 and T2. The L5 projectile with an initial velocity of 60 m/s reaches the peak compression displacement after T4. After the rebound (unloading) stage, the compression of chest displacement decreased, but did not show significant recovery at 5 ms.

5 Digital Model Data Processing and Result Analysis

The finite element model of the human thorax for non-lethal blunt ballistic impact damage assessment was validated by Bir biomechanical corridor and viscosity criteria. The viscosity criterion, is based on the dynamic deflection of the rib cage and is determined by the viscous response, which is the product of the rib cage deflection (compression) and the rib cage deflection rate normalized by the thickness of the thorax wall. The viscosity criterion VCmax is defined as the maximum viscous response of the thoracic cavity. Bir et al. [10] showed that VCmax could be used to assess the risk of non-lethal blunt ballistic impact damage. Its formula is shown in Eqs. (1) and (2).

$$Visous\ reponse = \left(v(t) \times C(t)\right) \tag{1}$$

$$VC\max = \left(v(t) \times C(t)\right)\max \tag{2}$$

Where, the mean compression *C(t)* is expressed as the ratio between the displacement *d(t)* measured at the impact point and the length $d_0 = 0.236$ m. *V(t)* is the velocity of the impact point. According to the AEP-99, test results need to be filtered and normalized by CFC1000. The digital filtering process adopts ISO 648:2015, SAEJ211, using a four-order Butterworth low-pass filter (CFC). Displacement-time curves under three experimental conditions of 20 m/s, 40 m/s, 60 m/s and VC-time curves are shown in Fig. 5 and Table 2.

The thorax compression of the dummy under three experimental conditions of 20 m/s, 40 m/s and 60 m/s presents an upward trend of first fast, then slow and then decline. In the early stage of impact (t = 1.22 ms under 20 m/s, t = 0.80 ms under 40 m/s, and t = 0.67 ms under 60 m/s), the thorax compression growth rate of the dummy is fast and close to the upper limit of displacement corridor. In the middle stage of impact (t = 3.37 ms under 20 m/s, t = 1.84 ms under 40 m/s, and t = 4.40 ms under 60 m/s), the thorax compression growth rate of the dummy is

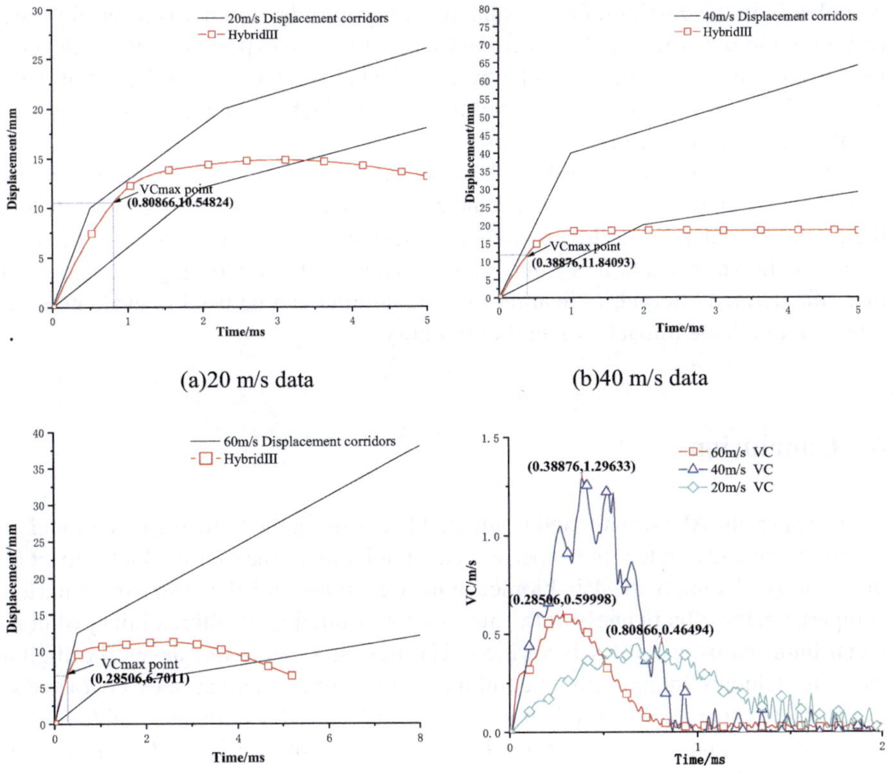

(a)20 m/s data (b)40 m/s data

(c)60 m/s data (d)VC data

Fig. 5 L5 projectile data at 20, 40, 60 m/s. (**a**) 20 m/s data (**b**) 40 m/s data (**c**) 60 m/s data (**d**) VC data

Table 2 Impact case and VCmax comparing with PMHS

Case	Projectile	Velocity (m/s)	*VC*max corridor (m/s)	Simulated result	
				*VC*max (m/s)	$t_{VC\max}$ (ms)
1	140 g L5	20	(0.24, 0.51)	0.464 94	0.808 66
2	140 g L5	40	(0.65, 2.35)	1.296 33	0.388 76
3	30 g L5	60	(0.14, 0.60)	0.599 98	0.285 06

slow, and the thorax compression peaks are generated, whose peak points are (3.01188, 14.84333) (1.51889, 18.49585) (2.48309, 11.21813), and then a downward trend, close to the lower limit of the displacement corridor; At the later stage of impact, the thorax compression of the dummy is lower than the lower limit of displacement corridor and tends to zero. From the whole impact process, it is found that the thorax compression displacement-time of the dummy is not completely in the corridor. On the one hand, because the dummy is a certain rigid response. On the other hand, according to the cadaveric experiment, the cadaveric bodies used in the experiment lack physiological characteristics, and most of them are over 70 years old, and the skin and tissue aging lack tolerance, so the displacement range continues to rise.

The point coordinates (0.80866, 10.54824), (0.38876, 11.84093) and (0.28506, 6.7011) correspond to the point coordinates of VCmax in its working condition, and the point coordinates and the previous displacement-time curves are all in the corridor. As shown in Table 2, VCmax values under the three working conditions all meet the defined interval, indicating that the dummy can be used to evaluate non-fatal blunt ballistic impact human thorax injury.

6 Conclusion

In this paper, the AEP-99 method is adopted to construct the finite element model of L5 projectile and conduct the experimental simulation of the thorax blunt impact of the dummy. Through the Bir biomechanical corridor and the viscosity criterion comparison, the effectiveness of the dummy for evaluating the thorax injury of non-lethal blunt ballistic impact is verified. The next step can be to develop a digital target for Chinese human thorax combined with Chinese human body characteristics, so as to promote the establishment of a standardized document suitable for the assessment of human thorax injury by blunt impact of non-lethal kinetic energy projectiles in China.

References

1. Bir C (2015) Ballistic injury biomechanics. In: Yoganandan N, Nahum A, Melvin J (eds) Accidental injury. Springer, New York. https://doi.org/10.1007/978-1-4939-1732-7_29
2. Nsiampa N (2016) Numerical assessment of non-lethal projectile thoracic impacts. Royal Military Academy & University of liege, Brussels. https://www.proquest.com/openview/f0fe623698f16ec26d9f02864b8b1816/1?pq-origsite=gscholar&cbl=2026366&diss=y
3. Bracq A, Delille R, Bourel B et al (2019) Numerical recreation of field cases on a biofidelic human FE model involving deformable less-lethal projectiles. Hum Factors Mech Eng Def Saf 3:5. https://doi.org/10.1007/s41314-019-0022-8
4. Wang S, Zhan RJ, Duan XY (2020) Finite element method evaluation on chest blunt injury by rubber projectile. J Syst Simul 32(9):1762–1770. https://doi.org/10.16182/j.issn1004731x.joss.19-0119
5. Lu, Qin, Li HT, B, WB (2023) Numerical simulation of non-lethal kinetic energy projectile impact on human torso model. J Phys Conf Series Hunan 2478. https://doi.org/10.1088/1742-6596/2478/12/122057
6. Qi W, Zhuang HW, Zhao FD (2017) Response testing method of hybrid III thorax subjected to non-lethal kinetic energy impact. Ordnance Mater Sci Eng 40(04):20–26. https://doi.org/10.14024/j.cnki.1004-244x.20170705.001

7. Zhao FD, Chen CM, Bao HT et al (2022) Numerical simulation of non-lethal kinetic p rojectiles blunt impact on dummy chest. J Ballist 34(4):30–37. https://doi.org/10.12115/j.issn.1004-499X(2022)04-005
8. Chen, C.M., Zhao, F.D., Ma, Z.M. (2023) Numerical analysis of thorax injury caused by the blunt impact of SIR-X sponge grenade. Hum Factors Mech Eng Def Saf, 7, 7(2). doi:https://doi.org/10.1007/s41314-023-00057-4
9. Bir CA (2000) The evaluation of blunt ballistic impacts of the thorax. Wayne State University 9966125. https://www.proquest.com/openview/e1044a1b9c93bd66c533bde53db5991d/1?pq-origsite=gscholar&cbl=18750&diss=y
10. Robbe C, Papy A, Nsiampa N et al (2023) NATO standardized method for assessing the thoracic impact of kinetic energy non-lethal weapons. Hum Factors Mech Eng Def Saf 7(1):7. https://doi.org/10.1007/s41314-023-00060-9
11. NATO (2021) STANREC 4744 edition 5 - aep 99 ed: b ver 1 thorax injury risk assessment of non-lethal projectiles—no classification, fully accessible at www.nato.nso.Int.

Open Access This chapter is licensed under the terms of the Creative Commons Attribution-NonCommercial-NoDerivatives 4.0 International License (http://creativecommons.org/licenses/by-nc-nd/4.0/), which permits any noncommercial use, sharing, distribution and reproduction in any medium or format, as long as you give appropriate credit to the original author(s) and the source, provide a link to the Creative Commons license and indicate if you modified the licensed material. You do not have permission under this license to share adapted material derived from this chapter or parts of it.

The images or other third party material in this chapter are included in the chapter's Creative Commons license, unless indicated otherwise in a credit line to the material. If material is not included in the chapter's Creative Commons license and your intended use is not permitted by statutory regulation or exceeds the permitted use, you will need to obtain permission directly from the copyright holder.

Thermal Simulation of Small Caliber Composite Tube

Yihui Liu, Kedong Zhou, Lei He, Di Zhang, and Haiyue Ren

Abstract Composite tubes offer significant weight reduction advantages over steel tubes, but their heat dissipation performance is inferior. To research the heat transfer characteristics and influencing factors of a small caliber composite tube, a one-dimensional unsteady-state heat transfer model was developed using the finite volume method, and its accuracy was proved by experimental data. This model was utilized to numerically simulate the heat transfer process of a small caliber composite tube, revealing the radial distribution and variation of temperature within the tube. The study results indicate that the heat dissipation performance of the small caliber composite tube is inferior to the steel tube in same size, failing to meet the specification of continuous heating of 120 cycles. Under the heating specification of 30 cycles, the highest temperatures were 478.8 °C at the inner surface of the liner, 223.8 °C within the composite jacket, and 227.1 °C at the outer surface of the composite tube. The composite jacket exhibiting a lag in temperature change compared to the steel liner. These findings provide guidance for the design of composite tubes.

Keywords Composite · Tube · Heat transfer · Finite volume method · Numerical simulation

1 Introduction

In order to withstand the high temperature and high-pressure loads during operation, steel tubes are often relatively heavy and bulky, and carbon fiber reinforced polymer matrix composites (CFRPs) are known for their high specific strength [1]. The density of CFRPs ranges between 1.5 g/cm^3 and 1.9 g/cm^3, whereas steel has a density of approximately 7.8 g/cm^3. Applying CFRPs to tubes to form composite

Y. Liu · K. Zhou (✉) · L. He · D. Zhang · H. Ren
School of Mechanical Engineering, Nanjing University of Science and Technology, Nanjing, China
e-mail: zkd81151@njust.edu.cn

© The Author(s) 2026

Z. Zhang, P. Mercorelli (eds.), *Mechanical Design and Simulation: Insights and Innovations*, Lecture Notes in Mechanical Engineering,
https://doi.org/10.1007/978-981-95-2405-1_21

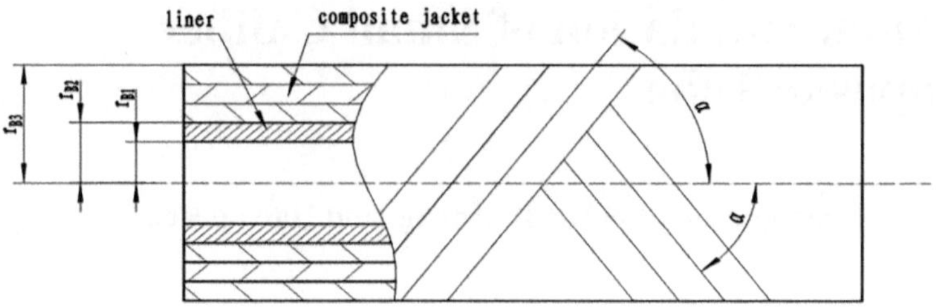

Fig. 1 Schematic diagram of composite tube

tubes offers significant weight reduction advantages compared to steel tubes [2]. A typical structure of composite tubes is shown in Fig. 1, with a liner as the inner layer and multiple layers of unidirectional composites at various angles as the outer layer.

Although CFRPs have high specific strength, their resin matrix typically exhibits poor high-temperature resistance. Therefore, studying the heat transfer in composite tubes is particularly important. Scholars both domestic and international have conducted research on the heat transfer of large caliber composite tubes [3–5], but there is relatively less research specifically on small caliber composite tubes [6]. Previous studies on tube heat transfer have often employed finite difference method (FDM) [7–12]. While FDM is advantageous in terms of mathematical properties, the derivation process may lack clarity in physical concepts and cannot guarantee the conservation properties of the resulting difference equations. In contrast, finite volume methods (FVM) effectively address these shortcomings and provide discrete equations with second-order accuracy [13]. Therefore, this study employs the FVM to develop a one-dimensional heat transfer program suitable for composite tubes, investigating the transient heat transfer during the heating process.

2 One-Dimensional Heat Transfer Model of Composite Tube

2.1 Assumptions

To ensure the feasibility of calculations, certain assumptions are made for the heat transfer model:

1. Ignore the influence of structure and coatings on the inner surface of the tube, simplifying the tube to a thick-walled cylinder with a smooth inner surface.
2. Due to the significantly larger radial temperature gradients compared to axial and circumferential gradients, simplify the model to a one-dimensional radial heat transfer model in cylindrical coordinates.
3. Disregard the impact of temperature changes on the heating process, assuming the heating process of each cycle is consistently identical and independent of each other.

4. Neglect nonlinear factors in the thermal properties of materials, assuming the thermal properties of the tube material remain constant.

2.2 Control Equation

According to reference [13], the one-dimensional transient heat transfer control equation in cylindrical coordinates is:

$$\rho_i c_i \frac{\partial T}{\partial t} = \frac{1}{r}\frac{\partial}{\partial r}\left(r\lambda_i \frac{\partial T}{\partial r}\right) \tag{1}$$

where T is the temperature at a point in the tube wall, t is time, r is the radial coordinate. λ_i, ρ_i, and c_i denote the thermal conductivity, density, and specific heat of the tube materials, respectively. When $r_{B1} \le r \le r_{B2}$, the subscript i is s for steel; when $r_{B2} < r \le r_{B3}$, the subscript i is c for composite. r_{B1}, r_{B2}, and r_{B3} denote the inner radius of the liner (inner surface of the composite tube), outer radius of the liner (inner surface of the composite jacket), and outer radius of the composite tube (outer surface of the composite jacket), respectively.

2.3 Boundary Conditions

During the heat transfer process in the tube, the time domain is divided into two periods: the period of high-temperature gas flow (including the heating period and the after-effect period) and the air cooling period (from the end of the after-effect period until just before the next heat cycle, or from the end of the after-effect period after the last heat cycle until the end of heat transfer). During the high-temperature gas flow period, the internal fluid in the tube is high-temperature, high-pressure, high-speed gas, exhibiting turbulent flow and forced convection heat transfer with the inner surface of the tube. Since the tube interior is a closed space, radiation heat transfer is neglected. During the air cooling period, the high-temperature gas has been expelled, and the internal fluid is air at a temperature and pressure consistent with the environment, undergoing natural convection heat transfer with the inner surface of the tube. The outer surface of the tube exchanges heat with the surrounding air through a combination of natural convection and radiation heat transfer. Based on the above analysis, both the internal and external boundaries of the tube heat transfer correspond to third-type boundary conditions.

Internal Boundary Conditions The internal boundary conditions can be expressed as:

$$\lambda_s \left.\frac{\partial T}{\partial t}\right|_{r=r_{B1}} + \eta_1 h_{1j}\left(T_j - T_{B1}\right) = 0 \tag{2}$$

where h_{1j} is the convective heat transfer coefficient of the high-temperature gas to the inner surface of the composite tube, η_1 is a correction factor for the composite heat transfer coefficient between the inner liner inner surface and the high-temperature gas, $\eta_1 = 2$ according to reference [14]. T_j is the fluid temperature inside the tube, with hg as the subscript during the period of high-temperature gas flow, and a during the air cooling period. T_{B1} is the temperature of the tube (liner) inner surface.

The formula for calculating the forced convective heat transfer coefficient at the inner boundary during the period of high-temperature gas flow is:

$$h_{1hg} = 0.023 \times \frac{\lambda_{hg}}{2r_{B1}} \left(\frac{2u_{hg}\rho_{hg}r_{B1}}{\mu_{hg}}\right)^{0.8} \left(\frac{c_{phg}\mu_{hg}}{\lambda_{hg}}\right)^{0.4} \tag{3}$$

where λ_{hg}, u_{hg}, ρ_{hg}, μ_{hg}, and c_{phg} represent the thermal conductivity, velocity, density, dynamic viscosity, and specific heat of the high-temperature gas, respectively.

The formula for calculating the natural convective heat transfer coefficient at the inner boundary during the air cooling period is:

$$h_{1a} = 0.53 \times \frac{\lambda_a}{2r_{B1}} \left[\frac{g\alpha_v \left(2r_{B1}\right)^3 \left(T_{B1} - T_a\right) c_{pa}\mu_a}{v_a^2 \lambda_a}\right]^{0.25} \tag{4}$$

where λ_a, α_v, T_a, c_{pa}, μ_a, and ν_a denote the thermal conductivity, volumetric expansion coefficient, temperature, specific heat, dynamic viscosity, and kinematic viscosity of air, respectively. g is the acceleration due to gravity.

Intermediate Boundary Conditions With the interface between the liner and the composite jacket as the intermediate boundary, the intermediate boundary conditions can be represented as:

$$\lambda_i \left.\frac{\partial T}{\partial r}\right|_{r=r_{B2}} + h_2\left(T_{B2s} - T_{B2c}\right) = 0 \tag{5}$$

where h_2 is the interfacial thermal contact conductance between the liner and the composite jacket. According to reference [15], h_2 = 14310.5 W/(m^2•K), T_{B2s} is the temperature on the liner side of the interface between the liner and the composite jacket, while T_{B2c} is the temperature on the composite jacket side of the interface.

External Boundary Conditions The external boundary conditions can be expressed as:

$$\lambda_c \left.\frac{\partial T}{\partial r}\right|_{r=r_{B3}} - \eta_3 h_3 (T_a - T_{B3}) = 0 \tag{6}$$

where h_3 is the composite heat transfer coefficient of the tube outer surface, η_3 is the correction factor for the composite heat transfer coefficient between the tube outer surface and the surrounding air, and according to reference [14], $\eta_3 = 1.65$. T_{B3} is the temperature of the tube outer surface (outer surface of the composite jacket).

The external boundary always involves natural convection heat transfer and radiation heat transfer in a large space. The heat transfer coefficient is calculated by:

$$\begin{cases} h_{3f} = 0.53 \times \dfrac{\lambda_a}{2r_{B3}} \left[\dfrac{g\alpha_v (2r_{B3})^3 (T_{B3} - T_a) c_{pa} \mu_a}{v_a^2 \lambda_a} \right]^{0.25} \\ h_{3r} = \dfrac{\varepsilon_c C_0 \varepsilon_a \left[\left(\dfrac{T_{B3}}{100} \right)^4 - \left(\dfrac{T_a}{100} \right)^4 \right]}{T_{B3} - T_a} \\ h_3 = h_{3f} + h_{3r} \end{cases} \tag{7}$$

where h_{3f} and h_{3r} are the convective heat transfer coefficient and the radial heat transfer coefficient of the tube outer surface, respectively. ε_c is the emissivity of the tube outer surface, C_0 is the Stefan-Boltzmann constant ($C_0 = 5.67$ W/(m^2•K^4)), and ε_a represents the emissivity factor of air.

3 Numerical Heat Transfer Model for Composite Tubes

3.1 Discretization of Control Equation

In cylindrical coordinates, considering the one-radian sector of the *r*-*θ* plane, the outer node method (Method A) is employed to discretize the computational domain along the radial direction (*r*-direction), as shown in Fig. 2. Within the interior node region (excluding boundaries and the interface between the liner and composite jacket), the implicit finite volume method (IFVM) is used to discretize Eq. (1). The discretized equation is then formulated as:

$$A_n T_{n-1}^{m+1} + B_n T_n^{m+1} + C_n T_{n+1}^{m+1} = D_n \tag{8}$$

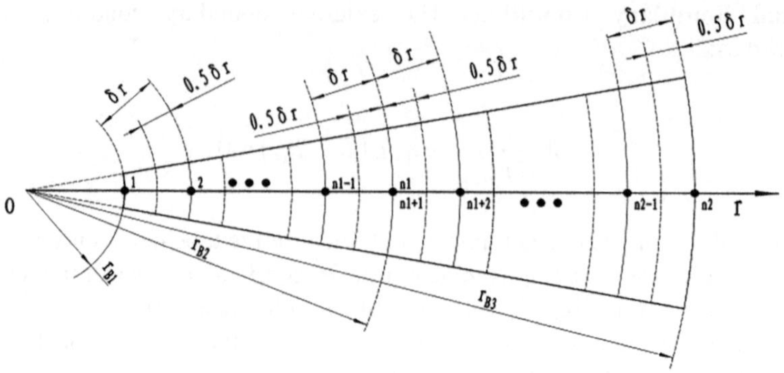

Fig. 2 Discretization of the spatial domain

$$
\begin{cases}
A_n = -\dfrac{(r_n - 0.5\delta r)}{\delta r} \\
B_n = \dfrac{(r_n - 0.5\delta r)}{\delta r} + \dfrac{(r_n + 0.5\delta r)}{\delta r} + \dfrac{r_n \delta r}{\alpha_i dt} \\
C_n = -\dfrac{(r_n + 0.5\delta r)}{\delta r} \\
D_n = \dfrac{r_n \delta r}{\alpha_i dt} T_n^m
\end{cases} \tag{9}
$$

where subscript n denotes the spatial domain node number, superscript m indicates the time step number, r_n represents the distance from the n-th node to the axis of the tube, $1 < n < n_1$ and $n_1 + 1 < n < n_2$. n_1 is the node at the liner side interface between the liner and the composite jacket, $n_1 + 1$ is the node at the composite jacket side interface, and n_2 is the node at the outer surface of the tube. δ_r is the distance between adjacent nodes, α_i is the thermal diffusivity coefficient of the material, defined as $\alpha_i = \lambda_i/(\rho_i \cdot c_i)$. For $1 \le n \le n_1$, subscript i is s (steel), and for $n_1 + 1 \le n \le n_2$, subscript i is c (composite).

3.2 Discretization of Boundary Conditions

Establishing thermal equilibrium equations at the boundaries, using the IFVM, we get the discretized formats for the inner boundary conditions, middle boundary conditions, and outer boundary conditions as:

$$\begin{cases} B_1 T_1^{m+1} + C_1 T_2^{m+1} = D_1 \\ A_{n1} T_{n1-1}^{m+1} + B_{n1} T_{n1}^{m+1} + C_{n1} T_{n1+1}^{m+1} = D_{n1} \\ A_{n1+1} T_{n1}^{m+1} + B_{n1+1} T_{n1+1}^{m+1} + C_{n1+1} T_{n1+2}^{m+1} = D_{n1+1} \\ A_{n2} T_{n2-1}^{m+1} + B_{n2} T_{n2}^{m+1} = D_{n2} \end{cases} \tag{10}$$

where:

$$\begin{cases} B_1 = \dfrac{(r_{B1} + 0.5\delta r)}{\delta r} + \dfrac{0.5 r_{B1} \delta r}{\alpha_s dt} + \dfrac{\eta_1 h_{1j} r_{B1}}{\lambda_s} \\ C_1 = -\dfrac{(r_{B1} + 0.5\delta r)}{\delta r} \\ D_1 = \dfrac{0.5 r_{B1} \delta r}{\alpha_s dt} T_1^m + \dfrac{\eta_1 h_{1j} r_{B1}}{\lambda_s} T_j \end{cases} \tag{11}$$

$$\begin{cases} A_{n1} = -\dfrac{(r_{B2} - 0.5\delta r)}{\delta r} \\ B_{n1} = \dfrac{(r_{B2} + 0.5\delta r)}{\delta r} + \dfrac{0.5 r_{B2} \delta r}{\alpha_s dt} + \dfrac{h_2 r_{B2}}{\lambda_s} \\ C_{n1} = -\dfrac{h_2 r_{B2}}{\lambda_s} \\ D_{n1} = \dfrac{0.5 r_{B2} \delta r}{\alpha_s dt} T_{n1}^m \end{cases} \tag{12}$$

$$\begin{cases} A_{n1+1} = -\dfrac{h_2 r_{B2}}{\lambda_c} \\ B_{n1+1} = \dfrac{(r_{B2} + 0.5\delta r)}{\delta r} + \dfrac{0.5 r_{B2} \delta r}{\alpha_c dt} + \dfrac{h_2 r_{B2}}{\lambda_c} \\ C_{n1+1} = -\dfrac{(r_{B2} + 0.5\delta r)}{\delta r} \\ D_{n1+1} = \dfrac{0.5 r_{B2} \delta r}{\alpha_c dt} T_{n1+1}^m \end{cases} \tag{13}$$

$$\begin{cases} A_{n2} = -\dfrac{(r_{B3} - 0.5\delta r)}{\delta r} \\ B_{n2} = \dfrac{(r_{B3} - 0.5\delta r)}{\delta r} + \dfrac{0.5 r_{B3} \delta r}{\alpha_c dt} + \dfrac{\eta_3 h_3 r_{B3}}{\lambda_c} \\ D_{n2} = \dfrac{0.5 r_{B3} \delta r}{\alpha_c dt} T_{n2}^m + \dfrac{\eta_3 h_3 r_{B3}}{\lambda_c} T_a \end{cases} \tag{14}$$

4 Calculation Results and Analysis

4.1 Model Validation

To verify the correctness of the model in this study, heating tests were conducted using a small caliber launch device. A thermal imaging camera was used to record the temperature changes of the tube outer surface during heating and air cooling periods. The heating specification was as follows: with a heating interval of 1 s between adjacent heating cycles. Heating cycles were in groups of 30 cycles, with a 10-s break between adjacent groups, totaling 120 cycles. After heat process, the tube was left to cool in ambient conditions (27 °C) for 180 s. The tests were conducted under standard atmospheric pressure.

At a cross-section located 338 mm from the nozzle of the tube, designated as Section 1, temperature variations of the outer surface during heating and air cooling periods were measured. Figure 3 shows the infrared thermal imaging image of the tube outer surface during heating. Figure 4 displays the experimental and calculated temperature curves of the outer surface at Section 1 of the tube during the heating test of a small caliber launch device. Table 1 presents a comparison between experimental and calculated temperature values at selected time points for Section 1 of the tube.

Comparing the calculated values with the measured ones, it was found that the maximum relative error was less than 5%, indicating a high degree of agreement. This confirms the reliability of the model used in this study.

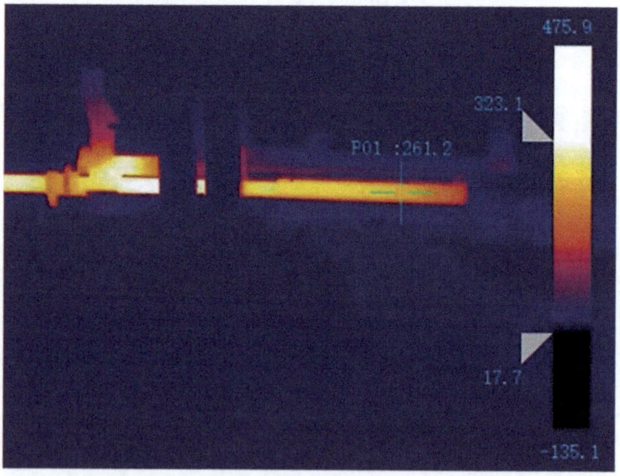

Fig. 3 Infrared thermal imaging image of the tube

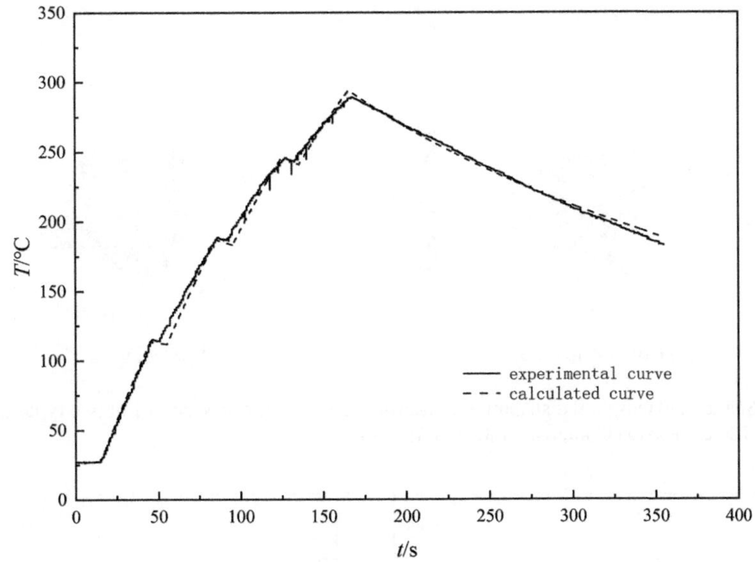

Fig. 4 Temperature curves of the outer surface at Section 1

Table 1 Comparison of calculated and measured temperatures

Moment	Calculated value/°C	Experimental value/°C	Relative error/%
30 cycles ended	113.2	114.6	−1.2
60 cycles ended	186.6	188.1	−0.8
90 cycles ended	246.6	245.4	0.5
120 cycles ended	293.7	288.3	1.9
Air cooling for 180 s	192.1	187.9	2.2

4.2 Temperature Field in a Small Caliber Composite Tube

According to the same heating specification and environmental factors, based on the one-dimensional transient heat transfer model of the composite tube mentioned earlier, the transient temperature field at Section 1 of a small caliber composite tube during heating was calculated. The spatial and temporal distribution of Section 1 is shown in Fig. 5a. For comparison, the transient temperature field at Section 1 of a steel tube under the same parameters was also calculated, depicted in Fig. 5b. A comparison reveals that the temperature gradient in the steel tube is continuous radially, whereas in the composite tube, there is a noticeable discontinuity at the interface between the liner and the composite jacket (at $r = 5$ mm), caused by differences in material properties on either side of the interface.

Extracting temperatures from the inner and outer surfaces of two types of tubes, the curves shown in Fig. 6a, b were obtained. From Fig. 6a, during the heat transfer process, the maximum temperatures on the inner surfaces of the composite tube and

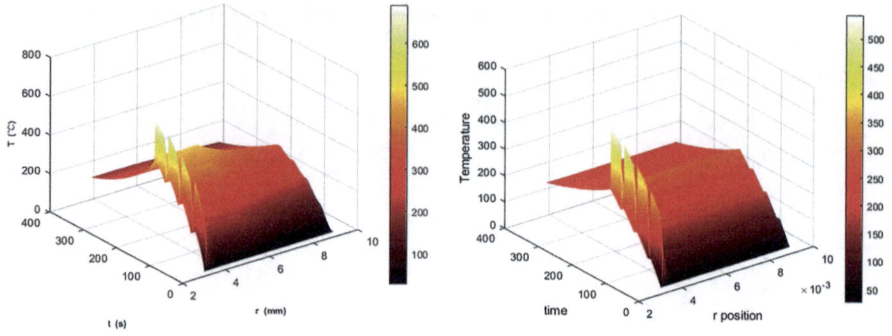

(a) Composite tube (b) Steel tube

Fig. 5 Spatial and temporal distribution of transient temperature at Section 1 of two types of tubes for heat 120 cycles. (**a**) Composite tube (**b**) Steel tube

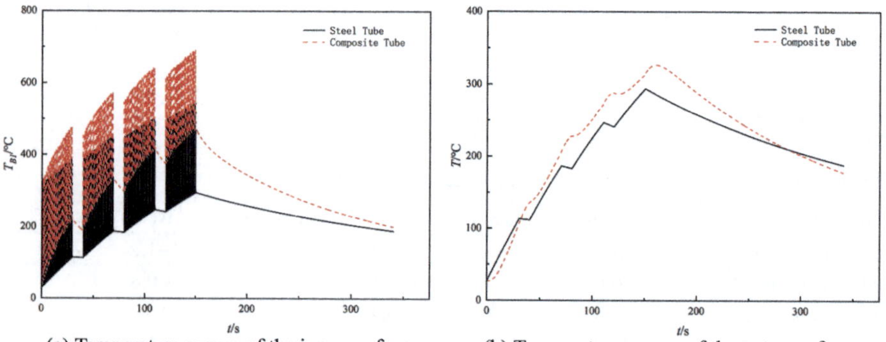

(a) Temperature curves of the inner surface (b) Temperature curves of the outer surface

Fig. 6 Temperature curves of the inner and outer surfaces of two types of tubes. (**a**) Temperature curves of the inner surface (**b**) Temperature curves of the outer surface

steel tube are 691.9 °C and 542.5 °C, respectively. After air cooling for 180 s, these temperatures decrease to 200.5 °C and 188.3 °C for the composite tube and steel tube inner surfaces, respectively. From Fig. 6a, b, it can be observed that during the heat transfer process, the maximum temperatures on the outer surfaces of the composite tube and steel tube are 326.7 °C and 293.7 °C, respectively. After air cooling for 180 s, these temperatures decrease to 178 °C and 188.1 °C for the composite tube and steel tube outer surfaces, respectively.

Comparison reveals that during the heat transfer process, the temperature on the inner surface of the composite tube consistently exceeds that of the steel tube. The temperature on the outer surface of the composite tube is lower than that of the steel tube for the first 30 heat cycles, surpasses it after the 30th heat cycle, and remains lower than the steel tube after air cooling for 128 s following the end of heating. The rise and fall of the temperature on the outer surface of the composite tube lag behind that of the steel tube.

These phenomena are attributed to the significantly lower thermal diffusion coefficient of the CFRPs compared to steel, which results in heat being trapped in the inner area of the tube and slow transmission to the outer area. Thus, throughout the heat transfer process, the temperature on the inner surface of the composite tube remains higher than that of the steel tube, and the temperature on the outer surface of the composite tube both rises and falls later than that of the steel tube. In the first 30 heat cycles, heat remains trapped in the inner area of the liner, causing the temperature on the outer surface of the composite tube to be lower than that of the steel tube. As the heating continues, the accumulated heat of the liner keeps increasing, and the temperature continues to rise. The accumulated heat and the temperature of the liner become much higher than those in the inner layer of the steel tube in same size. Additionally, since the thermal diffusion coefficient of the composite jacket is much lower than that of steel, by the time the 30th heat cycle is ended, the temperature on the outer surface of the composite tube is actually higher than that of the steel tube. Additionally, higher surface temperatures enhance heat exchange with the air, causing the temperatures of both the inner and outer surfaces of the composite tube to decrease faster than those of the steel tube. Therefore, after 128 s of air cooling, the temperature of the outer surface of the composite tube is lower than that of the steel tube.

Therefore, while applying composite materials to tubes is advantageous for reducing weight, it can lead to decreased heat dissipation during heating, resulting in heat accumulation. Furthermore, during heating, the highest temperature reached by the composite jacket is 473.2 °C, which exceeds the glass transition temperature (260 °C) of the resin. This can cause phase changes or even thermal decomposition in the composite resin matrix, leading to severe degradation of mechanical performance or damage, which is unacceptable. Consequently, the following sections will focus on studying the temperature field of composite tubes with a heating limit not exceeding 30 cycles.

The new heating specification is as follows: with a heating interval of 1 s between adjacent heat cycles, totaling 30 cycles, followed by air cooling for 60 s after heating. According to the new heating specification, the temperature-time curves for four typical positions on Section 1 of the composite tube are shown in Fig. 7. Here, T_{B1} represents the temperature of the inner surface of the tube (liner inner surface), T_{B2s} is the temperature at the steel-composite interface on the steel side (liner outer surface), T_{B2c} is the temperature at the steel-composite interface on the composite side (composite jacket inner surface), and T_{B3} is the temperature of the outer surface of the composite tube (composite jacket outer surface).

According to Fig. 7, during the heat transfer process in the composite tube, T_{B1} reaches a maximum of 478.8 °C, T_{B2s} reaches a maximum of 227.1 °C, T_{B2c} reaches a maximum of 223.8 °C, and T_{B3} reaches a maximum of 146 °C. As the number of heat cycles increases, T_{B1}, T_{B2s}, T_{B2c}, and T_{B3} show an increasing trend generally. However, compared to the slow increase in T_{B3}, T_{B1}, T_{B2s} and T_{B2c} show alternating increases and decreases with each high-temperature gas flow period and air cooling period, indicating that the regions affected by heat shock are closer to the inner area of the tube. Furthermore, T_{B3} continues to slowly rise after heating ends, reaching

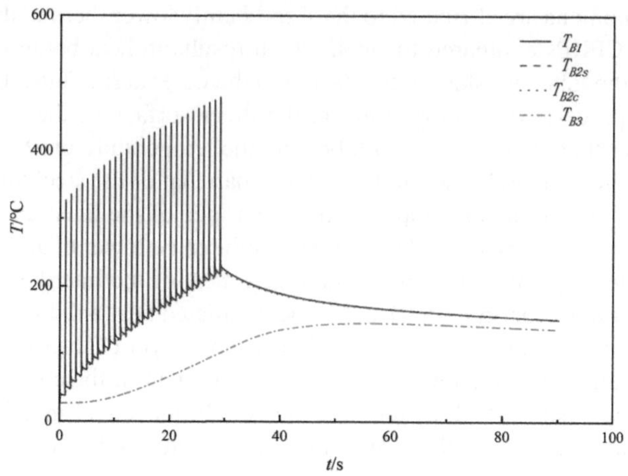

Fig. 7 The temperature-time curves for four typical positions on Section 1

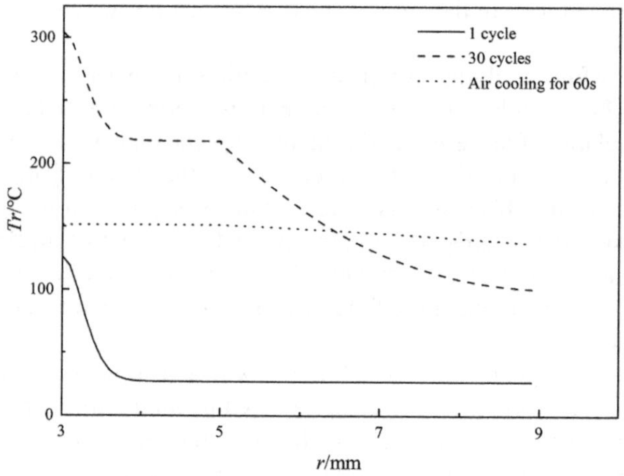

Fig. 8 The temperature distribution along the radius at three typical moments for Section 1

146 °C and then slowly decreasing, showing a noticeable lag effect compared to the inner area of the tube. The thermal diffusivity of the composite jacket is smaller than that of liner, resulting in slower heat propagation within the composite jacket.

Extract the temperature distribution along the radius at three typical moments during the heat transfer process of the composite tube, as shown in Fig. 8. At the end of the air cooling period after the first heat cycle, T_{B1} is 126.3 °C, with significant temperature gradients observed only within approximately 1 mm from the inner surface of liner towards the outer regions. Beyond this, the temperature remains relatively unchanged, indicating that during single heat cycle, the tube model can be simplified to that of a semi-infinite flat plate. By the end of the air cooling period after the 30th heat cycle, T_{B1}, T_{B2s}, T_{B2c}, and T_{B3} reach temperatures of 304.3 °C, 217.6 °C, 215.4 °C, and 100.9 °C respectively. At this moment, not only does the composite tube maintain significant temperature gradients near the liner inner surface, but large temperature gradients also appear within the composite jacket. The temperature difference between the inner and outer surfaces of the composite jacket is 114.5 °C, which exceeds the 86.7 °C difference between the inner and outer surfaces of the liner. This is due to the continuous transfer of heat from the liner to the composite jacket during successive heating, causing the composite jacket to heat up and develop temperature gradients. Additionally, because the thermal diffusivity of the composite material is much lower than that of steel, the temperature difference between the inner and outer surfaces of the composite jacket is greater than that in the liner. After air cooling for 60 s, the overall temperature gradient of the tube becomes more gradual, with a very small temperature difference of approximately 0.005 °C between the inner and outer surfaces of the liner. However, due to the lower thermal diffusivity of the composite compared to steel, the temperature difference between the inner and outer surfaces of the composite jacket remains significant, around 14.3 °C.

5 Conclusion

A one-dimensional transient heat transfer model for composite tubes is established and verified in this paper. Then the model was used to research the temperature field of a small caliber composite tube and the following conclusions were obtained:

1. During the transient heat transfer process, the maximum temperature of the composite tube exceeds that of the steel tube. The composite tube exhibits delayed heating and cooling compared to the steel tube, indicating poorer heat dissipation capabilities than a steel tube of the same size. The composite jacket causes heat accumulation.
2. Under the heating specifications of 1-s intervals for adjacent heat cycles and continuous heating of 120 cycles, the highest temperature in the composite jacket exceeds its glass transition temperature. Therefore, the composite tube cannot meet these heating specifications and is unsuitable for the scenario requiring high heating rate and continuous multi-cycle heating.
3. During continuous heating, the regions affected by heat shock are closer to the inner surface of the tube. During the heat transfer process, the temperature rise and fall of the composite jacket lag behind those of the liner. After continuous heating, the temperature gradient within the composite jacket exceeds that of the liner, and the temperature difference between the inner and outer surfaces of the composite jacket is greater than that of the liner.

Acknowledgments This work was supported by Postgraduate Research & Practice Innovation Program of Jiangsu Province (number KYCX24_0708).

References

1. Zhang J, Lin G, Vaidya U, Wang H (2023) Past, present and future prospective of global carbon fibre composite developments and applications. Compos Part B Eng 250. https://doi.org/10.1016/j.compositesb.2022.110463
2. Siengchin S (2023) A review on lightweight materials for defence applications: present and future developments. Def Technol 24:1–17. https://doi.org/10.1016/j.dt.2023.02.025
3. Littlefield A, Hyland E, Keating J (2007) 120MM prestressed thermoplastic composite overwrapped gun tube. In: SAMPE '07: M and P—From coast to coast and around the world, June 3, 2007–June 7, 2007, Baltimore, MD, USA, vol. 52: Soc. for the Advancement of Material and Process Engineering, in International SAMPE Symposium and Exhibition (Proceedings), p. SAMPE Baltimore/Washington Chapter
4. Littlefield AG, Hyland EJ (2009) Development and testing of prestressed carbon fiber composite overwrapped gun tubes. In: ICCM-17—17th International Conference on Composite Materials, 2009: International Committee on Composite Materials. ICCM International Conferences on Composite Materials
5. Chen LM, Qian LF, Hu HS (2005) Numerical heat transfer analysis and experiment of composite material barrel. In: ISTM/2005: 6th international symposium on test and measurement, Vols 1–9, conference proceedings, vol 25(3), pp 4526–4529. https://doi.org/10.3969/j.issn.1673-9728.2005.03.029
6. Wu Q, He L, Zeng X, Zhou K, Li J (2010) Numerical simulation and experimental study on temperature field of segment enwound composite gun barrel. J Ballistics 22(4):94–96. [Online]. Available: https://d.wanfangdata.com.cn/periodical/Ch9QZXJpb2RpY2FsQ0hJTmV3UzIwMjQxMTA1MTcxMzA0Eg1kZHhiMj AxMDA0MDIzGghkMXNxc2oxcQ%3D%3D
7. Chen S, Wang X, Shi Y, Xu C (2024) Transient heat transfer study of small-caliber barrel in continuous fire. Int J Therm Sci 196. https://doi.org/10.1016/j.ijthermalsci.2023.108737
8. Abaci WB, Hristov N, Ahmed NZ, Jerkovic D, Drakulic M (2022) Determination of the gun barrel walls temperature distribution and its experimental validation during multiple-shots firing process. Int J Therm Sci 179. https://doi.org/10.1016/j.ijthermalsci.2022.107667
9. Cialkowski M, Joachimiak M, Mierzwiczak M, Frckowiak A, Olejnik A, Kozakiewicz A (2024) Cauchy type nonlinear inverse problem in a two-layer cylindrical area. https://doi.org/10.1080/01457632.2024.2416285
10. Wang Y, Ding S, Zhang X (2023) A novel structure to inhibit barrel erosion induced by thermal effects in a propulsion system. Int Commun Heat Mass Transfer 147. https://doi.org/10.1016/j.icheatmasstransfer.2023.106991
11. Zhang W, Yu Y (2023) Analysis of gas-eroding barrel characteristics based on fluid-solid interaction. Explos Shock Waves 43(3):136–146. https://doi.org/10.11883/bzycj-2022-0390
12. Yagla JJ, Anderson GA (2022) Barrel heating in hypervelocity powder guns. In: 32nd International Symposium on Ballistics, BALLISTICS 2022, May 9, 2022–May 13, 2022, Reno, NV, USA, 2022, vol. 1: DEStech Publications, in Proceedings—32nd International Symposium on Ballistics, BALLISTICS 2022, pp 563–574. [Online]. Available: https://doi.org/10.12783/ballistics22/36093
13. Tao W (2001) Numerical heat transfer, 2nd edn. Xi'an Jiaotong University Press, Xi'an, p 566
14. Feng G, Zhou K, He L, Zhang Y, Qiao Z, Huang X (2017) Correction for the composite heat transfer cofficient of gun barrel heat transfer. J Eng Thermophys 38(2):296–302. [Online]. Available: https://d.wanfangdata.com.cn/periodical/Ch9QZXJpb2RpY2FsQ0hJTmV3UzIwMjQxMTA1MTcxMzA0EhBn Y3J3bHhiMjAxNzAyMDEyGghjd3B5YzY2dA%3D%3D
15. Chen L, Qian L (2008) Contact thermal resistance of composite materials barrel thermal performance analysis. Compos Sci Eng 5:28–33. https://doi.org/10.3969/j.issn.1003-0999.2008.05.006

Open Access This chapter is licensed under the terms of the Creative Commons Attribution-NonCommercial-NoDerivatives 4.0 International License (http://creativecommons.org/licenses/by-nc-nd/4.0/), which permits any noncommercial use, sharing, distribution and reproduction in any medium or format, as long as you give appropriate credit to the original author(s) and the source, provide a link to the Creative Commons license and indicate if you modified the licensed material. You do not have permission under this license to share adapted material derived from this chapter or parts of it.

The images or other third party material in this chapter are included in the chapter's Creative Commons license, unless indicated otherwise in a credit line to the material. If material is not included in the chapter's Creative Commons license and your intended use is not permitted by statutory regulation or exceeds the permitted use, you will need to obtain permission directly from the copyright holder.

Study on Wave Impact Characteristics of Three-Dimensional Wedge in Water

Dongying Li, Jing Yang, Lujun Zhao, and Tengjiao Sang

Abstract The large amphibious aircraft is typical of the hull type configuration, and its landing area can be simplified as a three-dimensional wedge. Because the amphibious aircraft has the necessary characteristics of taking off and landing on water, the impact characteristics of the three-dimensional wedge in the process of wave landing are studied, which provides guidance for the design and selection of the bottom of the amphibious aircraft. The theoretical wave height of wave is obtained by theoretical calculation, and the effectiveness of wave simulation in CFD is verified by comparison. Meanwhile, the influence of different bottom oblique angles and different wavelengths on the impact load of water wedge is studied. The calculation results show that the impact load of the wedge increases with the increase of the wavelength, and the water load will be saturated due to the limitation of the infiltration area. The increase of the oblique angle of the bottom of the ship makes the water load of the collision wave gradually decrease. It is suggested that the oblique angle of the bottom of the ship should be appropriately increased to reduce the collision load.

Keywords Amphibious aircraft · 3-D wedge · Water impact load · CFD

1 Introduction

The shape of amphibious aircraft combines the aerodynamic and hydrodynamic advantages of traditional land-based aircraft and streamlined hull. It is required to have the basic ability of taking off and landing, water skiing and maneuvering on the

D. Li (✉) · J. Yang · L. Zhao
AVIC Special Vehicle Research Institute, Jingmen, China

T. Sang
AVIC Special Vehicle Research Institute, Jingmen, China

Key Aviation Scientific and Technological Laboratory of High-Speed Hydrodynamics, Jingmen, China

© The Author(s) 2026

Z. Zhang, P. Mercorelli (eds.), *Mechanical Design and Simulation: Insights and Innovations*, Lecture Notes in Mechanical Engineering,
https://doi.org/10.1007/978-981-95-2405-1_22

water surface. To enhance the stability of water surface operation, the wings of large amphibious aircraft are mostly fixed wings, and the bottom of the fuselage is mostly hull. When the seaplane is running on the water surface, it is mainly affected by the hydrodynamic force. Aiming at the problems of water impact, take-off overload and hydroplaning stability of the seaplane during take-off and landing on the water surface, domestic and foreign scholars have carried out a lot of theoretical calculation analysis and experimental research. The water entry impact problem is a complex nonlinear problem. The problem of water entry impact is a kind of complex nonlinear problem. Wanger [1] linearized the free surface and the Bernoulli equation to approximate the physical surface as a plate, but there are still difficulties in the boundary value nonlinear problem. With the idea of asymptotic approximation, Cumberbatch [2] obtained a similar solution, which can be used to solve the impact problem of wedge flow. After introducing a small amount, Mackie [3] realized the linearization of the wet surface, which promoted the development of the numerical solution of the water entry impact problem. With the rapid development of computer technology and experimental testing technology, the numerical boundary solution and experimental verification of the water entry impact problem are no longer difficult.

The landing structure of the bottom of amphibious aircraft is the main factor affecting the surface performance of the aircraft [4]. As a typical structural form of the hull amphibious aircraft, it is still of great practical value to study the influence of water-entry load response characteristics of wedge. Wu Qiaorui [5] analyzed the distribution characteristics of slamming pressure of wedge shape with three shapes, linear, concave and convex, and verified the characteristics of wedge water slamming numerical by simulation and experimental. Chen Cheng studied the influence of the wedge angle on the inflow field and cavitation [6]. Considering the wave environment of the water surface, the slamming load and wave load generated by the wedge during the water entry impact will significantly increase the water entry pressure [7]. The initial state of the water entry of the structure, such as the structural mass, the water entry position, speed, attitude angle also has a significant influence on the impact load of the structure [8]. In fact, the influence of the shape of the water entry structure on the impact load can't be ignored. Through the experimental study, it is found that the maximum impact acceleration of the arc body is larger than that of the wedge body, and the response time is shorter [9]. Not limited to the study of water entry of rigid bodies, the impact of water entry of elastomers has also set off a boom. In terms of theoretical analysis, the MLM impact load model can be used to analyze the impact load at various locations on the surface of the wedge shape [10]. In the numerical simulation method, the algorithm based on CFD and FEA coupling [11] and the algorithm based on FEM and SPH coupling [12] are common, but the computing resources will be greatly increased. Mou Haolei et al. [13] studied the damage morphology of the structure through the water impact test of the fuselage section. In this paper, the calculation cost is considered comprehensively, and the typical cabin section of the amphibious aircraft is simplified as a wedge with an inclined angle of rise. The method of combining theory with numerical analysis is used to reproduce the change of wave characteristic parameters, and the effectiveness of simulation calculation is verified. At the same time, the simulation numerical analysis of the rigid body 's water impact process is carried out. By analyzing the results of wedge impact wave load under different parameters, it provides data reference for the design and selection of amphibious aircraft.

2 Numerical Calculation Method

Under macroscopic conditions, the typical size of the structure is much larger than the distance between atoms. The material of the object can be regarded as a continuum, and the mechanical properties between the control fluid and the solid can be established through the conservation of mass, energy and momentum. To describe the physical phenomena under real conditions as a solvable mathematical model, it is necessary to determine the constitutive equation of the physical model and consider the boundary conditions, so that the solution of the differential equation of the mathematical model is closed.

In Star ccm+ software, the continuous equation is transformed into algebraic equation by discrete thinking. Firstly, the continuous domain with physical meaning is discretized into a finite number of elements or grids. Store the unknown variables to the node, face center, body center and other specific positions of the unit. Then, the integral form or integral weak form of the differential equation is used to discretize the space, and the algebraic equations to be solved are generated for each discretized time step. Finally, the unknown variables can be obtained by solving these large discrete nonlinear equations.

It's necessary to specify the actual boundary conditions on the solution domain to solve the mathematical model. The boundary conditions of Newtonian fluid motion mainly include the entry and outflow of fluid, the restriction and obstruction of fluid. And symmetric boundary or periodic boundary can be used to simplify the modeling and solving process of the calculation model. Usually, the velocity distribution and fluid properties are known in the velocity inlet boundary, which can be used to calculate the volume, momentum and energy fluxes at the inlet. The outlet boundary of the pressure is the outflow condition of the working pressure, which can be considered as the static pressure of the environment. The wall boundary condition refers to the interface between fluid and solid, on which fluid and structure are impermeable to each other. The symmetrical plane boundary condition simulates the symmetrical state of geometry and fluid in the model. The symmetrical boundary can be regarded as a slip wall with shear stress of 0.

In order to solve the problem of wave impact on three-dimensional wedge, the CFD software Star-ccm+ is used to calculate the fluid response. The finite volume method is used to discretize the space, and the N-S equation is used to simulate the control state of the flow field. Considering the incompressibility of the fluid, the K-ω turbulence model and the all-y+ wall treatment method are used to simulate the gas-liquid two-phase interaction to obtain the free surface morphology change. The volume of fluid method (VOF) is used to identify the free flow and interaction of the liquid surface.

3 Analysis of Wedge Hitting Waves

3.1 Simulation Model

In this paper, Star ccm+ is used as the calculation platform to establish the CFD calculation model of the water impact wave of the amphibious aircraft wedge hull. It is considered that the outer surface of the structure has a large rigidity during the water impact process. Therefore, the coupling effect of the deformation of the structure on the fluid is ignored.

The typical cabin section of an amphibious aircraft with a certain oblique angle is shown in Fig. 1. Under a certain scale ratio, the length, width and height of the structure are about 600 mm^3 356.25 mm^3 231.25 mm. The structure of the cabin section is simplified as a wedge. The initial water entry attitude angle of the fixed wedge is 7°, and the lowest point is located at the average water surface. The process of the wedge entering the water and impacting the wave at a speed of 13.75 m/s is simulated and calculated.

The boundary conditions of the given model are shown in Fig. 2. The upstream, top and bottom of the basin are velocity inlets, both sides are symmetrical surfaces, and the

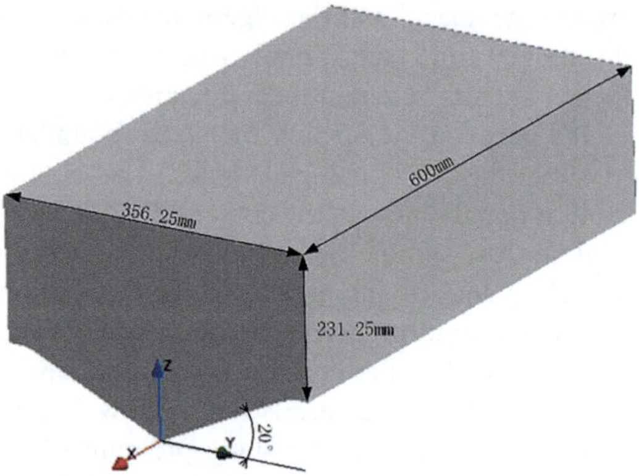

Fig. 1 Geometric model of wedge

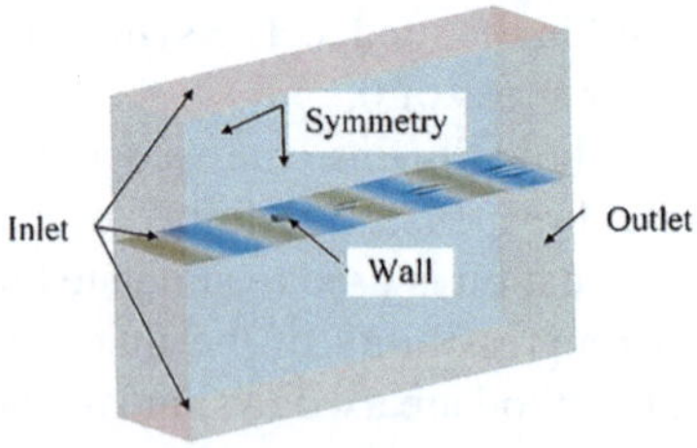

(a) Boundary conditions

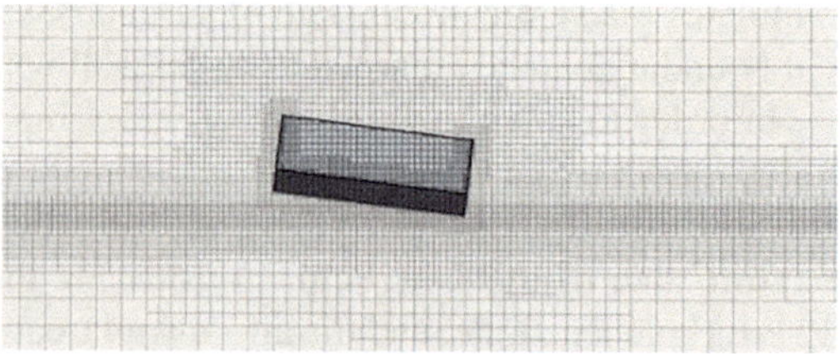

(b) Grid model of wedge

Fig. 2 FD model of wedge (**a**) Boundary conditions (**b**) Grid model of wedge downstream is pressure outlet. The grid size of the whole outer domain of the model is 20 mm, and the grid size of the transition region is 10 mm. In order to accurately capture the motion state of the free liquid surface, the grid size along the wavelength direction is 20 mm, the grid size in the wave height direction is 5 mm, and the rigid wall grid is controlled to 5 mm. The number of total calculation grid unit is 4,445,685.

3.2 Wave Theory Calculation

Assuming that the wave motion is an ideal and incompressible fluid irrotational motion, the velocity potential function Φ satisfies the Laplace equation (1):

$$\frac{\partial^2 \phi}{\partial x^2} + \frac{\partial^2 \phi}{\partial z^2} = 0 \tag{1}$$

Once the velocity Φ is determined, the fluid pressure distribution P can be calculated by introducing the Cauchy-Lagrange integral as Eq. (2) (g is the acceleration of gravity).

$$\frac{v^2}{2} + \frac{P}{\rho} + gz + \frac{\partial \Phi}{\partial t} = \mathrm{C}(\mathrm{t}) \tag{2}$$

There are three boundary conditions for wave motion: the bottom normal velocity v is 0 (non-penetrating condition), the relative pressure on the free surface is 0, and the water point on the free surface always satisfies the wave surface motion. The micro-amplitude wave theory under linear boundary conditions is often used to reduce the complexity of the problem. Assuming that the particle moves slowly, that is, v is 0, it is obtained that as Eq. (3):

$$\varsigma = A\cos(kx - \omega t) \tag{3}$$

The wave number k is obtained from the wavelength L by the equation$k = \frac{2\pi}{L}$. In the condition of infinite water depth, there is a dispersion relation: $\omega^2 = gk$. Thus, the circular frequency ω and encounter frequency f are obtained: $f = \frac{\omega}{2\pi}$.

In order to determine the accuracy of wave simulation, the wave height H is specified as 60 mm in the simulation model, and the wavelengths L are 4 m, 8 m and 10 m respectively. The wave height and frequency under different wave conditions are collected and compared with the above theoretical calculation results. The curve of wave height with time is shown in Fig. 3. It can be seen from the figure that, the amplitude and frequency obtained by simulation monitoring are equivalent to the theoretical calculation results.

The average values of the stable wave height and frequency in the statistical simulation are compared with the theoretical calculation results, and the errors are compared and analyzed as Table 1. It can be seen from the table that the simulation

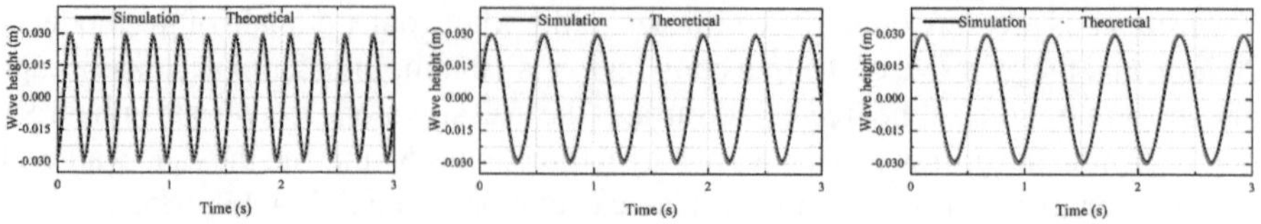

(a) L=4m (b) L=8m (c) L=10m

Fig. 3 The difference of wave height between theory and simulation. (**a**) L = 4 m (**b**) L = 8 m (**c**) L = 10 m

Table 1 Wave height and frequency

	Wave height (m)			Frequency (Hz)	
Wavelengths (m)	Theoretical	Simulation	Deviation	Theoretical	Simulation
4	0.06	0.0547	−4.33	25.52	25.52
8	0.06	0.0583	−2.77	13.58	13.58
10	0.06	0.0589	−1.83	11.2	11.2

frequency are consistent with the theoretical values. It shows that the wave input is correct. Since the simulation calculation is based on the N-S equation to solve the fluid motion state, the solution process is dependent on the discrete advantages and disadvantages of the computational domain and the truncation of the numerical iteration. Therefore, the damping dissipation of the simulation calculation leads to the simulation wave height being slightly smaller than the theoretical wave height. But the maximum deviation doesn't exceed 4.33%. The method used in this paper is accurate and reliable, and it has little effect on the simulation results of fluid motion.

4 Influencing Factors of Wedge Impact Wave Load

To explore the influence of oblique angle and wavelength on the impact wave load, considering the wave conditions of wavelength L of 4 m, 8 m and 10 m, the impact wave load of wedges with oblique angle of 18°, 20°, 23° and 25° was analyzed. As shown in Figs. 4 and 5, when the impact wave wavelength is 4 m, the load generated by the wedge at the initial impact wave is greater than other wave conditions. Therefore, the strength assessment of the structure needs to pay special attention to the dynamic impact force at the moment of impact.

The average value of the stable load under each working condition is extracted as shown in Fig. 6 below. The water load in the water-entry direction plays a dominant role in the process of water-entry impact wave of symmetrical structure. When the inclined angle of the wedge increases from 188 to 258, the proportion of the water load component along the water entry direction decreases, so the water load gradually decreases. The resultant force of the water load decreases by up to 56%, and the bending moment decreases by up to 58%. However, with the increase of the oblique angle, the infiltration area of the bottom of the ship increases, and the wave

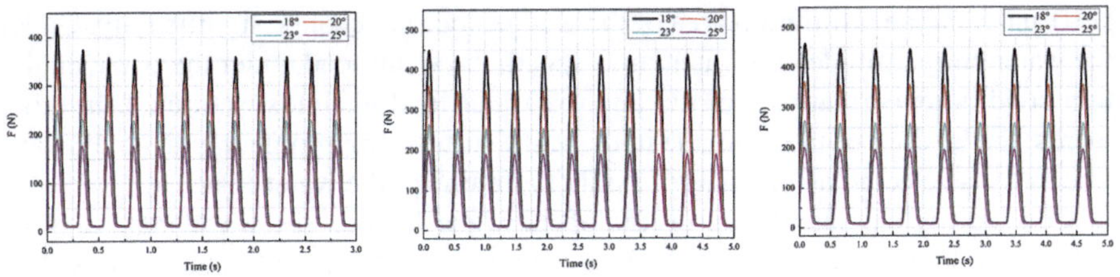

(a) L=4m (b) L=8m (c) L=10m

Fig. 4 The curve of total water load. (**a**) L = 4 m (**b**) L = 8 m (**c**) L = 10 m

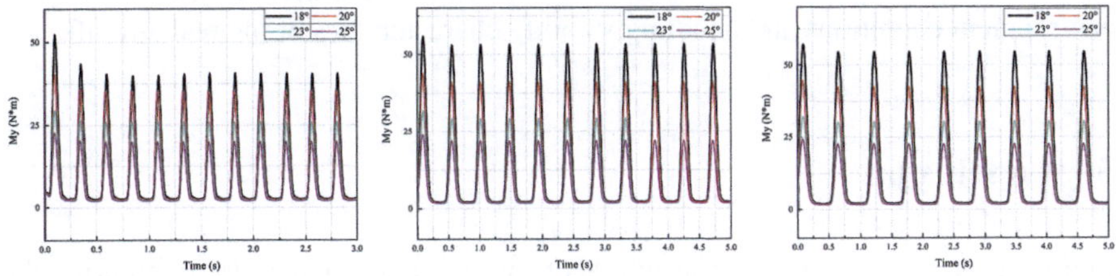

(a) L=4m (b) L=8m (c) L=10m

Fig. 5 The curve of total moment. (**a**) L = 4 m (**b**) L = 8 m (**c**) L = 10 m

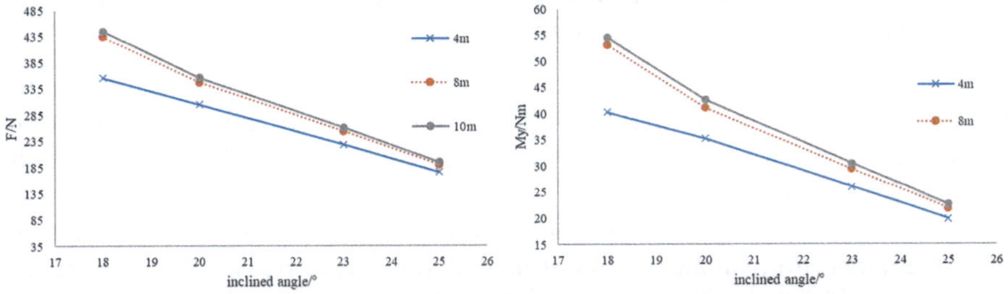

(a) Water load (b) Longitudinal moment

Fig. 6 The curve of water load variation with inclined angle. (**a**) Water load (**b**) Longitudinal moment

resistance and friction resistance of the water surface will increase, which will reduce the navigation performance to a certain extent. Therefore, in the selection of the bottom of the amphibious aircraft, the influence of the oblique angle on the navigation resistance, maneuverability and seakeeping can be considered comprehensively, and the oblique angle of the bottom can be increased appropriately. And when landing in a large wavelength wave environment, the effect of the increase of the inclined angle on the load reduction is more obvious.

When the wavelength increases from 4 to 10 m, as the wavelength increases, the infiltration area of the wedge gradually increases, and the wave impact load of the wedge increases gradually, the resultant force of the water impact load increases by 24.9% and the bending moment increases by 35.2% at most. When the wavelength is 10 and 8 m, the infiltration area of the wedge does not change significantly. Under the water entry attitude and water entry speed, the water load under the two calculation conditions is relatively close, and the water load is almost saturated. In other words, the landing of the seaplane in the wave surface environment with small wave steepness is selected as much as possible to minimize the water load.

Considering the calculation results of the water load by the wavelength and the oblique angle of the bottom, the load of the wedge with large oblique angle will be smaller in the still water surface, but the lateral stability and the surface rotation performance of the seaplane will be greatly weakened, and it is difficult to find the water condition of the still water surface. Therefore, the design and selection of the length and the oblique angle of the bottom should be carried out according to the service environment conditions and handling stability characteristics of the real aircraft.

5 Conclusion

In this paper, the process of water impact of wedge is simulated. By comparing the simulation results of waves with the theoretical results, the effectiveness of the simulation of water impact process is ensured. The influence of oblique angle and wavelength of ship bottom on the wave load is analyzed, which provides a data reference for the selection of the bottom oblique angle of the seaplane. The following regular conclusions are obtained through calculation:

1. The deviation between the wave height and the theoretical wave height in the simulation calculation model is less than 4.33%, which verifies the validity of the grid division in the simulation calculation model and the reliability of the calculation results.
2. With the increase of the oblique angle of the wedge, the water load and the longitudinal bending moment of the structure gradually decrease. In the design and selection of the aircraft, it can be considered to increase the oblique angle appropriately to reduce the risk of excessive impact load.
3. With the increase of the wavelength, the water load of the structure increases gradually. However, since the infiltration area does not increase infinitely with the wavelength, the water load of the structure will reach saturation under certain wavelength conditions at a specific water entry attitude and speed.

In this paper, the bottom of the surface vehicle is simplified as a wedge, and the dynamic conditions can't be considered in the process of simulating the collision wave. In the future, the collision wave model with full power of the seaplane can be established, and the lateral stability and rotation performance can be taken into account to obtain a more stringent design and selection range of the bottom oblique angle.

References

1. Wagner H (1932) Uber Stoss-und Gleitvorgange an der Oberflache von Flussigkeiten. Z Angew Math Mech 12(4):193–215. https://doi.org/10.1002/zamm.19320120402
2. Cunberbatch E (1960) The impact of a water wedge on a wall. J Fluid Mech 7:353–374. https://doi.org/10.1017/S002211206000013X
3. Mackie AG (1962) A linearized theory of the water entry problem. He Q J Mech Appl Math 15(2):137–151. https://doi.org/10.1093/qjmam/15.2.137
4. Zhang YJ, Cui B, Wang MZ et al (2023) Research progress of water landing test and theoretical analysis methods for amphibious aircraft. Acta Aeronaut Astron Sin 44(21):528665. https://doi.org/10.7527/S1000-6893.2023.28665
5. Wu QR, Zhang Z, Chen MH et al (2023) Research on water impact of curved wedge with cushioning effect. Shipbuild China 64(1):87–98. CNKI: CDMD: 2.1016.913373
6. Chen C, Shi WK, Shen YM et al (2023) Study on effect of dead-rise angle on evolution of multiphase Interface in wedge water entry problems. Chin J Hydrodyn 38(5):663–668. https://doi.org/10.16076/j.cnki.cjhd.2023.05.001
7. Zhou L, Li ZH, Sun TZ et al (2019) Experimental study on free fall slamming of wedge-shaped objects under still water and waves conditions. The Ocean Engineering 37(1):1–11. CNKI: SUN: HYGC.0.2019-01-001
8. Chen SY (2022) Numerical analysis of water impact characteristics of structure in wave environment. Nanjing University of Aeronautics and Astronautics. https://doi.org/10.27239/d.cnki.gnhhu.2022.000229
9. Zhang YQ, Bai ZN, Zeng X et al (2020) Study of water impact response of wedge- and arc-shaped structures. J Ship Mech 24(3):400–408. https://doi.org/10.3969/j.issn.1007-7294.2020.03.014
10. Wang YW, Zheng C, Wu WG et al (2021) On slamming load and structural response of a flexible wedge via analytical methods and numerical simulations. Explos Shock Waves 41(11):1–16. https://doi.org/10.11883/bzycj-2020-0276
11. Sun Z, Liu GJ, Li H (2021) Coupling solution of three-dimensional elastic wedge-shaped body entering water and slamming under wave conditions. In: Proceedings of the 30th National Hydrodynamics Symposium. https://doi.org/10.26914/c.cnkihy.2020.037486
12. Tian MY (2019) Study and implementation of the SPH-FEM coupled method for water entry process of structures[D]. Dalian University of Technology. https://doi.org/10.26991/d.cnki.gdllu.2022.003092
13. Mu HL, Gao F, Wang ZL et al (2024) Vertical water-entry impact characteristics of typical fuselage section of civil aircraft. J Vib Shock 43(1):297–307. https://doi.org/10.13465/j.cnki.jvs.2024.01.035

Open Access This chapter is licensed under the terms of the Creative Commons Attribution-NonCommercial-NoDerivatives 4.0 International License (http://creativecommons.org/licenses/by-nc-nd/4.0/), which permits any noncommercial use, sharing, distribution and reproduction in any medium or format, as long as you give appropriate credit to the original author(s) and the source, provide a link to the Creative Commons license and indicate if you modified the licensed material. You do not have permission under this license to share adapted material derived from this chapter or parts of it.

The images or other third party material in this chapter are included in the chapter's Creative Commons license, unless indicated otherwise in a credit line to the material. If material is not included in the chapter's Creative Commons license and your intended use is not permitted by statutory regulation or exceeds the permitted use, you will need to obtain permission directly from the copyright holder.

References

[illegible]

Satellite Truss Structure-Processing Equipment-Processing Technology Integrated Design System Based on Collaborative Optimization

Jianglong Liu, Congfa Zhang, Zhicheng Zhu, Jiaqi Li, and Muwei Shu

Abstract With the continuous improvement of satellite communication performance and the increasing size of satellite antenna, the traditional ground manufacturing method is facing challenges, and the on-orbit manufacturing technology is gradually becoming a trend. Multidisciplinary optimization design technology provides support for on-orbit manufacturing. The existing method adopts serial optimization, which easily leads to local optimization. In this paper, a satellite truss structure-equipment-process integrated design system is developed based on collaborative optimization method. It decouples the variables among the three subdisciplines of geometry structure, processing equipment and process parameters, and improves the ability to deal with high-dimensional optimization problems. Meanwhile, the introduction of proxy model method improves the design efficiency of satellite truss Additionally, workflows are built based on Isight software to enable the joint invocation of various third-party software tools. Finally, a visual GUI interface is developed using QT, allowing for rapid modification of constraint conditions and target variables, and realizing the visualization of optimization solutions.

Keywords Satellite truss · Collaborative optimization · Surrogate model · Multidisciplinary optimization design

1 Introduction

In recent years, as the precision requirements of satellites have increased, the demand for larger satellite antenna sizes has grown. Traditional on-site additive manufacturing of composite trusses can no longer meet the needs for large-scale, high-precision, and complex structural components [1, 2]. At this point, the

J. Liu · Z. Zhu · J. Li · M. Shu (✉)
Beijing Institute of Technology, Beijing, China

C. Zhang
Institute of Spacecraft System Engineering, Beijing, China

© The Author(s) 2026

Z. Zhang, P. Mercorelli (eds.), *Mechanical Design and Simulation: Insights and Innovations*, Lecture Notes in Mechanical Engineering,
https://doi.org/10.1007/978-981-95-2405-1_23

emergence of in-orbit additive manufacturing technology has broken through the bottlenecks of traditional on-site manufacturing [3–7].The key challenge in the manufacturing process lies in the comprehensive consideration of geometric structure, processing equipment, and processing technology parameters across the three disciplines.

In terms of geometric structure optimization, Tethers Unlimited, Inc. (TUI) developed the Trusselator technology [8–10], which assembles trusses after partial extrusion forming. NASA's experiments have shown that the mass of the trusses is reduced by 2–5 times compared to previous truss structures [8] .Tae-Hyun Kim et al. [11] proposed a deployable truss structure that retains the advantages of traditional truss structures while achieving higher packaging efficiency. Through dynamic model analysis and prototype testing, the feasibility of the structure was verified. Lu et al. [12] proposed an optimal design method for deployable antenna truss structures based on dynamic characteristic constraints. By conducting detailed dynamic analyses and structural optimization, lightweight and high-performance designs were achieved. Suh et al. [13] proposed a concept of a two-dimensional modular deployable truss, which exhibits excellent flexibility and high packaging efficiency. Experimental results demonstrated that the dynamic model accurately describes the behavior of the proposed deployable system. Li et al. [14], based on the finite element method, derived the thermo-dynamic coupling finite element dynamic equation and the fundamental equation for the solar radiation pressure field of a circular truss structure. The analysis revealed that the effect of solar pressure on deformation is minimal, providing guidance for the optimized design of large truss structures for space applications. Siriguleng et al. [15] designed and manufactured a scaled experimental model of a circular truss structure and analyzed its vibration modes through experiments. Li et al. [16] designed a truss structure that can be continuously built in space using strip materials, which improved the structural stiffness and manufacturing efficiency. YAO et al. [17] proposed a new rapid modeling and simulation method. Through software development and algorithm optimization, experiments showed significant improvements in the efficiency and quality of satellite truss structure design. Aborehab et al. [18] demonstrated, through a case paper, how to reduce the structural mass of small satellites while ensuring structural integrity and performance, with the design validated through finite element analysis. Jawad et al. [19] applied the artificial bee colony algorithm for the size and layout optimization of truss structures, validating the method's effectiveness and superiority through multiple test cases. El Bouzouiki et al. [20] proposed a new non-uniform cellular automata (CA) algorithm for topology and size optimization of truss structures, achieving lightweight designs while ensuring structural stability and meeting constraints. Kaveh et al. [21] introduced an improved IGWO algorithm, which showed higher efficiency and accuracy in solving structural optimization problems compared to the original GWO algorithm. Zuo et al. [22] proposed an effective hybrid optimization method for handling truss structure optimization problems with frequency constraints, and its effectiveness was verified through optimization examples of trusses at different scales.

In terms of processing equipment optimization, Kringer et al. [23] developed a method using robots to directly extrude photopolymers (DREPP) for the cost-effective manufacturing of space structures. Vélez et al. [24] proposed an alternative machine capable of continuously moving along the object being manufactured. This machine can produce objects larger than itself and supports an infinite manufacturing cycle, including printing, re-anchoring to a new site, and printing again.

In terms of processing technology optimization, Bhundiya et al. [25] designed methods by evaluating different materials and manufacturing processes, with a particular emphasis on deformation processing as an efficient and low-energy space manufacturing method. This approach enables rapid construction of structures while maintaining the required precision and performance. Liu et al. [26] proposed a new free-hanging 3D printing method, which significantly improves the printing accuracy of structures compared to traditional paths. Liu et al. [27] introduced a novel truss structure layout optimization method that considers modular constraints. By using MILP and heuristic algorithms, the method improves design efficiency and reduces material usage, with experiments demonstrating its feasibility. Yan et al. [28] proposed an integrated winding and splicing forming method based on thermoplastic prepreg tapes. The paper explored the forming process, developed a forming prototype, and optimized process parameters. Results showed that this method offers high manufacturing efficiency, with equipment that is easy to miniaturize, laying the foundation for the continuous in-orbit manufacturing of high aspect ratio thermoplastic tubes. Yan et al. [29] also determined the influence of relevant parameters on the forming process through numerical simulations and experimental verification. They proposed a crimping and pultrusion process for truss rods using carbon fiber (CF)/polyether-ether-ketone (PEEK) prepreg tapes as raw materials. This process features high efficiency, low energy consumption, and a high degree of integration.

In summary, for the current in-orbit manufacturing of satellite trusses, apart from NASA's research, which simultaneously considers the three disciplines of geometric structure, processing equipment, and process parameters, few studies address all three disciplines concurrently. Moreover, NASA's optimization approach follows a serial design scheme. This method can obtain a feasible solution, but it is probably not the optimal solution, because it ignores a large number of feasible regions in the design space. This paper proposes an integrated design for truss structures using a parallel optimization approach, treating the entire problem as a multidisciplinary optimization problem. By considering the collaborative optimization among the three disciplines—geometric structure, processing equipment, and process parameters—parallel optimization is achieved, significantly reducing the overall optimization time. Additionally, a visual interface is developed using QT, enabling the rapid generation of satellite truss designs that meet specified requirements.

In terms of optimization speed, compared to traditional single-discipline optimization, the multidisciplinary parallel optimization proposed in this study significantly improves efficiency for two main reasons: firstly, it adopts a collaborative optimization (CO) algorithm with a parallel optimization structure, greatly enhancing efficiency; secondly, it utilizes a neural network surrogate model to replace traditional simulation software, further reducing the optimization time required.

In terms of optimization performance and reliability, the advantages of CO in solving multidisciplinary optimization problems have been validated in fields such as improving the endurance of underwater gliders [30] and optimizing the design of bionic robotic fish systems [31] . Similarly, in this paper, CO was used to achieve parallel optimization of three disciplines, resulting in an optimal solution that meets engineering requirements across all three disciplines.

2 Definition of the Integrated Design Problem for In-Orbit Manufactured Truss Structures: Structure-Processing Equipment-Processing Technology

2.1 The Structure of Composite Trusses

The typical configuration of on-orbit additive manufacturing products for composite material trusses is shown in Fig. 1. The typical configuration consists of a truss structure composed of longitudinal beams and diagonal beams. During the entire optimization design process, factors such as mechanical performance and weight are comprehensively considered for optimization. However, during the modification of the optimization scheme, the process involves numerous disciplines, such as geometric structure, processing equipment, and process parameters. The design variables across these disciplines are highly dimensional and strongly coupled.

Based on these issues, this paper transforms the entire optimization problem into a multidisciplinary optimization problem, decouples variables, and adopts collaborative optimization between system-level and discipline-level. To address the long simulation optimization cycles and the unclear mechanisms of traditional methods, surrogate models are used to accelerate simulation speed, achieving high accuracy in predicting performance indicators such as stress, strain, and strength. Due to the lack of simulation software tools and the need to integrate third-party CAD, CAE, and other software, the simulation model of this system is developed using Isight software. Finally, addressing the visualization issues of the system, a

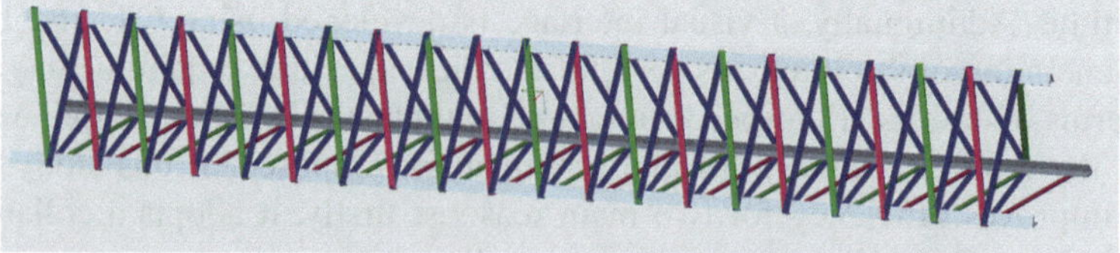

Fig. 1 Schematic diagram of the geometric structure of an on-orbit manufactured composite truss

human-machine interaction interface is developed based on the QT platform. This interface assists users in quickly modifying constraints and variables, performing simulation optimization, and adjusting parameter values. It also enables the visualization of the optimized design scheme. The overall technical framework is illustrated in Fig. 2.

Geometric structural optimization design involves continuously optimizing geometric parameters to achieve the lightest total mass of the truss under optimized topological structures and given load conditions, ultimately outputting the optimal geometric parameters. The primary geometric parameter is the width of the longitudinal beams, optimized using evolutionary algorithms and other heuristic methods. During the optimization process, it is necessary to meet various constraints, including stress constraints, displacement constraints, and frequency constraints. Stress and displacement constraints can be calculated using data-driven predictive models.

Optimization of processing equipment occurs after the design, analysis, and validation to determine the required truss geometry. This involves designing a processing equipment system (truss machine) capable of manufacturing the specified truss. The truss machine selects the most suitable truss machine units through a series of optimized parameter configurations.

Process technology optimization involves the performance of materials, processing temperatures, speeds, etc. These parameters directly affect the final product, i.e., the final quality and performance of the in-orbit manufactured composite truss. Currently, there are two methods for realizing the in-orbit manufacturing of composite trusses. The first method is cold bending forming of metal pipe fittings, used to construct large-scale truss structures. This method is mainly aimed at the needs of high-stiffness platforms and is currently under independent research and development to support physical manufacturing. The second method is based on the truss rod forming technology using pre-impregnated tape, following the technical route of NASA. This method involves feed mechanisms, main beam pulling mechanisms,

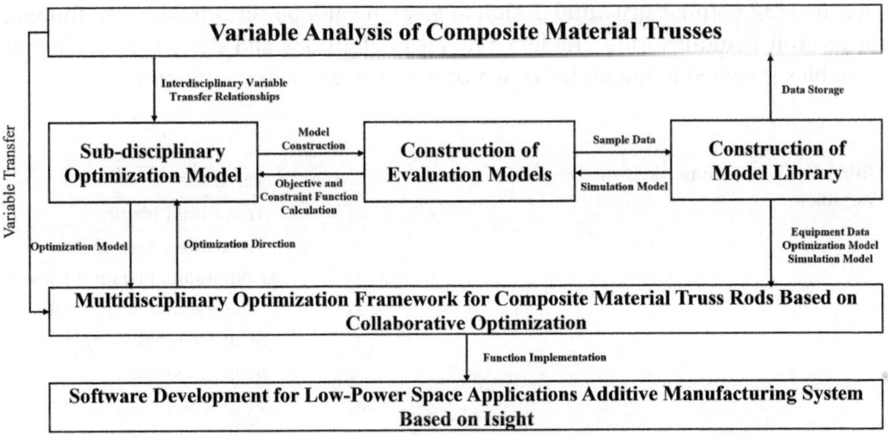

Fig. 2 Technical framework diagram for the integrated design of composite material trusses

diagonal beam winding mechanisms, and welding and guiding mechanisms to directly form small truss rods, which are then used to assemble larger trusses. This method is easy to manufacture, highly efficient, and applicable for ultra-large truss structures, although it has low stiffness. The needs of this work favor ease of manufacturing, hence lower stiffness requirements are acceptable, making pre-impregnated tape the focus of this project.

The definition of the integrated design problem for in-orbit manufactured trusses—structure, process, and processing equipment—entails an analysis of the geometric structure, processing equipment, and process parameters of the composite truss. This involves defining the problem considering the multitude of parameters that exist within the environment of in-orbit construction, as well as the product, processes, and equipment used. In different environments and for different products, various configurations, processes, and equipment can be selected to achieve construction in orbit.

2.2 *Optimization Model for In-Orbit Manufacture of Composite Trusses*

After taking into consideration factors such as load-bearing structure design, truss geometric structure design, processing equipment, and manufacturing processes, we have organized the variables corresponding to each of the three sub-disciplines into separate discipline design models, as detailed below:

Truss Geometric Structure Optimization The focus of truss geometric structure optimization is the spatial configuration of the structure, including the position of nodes, the length of beams, and the angles of connections, aimed at enhancing overall performance by improving the geometric shape. Optimizing variables such as helix angle, total length of the truss, and cross-sectional dimensions is a crucial step in achieving optimal structural design to meet the unique zero-gravity requirements of in-orbit manufacturing. Table 1 provides definitions and design ranges for the variables involved in the optimization of the truss geometric structure.

Table 1 Truss Geometric Structure Optimization Variables

Symbol	Variables
x_1	Truss Total length
x_2	Truss cross-section size
x_3	Longitudinal beam thickness
x_4	Longitudinal beam width
x_5	Strip pitch/helix angle
x_6	Strip thickness
x_7	Strip width

Process Parameter Optimization Process parameter optimization focuses on specific operational parameters in the manufacturing process, such as temperature, pressure, and speed, which directly impact the final product—the quality and performance of the in-orbit manufactured composite truss. In truss manufacturing, the optimization of main beam process parameters, diagonal beam process parameters, and welding process parameters is critical. Properly set process parameters not only enhance the mechanical performance and durability of the structure but also improve material utilization and reduce costs. Definitions and range designs for each variable in process parameter optimization are shown in Table 2.

Processing Technology Optimization Configuration Processing equipment optimization configuration primarily involves optimizing the equipment and machinery used in the in-orbit manufacturing process of oversized three-dimensional structures. This includes the feed mechanism, main beam pulling and pressing mechanisms, diagonal beam winding mechanisms, and welding and guiding mechanisms. By optimizing the settings of these devices, such as speed, accuracy, and force, manufacturing efficiency can be improved, material waste can be reduced, and product quality can be ensured. Additionally, the corresponding product performance is influenced by the width of the longitudinal beams in the truss geometric structure parameters. The optimization goal of this discipline is to achieve optimal performance and cost-effectiveness in the in-orbit manufacturing process through proper configuration and adjustment of processing equipment. Definitions and range designs for each variable in the processing equipment optimization configuration are shown in Table 3.

Table 2 Description table for variables in process parameter optimization

Symbol	Variables
x_8	Traction speed
x_9	Forming pressure
x_{10}	Heating/cooling temperature
x_{11}	Winding parameters
x_4	(Main beam) longitudinal beam width

Table 3 Description table for variables in processing equipment optimization configuration

Symbol	Variables
x_{12}	Motor parameters (power, size, volume)
x_{13}	Mold parameters (length, width, corresponding beam thickness)
x_{14}	Roller parameters (quantity, position)
x_{15}	Heating device parameters
x_4	(Main beam) longitudinal beam width

3 Integrated Design Method for Composite Material Truss Structure-Processing Equipment-Processing Technology Based on Collaborative Optimization

3.1 Subdiscipline Optimization Model Construction

To construct an integrated multidisciplinary optimization model for the three-dimensional structure of in-orbit manufactured composite trusses, it is necessary to comprehensively consider the three subdisciplines: truss geometric structure, processing equipment, and process parameters. Each discipline's optimization variables should be carefully analyzed and defined, with reasonable value ranges designed for each variable.

Geometric Structure Optimization For the geometric structure optimization discipline, a single-discipline optimization model is constructed based on the parameter transmission between the system and the discipline. The design problem is defined as follows:

Geometric structure optimization model:

$$\begin{cases} \min J_1(X) = (s_1 - x_1)^2 + (s_2 - x_2)^2 + (s_3 - x_3)^2 + (s_4 - x_4)^2 \\ +(s_5 - x_5)^2 + (s_6 - x_6)^2 + (s_7 - x_7)^2 + (s_{16} - y_1)^2 + (s_{17} - y_2)^2 \\ +(s_{18} - y_3)^2 + (s_{19} - y_4)^2 + (s_{23} - y_8)^2 \\ S.t. y_1 \ge m_1, y_2 \ge m_2, y_3 \ge m_3 \end{cases} \tag{1}$$

Where y_4 and y_8 are coupling variables.

The optimization focuses on improving the layout of the geometric structure to enhance structural efficiency and manufacturability. The formulation of the design optimization problem aims to minimize the total length of the truss while meeting structural performance requirements. The variable numbering and ranges are consistent with those in Table 1.

Process Parameter Optimization This section focuses on the process parameter optimization discipline, constructing a single-discipline optimization model based on the parameter transmission between the system and the discipline. The design problem is defined as follows:

Process parameter optimization model:

$$\begin{cases} \min J_2(X) = (s_4 - x_4)^2 + (s_8 - x_8)^2 + (s_9 - x_9)^2 + (s_{10} - x_{10})^2 \\ +(s_{11} - x_{11})^2 + (s_{19} - y_4)^2 + (s_{20} - y_5)^2 + (s_{21} - y_6)^2 \\ +(s_{16} - y_1)^2 + (s_{24} - y_9)^2 \\ S.t. y_4 \le m_4, y_5 \le m_5, y_6 \le m_6 \end{cases} \tag{2}$$

Where y_1 and y_9 are coupling variables.

The processing temperature and speed variables for the main beam process parameters, as well as for the diagonal beam process parameters, are directly related to the quality and performance of the structure, with their ranges determined based on the characteristics of the materials used and the structural requirements. The welding process parameters, such as welding temperature and speed, influence the stability and durability of structural connection points. The variable numbering and value ranges are consistent with those in Table 2.

Processing Technology Optimization For the load-bearing structure topology optimization discipline, a single-discipline optimization model is constructed based on the parameter transmission between the system and the discipline. The design problem is defined as follows:

Processing equipment optimization model:

$$\begin{cases} \min J_3(X) = (s_4 - x_4)^2 + (s_{12} - x_{12})^2 + (s_{13} - x_{13})^2 + (s_{14} - x_{14})^2 \\ \quad + (s_{15} - x_{15})^2 + (s_{22} - y_7)^2 + (s_{23} - y_8)^2 + (s_{24} - y_9)^2 \\ \quad S.t. \min(y_7, y_8, y_9) \end{cases} \tag{3}$$

The processing equipment variables are primarily intended to reasonably configure parameters, adjust processing equipment, and reduce costs. The variable numbering and value ranges are consistent with those in Table 3.

After organizing the parameter transmission between the system and subdisciplines and defining the optimization problems for each subdiscipline, the design variables, constraints, and optimization objectives of the subdiscipline surrogate models provide a parameter foundation. These serve as the driving parameters for the subdiscipline surrogate models.

3.2 Subdiscipline Surrogate Model Construction

Surrogate Model Based on Neural Network Deep Neural Networks (DNN) are computational models inspired by the mechanisms of resting and action potentials in natural neurons. There are many forms of neural networks, and the one used in this paper is the BP neural network.

After completing the parameter organization and optimization of the model, three surrogate models corresponding to each of the three sub-disciplines are built based on the BP neural network, and training and testing of the surrogate models are conducted. Specifically, the structural parameters of the surrogate models, such as the number of nodes in the input and output layers, the number of hidden layers, the number of nodes in the hidden layers, and the learning rate, are determined based on the sample data volume and design experience. After constructing the surrogate model, it is trained using sample data. Through iterative optimization, the model continuously adjusts and updates the weights until the parameters meet the required objectives, resulting in a high-precision model for solving the composite material truss in the design space. Finally, the model is tested, completing the construction of the sub-disciplinary surrogate models.

Geometric Structure Optimization Surrogate Model In Table 1 of Chapter "Simulation and Optimization of Temperature Field for Arc Additive Manufacturing Based on Cloud Computing", we defined the parameter variables of the geometric structure optimization discipline. In this section, I will use these as the input variables for the surrogate model, and define the output variables of the surrogate model, as shown in Table 4 below:

In this paper, the input layer of the geometric structure surrogate model consists of 7 nodes, representing the 7 inputs x_1, x_2, x_3, x_4, x_5, x_6, and x_7 . The hidden layer has 4 layers, with each layer containing 32 nodes. The output layer consists of 3 nodes, corresponding to the 3 outputs y_1, y_2 , and y_3 . The activation function is the sigmoid function, and the optimization function is gradient descent. The learning rate is set to 0.001, and the model is trained for 1000 epochs.

Process Parameter Optimization Surrogate Model The specific construction steps are similar to those of the geometric structure optimization surrogate model. After determining the input for the process parameter surrogate model, the specific hyperparameters of the surrogate model are determined based on existing data and the experience of engineers. The specific process parameters are shown in Table 5:

Table 4 Geometric structure optimization input-output table

Input	Symbol	Output	Symbol
Total truss length	x_1	Stress	y_1
Truss cross-sectional dimensions	x_2	Strain	y_2
(Main beam) longitudinal beam thickness	x_3	Mode	y_3
(Main beam) longitudinal beam width	x_4		
Strap pitch/pitch angle	x_5		
Strap thickness	x_6		
Strap width	x_7		

Table 5 Process parameter optimization input-output table

Input	Symbol	Output	Symbol
Towing speed	x_8	Strength	y_4
Forming pressure	x_9	Straightness	y_5
Heating/cooling temperature	x_{10}	Surface roughness	y_6
Winding parameters	x_{11}		
(Main beam) longitudinal beam width	x_4		

The input layer of the process parameter surrogate model consists of 5 nodes, representing the 5 inputs x_8, x_9, x_{10}, x_{11}, and x_4. The hidden layer has 4 layers, each with 32 nodes. The output layer consists of 3 nodes, corresponding to the 3 outputs y_4, y_5, and y_6. The activation function is the sigmoid function, and the optimization function is gradient descent. The learning rate is set to 0.001, and the model is trained for 1000 epochs.

Processing Technology Optimization Surrogate Model The construction steps are similar to those of the previous two sub-disciplines. After determining the input and output variables, as well as the model hyperparameters, the surrogate model is built. The specific parameters are shown in Table 6:

The input layer of the machining equipment surrogate model consists of 5 nodes, representing the 5 inputs x_{12}, x_2, x_{13}, x_{14}, x_{15}, as well as x_4. The hidden layer has 4 layers, each with 32 nodes. The output layer consists of 3 nodes, corresponding to the 3 outputs y_7, y_8, y_9. The activation function is the sigmoid function, and the optimization function is gradient descent. The learning rate is set to 0.001, and the model is trained for 1000 epochs.

3.3 *Collaborative Optimization*

In this paper, a method is proposed that uses a collaborative optimization algorithm to achieve parallel optimization, simultaneously optimizing the three disciplines, significantly improving the overall optimization efficiency. The Collaborative Optimization (CO) algorithm mainly decomposes the original multidisciplinary optimization problem into multiple subproblems, each corresponding to a discipline or subsystem. This transformation makes the problem easier to optimize and solve. In this paper, geometric structure, process parameters, and processing equipment are treated as three sub-disciplines, each with its corresponding optimizer and surrogate model for solving.

After defining the design variables, constraints, and optimization objectives in the optimization problems of the three sub-disciplines in Sects 2.1 and 2.2, and constructing the surrogate models, we simplify the original design problem as shown in Fig. 3.

Table 6 Processing equipment optimization input-output table

Input	Symbol	Output	Symbol
Motor parameters	x_{12}	Processing equipment dimensions	y_7
Mold parameters	x_{13}	Total equipment mass	y_8
Roller parameters	x_{14}	Energy consumption	y_9
Heating device parameters	x_{15}		
(Main beam) longitudinal beam width	x_4		

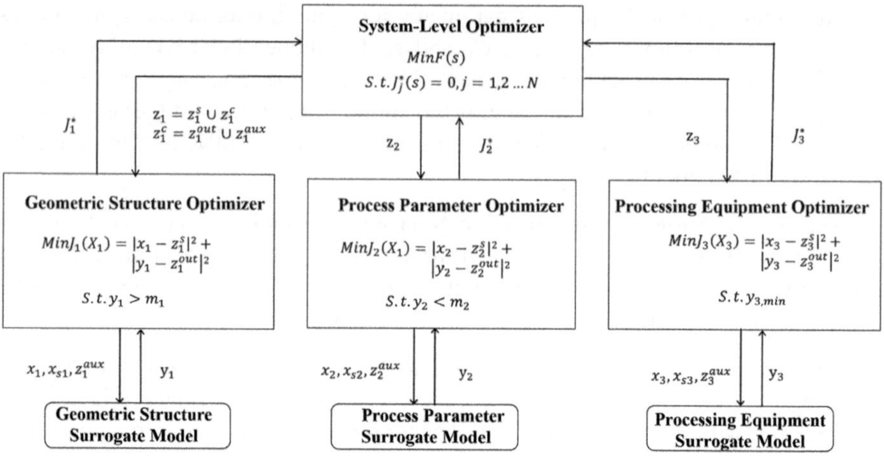

Fig. 3 Collaborative optimization algorithm framework diagram

Each discipline system-level variables$z_n = \left(z_n^s, z_n^c\right)$, z_n^sDesign variables of the sub-disciplines nn representing system-level requirements n Design variables; z_n^c represents the coupling variables in the discipline n that are required by the system level, $z_n^c = \left(z_n^{out}, z_n^{aux}\right)$, Where z_n^{out} represents the coupling variables that need to be optimized within the discipline, and z_n^{aux} represents the coupling variables determined by other disciplines, which remain unchanged during the optimization process of the discipline. In the process of subdiscipline optimization, the optimized variables are $(\boldsymbol{x}_n, \boldsymbol{y}_n)$,while$\left(z_n^s, z_n^c\right)$ are considered fixed values. The objective function is to minimize the consistency constraint J_n,that is, to minimize the difference between the subsystem-level design variables $\boldsymbol{x}_n$,coupling variables $\boldsymbol{y}_n$and the system-level transmitted variables $\boldsymbol{z}_n$.Coordinated optimization is performed between the system-level and discipline-level, ultimately achieving simulation results that meet the requirements.

In the entire optimization process,x_4 is the global variable,y_1, y_4, y_8, $y_9$are coupling variables. The final problem is simplified as follows, with the optimization algorithm used being the CO.

System-level model:

$$\begin{cases} minf(Y) = y_7 + y_8 + y_9 \\ S.t.J_1^*(X) \le \varepsilon \\ J_2^*(X) \le \varepsilon \\ J_3^*(X) \le \varepsilon \end{cases} \tag{4}$$

In the entire CO, the system-level optimizer and the discipline-level optimizers work together. The final goal of the collaborative optimization is to minimize the sum of y_7, y_8, y_9.

4 System Implementation Based on Isight

4.1 Isight Workflow Setup

This paper builds the workflow based on Isight software and implements the calling of various simulation software. The workflow in Isight is divided into the parameter reading section and the optimization section. The parameter reading section is mainly handled by the Data Exchanger component, which reads the parameters from the files generated by QT for configuration. The optimization part is built based on the optimizer and surrogate model, and the specific structure of the optimization part is shown in Fig. 3.

4.2 Integrated Software Development Based on QT

QT is a cross-platform application and user interface framework developed based on C++, allowing users to generate and create the GUI interface programs they need. In this platform, we mainly generate a txt file for parameter transfer. Users can intuitively click on the corresponding disciplines to set parameters, as well as modify and adjust the optimization objectives and constraints of the optimizer. By using the command-line client approach in the "Workflow Modification" button code, users can also select the corresponding zmf file to modify the Isight workflow.

4.3 Truss Structure Integrated Design System Implementation

After completing the Isight workflow and QT interface development, most of the work for the entire system has been completed. Finally, we connect the two parts by linking them together, mainly using a txt file to transfer parameters and complete the front-end and back-end connection. When QT generates the txt parameters, Isight will read these parameters before performing the simulation optimization and configure the model accordingly. This process is shown in Fig. 4.

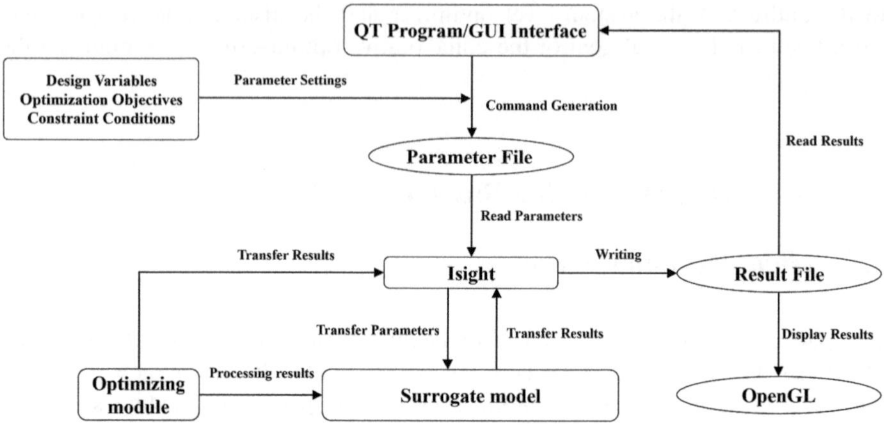

Fig. 4 Integrated system technical framework diagram

5 Conclusion

With the development of the aerospace field, optimizing a single discipline in the satellite truss design process is no longer sufficient to meet the current need for multidisciplinary collaboration and the requirement for rapid satellite truss solution generation. Therefore, this paper presents an integrated design system for composite material trusses, processes, and processing equipment.

1. A new design method is proposed, using collaborative optimization algorithms to treat the entire problem as a multidisciplinary optimization problem, achieving parallel optimization of three sub-disciplines.
2. To address the long simulation optimization cycles and the unclear mechanisms in traditional methods, surrogate models are used to speed up the simulation, achieving high-accuracy predictions for performance indicators such as stress, strain, and strength.
3. Based on the QT platform, a human-machine interface is developed, allowing users to quickly modify constraints, variables, perform simulation optimization, adjust parameter values, and visualize the optimized design solutions, thus completing the integration of the front-end and back-end.

This paper also has some limitations, the most significant regret being that, due to constraints in data sources, we used these data in internal experiments. This paper only introduces the framework of the relevant software and algorithms. In the future, we will validate our system and algorithms on more aerospace equipment to advance the development of on-orbit manufacturing technology.

References

1. Ferro CG, Varetti S, Maggiore P (2024) Experimental evaluation of mechanical compression properties of aluminum alloy lattice trusses for anti-ice system applications. Machines 12(6):404. https://doi.org/10.3390/machines12060404
2. Zhang D, Tian X, Zhou Y, Wang Q, Yan W, Akmal Zia A et al (2023) Spatial 3D printing of continuous fiber-reinforced composite multilayer truss structures with controllable structural performance. Polymers 15(21):4333. https://doi.org/10.3390/polym15214333
3. Sears PJ, Ho K (2018) Impact evaluation of in-space additive manufacturing and recycling technologies for on-orbit servicing. J Spacecr Rocket 55(6):1498–1508. https://doi.org/10.2514/1.A34367
4. Boschetto A, Bottini L, Macera L, Vatanparast S (2023) Additive manufacturing for lightweighting satellite platform. Appl Sci 13(5):2809. https://doi.org/10.3390/app13052809
5. Rinaldi M, Cecchini F, Pigliaru L, Ghidini T, Lumaca F, Nanni F (2020) Additive manufacturing of polyether ether ketone (PEEK) for space applications: a nanosat polymeric structure. Polymers 13(1):11. https://doi.org/10.3390/polym13010011
6. Liu J, Shi Y, Cui Y, Hou J, Zhang K, Huang A (2022) Research progress and prospect of on-orbit additive manufacturing technology. Chin Space Sci Technol 42(6):23–34. https://doi.org/10.16708/j.cnki.1000-758X.2022.0080
7. Yang X, Han J, Liu J, Zhang C, Liang J (2021) Review and prospects of development of space on-orbit manufacturing technology. J Astronaut 42(11):1343–1354. https://doi.org/10.3873/j.issn.1000-1328.2021.11.001
8. Hoyt RP, Cushing J, Slostad J, Jimmerson G (2014) Trusselator: on-orbit fabrication of high-performance composite truss structures. In: AIAA SPACE 2014 conference and exposition, p 4337. https://doi.org/10.2514/6.2014-4337
9. Hoyt RP, Slostad J, Cushing J, Moser T (2016) In-space manufacturing of Constructable™ long-baseline sensors using the Trusselator™ technology. In: AIAA space 2016, p 5244. https://doi.org/10.2514/6.2016-5244
10. Levedahl B, Hoyt RP, Silagy T, Gorges J, Britton N, Slostad J (2018) Trusselator™ technology for in-situ fabrication of solar array support structures. In: 2018 AIAA spacecraft structures conference, p 2203. https://doi.org/10.2514/6.2018-2203
11. Kim TH, Suh JE, Han JH (2021) Deployable truss structure with flat-form storability using scissor-like elements. Mech Mach Theory 159:104252. https://doi.org/10.1016/j.mechmachtheory.2020.104252
12. Dai L, Xiao R (2020) Optimal design and analysis of deployable antenna truss structure based on dynamic characteristics restraints. Aerosp Sci Technol 106:106086. https://doi.org/10.1016/j.ast.2020.106086
13. Suh JE, Jeong SY, Han JH (2019) A two-dimensional modular deployable truss structure with bistability. J Intell Mater Syst Struct 30(3):335–350. https://doi.org/10.1177/1045389X18812257
14. Li WJ, Wu YS, Wei QS, Wei C (2019, July) Multifield coupling dynamics of large truss structure for space applications. In: 2019 IEEE 10th International Conference on Mechanical and Aerospace Engineering (ICMAE). IEEE, pp 52–56. https://doi.org/10.1109/ICMAE.2019.00018
15. Siriguleng B, Zhang W, Liu T, Liu YZ (2020) Vibration modal experiments and modal interactions of a large space deployable antenna with carbon fiber material and ring-truss structure. Eng Struct 207:109932. https://doi.org/10.1016/j.engstruct.2020.109932
16. Li P, Ning H, Yan J, Xu B, Li H (2022) Design and mechanical characterisation of a large truss structure for continuous manufacturing in space. Materials 15(17):6025. https://doi.org/10.3390/ma15176025

17. Yao J, Zhang HY, Cheng Y et al (2022) Rapid modeling simulation of satellite truss structure based on simplified geometrical features of 3D CAD model and test verification. Spacecraft Environ Eng 39(5):515–521. https://doi.org/10.12126/see.2022.05.011
18. Aborehab A, Kassem M, Nemnem A, Kamel M, Kamel H (2020) Configuration design and modeling of an efficient small satellite structure. Eng Solid Mech 8(1):7–20. https://doi.org/10.1515/engsol-2020-0023
19. Jawad FK, Ozturk C, Dansheng W, Mahmood M, Al-Azzawi O, Al-Jemely A (2021) Sizing and layout optimization of truss structures with artificial bee colony algorithm. In: Structures, vol 30. Elsevier, pp 546–559. https://doi.org/10.1016/j.istruc.2021.02.014
20. El Bouzouiki M, Sedaghati R, Stiharu I (2021) A non-uniform cellular automata framework for topology and sizing optimization of truss structures subjected to stress and displacement constraints. Comput Struct 242:106394. https://doi.org/10.1016/j.compstruc.2020.106394
21. Kaveh A, Zakian P (2018) Improved GWO algorithm for optimal design of truss structures. Eng Comput 34:685–707. https://doi.org/10.1007/s00366-017-0547-3
22. Zuo W, Bai J, Li B (2014) A hybrid OC–GA approach for fast and global truss optimization with frequency constraints. Appl Soft Comput 14:528–535. https://doi.org/10.1016/j.asoc.2013.12.019
23. Kringer M, Böhrer C, Frey M, Pimpi J, Pietras M (2022) Direct Robotic Extrusion of Photopolymers (DREPP): influence of microgravity on an in-space manufacturing method. Front Space Technol 3:899242. https://doi.org/10.3389/fsyst.2022.899242
24. Vélez M, Toala E, Zagal JC (2020) Koala 3D: a continuous climbing 3D printer. Robot Comput Integr Manuf 64:101950. https://doi.org/10.1016/j.rcim.2020.101950
25. Bhundiya HG, Royer F, Cordero Z (2022) Engineering framework for assessing materials and processes for in-space manufacturing. J Mater Eng Perform 31(8):6045–6059. https://doi.org/10.1007/s11665-022-06458-5
26. Liu S, Li Y, Li N (2018) A novel free-hanging 3D printing method for continuous carbon fiber reinforced thermoplastic lattice truss core structures. Mater Des 137:235–244. https://doi.org/10.1016/j.matdes.2017.10.036
27. Liu Y, Wang Z, Lu H, Ye J, Zhao Y, Xie YM (2023, September) Layout optimization of truss structures with modular constraints. In: Structures, vol 55. Elsevier, pp 1460–1469. https://doi.org/10.1016/j.istruc.2023.06.004
28. Yan J, Geng C, Liang C, Wang C, Mu C, Duan Y, Zhang L (2023) Research on on-orbit roll forming technology and equipment for thermoplastic Prepreg tape. In: 2023 5th International Symposium on Robotics & Intelligent Manufacturing Technology (ISRIMT). IEEE, pp 196–199. https://doi.org/10.1109/ISRIMT57189.2023.10184971
29. Yan J, Li P, Geng C, Guo X, Zhang L (2024) Research and validation of CF/PEEK-based truss rod crimping and pultruding process for on-orbit isoform forming. Materials 17(10):2393. https://doi.org/10.3390/ma17102393
30. Wang S, Yang M, Niu W, Wang Y, Yang S, Zhang L, Deng J (2021) Multidisciplinary design optimization of underwater glider for improving endurance. Struct Multidiscip Optim 63:2835–2851. https://doi.org/10.1007/s00158-020-02488-6
31. Chen H, Li W, Cui W, Yang P, Chen L (2021) Multi-objective multidisciplinary design optimization of a robotic fish system. J Mar Sci Eng 9(5):478. https://doi.org/10.3390/jmse9050478

Open Access This chapter is licensed under the terms of the Creative Commons Attribution-NonCommercial-NoDerivatives 4.0 International License (http://creativecommons.org/licenses/by-nc-nd/4.0/), which permits any noncommercial use, sharing, distribution and reproduction in any medium or format, as long as you give appropriate credit to the original author(s) and the source, provide a link to the Creative Commons license and indicate if you modified the licensed material. You do not have permission under this license to share adapted material derived from this chapter or parts of it.

The images or other third party material in this chapter are included in the chapter's Creative Commons license, unless indicated otherwise in a credit line to the material. If material is not included in the chapter's Creative Commons license and your intended use is not permitted by statutory regulation or exceeds the permitted use, you will need to obtain permission directly from the copyright holder.

Simulation of Cargo Unloading Equipment

Tianrui Li

Abstract With the continuous development of the world economy, the import and export industry has entered a period of rapid development, in order to quickly carry out the flow of goods, there is a great demand for logistics transportation equipment, to develop a highly efficient professional container unloading and handling integrated equipment, in order to ensure the safety and reliability of cargo unloading equipment, It is necessary to use ANSYS analysis software to carry out static simulation of the equipment according to the actual situation, and observe whether its maximum deformation and allowable stress meet the requirements.

Keywords Cargo unloading equipment · Ansys · Statics simulation · Stress-strain

1 Introduction

With the continuous development of various types of trade, cargo boxes have become a necessary means of transport in the logistics industry, and are widely used in sea, air and land transportation, etc. However, the current logistics handling system is still not perfect enough, the level of information technology and automation is not high, and most of the unloading and handling work depends on manual work. Rapidly rising labor costs will push up logistics transportation costs, and then limit the development of domestic logistics field, in order to solve the problem, need to develop a high efficiency, professional automatic unloading equipment, cargo unloading equipment adopts the modular design principle, divided into grasping module, transfer module, lifting module, walking module and multi-sensing system, through the multi-sensing system to obtain information, The upper computer processes the information obtained by the multi-sensing system through the trajectory planning algorithm and intelligently and accurately controls the motion of the multi-modules. The module parts can be quickly adjusted according to different

T. Li (✉)
Changchun University of Science and Technology, Jilin Province, China

© The Author(s) 2026

Z. Zhang, P. Mercorelli (eds.), *Mechanical Design and Simulation: Insights and Innovations*, Lecture Notes in Mechanical Engineering,
https://doi.org/10.1007/978-981-95-2405-1_24

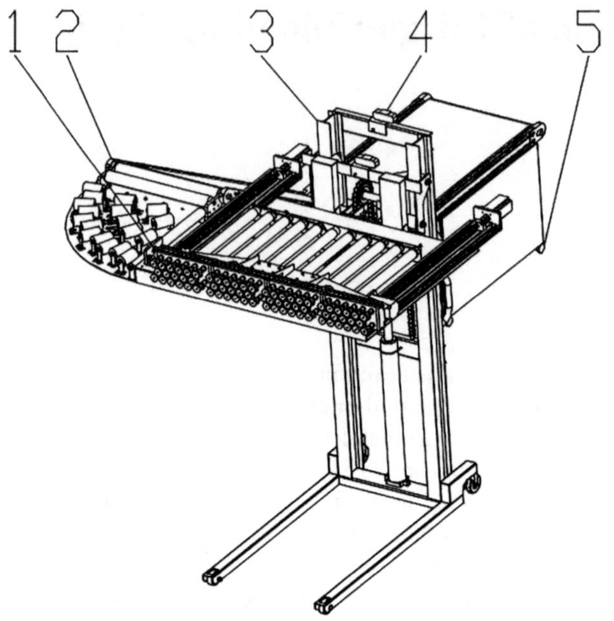

Fig. 1 Three-dimensional diagram of equipment

types of goods to be captured to realize the grasping of goods. The equipment model is shown as follows (see Fig. 1):

In the Fig. 1 represents the grasping module, 2 represents the transfer module, 3 represents the lifting module, 4 represents the multi-sensing system, and 5 represents the out-of-shape module. In order to verify the structural strength of the cargo unloading equipment, it is necessary to carry out static analysis of the whole machine to determine the peak position of the stress and deformation of the components. By analyzing the cloud image results, it can be verified whether the strength and stiffness of the components meet the established design requirements.

2 Static Analysis of the Whole Machine

The static analysis of the whole machine is to verify the anti-deformation ability of the equipment against external static forces [1], and analyze the deformation of the whole machine under the action of external static forces to determine whether the static stiffness of the whole machine meets the requirements. The size of the equipment intended to grab the goods is 430 × 210 × 270, and the weight of the goods is 0–30 kg. As shown in the figure (see Fig. 2), a total of 11 goods can be placed on the transfer module. When the lifting module lifts the transfer module to the top and places two goods, each weighing 30 kg, a total of 330 kg, the equipment has the

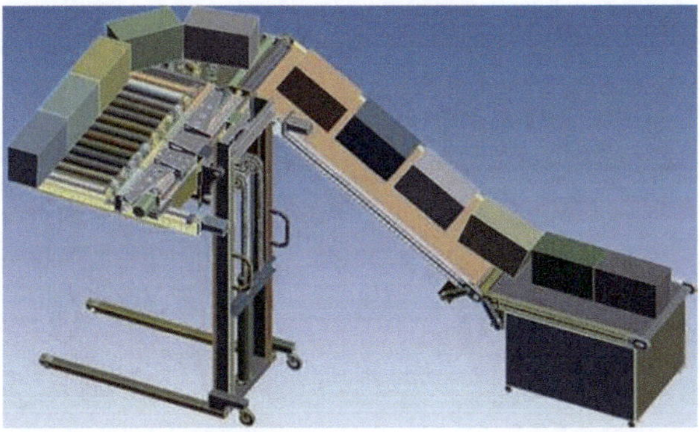

Fig. 2 Whole machine goods layout diagram

Table 1 Equipment material parameter table

Materials	Density	Young's modulus	Poisson's ratio	Yield limit
al6061	2.7 g/m^2	68.9 GPa	0.3	241 MPa
Steel	7.93 g/m^2	193 GPa	0.27	200 MPa

greatest force at this time, and it should be analyzed at this time. Meanwhile, static analysis should be carried out on the equipment without loading the goods. The two are compared to determine which parts of the equipment are stressed when adding goods. The structure of the equipment is relatively complex, and some small features that do not affect the analysis results of the characteristics of the equipment are ignored, such as, chamfer, small groove, sharp Angle and other features on each part [2]. If the manual grid division is relatively large, automatic grid division is chosen.

After the model construction is completed, the materials are allocated and the material, density, elastic modulus and Poisson ratio of each component of the whole machine are defined. Because alloy steel has good bearing capacity and impact performance [3], the lifting module and running module in the equipment are set as steel alloy steel. Because 6061 aluminum alloy is widely used in engineering equipment and other fields [4], the grasping module and transfer module are set to 6061 aluminum alloy and steel alloy steel (Table 1).

If manual grid division involves a large amount of engineering work, automatic grid division is selected [5], and the grid division situation is shown in the figure (see Fig. 3). After grid division, the number of units is 2.0652 million and the number of nodes is 5.5588 million [6].

Then "Bonded" contact was used to define the interaction of the whole machine model, and roller support was applied at the bottom of the model [7]. In order to simulate the stress of the equipment at full load, a heavy load was applied to the whole machine and a 3234N force was uniformly applied to the transfer module to simulate the pressure caused by goods on the equipment [8]. In order to simulate the stress situation of the equipment when goods were not loaded. Gravity is applied to the whole machine.

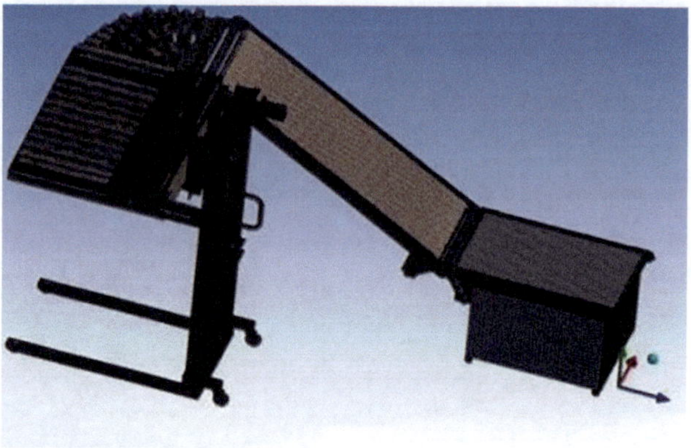

Fig. 3 Whole machine meshing

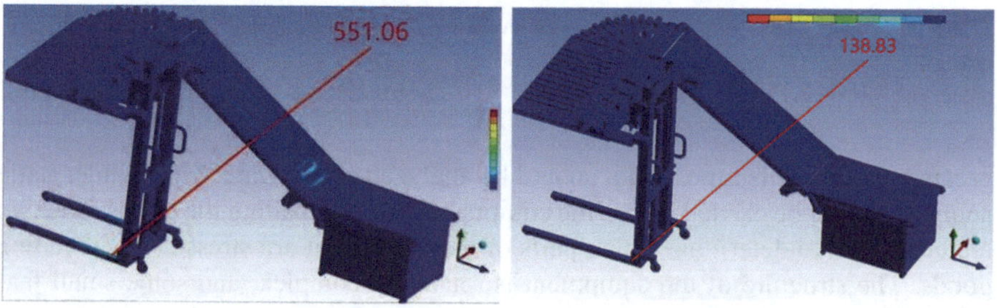

Fig. 4 Stress nephogram of whole machine

The stress cloud diagram of the whole machine is shown in the figure (see Fig. 4), the left side is the actual load when the load, the right side is the load when the load is not loaded, the maximum stress on the left side is 551.06 Mpa, the maximum stress on the right side is 110.12 Mpa, and the maximum stress is mainly at the bottom of the lifting module.

The maximum stress value of 551.06 Mpa is far less than the strength limit value of the material, so the structural design of the equipment is safe. It can be seen from the figure (see Fig. 4) that when the load capacity of the equipment rises, the stress of the whole machine is mainly concentrated in the bottom area of the lifting module and the trailing transmission part.

The deformation of the whole machine is shown in the figure (see Fig. 5), the left side is the actual load situation, the right side is the load situation when the goods are not loaded, Fig. 5 is the overall deformation diagram of the machine tool, the left side of the maximum deformation position is located on the trailing conveyor belt, the maximum deformation value is 7.8483 mm, the right side of the maximum deformation position is located at the junction of the bottom plate and the turntable, the maximum deformation value is 2.9695 mm. It can be seen from the figure that when the load capacity of the equipment rises, the total deformation of the whole

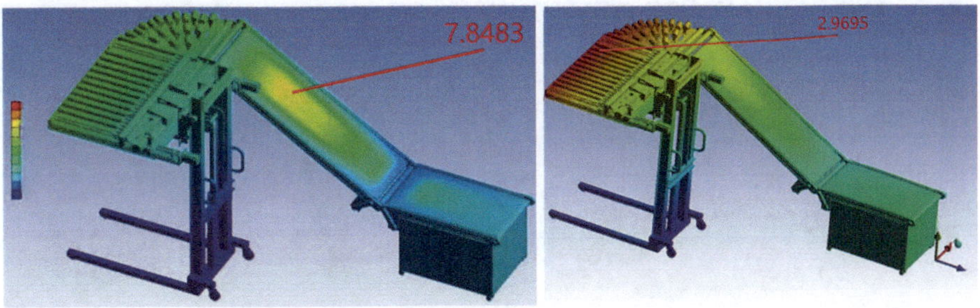

Fig. 5 Total deformation diagram of the whole machine

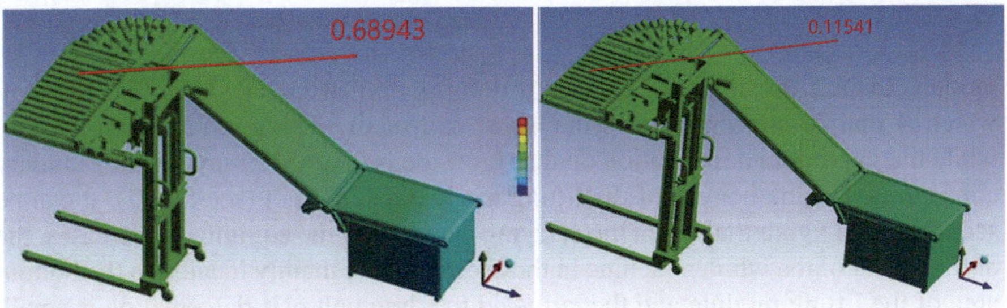

Fig. 6 Machine X direction deformation diagram

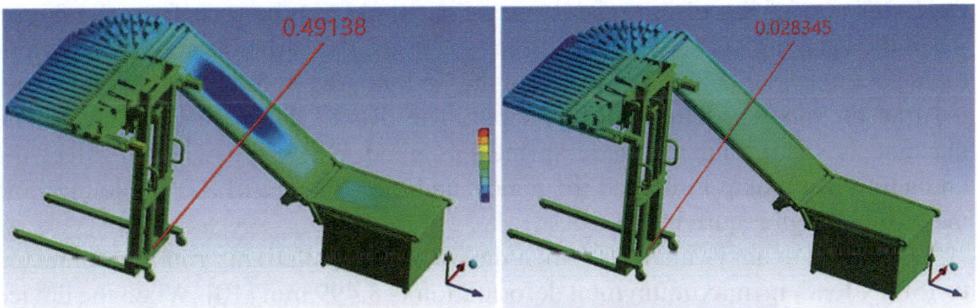

Fig. 7 Y-direction deformation diagram of the whole machine

machine is mainly located in the trailing transmission part. Figs. 6, 7 and 8 are the deformation diagrams of the equipment in the X, Y and Z directions. The maximum deformation of the equipment in the X direction is 0.68943 mm on the left side, which is mainly concentrated in the front end of the transfer module, while the maximum deformation on the right side is 0.11541 mm, which is mainly concentrated in the front of the transfer module (see Fig. 6). It can be seen from the figure that when the load capacity of the equipment increases, The X-direction deformation of the whole machine is mainly located in the front end area of the transfer

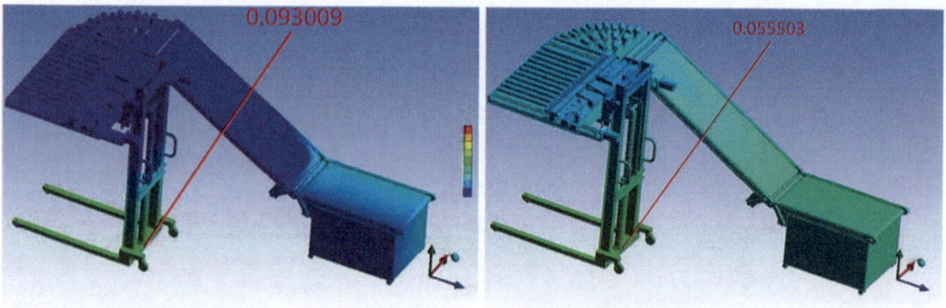

Fig. 8 Complete machine Z direction deformation diagram

module. In the Y direction, the maximum deformation on the left side is 0.49138 mm, which is mainly concentrated in the lower half of the lifting module and the tail, while the maximum deformation on the right side is 0.028345 mm, which is mainly concentrated in the bottom of the lifting module and the tail (see Fig. 7). It can be seen from the figure that when the carrying capacity of the equipment increases, the deformation of the whole machine in the Y direction is mainly located in the bottom area of the lifting module and the tail. And the drag belt part deformation is small. The maximum deformation on the left side of the equipment in the Z direction is 0.093009 mm, mainly concentrated in the lower part of the lifting module, and the maximum deformation on the right side is 0.055503 mm, mainly concentrated in the lower part of the lifting module (see Fig. 8). As can be seen from the figure, when the carrying capacity of the equipment rises, the Z-direction deformation of the whole machine is mainly located in the bottom area of the lifting module. And the transfer module deformation is small. The overall deformation of the whole machine is small, and the static stiffness is good [9], which will not affect the unloading equipment to unload the goods, and the structure of the whole machine meets the design requirements.

According to ansys analysis, when the joint type palletizing robot is subjected to 1500N load, its maximum total deformation is 8.499 mm [10]. When the device is subjected to 3234N, its maximum deformation is 7.8483N. Compared with the joint type palletizing robot, this device has better weighing and smaller deformation.

3 Conclusion

In this paper, through the finite element analysis of the whole equipment, the stress and strain of the whole equipment is determined. In order to ensure the stability, safety and reliability of the cargo unloading equipment, the following measures need to be applied to the main stress corresponding area when the load capacity of the equipment rises, as analyzed above, to disperse the stress of the equipment and reduce the deformation of the equipment. Specific measures include: Ensure the machining accuracy of the equipment to prevent pre-deformation caused by processing; Increase the material wall thickness of the equipment to improve the stiffness of the equipment; Support bars are added to the bottom of the transfer module and lifting module to disperse the stress; Change the material from 6060al to a higher strength material; Adjust the structure design, the front section of the transfer module and the bottom of the lifting module are designed as a circular arc to reduce the concentrated stress; Increase the number of wheels on the out-of-shape module to reduce the contact pressure.

References

1. Liang W, Li X, Meng C, Lin Y, Zeng L, Lan X, Yan C (2024) Harmonic response analysis of GIS bus shell and optimization strategy of vibration source based on ANSYS. Guangxi Electric Power 02:57–66. https://doi.org/10.16427/j.cnki.issn1671-8380.2024.02.009
2. Sun P, Miao T, Li Y, Zhao H, Zhou H, Tang S (2024) Simulation analysis of testing machine beam column structure based on ANSYS. Eng Test (04):52–56. https://doi.org/10.3969/j.issn.1674-3407.2024.04.017
3. Dong Z, Yang J, Fan F, Wang L (2024) Study on stress corrosion properties of high strength alloy steel. Dev App Mater (03):56–60. https://doi.org/10.19515/j.cnki.1003-1545.2024.03.007
4. Dong Q, Wan J, Peng G, Hu L, Luo Y, Peng Y-Z, Zhao Q-c (2024) Preparation and properties of superhydrophobic self-cleaning surface of 6061 aluminum alloy. J Xuzhou Eng Inst (Natural Science Edition) (04):55–61. https://doi.org/10.15873/j.cnki.jxit.000588
5. He P, Zhang D, Li J, Wang Y, He J, He W (2024) Design and simulation of bamboo basket breaking tool based on ANSYS. J Chin Agric Mech (05):116–121 + 271. https://doi.org/10.13733/j.jcam.issn.2095-5553.2024.05.018
6. Li H, Hao J, Ding H, Zhou Q (2024) Finite element static strength and modal analysis of vibrating screen frame based on ANSYS. Food Process 05:50–54. https://doi.org/10.20170/j.cnki.lsjg.1007-6395.20240511
7. Xiong Y, Liu Y, Cheng W, Gong Q, Chen B (2020) Static analysis of uniform tangential plane force of rotating shaft parts based on ANSYS. Modern Manuf Technol Equip 03:63–65. https://doi.org/10.16107/j.cnki.mmte.2020.0240
8. Teng F, Li Y (2024) Structural design of bionic knee joint robot. J Changchun Univ (10):7–12. http://kns.cnki.net/kcms/detail/22.1283.G4.20241219.1533.004.html
9. Sun W, Hu J, Li M, Yang J, Zhang Q, Xiang J, Lan W (2022) Static analysis of the main transmission components of the lock based on ANSYS. Agric Equip Vehicle Eng (12):158–163. https://doi.org/10.3969/j.issn.1673-3142.2022.12.031
10. Xuguang X (2024) Research on structure design simulation and joint control of palletizing robot. Master's Thesis. North University of China. https://link.cnki.net/doi/10.27470/d.cnki.ghbgc.2024.000829

Open Access This chapter is licensed under the terms of the Creative Commons Attribution-NonCommercial-NoDerivatives 4.0 International License (http://creativecommons.org/licenses/by-nc-nd/4.0/), which permits any noncommercial use, sharing, distribution and reproduction in any medium or format, as long as you give appropriate credit to the original author(s) and the source, provide a link to the Creative Commons license and indicate if you modified the licensed material. You do not have permission under this license to share adapted material derived from this chapter or parts of it.

The images or other third party material in this chapter are included in the chapter's Creative Commons license, unless indicated otherwise in a credit line to the material. If material is not included in the chapter's Creative Commons license and your intended use is not permitted by statutory regulation or exceeds the permitted use, you will need to obtain permission directly from the copyright holder.

Long Time Simulation of Flow Dynamics Based on Transfer Learning and Fourier Neural Operator

Tianhao Ren, Congfa Zhang, and Zhicheng Zhu

Abstract The use of deep learning for fluid dynamics simulation has become a hot topic. However, whether physics informed neural network or neural operator are used to solve the long time-dependent partial differential equation, the accuracy of the predicted solutions inevitably decreases gradually with time. We propose a transfer learning enhanced Fourier neural operator to improving accuracy and stability of Fourier neural operator for the long term prediction of partial differential equations. In the experiments of 2D cavity flow, our method reduces the error at the 1000th time step by 87.83% compared to the traditional Fourier neural operator in autoregressive prediction.

Keywords Computational fluid dynamics · Fourier neural operator · Transfer learning · Direct numerical simulation

1 Introduction

Computational Fluid Dynamics (CFD) is a discipline that uses numerical methods and computer technology to simulate fluid flow, heat transfer and related physical phenomena. It simulates the dynamic properties of complex fluid systems by solving the Navier-Stokes equations (N-S equations), which are the fundamental equations controlling fluid behaviour, in combination with discretisation techniques such as finite difference, finite volume or finite element methods. Although significant progress has been made in the past with traditional CFD methods, noise in the data and mesh generation continue to be a challenge that affects the accuracy and cost of simulation calculations, and thus deep learning has been introduced into the field of fluid mechanics [1].

T. Ren · Z. Zhu (✉)
Beijing Institute of Technology, Beijing, China

C. Zhang
Institute of Spacecraft System Engineering, Beijing, China

© The Author(s) 2026

Z. Zhang, P. Mercorelli (eds.), *Mechanical Design and Simulation: Insights and Innovations*, Lecture Notes in Mechanical Engineering,
https://doi.org/10.1007/978-981-95-2405-1_25

Currently, there are two main deep learning methods applied to the field of fluid mechanics. One is Physical Information Neural Network (PINN). In 2019 Raissi [2] et al. proposed PINN, which fuses the partial differential equation (PDE) into the loss function to solve the nonlinear problem as well as the problem of too small sample size. In 2020 Mao [3] et al. firstly investigated the use of PINN to approximately simulate the high-speed aerodynamic flow of the Euler equations. In the same year Rao et al. [4] proposed a mixed-variable scheme of PINN for fluid dynamics and applied it to simulate steady-state and transient laminar flows at low Reynolds numbers. In order to overcome some limitations of incompressible laminar and turbulent flows in simulation. Jin et al. [5] proposed a Navier-Stokes flow network (NSFnets), which is a solver that does not require mesh generation and at the same time is able to simulate both laminar and turbulent problems very well. Eivazi et al. [6] used the PINN to successfully solve incompressible turbulent flows with the Reynolds Mean Navier-Stokes equations for incompressible turbulence. Eivazi et al. [7] applied PINN to super-resolve the flow field data from a limited set of noisy measurements in both time and space, thus demonstrating that PINN is sufficiently capable in the context of data augmentation of hydrodynamics experiments.

To improve the speed of solving NS equations, another approach is to generalise neural networks to neural operators that learn mappings between function spaces [8]. For PDEs, neural operators directly learn the mapping from any function parameter dependence to the solution. Thus, they learn the entire family of partial differential equations, unlike classical methods that solve only one instance of an equation. The earliest proposed neural operator is DeepONet [9], which is able to learn mappings between function spaces and successfully solves many parametric PDEs. Based on this, Li et al. [10] proposed Fourier neural operator (FNO) by using Fourier transform as a graph kernel and obtained encouraging results in solving N-S equations. Wang et al. [11] trained DeepONet in a physics-aware manner, thus improving the sample efficiency.

FNO have been widely used in flow simulation and prediction due to their unique learning ability. 2022 Wen et al. [12] proposed U-FNO to simulate multiphase flow based on FNO, Li [13] et al. extended FNO to simulate three-dimensional turbulent eddy simulation, and Li [14] et al. enhanced FNO with implicit U-Net for long-term prediction of turbulence. FNO has been gradually applied to the fields of airflow dynamics of aircraft, airflow dynamics of automobile engines, and fluid analysis of underground oil reservoirs [15–17].

However, whether PINN or neural operators are used to solve the long time-dependent PDE system, the accuracy of the predicted solutions inevitably decreases gradually over time, and there are two main reasons here: (1) the trained model may accumulate errors in the long time prediction, which leads to the inaccurate results; (2) during the time evolution of the PDE, the functions of the propagator inputs and outputs may vary with the time, affecting the prediction results. To address the above challenges, Liu [18] et al. first proposed a DeepONet-based recursive propagator for learning evolutionary equations, Lyu [19] et al. improved the model accuracy by combining multi-fidelity data under a transfer learning paradigm, Xu et al.

[20] proposed a new transfer learning-based DeepONet method based on physical information to improve the accuracy of long-term prediction solutions. Inspired by this, we try to apply transferlearning to FNO and supplement it with Direct Numerical Simulation(DNS)to enhance the speed and accuracy of FNO for long-term prediction of partial differential equations.

We propose a new transfer learning enhanced FNO for predicting time-dependent partial differential equations. We pre-train a propagator for learning partial differential equations, and during the prediction of subsequent times, whenever the error metric reaches a certain level, the model automatically calls DNS to generate a small amount of accurate data, and re-trains the network parameters to continuously update the propagator through transfer learning. The resulting time-varying FNO has the following advantages over ordinary FNO: (1) The FNO after transfer learning is able to make accurate long-term predictions of the solutions of time-dependent partial differential equations; (2) The combination of DNS and FNO can accelerate the prediction process and improve computational efficiency.

The experimental results show that our method is applicable to the N-S equation, which improves the error accumulation of the conventional FNO when performing long-time flow field simulations. When autoregressive prediction is performed, the error at the 1000th time step is reduced by 87.83% (residual of PDE).

2 Methods

2.1 Fourier Neural Operators

The Fourier transform is a commonly used spectral method for solving partial differential equations, which converts a partial differential equation from the spatial domain to the frequency domain, solves the equation in the frequency domain, and then obtains the solution of the original equation through the Fourier inverse transform. Since multiplication in the frequency domain is equivalent to differential computation in the spatial domain, the Fourier transform can greatly accelerate the solution of partial differential equations.

FNO is a neural network architecture inspired by the Fourier Transform [11], which is capable of learning mappings between infinite dimensional functions. The main idea of FNO is to add Fourier layers to alleviate the computational cost of performing integral operations of neural operators by utilising FFTs in order to accelerate neural network computation.

As shown in Fig. 1, the overall structure of the FNO mainly contains the input function a(x), neural network P, Fourier layer, neural network Q, and output function u(x). Firstly, the input function a(x) is projected to the higher dimensional channel space through the neural network P, and then the output is projected back to the solid space by the neural network Q to obtain the output function u(x) after processing by the Fourier layer. Among them, the Fourier layer comprises a fast Fourier transform, a linear transform, a Fourier inverse transform, a convolution layer, and

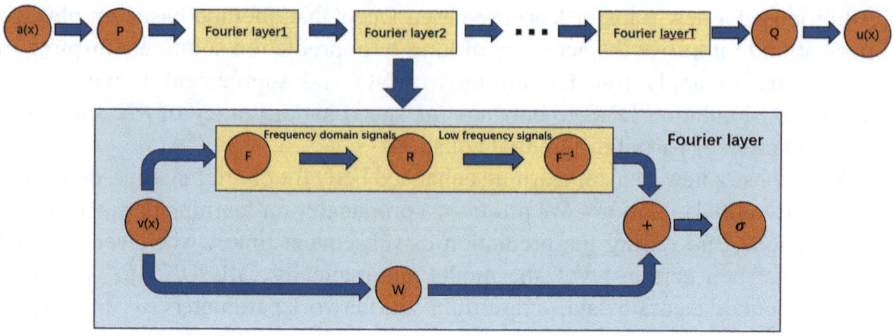

Fig. 1 Fourier neural operator

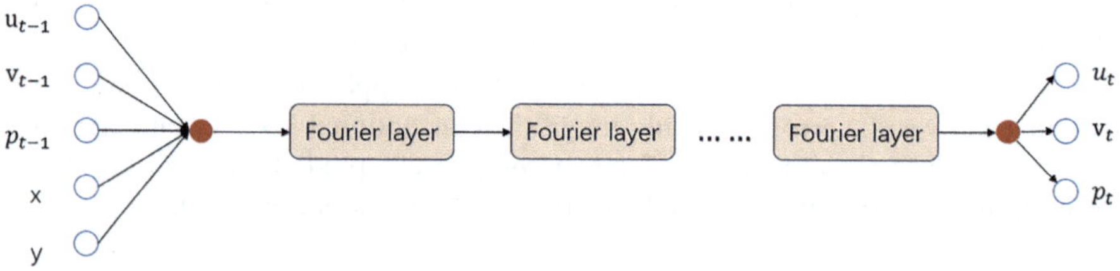

Fig. 2 Structure of the FNO solver

a nonlinear activation function, and in the Fourier layer, the input function v(x) on the one hand enters the frequency domain through the Fourier transform F. At this time, the linear transform R will carry out a high-frequency waveform thereof, and then subsequently undergoes the Fourier inverse transform processing. On the other hand, v(x) is linearly transformed by the convolution layer W, and the result is added with the above Fourier inverse transform result, and finally the activation function completes the nonlinear transformation.

Defining the Solver FNO has similar timing prediction ability as RNN and is often applied to flow field simulation and prediction. In this paper, we adopt the typical structure of FNO, keep the four-layer Fourier layer structure unchanged, and extend the input and output channels. For example, for a two-dimensional cavity flow, we use the velocity u_{t-1}, v_{t-1} and pressure p_{t-1} and the position x, y at the previous moment as the inputs, and predict the velocity u_t, v_t and pressure p_t at the next moment, and the specific structure is shown in Fig. 2.

2.2 Transfer Learning

Transfer learning is an important current technique in the field of machine learning that allows a pre-trained model to be adapted or migrated to a new and relevant task. The core idea of this approach is to use the knowledge of a well-trained model on a large dataset to improve the efficiency and performance of training on another relevant but different dataset. The process is divided into two main steps: the first step is pre-training on the source task with a large amount of data, and the second step is to apply the knowledge learnt from the source task, including feature relationships, to the target task.

Transfer learning can greatly reduce the annotation data requirements and training costs, while improving the learning efficiency and accuracy of the model, and has been widely used in small datasets, cross-domain and complex high-dimensional problems. However, how to effectively implement transfer learning, select appropriate source tasks and deal with domain differences are still hot research topics in the field. In this paper, we mainly adopt a fine-tuned transfer learning strategy by periodically retraining the last few layers of parameters of the pre-trained model for more accurate long-term prediction.

2.3 Transfer Learning Enhanced Fourier Neural Operator (TL-FNO)

Transfer learning is an approach in machine learning that allows a model to apply knowledge learnt from one task to another related task. This approach is particularly useful in situations where data is scarce, as it reduces the need for large amounts of labelled data, and transfer learning has become a hot topic in the field of deep learning and artificial intelligence. We mainly use a fine-tuned learning strategy: firstly, we train FNO with a large amount of data to get a pre-trained FNO model; then, we use the physical loss function as the model prediction index, we find that when the model predicts about 200 steps, the physical loss function will exceed 0.2, so we set the iteration number, and we set the number of iterations. Therefore, we set the number of iterations to call DNS solver after every 200 steps of FNO iteration, and DNS will continue to generate 50 steps of new data on the basis of the original prediction data; finally, we use the new data to re-train the model, and fine-tune the parameters of the last layer of the network, so as to make it better for long term prediction, and the specific flow of the transfer learning is shown in Fig. 3.

3 Experiments and Results

We verify the validity of the TL-FNO using a classical test case of a roof-driven cavity.

The roof-driven square-cavity flow is a benchmark case for validating computational methods for fluid dynamics problems. The model consists of a two-dimensional empty square cavity as shown in Fig. 4, where the fluid has tangential velocity at the top of the square cavity and zero velocity on the remaining three walls, and is characterised by a large vortex in the centre of the cavity and a number of small vortices in the corners.

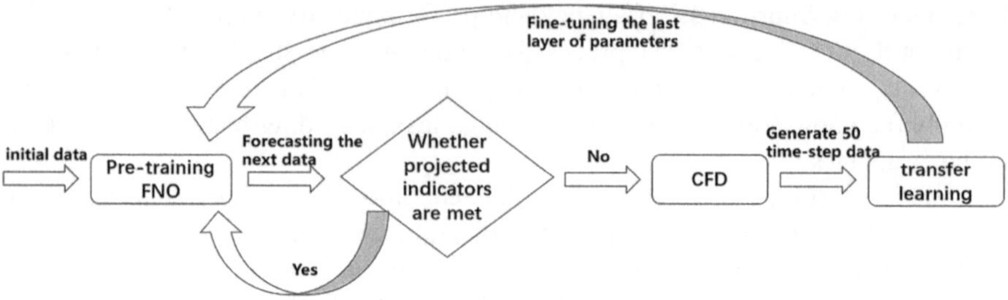

Fig. 3 Architecture of transfer learning enhanced Fourier neural operator

Fig. 4 Top cover driven square cavity flow

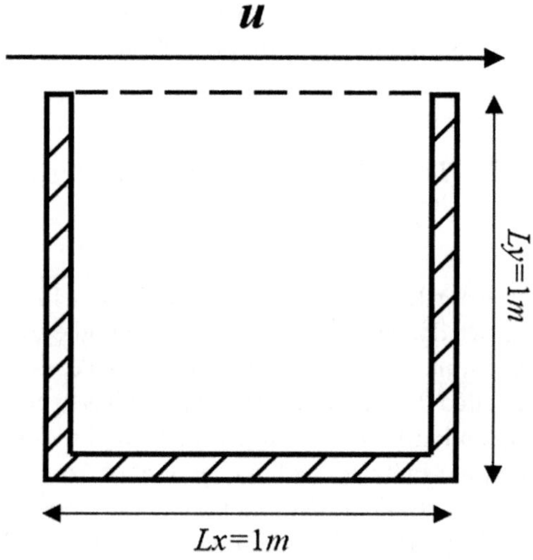

We divide the solution domain into a 100*100 rectangular grid, set the tangential velocity u = 1 and pressure p = 0 at the top boundary, and set the velocity and pressure gradient to zero at the left, right and bottom boundaries. Using the numerical simulation experiment DNS to generate 1000 sets of data, the first 500 sets of data are used as the training set and the last 500 sets of data are used as the test set.

Firstly, we use the 2D FNO as a pre-training model, set the optimizer as Adam, the learning rate as 0.001, and the loss function as L2 paradigm, and after 1000 epochs, the change curve of the loss function is shown in Fig. 5.

Immediately after that we use the pre-trained model for prediction. In each iteration cycle, FNO performs 200 prediction iterations and then calls the numerical simulation module (DNS), which reads the velocity prediction from the previous time step as the intermediate velocity, then solves the Poisson equation for the pressure to obtain the real pressure field, then corrects the intermediate velocity with the real pressure data to obtain the real velocity data, and so on, to generate 50 new steps of data. Finally, we use the new data to perform transfer learning for the model. The model reads the new data for 50 time steps, and is trained with 200 epochs of fine-tuning using Adam as the optimizer, 1e-5 as the learning rate, the last layer

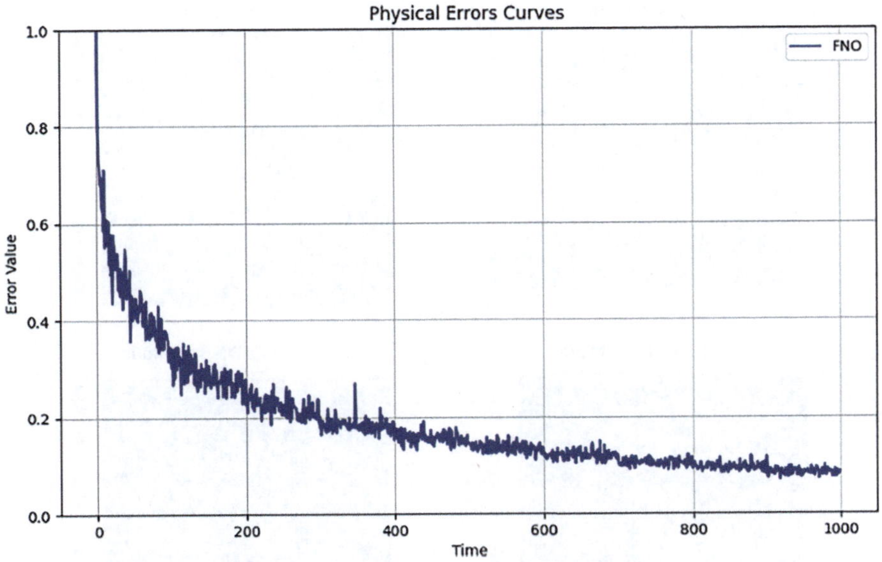

Fig. 5 Loss function variation curves during the pre-training process

parameter as the optimisation object, and the L2 paradigm as the loss function. In order to demonstrate the improvement effect of our method for the long-term prediction of FNO, we conducted a comparative analysis of the prediction results before and after the migration learning, and plotted the prediction cloud diagrams at the 500th and 1000th time steps under the four experimental conditions as shown in Fig. 6, and the corresponding physical loss function curves as shown in Fig. 7, respectively.

It can be seen from Fig. 6 that without transfer learning, the prediction results of FNO start to diverge around t = 500, while FNO-DNS, TL-FNO and TL-FNO-DNS still have good prediction results at t = 1000.

It can be seen from Fig. 7 that the physical loss function continues to grow when FNO alone is used for prediction, reaching about 0.8 at 1000 steps. While adding DNS numerical simulation and transfer learning respectively to the prediction process of FNO are both able to reduce the physical error to a certain extent. Among them, DNS can control the physical error to about 0.2 because it can eliminate the cumulative error of long-term prediction, while transfer learning can control the physical error to less than 0.1 because it can optimise the model periodically. And the long-term prediction effect is optimal when both of them work at the same time.

Finally, we briefly analyse the experimental results and draw conclusions:

1. Online transfer learning can dynamically update the model to adapt to changing time-step prediction needs, and transfer learning can efficiently update the parameters while retaining the learnt knowledge, which significantly reduces the error and improves the stability of long-term prediction.

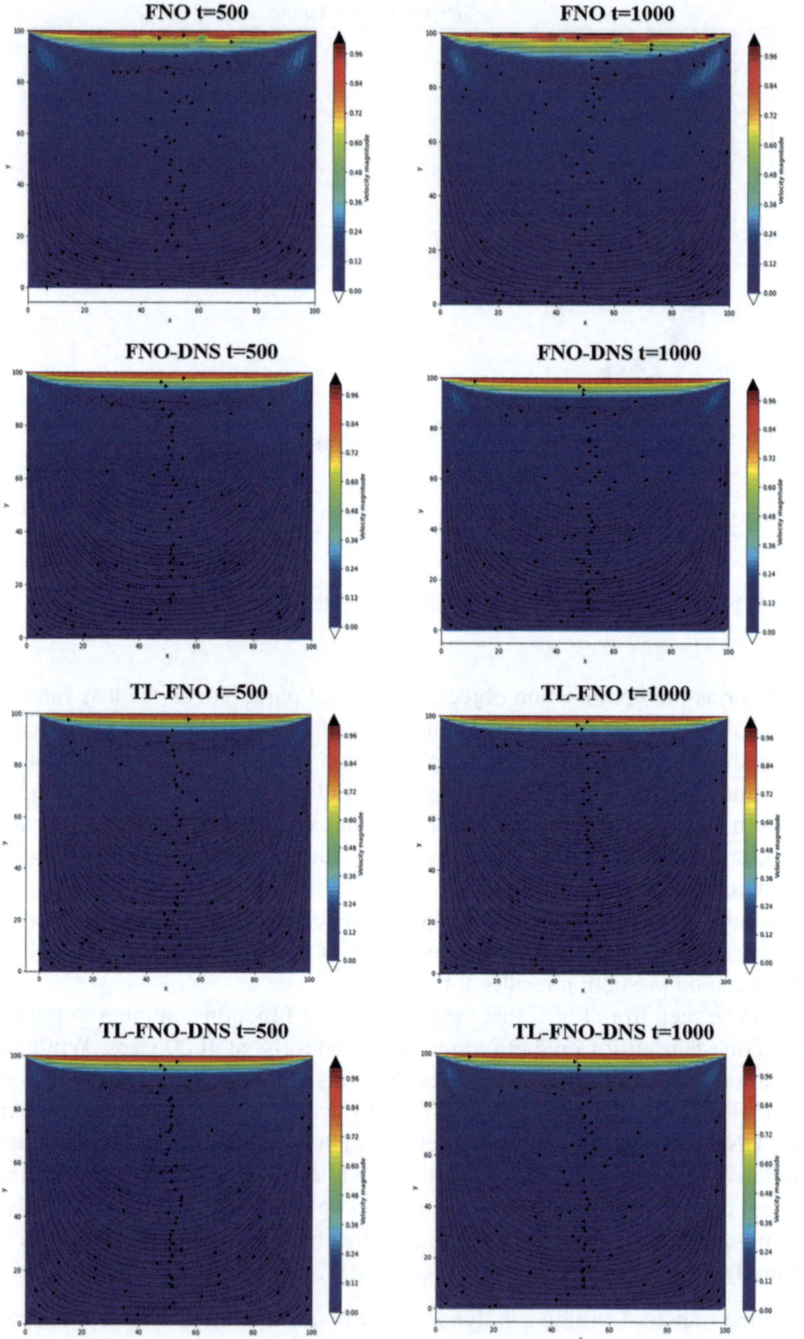

Fig. 6 Clouds of predicted results for the four experimental conditions

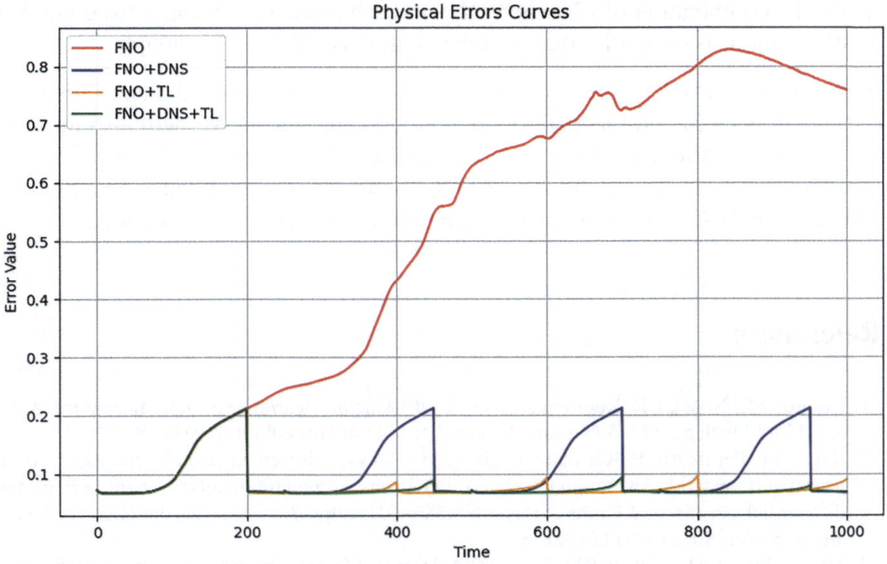

Fig. 7 Physical loss function curves for four experimental conditions

2. The data generated by DNS meets the constraints of physical conditions, such as Navier-Stokes equations and boundary conditions, and is therefore closer to reality. Therefore, the addition of DNS to the FNO model prediction process can effectively block the error propagation chain and reduce the cumulative error.
3. The experimental results show that both transfer learning and DNS can reduce the physical error. The common point of the two is that both add physical constraints to the model to a certain extent, which reduces the chain effect of error propagation, while compared with DNS, migration learning uses accurate data to correct the model for errors step by step, so the improvement effect of migration learning is better.

4 Conclusion

The application of deep learning in DNS has become a current and important trend. In this work, we combine FNO, and transfer learning, to predict the results with FNO, to generate accurate data with DNS, and transfer learning is used to fine-tune the model periodically to improve the long-term prediction accuracy of the model. When autoregressive prediction was performed, the error at the 1000th time step was reduced by 87.83%.

However, our work has some shortcomings:

1. our method is currently not validated on hydrodynamic problems in three dimensions;

2. for the combination of FNO and DNS in each prediction cycle, a fixed number of iterations is currently used, and the method is not flexible enough.

In the future, we will continue to work in these two directions and plan to verify its effectiveness on 3D problems through 3D fluid experiments. We will also improve the migration learning method, with the expected effect that when the error reaches a certain threshold during the model prediction process, the model will automatically call the DNS-generated data for fine-tuning of the migration learning.

References

1. Brunton SL, Noack BR, Koumoutsakos P (2020) Machine learning for fluid mechanics. Annu Rev Fluid Mech 52:477–508. https://doi.org/10.1146/annurev-fluid-010719-060214
2. Raissi M, Perdikaris P, Karniadakis GE (2019) Physics-informed neural networks: a deep learning framework for solving forward and inverse problems involving nonlinear partial differential equations. J Comput Phys 378:686–707. https://www.sciencedirect.com/science/article/abs/pii/S0021999118307125
3. Mao Z, Jagtap AD, Karniadakis GE (2020) Physics-informed neural networks for high-speed flows. Comput Methods Appl Mech Eng 360:112789. https://www.sciencedirect.com/science/article/abs/pii/S0045782519306814
4. Rao C, Sun H, Liu Y (2020) Physics-informed deep learning for incompressible laminar flows. Theor Appl Mech Lett 10(3):207–212. https://www.sciencedirect.com/science/article/pii/S2095034920300350
5. Jin X, Cai S, Li H et al (2021) NSFnets (Navier-Stokes flow nets): physics-informed neural networks for the incompressible Navier-Stokes equations. J Comput Phys 426:109951. https://www.sciencedirect.com/science/article/abs/pii/S002199912 0307257
6. Eivazi H, Tahani M, Schlatter P et al (2022) Physics-informed neural networks for solving Reynolds-averaged Navier–Stokes equations. Phys Fluids 34(7) https://pubs.aip.org/aip/pof/article/34/7/075117/2847279
7. Eivazi H, Wang Y, Vinuesa R (2024) Physics-informed deep-learning applications to experimental fluid mechanics. Meas Sci Technol 35(7):075303. https://iopscience.iop.org/article/10.1088/1361-6501/ad3fd3/meta
8. Kovachki N, Zongyi Li N, Burigede Liu C, Azizzadenesheli K, Kaushik Bhattacharya N, Stuart A, Anandkumar A, Editor C, Rosasco L, Li Z, Liu B, Bhattacharya K (2023) Neural operator: learning maps between function spaces with applications to PDEs. J Mach Learn Res 24(1):4061–4157. https://dl.acm.org/doi/10.5555/3648699.3648788
9. Lu L, Jin P, Pang G et al (2021) Learning nonlinear operators via DeepONet based on the universal approximation theorem of operators. Nat Mach Intell 3(3):218–229. https://www.nature.com/articles/s42256-021-00302-5
10. Li Z, Kovachki N, Azizzadenesheli K et al (2020) Fourier neural operator for parametric partial differential equations. arXiv preprint arXiv: 2010.08895. https://arxiv.org/abs/2010.08895
11. Wang S, Wang H, Perdikaris P (2021) Learning the solution operator of parametric partial differential equations with physics-informed DeepONets. Sci Adv 7(40):eabi8605. https://www.science.org/doi/full/10.1126/sciadv.abi8605
12. Wen G, Li Z, Azizzadenesheli K et al (2022) U-FNO—an enhanced Fourier neural operator-based deep-learning model for multiphase flow. Adv Water Resour 163:104180. https://www.sciencedirect.com/science/article/abs/pii/S0309170822000562
13. Li Z, Peng W, Yuan Z et al (2022) Fourier neural operator approach to large eddy simulation of three-dimensional turbulence. Theor Appl Mech Lett 12(6):100389. https://www.sciencedirect.com/science/article/pii/S2095034922000691

14. Li Z, Peng W, Yuan Z et al (2023) Long-term predictions of turbulence by implicit U-Net enhanced Fourier neural operator. Phys Fluids 35(7) https://pubs.aip.org/aip/pof/article-abstract/35/7/075145/2903750/Long-term-predictions-of-turbulence-by-implicit-U?redirectedFrom=fulltext
15. Meng D, Zhu Y, Wang J et al (2024) Koopman neural operator approach to fast flow prediction of airfoil transonic buffet. Phys Fluids 36(7) https://pubs.aip.org/aip/pof/article-abstract/36/7/075182/3304126/Koopman-neural-operator-approach-to-fast-flow?redirectedFrom=fulltext
16. Kumar V, Goswami S, Smith D et al (2024) Real-time prediction of gas flow dynamics in diesel engines using a deep neural operator framework[J]. Appl Intell 54(1):14–34. https://link.springer.com/article/10.1007/s10489-023-05178-z
17. Zhang K, Zuo Y, Zhao H et al (2022) Fourier neural operator for solving subsurface oil/water two-phase flow partial differential equation. SPE J 27(03):1815–1830. https://onepetro.org/SJ/article-abstract/27/03/1815/480880/Fourier-Neural-Operator-for-Solving-Subsurface-Oil?redirectedFrom=fulltext
18. Liu L, Cai W (2022) DeepPropNet--a recursive deep propagator neural network for learning evolution PDE Operators. arXiv preprint arXiv:2202.13429. https://arxiv.org/abs/2202.13429
19. Lyu Y, Zhao X, Gong Z et al (2023) Multi-fidelity prediction of fluid flow based on transfer learning using Fouriegr neural operator. Phys Fluids 35(7) https://pubs.aip.org/aip/pof/article-abstract/35/7/077118/2903201/Multi-fidelity-prediction-of-fluid-flow-based-on?redirectedFrom=fulltext
20. Xu W, Lu Y, Wang L (2023) Transfer learning enhanced deeponet for long-time prediction of evolution equations. In: Proceedings of the AAAI conference on artificial intelligence, vol 37(9), pp 10629–10636. https://ojs.aaai.org/index.php/AAAI/article/view/26262

Open Access This chapter is licensed under the terms of the Creative Commons Attribution-NonCommercial-NoDerivatives 4.0 International License (http://creativecommons.org/licenses/by-nc-nd/4.0/), which permits any noncommercial use, sharing, distribution and reproduction in any medium or format, as long as you give appropriate credit to the original author(s) and the source, provide a link to the Creative Commons license and indicate if you modified the licensed material. You do not have permission under this license to share adapted material derived from this chapter or parts of it.

The images or other third party material in this chapter are included in the chapter's Creative Commons license, unless indicated otherwise in a credit line to the material. If material is not included in the chapter's Creative Commons license and your intended use is not permitted by statutory regulation or exceeds the permitted use, you will need to obtain permission directly from the copyright holder.

An Interval Optimization Algorithm for Condition-Based Maintenance with Multi-methods Based on SA-BP Neural Network

Weigang Zhang, Junxian Xiao, and Yuxiong Zhang

Abstract To tackle the problem of low solving efficiency in interval optimization for multi-method condition-based maintenance, an algorithm based on fitting and random global search was put forward. Firstly, based on the traditional multi-method condition-based maintenance model, a maintenance strategy considering imperfect detection was constructed, in which the detection interval of each method was taken as the decision variable, the fault detection in time was taken as the constraint condition of the safety during the operation of the system, and the minimum expected total cost within the maximum time was taken as the decision goal. Secondly, simulated annealing algorithm was used to optimize the above model to obtain the best detection strategy. In addition, a prediction algorithm based on BP neural network is designed, and the mapping relationship between detection interval and detection cost and reliability is established, which is used to replace the recalculation part of the simulated annealing algorithm in each optimization. Finally, the proposed algorithm is verified and the accuracy of the algorithm is analyzed by taking a system degradation as an example. The analysis shows that the proposed method has a very important reference value for improving the computational efficiency of multi-method detection optimization problem under combinatorial optimization and optimizing the detection interval of multi-method detection.

Keywords Multi-method detection · Condition-based maintenance · BP neural network · Simulated annealing algorithm

---

W. Zhang
Civil Aviation University of China, College of Transportation Science and Engineering, Tianjin, China

J. Xiao (✉) · Y. Zhang
Civil Aviation University of China, College of Aeronautical Engineering, Tianjin, China

© The Author(s) 2026

Z. Zhang, P. Mercorelli (eds.), *Mechanical Design and Simulation: Insights and Innovations*, Lecture Notes in Mechanical Engineering,
https://doi.org/10.1007/978-981-95-2405-1_26

1 Introduction

In the fields of engineering and military, such as aircraft, ships, and nuclear power plants, equipment with high reliability undertakes critical missions [1]. To ensure the performance of the next mission, maintenance is required [2]. The multi-method detection condition-based maintenance strategy combines various detection means, which can timely discover system abnormalities and avoid serious consequences. Compared with single-method detection, it can reduce the total maintenance cost [3].

The optimization model of multi-method inspection and maintenance interval was first proposed by Wang Wenbin [3] to solve the problem of multiple nested inspections in production plants, and the branch and bound algorithm was used to solve it to determine the optimal interval of all inspections. On this basis, many scholars at home and abroad have carried out extended research on multi-method detection strategies. CAI Jing [4] considered the impact of different inspection levels on maintenance costs and decision making, studied the update probability of aircraft structure, and then studied the initiation and propagation of cracks by Monte Carlo simulation. Mohamad M [5] considered the failures of complex infrastructures caused by environmental or operational conditions. It is assumed that one of multiple detection methods can be selected at each detection time. Two mixed-integer nonlinear programming models are introduced to determine the optimal inspection strategy that maximizes the system reliability within its service life. Wang Wenbin [6] proposed a multi-spot inspection strategy based on the concept of three-stage fault delay time, and obtained the optimal multi-spot inspection strategy through numerical calculation when the cost of small inspection was less than the threshold. The above research shows that, for the problem of multi-method combination optimization, although different computational methods have been adopted, the computational efficiency is low and it takes a long time to find the optimal solution.

There are few studies on the efficiency optimization problem of multi-method condition-based maintenance combination solution, but researchers have extensively studied the combination solution optimization in other fields. Han Zhiwu [7] et al., in order to realize the optimization design of the wear reduction performance of the circular saw blade of Salix salix flat-stubble, obtained the wear degradation data under different sawtooth structures through field flat-stubble experiments, established a BP neural network model to predict the wear amount of the circular saw blade, and then combined with GA algorithm to find the optimal parameters of the sawtooth structure of the circular saw blade with the minimum wear amount as the optimization goal. It provides technical support for improving the work efficiency of Salix salix flat stubble. In order to improve the static characteristics of disc porous hydrostatic thrust bearings, Yan Ruzhong et al. [8] used mind evolution algorithm to optimize Back Propagation (BP) neural network to create a prediction model for the static characteristics of disc porous hydrostatic thrust bearings, and used genetic algorithm to perform multi-objective optimization on the combination of porous parameters of bearings to realize the rapid optimization design of bearing parameters. Ji Shanling et al. [9] proposed to combine multi-task Gaussian process and adaptive active learning for VM modeling in CVD systems. Subsequently, active learning methods based on different sampling criteria were improved to solve the problem of limited training data.

Therefore, based on the condition-based maintenance strategy of multi-method combination based on imperfect inspection, this paper constructs a prediction heuristic algorithm to solve the multi-method interval optimization model. Then, the discrete variables were substituted into the fitting formula and simulated annealing algorithm was used to optimize. Finally, the accuracy and efficiency of the proposed algorithm were analyzed, so as to provide reference for the design and optimization of condition-based maintenance activities under multi-method detection.

2 Problem Description

This paper studies components or equipment under a single failure mode. As shown in Fig. 1, their degradation experiences three states: normal, potential failure, and functional failure. When the system state is identified by multi-method detection, different detection methods correspond to different potential failure points P_i.

In the figure, F is the functional failure point of a component, that is, the condition that the component (or the equipment containing the component) fails to complete the specified function. $T_p = \min\left[T_{V_i,n}\right]$ is the nearest detection plan in the multi-method detection strategy, where $T_{V_i,n}$ is one of the detection methods $V_i(i = 1,2,\ldots,n)$, t_{V_i} is the detection interval and the detection time of the nth detection of the device. In $T_{V_1,1} < T_{V_2,1} < \ldots < T_{V_n,1}$, for example, $\min\left[T_{V_1,1}, T_{V_2,1}, \ldots, T_{V_n,1}\right]$ after $T_1 = T_{V_1,1}$, and then update testing plan, $T_{V_i,n}$ relationship may become $T_{V_2,1} < T_{V_1,2} < T_{V_3,1} < \ldots$, $\min\left[T_{V_1,2}, T_{V_2,1}, \ldots, T_{V_n,1}\right]$ after $T_1 = T_{V_2,1}$, and so on.

In order to describe the corresponding relationship between multiple detection methods and potential faults, let t_{P_i} be the occurrence time of the i-th potential fault

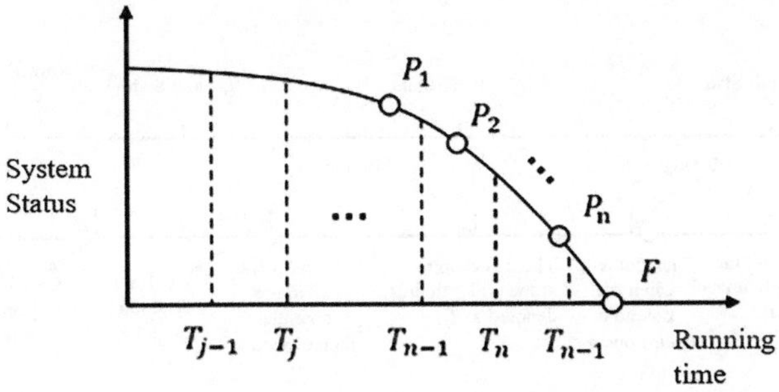

Fig. 1 Different potential failure points of the same failure mode before functional failure

point, and t_f be the occurrence time of equipment function failure, Then the relationship between the two is as follows:

$$T_{V_i,n} > t_{p_i}, \tag{1}$$

That is, the ith detection method can detect and identify the i th to n th potential faults.

The device status is identified through the above relationship, taking into account the error probability θ_{V_i} that may be caused by imperfect detection. If it is detected that the system has developed from P_i to the H_i fault state of P_{i+1}, preventive maintenance is carried out; if the detection misses a potential fault, causing it to evolve into a functional fault, corrective maintenance is carried out. The following analyzes the update of these two states.

H_i failure status update. When $T_{j-1} < t_{P_1} < T_j \cap t_{P_i} < T_n < t_{P_{i+1}} \cap t_{P_i} < T_{V_i,n}$, that is, the detection can detect that the equipment is in the potential failure state H_i, and preventive maintenance is carried out. In this case, assume that the multiple detection strategy $S = [V_1, V_2, \ldots, V_n]$ has undergone $[n, m, \ldots, l]$ maintenance activities since the last maintenance, and there may be missed detections, as shown in Fig. 2 below. Then the maintenance cost in this case can be expressed as follows:

$$C\left(t_{V_1}, t_{V_2}, \ldots, t_{V_i}\right) = n \times c_{V_1} + m \times c_{V_2} + \ldots + l \times c_{V_n} + c_{H_i} \tag{2}$$

In the formula, c_{V_i} is the single detection cost when the V_i detection method is adopted; c_{H_i} is the cost required to repair the faults found during the inspection and restore the equipment to a new state.

F failure status update. When $T_{j-1} < t_{P_1} < T_j \cap t_f \le T_n$, that is, the component experiences a functional failure between the detections at T_{n-1} and T_n, and corrective maintenance is carried out. In this case, assume that the multiple detection strategy $S = [V_1, V_2, \ldots, V_n]$ has undergone $[n-1, m-1, \ldots, l-1]$ maintenance activities since the last maintenance, and there will be one failure replacement activity, and there may be missed detections, as shown in in Fig. 3 below. Then the maintenance cost in this case can be expressed as follows:

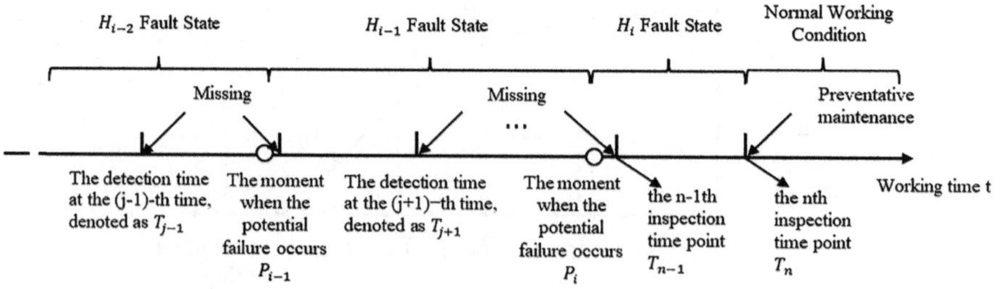

Fig. 2 Multiple methods detect H_i fault state updates

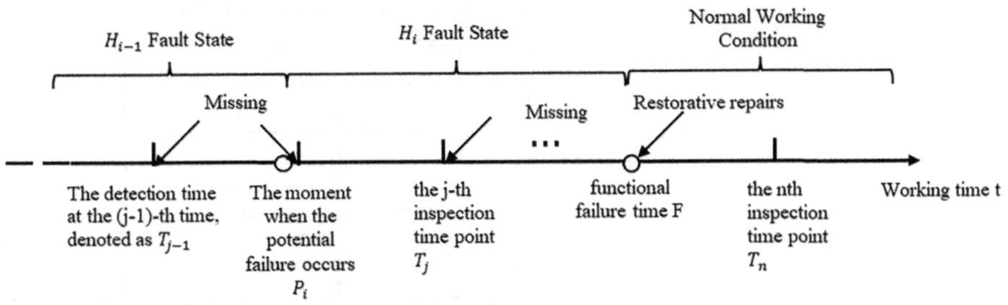

Fig. 3 F fault state update under multi-method detection

$$C\left(t_{V_1},t_{V_2},\ldots,t_{V_i}\right)=(n-1)\times c_{V_1}+\ldots+(l-1)\times c_{V_n}+c_f \tag{3}$$

To sum up, taking the detection intervals of each method as optimization variables, the total maintenance cost as the optimization objective, and the reliability as the constraint condition, the interval optimization model based on multi-method condition-based maintenance is established as follows:

$$\begin{cases} \min.\overline{C\left(t_{V_1},t_{V_2},\ldots,t_{V_i}\right)} = \left.\sum_{i=1}^{N}\sum_{i=1}^{k} C\left(t_{V_1},t_{V_2},\ldots,t_{V_i}\right)\right/ N \\ s.t.\begin{cases} R\left(t_{V_1},t_{V_2},\ldots,t_{V_i}\right)=1-\left[\left.\left[COUNT\left(T_{j-1}<t_{p_1}<T_j\cap t_f\geq T_n\right)\right]\right/N\right]\geq R_0 \\ T_{V_i,n}>t_{P_i} \end{cases} \end{cases} \tag{4}$$

where: $\overline{C\left(t_{V_1},t_{V_2},\ldots,t_{V_i}\right)}$ is the total unit maintenance cost of the component; N is the number of simulations; k is the number of detection devices; $R\left(t_{V_1},t_{V_2},\ldots,t_{V_i}\right)$ is the reliability of the equipment; COUNT is the counting operator.

3 Inspection Interval Optimization Algorithm Based on BP Neural Network and Simulated Annealing

For the linear problem in the model, BP neural network can efficiently output accurate prediction values through training, and establish a complex relationship model between input and output. However, it is not suitable for optimization and solution. The simulated annealing algorithm (SA) is an efficient global optimization method that can avoid the local optimal trap and find the global optimal solution [10, 11]. However, its optimization speed is relatively slow. In this paper, by combining the advantages of both, the predictive ability of the BP neural network and the optimization ability of SA are utilized to find the optimal solution for the multi-method

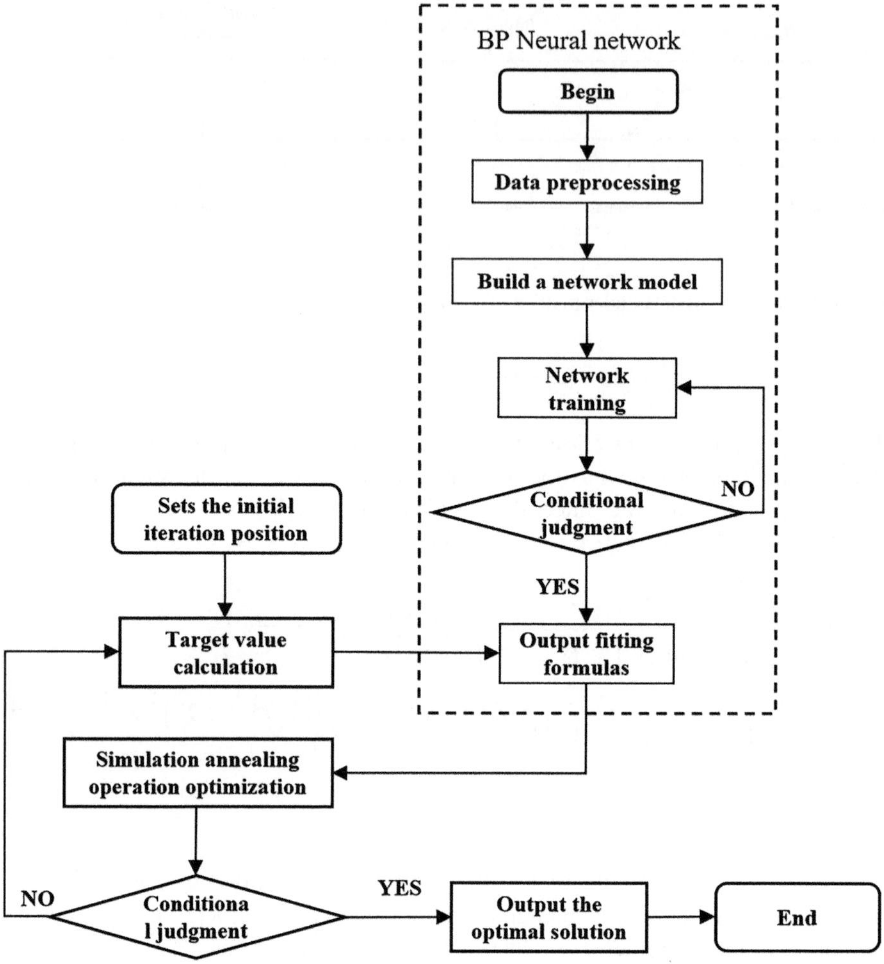

Fig. 4 Flow chart of the BP-GA-algorithm

detection strategy. The process of the BP-SA algorithm is divided into two steps: BP modeling and SA optimization, as shown in Fig. 4.

3.1 Generation of Training Data

In order to test the accuracy of the BP network, a large amount of data is needed for training to identify the mapping relationship and verify the accuracy. The discrete simulation method using the enumeration algorithm is adopted to provide valid data. Based on the practical application in engineering, taking the detection of an equipment system by the combination of three detection methods as an example, it is assumed that the smaller the deviation of the detected potential fault point from the normal working state, the lower the update cost but the higher the detection cost. According to the relevant parameters (Table 1), 8379 groups of sample data are obtained through simulation (Table 2). Among them, the unit of cost is yuan, and the unit of time is hour.

3.2 The Construction of BP Neural Network

Before optimizing the multi-method detection intervals, it is necessary to establish a relational model between the detection intervals, the total cost, and the reliability, and train the BP neural network with discrete simulation data. The network structure consists of 3 input units, 10 hidden layer units (determined based on the R coefficient analysis, as shown in the Fig. 5, When the number of hidden layer nodes is 10, the coefficient of determination no longer increases significantly. Therefore, the number of hidden layers is chosen to be 10), and 2 output units (total cost and reliability). The Levenberg-Marquardt algorithm (trainlm) is selected for backpropagation training to complete the construction of the BP neural network.

3.3 Simulated Annealing Algorithm Optimization

During the optimization process, the core idea of simulated annealing is to control the scope of the search process by gradually reducing the "temperature". In this paper, the exponential annealing method is used for temperature decrease. The temperature decay function is:

Table 1 Check methods and update parameters

c_{V_1}	c_{V_2}	c_{V_3}	θ_{V_1}	θ_{V_2}	θ_{V_3}	c_{H_1}	c_{H_2}	c_{H_3}	c_f
900	90	10	95	75	55	1000	3000	6000	20,000

Table 2 Multi-method detection of data samples

Samples	t_{V_1}	t_{V_2}	t_{V_3}	$\overline{C}$	R
1	145	115	18	887,095	0.9855
2	195	80	28	991032.9	0.9296
3	160	80	22	933787.2	0.9601
4	190	125	28	1,080,980	0.8869
⋮	⋮	⋮	⋮	⋮	⋮
8379	200	60	28	957634.2	0.9524

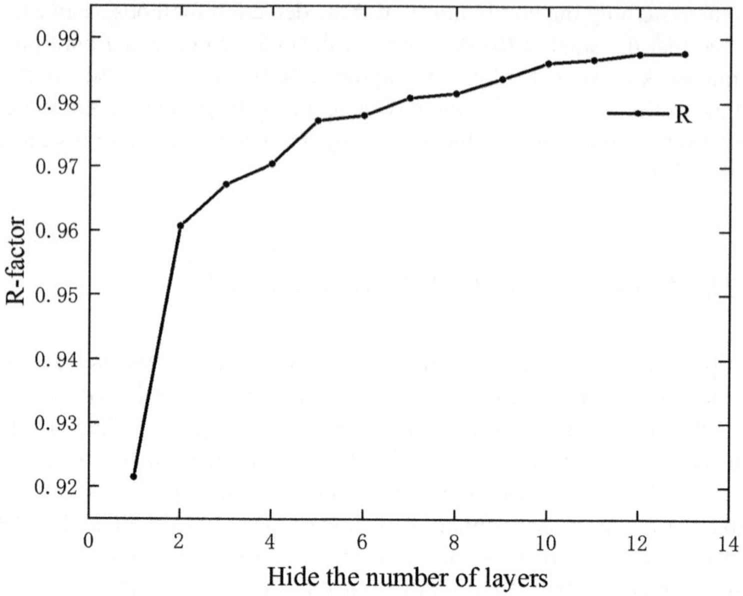

Fig. 5 Plot of the change of R factor

$$T_{k+1} = \alpha T_k, \tag{5}$$

Where: T_{k+1} is the system temperature at the k + 1-th iteration; α is the cooling coefficient; T_k is the system temperature at the k-th iteration.

In the simulated annealing algorithm, in order to avoid the algorithm falling into local optimal solutions, the Metropolis criterion is applied to determine whether to replace the current individual with a new individual. The Metropolis criterion is as follows:

$$P = \begin{cases} 1, & \eta_{new} \geq \eta_{current} \\ \exp\left(\dfrac{\eta_{new} - \eta_{current}}{T_k}\right), & \eta_{new} < \eta_{current} \end{cases} \tag{6}$$

Where P is the probability that the current individual is replaced by the new individual; η_{new} represents the efficiency of the newly generated individual; $\eta_{current}$ represents the efficiency of the current individual.

Therefore, when the new solution is better, the new solution is directly accepted. When the new solution is worse, it is accepted with a small probability, so as to maintain the diversity of exploration. Especially at high temperatures, it is easier to accept. As the temperature decreases, the probability of accepting worse solutions gradually decreases, and it gradually converges to the optimal solution.

4 Analysis of Result

4.1 Analysis of Prediction Effect

Based on the above BP neural network model, parameter adjustment, training and solving are carried out on the data set obtained by discrete solving of multi-method detection. The R coefficients of 1258 groups of test set data before and after the optimization of the BP neural network model are shown in Fig. 6. Meanwhile, in order to obtain the prediction effect of the model, the mean squared error (MSE) is selected as the evaluation index to evaluate the prediction performance of the test set.

According to the above figure and the prediction data, the MSE of the prediction model for predicting the total cost and reliability are 8.187×10^{-4}; 2.447×10^{-4} respectively, and its R coefficient approaches 1. Therefore, the BP neural network model can be used to predict the total cost and reliability of the multi-method detection strategy, and its prediction accuracy is very high.

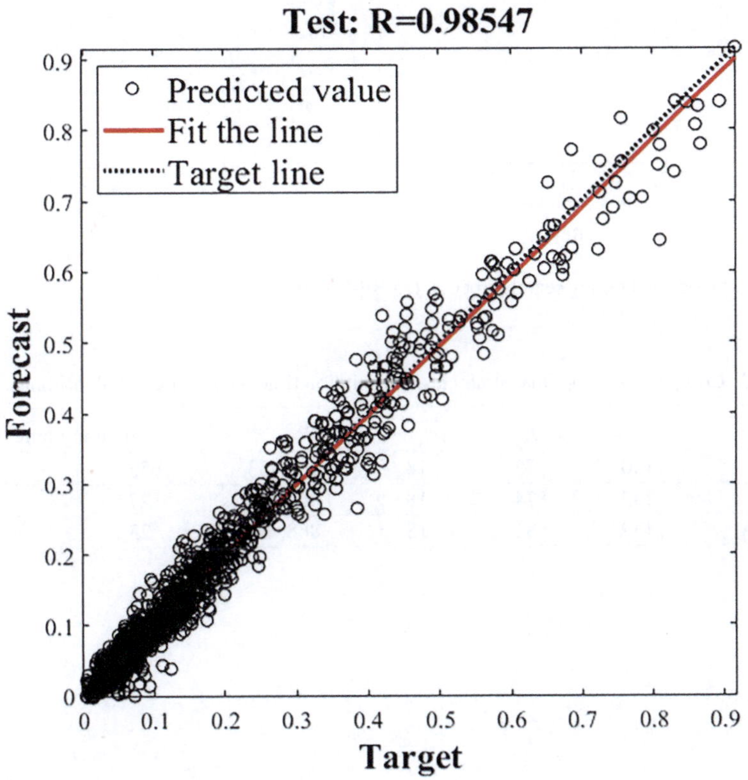

Fig. 6 Test set predictions

4.2 Analysis of Optimization Effect

In order to verify the accuracy and efficiency of the BP neural network—SA algorithm, comparisons are made with the enumeration algorithm and the traditional SA algorithm respectively. The search processes of the BP neural network—SA algorithm and the SA algorithm are shown in Fig. 7, and the optimal strategies and solution times of each algorithm are shown in Table 3.

Through comparison, it is found that the optimal solution result of the BP neural network—SA algorithm is basically consistent with that of the enumeration algorithm. In terms of the efficiency of searching for solutions, the convergence speed of the BP neural network—SA algorithm is significantly faster than that of the traditional SA algorithm and EA algorithm.

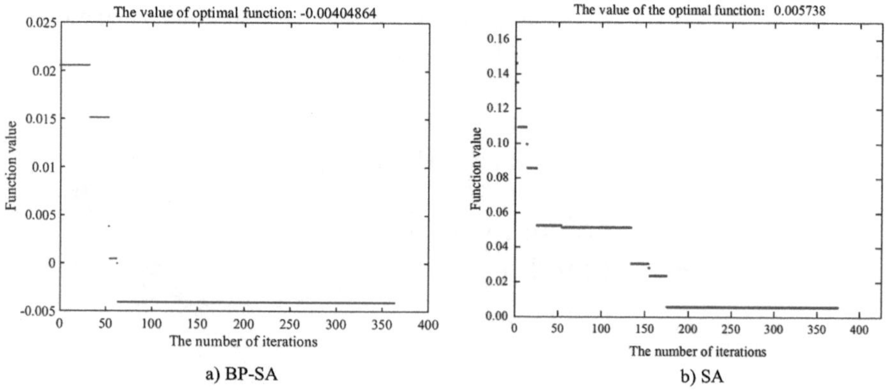

Fig. 7 The optimal policy search process. (**a**) BP-SA (**b**) SA

Table 3 Comparison of optimal strategies and solution time under different algorithms

	t_{V_1}	t_{V_2}	t_{V_3}	$\overline{C}$	Computing time /*s*
EA	140	75	18	864,753	653
SA	142	74	18	869,715	127
BP-SA	139	81	18	868,998	23

5 Conclusions

In order to solve the problem of low efficiency of multi-method condition-based maintenance, this paper proposes a solution algorithm combining BP neural network and simulated annealing to solve the above problems. The conclusions are as follows:

1. When the new algorithm combines and optimizes multiple methods with intervals, the optimal detection strategy is basically the same as that of the enumeration algorithm and the traditional simulated annealing algorithm.
2. In terms of computational efficiency, the efficiency of the new algorithm is 96.47% higher than that of the enumeration algorithm, and 81.89% higher than that of the traditional simulated annealing algorithm. The computational efficiency has been significantly improved.

References

1. Liu Y, Chen Y, Jiang T (2018) On sequence planning for selective maintenance of multi-state systems under stochastic maintenance durations. Eur J Oper Res 268(1):113–127. https://doi.org/10.1016/j.ejor.2017.12.036
2. Zhang ZM, Yan W, He SG et al (2024) Selective maintenance strategy optimization for multi-state series-parallel systems considering structural dependence under imperfect repair. Ind Eng Manag 29(04):139–147. https://doi.org/10.19495/j.cnki.1007-5429.2024.04.014
3. Wang WB (2000) A model of multiple nested inspections at different intervals. Comput Operat Res 27(6):539–558., ISSN 0305-0548. https://doi.org/10.1016/S0305-0548(99)00046-5
4. Cal J, Zuo H-f, Xu Y-M (2010) Maintenance cost analysis under different inspection levels for aircraft structure. In: 2010 Prognostics and system health management conference, Macao, China, pp 1–7. https://doi.org/10.1109/PHM.2010.5414568
5. Mohamad M, Alaa E, Kamran S et al (2017) A delay time model with multiple defect types and multiple inspection methods. IEEE Trans Reliab 66(4):1073–1084. https://doi.org/10.1109/TR.2017.2754519
6. Wang WB, Zhao F, Peng R (2014) Modeling of the optimal multiple inspection policy based on a three-stage failure process. Syst Eng Theory Pract 34(01):223–232. https://link.cnki.net/urlid/11.2267.N.20130814.1046.037
7. Han ZW, Liu ZG, Chang TT et al (2024) Optimization of wear reduction of Salix Cheilophila cutting tools--based on PSO-BP neural network combined with GA algorithm [J/OL]. J Agric Mech Res:1–7. https://doi.org/10.13427/j.issn.1003-188X.2025.08.032
8. Yan RZ, Yu Z (2024) Parameter optimization of porous bearing based on MEA-BP neural network. Mach Tool Hydraul 52(16):177–182
9. Ji SL, Dai M, Wen HY et al (2022) An improved virtual metrology method in chemical vapor deposition systems via multitask Gaussian processes and adaptive active learning. Int J Adv Manuf Technol 122(7–8):3149–3159
10. Li HB, Feng GS, Zhu Q et al (2023) Thermal layout optimization of power components based on simulated annealing algorithm. Electr Technol (13):13–16. https://doi.org/10.19768/j.cnki.dgjs.2023.13.004
11. Wang J (2023) Research on optimization technology of civil aviation transportation route based on simulated annealing algorithm. Electron Design Eng 31(20):77–81. https://doi.org/10.14022/j.issn1674-6236.2023.20.017

Open Access This chapter is licensed under the terms of the Creative Commons Attribution-NonCommercial-NoDerivatives 4.0 International License (http://creativecommons.org/licenses/by-nc-nd/4.0/), which permits any noncommercial use, sharing, distribution and reproduction in any medium or format, as long as you give appropriate credit to the original author(s) and the source, provide a link to the Creative Commons license and indicate if you modified the licensed material. You do not have permission under this license to share adapted material derived from this chapter or parts of it.

The images or other third party material in this chapter are included in the chapter's Creative Commons license, unless indicated otherwise in a credit line to the material. If material is not included in the chapter's Creative Commons license and your intended use is not permitted by statutory regulation or exceeds the permitted use, you will need to obtain permission directly from the copyright holder.

Simulation Analysis of Gas Pressure Energy Power Generation Test System Based on Gas-Liquid Conversion

Chenxu Pan and Tian Ji

Abstract The power bond diagram simulation can intuitively show the complex running state of the system. This paper takes a gas regulating station in Leshan as an example, according to the workflow of the gas regulating station, the gas is transferred from high pressure to low pressure during the transmission process, so the pressure difference energy generated in the pressure regulation process can be used in other fields. Taking the gas pressure energy power generation test system based on gas-liquid conversion as the research object, the AMESim software is used to build the model through the power bond diagram, the parameters are set according to the working condition of the pressure regulating station, and the test system is analyzed by gas-liquid simulation and underwent further experimental verification. The results show that in different gas-flow rate, adjusting the pressure and temperature of hydraulic device can effectively ensure the system continuously efficiently. Combining with the simulation results and experimental comparison data can intuitively show the system improves the power, reduces the energy loss, is feasible in the field of gas pressure power generation.

Keywords Power bond diagram · Gas-liquid conversion · Pressure energy · Pressure difference power generation

1 Introduction

Gas pressure energy power generation is a technology that converts the differential pressure energy generated in the process of gas pressure reduction into electric energy, which can effectively improve the efficiency of energy use [1, 2]. At present, the gas pressure energy is mainly used through the collection device or saving device at home and abroad, but there are problems such as complex structure and

C. Pan · T. Ji (✉)
Dalian Polytechnic University, Dalian, China
e-mail: jitian@dlpu.edu.cn

© The Author(s) 2026

Z. Zhang, P. Mercorelli (eds.), *Mechanical Design and Simulation: Insights and Innovations*, Lecture Notes in Mechanical Engineering,
https://doi.org/10.1007/978-981-95-2405-1_27

energy loss [3, 4]. Taking the working condition data and pressure regulation process of a gas pressure regulating station in Leshan as an example, the paper designs a gas pressure energy power generation test system based on the method of gas-liquid conversion. The test system directly connects the piston rod of the cylinder and the hydraulic cylinder, and there is no other device in the middle. The cylinder directly drives the hydraulic cylinder to do reciprocating linear motion. The pressure difference energy generated by gas pressure reduction in the pneumatic device is converted into hydraulic oil energy through the hydraulic device, and then the hydraulic oil energy is converted into electric energy. Compared with other existing systems, the system simplifies the internal structure, solves the problems of energy loss caused by many intermediate components and complex pipelines, and improves the overall energy conversion efficiency of the system. The power bond diagram establishes the simulation model of the test system, conducts gas-liquid simulation and combines with the experiment to prove its feasibility.

2 Principle and Research Method of Power Bond Diagram

Power bond diagram is a simulation modeling tool for mathematical equation operation using conservation of energy and transmitted signal causality [5, 6]. Different power bond diagrams correspond to one mathematical equation, so power bond diagrams can establish equations of state of a variety of physical systems to perform more complex operations and operational analysis [7, 8]. Therefore, the power bond diagram is highly readable during the analysis process, making the analysis results more intuitive [9, 10].

At present, AMESim software is one of the most suitable for power bonding software. The power bond diagram in AMESim software is used to model and simulate the gas pressure energy power generation test system based on the gas-liquid conversion principle, and the system is tested. Through simulation analysis and experimental results, the feasibility of the test system is studied.

3 Test System Design Scheme

The working condition of the gas regulating station is to reduce the pressure of the gas from 0.8 to 0.4 MPa, so the gas source input pressure of the test system should be 0.8 MPa and the output pressure is 0.4 MPa. The simulation model of the gas pressure energy power generation test system based on gas-liquid conversion established by using the power bond diagram in the AMESim software is shown in Fig. 1. The system includes pneumatic device and hydraulic device, and the differential pressure energy generated by gas pressure reduction is converted into hydraulic oil energy through the hydraulic device, and the hydraulic motor drives the generator to realize the pressure difference power generation.

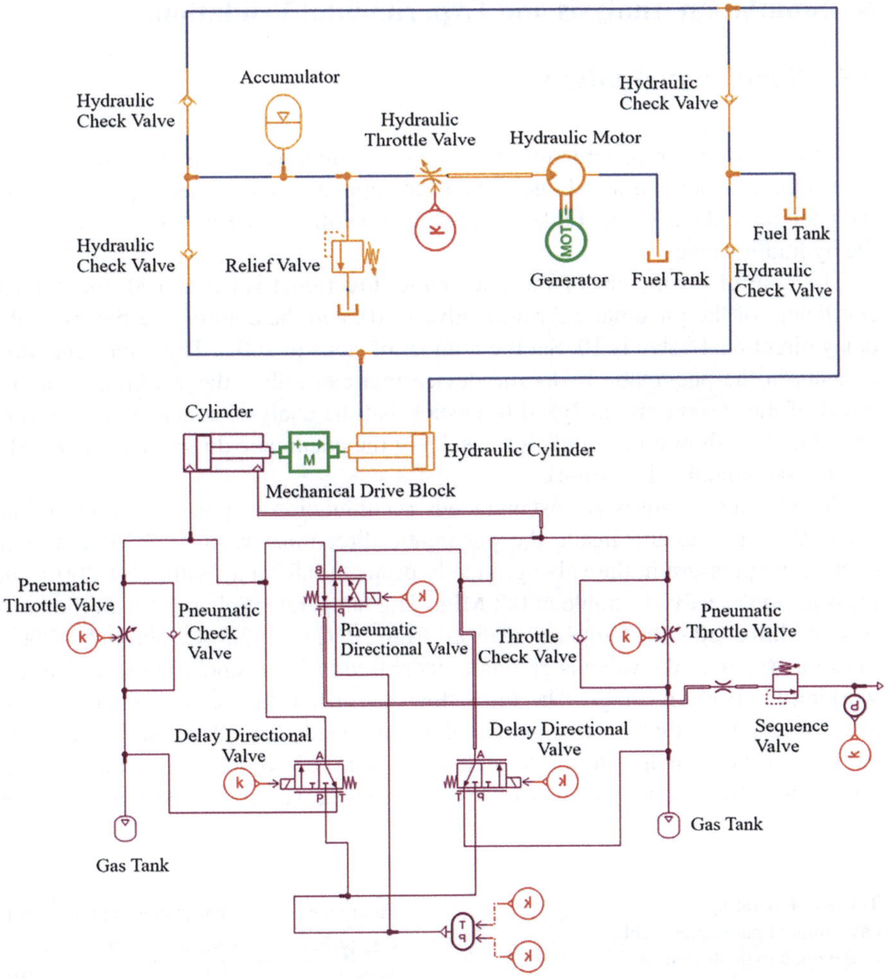

Fig. 1 Simulation model diagram of the gas pressure energy power generation test system

The working process of the test system is as follows: 0.8 MPa gas reaches the pneumatic directional valve, the pneumatic directional valve switches to the right position, and the cylinder piston goes right. Under the delay inflation of the pneumatic check valve, the delay directional valve switches to the left position, the pneumatic directional valve switches to the left position, the cylinder piston goes left, and reciprocating movement. The gas discharged from the cylinder reaches the sequence valve after the pneumatic directional valve, and is directly transported through the sequence valve into the pipeline. The output gas pressure is 0.4 MPa. The piston rod of the cylinder is connected to the piston rod of the hydraulic cylinder. The cylinder drives the hydraulic cylinder to reciprocate motion, and the hydraulic oil pushes the hydraulic motor to rotate, thus driving the generator to generate electricity.

4 Simulation Analysis and Experimental Validation

4.1 Simulation Analysis

According to the working conditions of the gas regulating station, the parameters of the simulation model is set. Table 1 shows the operating environment parameters of the pneumatic device, and Table 2 shows the operating environment parameters of the hydraulic device.

The control coefficient of the pneumatic directional valve is 100, the control coefficient of the pneumatic throttle valve is 10, and the control coefficient of the delay directional valve is 10. Set the number of steps to 200 s. The important components in the pneumatic hydraulic device that can reflect the working state and effect of the system are analyzed to ensure that the analysis conclusion is reasonable. Figure 2 shows the simulation results of the pneumatic directional valve (left) and the sequential valve (right).

The Gas input causes an instantaneous accumulation of pressure. In the initial stage, the gas pressure inside the pneumatic directional valve is 3 MPa. Within 0–50 s, the pressure in the valve gradually drops to 0.8 MPa. Within 50–200 s, the pressure in the valve is stable at 0.8 MPa. The temperature changes with the pressure. Within 0–50 s, the temperature drops from 20 °C to −42.5 °C. Within 50–200 s, the temperature in the valve is gradually increased to 15 °C and is stable in the safe working temperature range. The mass flow rate inside the valve is maintained at about 0.12 g/s. Within 0–30 s, the gas-flow rate inside the valve changes dramatically, with the enthalpy flow rate increasing from 1.3 to 5.2 J/s. As the pressure inside the valve gradually decreases to a stable level, within 30–200 s, the enthalpy

Table 1 Operating environment parameters table of the pneumatic device

Parameter	Parameter values	Unit
Gas type	Air	/
Input pressure	0.8	MPa
Output pressure	0.4	MPa
Gas-flow rate	0.02	m^3/s
Velocity of flow	1.8	m/s

Table 2 Operating environment parameters table of the hydraulic device

Parameter	Parameter values	Unit
Hydraulic oil type	Standard	/
Internet traffic	24	L/min
Initial oil pressure	10	MPa
Transmission load rate	0.9	/
Electric power generation	10	kW
Rotating speed of hydraulic motor	1500	r/min

flow rate drops from 5.2 to 1.1 g/s and gradually stabilizes, indicating that the gas pressure transmission is stable.

The sequence valve acts as a pressure stabilizer. So in the initial stage, the internal pressure of the sequence valve is consistent with the pneumatic directional valve, and its pressure value is 3 MPa. Due to the gas depressurization, the gas input and output pressure value of the sequence valve are 0.4 MPa, which proves that the pneumatic device completes the gas depressurization of the test system.

The hydraulic motor simulation results are shown in Fig. 3. The initial working pressure of the hydraulic motor is 10 MPa. Its pressure rises to 10.8 MPa at 45 s and remains relatively stable. At 86 s, its pressure gradually rises to 11.6 MPa. Its temperature rises steadily to 37.3 °C, which is within the safe operating temperature range of the hydraulic device. The leakage flow rate remains stable at 0.002 L/min, indicating that there is a small amount of oil overflow phenomenon. The volumetric flow rate of the hydraulic motor is stable at around 23.8 L/min, indicating that the test system can ensure continuous and stable power generation.

4.2 *Experimental Verification*

Conduct experiments on the test system, observe its working effect, and compare the data with the power of a gas regulating station in Leshan to further analyze the performance of the system and verify its feasibility.

The gas input pressure of the test system is set the same as the gas regulating station with a pressure value of 0.8 MPa. The temperature and pressure values of the hydraulic device in the test system are recorded under different gas-flow rate conditions. The power generation equipment of the gas regulating station is equipped with a pressure collection device, which loses part of the pressure energy. Therefore, under the same gas-flow rate condition, by comparing the power values of the test system and the gas regulating station, the effect of using pressure energy to generate electricity can be intuitively reflected. Table 3 is the experimental data table.

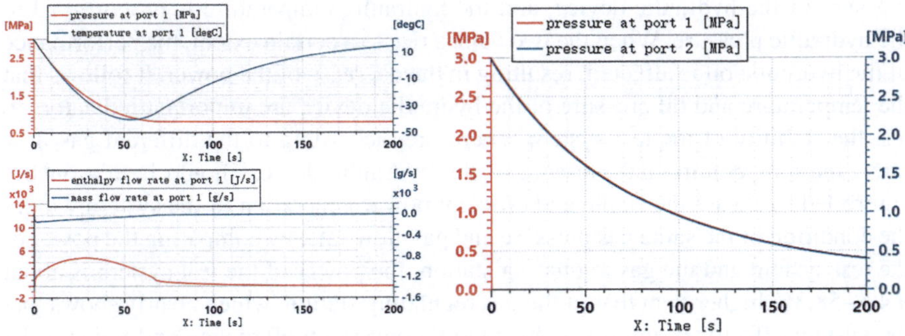

Fig. 2 Simulation results of the pneumatic directional valve (left) and the sequence valve (right)

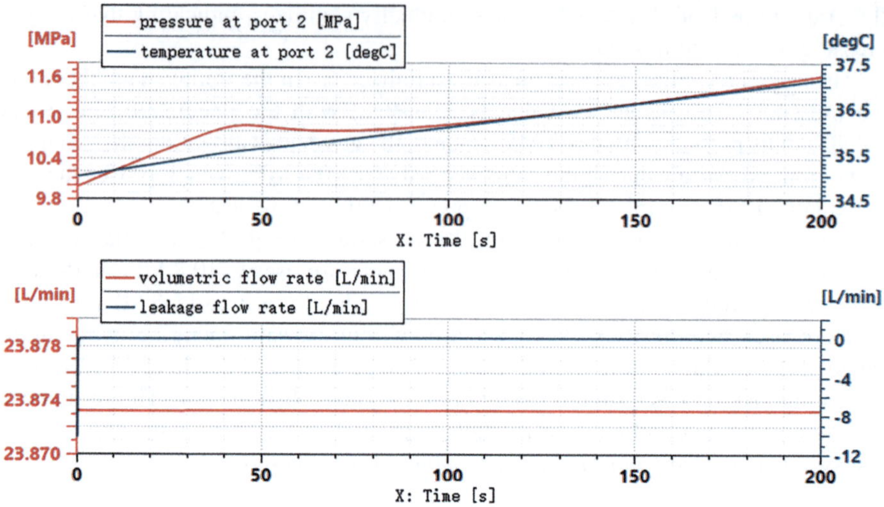

Fig. 3 Simulation results of the hydraulic motor

Table 3 Experimental data table

Gas-flow rate (Nm^3/h)	Power of the gas regulating station (kW)	Power of the test system (kW)	Hydraulic temperature (°C)	Hydraulic pressure (MPa)
60	4.8	5.4	46	10.6
		8.5	65	12
		5.1	78	14.2
100	5.3	6.2	52	12.1
		8.4	66	13.5
		5.6	72	15.1
150	6.5	7.9	56	12.5
		10	69	13.9
		6.8	74	15.2

It can be seen from Table 3 that the power is affected by the temperature and pressure of the hydraulic device, and the hydraulic temperature is proportional to the hydraulic pressure. When the two values rise to a certain extent, the performance of the hydraulic oil is affected, resulting in the decrease of the power. It follows that the temperature and oil pressure of the hydraulic device are important indicators to test the stability of the test system. Therefore, according to the different gas-flow rates, the temperature and pressure of the hydraulic device should be reasonably controlled to ensure the stable and efficient power generation of the system. Under the condition of the same gas pressure and gas-flow rate, by comparing the power of the test system and the gas regulating station, the power of the test system is about 9.4%–58.3% higher than that of the gas regulating station, which clearly shows that the system effectively improves the energy conversion efficiency and reduces the energy loss and waste.

The Simulation analysis and the experimental data analysis show that the test system has a stable and good power generation effect, and can effectively improve the power, which is feasible in the field of gas pressure energy power generation.

5 Research Conclusion

The paper simulates the gas pressure energy power generation test system based on gas-liquid conversion through the power bonding diagram, observes the running state and data changes of the main components in the test system, and analyzes it combined with the experimental verification. The results show that the direct connection between cylinder and hydraulic cylinder can reduce the number of other intermediate components and simplify the pipe structure, which can effectively reduce the loss of pressure energy and improve the energy conversion efficiency. At the same time, the pressure and temperature of the hydraulic device should be adjusted in time according to the change of gas-flow rate in the pneumatic device. The hydraulic device cooperates with the pneumatic device to ensure that the system is stably maintained in the best working condition and play the maximum performance. Under the same gas source environment, the power of the test system is higher than that of other existing equipments, and the test system can be generated steadily and continuously, which proves that the test system is feasible in the field of gas pressure energy power generation.

References

1. Lei L, Chang Y (2014) Recovery of pressure regulation pressure in front of natural gas pipeline gate station. China Petroleum and Chemical Standard and Quality 12:1. https://doi.org/10.3969/j.issn.1673-4076.2014.12.150
2. Li YY, Liu MM (2022) Discussion on the pressure energy utilization scheme of natural gas diversion station. China Petrol Chem Stand Qual 42(01):157–158. https://doi.org/10.3969/j.issn.1673-4076.2022.01.062
3. Zuo MX, Wei D, Xiong YXD (2025, 2025) Fuzzy PI model predictive control for gas pressure energy power generation system. Control Eng China:1–11. https://doi.org/10.14107/j.cnki.kzgc.20220512
4. Cheng SJ, Zhao L, Yang HHD (2022) Evaluation model and application of pressure energy generation potential of natural gas pipeline network. Chem Eng Equip 08:50–52. https://doi.org/10.19566/j.cnki.cn35-1285/tq.2022.08.129
5. Du XF (2018) Development and application of ice making technology for pressure energy power generation of natural gas pipeline network. Sci Technol Vision (14):25–26. https://doi.org/10.19694/j.cnki.issn2095-2457.2018.14.010
6. Qiu L, Zhang ZC, Wang KD (2022) IES optimization scheduling of pressure energy power generation of natural gas pipe network. Power Syst Technol 46(04):1457–1464. https://doi.org/10.13335/j.1000-3673.pst.2021.0557
7. Niu GX, Wang W (2025) Study on the dynamic characteristics of the pilot liquid control check valve. Coal Mine Machinery 46(01):53–55. https://doi.org/10.13436/j.mkjx.202501015

8. Wu QW, Guan D, Yang ZW (2024) Design of bionic feed energy damping seat based on power bond diagram theory. J Yangzhou Univ (Natural Science Edition) 27(02):1–8. https://doi.org/10.19411/j.1007-824x.2024.02.001
9. Hu NY, Yang HW, Zhu ZY (2024) Simulation study of inclined disc axial plunger pump based on AMESim. Pearl River Water Transp 23:11–14. https://doi.org/10.14125/j.cnki.zjsy.2024.23.011
10. Ruan ZD, Qiao ZF (2024) Modeling and simulation study of ceramic grinding and throwing force control actuator based on power bond diagram theory. J Tianjin Univ Technol 40(6):34–40. https://doi.org/10.3969/j.issn.1673-095X.2024.06.006

Open Access This chapter is licensed under the terms of the Creative Commons Attribution-NonCommercial-NoDerivatives 4.0 International License (http://creativecommons.org/licenses/by-nc-nd/4.0/), which permits any noncommercial use, sharing, distribution and reproduction in any medium or format, as long as you give appropriate credit to the original author(s) and the source, provide a link to the Creative Commons license and indicate if you modified the licensed material. You do not have permission under this license to share adapted material derived from this chapter or parts of it.

The images or other third party material in this chapter are included in the chapter's Creative Commons license, unless indicated otherwise in a credit line to the material. If material is not included in the chapter's Creative Commons license and your intended use is not permitted by statutory regulation or exceeds the permitted use, you will need to obtain permission directly from the copyright holder.

Numerical Simulation Study on the Spread of Droplet Virus in Ship Isolation Ward

Xuerun Jing and Guozeng Feng

Abstract This study zeroes in on a standard double bed ship isolation ward, employing Computational Fluid Dynamics (CFD) to probe airflow arrangements. It scrutinizes three ventilation tactics while altering the contaminant mass flow rate from human exhalation, assessing comfort and cross-infection risks to pinpoint the best ventilation setup. The findings reveal displacement ventilation's prowess in eliminating contaminants, thwarting cross-infections effectively. The diameter of contaminants exhaled by patients barely affects comfort but impacts indoor contaminant concentrations. This research is pivotal for curbing cross-infections in shipboard wards and offers crucial insights for isolation ward ventilation design on ships during respiratory droplet-borne viral disease outbreaks. It underscores the significance of proper ventilation in mitigating airborne virus transmission, providing a detailed analysis of airflow and contaminant dispersion via CFD. The results suggest displacement ventilation is an apt choice for shipboard isolation wards, offering energy efficiency and low noise. This study aids in devising effective infection control measures in healthcare settings, enhancing safety and comfort in shipboard isolation wards.

Keywords CFD · Cross-infection risk · Mixed ventilation · Displacement ventilation · Droplet transmission

1 Introduction

More than 85% of human time is spent indoors, such as in transportation, public places, residential areas, etc. [1]. Droplets are an important carrier of virus transmission. When infectious individuals speak, cough, or sneeze, infectious droplets are

X. Jing · G. Feng (✉)
School of Energy and Power, Jiangsu University of Science and Technology, Zhenjiang, China
e-mail: fengguozeng@just.edu.cn

© The Author(s) 2026

Z. Zhang, P. Mercorelli (eds.), *Mechanical Design and Simulation: Insights and Innovations*, Lecture Notes in Mechanical Engineering,
https://doi.org/10.1007/978-981-95-2405-1_28

produced from the nose and mouth and spread through the air. The risk of human cross-infection is related to exposure time and virus concentration, and the distance of virus transmission depends on the distance of airflow transmission. Reasonable airflow organization can reduce the diffusion of contaminants and improve indoor air quality.

With the continuous spread of the COVID-19, which is relevant in the context of outbreaks of diseases such as influenza, respiratory syncytial virus, and emerging infectious diseases where respiratory droplet transmission is a primary route of spread. Isolation plays a crucial role in preventing the spread of epidemics and ensuring the recovery of patients with respiratory diseases. To better understand the potential risks of COVID-19 transmission through respiratory droplets indoors and explore potential preventive measures [2]. This study innovatively applies displacement ventilation in marine isolation wards, highlighting its superiority over mixed ventilation and no ventilation. Unlike mixed ventilation, which can redistribute contaminants and increase cross-infection risks, displacement ventilation effectively removes pollutants and maintains a comfortable environment [3].

1.1 Model Setup

A typical double bed ship isolation ward, as shown in Fig. 1, measures 4 m × 3 m × 2.4 m (length × width × height). The replacement ventilation inlet, also serving as a mixed ventilation outlet, and the displacement ventilation outlet, as well as the mixed ventilation inlet, all have a size of 0.4 m × 0.3 m. The human body model is simplified into four parts: head, torso, arms, and legs. An integrated source is set up in the human mouth to simulate the release of viruses carried by droplets.

Omitting factors like internal items, structural heat transfer maintenance, and solar radiation, a simple geometric hexahedron room is established in numerical

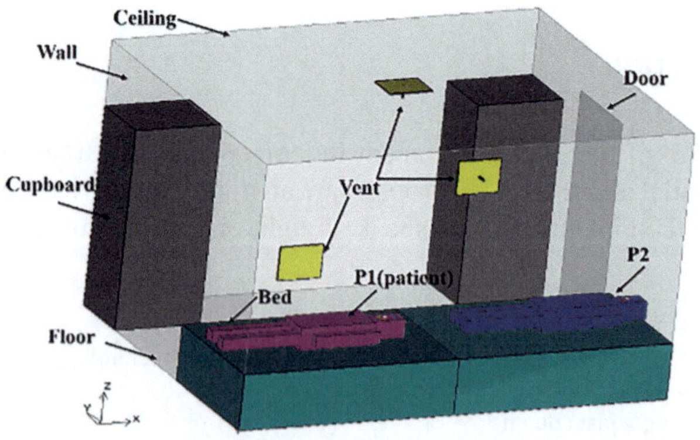

Fig. 1 Schematic of a double bed ship isolation ward

Table 1 Investigated cases

Cases	Air distribution	Diameter class (μm)	Mass flow rate (kg/s) [6]
Case 1	No ventilation	10	8.5E-11
Case 2		50	1.6E-9
Case 3	Mixed ventilation	10	8.5E-11
Case 4		50	1.6E-9
Case 5	Displacement ventilation	10	8.5E-11
Case 6		50	1.6E-9

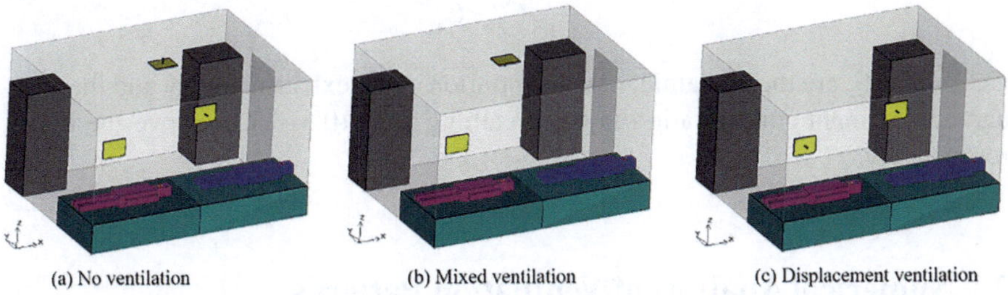

Fig. 2 Three types of ventilation patterns

simulation to reduce workload and speed up calculation. Two people lie in the room, one infected (P1). The impact on the other is simulated via droplets produced by P1's speech. Focusing on small (10 μm) and medium (50 μm) particle sizes, simulations are conducted on virus droplet diffusion under different ventilation patterns. The air supply rate is 12 ACH [4], supply temperature 23 °C, and return air outlet is outflow. The heat flow rate of P1 and P2 is 45 W/m^2 [5], with mouth size 0.02 m × 0.02 m. Walls are considered adiabatic. Simulated working and boundary conditions are in Table 1. Figure 2a–c shows the three ventilation patterns.

1.2 Numerical Simulation Method

Using the widely used RNG k-epsilon turbulence model in practical engineering for numerical simulation calculations [7]. This turbulence model is an improvement on the standard k-epsilon model, which can more accurately predict the distribution of the flow field. The general form of the control equation is given in Eq. (1) [8].

$$\frac{\partial}{\partial x_j}\left(u_j\varphi\right)=\frac{\partial}{\partial x_j}\left(\Gamma_{\varphi,eff}\frac{\partial\varphi}{\partial x_j}\right)+S_\varphi \tag{1}$$

Where, u_j is the velocity component in the j th. φ is the generic variable. $\Gamma_{\varphi,eff}$ is the effective exchange diffusion coefficient for variable. S_φ is the source term for the variable.

1.3 Evaluation of Cross-Infection Risk

Contaminant removal effectiveness (CRE) [9] is used to evaluate the system performance in the removal of contaminants in the breathing zone. The CRE depends on the airflow pattern, the contaminant sources 'properties and positions. The smaller the CRE, the worse the contaminant removal performance [10], this significantly increases the risk of cross-infection. CRE is expressed as:

$$\varepsilon = \frac{C_e - C_s}{C_p - C_s} \tag{2}$$

C_e, C_s, and C_p are the contaminant concentration at the exhaust, supply and the average contaminant concentration in the breathing zone (0.3–0.7 m above the floor), respectively.

2 Numerical Analysis of Ventilation Patterns

2.1 Velocity Distribution

Figure 3a–f illustrates the velocity distribution at the Y = 1.5 m plane. Displacement ventilation (Case 5 and Case 6) provides the most uniform airflow, suitable for resting. Mixed ventilation (Case 3 and Case 4) results in higher velocities and large vortices. No ventilation (Case 1 and Case 2) leads to low airflow velocities with high gradients around the patient.

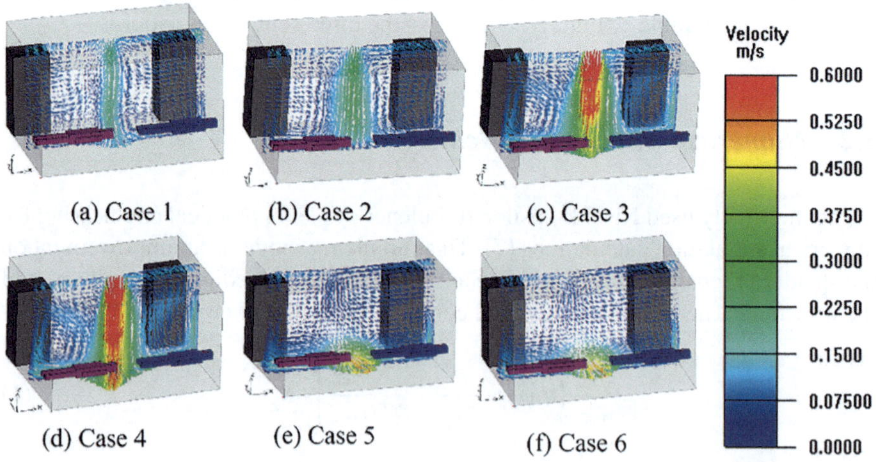

Fig. 3 Velocity vector distribution in different cases; (**a**) and (**b**) low velocity, high gradients around the patient, peaking at 0.1 m/s, (**c**) and (**d**) higher velocities, averaging 0.3 m/s, forming large vortices, (**e**) and (**f**) uniform flow, velocities between no ventilation and mixed ventilation, no vortices

2.2 Temperature Distribution

Figure 4a–f shows the temperature distribution in the isolation ward. Displacement ventilation (Case 5 and Case 6) offers the most comfortable temperature stratification, with an average temperature of 24.2 °C around the human body, dropping to 23 °C near the floor. Mixed ventilation (Case 3 and Case 4) improves the temperature field somewhat, with an average temperature of 25.5 °C around the human body. No ventilation (Case 1 and Case 2) results in higher temperatures, especially near the human body.

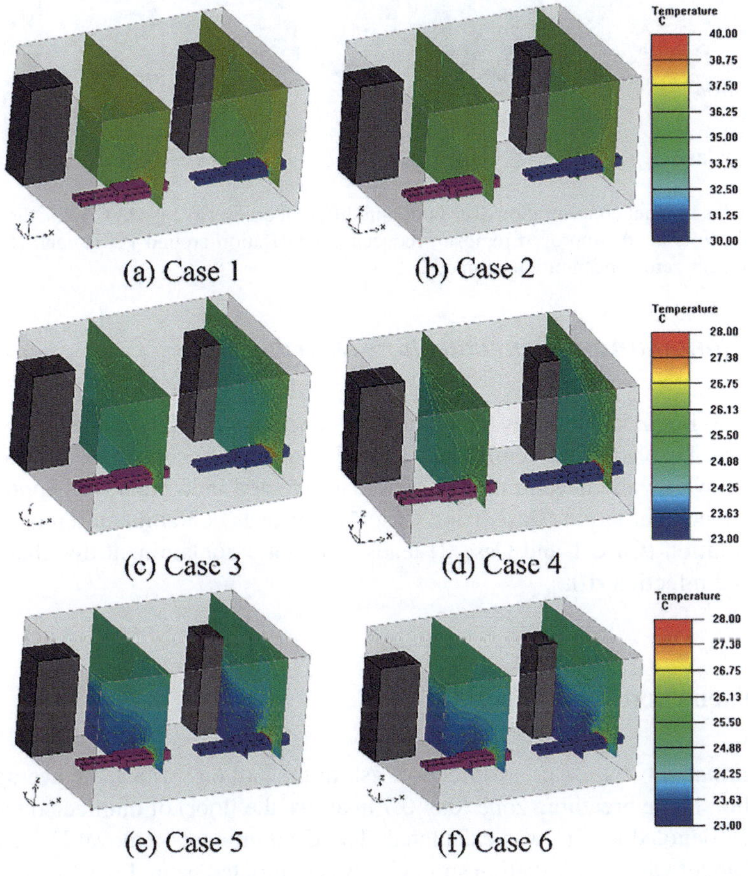

Fig. 4 Temperature distribution in different cases; (**a**) and (**b**) temperatures around 30 °C, warmer near the body, (**c**) and (**d**) average 25.5 °C around the body, warmer in the upper ward, (**e**) and (**f**) stratified, 24.2 °C around the body, 23 °C near the floor

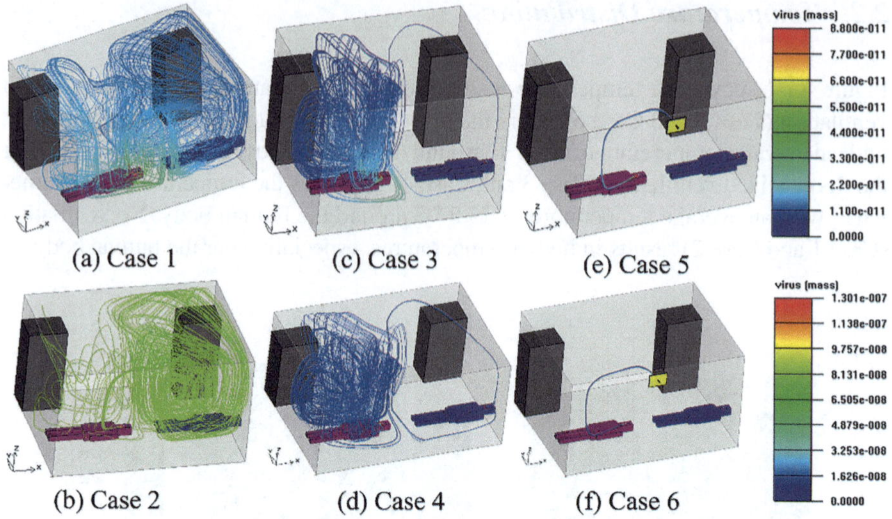

Fig. 5 Contaminant concentration distribution in different cases; (**a**) and (**b**) Chaotic distribution, high risk, (**c**) and (**d**) improved removal, reduced concentration around P2, (**e**) and (**f**) optimal removal, near-zero concentration around P2

2.3 Contaminant Concentration Distribution

Figure 5a–f depicts the virus (contaminant) concentration distribution in the isolation ward. Displacement ventilation (Case 5 and Case 6) shows the best contaminant removal, with concentrations near the uninfected individual (P2) approaching zero. Mixed ventilation (Case 3 and Case 4) improves contaminant removal, while no ventilation (Case 1 and Case 2) leads to chaotic contaminant distribution and increased infection risk.

3 Discussion

To quantitatively assess cross-infection risk in the isolation ward, the average concentration in the breathing zone (0.3–0.7 m above the floor) of uninfected individuals (P2), denoted as Cp, was calculated. The Contaminant Removal Effectiveness (CRE) under various ventilation strategies was computed using Eq. (2), with results shown in Fig. 6. At steady state, displacement ventilation (Case 5 and Case 6) shows CRE values exceeding 1, indicating complete contaminant discharge with airflow, optimizing removal and minimizing cross-infection risk. Mixed ventilation (Case 3 and Case 4) has CRE near 1, showing effective though not complete removal. Conversely, no ventilation (Case 1 and Case 2) has CRE significantly below 1, signifying poor removal efficiency and heightened cross-infection risk.

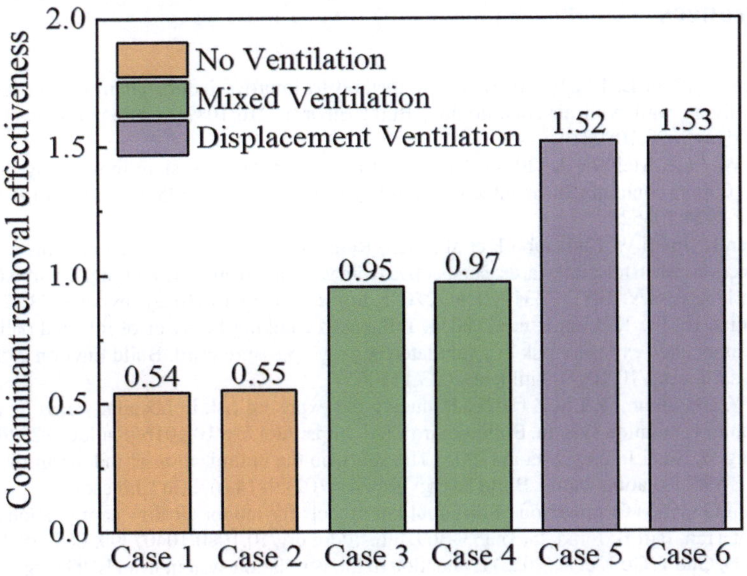

Fig. 6 Contaminant removal effectiveness in the breathing zone

In the simulations, anomalies were observed near the ventilation outlets, likely due to grid resolution and boundary conditions. The current model, a simplified hexahedron room, ignores internal items, structural heat transfer, and solar radiation, which may affect result accuracy. Future research should consider more complex models and dynamic conditions, and explore multi-field coupling to better understand airflow and contaminant distribution in marine isolation wards.

4 Conclusion

This study uses numerical simulation to investigate airflow distribution in a typical double-bed ship isolation ward under different ventilation patterns and varying contaminant mass flow rates. The following conclusions are drawn:

(1) The use of Computational Fluid Dynamics (CFD) for simulating isolation wards effectively predicts airflow distribution, making it a vital tool for preventing cross-infection from droplet-borne viruses. This method is applicable not only to marine isolation wards but also to hospital wards, aircraft cabins, and public transport vehicles. (2) Displacement ventilation is found to generate favorable temperature and velocity fields, enhancing human thermal comfort. Moreover, it achieves optimal contaminant removal effects, significantly reducing the risk of cross-infection. Therefore, displacement ventilation emerged as the optimal ventilation pattern.

References

1. Meng X, Yan L, Liu F (2022) A new method to improve indoor environment: combining the living wall with air-conditioning. Build Environ 216:108981. https://doi.org/10.1016/j.buildenv.2022.108981
2. Su W, Ai Z, Melikov A (2024) Potential of a bed ventilation system in reducing the risk of exposure to contaminants in infectious wards. J Build Eng:84108525. https://doi.org/10.1016/j.jobe.2024.108525
3. Chen R, Junqi W, Zhuangbo F et al (2023) Refined design of ventilation systems to mitigate infection risk in hospital wards: perspective from ventilation openings setting. Environ Pollution (Barking, Essex: 1987) 333122025-122025. https://doi.org/10.1016/j.envpol.2023.122025
4. Zhijian L, Zhe H, Lina H et al (2024) Influence of talking behavior of infected patients and the associated exposure risk in a ventilated negative-pressure ward. Build Environ 248111065. https://doi.org/10.1016/j.buildenv.2023.111065
5. Lu Y, Oladokun M, Lin Z (2020) Reducing the exposure risk in hospital wards by applying stratum ventilation system. Build Environ 183. https://doi.org/10.1016/j.buildenv.2020.107204
6. Deng W, Shen J, Tang X et al (2005) The study on the optimization of indoor air distribution for SARS isolation wards. Build Energy Environ (02):9–14+67. (in Chinese)
7. Chen Q (1995) Comparison of different k-ε models for indoor air flow computations. Numer Heat Tran, Part B: Fund 28(3):353–369. https://doi.org/10.1080/10407799508928838
8. Shi D, Guo Y, Gu X et al (2022) Evaluation of the ventilation system in an LNG cargo tank construction platform (CTCP)by the AHP–entropy weight method. Build Simul 15(7):1277–1294. https://doi.org/10.1007/s12273-021-0845-0
9. Cao G et al (2014) A review of the performance of different ventilation and airflow distribution systems in buildings. Build Environ 73:171–186. https://doi.org/10.1016/j.buildenv.2013.12.009
10. Novoselac A, Srebric J (2003) Comparison of air exchange efficiency and contaminant removal effectiveness as IAQ indices. Build Eng 109(2):11

Open Access This chapter is licensed under the terms of the Creative Commons Attribution-NonCommercial-NoDerivatives 4.0 International License (http://creativecommons.org/licenses/by-nc-nd/4.0/), which permits any noncommercial use, sharing, distribution and reproduction in any medium or format, as long as you give appropriate credit to the original author(s) and the source, provide a link to the Creative Commons license and indicate if you modified the licensed material. You do not have permission under this license to share adapted material derived from this chapter or parts of it.

The images or other third party material in this chapter are included in the chapter's Creative Commons license, unless indicated otherwise in a credit line to the material. If material is not included in the chapter's Creative Commons license and your intended use is not permitted by statutory regulation or exceeds the permitted use, you will need to obtain permission directly from the copyright holder.

BY NC ND

The Effects of Various Ventilation Modes and Patient Relative Positions on Airflow Distribution in a Two-Bed Isolation Ward

Xingyu Wan and Guozeng Feng

Abstract The COVID-19 pandemic presented substantial challenges to the healthcare systems in treating infected patients. The isolation ward is a crucial component in controlling the spread of respiratory diseases. A two-bed isolation ward is served as the subject of research, and the numerical simulation is used to analyze the indoor airflow distribution based on 12 cases with three ventilation modes and four patient relative postures. Three ventilation modes encompass stratum ventilation (SV), displacement ventilation (DV), and downward ventilation (DWV). Four patient relative postures involve supine (Sp-Sp), sitting (St-St), and side-lying in the same direction (Ly-Ly1) or back-to-back (Ly-Ly2). The results show that regarding the temperature distribution, and velocity distribution, air age distribution, DWV performs better than other ventilation modes. While regarding the contaminant concentration distribution, DV can better reduce the risk of cross-infection for patients and healthcare worker. This work can contribute valuable insights for the systematic design of wards in response to the emergent respiratory infectious diseases. It has significant practical implications for controlling the risk of cross-infection of respiratory diseases among personnel.

Keywords Two-bed isolation ward · Ventilation mode · Patient relative posture · Airflow distribution

1 Introduction

At the end of 2019, the outbreak of the COVID-19 pandemic caused by the severe acute respiratory syndrome coronavirus 2 (SARS-CoV-2) has drew global attention [1, 2]. Researchers find that aerosolized droplet is the significant mode of

X. Wan · G. Feng (✉)
School of Energy and Power, Jiangsu University of Science and Technology, Zhenjiang, China
e-mail: fengguozeng@just.edu.cn

© The Author(s) 2026

Z. Zhang, P. Mercorelli (eds.), *Mechanical Design and Simulation: Insights and Innovations*, Lecture Notes in Mechanical Engineering,
https://doi.org/10.1007/978-981-95-2405-1_29

transmission for SARS-CoV-2 [3, 4]. To minimize the cross-infection among personnel, adopting an efficient ventilation mode in wards is crucial [5]. Moreover, reference [6] indicates that the patient relative postures can influence the indoor airflow organization, playing a vital role in the distribution of indoor contaminant, thereby affecting the risk of cross-infection.

On one hand, numerous researches have explored the impact of various ventilation modes on airflow and contaminant distribution within wards. Lu et al. [7] examined the airflow distribution with mixed, displacement, stratified, and downward ventilation. Ren et al. [8] investigated the airflow distribution with displacement, stratum, and downward ventilation in hospital wards.

On the other hand, previous researches have examined the importance of patient relative postures. Chen et al. [9] studied the indoor airflow distribution in isolation wards, focusing on scenarios where patients are sitting, lying on their backs or their sides. Liu et al. [10] investigated the airflow distribution when patients are lying down and sitting. The results showed that the dispersion of droplets depends on patients posture and air change rate.

Therefore, a two-bed isolation ward is taken as the subject, and numerical simulation is employed to investigate the airflow distribution with 12 cases formed by three ventilation modes and four patient relative postures. This work provides valuable reference for the systematic design of isolation wards by changing the ventilation modes and patient relative postures. It has significant practical implications for controlling the risk of cross-infection of respiratory diseases among personnel.

2 Methodology

2.1 Geometry Model

A two-bed isolation ward is established, as shown in Fig. 1. The room has dimensions of 6 m × 3.5 m × 2.5 m (L × W × H) with two bedside tables, two lamps for lighting, two windows, and a cabinet placed near the corner of the wall. Two patients (denoted as P1 and P2) stay on the beds and one healthcare worker stands between the beds.

2.2 Mathematical Model

$$\frac{\partial}{\partial x_i}(\rho k u_i) = \frac{\partial}{\partial x_j}\left(\alpha_k \mu_{eff} \frac{\partial k}{\partial x_j}\right) + G_k + G_b - \rho\varepsilon - Y_M + S_k \tag{1}$$

$$\frac{\partial}{\partial x_i}(\rho \varepsilon u_i) = \frac{\partial}{\partial x_j}\left(\alpha_\varepsilon \mu_{eff} \frac{\partial \varepsilon}{\partial x_j}\right) + C_{1\varepsilon}\frac{\varepsilon}{k}(G_k + G_{3\varepsilon}G_b) - G_{2\varepsilon}\rho\frac{\varepsilon^2}{k} + S_\varepsilon \tag{2}$$

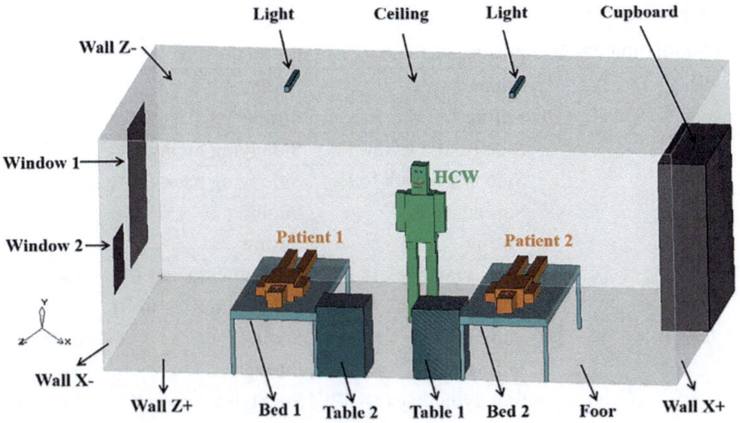

Fig. 1 Geometry model of two-bed isolation ward

Where G_k represents the turbulent Prandtl number for the k-equation; G_b denotes the turbulent kinetic energy generated by buoyancy; Y_M signifies the effect of compressible turbulence fluctuations on the total dissipation rate; $C_{1\varepsilon}$, $C_{2\varepsilon}$, $C_{3\varepsilon}$ are empirical constants; μ_{eff} is the effective turbulent viscosity coefficient; α_k is the inverse of the effective Prandtl number for turbulent kinetic energy; α_ε is the inverse of the effective Prandtl number for the dissipation rate; C_μ is a constant, which can be taken as 0.0845; S_k, S_ε are user-defined source terms.

2.3 Simulation Case

Combine the ventilation modes with the relative postures results in a total of 12 cases, as shown in Table 1.

2.4 Boundary Conditions Setting

Supply air inlets are set as velocity inlets with an air exchange rate of 12 times per hour. The supply air temperatures are set to 24 °C and the exhaust air outlets are set as outflow conditions. Ambient temperature is set at 31 °C, with the heat flux of patients and the healthcare worker at 45 W/m^2. The heat flux of lights is 150 W/m^2. The ceiling, walls, windows, and floor are all modeled as adiabatic, no-slip boundaries.

Yuce [11] achieved good research results using CO_2 as a tracer gas in the study of contaminant from the patient's mouth. Therefore, CO_2 is selected as the tracer gas to study the dispersion of contaminant. The mouth area of the patient is 97 mm^2, the

Table 1 Simulation cases

Case	Ventilation mode	posture of P1	posture of P2	Relative posture
Case1	SV	Supine	Supine	Sp-Sp
Case2		Sitting	Sitting	St-St
Case3		Side lying back to P2	Side lying facing P1	Ly-Ly1
Case4		Side lying back to P2	Side lying back to P1	Ly-Ly2
Case5	DV	Supine	Supine	Sp-Sp
Case6		Sitting	Sitting	St-St
Case7		Side lying back to P2	Side lying facing P1	Ly-Ly1
Case8		Side lying back to P2	Side lying back to P1	Ly-Ly2
Case9	DWV	Supine	Supine	Sp-Sp
Case10		Sitting	Sitting	St-St
Case11		Side lying back to P2	Side lying facing P1	Ly-Ly1
Case12		Side lying back to P2	Side lying back to P1	Ly-Ly2

exhaled CO_2 gas temperature is 35 °C, and the velocity is 0.89 m/s. The velocity direction is perpendicular to the patients' faces.

3 Analysis of Airflow Distribution

3.1 Temperature Distribution

Figure 2 presents the temperature distribution for different patient relative postures with various ventilation modes (X = 1.8 m and X = 4.2 m). It shows that when the ventilation mode is SV, the average temperature within the ward is the highest, followed with DV and DWV.

When SV is conducted in the ward, cold airflow is introduced through the supply vent. For DV mode, the cooled airflow introduced from the lower supply air inlets mixes with the thermal plume around the human body, causing the temperature to rise and the air to flow upward. When it comes to DWV, the average airflow temperature is lower compared to DV and SV.

3.2 Velocity Distribution

Figure 3 shows the velocity distribution for different patient relative postures with various ventilation modes (X = 1.8 m, X = 4.2 m). It shows that among three ventilation modes, when the patient posture is Sp-Sp, the airflow distribution is the most

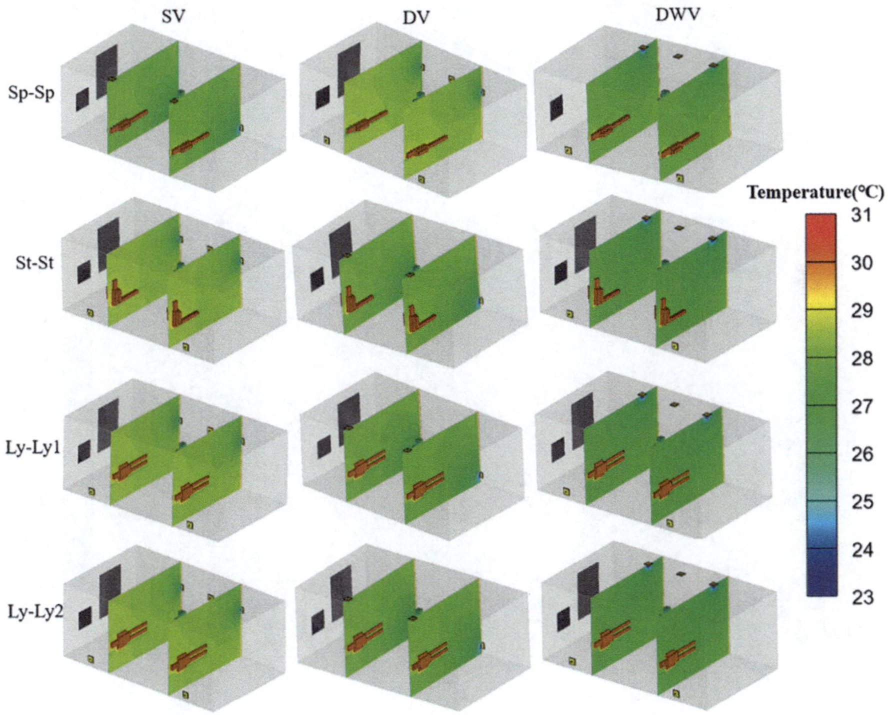

Fig. 2 Temperature distribution

uniform, but the airflow velocity is higher. When patients are all seated or lying on their sides, the airflow velocity shows a certain gradient change.

3.3 Air Age Distribution

Figure 4 shows the mean air age distribution for different relative postures with various ventilation modes (Z = 1.6 m). Comparing three ventilation modes, it can be observed that for DWV, the area near the head of healthcare personnel has the smallest air age. In contrast, this area has the highest air age for DV, while the SV mode yields values between the other two.

When SV is used in the ward, the area above the healthcare worker's head receives fresher airflow because of the air supply outlet being directly above them. For DV, the air age near the healthcare worker's head generally maintains around 265 s. When it comes to DWV, the air freshness near the healthcare worker's head is the highest.

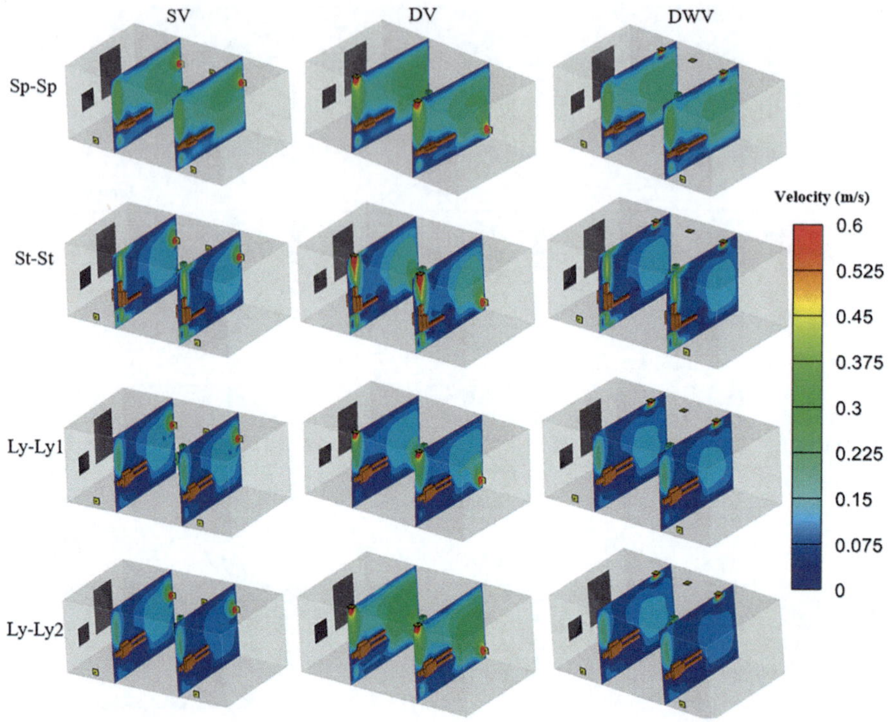

Fig. 3 Velocity distribution

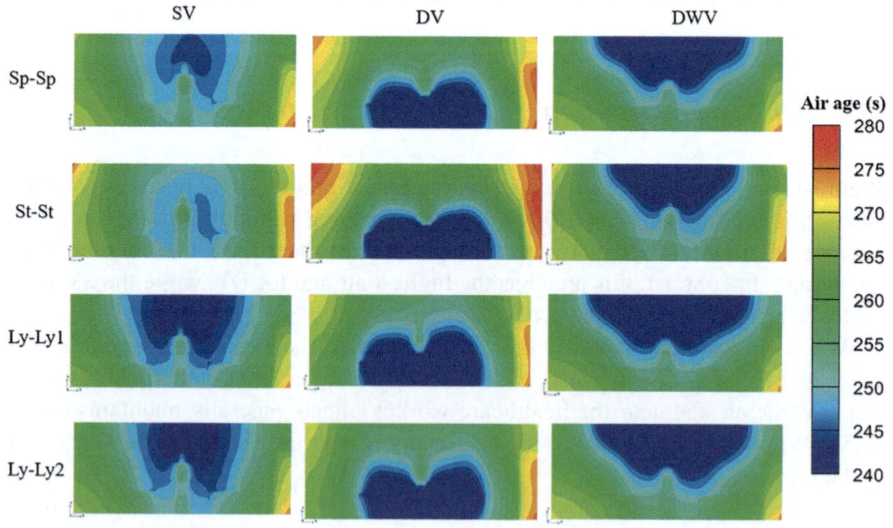

Fig. 4 Air age distribution

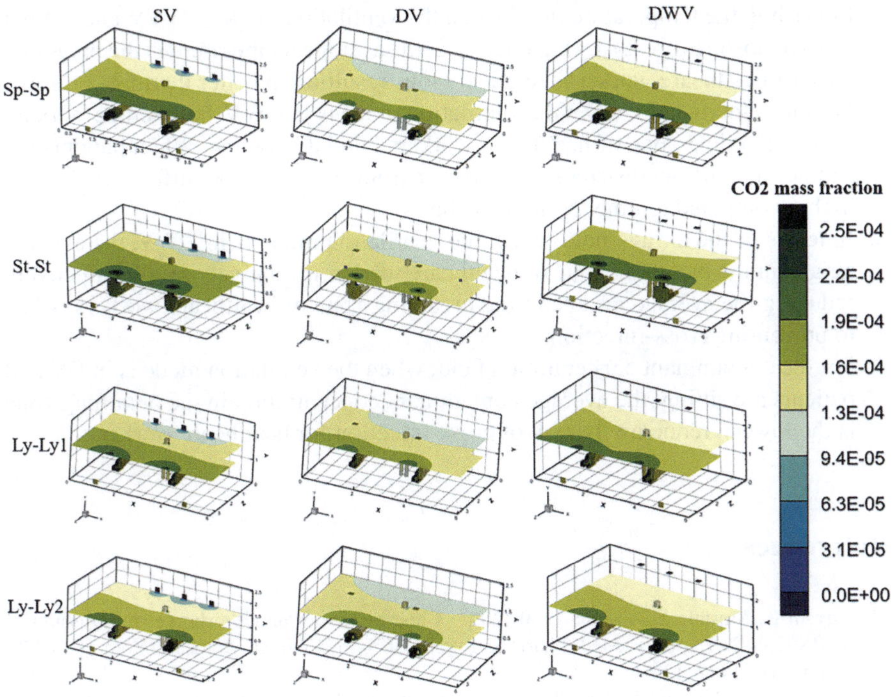

Fig. 5 Contaminant concentration distribution

3.4 Contaminant Concentration Distribution

Figure 5 shows the contaminant concentration distribution for different relative postures with various ventilation modes (Y = 1.5 m). It can be seen from the figure that the three ventilation modes all exhibit a stratification phenomenon in concentration, with the highest contaminant concentration near the patient's mouth (within a 0.5-m radius). As the distance from the patient's mouth increases, the contaminant concentration tends to decrease. When DV is used in the ward, the average contaminant concentration in the head area of the healthcare worker is the lowest, with a mass fraction of approximately 0.00013.

4 Conclusion

The work is set against the backdrop of the 2019 COVID-19 pandemic outbreak. A two-bed isolation ward is served as the subject of research, and the numerical simulation is used to analyze the airflow distribution based on 12 cases with three ventilation modes and four patient relative postures.

1. Regarding the temperature field, when the ventilation mode is DWV and patient relative postures are Ly-Ly1 and Ly-Ly2, the average temperature in the breathing zone of healthcare worker is relatively low, resulting in better thermal comfort.
2. For the velocity field, when the ventilation mode is in DWV, the average velocity in the breathing zone of healthcare workers is relatively low. When patients are supine, the airflow distribution is the most uniform, but the airflow velocity is higher compared to other relative postures.
3. In terms of the air age field, when the ventilation mode is in DWV, the average air age in the breathing zone of healthcare worker is relatively small. Effectively reducing the average air age in the breathing zone of healthcare workers is key to preventing cross-infection.
4. For the contaminant concentration field, when the ventilation mode is in DV and patients are sitting, the average contaminant concentration in the breathing zone is the lowest, reducing the risk of cross-infection for healthcare worker.

References

1. Atiyani R, Mustafa S, Alsari S et al (2021) Clearing the air about airborne transmission of SARS-CoV-2. Eur Rev Med Pharmacol Sci 25(21):6745–6766. https://doi.org/10.26355/eurrev_202111_27120
2. Ye ZW, Yuan S, Yuen KS et al (2020) Zoonotic origins of human coronaviruses. Int J Biol Sci 16(10):1686. https://doi.org/10.7150/ijbs.45472.eCollection2020
3. Li HY, Leong FY, Xu G et al (2021) Airborne dispersion of droplets during coughing: a physical model of viral transmission. Sci Rep 11(1):4617. https://doi.org/10.1038/s41598-021-84245-2
4. Liu K, Allahyari M, Salinas JS, Zgheib N, Balachandar S (2021) Peering inside a cough or sneeze to explain enhanced airborne transmission under dry weather. Sci Rep 11(1):9826. https://doi.org/10.1038/s41598-021-89078-7
5. Koch S, Khomenko S, Cirach M et al (2022) Impacts of changes in environmental exposures and health behaviours due to the COVID-19 pandemic on cardiovascular and mental health: a comparison of Barcelona, Vienna, and Stockholm. Environ Pollut 304:119124. https://doi.org/10.1016/j.envpol.2022.119124
6. Yang C, Yang X, Zhao B et al (2015) The ventilation needed to control thermal plume and particle dispersion from manikins in a unidirectional ventilated protective isolation room. Build Simul. Springer. https://doi.org/10.1007/s12273-014-0227-6
7. Lu Y, Oladokun M, Lin Z (2020) Reducing the exposure risk in hospital wards by applying stratum ventilation system. Build Environ 183:107204. https://doi.org/10.1016/j.buildenv.2020.107204
8. Ren C, Wang J, Feng Z et al (2023) Refined design of ventilation systems to mitigate infection risk in hospital wards: perspective from ventilation openings setting. Environ Pollut 333:122025. https://doi.org/10.1016/j.envpol.2023.122025
9. Chen M, Zhang Y, Li T (2023) Airborne spread of human coughing droplets with different postures and relative orientation in wards combined with peacetime and epidemic periods. Build Environ 245:110863. https://doi.org/10.1016/j.buildenv.2023.110863
10. Liu H, Liu Z, Wang Y et al (2022) Distribution of droplets/droplet nuclei from coughing and breathing of patients with different postures in a hospital isolation ward. Build Environ 225:109690. https://doi.org/10.1016/j.buildenv.2022.109690
11. Yuce BE, Aganovic A, Nielsen PV et al (2023) Analysis of parameters influencing pathogen concentration in a room with displacement ventilation using computational fluid dynamics and Taguchi methods. J Build Eng 80:108002. https://doi.org/10.1016/j.jobe.2023.108002

Open Access This chapter is licensed under the terms of the Creative Commons Attribution-NonCommercial-NoDerivatives 4.0 International License (http://creativecommons.org/licenses/by-nc-nd/4.0/), which permits any noncommercial use, sharing, distribution and reproduction in any medium or format, as long as you give appropriate credit to the original author(s) and the source, provide a link to the Creative Commons license and indicate if you modified the licensed material. You do not have permission under this license to share adapted material derived from this chapter or parts of it.

The images or other third party material in this chapter are included in the chapter's Creative Commons license, unless indicated otherwise in a credit line to the material. If material is not included in the chapter's Creative Commons license and your intended use is not permitted by statutory regulation or exceeds the permitted use, you will need to obtain permission directly from the copyright holder.

Simulation Study on the Mechanical Properties of Fish-Eye Terminal Insertion and Extraction

Li Wang, Anming Pan, Chen Qiang, Jiaxin Song, and Yuqing Zheng

Abstract With the development of automobile electronics and other industries, fish-eye terminals are widely used. Before producing fish-eye terminals, Engineers often use the simulation method to determine whether it can meet the specific mechanical performances or not during the insertion process and the crack is not allowed to occur in the local plastic deformation zones. Therefore, this paper takes a typical fish-eye terminal connector as the object, and uses two software of Hypermesh and LS-DYNA to carry out the insertion and extraction process simulation. By using the symmetry of the model to take half of the finite element analysis and modeling, the insertion and extraction force, sidewall pressure and elastic rebound curves with time are obtained respectively. In order to improve the prediction accuracy of simulation model, some problems also need to solve such as the need to align with the experimental data, to supply the material tensile curve and to detect physical surface flaws.

Keywords Fish-eye terminal · Insertion and extraction force · Finite element simulation

---

L. Wang
Cwb Automotive Electronics Co., Ltd, Zhejiang, China
e-mail: wang.li@cwb.com.cn

A. Pan · C. Qiang
Zhejiang Guanghe Intelligent Technology Co., Ltd, Zhejiang, China

J. Song · Y. Zheng (✉)
Huzhou University, Huzhou, China
e-mail: zhengyuqing@zjhu.edu.cn

© The Author(s) 2026

Z. Zhang, P. Mercorelli (eds.), *Mechanical Design and Simulation: Insights and Innovations*, Lecture Notes in Mechanical Engineering,
https://doi.org/10.1007/978-981-95-2405-1_30

1 Introduction

With the development of electronics, telecommunications, aerospace and other industries, the demand for equipment performance, system integration and signal transmission accuracy are increasing [1]. As a critical component, its reliability is essential and is one of the key indicators of a connector [2]. Thus, it is important to analyze the contact characteristics of electrical connector contacts for their reliability analysis [3]. Fish-eye hole shaped compatible pins (fish-eye terminals) are widely used in high-speed electrical connectors [4], and their solderless technology avoids soldering problems [5]. The application of fish-eye terminals provides highly reliable connectivity solutions and promotes the transformation of electronic technology [6]. However, before mass production, their mechanical properties need to be rigorously evaluated to verify their compliance with standards by simulating and analyzing the insertion and extraction forces, contact stresses, etc. [7], and the uneven distribution of stresses can trigger problems such as local overheating and signal attenuation [8].

The design and processing of the mould as well as the mass production need to follow strict prerequisites. Only if it is proven that the fish-eye terminal meets the mechanical property indicators and does not crack during insertion due to large deformations can it proceed to the mass production stage [9].

This paper takes a typical fish-eye terminal connector as the research object, simulates and analyzes the contact process between the fish-eye terminal and the PCB board and obtains the mechanical properties curve diagram. Based on the analysis results of the graphs, it can lay the foundation for the next step of whether the mass production can be done.

2 Finite Element Analysis of Fish-Eye Terminal Insertion and Removal Process

2.1 Geometric Modeling

Insertion force and withdrawal force are key indicators of the electrical performance of connectors, which are essential to ensure the stable transmission of electrical signals [10]. In order to investigate whether the electrical signal of the fish-eye terminal can be transmitted stably, a simulation model of the fish-eye terminal pressed into the copper-plated hole is established. A 1:1 3D model of the selected structure was created in Solidworks (Fig. 1). The thickness of the fish-eye terminal is 0.64 mm, the thickness of the PCB board is 1.6 mm, the inner diameter is 1.115 mm.

a. Fish-eye Terminal 3D Model b. PCB boards with copper plating

Fig. 1 3D model of fish-eye terminal, PCB board, and coated copper. (**a**) Fish-eye Terminal 3D Model (**b**) PCB boards with copper plating

2.2 Model Meshing

The finer the finite element meshing, the more accurate the results will be, but also the more time consuming. Therefore, when meshing, the critical parts can be subdivided and other parts coarsened. The specific model mesh is shown in Fig. 2.

2.3 Model Materials and Boundary Conditions

Table 1 lists the key parameters of the fish-eye connector.

When simulating the insertion and removal process of fish-eye terminals, certain parameters and conditions need to be set so that calculations can be performed in LSDYNA. The specific settings are as follows: Fix the bottom and sides of the PCB board in the XYZ direction. Since the fish-eye connector is modeled as one half of the whole, constraints in the Z direction are established on the symmetry plane. A reference point is selected and coupled at the end of the connector, and a downward displacement load in the Y direction is applied; the load displacement curve is shown in Fig. 3.

3 Simulation Results Extraction and Discussion

In order to improve the computational efficiency, the fish-eye terminals were placed 1.4 mm away from the PCB, thus reducing the runtime of the "empty-contact". Images of the mechanical properties of the fish-eye terminals were then extracted from a post-processing module along with the dynamic simulation process (Fig. 4).

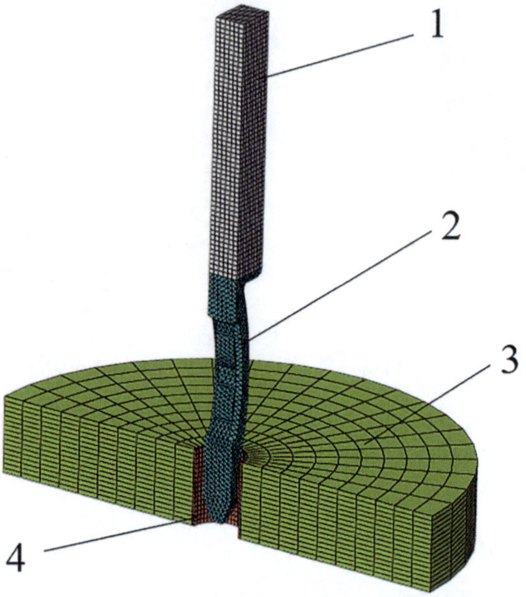

1. Fish-eye Terminal Tail 2. Fish-eye Terminal Front 3. PCB Board 4. Coated Copper

Fig. 2 Grid division. (1) Fish-Eye terminal tail (2) Fish-Eye terminal front (3) PCB board (4) Coated copper

Table 1 Fish-eye terminal material property table

Name	Material	Poisson's ratio	Density ton (mm^3)	Modulus of elasticity (Mpa)	Yield strength (Mpa)	Tangent modulus (Mpa_
Fish-eye terminal	C51900	0.34	8.8e-9	118,000	630	500
PCB	PCB-FR4	0.145	1.92e-9	21,500	206	500
Copper	Copper	0.31	8.9e-9	114,000	176	200

Precisely extracted graphs of insertion force, withdrawal force, sidewall pressure and rebound volume versus time are presented in Fig. 5.

In Fig. 5a it can be obtained that the maximum value of pressure exerted on the side wall reaches 148.496 N. In Fig. 5b, the pull-out force of the fish-eye terminal is

Fig. 3 Loading displacement graph

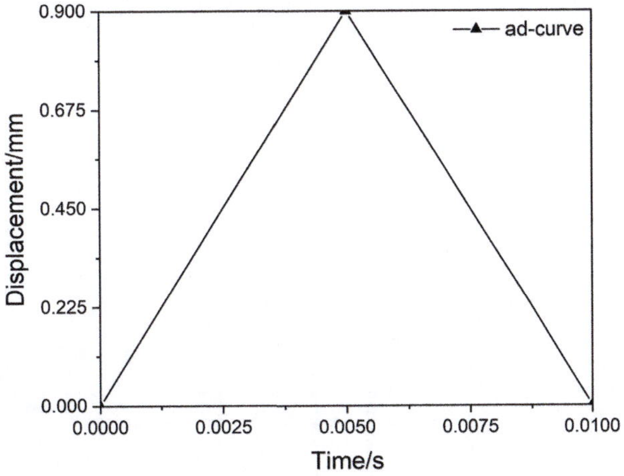

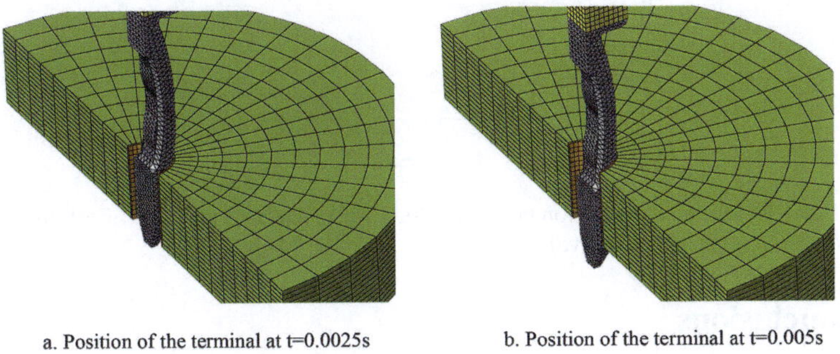

Fig. 4 Positions at different moments of the dynamic insertion and removal process. (**a**) Position of the terminal at t = 0.0025 s (**b**) Position of the terminal at t = 0.005 s

47.745 N which is greater than 35 N, while the insertion force is basically greater than 100 N, so the model structure needs to be further optimized and designed or replaced with new materials. In Fig. 5c, a small deformation occurs due to the small contact at the beginning. The subsequent sharp drop indicates a large elastic deformation of the terminal with a maximum rebound X of 0.009 mm.

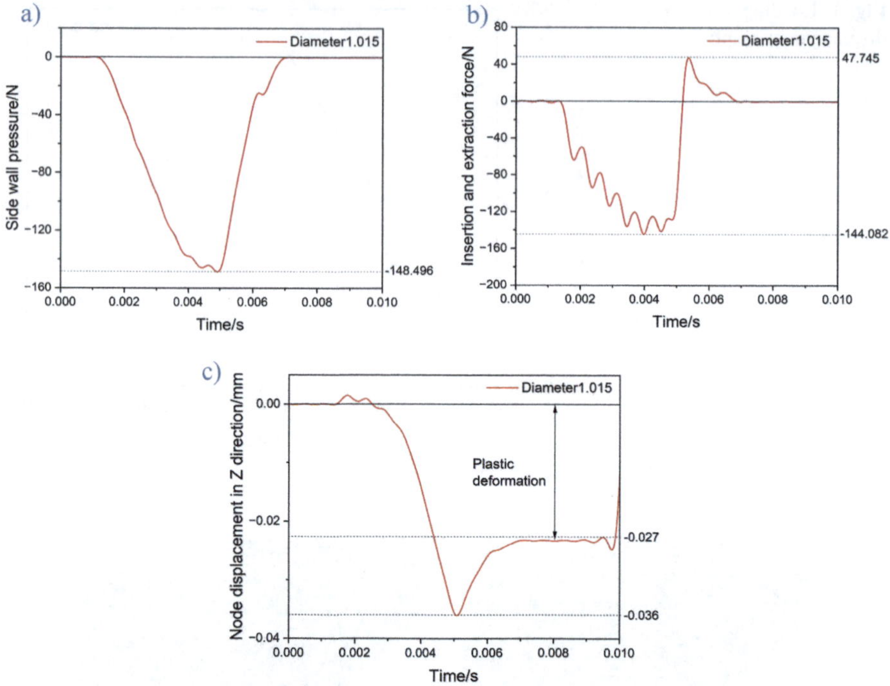

Fig. 5 Fish-eye terminal simulation curve. (**a**) Side wall pressure curve (**b**) Insertion and extraction force curve (**c**) Rebound curve)

4 Conclusions

In this paper, half of a typical fish-eye terminal block is taken as the research object. Hypermesh and LS-DYNA simulation techniques were used to study the mechanical properties of the connector, and the relevant mechanical performance curves were obtained.

Considering the computational efficiency, half of the models are modeled and analyzed in this paper. The change curves of insertion force and pull-out force etc. with time were obtained through hyper-mesh simulation. During the simulation process, the maximum value of pull-out force reaches 47.745 N greater than 35 N, which meets the established requirements, and the maximum rebound value of elastic deformation is 0.009 mm.

Since the simulation process is unstable, it needs to be successfully aligned with the experimental data to accurately predict whether the structural design of the terminal meets the design requirements.

References

1. Zefang FC (2022) Reliability study on electrical contact structure of high-speed electrical connectors. Huazhong University of Science and Technology
2. Song J, Shukla A, Probst R (2024) The state of health of electrical connectors. Machines 12(7):474–474
3. Zeng Y, Chen X (2025) Study on the stress of contact member of electrical connector with laminated fork structure. Automot Digest: 1–6
4. Xia Y (2019) Research on insertion and extraction force and reliability of fish-eye terminals of electrical connectors. Fail Anal Prev 14(05):289–293
5. Rongzhihang Information Technology (Suzhou) Co. Wave crest type hollow fish eye terminal connector: CN202221606552.3 (2022)
6. Xuesong M (2024) Research on structure optimization and reliability of fisheye terminal connector. Harbin Institute of Technol
7. Hou B, Sheng Y, Geng X et al (2024) Identification of material properties and simulation study of crimping characteristics of fisheye-type high-speed connectors. Elect Packag 24(03):25–32
8. Deng Z, Chen M, Lai W et al (2020) Effect of temperature inhomogeneity on current distribution in multichip parallel crimped IGBTs. China Electric Power 53(12):10–17
9. Yu B, Zhan S, Yin M, et al (2020) A fish-eye terminal and a design method of fish-eye terminal. Zhejiang Province:CN202010923768.1
10. Saputra MH, Nor MSN, Rijanto E et al (2025) Characterization of plugging and unplugging process for electric vehicle charging connectors based on force/torque measurements. Measurement 242(PB):115876–115876

Open Access This chapter is licensed under the terms of the Creative Commons Attribution-NonCommercial-NoDerivatives 4.0 International License (http://creativecommons.org/licenses/by-nc-nd/4.0/), which permits any noncommercial use, sharing, distribution and reproduction in any medium or format, as long as you give appropriate credit to the original author(s) and the source, provide a link to the Creative Commons license and indicate if you modified the licensed material. You do not have permission under this license to share adapted material derived from this chapter or parts of it.

The images or other third party material in this chapter are included in the chapter's Creative Commons license, unless indicated otherwise in a credit line to the material. If material is not included in the chapter's Creative Commons license and your intended use is not permitted by statutory regulation or exceeds the permitted use, you will need to obtain permission directly from the copyright holder.

A Domain Adaptation Hypergraph Residual Adversarial Network for Cross-Domain Fault Diagnosis

Junlan Hu, Jiaxing Zhu, Gaocai Fu, and Buyun Sheng

Abstract Deep learning-based fault diagnosis methods rely on consistent data distributions, but shifts in operational conditions often lead to distributional discrepancies between training and real-world data. Limited labeled samples for fine-tuning further hinder model performance. To address these challenges, we propose the Domain Adaptation Hypergraph Residual Adversarial Network (Res-DAHGAN), which integrates a hypergraph residual feature extractor, domain discriminator, and label classifier with a domain alignment mechanism. Res-DAHGAN enables cross-domain fault diagnosis without requiring labeled target domain data. Raw vibration signals are converted to frequency-domain spectra using the Fast Fourier Transform (FFT), and sample hypergraphs are constructed via the K-nearest Neighbor (KNN) algorithm. Both source and target domain samples are processed by Res-DAHGAN to extract domain-invariant features for fault identification. Results on two datasets show that Res-DAHGAN outperforms current state-of-the-art methods in accuracy, robustness, and generalization.

Keywords Rotating machinery · Fault diagnosis · Cross-domain · Domain adaptation · Hypergraph residual network

1 Introduction

Rotating machinery, such as bearings and gears, is crucial to modern manufacturing industries. The reliable operation of these components impacts production safety and efficiency, making fault diagnosis vital for improving equipment reliability and reducing maintenance costs [1]. Traditional fault diagnosis relies on costly and inefficient manual feature extraction. However, deep learning-based approaches have

J. Hu · J. Zhu · G. Fu · B. Sheng (✉)
School of Mechanical and Electrical Engineering, Wuhan University of Technology, Wuhan, China
e-mail: shengby@whut.edu.cn

© The Author(s) 2026

Z. Zhang, P. Mercorelli (eds.), *Mechanical Design and Simulation: Insights and Innovations*, Lecture Notes in Mechanical Engineering,
https://doi.org/10.1007/978-981-95-2405-1_31

become a leading trend due to their ability to automatically extract features and learn data patterns, enabling end-to-end fault diagnosis even in complex scenarios [2]. Deep learning models assume that training and testing data share the same distribution, a condition often unmet in practice. Operating under varying conditions (e.g., speed, load), rotating machinery causes significant distributional inconsistencies, leading to performance degradation when models trained on benchmark data are applied to target tasks. Fine-tuning with labeled target data is one potential solution, but gathering sufficient labeled samples across conditions is costly and often impractical [3].

Domain adaptation (DA) provides a promising solution to aligning source and target domain distributions, enabling knowledge transfer from the source to the target domain. DA methods are typically divided into adversarial-based and discrepancy-based approaches [4]. Adversarial methods, inspired by generative adversarial networks (GANs), use feature extractors to map both domains into a shared feature space. Adversarial training aims to obscure domain boundaries, fostering domain-invariant features. For example, Chen et al. [5] proposed the DATN method, using an asymmetric encoder to extract layered representations and minimize distributional differences through adversarial training. Liu et al. [6] introduced DADA, combining deep autoencoders with adversarial networks to address domain inconsistencies. However, adversarial-based DA methods often struggle with convergence and learning stable features, particularly when distribution gaps are large. Discrepancy-based methods, on the other hand, utilize distribution distance metrics to reduce domain discrepancies and improve feature transferability. For example, Zhang et al. [7] proposed DCDAN, incorporating MK-MMD into deep convolutional networks to minimize domain discrepancies via multi-kernel MMD. Li et al. [8] developed DBDANN, using multi-layer convolutions to extract transferable features and MMD to minimize discrepancies in both marginal and conditional distributions. Despite superior convergence, discrepancy-based methods require high-quality input features, as misalignment with irrelevant features impairs transfer performance. Ultimately, DA success depends on identifying shared feature representations across domains.

Traditionally, research has emphasized the data itself, often neglecting the intrinsic relationships that could identify more suitable domain-invariant features. Inspired by hypergraph neural networks, we propose Res-DAHGAN for unsupervised cross-domain fault diagnosis of rotating machinery. This method combines hypergraph residual networks with domain adaptation, enabling effective domain alignment and feature transfer.

The main contributions of this paper are as follows:

- A hypergraph-based approach is introduced to effectively capture the shared features across the source and target domains, utilizing a hypergraph residual network to learn domain-invariant representations.
- Domain adaptation is enhanced under limited labeled data conditions by employing domain adversarial training and feature distribution discrepancy metrics, facilitating effective domain alignment and seamless knowledge transfer without requiring labeled target data.

- We propose Res-DAHGAN, which outperforms existing domain adaptation techniques in accuracy, convergence, and generalization, achieving superior performance across 24 transfer tasks with exceptional robustness.

The remainder of the paper is organized as follows. Section 2 introduces the proposed Res-DAHGAN method, Sect. 3 provides the comparative experiments to evaluate the effectiveness and superiority of the method, and Sect. 4 concludes the paper.

2 The Proposed Res-DAHGAN Method

The fault diagnosis process based on Res-DAHGAN, shown in Fig. 1, includes three key modules: (1) data preprocessing and hypergraph construction, (2) hypergraph residual feature extractor, and (3) domain adaptation. These modules are integral for effective cross-domain knowledge transfer. The following sections describe each module and its role.

2.1 Data Preprocessing and Hypergraph Construction

The raw vibration data are standardized and segmented into equal-length windows of size *L*. The Fast Fourier Transform (FFT) is then applied to convert the time-domain signals into frequency-domain vectors, resulting in source domain samples $X_s \in \mathbb{R}^{f \times n_s}$ and target domain samples $X_t \in \mathbb{R}^{f \times n_t}$, where $f = L/2$, and n_s and n_t represent the number of samples in the source and target domains, respectively. These samples are represented as a hypergraph, a structure capable of capturing complex

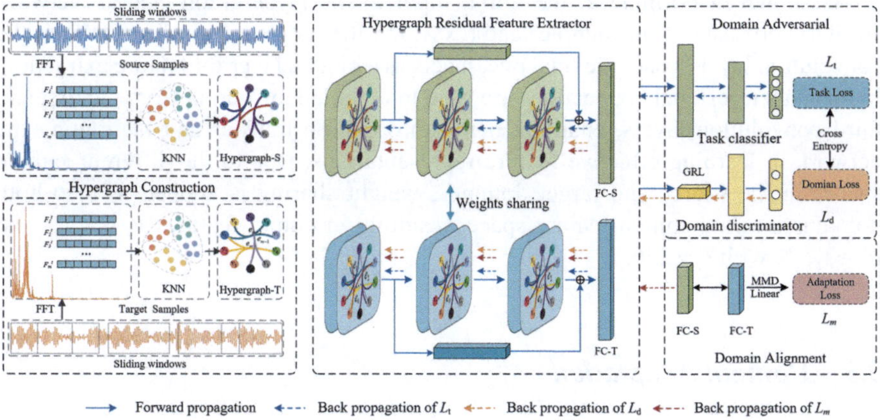

Fig. 1 Res-DAHGAN framework overview

relationships between nodes. Given a node set $V = \{v_1, \ldots, v_n\}$ and a hyperedge set $=\{e_1, \ldots e_n\}$, the incidence matrix $H \in R^{|V| \times |E|}$ is defined as follows:

$$H\left(v_i, e_j\right) = \begin{cases} 1, \; if \; v_i \in e_j \\ 0, \; if \; v_i \notin e_j \end{cases} \tag{1}$$

When $H(v_i, e_j) = 1$, node v_i is connected to other nodes via hyperedge e_j. K-nearest neighbors (KNN) is employed to establish associations between sample nodes by calculating distances. The KKK most similar nodes are connected, with Euclidean distance used for distance computation, and K set to 12. The source domain hypergraph $H_s \in \mathbb{R}^{n_s \times n_s}$ and target domain hypergraph $H_t \in \mathbb{R}^{n_t \times n_t}$ are then used as model inputs.

2.2 *Hypergraph Residual Feature Extractor*

As shown in the Fig. 1, the feature extractor is designed to capture high-order complex features from both the source and target domains. A hypergraph residual network, consisting of three layers of hypergraph convolution, is constructed. Spectral convolution theory is used to define the hypergraph convolution as follows:

$$X^{(l+1)} = \sigma\left(D_v^{-\frac{1}{2}} H W D_e^{-1} H^T D_v^{-\frac{1}{2}} X^{(l)} \theta^{(l)} \right) \tag{2}$$

Where $X^{(l)}$ and $X^{(l+1)}$ represent the input and output of the l-th layer, D_v and D_e are the node degree and hyperedge degree matrices of the hypergraph, respectively. W is the hyperedge weight, and $\theta^{(l)}$ are the learnable parameters. σ represents the activation function.

As shown in Formula 2, the initial input to the Hypergraph Neural Network (HGNN) consists of the sample features X and the incidence matrix H. When the convolution layer l is not the final one, it acts as a hidden layer for feature extraction. In this paper, residual connections are used to combine the output from the first and third convolution layers, enabling better information transfer and facilitating the network in learning effective feature representations. Due to the different sample spaces in the source and target domains, weight sharing is applied to map both domains into a common feature space, resulting in output features $F_s \in \mathbb{R}^{d \times n_s}$ and $F_t \in \mathbb{R}^{d \times n_t}$, with $d = 256$.

2.3 *Domain Adaptation*

Due to differences in operating conditions, the feature distributions between the source and target domains are not identical, making unsupervised cross-domain fault diagnosis a challenging task. The source domain $D_s = \left\{\left(x_i^s, y_i^s\right)\right\}$ contains both samples and labels, while the target domain $D_t = \{(x_i^t)\}$ consists only of unlabeled samples. The data distributions are given by $P(X_s) \neq P(X_t)$ and $(X_s | Y_s) \neq P(X_t | Y_t)$.

This paper utilizes domain adversarial and alignment methods to align feature distributions, as illustrated in Fig. 1. Domain adversarial involves three components: the feature extractor G_f, the task classifier G_y, and the domain discriminator G_d. The domain discriminator receives the extracted features and determines their domain origin in the adversarial phase. The task classifier handles fault classification, using only source domain features F_n due to the lack of target domain labels. The classification loss is computed via cross-entropy loss, as shown below:

$$L_t\left(\theta_f,\theta_y\right) = -\frac{1}{n_s}\sum_{x_i \in D_s} L\left(G_y\left(G_f\left(x_i,H_s,H_t\right)\right),y_i\right) \tag{3}$$

$$L_d\left(\theta_f,\theta_d\right) = -\frac{1}{n_s+n_t}\sum_{x_i \in D_s,D_t} L\left(G_d\left(G_f\left(x_i,H_s,H_t\right)\right)\right) \tag{4}$$

Here, L_t and L_d represent the task classification loss and domain discrimination loss for sample x_i. Adversarial training minimizes the task classification loss and maximizes the domain discrimination loss. The Gradient Reversal Layer (GRL) reverses the gradient direction of the domain discriminator to confuse its domain classification.

Additionally, we utilize linear Maximum Mean Discrepancy (MMD) to assess the domain distribution discrepancy, thereby facilitating effective feature alignment. The formula is as follows:

$$L_m\left(F_s,F_t\right) = \frac{1}{d}\frac{1}{n_s}F_s - \frac{1}{n_t}F_{tF}^2 \tag{5}$$

Where $F_s \in \mathbb{R}^{d \times n_s}$ and $F_t \in \mathbb{R}^{d \times n_t}$ denote the feature matrices of the source and target domains. Based on Eqs. (3), (4) and (5), the global loss is given by:

$$L_{all} = L_t + \mu L_d + \lambda L_m \tag{6}$$

Here, μ and λ are the weights for the domain discrimination loss and alignment loss, respectively, where higher values increase the influence of the corresponding module. The initial weights are set to 1 and 2, and the Adam optimizer is used to update them dynamically, enhancing convergence. Parameter settings for each layer are provided in Table 1.

3 Experimental Validation

This section evaluates the diagnostic performance of Res-DAHGAN through two experimental setups and ablation studies to assess key modules.

Table 1 Layer-wise network parameters of Res-DAHGAN

Module	Layer	Type	Activation	In-dim	Out-dim
Feature extractor	1-1	HGNN conv	ReLU	2048	1024
	1-2	HGNN conv	ReLU	1024	512
	1-3	HGNN conv	ReLU	512	256
	1-13	Linear	–	1024	256
Domain adversarial	2-1	Linear	LeakyReLU(0.2)	256	128
	2-2	Linear	LeakyReLU(0.2)	128	64
	2-3	Linear	–	64	C
Domain adaptation	3-1	GRL	–	256	256
	3-2	Linear	ReLU	256	128
	3-3	Linear	LeakyReLU(0.2)	128	2

3.1 Data Description

The KAT dataset [9], provided by the University of Paderborn, contains vibration signals from rolling bearings under various fault and operating conditions, sampled at 64 kHz. It includes four operating conditions: A: N09_M07_F10, B: N15_M07_F10, C: N15_M07_F04, and D: N15_M01_F10, where N is rotational speed, M is load, and F is radial force. Typical faults are selected for the experiments, as shown in Table 2.

3.2 Implementation Details

Experiments were conducted on a Windows 11 system with Intel Core i9-13900HX, GeForce RTX 4060, and 16 GB RAM. The initial learning rate was 0.001, weight decay 0.0005, and StepLR with a decay rate of 0.9 and step size of 200. Training ran for 300 iterations with a batch size of 64. The dataset was split 70% for training and 30% for testing, with dropout (0.5) to reduce overfitting. Results were averaged over five runs.

3.3 Comparison Approaches

To assess the effectiveness of Res-DAHGAN, we compared it with four state-of-the-art unsupervised domain adaptation methods:

- DANN: Domain adversarial learning with Gradient Reversal Layer (GRL), ResNet1D feature extractor, Adam optimizer (learning rate 0.001, batch size 64).
- DDC: Uses Maximum Mean Discrepancy (MMD) loss to align distributions, with ResNet1D and training parameters similar to DANN.

Table 2 Bearing Dataset Details under Different Operating Conditions

Condition	Name	Location	Description	label	Numbers
A B C D	K003	Normal	Normal	0	100
	KA01	Outer ring	EDM	1	100
	KA05	Outer ring	Electric engraver	2	100
	KA07	Outer ring	Drilling	3	100
	KI01	Inner ring	EDM	4	100
	KI08	Inner ring	Electric engraver	5	100

- MK-MMD: An extension of MMD with multiple kernel functions, using the same feature extractor and parameters as DDC.
- Deep-Coral: Aligns covariance matrices using the CORAL method, with ResNet1D as the feature extractor, Adam optimizer, and learning rate of 0.001.

3.4 Results and Discussion

Figure 2 shows the accuracy and loss curves for transfer task B-C. Res-DAHGAN outperforms DANN and DDC, with smoother, more stable curves. DANN and DDC show fluctuating accuracy and unstable loss curves, indicating poor optimization. MK-MMD and Deep-Coral show stable accuracy and gradual loss convergence, but their final performance is still inferior to Res-DAHGAN. These results confirm Res-DAHGAN's superior feature learning and domain adaptation in unsupervised domain adaptation.

Experimental results in Table 3 show Res-DAHGAN achieves an average accuracy of 96.20% across all tasks, outperforming other methods in all 12 transfer tasks. This indicates superior stability and accuracy. Other methods exhibit lower accuracy and more instability. Task difficulty influences performance: Tasks B-D and D-B have near-100% accuracy, suggesting minimal distribution differences between B and D. However, in the challenging Task C-A, Res-DAHGAN shows robust performance with 87.80% accuracy, outperforming all other methods, further confirming its adaptability and robustness in cross-domain tasks.

3.5 Ablation Experiment

Ablation studies were conducted to evaluate the contribution of key modules by removing the residual structure, domain adversarial module, and domain alignment module. Four variants of the DAHGAN framework—Res-DAHGNN, Res-HGAN, and Res-HGNN—were designed and evaluated across tasks, as shown in Fig. 3.

The results show that Res-DAHGAN achieves the highest performance (96.25% accuracy). Removing the residual structure (DAHGAN) reduces accuracy to

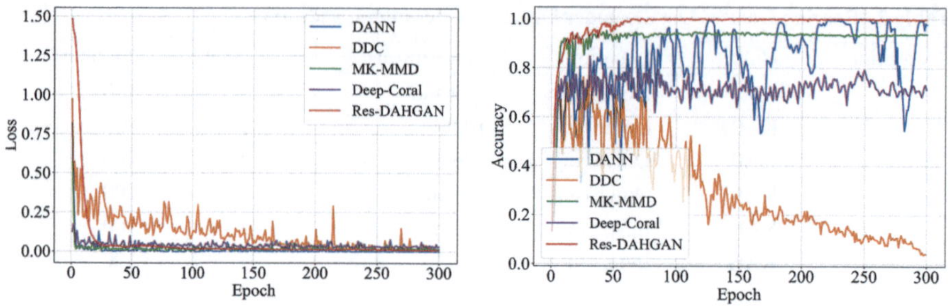

Fig. 2 Accuracy and loss curves of different methods for transfer task B-C

Table 3 Accuracy of different methods across all transfer tasks

Tasks	DANN	DDC	MK-MMD	Deep-Coral	Res-DAHGAN
A-B	66.83%	58.50%	60.33%	54.83%	98.70%
A-C	70.17%	78.50%	93.50%	59.50%	99.17%
A-D	90.33%	72.33%	80.67%	48.50%	98.67%
B-A	77.33%	71.67%	44.33%	37.17%	85.53%
B-C	81.83%	80.83%	91.00%	79.33%	99.63%
B-D	100.00%	100.00%	100.00%	98.00%	99.77%
C-A	71.00%	85.17%	21.83%	48.00%	87.80%
C-B	83.17%	81.50%	92.83%	71.17%	99.67%
C-D	95.17%	84.33%	96.67%	79.50%	99.70%
D-A	80.00%	64.67%	71.67%	44.00%	86.13%
D-B	100.00%	100.00%	100.00%	97.00%	99.83%
D-C	92.50%	89.33%	97.00%	76.83%	99.77%
Average	84.03%	80.57%	79.15%	66.15%	96.20%

94.32%, while omitting the domain adversarial module (Res-DAHGNN) drops accuracy to 93.67%. Removing the domain alignment module (Res-HGAN) causes a significant decrease to 85.63%. The Res-HGNN model, lacking both modules, shows a further drop to 74.89%. These findings confirm the essential role of the residual structure, domain adversarial, and domain alignment modules in improving Res-DAHGAN's performance in cross-domain fault diagnosis.

3.6 Application on Real-World Gear Fault Dataset

Real-world gear fault data were used to assess the method's practicality. Data were collected from a first-stage gearbox, with a 5 kHz sampling frequency [10]. Gear faults were categorized into four types: 1/4 crack, 1/2 crack, 3/4 crack, and healthy

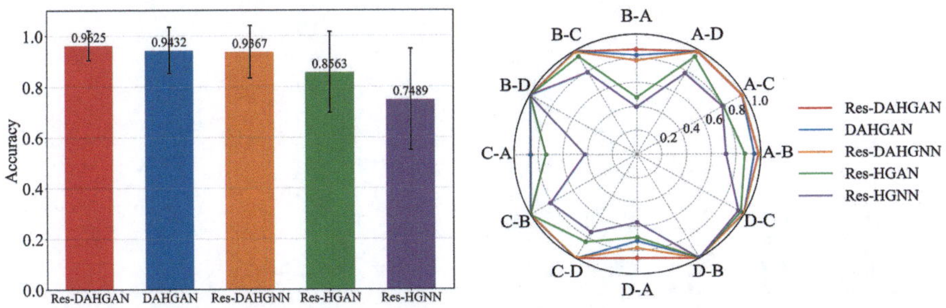

Fig. 3 Accuracy of transfer tasks for different method variants

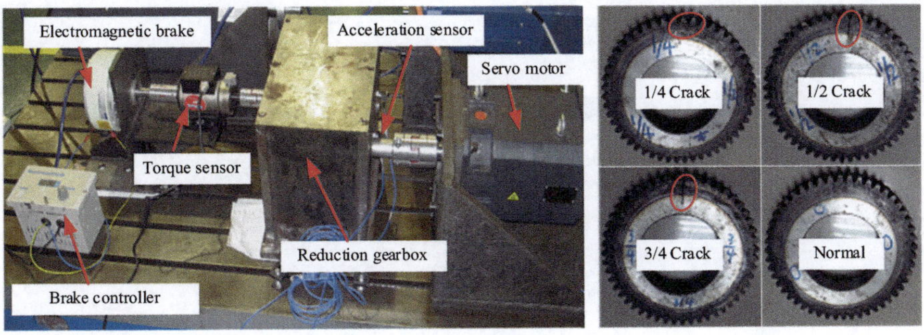

Fig. 4 Gear data acquisition device and gear fault types

gear, as shown in Fig. 4. The dataset includes data collected under four load conditions: 2 Nm (L1), 4 Nm (L2), 6 Nm (L3), and 8 Nm (L4). A total of 12 transfer tasks were evaluated, with 100 fault samples per condition, each consisting of 4096 data points.

Figure 5 displays the gearbox dataset results, where Res-DAHGAN achieves the highest average accuracy (0.9653), outpacing Deep-Coral (0.7644), DANN (0.6522), MKMMD (0.6527), and DDC (0.3496). Its low standard deviation highlights stability. The radar chart further shows its consistent effectiveness, with an average accuracy of 96.35% across 12 tasks. In the L2-L4 transfer task, Res-DAHGAN surpasses other methods, confirming its strong domain adaptation and real-world applicability.

The experiment compares all domain adaptation methods via t-SNE visualization for the L2-L4 transfer task, as shown in Fig. 6. Res-DAHGAN demonstrates superior domain separation, suggesting more effective domain adaptation between the source and target domains. Other methods exhibit greater variability in diagnostic accuracy, which can be attributed to limited sample sizes and domain distribution disparities, hindering convergence. In contrast, Res-DAHGAN exhibits stable and reliable performance, indicating its robust convergence properties, which are essential for accurate domain adaptation in this task.

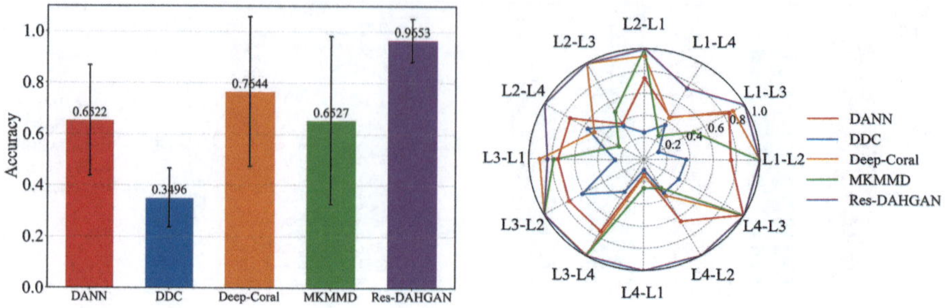

Fig. 5 The accuracy chart of gearbox dataset experiment

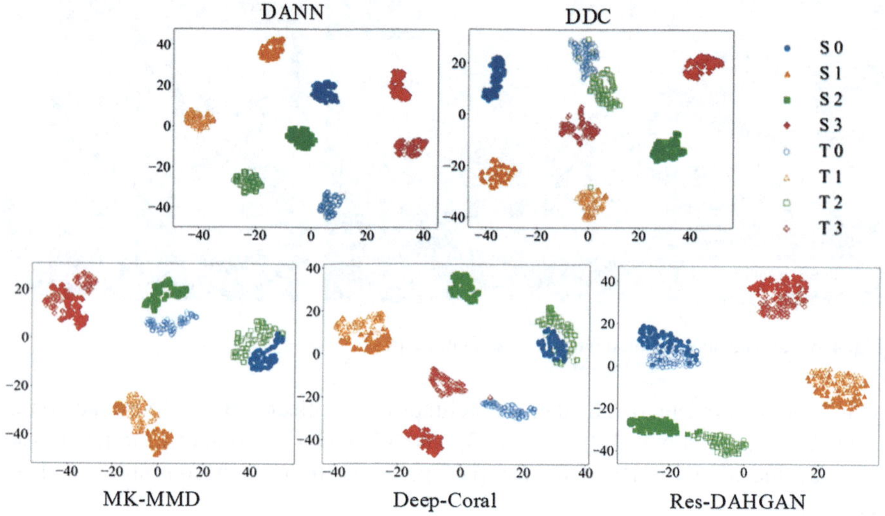

Fig. 6 t-SNE visualization of transferable features for L2-L4

4 Conclusion

This paper presents an unsupervised cross-domain fault diagnosis method, Res-DAHGAN, designed to overcome challenges such as data distribution inconsistency and the absence of labeled data in target domains during cross-condition fault diagnosis for rotating machinery. By integrating hypergraph residual networks with domain adaptation techniques, the proposed method facilitates effective feature transfer and alignment between source and target domains. Experimental results on rolling bearing and gear datasets validate the feasibility and superiority of Res-DAHGAN, demonstrating optimal diagnostic accuracy across 24 transfer tasks. Convergence analysis and feature visualization further corroborate the robustness and generalization capability of the method in complex cross-domain fault diagnosis tasks. However, the assumption of consistent label spaces between source and target domains may not hold in real-world industrial scenarios. Therefore, future work will focus on extending Res-DAHGAN to address the open-set transfer fault diagnosis challenge, which involves handling label space discrepancies between source and target domains.

References

1. Nath AG, Udmale SS, Singh SK (2021) Role of artificial intelligence in rotor fault diagnosis: a comprehensive review. Artif Intell Rev 54:2609–2668. https://doi.org/10.1007/s10462-020-09910-w
2. Zhang S, Zhang S, Wang B, Habetler TG (2020) Deep learning algorithms for bearing fault diagnostics—a comprehensive review. IEEE Access 8:29857–29881. https://doi.org/10.1109/ACCESS.2020.2972859
3. Misbah I, Lee CKM, Keung KL (2024) Fault diagnosis in rotating machines based on transfer learning: literature review. Knowl-Based Syst 283:111158. https://doi.org/10.1016/j.knosys.2023.111158
4. Jiang L, Lei W, Wang S et al (2023) A deep convolution multi-adversarial adaptation network with correlation alignment for fault diagnosis of rotating machinery under different working conditions. Eng Appl Artif Intell 126:107179. https://doi.org/10.1016/j.engappai.2023.107179
5. Chen Z, He G, Li J et al (2020) Domain adversarial transfer network for cross-domain fault diagnosis of rotary machinery. IEEE Trans Instrum Meas 69:8702–8712. https://doi.org/10.1109/TIM.2020.2995441
6. Liu Z-H, Lu B-L, Wei H-L et al (2021) Deep adversarial domain adaptation model for bearing fault diagnosis. IEEE Trans Syst Man Cybern Syst 51:4217–4226. https://doi.org/10.1109/TSMC.2019.2932000
7. Zhang Y, Ren Z, Zhou S (2020) A new deep convolutional domain adaptation network for bearing fault diagnosis under different working conditions. Shock Vib 2020:8850976. https://doi.org/10.1155/2020/8850976
8. Li Q, Tang B, Deng L et al (2020) Deep balanced domain adaptation neural networks for fault diagnosis of planetary gearboxes with limited labeled data. Measurement 156:107570. https://doi.org/10.1016/j.measurement.2020.107570
9. Lessmeier C, Kimotho J, Zimmer D, Sextro W (2016) Condition monitoring of bearing damage in electromechanical drive systems by using motor current signals of electric motors: a benchmark data set for data-driven classification
10. Yang C, Zhou K, Liu J (2022) SuperGraph: spatial-temporal graph-based feature extraction for rotating machinery diagnosis. IEEE Trans Ind Electron 69:4167–4176. https://doi.org/10.1109/TIE.2021.3075871

Open Access This chapter is licensed under the terms of the Creative Commons Attribution-NonCommercial-NoDerivatives 4.0 International License (http://creativecommons.org/licenses/by-nc-nd/4.0/), which permits any noncommercial use, sharing, distribution and reproduction in any medium or format, as long as you give appropriate credit to the original author(s) and the source, provide a link to the Creative Commons license and indicate if you modified the licensed material. You do not have permission under this license to share adapted material derived from this chapter or parts of it.

The images or other third party material in this chapter are included in the chapter's Creative Commons license, unless indicated otherwise in a credit line to the material. If material is not included in the chapter's Creative Commons license and your intended use is not permitted by statutory regulation or exceeds the permitted use, you will need to obtain permission directly from the copyright holder.

Wear Prediction for Spur Gear under Dynamic Load and Mixed Lubrication

Lang Li, Tengjiao Lin, and Bingkui Chen

Abstract The gear system in an aircraft engine operates under harsh conditions, subjected to high-speed and heavy-load situations. High sliding velocity and load lead to non-uniform wear on the gear surfaces. The tooth surface wear alter the tooth profile, affect the normal meshing of gear consequently. In order to reveal the non-uniform wear characteristics for the gear pairs, based on the timing-varying stiffness of meshing gear pair, a lumped parameter model of gear pair with nonlinear internal excitation is established. Based on the gear meshing contact and motion analysis, a parameterized model of the gear pair, incorporating nonlinear internal excitation, is constructed. Combined with the line contact mixed EHL model, the tooth surface contact temperature is calculated by considering the comprehensive friction coefficient and heat flux density. According to the classical Archard wear calculation theory, a simulation model of spur gear tooth surface wear considering mixed EHL is constructed to calculate the distribution of wear on the tooth profile.

Keywords Tooth surface Wear · EHL · Dynamic load

1 Introduction

The cylindrical gear transmission system in an aircraft engine inevitably experiences wear on the tooth surfaces due to high sliding velocities and loads during operation. Therefore, it is necessary to conduct numerical simulation and analysis of the wear state for spur gear pair considering dynamic load and mixed elastohydrodynamic lubrication (EHL) conditions. Considerable research on gear surface wear has already been conducted. Flodin et al. [1, 2] were among the first to use the Archard model to analyze the wear of both spur and helical gear, revealing the distribution pattern of gear wear across the tooth profile. Masjedi et al. [3–5] expounded

L. Li · T. Lin (✉) · B. Chen
State Key Laboratory of Mechanical Transmission, Chongqing University, Chongqing, China
e-mail: tjlin1968@162.com

© The Author(s) 2026

Z. Zhang, P. Mercorelli (eds.), *Mechanical Design and Simulation: Insights and Innovations*, Lecture Notes in Mechanical Engineering,
https://doi.org/10.1007/978-981-95-2405-1_32

the wear characteristics of cylindrical spur gears with different rough surface profiles under mixed EHL, predicting the wear rate, which has been widely recognized. Wang et al. [6] investigated the wear of heavy-duty helical gear considering line-contact mixed EHL, analyzing the wear patterns under different conditions. Considering the thermal EHL and point contact model, Xiao et al. [7] calculated the wear of heavy-duty herringbone gears in the start-stop stage. He et al. [8] considered the dynamic meshing forces obtained through finite element (FE) analysis, improving the traditional Archard model based on this. Based on contact analysis, Wallker et al. [9] constructed a multi-physical field transient wear calculation model and analyzed the wear amount of helical gears. In order to obtain the contact force for the calculation of gear wear, Pathak et al. [10] used the linear complementarity theory to the contact analysis of gears.

Numerous scholars have analyzed the tooth surface wear of various types of gears from the perspective of numerical computation. However, there has been limited research on wear calculations under mixed EHL conditions that take dynamic meshing forces into account. This paper focuses on the study of aircraft spur gear pairs, considering the dynamic load and lubrication at the same time, constructing a multi-degree-of-freedom gear dynamics model, calculating the dynamic tooth loads, establishing a gear line contact wear model under mixed lubrication conditions and dynamic load. This model is closer to the actual operation state of the gear, and the influence of more actual excitation on wear can be considered in the future.

2 Analysis of Dynamic Load for Tooth Surface

2.1 Load Distribution Ratio Between Teeth

In one cycle of gear meshing, during single-tooth engagement, the dynamic load is borne by a single tooth, and there is no issue of load distribution. The normal dynamic load is shared by them when the two teeth mesh. In consideration of the geometric backlash (δ) caused by errors and wear, a pair of parallel springs (k_1 and k_2) can be used to simulate the load distribution between two meshing gear pairs, as shown in Fig. 1. The relationship between the loads carried by each pair of gears is as follows:

$$\begin{cases} F_{\mathrm{n}} = F_{\mathrm{n1}} + F_{\mathrm{n2}} \\ F_{\mathrm{n1}} = \left[x(t) - e_1(t)\right]k_1(t) \\ F_{\mathrm{n2}} = \left[x(t) - e_2(t)\right]k_2(t) \end{cases} \tag{1}$$

Where, F_{n} refers to the total dynamic load; $F_{\mathrm{n}j}$ is the load shared by the teeth; $k_j(t)$ is the stiffness of the gear pair; $x(t)$ is the displacement between the gear pairs; $e_j(t)$ is

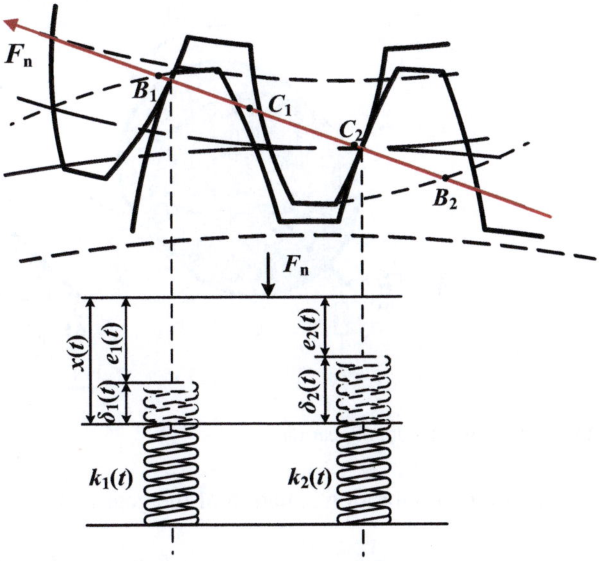

Fig. 1 Diagram of load distribution coefficient

the comprehensive error of the gear pair, primarily considering the combined error caused by wear; $j = 1, 2$ represent the first and second pairs of teeth.

The load distribution coefficient β and the calculation formulas for the loads F_{n1} and F_{n2} are as follows [5]:

$$\begin{cases} \beta = \dfrac{1}{k_1(t)+k_2(t)}\left[k_1(t) - \dfrac{\delta k_1(t)k_2(t)}{F_n}\right] \\ F_{n1} = \dfrac{k_1(t)(F_n - \delta k_2(t))}{k_1(t)+k_2(t)}, F_{n2} = \dfrac{k_2(t)(F_n + \delta k_1(t))}{k_1(t)+k_2(t)} \end{cases} \tag{2}$$

2.2 Dynamic Load Analysis of Tooth Surface

Figure 2 displays the 6-degree-of-freedom dynamic model of the spur gear pair. The time-varying meshing stiffness and damping of this model can be expressed by k_m and c_m. k_{x1}, k_{y1}, k_{x2}, k_{y2} are the support stiffnesses of bearings of the bull wheels (driving) and pinion (driven), respectively. c_{x1}, c_{y1}, c_{x2} and c_{y2} are the support damping of the bearings for the bull and pinion gears, respectively. $e(t)$ represent the static transmission error. The relative displacement δ_{12} and dynamic meshing force F_s of the meshing point in the meshing direction are as follows,

$$\begin{cases} \delta_{12} = (x_1 - x_2)\sin\alpha_n + (y_1 - y_2)\cos\alpha_n + r_{b1}\theta_1 - r_{b2}\theta_2 - e(t) \\ F_s = k_m\delta_{12} + c_m\dot{\delta}_{12} + F_c(t) \end{cases} \tag{3}$$

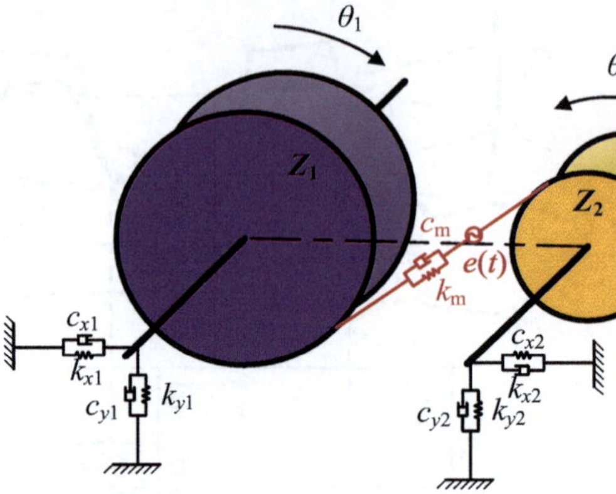

Fig. 2 Dynamic model of gear pair

The differential vibration equation of the gear pair is,

$$\begin{cases} m_1\ddot{x}_1 + c_{x1}\dot{x}_1 + k_{x1}x_1 + F_s \sin\alpha_n - F_{f1}\cos\alpha_n = 0 \\ m_1\ddot{x}_1 + c_{y1}\dot{y}_1 + k_{y1}y_1 + F_s \cos\alpha_n + F_{f1}\sin\alpha_n = 0 \\ J_1\ddot{\theta}_1 + r_{b1}F_s - T_{f1} = T_{in} \\ m_2\ddot{x}_2 + c_{x2}\dot{x}_2 + k_{x2}x_2 - F_s \sin\alpha_n + F_{f2}\cos\alpha_n = 0 \\ m_2\ddot{y}_2 + c_{y2}\dot{y}_2 + k_{y2}y_2 - F_s \cos\alpha_n - F_{f2}\sin\alpha_n = 0 \\ J_2\ddot{\theta}_2 - r_{b2}F_s + T_{f2} = -T_{out} \end{cases} \tag{4}$$

Where, m_k, J_k, F_{fk} and T_{fk} represent the mass, moment of inertia, frictional force of the gear pair and moment of friction, respectively. Here, $k = 1, 2$ corresponds to the bull and pinion gears, respectively. T_{in} and T_{out} represent the input and load torques.

3 The Wear Model under EHL

Figure 3 illustrates the equivalent cylindrical contact model at the contact point i for an involute spur gear, and defines the dimensionless function Λ as the ratio of the distance from P to i and distance from P to N_1, which can be expressed as,

$$\Lambda = \frac{\overline{Pi}}{\overline{N_1P}} = \frac{\tan\alpha_i}{\tan\alpha_n} - 1 \tag{5}$$

Where, α_i is the pressure angle of meshing point. α_n is base circle pressure angle.

Then the relationship between the curvature radii R_{1i} and R_{2i} for a meshing point i on the bull gear and pinion and the function Λ is as follows,

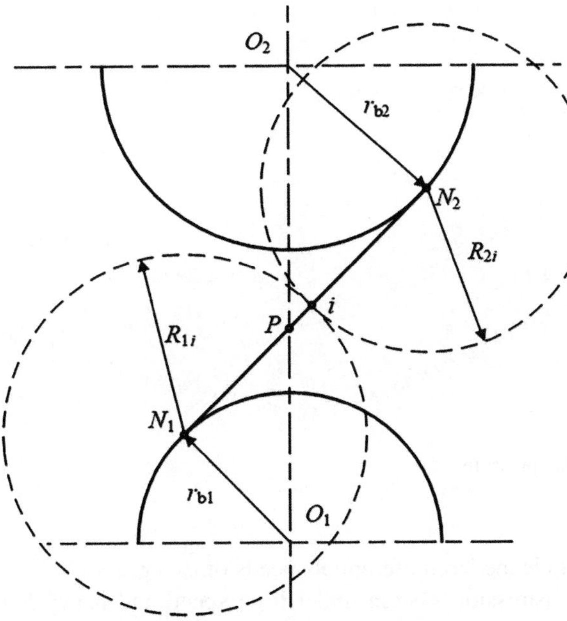

Fig. 3 Equivalent cylindrical contact model

$$R_{1i} = (\Lambda + 1) r_{b1} \tan\alpha_n, R_{2i} = (\Lambda - 1) r_{b1} \tan\alpha_n \tag{6}$$

Based on the hertz contact pressure calculation formula, the contact half-width a_h and the Hertz contact pressure p_h of the tooth surface can be expressed as [1],

$$a_h = \sqrt{\frac{8F_n R_{ei}}{\pi b E^*}}, p_h = \sqrt{\frac{\pi F_n E^*}{32 b R_{ei}}} \tag{7}$$

Where, F_n is the normal meshing force. R_{ei} represent the equivalent radius of curvature of the mesh node. b refers to width of the gear teeth. R_{ji} refers to the radius of curvature of gear meshing point. E^* represent the equivalent modulus of gear material. $j = 1, 2$, representing the first and second pairs of teeth. r_b represent the base circle radius.

Figure 4 shows the Hertzian contact pressure distribution of meshing gear pair.

According to gear meshing principle, the circumferential speed of the bull gear and pinion is different at the contact point. Therefore, there will be relative sliding at the meshing point on the two tooth surfaces. This calculation formula of the relative moving sliding distance S_1 and S_2 of two gears at the meshing points is [2],

$$S_1 = 2a_h \left|1 - u_{t2} / u_{t1}\right|, S_2 = 2a_h \left|1 - u_{t1} / u_{t2}\right| \tag{8}$$

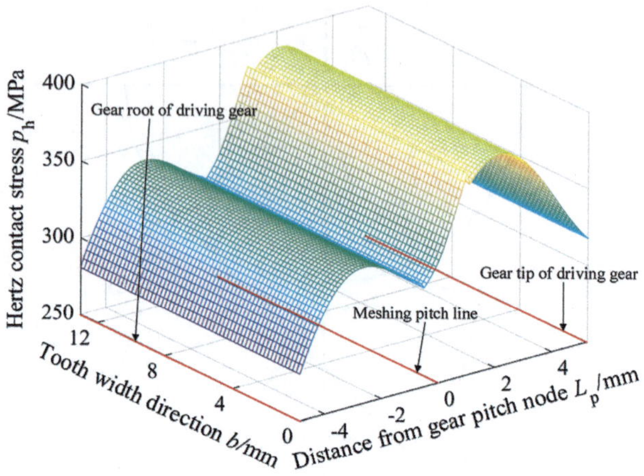

Fig. 4 Hertz contact pressure

Where, u_{t1} and u_{t2} is the circumferential speeds of two gears.

Spur gear transmission system under high speed and heavy load often works under mixed EHL. Consequently, it is necessary to establish a linear contact mixed EHL model when calculating wear. According to the research of Masjedi et al., the formulas for calculating the convex load ratio and dimensionless minimum oil film thickness of the linear contact mixed EHL model are as follows [4],

$$\begin{cases} L_a = 0.005W^{-0.408}U^{-0.088}G^{0.103}\ln\left(1+4470\sigma^{-6.015}V^{1.168}W^{0.485}U^{-3.741}G^{-2.898}\right) \\ H_{min} = \dfrac{h_{min}}{R_{ei}} = 1.652W^{-0.077}U^{0.716}G^{0.695}\left(1+0.026\sigma^{1.120}V^{0.185}W^{-0.312}U^{-0.809}G^{-0.977}\right) \end{cases} \tag{9}$$

Where, h_{min} is the minimum oil film thickness. L_a is the bearing ratio of micro-convex body. H_{min} refers to the dimensionless oil film thickness. U is the dimensionless rolling rate. W represent the dimensionless load parameter. V is dimension-free surface hardness. G is dimensionless elastic modulus. σ is dimensionless surface roughness.

During the operation of gears, due to the friction between micro-protrusions and the extrusion of lubricating oil film, a large amount of heat will be generated, resulting in an increase in the temperature rise at the meshing point of tooth profile. This temperature rise of the gear contact area is as follows according to the Green's function deduced by Tian and Kennedy [5],

$$\Delta T = \frac{2a_h q_m}{\sqrt{\pi}\left(K_1\sqrt{1+P_{e1}} + K_2\sqrt{1+P_{e2}}\right)} \tag{10}$$

Where, q_m refers to the average heat flux. K_1 and K_2 are the heat conduction coefficients of the big and small gears respectively. P_{e1} and P_{e2} are Peclet numbers of the bull gear and pinion respectively [5].

$$V_h = K\psi\left(\frac{La}{100}\right)\frac{FS}{H_d}, \psi = 1 - \exp\left[-\frac{\chi}{u_r t_0}\exp\left(-\frac{E_a}{R_g T_s}\right)\right] \tag{11}$$

Where, K represent the dimensionless wear coefficient. V_h is the unit wear. ψ is the oil film loss coefficient. χ is the diameter of oil film molecules in the adsorption area. T_0 refers to the basic time of molecular vibration in adsorption state. R_g refers to gas constant. E_a represent the adsorption heat of lubricating oil molecules on the tooth surface.

By discretizing the above formula, the formula for calculating the wear of a meshing point i on the gear tooth surface is obtained as follows,

$$\Delta h = \int_0^S k_h \psi \frac{L_a}{100} p_h dS \tag{12}$$

Where, Δh represent the wear depth of a meshing point. k_h is the wear coefficient.

4 Simulation and Result Analysis

The results of wear and cumulative wear of the bull gear under EHL is shown in Fig. 5. From the calculation results, for the bull gear, the wear amount is the largest at the tooth tip, which decreases nonlinearly from the top of the tooth to the mesh pitch node, and has a sudden change when the single and double teeth are alternately engaged, and the wear amount at the pitch node is the smallest; Correspondingly, for the pinion, the distribution law of wear from the tooth tip to the tooth root is consistent with that of the bull gear from tooth root to tooth tip.

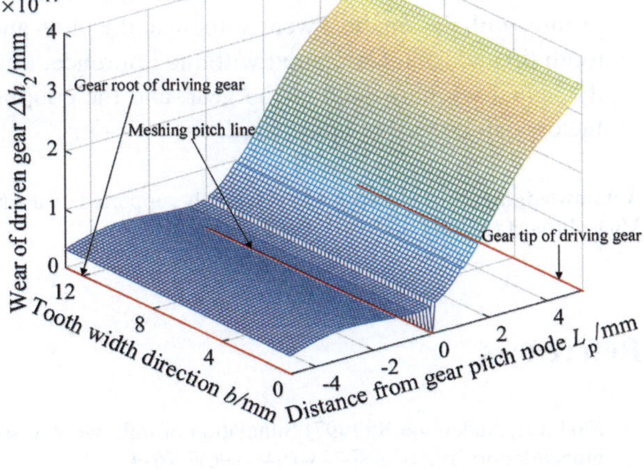

Fig. 5 Wear of the bull gear under EHL

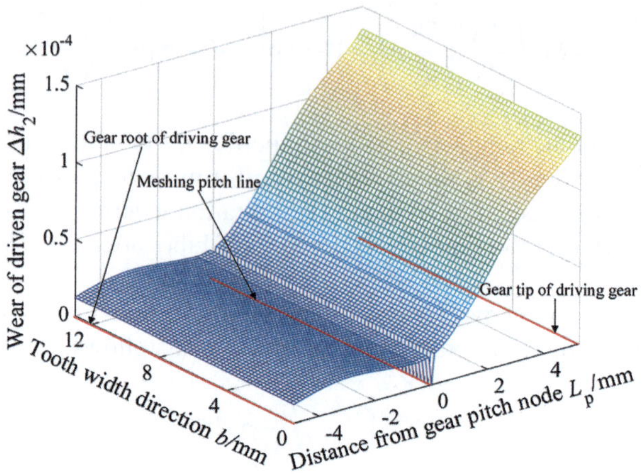

Fig. 6 Wear of the bull gear under dry friction

Compared with Fig. 6, the order of magnitude of the wear calculation results considering EHL is more realistic [5, 8].

5 Conclusion

1. The 6-degree-of-freedom gear dynamic model for spur gears was established, and it was found that the dynamic load on the mesh tooth fluctuates around the static meshing force, while there is a significant difference in the amplitude of the tooth surface Hertz contact pressure between the single-tooth meshing interval and the double-tooth meshing interval.
2. The tooth surface wear calculation method considering dynamic loads and mixed EHL was constructed, and numerical simulation was performed using Matlab software to solve the distribution of tooth surface wear. It is concluded that the tooth surface wear amount is non-uniformly distributed along the tooth profile, with the smallest wear amount at the pitch and larger wear amounts at the tooth tip and root. Compared with the reference, this model is more in line with the actual working state of the gear, and the coupling effect of wear and other factors can be further considered.

Acknowledgments This work was financially supported by the National Science and Technology Major Project (No: HT-J2019-IV-0001-0068).

References

1. Flodin A, Andersson S (1997) Simulation of mild wear in spur gears. Wear 207(1–2):16–23. https://doi.org/10.1016/S0043-16480(96)7467-4

2. Flodin A, Andersson S (2000) Simulation of mild wear in helical gears. Wear 241(2):123–128. https://doi.org/10.1016/S0043-1648(00)00384-7
3. Akbarzadeh S, Khonsari MM (2009) Prediction of steady state adhesive wear in spur gears using the EHL load sharing concept. ASME J Tribol 115(3):493–503. https://doi.org/10.1115/1.3075859
4. Masjedi M, Khonsari MM (2014) Theoretical and experimental investigation of traction coefficient in line-contact EHL of rough surfaces. Tribol Int 70:179–189. https://doi.org/10.1016/j.triboint.2013.10.004
5. Masjedi M, Khonsari MM (2015) On the prediction of steady-state wear rate in spur gears. Wear 342:234–243. https://doi.org/10.1016/j.wear.2015.08.010
6. Wang H, Zhou C, Lei Y et al (2019) An adhesive wear model for helical gears in line-contact mixed elastohydrodynamic lubrication. Wear 426:896–909. https://doi.org/10.1016/j.wear.2019.01.104
7. Xiao Z, Shi X, Wang X et al (2021) Lubrication analysis and wear mechanism of heavily loaded herringbone gears with profile modifications in full film and mixed lubrication point contacts. Wear 477:203790. https://doi.org/10.1016/j.wear.2021.203790
8. He Z, Hu Y, Zheng X et al (2022) a calculation method for tooth wear depth based on the finite element method that considers the dynamic mesh force. Machines 10:69. https://doi.org/10.3390/machines10020069
9. Walker J, Mohammadpour M, Theodossiades S et al (2022) A multi-physics transient wear model for helical gear pairs. Tribol Int 169:107463. https://doi.org/10.1016/j.triboint.2022.107463
10. Pathak M, Jayanti L, Rakshit S (2024) Linear complementarity formulations for contact analysis and wear prediction in gears. Mech Mach Theory 198:105669. https://doi.org/10.1016/j.mechmachtheory.2024.105669

Open Access This chapter is licensed under the terms of the Creative Commons Attribution-NonCommercial-NoDerivatives 4.0 International License (http://creativecommons.org/licenses/by-nc-nd/4.0/), which permits any noncommercial use, sharing, distribution and reproduction in any medium or format, as long as you give appropriate credit to the original author(s) and the source, provide a link to the Creative Commons license and indicate if you modified the licensed material. You do not have permission under this license to share adapted material derived from this chapter or parts of it.

The images or other third party material in this chapter are included in the chapter's Creative Commons license, unless indicated otherwise in a credit line to the material. If material is not included in the chapter's Creative Commons license and your intended use is not permitted by statutory regulation or exceeds the permitted use, you will need to obtain permission directly from the copyright holder.

Simulation Analysis of the Feeding Mechanism of a Self-Propelled Total Mixed Ration Preparation Machine

Baoxing Wang, Zheng Zhang, Zhenxing Zhao, Qian Bao, Changzhe Liu, and Huabin Qin

Abstract In response to the current situation where Chinese self-propelled total mixed ration preparation machines lack the reclaiming mechanism and parameter optimization, the model of the reclaiming mechanism was designed and optimized. EDEM was employed to conduct orthogonal experimental simulation analysis on four factors that affect the reclaiming effect of the reclaiming mechanism, namely the feed rate, revolution speed, spiral blade pitch, and blade length ratio. Eventually, the primary and secondary order of the influence of each factor on the feeding efficiency is as follows: feed rate, spiral blade pitch, blade length ratio, and revolution speed. And the primary and secondary order of the influence of each factor on the backflow rate is: spiral blade pitch, blade length ratio, revolution speed, and feed rate. Considering both the feeding efficiency and backflow rate, the optimal parameter combination is a feed rate of 4 m/min, a spiral blade pitch of 230 mm, a blade length ratio of the reclaiming cutter of 1.8, and a revolution speed of the reclaiming roller of 230 r/min. When selecting the best scheme, the feeding efficiency is 82.6 m^3/h, and the backflow rate is 38.93%.

Keywords Total mixed ration preparation machines · The feeding mechanism · Orthogonal test

1 Introduction

In the late 1960s, the total mixed ration feeding technology gradually began to be promoted in the United States, Britain, Israel, and other countries [1]. After more than 50 years of development, Europe, the United States, and other countries have

B. Wang (✉) · Z. Zhang · Z. Zhao · Q. Bao · C. Liu · H. Qin
Weifang Vocational College, Weifang, Shandong, China
e-mail: 2022020020@sdwfvc.edu.cn

© The Author(s) 2026

Z. Zhang, P. Mercorelli (eds.), *Mechanical Design and Simulation: Insights and Innovations*, Lecture Notes in Mechanical Engineering,
https://doi.org/10.1007/978-981-95-2405-1_33

become more mature in total mixed ration feeding technology, and the research content has become more comprehensive [2–4]. At present, a variety of models of total mixed ration preparation machines have been developed. The 12.1DL self-propelled total mixed ration preparation machine developed in France has a single vertical auger. The Triotrac self-propelled total mixed ration preparation machine developed by the Netherlands has a conveyor belt width of 1.85 m, which can realize stepless speed change control.

In recent years, China's research institutes, universities and enterprises have made some research results. For example, Yu Keqiang et al. of Northeast Agricultural University developed a rotary total mixed ration mixer [5], Li Long of Inner Mongolia Agricultural University developed a 9JQL-8.0 traction TMR feed mixer [6], and 9JLT-10 total mixed ration mixer developed by Heilongjiang Academy of Agricultural Machinery Engineering [7]. China's self-propelled total mixed ration preparation machine has been studied, but it lacks research on the feeding mechanism and optimization method, and has not overcome key technologies and has not formed industrialization.

Currently, in most ranches in China, a combination of multiple machines—such as silage loaders, conveyor belts, total mixed ration (TMR) mixers, and feed pushers—is adopted to complete the task of feeding cattle. The low degree of automation in the entire operation process restricts the rapid development of China's cattle breeding industry.

To promote the development of China's cattle breeding industry, this paper presents the development of a self-propelled total mixed ration (TMR) mixer. The feeding mechanism of the machine was designed precisely, and the discrete element simulation analysis and parameter optimization of the feeding process of the machine were carried out by EDEM software.

2 Structure of Self-Propelled Total Mixed Ration Preparation Machine

The 25 m^3 self-propelled total mixed ration preparation machine is mainly composed of a feeding mechanism, a stirring mechanism, a material throwing mechanism, and a transmission system. The overall structure of the machine is shown in Fig. 1.

The feeding mechanism of the self-propelled total mixed ration preparation machine is driven by a closed hydraulic transmission system, and the material is transported to the mixing and stirring mechanism through the feeding conveyor belt in the feeding mechanism. The mixing and stirring mechanism cuts and mixes the materials to ensure that the materials reach the ideal mixing state. After that, the prepared material is scattered to the feeding area by the spreading mechanism.

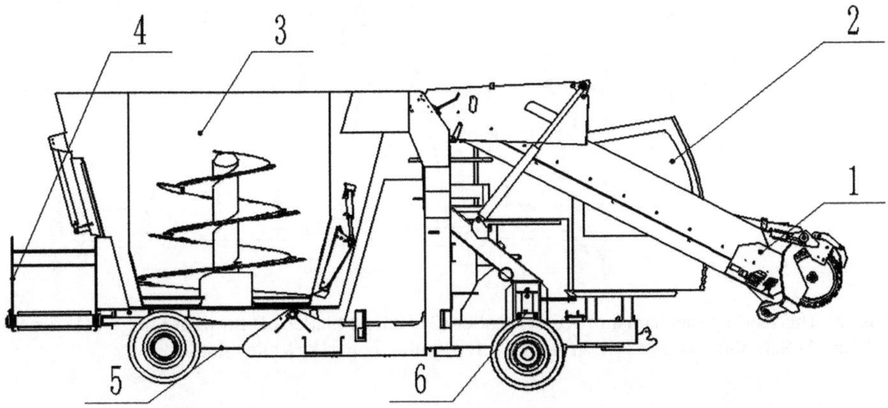

Fig. 1 Structure diagram of self-propelled total mixed ration preparation machine. (1) Feeding mechanism (2) Wheelhouse (3) A stirring mechanism (4) A material throwing mechanism (5) Chassis (6) Transmission system

3 The Structure of the Feeding Mechanism Part

The reclaimer of the self-propelled total mixed ration preparation machine mainly consists of a feeding part and a material conveying part. The structure of the feeding part mainly includes a roller, helical blades, straight knives, curved knives, a main shaft, a hydraulic motor and a roller shield. There is a reclaiming main shaft in the central part of the roller, and the hydraulic motor is connected to the reclaiming main shaft. Moreover, two sections of helical blades with opposite rotation directions are welded onto the circumference of the roller, and the straight knives and the curved knives are installed on the helical blades at intervals. Meanwhile, a notch is made in the middle of the rear part of the shield. The structure of the whole part is shown in Fig. 2.

The hydraulic motor drives the roller to rotate at a high speed, and the reclaiming blades (straight knives and curved knives) mounted on the spiral blade remove the silage from the silage pile. The reclaiming knife installed on the left-handed spiral blade and the right-handed spiral blade pushes the silage material along the axial direction of the roller to the middle position, and then throws the silage material onto the rear material conveyor belt. By adjusting the speed of the hydraulic motor, the feeding efficiency of the whole mechanism can be controlled.

The material conveying part is mainly composed of hydraulic motor, drive roll, passive roller, conveyor belt, conveyor belt shield and lifting hydraulic cylinder. The conveyor belt is equipped with a scraping plate to enhance the conveying capacity of the material. As shown in Fig. 3, the figure is the side view of the reclaiming mechanism, which can clearly reflect the conveying part.

The hydraulic motor drives the active roller to rotate at high speed, and the active roller drives the conveyor belt to move and throw the material into the rear tank. The

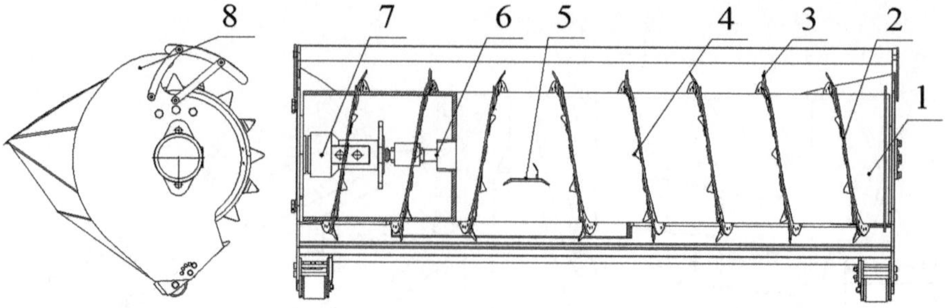

Fig. 2 The feeding mechanism part. (1) Roller (2) Helical blades (3) Straight knives (4) Curved knives (5) Scraping blade (6) Main shaft (7) Hydraulic motor (8) Roller shield

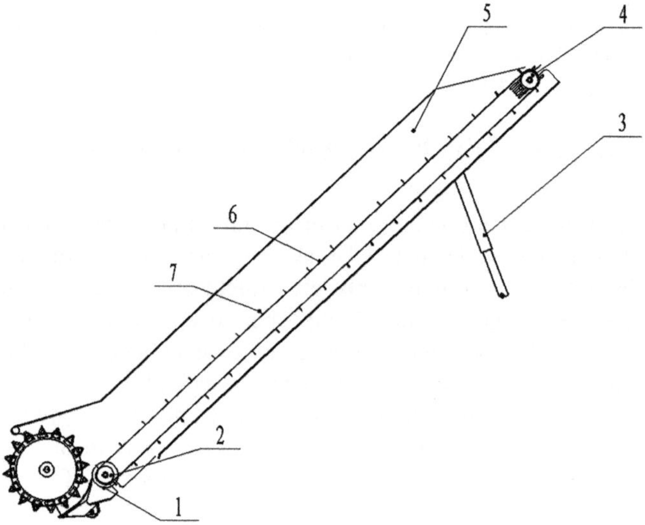

Fig. 3 Material conveying part. (1) Hydraulic motor (2)Drive roll (3) Lifting hydraulic cylinder (4) Passive roller (5) Conveyor belt shield (6) Conveyor belt (7) Scraping plate

scrapers set at intervals on the conveyor belt can prevent the silage from sliding and improve the conveying efficiency.

4 Simulation Analysis

4.1 Establish an EDEM Simulation Model

The discrete element software EDEM, a product of DEM Solutions, was established in 2002 and is headquartered in Edinburgh, UK. The software is the first particle discrete element method software that can be used in many fields around the world

[8]. For example, EDEM can be used for related research and analysis in agriculture, mining, pharmaceutical and raw material processing industries. In addition, EDEM can be coupled with a variety of software simulation analysis to achieve the study of complex problems.

The establishment of the correct particle model is the basis of the whole simulation analysis. To improve computational efficiency and reduce simulation time, the simulation analysis uses the main components of corn stover in silage, namely corn stover skin and pith, as the material. The average size of the actual material is statistically analyzed, and then the external dimensions are doubled, with spherical particles combined to simulate the external shapes of both. The particle model of the corn stover skin is 91 mm long, 10 mm wide, and 10 mm high, with a density of 287 kg/m^3; The particle model of the corn stover pith is 39 mm long, 38 mm wide, and 26 mm high, with a density of 153 kg/m^3. Material property parameters are referenced in Literature 10 [9].

4.2 *Test Scheme*

Based on the preliminary single-factor trial scheme, key factors affecting the feeding efficiency and backflow rate, such as the feed Rate, revolution speed, spiral blade pitch, and the length ratio of the feeding blade, were selected for orthogonal testing. Each factor has three levels, so an L9(3^4) orthogonal table, which is an equal-level orthogonal test table, was chosen for four factors and three levels. The trial scheme is shown in Table 1.

The mass of the material collected by the mechanism during the simulation time is analyzed statistically, and the volume of material collected per hour is calculated to determine the feeding efficiency. The mass of the material splashed out of the

Table 1 Test design and results

	Experimental factors				Targets of test	
Group number	Feed rate (m·min^{-1})	Revolution speed (r·min^{-1})	Spiral/ blade/pitch (mm)	Blade length ratio	feeding efficiency (m^3/h)	Backflow rate (%)
1	1	160	230	1	27.93	38.72
2	1	230	330	1.4	25.47	44.07
3	1	300	430	1.8	23.88	54.34
4	2.5	160	330	1.8	56.94	52.1
5	2.5	230	430	1	52.92	48.44
6	2.5	300	230	1.4	58.22	39.4
7	4	160	430	1.4	69.16	53.81
8	4	230	230	1.8	82.60	38.93
9	4	300	330	1	78.71	42.34

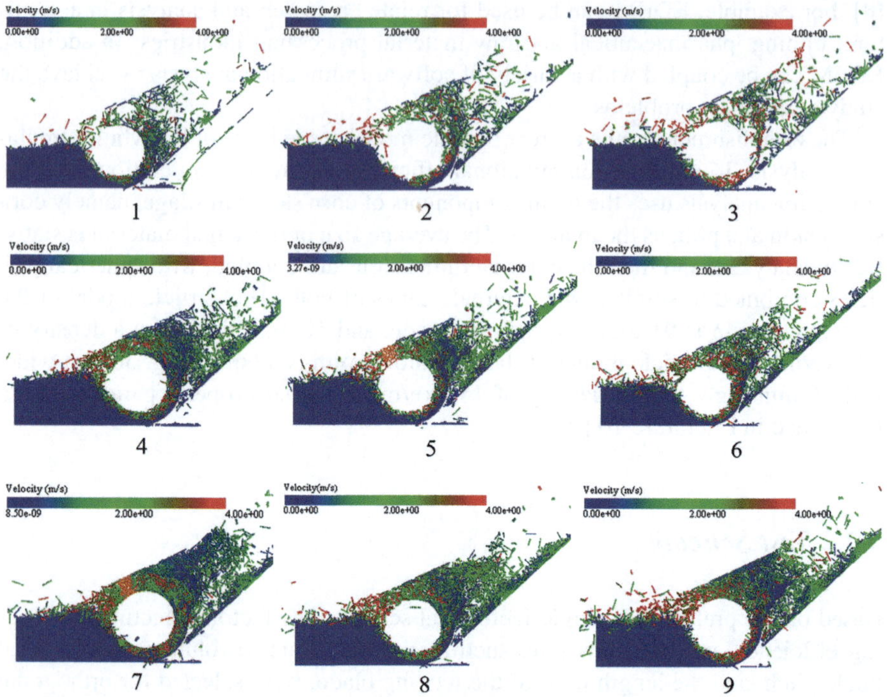

Fig. 4 Particle velocity distribution under different groups

roller per unit time is compared with the total mass of the material sucked into the roller per unit time, resulting in the backflow rate.

The particle velocity distribution in different groups is shown in Fig. 4. The figure can clearly reflect the velocity of material particles at different positions and the distribution of particles. From the diagram, it can be seen that the feeding amount of 1 ~ 3 groups is less, and the feeding amount of 7 ~ 9 groups is the largest, so it can be preliminarily determined that the forward speed has a great influence on the feeding efficiency.

4.3 Variance Analysis of Simulation Results

Analysis of variance (ANOVA) is an analysis method to distinguish the fluctuation of test index caused by the change of test factor level from the fluctuation of test index caused by test interference factors.

In order to measure the F ratio, it is necessary to use the critical value table of the F-distribution and the standard value. The critical value table of the F-distribution is a numerical table of the standard values under various confidence levels and degrees of freedom, which is compiled based on the principles of mathematical statistics. These numerical values can be obtained from Reference [10].

Table 2 Analysis of variance table

Targets of test	Source	Sum of squares	Degrees of freedom	Mean square	F	F_a	Sig.
The feeding efficiency	A	3955.62	2	1977.81	267.905	$F_{0.01}(2,4) = 18$	※※
	B	10.5	2	5.25	0.711	$F_{0.05}(2,4) = 6.94$	
	C	89.62	2	44.81	6.07	$F_{0.1}(2,4) = 4.32$	(※)
	D	19.03	2	9.515	1.289	$F_{0.25}(2,4) = 2$	
	Error-E	29.53	4	7.3825	—		
The backflow rate	A	3.97	2	1.985	1	$F_{0.01}(2,2) = 99$	
	B	29.85	2	14.925	7.519	$F_{0.05}(2,2) = 19$	[※]
	C	261.2	2	130.6	65.793	$F_{0.1}(2,2) = 9$	※
	D	41.98	2	20.99	10.574	$F_{0.25}(2,2) = 3$	(※)
	Error-E	3.97	2	1.985	—		

The significance levels of the influence of experimental factors on the indicators, from high to low, are indicated as follows: ※※, ※, and (※). The analysis of variance for the two experimental indicators, the feeding efficiency and the backflow rate, is shown in Table 2.

As can be seen from Table 2, The F-ratios of factors A, B, C, and D under each test index are F_A, F_B, F_C, and F_D, respectively. Under the indicator of feeding efficiency, the order of the F ratios is $F_A > Fc > F_D > F_B$. Therefore, the primary and secondary order of the influence of each experimental factor on the feeding efficiency is A, C, D, B. Combined with the influence degree of each experimental factor on the feeding efficiency, the relatively optimal scheme is $A_1C_{-1}D_1B_0$. Under the indicator of backflow rate, the order of the F ratios is $F_C > F_D > F_B > F_A$. Therefore, the primary and secondary order of the influence of each experimental factor on the backflow rate is: C, D, B, A. Combined with the influence degree of each factor on the backflow rate, the relatively optimal scheme is $C_{-1}D_{-1}B_0A_1$.

Since the importance of the two experimental indicators is in the order of feeding efficiency and backflow rate, considering comprehensively the influence degree of each factor level on the values of the two indicators and the relatively optimal schemes under each indicator, the optimal scheme selected is $A_1C_{-1}D_1B_0$, that is, the feed rate is 4 m/min, the spiral blade pitch is 230 mm, the blade length ratio of the reclaiming cutter is 1.8,and the revolution speed is 230 r/min. Under the optimal scheme, the feeding efficiency is 82.6 m^3/h and the backflow rate is 38.93%.

5 Conclusions

The drum-type reclaiming mechanism model for the self-propelled total mixed ration preparation machine was innovatively designed, and the orthogonal experimental simulation analysis of the feed rate, revolution speed, spiral blade pitch, and blade length ratio of the reclaiming mechanism was conducted using EDEM. It was found during the simulation process that, due to the existence of cutters on the reclaiming drum, some corn stover silage rotates along with the high-speed rotating reclaiming drum, which will thus affect the feeding efficiency. Analysis of variance was performed on the simulation test results, taking into account the influence degree of each experimental factor level on the two indicator values of feeding efficiency and backflow rate. The optimal scheme selected is $A_1C_{-1}D_1B_0$, that is, the feed rate is 4 m/min, the revolution speed is 230 r/min, the blade length ratio of the reclaiming cutter is 1.8, and the spiral blade pitch is 230 mm. Finally, Under the optimal scheme, the feeding efficiency reaches 82.6 m^3/h, while the backflow rate is 38.93%.

References

1. Schingoethe DJ. A 100-year review: Total mixed ration feeding of dairy cows . Dairy Sci 2017 Dec;100 (12):10143–10150. doi: https://doi.org/10.3168/jds.2017-12967. PMID: 29153159
2. Liu Z (2023) The advantages, existing problems, and future development suggestions of Total mixed ration preparation machines. Agric Mach (03):57–58. https://doi.org/10.16167/j.cnki.1000-9868.2023.03.030
3. Li Y, Duan X, Zhang W et al (2024) Research progress on the application of total mixed ration (TMR) in beef cattle. China Feed (14):5–9. https://doi.org/10.15906/j.cnki.cn11-2975/s.20241402
4. Yan X, Ma H, Li W et al (2022) Key technologies and application of Total mixed ration (TMR) feeding for beef cattle. Feed Res 45(24):137–141. https://doi.org/10.13557/j.cnki.issn1002-2813.2022.24.027
5. Yu K, Li L, He X et al (2015) Experimental design and principle analysis on paddle-wheel total mixed ration mixer. Trans Chin Soc Agric Mach 46(7):109–117,170. https://doi.org/10.6041/j.issn.1000-1298.2015.07.017
6. Long L (2012) 9JQL-8.0 traction type TMR (Total mixed ration) feed blender mixing dragon design and analysis [D]. Inner Mongolia Agricultural University, Hohhot. https://kns.cnki.net
7. Tian F, Chen Y, Song Z et al (2020) Design and experiment of self-propelled TMR preparation mixer for silage straw feed. Trans Chin Soc Agric Mach 51(3):106–114. https://doi.org/10.6041/j.issn.1000-1298.2020.03.012
8. Liang J (2024) Application of EDEM software in lightweighting of agricultural machinery. J Shijiazhuang Vocat Techn College 36(2):10–15. https://kns.cnki.net
9. Niu Q, Wang B, Dong H et al (2022) Feeding simulation and parameter optimization of self-propelled total mixed ration preparation machine. J Chin Agric Mech 43(10):99–106. https://doi.org/10.13733/j.jcam.issn.2095-5553.2022.10.015
10. Baoxing W (2022) Design and optimization of self-propelled Total mixed ration preparation machine [D]. Shandong Agric Univ. https://doi.org/10.27277/d.cnki.gsdnu.2022.000560

Open Access This chapter is licensed under the terms of the Creative Commons Attribution-NonCommercial-NoDerivatives 4.0 International License (http://creativecommons.org/licenses/by-nc-nd/4.0/), which permits any noncommercial use, sharing, distribution and reproduction in any medium or format, as long as you give appropriate credit to the original author(s) and the source, provide a link to the Creative Commons license and indicate if you modified the licensed material. You do not have permission under this license to share adapted material derived from this chapter or parts of it.

The images or other third party material in this chapter are included in the chapter's Creative Commons license, unless indicated otherwise in a credit line to the material. If material is not included in the chapter's Creative Commons license and your intended use is not permitted by statutory regulation or exceeds the permitted use, you will need to obtain permission directly from the copyright holder.

Collision Simulation Model Accuracy Prediction Based on BP Neural Network

Hailan Xu, Zhenfei Zhan, Yajun Luo, Changhao Li, and Hailiang Chen

Abstract Based on the car's frontal collision test, this article uses MADYMO software to digitally model the occupant's frontal collision. By adjusting the occupant's sitting angle, the AEB (Autonomous Emergency Braking) pre-braking conditions, collision conditions and collision conditions under the action of AEB are simulated. The simulation data is put into the BP (Back Propagation) neural network for training, in which the simulation data of AEB pre-braking conditions and the simulation data of collision conditions are used as input, and the simulation data of collision conditions under the action of AEB are used as output. The input and output data include seat belt force, Simulation data such as head acceleration, neck bending moment, chest compression, etc. Then, the AEB pre-braking condition test data and collision condition test data are put into the trained prediction model to predict the data of the collision condition under the action of AEB. Finally, the EEARTH (Enhanced Error Assessment of Response Time Histories) method is used to score the simulation data separately from the predicted data and experimental data, and the predicted scores and actual scores of various data are compared and analyzed. The results show that the built BP neural network model has good prediction effect and the accuracy prediction method is feasible.

Keywords MADYMO · Digital modeling · BP neural network · EEARTH method

H. Xu
China Merchants Testing Vehicle Technology Research Institute, Chongqing, China

Z. Zhan (✉) · Y. Luo · C. Li · H. Chen
School of Mechatronics and Vehicle Engineering, Chongqing Jiaotong University, Chongqing, China
e-mail: zhenfeizhan@cqjtu.edu.cn

© The Author(s) 2026

Z. Zhang, P. Mercorelli (eds.), *Mechanical Design and Simulation: Insights and Innovations*, Lecture Notes in Mechanical Engineering,
https://doi.org/10.1007/978-981-95-2405-1_34

1 Introduction

Vehicle collision testing is a crucial method for evaluating the structural strength of vehicles and the protection performance for occupants. However, actual collision tests are not only costly but also accompanied by expensive testing equipment and difficulties in data acquisition, especially under complex scenarios or extreme conditions. In recent years, with the development of simulation technology, establishing digital models to simulate vehicle collisions in a virtual environment has become an effective alternative. The accuracy of the simulation model directly affects the reliability of the simulation results. Insufficient model accuracy may lead to incorrect safety performance assessments. Therefore, improving the accuracy of collision simulation models and verifying their reliability have become important research topics in automotive safety. This paper proposes a collision simulation model prediction method based on BP neural network. The model is trained using the simulation data of AEB (Autonomous Emergency Braking) pre-braking conditions, the simulation data of collision conditions, and the simulation data of collision conditions under AEB. Then, the test data of AEB pre-braking conditions and the test data of collision conditions are used to predict the test data of collision conditions under AEB. Finally, the EEARTH method is used to analyze and score the test data and the prediction data to prove the reliability of the prediction method.

Researchers have conducted in-depth research on the reliability verification of complex simulation models with uncertain factors [1], multivariate static output variables [2], and multiple sub-models [3]. In the field of model validation, researchers have developed a range of methods, from classical statistical approaches to advanced techniques such as Bayesian networks, to address issues related to uncertain simulation models and to evaluate the overall performance of complex systems [4–12].

The first part of the text is a description of the establishment of the occupant frontal collision simulation model. The second part quantitatively verifies the collision simulation model based on the EEARTH method. The third part proposes a BP neural network to predict the model. Finally, the EEARTH method is used to achieve the accuracy prediction of the simulation model.

2 Occupant Frontal Collision Simulation Modeling

In the field of collision safety, occupant posture has a significant impact on the degree of injury sustained during a collision. Therefore, in the digital modeling process, accurately simulating the posture of the occupant is one of the key factors to improve the accuracy of the simulation model. This study uses MADYMO software to model and simulate the collision process, and considers the occupant's posture as a design variable. The simulation modeling process and the collision model are shown in Figs. 1 and 2.

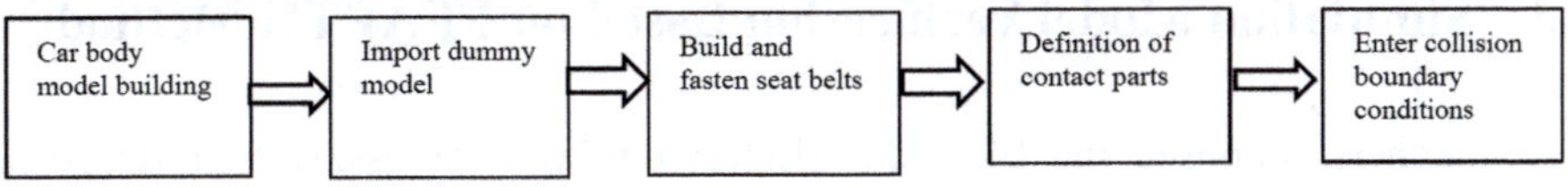

Fig. 1 Simulation model building process

Fig. 2 Import dummy model

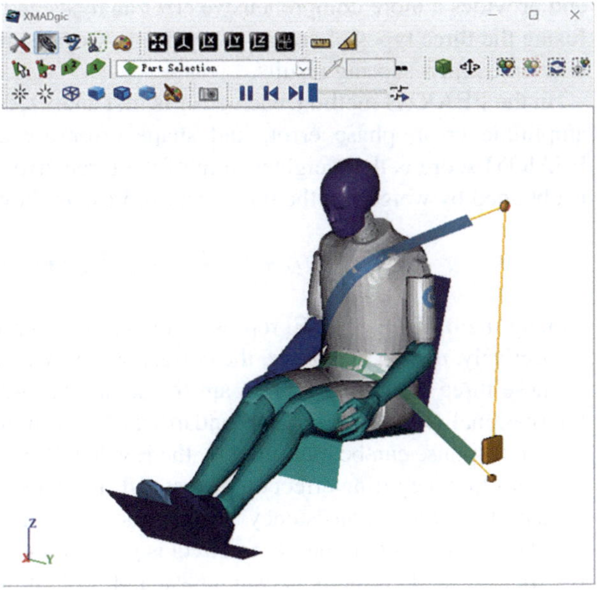

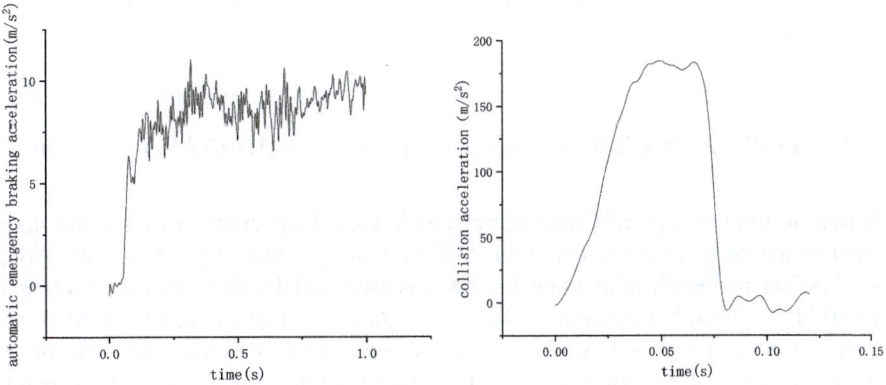

Fig. 3 Automatic emergency braking acceleration and collision acceleration

This study simplifies the action process of active safety devices into automatic emergency braking. Figure 3 shows the acceleration curve of the AEB stage and the acceleration curve of the collision. The relevant data is provided by China Merchants Vehicle Inspection Institute.

3 Simulation Model Verification Based on EEARTH Method

Researchers proposed the EEARTH (Enhanced Error Assessment of Response Time Histories) method to conduct a comprehensive evaluation of the dynamic response of the system. The EEARTH method not only inherits the advantages of traditional EARTH error analysis, but also further expands its scope of application and provides a more comprehensive error analysis and evaluation of the model. By fusing the three types of errors, a composite score is generated that is more consistent with expert scoring results.

In the EEARTH method, each error is first analyzed separately, and the scores of amplitude error, phase error, and shape error are calculated respectively. The EEARTH score is the weighted sum of the three error scores. The EEARTH score is obtained by weighting the three error scores, as shown in Formula 1.

$$E = \omega_p \cdot E_p + \omega_m \cdot E_m + \omega_s \cdot E_s \quad (1)$$

Among them, E_p, E_m and E_s represent the phase, amplitude and shape error scores respectively, ω_p, ω_m and $\omega_m$are the corresponding weighting factors, and the values of these three weighting factors are the same. According to the provisions of the International Organization for Standardization Committee, the fitting effect of the system response can be measured by the E value. When the E value exceeds 94%, it indicates that the fitting effect of the test and simulated response is very good, showing a high degree of consistency between the two; when the E value is between 80% and 94%, it means that the fitting effect is good and the correlation between the two is strong. If the E value is between 58% and 80%, it means that the fitting effect is general, and there may be certain errors and deviations; finally, when the E value is lower than 58%, it means that there is almost no correlation between the test and simulated response.

3.1 AEB Pre-braking Condition Simulation Model Verification

Based on the test data of China Merchants Vehicle Inspection Institute, this study verified the deceleration effect of the AEB system by simulating the movement of a sled. A 50th percentile adult male dummy was used, and the sled speed was increased to 50 km/h through a traction device. Subsequently, under the action of AEB, the sled was decelerated to 30 km/h. Figure 3 shows the acceleration time curve of the actual vehicle test. By collecting test data and simulation data, such as the dynamic response curve shown in the Fig. 4, it is analyzed that the shapes of the simulation curve and the test curve of each response quantity are roughly consistent.

The EEARTH method in this paper is implemented using MATLAB program. The error scores and EEARTH comprehensive scores of each response of the occupant during AEB pre-braking are shown in Table 1.

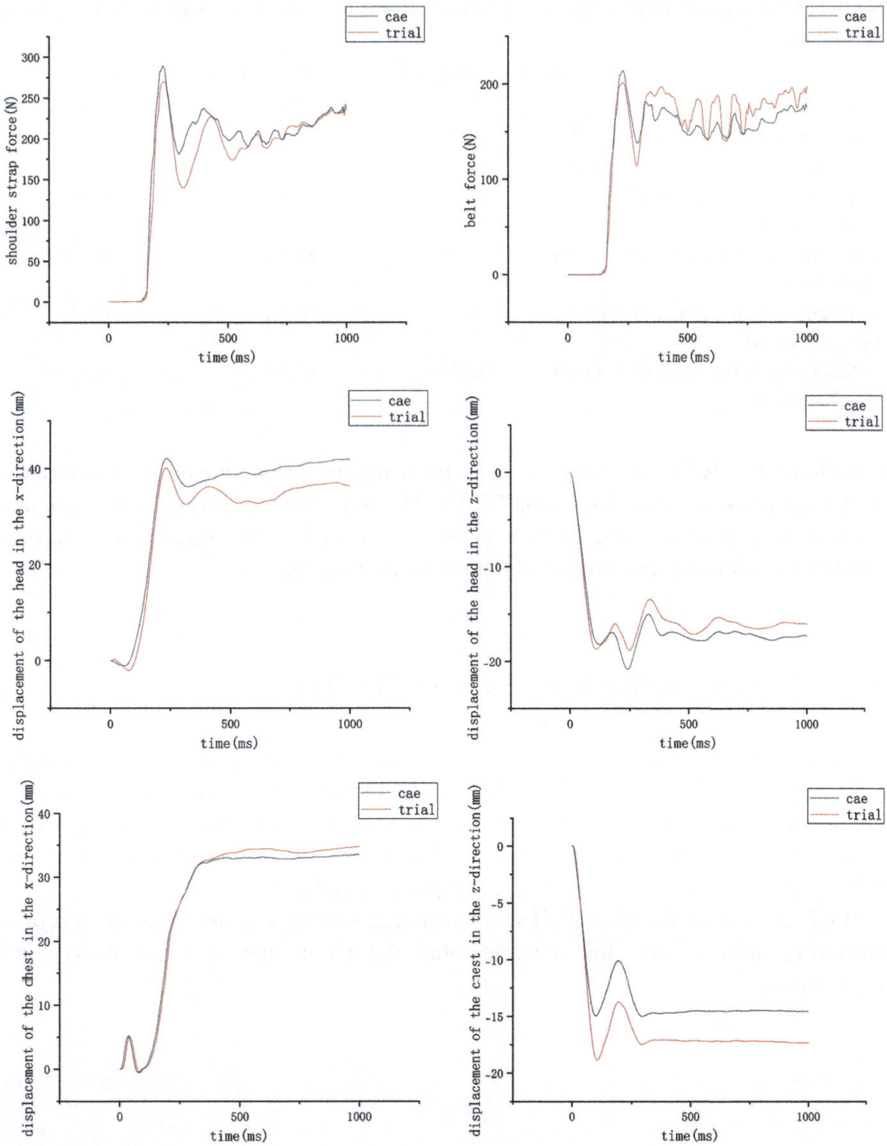

Fig. 4 Simulation curves and test curves

The results show that the EEARTH comprehensive score of the Z-direction displacement of the head is the highest, reaching 88.7%, which indicates that the response quantity has a very good fit between the test and the simulation and has a high correlation. However, the EEARTH comprehensive score of the lap belt force

Table 1 Characteristic error scores and EEARTH scores of each response quantity during AEB pre-braking

Response	Phase error score	Amplitude error score	Morphological error score	EEARTH score
Shoulder strap force (N)	95.1%	95.5%	62.9%	84.4%
Belt force (N)	97.5%	91.4%	47.8%	78.9%
x-direction displacement of the head (mm)	97.5%	88.8%	57.8%	81.4%
z-direction displacement of the head (mm)	95.0%	92.4%	78.7%	88.7%
x-direction displacement of the chest (mm)	95.0%	76.2%	70.5%	80.6%
z-direction displacement of the chest(mm)	95.0%	68.9%	85.5%	83.1%

is the lowest, only 78.9%. Although it is still within an acceptable range, it is weaker than other response quantities. The EEARTH comprehensive scores of the remaining response quantities are all over 80%, which further illustrates that there is a good fit between the test curve and the simulation curve.

3.2 Collision Simulation Model Verification

The test vehicle is first accelerated to 30 km/h by the traction device, and then the collision process lasts for 120 ms. The acceleration of the collision test uses the acceleration curve shown in Fig. 3. Figure 5 respectively show the comparison curves of the test and simulation results during the collision process, such as seat belt force, head acceleration and other response quantities.

Table 2 presents the EEARTH scores for each response quantity, with all scores clustering around 80. This indicates that the simulation and test curves are well-matched.

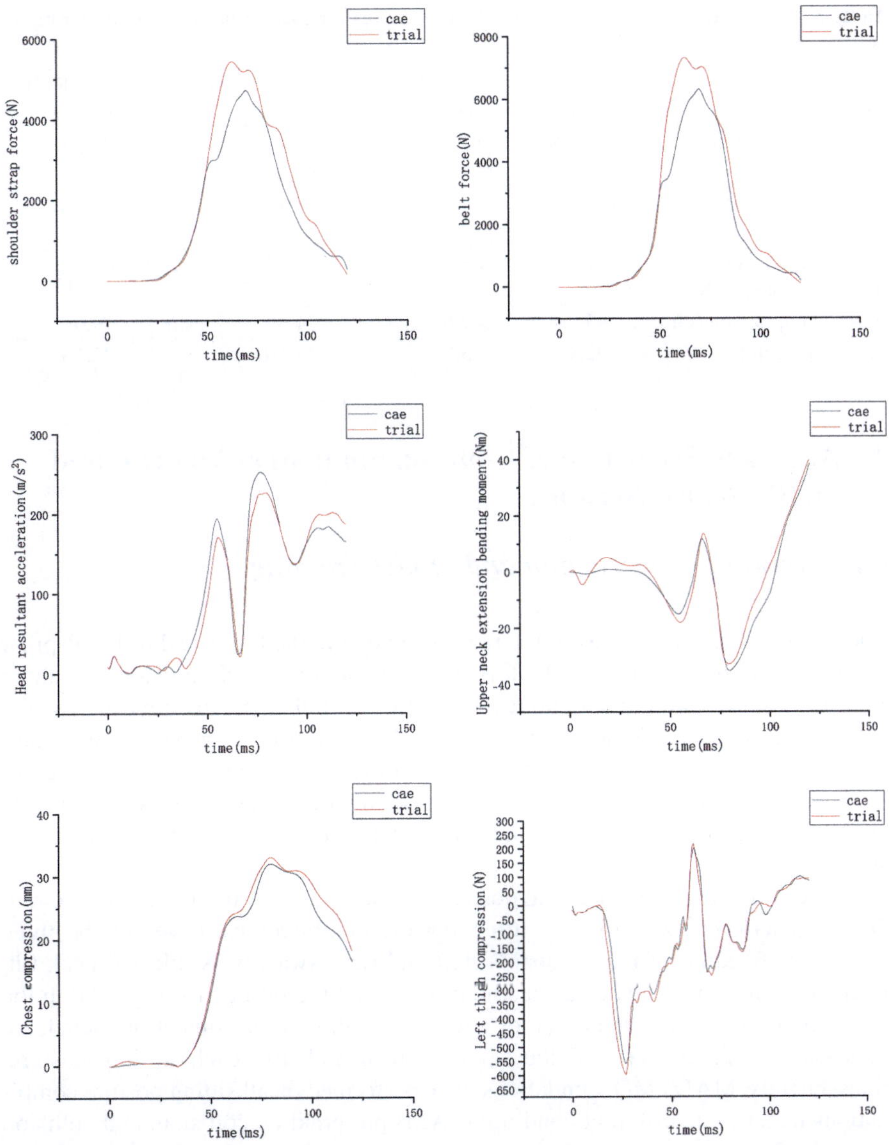

Fig. 5 Simulation curves and test curves

Table 2 Characteristic error scores and EEARTH scores for each response quantity during the collision process

Response	Phase error score	Amplitude error score	Morphological error score	EEARTH score
Shoulder strap force (N)	95.0%	92.1%	65.2%	84.1%
Belt force (N)	95.0%	90.6%	64.0%	83.2%
Head resultant acceleration (m/s^2)	95.0%	86.3%	74.4%	85.2%
Upper neck extension bending moment (N)	97.5%	66.7%	77.0%	80.4%
Chest compression (mm)	95.0%	69.6%	74.3%	79.6%
Left thigh compression (N)	92.5%	75.8%	48.1%	72.1%

4 Accuracy Prediction of Collision Simulation Model Based on BP Neural Network

4.1 Network Construction and Model Training

The acceleration-time curves of the reference point at the lower end of the B-pillar of the vehicle in the AEB pre-braking test, collision test and collision test under AEB intervention were input into the MADYMO simulation software respectively.

The collision acceleration curve under AEB intervention is generated by superimposing the AEB pre-braking curve on the collision input curve. Before the collision occurs, the AEB pre-braking system gradually decelerates the test bench from an initial speed of 50–30 km/h, and its acceleration variation curve over time is show in Fig. 6.

To explore the various possible postures of the occupants in the vehicle, we randomly generated 100 different sitting posture combinations, covering the three degrees of freedom of the occupants' left and right sway, waist tilt and neck tilt. Each combination simulates a specific posture that the occupants may take in the car. In order to improve the accuracy and applicability of the simulation model, we adjusted the parameters of the simulation model for each sitting posture. Subsequently, MADYMO simulations were performed on all sitting posture combinations under three different conditions: AEB pre-braking, collision, and collision under AEB intervention, and 100 sets of simulation data were collected. Each output file records the data results under these three conditions in detail. In the model training and validation stage, we selected the first 80 sets of data for training, and the remaining 20 sets of data were used as test sets to evaluate the performance and generalization ability of the model.

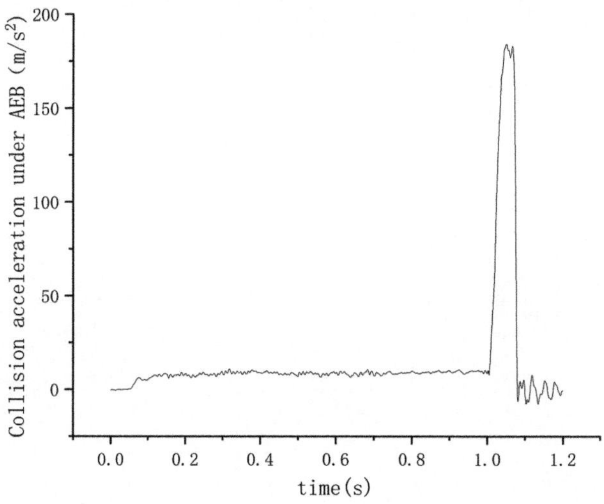

Fig. 6 Collision acceleration time history curve under AEB

4.2 Verification of Generalization Method

The model generalization idea proposed in this paper is to learn the characteristic laws between these conditions by training the simulation data of AEB pre-braking conditions, collision conditions and collision conditions under AEB, with the aim of predicting the experimental data of collision conditions under AEB. The specific implementation steps include inputting the AEB pre-braking condition and collision condition data obtained from the actual vehicle test into the trained prediction model, and then using these models to predict the experimental data of collision conditions under AEB. In order to verify the effectiveness of the generalization method, we use the EEARTH method to score the predicted data and simulation data, and compare the EEARTH predicted scores of each response quantity with the actual scores. This method can not only evaluate the generalization ability of the model, but also ensure the accuracy and reliability of the prediction results.

Figure 7 show the comparative analysis of experimental data, simulation data and predicted data of some dynamic response quantities.

The EEARTH model confirmation method is used to score the fit of the test and simulation data of each group of response quantities during the collision process with the AEB active safety device, and the predicted data and simulation data are scored at the same time. Table 3 lists the characteristic error scores and EEARTH

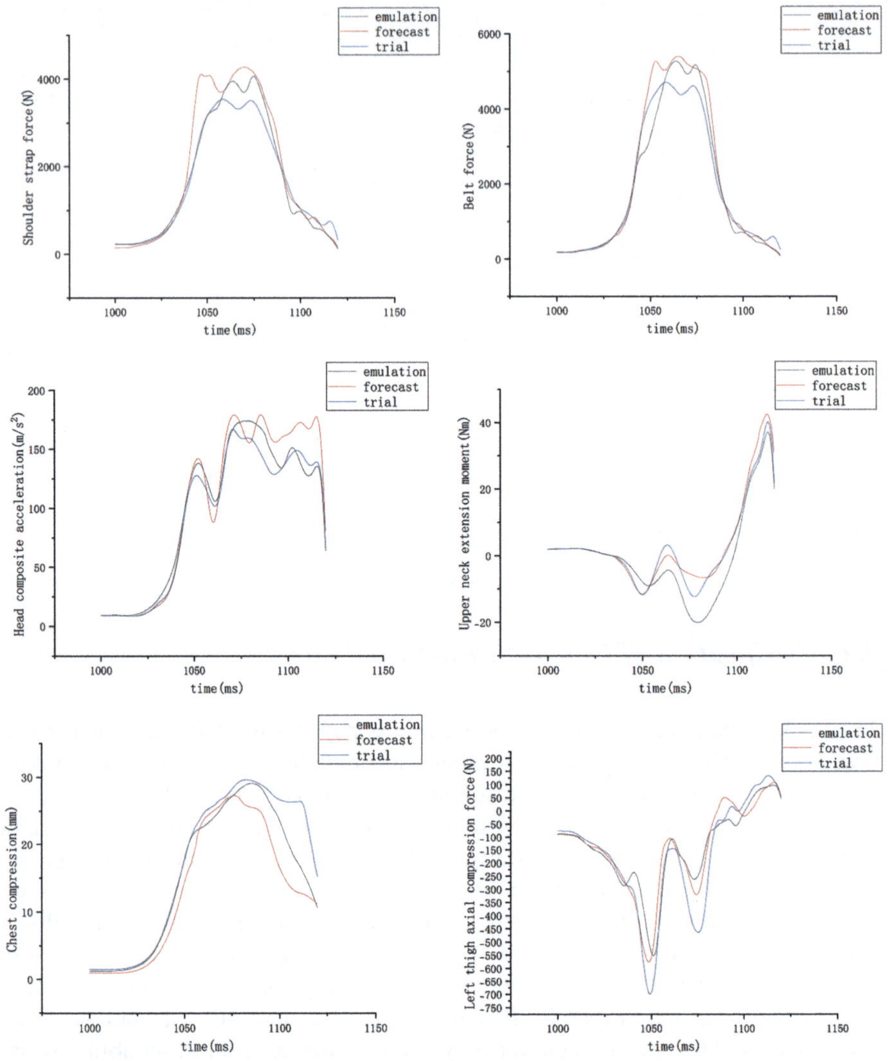

Fig. 7 Simulation curves, test curves and predicted curves

comprehensive scores of each response quantity of the experimental data and simulation data, reflecting the actual situation of the collision simulation model confirmation under the action of AEB. The evaluation results show that the EEARTH score of the synthetic acceleration of the head is the highest, reaching 89.9%, while the score of the axial compression force of the left thigh is relatively low, only 78.3%. In addition, the EEARTH scores of other response parameters are all over 80%, indicating that the curve fit between the experimental data and the simulation data is good.

Table 3 EEARTH scoring of experimental and simulation fitting of various response quantities during collision under AEB

Response	Phase error score	Amplitude error score	Morphological error score	EEARTH score
Shoulder strap force (N)	97.5%	92.6%	76.8%	88.9%
Belt force (N)	90.0%	91.7%	74.3%	85.3%
Head resultant acceleration (m/s^2)	97.5%	92.6%	79.6%	89.9%
Upper neck extension bending moment (N)	95.0%	71.4%	79.9%	82.1%
Chest compression (mm)	82.5%	98.0%	71.8%	84.1%
Left thigh compression (N)	95.0%	76.4%	63.5%	78.3%

Table 4 EEARTH score for prediction and simulation fitting of various response quantities during collision under AEB

Response	Phase error score	Amplitude error score	Morphological error score	EEARTH score
Shoulder strap force (N)	92.5%	93.5%	70.3%	85.4%
Belt force (N)	97.5%	92.7%	72.2%	87.5%
Head resultant acceleration (m/s^2)	95.0%	88.5%	64.5%	82.6%
Upper neck extension bending moment (N)	97.5%	45.7%	74.9%	72.7%
Chest compression (mm)	95.0%	70.4%	67.4%	77.6%
Left thigh compression (N)	92.5%	89.3%	71.6%	84.5%

Table 4 lists the error scores of each response quantity and the EEARTH comprehensive score of the predicted data and the simulation data, which reflect the prediction effect of the simulation model in the generalization process. It can be seen that the highest prediction score is the belt force, which reaches 87.5%, while the score of the upper neck extension moment is the lowest, which is 72.7%.

5 Conclusion

This paper proposes to use BP neural network to predict the collision simulation model. In order to ensure the reliability of the neural network input data, the EEARTH method is first used to verify the established collision simulation model. The results show that the simulation and experiment of each response quantity have a good fitting effect. Then the trained neural network is used to predict the collision test data under the action of AEB, and finally the simulation model accuracy prediction is completed using EEARTH. The results show that the established neural network prediction model has good prediction accuracy and the simulation model accuracy prediction method is effective. The proposed method not only provides a reliable and efficient way to predict the accuracy of collision simulation models, but also has significant implications for future research and practice in automotive safety.

References

1. Liu T, Niu Y, Chen G (2023) A general methodology for simulation model validation and model updating of the coordination mechanism with uncertainty [J]. J Mech Sci Technol 37(12):6271–6286
2. Kwag S, Gupta A, Dinh N (2018) Probabilistic risk assessment based model validation method using Bayesian network [J]. Reliab Eng Syst Safe 169:380–393
3. Ao D, Hu Z, Mahadevan S (2017) Design of validation experiments for life prediction models [J]. Reliab Eng Syst Safe 165:22–33
4. Li X (2024) A modular Bayesian approach to model calibration of low-fidelity simulation models [J]. Qual Reliab Eng Int 40(1):428–445
5. Vicino D, Wainer GA, Dalle O (2021) Uncertainty on discrete-event system simulation [J]. ACM Trans Model Comput Simul 32(1):1–27
6. Barzanooni R, Duddeck F (2024) Incorporating uncertainty into the validation of automotive crash simulation models for component analysis [J]. Int J Crashworth:1–8
7. Zhenfei Z (2011) Research on the theory and application of uncertain multivariate dynamic system model verification for automobile safety [D]. Shanghai Jiaotong University
8. Damblin G, Keller M, Barbillon P et al (2016) Bayesian model selection for the validation of computer codes [J]. Qual Reliab Eng Int 32(6):2043–2054
9. Hill RR, Rupp CC, Jones KM et al (2019) A case study in engineering model validation using new wavelet-based methods [J]. Qual Reliab Engng Int 35:815–823
10. Shenglin L (2021) Research on evaluation and optimization methods for complex simulation [D]. Harbin Inst Technol. https://doi.org/10.27061/d.cnki.ghgdu.2020.001421
11. Li W, Lin S, Qian X et al (2021) An evidence theory-based validation method for models with multivariate outputs and uncertainty [J]. Simulation 97(12):821–834
12. Qingmiao W (2021) Research on simplified model error evaluation method based on machine learning and its application in automobile safety [D]. Chongqing University. https://doi.org/10.27670/d.cnki.gcqdu.2019.000320

Open Access This chapter is licensed under the terms of the Creative Commons Attribution-NonCommercial-NoDerivatives 4.0 International License (http://creativecommons.org/licenses/by-nc-nd/4.0/), which permits any noncommercial use, sharing, distribution and reproduction in any medium or format, as long as you give appropriate credit to the original author(s) and the source, provide a link to the Creative Commons license and indicate if you modified the licensed material. You do not have permission under this license to share adapted material derived from this chapter or parts of it.

The images or other third party material in this chapter are included in the chapter's Creative Commons license, unless indicated otherwise in a credit line to the material. If material is not included in the chapter's Creative Commons license and your intended use is not permitted by statutory regulation or exceeds the permitted use, you will need to obtain permission directly from the copyright holder.

Simulation Analysis Study of the Air Spring Suspension System in the Cabins of Specialty Vehicles

Xiaoyong Liu, Yu Dai, and Wangqiang Xiao

Abstract Special vehicles are often used in transportation under harsh road conditions, with particularly severe vibration and impact environments. As an effective means of vibration control, the air spring damping system has been widely applied in the cab of special vehicles. Therefore, this article focuses on the simulation analysis of the air spring damping system in the cab of special vehicles. Firstly, this article introduces the basic principles and characteristics of the air spring damping system. Then, a finite element model of the air spring damping system is established using finite element software, and the vibration isolation effect of the air spring under different stiffnesses is studied in detail through simulation analysis. The research results show that the air spring can achieve a vibration reduction rate of more than 45.5% under various stiffness settings, especially when the stiffness is 88.8 N/mm (small stiffness), the vibration reduction rate can be up to 90.6%.

Keywords Vehicle driver cabin vibration damping control · Air spring suspension system · Simulation analysis

1 Introduction

Due to their unique functional design, special vehicles are often used in transportation environments where conditions are extremely harsh and road conditions are extremely severe. In these environments, the vehicle not only has to withstand the severe vibration brought about by the rugged road surface, but also may encounter sudden shocks and bumps, these vibrations and shocks will not only affect the stability and safety of the vehicle, but also cause serious damage to the driver's driving experience and the vehicle components [1], therefore, this paper adopts the air-spring-type vibration isolation system as the object of the study to analyze the effect of its vibration damping.

X. Liu (✉)
713th Research Institute of China State Shipbuilding Corporation Limited, Zhengzhou, China

Y. Dai · W. Xiao
School of Aerospace Engineering, Xiamen University, Xiamen, China

© The Author(s) 2026

Z. Zhang, P. Mercorelli (eds.), *Mechanical Design and Simulation: Insights and Innovations*, Lecture Notes in Mechanical Engineering,
https://doi.org/10.1007/978-981-95-2405-1_35

The researchers carried out a series of studies on the air spring suspension system, aiming to solve the problems of the system and improve its performance. The research focuses on system design optimization [2], simulation analysis of vibration damping effect [3], system stability study [4], and evaluation of system performance under different working conditions [5]. The researchers explored the dynamic response of the system under different conditions and sought the optimal vibration damping effect by adjusting the stiffness, pressure and other parameters of the air spring [6]. In addition, for practical application scenarios, the researchers also conducted field tests on the system [7, 8]. From the above studies, it can be found that the following deficiencies still exist: the further improvement of system stability needs to be strengthened, especially the performance under extreme working conditions needs to be optimized; the simulation and analysis studies of air spring suspension systems have not yet been fully developed [9–11]. As a result, the existing simulation and analysis research of air spring suspension system is not enough. For this reason, this paper proposes a multiple air-spring type suspension system and simulation analysis of its vibration damping effect is carried out, and the influence of design parameters on the vibration isolation system is discussed in detail. Finally, the vibration isolation effect of the vibration isolator is analyzed by the force transmission rate.

2 Special Vehicle Air Spring Damping System

Special vehicle air spring damping system dynamics model mainly consider the action part of the special vehicle, the suspension device and the vehicle body and so on. The vehicle body is regarded as a plane vibration rigid body, the tire is regarded as a spring, and the air spring damping system is regarded as a damper and a spring, so the 1/2 special vehicle air spring damping system dynamics model is shown in Fig. 1.

Fig. 1 Dynamic model of air spring damping system

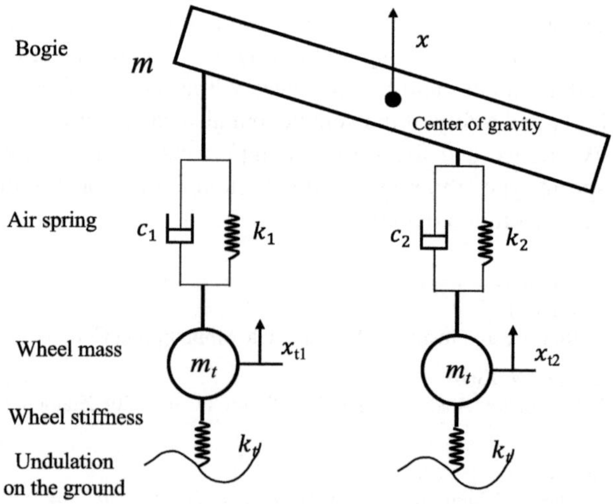

3 Modeling of Air Spring Damping System

3.1 Structure of Air Spring Damping System

This air spring damping system mainly consists of upper and lower cover plates and rubber zinan, as well as the gas filling inside, as shown in Fig. 2. The upper and lower cover plates play a supporting role, and their function is to install the air spring between the frame and the body.

3.2 Finite Element Modeling

Establish the three-dimensional model of the air spring, imported into the finite element software, in order to reduce the calculation, the air spring model is simplified accordingly, and its meshing, as shown in Fig. 3. Air spring finite element model is the most important thing is to model the rubber airbag, the stiffness of its upper and lower cover plate is far more than the rubber bag, it can be considered to be rigid, the impact in the simulation calculation is very small, you can ignore its deformation, the choice of rigid body Rigid unit for simulation. The upper cover plate is set as the reference point of the rigid body on the center axis of the plate, which is used to apply displacement, load and constraint. The elastic modulus of the upper and lower cover plates is 2×10^5pa, Poisson's ratio is 0.3, and density is 7.8×10^{-9}.

Rubber Airbag Setup Air springs have a unique airbag construction, which is made of a composite of rubber and a cord layer. In this case, the rubber, as the core component, is tightly bonded to the cord layer by a vulcanization process and exhibits significant mechanical anisotropy and geometrical nonlinear characteristics. The elastic properties of this rubber material are particularly outstanding, and its highly nonlinear character with very small volume compressibility allows it to be accurately described by a hyperelastic material model. In particular, the Mooney Rivlin

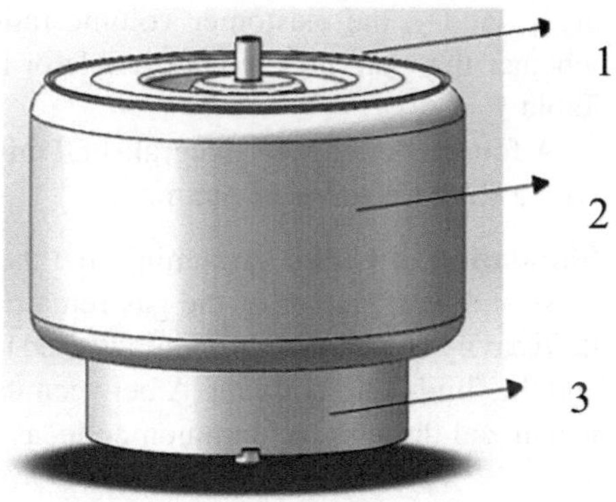

Fig. 2 Components of gas spring structure. (1) Top plate (2) Rubber airbag (3) Bottom cover

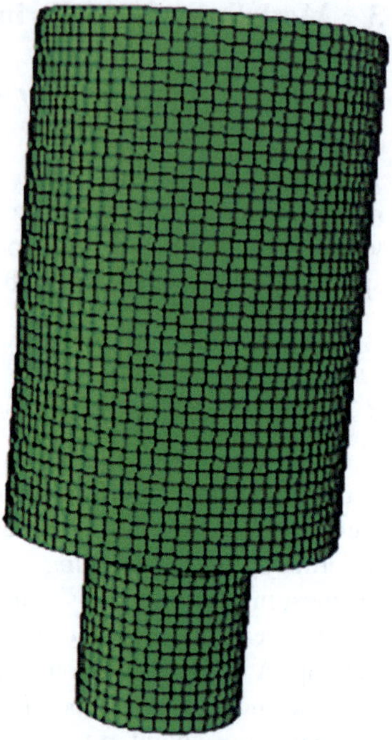

Fig. 3 Finite element model of air spring damping system

model exhibits excellent computational precision and accuracy for small and medium deformations and small strains, so that the mechanical properties can be described by the strain energy density function of the Mooney-Rivlin model:

$$U = C_{10}(I_1 - 3) + C_{01}(I_2 - 3) + \frac{1}{D_1}(J - 1)^2 \tag{1}$$

where C_{10}, C_{01} are the hyperelastic material constants, which are obtained from experimental fitting; I_1,$I_2$are the first and second deviations, which $I_1 = \lambda_1^2 + \lambda_2^2 + \lambda_3^2$, $I_2 = \lambda_1^{-2} + \lambda_2^{-2} + \lambda_3^{-2}$, and the principal tensile coefficients in the three directions areλ_1,$\lambda_2$andλ_3; the elastomer volume ratio is *J*; and the value of D1 depends on whether the material is compressible or not. The material parameters are shown in Table 1.

A four-node reduced integral shell unit SR4 is used to simulate the rubber layer with a shell thickness of 3 mm.

Simulation of Gases Assuming that the gas in the cavity is an ideal gas, and the mass and temperature of the gas remain constant during the working process, the fluid cavity (fluid cavity) module is used to define the closed gas in the airbag cavity, and the fluid-solid coupling A between the gas unit and the inner wall of the airbag is realized through the common node, as shown in Fig. 4.

Table 1 Mooney-Rivilin model

Model parameters	Parameter values
C_{10}	0.4825(MPa)
C_{01}	0.1206(MPa)
D_1	0
Temperature	25 °C
Density	1e-9 kg/m

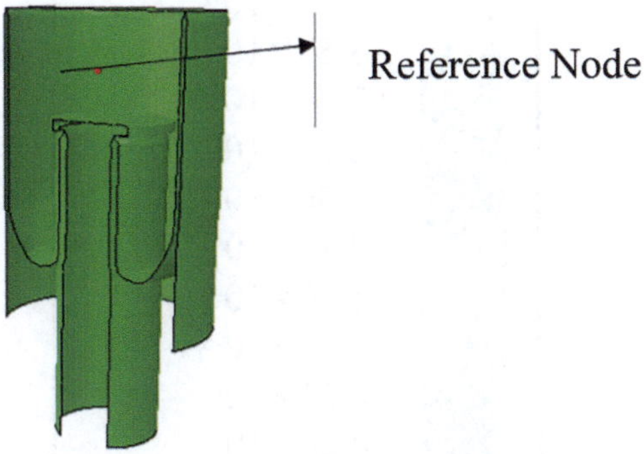

Fig. 4 Cavity reference node

Contact and Boundary Conditions The gas spring undergoes tensile and compressive deformation during operation, and its rubberized air bladder will interact with the upper and lower covers. In order to solve this contact problem between the deformed body and the contacting body, we adopt the Lagrange method. In simulating the contact between the rubber airbag and the upper and lower cover plates, we chose the tie connection to simulate this contact relationship. Given the relatively low stiffness of the rubber airbag, we set it as the passive body, while the upper and lower cover plates act as the active body. When the shape of the rubber airbag changes during inflation, the parts that were not previously in contact with the upper and lower cover plates begin to generate contact. In this case, we set the contact between the rubber airbag and the upper and lower cover plates as a no-slip state to ensure the stability and reliability of the contact surface.

Load Step Setting According to the test method of GB/T 13061-1991 "Air Spring Rubber Airbag for Vehicle Suspension", 2 analysis steps are set up in the finite element software: Step-1 constrains 6 degrees of freedom of the upper and lower cover plates, and fills 0.1Mpa gas into the airbag; Step-2 releases the vertical displacement constraints of the upper cover plate, and the rest of the constraints of the upper and lower cover plates are kept unchanged, and ±5mm vertical displacement load is applied to the upper cover plate, this analysis step can be interpreted as the stretching and compression process of upward and downward vibration of air spring. The vertical displacement load is applied to the upper cover plate, and this analysis step can be interpreted as the stretching and compression process of the air spring that vibrates upward and downward of the car body, and the tensile and compressive

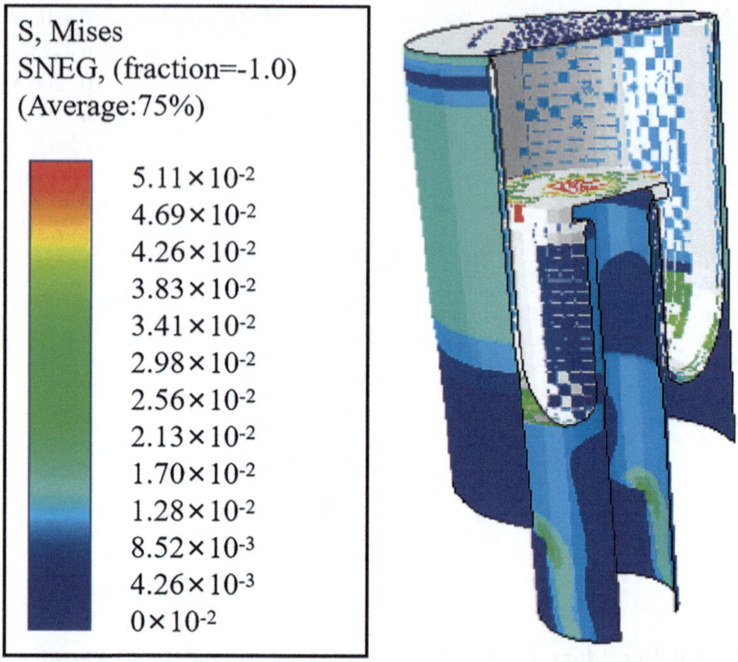

Fig. 5 Gas spring inflation deformation

stress cloud diagram of the air spring is shown in Fig. 5 after the inflation of the air spring.

4 Simulation Results of Air Spring Damping System

4.1 Effect of Different Initial Air Pressure on Vertical Static Stiffness

Different amount of gas filled in the gas pocket of the gas spring affects its pressure bearing capacity. Different initial air pressure is applied to the gas spring, the amplitude is set to 5 mm, the rest of the parameters remain unchanged, and the same analysis steps are set to obtain the results as shown in Fig. 6.

From the above figure, the vertical static stiffness of the air spring damping system can be calculated when the air spring is damped at different initial air pressures, as shown in Table 2.

From Fig. 6 and Table 2, it can be seen that under the condition of other conditions being the same, with the increase of the initial air pressure, the support reaction force on the air spring is increasing, and increasing the initial pressure

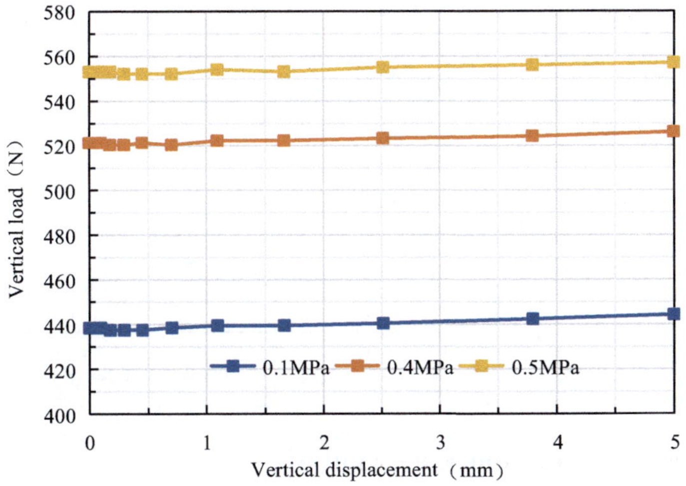

Fig. 6 Plot of vertical stiffness for different initial air pressures

Table 2 Sagittal stiffness at different initial air pressures

Initial internal pressure (MPa)	0.1	0.4	0.5
Vertical load (N)	444.039	526.113	557.936
Vertical stiffness (N/mm)	88.8	105.2	111.2

of the compressed gas inside the air bag can also improve the vertical stiffness of the air spring. Therefore, changing the initial gas pressure of the air spring, i.e. changing the initial load of the air spring, can significantly change the vertical stiffness of the air spring, thus changing the vibration isolation rate of the air spring.

The amplitude of the gas spring is the vibration amplitude of the gas spring system in the vertical direction, the performance under this vibration amplitude as the benchmark for evaluating the performance of the gas spring, the amplitude affects the load carrying capacity of the gas spring, also affects the vertical stiffness of the gas spring. The amplitude affects the load carrying capacity of the air spring and also the vertical stiffness of the air spring. When the other conditions are the same, different amplitude displacements are applied to the air spring, with amplitudes of 1 mm, 2 mm, 3 mm, 4 mm, 5 mm, 6 mm. the air spring displacements-branching reaction forces under different displacement loads are shown in Fig. 7.

From the above figure, it can be seen that when the vertical displacement becomes larger, the static stiffness of the air spring damping system also becomes larger and exhibits an approximately linear increase, while other conditions remain consistent.

Fig. 7 Plot of vertical static stiffness for different amplitudes

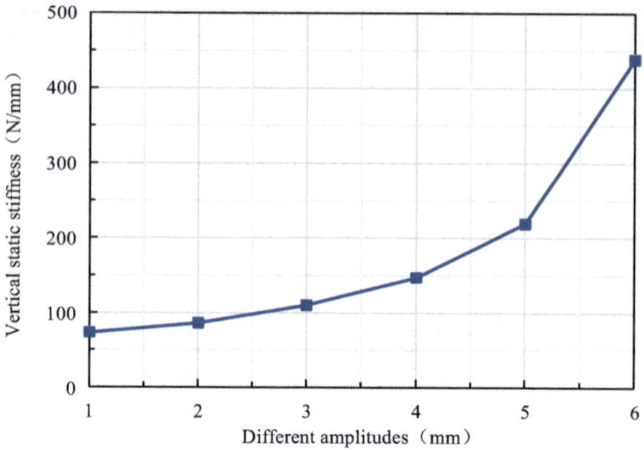

5 Force Transmission Rate Analysis

Vibration often has a detrimental effect on the performance of machines, instruments, and equipment. In order to eliminate this effect, vibration isolation is an effective measure. The force transmitted to the foundation by the vibration isolation device has two parts: the force transmitted to the foundation by the spring and the force transmitted to the foundation by the damping.

Then the index force transfer coefficient for evaluating the effectiveness of positive vibration isolation:

$$T_F = \frac{F_T}{F} = \frac{\sqrt{1+(2\xi r)^2}}{\sqrt{(1-r^2)^2+(2\xi r)^2}} \tag{2}$$

In Eq. (2), $\xi = \frac{c}{2m\omega_n}$ is the damping ratio, $r = \frac{\omega}{\omega_n}$; c is the damping coefficient, $\omega_n = \sqrt{\frac{k}{m}}$ is the intrinsic frequency, and k is the system stiffness.

Taking the excitation frequency ω = 1.3 Hz, m = 611.25 kg, c = 0.45KN*s/m, and taking k = 88.8KN/m (105.2 KN/m or 111.2KN/m) when substituting into Eq. (2), the force transfer coefficient can be obtained as shown in Table 3.

From Table 3, it can be seen that as the stiffness of the air spring increases, the transmission rate of the air spring increases in the working process, and the vibration isolation effect is worse. When the stiffness of the air spring is 88.8 N/mm, the force transfer rate is 9.4%, and when the stiffness of the air spring is increased to 220.4 N/mm, the force transfer rate is 54.5%, and the vibration isolation rate is reduced by about 5 times. Therefore, changing the stiffness of air springs by choosing different initial air pressures, different loads and different amplitudes is particularly important for the vibration isolation effect of air springs.

Table 3 Force transfer rates of gas springs with different stiffnesses

Stiffness (N/mm)	88.8	105.5	111.2	114.2	220.4
Force transfer rate T_F	9.4%	11.3%	23.0%	30.9%	54.5%

6 Conclusion

This paper establishes a finite element model of air spring damping system, analyzes the influence of different factors on the stiffness of air springs, and takes the force transfer coefficient as an index to analyze the vibration isolation effect of air springs under different selections, and obtains the following conclusions:

1. Changing the initial air pressure of the air spring, i.e., changing the load carrying capacity, can significantly change the vertical stiffness of the air spring, and thus change the force transmission rate, while other conditions remain consistent.
2. All other conditions being equal, as the vertical displacement becomes larger, the static stiffness of the air spring damping system also becomes larger.
3. The stiffness characteristics of the air spring is an important indicator of its performance, but also an important indicator to judge the vibration isolation performance of the cab, according to the force transfer coefficient can be seen, with the increase of stiffness, the force transfer coefficient TF increases, the cab vibration isolation performance is reduced.
4. From the simulation analysis, different stiffness of the air spring vibration damping rate can reach more than 45.5%, when the selected stiffness is 88.8 N/mm (stiffness is small), the vibration damping rate of the air spring can reach 90.6%.

References

1. Lei Z, Xuetao S, Jie C et al (2024) Research Progress in Reliability Analysis and Optimal Design of Automotive Structures [J]. China Mechanical Engineering. 35(11):1948–1962+1970. https://kns.cnki.net/kcms2/article/abstract?v=d24vdHNzaZj_1-YrEHbMtFlriRso23FLB-deikilwE8xjS3S0fM0jB4WzkUfBnmbRQkXsWMucn31MdU3PayznUYz_bZ6WUHrqko3mhVi-n2zCr-rwX76sCmQd6s22a3ByPxdGr52wkZ2tfIK4xr2MdZQZZ6JwNOLym691hdu6WBmJgXbi0djan9IJKJgJx0u&uniplatform=NZKPT&language=CHS
2. Wenji L (2023) Theoretical research and finite element analysis of hyperbolic Air Springs [J]. Environ Technol 41(12):155–158+181. https://kns.cnki.net/kcms2/article/abstract?v=Nyg97wmOeE4VjUIbzbber2MTrd5WwoFj2-vJ6kmDWV2IZhF5yI6B-6YbwXBz03hAVDvFpORd57wiVeL3VCoCEzKxE0semMhCje8saYcKadclIP93JDCe7QsCpQxS1T2wkeC-Ui_ng2xv9s-JpAY3gHgDlX3SFArDUiR5qEeiqgCWKuy SFFCM8hnm5NPdfWwHqz7uxuhrlxhs=&uniplatform=NZKPT&language=CHS
3. Taichi W, Masahiko A (2020) Vibration control of air spring type tilting train. The Proceedings of the International Conference on Motion and Vibration Control 2020.15(0):10105–10115. https://www.jstage.jst.go.jp/browse/jsmeintmovic/_pubinfo/-char/en

4. Cong P, Magee RA, Zhang T (2019) Efficient calculation of the hydrodynamic coefficients and dynamic stiffness of an air-spring type vibration absorber. Ocean Eng 192:106550–106550. https://doi.org/10.1016/j.oceaneng.2019.106550
5. Song H, Lin J, Liu W (2023) Research on the sealing performance of the upper sub-mouth of the pressure self-sealing air spring. Railway Vehicle 61(04):99–103. https://kns.cnki.net/kcms2/article/abstract?v=Nyg97wmOeE6RNmdIjx9fd2H25ehNyABFwuSiEKXTcJabm_wHkJADqUKlclEo0fHYoMxGk1-thT-n1XakyoEPd_KPIYFGbt7my-0I25xeTxWpkx9oVKv7qFjaE52BBM4wPl56J3_TWyFHD7NuliGpAQIlGWrbqFjIi1-mzgl1RTq-lwPSbFV8X1Qe8slAkHj9&uniplatform=NZKPT&language=CHS
6. Du J, Dong G, Han J et al (2024) Research on welding fatigue characteristics of fully floating air spring cab. Mech Sci Technol Aerospace Eng:1–12. https://doi.org/10.13433/j.cnki.1003-8728.20230249
7. Yin C, Zhang Y, Geng G et al (2023) Simulation and optimal matching research on the suspension system of commercial vehicle cab based on ride comfort. J Mach Des 40(10):128–135. https://doi.org/10.13841/j.cnki.jxsj.2023.10.021
8. Yin C, Luo D, Yue S (2023) Research on vibration performance of the suspension system of commercial vehicle cab-seat. J Mach Des 40(09):99–107. https://doi.org/10.13841/j.cnki.jxsj.2023.09.016
9. Editorial Department of China Journal of Highway and Transport (2023) Academic research overview of automotive engineering in China 2023. China J Highw Transp 36(11):1–192. https://doi.org/10.19721/j.cnki.1001-7372.2023.11.001
10. Jianghua F, Yakun Y, Jintao S et al (2023) Review of research on magnetorheological suspension for automotive powertrain. J Chongqing Univ Sci Technol (Natural Science Edition) 37(06):29–38. https://kns.cnki.net/kcms2/article/abstract?v=d24vdHNzaZijBdpjIIHVXp8FYSu0YSevdxd3W9wLJL15Rbzm5GJNJR CeER-06iL9_Nz9Fy25oki69sr7r6UH4iO3kAd21lDWFKnru2pJTxkBYILQQyxxPXMo7zQkg04pC5ejxpT9G_41H3VUbhqGoeCcwEl0DWscFctaG4_PTDdJDl3FeWHPwY6YNquzS3Wu&uniplatform=NZKPT&language=CHS
11. Wang L, Huang Z, Li H et al (2023) Simulation analysis and optimization of an air suspension system for a light commercial vehicle. J Mech Strength 45(05):1159–1165. https://doi.org/10.16579/j.issn.1001.9669.2023.05.020

Open Access This chapter is licensed under the terms of the Creative Commons Attribution-NonCommercial-NoDerivatives 4.0 International License (http://creativecommons.org/licenses/by-nc-nd/4.0/), which permits any noncommercial use, sharing, distribution and reproduction in any medium or format, as long as you give appropriate credit to the original author(s) and the source, provide a link to the Creative Commons license and indicate if you modified the licensed material. You do not have permission under this license to share adapted material derived from this chapter or parts of it.

The images or other third party material in this chapter are included in the chapter's Creative Commons license, unless indicated otherwise in a credit line to the material. If material is not included in the chapter's Creative Commons license and your intended use is not permitted by statutory regulation or exceeds the permitted use, you will need to obtain permission directly from the copyright holder.

Simulation and Analysis of Elastic Flow Lubrication of Three-Stage Planetary Gear System

Xiaojun Yin, Wen Liu, Xiansheng Pan, and Ziwen Li

Abstract The structure of planetary gear system is complex, and it is difficult to predict the flow of oil by empirical formula, so it is necessary to simulate and analysis the transient flow field of the gear system with immersion oil lubrication, to study the oil distribution law and the convective heat transfer characteristics of the gears. Based on the actual structure of the planetary gear system, this paper establishes a simulation model of the planetary gear system with immersion oil lubrication, reveals the law of oil movement, and analysis the convective heat transfer characteristics of the planetary wheel system gears in the immersion oil lubrication. The results show that the planetary gears convective heat transfer characteristics, in the mesh area oil distribution, convective heat transfer coefficient is large, lubrication effect is good; planetary gear system gears convective heat transfer coefficient presents the first level of the largest, the second level of the second, the third level of the smallest rule of change.

Keywords Planetary gear · Convective heat transfer · Heat flow density

1 Introduction

Planetary gear systems are often utilized in low-speed, heavy-duty, high power density, and large ratio applications. The lubrication method of closed gear systems is usually oil immersion lubrication, and higher temperatures can cause thermal deformation of the gears, which affects the service performance of the gears. At present, there have been a large number of studies on gear immersion lubrication and convective heat transfer. Li et al. [1, 2] established a numerical analysis model for spur gear oil injection lubrication, which revealed that the heat on the gear surface is conducted by the lubricating oil film; Ouyang et al. [3] established a mathematical

X. Yin · W. Liu (✉) · X. Pan · Z. Li
State Key Laboratory of Mechanical Transmission, Chongqing University, Chongqing, China
e-mail: Liuwen@cqu.edu.cn

© The Author(s) 2026

Z. Zhang, P. Mercorelli (eds.), *Mechanical Design and Simulation: Insights and Innovations*, Lecture Notes in Mechanical Engineering,
https://doi.org/10.1007/978-981-95-2405-1_36

model for oil injection lubrication by taking into consideration the layout of the nozzles, the injection parameters and the geometrical parameters of the face gears; Guo et al. [4, 5] based on the principle of gear meshing and elastic flow lubrication theory to solve the gear meshing lubricant quantity, and studied the gear parameters and working condition parameters on the required lubricant quantity; Zhu et al. [6, 7] to establish the gear box process simulation model, calculate the convective heat transfer coefficient at different working condition parameters and oil temperature.

This paper takes the planetary gear transmission system of shield machine as the research object, establishes the simulation model of planetary gear system with immersion oil lubrication, reveals the oil motion law, analyzes the convective heat transfer characteristics of planetary wheel system gears during immersion oil lubrication, and provides the basis for the research of oil lubrication of the gear system, steady state temperature field and so on.

2 Simulation of Planetary Wheel System Immersion Oil Lubrication

2.1 Basic Theory of Flow Field Simulation

1. fundamental conservation equation

The total mass of the fluid micronutrient is *ρdxdydz*, its mass increment per unit time is *(∂ρ/∂t)dxdydz*, The conservation equation is as follows [8]:

$$\partial \rho / \partial t + \partial\left(\rho v_x\right) / \partial x + \partial\left(\rho v_y\right) / \partial y + \partial\left(\rho v_z\right) / \partial z = 0 \tag{1}$$

When analyzing an incompressible fluid, the mass conservation equation is expressed as:

$$\partial v_x / \partial x + \partial v_y / \partial y + \partial v_z / \partial z = 0 \tag{2}$$

The mass of the fluid micrometric element is ρdxdydz, the acceleration in the y-direction is $a_y = \partial v_y / \partial t + \left(\vec{V} \cdot \nabla\right) v_y$,∇is the Hamiltonian operator, $\vec{V} \cdot \nabla$is the dispersion, and the momentum conservation equation is as follows:

$$\begin{cases} \partial\left(\rho v_x\right) / \partial t + \nabla \cdot \left(\rho v_x \vec{V}\right) = \rho F_x - \partial \rho / \partial x + \partial \tau_{xx} / \partial x + \partial \tau_{yx} / \partial y + \partial \tau_{zx} / \partial z \\ \partial\left(\rho v_y\right) / \partial t + \nabla \cdot \left(\rho v_y \vec{V}\right) = \rho F_y - \partial \rho / \partial y + \partial \tau_{xy} / \partial x + \partial \tau_{yy} / \partial y + \partial \tau_{zy} / \partial z \\ \partial\left(\rho v_z\right) / \partial t + \nabla \cdot \left(\rho v_z \vec{V}\right) = \rho F_z - \partial \rho / \partial z + \partial \tau_{xz} / \partial x + \partial \tau_{yz} / \partial y + \partial \tau_{zz} / \partial z \end{cases} \tag{3}$$

where: *Fx*, *Fy*, *Fz* are the external volumetric forces in the *x*, *y*, *z* directions of the volumetric force vector F on the fluid microelement, respectively; τ_{xx}, τ_{xy}, τ_{xz}, τ_{yy}, τ_{yz}, τ_{zz} is the component of the viscous stress τ.

The equation for conservation of energy is given below:

$$\begin{aligned}&\frac{\partial}{\partial t}\left[\rho\left(e+\frac{V^2}{2}\right)\right]+\nabla\cdot\left[\rho\left(e+\frac{V^2}{2}\right)\vec{\mathrm{V}}\right]=\frac{\partial(v_z\tau_{xz})}{\partial x}+\frac{\partial(v_z\tau_{yz})}{\partial y}+\frac{\partial(v_z\tau_{zz})}{\partial z}+\\&\frac{\partial}{\partial x}\left(\lambda\frac{\partial T}{\partial x}\right)+\frac{\partial}{\partial y}\left(\lambda\frac{\partial T}{\partial y}\right)+\frac{\partial}{\partial z}\left(\lambda\frac{\partial T}{\partial z}\right)-\left(\frac{\partial(v_xp)}{\partial x}+\frac{\partial(v_yp)}{\partial y}+\frac{\partial(v_zp)}{\partial z}\right)+\\&\frac{\partial(v_x\tau_{xx})}{\partial x}+\frac{\partial(v_x\tau_{yx})}{\partial y}+\frac{\partial(v_x\tau_{zx})}{\partial z}+\frac{\partial(v_y\tau_{xy})}{\partial x}+\frac{\partial(v_y\tau_{yy})}{\partial y}+\frac{\partial(v_y\tau_{zy})}{\partial z}+\rho\mathbf{F}\cdot\vec{\mathrm{V}}\rho\dot{q}\end{aligned}\tag{4}$$

where V is the volume per unit mass of the fluid microelement; e is the internal energy per unit fluid microelement;$\dot{q}$ is the volumetric heating rate per unit fluid microelement; λ is the thermal conductivity of the fluid microelement; and T is the body temperature of the fluid microelement.

2. computational theory of turbulence modeling

   In order to consider the rotational effect of the fluid and to improve the accuracy of the calculation of the flow field in strong cyclonic flow and flow field with large velocity gradient, this paper adopts the RNG k-ε model for the simulation of the flow field of the planetary wheel system [9]:

$$\frac{\partial(\rho k)}{\partial t}+\frac{\partial(\rho v_i k)}{\partial x_i}=\frac{\partial}{\partial x_l}\left(\alpha_k\mu_e\frac{\partial k}{\partial x_l}\right)+G_k+G_b-\rho\varepsilon-Y_m-\rho\varepsilon \tag{5}$$

$$\frac{\partial(\rho\varepsilon)}{\partial t}+\frac{\partial(\rho v_i\varepsilon)}{\partial x_i}=\frac{\partial}{\partial x_l}\left(\alpha_\varepsilon\mu_e\frac{\partial k}{\partial x_l}\right)+\frac{C_{1\varepsilon}\varepsilon}{k}(G_k+C_{3\varepsilon}G_b)-C_{2\varepsilon}\rho\frac{\varepsilon^2}{k}-R_\varepsilon \tag{6}$$

$$R_\varepsilon=\frac{C_\mu\rho\eta^3\left(1-\frac{\eta}{\eta_0}\right)\varepsilon^3}{1+\beta\eta^3}\cdot\frac{\varepsilon^3}{k} \tag{7}$$

Where α_k, α_ε are the Platt's constants corresponding to the pulsation kinetic energy k and the energy dissipation rate ε; μ_e is the effective turbulent viscosity; η is the fraction of turbulence time in the mean flow time of the fluid; η_0 is the typical value of η in shear fluids; C_μ and β are constants.

For an unpressurizable fluid, the turbulent transient control equations are:

$$\begin{cases}\vec{\mathrm{V}}\cdot\nabla=0\\ \frac{\partial v_x}{\partial t}+\nabla\cdot(v_x\mathbf{V})+\frac{1}{\rho}\frac{\partial p}{\partial x}=\nu\nabla\cdot grad(v_x)\\ \frac{\partial v_y}{\partial t}+\nabla\cdot(v_y\mathbf{V})+\frac{1}{\rho}\frac{\partial p}{\partial y}=\nu\nabla\cdot grad(v_y)\\ \frac{\partial v_z}{\partial t}+\nabla\cdot(v_z\mathbf{V})+\frac{1}{\rho}\frac{\partial p}{\partial z}=\nu\nabla\cdot grad(v_z)\end{cases}\tag{8}$$

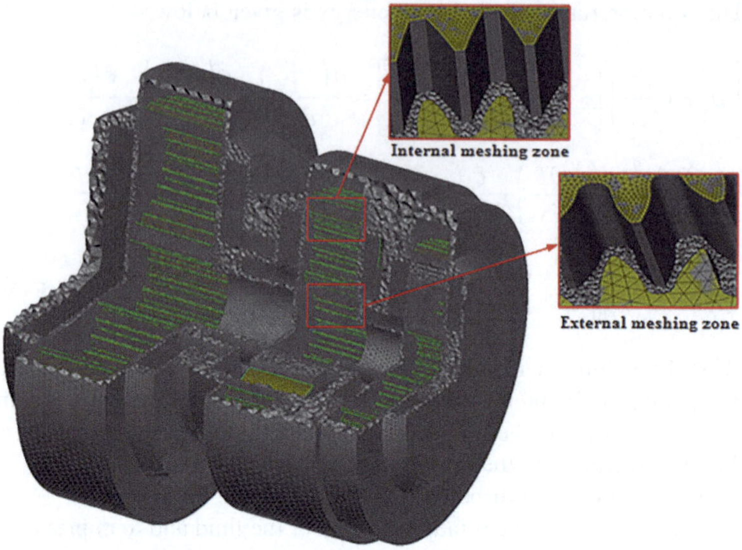

Fig. 1 Grid model of planetary gear reducer for TBM

2.2 Gas-Liquid Two-Phase Flow Simulation Model

After establishing the 3D model of the planetary gear train, it is imported into ANSA software for pre-processing. The surface of the gears, planetary racks and the inner gear ring is swept to generate a more regular unstructured mesh (tetrahedral mesh), and then the local mesh encryption is performed for the gear tooth surface, and the gear endface, the planetary racks and the box is generated as the fluid mesh with a growth rate of 1.05, the surface mesh of the gears, and the total number of face meshes is 3.4 million.

The fluid domain is divided into CFD body mesh with a growth rate of 1.1, and the fluid domain mesh of the three-stage planetary wheel system is divided into a total of 30 million, as shown in Fig. 1.

3 Flow Field Characterization

Import the pre-processed model into Fluent for boundary condition setting, select Pressure-Based solver (Pressure-Based), set the multiphase flow model as Volume of fluid, and select Implicit algorithm (Implicit), open the energy equation, set the turbulence model as RNG k-ε model, and the lubricating oil as #220 gear oil.

1. Velocity distribution in the fluid domain

   The velocity distribution cloud and velocity distribution vector plots of the fluid domain calculation results are shown in Figs. 2 and 3.

Fig. 2 Velocity distribution cloud

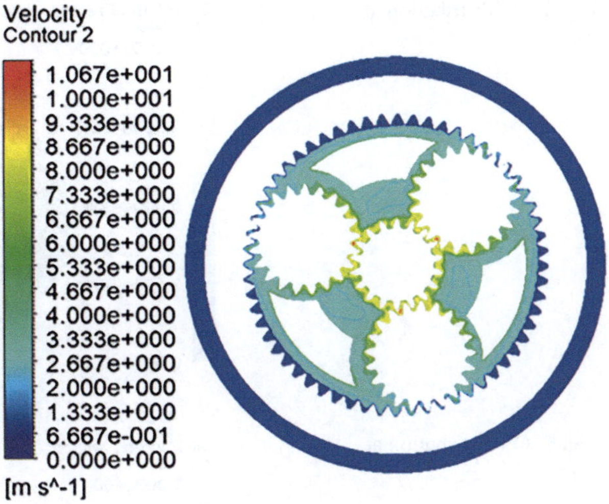

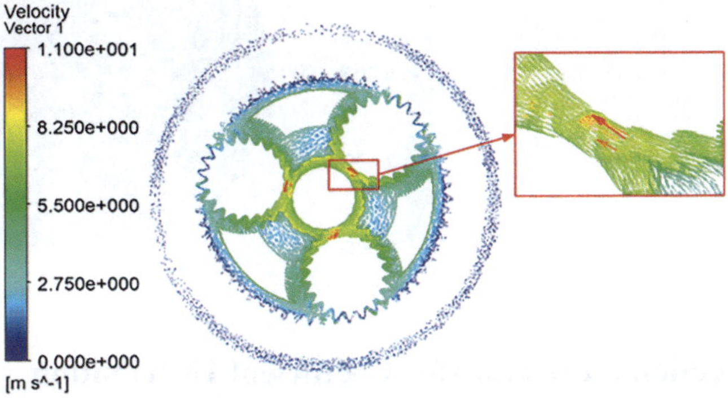

Fig. 3 Vector diagram of velocity distribution

As can be seen from the figure, the first stage of the gear system fluid domain maximum flow velocity of 10.67 m/s, located in the meshing area, this is due to the meshing area of the fluid is not only affected by the gear linear velocity, but also by the meshing of the tooth surface of the extrusion between the role, resulting in the meshing area of the fluid velocity is higher than that in other areas of the fluid velocity. The fluid domain velocity near the tooth surface is about 7 m/s, which is closer to the linear velocity of the tooth surface.

2. Oil distribution

When the gear rotates, the lubricant enters the gear tooth meshing area, most of the lubricant will be attached to the tooth surface, and then squeezed by the gear teeth, the lubricant will be spread along the tooth direction, and the excess lubricant on the tooth surface will be overflowed to the end face. The oil distribution at 0.4 s and 2.4 s is shown in Figs. 4 and 5.

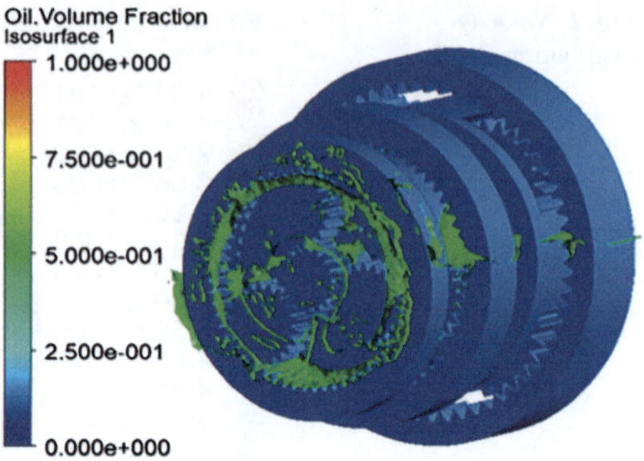

Fig. 4 Oil distribution at 0.4 s

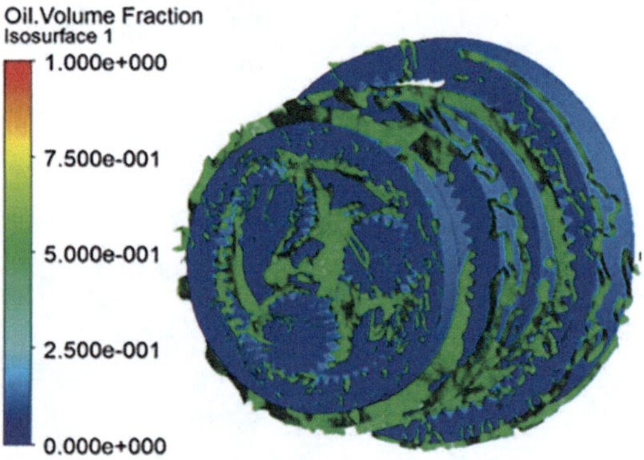

Fig. 5 Oil distribution at 2.4 s

4 Convection Heat Transfer Coefficient Distribution

After the oil-immersed lubrication flow field of the planetary wheel system of the shield machine tends to stabilize, the convective heat transfer coefficient of each gear of the first-stage planetary wheel system is extracted from the cloud diagram, as shown in Fig. 6, the distribution law of the convective heat transfer characteristics is close to the distribution law of the oil and gas ratio, and the convective heat transfer coefficient is large in the position where there is a lot of oil distribution.

Shield machine three-stage planetary wheel system gear teeth average convective heat transfer coefficient as shown in Table 1.

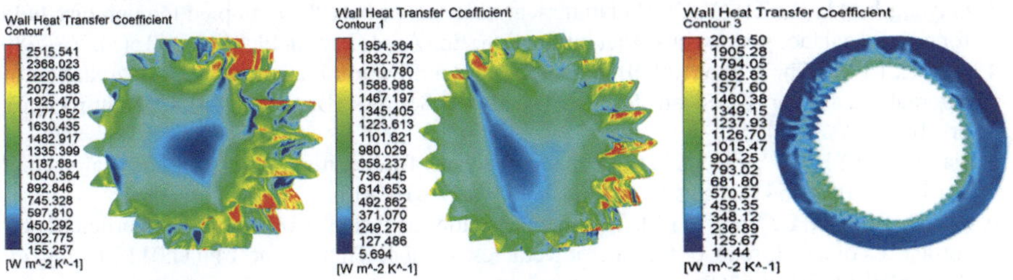

(a) Sun (b) Planet (c) Ring

Fig. 6 First stage gear convective heat transfer coefficient diagram. (**a**) Sun (**b**) Planet (**c**) Ring

Table 1 Convective heat transfer coefficient α/(W•m^{-2}•K^{-1})

	Sun	Planet 1	Planet 2	Planet 3	Planet 4	Ring
First level	1256.6	843.9	840.4	839.8	–	587.0
Second level	566.5	372.8	376.1	378.2	–	261.6
Third level	351.3	148.6	151.9	147.2	145.7	133.0

5 Conclusion

1. After the pre-processing, the mesh model is imported into Fluent to set the boundary conditions, and the UDF function is used to control the rotation of the planetary wheel system by choosing the pressure-based solver, VOF multiphase flow model and RNG k-ε model with the proportion of the oil immersion depth of 50%.
2. Analyze the oil-air ratio distribution, oil motion law and convective heat transfer coefficient of the planetary wheel system of shield machine based on immersion oil lubrication. The convective heat transfer coefficient of each gear tooth of the planetary wheel system shows the maximum rule of the first level, the second level is the second, and the third level is the smallest. The conclusion is basically consistent with the law described in the literature [10], indicating that the study is basically correct.

References

1. Li KY, Chen GD, Liu D (2016) Study of the influence of lubrication parameters on gear lubrication properties and efficiency. Ind Lubr Tribol 68(6):647–657. https://doi.org/10.1108/ILT-06-2015-0089
2. Kromer C, Cordes L, Keller MC et al (2019) Analytical solution to the heat transfer in fling-off cooling of spur gears. J Heat Transf-Trans Asme 141(8):082103. https://doi.org/10.1115/1.4043894

3. Ouyang B, Ma FY, Dai Y (2021) Numerical analysis on heat-flow-coupled temperature field for orthogonal face gears with oil-jet lubrication. Eng Appl Comput Fluid Mech 15(1):762–780
4. Guo R, Li YT, Shi Y et al (2020) Research on identification method of wear degradation of external gear pump based on flow field analysis. Sensors (Basel) 20(14):4058. https://doi.org/10.3390/s20144058
5. Jia FC, Lei YL, Fu Y et al (2020) Demand analysis of lubricating oil in spur gear pairs. Appl Sci-Basel 10(16):5417. https://doi.org/10.3390/app10165417
6. Zhu B, Wang X, Luo L et al (2022) Influence of lubricant supply on thermal and efficient performances of a gear reducer for electric vehicles. J Tribol-Trans Asme 144(1):011202. https://doi.org/10.1115/1.4052681
7. Shen LJ, Zhu YM, Shao S et al (2023) Research on splash lubrication characteristics of a spiral bevel gearbox based on the MPS method. Lubricants 11(12):520. https://doi.org/10.3390/lubricants11120520
8. Wang SS, Wang SK, Wang J et al (2020) Temperature field simulation and experimental study of anti-backlash single-roller enveloping hourglass worm gear. Chin J Mech Eng 33(1):59. https://doi.org/10.1186/s10033-020-00475-x
9. Li W, Tian Y (2017) Unsteady-state temperature field and sensitivity analysis of gear transmission. Tribol Int 116:229–243. https://doi.org/10.1016/j.triboint.2017.07.019
10. Yuhao X, Tengjiao L, Wen L (2022) Characteristics of internal flow field and convection heat transfer in planetary gear reducer of shield machine. J Phys Conf Ser 2383(1):012023. https://doi.org/10.1088/1742-6596/2383/1/012023

Open Access This chapter is licensed under the terms of the Creative Commons Attribution-NonCommercial-NoDerivatives 4.0 International License (http://creativecommons.org/licenses/by-nc-nd/4.0/), which permits any noncommercial use, sharing, distribution and reproduction in any medium or format, as long as you give appropriate credit to the original author(s) and the source, provide a link to the Creative Commons license and indicate if you modified the licensed material. You do not have permission under this license to share adapted material derived from this chapter or parts of it.

The images or other third party material in this chapter are included in the chapter's Creative Commons license, unless indicated otherwise in a credit line to the material. If material is not included in the chapter's Creative Commons license and your intended use is not permitted by statutory regulation or exceeds the permitted use, you will need to obtain permission directly from the copyright holder.

Simulation Calculations of TVMS of Aerospace Bevel Gears Taking Temperature into Account

Lei Ye, Tengjiao Lin, and Bingkui Chen

Abstract Aiming at the temperature effect problem of aerospace bevel gears under high-speed and high-load conditions, a time-varying meshing stiffness (TVMS) simulation calculation method is proposed to take into account the influence of temperature. Based on the theory of frictional heat generation, the formula of heat flow density on the tooth surface is deduced, and the relative sliding velocity of the curved bevel gear is simplified by combining with the equivalent cylindrical gear parameters to complete the analysis of tooth contact pressure; the gear pair's steady-state thermal field is solved via loading the boundary conditions through ANSYS APDL program. The static contact finite element (FE) method is further used to apply the temperature field results as thermal loads, extract the normal contact force and integrated deformation of the meshing tooth surface, and calculate the TVMS. The results show that the increase in temperature leads to thermal expansion in gears, contact regions increases, the integrated deformation decreases, and the TVMS is significantly increased. This work establishes theoretical foundations for thermomechanical coupling design and performance optimization of aerospace bevel gears.

Keywords TVMS · Heat flow density · Convective heat transfer

1 Introduction

As pivotal components in the aerospace power transmission mechanisms, curved bevel gears are not only subjected to complex mechanical loads, but also face the challenges brought by temperature variations. In particular, the effect of temperature on the TVMS cannot be ignored. Therefore, it is necessary to study the TVMS of curved-tooth bevel gears considering the temperature effect. Current research on

L. Ye · T. Lin (✉) · B. Chen
State Key Laboratory of Mechanical Transmission, Chongqing University, Chongqing, China
e-mail: tjlin1968@162.com

© The Author(s) 2026

Z. Zhang, P. Mercorelli (eds.), *Mechanical Design and Simulation: Insights and Innovations*, Lecture Notes in Mechanical Engineering,
https://doi.org/10.1007/978-981-95-2405-1_37

gear TVMS includes: Huang et al. [1] established an analytical model of TVMS of variable thickness gears based on the principle of load-contact FE analysis; Liu et al. [2] enhanced helical gear TVMS accuracy using energy potential principles, which improved the calculation accuracy; Huang et al. [3] established a solution model for TVMS under elastic flow lubrication, and analyzed the effects of elastic flow lubrication and lamellar coupling on the mesh characteristics; Pei et al. [4] developed a TVMS model considering lubrication and friction nonlinearities, which provided a theoretical support for the performance optimization of involute straight-toothed cylindrical gears; Wen et al. [5] proposed a solution model for gear TVMS of helical/bevel gears by using the slice method and analyzed its effect through parameter adjustment; Chakroun et al. [6] applied the local slope method to the analysis of bevel gear TVMS and verified the feasibility of the method. Although these studies have achieved fruitful results in the calculation of TVMS of different types of gears, studies on TVMS of curved-tooth bevel gears under the influence of temperature is still relatively limited.

This study focuses on the effect of temperature on TVMS in curved bevel gears. Steady-state thermal fields are determined by calculating the flux density and convection coefficients. The temperature results are treated as thermal loads in the FE model, and the TVMS is calculated by extracting the normal contact force (F_n) and gear pair deformation (δ_n) during meshing. The results show that the temperature-induced thermal expansion increases the contact area, reduces the deformation, and enhances the TVMS, which provides important insights for thermo-mechanical design optimization.

2 Calculation of Heat Flow Density of Curved Bevel Gear

Based on the theory of friction heat generation, assuming that the instantaneous heat fluxes of the master and slave wheels of the gear pair in the meshing process are φ_i (i = 1, 2), then:

$$\begin{cases} \varphi_1 = \beta_a \varphi_T \\ \varphi_2 = (1 - \beta_a)\varphi_T \end{cases} \tag{1}$$

Where, β_a is the heat distribution coefficient; φ_T is the total frictional heat flow.

The expression for the total frictional heat flux φ_T of the gear pair is given by [7]:

$$\varphi_T = \eta \mu \sigma_{HP} V_s \tag{2}$$

Where, η represents thermal conversion efficiency; μ is the tooth friction coefficient; σ_{HP} is the tooth contact pressure; and V_s is tooth sliding velocity.

The resulting expression for the heat flow density on the tooth surface is:

$$q = \frac{t_z}{t_c} \varphi_i \tag{3}$$

Where, t_z is the single tooth meshing period; t_c is the time required for 1 week of gear rotation.

It can be seen that the heat flow density of the tooth surface is determined by the relative sliding speed, the tooth contact pressure and the comprehensive friction coefficient of the tooth surface. Considering the complex shape of the teeth of the curved bevel gear, it can be simplified by deriving the corresponding equivalent cylindrical helical gear parameters. The following quantitative relationship exists between the number of bevel gear teeth and its equivalent cylindrical gear teeth:

$$z_{vi} = \frac{z_i}{\cos\delta_i} \tag{4}$$

Where, z_{vi} is the number of equivalent cylindrical gear teeth; z_i is the number of bevel gear teeth; δ_i is the split cone angle.

Based on the gear meshing principle, the normal partial velocity V_n of the master and slave wheels at the meshing point A in the process of spur gear meshing is equal, while the tangential partial velocities V_{t1} and V_{t2} are different, and there is a relative sliding phenomenon on the meshing surface. Assuming that the transmission ratio is i, V_{t1}, V_{t2} can be expressed as respectively:

$$\begin{cases} V_{t1} = R_1\omega_1 = \dfrac{\pi n_1}{30}\left(1 + \dfrac{\overline{PA}}{\overline{N_1P}}\right) r_{b1}\tan\alpha \\ V_{t2} = R_2\omega_2 = \dfrac{\pi n_2}{30}\left(\dfrac{\overline{PA}}{\overline{N_1P}} - i\right) r_{b1}\tan\alpha \end{cases} \tag{5}$$

So the relative sliding speed of the teeth is:

$$V_s = |V_{t1} - V_{t2}| \tag{6}$$

The equivalent gear of a straight-toothed bevel gear is a straight-toothed cylindrical gear with index circle $R\cos\delta$ and number of teeth $z/\cos\delta$. And the curved bevel gear is based on the straight bevel gear with helix angle added, so the equivalent gear of the arc bevel gear can be expressed by the helical cylindrical gear with the number of teeth $z/\cos\delta$ and the helix angle β. The speed of the equivalent helical gear of the curved bevel gear at any point on its end face meshing line and the relative sliding speed can be expressed as:

$$\begin{cases} V_{vt1} = R_{vn1}\omega_1\cos\beta = R_{vt1}\omega_1 \\ V_{vt2} = R_{vn2}\omega_2\cos\beta = R_{vt2}\omega_2 \end{cases} \tag{7}$$

$$V_s' = |V_{vt1} - V_{vt2}| \tag{8}$$

In this section, ANSYS hydrostatic simulation is used to analyze the contact pressure on the teeth of a curved bevel gear. The model has 2,610,270 cells and 313,302

Fig. 1 Contact pressure calculation model

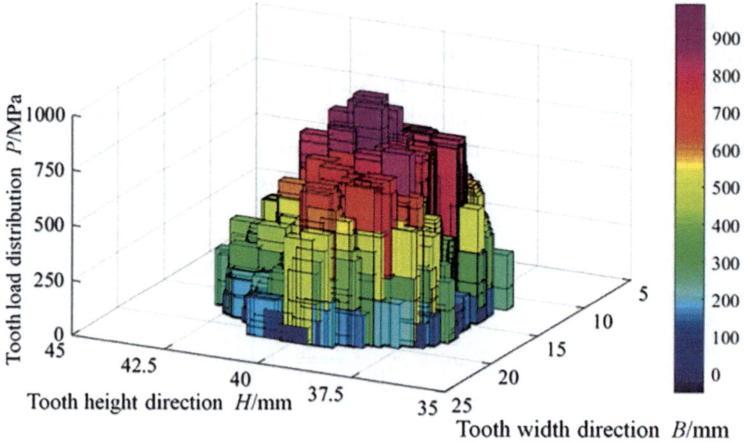

Fig. 2 Hertz contact pressure

nodes as shown in Fig. 1. Axial and radial constraints are applied to the master wheel while the driven wheel is fixed. An APDL program is written to simulate the meshing process and extract the contact pressure to generate the Hertzian contact pressure distribution as shown in Fig. 2.

In order to simplify the calculation, the tooth surface friction coefficient in this paper takes the value of 0.05 [8]. According to the calculated relative sliding speed and tooth contact pressure, combined with the tooth friction coefficient, the heat flow density of the bevel gear is obtained from Eq. (3) as shown in Figs. 3 and 4.

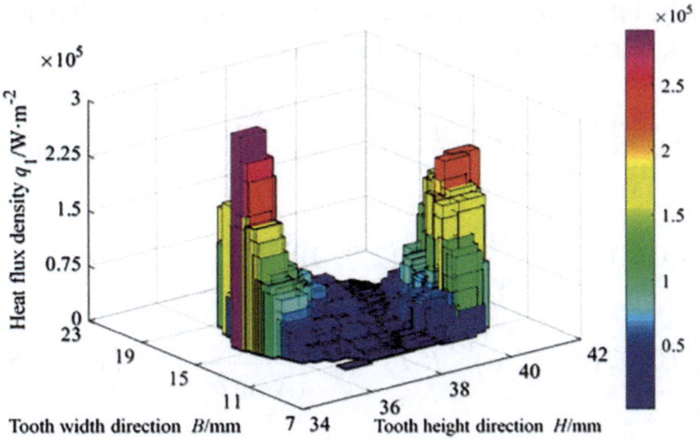

Fig. 3 Active gear heat flow density

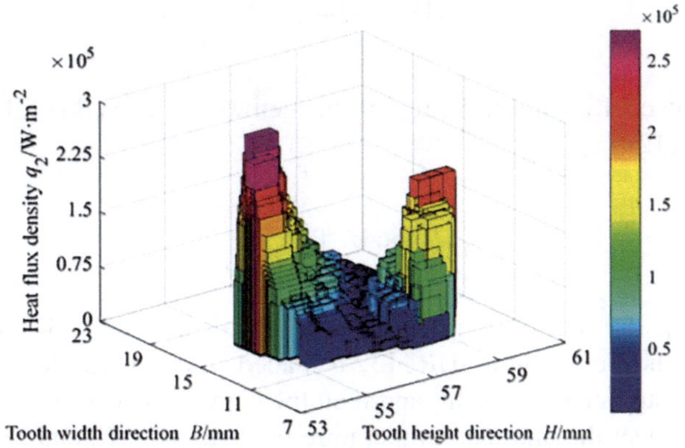

Fig. 4 Slave gear heat flow density

3 Steady-State Temperature Field Analysis

Utilizing principle of disc convection heat transfer [9], the curved bevel gear is equivalent to a rotating cone, then its end face flow heat transfer coefficient expression is:

$$h_{ez} = N_u K_t \left(\frac{\omega}{v \sin \alpha_e} \right)^{1/2} \tag{9}$$

Depending on the lubricant flow state, the flow convection coefficients of the curved bevel gear end face can be further expressed as:

$$h_{ez}=\begin{cases}2K_t\left[\dfrac{1.02P_r(m_r+2)}{60}\right]^{1/3}\left(\dfrac{\omega}{v\sin\alpha_e}\right)^{1/2} & \left(R_e\le 2.0\times10^5\right)\\ 1.0\times10^{-19}K_t\left(\dfrac{\omega}{v}\right)^4 r_i^7 & \left(2.0\times10^5<R_e<2.5\times10^5\right)\\ 0.0197(m_r+2.6)^{1/5}K_tP_r^{3/5}R_{ei}^{4/5}\left(\dfrac{\omega}{v\sin\alpha_e}\right)^{1/2} & \left(R_e>2.5\times10^5\right)\end{cases} \tag{10}$$

The convection coefficients of the curved bevel gear tooth-to-face flow is determined by Eq. (11), it can be expressed as [10]:

$$h_{cz}=1.418\left(\frac{vc_0\rho_0}{K_tz}\right)^{1/4}\frac{\sqrt{K_tvc_0\omega}}{2\pi} \tag{11}$$

The convection coefficients at the top of the teeth of the curved bevel gear is determined by Eq. (12):

$$h_d=0.664K_t\mathrm{P}_r^{1/3}\left(\frac{\omega}{v}\right)^{1/2} \tag{12}$$

Where, c_0 and ρ_0 are the specific heat capacity and density of the lubricant.

The surface heat effect unit SURF152 is loaded with the heat flow density on the meshing tooth surface of the gear, and then the convection coefficients are applied to the gear surface; the ambient temperature is set to 110 °C, and the steady-state temperature field of the curved bevel gear pair is solved as shown in Figs. 5 and 6.

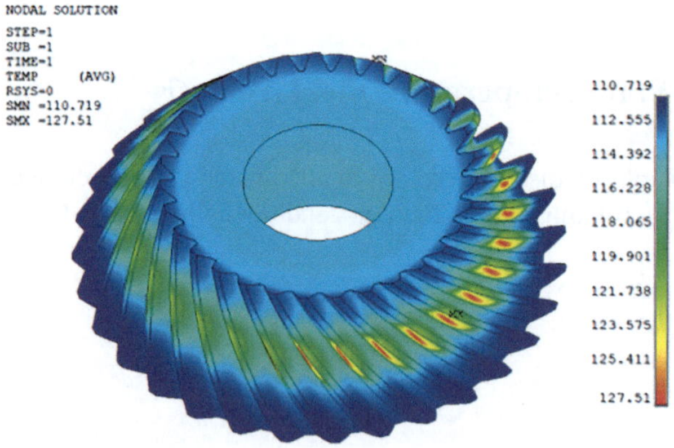

Fig. 5 Active gear body temperature

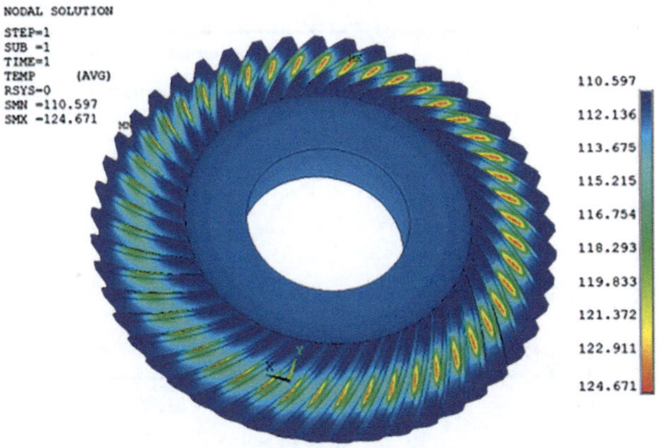

Fig. 6 Slave gear body temperature

4 TVMS Taking into Account Temperature

When the gear body temperature increases, the gear due to thermal deformation leads to the meshing tooth surface deviation from the theoretical position, so that the gear meshing stiffness is affected. The TVMS of the gear pair is calculated as:

$$k_n = \frac{F_n}{\delta_n} \tag{13}$$

Where, F_n is the normal contact force on the meshing tooth surface; δ_n is the integrated deformation of the gear pair.

In this section, the results of temperature calculation in the previous section are applied to all nodes of the model with thermal loads; the simulated values of the normal contact force F_n on the meshing tooth surface and the integrated deformation δ_n of the gear pair are extracted during the meshing process, and the TVMS taking into account the effect of temperature is obtained by substituting into Eq. (13), as shown in Figs. 7 and 8. It can be analyzed that the thermal expansion effect occurs in the gear teeth after the temperature increases, leading to an increase in the contact area of the gear teeth in the meshing process, which makes the integrated deformation δ_n of the nodes on the contact surface decrease, and thus improves the TVMS.

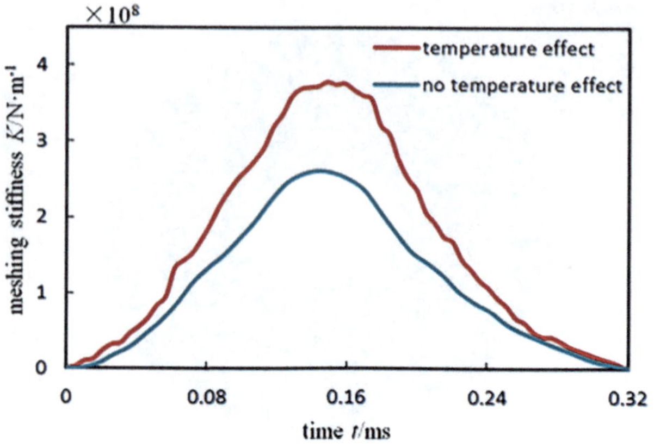

Fig. 7 Single tooth meshing stiffness

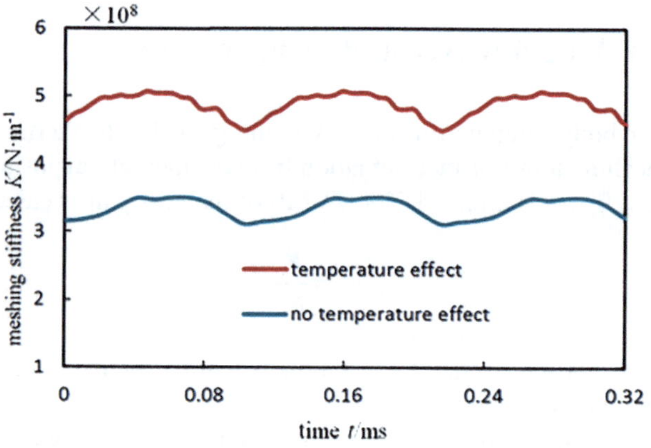

Fig. 8 Combined meshing stiffness

5 Conclusion

1. The steady state temperature field simulation analysis was completed. The highest peak temperature of the gear pair occurs at the top of the active gear tooth near the pitch line. The maximum temperature of the gear pair is 127.51 °C, and the temperature rise is 27.51 °C.
2. The TVMS of the curved bevel gears considering the effect of temperature is obtained, and the results show that the increase in temperature leads to thermal expansion of the gear teeth, the contact area increases, the integrated deformation decreases, and the TVMS is significantly increased.

Acknowledgments This work was financially supported by the National Science and Technology Major Project (No: HT-J2019-IV-0001-0068).

References

1. Huang Y, Du X, Zhu C et al (2019) Mesh stiffness analysis of beveloid gears for the rotating vector transmission. J Mech Sci Technol 33(8):3943–3953. https://doi.org/10.1007/s12206-019-0739-6
2. Liu Z, Sun Y, Zhou C et al (2023) Helical gear mesh stiffness improvement algorithm and its influence factors analysis. J Mech Transm 47(3):39–48. https://doi.org/10.16578/j.issn.1004.2539.2023.03.006
3. Huang W, Ma H, Zhao Z et al (2023) An iterative model for mesh stiffness of spur gears considering slice coupling under elastohydrodynamic lubrication. J Cent South Univ 30(10):3414–3434. https://doi.org/10.1007/s11771-023-5419-0
4. Pei X, Nie J, Guo H et al (2024) An involute gear pair meshing stiffness model considering time-varying friction under mixed lubrication. Nonlinear Dyn 56(12):1–19. https://doi.org/10.1007/s11071-024-10056-5
5. Wen J, Yao H, Yan Q et al (2023) Research on time-varying meshing stiffness of marine beveloid gear system. Mathematics 11(23):3–26. https://doi.org/10.3390/math11234774
6. Chakroun E, Battarra M, Mucchi E (2025) The bevel local slope approach: a method for mesh stiffness estimation in spur, helical and spiral bevel gears. Simul Modell Pract Theory 139:103047. https://doi.org/10.1016/j.simpat.2024.103047
7. Ru X, Wang G (2016) Study on contact characteristics and trimming of helical gears considering thermoelastic coupling. Mech Des Manuf 11:101–104. https://doi.org/10.19356/j.cnki.1001-3997.2016.11.025
8. Wang Y, Tang W, Chen Y et al (2017) Investigation into the meshing friction heat generation and transient thermal characteristics of spiral bevel gears. Appl Therm Eng 119:245–253. https://doi.org/10.1016/j.applthermaleng.2017.03.071
9. Gardon G, Astarita T, Carlomagno G (2007) Infrared heat transfer measurements on a rotating disk. Opt Diagn Eng 1(2):1–7. https://doi.org/10.1155/S1023621X97000018
10. Patir N, Cheng H (1979) Prediction of the bulk temperature in spur gears based on finite element temperature analysis. ASLE Trans 22(1):25–36. https://doi.org/10.1080/05698197908982899

Open Access This chapter is licensed under the terms of the Creative Commons Attribution-NonCommercial-NoDerivatives 4.0 International License (http://creativecommons.org/licenses/by-nc-nd/4.0/), which permits any noncommercial use, sharing, distribution and reproduction in any medium or format, as long as you give appropriate credit to the original author(s) and the source, provide a link to the Creative Commons license and indicate if you modified the licensed material. You do not have permission under this license to share adapted material derived from this chapter or parts of it.

The images or other third party material in this chapter are included in the chapter's Creative Commons license, unless indicated otherwise in a credit line to the material. If material is not included in the chapter's Creative Commons license and your intended use is not permitted by statutory regulation or exceeds the permitted use, you will need to obtain permission directly from the copyright holder.

Analysis and Verification of Heat Source Model for Laser Quenching on 45 Steel Surface

Jun Xu, Bucheng Zhang, Gangyan Li, Yutong Guo, and Rui Sheng

Abstract The heat transfer process of medium carbon steel laser quenching has complex unsteady heat transfer characteristics, which are difficult to be accurately described by adopting traditional heat conduction models. The unsteady heat transfer control equation of medium carbon steel laser quenching considering the above influences is deduced, the heat conduction and temperature field variation laws of this process are explored, and a simulation model of unsteady heat transfer characteristics of medium carbon steel laser quenching is established. A laser quenching test system for medium carbon steel and measuring the temperature data during the actual quenching process is established to verify the reliability of the simulation method. The prosed heat source model for laser quenching on 45 steel surface has been proved to be effective by comparing the simulation results with the test ones.

Keywords Laser quenching · Solid-state phase transformation · Unsteady-state heat transfer

1 Introduction

In the fields of materials science and surface engineering, laser quenching technology has attracted extensive attention due to its significant ability to enhance the surface hardness and wear resistance of metal materials. In numerical simulation and simulation analysis of laser quenching for different materials, Xidong Zhu et al. [1] conducted numerical simulation and experimental verification on the influence of laser quenching on the surface properties of 45 steel, and used ANSYS software to simulate and analyze the influence of process parameters such as laser power and scanning speed on the temperature field, and predicted the depth and width of the quench-hardened layer. Mingzhi et al. [2] conducted detailed simulation analysis of

J. Xu · B. Zhang · G. Li (✉) · Y. Guo · R. Sheng
School of Mechanical & Electric Engineering, Wuhan University of Technology, Wuhan, China

© The Author(s) 2026

Z. Zhang, P. Mercorelli (eds.), *Mechanical Design and Simulation: Insights and Innovations*, Lecture Notes in Mechanical Engineering,
https://doi.org/10.1007/978-981-95-2405-1_38

the laser quenching process of H13 steel using COMSOL finite element software, and discussed the influence of laser power on the surface modification layer performance of H13 steel. The significance of this research lies in the further improvement of the simulation analysis method for the non-steady heat transfer characteristics of laser quenching of medium-carbon steel. Qi et al. [3] obtained the ablation thermal response under a thermal load by establishing a thermal property model that varied with the temperature and time. The thermal property model did not distinguish between the phase transformations during the heating and cooling processes and did not consider the effect of the phase transformation from Austenite to Martensite on the thermal properties. Chao et al. [4] studied the thermal expansion behavior and corresponding phase transformation of laser powder bed melted martensitic steel in various directions during continuous heating. However, they only selected a set of physical property parameters from the American Society of Metals Handbook, which did not reflect the continuity of the material thermal property parameters that change with temperature. Li et al. [5] proposed a numerical simulation method for residual stresses in titanium alloy welding by considering solid-state phase transformations. The Johnson Mehl Avrami Kolmogorov and Koistinen–Marburger equations were used to describe the diffusion and non-diffusion transformations during the phase transformation process of titanium alloys [6–13], providing an important reference for the analysis of the thermal property parameters in the laser quenching of medium-carbon steel. Compared with existing studies, this method comprehensively considers the changes in material physical properties caused by the non-steady heat transfer characteristics. This not only includes the phase transformation from ferrite to austenite during heating, but also takes into account the influence of the conversion between different phases during heating and cooling on the simulation results of the laser quenching temperature field, as well as the effects of boundary conditions such as air convection. This simulation method not only enhances the practicability of the theoretical model but also provides important theoretical support and optimization guidance for the industrial application of laser quenching technology, helping to improve processing quality, efficiency, and economy. In addition, the results of this research can be extended to other heat treatment processes and material processing fields, and have broad application prospects and academic value.

2 Methodology

To solve for the energy distribution in the heat source loading area on the upper surface, it is necessary to obtain the temperature variation in the heat source loading area on the upper surface of the sample after the laser starts loading [14]. This experiment utilized a medium carbon steel laser quenching test system and standard test methods to obtain the temperature field test results under a laser power of 1050 W. The relationship graph between the coordinate and the temperature is shown in Fig. 1.

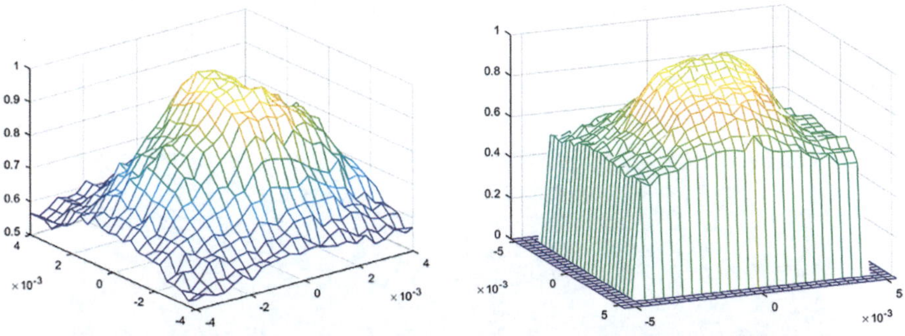

(a) MATLAB solution results (b)Schematic diagram

Fig. 1 Energy distribution of laser heat source. (**a**) MATLAB solution results. (**b**) Schematic diagram

In order to obtain simulation results that are closer to the actual situation, it is necessary to correct the energy distribution of the laser heat source obtained by solving with MATLAB. According to the actual working conditions of laser quenching, the size of the laser heat source model is defined. That is, the energy distribution of the heat source beyond this size range is all 0. Therefore, the edge of the laser heat source energy distribution is set to 0. The weighted super-Gaussian function [15] as shown in Eq. (1) is adopted as the basic form of the fitting function. After the function fitting, the energy distribution function of the laser heat source is obtained as shown in Eq. (2).

$$I = a\cdot\exp\left[\frac{b\cdot(-2)\left(x^{n}+y^{n}\right)}{(-0.004)^{n}}\right]+c\cdot\exp\left[\frac{d\cdot(-2)\left(x^{50}+y^{50}\right)}{(-0.004)^{50}}\right] \tag{1}$$

$$I = 0.483\cdot\exp\left[\frac{0.947\cdot(-2)\left(x^{4}+y^{4}\right)}{(-0.004)^{4}}\right]+0.517\cdot\exp\left[\frac{0.22\cdot(-2)\left(x^{50}+y^{50}\right)}{(-0.004)^{50}}\right] \tag{2}$$

The chemical compositions of the test pieces are listed in Table 1.

3 Results and Discussion

During quenching, the laser is focused on a small area of the workpiece surface and scanned along a specific path. The temperature change in the heated area was severe and there was a significant temperature gradient, whereas the temperature change in the area far away from the quenching path was relatively mild. Therefore, mesh refinement was performed on the laser quenching path and its surrounding areas, and the quenching path and surrounding areas on the workpiece surface were refined. A 0.5 mm hexahedral mesh was used to accurately capture the temperature

Table 1 Chemical composition of experimental material (%)

C	Cu	Si	Mn	P	S	Cr	Ni	Fe
0.47	0.01	0.29	0.55	0.016	0.007	0.02	0	98.637

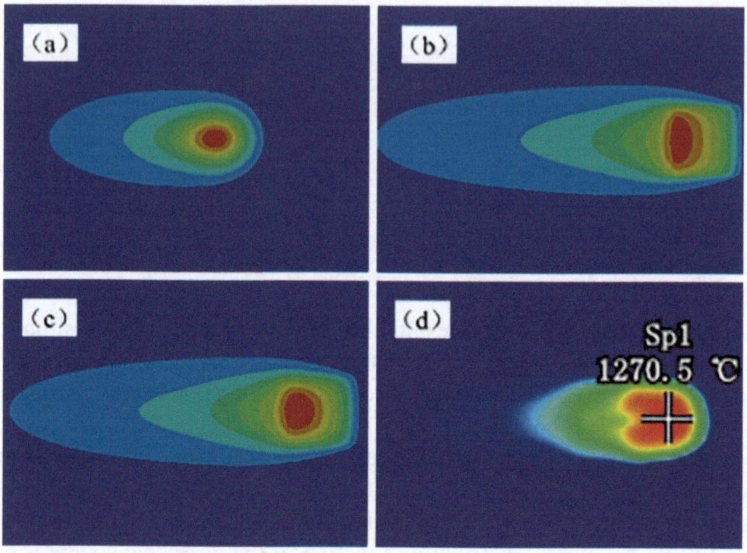

Fig. 2 Cloud diagrams of temperature field distribution for different heat source models

changes in this area. A 1 mm thick hexahedral mesh was used to ensure the computational efficiency of the remaining parts of the workpiece. The temperature field of the model was solved and the data were extracted during the laser quenching process at a power of 2100 W. To verify the accuracy of the heat source modeling method based on experiments and numerical analysis proposed, two common empirical heat source models, Gaussian distribution and uniform distribution, are selected for comparative analysis. By comparing the deviation between the temperature field simulation results under different heat source models and the experimental measurement data, especially the temperature error at the characteristic point A near the heat source application boundary, the accuracy and applicability of different heat source models can be evaluated. Figure 2 shows the temperature field distribution cloud diagrams of the three heat source models and the experiment at the same moment. It can be seen that the temperature distribution obtained by the proposed heat source model is closer to the experimental results.

Under the same absorption rate, the influence of the heat source energy function on the boundary where the heat source is applied is the most significant. To quantitatively evaluate the accuracy of different heat source models, the simulation results of the temperature at the characteristic point A on the boundary where the heat source is applied at different times were extracted and compared with the experimental measurement results. To ensure the validity of the comparison results, the

Table 2 Simulation results of temperature at feature point A

Moment(s)		0	0.05	0.1
Experimental temperature (°C)		526.9	519.64	478.37
Gaussian distribution	Simulated temperature (°C)	163.8	163.18	160
	Relative error	−68.91%	−68.60%	−66.55%
Uniform distribution	Simulated temperature (°C)	674.09	549.37	480.11
	Relative error	27.94%	5.72%	0.36%
New model	Simulated temperature (°C)	536.77	536.13	452.26
	Relative error	1.87%	3.17%	−5.46%

moment when point C reaches the highest temperature was taken as the 0 moment, and three moments of 0, 0.05 and 0.1 s were selected for comparison and analysis. The results are shown in Table 2.

The results show that the temperature prediction errors at point C for the Gaussian distribution and uniform distribution heat source models are relatively large, with the maximum relative errors reaching 68.91% and 27.94% respectively. In contrast, the prediction results of the heat source model proposed are highly consistent with the experimental values, with the relative error controlled within 6%. These results verify the accuracy of the heat source modeling method proposed, indicating that it can describe the temperature distribution near the boundary more precisely. The Gaussian heat source model and the uniform heat source model adopt simplified processing methods when describing the laser energy distribution, which leads to significant prediction errors at the boundary where the heat source is applied. The Gaussian distribution assumes that the laser energy reaches its maximum at the center of the spot and decreases exponentially with the distance from the center. Due to the assumption of a lower energy distribution at the edge of the spot, the actual energy input in these areas may be underestimated, resulting in lower temperature prediction values. On the other hand, the uniform heat source model may introduce sudden changes in energy input at the edge of the spot, leading to higher temperature prediction values. Compared with the Gaussian heat source model and the uniform heat source model, the heat source model proposed demonstrates a more accurate and precise temperature prediction capability near the boundary. The fitted heat source function can more accurately reflect the actual laser energy distribution on the material surface, including the energy change gradient at the edge of the spot. Using the fitted heat source function as the boundary condition for the temperature field simulation can significantly improve the prediction accuracy of key parameters such as the heat-affected zone boundary and the quenched layer thickness. In addition, the heat source model proposed has strong applicability and flexibility. By obtaining temperature field test data under different laser quenching conditions and combining it with numerical analysis methods for heat source function fitting, heat source models suitable for different laser types, laser parameters, and material systems can be constructed. This heat source model construction method based on experimental data can effectively overcome the limitations of traditional empirical models and improve the universality and accuracy of laser quenching temperature field simulation.

4 Conclusions

The heat source function solution method based on experiments and numerical analysis can precisely describe the energy distribution of the laser; the material physical parameters take into account the influence of temperature and phase transformation, significantly improving the prediction accuracy of the maximum temperature and the cooling rate in the phase transformation temperature range. Through the comparative analysis with Gaussian distribution and uniform distribution heat source models, the superiority of the proposed heat source modeling method in describing the spatial energy distribution of laser heat sources and predicting the temperature field near the heat source boundary has been verified. This method provides new ideas and means for further improving the accuracy of temperature field simulation of laser quenching of medium carbon steel. The model established can accurately predict the evolution of the temperature field, laying a foundation for further optimization of the laser quenching process and performance prediction of medium carbon steel.

References

1. Zhu X (2021) Research on the process basis of laser quenching on the surface properties of 45 steel. Xinjiang University, Ürümqi. https://doi.org/10.27429/d.cnki.gxjdu.2021.000618
2. Zhi M, Fangping Y, Jinhua L (2023) Simulation analysis and experimental research on laser quenching of H13 steel surface. Manuf Technol Mach Tool 11:39–46. https://doi.org/10.19287/j.mtmt.1005-2402.2023.11.006
3. Qi Z, Xu S, Yajun Y et al (2024) Cooling curve characteristics and particle swarm optimization algorithm based on optimization casting parameters in thermal and physical properties of the. J Spec Cast Nonferrous Alloys 44(2):165–169. https://doi.org/10.15980/j.t.ZZZ.2024.02.004
4. Yu Chao B, Cuiling Z, Jiayi Z, Hao W et al (2022) Abnormal thermal expansion behaviour and phase transition of laser powder bed fusion maraging steel with different thermal histories during continuous heating. Addit Manuf 53:102712. https://doi.org/10.1016/j.addma.2022.102712
5. Yu L, Quan WZ, Wen T (2023) Numerical simulation method of welding residual stress in titanium alloy considering solid phase transformation. Ship Hydrodyn 27(1):109–120. https://doi.org/10.3969/j.issn.1007-7294.2023.01.01
6. Lusquinos F, Conde JC, Bonss S et al (2007) Theoretical and experimental analysis of high power diode laser (HPDL) hardening of AISI 1045 steel. Appl Surf Sci 254(4):948–954. https://doi.org/10.1016/j.apsusc.2007.07.200
7. Tricarico L, Ancona A, Palumbo G et al (2017) Discrete spot laser hardening and remelting with a high-brilliance source for surface structuring of a hypereutectoid steel. Mater Des 115:194–202. https://doi.org/10.1016/j.matdes.2016.10.079
8. Chen G (2023) Research on simulation and adaptive monitoring technology of laser quenching characteristics of QT600-3 ductile iron. University of Electronic Science and Technology of China, Chengdu. https://doi.org/10.27005/d.cnki.gdzku.2023.004912
9. Xu X (2022) Research on numerical simulation and temperature control technology of laser phase transformation hardening. Shandong University, Jinan. https://doi.org/10.27272/d.cnki.gshdu.2022.004419
10. Martinovs A, Polukoshko S, Zaicevs E et al (2020) Laser hardening process optimization using FEM. Eng Rural Dev Jelgava 20:1500–1508. https://doi.org/10.22616/ERDev.2020.19.TF372

11. Tobar MJ, Álvarez C, Amado JM et al (2006) Laser transformation hardening of a tool steel: simulation-based parameter optimization and experimental results. Surf Coat Technol 200(22–23):6362–6367. https://doi.org/10.1016/j.surfcoat.2005.11.067
12. Huiying D (2023) Research on laser heat treatment experiment and numerical simulation of GGG70L ductile iron. Shandong University, Jinan. https://doi.org/10.27272/d.cnki.gshdu.2023.001949
13. Lee S, Lee Y (2009) Latent heat of martensitic transformation in a medium-carbon low-alloy steel. Scr Mater 60(11):10161019. https://doi.org/10.1016/j.scriptamat.2009.02.042
14. Kang SH, Im YT (2007) Thermo-plastic finite element analysis of quenching process of carbon steel. J Mater Process Technol 192:381–390. https://doi.org/10.1016/j.jmatprotec.2007.04.101
15. Zheng M, Junjie L et al (2023) Laser quenching temperature field and phase transformation hardened zone of 45 steel. Heat Treat Metals 48(9):258–264. https://doi.org/10.1016/j.mechmat.2023.09.006

Open Access This chapter is licensed under the terms of the Creative Commons Attribution-NonCommercial-NoDerivatives 4.0 International License (http://creativecommons.org/licenses/by-nc-nd/4.0/), which permits any noncommercial use, sharing, distribution and reproduction in any medium or format, as long as you give appropriate credit to the original author(s) and the source, provide a link to the Creative Commons license and indicate if you modified the licensed material. You do not have permission under this license to share adapted material derived from this chapter or parts of it.

The images or other third party material in this chapter are included in the chapter's Creative Commons license, unless indicated otherwise in a credit line to the material. If material is not included in the chapter's Creative Commons license and your intended use is not permitted by statutory regulation or exceeds the permitted use, you will need to obtain permission directly from the copyright holder.

Simulation Analysis Model for Hardness in Laser Quenching of Medium Carbon Steel

Chenxu Hu, Yutong Guo, Rui Sheng, Bucheng Zhang, Yanglin Chen, and Gangyan Li

Abstract This article focuses on the simulation analysis of hardness in laser quenching of medium carbon steel. Firstly, the theoretical calculations of hardness were analyzed. Subsequently, a hardness testing system for medium carbon steel undergoing laser quenching is established. Finally, a simulation analysis model is constructed based on the COMSOL Multiphysics simulation platform. The accuracy of the simulation analysis model for hardness in laser quenching of medium carbon steel is verified by comparing it with experimental results. The results indicate that the simulation errors for hardness are relatively small, whereas the errors for the size of the hardened layer are more significant.

Keywords Laser quenching · Hardness · Phase transformation · Simulation analysis · Hardened layer

1 Introduction

Laser quenching technology uses high-energy laser beams to rapidly heat material surfaces beyond the austenitizing transformation temperature. As the surface cools, martensite forms, creating a hardened layer via this process [1]. This treatment enhances surface hardness, wear resistance, fatigue resistance, and oxidation resistance [2]. The process also allows for division into three layers: the hardened layer, the transition layer, and the base material [3].

C. Hu · Y. Guo · R. Sheng · B. Zhang · Y. Chen · G. Li (✉)
School of Mechanical & Electric Engineering, Wuhan University of Technology, Wuhan, China
e-mail: gangyanli@whut.edu.cn

© The Author(s) 2026

Z. Zhang, P. Mercorelli (eds.), *Mechanical Design and Simulation: Insights and Innovations*, Lecture Notes in Mechanical Engineering,
https://doi.org/10.1007/978-981-95-2405-1_39

Liverani et al. [4] conducted a numerical study using a semiconductor laser to process AISI 9810 steel gears through laser heat treatment, investigating the effects of quenching layer depth and residual stress states. Martinez et al. [5] developed a semi-empirical model-based tool to predict temperature fields and quenching layer thickness in laser surface treatments. Chen Kaiye [6] investigated temperature, stress fields, microstructural changes, and hardness evolution during traditional laser quenching and galvanometer scanning laser quenching of carbon steel using numerical simulation and experiments.

Hardness detection requires specialized equipment, and internal hardness testing is particularly challenging. While simulation analysis has numerous advantages to compensate for the shortcomings of hardness analysis in simulations, this study employs 45 steel as a research object to establish a middle carbon steel laser quenching hardness calculation model, thereby minimizing research costs and enhancing efficiency.

2 Hardness Calculation Principles for Laser Quenching of Medium Carbon Steel

Similar to traditional quenching, the hardened layer formed after laser quenching consists of martensite and retained austenite [7]. The calculation model of martensite, ferrite and pearlite is given by Eqs. (1) and (2) [8]:

$$HV_M = 127 + 949C + 27Si + 11Mn + 8Ni + 16Cr + 21\ln(V_r) \quad (1)$$

$$\begin{aligned} HV_{F-P} &= 42 + 223C + 53Si + 30Mn + 12.6Ni + 7Cr \\ &+ 19Mo + (10 - 19Si + 4Ni + 8Cr + 130V)\ln(V_r) \end{aligned} \quad (2)$$

Where: HV_M is the Vickers hardness of martensite, $HV_{F\text{-}P}$ is the Vickers hardness of ferrite and pearlite, V_r is the cooling rate at 700 °C, C, Si, Mn, Ni, and Cr are the chemical element contents in steel.

The hardness of mixed microstructures is calculated using a weighted sum, as shown in Eq. (3):

$$HV = HV_{F-P}\left(\xi^F + \xi^P\right) + HV_M \xi^M \quad (3)$$

Where: ξ^F, ξ^P and ξ^M are the volume fractions of ferrite, pearlite, and martensite, respectively.

The calculation model involves the microstructural volume fraction and temperature. Therefore, simultaneously, we also need to analyze the phase transformation and heat transfer processes during the laser quenching of middle carbon steel.

2.1 Analysis of Solid-State Phase Transformation in Laser Quenching of Medium Carbon Steel

During the heating phase, the austenite transition between ferrite and pearlite can be described by the Johnson-Mehl-Avrami-Kolmogorov (JMAK) model [9], the volume fraction of austenite during laser quenching of medium carbon steel can be expressed as Eq. (4).

$$f_A(T) = \left[1 - \exp\left(-b(T)t^n\right)\right] \cdot F_A(T) \tag{4}$$

Where: $f_A(T)$ is the volume fraction of austenite, $F_A(T)$ is the equilibrium volume fraction of austenite, T is the temperature, t is the time, b and n are empirical parameters. For austenite transformation, n is taken as 4, b is determined based on the nucleation and growth rate.

During the cooling phase, the transformation of austenite into martensite is displacive and can be described by the Koistinen-Marburger model. The volume fraction of martensite can be given by Eq. (5).

$$f_M(T) = 1 - \exp\left(-\beta\left(M_s - T\right)\right) \tag{5}$$

where: f_M is the volume fraction of martensite, β is the Koistinen-Marburger model coefficient, taken as 0.011(1/°C), M_s is the martensite start temperature.

2.2 Analysis of Heat Transfer Process in Laser Quenching of Medium Carbon Steel

The heat transfer process [10] of laser quenching can be expressed by Eq. (6).

$$\rho c(T)\frac{\partial T}{\partial t} = k(T)\nabla^2 T + S \tag{6}$$

Where: T is the temperature, t is the time variable, ρ is the density, c is the specific heat capacity, k is the thermal conductivity, S is the source term. The energy losses due to convection and radiation [11] can be expressed as Eq. (7):

$$q_{lost} = h\left(T - T_W\right) + \sigma\varepsilon\left(T^4 - T_W^4\right) \tag{7}$$

Where: T is the surface temperature, T_W is the ambient temperature, σ is the Stefan-Boltzmann constant, with a value of 5.67×10^{-8} W/(m^2·K^4), ε is the emissivity of the specimen surface, h is the convective heat transfer coefficient of the specimen surface.

3 Hardness Testing System for Laser Quenching of Medium Carbon Steel

The test specimen used in this study is a 45 steel block with dimensions of 100 mm × 30 mm × 10 mm, as shown in Fig. 1. The chemical composition of the 45 steel is provided in Table 1. A laser quenching test system for medium carbon steel was established, as shown in Fig. 2.

Fig. 1 45 steel specimen

Table 1 Chemical composition of 45 steel (%)

C	Cu	Si	Mn	P	S	Cr	Ni	Fe
0.47	0.01	0.29	0.55	0.016	0.007	0.02	0	98.637

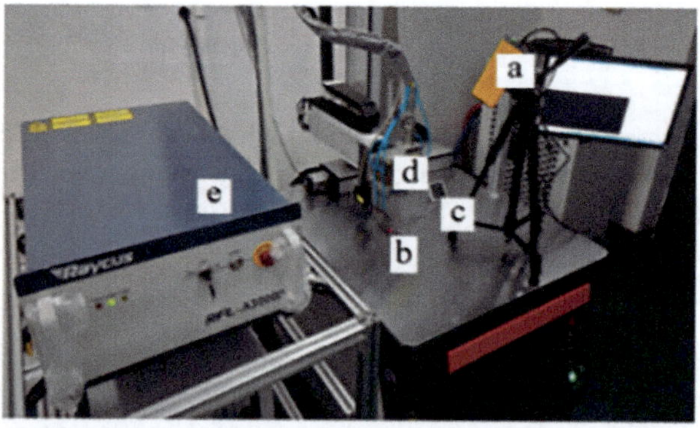

(a) Thermal imager, (b) Specimen, (c) Hy-gro-thermograph, (d) Lens of laser, (e) Laser gener-ator

Fig. 2 Laser quenching test bench. (**a**) Thermal imager, (**b**) specimen, (**c**) hygro-thermograph, (**d**) lens of laser, (**e**) laser generator

4 Simulation Analysis Model for Hardness in Laser Quenching of Medium Carbon Steel

Using the COMSOL Multiphysics 6.2 platform, a simulation analysis model for solid heat transfer and solid-state phase transformation during laser quenching of medium carbon steel was established.

4.1 Material Properties for Simulation Analysis of Laser Quenching Hardness

The overall material properties are calculated as a weighted average of the properties of individual phases based on their volume fractions. The specific heat and thermal conductivity of the material were calculated using JmatPro software, as shown in Figs. 3 and 4.

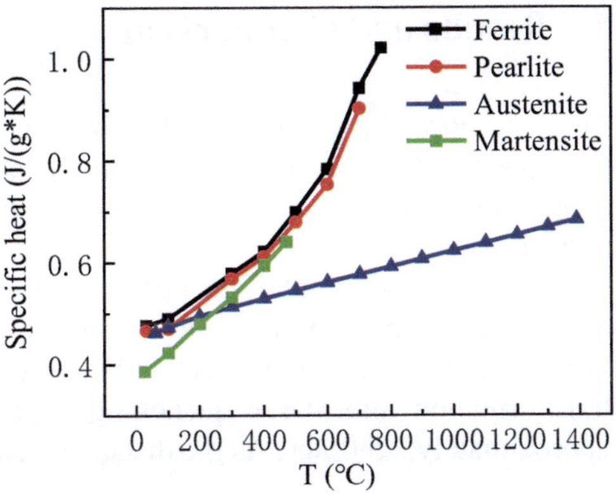

Fig. 3 Specific heat of each phase

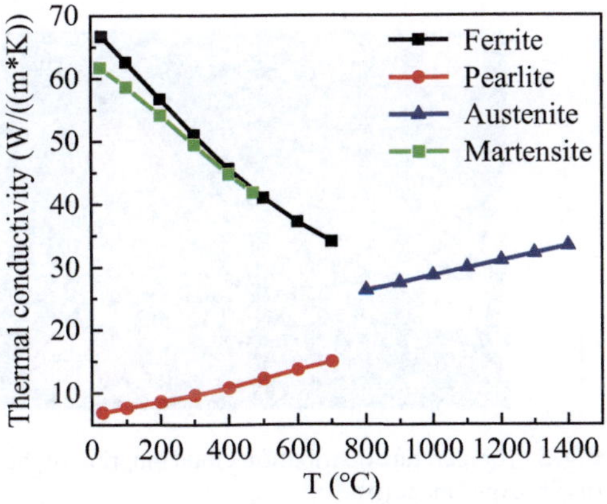

Fig. 4 Thermal conductivity of each phase

4.2 *Heat Source Model for Simulation Analysis of Laser Quenching Hardness*

During laser quenching, the energy density of the heat source can be represented by Gaussian function [12]. The heat source model of the semiconductor laser used in this paper is more consistent with the Ultra Gaussian heat source model, which can be expressed by Eq. (8):

$$q_{laser} = \alpha \frac{P}{A} \exp\left(\frac{(-1/2)(x^4 + y^4)}{(-0.004)^4}\right) \tag{8}$$

where: α is the absorptivity of the specimen surface to the laser, P is the laser power, set to 2200 W, A is the area of the laser spot on the specimen surface, with x, y being the coordinates on the respective axes, 0.004 is the width of the laser spot, representing the calculated size of the laser-irradiated region.

5 Results and Comparison

Figure 5 shows the temperature distribution cloud diagram when the center point of the upper surface reaches its maximum temperature. It can be observed that a "comet-like" tail remains in the region scanned by the laser spot on the specimen surface.

From the center point of the upper surface were extracted and compared with the experimentally measured hardness results, as shown in Fig. 6. The results show that in the region close to the laser-irradiated area, the 45 steel undergoes martensitic transformation after laser quenching, significantly increasing the hardness by approximately 2.5 times. As the distance from the laser-irradiated region increases,

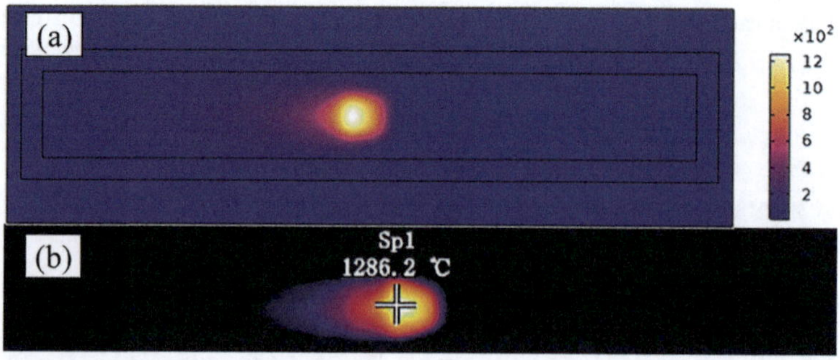

Fig. 5 Temperature distribution cloud diagram of the center point on the upper surface: simulation (**a**) vs. experiment (**b**)

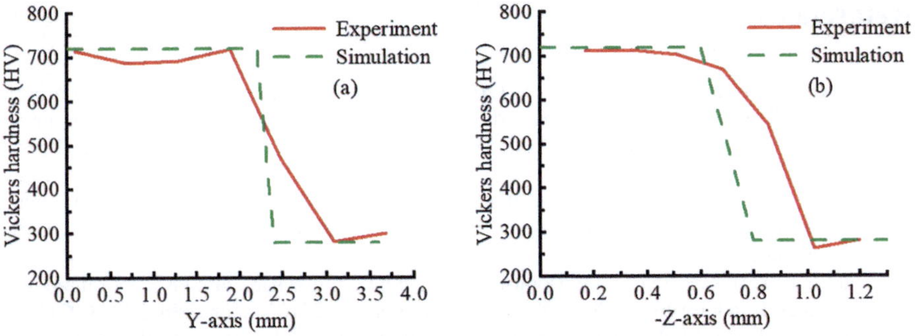

Fig. 6 Experimental and simulated Vickers hardness variation along the Y-axis (**a**) and the Z-axis (**b**) at the center point of the upper surface

Table 2 Comparison of experimental and simulation results for laser quenching at 2200 W and 1 m/min

		Experiment	Simulation	Relative error (%)
Center point	Maximum temperature (°C)	1286	1264	−1.7
	Vickers hardness (HV)	694	720	3.7
Hardened layer	Width (mm)	1.894	2.2	16.1
	Depth (mm)	0.685	0.6	−12.4

the hardness of 45 steel decreases sharply, and the transition layer spacing is small. The hardness of the base layer remains largely unchanged.

The experimental results, simulation results, and their relative errors are listed in Table 2. For temperature and hardness values, the relative error between the simulation and experimental results is less than 5%, indicating a good fit with the experimental data. However, for the dimensions of the hardened layer, the relative error is around 15%, which is relatively large.

6 Conclusion

This study is based on the COMSOL Multiphysics simulation platform to analyze the laser quenching process of middle carbon steel through numerical simulations and experimentally verifies the accuracy of the simulation model. The results indicate that while the relative errors in the maximum temperature and hardness at the center of the top surface between simulation and experiment are relatively small, there is a larger relative error in the thickness of the hardened layer. This discrepancy may be attributed to the transition layer being relatively narrow, making it challenging to precisely capture its boundaries during hardness measurements. Additionally, differences between the heat source model used in simulations and the laser generator's actual output lead to significant discrepancies between the two results.

References

1. Yanpeng Z, Weimin W, Feng S et al (2025) Effect of laser quenching on corrosion properties of Ti6Al4V-Cu alloy fabricated by wire-powder cooperative arc additive manufacturing. Metal Work (Heat Process):1–7. http://kns.cnki.net/kcms/detail/11.5627.TH.20250124.1126.006.html
2. Xia D (2025) Application research of laser surface modification technology in improving tool wear resistance. Agric Mach Use Mainten (2):84–86+90. https://doi.org/10.14031/j.cnki.njwx.2025.02.019
3. Zhuo H (2024) Research on ultrasonic testing technology for hardened layer morphology of alloy steel. Jilin University, Changchun. https://doi.org/10.27162/d.cnki.gjlin.2024.002118
4. Liverani E, Lutey AHA, Ascari A et al (2016) A complete residual stress model for laser surface hardening of complex medium carbon steel components. Surf Coat Technol 302:100–106
5. Martínez S, Ukar E, Tabernero I, Lamikiz A (2012) Development of a laser hardening simulation method including material characterization for rapid heating rates. In: Materials science forum, vol 713. Trans Tech Publications Ltd., Wollerau, pp 49–54
6. Kaiye C, Ye W, Guolong W et al (2020) Simulation and experimental research on temperature characteristics of laser quenching based on scanning galvanometer. Surf Technol 49(5):251–258. https://doi.org/10.16490/j.cnki.issn.1001-3660.2020.05.030
7. Wen F (2016) Simulation research on laser quenching of 45 steel. Northeastern University, Boston
8. Jixun W, Guangyu Z, Yumin X (1994) Application technology of steel hardenability. Mechanical Industry Press, Beijing
9. Youmin R (2017) Research on influence mechanism of phase transformation and precipitation on thermo-elastic-plastic behavior and weld mechanical properties in laser welding of high-strength steel. Huazhong University of Science and Technology, Wuhan
10. Luo S, Wang G, Ma X, Zheng L, Wang R (2022) Simulation of flame spread of titanium-alloy sheet under effect of titanium droplet. Acta Aeronaut Astronaut Sin 43(9):425652. https://hkxb.buaa.edu.cn/CN/Y2022/V43/I9/425652
11. Lifeng W, Weimin L, Yang Y et al (2023) Research status of laser quenching of steel. Metal Working (Heat Processing):20–31. https://doi.org/10.3969/j.issn.1674-165X.2023.05.003
12. Xidong Z, Xieeryazidan A, Changqi C (2024) Numerical simulation of 45 steel surface laser quenching based on ANSYS. Hot Work Technol 53(22):51–55+61. https://doi.org/10.14158/j.cnki.1001-3814.20202583

Open Access This chapter is licensed under the terms of the Creative Commons Attribution-NonCommercial-NoDerivatives 4.0 International License (http://creativecommons.org/licenses/by-nc-nd/4.0/), which permits any noncommercial use, sharing, distribution and reproduction in any medium or format, as long as you give appropriate credit to the original author(s) and the source, provide a link to the Creative Commons license and indicate if you modified the licensed material. You do not have permission under this license to share adapted material derived from this chapter or parts of it.

The images or other third party material in this chapter are included in the chapter's Creative Commons license, unless indicated otherwise in a credit line to the material. If material is not included in the chapter's Creative Commons license and your intended use is not permitted by statutory regulation or exceeds the permitted use, you will need to obtain permission directly from the copyright holder.

Numerical Simulation of Liquid Sloshing in Type B Liquid Cargo Tanks

Si-Cong Wang, Dong-Liang Fu, Yi Zhang, Bo Zhang, Bi-Wen Chen, and Gang Wu

Abstract This paper utilizes the Liquefied Natural Gas (LNG) ship B-type liquid cargo tank as the primary research object to investigate the extent of liquid sloshing within the liquid cargo tank under the emergency stop braking condition, considering varying filling rates. In this multiphase flow problem, numerical simulation is carried out by using the Volume of Fluid (VOF) method to analyze the magnitude of the impact of liquid sloshing on the wall surface of the liquid tank in the B-type liquid cargo tank of the LNG carrier under the emergency stop braking condition, as well as the change of the impact of the liquid sloshing on the wall surface of the liquid tank under the conditions of different filling rates. The findings indicate that, under the emergency stop braking condition, the impact force of the fluid in the front head of the liquid cargo tank is significantly greater than that in the rear head. Additionally, the impact force of the liquid sloshing on the wall surface of the liquid tank under different filling rates exhibits a tendency to first increase and then decrease. The results of this research provide valuable guidance and reference for the design and calculation of the B-type liquid cargo tanks of the LNG carriers.

Keywords Liquid cargo tank · Liquid sloshing · Mulitiphase · Filling rate · Impact force

1 Introduction

Marine LNG storage tanks are mobile vessels, and if they are underfilled, the motion of the ship will induce the shaking of the liquid in the tanks [1]. Liquid sloshing is a complex, nonlinear fluid motion. The dynamic interaction between liquid sloshing

---

S.-C. Wang · G. Wu (✉)
Merchant Marine College, Shanghai Maritime University, Shanghai, China
e-mail: wugang@shmtu.edu.cn

D.-L. Fu · Y. Zhang · B. Zhang · B.-W. Chen
Shanghai Marine Equipment Research Institute, Shanghai, China

© The Author(s) 2026

Z. Zhang, P. Mercorelli (eds.), *Mechanical Design and Simulation: Insights and Innovations*, Lecture Notes in Mechanical Engineering,
https://doi.org/10.1007/978-981-95-2405-1_40

within cargo tanks has been shown to have a significant impact on the stability of a ship's voyage [2]. Common shaking phenomena in the liquid tanks include standing waves, traveling waves, hydrodynamic leaps, and combined waves, as well as strong nonlinear phenomena such as vortices and splashes. When liquid shaking occurs in the liquid cargo tank, the free liquid surface impacts the tank wall, generating a significant impact force [3–7]. Researchers domestically and internationally have extensively investigated this subject, primarily focusing on three categories of research: theoretical analysis, numerical simulation, and experimental research methods. However, due to limitations imposed by various factors, the theoretical analysis method often falls short in adequately addressing and resolving practical challenges. The experimental research method, while the most intuitive, is often limited by the necessity of significant manpower and material resources, which restricts its applicability to more limited conditions. The China Classification Society has conducted experimental research and formulated assessment guidelines [8], and numerous scholars have also conducted experimental research on the swaying loads effect. The advent of sophisticated computer technology has elevated numerical simulation to a prominent research method, facilitating the exploration of numerous problems, including liquid sloshing. Chen et al. [9] employed the finite element method to assess the structural integrity of an autonomous B-type liquid cargo tank. Xie et al. [10] employed the VOF method to numerically simulate the filling condition of LNG-FSRU under sloshing conditions and investigated the perturbation change process of a liquid tank filled with LNG in different stratification states. He et al. [11] and others used the CFD method to numerically simulate the shaking of C-type LNG.

2 Simulation Models Verification and Research Scheme

2.1 Establishment of Numerical Simulation Model

The liquid cargo tank model has been designed according to the specifications of the new independent LNG ship B-type liquid cargo tank of Hudong-Zhonghua Shipbuilding (Group) Co. The liquid cargo tank has a length of 3.79 m, a width of 4 m, and a height of 2.675 m. The geometrical model was created using SOLIDWORKS software, and subsequently, the SOLIDWORKS software was employed to generate the geometrical model. The model was then imported into ICEM to delineate the fluid mesh and encrypt the boundary layer. It is assumed that the tank is rigid; that is to say, the motion and deformation generated by the tank under the action of fluid load are not considered. Figure 1 shows the mesh model of the liquid cargo tank. The entire model is segmented by a tetrahedral mesh with a mesh size of 0.01 m and a mesh number of 53,718.

The VOF method is employed in this study to solve for the sloshing of the liquid within the liquid cargo hold, and the configuration of the solver is delineated in Table 1. For the residuals of the continuity, momentum, volume fraction, and turbulence equations, the convergence criterion was set to 1×10^{-4}. The transient

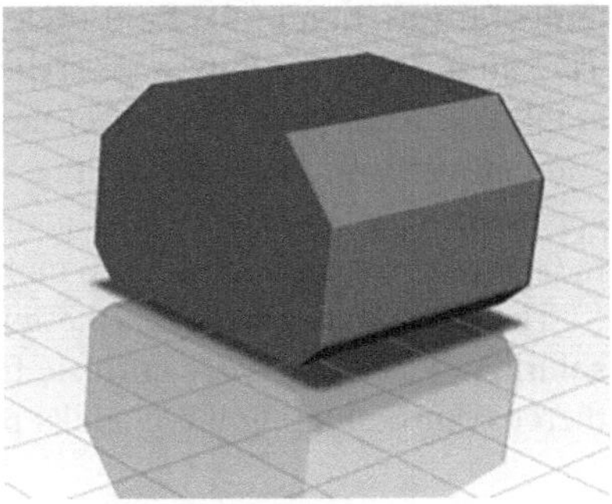

Fig. 1 Mesh model of liquid cargo tanks

Table 1 Solver settings

Parameters	Options
Computational model	3D double precision
Turbulence model	Standard k-epsilon model
Physical model	VOF model
Calculation method	Pressure–velocity coupled SIMPLE algorithm
Pressure discrete format	Body force weighted algorithm
Time-discrete format	First-order implicit format
Time step	0.005 s

calculations were executed at a time step of 5×10^{-3} s. The ambient pressure was set to the atmospheric pressure of 1.013×10^5 Pa.

2.2 Verification of Numerical Simulation Models

In this paper, the analytical solution of the first-order intrinsic frequency of the liquid cargo hold model is compared with that calculated by numerical simulation. This comparison is used to verify the accuracy of the simulation model. The formula for calculating the intrinsic frequency of the liquid cargo hold is as follows [6]:

$$f = \frac{\sqrt{\frac{g\pi}{b}\tanh\left(\frac{\pi h}{b}\right)}}{2\pi} \quad (1)$$

Where g is the acceleration of gravity, b is the effective sway length, and h is the height of the filled liquid.

In the context of an emergency braking scenario, characterized by a 50% filling rate, the internal fluid exhibits oscillatory behavior, exhibiting an intrinsic frequency at t = 0, coinciding with the cessation of movement of the liquid cargo tank. The temporal variation in the fluid force exerted on the front head is depicted in Fig. 2. The fluid force exhibits a periodic fluctuation, with a concomitant decrease in amplitude over time. The spectrum analysis of fluid shaking in the liquid cargo compartment, as depicted in Fig. 3, reveals a first-order shaking frequency of approximately 0.4 Hz. This frequency is calculated using Eq. (1) and is equivalent to 0.404 Hz for the liquid cargo tank at a 50% filling rate.

The discrepancy between the numerical simulation solution and the analytical solution of the first-order intrinsic frequency of the simulation model is negligible, thereby substantiating the model's precision.

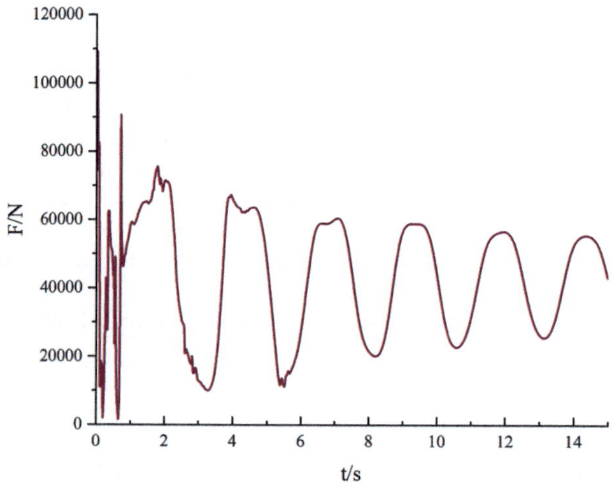

Fig. 2 The force of the fluid on the anterior head over time

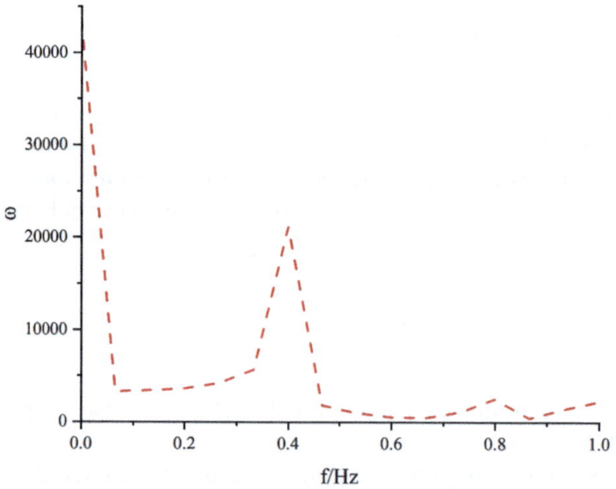

Fig. 3 The amplitude-frequency characteristic profile after the FFT transformation

Table 2 Table of movement states at each time point

t/s	State of motion
0 s ago	Moving at a constant speed of 19.6 m/s
0 to 1 s	Decelerating with an acceleration of 19.6 m/s^2
After 1 s	The liquid cargo tank is stationary

2.3 Research Methods

In order to investigate the fluid shaking and the magnitude of the fluid impact on the walls of the B-type liquid cargo tanks under the emergency stopping condition, a VOF model is used in Fluent software to monitor the free liquid level in the tanks. Three different filling rates of 20%, 50% and 80% were set. Water and air were used as the fluid medium in the model without energy changes, and air was considered as an ideal gas, and the initial densities of water and air were set to 998.2 and 1.225 kg/m^3, respectively. And the front and rear surfaces of the liquid cargo tank are defined as force monitoring points. The motion model of the emergency stopping experiment is shown in Table 2.

3 Results and Discussion

As illustrated in Fig. 4, consider a filling rate of 50% at t = 0 s, representing the initial state of uniform motion prior to braking. At t = 0.38 s, the tank undergoes a fluid impact due to inertia, as depicted in the figure. It is evident that the impact force attains its maximum at t = 0.38 s, with a value of Fmax = 394909.41 N. Due to the inertia, the fluid exhibits a propensity to move forward, however, it is impeded by the front head of the liquid cargo compartment. Due to inertia, the fluid exhibits a propensity to move forward, but is impeded by the front head of the liquid cargo compartment, consequently forcing it to move upward along the wall of the front head. Consequently, at t = 0.38 s, the volumetric force of the fluid is applied to the front head in its entirety, reaching the maximum impact force. At t = 0.87 s, the deceleration movement comes to a conclusion. The fluid flows back to the rear head, and the impact force of the fluid on the rear head reaches a maximum value of 222628.51 N. It is evident that the maximum value of the impact force on the front head is considerably larger than that on the rear head. The deceleration process concludes after 1 s, and the fluid oscillates back and forth in the liquid cargo compartment before eventually returning to a stationary state.

The analysis of Fig. 5 is as follows: As illustrated in the Fig. 4, at a filling rate of 20%, the volume occupied by the fluid is minimal, and the time for the volumetric force of the fluid to be applied to the front head is delayed, reaching its maximum at 0.53 s, and the maximum impact force has been measured to be

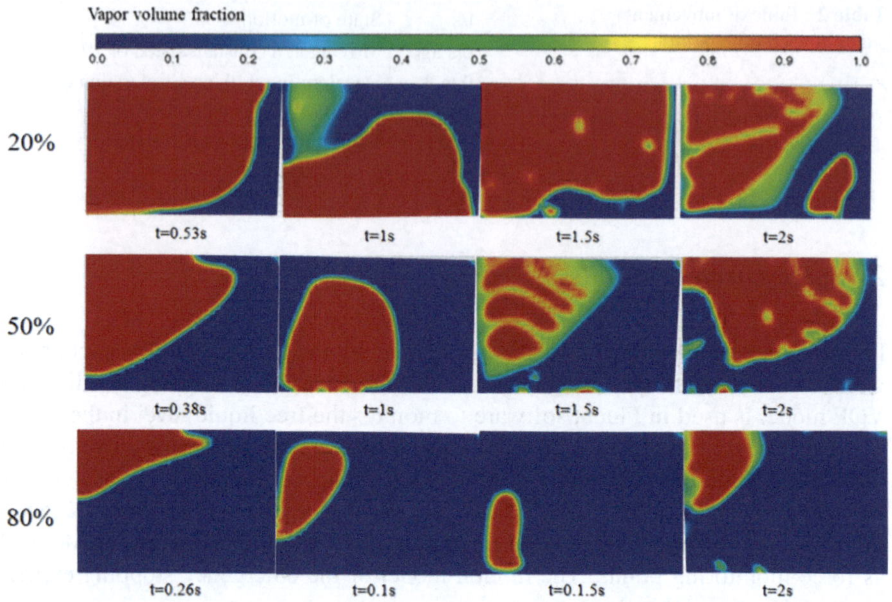

Fig. 4 Cloud diagram of liquid sloshing process under different filling rates

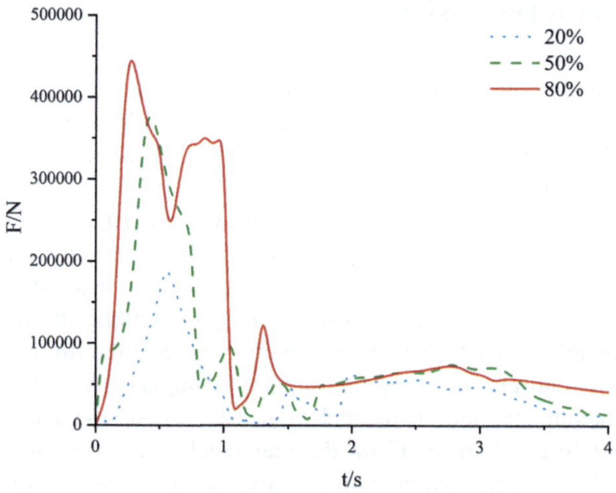

Fig. 5 The impact force of the liquid cargo compartment changes over time

194,833.3 N. Conversely, at a filling rate of 80%, the volume occupied by the fluid is substantial, and the time to reach the maximum impact force is reduced, reaching a maximum at 0.26 s, and the maximum impact force has been measured to be 455314.12 N. Consequently, between 20% and 80% filling rate, the maximum impact force on the liquid cargo ship increases with increasing filling rate, and the time of occurrence of the maximum impact force shortens with increasing filling rate. This indicates that the larger the volume of liquid occupying the liquid cargo hold, the smaller the area leading to liquid activity, and the earlier the time for the fluid to apply the full volumetric force to the front head.

4 Conclusion

This paper utilizes the VOF fluid volume method to simulate the shaking process of the B-type liquid cargo tank of an LNG ship under the emergency stop braking condition. The primary focus of this study is the discussion of the impact force on the bulkhead caused by liquid shaking in the B-type liquid cargo tank under the emergency stop braking condition, as well as the change rule of the impact force on the bulkhead caused by liquid shaking under the condition of different charging rates.

References

1. Zhimei L, Shiming F (2010) Research progress of ship liquid tank swinging. Shanghai Shipbuild (2):14–16+30
2. Zhang Z, Tao A, Wu Q et al (2023) Review on the progress and issues in liquid tank sloshing of ships. China Ocean Eng 37(5):709–724
3. Li C, Zhao Z, Ding J et al (2024) Numerical study of thermodynamic and hydrodynamics in a type B LNG cargo tank during sloshing excitation. Appl Therm Eng 247:123048
4. Wen D, Yi Y, Zhengyi Z et al (2019) Force analysis of the support structure of type B independent liquid cargo compartment of LNG vessel under swinging load. Ship Sci Technol 41(15):17–22
5. Wenfu L, Susu Z, Xu Y et al (2024) Numerical simulation and test validation of the coupling motion of fluid compartment motion and fluid sway. Ship Mech 28(10):1516–1525
6. Zhijun W, Qianjin Y, Shilun R et al (2012) Study on the test mechanism of the swinging impact load. Ship Mech 16(8):885–892
7. Wenfu L, Hongxiang X, Wenyong T et al (2015) Numerical analysis of swinging load of independent B LNG prism chamber. Ship Eng 37(7):22–25+72. https://doi.org/10.13788/j.cnki.cbgc.2015.07.022
8. CCS (2013) Code for natural gas fuel-powered vessels. People's Communications Press, Beijing
9. Qian C, Jun L, Wenyong T (2015) Study on the swinging strength of independent type B LNG hull. Ship Sci Technol 37(9):12–15
10. Mengbin X, Hanhua Z, Hao M (2022) Numerical simulation of LNG-FSRU injection condition under sway condition. China Navig 45(2):48–55
11. Xiaocong H, Rong H (2016) Numerical analysis of liquid swinging in marine LNG storage tanks. Ship Sea Eng 45(3):12–17

Open Access This chapter is licensed under the terms of the Creative Commons Attribution-NonCommercial-NoDerivatives 4.0 International License (http://creativecommons.org/licenses/by-nc-nd/4.0/), which permits any noncommercial use, sharing, distribution and reproduction in any medium or format, as long as you give appropriate credit to the original author(s) and the source, provide a link to the Creative Commons license and indicate if you modified the licensed material. You do not have permission under this license to share adapted material derived from this chapter or parts of it.

The images or other third party material in this chapter are included in the chapter's Creative Commons license, unless indicated otherwise in a credit line to the material. If material is not included in the chapter's Creative Commons license and your intended use is not permitted by statutory regulation or exceeds the permitted use, you will need to obtain permission directly from the copyright holder.

Research on Optimization of Vibration Dispersion for Batch Centrifugal Pumps Based on the Key Process Parameter

Wanbo Yu, Zhenlai Ma, Tao Sun, Guoping Li, and Xinxin Zhao

Abstract Currently, after the batch production and use of the centrifugal pump, there is an obvious phenomenon of vibration dispersion. Aiming at the convergence of vibration and noise control of batch centrifugal pump products, this paper carries out the research from the point of view of process. From the perspective of location and optimization of the key process parameters affecting vibration and noise, focusing on the main excitation sources of centrifugal pumps. Taking the clearance between the wear ring and the impeller as the research object, comprehensive consideration of the influence of error of wear ring clearance on flow-induced vibration of pump, utilizing three-dimensional dimensional chain modeling and analysis to screen out the key process parameters that affect vibration. Based on mathematical statistics, fluid numerical simulation and fluid–structure interaction, the impact of key process parameters on error of the wear-ring clearance, which the dispersion level of vibration is obtained. Finally, an experimental control group is established to verify the key process parameters.

Keywords Centrifugal pump · Clearance of wearing ring · Characteristic vibration · Vibration dispersion · Process optimization

1 Introduction

Centrifugal pumps are widely used in various fields such as energy and power, metallurgy, and petrochemical engineering as fluid transportation equipment. Which stability and reliability have a significant impact on industrial systems. As the most widely used general machinery and equipment, excessive vibration is a common

W. Yu (✉) · Z. Ma · G. Li · X. Zhao
Shanghai Marine Equipment Research Institute, Shanghai, China

T. Sun
School of Mechatronic Engineering and Automation, Shanghai University, Shanghai, China

© The Author(s) 2026

Z. Zhang, P. Mercorelli (eds.), *Mechanical Design and Simulation: Insights and Innovations*, Lecture Notes in Mechanical Engineering,
https://doi.org/10.1007/978-981-95-2405-1_41

failure of centrifugal pumps [1]. Current research on vibration optimization of centrifugal pumps mainly focuses on optimizing the structural design of centrifugal pumps through experiments and numerical simulations to reduce the excitation forces causing vibration from various aspects. For example, Yuan [2] proposed two types of novel complex impellers that have a positive effect on reducing centrifugal pump vibration and concluded that complex impellers are beneficial for reducing pump vibration. Zeng et al. [3] proposed installing splitter blades on the impeller to effectively reduce fluid-induced vibration.

From a design perspective, the vibration design level of centrifugal pumps is advanced, and the means of structural optimization are also becoming more sophisticated. However, the vibration performance of batch products is difficult to meet design requirements, and the vibration levels among products of the same batch are also inconsistent with large dispersion levels [4], making it difficult to effectively control the vibration of the equipment. This is due to errors in the processing and assembly of centrifugal pumps, which have been particularly serious for a long time in the past. And this issue has long plagued production processing and debugging personnel. As structural design optimization reaches a bottleneck, some scholars have realized the importance of process during research. Researchers have found that multiple key parameters during the assembly process, such as bearing oil clearance, the fit between the bearing and the bearing seat, rotor dynamic balance, and tongue clearance, all affect the vibration of centrifugal pump [5–7].

However, current research has not provided targeted process optimization suggestions for centrifugal pump sets based on vibration, relying more on repeated human attempts or monotonically improving accuracy to achieve process optimization. However, due to processing conditions and cost limitations, the accuracy grade of equipment parts cannot be infinitely increased, and blindly improving part accuracy often fails to yield desirable results. Therefore, it is necessary to screen out process parameters that affect vibration, sort out the influence mechanism of process parameters, and ultimately form a direction for process optimization targeting vibration.

2 Research Design

This paper takes a certain type of centrifugal pump as the research object, with its model shown in Fig. 1:

From the perspective of assembly, the assembly sequence is as shown in Fig. 3: (1) Shaft, (2) Bearing cover (3) Bearing, (4) Bearing cover, (5) End cover, (6) Bracket, (7) Pump body, (8) Mechanical seal, (9) Bushing, (10) Impeller, (11) Wear Ring, (12) Pump cover.

Therefore, this paper will select key vibration excitation sources and establish a mutual mapping relationship between process parameters, excitation sources, and vibration characteristics to screen out key process parameters.

Currently, with the further improvement of the processing accuracy of centrifugal pump rotors, the influence of fluid excitation on the vibration characteristics of

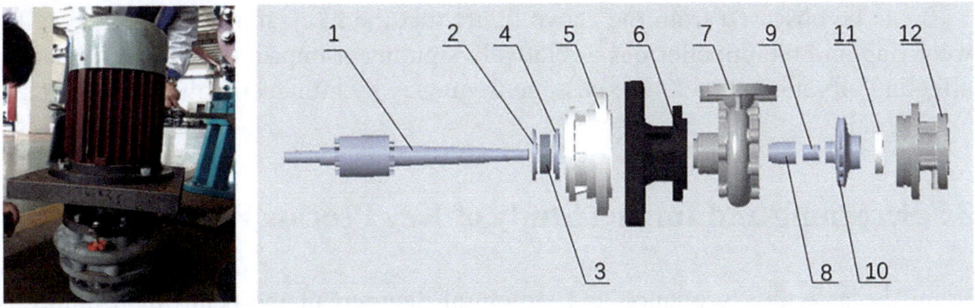

Fig. 1 Physical and virtual prototype models of the centrifugal pump

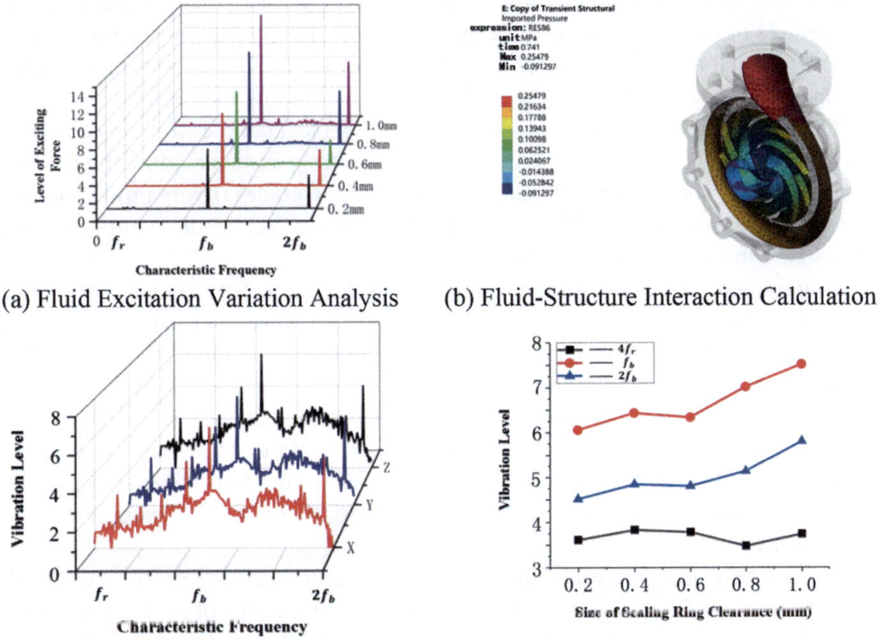

(a) Fluid Excitation Variation Analysis

(b) Fluid-Structure Interaction Calculation

(c) Vibration Levels under Fluid Excitation

(d) Analysis of Characteristic Vibration Changes

Fig. 2 Changes in equipment vibration under the influence of changes in wear ring clearance. (**a**) Fluid excitation variation analysis. (**b**) Fluid–structure interaction calculation. (**c**) Vibration levels under fluid excitation. (**d**) Analysis of characteristic vibration changes

pump sets has become increasingly evident. Some scholars have found that it is very important to improve the fluid excitation of the pump by controlling the wear clearance, which is crucial for the vibration optimization of the pump [8]. Therefore, this paper establishes a whole-machine fluid model, combines numerical simulation and fluid–structure interaction, and initially analyzes the excitation and overall vibration levels under different wear ring clearances, as shown in Fig. 2. The Vibration Level in the figure is a dimensionless parameter used to compare the strength of vibrations.

It can be observed from the above figure that the fit clearance size between the wear ring and the impeller has a relatively significant impact on vibration, mainly affecting vibration at the blade passage frequency by influencing fluid excitation.

3 Screening and Impact Study of Key Process Parameters

From the assembly sequence and structural diagram of the centrifugal pump as Fig. 1, the fit between the wear ring and the impeller is affected by the entire machine's assembly process, where the size and geometric tolerances of all components will directly or indirectly affect the size of the wear-ring clearance. From the perspective of process control, since the formation of the wear ring clearance is not directly formed through assembly, after batch production, its actual value is easy to deviate from the design value and is difficult to control and measure.

Therefore, it is necessary to clarify the dispersion level of the wear ring clearance under the current process state, locate the key process parameters affecting the wear ring clearance, sort out the influence of the accuracy grades of key process parameters on vibration, and verify the vibration control effect of the equipment after optimizing the key process parameters.

To complete the above research, it is not sufficient to focus solely on a single tolerance. Instead, it is necessary to consider the size and geometric tolerances of all components from the perspective of the entire machine assembly to form a three-dimensional dynamic relationship. Therefore, facing the wear ring clearance, a three-dimensional dimensional chain needs to be formed and solved. First, the contribution of process parameters to the structural error of the ring wear ring clearance is completed, and the impact on the structural and vibration dispersion levels is analyzed in combination with different combinations of process parameter accuracy grades.

According to the assembly sequence of the centrifugal pump, by judging the size, geometric tolerances, and mutual assembly relationships of each component, a logically closed chain for the size chain of the ring wear ring clearance is formed as shown in Fig. 3:

Based on the above, set the size and geometric tolerance ranges of each component according to the design drawings, and combine the three-dimensional dimensional chain model to solve for the size distribution of the wear ring clearance under this process level and the process parameters that have a greater impact on the size of the wear ring clearance, as shown in and Fig. 4.

The above process parameters are analyzed by discrete analysis of the gap size of the wear ring. And the results are shown in Fig. 5.

It can be seen from Fig. 5 that when the above process parameters are controlled and the accuracy is improved, the size discretization and error of the wear-ring clearance are effectively controlled.

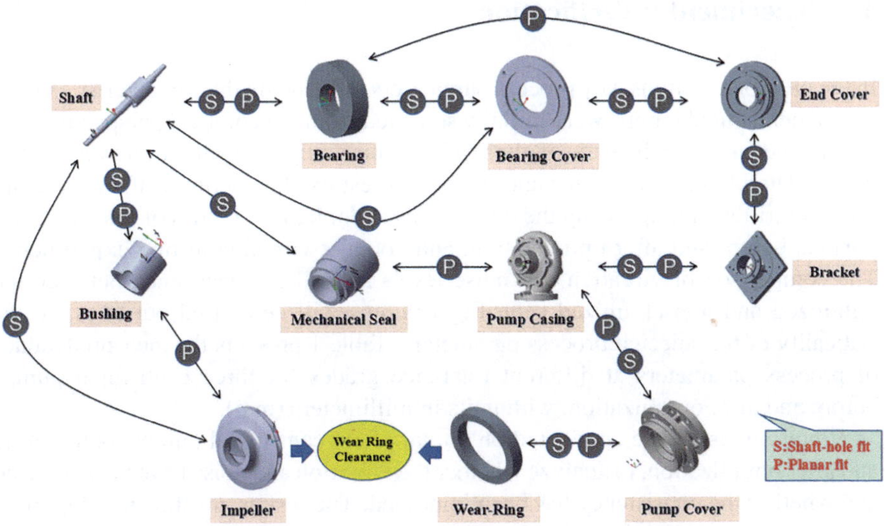

Fig. 3 Logically closed chain for the size chain of the wear ring clearance

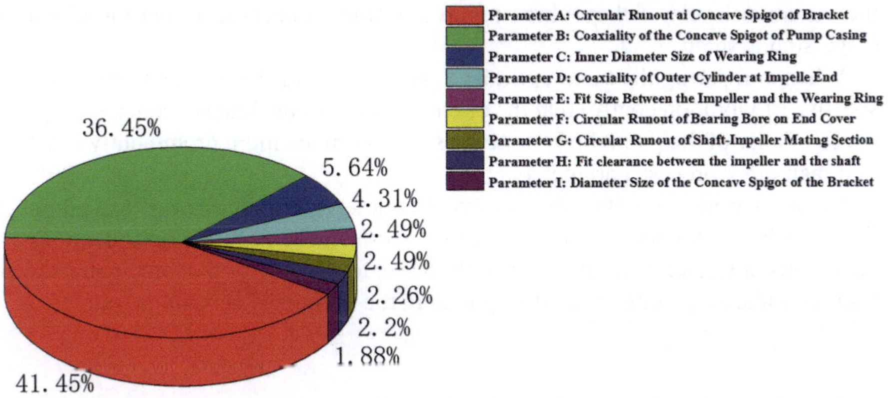

Fig. 4 Three-dimensional tolerance solution results for the wear ring clearance

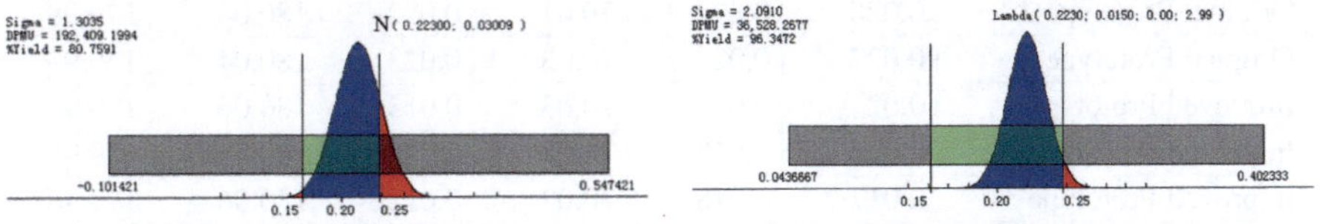

(a)Size distribution in the original state of process parameter

(b)After optimizing the process parameters with greater contribution

Fig. 5 The optimal distribution level of the wear ring clearance size under different process levels. (**a**) Size distribution in the original state of process parameter. (**b**) After optimizing the process parameters with greater contribution

4 Experimental Verification

In the previous sections, key process parameters influencing the characteristic vibration of centrifugal pumps were initially screened from a theoretical perspective, and the manner and mechanism of their impact on equipment vibration were determined. Therefore, based on the above analysis results, this section will manufacture three centrifugal pumps using the dimensions and tolerance grades of relevant components before and after optimization, and conduct vibration testing experiments. The comparison of vibration and noise levels as well as consistency between the optimized and non-optimized centrifugal pumps will be carried out to prove the criticality of the selected process parameters. Table 1 presents the measured values of process parameters at different tolerance grades for three centrifugal pumps before and after optimization, with units in millimeters (mm).

Vibration tests were conducted on a total of six centrifugal pumps, both before and after optimization, to analyze whether the vibration and noise have been reduced and whether the consistency has been improved. The results are shown in Fig. 6.

It can be observed from the figures that the vibration acceleration of the three optimized centrifugal pumps has been significantly reduced, especially near the blade passing frequency, where there is a notable decrease in vibration. Moreover, the vibration levels of these three pump are more concentrated and the vibration dispersion is small.

When comparing the three optimized pumps, it is further verified from the perspective of vibration control that the circular runout of the bracket concave stop and the coaxiality of the pump body concave stop contribute more prominently to vibration than other process parameters.

Before optimization, the vibration accelerations of the three centrifugal pumps at the same frequency had relatively large deviations, whereas after optimization, the deviations in vibration acceleration at the same frequency for the three pumps were smaller, indicating better consistency among the optimized centrifugal pumps.

Table 1 Comparative experiment parameter table

	A	B	C	D	E	F
Original Prototype 1	0.029	0.028	70.02	0.023	180.03	179.98
Original Prototype 2	0.028	0.027	70.01	0.022	180.04	179.99
Original Prototype 3	0.027	0.027	70.03	0.023	180.04	179.99
Improved Prototype 1	0.02	0.02	70.03	0.013	180.06	179.97
Improved Prototype 2	0.018	0.018	70.01	0.013	180.05	179.98
Improved Prototype 3	0.017	0.018	70.02	0.012	180.06	179.96

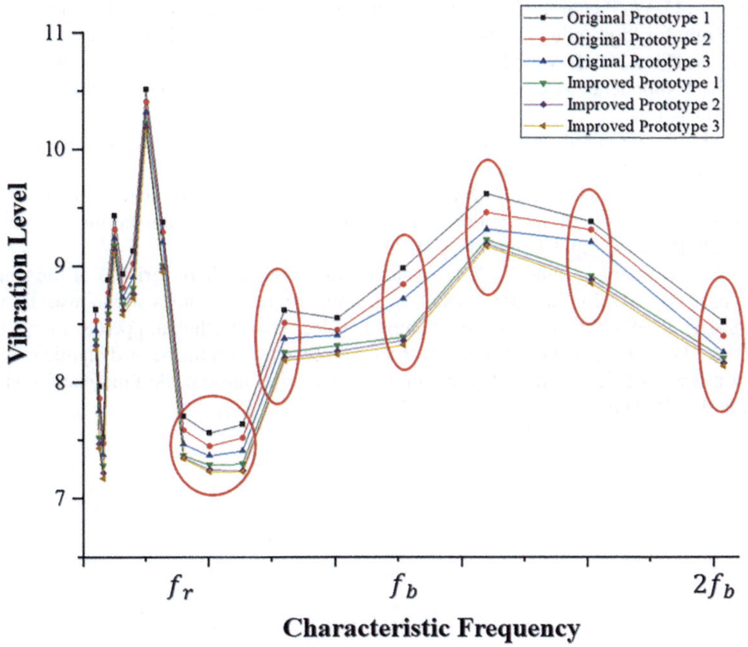

Fig. 6 One-third octave band spectrum of vibration acceleration

5 Conclusion

This paper focuses on the vibration and noise control of centrifugal pumps, centering on the main excitation sources and process control challenges of centrifugal pumps. Taking the fit clearance between the wear ring and the impeller as the research object, from the perspective of process optimization, three-dimensional dimensional chain modeling and analysis were utilized to screen the process parameters that have a significant impact on the wear fit clearance in the assembly of centrifugal pumps. Based on mathematical statistical methods, fluid numerical simulations, and fluid–structure interaction calculations, the influence levels of various key process parameters on the size of the wear ring clearance and vibration dispersion were obtained. Finally, through the establishment of experimental control groups, the key process parameters were validated, and preliminary optimization suggestions were proposed.

References

1. Dai C, Guo C, Chen YP et al (2021) Analysis of the influence of different bionic structures on the noise reduction performance of the centrifugal pump. Sensors 21(3):886–901
2. Yuan Y, Yuan S, Tang L (2019) Investigation on the effect of complex impeller on vibration characteristics for a high-speed centrifugal pump. Proc Inst Mech Eng A J Power Energy 234(5):611–624

3. Zeng G, Li Q, Wu P et al (2020) Investigation of the impact of splitter blades on a low specific speed pump for fluid-induced vibration. J Mech Sci Technol 34(7):2883–2893
4. Awais AK, Hussein MH, Emad SA et al (2015) Computer-aided process planning in prismatic shape die components based on standard for the exchange of product model data. Adv Mech Eng 7(11):1–11
5. Sun T, Shu P, Yu Q et al (2022) Influence of bearing process parameters on vibration of centrifugal pump. J Braz Soc Mech Sci Eng 44(9):1–12
6. Shi Z, Sun T, Ma Z et al (2022) Machining accuracy allocation of shaft system considering vibration and cost. In: 4th international conference on mechanical engineering and automation (MEA). IOP Publishing, Chongqing, p 012053
7. Sun T, Yao Y, Zhang J, Yang L (2021) Analysis of vibration characteristics of the centrifugal pump base on CFD. In: 7th international symposium on mechatronics and industrial informatics (ISMII). IEEE CS-Conference Publishing Services (CPS), Zhuhai, pp 168–171
8. Zhou W, Ma J, Ma Z et al (2024) Fluid-structure interaction on the rotor-dynamic characteristics of a low-specific-speed centrifugal pump considering multi-scale fluid excitation effects. Phys Fluids 36(11):117157

Open Access This chapter is licensed under the terms of the Creative Commons Attribution-NonCommercial-NoDerivatives 4.0 International License (http://creativecommons.org/licenses/by-nc-nd/4.0/), which permits any noncommercial use, sharing, distribution and reproduction in any medium or format, as long as you give appropriate credit to the original author(s) and the source, provide a link to the Creative Commons license and indicate if you modified the licensed material. You do not have permission under this license to share adapted material derived from this chapter or parts of it.

The images or other third party material in this chapter are included in the chapter's Creative Commons license, unless indicated otherwise in a credit line to the material. If material is not included in the chapter's Creative Commons license and your intended use is not permitted by statutory regulation or exceeds the permitted use, you will need to obtain permission directly from the copyright holder.

Innovations in Materials and Manufacturing Processes

An Impact Fatigue Life Prediction Model of Metal Materials Based on Damage Mechanics

Shumao Cheng and Lei Yan

Abstract This article proposes a metal material impact fatigue life prediction model based on damage evolution laws. This model describes the damage evolution process of metal materials as a gradually increasing process over time, and by monitoring the changes in damage parameters, the lifespan of materials can be predicted. Meanwhile, considering the influence of the microstructure of metal materials on their fatigue performance, this paper also introduces the microscale characteristics of material parameters, further improving the accuracy and precision of the model. By comparing and verifying with relevant experimental results, it is shown that the combination of continuous damage mechanics and finite element analysis method can effectively evaluate the impact damage and subsequent fatigue life of metal structures.

Keywords Damage mechanics · Metal · Impact · Life prediction

1 Introduction

Deterministic life cannot fully reflect the probabilistic statistical characteristics of fatigue life, such as probability distribution types and standard deviations, kurtosis, and skewness. Probabilistic fatigue life should be able to characterize its probability distribution and corresponding distribution parameters. At present, most studies are based on the improvement of these models, such as Bai Enjun, who equates the fatigue life of different levels of stress to the highest level of stress, achieving the fitting of p-S-N curves under small sample sizes. Liu et al. predicted the fatigue life at the bolt holes of the turbine disc using an improved average stress formula. Luo et al. constructed a functional function for bridge fatigue damage based on the S-N curve and Miner's criterion, which is used for fatigue reliability evaluation of

S. Cheng · L. Yan (✉)
Tianjin Huayuan Times Metal Products Co., Ltd., Tianjin, China

© The Author(s) 2026

Z. Zhang, P. Mercorelli (eds.), *Mechanical Design and Simulation: Insights and Innovations*, Lecture Notes in Mechanical Engineering,
https://doi.org/10.1007/978-981-95-2405-1_42

simply supported beam bridges. Ma Yu'e established S-N curves through fatigue experiments on glass fiber aluminum alloy laminates (FMLs), providing material properties and information for predicting their fatigue life. Doudard et al. proposed a probabilistic dual scale model under the framework of the weakest link theory for predicting the dispersion of S-N curves. Due to the complexity of impact fatigue testing methods and material impact fatigue performance description, there is currently no systematic theory, standard testing methods, and rich experimental data to guide material selection and structural design of components subjected to repeated impacts. This article relies on damage mechanics and finite element analysis methods to explore the fracture mechanism and behavior of metal materials, establish a prediction model for the impact fatigue life of metal materials, and evaluate and correct the accuracy of the model through experimental verification [1].

2 Deformation Characteristics of Metal Materials in Impact Fatigue

When studying the impact fatigue deformation characteristics of metal materials, the first step is to study the deformation characteristics of the material under a single impact load, and then study the deformation characteristics of the material under an impact fatigue load. The following will be elaborated in sequence [2].

2.1 Single Impact Load

Compared with nonimpact stress, the duration of impact stress under a single impact load is very short, only about 0.1–1% of the nonimpact fatigue stress under one cycle. The typical stress characteristics of metal materials under tensile impact load are shown in Fig. 1.

Fig. 1 Stress variation during a single impact

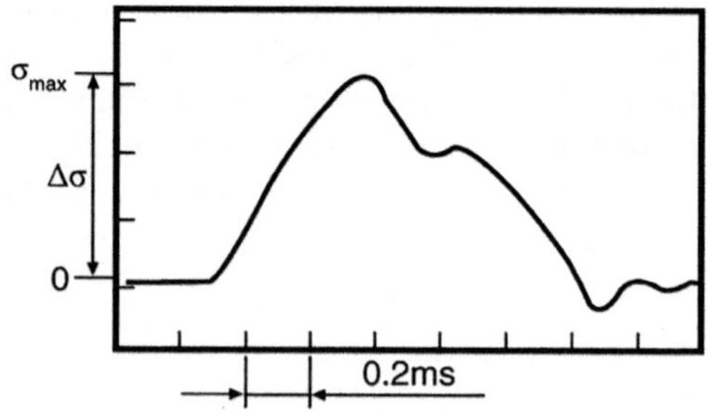

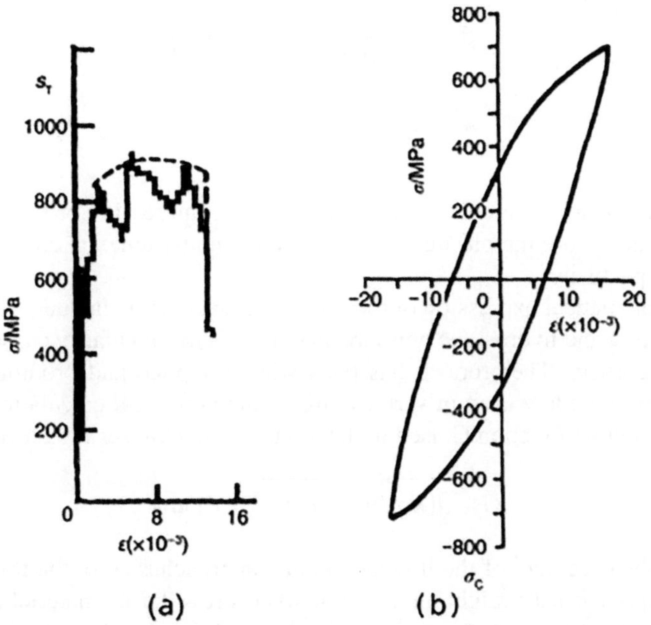

Fig. 2 Stress–strain curve of carbon steel under low cycle impact fatigue load and hysteresis loop under nonimpact low cycle fatigue load. (**a**) Impact fatigue, (**b**) non impact fatigue

2.2 *Impact Fatigue Load*

Due to the higher dynamic yield strength of medium carbon steel under impact fatigue loading (see Fig. 2), the same maximum stress intensity factor is applied K_{max}. The plastic deformation zone at the crack tip under impact fatigue is smaller than that under nonimpact fatigue (see Fig. 2). Multi-system slip under impact fatigue is also more complex than under nonimpact fatigue. The measurement results of dislocation density at the fatigue crack tip also indicate that under the same conditions K_{max}, horizontally, the dislocation density of nonimpact fatigue is greater than that of impact fatigue. This trend can be attributed to the susceptibility of multiple system slips under nonimpact fatigue [3].

3 Theoretical Model

3.1 *Plastic Damage Model*

The uniaxial effective tensile stress and compressive stress are used as the basis for determining the yield and failure surfaces:

$$\bar{\sigma}_t = \frac{\sigma_t}{(1-d_t)} = E_0 - \left(\varepsilon - \varepsilon_t^{pl}\right) \tag{1}$$

$$\bar{\sigma}_c = \frac{\sigma_c}{(1-d_c)} = E_0 - \left(\varepsilon - \varepsilon_c^{pl}\right) \tag{2}$$

where d_c and d_t are plastic damage factors during compression or tension, respectively; ε_c^{pl} and ε_t^{pl} are the plastic strain of concrete under uniaxial compression and tension, respectively.

The mathematical expression of the D-P strength criterion includes intermediate principal stress and hydrostatic pressure, and there is no singularity caused by sharp edges and corners. Therefore, it has been widely applied and promoted after its proposal and is widely used in various large-scale numerical calculation software. The flow potential function G used in the model is the Drucker Prager function, i.e.

$$G = \sqrt{(\kappa\sigma_0 \tan\psi)^2 + \bar{q}^2} - \bar{p}\tan\psi \tag{3}$$

where the displacement of the flow potential κ approaches zero, the flow potential function approaches a straight line; σ_0 is uniaxial stress during material failure; ψ is under high lateral pressure $\bar{p} - \bar{q}$ shear dilation angle on the plane; $\bar{p}$ is the average hydrostatic pressure; $\bar{q}$ is the average equivalent effective stress.

3.2 *Fatigue Damage Model*

Chaboche and Lesnes proposed a nonlinear uniaxial stress fatigue damage model:

$$\dot{D} = \frac{dD}{dN} = \left[1-(1-D)^{\beta+1}\right]^{\alpha}\left[\frac{\sigma^{\alpha}}{M(\sigma_m)(1-D)}\right]^{\beta} \tag{4}$$

where N is the number of cycles; σ_a and σ_m are the stress amplitude and average stress of fatigue load, respectively; β is the material constant; α and $M(\sigma_m)$ are as follows:

$$\alpha = 1 - a\frac{\sigma_a - \sigma_0}{\sigma_b - \sigma_a} \tag{5}$$

$$M(\sigma_m) = M_0(1 - b\sigma_m) \tag{6}$$

where σ_b is the ultimate tensile strength of the material; σ_0 is the fatigue limit of the material; a, M_0 are both material constants. For metallic materials, Dattoma et al. obtained a = 0.0801; $\langle\sigma\rangle$ is Macquarie brackets (Macaulay brackets).

Lin Youzhi and others revised the above model based on the theory of distortion energy density to consider the situation of multi-axial stress.

4 Prediction of Impact Fatigue Life of Aluminum Alloy

4.1 2A12 Impact and Fatigue Testing of Aluminum Alloy Plates

The shape and dimensions of the test piece are shown in Fig. 3. Using a drop hammer device to produce impact pit defects, a pit with a radius of r = 3 mm and a depth of d = 0.4 mm was obtained through calibration. Then, a high-frequency fatigue machine was used to conduct two sets of fatigue tests on the sample with pit defects, with a stress ratio of $R = -1$ and a maximum nominal stress of 200 and 230 MPa, respectively. The test results are listed in Table 1 [4].

Figure 4 shows the cross-sectional view of the 2A12 test piece with a 0.4 mm impact pit defect after fatigue failure.

The cyclic action of impact load forms impact fatigue, and impact fatigue load has impact load characteristics. Due to the short-term and strong load characteristics of impact loads, the interior of metal materials is in a dynamic equilibrium process

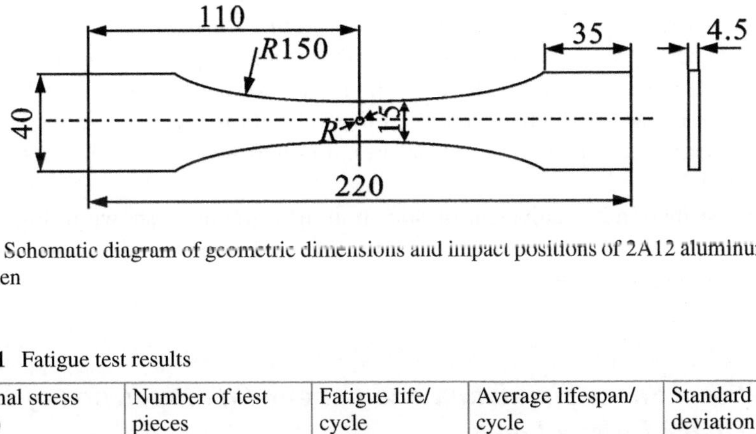

Fig. 3 Schematic diagram of geometric dimensions and impact positions of 2A12 aluminum alloy specimen

Table 1 Fatigue test results

Nominal stress (MPa)	Number of test pieces	Fatigue life/ cycle	Average lifespan/ cycle	Standard deviation
230	3	56,100	72,833	11,833
		81,300		
		81,100		
200	5	220,200	252,340	34,967
		232,000		
		287,500		
		220,500		
		301,500		

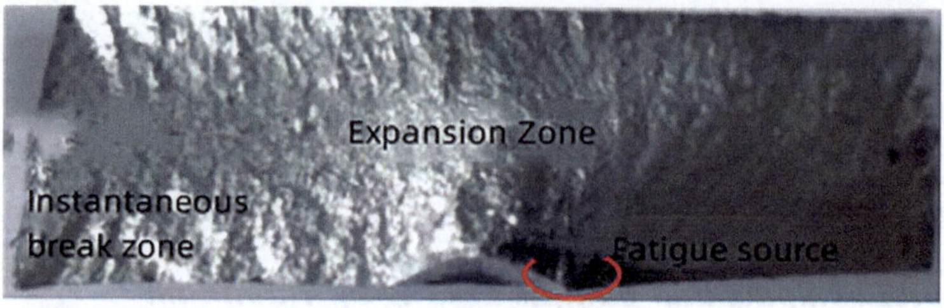

Fig. 4 The fatigue fracture surface of the 2A12 aluminum alloy test piece

that rapidly changes over time. The mechanical properties of the material are determined by both external loads and internal responses. The dynamic response behavior of metal materials determines their unique impact fatigue characteristics. The rheological stress of the material is significantly higher than that of conventional fatigue, the cyclic hardening ability and fatigue strength are lower than those of conventional fatigue, the crack propagation rate is higher, and the fatigue life is relatively lower [5].

4.2 *Finite Element Simulation of the Impact Process*

We establish a finite element model corresponding to fatigue testing using ABAQUS software, as shown in Fig. 5, mechanical performance parameters are shown in Table 2. When the impact energy is 2.64 J, the simulated pit depth obtained is consistent with the experiment. There is significant local deformation and residual stress at the impact crater, therefore the stress relaxation effect has been considered [6]. The distribution of their local deformation and residual stress are shown in Figs. 6 and 7.

The distribution of equivalent plastic strain near the pit is shown in Fig. 8. The equivalent plastic strain is mainly concentrated inside the pit and reaches its maximum at the bottom of the pit.

4.3 *Local Stress Simulation and Fatigue Life Prediction of Pits Under Fatigue Loads*

By substituting the equivalent stress, equivalent plastic strain, and other data during the impact process of Points A and B into Formula (4), the initial plastic damage of the dangerous point can be obtained, as shown in Table 3. The material parameters in the plastic damage model and fatigue damage model refer to Chen et al.'s research, and the relevant parameter values are shown in Table 4. The comparison of the fatigue test results and the prediction results of the injury model is shown in Table 5.

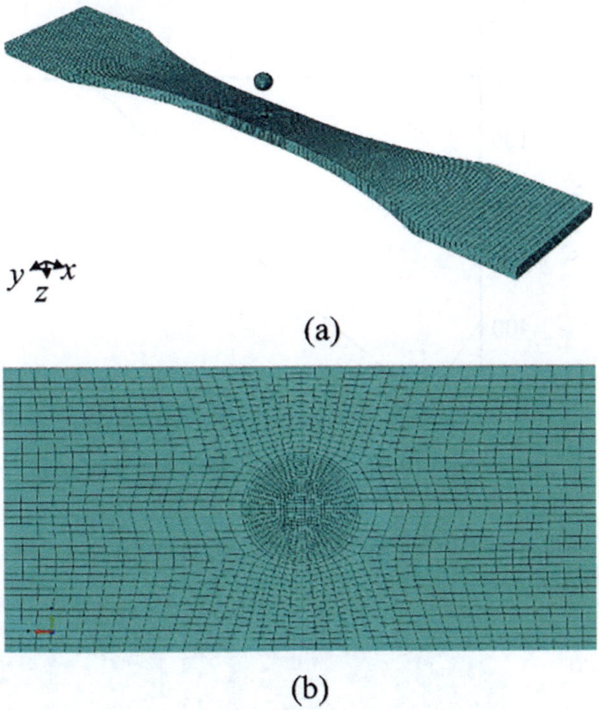

Fig. 5 Finite element model for impact simulation of 2A12 aluminum alloy specimen. (**a**) Whole grid, (**b**) local grid

Table 2 2A12 mechanical property parameters of aluminum alloy

E (MPa)	ν	σ_y(MPa)	σ_b(MPa)	ρ(kg · m^{-3})
68,000	0.33	410	547	2780

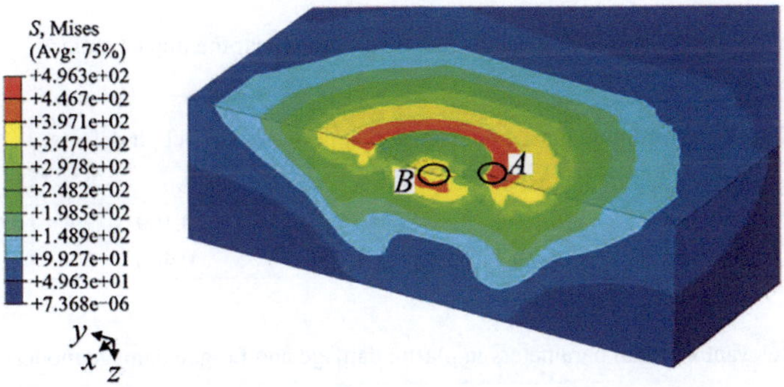

Fig. 6 Cloud map of residual stress distribution near the impact crater

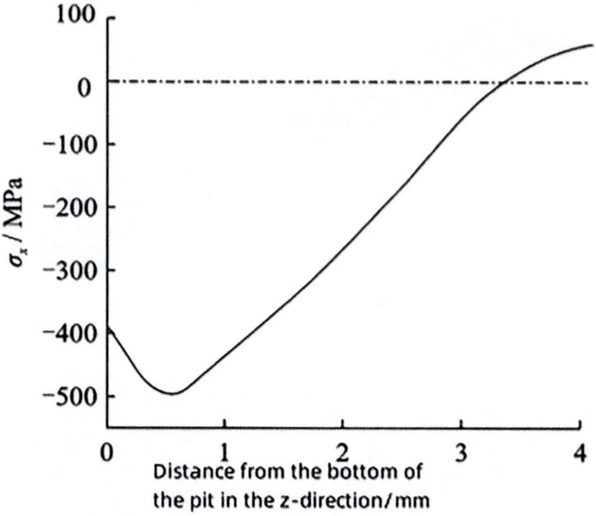

Fig. 7 The distribution curve of residual stress σ × along the z-direction (thickness direction) at the bottom of the pit

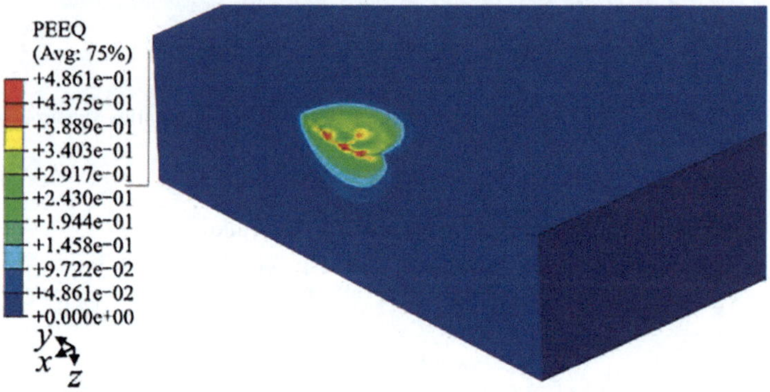

Fig. 8 Cloud map of equivalent plastic strain distribution near the impact crater

Table 3 Calculation of initial plastic damage at risk points of impact pit defects

Dangerous points	σ_{eq}(MPa)	R_v	Δp	D_0
A	496.196	0.887	0.109	0.094
B	384.606	1.409	0.486	0.362

Table 4 Relevant material parameters in plastic damage and fatigue damage models

S_0(MPa)	S_0	σ_0(MPa)	α	β	M_0	b
1.69	3	137	0.0801	1.95	114,339	0.003

Table 5 Comparison between fatigue test results and damage model predictions

Nominal stress (MPa)	Test average life/cycle	Predicting lifespan/cycle	Error (%)
200	252,340	309,193	22.53
230	72,833	83,108	14.11

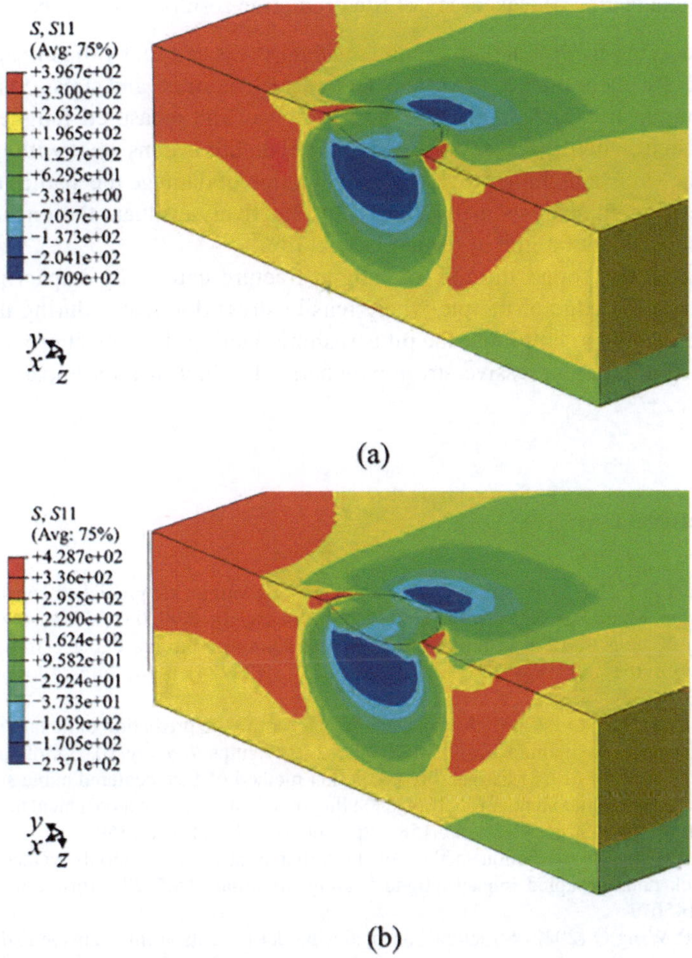

Fig. 9 Stress distribution of impact pits under subsequent fatigue maximum tensile load. (**a**) Stress nephogram under consequent load of 200 MPa. (**b**) Stress nephogram under consequent load of 230 MPa

Figure 9 shows the stress distribution of the impact pit under the subsequent fatigue maximum tensile load.

By comparing the theoretical predicted values with the experimental results, it can be seen that the predicted life based on the continuous damage mechanics model is consistent with the experimental results.

5 Conclusion

Based on the theory of continuous damage mechanics, we research the impact of 2A12 aluminum alloy.

The impact damage and subsequent fatigue life were evaluated, and combined with fatigue testing, the accuracy of life prediction results was verified.

1. Considering the initial damage and residual stress caused by the impact comprehensively, the influence of various factors such as stress and stress concentration on fatigue life, and prediction, the location of the measured fatigue source is consistent with the test results, and the fatigue life is consistent with the test.

   The value matching proves the combination of damage mechanics theory and finite element analysis method. We can effectively evaluate the impact damage and subsequent fatigue of metal material life.
2. Research has found the risk of fatigue fracture caused by impact pit defects points at the edge of the pit, where tensile stress dominates during the loading process, and the bottom of the pit is rushing. During the impact process, significant residual compressive stress is generated, which has an impact on fatigue failure.

References

1. Peng Y, Liu Y, Li H et al (2022) Research on low cycle fatigue life prediction considering average strain. Mater Res Express 9:016521. https://doi.org/10.1088/2053-1591/ac4b4d
2. Huang X, Sun B, Li Z (2020) Multi-scale modeling of fatigue damage in a metal wire film with the thickness effect. J Mater Res 35(23–24):1–10. https://doi.org/10.1557/jmr.2020.307
3. Hong H, Wang J, Li P (2021) A revised approach for the life prediction of metal materials fabricated by additive manufacturing. Mechanika 27:187. https://doi.org/10.5755/j02.mech.28216
4. Wang Q, Geng P (2022) Fatigue life prediction method of face-centered cubic single-crystal metals under multiaxial nonproportional loading based on structural mechanical model. Fatigue Fract Eng Mater Struct 45(1):133–158. https://doi.org/10.1111/ffe.13590
5. Mi C (2022) Life prediction method of dissimilar lightweight materials welded joints with precrack under coupled impact-fatigue loading. Materials 15:5077. https://doi.org/10.3390/ma15145077
6. Geng P, Wang Q (2022) Structural mechanics model for anisotropic damage and fatigue life prediction method of face-centered cubic metal materials. Int J Fatigue 156:106657

Open Access This chapter is licensed under the terms of the Creative Commons Attribution-NonCommercial-NoDerivatives 4.0 International License (http://creativecommons.org/licenses/by-nc-nd/4.0/), which permits any noncommercial use, sharing, distribution and reproduction in any medium or format, as long as you give appropriate credit to the original author(s) and the source, provide a link to the Creative Commons license and indicate if you modified the licensed material. You do not have permission under this license to share adapted material derived from this chapter or parts of it.

The images or other third party material in this chapter are included in the chapter's Creative Commons license, unless indicated otherwise in a credit line to the material. If material is not included in the chapter's Creative Commons license and your intended use is not permitted by statutory regulation or exceeds the permitted use, you will need to obtain permission directly from the copyright holder.

Research Progress on Biaxial Tensile Testing

Chang Liu, Wei Huang, Zhiqiang Chang, Fujun He, Yake Wu, and Chao Li

Abstract Since the 1970s, part of the solid rocket motor (SRM) has experienced disintegration when the tactical missile is launched at a low temperature of −40 °C. Analyses of the emission of a large number of solid missiles revealed that the destruction of the structural integrity of the solid propellant grain was the major contributor to SRM failure. Due to the coupled effects of low temperature and the ignition pressure load, the structure integrity of solid propellant grain is aggravated. Therefore, improving the structure integrity of propellant grain during the ignition of SRM for tactical missiles at low temperatures is one of the key problems to tackle during the research and production of weapons. As previous reports show, the mechanical behaviors of solid propellants at low temperatures and strain rates from 1 to 102 s^{-1} are very important to effectively analyze the structure integrity of propellant grain for SRM of the tactical missiles during ignition at low temperatures. However, the mechanical behaviors of thermal aging solid propellants at low temperatures under dynamic loading are rarely reported. It is urgent to carry out research in this regard through new mechanical test methods. In this paper, we analyze the research progress of biaxial tensile testing schemes for different materials such as metals and composite materials. It was found that there are currently many biaxial tensile tests on rubber and composite materials, which can provide effective references for other materials.

Keywords Research progress · Biaxial tensile · Testing

1 Introduction

Since the 1970s, part of the solid rocket motor (SRM) has witnessed the disintegration phenomenon when the tactical missile was launched at a low temperature of −40 °C. Through analysis of the emission of a large number of solid missiles, it was

C. Liu (✉) · W. Huang · Z. Chang · F. He · Y. Wu · C. Li
Qingzhou Hi-Tech Institute, Qingzhou, China

© The Author(s) 2026

Z. Zhang, P. Mercorelli (eds.), *Mechanical Design and Simulation: Insights and Innovations*, Lecture Notes in Mechanical Engineering,
https://doi.org/10.1007/978-981-95-2405-1_43

found that structural disintegration of the solid propellant gain contributes considerably to the failure of SRMs. Due to the coupled effects of low temperature and the ignition pressure load, the structure integrity of solid propellant grain is aggravated. Therefore, in the research and production of weapons, it is necessary to enhance the structural integrity of the propellant grain during low-temperature ignition of SRMs in tactical missiles. Previous studies have revealed that controlling the mechanical behaviors of the solid propellants at a low temperature and a strain rate within a range between 1 and 102 s^{-1} is a solution to effective analysis of the structural integrity. However, reports on the mechanical behavior of the thermal aging solid propellant at low temperatures under dynamic loading are rare or even lacking. It is urgent to carry out studies in this regard via new mechanical test methods.

Given what is specified above, the tri-component Hydroxyl-Terminated Polybutadiene (HTPB) propellant used in the SRM of a typical tactical missile was chosen as the research object in this thesis. Then, a method combining experimental investigation and theoretical analysis was applied to systematically study the mechanical behaviors of the thermal aging propellant under low-temperature dynamic loading conditions, which contains dynamic biaxial tensile and compressive mechanical properties at low temperatures, microscopic damage, and the strength criterion. The fault envelope of thermal aging HTPB propellant was established according to the twin shear strength theory.

At present, both domestically and internationally, mechanical biaxial tensile testing is a popular method for the investigation of the mechanical characteristics of materials. Regarding biaxial mechanical property experimental research, according to material properties, it mainly includes metal mechanical property tests and composite material mechanical property tests. This article mainly summarizes the mechanical property tests of two types of materials and points out the shortcomings in current research and the need for further focused research.

2 Multi-axis Mechanical Performance Test of Propellant

Wang Zhejun [1] simulated the stress ratio relationship between horizontal and vertical directions when Flat noodles propellant is stretched, as shown in Fig. 1, through finite element software, and found that Flat noodles can perform approximately 1:2 biaxial stretching. Based on the domestic and foreign experiments, it is shown that flat noodles propellant can effectively simulate the stress state of real grain. Therefore, Wang Zhejun obtained the same change rule as the uniaxial tensile test by relying on the high strain rate testing machine to test the flat noodles test piece, as shown in Fig. 2.

At present, biaxial tensile research mainly relies on cross-shaped propellants. Zhang Lihua employed a uniaxial testing machine and self-developed the testing equipment to perform biaxial tensile tests on the propellants, which verified the anisotropic characteristics of solid propellants. To conduct a true biaxial tensile test, Qiang Hongfu et al. thinned the central region of a cross-shaped test piece with a

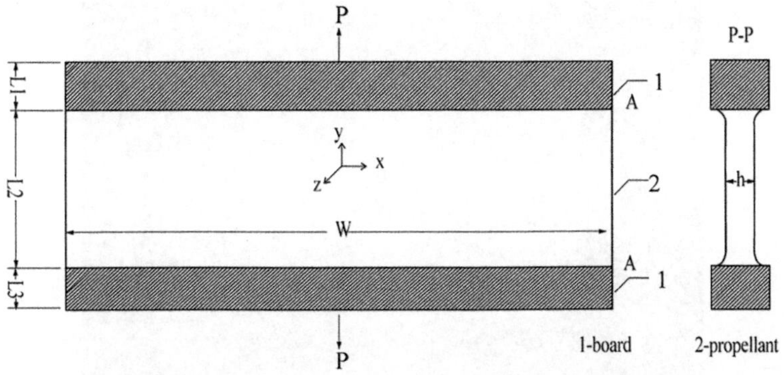

Fig. 1 Schematic diagram of flat noodles propellant

Fig. 2 Wang Zhejun et al. carried out flat noodles test

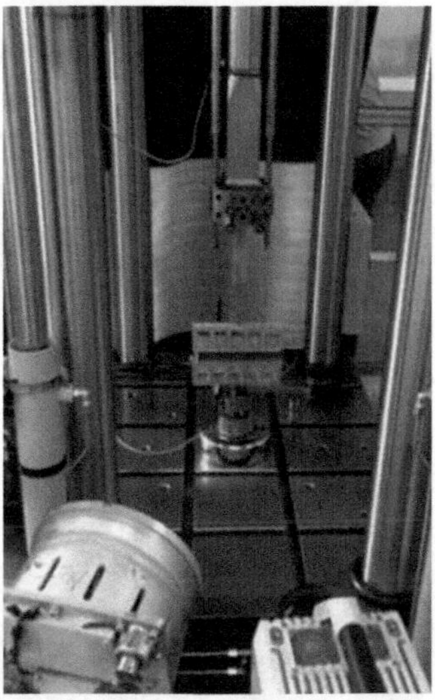

biaxial tensile machine in Fig. 3. Subsequently, Jia Yonggang et al. conducted biaxial tests by improving the bathtub-type propellant designed by Qiang Hongfu and combining it with a cross-shaped test piece with grooves on the arm in Fig. 4.

Foreign researcher Jalocha believes that the experimental scheme adopted by Jia Yonggang is not very reliable. Therefore, he redesigned the cross-shaped test piece by canceling the slotted design and expanding the radius of the propellant arc area, finally achieving the biaxial tensile test research in Fig. 5.

Zhang et al. conducted biaxial tensile, biaxial compression, and tensile compression performance tests on bathtub-shaped specimens, rectangular specimens, and similar dumbbell-shaped specimens using a biaxial testing machine.

To simulate the effects of the pressure environment in SRM of propellant columns, Liu et al. [2] conducted mechanical experiments on solid propellants by adding confining pressure devices and applying pressure loads. The action of pressure

Fig. 3 Qiang Hongfu et al. experimental system

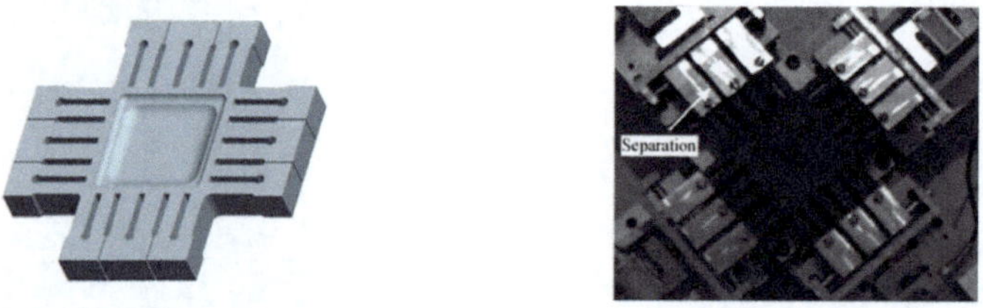

(a) Optimization of the cross-test piece (b) Test piece and machine loading

Fig. 4 Jia Yonggang et al. experimental system. (**a**) Optimization of the cross-test piece. (**b**) Test piece and machine loading

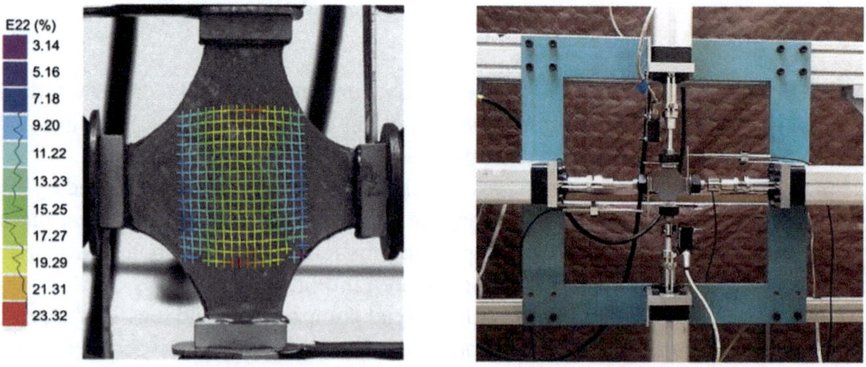

(a) Improve cross propellant (b) Test piece and device system

Fig. 5 Jalocha et al. experimental setup. (**a**) Improve cross propellant. (**b**) Test piece and device system makes it difficult to observe the damage and increase the strength of the propellant. This agrees with the conclusions reached by He Tieshan et al. in simulating the ignition and pressure building of NEPE propellant in a high-pressure testing machine. Zhang Ya et al. conducted mechanical tests under confining pressure on HTPB propellant specimens of different ages and found that when pressure exists, the propellant has higher mechanical tensile strength. In the confining pressure environment, the propellant can be stored better.

Regarding triaxial mechanical performance tests, both domestically and internationally, tests are mainly conducted within the quasi-static ($<1\ s^{-1}$) range. Jones designed thick disc specimens for research and conducted radial compression tests using a compression testing machine to simulate their stress state from multiple perspectives. In fact, during the SRM operation, the multi-axis specimen presents a stress state that has a higher resemblance with the real situation of the internal grain. Therefore, foreign researcher Sanal Kumar provided a research circular tube test piece to better simulate the stress state of the grain during ignition and compression. Relaxation tests were designed for different stress states. The tests showed that at the beginning of the test, the stress composition was not enough to change the relaxation modulus of the propellant. As the test continued to be loaded under tension, the relaxation modulus increased with the dimension of stress. Under the same test conditions, the relaxation modulus of the triaxial test was the highest.

3 Research Progress on Influencing Factors of Mechanical Properties of Propellants

1. Temperature is the influencing factor that causes changes in the mechanical properties of HTPB propellants. Pang et al. found that the initial modulus and maximum tensile strength of solid propellants are inversely proportional to temperature, but there is no significant proportional relationship between the maximum elongation. Lai Jianwei and Wang Xiaoying [3] both conducted low-temperature and low-temperature recovery tensile tests at room temperature to study the effects of low temperatures on propellants. Through low-temperature tensile tests, Wang Zhejun studied the performance changes of propellants under uniaxial and biaxial tensile tests. Shekhar compared the changes in propellant performance caused by various temperatures and found through experiments that low temperatures increase the maximum tensile strength of propellants.
2. In terms of strain rate, the experiment mainly relies on changing the tensile or compression rate of the test piece. Lai Jianwei et al. compared the mechanical properties of the propellant at different compression rates by changing its compression rate and found that the magnitude of the compression rate can change the strength of the propellant. Via mechanical tests on the modified double-base solid propellants under varied strain rates, Sun et al. established constitutive equations based on the mechanical performance parameters. Sun Chaoxiang [4,

5] conducted research on solid propellants by changing the strain rate and found that high strain rates have a greater impact.
3. Solid propellants face aging problems from the moment they are produced. To study the changes in mechanical properties of propellants after aging, natural aging and thermal aging tests are mainly conducted in current experiments. Currently, domestic and foreign research mainly relies on the Time–Temperature Superposition Principle (TTSP) to conduct high-temperature thermal aging tests on test specimens. Kivity et al. analyzed the intrinsic causes of propellant aging from a mechanistic perspective. Zhang Ya and Wang Zhejun [6] conducted experiments on propellants with different aging times and found that the strength of aged propellants increased compared to those without aging.
4. To simulate the real force state of propellants during SRM operation, researchers at home and abroad have proposed experimental schemes from one-dimensional to multi-dimensional for research. For uniaxial tensile testing, JANNAF standards and GJB 770B-2005 standards have been introduced as references, but there is still a lack of standards that can be referenced for multiaxial mechanical testing. To better study the mechanical properties of propellants under multidimensional stress states, Zhang Ya conducted biaxial tensile and biaxial compression tests using a biaxial testing machine. Compared with his uniaxial tensile and compression studies, it was found that the biaxial tensile strength increased, but the elongation rate decreased; Wang Zhejun also found that biaxial will enhance the maximum tensile strength of propellant by carrying out flat noodles test pieces, and found that solid propellant will be constrained in two directions under biaxial conditions, so its elongation will decrease.

4 Conclusion

1. At present, only testing standards for single-axis quasi-static conditions have been issued domestically and internationally. Due to device limitations, testing under dynamic loading is still very rare, especially for biaxial high strain rate loading tests. Although Qiang Hongfu and others have conducted dynamic single-axis biaxial tensile tests, there is still insufficient understanding of the mechanical properties of real propellants.
2. There is a lack of multi-factor analysis. Currently, only single-factor mechanical performance tests such as temperature, strain rate, stress state, or aging have been conducted, and multi-factor loading tests for HTPB propellants under biaxial tension after low temperature, high strain rate, and thermal aging have not been carried out yet.

References

1. Wang Z-J (2016) Experimental and theoretical research on mechanical behaviors of HTPB propellant at low temperatures under dynamic loading. Xi'an Hi-Tech Institute, Xi'an
2. Liu C, Qiang H-F, Wang Z-J et al (2022) Strength criterion of aged HTPB propellant at low temperature under dynamic loading. J Propuls Technol 39(11):2581–2587
3. Xiaoying W, Yue W, Xinmei Y et al (2016) Research on low temperature mechanical performance of NEPE propellant. Chem Propell Polym Mater 14(5):64–67
4. Sun C-X, Ju Y-T, Zheng J et al (2023) Experimental research on the compressive properties of modified double base propellant at low and high strain rates. Acta Armamentarii 34(6):698–703
5. Sun C, Xu J, Chen X et al (2021) Strain rate and temperature dependence of the compressive behavior of a composite modified double-base propellant. Mech Mater 89:35–46
6. Wang Z, Qiang H, Wang G, Huang Q (2021) Tensile mechanical properties and constitutive model for HTPB propellant at low temperature and high strain rate. J Appl Polym Sci 132(24):42104

Open Access This chapter is licensed under the terms of the Creative Commons Attribution-NonCommercial-NoDerivatives 4.0 International License (http://creativecommons.org/licenses/by-nc-nd/4.0/), which permits any noncommercial use, sharing, distribution and reproduction in any medium or format, as long as you give appropriate credit to the original author(s) and the source, provide a link to the Creative Commons license and indicate if you modified the licensed material. You do not have permission under this license to share adapted material derived from this chapter or parts of it.

The images or other third party material in this chapter are included in the chapter's Creative Commons license, unless indicated otherwise in a credit line to the material. If material is not included in the chapter's Creative Commons license and your intended use is not permitted by statutory regulation or exceeds the permitted use, you will need to obtain permission directly from the copyright holder.

References

Study on the Low-Temperature Deformation Mechanisms of Dual-Phase Titanium Alloys Under Nanoindentation

Jiayi Li, Gaoyang Liu, Xinhai Yao, and Dong He

Abstract To investigate the influence of temperature on dual-phase titanium alloys' deformation behavior and mechanisms during nanoscale indentation, molecular dynamics simulations are conducted. The atomic strains, deformation behavior, dislocation types, and lengths of the dual-phase titanium alloy are systematically analyzed during low-temperature nanoscale indentation. The results show that within the temperature range of 0.01–70 K, the decreases in critical load of the dual-phase titanium alloy with increasing temperature indicates that plastic deformation is most likely to occur at 70 K. Near the indentation region, the distribution of atoms with high stress and strain becomes more concentrated. At 20 K, {10−11} twinning is observed in the dual-phase titanium alloy structures, suggesting that twinning is more easily activated at low temperatures. Furthermore, the dual-phase titanium alloys are dominated by 1/3<1−100> partial dislocations and mixed dislocations, and the total dislocation length increased with temperature, indicating that slip deformation is the predominant deformation mechanism at 70 K.

Keywords Titanium alloys · Molecular dynamics · Ultra-low temperature · Plastic deformation

1 Introduction

Titanium and its alloys are widely used in fields such as automotive and aerospace due to their high specific strength, relatively low density, good fracture toughness, and corrosion resistance [1]. It is well known that different deformation mechanisms occur in metals at different temperatures [2]. Generally, as temperature decreases, alloys tend to exhibit increased strength but decreased ductility. The $\alpha + \beta$ dual-phase titanium alloy Ti6Al4V has been widely used in low-temperature

J. Li (✉) · G. Liu · X. Yao · D. He
North China University of Technology, Beijing, China

© The Author(s) 2026

Z. Zhang, P. Mercorelli (eds.), *Mechanical Design and Simulation: Insights and Innovations*, Lecture Notes in Mechanical Engineering,
https://doi.org/10.1007/978-981-95-2405-1_44

environments [3]. To ensure that titanium alloys can be used at low temperatures, effective control of multiple deformation mechanisms to simultaneously enhance low-temperature strength and plasticity becomes crucial.

He and Zheng [4] employed molecular dynamics (MD) simulation to analyze the influence of the $\gamma/\alpha 2$ interface on the deformation behavior of TiAl alloys during nanoindentation. The results revealed that the interface acted as a barrier to dislocation motion, enhancing the strength of TiAl alloys. Chen et al. [5] utilized MD simulation to investigate the influence of temperature on the deformation mechanisms of TiAl/Ti3Al composite materials and discovered that the phase interfaces can promote phase formation. Song and Liu [6], through MD simulations, analyzed the wear mechanisms of high-Nb TiAl alloys under different displacements and loads. The results indicated that the gradient twin-like layer interfaces hindered twin deformation and dislocation motion, thereby improving the ability of the alloy to resist wear.

Atomic-level computational simulations are capable of observing atomic arrangements and are extensively employed to investigate the development of microstructures. Molecular dynamics computational methods are important tools for studying the deformation mechanisms of titanium alloys. Molecular dynamics simulations can precisely monitor the dynamic migration and phase transformation processes of materials, along with the nucleation and progression of internal dislocations. To address the issue of plastic instability in titanium alloys at extremely low temperatures, this study utilizes molecular dynamics simulations to investigate the behavior of plastic deformation and mechanisms of titanium alloys at ultra-low temperatures.

2 Methodology

The molecular dynamics analytical method involves integrating motion equations to track the trajectories of individual atoms in a multiparticle system and study its physical properties. Given the equations of motion, numerical integration is performed with an appropriate time step until a specified mechanical convergence criterion is reached, resulting in final position coordinates. Then, using statistical mechanics methods, the simulated particle positions, velocities, and accelerations are used to calculate and obtain thermodynamic and dynamic properties.

Molecular dynamics simulations were conducted utilizing the LAMMPS [7] software package and the atomic data acquired from the simulations were subsequently analyzed using the OVITO visualization software [8]. A dual-phase titanium alloy model, depicted in Fig. 1, was established for nanoindentation calculations using the lattice constants of a = 2.9525 Å and c = 4.6944 Å for the hexagonal close-packed (hcp) structure, and a lattice constant of 3.32 Å for the bcc structure. The model was developed based on a hybrid potential function. The red atoms correspond to the hcp structure, while the blue atoms correspond to the bcc structure, and the green atoms represent the faced-centered cubic (fcc) structure.

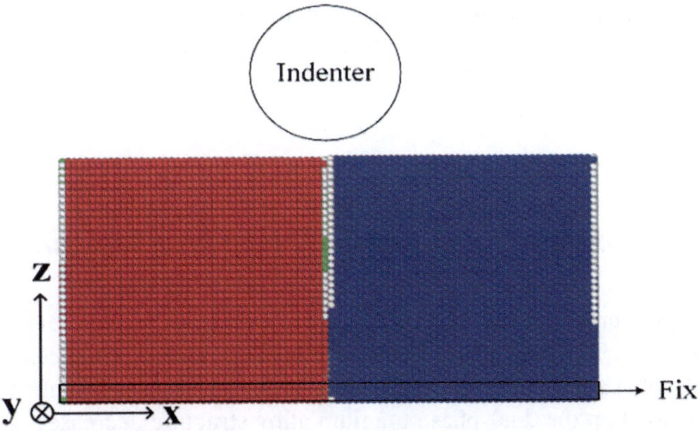

Fig. 1 The dual-phase titanium alloy nanoindentation model

The crystal orientations of the α-phase in the x, y, and z axes are [0001], [1−100], and [11−20], while the crystal orientations of the β-phase in the x, y, and z axes are [110], [−112], and [1−11]. The dimensions of the dual-phase model are 206 Å in length, 219 Å in width, and 103 Å in height. In the α phase, random additions of Al atoms comprising a weight percentage of 0.1 and V atoms comprising a weight percentage of 0.02 were incorporated based on hcp-Ti. In the β phase, random additions of Al atoms with a mass fraction of 0.03 and V atoms with a mass fraction of 0.25 were incorporated based on bcc-Ti. The final dual-phase model had an Al mass fraction of 0.06 and a V mass fraction of 0.04. A spherical virtual indenter with a radius of 25 Å was located 5 Å away from the model. To prevent rigid body motion during the nanoindentation process, the atoms at the bottom were fixed as the boundary layer with a thickness of three atomic layers, equivalent to 8.85 Å.

Initially, the NPT relaxation technique was employed to stabilize the model over 20 ps to attain system equilibrium. During the nanoindentation process, periodic boundary conditions were imposed along the x and y axes, whereas open boundary conditions were employed along the z-axis. A spherical indenter moved downward at a consistent speed of 40 m/s, applying compression in the z-direction. The step size for the nanoindentation process was set to 60 ps, with a maximum indentation depth of 19 Å.

The function responsible for defining the interaction among atoms is referred to as a potential function. According to the principle of superposition, the collective potential energy of a system can be formulated as a function of the particle coordinates [9]:

$$U(R)=\sum_{k=1}^{N}U_1(R_k)+\sum_{k<l}^{N}U_2(R_k,R_l)+\sum_{k<l<m}^{N}U_3(R_k,R_l,R_m)+\ldots \quad (1)$$

In the equation, U(R) represents the total potential energy, which can be decomposed into multiple contributions. The material studied in this paper is titanium alloy, where both the Embedded Atom Method (EAM) potential and the Modified

Embedded Atom Method (MEAM) potential are employed to describe the interactions between titanium elements and alloying elements [10].

3 Results and Discussions

3.1 The Deformation Behavior of Dual-Phase Titanium Alloys

The load-depth curves of dual-phase titanium alloys under varying temperatures are shown in Fig. 2a. The dual-phase titanium alloy structure experiences a maximum load of 391 nN at a temperature of 0.01 K. Within the temperature interval from 20 to 70 K, the load on the dual-phase titanium alloy structure decreases with increasing temperature, reaching a minimum load of 200 nN at 70 K.

The critical load of dual-phase titanium alloys at different temperatures is shown in Fig. 2b, and it diminishes as the temperature rises. At a temperature of 0.01 K, the dual-phase titanium alloy exhibits the maximum critical load of 19.91 nN, while at 70 K, the lowest critical load of 5.81 nN is observed, indicating the highest susceptibility to plastic deformation.

Different temperature-dependent atomic strain distributions in dual-phase titanium alloys after nanoindentation are illustrated in Fig. 3. The findings indicate that the high-strain regions of the α phase remain near the indentation zone, while the high-strain regions of the β phase start to spread towards the peripheral regions as the temperature increases. The area of diffusion expands gradually with rising temperature.

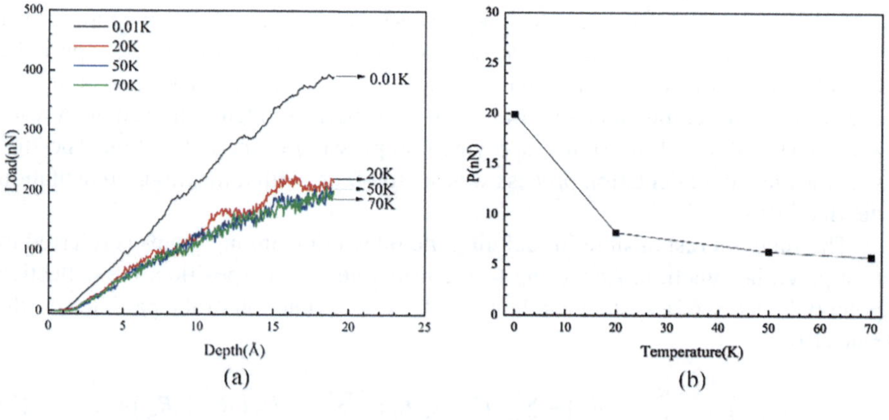

Fig. 2 The deformative characteristics of dual-phase titanium alloys at different temperatures. (**a**) The load-depth curves. (**b**) Critical load

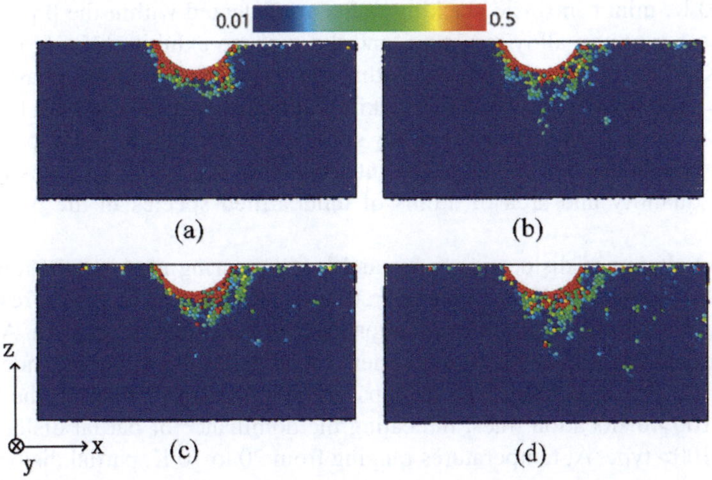

Fig. 3 The atomic strains of the dual-phase titanium alloy at different temperatures. (**a**) 0.01 K, (**b**) 20 K, (**c**) 50 K, (**d**) 70 K

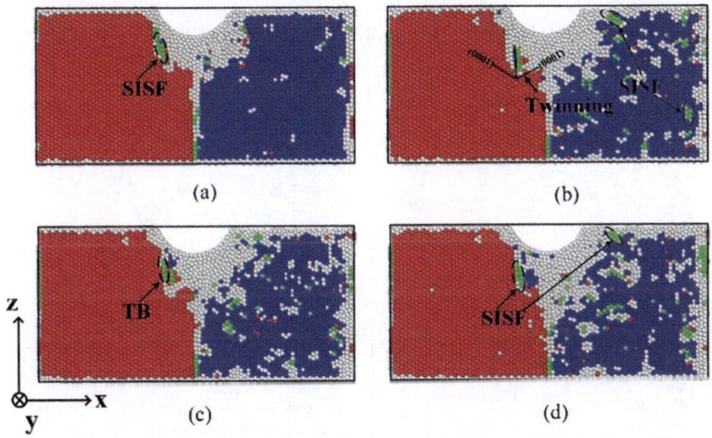

Fig. 4 The defect evolution in the dual-phase titanium alloy at different temperatures. (**a**) 0.01 K, (**b**) 20 K, (**c**) 50 K, (**d**) 70 K

3.2 The Deformation Mechanisms of Dual-Phase Titanium Alloys

Figure 4 illustrates the defect evolution in dual-phase titanium alloys at different temperatures. At 0.01 K, intrinsic stacking faults are observed in the α phase of the dual-phase titanium alloy structure, while a small number of atoms of unidentified species are present in the β phase.

At 20 K, minor intrinsic stacking faults are detected within the β phase of the dual-phase titanium alloy structure, and the α phase exhibits {10−11} twinning. At 50 K, grain boundaries appear within the α phase of the dual-phase titanium alloy structure. At 70 K, intrinsic stacking faults are observed in both the α and β phases of the dual-phase titanium alloy structure. With increasing temperature, the number of unidentified atoms at the interface between the two phases increases, and the quantity and area of atoms of unidentified species in the β phase also increase.

The evolution of dislocation types and lengths during nanoindentation in dual-phase titanium alloys at varying temperatures is depicted in Fig. 5. The results indicate that the overall length of dislocation lines grows with temperature. At 0.01 K, the cumulative length of dislocation lines is relatively short when the indenter penetrates to 0–30 ps. However, after 30 ps, there is a sharp increase in the length of 1/3<1−100> dislocation lines, indicating the dominance of partial dislocations of 1/3<1−100> type. At temperatures ranging from 20 to 70 K, partial dislocations of

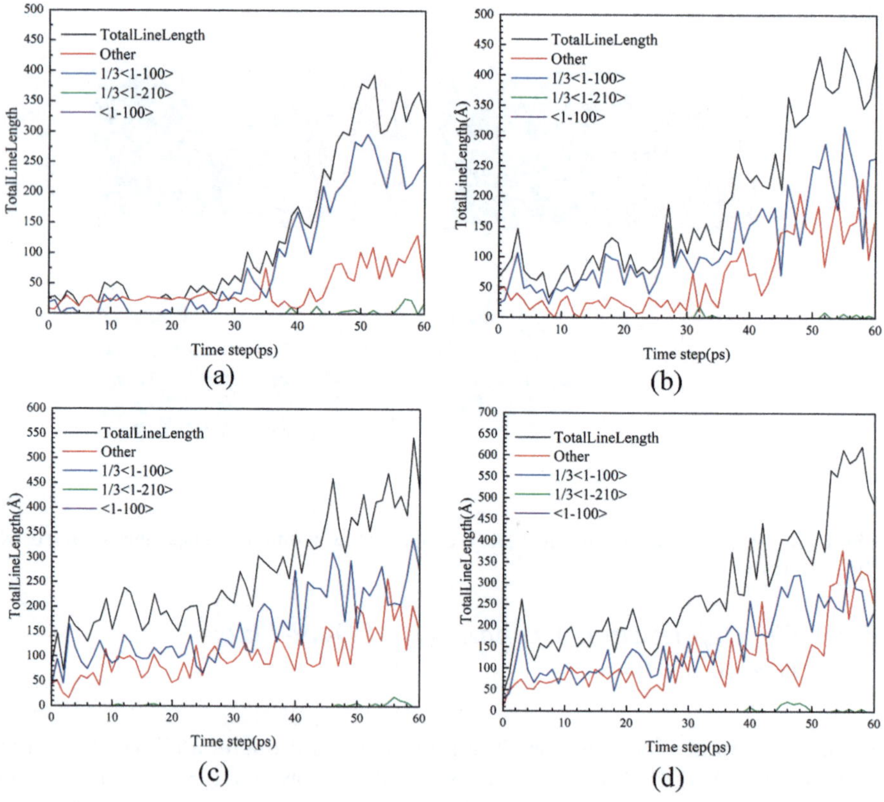

Fig. 5 The variation in dislocation types and lengths in the dual-phase titanium alloy at different temperatures

1/3<1−100> type and mixed dislocations are dominant, and with increasing temperature, the proportion of mixed dislocation line lengths gradually increases. At 70 K, the lengths of 1/3<1−100> dislocation lines and mixed dislocation lines are essentially the same.

4 Conclusions

This research utilized molecular dynamics simulations to explore the nanoscale indentation process of dual-phase titanium alloys at low temperatures. The behavior under deformation and mechanisms of dual-phase titanium alloys in the course of low-temperature nanoindentation were systematically examined. The results showed that the dual-phase titanium alloys exhibit a decrease in critical load with increasing temperature. This implies that plastic deformation is most likely to occur at 70 K, as it is the temperature at which the materials are most susceptible to undergo plastic deformation. As the region near the indentation area is approached, there is an increased density of atoms experiencing high stress and strain, and the area of the high-strain region in the dual-phase titanium alloy structure expands with increasing temperature. At a temperature of 20 K, the dual-phase titanium alloys exhibit the presence of {10−11} twinning, indicating that the mechanism of plastic deformation in the material is predominantly governed by twinning at lower temperatures. The dominant dislocation types are 1/3<1−100> partial dislocations and mixed dislocations. Moreover, the overall length of dislocations increases with the temperature rise. This indicates that at 70 K, the deformation mechanism of titanium alloys tends to favor slip deformation.

References

1. Yu Y, Jiang P (2014) Recent advances in the development and application of cryogenic titanium alloys. Mater Dev Appl 29(6):118–122
2. Liu W, Wang N, Chen Y (2022) Damping behavior of typical titanium alloys by varied frequency micro harmonic vibration at cryogenic temperatures. J Mater Res Technol 21:3746–3755
3. Yin X, Wang H, Wang K (2024) Research on mechanical properties of Ti-6Al-4V ELI titanium alloy in wide temperature range (−253–350 °C). Therm Process Tech 53(4):119–122+126
4. He T, Zheng M (2024) Molecular dynamics for nanoindentation in the $\gamma/\alpha 2$ interfacial strengthening mechanism of a duplex full lamellar TiAl alloy. Mater Today Commun 38:107965
5. Chen Y, Li J, Liu B (2022) Unraveling hot deformation behavior and microstructure evolution of nanolamellar TiAl/Ti3Al composites. Intermetallics 150:107685
6. Song Q, Liu L (2022) Fretting wear behaviors and mechanism of a high Nb–TiAl alloy with full lamellar microstructure at ambient temperature. J Mater Sci 57(32):15636–15650
7. Plimpton S (1995) Fast parallel algorithms for short-range molecular dynamics. J Comput Phys 117(1):1–19

8. Stukowski A (2010) Visualization and analysis of atomistic simulation data with OVITO-the open visualization tool. Model Simul Mater Sci Eng 18:015012
9. Griebel M, Zumbusch G, Knapek S (2007) Numerical simulation in molecular dynamics numerics, algorithms, parallelization, applications. Springer, Berlin
10. Liu X, Xu L, Zhang S (2020) Molecular dynamics simulation of Ti-6AL-4V diffusion bonding behavior under different process parameters. Mater Technol 54:365–372

Open Access This chapter is licensed under the terms of the Creative Commons Attribution-NonCommercial-NoDerivatives 4.0 International License (http://creativecommons.org/licenses/by-nc-nd/4.0/), which permits any noncommercial use, sharing, distribution and reproduction in any medium or format, as long as you give appropriate credit to the original author(s) and the source, provide a link to the Creative Commons license and indicate if you modified the licensed material. You do not have permission under this license to share adapted material derived from this chapter or parts of it.

The images or other third party material in this chapter are included in the chapter's Creative Commons license, unless indicated otherwise in a credit line to the material. If material is not included in the chapter's Creative Commons license and your intended use is not permitted by statutory regulation or exceeds the permitted use, you will need to obtain permission directly from the copyright holder.

Thermal Cure Coupling Optimization Analysis of Integrated Thermal Protection System Connector

Bin Wang, Dongqiang Zhao, Yu Ning, Cunxian Wang, Peng Du, and Feng Xiong

Abstract This study focuses on the heat conduction characteristics and stress distribution patterns of the Z-shaped connecting bracket in thermal protection structures. A two-dimensional heat conduction model is established, considering the thermal radiation interaction between the outer panel and the external environment. By simulating heat conduction under different insulation layer thicknesses, the relationship between the maximum temperature of the base panel and insulation layer thickness is revealed. The model's accuracy is verified through comparisons with a three-dimensional model and actual temperature data. Furthermore, a 1/8 scale model of the thermal protection structure is established to analyze the impact of pressure loads on bracket stress, identifying the main causes of high-stress areas. The temperature gradient in the structure's thickness direction is found to be a key factor causing higher stress on the bracket. This research proposes a novel numerical approach for the design and optimization of thermal protection structures in aerospace vehicles.

Keywords Connector · Z-shaped · Stress · Thermal analysis

1 Introduction

In the exploration of outer space, reusable launch vehicles (RLVs) play a crucial role due to their multifunctionality [1], making them a significant focus of high-tech development in several developed countries. Since the 1980s, countries around the world have been actively researching RLVs [2]. Although some countries have

B. Wang · D. Zhao · Y. Ning (✉) · P. Du · F. Xiong
AVIC the First Aircraft Institute, Xi'an, China

C. Wang
AVIC the First Aircraft Institute, Xi'an, China

Northwestern Polytechnical University, Xi'an, China

© The Author(s) 2026
Z. Zhang, P. Mercorelli (eds.), *Mechanical Design and Simulation: Insights and Innovations*, Lecture Notes in Mechanical Engineering,
https://doi.org/10.1007/978-981-95-2405-1_45

temporarily suspended certain aerospace plane projects in recent years due to funding issues, it is undeniable that reusable aerospace planes indeed hold the potential to significantly reduce launch costs on a large scale in future developments [3].

As one of the most critical components of spacecraft, the thermal protection structure shields the internal components from the high temperatures encountered during reentry into the Earth's atmosphere [4]. The thermal protection system faces complex loads during atmospheric reentry, including thermal, acoustic, and vibrational challenges, which trouble researchers in the design of aircraft thermal protection structures [5]. The issues related to high-temperature stresses induced by thermal flux loads are particularly significant [6].

Thermal protection structures for aerospace vehicles primarily consist of three main types: passive, semi-passive, and active [7]. The latter two types can withstand high temperatures but involve complex structures and high manufacturing and maintenance costs, so they are less commonly used or limited to high-temperature areas like the nose of the spacecraft. In contrast, passive thermal protection systems are simpler in structure and easier to manufacture and maintain [8]. Among passive systems, the tile-based structure with insulation layers is particularly favored by designers [9]. This type of structure effectively isolates high temperatures from penetrating the internal structure of the aerospace vehicle [10].

Passive thermal protection systems, specifically those using insulation layer-based structures, consist of a thermal structure on the outer layer and a cold structure within the airframe frame. The connection between the cold and thermal structures often experiences stress concentration due to thermal mismatch and temperature gradients. In severe cases, this can adversely affect the safety and lifespan of the spacecraft. Historically, there have been two catastrophic accidents involving aerospace vehicles, resulting in loss of life and destruction of the craft. Subsequent investigations revealed a direct correlation between these accidents and the thermal protection systems along with their associated connecting structures.

The Z-shaped bracket, as a relatively simple connection between external panels and internal structures, has been studied extensively by scholars, as shown in Fig. 1. Blosser et al. applied the Z-shaped bracket in metal thermal protection systems and

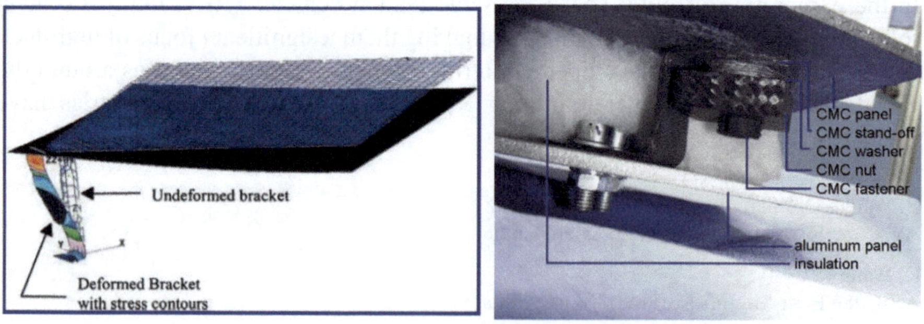

(a) (b)

Fig. 1 Z-shaped bracket in thermal protection structure

investigated its deformation under thermal loads [11]. Böhrk and Beyermann utilized ceramic matrix composite materials for Z-shaped brackets and employed composite material bolts to mitigate thermal mismatch issues [12]. Z-shaped brackets are convenient to disassemble, straightforward to apply, and are considered a favorable choice for the connection between internal and external structures in removable thermal protection systems. However, systematic studies on the structural dimensions of Z-shaped brackets are lacking.

In the present work, the heat conduction characteristics and stress distribution patterns of the Z-shaped connecting bracket in thermal protection structures were investigated through simulation. In Sect. 2, the thermal flux load on aerospace vehicles during reentry into the Earth's atmosphere is addressed briefly. In Sect. 3, a two-dimensional heat conduction model, considering the thermal radiation interaction between the outer panel and the external environment, is established to analyze the effect of insulation layer thickness on the maximum temperature of the base panel. Then in Sect. 4, an optimization analysis of the design of the bracket structure is performed. Conclusions are summarized in Sect. 5.

2 Thermal Flux Load

Most aerospace vehicles experience significant thermal loads during reentry into the Earth's atmosphere, exemplified by a cylindrical mono-stage orbiter with wings in a certain region of the United States [8], as shown in Fig. 2. The highest temperatures on the nose and leading edges of the wings reach up to 1422 K (2100 °F). This study focuses on the windward surface of the aircraft subjected to extensive high thermal flux loads, with maximum temperatures reaching 1033 K (1400 °F). The

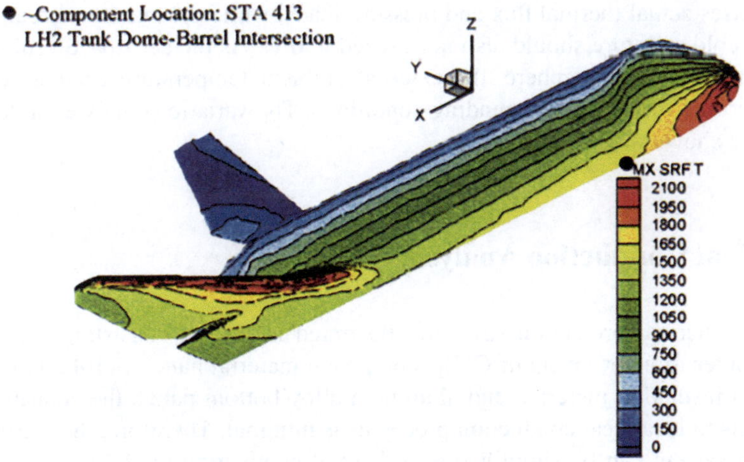

Fig. 2 Distribution of maximum temperatures in aerospace vehicles [8]

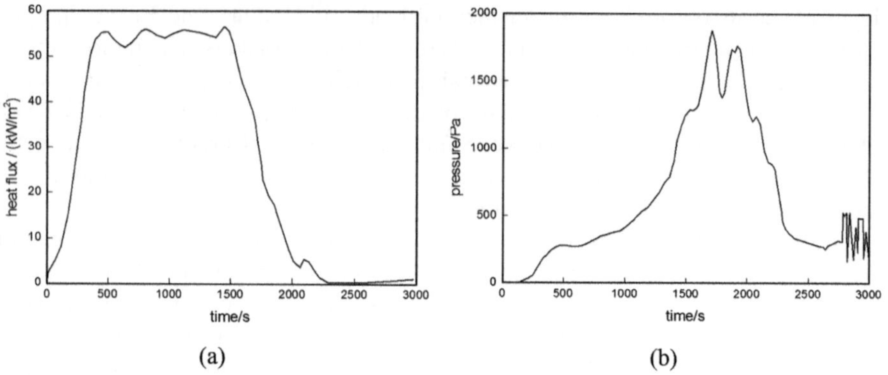

Fig. 3 Windward surface thermal flux load and pressure load

Table 1 External temperature variation during aerospace vehicle reentry into the atmosphere

Time (s)	0	450	1575	2175	3000
Ambient temperature (K)	213	243	273	291	295

relationship between thermal flux load and time during reentry is depicted in Fig. 3a, while the relationship between pressure load and time is shown in Fig. 3b. In Fig. 3b, pressure refers to the difference between internal and external pressures, with positive values indicating that internal pressure exceeds external pressure. The internal components of spacecraft are commonly made from aluminum alloys. Therefore, the maximum temperature at the bottommost layer of the thermal protection structure should not exceed the highest operational temperature of aluminum alloys, which is typically 450 K [13]. Hence, the external load on the thermal structure comprises actual thermal flux and pressure loads, while the maximum temperature of the cold structure should also not exceed 450 K. As the aerospace vehicle reenters the Earth's atmosphere, the external ambient temperature changes with the spacecraft's speed and surrounding conditions. The variations in external temperature are shown in Table 1.

3 Heat Conduction Analysis

For the thermal protection structure illustrated in Fig. 1, consisting sequentially from outer to inner layers of C/SiC composite material panel, Saffil alumina fiber aerogel insulation material, and aluminum alloy bottom panel, the contribution of brackets to local heat conduction processes is minimal. Therefore, the thermal protection system can be simplified to a thermal conduction model of "outer panel-insulation material-bottom panel". Using the windward surface thermal flux load of

Table 2 Mechanical properties of C/SiC material

E_1 (GPa)	E_2 (GPa)	E_3 (GPa)	μ_{12}	μ_{13}	μ_{23}	G_{12} (GPa)	G_{13} (GPa)	G_{23} (GPa)
118	118	20.6	0.27	0.07	0.07	21	10.8	10.8

Table 3 Thermal properties of C/SiC composite material

		Conductivity (W/(m · K))			Mean linear expansion (10^{-6} K^{-1})		
Temp (K)	Specific heat (J/kg · K)	X	Y	Z	X	Y	Z
293	700	1.25	1.25				
373	710				1.33	1.33	
473	850			1.19	2.42	2.42	1.09
573	970	2.67	2.67	1.00	2.83	2.83	1.19
673	1050			0.87	3.16	3.16	1.27
773	1100			0.83	3.01	3.01	1.32
873	1150	2.13	2.13	0.85	2.99	2.99	1.43
973	1220			0.86	3.12	3.12	1.56
1073	1280			0.87	3.18	3.18	1.68
1173	1300			0.88	3.29	3.29	1.80
1273	1290	2.04	2.04	0.89	3.36	3.36	1.93
1373					3.47	3.47	
1473		2.62	2.62		3.44	3.44	

the aerospace vehicle and the maximum allowable temperature of internal structures as conditions, the minimum thickness of the insulation material can be calculated to determine the thickness parameters of the thermal protection structure.

Saffil alumina fiber can withstand temperatures up to 1600 °C, with a density of 96 kg/m^3, specific heat capacity of 1000 J/kg · K, and thermal conductivity of 0.08 W/(m · K) [14], disregarding its structural rigidity. Aluminum alloy has an elastic modulus of 70 GPa, thermal conductivity of 100 W/(m · K), and thermal expansion coefficient of 2.3×10^{-5} K^{-1} [15]. The 2D woven C/SiC composite material, with a density of 2.05×10^3 kg/m^3 between room temperature and 1100 K, exhibits minimal changes in elastic modulus and strength with temperature [16]. Hence, mechanical parameters of C/SiC composite materials in this study are considered temperature-independent, sourced in [17], as listed in Table 2. With an in-plane compressive strength of 409 MPa, tensile strength of 260 MPa, and shear strength of 120 MPa, thermal-related parameters are sourced from [18], as listed in Table 3.

As temperature increases, the heat flux density due to solid conduction experiences only a slight increase, whereas thermal radiation increases significantly as it is proportional to the fourth power of temperature difference [19]. When temperatures exceed 850 °C, thermal radiation becomes the dominant factor. Since the outer panel of the spacecraft remains at high temperatures for most of the reentry phase, convective heat exchange between the outer panel and the external environment, as

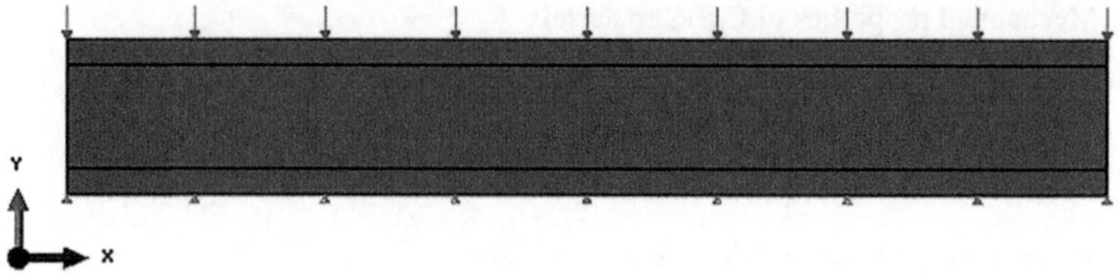

Fig. 4 2D thermal conduction model

well as heat conduction, are minimal compared to thermal radiation. Therefore, only thermal radiation is considered in the heat exchange between the outer panel and the external environment.

A 2D model is established, as shown in Fig. 4, with the Y-axis direction representing the direction from the inside to the outside of the insulation structure. The outermost layer is C/SiC composite material with a thickness of a = 5 mm, followed by an aerogel insulation layer with thickness b, and the innermost layer is an aluminum alloy bottom panel with thickness c = 5 mm. Displacement in the Z-direction and rotational degrees of freedom around the X and Y axes are constrained. Since the model only considers heat conduction, no additional displacement constraints are necessary.

The external surface emissivity is 0.8, and the Boltzmann constant is 5.67×10^{-8}. External environmental temperature variations are detailed in Table 1. The thermal flux load is depicted in Fig. 3a and is applied to the outermost surface. Despite the low heat flux density of up to 3000 s, the outermost layer may still exhibit temperatures higher than those of the internal structure due to temperature gradients. Consequently, heat conduction from the outermost layer inward leads to an increase in the temperature of the aluminum alloy bottom panel. Therefore, the calculation time for this 2D thermal conduction model is set to 5000 s.

By varying the thickness b of the insulation layer and monitoring the temperature changes on the outer surface of the C/SiC composite material layer and the bottommost temperature of the aluminum alloy panel, the temperature variations of the outer and bottom layers are depicted in Fig. 5. From the figures, it is observed that the thickness of the insulation layer has a minimal impact on the maximum temperature of the outermost layer, which reaches 1000 K at 550 s and maintains around 1050 K for the next 600 s before gradually decreasing to below 450 K after 3000 s. The temperature of the bottom panel varies with the thickness of the insulation layer. The decrease in temperature after 3000 s for different thickness structures of the outermost layer also slows down with increasing insulation layer thickness, attributed to the insulating effect of the insulation layer.

The relationship between the maximum temperature of the bottom panel and the thickness *b* of the insulation layer is illustrated in Fig. 5. The maximum temperature of the bottom panel decreases with the increasing thickness of the insulation layer, and the rate of decrease diminishes as the thickness increases. Therefore, excessively increasing the thickness of the insulation layer to reduce the maximum temperature of the bottom panel may not yield optimal cost-effectiveness. When the

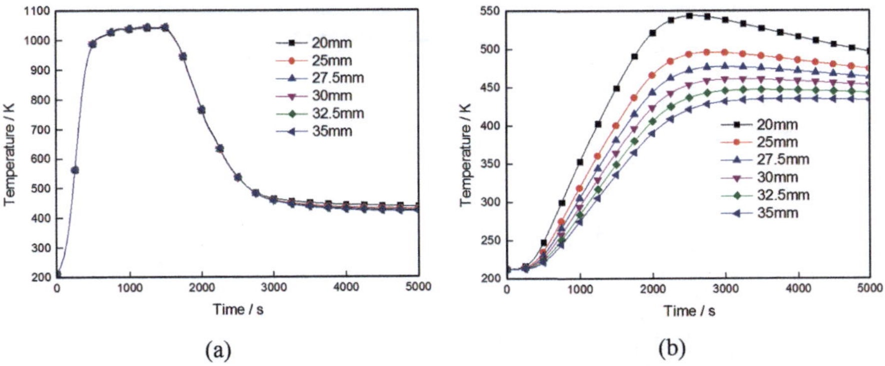

Fig. 5 Temperature variation of outermost layer and bottom panel in 2D model

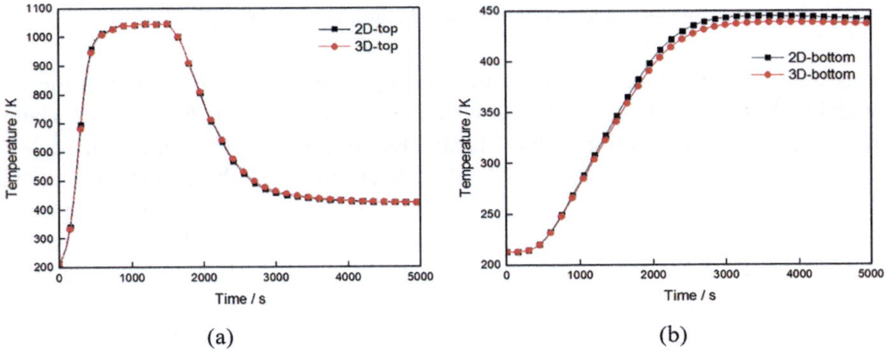

Fig. 6 Temperature comparison between 2D and 3D models

insulation layer thickness is increased to 32.5 mm, the maximum temperature of the bottom panel is 447 K, satisfying the temperature requirements for the internal structure.

Due to the inability of the 2D model to consider the influence of heat conduction perpendicular to the plane, this study uses a structure with an insulation layer thickness of 33 mm as an example. A comparison between the results of 3D and 2D models is conducted to validate the correctness of the 2D model through numerical simulation in 3D. A 3D model is established with an insulation layer thickness of 33 mm, and both the C/SiC outer panel and the aluminum alloy bottom panel have a thickness of 5 mm. The model is without boundary constraints, and the calculation time is set to 5000 s. The temperatures of the outer surface of the outer panel and the aluminum alloy bottom panel extracted from the 3D model are compared with those from the 2D model, as shown in Fig. 6.

The comparison shows that the highest temperatures and temperature variation trends of the outermost layer in the 3D model are consistent with those of the 2D model. The temperature trends of the aluminum alloy bottom panel are also

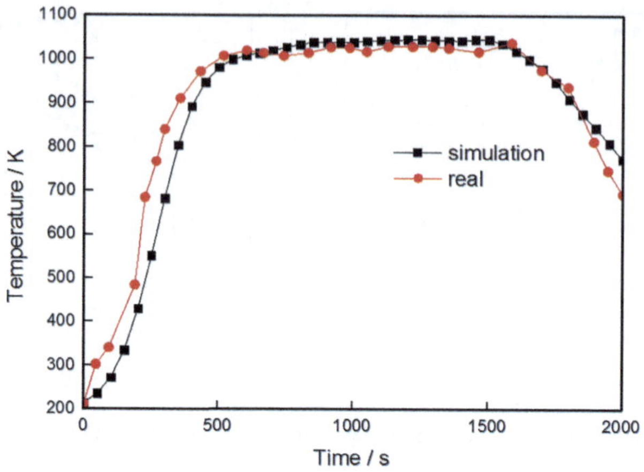

Fig. 7 Comparison of calculated results with actual temperature values

consistent, with the highest temperature in the 2D model at 445 K and in the 3D model at 439 K, a difference of 6 K, which is within 2% of the temperature value. The temperature distribution at 3000 s extracted from both the 2D and 3D models is shown in Fig. 7, with a difference of only 1 K at this time. The temperature of the bottom panel is 445 K in the 2D model and 438 K in the 3D model, a difference of 7 K, which also remains within 2% of the temperature value. Therefore, it is concluded that the 2D model can effectively represent the thermal conduction of the 3D structure and the temperature variation of the bottom panel.

Based on the previous analysis, it is evident that after 2000 s of reentry into the atmosphere, the temperature variations of the outer layer differ with changes in the insulation layer thickness. The choice of insulation material also affects the temperature variations after 2000 s. Therefore, the temperature variations of the outer surface of the C/SiC layer in the 3D model are compared with the temperature variations on the windward surface of a reentry vehicle as reported in [12]. Figure 7 illustrates this comparison, showing that the calculated results align with the actual trends observed. The maximum temperature values and their durations are consistent, demonstrating the accuracy of the computational approach in this study and its correspondence with the actual temperature variations on the outer surface of the thermal structure.

4 Optimization Analysis of Bracket Structure Form

Due to the discrepancy in temperatures between the outer panel and the internal structure, there exists a thermal expansion mismatch. If the stiffness of the bracket structure perpendicular to the thickness direction is too high, it may resist the deformation of the outer panel excessively, potentially leading to excessive deformation

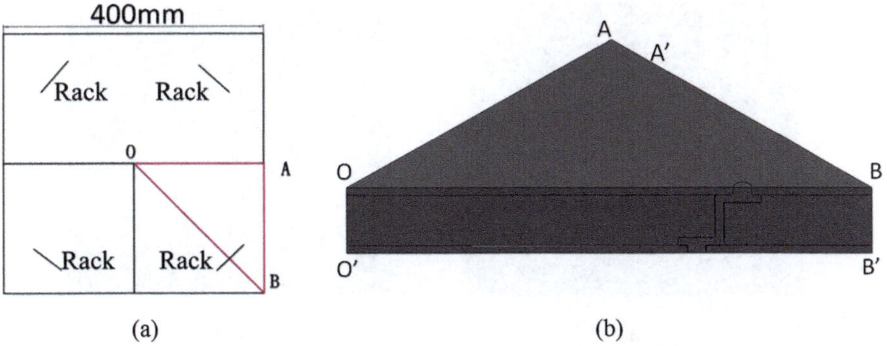

Fig. 8 Schematic diagram of bracket positions

or even plastic deformation. To ensure a certain amount of deformation tolerance for the outer panel in the perpendicular direction, the bracket structure is designed based on the distribution position of brackets described in [5], as shown in Fig. 8.

The dimensions of the single thermal protection structure are 400 mm per side, and the brackets are perpendicular to the diagonal. Given the symmetry of boundary conditions, loads, and material properties along the axes of symmetry of the square structure, a 1/8 model is established, as depicted by the red line segment OAB in Fig. 8b. Here, Points A′ and B′ are located on the underside of the bottom panel, directly below Point A.

The strength of the bolts has been extensively studied in previous sections. In this chapter, the upper and lower ends of the brackets are simplified to cylindrical bodies with a diameter of 10 mm, integrated into the bracket structure. The connection between the bolts and the panels as well as the brackets, is tight; therefore, the edges of the cylindrical bodies are tied to the holes, ensuring that the upper and lower parts of the brackets deform in line with the adjacent panels. The Saffil alumina fiber gel filled between the inner and outer panels is a flexible material, and its stiffness and strength are disregarded in this analysis. Hence, the interaction between the gel and the panels is simplified to a tied connection to facilitate calculation. Within the structure, all other components are treated as contact interactions, with normal direction rigidity and a friction coefficient of 0.2. Thermal impedance between components is not considered. Symmetrical constraints are applied to the OO′B′B and OO′A′A planes, along with constraints on the O′A′B′ plane in the thickness direction displacement and perpendicular displacement of Points O and O′. Thermal flux and pressure loads are applied on the OAB plane.

This study designs four types of brackets based on the Z-shaped structure, as shown in Fig. 9, aiming to investigate the influence of different load paths on the maximum stress of the brackets. By comparing the computational results, the optimal structure is selected. The boundary conditions and load settings are consistent with those described earlier, maintaining a vertical distance of 80 mm between the vertical part of the bracket and BB′. Among them, Structure 1, as shown in the right

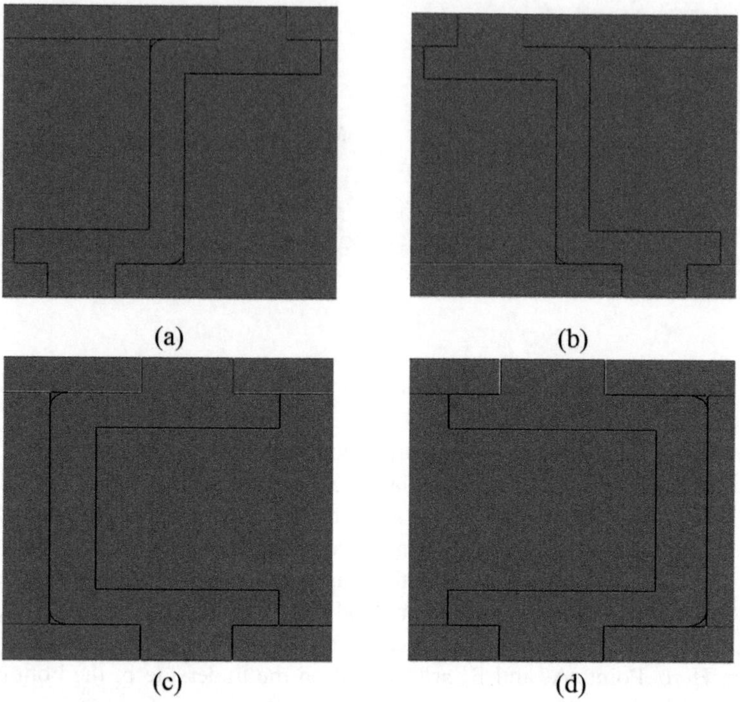

Fig. 9 Four types of support structures

diagram of Fig. 8, has a bracket mesh size of 1 mm, a thickness of 5 mm, and a distance of 33 mm between the outer panel and the bottom panel.

Four types of support structures' stress distribution at maximum stress are shown in Fig. 10. Due to the tie connection between the cylindrical bodies replacing bolts and the threaded holes on the panels, thermal expansion is restricted, causing stress to be misrepresented here. However, this does not affect the structural integrity of the supports themselves. Hence, the stress contours conceal the cylindrical bodies and the adjacent layer of elements. From the figure, it can be observed that the high-stress regions for all four support structures are located at the corner joints near the base, with Structure Three exhibiting the lowest stress and Structure Four experiencing the highest. The variation of maximum stress over time on the supports is depicted in Fig. 11. The curve "Structure-3-2" represents the change in maximum stress on the support structure without additional pressure load for Structure (c). The trends in maximum stress variation on the supports are similar in the initial 500 s, followed by a gradual increase to reach peak values, maintaining relatively high levels until around 1700 s, after which the stress begins to decline.

From the figure, it can be observed that the influence of pressure load on the maximum stress on the supports is relatively small. The primary reason for the presence of high-stress zones on the supports is that the temperature of the outer panel is higher than that of the bottom panel, causing inconsistent deformation in the

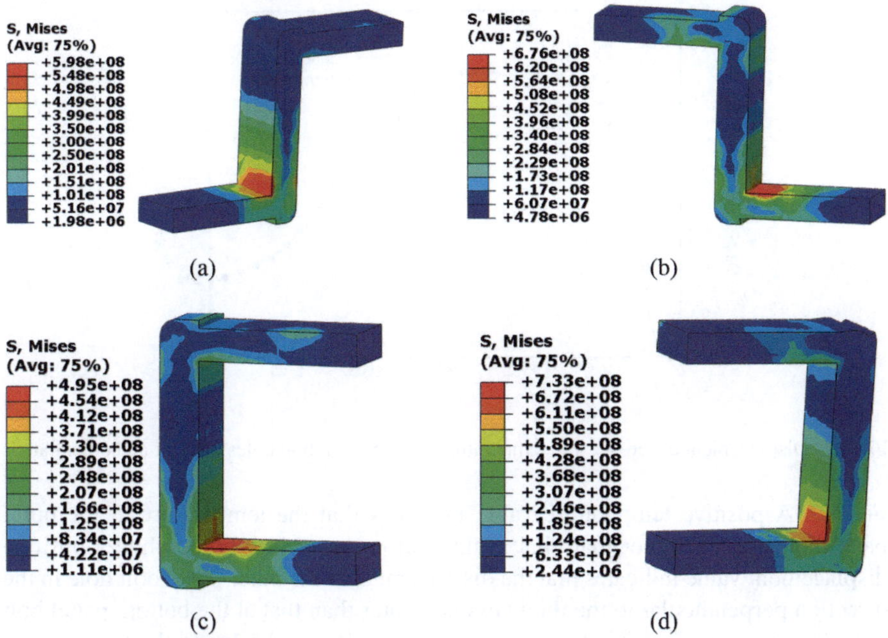

Fig. 10 Stress distribution of four types of supports at maximum stress

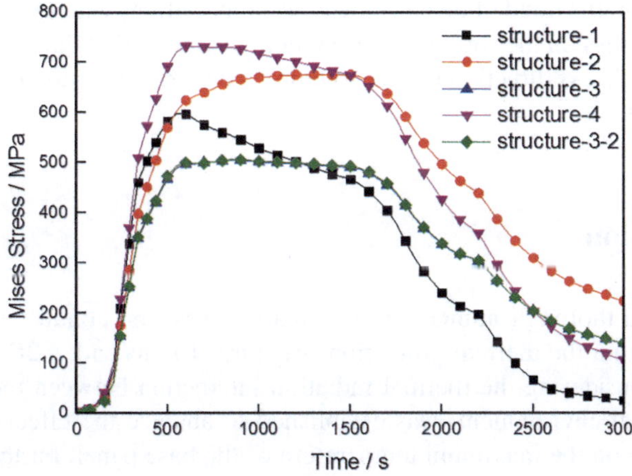

Fig. 11 Variation of maximum stress on the support structure

thickness direction of the structure. As the supports undergo thermal expansion, they resist this inconsistency by bearing complex forces such as bending and compression, resulting in the formation of high-stress zones.

The temperature difference and displacement difference perpendicular to the thickness direction at the bolt holes of the inner and outer panels are shown in

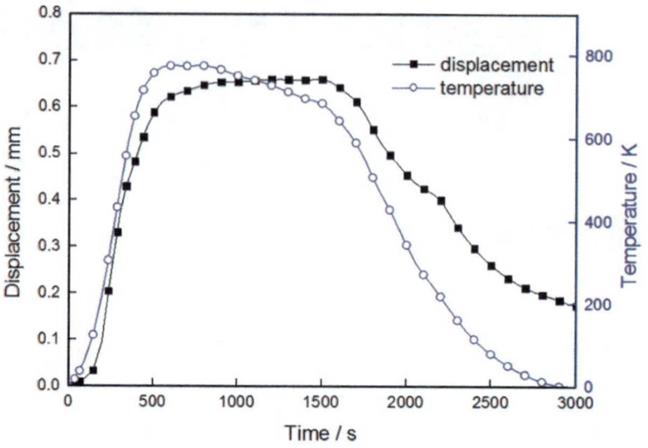

Fig. 12 Displacement difference and temperature difference at bolt holes of inner and outer panels

Fig. 12. A positive temperature value indicates that the temperature at the outer panel bolt hole is higher than that at the bottom panel bolt hole, while a positive displacement value indicates that the displacement at the outer panel bolt hole in the direction perpendicular to the thickness is greater than that at the bottom panel bolt hole. The appearance of displacement difference is mainly due to the temperature gradient in the thickness direction of the structure. The trend of displacement difference in the figure is consistent with the temperature difference, confirming this point. Moreover, the trend of displacement difference aligns with the variation trend of maximum stress on the supports, further demonstrating that the temperature gradient in the thickness direction of the structure is the primary cause of higher stress on the supports.

5 Conclusion

In this study, a thorough numerical investigation was first conducted on the heat conduction within the thermal protection structure. To this end, a 2D heat conduction model, considering the thermal radiation interaction between the outer panel and the external environment, was established to analyze the effect of insulation layer thickness on the maximum temperature of the base panel. Furthermore, verification of the 2D model was conducted by comparing its numerical results with both the outcomes of a 3D model and actual temperature data. Results revealed that the insulation layer's thickness has a pronounced influence on the temperature of the bottom panel, and specifically, a thickness of 32.5 mm successfully fulfills the temperature requirements for the internal structure.

Secondly, a comprehensive optimization analysis is conducted on the design of the bracket structure utilizing simulations. Four distinct Z-shaped designs were modeled to analyze their stress distribution patterns. Upon examination, high-stress zones were pinpointed at the root corners, exhibiting variations in stress intensities across different designs. Notably, the temperature gradient along the thickness of the structure emerged as the primary contributor to elevated stress levels, offering crucial insights for the design and optimization of thermal protection structures.

The present research provides theoretical support and practical reference for the design and optimization of thermal protection structures in aerospace vehicles. To enhance the support structure's overall performance, future research endeavors might involve a more profound exploration of the impact that additional environmental variables, particularly wind speed and humidity, exert on heat conduction mechanisms and the intricate patterns of stress distribution.

Acknowledgments This work was financially supported by the National Natural Science Foundation of China (Grant Nos. 12002264, 12141203), the project of Shaanxi Province Key Research and Development Plan (Grant No. 2024GH-ZDXM-30), and Fundamental Research Funds for the Central Universities (Grant No. D5000230140).

References

1. Wang X, Xu X, Yang QC (2024) Numerical analysis on the thermal environment of reusable launch vehicle during supersonic retropropulsion. Int J Therm Sci 198:108857
2. Ma ZH (2004) Performance analysis and research on thermal protection system of reusable launch vehicles (in Chinese). Northwestern Polytechnical University Press, Xi'an
3. Balesdent M, Brevault L, Paluch B (2023) Multidisciplinary design and optimization of winged architectures for reusable launch vehicles. Acta Astronaut 211:97–115
4. Blosser ML (2003) Thermal protection systems for reusable launch vehicles. In: 14th annual thermal and fluids analysis workshop, Hampton, pp 182–188
5. Zhang CY, Yu P, Peng HF (2024) Prediction of thermal contact resistance for reusable heat-pipe cooled thermal protection system based on an inverse thermos-mechanical coupling method. Renew Energy 227:120541
6. Izaguirre I, Loewenhoff T, Prado DJ (2023) Thermal fatigue response of W-Eurofer brazed joints by the application of high heat flux loads. J Mater Process Technol 319:118056
7. Zhang SL, Zhang ZY, Wang Y (2024) Design and performance verification of thermal protection structure of high-temperature probe in aeroengine. Appl Therm Eng 253:123745
8. Brociek R, Hetmaniok E, Napoli C (2024) Identification of aerothermal heating for thermal protection systems taking into account the thermal resistance between layers. Int J Heat Mass Transf 218:124772
9. Zhang XL, Liang MK, Sun FH (2024) Thermal insulating performance of thermal protection system with sandwich structure based on Al_2O_3/SiC composites aerogels. Ceram Int 50(18B):33946–33952
10. Sergis A, Hardalupas Y, Flinders K (2023) A quantitative study on the thermal performance of self-modified heat transfer surfaces in high heat flux flow systems. Int J Heat Mass Transfer 215:124525
11. Blosser ML, Chen RR, Schmidt IH (2002) Advanced metallic thermal protection system development. In: 40th aerospace sciences meeting & exhibit, Reno, pp 14–17
12. Böhrk H, Beyermann U (2010) Secure tightening of a CMC fastener for the heat shield of re-entry vehicles. Compos Struct 92(1):107–112
13. Blosser ML (2000) Advanced metallic thermal protection systems for reusable launch vehicles. University of Virginia Press, Hampton

14. Tsirogiannis EC, Psarommatis F, Prospathopoulos A (2024) Composite armor philosophy (CAP): holistic design methodology of multi-layered composite protection systems for armored vehicles. Defence Technol 14:7–29
15. Gale WF, Totemerier TC (2003) Smithells metals reference book, 8th edn. Elsevier, Amsterdarn
16. Tongji University (2007) Design specification for aluminum alloy structures (in Chinese)
17. Zhang C, Guan TH, Ren TF (2022) Influences of SiC infiltration and coating on compressive mechanical behaviors of 2D C/SiC composites up to 1,600°C at wide-ranging strain rates. J Eur Ceram Soc 42(9):3787–3801
18. Zhang BF (2010) A study on the intrinsic and service performance of C/SiC composite materials with precursor transformation. University of National Defense Technology Press, Changsha
19. Blosser ML, Poteet CC, Chen RR (2004) Development of advanced metallic-thermal-protection system prototype hardware. J Spacecr Rockets 41(2):183–194

Open Access This chapter is licensed under the terms of the Creative Commons Attribution-NonCommercial-NoDerivatives 4.0 International License (http://creativecommons.org/licenses/by-nc-nd/4.0/), which permits any noncommercial use, sharing, distribution and reproduction in any medium or format, as long as you give appropriate credit to the original author(s) and the source, provide a link to the Creative Commons license and indicate if you modified the licensed material. You do not have permission under this license to share adapted material derived from this chapter or parts of it.

The images or other third party material in this chapter are included in the chapter's Creative Commons license, unless indicated otherwise in a credit line to the material. If material is not included in the chapter's Creative Commons license and your intended use is not permitted by statutory regulation or exceeds the permitted use, you will need to obtain permission directly from the copyright holder.

Research on Digital Process Design of Display and Control Equipment Based on Teamcenter

Mingchun Zhou

Abstract To enhance the guidance of 3D models for actual production and realize the process digitalization of display and control equipment, a digital design architecture of display and control equipment based on Teamcenter is proposed to analyze process design planning and BOM construction, which can effectively ensure the consistency of product physical objects and design drawings and enhance the stability of product process quality. In display control equipment processing, assembly example verification aims to improve the process of production practice guidance ability.

Keywords Teamcenter · Digital process design · Display and control equipment · BOM

1 Introduction

The display and control equipment can realize the needs of information processing, operation, control, and human-computer interaction. It has rich functions, strong integration, and many wiring harness connection points, and the overall hardware platform is more complex. The equipment consists of a large number of parts, complex modeling, specifications, and models, cumbersome processing, heavy processing technology design, subject to external space restrictions and product size requirements, complex assembly relations, and limited wiring harness space layout, so assembly process design is more difficult. At present, the process design of display and control equipment is still mainly based on two-dimensional engineering drawings and combined with technical requirements and other text descriptions, the degree of digitization is low, the intuitiveness is poor, and the manufacturing intention of the drawing itself cannot be quickly conveyed, and the technical level of the operator is high, which is easy to cause rework and modification in the later period,

M. Zhou (✉)
Jiangsu Automation Research Institute, Lianyungang, Jiangsu, China

© The Author(s) 2026

Z. Zhang, P. Mercorelli (eds.), *Mechanical Design and Simulation: Insights and Innovations*, Lecture Notes in Mechanical Engineering,
https://doi.org/10.1007/978-981-95-2405-1_46

directly affecting the production schedule of the product and increasing the production cost. With the development of technology, digital design based on 3D models has become the development trend in the manufacturing industry. Aerospace enterprises are relatively advanced in digital construction and can realize 3D digital design and management of flight products to a certain extent [1–3]. However, the digital development of display and control equipment in the shipping industry started late and has not yet realized the digital process design of the whole process. Compared with the production of fine management, short cycle delivery requirements are far away, and cannot meet the needs of use. Therefore, this paper optimizes and improves the digital process design of a certain display and control equipment based on Teamcenter, which can effectively ensure the consistency of product physical objects and design drawings, enhance the stability of product process quality, strengthen the guiding role of technology in the manufacturing process, and improve batch production efficiency.

2 Digital Process Design Data Integration Based on Teamcenter

The PLM (Product Lifecycle Management) system used now is the Siemens Teamcenter system, which has been deeply used in various design processes and production links. The system can effectively realize the design model, process information, flow information, and other product design and production process data, through the integrated interface, to realize the digital design and manufacturing of products.

2.1 *System Architecture Design*

To improve the quality of product process design, a digital process design framework based on Teamcenter is proposed, as shown in Fig. 1, including an integrated interface, interactive application, business function, and basic data. Through the Teamcenter system framework, we achieve digital design and process comprehensive control, to achieve integration with NX/CAD and other software, MDM/MES/ERP, and other systems. The interactive application provides users with a visual operation window to complete process design and simulation. A business function is the business component that realizes the system function, including BOM construction, process design, three-dimensional annotation, document management, information summary, etc. Basic data includes tooling, equipment, and material information, including process knowledge, process parameters, etc., to facilitate the implementation of process invocation [4].

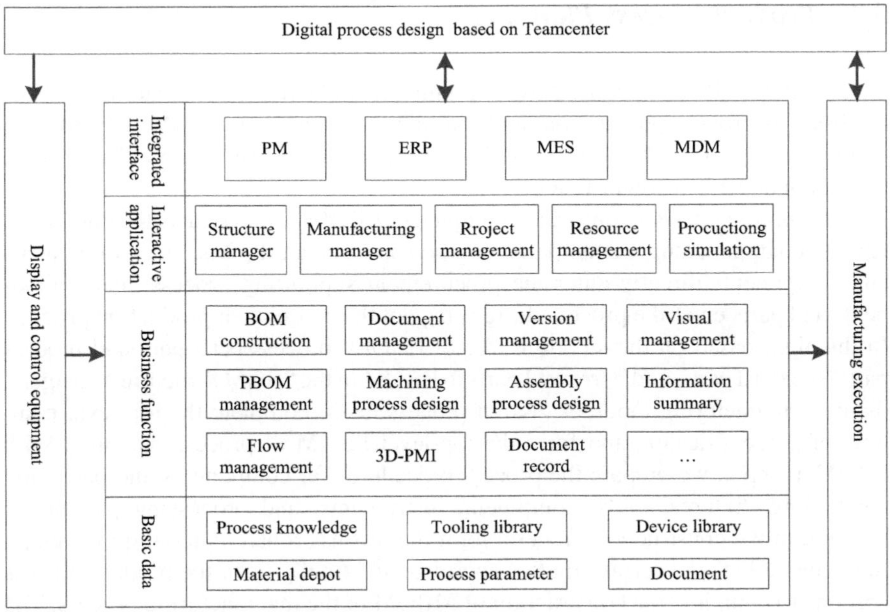

Fig. 1 Digital process design framework based on Teamcenter

2.2 *Process Resource Database*

Process resource library is an important part of digital process design, which is used to store and call process resources uniformly. Based on the classification management of Teamcenter, a classification library of materials, equipment, tooling, tools, instruments, measuring instruments, typical processes, process templates, process parameters, etc., has been constructed. These resources are stored in each classification database and have corresponding unique codes, which can be selected by classification query or search in the resource database.

3 Digital Process Design Process of Display and Control Equipment

The digital process design of display and control equipment is based on the three-dimensional model of the product and its dependency relationship as the path to carry out the process work planning, parts processing, and assembly design while combining the process simulation and simulation to verify the production process.

3.1 Product Process Planning

Process BOM (PBOM) is the core of product manufacturing and data integration, the basis of process preparation, and the link between design BOM (EBPM) and manufacturing BOM (MBOM). EBOM is the database of PBOM, and PBOM is the derivation and extension of EBOM data.

After product design approval is completed, EBOM is released as the data basis. EBOM contains components, parts, components, standard parts, etc. Components and components directly enter the procurement supporting process, and components and parts enter the process design stage of the subsequent production process. In this link, based on the structure tree of EBOM, process personnel add process parts, standard parts, and borrowed parts to EBOM in the "BOM Structure Manager", divide manufacturing modules, assign process tasks, complete the top-level planning of process design, and form the top-level PBOM of process planning. With PBOM as input, we prepare the process procedures for components and parts, add the tooling, fixtures, tools, equipment, instruments, and processing parameters required in the construction process, form the process route, material quota, material demand list, and release it after approval, and finally form the product process baseline to complete the construction of MBOM in the manufacturing process. The BOM construction and transformation process is shown in Fig. 2 [5].

3.2 Product Process Design

According to the traditional two-dimensional engineering drawing process design mode, the two-three-dimensional hybrid digital design and manufacturing mode of two-dimensional engineering drawing + three-dimensional model is formed to realize the three-dimensional process design based on MBD. The realization method of "table + three-dimensional model" is adopted, which not only takes into account the production and application requirements of existing two-dimensional drawings but also realizes the three-dimensional numerical control machining. To adapt to the production status of multi-variety and small-batch product models, it can meet the production needs of short-cycle delivery [6–8].

For the process design of parts, the process personnel first confirm the raw material model and material quota according to the model data, and formulate the processing flow according to the characteristics of the parts, such as "heat treatment—number milling—electroplating—spraying", and add process parameters, tooling equipment, tools and measuring tools, operating requirements, precautions, inspection requirements, etc., according to the operational requirements of each process. NX interface is used to call the design model for correction, and NXWave technology, synchronous modeling technology, and PMI 3D labeling technology are used to correct the model, forming the 3D process median difference

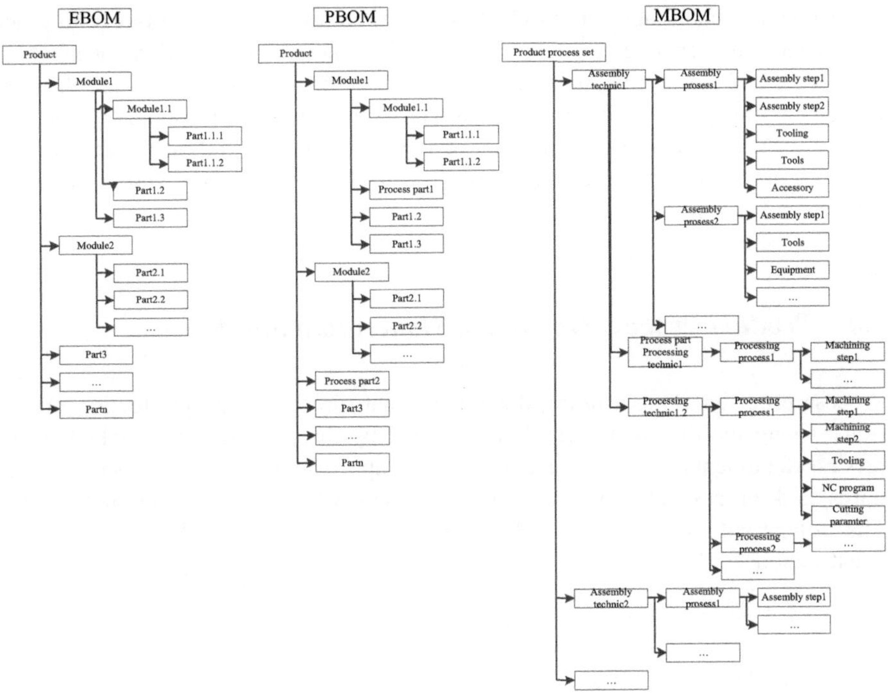

Fig. 2 BOM construction and transformation

model required by each corresponding process. Based on the process procedures and the middle difference model, the programmer uses the NX CAM module to prepare the CNC machining tool track file under the process, generates the NC code program, attaches it to the corresponding process, and simulates the processing program, and sends it to the corresponding machine tool after confirming that it is correct.

For the process design of components, the assembly steps are first divided and the assembly process procedures are designed, including the assembly method, assembly object (middleware and consumption materials), assembly tools and equipment, assembly order, and other information. If necessary, the process will be refined into work steps to further clarify the details of the work content [9].

3.3 *Product Processing Simulation*

Part machining simulation and optimization is based on a virtual machining environment, CAM programming foundation, machine tool model library, tool model library, cutting parameter library, and other processing resources. We select the corresponding manufacturing model, tool type, cutting speed, feed rate, and other parameters, and generate the tool path. Through the simulation function, the machining process of the tool is simulated according to the generated path, and the tool movement trajectory, cutting state, and possible interference and collision are observed. According to the simulation results, the tool path and machining parameters are optimized and adjusted to improve machining efficiency and avoid interference and collision [10, 11].

3.4 Product Process Document Generation and Release

Process procedures are guiding documents that directly guide production operations. Using the document template provided by the platform, a complete set of process documents is formed and signed after approval, including a process procedures card, process drawings card, process details card, process material card, etc. The integrated interface to ERP, MES, etc., directly guides the operator construction.

4 Application Cases

Taking a certain type of display and control equipment as an example, as shown in Fig. 3, the digital process design is carried out based on the digital process design platform of display and control equipment.

Through statistical analysis, using a digital process design platform, the process design efficiency of display and control equipment has been greatly improved, the process documents are more standardized and more guiding, and the final assembly efficiency of equipment has been increased by 28%, and the time saving of some components is shown in Table 1.

Fig. 3 Digital process design process of a display and control equipment. (**a**) EBOM of a display and control device, view the composition structure in the structure manager, and adjust the PBOM as needed; (**b**) Preparation of process procedures based on PBOM, completion of the process sequence, material quota, tooling and fixtures, and other resources, and construction of MBOM to guide production; (**c**) Addition of a three-dimensional mean difference model to the processing process and completion of PMI labeling; (**d**) Program and process program simulation based on NX CAM to verify program correctness; (**e**) Division of the assembly nodes and improvement of the assembly process and resource summary; (**f**) Fine chemical steps according to needs form a "table + three-dimensional model" process rules and issue to ERP and MES systems after publication

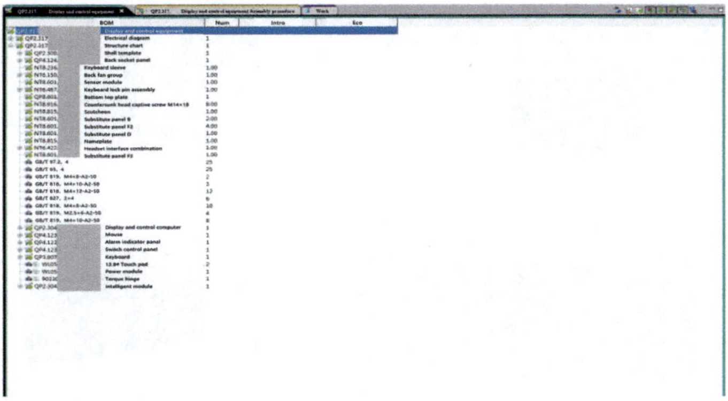

a

b

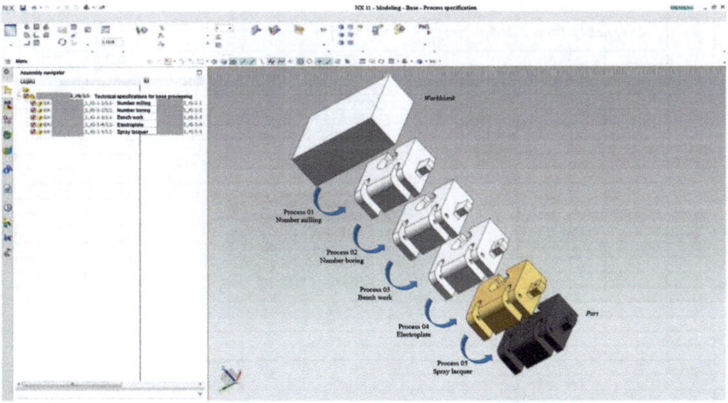

c

(continued)

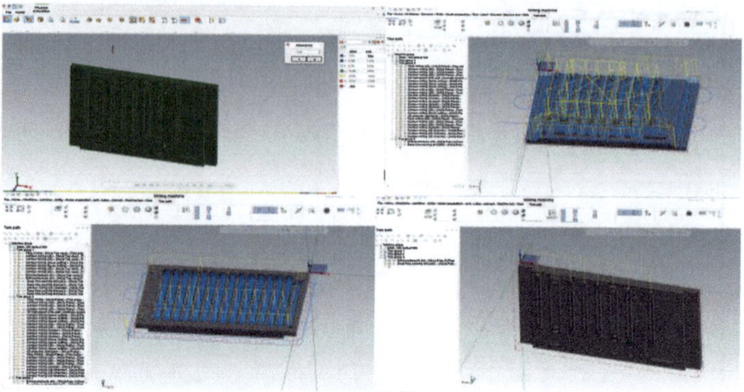

d

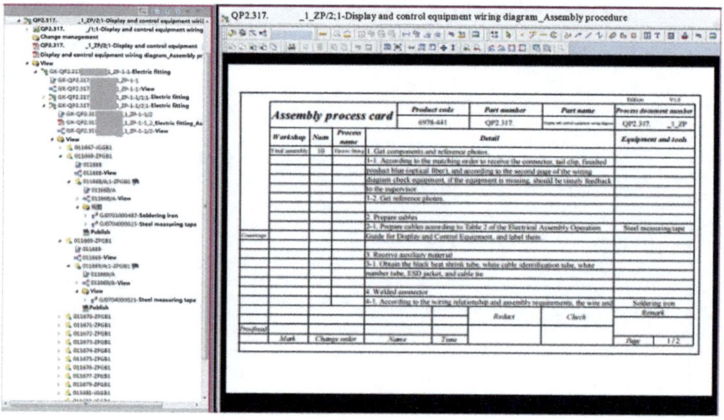

e

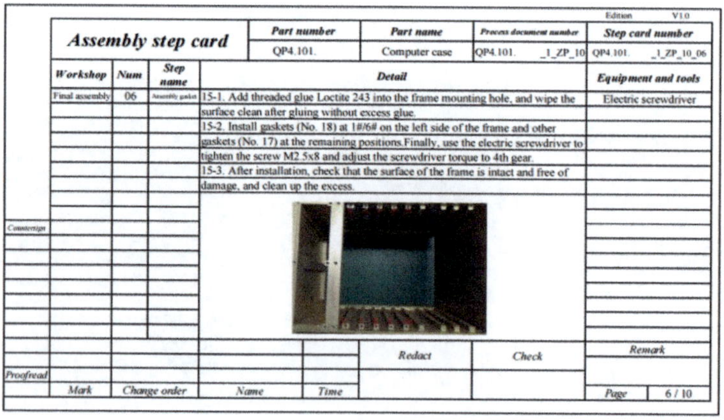

f

Fig. 3 (continued)

Table 1 Production time comparison

Category	Before (h)	After (h)	Time-saving rate (%)
Front-panel	4	1.2	70.0
Locking shaft	4.8	1.56	67.5
Lock	0.9	0.4	55.6
Case frame	24	12	50.0
Printed board	24	13.6	43.3
Transition plate	108	67.2	37.8
Display mounting board	960	720	25.0
Display bracket mounting seat	4224	3192	24.4
Equipment assembly	522	408	21.8

5 Conclusion

Digital process design and manufacturing of display and control equipment is an important way to develop high-quality products in the new era and an important means to achieve rapid delivery. Based on Teamcenter, this paper builds a digital process design platform for display and control equipment, analyzes the transformation from EBOM to PBOM to MBOM, analyzes the digital process design process of parts processing and whole machine assembly of display and control equipment, and takes parts and assembly of a display and control equipment as an example to illustrate its application. Through the application of a structured system, the digital level of process design is improved, which can better meet the needs of product design and manufacturing, and has certain promotion significance in the Marine electronic equipment industry. In the future, based on the existing platform and combined with MBD, further engineering research can be carried out on assembly simulation, and product rationality review and operation process simulation can be realized through full factor simulation, to further guarantee product production quality.

References

1. Shao CY, Ji JP, Guo Y, Liu SW (2023) Research and application of digital technology based on 3D aircraft models. Mech Eng (4):119–122
2. Xin ZY (2023) Research on simulation technology based on aircraft digital assembly process. Technol Innov Appl (19):114–117. https://doi.org/10.19981/j.CN23-1581/G3.2023.19.026
3. Guan XQ (2020) 3D digital assembly process system analysis. China Metal Bull (1):288–290
4. Liu YJ, Li WH, Pu Y, Yang Q, Liu L, Ying JQ, Wang ZP (2022) Research on 3D digital assembly process design technology of projectile and arrow products based on Teamcenter. Mach Tool Hydraul (1):117–120
5. Liu Y (2023) Research on three-dimensional digital construction process expression based on master's degree (Dissertation, Nanchang University). Master of https://link.cnki.net/doi/10.27232/d.cnki.gnchu.2023.003361

6. Xu B (2023) Application of digital assembly technology in electronic equipment production. China Steel Focus (7):18–20
7. Li D, Sun B (2021) Exploration of a new model of integrated R&D of design process based on Teamcenter. New Technol New Process (1):16–22. https://doi.org/10.16635/j.carolcarrollnki.1003-5311.2021.01.004
8. Cao J (2023) Development and research of a structured process preparation system based on Teamcenter. Mech Electr Inf (17):85–88. https://doi.org/10.19514/j.carolcarrollnkicn32-1628/tm.2023.17.023
9. Lou Q, Jiao GM, Li WS, Lin ZL, Zhao Z, Ye R, Zhang L (2022) Structured process design process based on MBD model. Inf Record Mater (6):210–214. https://doi.org/10.16009/j.carolcarrollnkicn13-1295/tq.2022.06.023
10. Liu H, Zeng H, Yang L (2022) Digital simulation machining of 3D solid parts based on CAM. J Gansu Normal Coll (2):21–26
11. Meng Q, Liu JH, Tang CT, Zhou ZY (2014) Research and application of process-based structured assembly process design technology for complex products. Comput Integr Manuf Syst (2):301–312. https://doi.org/10.13196/j.carolcarrollims.2014.02mengqian.0301.12.2014029

Open Access This chapter is licensed under the terms of the Creative Commons Attribution-NonCommercial-NoDerivatives 4.0 International License (http://creativecommons.org/licenses/by-nc-nd/4.0/), which permits any noncommercial use, sharing, distribution and reproduction in any medium or format, as long as you give appropriate credit to the original author(s) and the source, provide a link to the Creative Commons license and indicate if you modified the licensed material. You do not have permission under this license to share adapted material derived from this chapter or parts of it.

The images or other third party material in this chapter are included in the chapter's Creative Commons license, unless indicated otherwise in a credit line to the material. If material is not included in the chapter's Creative Commons license and your intended use is not permitted by statutory regulation or exceeds the permitted use, you will need to obtain permission directly from the copyright holder.

Study on the Tensile Mechanical Properties of Residual Film in Non Ridge Direction

Dingguo Zhang, Zhe Song, and Yaxin Zhang

Abstract Aiming at the issue of the bottleneck in enhancing the picking rate, a methodology is devised to boost the retrieval rate of mechanical residual film collectors, leveraging the mechanical property traits of the working target. The 0.01 mm thick plastic film prevalently utilized in the semi-arid regions of Northeast China was chosen as the specimen for investigation. Specialized testing modalities were employed to conduct a comprehensive and detailed examination of the mechanical attributes of the residual film across diverse orientations. In the course of the experiment, it was ascertained that the mechanical characteristics of the film were conspicuously influenced by the pre-tension during sowing processes. Notably, the residual film attains its peak tensile strength at an angle of around 40.6° relative to the ridge direction. This crucial datum forms the central foundation for augmenting the picking rate of residual film pickers and can serve as a guideline for the design, structural refinement, and operational parameter optimization of such apparatus.

Keywords Agricultural engineering · Residual film tensile strength · Tensile strength coefficient · Non monopolistic direction

D. Zhang
School of Biological Sciences and Technology, Yili Normal University, Yining, China

Xinjiang Key Laboratory of Lavender Conservation and Utilization, Yining, China

Z. Song
School of Automation and Electrical Engineering, Inner Mongolia University of Science and Technology, Baotou, China

Y. Zhang (✉)
Xinjiang Key Laboratory of Lavender Conservation and Utilization, Yining, China

School of Physics and Intelligent Manufacturing Engineering, Chifeng University, Chifeng, China

© The Author(s) 2026

Z. Zhang, P. Mercorelli (eds.), *Mechanical Design and Simulation: Insights and Innovations*, Lecture Notes in Mechanical Engineering,
https://doi.org/10.1007/978-981-95-2405-1_47

1 Introduction

The mechanical attributes of residual film are of vital significance in residual film collection activities and exert a non-negligible influence on the retrieval rate of residual film collectors. In the practical process of residual film retrieval, it was discovered through in-depth probing that there are substantial disparities in the mechanical characteristics manifested by residual films in different directions. Additionally, as time elapses, the mechanical properties of residual film will experience remarkable alterations. Among them, the mechanical features chiefly affecting residual film collection work are mainly manifested in tensile and tearing properties [1, 2].

During the utilization of plastic film, a multitude of small cracks will gradually emerge, and the orientation of these cracks is highly random. When the residual film is torn, its tearing direction frequently follows the path of crack propagation. Based on this feature, from a statistical viewpoint, the tearing performance of residual film does not have a pronounced impact on the picking rate of residual film, which has been corroborated by relevant research literature [3–5]. Conversely, the tensile performance of the residual film becomes the central mechanical property factor influencing the residual film picking rate. Further research has indicated that the tensile properties of plastic films are restricted by elements such as the angle between the picking direction and the residual film laying direction, as well as the laying duration.

In light of this, an in-depth exploration of the mechanical properties of spineless residual films is undeniably of paramount value and significance for enhancing the residual film pickup rate. The associated research findings will furnish a robust and dependable theoretical underpinning for the innovation of design philosophies, the optimization and refinement of structural arrangements, and the scientific and rational configuration of operational parameters for residual film pickers. This will actively propel the advancement of residual film recycling technology and make a substantial contribution to safeguarding the agricultural ecological milieu.

1.1 Sample Production

According to the plastic film standard (GB13735-1992 "Polyethylene Blown Plastic Agricultural Ground Cover Film") 26, a 0.01 mm thick plastic film widely used in the central and eastern parts of Inner Mongolia was selected for testing. Make tensile specimens and mark them, ensuring that there are no visible damages on the surface and smooth edges without burrs. The standard dimensions of A-type transverse and longitudinal tensile specimens are shown in Fig. 1. The longitudinal and transverse specimens are taken with a width of 20 mm and a length of 150 mm, and the distance between the two marked lines in the middle part of the clamp is 100 mm.

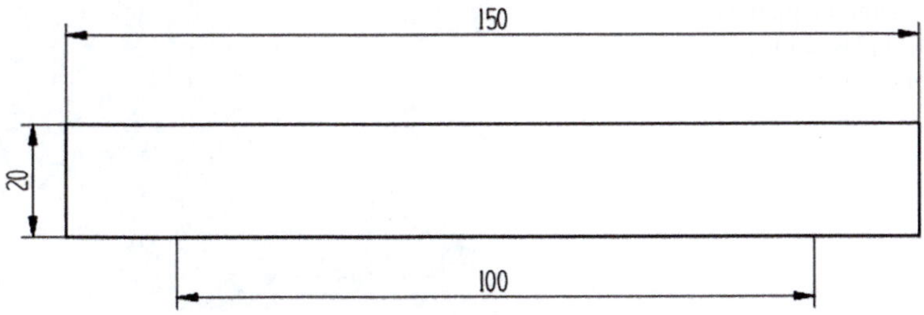

Fig. 1 A-type transverse and longitudinal tensile specimen

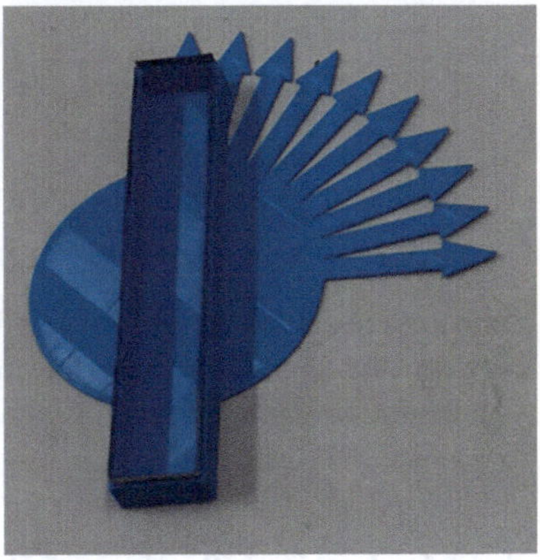

Fig. 2 Residual plastic film sampler with inclination direction

1.2 Sampler Design

In accordance with the plastic film standard (GB13735-1992 "Polyethylene Blown Plastic Agricultural Ground Cover Film"), a 0.01 mm thick plastic film commonly employed in the central and eastern regions of Inner Mongolia was chosen for testing. Tensile specimens were fabricated and labeled, with the surface being free from visible defects and the edges being smooth without burrs. The standard dimensions of A-type transverse and longitudinal tensile specimens are illustrated in Fig. 2. The longitudinal and transverse specimens were obtained with a width of 20 mm and a length of 150 mm, and the distance between the two marked lines in the central part of the clamp was set at 100 mm.

1.3 Test Piece Clamping

The mechanical stretching apparatus selected is the JHY-5000 single-arm mechanical testing machine manufactured by Xiamen Jinheyuan Technology Co., Ltd. During the experiment, in order to guarantee that the residual film specimen is free

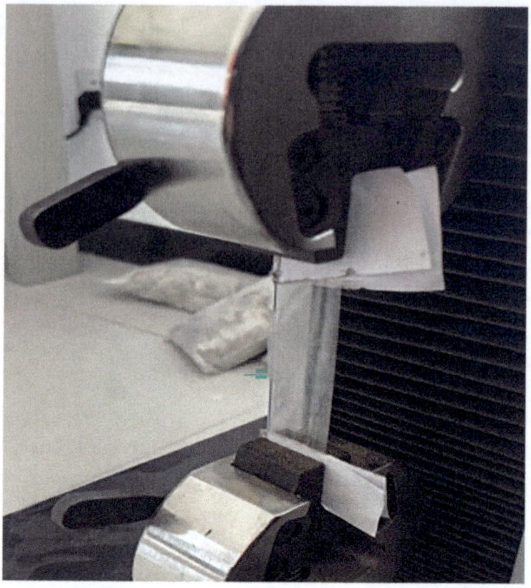

Fig. 3 Clamping method of residual plastic film tensile specimens

from torsion and has a flat surface when clamped, a pair of origami strips are inserted between the sample and the fixture, as depicted in Fig. 3. The stretching clamp secures the origami, and the origami immobilizes the residual film via friction.

2 Mechanical Testing of Residual Film Under Different Directions of Tension

2.1 Materials and Methods

The experimental field is situated in Yikenzhong Township, Ningcheng County, Chifeng City, which represents a typical semi-arid agricultural region in Northeast China. Seven levels of data are selected at angles of 0°, 15°, 30°, 45°, 60°, 75°, and 90° relative to the ridge planting direction. Eight samples are chosen for each level for testing, and the average tensile strength and elongation of the samples are computed based on each set of valid data.

2.2 Results

The tensile mechanical properties of residual plastic film at various angles from the ridge planting direction are presented in Fig. 4.

The tensile strength and elongation rate of residual plastic film at different angles with respect to the ridge direction are shown in Tables 1 and 2.

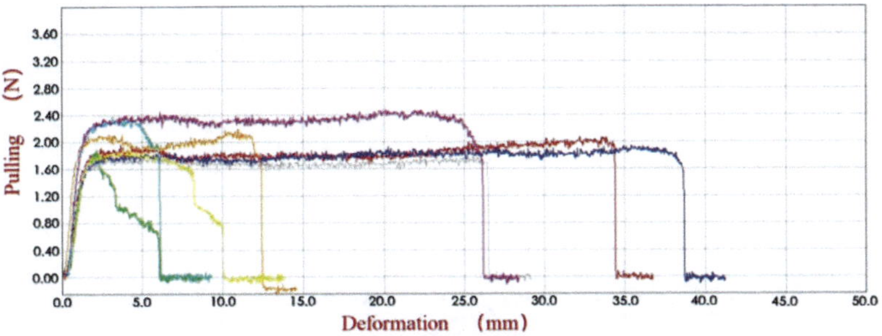

At a 0 degree angle to the direction of ridge cultivation

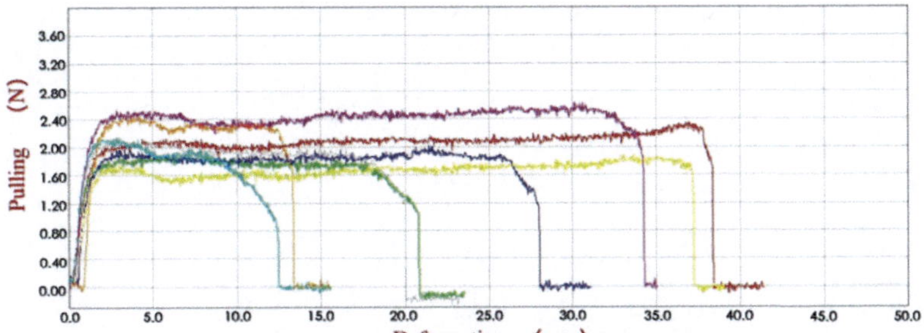

At a 15 degree angle to the direction of ridge cultivation

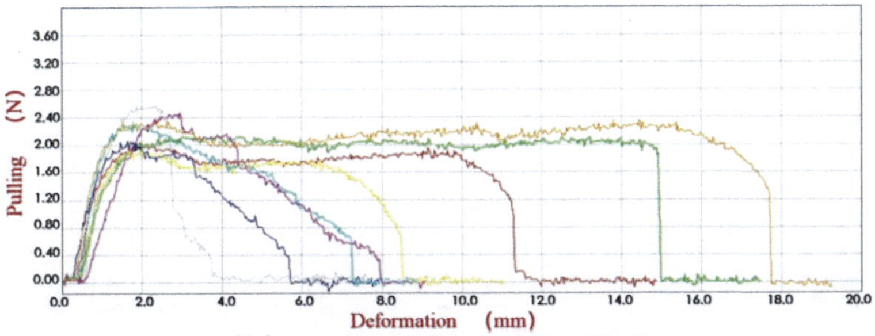

At a 30 degree angle to the direction of ridge cultivation

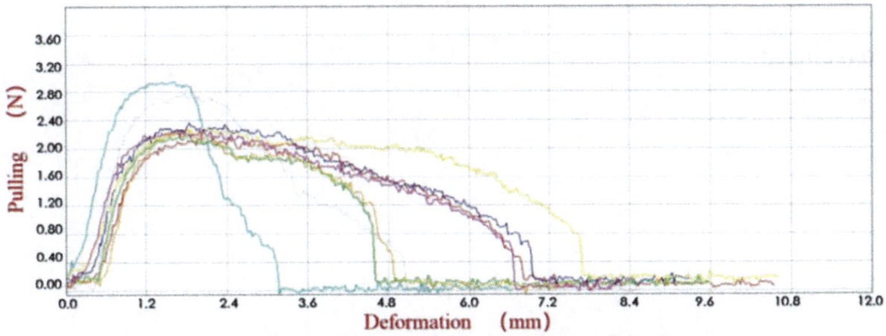

At a 45 degree angle to the direction of ridge cultivation

Fig. 4 Tensile curves of residual plastic film in different directions

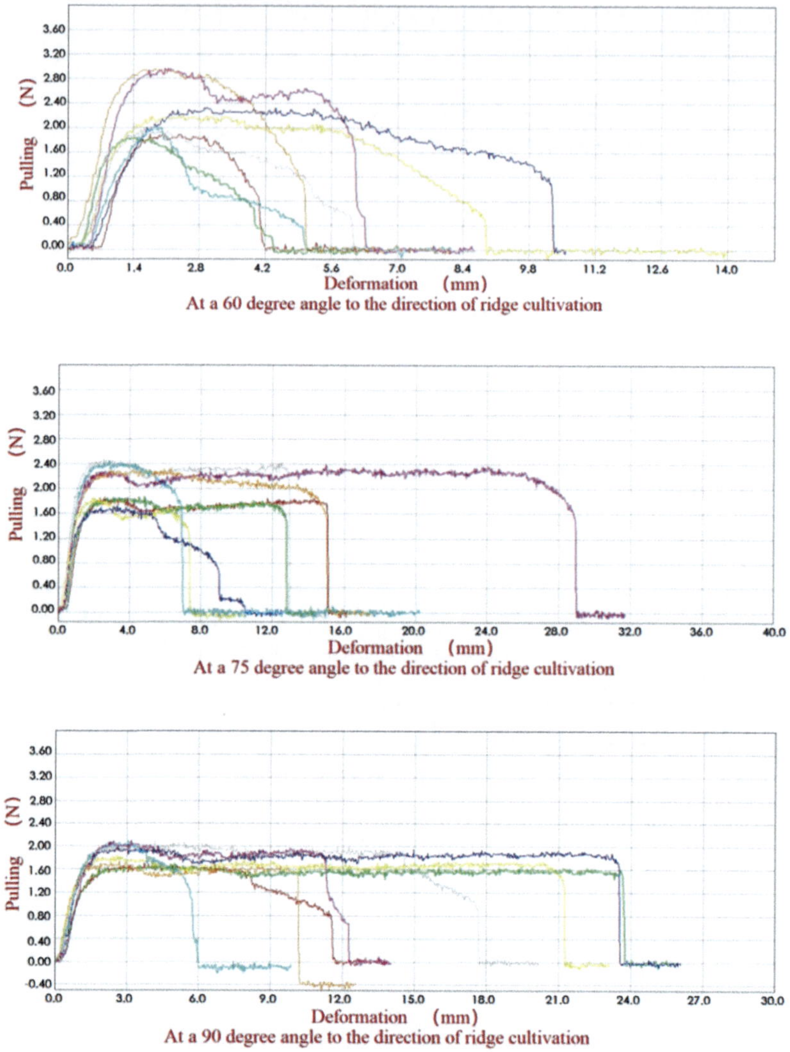

Fig. 4 (continued)

Table 1 Tensile strength of residual plastic film at different angles to ridge direction

Angle with ridge planting direction (°)	Tensile strength (MPa)								Average (MPa)
0	10.4	9.4	9.7	9.6	9	11.9	12.4	11	10.425
15	11.7	9.4	10.1	9.4	10.5	10.7	13.2	12.3	10.9125
30	10.1	10.6	10.3	9.8	12.9	11.7	12.3	11.9	11.2
45	10.8	11	11.8	11.5	13.9	14.8	11.4	11.3	12.0625
60	9.5	9.2	11.6	11.1	9.9	10.1	14.8	14.8	11.375
75	9.3	9.4	8.6	9.3	12.4	12.2	12	11.6	10.6
90	8.5	8.5	9.9	9.2	10.4	10.5	10.5	8.6	9.5125

Table 2 Elongation rate of residual plastic film at different angles to ridge direction

Angle with ridge planting direction (°)	Elongation rate (%)								Average (%)
0	36.8	8.7	41.2	13.8	29.1	9.3	28.3	14.5	22.71
15	41.4	23.5	31.1	39	23.2	15.6	35	15.6	28.05
30	14.9	17.5	7.9	11	7.6	8.9	9.1	19.2	12.01
45	10.5	9.6	9.3	10.6	8.5	6.6	8.3	7.5	8.86
60	6.6	6.4	10.6	14.1	8.7	8.1	8.6	8.1	8.90
75	16.2	14.4	12.5	10.4	14.8	20.2	31.7	17.5	17.21
90	13.9	26	23.1	20.2	9.8	14.0	12.5	26.1	18.20

3 Statistical Analysis

Given the complex working conditions of plastic film coverage, the mechanical characteristics of residual plastic film possess a high degree of randomness. The tensile test results of 0.01 mm thick residual plastic film in the central and eastern parts of Inner Mongolia at different angles to the ridge planting direction were analyzed using Design Expert software, as illustrated in Table 3. The P-value of the non-ridge tensile strength of residual plastic film is 0.0366; the P-value of the ridge elongation test of residual plastic film is less than 0.0001, and a P-value less than 0.05 implies a significant statistical difference.

The relationship between the tensile strength of 0.01 mm thick residual plastic film and the angle of the ridge planting direction is depicted in Fig. 5.

The tensile strength of residual plastic film presents a tendency of initially rising and subsequently declining as the angle with the ridge planting direction augments. The tensile strength of the residual film either along the ridge planting direction or in the perpendicular direction to it is not at its peak. The maximal tensile strength of residual plastic film emerges in the direction approximately 45° relative to the ridge planting direction.

The correlation between the elongation and the angle formed by the ridge direction of a 0.01 mm thick residual plastic film is illustrated in Fig. 5. Evidently, the elongation of the residual plastic film displays a U-shaped pattern as the angle enlarges, attaining its minimum value around 45°. Further investigations have

Table 3 Statistical analysis of tensile test results with different angles between residual plastic film and ridge planting direction

Dependents	Sum of squares	df	Mean square	F-value	P-value	
Tensile strength	0.712	6	0.1187	2.47	0.0366	Significant
Elongation rate	2492.76	6	415.46	7.71	<0.0001	Significant

Fig. 5 The relationship between the elongation of residual film and the ridge angle

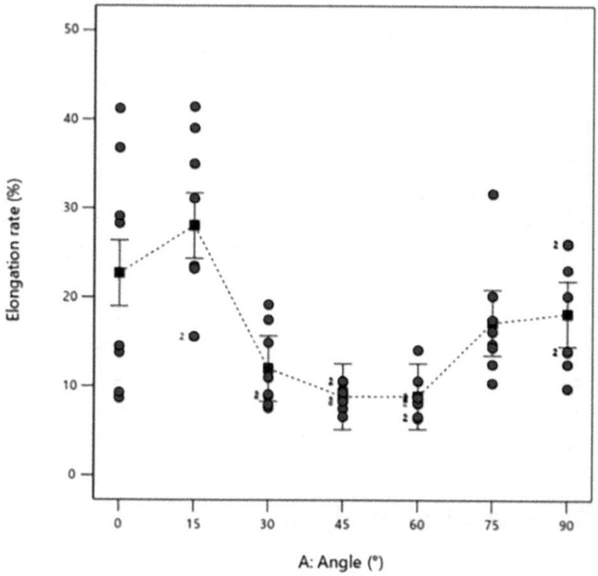

uncovered a negative association between the tensile strength and elongation of residual plastic film, i.e., the lower the elongation, the greater the tensile strength. It is remarkable that the maximum tensile strength and minimum elongation of residual plastic film do not occur at the position conventionally assumed to be in line with the ridge direction but at a 45° angle between the residual plastic film and the ridge direction, which challenges traditional perception.

4 Conclusion

In the central and eastern regions of Inner Mongolia, comprehensive tensile strength and elongation tests were carried out on 0.01 mm thick residual films which are commonly used. During the testing, detailed measurements were made at different angles relative to the ridge direction to deeply investigate the mechanical properties of the residual films. The direction of the maximum tensile strength of the residual film is not constant but is closely related to the stress the film endures during sowing. After numerous rigorous experiments, it was discovered that under specific experimental circumstances and conditions, the residual film had the highest tensile strength when the testing direction was at a 40.6-degree angle to the ridge direction. This crucial data serves as an important foundation for the optimization design of the residual film picker. In the design process of the residual film picker, after comprehensive consideration and selection, the swing adjustment mechanism with a deviation of ±15° from the ridge direction was ultimately chosen as the core picking mechanism. Thus, in practical application, operators can flexibly adjust the direction of the strong tensile strength of the residual film to ensure that the picking mechanism can better conform to the characteristics of the residual film, thereby remarkably enhancing the picking rate of the residual film, reducing farmland residual film pollution, and facilitating the sustainable development of agriculture.

Acknowledgements This research is funded by the Intelligence Aid Xinjiang Innovation and Expansion Talent Initiative, the 2024 Open Project of Xinjiang Lavender Resource Protection and Utilization Key Laboratory (LCUY2407) at Yili Normal University, and the Youth Science and Technology Talent Support Scheme (NJYT22117) of the Inner Mongolia Autonomous Region Higher Education.

References

1. Huang SC, Chen JT (2022) Effects of different colors of plastic film mulching on growth and yield of sweetcorn. Mod Agric Sci Technol:11–14
2. Bu LD et al (2014) Attainable yield achieved for plastic film-mulched maize in response to nitrogen deficit. Eur J Agron 55:53–62
3. Wang S et al (2016) Effect of split application of nitrogen on nitrous oxide emissions from plastic mulching maize in the semiarid Loess Plateau. Agric Ecosyst Environ 220:21–27
4. Cuello JP et al (2015) Impact of plastic film mulching on increasing greenhouse gas emissions in temperate upland soil during maize cultivation. Appl Soil Ecol 91:48–57
5. Jiye B et al (2018) The effect of plastic film covering on crop yield. J Agric Eng 11:172–175

Open Access This chapter is licensed under the terms of the Creative Commons Attribution-NonCommercial-NoDerivatives 4.0 International License (http://creativecommons.org/licenses/by-nc-nd/4.0/), which permits any noncommercial use, sharing, distribution and reproduction in any medium or format, as long as you give appropriate credit to the original author(s) and the source, provide a link to the Creative Commons license and indicate if you modified the licensed material. You do not have permission under this license to share adapted material derived from this chapter or parts of it.

The images or other third party material in this chapter are included in the chapter's Creative Commons license, unless indicated otherwise in a credit line to the material. If material is not included in the chapter's Creative Commons license and your intended use is not permitted by statutory regulation or exceeds the permitted use, you will need to obtain permission directly from the copyright holder.

Study on Performance of Hybrid Turbofan Variable-Cycle Engine with Intermediate Series Fan

Zongxin Yang

Abstract In this paper, the cycle analysis and installation characteristics of the engine are analyzed, and the variable cycle mechanical structure is designed. The intermediate series fan variable cycle engine moves the fan of the traditional large bypass ratio turbofan engine back to the low and high-pressure compressor connection, so it becomes the intermediate series fan. The fan adopts CDFS (core engine fan) technology and has the characteristics of high speed and a large boost ratio. In front of the fan, the auxiliary air intake of the external culvert is designed, and the intake regulating device is installed to control the airflow of the external culvert and adjust the bypass ratio. The adjustment structure refers to the camera aperture adjustment mechanism and has the characteristics of a continuous adjustable opening and a stable intermediate process. The movement direction of blade adjustment is perpendicular to the direction of incoming flow, and the influence of incoming flow impact on blade movement can be minimized in the adjustment process. It is driven by a hydraulic actuator cylinder and has the advantages of a simple structure and easy realization. A mixed turbofan variable-cycle engine with an intermediate series fan can provide thrust for civil airliners with a flight speed of Mach 0–3 and has good economic performance. The proposed scheme has certain reference significance and value for the development of the next-generation supersonic airliner and the development of a military variable cycle engine.

Keywords Bypass ratio · Variable cycle engine · Series fan

1 Introduction

Turbojet engines and turbofan engines are the types of engines used by current mainstream supersonic aircraft. Under subsonic flight conditions, the fuel consumption of gas turbine engines increases with the decrease in bypass ratio, so turbofan

Z. Yang (✉)
Air Force Engineering University, Xi'an, China

© The Author(s) 2026

Z. Zhang, P. Mercorelli (eds.), *Mechanical Design and Simulation: Insights and Innovations*, Lecture Notes in Mechanical Engineering,
https://doi.org/10.1007/978-981-95-2405-1_48

engines can show better fuel consumption performance at low-speed cruise. Under supersonic flight conditions, the fuel consumption of the engine will increase with the increase in bypass ratio. At the same time, the bypass ratio of the turbofan engine will increase with the increase in Mach number. These two factors lead to the deterioration of the fuel consumption performance of turbofan engines during supersonic cruise. Reducing engine fuel consumption and aircraft fuel consumption is of great significance for improving aircraft range and expanding economic benefits. Variable-cycle engine is a gas turbine engine that can fly in a wider envelope by changing the working cycle through the adjustment of variable geometry components. Variable-cycle engine is essentially a kind of aero-engine that can change the bypass ratio by adjusting the geometric components to meet the performance requirements of thrust and fuel consumption in different flight states. There is still a certain gap between domestic and foreign variable-cycle engine technology, which mainly focuses on performance analysis and simulation of variable-cycle engines. Zhou et al. established a mathematical simulation model for the overall performance of a dual-bypass variable-cycle engine and analyzed the control laws of ten adjustable variables of a variable-cycle engine under typical flight conditions and their effects on installation performance and stability [1]. In the Nanjing University of Aeronautics and Astronautics, a nonlinear component-level model of a dual external bypass variable cycle engine was established, which can simulate the action/steady state and mode switching under the single and double bypass operating mode. The influence of adjustable components on the performance of the whole engine was analyzed, a joint adjustment scheme of multiple components was designed, and a preliminary open-loop adjustment scheme for mode switching was presented [2]. Jia from Northwestern Polytechnical University established the whole engine models of a dual external variable cycle engine and triple-external adaptive cycle engine with key components, studied the flight/engine integration and installation performance, and proposed a new method for optimal design of steady-state control rules [3]. Wang et al. [4], in order to study the variable geometry performance benefits of a single external variable cycle engine, established a simulation model of the overall performance of a single external variable cycle engine, and verified the calculation accuracy of the simulation model through a numerical example. Based on the performance calculation and simulation model of the adaptive cycle engine, Jiang et al. [5] combined the mode conversion and acceleration and deceleration process to realize the rapid thrust change process on the basis of relevant research. According to the mode conversion and acceleration and deceleration control law design, the performance analysis of the transition process of rapid thrust change under typical calculation examples was carried out and the feasibility of the control law design method was verified. Hou [6] summarized the research progress of the most promising combined power, especially the rocket ram combined power, analyzed the technological development characteristics and breakthroughs, and put forward the main development direction of the rocket ram combined engine in the future and the development suggestions of combined power technology. Ren et al. [7] proposed a new conceptual structure for a turboshaft-turbofan variable cycle (TSFVCE) engine for rotary wing VTOL high-speed cruise aircraft, which can operate in both turboshaft and turbofan modes. Wang et al. [8] carried out research on the advance/launch matching problem of adaptive cycle engines and proposed to use the FLADE (Fan on Blade) component unique to adaptive cycle engines to achieve advance/launch matching under sub-supersonic cruise tasks. Yu et al. [9] discussed the development history, layout position, and structural form of the rear bypass ejector of the variable cycle engine, analyzed the structural and functional characteristics of the rear bypass ejector, and pointed out the influence on the aero-thermal performance and structural size of the afterburner, as well as the influence on engine thrust and fuel consumption. Chen et al. [10] studied the starting control law of double-external-cultural-cycle engines. Based on the component-level modeling method, taking into account variable geometry characteristic processing of adjustable components and low-speed characteristic extrapolation of rotating components, they adopted the flow balance method to construct common working equations and established the mathematical starting model of variable cycle engines.

2 Intermediate Series Fan Variable Cycle Engine Structure

The design scheme is a type of variable cycle engine with CDFS (core engine fan), which is composed of a compressor, series fan, double ring combustion chamber, turbine, and exhaust device, and is equipped with flow channel regulation and combustion regulation device, combined with the characteristics of turbojet engine and large bypass ratio turbofan engine, and has the characteristics of large thrust and low fuel consumption. The core geometric regulating component of the intermediate series fan variable cycle engine is the air intake regulating device of the auxiliary intake valve, which can continuously and stably control the opening of the auxiliary intake port of the engine outer bypass, thereby continuously changing the airflow of the outer bypass, increasing or reducing the engine bypass ratio, changing the cycle state, and realizing the purpose of the engine variable cycle. The regulating device is installed in front of the fan and is integrated with the engine compartment. The MR-01 low-pressure compressor blade does not compress the external bypass gas but adds a series fan in the middle of the high and low-pressure compressor to compress the external bypass air, and the series fan is connected to the high-pressure rotor, driven by the output power of the high-pressure turbine. In addition, the scheme adds a variable geometry engine casing before the series fan, and combined with the series fan, adjusts the intake volume of the external bypass by controlling the opening of the adjustable structure of the casing to change the engine bypass ratio, so as to achieve variable cycle, as shown in Fig. 1.

In the small bypass ratio mode, the airflow is divided into two streams after entering the engine inlet. One stream flows into the inner passage, is compressed by the compressor, enters the combustion chamber, burns, and expands in the turbine to work; the other stream enters the outer bypass, is compressed by the fan, and mixes with the first stream at the turbine outlet interface to enter the afterburner. In the high bypass ratio mode, the airflow is divided into three streams, one stream

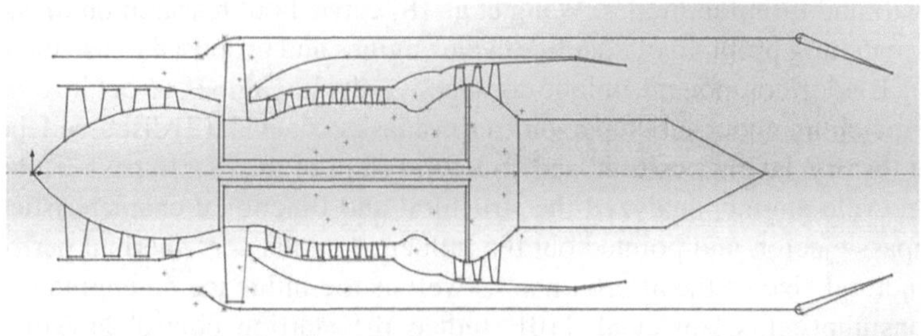

(a) Schematic diagram of the principle of intermediate series fan engine.

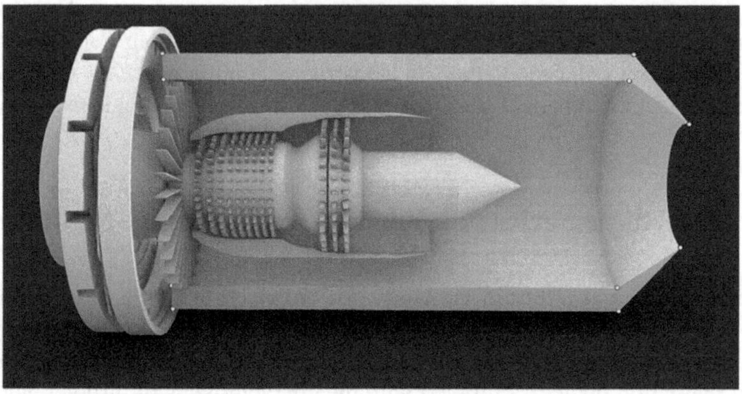

(b) Three-dimensional structure diagram.

Fig. 1 Schematic diagram of intermediate series fan engine. (**a**) Schematic diagram of the principle of intermediate series fan engine. (**b**) Three-dimensional structure diagram

flows into the inner passage from the engine inlet, and the second stream flows into the outer passage from the engine inlet. The third stream flows into the outer passage through the auxiliary intake valve at the front of the fan and mixes with the second stream. After CDFS compression, it mixes with the first stream after combustion expansion, works, and flows into the afterburner together. Finally, it flows out through the tail nozzle.

Compared with the conventional turbofan engine with double bypass, the external bypass air flow rate is adjustable in this scheme, and the bypass ratio is increased by increasing the external bypass air flow rate, reducing fuel consumption, and meeting the thrust performance requirements of the conventional turbofan engine.

3 Estimation of Propulsion System Characteristics

In this paper, the performance of the intermediate tandem engine is estimated to a certain extent. Due to the lack of specific aircraft flight tasks, the component characteristics cannot be completely determined for the time being, and many statistical laws and existing engine statistical estimation models are referred to in the estimation process.

3.1 Installation Thrust Performance Estimation

In terms of propulsion system performance, the installed thrust of the engine is different from the non-installed thrust due to the installation losses of the engine, such as the additional resistance of the intake system, the overflow resistance of the intake system, the boundary layer discharge resistance of the intake system, the friction resistance of the rear body of the exhaust system and the pressure differential resistance. This must also take into account changes in the total intake pressure recovery coefficient and nozzle thrust coefficient, power extraction, external pilot air, intake resistance, and nozzle resistance. The following Eq. (1) expresses the relationship between installed and non-installed thrust:

$$F_A = F - F_R - X_{IN} - X_{OUT} \tag{1}$$

where F_A: installed thrust; F: uninstalled thrust; FR: Compressor extraction power; X_{IN}: Intake system resistance; X_{OUT}: Exhaust system resistance.

3.2 Estimation of Engine Size and Weight

The weight and size of the propulsion system have a great impact on the performance of the aircraft, and the weight of the propulsion system in modern aircraft accounts for 7–14% of the total weight of the aircraft. The increase in propulsion system weight and impact on aircraft performance also affect life cycle costs, while the size of the propulsion system directly affects aircraft drag. Therefore, in the overall design of the propulsion system, the weight and size of the propulsion system must be estimated. However, in the project design stage, the structure of the propulsion system has not yet been determined, and its dimensions and weight can only be calculated by statistical methods.

3.2.1 Engine Size Estimation

One of the main parameters that affect the resistance and weight of the aircraft structure is the upwind area of the engine, that is, the maximum cross-sectional area of the engine. The maximum cross-sectional area of the intermediate series fan engine is located in the front section of the series fan, so the maximum cross-sectional area of the engine can be obtained by estimating the engine inlet area and the space occupied by the adjustable mechanical structure According to the results of cycle analysis, the front section diameter of the engine series fan is about 1.29 m. Considering the installation volume of the adjustable structure, the maximum diameter of the engine windward side is estimated to be 1.5 m. Engine length is generally determined by statistical analysis of available engine technical data.

$$L_e = D_{e\max} \cdot f\left(L / D_{e\max}\right) \tag{2}$$

where L_e: Engine length; D_{emax}: Maximum engine windward diameter; f (L/D_{emax}): Statistical proportion coefficient. Figure 2 shows a simple example of a statistical curve for f (L/Demax).

3.2.2 Engine Size Estimation

Engine-delivered weight does not include engine accessories. The delivered weight of the engine is related to many factors such as engine thrust, design point cycle parameters, structural form, materials used, and the technical level at that time, so it is very difficult to get an accurate analytical equation for calculating the weight of the engine. At present, there are two widely used estimation methods:

1. Calculating engine weight by parts. This calculation method can obtain higher precision of engine weight, but because there are many engine parts, it is difficult to know the weight of engine parts in the design stage. Therefore, this method is not suitable for the overall engine design stage.
2. Using available engine statistics to estimate engine weight. There are many forms and methods of engine weight estimation models established by statistical methods. Moreover, the engine uses involved are different, the focus considered

Fig. 2 Statistical curve

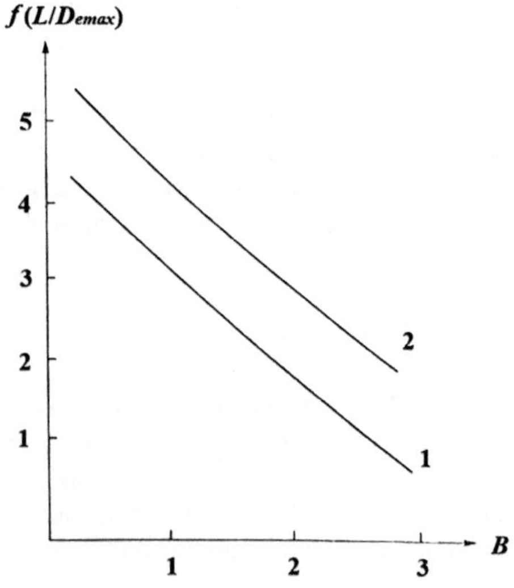

in the construction of the weight estimation structure is different, and the estimation model obtained is naturally different.

Estimation model:

$$W_e = C_t \cdot AFTF \cdot W_{aH} \cdot W_s \tag{3}$$

where C_t is the engine technical level parameter. For the value of the development technology level of different countries and the new technology level adopted in the engine, $C_t = 0.825$; AFTF is the correction coefficient of turbine front temperature, and AFTF can be calculated according to the following equation:

$$AFTF = 0.6867 - 0.0347\left(T_4 - 1850\right)/150 \tag{4}$$

where W_{aH} is the design point engine internal airflow; W_s is the engine weight per unit flow and is determined by statistical method. According to the above model estimation and taking into account the weight of the mechanical regulating mechanism of the outer bypass, $W_e = 2676$ kg.

3.3 *Estimation of Installed Fuel Consumption*

According to empirical estimates, for different engine uses, the installed fuel consumption rate is 5–10% higher than the non-installed fuel consumption rate. In the estimation of installation fuel consumption in this paper, this value is taken as 8%. First of all, according to the calculated fuel consumption value of 0.54 kg/(daN·Hr) at the engine design point, the unit fuel consumption rate at takeoff is first simply estimated, and the estimated result is 0.62 kg/(daN·Hr), which can still meet the target requirements.

With reference to the fuel consumption ratio of CJ828 in each flight stage as the weight to estimate the fuel consumption of the intermediate tandem fan engine installation, and the flight altitude of CJ828 combined with the variable cycle characteristics of the engine, the best flight speed and bypass ratio are selected at a given altitude, so as to determine the unit fuel consumption of the engine at this flight altitude. Finally, the weight is brought in to estimate the installation fuel consumption by weighted average. The planned mission is shown in Table 1:

According to the above estimation method, the estimated result is 1.02 (kg/(daN·Hr)), which is a bit too high for a civil airliner. However, it should be noted that when making the above estimation, the gas temperature in front of the turbine is always 1930 K, which will have a great impact on the fuel consumption of the engine. The gas temperature in front of the turbine is not always 1930 K. In the cruise and glide stages of the aircraft, the economic performance of the engine is to be pursued to the greatest degree. The thrust value shown in Table 1 is not needed at this time, and these values are much higher than the actual flight process, resulting in much higher fuel consumption. It is found that the fuel consumption rate in the

Table 1 Thinking about the mission

Flight phase	Flight altitude (m)	Flight speed (Ma)	Thrust (kN)	Bypass ratio	Unit fuel consumption (kg/(daN·Hr))	Weight	Average unit fuel consumption (kg/(daN·Hr))
Take-off phase	0–450	0.4	122.4	4	0.62	0.03	1.02
Climb phase	450–10,000	0.6	112.0	3	0.84	0.08	
Cruise stage 1	10,000	0.7	97.2	4	0.80	0.25	
Stair climb	10,000–12,000	0.8	105.2	3.5	0.82	0.01	
Cruise stage 2	12,000	0.8	95.4	4	0.82	0.2	
Glider stage	12,000–4500	0.6	112.0	3	0.84	0.02	
Switching section	4500	0.6	94.7	4	0.79	0.09	
Stand-by stage	450	0.4	122.4	4	0.62	0.07	
Descent phase	450–0	0.4	122.4	4	0.62	0.015	

three stages of cruise, glide, and transition is much lower than the value in the table, which can reduce the unit fuel consumption rate by 40–50%, so the final calculation result is corrected, and the correction result is 0.45 (kg/(daN·Hr)), which is in line with the current engine performance requirements of civil aircraft. If the aircraft is flying supersonic in the cruise phase, the fuel consumption at maximum speed is 1.19 (kg/(daN·Hr)), the speed is Mach 3, the speed is increased by 275%, and the fuel consumption is increased by 46.9%. The data show that the engine solution will make the next generation of supersonic passenger aircraft speed and economic performance improve significantly.

4 Conclusion

In this paper, the next-generation engine design of supersonic passenger aircraft is proposed. The MR-01 proposal provides enough thrust for a supersonic airliner to achieve sea-level take-off and supersonic cruise without an afterburner. Its simple variable geometry channel adjustment mode greatly reduces the difficulty of design and production of variable cycle turbofan engines, and the stable and continuous airflow control mode makes it highly feasible. The proposal of MR-01 provides some reference value for the research of variable cycle engines.

References

1. Hong Z, Zhanxue W, Zengwen L et al (2014) Study on variable geometry characteristics of the dual external variable-cycle engine. Acta Aeronaut Astron Sin 35(8):2126–2135
2. Xuezhong G (2012) Research on modeling and control law of variable cycle engine. Nanjing University of Aeronautics and Astronautics, Nanjing
3. Linyuan J (2017) Research on control law design method of variable cycle engine. Northwestern Polytechnical University, Xi'an
4. Mu Y-w, Wang F-m, Zhu D (2024) Single change cycle engine outside culvert geometry simulation. J Aircraft Engine 50(2):52–57. https://doi.org/10.13477/j.carolcarrollnkiaeroengine.2024.02.006
5. Yongrui J, Yihao X, Jiu Z et al (2024) Adaptive cycle engine thrust quickly changed a preliminary study on the transient performance. J Propuls Technol (6):26–36. https://doi.org/10.13675/j.carolcarrollnkiTJJS.2307012
6. Hou X (2023) Research progress of combined cycle engine technology. Acta Aeronaut Astron Sin 44(21):35–47
7. Cheng R, Linyuan J, Xiankun B et al (2024) Vortex axis—turbofan variable cycle engine design and performance matching study. J Propul Technol (5):31–40. https://doi.org/10.13675/j.carolcarrollnkiTJJS.2212021
8. Yifan W, Haoying C, Haibo Z (2019) Adaptive cycle engine feed/fire matching for cruise mission. Acta Aeronaut Sin 45(2):109–120. (in Chinese)
9. Haitao Y, Lei S, Yao F et al (2024) Variable cycle engine bypass after the research status of ejector and the key technology. J Aircraft Engine 50(3):1–8. https://doi.org/10.13477/j.carolcarrollnkiaeroengine.2024.03.001
10. Li C, Xuechong G, Min C (2024) Variable cycle engine start modeling and control law study. J Air Power 33(2):253–261. https://doi.org/10.13224/j.carolcarrollnkijasp.20210470

Open Access This chapter is licensed under the terms of the Creative Commons Attribution-NonCommercial-NoDerivatives 4.0 International License (http://creativecommons.org/licenses/by-nc-nd/4.0/), which permits any noncommercial use, sharing, distribution and reproduction in any medium or format, as long as you give appropriate credit to the original author(s) and the source, provide a link to the Creative Commons license and indicate if you modified the licensed material. You do not have permission under this license to share adapted material derived from this chapter or parts of it.

The images or other third party material in this chapter are included in the chapter's Creative Commons license, unless indicated otherwise in a credit line to the material. If material is not included in the chapter's Creative Commons license and your intended use is not permitted by statutory regulation or exceeds the permitted use, you will need to obtain permission directly from the copyright holder.

Design and Development of Underwater Pump-Suction Foreign Body Salvage Device for Nuclear Power Plant

Yang Chen, Yu Du, Junlong Xu, Maocheng Hong, and Jiawei You

Abstract Foreign matter entering the reactor system of a nuclear power plant will lead to the deterioration of the performance of the system equipment, make the operation of the equipment abnormal, and will lead to equipment damage and even affect the safety of the reactor in serious cases. The environment of the primary circuit of the nuclear power plant is radioactive, and when foreign matter enters the primary circuit, the traditional salvage method can not meet the need for radiation protection. In this paper, a foreign body salvage device is designed, which uses water as a shielding layer to retrieve underwater foreign bodies by pump-suctioning and realizes the convenient salvage of foreign bodies in the radioactive environment.

Keywords Reactor · Foreign matter · Salvage · Pump-suction

1 Introduction

There are three main sources of foreign bodies in nuclear power plants: first, foreign bodies that do not belong to the system and equipment of nuclear power plants; The second is the foreign matter that originally fell off the equipment of the nuclear power plant; Third, due to the operation of the nuclear power plant, the sediment of the working medium dissolved matter formed by precipitation [1]. The impact of foreign matter on nuclear power plants is mainly reflected in the following three points: First, security threats. If foreign matter in a nuclear power plant enters the core or primary circuit system, it may lead to damage to the fuel assembly cladding, increase the radiation level, and even cause the spread of radioactive pollution, which directly threatens the safe and stable operation of the nuclear power plant. Foreign objects can also damage system equipment, such as filters, chemical

Y. Chen (✉) · Y. Du · J. Xu · M. Hong · J. You
CGNPC Inspection Technology Co. Ltd., Suzhou, Jiangsu, China

National Engineering Research Center for Nuclear Power Plant Safety & Reliability, Suzhou, Jiangsu, China

© The Author(s) 2026

Z. Zhang, P. Mercorelli (eds.), *Mechanical Design and Simulation: Insights and Innovations*, Lecture Notes in Mechanical Engineering,
https://doi.org/10.1007/978-981-95-2405-1_49

processors, etc., resulting in equipment failure and further aggravating safety risks. Second, operation obstacles. Foreign objects can block the flow of coolant and cause the nuclear reactor to overheat, affecting the normal operation of the nuclear power plant. Foreign matter attached to the surface of the fuel rods can cause the fuel rods to overheat and even cause the fuel rods to fail, increasing the risk of nuclear reactors. Third, Environmental pollution. The environment around a nuclear power plant is sensitive, and if a leak is caused by a foreign object, it may contaminate the surrounding environment, causing harm to the ecosystem and human health.

The common anti-foreign body measures in nuclear power plants are to establish a strict anti-foreign body mechanism, such as establishing a foreign body isolation area, controlling the number of openings in the system, and recording the number and status of items brought in. This kind of method can play an effective role in restricting foreign bodies, but it can not completely solve the consequences of foreign bodies caused by foreign bodies. And can not cope with the system equipment shedding and working fluid deposition of foreign bodies. However, the traditional salvage method can not meet the requirements of a nuclear power plant for foreign body salvage radiation protection [2]. In this paper, the foreign matter in the primary pressure vessel of the nuclear power plant is studied, and an underwater pump-suction device is developed to recover foreign matter with water as a shielding layer.

2 Application Objects and Research Status

2.1 Application Objects

In this paper, 14 foreign body incidents that occurred in 18 nuclear power units of 4 domestic nuclear power plants from 2020 to 2023 were investigated. The findings were as follows: (1) According to the source of foreign body, 13 foreign body incidents were caused by foreign body introduction, and 1 foreign body was deposited by system equipment; (2) According to the weight analysis of foreign bodies, the weight of foreign bodies in the 14 cases was all less than 1 kg; (3) According to the location analysis of foreign bodies, 10 foreign bodies occurred in open containers, and 4 foreign bodies occurred in open pipes. Based on the above findings, the use scenario designed in this paper is the steel pressure vessel in the radioactive environment of the nuclear power plant, the reactor pressure vessel, the reactor main pipeline, and the spent fuel pool are the main use scenarios [3], and the foreign matter with a total weight of less than 1 kg is the application object.

2.2 Research Status

Other common forms of foreign body grasping equipment in China mainly include air claw type and multi-functional special type, as shown in Fig. 1. The air claw salvage tool simulates hand grip to grasp foreign bodies. Its advantages are simple

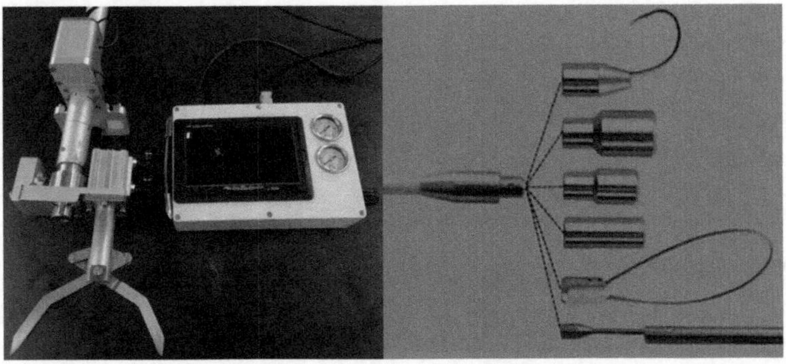

Fig. 1 Air claw fishing tool (left) and multi-function fishing tool (right)

to grasp foreign bodies, and foreign bodies are not easy to fall after grasp. The disadvantages are that it is difficult to grasp foreign bodies in thin gaskets, can only grasp foreign bodies in the vertical direction, and can not grasp foreign bodies on the side. In addition, most of the deposited foreign bodies formed by the precipitation of working fluid have a soft surface texture, which is prone to causing foreign bodies to be crushed during the grasping process, forming more small foreign bodies, so the air claw type cannot cope with such foreign bodies. The multi-functional special fishing tool is equipped with multiple grasping joints, such as fishhook type, strong magnetic type, and lasso type, which has the advantage of being suitable for different grasping scenarios. The disadvantage is that when applied to a radioactive environment, it will cause radioactive contamination of multiple grasping joints, and in the process of replacing grasping joints, personnel need to directly contact the grasping joints that may be contaminated by radioactivity, and the radioactive exposure risk of personnel is high [4].

3 Equipment Design

3.1 Overall Design

The overall structure of the underwater foreign body salvage device is shown in Fig. 2. The underwater foreign body salvage device adopts the principle of pump-suctioning, and the foreign body can be sucked into the collection pipe when the collecting mouth is close to the underwater foreign body. The collection port is divided into straight pipe type and bent pipe type, which can pump-suction vertical direction (in the container) and lateral foreign matter (in the pipeline), respectively. Foreign body salvage and collection are completed underwater to avoid frequent water discharge. The equipment is designed to salvage foreign bodies with diameters of 0.5–50 mm and weights less than 1 kg in reactor pressure vessels and spent fuel pools.

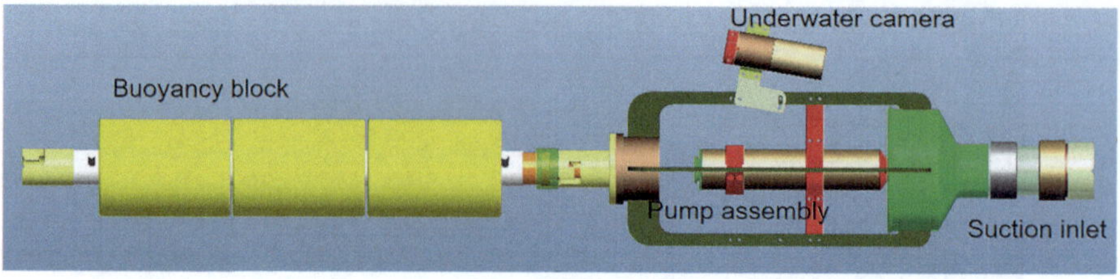

Fig. 2 Overall design

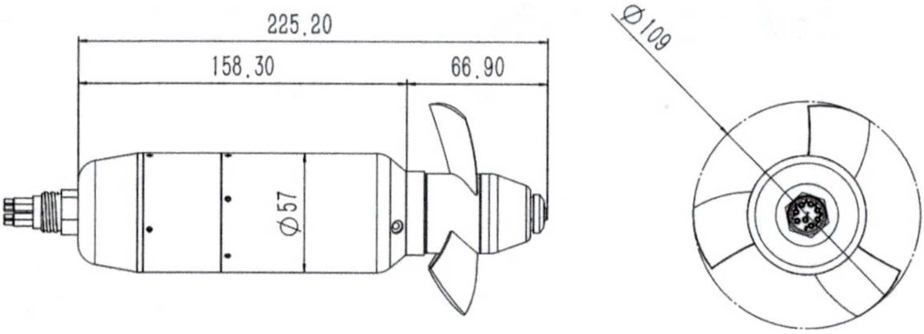

Fig. 3 Pump-suctioning size

Table 1 Main parameters of pump-suctioning assembly

Performance index	Parameter	Performance index	Parameter
Input power	580 W	Working depth	0–850 m
Rated voltage	48VDC	Weight	≤1.7 kg
Rated thrust	≥10 kg	Blade material	Stainless steel
Seal construction	Magnetically coupled rotary seal	Signal power supply	12 V ± 5%

3.2 *Design of Pump-Suction Assembly*

The salvage device realizes the capture and release of foreign bodies through the positive and negative rotation of the pump-suction assembly. Its geometric dimensions are shown in Fig. 3, and its main parameters are shown in Table 1. The rear end of the pump-suction assembly is designed with a filter screen with a diameter of 0.5 mm to ensure the fishing accuracy of the equipment. Since the pump-suction assembly is the core part of the equipment and it works in the underwater radioactive environment for a long time, the pump-suction assembly is designed to be sealed and radiation resistant. The outer surface of the main body of the pump-suction component is made of integrated aluminum alloy, and the outer shell, the blade end, and the tail end are sealed by an O-ring (outer seal). The inner surface of the main body of the pump-suction assembly is lined with a thin lead plate as the overall shielding material of the pump-suction assembly. The pump-suction motor, controller, encoder, and other components are placed inside the main body of the pump-suction assembly, and circular lead covers are designed at both ends to seal

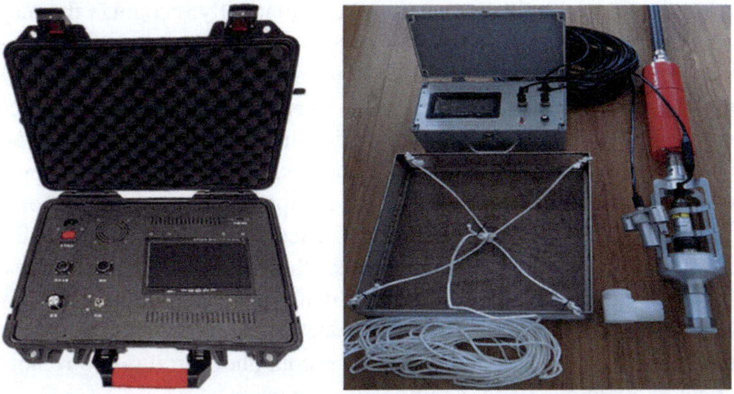

Fig. 4 Control box and auxiliary tooling

the main components of the pump-suction assembly (inner seal). The circular lead cover at the blade end is provided with a motor through the hole, and the through hole is sealed by a stuffing box. The circular lead cover at the tail end is provided with a wiring control port, and the wiring hole is provided with a static seal. The pump-suction motor shaft is connected to the blade shaft by a coupling placed at the blade end. The seal of the blade shaft is a magnetically coupled dynamic seal to ensure the sealing performance under long-term motion. The blades are made of stainless steel to ensure their long-term durability in underwater environments. The joint at the tail end is the only non-metal part of the pump-suction assembly that has been exposed to radioactivity for a long time [5]. According to research, the radiation resistance of polytetrafluoroethylene is good, and the tolerance dose is up to 100 Mrad. Its dielectric constant and dielectric loss are low, and the electromagnetic shielding effect is good. Therefore, the end joint shell uses its shell material to choose polytetrafluoroethylene. In addition to the O-ring static sealing at the joint, the glue-filling process is also used to ensure the waterproof performance of the joint.

3.3 *Design of Auxiliary Tooling*

The design of auxiliary tooling includes the collection frame, control box, and installation tool, as shown in Fig. 4. The underwater action of the equipment is controlled through the control box. One side of the panel of the control box is the display area, which is used to display the environment picture captured by the underwater environment camera and locate the relative position of the suction inlet and foreign matter, so as to facilitate suction operation. The control box is equipped with foreign object recognition software, which can identify common nuclear power plant foreign objects such as bolts and nuts. When such foreign objects are found in the environment camera, the software will automatically recognize them and give a prompt on the top right of the display screen. The underwater camera adopts an irradiation-resistant design, and its maximum irradiation dose rate is not less than 20 Gy/h. The other side of the panel is the interface and pump-suction suction control area. The underwater camera connection interface and pump-suction suction assembly interface are provided in this area, both of which are used to control the salvage and release of foreign bodies of the equipment and adjust the suction output of the equipment. The collection frame is used to collect foreign bodies. When retrieving highly radioactive foreign bodies, the collection frame is placed in the water by the aerial crane. After the foreign matter is placed in the collection frame, the on-site personnel shall evacuate and place the foreign matter together with the collection frame into the prepared shielding container through the remote control aerial crane to avoid personnel's radioactive exposure.

3.4 Device Installation and On-Site Implementation

At the beginning of foreign body salvage, we first connect the video cable and submersible motor cable to the control box, connect the power supply of the control box, and then click the pump-suction assembly to confirm the successful connection of the equipment. The buoyancy block is connected to the salvage equipment to reduce the gravity of the equipment in the water, as shown in Fig. 5. The carbon fiber long rod is connected to the buoyancy block, and the number of long rods is determined according to the water depth. The length of each long rod is 2 m. Meanwhile, the foreign body collection frame is put into the pool through the safety

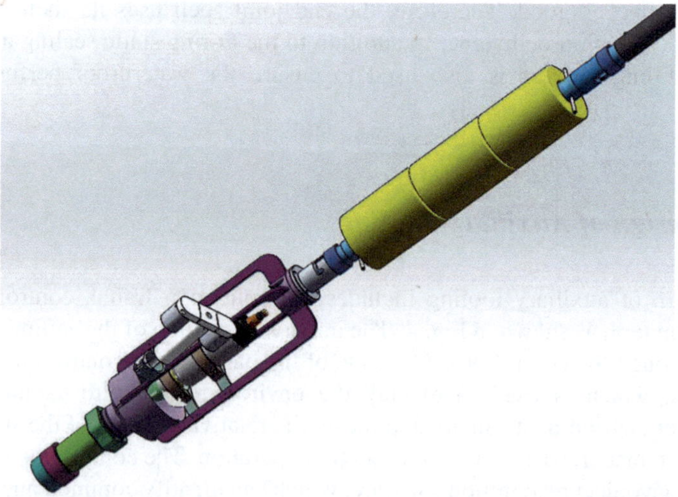

Fig. 5 Use renderings

rope. We turn on the power switch, determine the position of the foreign body through the camera, open the pump-suction assembly, and gradually adjust the motor speed, so that the foreign body pump-suction is sucked into the suction inlet and attached to the filter screen, and the filter screen can avoid foreign body with a diameter greater than 0.5 mm into the blade wheel. We move the collection tube to the foreign body collection frame, stop or reverse the motor, and discharge the foreign body into the collection frame. After all foreign bodies are recovered, we turn off the pump-suction suction assembly, lift the foreign body salvage device and the foreign body collection frame off the water surface, and recover the equipment.

4 Performance Test

The positive and negative thrust of the pump-suction assembly is the key to realizing the salvage capability of the equipment. Because the water flow disturbance will affect the fishing capacity of the pump-suction assembly, when the water flow is opposite to the pump-suction direction, the pump-suction capacity will be weakened significantly. Therefore, this paper conducts rotational speed-thrust test and power-thrust test for the pump-suction assembly under different flow rates (0, 1, 2, 3, 4 m/s). The test conditions are as follows: with the test water depth set at 1 m underwater, we take inhalation as the positive direction, and the test position is close to the suction inlet. The test results are shown in Fig. 6. According to the test results, as the speed increases, the output of the pump-suction assembly also increases, and the higher the speed, the greater the effect of the speed on the pump-suction output. The higher the input power of the pump-suction assembly, the higher the output of the pump-suction assembly, but the higher the power, the output increase caused by the increase of power is significantly reduced. From the test result diagram, taking the test curve in still water as an example, it can be found that when the power of the

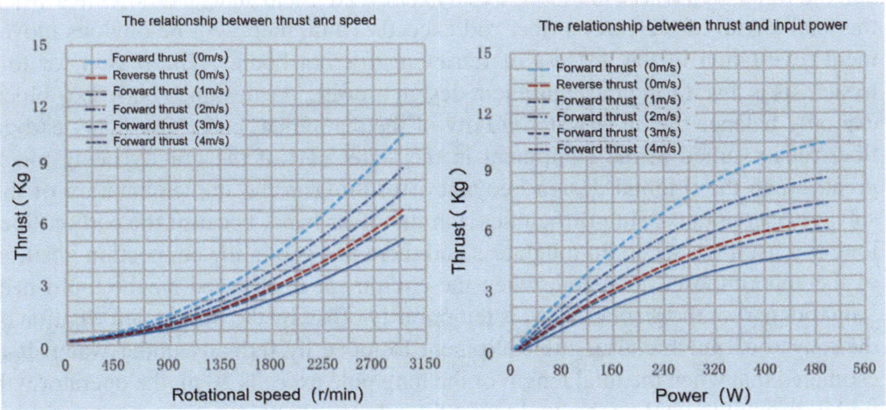

Fig. 6 Thrust speed curve (left) and thrust input power curve (right)

pump-suction component exceeds 400 W, as the power is increased by 10%, the pump-suction output is increased by less than 8% and the value is less than 1 kg. In this case, the influence of power on the pump-suction output is significantly reduced, and the pump-suction efficiency is significantly reduced. At this time, the corresponding pump-suction output is about 9.75 N, and the speed is about 2800 r/min. When the operating speed of the pump-suction assembly is above 2250 r/min, the output of the pump-suction assembly begins to increase by more than 10% when the speed increases by 10%, and the output of the pump-suction assembly begins to be significantly affected by the speed increase. In this case, the output of the pump-suction group can meet the requirements of conventional fishing when the output of the pump-suction group is above 6 kg, and the corresponding power is 240 W. Therefore, the operating efficiency of the pump-suction assembly is the highest when the operating speed of the pump-suction assembly is 2250–2800 r/min and the input power is 240–400 W. In the actual operation test, influenced by many factors such as water flow rate and uneven adsorption, combined with historical experience, when the operating speed of the pump-suction component is controlled between 2400 and 2700 r/min, the input power can be stable within the maximum operating efficiency interval, which is the actual optimal operating speed interval.

5 Engineering Applications

According to the test results under laboratory conditions, the equipment will be applied in domestic nuclear power plant A in 2023. This equipment can meet the requirements for effective salvage of foreign bodies within the design range in both laboratory environments and on-site use of nuclear power plants. The actual salvage situation is shown in Fig. 7. However, in the actual application process, it is found that the salvage operation in the reactor pressure vessel of a nuclear power plant is inconvenient and inefficient due to the large height drop between the operation area and the water area where the foreign body is located. Field analysis shows that when the total length of the carbon fiber rod exceeds 10 m, there will be obvious movement retardation in the process of retrieving foreign bodies. The reason for this hysteresis is that during the equipment design process, the designed buoyancy block can only balance the underwater gravity of the equipment body. Too many carbon fiber rods resulted in the equipment in the water part of the gravity being much greater than the original design buoyancy, resulting in the implementation of the salvage operators causing difficulties. On the other hand, because the carbon fiber long rod itself has a certain toughness and there is a gap in the connection position of the superimposed long rod, when the equipment is operated by the multi-heel long rod, part of the displacement is temporarily offset by the plastic deformation of the long rod, which causes the equipment to move hysterically under water. It is estimated that when the total length of the long pole exceeds 20 m, the operator will find it difficult to operate the equipment for salvage operations.

Fig. 7 Foreign body salvage effect

Fig. 8 Bridge tool effect drawing (left) and physical drawing

In view of the above situation, the equipment has developed a bridge tool, as shown in Fig. 8, which is used for foreign body salvage when the height drop is more than 10 m. The tool is usually used in conjunction with the bridge over the pool of the reactor pressure vessel in a nuclear power plant. The tool realizes the positioning in the north-south direction and the east-west direction by moving the pulley and the bridge, respectively, and realizes the positioning fine-tuning by the linear guide rail. The vertical height positioning is adjusted by the carbon fiber long rod, and the vertical height is fine-tuned by the worm gear. Through the pulley rolling on the bridge tool and the motor movement of the bridge instead of the original manual operation positioning, the restraint of the carbon fiber long rod is added, and the fishing efficiency of the fishing equipment in response to the high drop operation has been significantly improved. At the beginning of 2024, when the foreign body salvage work in the reactor pressure vessel pool of a domestic nuclear power plant B was carried out (the actual height drop is about 15 m), by using the bridge tool, the actual construction period was saved by 1 man-hour/person, the work efficiency was improved by nearly 25%, and the collective radiation dose of personnel was reduced by 1 mSv in a single salvage.

6 Conclusion

In this paper, a foreign body salvage device is designed to achieve the salvage of foreign bodies inside containers and pipelines under the condition of meeting the requirements of personnel's reflective protection. The salvage capability can meet the salvage requirements of common radioactive foreign bodies in nuclear power plants and provide a special bridge tool to improve the applicability of the equipment in nuclear power plants when the salvage operation is carried out with a large height drop. This study provides a useful solution for the disposal of underwater foreign bodies in nuclear power plants. However, in practical application, on the one hand, there are many uncertain factors in the process of foreign body salvage, such as the type and location of the foreign body, which can easily affect the efficiency of foreign body retrieval. On the other hand, even if there is water as a shielding layer during the actual operation, there is still a small amount of radioactivity in the working environment of personnel. Therefore, improving the remote control of equipment to limit the exposure dose of personnel and improve the grasping efficiency of equipment is the future research direction of this subject.

References

1. Xie W, Zhao X, Chen H, Chen Q, Liu S (2023) Radiation protection for foreign body salvage work in primary of nuclear power plant. J Radiat Res Radiat Process 41(3):104–110
2. Chen X, Li W, Bai L (2023) Foreign material management during NPP construction. Nucl Saf 22(1):7–10
3. Zhang G (2017) The study on analysis and improvement of abnormal cause control rod drop in PWR nuclear power station. Nucl Electron Detect Technol 37(7):683–687
4. Wang W, Zhang X, Wang W, Guan X, Yan S (2023) Radiation risk control for salvaging foreign object in the reactor pool of nuclear power plants. Radiat Prot Bull 43(3–4):51–56
5. Pei S, Guo Y, Huang S, Wang C, Mao J (2024) Analysis and optimization of the blocking plate in nuclear power plant based on the finite element method. Nucl Sci Eng 44(2):350–359

Open Access This chapter is licensed under the terms of the Creative Commons Attribution-NonCommercial-NoDerivatives 4.0 International License (http://creativecommons.org/licenses/by-nc-nd/4.0/), which permits any noncommercial use, sharing, distribution and reproduction in any medium or format, as long as you give appropriate credit to the original author(s) and the source, provide a link to the Creative Commons license and indicate if you modified the licensed material. You do not have permission under this license to share adapted material derived from this chapter or parts of it.

The images or other third party material in this chapter are included in the chapter's Creative Commons license, unless indicated otherwise in a credit line to the material. If material is not included in the chapter's Creative Commons license and your intended use is not permitted by statutory regulation or exceeds the permitted use, you will need to obtain permission directly from the copyright holder.

Development and Optimal Machining Path Determination of Incremental Forming Equipment for 120° Sidewall Machining Angles

Zihang Liu and Chun Xu

Abstract In response to the limitations of incremental forming technology in the manufacture of thin-walled parts with large forming angles, this paper presents the independent development of an incremental forming equipment capable of multi-angle processing. The equipment achieves independent control over the two rotational degrees of freedom of the forming tool through the coordinated operation of the motor and the electromagnetic clutch, thereby enabling multi-angle plastic forming of metal sheets. By employing a strategically designed "pyramid" incremental forming path, the equipment has successfully processed thin-walled parts with forming angles exceeding 90°.

Keywords Incremental forming · Multi-angle machining · Equipment development · Large angle part forming

1 Introduction

Incremental forming processes, which do not require mold, offer high flexibility and are well-suited for the production of small batches of parts, thus becoming a research focus both domestically and internationally [1–3]. However, the application of incremental forming equipment still encounters a technical bottleneck in machining thin-walled metal parts with side wall angles exceeding 90°, which restricts its further large-scale application. Researchers have conducted relevant studies: Zhang

Zihang Liu and Chun Xu are co-first author.

Z. Liu
School of Mechanical and Engineering, Shanghai Institute of Technology, Shanghai, China

C. Xu (✉)
School of Materials Science and Engineering, Shanghai Institute of Technology, Shanghai, China

© The Author(s) 2026

Z. Zhang, P. Mercorelli (eds.), *Mechanical Design and Simulation: Insights and Innovations*, Lecture Notes in Mechanical Engineering,
https://doi.org/10.1007/978-981-95-2405-1_50

et al. [4] achieved the machining of conical parts with a 70° side wall through a spiral machining path; Zhu et al. [5] using a square punch and an alternating composite forming method, were only able to process box-shaped parts with 90° straight walls; Li et al. [6] based on the technology of three-axis and five-axis CNC progressive composite forming, although capable of processing parts with angles greater than 90°, only formed a small-sized conical protrusion in a specific area of the square pyramid, making it difficult to achieve all-around large-angle machining of parts. Therefore, addressing the challenge of machining complex thin-walled parts with side walls greater than 90° remains a key research focus for progressive forming equipment.

The principle of incremental forming is essentially the superimposed combination of layer-by-layer manufacturing. It involves the discretization of complex three-dimensional shapes into orderly two-dimensional sections along a specific axis. The forming tool follows a predetermined planar motion trajectory, starting from point processing and gradually accumulating into line processing, to carry out layer-by-layer forming of the metal sheet. This continuous accumulation process achieves the forming of three-dimensional shape parts from the sheet metal [7–11]. For the processing of side walls with angles greater than 90°, the tool head is required to be tilted at a certain angle. However, the tool head is typically limited by its fixed installation, especially affected by the angle of reaction force from the sheet metal deformation, and its tilting angle must not exceed a critical value; otherwise, the tool head may exhibit vibration phenomena, and in severe cases, it can lead to lateral bending of the tool head and fixtures.

This paper presents the independent development of an incremental forming equipment capable of multi-angle machining, which improves the tool mounting method and eliminates machining vibrations. Moreover, by optimizing the machining path, it achieves the forming of parts with a 120° angle of deformation.

2 Experimental Equipment

This device is a compact five-axis incremental forming machine, utilizing a dual-head five-axis configuration. In addition to employing the most basic screw transmission method to achieve the XYZ three-axis motion of the tool, its forming tool can realize dual degrees of freedom, including rotation around the Z-axis and oscillation around the Y-axis (or X-axis). In the design of traditional CNC machines, it is often necessary to use two motors to control the single rotational degree of freedom of the tool to achieve rotation in the aforementioned two directions. However, for the small device designed in this paper, the use of multiple motors would not only occupy a large space but also lead to an excessively long cantilever structure at the tool, thereby reducing the stability of the processing. Therefore, this device achieves independent control of the tool's two rotational degrees of freedom through the engagement and disengagement mechanism of an electromagnetic clutch. This design not only enables the forming tool to perform multi-angle processing of metal sheets but also enhances the stability of the processing process. The design principle is detailed in Fig. 1.

The motor output shaft is connected to the input end of the electromagnetic clutch through a key connection at one end, and also achieves circumferential fixation with the horizontal bevel gear through a key connection at the other end, thereby enabling synchronous rotation of the three components. When the electromagnetic clutch is in the power-off state, its input and output ends are separated, and the motor output shaft drives the clutch input end and the horizontal bevel gear to rotate, achieving the tool head's oscillation around the Y (or X) axis through the bevel gear meshing transmission. When the electromagnetic clutch is energized and closed, the input and output ends of the clutch are tightly connected, and the entire clutch achieves synchronous rotation with the motor output shaft through a key connection, thereby driving the tool to rotate as a whole around the Z axis.

As shown in Fig. 2, although the use of one motor is reduced, the total length H of the forming tool is still excessively long. If it is directly suspended vertically below the beam, a large cantilever structure will inevitably be formed, reducing the stability of the machining process. Therefore, a "riding" structural design is adopted: the main structure of the tool assembly stably "rides" between the screw rod and the guide rail, with the screw rod positioned above the guide rail and

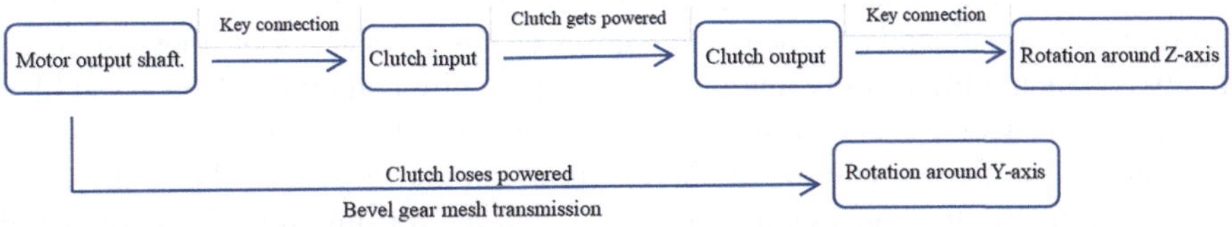

Fig. 1 Schematic diagram of the multi-angle machining tool structure

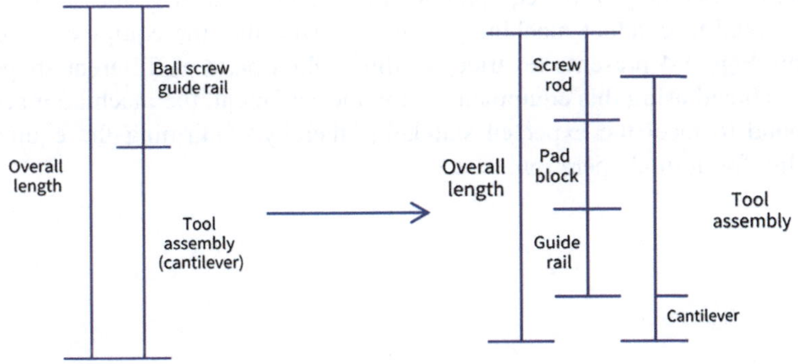

Fig. 2 Design principle of the "riding" structure of the forming tool

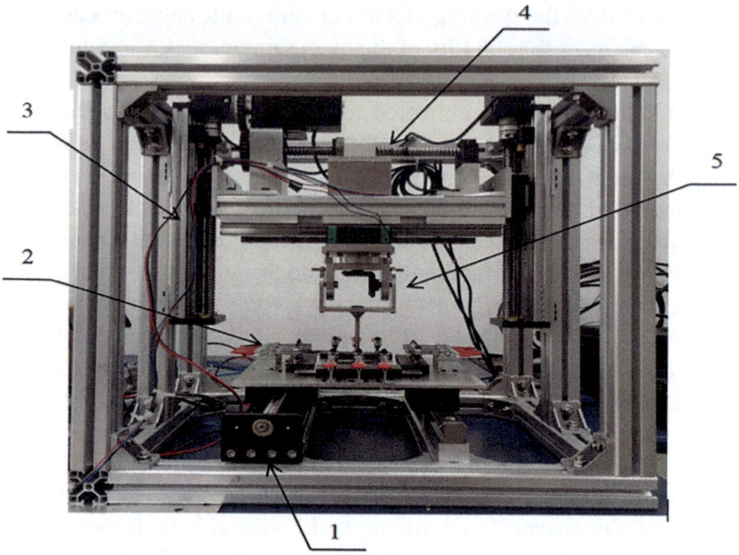

Fig. 3 Actual device photo: 1—X-axis; 2—fixture; 3—Z-axis; 4—Y-axis; 5—Forming tool

connected to the top of the tool assembly through the nut seat on it. Meanwhile, the slider on the guide rail is connected to the bottom of the tool assembly. This structural layout significantly reduces the cantilever length of the tool, thereby effectively enhancing the rigidity and stability of the entire equipment. The assembled equipment structure is shown in Fig. 3. The equipment consists of five parts: the X-direction motion structure, the Y-direction motion structure, the Z-direction motion structure, the sheet clamping device, and the forming tool. In incremental forming, positive forming cannot be separated from mold support, making it difficult to achieve truly flexible forming. Therefore, this equipment adopts a negative forming method.

After the assembly of the equipment structure and control system is completed, it is essential to conduct machining tests to ensure that the equipment functions properly. Figure 4 presents a variety of thin-walled parts of different shapes that were machined using this equipment. Upon measurement, the machining accuracy was found to meet the expected standards, thereby confirming the equipment's capability for normal operation.

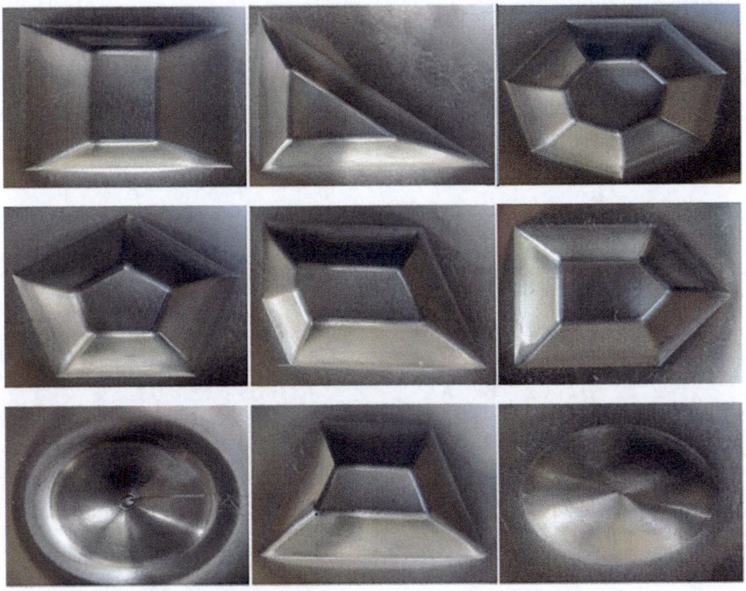

Fig. 4 Conventional thin-walled parts

3 Experimental Material

The experimental material used in this paper is commercially available O-temper 1060 series aluminum alloy sheets, with dimensions of 150 × 150 × 0.5 mm.

4 Experimental Plan

4.1 90° Sidewall Machining Path

The 90° side wall forming path is shown in Fig. 5: The tool starts from the preset starting point A1, following the inclined path from A1 to A2, and then from A3 to A4, to complete the first pass of the forming operation. After completing this pass, the tool returns to the starting point A1, and along the vertical path from A1 to B1, performs the second pass of vertical machining on the side wall of the part to form a vertically oriented side wall. Subsequently, the tool continues along the inclined path from B1 to B2, and then from B3 to B4, to execute the second pass of the forming operation, repeating the aforementioned forming process until the side wall of the part is completely transformed from the initial concave shape to the desired straight wall shape.

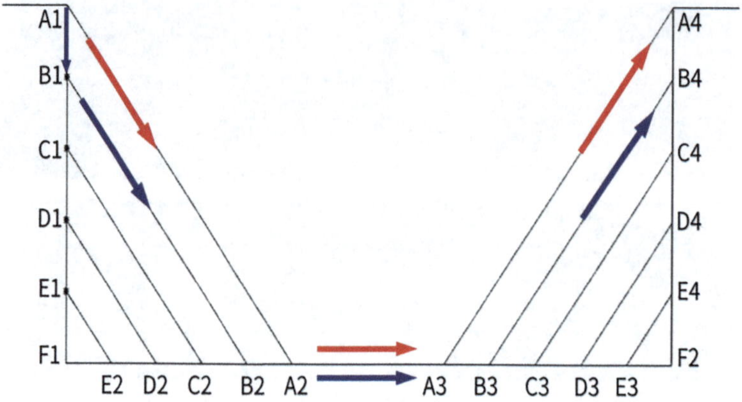

Fig. 5 Forming path of straight-walled parts

Fig. 6 "Pyramid"-style expansion forming path

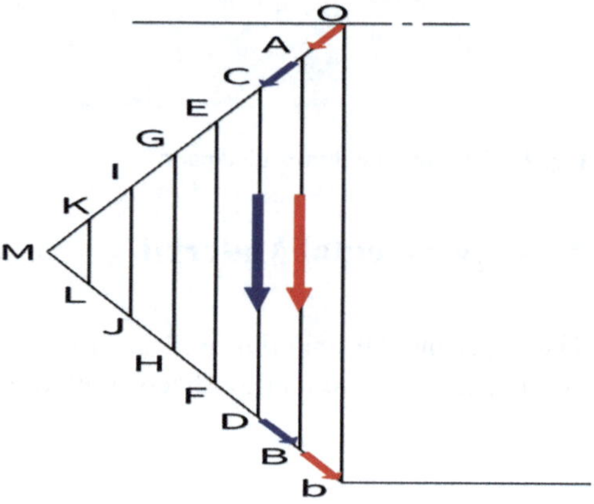

4.2 Greater Than 90° Sidewall Machining

The forming path is shown in Fig. 6. Employing a "pyramid-like" layer-by-layer expansion path strategy for further large-angle forming of straight-walled parts. The processing tool completes the first large-angle expansion of the straight-walled part along the path from the starting point O to OA-AB-Bb. Afterward, the tool head returns to point A and performs the second expansion forming of the straight-walled part along the path AC-CD-DB. This process will be repeated until the expansion forming of the entire part is completed. Similar to the forming principle of straight-walled parts, this method achieves consistency in horizontal spacing between adjacent expansions through the "pyramid"-like layer-by-layer expansion forming. During a single layer expansion, the plastic deformation experienced by the metal at any position is relatively consistent in magnitude and direction, ensuring the uniformity of the deformation process. This deformation characteristic helps to achieve a uniform thickness distribution of the sheet metal side wall during the forming process, significantly reducing the risk of material rupture.

Traditional incremental forming equipment is typically modified from three-axis CNC machines, and due to the fixed nature of their forming tools, they can only process thin-walled parts with forming angles less than 90°, resulting in concave pit shapes. However, the equipment independently developed in this study has overcome this limitation, enabling the forming tool to tilt at multiple angles. Based on the straight wall piece, this equipment can further achieve a large-angle expansion, thereby successfully processing thin-walled parts with forming angles greater than 90° in a concave style.

5 Experimental Results and Analysis

Figure 7a, b present the thin-walled parts with a 120° forming angle obtained from the experiment in this study. The surface of the parts is smooth and even, with no apparent machining vibration marks. Figure 7c is a cross-sectional view taken along the center line of the part, and it can be observed that the thickness of the part's side wall gradually decreases with the increase in forming depth. This variation in thickness is consistent with the thickness change observed in stamping formed sheet metal. To visually observe the trend of changes in the thickness of the part's side wall, Fig. 8 displays the image of this cross-section under a microscope. In the initial unprocessed area, due to the entanglement effect of the metal in the processing area, a noticeable elastic deformation has occurred. As the forming depth increases, the thickness of the upper half of the processed area gradually decreases, especially at the corner, where the thinning is most pronounced. In contrast, the thickness of the lower half of the processed area is similar to that at the corner and exhibits a more uniform thickness distribution. Measurements revealed that the minimum wall thickness is located in the lower processed area, with a value of 0.1156 mm, and compared to the initial thickness of the sheet metal of 0.5 mm, the maximum thinning rate reached 76.88%.

Incremental forming is a process that combines elastic and plastic deformation. During the forming process, the metal sheet first undergoes elastic deformation, which increases as the forming depth increases. When the forming depth reaches a

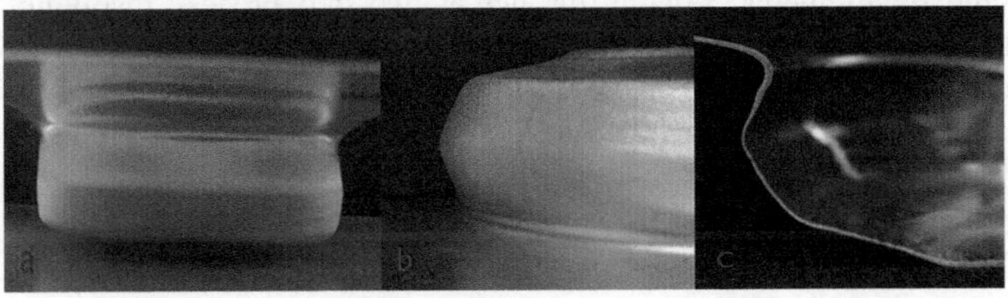

Fig. 7 Thin-walled part with a forming angle of 120°

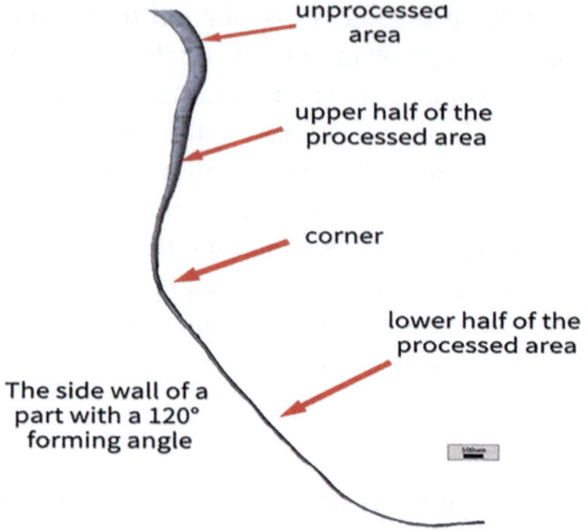

Fig. 8 Cross-sectional view of the side wall

certain value, the sheet reaches its yield limit, resulting in plastic deformation. After the processing is completed, when the forces applied by the forming tool head and the sheet clamping device are removed, the stored elastic deformation energy within the sheet is released, causing the part to spring back. Since spring back is inevitable in the incremental forming process, the measurement results show that the actual forming angle of the part's side wall is 118°, slightly lower than the expected 120°.

6 Conclusion

This paper successfully developed a compact five-axis incremental forming machine, which is primarily composed of five parts: the XYZ three-axis motion structure, the sheet clamping device, and the forming tool. By utilizing the engagement and disengagement of an electromagnetic clutch, it achieves independent control of the two rotational degrees of freedom of the forming tool with a single motor, enabling the tool to perform multi-angle processing and forming of metal sheets. Additionally, the "riding-style" structural design shortens the cantilever length of the tool, thereby improving the vibration issues caused by an excessively long cantilever during processing. After the development of the equipment, various shapes of thin-walled parts were processed and tested, verifying that the equipment meets the usage requirements. Finally, through a two-step process involving vertical machining paths and a "pyramid"-style layer-by-layer expansion forming path, the machine achieved the processing and forming of parts with a 120° side wall forming angle, significantly expanding the forming range of incremental forming.

References

1. Najm SM, Paniti I, Trzepieciński T et al (2021) Parametric effects of single point incremental forming on hardness of AA1100 aluminium alloy sheets. Materials 14(23):7263
2. Cheng Z, Li Y, Li J et al (2022) Ultrasonic assisted incremental sheet forming: constitutive modeling and deformation analysis. J Mater Process Technol 299:117365
3. Abdel-Nasser Y, Ma N, Rashed S et al (2023) Numerical modelling for efficient analysis of large size multi-stage incremental sheet forming. J Manuf Mater Process 8(1):3
4. Zhang C, Lu SH, Li ZF et al (2018) Study on multi-pass incremental forming technology of deep cone parts with large angle. Hot Work Technol 47(21):135–137
5. Zhu ZS, Shi WX, Wang H et al (2022) Experimental study on wall thickness distribution in multi-pass incremental forming of TB8 titanium alloy vertical walls square box. J Netshape Form Eng 14(5):8–13
6. Zhu H, Li JL, Liu YB et al (2016) CNC incremental compound forming based on three-axis and five-axis. Forging Stamp Technol 41(8):18–23
7. Hussain G, Gao L, Dar NU (2007) An experimental study on some formability evaluation methods in negative incremental forming. J Mater Process Technol 186(1):45–53
8. Tisza M (2012) General overview of sheet incremental forming. J Achiev Mater Manuf Eng 55(1):113–120
9. Mo JH, Han F (2008) State of the arts and latest research on incremental sheet NC forming technology. China Mech Eng 19(4):491–497
10. Żaba K, Puchlerska S, Kuczek Ł et al (2022) Effect of step size on the formability of Al/Cu bimetallic sheets in single point incremental sheet forming. Materials 16(1):367
11. Yu JH, Jung KS, Murugesan M et al (2022) Study on the incremental sheet metal forming process using a metal foam as a die. Int J Mater Form 15(6):71

Open Access This chapter is licensed under the terms of the Creative Commons Attribution-NonCommercial-NoDerivatives 4.0 International License (http://creativecommons.org/licenses/by-nc-nd/4.0/), which permits any noncommercial use, sharing, distribution and reproduction in any medium or format, as long as you give appropriate credit to the original author(s) and the source, provide a link to the Creative Commons license and indicate if you modified the licensed material. You do not have permission under this license to share adapted material derived from this chapter or parts of it.

The images or other third party material in this chapter are included in the chapter's Creative Commons license, unless indicated otherwise in a credit line to the material. If material is not included in the chapter's Creative Commons license and your intended use is not permitted by statutory regulation or exceeds the permitted use, you will need to obtain permission directly from the copyright holder.

Numerical Analysis of Flow Performance and Design of Gear Pump with Unload Groove

Yi Wang, Bing Han, and Jianyao Wang

Abstract To improve the problem of oil trapping during the operation of external gear pumps, studies on the design of external gear pump teeth and unload groove and the analysis of the effect of their flow characteristics have been carried out. An external gear pump with a new type of unload groove was designed with reference to the general design specification. Based on the computational fluid dynamics (CFD) method, a numerical simulation model of the internal flow field of the gear pump was established, and the simulation analysis of the internal flow characteristics and pressure distribution of the gear pump was carried out to obtain the influence law of unload groove on the performance of the gear pump. The simulation results showed that the unload groove gear pump had a better flow rate, higher capacity efficiency, and less radial pressure than the Default gear pump under the same working conditions and design parameters. The study showed that the external gear pump with unload groove designed in this paper has significantly improved the problems caused by the oil trapping problem compared with the default gear pump.

Keywords External gear pump · CFD · Unload groove · Optimization design

1 Introduction

As design and manufacturing technologies continue to advance, gear pumps have gradually evolved into a significant power equipment in today's context. They are highly regarded for their simplicity in structure, compact size, lightweight, and low sensitivity to contamination, holding a significant share in hydraulic transmission

Y. Wang · J. Wang (✉)
School of Ocean and Civil Engineering, Shanghai Jiao Tong University, Shanghai, China
e-mail: kingone@sjtu.edu.cn; wjy2011@sjtu.edu.cn

B. Han
National Engineering Research Center of Ship & Shipping, Control System, Shanghai, China

© The Author(s) 2026

Z. Zhang, P. Mercorelli (eds.), *Mechanical Design and Simulation: Insights and Innovations*, Lecture Notes in Mechanical Engineering,
https://doi.org/10.1007/978-981-95-2405-1_51

and fluid control technology [1]. They find extensive applications in fields such as marine engineering and shipbuilding [2].

Gear pumps can be classified into two categories: internal meshing gear pumps and external meshing gear pumps. Compared to internal meshing gear pumps, external meshing gear pumps have a simpler structure, smaller size, ease of manufacturing, strong self-priming capability, and lower costs for both maintenance and production. However, external gear pumps still experience issues related to oil trapping [3], resulting in problems such as radial hydraulic imbalance of gears, severe bearing wear, significant flow pulsations, and high noise levels [4]. Therefore, the design and performance analysis of high-performance external meshing gear pumps have remained a focus of research [5].

To address the drawbacks of external meshing gear pumps, Zhang et al. [6] conducted simulations of the internal flow field of gear pumps using Fluent software. They identified the characteristics of oil trapping in gear pumps and analyzed the factors influencing it, as well as the issues related to internal leakage and gear stress. Through an analysis of the flow field characteristics of gear pumps, they elucidated the principles, characteristics, affected regions, and causes of oil trapping, providing a theoretical basis for the research and optimization design of external meshing gear pumps. In the effort to improve oil trapping in gear pumps, the design of unloading slots in gear pumps has received extensive research attention [7, 8]. These slots, situated on both sides of the gear meshing area, are designed to release the high-pressure oil formed during gear meshing, thereby mitigating the oil trapping phenomenon. Guo et al. [9] obtained the expression for the unloading area using the geometric pattern expansion method, providing a theoretical reference for unloading slot design. Liu et al. [10] conducted comprehensive design of unloading slots and numerical simulation of flow fields, demonstrating that the structurally designed unloading slots can effectively alleviate the oil trapping pressure in the gear meshing area, thus improving the oil trapping issue in gear pumps. It is evident that a well-designed unloading slot has a positive significance in addressing the oil trapping problem in gear pumps [11]. However, traditional unloading slot design methods are largely empirical and can only mitigate the oil trapping phenomenon to some extent. Moreover, the transient oil pressure remains relatively high, and improper design parameters can lead to issues such as reduced flow rates in gear pumps.

This paper aims to propose a design approach for an external gear pump and its unloading grooves based on task requirements. It establishes a numerical simulation model of the internal flow field of the gear pump using CFD methods. The paper conducts simulation analyses on the internal flow characteristics and pressure distribution of different types of gear pumps, and verifies that gear pumps with unloading grooves exhibit superior performance.

2 Gear Pump Model and Parameter Design

2.1 Pump Parameter Design

The research in this paper focuses on external gear pumps. To begin with, it is essential to design the functional parameters of the gear pump according to the task requirements. During the gear pump design process, precise determination of key parameters is imperative, including “required power”, “pump shaft input power”, “motor output power”, “motor speed”, “pump drive shaft speed”, and “displacement”. The accurate determination of these parameters significantly impacts the final performance of the gear pump. The design requirements for the gear pump studied in this paper are as follows: pressure differential $\Delta p = 0.1$ MPa, flow rate $Q = 1000$ mL/min, and motor current $I < 1$ A.

According to design specifications, the design parameters of the gear pump under study are as follows, as shown in Table 1.

In the calculation of the required power P_{liquid}, Δp represents the design pressure differential, and Q represents the design flow rate. In the calculation of the pump shaft input power P_{shaft}, η_r is the volumetric efficiency, which describes the energy losses due to liquid leakage and compression, and is assumed to be 0.8. η_m is the mechanical efficiency, which describes the friction losses between components or between components and the fluid, and is assumed to be 0.75. In the calculation of

Table 1 Gear pump design parameters calculation results

Parameter name and symbol	Method of calculation	Calculation results
Pump shaft input power P_{shaft}	$\frac{P_{liquid}}{\eta_r \eta_m}$	2.78 W
Motor output power P_{motor}	$\frac{P_{shaft}}{\eta_t}$	2.74 W
Motor speed n_{motor}	Motor performance curve	7100 rpm
Pump drive shaft speed n_{shaft}	$\frac{n_{motor}}{5}$	1420 rpm
Displacement q	$\frac{Q_t}{n_{shaft}} = \frac{Q}{\eta_r n_{shaft}}$	0.88 mL/r
Pump shaft input power P_{shaft}	$\frac{P_{liquid}}{\eta_r \eta_m}$	2.78 W
Motor output power P_{motor}	$\frac{P_{shaft}}{\eta_t}$	2.74 W
Motor speed n_{motor}	Motor performance curve	7100 rpm
Pump drive shaft speed n_{shaft}	$\frac{n_{motor}}{5}$	1420 rpm
Displacement q	$\frac{Q_t}{n_{shaft}} = \frac{Q}{\eta_r n_{shaft}}$	0.88 mL/r

motor output power P_{motor}, η_t represents the transmission efficiency of the driving mechanism and is assumed to be 0.9. The calculation of motor speed is based on the motor performance curve, as illustrated in Fig. 1. From the graph, the motor operates at the following parameters: output power of 4 W, output torque of approximately 5.5 mN · m, motor speed of approximately 7100 rpm, and input current of approximately 0.48 A, meeting the requirement of a motor operating current below 1 A. In the calculation of pump drive shaft speed n_{shaft}, n_{motor} represents the motor speed, and the transmission ratio between the motor and pump shaft is 5:1. The calculation of displacement q is based on the theoretical flow rate of the pump Q_t and the actual flow rate Q.

After the design is completed, it is necessary to verify the motor's output torque. The input torque of the pump shaft can be calculated as follows:

$$M_{shaft} = \frac{50}{\pi}\Delta pq\frac{1}{\eta_m}\times 10^{-3} = 18.68\left(\text{mN}\cdot\text{m}\right) \tag{1}$$

Where Δp represents the design pressure differential in bars.

This results in the motor's output torque as follows:

$$M_{motor} = \frac{M_{shaft}}{\eta_t}\cdot\frac{1}{5} = \frac{18.68}{0.9\times 5} = 4.15 < 5.5\left(\text{mN}\cdot\text{m}\right) \tag{2}$$

This complies with the design verification requirements.

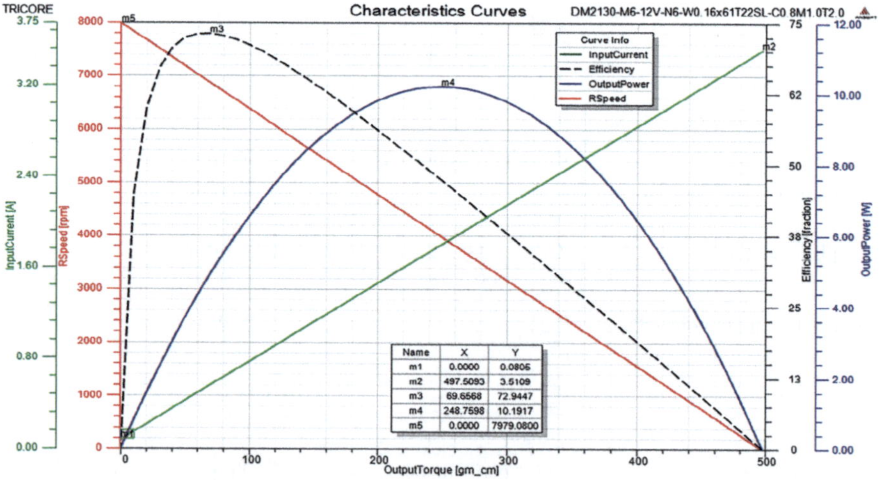

Fig. 1 Motor performance graphs

2.2 Unload Groove Design

In the design of external gear pumps, to ensure smooth operation, gear overlap generally exceeds 1. This leads to the occurrence of oil trapping during the pump's operation. Oil trapping, during a specific meshing period of the gears, results in the formation of a closed chamber between the two engaging teeth, which shrinks and expands as the gears rotate. When the chamber shrinks, the liquid inside is pressurized, causing a sharp increase in pressure, exerting a significant load on the gear shaft and bearings, leading to gear pump vibration, noise, and reduced operational stability and lifespan. When the chamber expands, the internal pressure drops, creating a vacuum, which can lead to cavitation and vibration noise issues.

In this design, double rectangular unload grooves are symmetrically arranged relative to the gear centerline. The positions of symmetrically arranged double unload grooves should satisfy the following conditions:

(a) When the trapped oil volume begins to decrease, this volume should be connected to the pressure oil chamber.
(b) When the trapped oil volume is at its minimum, it should not be connected to either the pressure oil chamber or the suction oil chamber.
(c) When the trapped oil volume begins to increase, it should be connected to the suction oil chamber.

Figure 2 illustrates the structure of symmetric double rectangular unload grooves with gear tooth clearance. In the figure, the trapped oil volume V_B is at its minimum, and the edges of the two unload grooves align precisely with engagement points D and E. The distance between the two unload grooves ω should ensure that the trapped oil volume V_B remains connected to the pressure oil chamber until it reaches its minimum

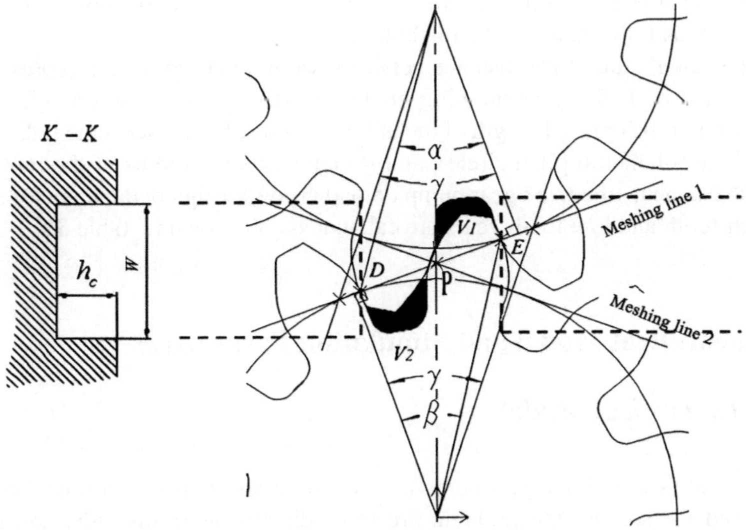

Fig. 2 Symmetrical double rectangular discharge chute

Table 2 Unload grooves design parameters

Parameter name and symbol	Method of calculation	Calculation results (mm)
Distance between two unload grooves a	$\pi m cos^2\alpha_n$	2.2
Distance between high-pressure side unload groove edge and gear centerline a_g	$\frac{1}{2}(\pi m cos\alpha_n - b tan\beta) \cdot cos\ \alpha_n$	0.116
Distance between low-pressure side unload groove edge and gear centerline a_d	$\frac{1}{2}(\pi m cos\alpha_n + b tan\beta) \cdot cos\ \alpha_n$	2.104
Optimal unload groove length c	$2\left(R - \sqrt{R_f^2 - \left(\frac{a}{2}\right)^2}\right)$	2.21
Unload groove depth h	$h \geq 0.8$ mm	0.8

Table 3 Design parameters of unload grooves for different gear pumps

Parameter name and symbol	Unit	High-tooth gear	Full-tooth gear	Low-tooth gear
Unload groove spacing a	mm	2.2		
Unload groove length C	mm	2.85	2.21	1.79
Unload groove depth h	mm	0.8		

position. When V_B reaches the minimum position, the trapped oil volume V_B should not be connected to either the pressure oil chamber or the suction oil chamber. After passing the minimum position, V_B should remain connected to the suction oil chamber.

Therefore, the dimension of a has strict requirements. If a is too large, the oil trapping phenomenon cannot be completely eliminated, and if it is too small, it may cause leakage by connecting the suction oil chamber and the pressure oil chamber, reducing the efficiency of the gear pump. The final design parameters for the unload grooves are as follows, as shown in Table 2.

In the calculation of the distance between two unload grooves, m represents the module, and α_n is the gear meshing angle (standard gear pressure angle). In the calculation of the optimal length of the unload groove, R represents the radius of the pitch circle (diametral pitch circle), and R_f represents the radius of the root circle. Using the same method, the gear pump unload groove design parameters for mounting high-tooth and low-tooth gears are calculated and shown in Table 3.

3 Theoretical Model and Simulation Parameters

3.1 *Theoretical Model*

The internal flow within a gear pump involves high-speed turbulent flow. The equations used for solving the problem are three-dimensional, unsteady, and incompressible mass and momentum conservation equations [12]:

$$\frac{\partial \rho}{\partial t}+\frac{\partial\left(\rho \overline{u_j}\right)}{\partial x_j}=0 \tag{3}$$

$$\rho\frac{\partial \overline{u_i}}{\partial t}+\rho\overline{u_j}\frac{\partial \overline{u_i}}{\partial x_j}=\rho f_i-\frac{\partial \overline{p}}{\partial x_i}+\mu\frac{\partial^2 \overline{u_i}}{\partial x_j \partial x_j}-\rho\frac{\partial \overline{u_i' u_j'}}{\partial x_j} \tag{4}$$

For simulating the turbulent flow, the RNG $k-\varepsilon$ model is chosen due to its accuracy. The RNG model equations for non-steady flow simulation are as follows:

$$\rho\frac{Dk}{Dt}=\frac{\partial}{\partial x_j}\left(\alpha_k u_{eff}\frac{\partial_k}{\partial x_j}\right)+2u_t\overline{S_{ij}}\frac{\partial \overline{u_i}}{\partial x_j}-\rho\varepsilon \tag{5}$$

$$\rho\frac{D\varepsilon}{Dt}=\frac{\partial}{\partial x_j}\left(\alpha_\varepsilon u_{eff}\frac{\partial_\varepsilon}{\partial x_j}\right)+2C_{1\varepsilon}\frac{\varepsilon}{k}v_t\overline{S_{ij}}\frac{\partial \overline{u_i}}{\partial x_j}-C_{2\varepsilon}\rho\frac{\varepsilon^2}{k}-R \tag{6}$$

Where: $C_{1\varepsilon}$ and $C_{2\varepsilon}$ represent empirical constants; k represents turbulent kinetic energy; ε represents turbulent dissipation rate; α_k and α_ε represent turbulent flow dissipation rates.

3.2 Simulation Parameters and Settings

To validate the performance of the designed unload grooves, two commonly used gear pump types are selected for comparative simulation analysis: the standard gear pump and the inlet-outlet gear pump. The models for each type of gear pump are shown in Fig. 3.

After importing the simulation models into PumpLinx, simulation parameters are set based on common working conditions for ship gear pumps. The working fluid is water, and both inlet and outlet boundary conditions are set as pressure boundaries. The ambient temperature is 300 K, and the time step for a single driven wheel is set to 30 s. The gear rotation speed is 1420 rpm.

Since the flow inside the gear pump is three-dimensional and unsteady, using a standard grid for calculations would result in short time steps due to the narrow gap

a) conventional gear pump b) inlet and outlet gear pump c) unload groove gear pump

Fig. 3 Various gear pump models. (**a**) Conventional gear pump, (**b**) inlet and outlet gear pump, (**c**) unload groove gear pump

a) conventional gear pump b) inlet and outlet gear pump c) unload groove gear pump

Fig. 4 Simulation model meshing for different types of pump models. (**a**) Conventional gear pump, (**b**) inlet and outlet gear pump, (**c**) unload groove gear pump

between gears, increasing simulation costs. To address this issue, PumpLinx utilizes the Geometry Conformal Adaptive Binary Tree (CAB) algorithm. The software generates Cartesian grids for inlet, outlet, and unload grooves, and generates structured grids around the gears, effectively improving computational efficiency. The meshing for this simulation is done using PumpLinx's built-in automated grid generator, creating high-quality Cartesian hexahedral grids as shown in Fig. 4.

4 Simulation Results and Analysis

During the operation of a gear pump, its performance differences are reflected in several parameters, such as the inlet and outlet flow rates and pressure, which can reflect the pump's conveying performance, and the volumetric efficiency, which can reflect the cavitation condition and flow quality within the gear pump. Therefore, this section will focus on the simulation results of these parameters to explore the performance differences among different types of gear pumps.

To explore the performance of the designed unload groove gear pump, this study conducted comparative simulations for three types of gear pumps: standard gear pump, inlet-outlet gear pump, and unload groove gear pump, all with the same gear shape and under the same operating conditions. The simulation results are analyzed to evaluate the performance differences among different gear pump types.

After mesh generation and simulation parameter settings, internal flow characteristics of the standard gear pump, inlet-outlet gear pump, and unload groove gear pump were simulated using the PumpLinx software. The final simulation results are presented below. Figure 5 shows the pressure distribution simulation results for the three types of gear pumps, Fig. 6 shows the flow rate distribution simulation results, and Fig. 7 includes charts showing changes in various parameters during the simulation.

From Fig. 5, it is evident that in a gear pump, the contact area between the two gears is the high-pressure region in the pump. The simulation results show that the liquid pressure in this region in the unload groove gear pump is significantly lower

a) conventional gear pump b) inlet and outlet gear pump c) unload groove gear pump

Fig. 5 Gear pump pressure distribution simulation results. (**a**) Conventional gear pump, (**b**) inlet and outlet gear pump, (**c**) unload groove gear pump

a) conventional gear pump b) inlet and outlet gear pump c) unload groove gear pump

Fig. 6 Gear pump flow rate distribution simulation result. (**a**) Conventional gear pump, (**b**) inlet and outlet gear pump, (**c**) unload groove gear pump

than in the other two types of gear pumps, indicating an improvement in mitigating oil trapping in this area. Figure 7a–c show that for all three gear pump types, the unload groove gear pump has significantly smaller variations in inlet and outlet flow rates and pressure.

Additionally, based on gear pump design parameters, the ideal fluid discharge for the gear pump can be calculated using Eq. (7), and ideal flow rates can be calculated using Eq. (8). These calculated values serve as the basis for calculating the volumetric efficiency of the gear pump for each type and assessing the flow field quality.

$$q_{ideal} = 4\pi R\left(R_a - R\right)B \times 10^{-3} = 2h_{ac}\pi Zm^2 B \times 10^{-3} \tag{7}$$

$$Q_{ideal} = q_{ideal} \times n_{shaft} \tag{8}$$

Finally, the simulation results for the three types of gear pumps are summarized in Table 4.

It is clear that the unload groove gear pump outperforms the other two types of gear pumps in terms of inlet and outlet flow rates, gear shaft loads, and volumetric efficiency. The unload groove type also exhibits smoother operation with smaller peak values in its performance curves. Therefore, compared to the other two types, the unload groove gear pump demonstrates superior performance, lower pressure, and reduced noise.

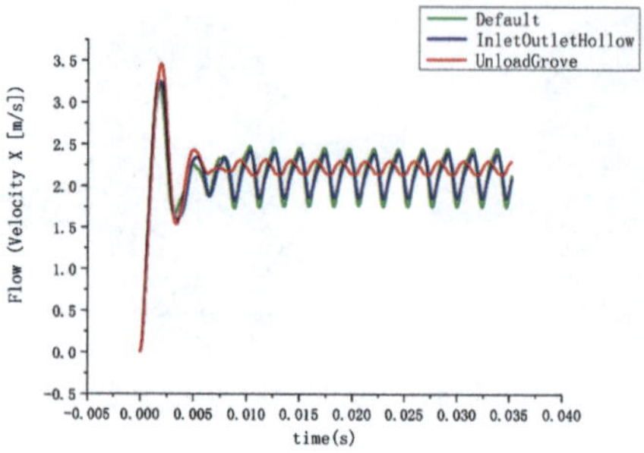

a) Inlet flow rate variation

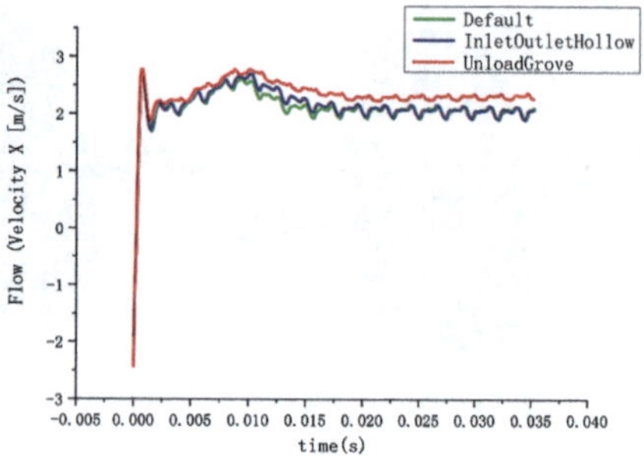

b) Outlet flow rate variation

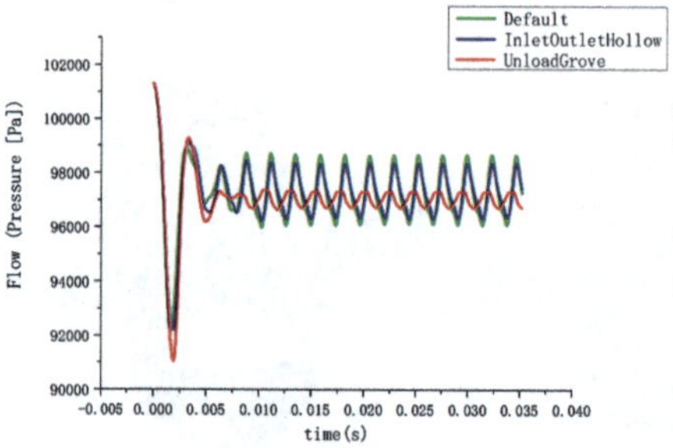

c) Inlet pressure variation

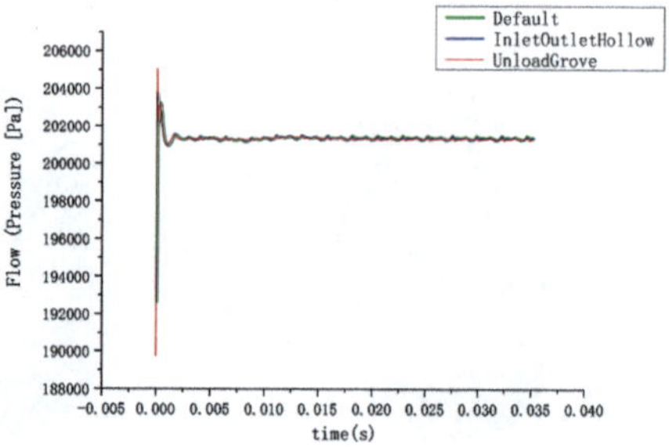

d) Outlet pressure variation

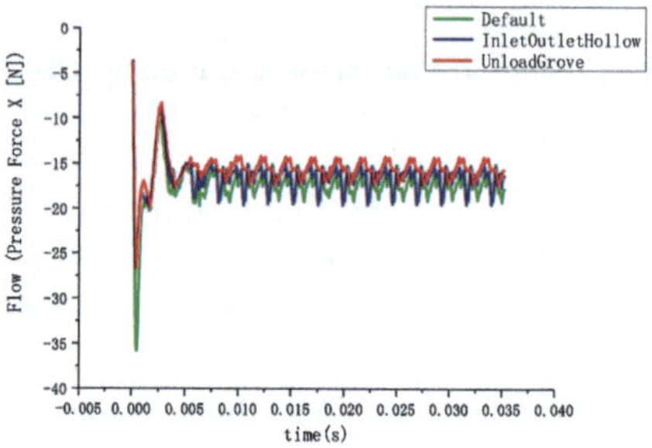

e) X-direction driving force variation

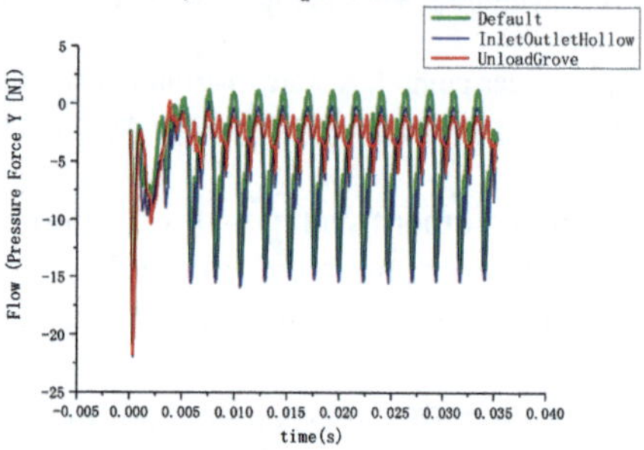

f) Y-direction driving force variation

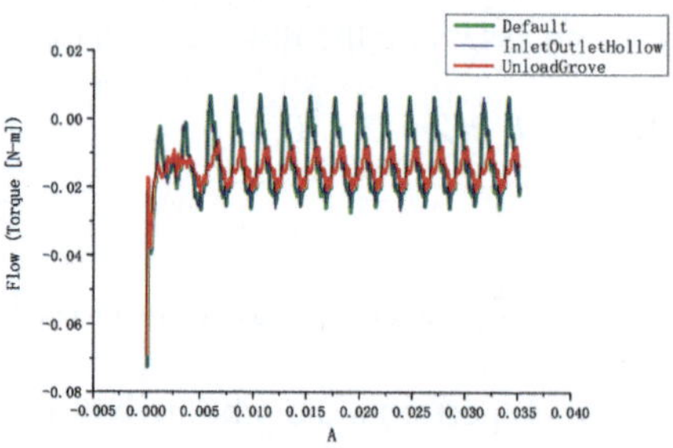

g) Gear driving torque

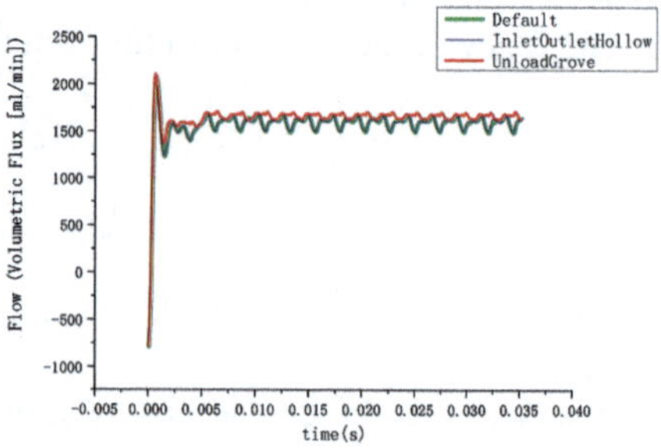

h) Flow rate variation

Fig. 7 Comparison of simulation results of different gear pumps. (**a**) Inlet flow rate variation. (**b**) Outlet flow rate variation. (**c**) Inlet pressure variation. (**d**) Outlet pressure variation. (**e**) X-direction driving force variation. (**f**) Y-direction driving force variation. (**g**) Gear driving torque. (**h**) Flow rate variation

Table 4 Simulation results for different type of gear pump

Simulation parameters	Conventional gear pump	Inlet and outlet gear pump	Unload groove gear pump
Inlet flow rate (m/s)	2.1	2.1	2.2
Outlet flow rate (m/s)	2.2	2.2	2.4
Inlet pressure (Pa)	97,500	97,500	97,000
Outlet pressure (Pa)	201,500	201,500	201,500
X-direction gear driving force (N)	17	17	15
Y-direction gear driving force (N)	7	7	3
Gear driving torque (N-m)	0.015	0.015	0.015
Flow rate (mL/min)	1500	1500	1600
Volumetric efficiency	86.57%	86.57%	92.35%

5 Conclusion

This study explored a method for assisting the design of gear pumps by analyzing the internal flow field through simulations. It resulted in the design of a high-performance unload groove gear pump. The study first performed the basic parameter design for a gear pump based on given performance requirements. Furthermore, it proposed a novel gear pump unload groove design to address the problem of oil trapping in external gear pumps. Standard turbulent flow modeling was used to characterize the flow behavior within the gear pump, and simulations were conducted using PumpLinx software to analyze different types of gear pumps. The key findings are as follows:

1. A novel unload groove gear pump was designed based on the given performance requirements.
2. Through simulation and analysis of the internal flow fields of different types of gear pumps operating under the same conditions, it was found that the unload groove gear pump outperformed the other two types of gear pumps in terms of inlet and outlet flow rates, gear shaft load, flow rate, and volumetric efficiency.

Acknowledgments This work was supported by the National Natural Science Foundation of China (12202265) and the Natural Science Foundation of Fujian Province of China (2022J01131710).

References

1. Castilla R et al (2009) Experimental study of the shaft motion in the journal bearing of a gear pump. J Eng Gas Turbines Power 131(5):52502
2. Husár Š (2016) Gear of hydraulic pumps for ship equipment. NAŠE MORE 63(3):108–114
3. Yue W et al (2020) The performance simulation for a gear pump by trapped oil model. In: 2020 11th international conference on mechanical and aerospace engineering (ICMAE), pp 223–227

4. Wang S, Sakurai H, Kasarekar A (2011) The optimal design in external gear pumps and motors. IEEE ASME Trans Mechatron 16(5):945–952
5. Yin DY et al (2014) A new way to eliminate the phenomenon of trapping oil in gear pumps. Appl Mech Mater 628:181–185. (Materials Science, Mechanical Engineering and Applied Research)
6. Zhang J, Mao Z, Yang G (2014) The simulation and analysis of the flow field for external helical gear pump. Chin Hydraul Pneum 270(2):10–13
7. Gamez-Montero PJ, Garcia-Vilchez M, Raush G et al (2012) Teeth clearance and relief grooves effects in a trochoidal-gear pump using new modules of GeroLAB. J Mech Des 134(5):054502
8. Liu H et al (2022) A new type unloading groove design and simulation for engine fuel gear pump. In: 2022 13th international conference on mechanical and aerospace engineering (ICMAE), pp 86–90
9. Guo S, Chen D (2020) Calculation of unloading area of internal gear pump and optimization. Math Probl Eng 2020:1–9
10. Liu Z, Hu Y, Cui L et al (2020) Design and inner flow simulation for a gear pump with relief groove. Chin Hydraul Pneum 349(9):100–107
11. An S, Li L (2016) Optimization design of relief grooves in external gear pump inside hydraulic automatic transmission. Mach Des Manuf 304(6):126–129
12. Bai K, Ma X, Li Y et al (2021) Numerical simulations of internal flow field and mass flow characteristic of gear pumps in torpedo based on PumpLinx. Ship Sci Technol 43(5):56–59

Open Access This chapter is licensed under the terms of the Creative Commons Attribution-NonCommercial-NoDerivatives 4.0 International License (http://creativecommons.org/licenses/by-nc-nd/4.0/), which permits any noncommercial use, sharing, distribution and reproduction in any medium or format, as long as you give appropriate credit to the original author(s) and the source, provide a link to the Creative Commons license and indicate if you modified the licensed material. You do not have permission under this license to share adapted material derived from this chapter or parts of it.

The images or other third party material in this chapter are included in the chapter's Creative Commons license, unless indicated otherwise in a credit line to the material. If material is not included in the chapter's Creative Commons license and your intended use is not permitted by statutory regulation or exceeds the permitted use, you will need to obtain permission directly from the copyright holder.

Research on Topological Grinding of Circumferentially Arranged Fish Scale Structured Surfaces on Inner Cylindrical Surfaces

Yuanzhi Qi, Xingshan Li, and Yushan Lyu

Abstract Due to the excellent drag-reducing performance of fish scale surfaces in fluid environments, this study aims to grind a circumferentially arranged fish scale-structured surface on the inner cylindrical surface of a workpiece. Based on the characteristics of traveling wave structures and scale morphology, a fish scale surface model with circumferential arrangement on the inner cylinder was established, a topological mapping matrix was constructed, and a structured grinding wheel was designed. The effects of different speed ratios and grinding depths on the surface morphology were also investigated. The simulation results indicate that the designed grinding wheel can process a circumferentially ordered fish scale microstructure, and the grinding parameters have a significant regulatory effect on the texture morphology.

Keywords Inner cylindrical grinding · Fish scale surfaces · Structured grinding wheel · Abrasive clusters

1 Introduction

In recent years, fish scale surfaces with traveling wave characteristics have garnered significant attention [1, 2] due to their excellent drag-reducing performance in fluid environments. The key to achieving this drag reduction effect lies in the successful fabrication of surfaces with similar structural characteristics. Scholars both domestically and internationally [3–9] have proposed a series of innovative manufacturing techniques. For example, Piotr [3] used a diamond dresser to create millimeter-scale structured surfaces on conventional ceramic grinding wheels; Guo et al. [4] applied laser processing technology to produce structured grinding wheels for the surface machining of hard and brittle material workpieces. However, there is limited research on the grinding of inner cylindrical surfaces in the existing grinding

Y. Qi · X. Li (✉) · Y. Lyu
School of Mechanical Engineering, Shenyang Ligong University, Shenyang, Liaoning, China

© The Author(s) 2026

Z. Zhang, P. Mercorelli (eds.), *Mechanical Design and Simulation: Insights and Innovations*, Lecture Notes in Mechanical Engineering,
https://doi.org/10.1007/978-981-95-2405-1_52

studies, and the methods for designing grinding wheels based on workpiece surface morphological characteristics are also scarce.

In this study, a grinding wheel and grinding process suitable for the fish scale structured inner cylindrical surface were designed using a reverse solving method. The feasibility of the design was verified through simulation experiments, which also revealed the impact of grinding parameters on the texture characteristics of the inner fish scale surface.

2 Fish Scale Surface Characteristic and Model Establishment

2.1 Fish Scale Unit Contour and Traveling Wave Structural Characteristics

Studies have shown that the surfaces of different types of fish scales [10] have common characteristics: overlapping scales, longitudinal staggered arrangement, and each scale consisting of four elliptical arcs (see Fig. 1). Additionally, Fig. 2

Fig. 1 Surface structure of fish scales

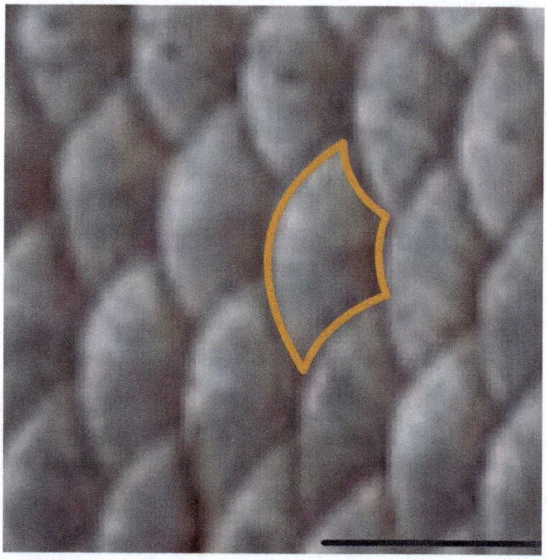

Fig. 2 Schematic diagram of the traveling wave structure

illustrates a schematic of the traveling wave indentations formed on a solid surface when in contact with fluid. The fish scale surface can be considered as a special type of traveling wave structured surface.

2.2 *The Model Establishment of the Inner Circular Fish-Scale Surface*

Based on the contour analysis of the fish scale units, the schematic diagram of the inner cylindrical fish scale surface shown in Fig. 3 was designed. The length $\boldsymbol{l}_w$, width $\boldsymbol{w}_w$, and depth $\boldsymbol{d}_w$ of the fish scale units are evenly divided. The length direction of the fish scale unit is divided into $\boldsymbol{a}$ equal units, and the width direction is divided into $\boldsymbol{b}$ equal units. By determining the position coordinates of the fish scale unit, the coordinate function in the depth direction can be obtained: $\boldsymbol{d}_w = \boldsymbol{f}_w(\boldsymbol{l}_w, \boldsymbol{w}_w)$. Therefore, the feature sets in the length, width, and depth directions of the fish scale units are:

$$\boldsymbol{L}_w = \begin{bmatrix} l_{w1} & l_{w2} & \cdots & l_{w(a-1)} & l_{wa} \end{bmatrix}^T, \quad \boldsymbol{W}_w = \begin{bmatrix} w_{w1} & w_{w2} & \cdots & w_{w(b-1)} & w_{wb} \end{bmatrix}^T$$

$$\boldsymbol{D}_w = \begin{bmatrix} d_{w11} & d_{w12} & \cdots & d_{w1(a-1)} & d_{w1a} \\ d_{w21} & d_{w22} & \cdots & d_{w2(a-1)} & d_{w2a} \\ \vdots & \vdots & \ddots & \vdots & \vdots \\ d_{w(b-1)1} & d_{w(b-1)2} & \cdots & d_{w(b-1)(a-1)} & d_{w(b-1)a} \\ d_{wb1} & d_{wb2} & \cdots & d_{wb(a-1)} & d_{wba} \end{bmatrix} \tag{1}$$

Thus, the topological feature vector $\boldsymbol{M}_w$ of the fish-scale unit can be expressed as:

$$\boldsymbol{M}_w = \left[\boldsymbol{L}_w\, \boldsymbol{W}_w\, \boldsymbol{D}_w\, 1\right]^T \tag{2}$$

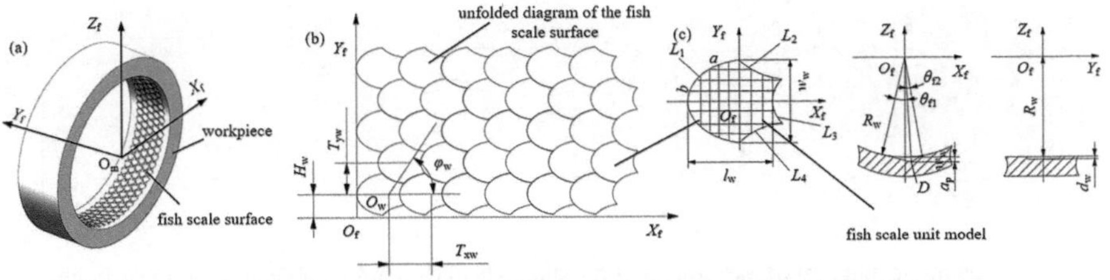

(a) Schematic diagram of the inner cylindrical surface of the workpiece (b) Unfolded diagram of the fish scale surface (c) Three-view diagram of the fish scale unit

Fig. 3 Schematic diagram of the arrangement of inner cylindrical fish scale units. (**a**) Schematic diagram of the inner cylindrical surface of the workpiece. (**b**) Unfolded diagram of the fish scale surface. (**c**) Three-view diagram of the fish scale unit

Based on the arrangement properties of the inner cylindrical fish scale structured surface, it is known that it is primarily related to the arrangement period $\boldsymbol{T}_{\mathbf{xw}}$ in the $\boldsymbol{X}_{\mathbf{f}}$ direction, the phase shift $\boldsymbol{\varphi}_{\mathbf{w}}$ when staggered, and the arrangement period $\boldsymbol{T}_{\mathbf{yw}}$ in the $\boldsymbol{Y}_{\mathbf{f}}$ direction. Therefore, the arrangement vector $\boldsymbol{T}_{\mathbf{fw}}$ of the inner cylindrical fish scale units can be established:

$$\boldsymbol{T}_{fw} = \left[\boldsymbol{T}_{xw}\ \boldsymbol{T}_{yw}\ \boldsymbol{\varphi}_{w}\right]^{T} \tag{3}$$

2.3 Structured Grinding Wheel Design

Based on the principles of inner cylindrical grinding and the topological mapping method, combined with the inner cylindrical fish scale structural unit model and its arrangement pattern, a schematic diagram of the grinding wheel fish scale abrasive cluster contour arranged circumferentially, as shown in Fig. 4, was designed. The matrices for the unit length $\boldsymbol{L}_{\mathbf{s}}$, unit width $\boldsymbol{W}_{\mathbf{s}}$, and unit depth $\boldsymbol{D}_{\mathbf{s}}$ of the abrasive cluster are as follows:

$$\boldsymbol{L}_{s} = \left[l_{s1}\ \ l_{s2}\ \ \cdots\ \ l_{s(a-1)}\ \ l_{sa}\right]^{T},\ \boldsymbol{W}_{s} = \left[w_{s1}\ \ w_{s2}\ \ \cdots\ \ w_{s(b-1)}\ \ w_{sb}\right]^{T}$$

$$\boldsymbol{D}_{s} = \begin{bmatrix} d_{s11} & d_{s12} & \cdots & d_{s1(a-1)} & d_{s1a} \\ d_{s21} & d_{s22} & \cdots & d_{s2(a-1)} & d_{s2a} \\ \vdots & \vdots & \ddots & \vdots & \vdots \\ d_{s(b-1)1} & d_{s(b-1)2} & \cdots & d_{s(b-1)(a-1)} & d_{s(b-1)a} \\ d_{sb1} & d_{sb2} & \cdots & d_{sb(a-1)} & d_{sba} \end{bmatrix} \tag{4}$$

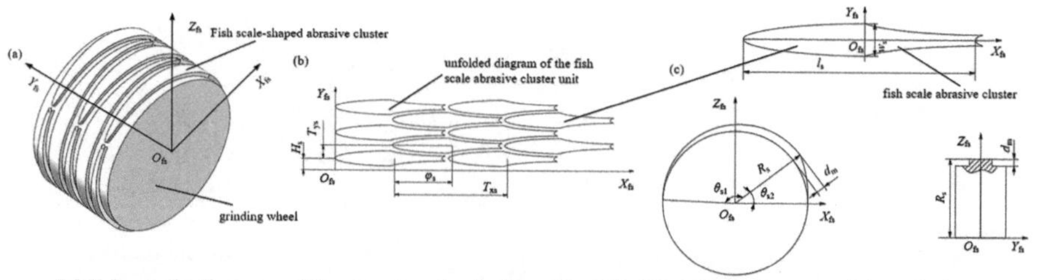

(a) Schematic diagram of the structured grinding wheel (b) Unfolded diagram of the grinding wheel surface (c) Three-view diagram of the abrasive cluster unit

Fig. 4 Schematic diagram of the fish scale structured grinding wheel. (**a**) Schematic diagram of the structured grinding wheel. (**b**) Unfolded diagram of the grinding wheel surface. (**c**) Three-view diagram of the abrasive cluster unit

Thus, the topological feature vector $\boldsymbol{M}_s$ of the grinding wheel abrasive cluster unit, the topological transformation matrix $\boldsymbol{T}_{fs}$ for the arrangement feature parameters of the grinding wheel abrasive cluster:

$$\boldsymbol{M}_s = \begin{bmatrix} \boldsymbol{L}_s & \boldsymbol{W}_s & \boldsymbol{D}_s & 1 \end{bmatrix}^T, \quad \boldsymbol{T}_{fs} = \begin{bmatrix} \boldsymbol{T}_{xs} & \boldsymbol{T}_{ys} & \varphi_s \end{bmatrix}^T \tag{5}$$

The space occupied by the grinding wheel abrasive cluster unit can be regarded as a topological space $\boldsymbol{M}_s$, and the space occupied by the workpiece fish scale unit also constitutes a topological space $\boldsymbol{M}_w$. From a topological perspective, there exists a mapping relationship between them, namely: $\boldsymbol{M}_w \rightarrow \boldsymbol{M}_s$. Figure 5 shows a schematic diagram of grinding with a structured grinding wheel. In actual grinding motion, the grinding wheel only has feed motion in the circumferential direction, so the topological feature length $\boldsymbol{l}_s$ is stretched in the circumferential direction, while the width $\boldsymbol{w}_s$ remains unchanged. Based on the point set topology theory and grinding principles, the following topological mapping equations are established:

$$\boldsymbol{L}_w = \begin{bmatrix} \frac{\omega_s}{\omega_w} & \cdots & 0 \\ \vdots & \ddots & \vdots \\ 0 & \cdots & \frac{\omega_s}{\omega_w} \end{bmatrix}^{-1} \boldsymbol{L}_s + \begin{bmatrix} \frac{\tau\omega_s}{a\omega_w} & \cdots & 0 \\ \vdots & \ddots & \vdots \\ 0 & \cdots & -\frac{\tau\omega_s}{a\omega_w} \end{bmatrix}^{-1}, \quad \boldsymbol{W}_w = \begin{bmatrix} 1 & \cdots & 0 \\ \vdots & \ddots & \vdots \\ 0 & \cdots & 1 \end{bmatrix}^{-1} \boldsymbol{W}_s$$

$$\boldsymbol{M}_w = \begin{bmatrix} \frac{\omega_s}{\omega_w} & 0 & 0 & -\tau\frac{\omega_s}{\omega_w} \\ 0 & 1 & 0 & 0 \\ 0 & 0 & -1 & 0 \\ 0 & 0 & 0 & 1 \end{bmatrix}^{-1} \boldsymbol{M}_s, \quad \boldsymbol{T}_{fw} = \begin{bmatrix} \frac{\omega_s}{\omega_w} & 0 & 0 \\ 0 & 1 & 0 \\ 0 & 0 & \frac{\omega_s}{\omega_w} \end{bmatrix}^{-1} \boldsymbol{T}_{fs} \tag{6}$$

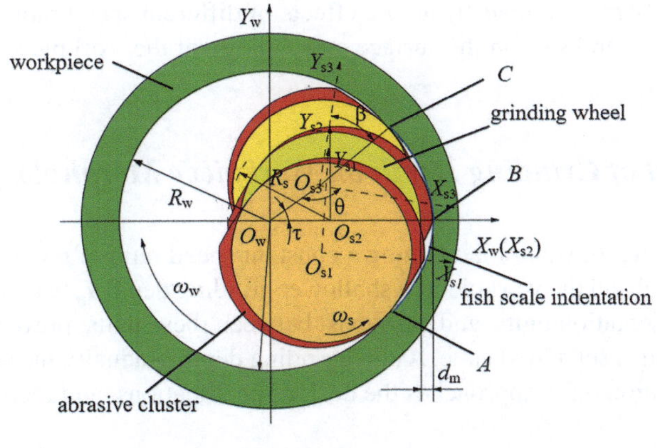

Fig. 5 Schematic diagram of grinding with a grinding wheel on the inner cylindrical surface

In these equations, ω_S represents the angular velocity of the grinding wheel, and ω_w represents the angular velocity of the workpiece.

3 Grinding Process Simulation Analysis

3.1 Simulation Conditions and Methods

According to the grinding motion principle shown in Fig. 5, a structured grinding wheel model is first established using simulation software. By inputting grinding parameters, a motion simulation of the fish scale-structured surface is conducted to obtain the coordinates of the inner cylindrical surface of the workpiece. The model of the inner cylindrical surface is then evaluated to check for interference between the coordinate points and the surface. If interference is detected, the coordinate points are assigned to the nearest grid points, resulting in a complete fish scale-structured surface morphology. Due to potential deviations between the simulation results and the designed dimensions, let ξ represent the magnitude of the deviation, which is controlled within 1.25%.

$$\xi = \left| \frac{l_w}{w_w} - \frac{6.28}{5} \right| \times 100\% \tag{7}$$

A workpiece with an inner diameter of Φ40 mm, an outer diameter of Φ60 mm, and a width of 25 mm was selected. Each fish scale indentation unit was designed with a unit length of 6.28 mm, a width of 5 mm, and a depth of 20 μm. Given the workpiece feed speed, for the stability of the grinding wheel during the grinding process and the aforementioned topological mapping relationship, the contour of the fish scale abrasive cluster unit was designed in the shape of an elliptical arc, with a major axis of 47 mm and a minor axis of 5 mm. The fish scale abrasive clusters are arranged in 2 units circumferentially and 5 rows axially. Given the workpiece feed speed v_w = 3.14 m/min, investigate the effects of different speed ratios ω_S / ω_-, i.e., p and grinding depths a_p on the surface morphology of the workpiece.

3.2 Effect of Grinding Depth on Workpiece Morphology

As shown in Fig. 6, when maintaining a constant speed ratio of p = 10 between the grinding wheel and the workpiece, a shallower grinding depth a_p (e.g.,10 μm) results in smaller indentation units, and gaps exist between these units, preventing overlap, thus displaying a separated state. As the grinding depth gradually increases, the size of the indentation units approaches the design specifications, and their arrangement

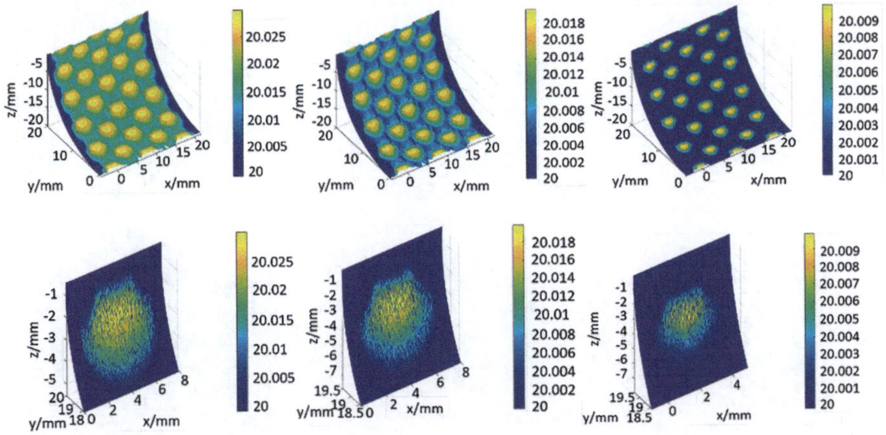

(a) v_w=3.14m/min, p=10, a_p=30μm; (b) v_w=3.14m/min, p=10, a_p=20μm; (c) v_w=3.14m/min, p=10, a_p=10μm.

Fig. 6 Simulation diagrams of the inner cylindrical surface of the workpiece ground at different depths of cut. (**a**) v_w = 3.14 m/min, p = 10, a_p = 30 μm; (**b**) v_w = 3.14 m/min, p = 10, a_p = 20 μm; (**c**) v_w = 3.14 m/min, p = 10, a_p = 10 μm

changes, transitioning from a separated state to an adjacent state, and eventually, at greater grinding depths a_p (e.g.,30 μm), the units start to intersect or overlap.

.

3.3 *Effect of Speed Ratio on Workpiece Morphology*

As shown in Fig. 7, with the increase in speed ratio p, both the length of the fish scale units and their circumferential arrangement undergo significant changes. Specifically, when the p is relatively low (e.g., 5), the fish scale units are longer, and the units are separated from each other, displaying a separated state. As the speed ratio p increases, the length of the fish scale units gradually shortens, transitioning from a separated state to an adjacent state (e.g., 10) and finally to an intersecting state (e.g., 15). A higher speed ratio leads to overlapping between the indentation units, reflecting an intersecting arrangement characteristic. Additionally, the number of fish scale units increases as the speed ratio increases, significantly enhancing the circumferential arrangement density. The characteristic values are shown in Table 1.

.

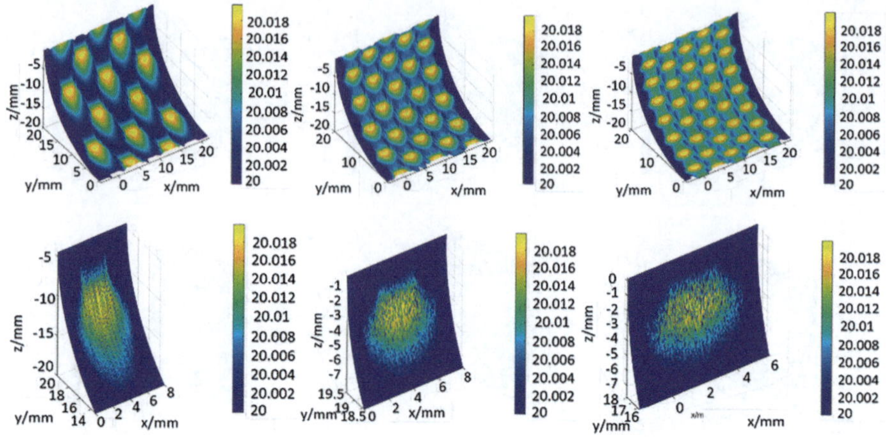

(a) v_w=3.14m/min, p=5, a_p=20μm; (b) v_w=3.14m/min, p=10, a_p=20μm; (c) v_w=3.14m/min, p=15, a_p=20μm.

Fig. 7 Simulation diagrams of the inner cylindrical surface of the workpiece ground at different spindle speeds. (**a**) v_w = 3.14 m/min, p = 5, a_p = 20 μm; (**b**) v_w = 3.14 m/min, p = 10, a_p = 20 μm; (**c**) v_w = 3.14 m/min, p = 15, a_p = 20 μm

Table 1 Comparison between the simulated pit parameters and the designed pit parameters

Grinding depth a_p (μm)	Speed ratio p	Length l_w (mm)	Width w_w (mm)	Depth d_w (mm)	Deviation value ξ (%)
10	10	4.22	4.05	10	20.6
20	5	12.44	5	20	123.2
20	10	6.23	5	20	1.01
20	15	4.07	5	20	44.2
30	10	6.23	5	30	1.01

4 Conclusion

By analyzing the contour of fish scale units and based on the principles of inner cylindrical grinding and point set topology theory, a circumferentially arranged fish scale-structured grinding wheel was designed and validated through simulation. The following conclusions were drawn: the topological mapping method enables high-precision grinding of fish scale surfaces with errors controlled within 1.25%; different grinding parameters affect the morphology and arrangement of fish scale units (separated, adjacent, intersecting); due to the randomness of abrasive particle size, increasing abrasive density is necessary to ensure processing quality.

Acknowledgements The authors gratefully acknowledge the support for this work from Chinese National Science Foundation (No. 51875368).

References

1. Browning A, Ortiz C, Boyce CM (2013) Mechanics of composite elasmoid fish scale assemblies and their bioinspired analogues. J Mech Behav Biomed Mater 19:75–86. https://doi.org/10.1016/j.jmbbm.2012.11.003
2. Wanting R, Haifeng Z, Tengjiao Z et al (2020) Drag reduction using lubricant-impregnated anisotropic slippery surfaces inspired by bionic fish scale surfaces containing micro-/nano-structured arrays. Adv Eng Mater 23(1). https://doi.org/10.1002/adem.202000821
3. Piotr S (2009) Regular surface texture generated by special grinding process. J Manuf Sci Eng 131(1):011015. https://doi.org/10.1115/1.3070511
4. Guo B, Zhao Q (2015) On-machine dry electric discharge truing of diamond wheels for micro-structured surfaces grinding. Int J Mach Tools Manuf 88:62–70. https://doi.org/10.1016/j.ijmachtools.2014.09.011
5. Islam MM, Kim H, Han SD et al (2016) Convex diamond patterns by grinding with a wheel which is dressed by a rounded tool. J Mech Sci Technol 30(4):1865–1873. https://doi.org/10.1007/s12206-016-0344-x
6. Kim H, Ko JT (2015) Verification of simulation of surface texturing on planar surface by grinding. Int J Precis Eng Manuf 16(2):225–231. https://doi.org/10.1007/s12541-015-0030-4
7. Silva DJE, Bottene CA, Oliveira DGFJ et al (2016) Grinding process for profiled texturing. CIRP Ann Manuf Technol 65(1):337–340. https://doi.org/10.1016/j.cirp.2016.04.116
8. Denkena B, Leon L, Wang B (2009) Grinding of microstructured functional surfaces: a novel strategy for dressing of microprofiles. Prod Eng 3(1):41–48. https://doi.org/10.1007/s11740-008-0134-0
9. Chen C, Tang J, Chen H et al (2018) An active manufacturing method of surface micro structure based on ordered grinding wheel and ultrasonic-assisted grinding. Int J Adv Manuf Technol 97(5–8):1627–1635. https://doi.org/10.1007/s00170-018-2044-4
10. Muthuramalingam M, Puckert KD, Rist U et al (2020) Transition delay using biomimetic fish scale arrays. Sci Rep 10(1):14534. https://doi.org/10.1038/s41598-020-71434-8

Open Access This chapter is licensed under the terms of the Creative Commons Attribution-NonCommercial-NoDerivatives 4.0 International License (http://creativecommons.org/licenses/by-nc-nd/4.0/), which permits any noncommercial use, sharing, distribution and reproduction in any medium or format, as long as you give appropriate credit to the original author(s) and the source, provide a link to the Creative Commons license and indicate if you modified the licensed material. You do not have permission under this license to share adapted material derived from this chapter or parts of it.

The images or other third party material in this chapter are included in the chapter's Creative Commons license, unless indicated otherwise in a credit line to the material. If material is not included in the chapter's Creative Commons license and your intended use is not permitted by statutory regulation or exceeds the permitted use, you will need to obtain permission directly from the copyright holder.

ClayArtisanX: Bridging Craftsmanship and Digital Fabrication in Ceramic 3D Printing

Yanhong Jia, Yanzhen Wu, Yanchen Li, and Huirong Le

Abstract The advancement of extrusion-based clay 3D printing has sparked a wave of artistic exploration in ceramics. Emerging techniques and tools have empowered designers to evolve beyond creators, taking on roles as innovators in material science, manufacturing processes, and human-machine collaboration. However, mainstream slicing software primarily caters to engineering and manufacturing needs, restricting designers' creative freedom with its rigid layer-by-layer slicing approach. This study examines the role of designers in clay 3D printing and introduces a novel design tool that seamlessly integrates craftsmanship with digital fabrication. By conducting expert user interviews, the study identifies key challenges faced by practitioners and develops a comprehensive design-to-manufacturing workflow. This workflow is implemented via a digital plugin for modeling, slicing, and G-code generation, providing designers with a flexible and efficient solution to bridge the gap between traditional craftsmanship and digital fabrication.

Keywords Digital fabrication · 3D printing · Clay · Design tools

1 Introduction

Computer-aided design and digital fabrication have profoundly transformed creative methods and expanded the design possibilities for artists and designers [1]. The concept of digital craftsmanship can be traced back to McCullough's seminal

Y. Jia · Y. Wu · H. Le (✉)
Department of Mechanical Engineering, Tsinghua University, Beijing, P. R. China

The Future Laboratory, Tsinghua University, Beijing, P. R. China
e-mail: Lehr@mail.tsinghua.edu.cn

Y. Li
The Future Laboratory, Tsinghua University, Beijing, P. R. China

Academy of Arts and Design, Tsinghua University, Beijing, P. R. China

© The Author(s) 2026

Z. Zhang, P. Mercorelli (eds.), *Mechanical Design and Simulation: Insights and Innovations*, Lecture Notes in Mechanical Engineering,
https://doi.org/10.1007/978-981-95-2405-1_53

work, *Abstracting Craft* (1998), which proposed studying digital design practices as a form of craftsmanship. As computational and fabrication technologies have advanced, the definition of digital craftsmanship has continually evolved [2]. The term "Digital" introduces innovative computational and fabrication techniques, while "Crafting," as a practice rooted in handmade production, embodies materials, culture, and traditional design methodologies [3–5].

In the realm of clay 3D printing, which uniquely blends craftsmanship with digital fabrication, many practitioners aim to bridge the gap by integrating traditional craftsmanship techniques with digital manufacturing technologies [6]. The Belgian design studio pioneered clay 3D printing in 2009 by adapting a desktop FDM printer to develop the first-generation clay extruder [7]. In collaboration with UK artist Jonathan Keep, an open-source community for clay 3D printers was established and maintained. This initiative supported artists and designers in constructing their 3D printers while fostering the sharing of ideas and creations within the community. Examples of clay 3D printing works and printers are shown in Fig. 1.

However, despite the availability of fully open-source equipment, clay 3D printing continues to face numerous challenges and uncertainties, which significantly dampens enthusiasm for creative exploration. Designers, in particular, grapple with dual challenges in both engineering and design (Fig. 2).

1.1 Material Challenges

Clay materials, characterized by their larger particle sizes, offer lower 3D printing precision compared to other materials such as plastics and metals, often resulting in visible layer marks [8]. Additionally, as a natural material, clay possesses high moisture content and significant rheological properties, which can lead to deformation and collapse during the printing process, resulting in frequent failures. Furthermore, clay undergoes substantial shrinkage due to its high moisture content. The solid-phase content of clay used in 3D printing typically ranges from 42% to 49% by volume, with shrinkage reaching up to 20% during drying and sintering [9]. The inherent anisotropy of the layer-by-layer 3D printing process further exacerbates issues such as cracking and warping. Addressing these challenges necessitates a comprehensive consideration of design, manufacturing, and material factors to minimize deformation, collapse, cracking, and warping in clay 3D printing.

1.2 Design Challenges

Designers frequently spend significant time exploring the capabilities and limitations of materials and manufacturing platforms, which hinders efficient design iteration and optimization. Most existing modeling and slicing software is not specifically tailored for clay materials, making it challenging for designers to

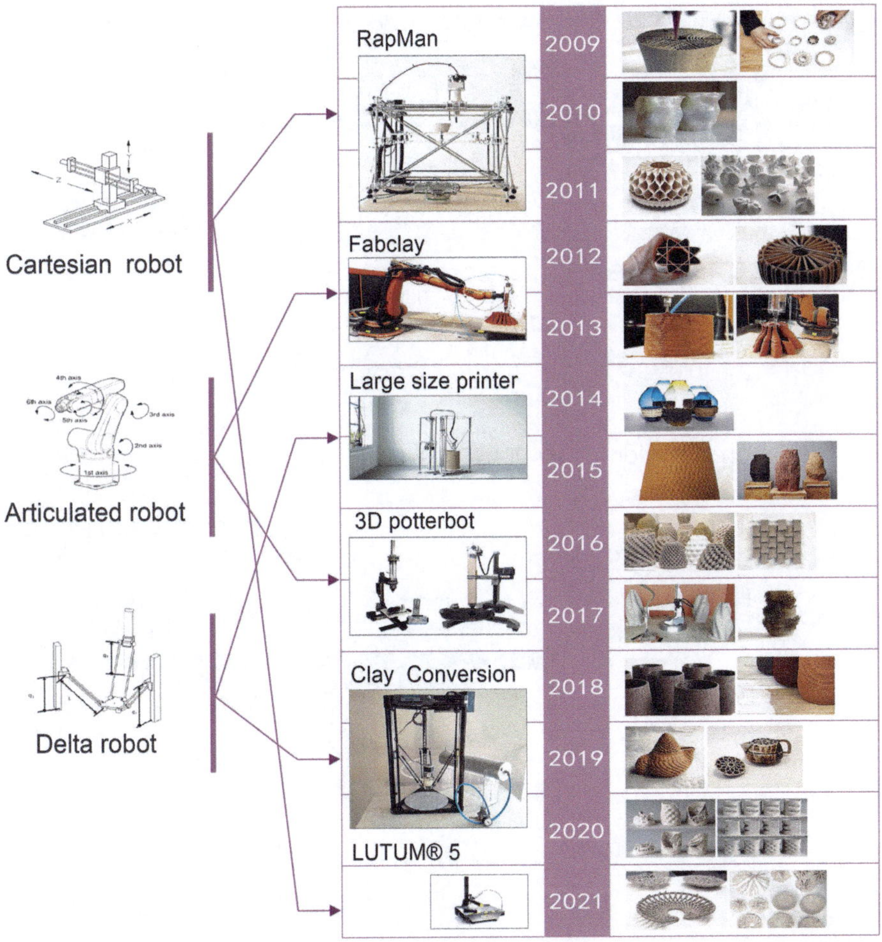

Fig. 1 Artistic works created by using clay 3D printing

achieve successful prints on clay 3D printers using parameters typically optimized for traditional materials like plastics and metals [10]. This lack of specialized software severely impacts the creative design process, limiting the full potential of clay 3D printing.

To tackle these equipment and material challenges, researchers have increasingly focused on solutions from both design and software perspectives. A critical area of exploration is understanding the role of designers in clay 3D printing and developing design tools that enable a seamless integration of traditional craftsmanship with modern manufacturing techniques [11, 12].

Through expert user experiments, this study identifies the key challenges faced by clay 3D printing practitioners and introduces an integrated design-to-manufacturing workflow implemented via a digital plugin for modeling, slicing,

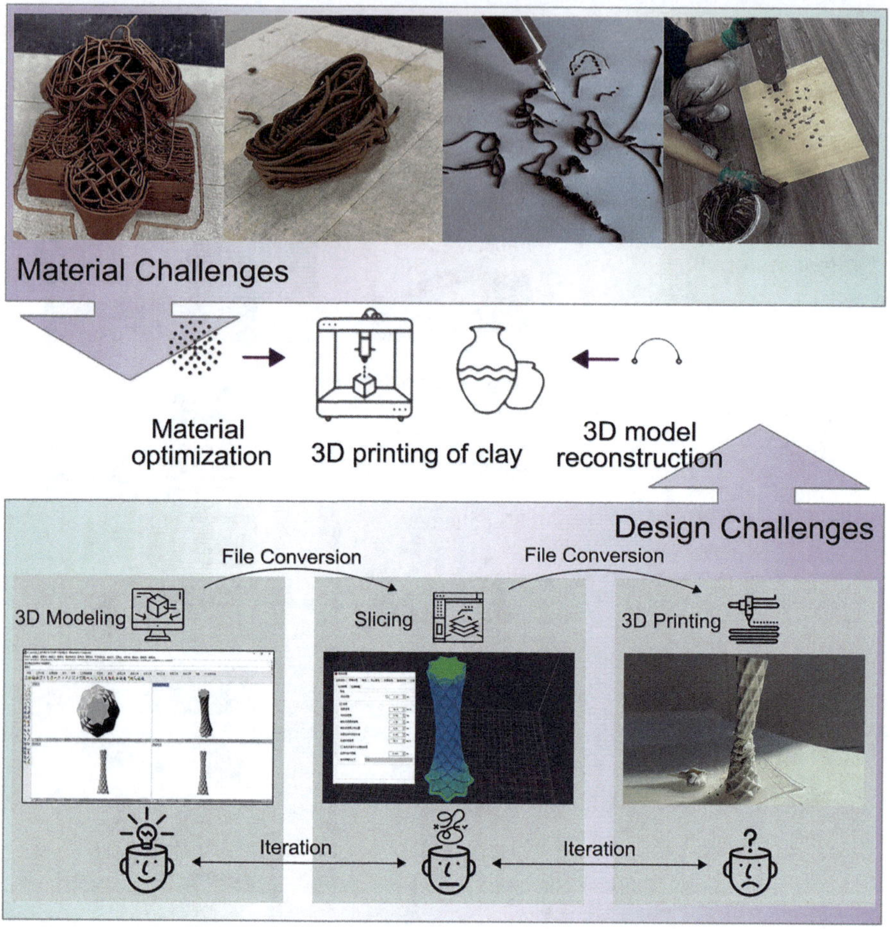

Fig. 2 Material and design challenges in 3D printing of clay

and G-code generation. This tool offers designers a more flexible and efficient solution, effectively bridging the gap between traditional craftsmanship and digital fabrication.

2 Materials and Methods

2.1 Clay Preparation

This section investigates the printability of traditional clay materials for 3D printing. The consistency of clay was first evaluated by measuring the extrusion force required to extrude the same volume of five common clay types: China clay, Ball

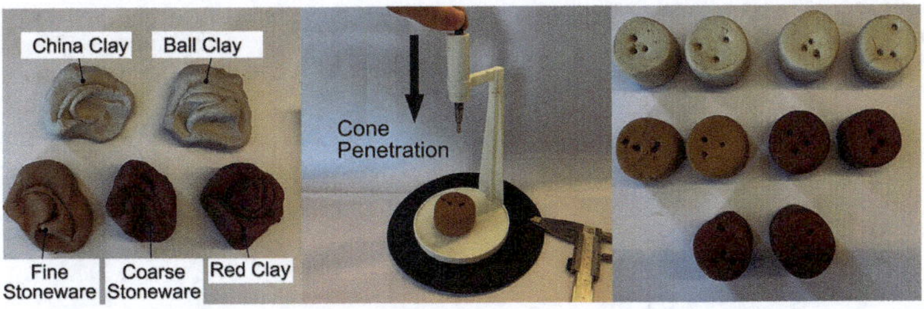

Fig. 3 Consistency measurement of five types of clay

clay, Fine stoneware, Coarse stoneware, and Red clay. This was done at a uniform rate through a narrow nozzle under consistent conditions. Next, the effects of varying air compressor pressures on extrusion speed were analyzed for different nozzle diameters. These experiments established baseline parameters, including clay consistency, water content, extrusion force, air pressure, and nozzle diameter, to facilitate smooth and reliable extrusion. Based on the findings, the slicing data in the digital plugin was optimized to align printing parameters with the material properties of the clay, significantly improving print success rates. This process is illustrated in Fig. 3.

2.2 *ClayArtisanX: Digital Plugin Development*

For the most common XYZ-type 3D printers, the typical workflow begins with the designer creating a 3D model, which is then exported in STL file format to third-party slicing software. The software slices the model and generates a G-code that the 3D printer can interpret. Finally, the G-code is imported into the 3D printer to initiate the printing process. As illustrated in Fig. 2, the file undergoes two key conversions: first, the model file is transformed into a slicing-compatible format, and second, it is converted into machine-readable G-code. Each stage, including modeling, slicing, and printing, often requires iterative adjustments due to discrepancies between the design and the printed result, which arise from indirect interactions and complex operations. This entire workflow reflects the transition from design to manufacturing, where each conversion and iteration step can introduce deviations from the intended outcome.

To address these issues, this paper proposes a workflow that enables direct control over G-code generation within the modeling software, as shown in Fig. 4. By customizing the Grasshopper plugin within Rhino, the model file is converted directly into machine-readable printing paths, facilitating seamless integration between design and manufacturing. This visual programming-driven approach introduces a novel method of digital craftsmanship, integrating material properties, geometric shapes, and digital fabrication. The proposed workflow not only enhances

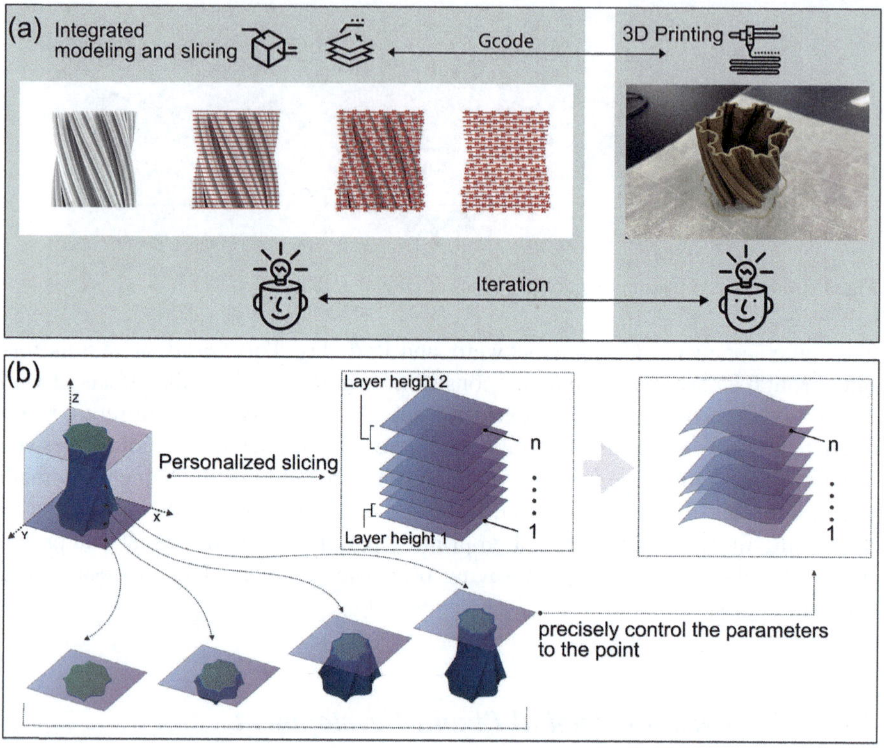

Fig. 4 (**a**) ClayArtisanX workflow; (**b**) Special slice illustration

design flexibility but also establishes a more intuitive and direct connection between the design and production processes.

2.3 *Printing Platform*

Printing tests were carried out by using a clay 3D printer with a traditional XYZ structure, utilizing Liquid Deposition Modeling (LDM) technology. The material storage system applied continuous pressure to a suspended clay tank via an air pump. The extrusion system featured a wear-resistant screw rod that ensured stable and controllable extrusion, allowing for the continuous extrusion of up to 2 kg of clay in a single session. The printer supported a build volume of 300 × 300 × 400 mm. The equipment setup is illustrated in Fig. 5.

Fig. 5 Clay 3D printer setup

3 Experiments and Results

3.1 User Experiments

Using a personas-based approach [13], this study categorizes users by integrating their professional backgrounds, needs, and preferences into distinct persona characteristics. This approach establishes a unified framework for understanding user roles, facilitating more focused and effective participant interviews.

Eight expert users participated in this experiment, with their professional fields represented by the text within the white circles in Fig. 6. Based on their backgrounds, the users were classified into three main categories:

Persona A: Ceramic experts with an interest in new technologies, aiming to explore the artistic potential of ceramic materials.

Persona B: Engineering experts experienced in 3D printing, with a focus on researching 3D printer control systems and key forming technologies.

Persona C: Design experts with an interest in human-computer interaction, aiming to investigate the integration of humans, machines, and materials in digital craftsmanship.

Based on these three personas, clay 3D printing tests were conducted, and case analyses were performed on selected representative works. The user experiment setup included a computer with the ClayArtisanX plugin installed and a 3D printer. Multiple cameras were positioned at various angles to capture the experimental process. The setup is depicted in Fig. 7.

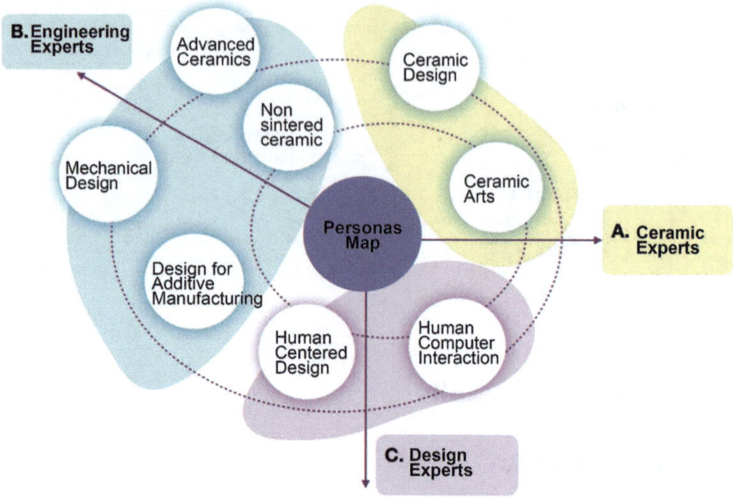

Fig. 6 Personas map

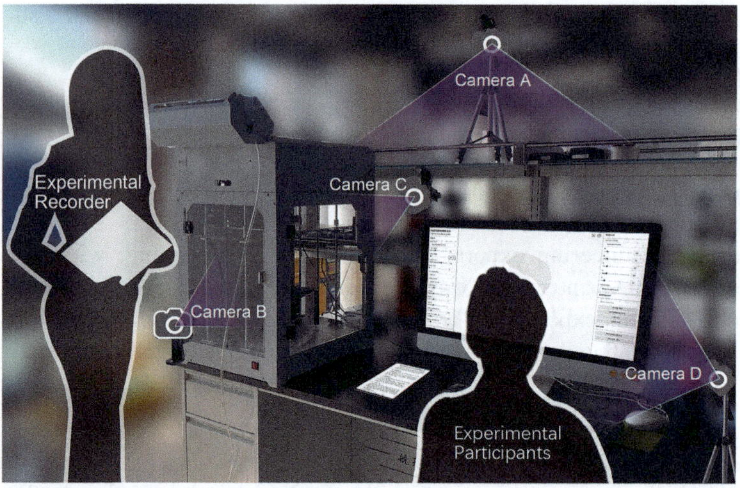

Fig. 7 User experiment setup

3.2 3D Printed Works and Design Analysis

The collection of works is presented in Fig. 8a. Following the three user personas, eight participants conducted clay 3D printing tests using the ClayArtisanX plugin. An overview of their printed works is displayed, with one representative work selected from each persona for detailed case analysis.

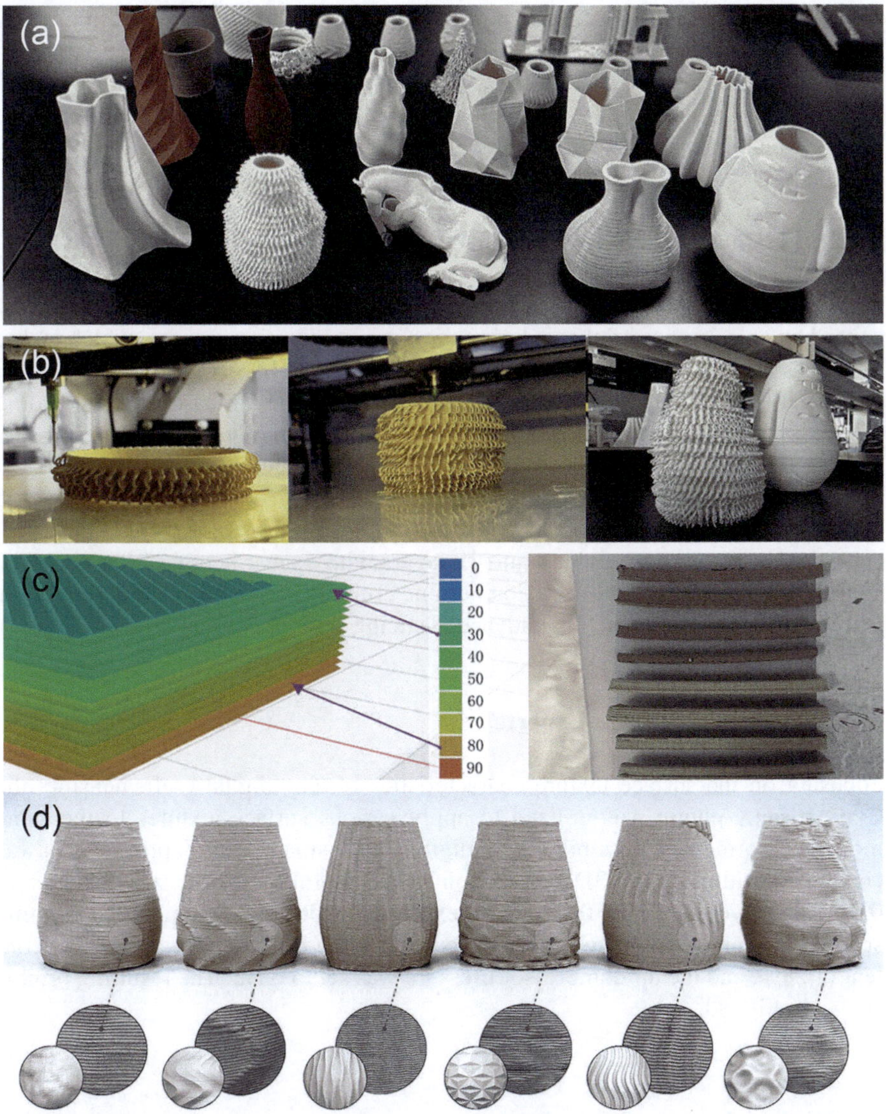

Fig. 8 (**a**) Overview of clay 3D printed works; (**b**) Special printing path planning; (**c**) Functional gradient design; (**d**) Parametric generation of various surface textures

3.2.1 Persona A: Ceramic Experts

The participant began by downloading an open-source 3D model from the internet, printing it with clay, and firing it to create a ceramic object. However, the smooth surface failed to effectively capture the essence of the original design. To address this issue, the author utilized the ClayArtisanX plugin to create a customized printing path that highlights the digital craftsmanship unique to clay 3D printing. By designing a distinctive slicing path, a layered, curled surface was achieved, as shown in Fig. 8b. This specialized path allowed for a natural drooping effect during printing, resulting in intricate surface details and a "fuzzy" texture that enhanced the realism and dimensionality of the piece.

3.2.2 Persona B: Engineering Experts

A significant challenge in clay 3D printing is edge curling and warping, primarily caused by anisotropic shrinkage during the ceramic sintering process, which makes the edges susceptible to deformation or even cracking [14]. To mitigate this issue, adjustments were made to the printing parameters of the flat base layer, including speed, layer height, and ventilation, to minimize deformation during sintering and reduce edge warping, as illustrated in Fig. 8c. Through multiple rounds of testing, it was determined that a clay mixture containing 28% water, combined with optimized printing speeds, significantly reduced base layer warping. During the slicing process, the base layer was divided into 12 layers with a layer height of 0.6 mm. The ClayArtisanX plugin was utilized to fine-tune printing parameters within the G-code generation component, enabling advanced slicing settings.

3.2.3 Persona C: Design Experts

Focusing on the surface texture characteristics of clay digital craftsmanship, the ClayArtisanX plugin was utilized to apply various surface textures to the same model, exploring new possibilities for digital expression in clay. A printing test was conducted with the clay 3D printer configured as follows: a nozzle diameter of 0.8 mm, a layer height of 0.6 mm, an extrusion width of 1.0 mm, and a printing speed of 40 mm/s. The clay material used was white kaolin clay with a water content of 28%, and the air compressor pressure was set to 6 bar. The resulting print is shown in Fig. 8d.

4 Discussions and Limitations

Observations and case analyses of the practical works highlight certain limitations, as they primarily remain at the forming stage with minimal post-processing. This underscores the significant potential and opportunities for advancing digital craftsmanship in clay. For designers working with digital tools, the greatest challenge lies in leveraging technology to craft a unique visual language without being constrained by algorithmic limitations [15]. While isotropic clay materials were used in the tests and prints conducted in this study, clay 3D printing has the potential to produce structures with anisotropic characteristics or varying functional gradients by carefully controlling deposition parameters. Tools that enable structured parameter control across different regions to achieve multifunctional structures are particularly valuable for researchers. Although the ClayArtisanX plugin developed in this study addresses both manufacturing and design, future work should emphasize the customization of material properties to further enhance its applicability.

5 Conclusion

This open-ended exploration validates the usability and practicality of the proposed design tools while offering a fresh perspective on personalized customization in clay 3D printing. As additive manufacturing technology continues to advance, the variety of printable clay types is expected to increase, presenting both new opportunities and challenges for the evolution of digital clay craftsmanship. With the growing range of printable clay, compatible glazes and decorative techniques may emerge, expanding the possibilities for post-processing 3D printed works. Looking ahead, digital clay craftsmanship has the potential to evolve into digital ceramic craftsmanship, enabling the creation of even more remarkable and accessible 3D-printed pieces for the public. Future design practices should explore innovative post-processing methods and techniques to fully realize the creative potential of digital clay craftsmanship.

Acknowledgments This work was supported by the Tsinghua University Initiative Scientific Research Program-International Cooperation Special Project (20233080011), Beijing Natural Science Foundation (3244042), and China National Postdoctoral Program for Innovative Talents (BX20230185). The authors are also grateful for the support of colleagues at the Future Lab of Tsinghua University.

References

1. Anton AG (2022) Computational practice and perspectives in the arts. Northwestern University, Evanston, IL
2. Oxman N (2007) Digital craft: fabrication-based design in the age of digital production. In: Workshop Proceedings for Ubicomp 2007: International Conference on Ubiquitous Computing, pp 534–538
3. Marshall J (2024) Hand thought: hybrid practices and a digital craft ethos. In: Craft and design practice from an embodied perspective. Routledge, London, pp 109–123. ISBN: 9781003328018
4. Liu G, Shi Q, Yao Y et al (2024) Learning from hybrid craft: investigating and reflecting on innovating and enlivening traditional craft through literature review. In: Proceedings of the 2024 CHI Conference on Human Factors in Computing Systems, pp 1–19. https://doi.org/10.1145/3613904.3642205
5. Rachmad YE (2021) Future craft: the art of digital business. Digital Book Humanities Publishing, Professional Book Providence Special Issue

6. Yin X, Guo C, Sun B, Chen H, Wang H, Li A (2023) The state of the art in digital construction of clay buildings: reviews of existing practices and recommendations for future development. Buildings 13(9):2. https://doi.org/10.3390/buildings13092381
7. Unfold (2021) Ceramic 3D Printing. Unfold [Online]. https://unfold.be/pages/ceramic-3d-printing. Accessed 17 Nov 2024
8. Donyadari R, Ortiz BF, Khondoker MAH (2024) Optimization of printing parameters for extrusion 3D printing of ceramic clay. Eng Proc 76(1):47. https://doi.org/10.3390/engproc2024076047
9. Madrid JA, Ortega GS, Carabaño JG, Olsson NOE, Ríos JAT (2023) 3D claying: 3D printing and recycling clay. Crystals 13(3):375. https://doi.org/10.3390/cryst13030375
10. Çapunaman ÖB, Iseman EK, Gürsoy B (2023) Material-in-the-loop fabrication: a vision-based adaptive clay 3D printing workflow on indeterminate sand surfaces. In: Computer-aided architectural design. INTERCONNECTIONS: co-computing beyond boundaries, CAAD futures 2023, vol 1. Communications in Computer and Information Science, Springer, Cham, p 819. https://doi.org/10.1007/978-3-031-37189-9_27
11. Bourgault S, Wiley P, Farber A, Jacobs J (2023) CoilCAM: enabling parametric design for Clay 3D printing through an action-oriented toolpath programming system. In: Proceedings of the 2023 CHI Conference on Human Factors in Computing Systems (CHI '23). ACM, New York, NY, pp 1–16. https://doi.org/10.1145/3544548.3580745
12. Jia Y (2021) Applications of computer-aided manufacturing in ceramic 3D-printing-path design. In: 2021 2nd international conference on intelligent design (ICID), Xi'an, China, pp 339–342. https://doi.org/10.1109/ICID54526.2021.00074
13. Pruitt J, Adlin T (2010) The persona lifecycle: keeping people in mind throughout product design. Elsevier, Amsterdam. https://doi.org/10.1016/B978-0-12-566251-2.X5000-X
14. Manière C, Harnois C, Marinel S (2021) 3D printing of porcelain: finite element simulation of anisotropic sintering. Int J Adv Manuf Technol 116:3, 263–3, 275. https://doi.org/10.1007/s00170-021-07304-y
15. Bunnell K (2004) Craft and digital technology. In: The World Crafts Council 40th Anniversary Conference

Open Access This chapter is licensed under the terms of the Creative Commons Attribution-NonCommercial-NoDerivatives 4.0 International License (http://creativecommons.org/licenses/by-nc-nd/4.0/), which permits any noncommercial use, sharing, distribution and reproduction in any medium or format, as long as you give appropriate credit to the original author(s) and the source, provide a link to the Creative Commons license and indicate if you modified the licensed material. You do not have permission under this license to share adapted material derived from this chapter or parts of it.

The images or other third party material in this chapter are included in the chapter's Creative Commons license, unless indicated otherwise in a credit line to the material. If material is not included in the chapter's Creative Commons license and your intended use is not permitted by statutory regulation or exceeds the permitted use, you will need to obtain permission directly from the copyright holder.

Research on the Design of Conjugate Cam Shedding Based on High-Speed Rapier Loom

Quan Zhang, Heng Chen, Xinwei Ye, and Ying Fu

Abstract Cam shedding is a kind of shedding form commonly used in modern high speed loom, which has the characteristics of high motion precision and good impact resistance. In this paper, conjugate cam is applied to the shedding mechanism design of rapier loom. By analyzing the principle of cam mechanism, studying the motion characteristics of the follower, calculating the key parameters such as pressure Angle and curvature radius, the conjugate cam shedding mechanism of high-speed rapier loom is designed. In the Matlab environment, the main cam theoretical contour design program was written, the reasonable cam pressure Angle and roller radius were determined, the three-dimensional model of conjugate cam mechanism was established by using SolidWorks, and the Motion law of pendulum rod, two-arm rod and heald frame was verified by the simulation of motion module. At the same time, the conjugate cam and heald frame were analyzed by finite element statics. The simulation results show that the angular displacement, angular velocity and angular acceleration of the swinging rod and the two-arm rod are consistent with the theoretical profile expectation of Matlab, and conform to the motion law of the rapier loom heald frame. In the process of motion of the heald frame, the conjugate cam and the roller can keep continuous contact with each other, which has good motion stability and mechanism performance. It provides strong support for the design innovation and practical application of rapier loom shedding mechanism.

Keywords Rapier loom · Conjugate Cam · Shedding mechanism · Simulation · Frame

Q. Zhang · H. Chen (✉) · X. Ye
Xi Jing University, Xian, China

Y. Fu
Shaanxi Brilliance Petroleum Technology Co., Xian, China

© The Author(s) 2026

Z. Zhang, P. Mercorelli (eds.), *Mechanical Design and Simulation: Insights and Innovations*, Lecture Notes in Mechanical Engineering,
https://doi.org/10.1007/978-981-95-2405-1_54

1 Introduction

Currently, Rongyi Zhentao, Zhao Shihai and Yuan Ruwang investigated the influence of fabric organisation laws on the high-speed conjugate cam opening process. By analysing the motion law of the follower and the resting time of the heald frame, the role of these factors on the performance of the cam opening process was explored. The results show that the best cam performance is achieved in the opening process with the opening rest angle removed when the follower follows the simple harmonic motion law. In addition, this paper devised a matching relationship between the fabric organisation laws and the cam contour lines within and between groups of fabric loops, which enabled significant optimisation of the cam performance in practical machining [1]. Wei Z, Wang Z, LiD study deals with the rigid-flex coupled dynamics modelling and dynamic accuracy analysis of planar cam four-link mechanism with multiple gap joints. The research of Wei Zhe, Wang Zhe and Li De is aimed at the rigid-flexible coupling dynamic modeling and dynamic accuracy analysis of the multi-clearance sub-planar cam four-bar linkage mechanism. By integrating the gap collision model, finite element model and rigid-flexure coupled dynamics model, it is applied to the cam-link combined multibody mechanism. Within the speed range of 600–800 rpm, it is found that the dynamic performance of the system is less affected by the speed. Meanwhile, through the comparative analysis of various materials, the results show that the selection of carbon fibre composite material has the least influence on the motion accuracy of the system [2]. Haifei Qiu, Chunfeng Li, and Ming Chen of Xijing College conducted a research on the design of conjugate cam opening for external high-speed weaving machine. In order to reduce the wear and contact stresses between the cam and the rotor, a new external conjugate cam opening system based on biaxial structure was constructed on the basis of the mechanistic equivalence transformation. The programming design and dynamic calculation of the theoretical cam profile and its key performance parameters (α, R, ρ) are realised in MATLAB environment. Meanwhile, a functional prototype simulation model of the single-page heald frame was established in ADAMS/View, and the dynamics of the mechanism was analysed. These works provide a reference basis for the design and performance improvement of high-speed cam opening system [3].

2 Structural Modelling and Assembly Design of Main and Secondary Cams

2.1 *Structural Modelling of Primary and Secondary Cams*

The theoretical profile of the main cam is inverted by MATLAB, and the theoretical profile of the main cam minus the roller radius is the actual profile of the main cam. And then the actual contour of the main cam into SolidWorks to generate a

Fig. 1 Cam section

Fig. 2 Initial 3D model of conjugate cam mechanism

conjugate cam main cam three-dimensional model, the main cam as shown in Fig. 1; Known conjugate cam cam axis and pendulum axis distance of 180 mm, the establishment of a simple base model, the base is 430 mm long, 340 mm wide, 5 mm thick, in the base to play a hole, the diameter of 50 mm, the camshaft is fixed in the base, and the diameter of 50 mm hole to impose the hinge with; Establishment of the roller connector three-dimensional model, the known diameter of the roller is 77 mm, the thickness is 20 mm; Pendulum length of 120 mm, the main and secondary pendulum angle of 130°, the two pendulum combined into a roller connector, the rollers were fixed on the roller connector, and finally its assembly. The initial 3D model of the conjugate cam mechanism is shown in Fig. 2. The parameters of the conjugate cam are shown in Table 1.

2.2 *Auxiliary Cam Profile Design*

First, various parameters of the main cam and the main cam are inversely calculated through MATLAB. Then, through the assembly function of SolidWorks, the main cam, the rocker arm, and the roller are reasonably assembled to form a conjugate

Table 1 Parameters of conjugate cam parts

Name	Lengths/mm	Breadth/mm	Thicknesses/mm
Foundation	430	340	5
Rollers	77	77	20
Pendulum	120	50	20

Fig. 3 Add results

cam. Finally, through the trajectory tracking command in the SolidWorks motion analysis, first select the displacement/velocity/acceleration option, then select the tracking path command, and finally select the main cam surface and the roller center as the tracking point to inversely calculate the theoretical profile of the auxiliary cam. The actual profile of the auxiliary cam can be obtained by subtracting the roller radius from the theoretical profile of the auxiliary cam. The main and auxiliary cams and the roller connectors are assembled to obtain the conjugate cam [4].

Use the motion simulation function in SolidWorks to add a motor to the camshaft, select the camshaft, enter the speed of 6RPM, and set the animation duration to 10 s, that is, the conjugate cam mechanism completes one rotation.

Click the Results and Graphics command, select the roller center point and the conjugate cam secondary cam surface, select displacement, velocity, and acceleration in the selection category; select tracking path in the selection subcategory. As shown in Fig. 3, add the results.

Click the calculation motion example to generate the auxiliary cam theoretical profile curve, as shown in Fig. 4, which is the tracking and generation state of the auxiliary cam theoretical profile curve. As shown in Fig. 5, the auxiliary cam theoretical profile curve is different from the actual profile curve by a roller radius. You need to use the equidistant curve to subtract a roller radius, and then form an actual profile curve. Finally, use the stretch command in SolidWorks to stretch the actual profile curve to get the auxiliary cam. Using the Template [5].

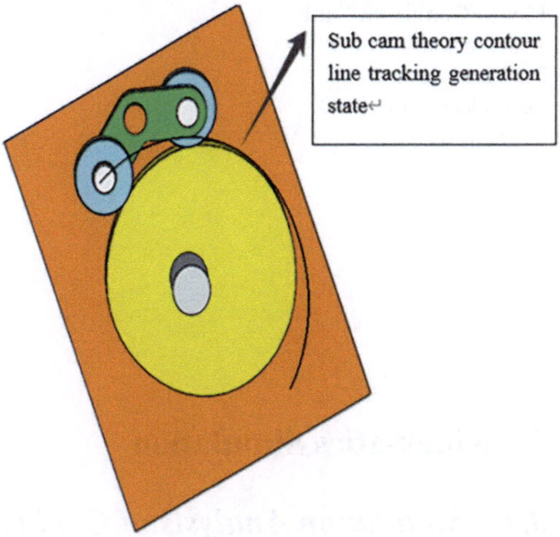

Fig. 4 Auxiliary cam theoretical profile tracking generation state

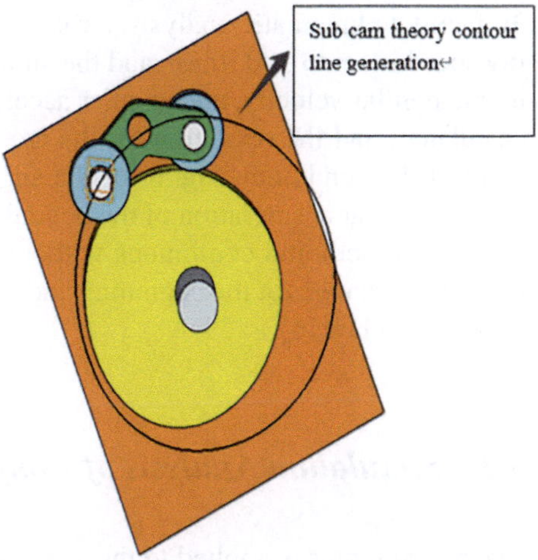

Fig. 5 Theoretical profile curve of auxiliary cam

2.3 Conjugate Cam Mechanism Assembly

The design and manufacture of the conjugate cam structure assembly need to take into account the shape of the cam, the size and material of the roller, the size and material of the rocker, etc., to ensure accurate and reliable transmission effect. Through reasonable design and material selection, the purpose of reducing friction, increasing stability, improving transmission efficiency and extending service life can be achieved.

The conjugate cam designed in this paper is mainly composed of a main cam, a secondary cam, two rollers, and a roller connector composed of two rockers, as shown in Fig. 6.

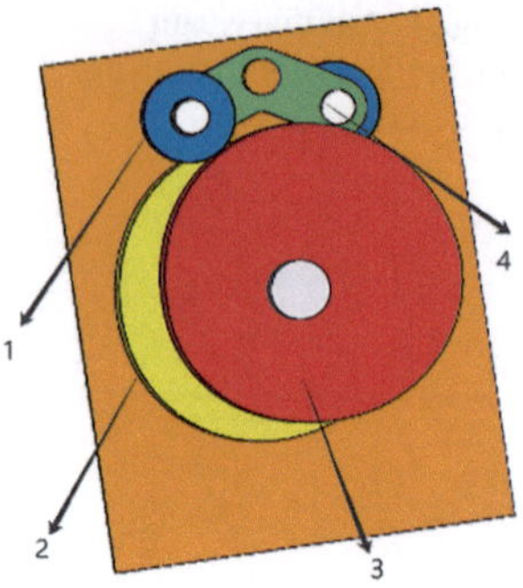

Fig. 6 Conjugate Cam Assembly. 1—Roller; 2—Main Cam; 3—Secondary cam; 4—Swing rod

3 Kinematics Simulation

3.1 *Simulation Analysis of Conjugate Cam Rocker*

To facilitate kinematic analysis, a rotary motor is applied to the camshaft, the driving speed is set to 300 r/min, and the simulation time is 0.2 s. The angular displacement, angular velocity, and angular acceleration of the pendulum are simulated and calculated, and the operation results are shown in Fig. 7 for the angular displacement of the pendulum, Fig. 8 for the angular velocity of the pendulum, and Fig. 9 for the angular acceleration of the pendulum. It can be seen from the results that the curve is smooth and continuous without strong impact, but the acceleration curve has abrupt changes at the beginning and end, which conforms to the cosine acceleration motion law [6].

3.2 *Simulation Analysis of Conjugate Cam Rocker*

The rotary motor is applied to the gear shaft. Since the gear ratio is 2:1, the driving speed is set to 600 r/min and the simulation time is 0.2 s. The angular displacement, angular velocity and angular acceleration of the two-arm lever are simulated and calculated. The operation results are shown in Figs. 10, 11 and 12. The curve is smooth and continuous without strong impact, but the acceleration curve has abrupt changes at the beginning and end, which conforms to the cosine acceleration motion law [7].

3.3 *Heald Frame Simulation Analysis*

A rotary motor is applied to the gear shaft. Since the gear ratio is 2:1, the drive speed is set to 600 r/min and the simulation time is 0.2 s. The linear displacement, linear velocity and linear acceleration of the heald frame are simulated and calculated, and

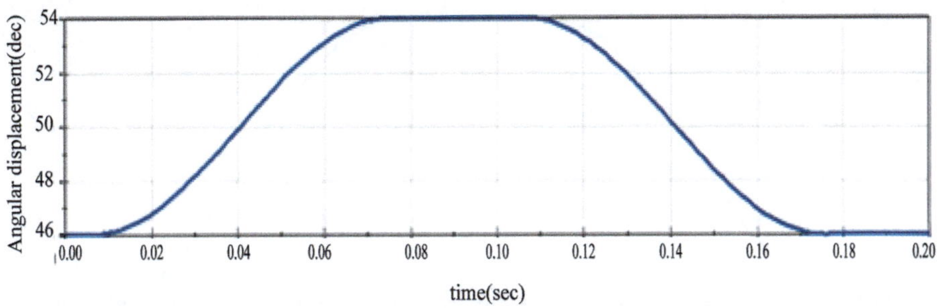

Fig. 7 Angular displacement of the pendulum

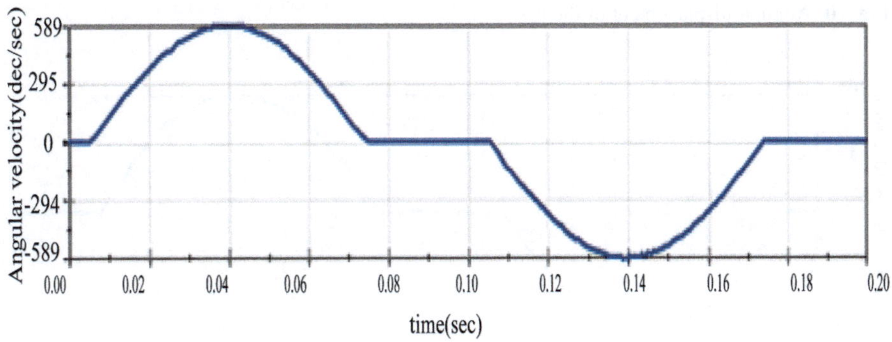

Fig. 8 Angular velocity of pendulum

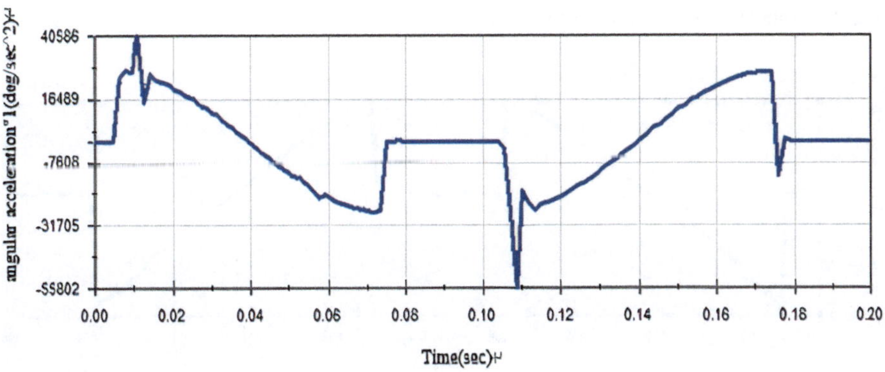

Fig. 9 Angular acceleration of the pendulum

the operating results are shown in Fig. 13 for the linear displacement of the heald frame, Fig. 14 for the linear velocity of the heald frame, and Fig. 15 for the linear acceleration of the heald frame. From the results, it can be seen that the curve is smooth and continuous without a strong impact, but the acceleration curve has mutations at the beginning and end, so it conforms to the law of cosine acceleration

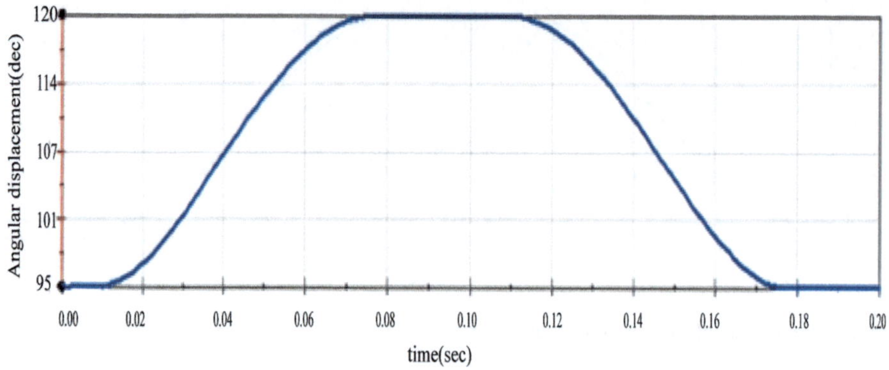

Fig. 10 Angular displacement of the two arm

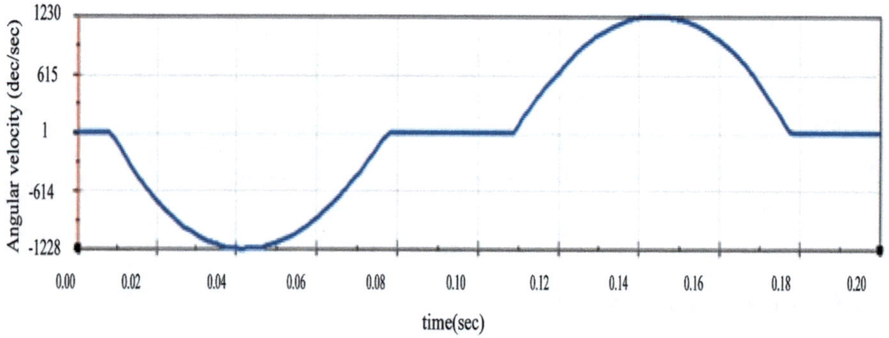

Fig. 11 Angular velocity of the second arm

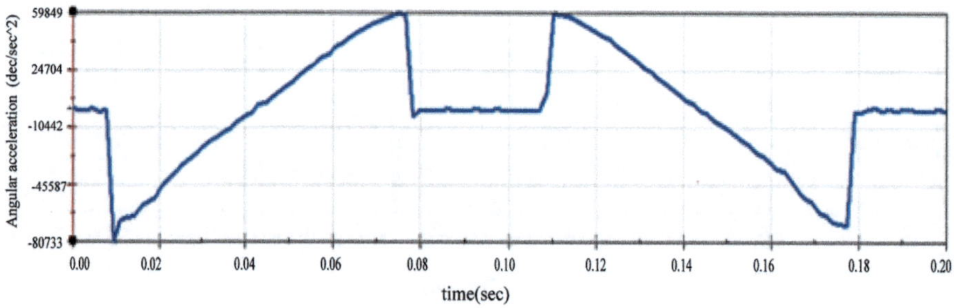

Fig. 12 Angular acceleration of the two-arm

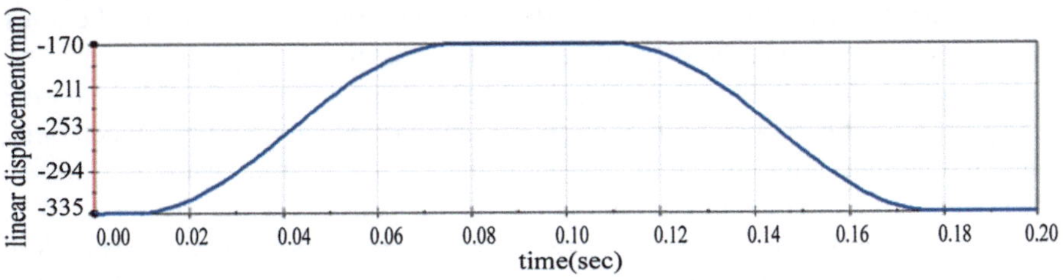

Fig. 13 Heald frame linear displacement

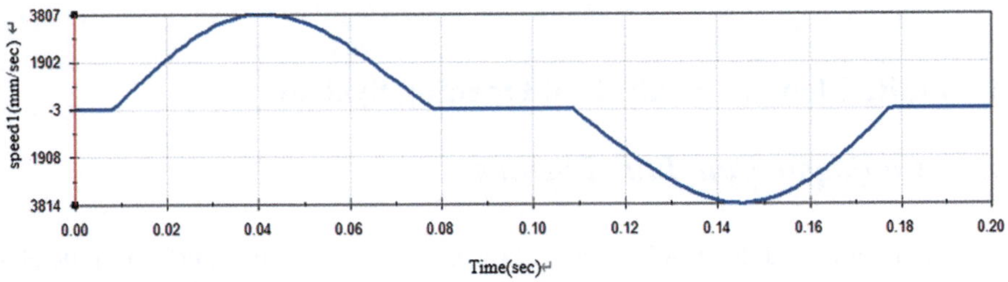

Fig. 14 Heald frame linear speed

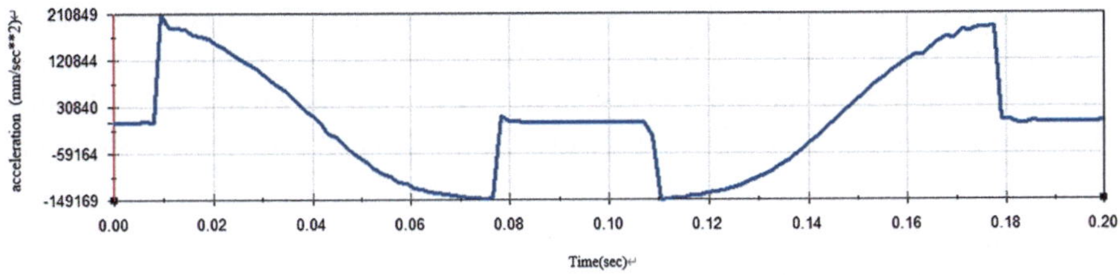

Fig. 15 Heald frame linear acceleration

motion. And in general, it can be divided into four stages, namely "push stroke-far rest-return stroke-near rest", which conforms to the segmented characteristics of the cam theoretical profile [8].

3.4 Simulation Results Analysis

By comparison, when the camshaft rotates one circle in the time range of 0–0.2 s, the motion law curves of the rocker, heald frame and two-arm lever have the same changing trend, which can be generally divided into four stages, namely "push stroke-far rest-return stroke-near rest", which conforms to the segmented characteristics of the cam theoretical profile. The conjugate cam opening mechanism after inverse analysis can open at a higher speed than the opening mechanism manufactured by direct mapping under the same working conditions, and the transmission is smoother and the dynamic performance is greatly improved.

Comparing the two-arm lever, heald frame and rocker lever, it can be seen that the speed is in the opposite direction to the angular velocity, and the acceleration curve has certain mutations at the beginning and end, indicating that the cam opening system has a flexible impact when starting and stopping, which conforms to the cosine acceleration motion law, indicating that the design is reasonable. Therefore, from the simulation results, the lifting law of the heald frame conforms to the motion law of the follower, indicating that the design of the conjugate cam opening system is reasonable [9].

4 Finite Element Analysis of Opening Systems

4.1 Conjugate Cam Initial Setting

The simulation module of SolidWorks software is mainly used to perform finite element analysis on the conjugate cam mechanism. The example added to the conjugate cam mechanism is static stress analysis. The material is set to alloy steel. The elastic modulus of alloy steel is 2.1 × 1011 N/m^2, the Poisson's ratio is 0.28, the maximum yield strength that the material can withstand is 2.95 × 108 N/m^2, and the mass density of the material is 7800 kg/m^3. These are some parameters of the materials used for the conjugate cam. The specific settings are shown in Fig. 16. Finally, the conjugate cam will be subjected to finite element analysis to obtain stress and

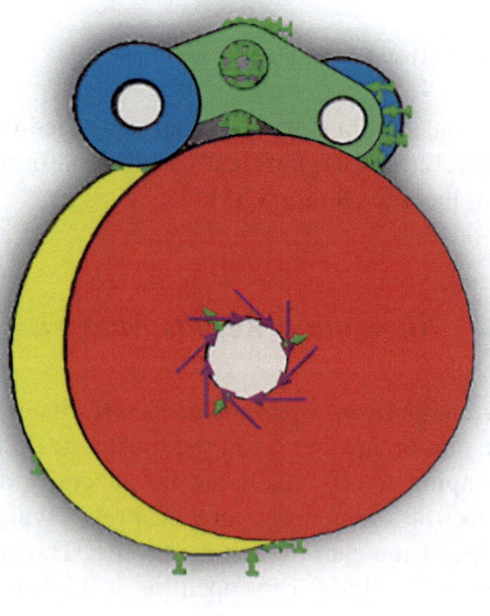

Fig. 16 Adding Fixtures and External Loads

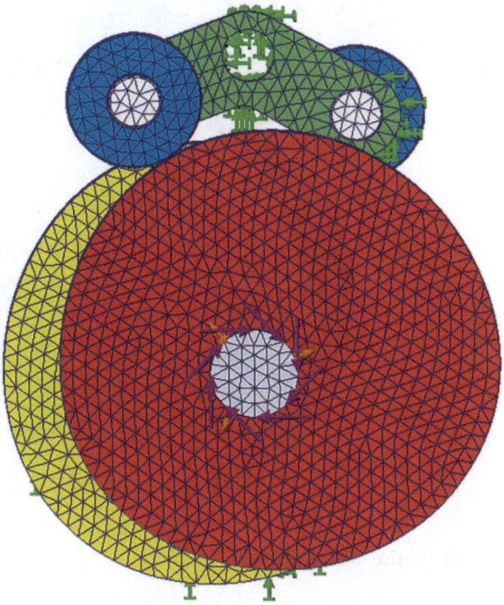

Fig. 17 Finite element meshing

displacement results. Then, by comparing the maximum yield strength of the material, the rationality of the conjugate cam design can be determined.

4.2 Conjugate Cam Meshing

The conjugate cam is meshed and the mesh density is set to good. There are 47,595 cells and 74,924 nodes. The finite element mesh division is shown in Fig. 17.

4.3 Stress and Displacement Analysis of Conjugate Cams

The finite element analysis results are shown in Figs. 18 and 19, where Fig. 18 is the finite element stress result. It can be seen from the figure that the maximum stress result is1.804 × 107 N/m^2, which is much smaller than the maximum yield strength of alloy steel 2.95 × 108 N/m^2, Fig. 19 is the finite element displacement result, and the maximum displacement is 2.222 × 10^{-2} mm. The results show that the displacement is very small and the impact on the mechanism is almost negligible [10].

The finite element analysis results of the conjugate cam mechanism show that, under a torque of 180 N m, compared with the material properties of alloy steel, both the stress and displacement results meet the requirements, indicating that the strength and stiffness of the conjugate cam mechanism are sufficient, so the design of the conjugate cam mechanism is reasonable.

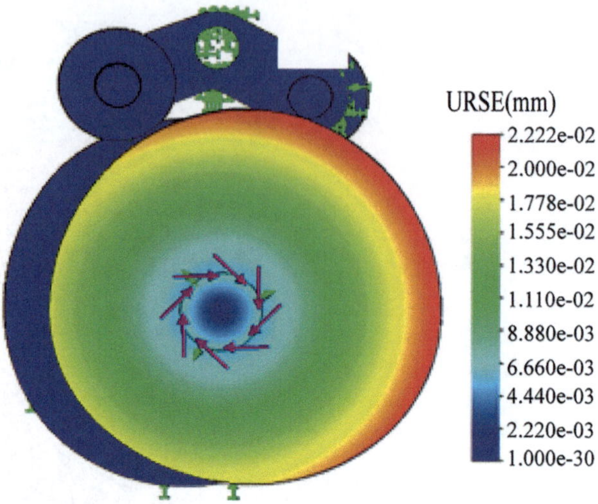

Fig. 18 Finite element stress results

Fig. 19 Finite element displacement results

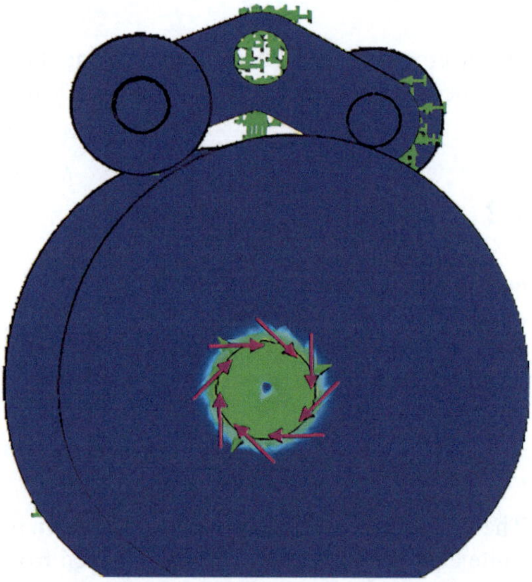

5 Conclusion

The theoretical contour line of the master cam was inverted by Matlab, and the pendulum angular displacement curve and the master cam pressure angle curve were calculated. According to the law of motion of the follower, the cosine acceleration law of motion is used, and the pressure angle is finally calibrated to ensure the working life of the conjugate cam mechanism. The CAD theoretical contour line of the main cam is imported into SolidWorks software to derive the main cam, and the theoretical contour line of the slave cam is inverted by the tracking path in Motion analysis to finally obtain the conjugate cam. The design and modelling of the drive mechanism parts and heald frame components in the conjugate cam opening mechanism of the high-speed rapier loom are completed in SolidWorks, and the parts are assembled to finally complete the conjugate cam opening mechanism. The kinematic simulation of the conjugate cam mechanism and conjugate cam opening mechanism was carried out using the Motion analysis module of SolidWorks software. By analysing the motion results of the pendulum bar, the two-arm bar and the heald frame, it is concluded that the conjugate cam mechanism and the conjugate cam opening mechanism are reasonably designed. Using the Simulation module of SolidWorks software, the finite element analysis of the conjugate cam and the heald frame is completed, and the results show that the stresses and displacements are within a reasonable range, which verifies the reasonable design of the conjugate cam mechanism and the heald frame.

Acknowledgments This work was financially supported by Xi'an Science and Technology Plan Project Fund (23NYGG0075) fund.

References

1. Zhentao R, Shihai Z, Ruwang Y (2023) Influence of fabric organisation pattern on high-speed conjugate cam opening process. Wool Textile Sci Technol 51(01):10–12
2. Wei Z, Wang Z, Li D et al (2024) Rigid-flex coupled dynamics modelling and dynamic accuracy analysis of planar cam four-link mechanism with multiple clearance joints. J Inst Mech Eng Pt K J Multibody Dynam 25(03):11–18
3. Qiu HF, Li CF, Chen M et al (2022) Design of a conjugate cam opening for external high-speed weaving machine. Mech Des 39(11):3–5
4. Zhengxing G, Shunqi M et al (2023) Research progress of conjugate cam weft beating mechanism for weaving machine: https://elibrary.ru/vxxqfc. Grand Altai Res Educ 2(20):106–113
5. Shi N, Liu Q, Li Y (2023) Inversion and construction of working contour surface of arc indexing cam. J Ordnance Equip Eng 44(6):8–10
6. Yizhentao R, Shihai Z, Ruwang Y (2023) Influence of fabric weave pattern on high-speed conjugate cam opening process. Wool Textile J 51(1):9–12
7. Zhang X, Fang L, Chen Q et al (2024) Optimisation of double weaving shaft weaving process for small jacquard fabric. Cotton Textile Technol 52(630):10–16
8. Yang XU, Xianbo YIN, Guosheng XIE et al (2024) Technological advance in weaving machinery from ITMA ASIA+ CITME 2022. Cotton Textile Technol 52(631):5–17
9. Lee MC, Choi DI, Ji JF et al (2012) Design method of conjugate cam opening mechanism for rapier loom based on dynamic characteristics. J Xuzhou Eng Coll (Nat Sci Ed) 27(03):11–15
10. Jing Z, Zhijia L, Y U (2024) Research and development of integrated weaving forming covering blanket fabric with flocculus. Cotton Textile Technol 52(631):4–12

Open Access This chapter is licensed under the terms of the Creative Commons Attribution-NonCommercial-NoDerivatives 4.0 International License (http://creativecommons.org/licenses/by-nc-nd/4.0/), which permits any noncommercial use, sharing, distribution and reproduction in any medium or format, as long as you give appropriate credit to the original author(s) and the source, provide a link to the Creative Commons license and indicate if you modified the licensed material. You do not have permission under this license to share adapted material derived from this chapter or parts of it.

The images or other third party material in this chapter are included in the chapter's Creative Commons license, unless indicated otherwise in a credit line to the material. If material is not included in the chapter's Creative Commons license and your intended use is not permitted by statutory regulation or exceeds the permitted use, you will need to obtain permission directly from the copyright holder.

A Review of Research on Multilayered Vessels

Zhengyan Xiao, Xuesheng Wang, and Zhibin Su

Abstract Multilayered vessels have been widely applied in engineering fields due to their unique layered characteristics, attracting considerable attention from researchers. Significant progress has been made in exploring key aspects. However, challenges remain regarding the universality of theoretical models, the precision of numerical simulations, and the comprehensiveness of experimental validations. This review systematically summarizes recent advances in this domain, providing a detailed chronological analysis of key research areas. Furthermore, the study identifies existing limitations and proposes potential directions for future research, aiming to offer a theoretical basis for optimizing the design and improving the performance of multi-layered containers.

Keywords Multilayered vessel · Prestressing · Thermal stress · Interlayer gap

1 Introduction

In 1930, A. O. Smith Corporation in the United States first proposed a multilayered wrapped cylinder structure processed by tensioning steel plates with steel wire, as shown in Fig. 1. This innovation not only met the industrial demands for high-pressure vessels at that time but also laid the foundation for the subsequent design and development of multilayered vessels.

In the late 1970s, Germany's Krupp Corporation introduced improvements to the multilayered structure developed by A. O. Smith Corporation, creating an integrated multilayered vessel, as shown in Fig. 2. In this design, all longitudinal and

Z. Xiao · X. Wang (✉)
East China University of Science and Technology, Shanghai, China
e-mail: wangxs@ecust.edu.cn

Z. Su
Shanghai Boiler Works Co., Ltd, Shanghai, China

© The Author(s) 2026

Z. Zhang, P. Mercorelli (eds.), *Mechanical Design and Simulation: Insights and Innovations*, Lecture Notes in Mechanical Engineering,
https://doi.org/10.1007/978-981-95-2405-1_55

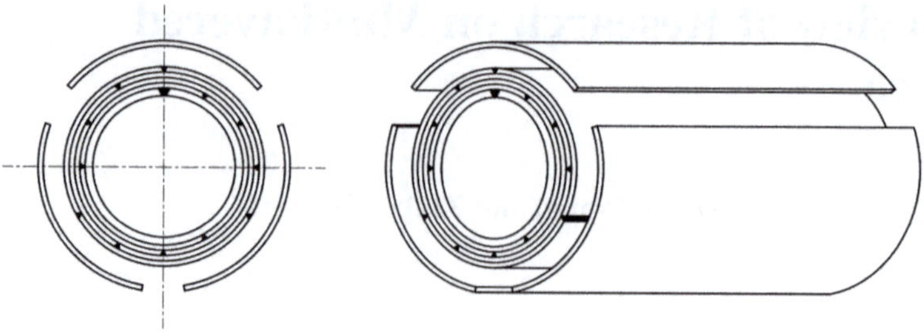

Fig. 1 Multilayered cylinder structure of the A. O. Smith

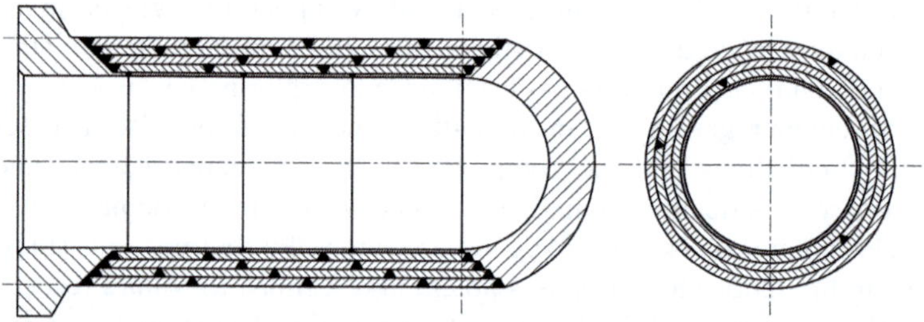

Fig. 2 Multilayered cylinder structure of the Krupp

(a) Ammonia Tower (b) High Pressure Extractor (c) H_2 Storage Tank

Fig. 3 Practical application of multilayered vessels

circumferential weld seams were distributed separately, with the circumferential weld seams of adjacent layers offset axially by a certain distance and the longitudinal weld seams of adjacent layers staggered at specific angles in the circumferential direction. The integrated multilayered vessel developed by Krupp Corporation demonstrated a range of comprehensive advantages, establishing itself as an ideal structure for multi-layer composite high-pressure vessels.

Since the advent of multilayered vessels, they have been widely applied in industries such as hydrogenation reactors, ammonia synthesis towers, and supercritical CO_2 extraction. Notably, China's first high-pressure vessel also adopted this structure. With the advancement of energy utilization, multi-layer wrapped vessels, recognized as a safe and reliable design, are now being extensively promoted for large-capacity high-pressure gaseous hydrogen storage, as shown in Fig. 3.

Due to the layered characteristics of multilayered vessels, numerous scholars have conducted studies on aspects such as the contact states between layers, stress states involving gaps, prestressing conditions, self-reinforcement or overpressure treatment, and thermal stress. However, studies on the fatigue issues of multilayered vessels are currently scarce. To identify the shortcomings of existing studies and establish the next steps for research, it is necessary to summarize and analyze previous work. This paper will present important past research in various areas in chronological order and highlight the limitations of current research, while also proposing directions for further investigation.

2 Research on Contact States Between Layers

In 1989, Wu et al. [1] divided the adjacent layers of multilayered vessels along the circumferential symmetrical plane of the cylindrical longitudinal weld seams, resulting in the friction zone and Lame zone. Based on the mechanical continuity conditions and stress continuity at the interface, they derived the calculation formula for the inter-layer friction angle of each layer, as shown in Eq. (1):

$$\alpha_j = \frac{1}{2f}\ln\left(\frac{R_{j+1}}{R_j}\cdot\frac{R_{n+1}^2 - R_j^2}{R_{n+1}^2 - R_{j+1}^2}\right) \tag{1}$$

From the Eq. (1), it can be deduced that to avoid the overlapping of the friction zones between adjacent layers, the angle by which the longitudinal weld seams of adjacent layers should be offset in the circumferential direction must be greater than twice the value of α_{n-1}. Therefore, the condition is:

$$\beta \geqslant \beta_{\min} = 2\alpha_{n-1} = \frac{1}{f}\ln\left(\frac{R_n}{R_{n-2}}\cdot\frac{R_{n+1}^2 - R_{n-1}^2}{R_{n+1}^2 - R_n^2}\right) \tag{2}$$

Subsequently, in 1993, Wu et al. [2], building upon the research on circumferential friction forces, considered the influence of axial friction forces between the layers. The study proposed that, to ensure the axial friction force between the layers can effectively carry the axial load, there must be a continuous contact region of a certain length between adjacent layers. The minimum contact length is given by Eq. (3):

$$l_{\min} = \lambda\pi R_0^2\left(k^2 - 1\right) \left\{2f\left[\sum_{j=0}^{n-1}\left(k_{j+1}^2 - 1\right)\left(\pi - \alpha_j - \alpha_{j+1}\right)R_{j+1} + \sum_{j=1}^{n-1}\frac{R_{j+1} - R_j}{f}\left(1 + k_j \cdot k_{j+1}\right)\right]\right\} \tag{3}$$

In 2023, Alaydin et al. [3] investigated the interaction of multilayered cylinders under friction and proposed a novel isogeometric Kirchhoff–Love (KL) shell computational method for multilayered structures. Their study revealed that interlayer friction increases the bending stiffness of the structure and induces significant hysteresis behavior.

3 Research on Stress States Involving Gaps

In 1968, Pimshtein [4], based on the uniform gap model as shown in Fig. 4, presented a formula for the calculation of the three principal stresses under the condition of gaps between the layers. However, the detailed derivation process was not provided in the literature. In the following year, Pimshtein [5], based on elastic-plastic theory, analyzed the stress state of multilayered cylinders with gaps. However, his conclusions were only applicable to the ideal situation where the gaps were completely eliminated under internal pressure.

In 1985, J. Rasty and M. Sabbaghian [6], building on the stress formulas derived by previous researchers based on the uniform gap assumption for layers, proposed a new gap model. The gap was approximated by two circular arcs with different radius: the inner arc radius R_g, the angular span of the gap arc corresponding to an angle of 2θ, and the maximum radial gap h as shown in Fig. 5. They also derived the additional strain required to eliminate such gaps, as given by Eq. (4):

$$\varepsilon_\theta = \frac{h\theta}{\pi R_g}\left[1-\frac{2+\frac{h}{R_g}}{2\left(1+\frac{h}{R_g}-\cos\theta\right)}\right]+\frac{1}{\pi}\left[1+\frac{h}{R_g}-\frac{\left(\frac{h}{R_g}\right)\left(2+\frac{h}{R_g}\right)}{2\left(1+\frac{h}{R_g}-\cos\theta\right)}\right] \times \arcsin\left[\frac{\frac{h}{R_g}\left(2+\frac{h}{R_g}\right)\sin\theta}{2\left(1+\frac{h}{R_g}-\cos\theta\right)\left(1+\frac{h}{R_g}\right)-\frac{h}{R_g}\left(2+\frac{h}{R_g}\right)}\right] \tag{4}$$

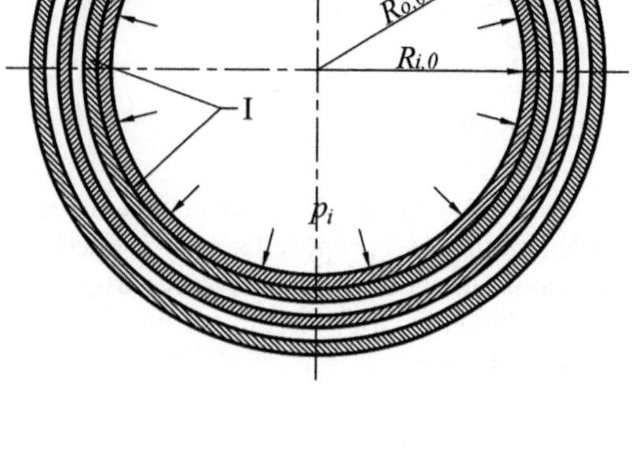

Fig. 4 Interlayer gap model

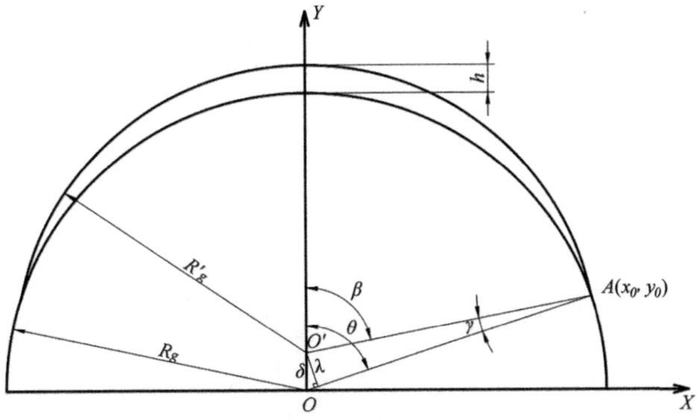

Fig. 5 Typical geometry of a gap

4 Research on Prestressing

In 2011, Zhen and Jiang [7] conducted research on the prestressing of the laminates and the clamping force of the hydraulic clamping device. They proposed a more concise formula for calculating the prestressing, as shown in the Eq. (5). They also carried out an overall clamping experiment on an inner cylinder with an inner diameter of 800 mm and a thickness of 30 mm, which was wrapped with a 10 mm thick outer steel plate.

$$\sigma_{\theta mc} = \sum_{m=1}^{n} \left(-p_{mc} \cdot \frac{2K_m^2}{K_m^2 - 1} \right) \tag{5}$$

The experimental results showed that the hydraulic clamping device effectively generates initial prestressing at the interfaces between the layers and the inner cylinder. Moreover, the magnitude of the clamping force should not be high enough to cause plastic yielding of the laminates during the processing.

In 2022, Wang et al. [8] derives the stress calculation formulas for the inner cylinder, considering the prestress of the shell, based on the Lamé equation and different strength theories. Furthermore, an optimization design model is proposed, aiming to minimize the material cost of the shell while accounting for the prestress and the use of different materials for the inner cylinder. The model takes the thickness of the inner cylinder, the number of lamination layers, and the thickness of each layer as variables.

In 2023, Wang et al. [9] derives the stress calculation expressions for the multilayered high-pressure vessel under internal pressure, both with and without prestress, as well as the formulas for calculating the ultimate load. Numerical simulations were conducted for validation. The results show that the theoretical and numerical simulation results are in good agreement. The application of prestress significantly reduces the working stress level of the inner cylinder and effectively improves the stress state of the shell under elastic conditions, thereby enhancing the elastic load-bearing capacity of the shell.

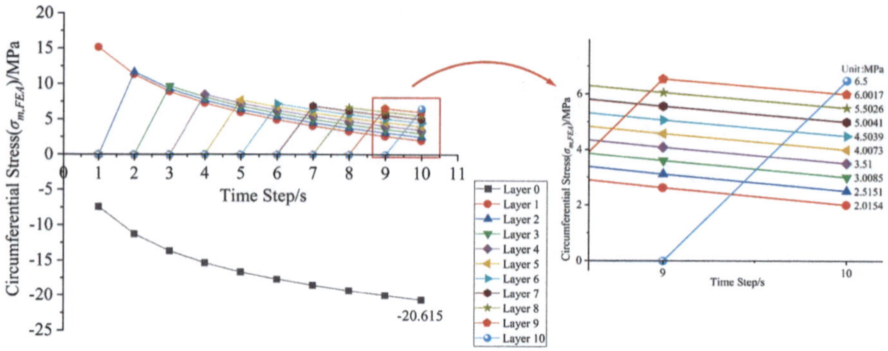

Fig. 6 FEA verification

In 2024, Xiao et al. [10], based on the ideal model without interlayer gaps and building on the work of Zhen [7], proposed a new method for calculating prestressing, as Eq. (6):

$$\sigma_m^{'} = \sigma_m - \sum_{u=m+1}^{n}\left(\sigma_u \cdot \frac{K_{m-1}^2+1}{K_{u-1}^2-1} \cdot \frac{k_u^2-1}{k_u^2+1} \cdot \prod_{v=m}^{u-1} k_v^2\right) \tag{6}$$

They employed a finite element analysis approach with multiple loading steps to validate the accuracy of their calculation method, as shown in Fig. 6. Additionally, they provided the limiting conditions for prestressing under small deflection instability conditions.

5 Research on Self-Reinforcement or Overpressure Treatment

In 2018, Rahman Seifi [11] investigated the maximum working pressure of shrink-fitted multi-layer containers after self-reinforcement treatment using a three-layer model that considered the Bauschinger effect and reverse yielding conditions. The results demonstrated that, compared to conventional single-layer type, the load-bearing capacity of multilayered vessels could be increased by up to two times through appropriately arranged interference fits. The findings aligned closely with the conclusions of E.-Y. Lee and N. Kumar.

In 2024, Khodayari [12] investigated the burst pressure of double-layered cylinders with internal surface cracks. The author conducted experimental and finite element analysis to study the effects of various parameters, such as crack geometry, percentage of autofrettage, and radial interference values, on the distribution of the J-integral at the crack tip and the burst pressure. This study provides a systematic investigation into the influence of cracks on the burst pressure of multilayered cylinders, offering theoretical insights and engineering guidance to enhance the design safety and service life of thick-walled cylinders.

6 Research on Thermal Stress

In 1990, Zukhova V. and Pimshtein P [13]. examined multi-layered cylinders subjected to internal pressure and internal heating. The study highlighted that the heat transfer characteristics of multilayered cylinders include the thermal resistance of the layers, the contact thermal resistance between layers, and the surface heat transfer resistance. Based on this, the steady-state heat flux can be calculated using Eq. (7):

$$q_l = \frac{2\pi\left(T_{\text{me}}^{\text{in}} - T_{\text{me}}^{\text{ou}}\right)}{\dfrac{1}{K_{\text{in}} r_{\text{in}}} + \sum_{i=1}^{n} \dfrac{1}{\lambda_i} \ln \dfrac{r_i}{r_{i-1}} + \sum_{i=1}^{n-1} R_i + \dfrac{1}{K_{\text{ou}} r_{\text{ou}}}} \tag{7}$$

In 2021, Sim et al. [14] studied the stress distribution in multi-layer hollow spherical pressure vessels under thermomechanical loads and proposed an analytical solution based on a recursive algorithm. The results show that the proposed recursive analytical method can effectively estimate the thermomechanical stress distribution in multi-layer hollow spherical pressure vessels, with the calculated results being in high agreement with those obtained from FEA, and the error being within 1%, as shown in Fig. 7. In the same year, Zhou et al. [15] focuses on the analytical modeling of steady-state thermal conduction in a multilayer cylinder with circumferentially-varying convective heat transfer boundary conditions.

In 2023, Jerzy et al. [16] investigated the steady-state temperature distribution in multilayer elliptical cylinders and proposed a combined analytical-numerical method to solve this problem. The results indicate that the temperature distribution

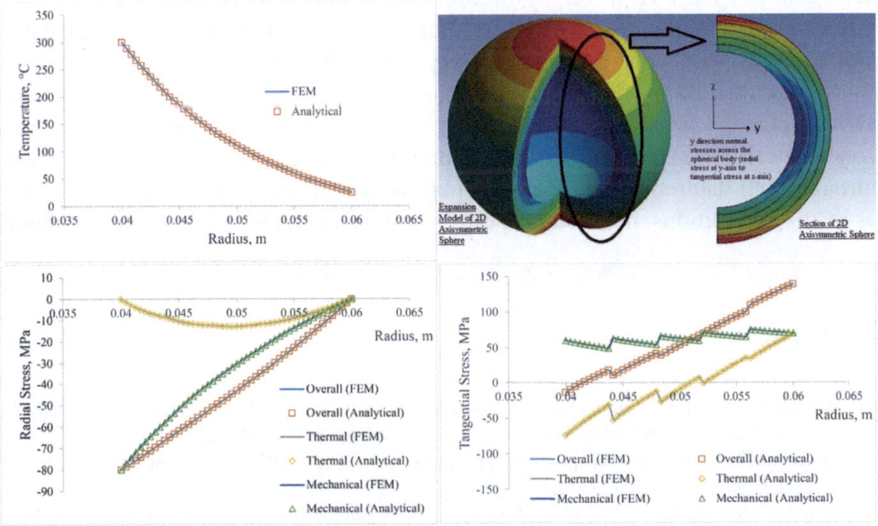

Fig. 7 Comparison of theoretical and finite element calculations [14]

within the insulating layer is approximately linear. In test systems with significantly different thermal conductivities, the temperature distribution exhibits a piecewise continuous zigzag pattern. Furthermore, the interfaces between the layers of the multilayer elliptical structure are not isothermal.

In 2024, Tokovyy et al. [17] proposed a method for solving the plane axisymmetric problem of multilayer hollow cylinders under elastic and thermoelastic conditions, applicable to cases where the material properties of each layer have an arbitrary distribution in the radial direction. The core idea of the method is to represent the distribution of material properties throughout the structure as a gradually varying function, and solve the corresponding compatibility equations using a direct integration approach, thereby obtaining the strain relationships. This method avoids the complexity of using generalized derivatives.

7 Conclusions

Currently, the calculation of prestressing primarily considers the effects of weld shrinkage and initial wrapping force, typically assuming an ideal multilayered cylinder without gaps. However, due to the surface roughness and the ellipticity from rolling, gaps between the layers significantly affect the distribution and magnitude of prestressing. Nevertheless, during manufacturing, the prestressing of the layers accumulates, leading to large compressive stresses in the inner cylinder and an increased risk of instability. Therefore, a safe and reliable prestressing calculation method is necessary to limit the overall prestress.

Additionally, existing stress calculation formulas for multi-layer cylinders with gaps assume small gaps and overlook the impact of gap elimination on the cylinder's diameter ratio. However, as multilayered vessels increase in pressure and size, the accumulation of gaps may lead to significant errors. Thus, more accurate calculation methods are required to meet safety standards.

At present, there is a lack of research on the fatigue performance of multilayered vessels, posing certain risks in evaluating their operational reliability under fatigue conditions. Future studies should focus on the fatigue performance of these vessels through both theoretical and experimental investigations, considering factors such as loading conditions, material properties, and the characteristics of internal media.

References

1. Wu J, Zhu X, Chen G (1989) Research on the leak-only, non-explosive behavior of a new type of integral multi-layer high-pressure vessel. J Press Vessel (China) 2:25–30
2. Wu J (1993) Discussion on the design method of integral multi-layer clamped high-pressure vessels. Chem Eng Mach 20(1):33–38
3. Alaydin MD, Bazilevs Y (2023) Multilayer shells interacting through friction. J Appl Mech 90:121012. https://doi.org/10.1115/1.4062139

4. Pimshtein P (1968) Strength of multilayer high pressure vessels. Chem Petrol Eng 4(7):574–578. https://doi.org/10.1007/BF01136754
5. Pimshtein P, Tatarinov V (1969) Elastoplastic operation of multilayer cylinders with gaps. Chem Petrol Eng 5(7):523–528. https://doi.org/10.1007/BF01832409
6. Rasty J, Sabbaghian M (1985) Effect of imperfect contact between adjacent layers on the integrity of wrapped vessels. J Press Vessel 98:167–176. https://doi.org/10.1115/1.3265596
7. Liang Z, Nan J, Sijia L (2011) Experimental study of integrated multilayer clamping high pressure vessel. J Press Vessel Technol 133(6):061206. https://doi.org/10.1115/1.4004568
8. Wang T, Qin G, Qian C (2022) Optimization design of multilayer wrapped high pressure vessel cylinder. Energy Conserv Technol (China) 40(05):387–391+402
9. Wang T, Qian C, Wu Z et al (2023) Analysis of load-bearing capacity of multi-layer wrapped high pressure vessels. J Mech Eng (China) 59(14):169–178
10. Xiao Z, Wang X, Wang H (2024) Prestressing estimation for multilayer clamping high pressure vessel laminates. J Press Vessel Technol 146(5):051703. https://doi.org/10.1115/1.4066003
11. Seifi R (2018) Maximizing working pressure of autofrettaged three layer compound cylinders with considering Bauschinger effect and reverse yielding. Meccanica 53(10):2485–2501. https://doi.org/10.1007/s11012-018-0834-2
12. Khodayari MS, Seifi R (2024) Bursting pressure of the autofrettaged and interferenced double-layer cylinder with an internal surface crack. J Fail Anal Prev 24(2):721–735. https://doi.org/10.1007/s11668-024-01876-z
13. Zukhova V, Pimshtein P (1990) Thermally stressed state of a laminated cylinder exposed to internal pressure and steady-state external heating. Sov Appl Mech 25(8):808–813
14. Sim LC, Yeo WH, Purbolaksono J et al (2021) Analytical solution of thermo-mechanical stresses of multi-layered hollow spherical pressure vessel. Int J Press Vessel Pip 191:104355. https://doi.org/10.1016/j.ijpvp.2021.104355
15. Zhou L, Parhizi M, Jain A (2021) Temperature distribution in a multi-layer cylinder with circumferentially-varying convective heat transfer boundary conditions. Int J Therm Sci 160:106673. https://doi.org/10.1016/j.ijthermalsci.2020.106673
16. Gołębiowski J, Zaręba M (2023) The stationary thermal field in a multilayer elliptic cylinder. Appl Sci 13(4):2354. https://doi.org/10.3390/app13042354
17. Tokovyy YV (2024) Elastic and thermoelastic response of multilayer inhomogeneous hollow cylinders. Mech Adv Mater Struct. https://doi.org/10.1080/15376494.2023.2186548

Open Access This chapter is licensed under the terms of the Creative Commons Attribution-NonCommercial-NoDerivatives 4.0 International License (http://creativecommons.org/licenses/by-nc-nd/4.0/), which permits any noncommercial use, sharing, distribution and reproduction in any medium or format, as long as you give appropriate credit to the original author(s) and the source, provide a link to the Creative Commons license and indicate if you modified the licensed material. You do not have permission under this license to share adapted material derived from this chapter or parts of it.

The images or other third party material in this chapter are included in the chapter's Creative Commons license, unless indicated otherwise in a credit line to the material. If material is not included in the chapter's Creative Commons license and your intended use is not permitted by statutory regulation or exceeds the permitted use, you will need to obtain permission directly from the copyright holder.

Study on the Deformation Characteristics of Oil and Gas Pipelines Subjected to Landslides

Jun Zhong, Hai Yu, and Chao Li

Abstract With the growing demand for energy, oil and gas pipelines play a vital role in the transportation of oil and gas. The terrain of our country is complex, and landslide is one of the main geological disasters causing damage to oil and gas pipelines. In this paper, a full-scale model test with small pipe diameters was carried out by controlling the slope gradient and displacement through a model box loading device, revealing the deformation law of oil and gas pipelines under transverse landslide. The research results indicated that under the action of transverse soil displacement, the increment of pipeline strain increases with the increase of displacement; obvious local buckling phenomenon occurs at the mid-span of the pipeline, and the overall failure form is symmetrical bending. The ultimate compressive strain value of the test is basically consistent with the calculated value in the specification, but the test value when the pipeline buckles and fails is significantly larger than the numerical simulation result. Finally, the friction coefficient between the pipeline and soil in the finite element model was corrected based on the scaled test data. By comparing the pipeline axial strain data from the scaled landslide test with that of the full-scale finite element model, the reliability of the finite element model was verified. This is of great significance for ensuring the safe operation of oil and gas pipelines and reducing losses caused by geological disasters.

Keywords Pipelines · Landslide · Transverse displacement · Deformation · Model experiments

J. Zhong (✉) · C. Li
China Petroleum Pipeline Engineering Corporation, Langfang, China

H. Yu
Construction Project Management Branch of China National Petroleum Pipeline Network Group Co., Ltd, Langfang, China

© The Author(s) 2026

Z. Zhang, P. Mercorelli (eds.), *Mechanical Design and Simulation: Insights and Innovations*, Lecture Notes in Mechanical Engineering,
https://doi.org/10.1007/978-981-95-2405-1_56

1 Introduction

The deformation test and effect study of oil and gas pipelines under the action of transverse landslide is a complex and important topic, which involves the threat of landslide geological disasters to the safe operation of oil and gas pipelines. Theoretical analysis and numerical simulation are the basic methods for studying the interaction between landslides and pipelines [1–3]. Theoretical analysis involves numerous assumptions, and its results are only applicable to materials in the elastic stage, making it difficult to apply to pipelines passing through large deformation sections such as landslides. Numerical simulations often employ soil spring models or solid contact models, which involve significant simplifications and can lead to grid distortion during large deformation calculations, reducing computational accuracy. Therefore, it is an effective method to conduct model tests on the interaction between pipe and soil in the moving section of the slope using large-scale landslide models [4–6]. At present, most of the model tests simulate the pipe-soil interaction under strike-slip faults through soil boxes, resulting in the lack of slope solid calculation models. For example, Trautmann and O'Rourke conducted a series of full-scale tests on pipe soil interaction using a soil box with a length, width, and height of 1.2 × 2.30 × 1.20 m for pipelines with diameters of approximately 110 mm and 220 mm, and provided a calculation method for pipe soil interaction under dense sand conditions according to ASCE specifications. Ashrafy used 1.5 × 1.5 × 8 m soil box models and steel pipes with diameters of 113 mm and 218 mm, simulating the interaction between pipe-soil under strike-slip faults. The results show that the maximum load and initial stiffness of the soil on the pipeline are greater than those recommended in the ASCE guidelines [7–10]. This paper focuses on the characteristics of oil and gas pipelines crossing landslides. Utilizing a large-scale landslide model test platform, it conducts model tests on the interaction between the pipeline and soil in the moving section of the slope. The aim is to reveal the deformation and failure patterns of pipelines during landslide occurrence, and provide a scientific basis for establishing safety design criteria for pipelines in areas prone to landslides in mountainous regions.

2 Model Test of Compacted Sand

2.1 Model Experimental Scheme

Reasonably designing soil and pipeline models, conducting indoor physical model tests, and utilizing model test methods to simulate pipe-soil interaction are effective ways to reveal the deformation and failure patterns of pipelines during landslide processes. According to the size of the model box of the test platform, the design slope size is 4.2 m (length) × 1.05 m (width) × 0.9 m (height), and the slope angle is set to 30°. The pipeline is buried at the slope shoulder position, and the buried depth at the top of the pipe is 0.4 m. The total length of the pipeline is about 1.5 m, the length in the soil is 105 cm, the thickness of the model box is 2.5 cm, and the suspended part of the pipe extends from each end of the pipe is 20 cm. The steel pipe selected for this test pipeline is Q235 galvanized seamless steel pipe, and the 40 mm pipe diameter is selected for the test with an increased pipe diameter, which needs to be re-opened in the side wall of the model box.

2.2 *Loading Process and Data Collection*

After the model is made, the model is left to stand for 48 h, and the loading test is carried out after the whole slope is consolidated and stabilized. Throughout the process, the data is recorded using an acquirer. After the model is stabilized, the graded slope is lifted, the slope of each climb is 5°, and the model is stationary after each climb, and the angel model is re-stabilized after the climb. Before and after the trial, the data were recorded. When the slope rises to 30°, displacement loading is performed. Due to the lack of precise control means of the loading device, the loading adopts a one-time loading until the displacement load causes the pipe to be destroyed. During the whole test, the data is continuously recorded, and a multi-camera camera is used to record the deformation process of the entire slope for analysis.

2.3 *Analysis of Pipeline Calibration Test Process and Results*

For the pipeline selected for the model test, in order to obtain its mechanical parameters and performance indicators, the same steel pipe needs to be used for calibration test. The strain gauge is laid on the steel pipe, and then the three-point bending test of concentrated load is carried out. The initial preload is 0.5 Mpa, and then 0.2 Mpa is loaded each time, and the deflection value and strain value are recorded after 5 min after each stage of loading is stopped and stabilized for 5 min, and then the next stage of loading is carried out until the model enters the plastic failure stage.

After the calibration test, the relationship between the strain value, deflection value and loading strength obtained by the test was processed by using the principle of material mechanics formula, and the stress-strain curve of the pipe was obtained (Fig. 1).

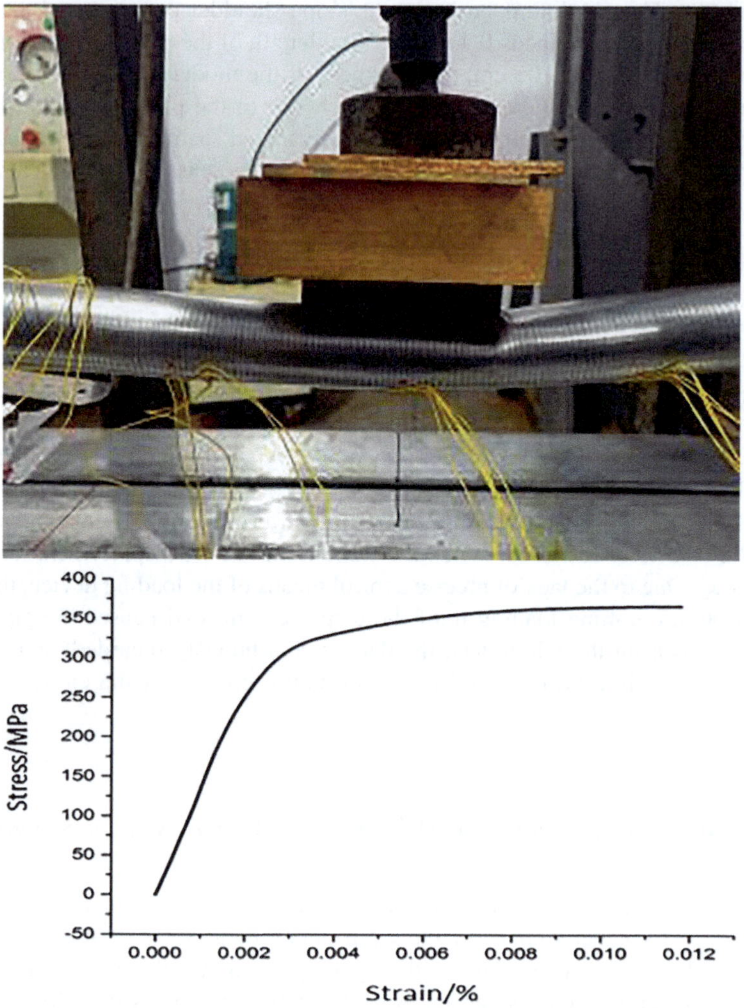

Fig. 1 Pipeline calibration test and the stress-strain curve of the test pipeline

2.4 *Pipeline Strain Analysis*

With the promotion of the lower jack of the model box, the angle of the bottom surface of the whole slope increases from 0° to 30°, and the load on the pipeline changes, which is manifested as the increase of slope thrust, and the strain data of each section responds obviously. The sliding side of the pipeline is in a state of compression, and most of the strain gauges of the pipeline are shown as negative values, and their values increase with the increase of slope, and the mid-span strain can reach about −1000 at 30°. The back-slip side of the pipe is in tension, and most of the strain gauges show positive values, which also increase with the increase of slope, and the mid-span strain can reach about +1000 at 30° (Fig. 2). When the slope of the model box is raised to 30°, the steel plate behind the model box is pushed by the hydraulic servo controller to move forward to push the slope. In the process of moving, the slope interacts with the pipeline, and the pipeline changes significantly from the elastic deformation stage to the plastic deformation stage and then to yield failure. In the test process, it is found that the slope movement is more integral, and the displacement of the thrust plate behind the model box is regarded as the displacement of the landslide as a whole, and the change of the strain values of different cross-sections in the process of displacement loading is plotted as a variable.

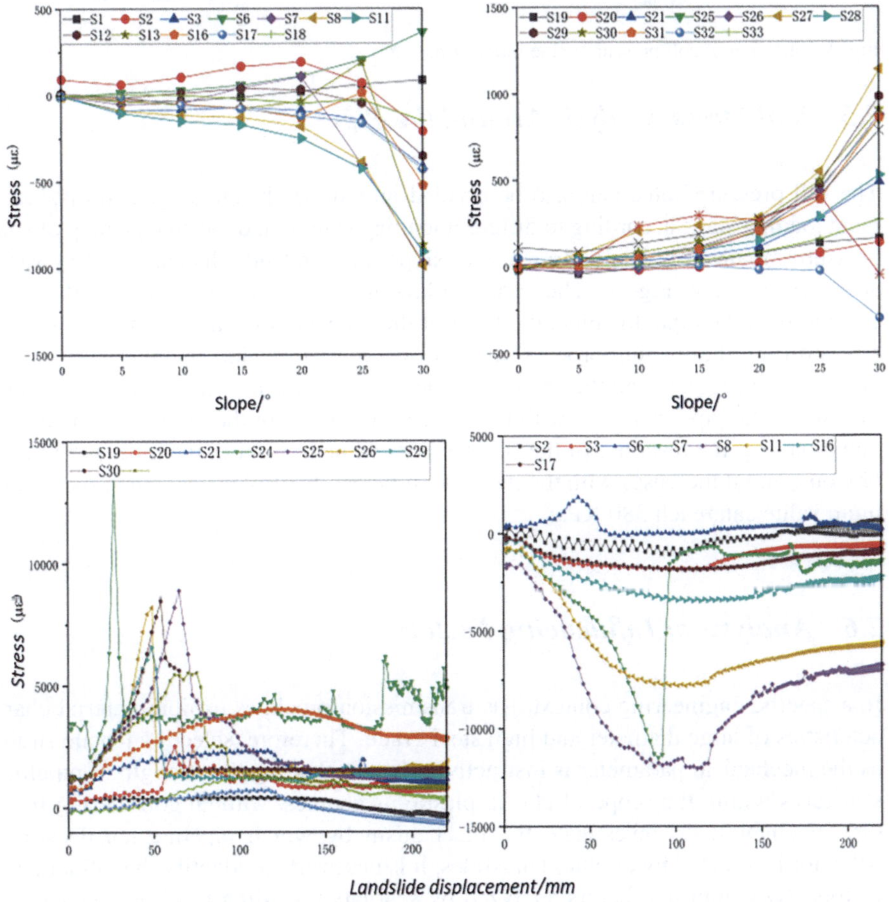

Fig. 2 Variation of strain values at different slopes

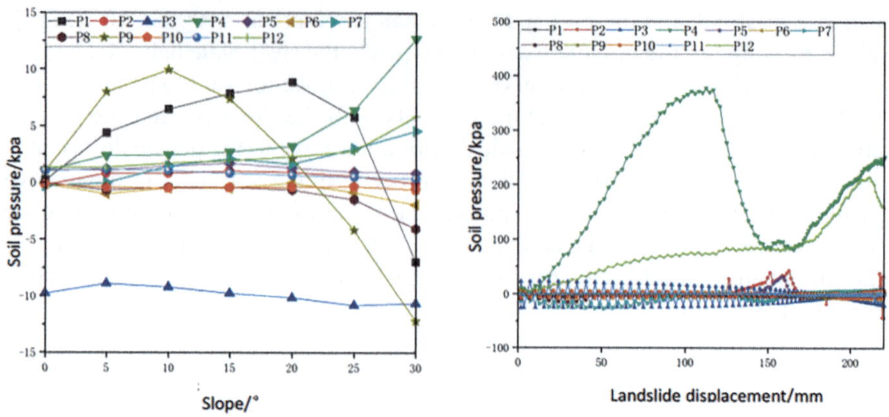

Fig. 3 Variation of soli pressure value monitoring data

2.5 *Soil Stress Analysis Around the Pipe*

The soil pressure value can also be divided into slope change stage and displacement loading stage according to different loading stages, and the change map of soil pressure monitoring data is drawn with slope and landslide displacement as variables respectively (Fig. 3). The variation law of the soil pressure value in the four directions of the pipe, the pipe, the front of the pipe and the back of the pipe is different during the loading process, and the soil pressure value changes little in the process of ascending, and the obvious is that the soil pressure value on the slope side in front of the pipe increases with the increase of slope. In the process of displacement loading, the response of the soil pressure value in front of the pipe is the most obvious, and it increases with the increase of landslide displacement, and the maximum value can reach 380 KPa.

2.6 *Analysis of Influencing Factors*

In a specific engineering context, gas transmission pipelines exhibit material characteristics of large diameter and high steel grade. Therefore, selecting tensile strain as the mechanical parameter is instructive when evaluating the strength of pipeline structures within the scope of elastic-plastic mechanics, without considering fracture mechanics. To investigate the mechanism by which pipeline tensile strain behavior is affected by circular landslides, it is necessary to identify the influencing factors. By compiling factors analyzed by scholars regarding pipeline mechanical parameters under landslide conditions, it was discovered that the primary influencing factors encompass pipeline parameters and landslide parameters. Firstly, by identifying the factors that influence the calculation of pipeline strain in the slope movement section, all types of influencing factors are determined. Subsequently, a dimensionless analysis is conducted for each type of factor to establish the dimensionless expression form of the final strain calculation model. Then, using the control variable method, the study focuses on the impact of individual factors on pipeline strain, specifically examining their mechanical behavior. By observing the trends in change, a functional expression for the influence of each factor on pipeline strain is determined. Furthermore, the coupling effects among these factors are discussed. The strain calculation model's effectiveness is tested, and a new working condition is devised for further testing. In this study, we focus on eight commonly used pipe diameter parameters in engineering projects: D219, D406, D610, D813, D914, D1016, D1219, and D1422. For each of these diameters, we conducted corresponding mechanical behavior research. Excluding factors other than the diameter-to-thickness ratio, the results are presented in Fig. 4. As evident from the figure, within the same pipe diameter range, as D increases and grows linearly within the D/t ratio, the impact of other influencing factors on the strain trend remains relatively unchanged.

Under the influence of different aspect ratios, the maximum tensile strain exhibits a linear relationship with the depth-to-diameter ratio. When the diameter-to-thickness ratio of the pipeline is small, meaning the wall thickness is relatively

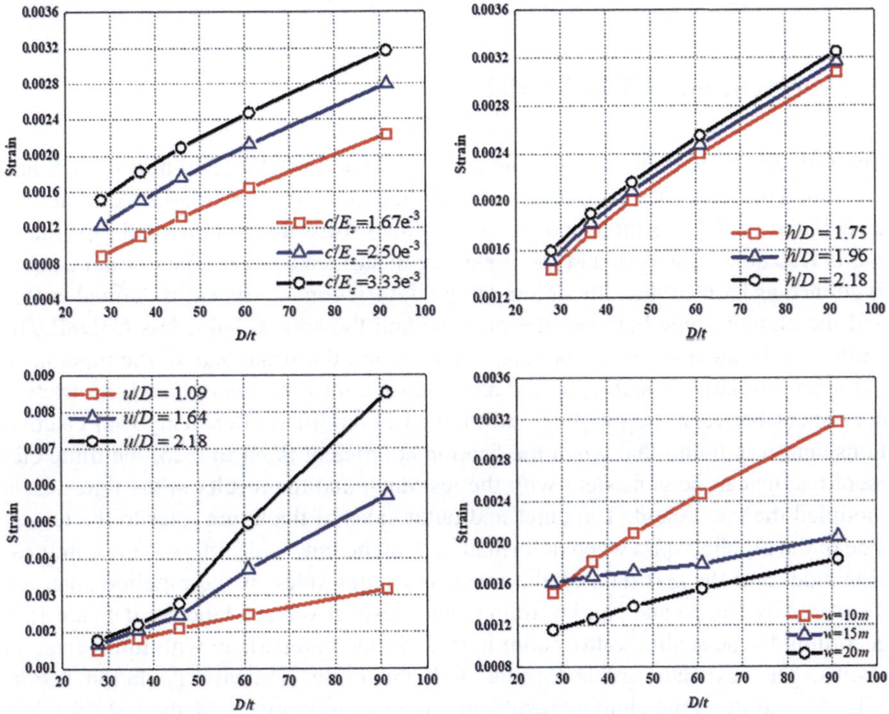

Fig. 4 Study on horizontal mechanical behavior of other influencing factors 3 under different diameter thickness ratios

large, the growth rate of the pipeline's maximum tensile strain initially increases and then decreases as the maximum landslide displacement increases. When the diameter-to-thickness ratio is large, the growth rate of the maximum tensile strain remains relatively constant and exceeds that of the smaller ratio. Specifically, with a diameter-to-thickness ratio of 25.4, when the maximum landslide displacement increases fivefold, the maximum tensile strain only rises by 40%. However, at a ratio of approximately 76.2 (three times the original ratio), a fivefold increase in maximum landslide displacement leads to a 304% increase in maximum tensile strain. Landslide displacement, a natural disaster, is difficult to intervene with human efforts, but increasing the wall thickness by three times can significantly reduce the maximum tensile strain. Therefore, this measure is highly effective in mitigating the axial tensile impact on pipelines caused by large soil displacements due to landslides. In practice, the maximum tensile strain initially increases and then decreases as the aspect ratio increases. This phenomenon occurs because, when the landslide width is small, the maximum tensile strain occurs at the junction between sliding and non-sliding soils; as the landslide width increases, the maximum tensile strain occurs at the center of the sliding soil. In engineering qualitative evaluations, a quick conclusion can be drawn: under the same other conditions, the larger the width-to-diameter ratio, the smaller the maximum tensile strain produced by the pipeline.

3 Verification of Test Model

The full-size large-diameter pipeline can be selected as OD 1219 mm × 18.4, and the scaled test diameter is 40 mm, so the similarity ratio is about 30:1. A numerical calculation model is established using ABAQUS software, and the steel pipe material is X80 steel. The contact type is surface to surface contact. The normal contact is defined as hard contact, the initial tangential friction coefficient is defined as 0.3, and the element type between the pipeline and the soil is defined as C3D8R.The strain data in the test data is the most reliable, and the strain data of the pipeline at 3/4 span is relatively stable, so the axial strain data is compared, and the friction coefficient between the pipeline and the soil is modified after many trial calculations, and it is found that when the friction coefficient is set to 0.25, the finite element results can be consistent with the test data, and the results in the figure have modified the coefficient. The outer and outer sides of the figure refer to the tensile side of the pipeline span when deformation, and the inner side refers to the compression side, and the test data of the inner and outer sides of the pipeline span are measured by No. 34 and No. 16 strain gauges respectively, and the test data are 1:30 according to the scaling ratio Zoom in to facilitate comparison with finite element results. The test data are consistent with the finite element calculation results (Fig. 5), and the finite element results are more conservative than the test data, and the material, boundary conditions, diameter-thickness ratios and other local dimensions between the two do not fully meet the similarity ratio, which affects the results to a certain extent.

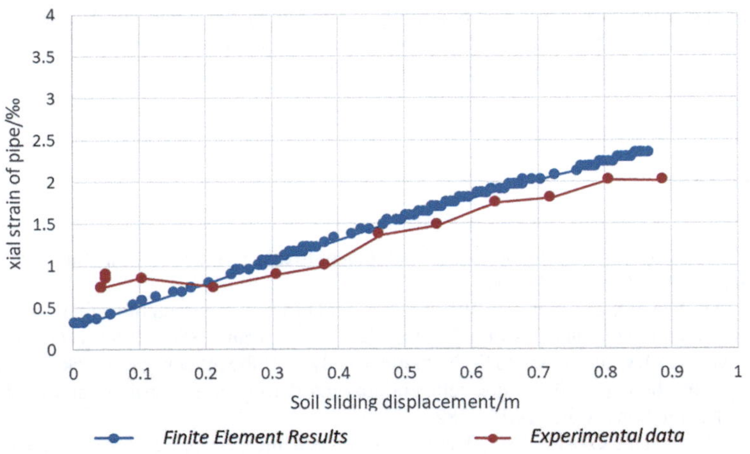

Fig. 5 Comparison of strain on the middle and outer sides of the pipe span

4 Conclusion

The reliability of the finite element model was verified by comparing the axial strain data of the pipeline between the landslide scale test and the original size finite element model. The friction coefficient between the pipe and soil in the finite element model was corrected through scaled test data. The conclusions of this study are as follows:

1. Under the action of transverse soil displacement, obvious local buckling phenomenon occurs at the mid-span of the pipeline, and the overall failure form is symmetrical bending, the increment of pipeline strain increases with the increase of displacement. In the engineering design stage, larger wall thickness or strain design method can be used to increase the strain capacity of the pipeline.
2. The ultimate compressive strain value of the test is basically consistent with the calculated value in the specification, but the test value when the pipeline buckles and fails is significantly larger than the numerical simulation result. When using numerical tools to calculate, the coefficient should be increased to make the design scheme safer.
3. The pipeline should try to avoid passing through sections with a slope greater than 20° at the design stage, and necessary protective measures should be taken to protect the operation safety of the pipeline. A study of pipeline protection measures should be carried out in the future, such as the installation of cushioning devices in areas where lateral landslides may occur. The buffer device is capable of effectively absorbing the displacement of the landslide. At the same time, it lowers the strain value of the pipeline and strengthens the pipeline's ability to resist the deformation induced by the landslide, thus ensuring the safe operation of the pipeline.

Acknowledgments This work was financially supported by the funds of "Research on Safety Design Technology of Oil and Gas Pipelines in Mountainous Areas" (GK22-A-2) and "Pipeline Integrity Management in Strain Design Areas" (20221256126).

References

1. Xin YP (2020) Current situation and development trend of oil and gas pipeline technology in China. Nat Gas Oil 38(2):26–31
2. Guo SD, Wang Q, Lin Y et al (2019) Analysis and prevention of geological hazards along the China-Myanmar Oil and Gas Pipeline. Oil Gas Storage Transp 38(9):1059–1064, 1071
3. Shan W, Ma M, Guo Y et al (2022) Numerical analysis of the influence of block stone embankment filling height on the water, temperature, and deformation distributions of subgrade in permafrost regions. Water 14(9):1382
4. Wei S, Tao Y, Ying G et al (2022) Thermohyrdromechanical coupling analysis and engineering verification of gravel pile groups for strengthening permafrost marshland highway foundations. Bull Eng Geol Environ 81(7):294
5. Xian Y, Wei X, Zhou H et al (2022) Snowmelt—triggered reactivation of a loess landslide in Yili, Xinjiang, China: mode and mechanism. Landslides 19(8):1843–1860
6. Wang L, Song D, Zhou G et al (2022) Debris flow overflowing flexible barrier: physical process and drag load characteristics. Landslides 19(8):1881–1896
7. Chiarelli D, Galizzi M, Bocchiola D et al (2023) Modeling snowmelt influence on shallow landslides in Tartano valley, Italian Alps. Sci Total Environ 856(1):158772
8. Zhao XZ (2014) Strain response analysis of oil and gas pipelines subject to lateral landslide. J Nat Hazards 23(4):250–256
9. Yang C, Chao W, Weng M et al (2022) Outburst debris flow of Yusui Stream caused by alarge—scale Silabaku landslide, Southern Taiwan. Landslides 19(7):1807–1811
10. Wan X, Pei W, Lu J et al (2022) Prediction of the unfrozen water content in soils based on premelting theory. J Hydrol 608:127505

Open Access This chapter is licensed under the terms of the Creative Commons Attribution-NonCommercial-NoDerivatives 4.0 International License (http://creativecommons.org/licenses/by-nc-nd/4.0/), which permits any noncommercial use, sharing, distribution and reproduction in any medium or format, as long as you give appropriate credit to the original author(s) and the source, provide a link to the Creative Commons license and indicate if you modified the licensed material. You do not have permission under this license to share adapted material derived from this chapter or parts of it.

The images or other third party material in this chapter are included in the chapter's Creative Commons license, unless indicated otherwise in a credit line to the material. If material is not included in the chapter's Creative Commons license and your intended use is not permitted by statutory regulation or exceeds the permitted use, you will need to obtain permission directly from the copyright holder.

Design and Research on New Type of Balance Valve Suitable for Miter Gate Hoist of Ship Lock

Fangchao Yan, Tianwei Huang, and Hanzhao Li

Abstract The miter gate hoist of the ship lock is the core component of the ship lock operation, and its opening and closing operation requires the gates on both sides to be synchronized and adapt to the external load changes caused by undercurrent, over-irrigation and over-discharge. A new type of pilot plug-in balance valve, designed for a 1:50 external load ratio, has been developed. We have constructed an AMESim model for the new pilot plug-in balance valve and hydraulic gate system, and conducted simulations to study its dynamic characteristics and anti-interference capabilities. The results indicate that the new type of pilot plug-in balance valve exhibits good dynamic performance. It can adapt to changes in external loads and is insensitive to them. The hydraulic hoist will not exhibit hydraulic cylinder crawling or herringbone gate drift under external loads.

Keywords New pilot-injected balance valve · Dynamic behavior · AMESim model · Sliding phenomenon · Drift phenomenon

1 Introduction

The performance of the balancing valve directly affects the reliability and stability of the entire hydraulic system operation. Unlike hydraulic balance valves commonly used in crane engineering machinery, ship lock inclined gate hoisting machinery is usually a double-acting, center-supported, bi-directional swing, horizontal installation hydraulic direct-link hoisting machinery that mainly bears horizontal loads [1, 2].

The ship lock miter gate equipment is a large inertia load, the miter gate operation process requires smooth action, small impact, and the operation of the double miter gate coordination and synchronization, a lock miter gate is often subject to changes in the opening and closing of the variable load, the hydraulic hoist

F. Yan · T. Huang (✉) · H. Li
Three Gorges Navigation Authority, Hubei Yichang, China

© The Author(s) 2026

Z. Zhang, P. Mercorelli (eds.), *Mechanical Design and Simulation: Insights and Innovations*, Lecture Notes in Mechanical Engineering,
https://doi.org/10.1007/978-981-95-2405-1_57

crawling, shaking and miter gate drift and other problems [3, 4]. The reverse load is large when the hermitage door is open, and the operation process requires the cylinder to be stable, the water level changes regularly throughout the year, and the repeated changes of the external load put forward higher requirements for the performance of the balance valve. At present, the high-end balance valve on the market basically relies on foreign products, and faces problems such as difficulties in purchasing spare parts and uncontrollable delivery [5, 6]. At the same time, the domestic balance valve can not meet the requirements of stability and fast response, and has the characteristics of small flow and high pressure loss [7, 8]. In practical application, two or more balancing valves are used in parallel to meet the large flow demand of the miter gate hoist with high operating pressure, so it is urgent to develop and design a new balancing valve suitable for the miter gate hoist.

In order to solve the problem of crawling and drifting of the hydraulic hoist of the miter gate of the ship lock, a new type of pilot plug-in balance valve adapted to the operating condition of the ship lock is studied and designed.

2 The Hydraulic Hoist for the Miter Gate of the Ship Lock

2.1 System Principle

The normal operation of the hydraulic hoist for the miter gate of the ship lock is the key to ensuring the safe and efficient operation of the ship lock. The hoist has a large scale and complex structure. A high pressure and large flow stepless speed control circuit for the miter gate hoist of ship lock is formed by the combination of the high pressure and large flow piston pump and the integrated control structure of the two-way plug valve [2].

At present, the hydraulic hoist of the miter gate of a ship lock cannot adapt to the large change of water level, including the crawling of the miter gate hoist and the drifting of the miter gate closing [3]. The crawling phenomenon of the hoist gate refers to the alternating movement of the hydraulic cylinder during movement, where the displacement jumps and stops, or moves faster and slower. Once the operation of the hoist miter gate of the ship lock appears crawling and drifting, it will cause the miter gate to be eventually mislocated, the water is unstable, the closing signal cannot be triggered, and the automatic operation process of the ship lock will be interrupted. In serious cases, it will also lead to the guide wheel card jamming, damage the metal structure of the miter gate, reduce the efficiency of safe operation, and bring about the risk of safe operation.

In this paper, a new type of pilot-injected balance valve is designed to adapt to the operating conditions of ship locks, in response to the failure phenomenon of the balancing valve due to the failure of the balancing valve to hold the speed during the fluctuation of the outer load of the sluice gates. The proposed high flow balance valve for ship locks can directly increase the diameter of the main valve component by enlarging the main valve core, thereby increasing the flow of the high flow balance valve for ship locks. It has the characteristics of easy adjustment and can better meet the high flow demand of ship lock opening and closing machines with high operating pressure. It also has the feature of low pressure loss.

The Principle of a New Type of Pilot Plug-in Balance Valve The structure of a new type of pilot plug-in balance valve is shown in Fig. 1. The control mode is pilot type, the connection and installation mode is cartridge type, the main spool is innovatively designed auxiliary spool on the basis of the spool structure of two-way cartridge valve. The valve body has four oil ports, namely load port B, port A connected to the electromagnetic directional valve, pilot control port X on the end cap, and pilot drain port L located on the control piston valve body. A. B ports are all located on the main valve body. The tail of the pilot valve core is machined with threads, which can be used for installing overcompensation damping. A quick closing valve port is designed between the pilot valve core and the reset spring. The quick closing valve port is in a normally open state when the pilot valve core is not open, and in a normally closed state when the pilot valve core is open. The pilot valve core and the control piston are in a coaxial position, and there is elastic contact between the control piston and the pilot valve core. The large end of the control piston is designed with threaded holes for installing bypass damping. The control piston is reset by a reset spring. The control piston part has been designed to be detachable, allowing for the replacement of reset springs with different stiffness.

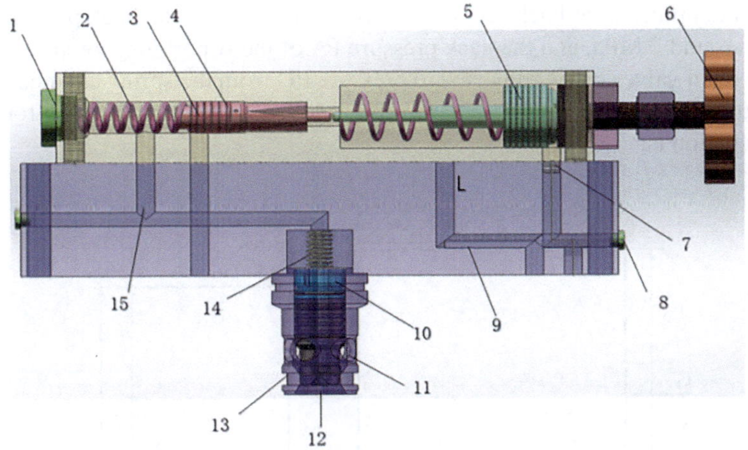

Fig. 1 Structure of a new type of pilot plug-in balance valve. 1—Plug; 2—Reset spring; 3—Pilot valve core; 4—Pilot valve port; 5—Control piston; 6—Adjusting screw; 7—Pilot inlet damping; 8—Pilot port; 9—Recycle port; 10—Auxiliary valve core; 11—Load port (B port); 12—A port; 13—Main valve sleeve; 14—Main valve spring; 15—Overcompensation damping

The pilot control oil port is located at the right end of the valve body, and inlet damping is installed on the sealing surfaces of the upper and lower valve bodies.

Structural innovation has been carried out on the main valve core of plug-in valve, with the addition of a throttling groove at the front end to achieve throttling effect. The throttling groove is evenly distributed at the front end of the main valve core. The auxiliary valve core is connected to the main valve core, and a feedback valve port is designed on the auxiliary valve core to enhance the overcompensation effect. Eight small holes with a diameter of 1 mm are designed on the inner side of the shoulder as flow channels. The feedback valve port communicates with the hydraulic oil at port B, and after a certain pressure loss, it enters the spring chamber. The hydraulic oil at load port B acts on the shoulder, and its force is consistent with the opening direction of the main valve port. After the main valve is opened, the B port and A port are connected, and the overcurrent area of the main valve port determines the size of the main valve flow.

3 The Mathematical Modeling of Balance Valve

Balance valve static characteristics are mainly analyzed when the balance of the valve components movement to reach equilibrium state when the main spool and the pilot part of the force.

The force analysis of the main valve core is shown in Fig. 2. The A port of the main valve is connected to the reversing valve. When testing the single characteristic and the influence of back pressure on the balance valve, the back pressure pressure is around 2 MPa, and the back pressure PA of the A port is ignored.

The main valve core is subjected to pressure PD from the spring chamber on A3, reset spring force Ft, load port pressure PB, and hydraulic force Ff. The force balance equation is:

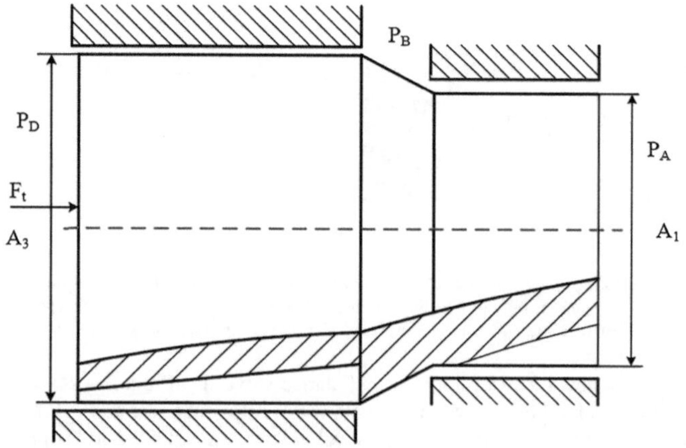

Fig. 2 Force analysis of the main valve

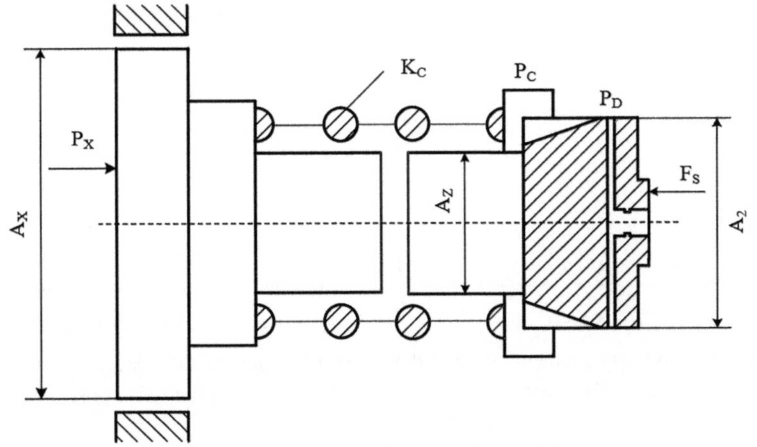

Fig. 3 Analysis of the overall force on the pilot valve and control piston

$$P_B\left(A_3 - A_1\right) + P_A A_1 - P_D A_3 - F_t - F_f = 0 \tag{1}$$

In the formula, A_3 is the big end area of the main valve core, and A_1 is the small end area of the main valve core

The force of pilot valve core and control piston is shown in Fig. 3. The pilot valve core and control piston can be regarded as a whole for force analysis.

After simplification, this whole is subjected to the pilot chamber pressure PD, the reset spring force FS, the main valve spring chamber pressure PC, and the spring force Fr of the control spring. The static pressure balance equation is as follows:

$$P_X A_x + P_c\left(A_2 - A_z\right) - P_D A_2 - F_s - F_c = 0 \tag{2}$$

The spring force is expressed as:

$$F_s = K_s\left(x_s + x_{ms}\right) + F_{s0} \tag{3}$$

$$F_c = K_c x_p + F_{c0} \tag{4}$$

4 Simulation Analysis of Balance Valve for Ship Lock Herringbone Gate Opening and Closing Machine

4.1 Modeling of Valve Ports for a New Type of Pilot Plug-in Balance Valve

In order to accurately and effectively build simulation models, it is necessary to obtain the flow area displacement curves of each valve port. Draw a 3D model based on the design dimensions, calculate the flow area data of each key valve port

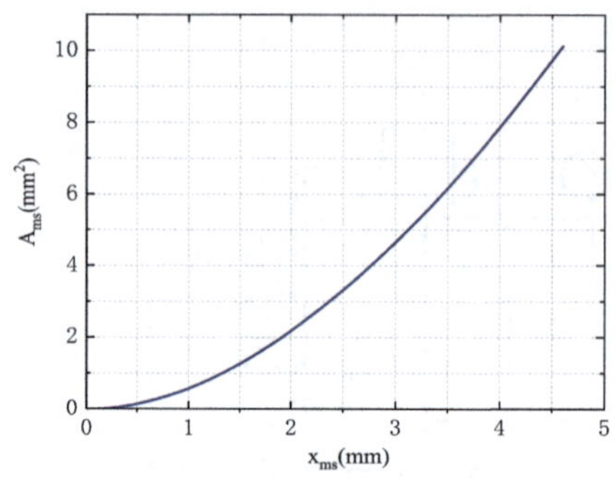

(a)Flow area and displacement curve of the main valve port

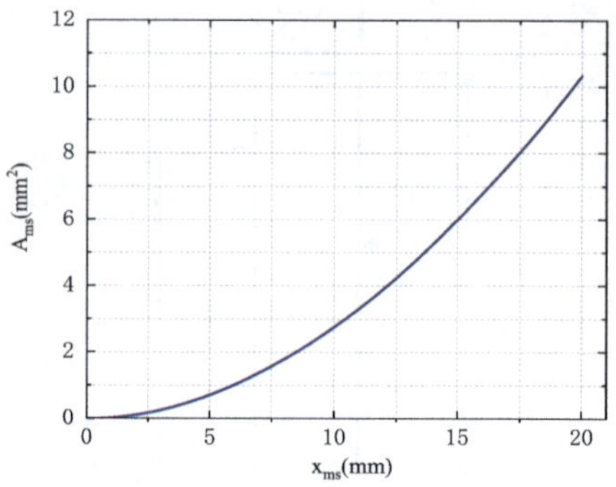

(b)Flow area and displacement curve of the pilot valve port

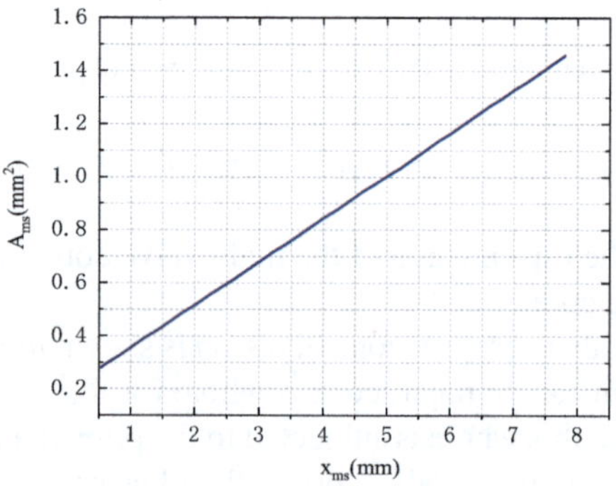

(c)Flow area and displacement curve of the feedback valve port

Fig. 4 Overflow area curve of each valve port

according to the actual dimensions, and use Origin to draw the flow area displacement curve of the valve port. The main valve port, feedback valve port, and pilot valve core of the new type of pilot plug-in balance valve are all irregular valve ports. The flow area displacement curves of each valve port were calculated and shown in Fig. 4.

4.2 Simulation Model of New Pilot Plug-in Balance Valve AMESim

Based on the working principle and three-dimensional model dimensions of the new pilot plug-in balance valve, the HCD simulation model of the new pilot plug-in balance valve was established using the HCD library in AMESim software, as shown in Fig. 5.

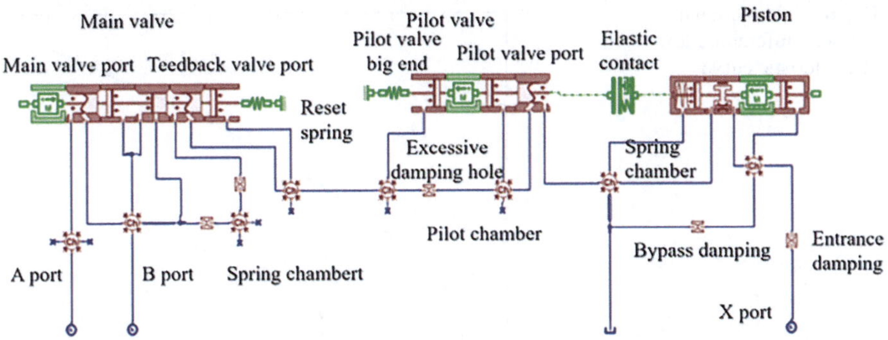

Fig. 5 Balance valve simulation AMESim model

Table 1 Main parameter of balance valve simulation model

Parameter variables	Numerical value
Hydraulic oil density kg/m^3	860
Hydraulic oil kinematic viscosity cP	46
Hydraulic oil volume Modulus of elasticity MPa	1700
Hydraulic oil discharge coefficient	0.7
Main valve end diameter mm	13.5
Main valve big end diameter mm	17
Main valve core spring N/mm	10
Main valve core quality g	100
Pilot valve core big end mm	14
Pilot valve core small end mm	(12, 3)
Pilot valve core reset spring N/mm	10
Pilot valve core quality g	120
Piston big end mm	28
Piston small end mm	3
Control spring N/mm	450

Static characteristics analysis of a new balance valve

The new type of pilot plug-in balance valve HCD model mainly consists of a main valve core, an auxiliary valve core, a pilot valve core, a control piston, an overflow valve, a damping hole, a feedback spring, a reset spring, a pressure source, etc.

The working principle of a balance valve is complex, and there are various influencing factors. In order to facilitate modeling and identify important influencing parameters, it is necessary to ignore certain external factors and make certain idealized assumptions. In an ideal state, simplified dynamic equations for each moving component and chamber are established based on Newton's second law and flow continuity equation, and the main parameters of the balance valve simulation model are obtained, as shown in Table 1.

Fig. 6 Unidirectional pressure differential flow characteristic curve

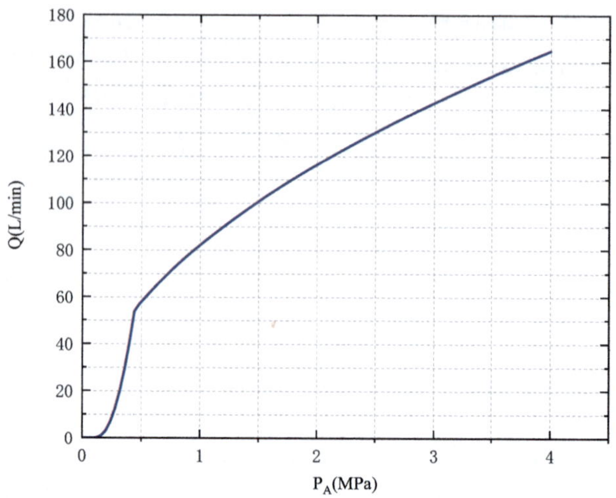

Single Characteristic When simulating a single characteristic, control port X is not set to control pressure (Px = 0 MPa), port B (load port) is connected to an oil tank or the pressure source is set to 0 MPa, and the pressure at port A is set to linearly change from 0–4 MPa. When drawing a one-way function diagram, the pressure at port A is taken as the x-axis, and the flow rate at port B is taken as the y-axis. The one-way pressure difference flow rate characteristic curve is drawn as shown in Fig. 6.

We can see that the one-way opening pressure of the balance valve is around 0.1 MPa to 0.2 MPa. The maximum flow rate of the balance valve main valve reaches 165 L/min, which is relatively small due to the small flow area design of the main valve port. The pressure difference at the main valve port is between 0.1 MPa and 0.2 MPa, indicating that the one-way opening pressure of the balance valve is low, energy consumption is low, and the throttling effect is good.

Flow Control Characteristic Curve The flow control characteristics of a balance valve refer to the relationship between the main valve flow rate and the pilot control pressure when the pilot control port pressure changes linearly when the load port pressure is at different constant values. The simulation of flow control characteristics is divided into ideal no back pressure and back pressure.

Under ideal no back pressure conditions, that is, when the pressure at port A is 0 MPa, the pressure at load port B is set to 5 MPa, 7 MPa, 9 MPa, and 11 MPa respectively, and the pressure at control port Px is set to linearly change from 0 to 5 MPa. The relationship between the main valve flow rate Q and the pilot pressure Px is shown in Fig. 7.

As shown in Fig. 7, when the load port pressure is constant, the flow control characteristics are divided into three stages. When the pilot pressure is lower than a certain opening pressure, the main valve does not open, and the flow rate through

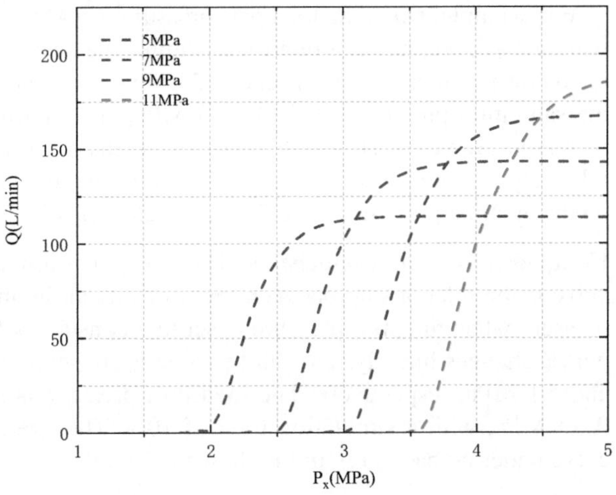

Fig. 7 Ideal No Back Pressure Flow Characteristic Curve

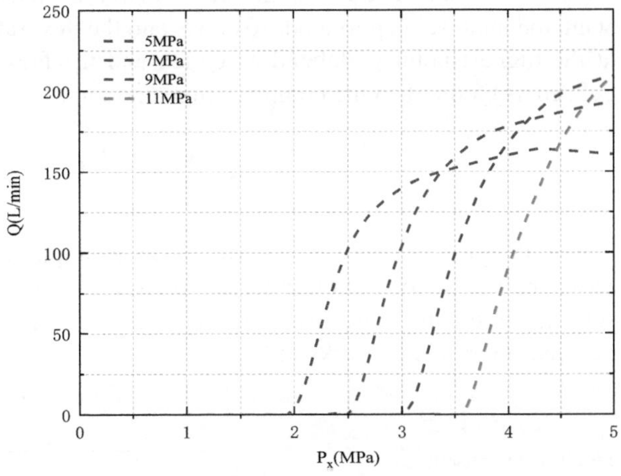

Fig. 8 Flow characteristic curve under back pressure

the main valve is 0 L/min; In the second stage, when the pressure reaches between the opening pressure and the saturation pressure, the flow rate through the main valve increases linearly with the pilot control pressure. In the third stage, when the pressure at the pilot port reaches saturation pressure, the flow rate of the main valve remains unchanged. From the simulation results, it can be seen that the control range of the new pilot plug-in balance valve can reach 1.55 MPa to 4.5 MPa, with a large pressure regulation range and good balance valve regulation performance.

If the A port is set with a back pressure of 1.2 MPa, simulate the flow control characteristics again with the same ideal no back pressure setting, and obtain the relationship between the main valve flow Q and the pilot control pressure Px as shown in Fig. 8. From the above figure, it can be seen that applying a back pressure of 1.2 MPa significantly increases the saturation flow rate of each load pressure of the balance valve.

For example, when the load port pressure is 5 Mpa, the ideal condition flow rate without back pressure is around 115 L/min. After setting the back pressure, the saturated flow rate increased to around 165 L/min due to the influence applied. Similarly, the saturation pressures of 7 MPa, 9 MPa, and 11 MPa at the load port have all increased. However, as the load pressure increases, the saturation flow rate decreases relatively. It can also be seen from the figure that the back pressure has little effect on the opening pressure of the flow characteristics.

Compensation Characteristics The compensation characteristic of a balance valve is the relationship between the flow rate at the main valve port and the load pressure when the pilot pressure is set to a constant value and the load pressure at port B changes linearly. The pilot pressures are set at 4.0 MPa, 4.3 MPa, 4.8 MPa, and 5.1 MPa, respectively. The load port pressure is set to linearly change from 0–15 MPa, with a simulation time of 10 s. The overcompensation characteristic curve under no back pressure is shown in Fig. 9.

When the pilot pressure is constant, the curve basically satisfies the 1/2 power of the flow formula. As the load gradually increases, the pilot pressure remains constant, the main valve port tends to close, and the flow rate decreases. If the flow rate at the highest point is defined as Qmax and the flow rate at the lowest point is defined as Qmin, then the compensation amount is

$$\xi = \frac{Q_{\max} - Q_{\min}}{Q_{\max}} \tag{5}$$

Taking the pilot pressure of 4.3 MPa as an example, the compensation amount is 85%, and the compensation inflection point is around 12.4 MPa. As the pilot pressure increases, the compensation decreases and the inflection point of the curve becomes increasingly backward.

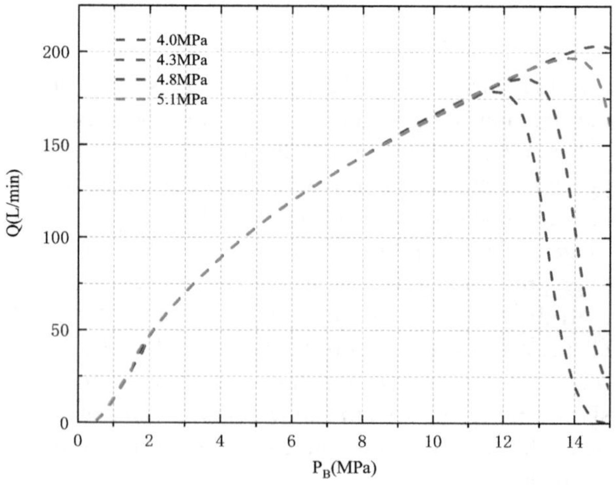

Fig. 9 Compensation characteristic curve without back pressure

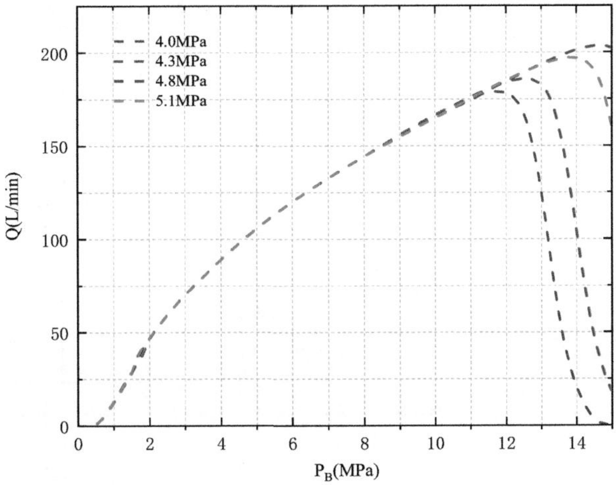

Fig. 10 Compensation characteristics at a back pressure of 2 MPa

Table 2 Parameter of system

Parameter variables	Numerical value
Hydraulic cylinder diameter mm	122
Hydraulic cylinder piston diameter mm	80
Effective displacement of hydraulic cylinder m	2.5
Total mass of hydraulic cylinder kg	200
The displacement of variable displacement pump L/min	30
Motor speed r/min	1500

From Fig. 10, We can see that when the A port is designed with a back pressure of 2 MPa, all other setting conditions remain unchanged, and the compensation characteristics under back pressure can be obtained. Similarly, the compensation amount can be calculated as 87% when the pilot pressure is 4.3 MPa under back pressure. The turning point of compensation has been slightly advanced.

Dynamic Characteristics Analysis of the New Type of Balance Valve To verify that the dynamic characteristics of the new type of balance valve can adapt to the operating conditions of the ship lock's herringbone gate opening and closing machine, it is necessary to connect the loading cylinder. Taking the herringbone gate opening condition as an example, an AMESim model of the hydraulic opening and closing machine system is built. During testing, the external load is reduced by 1:50, which means the external load is reduced by 1/50 of its original size. The same hydraulic cylinder parameters are also reduced in the same proportion. The system parameters are shown in the Table 2.

The load force acting on the hydraulic cylinder under negative load is opposite to the direction of movement of the hydraulic cylinder. When subjected to forward

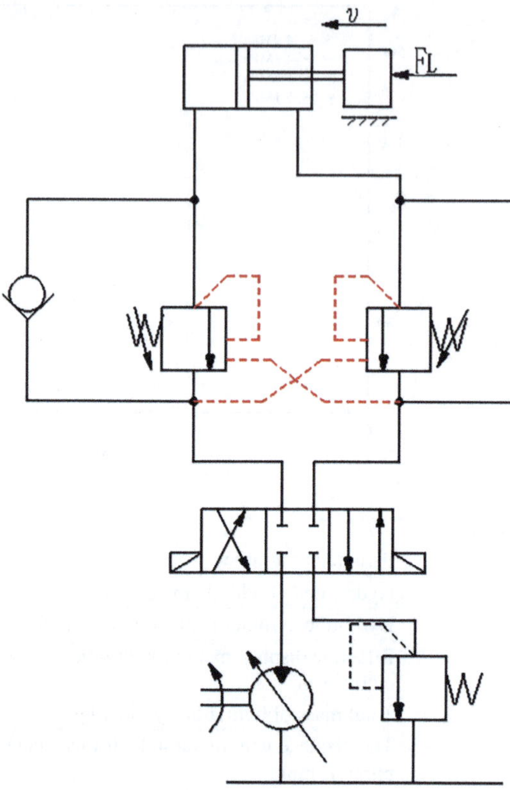

Fig. 11 Balance circuit and the relationship between negative load speed and force

load, the load force on the hydraulic cylinder is in the same direction as the speed. The schematic diagram of positive load is shown in Fig. 11.

The stability of the opening and closing operation of the ship lock's herringbone gate mainly relies on the bidirectional load balancing valve type balancing circuit in the hydraulic system of the hydraulic opening and closing machine. Based on the principle of the balancing circuit and the structure of the new plug-in balancing valve, a simulation model of the bidirectional load balancing valve balancing circuit was built using the hydraulic application library, mechanical library, signal library, and HCD library in AMESim, as shown in Fig. 12.

To analyze the impact of sudden load changes on hydraulic cylinder speed, the external load is set as shown in Fig. 13, with a simulation time of 20 s and a time interval of 0.1 s. The external load is a positive load, which is a tensile force for the hydraulic cylinder.

The hydraulic cylinder speed variation is shown in Fig. 14. The balance valve reaches a steady state after adjusting the hydraulic cylinder speed for about 1 s. The initial overshoot of the hydraulic cylinder speed is 5.6%. The operating speed of the hydraulic cylinder is 72 mm/s. From this, it can be seen that when the external load

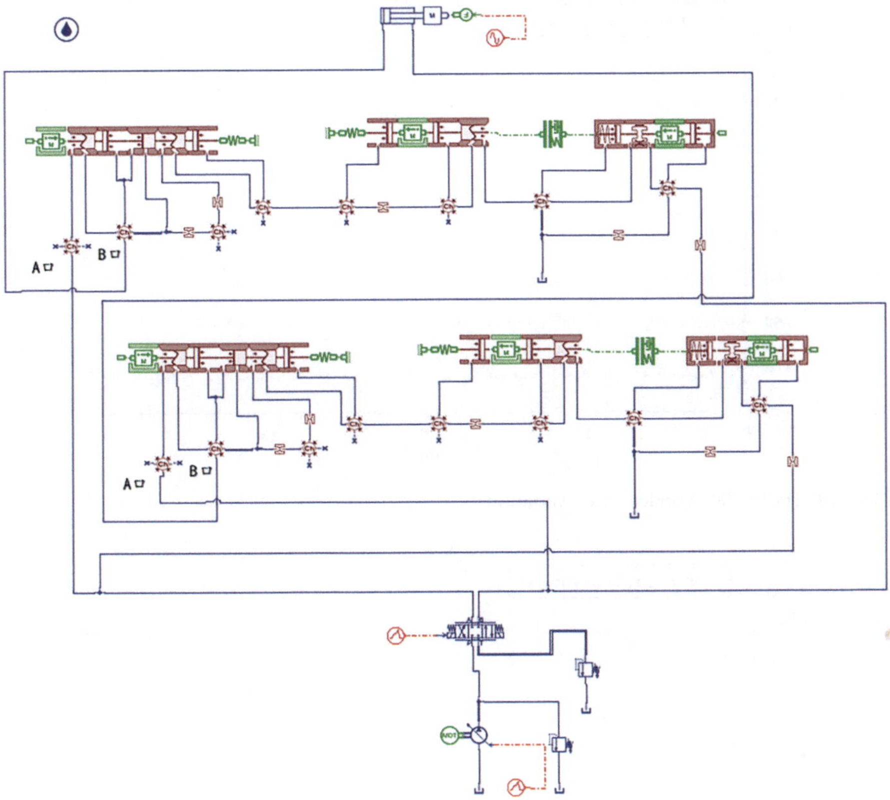

Fig. 12 Simplified balance system of hydraulic system for ship lock hydraulic opening and closing machine

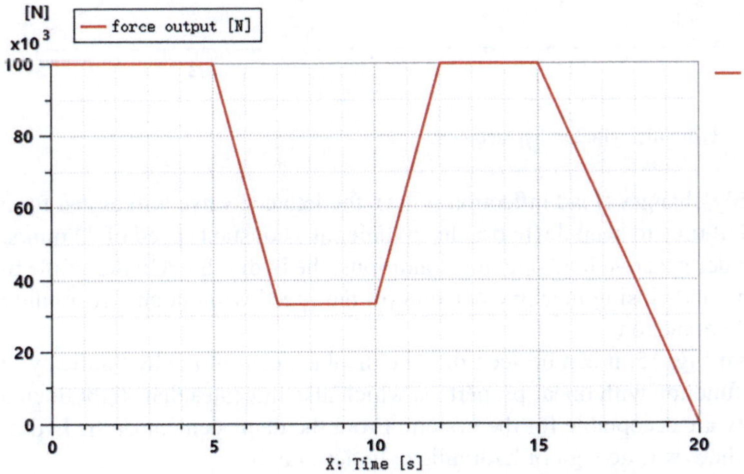

Fig. 13 Changes in external load

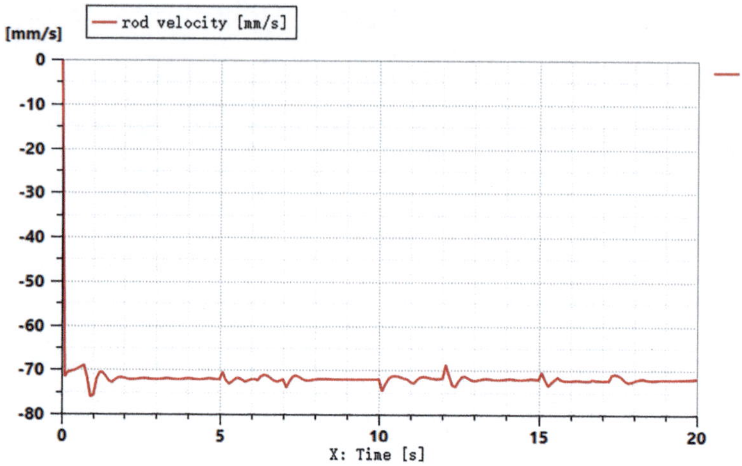

Fig. 14 Hydraulic cylinder speed variation

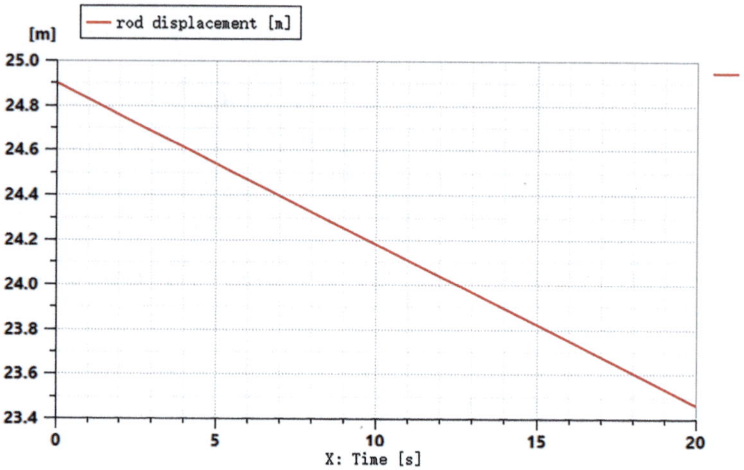

Fig. 15 Hydraulic cylinder displacement

suddenly changes (load inflection point), the balance valve will adjust for a certain period of time to keep the hydraulic cylinder at a constant speed of 72 mm/s. For the gate under negative load opening conditions, the hydraulic cylinder of the hydraulic opening and closing machine can control the speed without any hydraulic cylinder drift phenomenon.

From Fig. 15, it can be seen that the displacement of the hydraulic cylinder is a linear function with linear properties, which also indicates that slight fluctuations in velocity are acceptable for the system. From the displacement of the hydraulic cylinder, there was no sign of hydraulic cylinder crawling.

5 Adaptability Analysis of Balance Valve to Different Reverse Head Operating Conditions of Miter Gate

5.1 Submerged Water Depth of 20 m and Reverse Head Condition of 0.4 m for the Miter Gate

In order to verify the adaptability of the balance valve during the opening of the trapdoor at different submergence depths and reverse water heads. The reverse water head generated when the downstream water level of the initial gate is higher than the upstream water level causes the gate to bear a reverse load, resulting in a negative dynamic water resistance torque. As the gate opens, it gradually becomes a positive dynamic water resistance torque. The hydrodynamic resistance moment shows a saddle shaped curve with two large ends and a small middle as the gate opening time changes, and the maximum value appears in the early stage of opening. In the middle and later stages of opening the door, the operation of the zigzag gate is relatively stable, and the dynamic water resistance torque value is basically not affected by the height of the reverse water head, stabilizing within a lower range. The load variation curve is shown in Fig. 16.

From Fig. 17, it can be seen that during the initial stage of reverse head external load loading, the hydraulic cylinder speed fluctuates to a certain extent, reaching the design speed of 72 mm/s and maintaining stability.

From Fig. 18, it can be seen that the pressure in the rodless chamber of the herringbone door fluctuates with the change of external load, while the pressure in the rodless chamber fluctuates slightly and remains basically constant, with a pressure of 5.5 MPa, indicating that there is no crawling phenomenon.

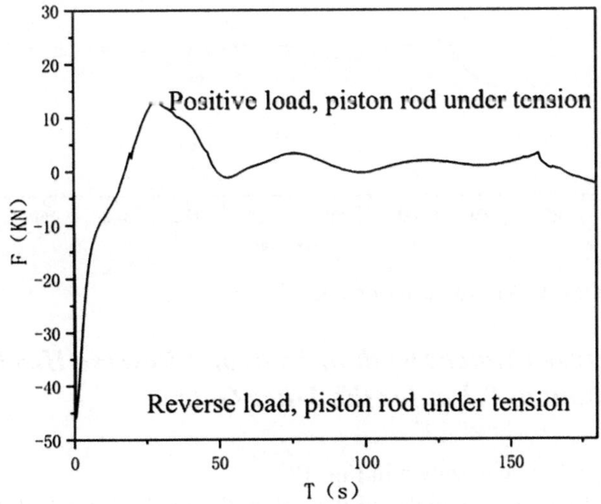

Fig. 16 External load

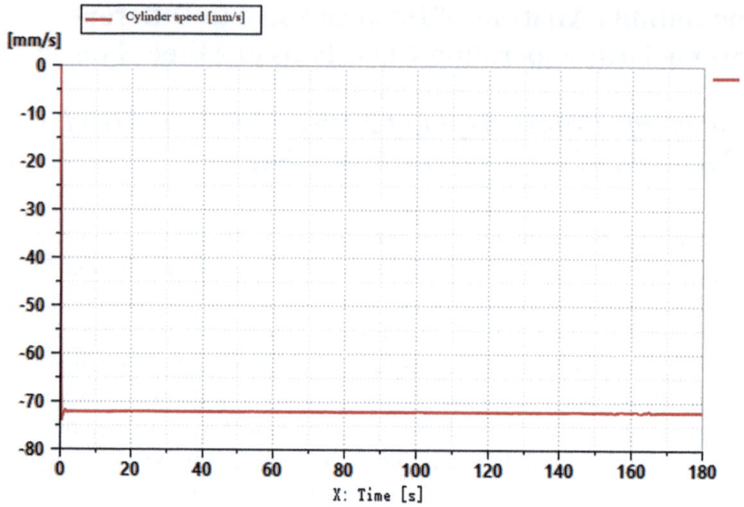

Fig. 17 Hydraulic cylinder speed

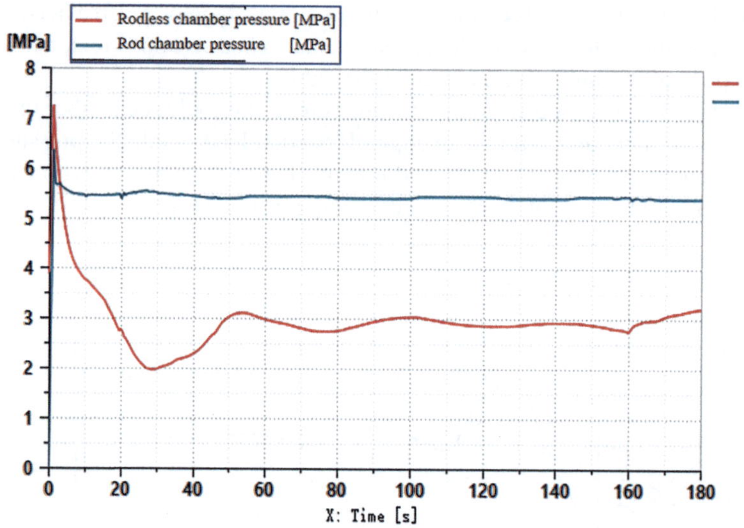

Fig. 18 Hydraulic cylinder two chamber pressure

5.2 Submerged Water Depth of 35 m and Reverse Head Condition of 0.3 m for the Miter Gate

The external load force is shown in Fig. 19.

From Fig. 20, it can be analyzed that when the load is switched, the hydraulic cylinder speed can still maintain the design speed of 72 mm/s, which is basically constant.

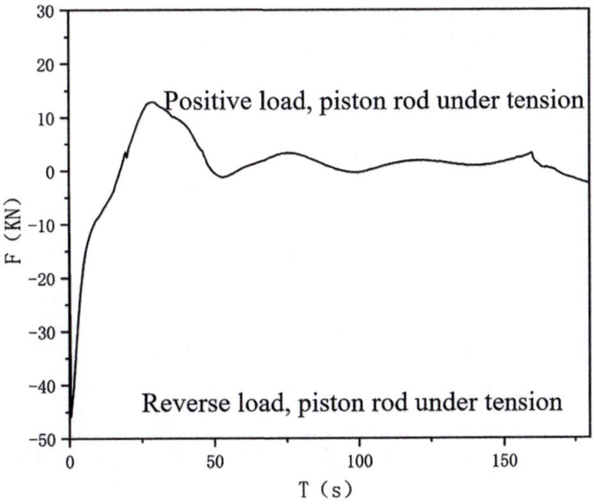

Fig. 19 External load force

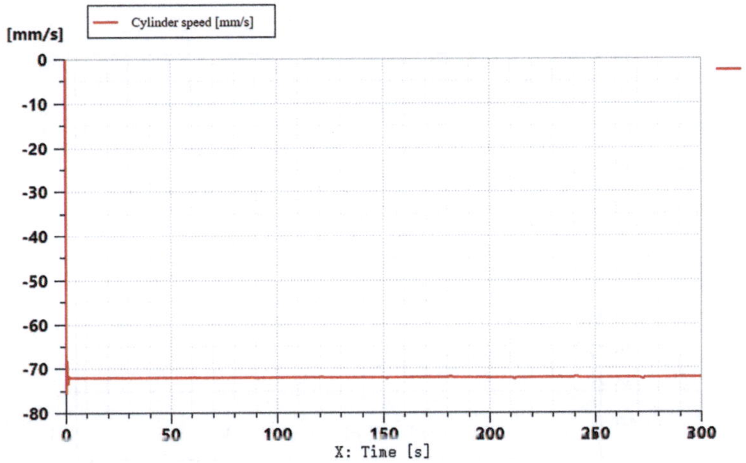

Fig. 20 Hydraulic cylinder speed

As shown in Fig. 21, when the external load fluctuates between positive and reverse load, the pressure in the rodless chamber also fluctuates with the external load, while the pressure in the rodless chamber remains basically unchanged, further indicating that the gate will not experience crawling phenomenon.

5.3 Submerged Water Depth of 35 m and Reverse Head Condition of 0.4 m for the Miter Gate

The external load force changes under reverse head conditions with a submerged depth of 35 m and 0.4 m are shown in Fig. 22.

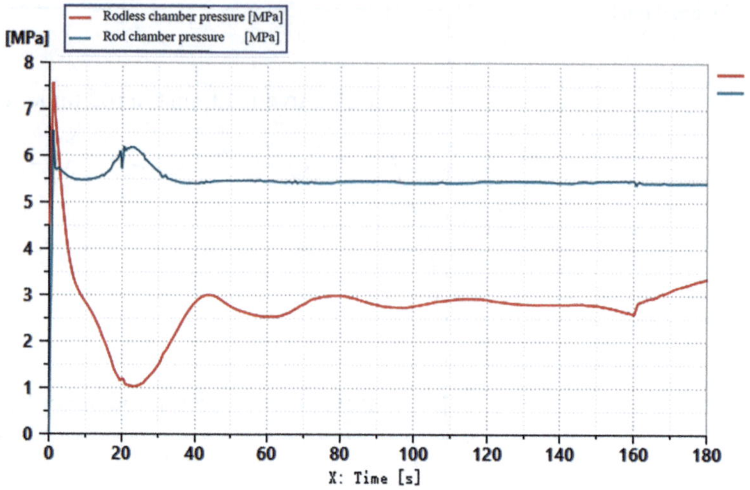

Fig. 21 Hydraulic cylinder two chamber pressure

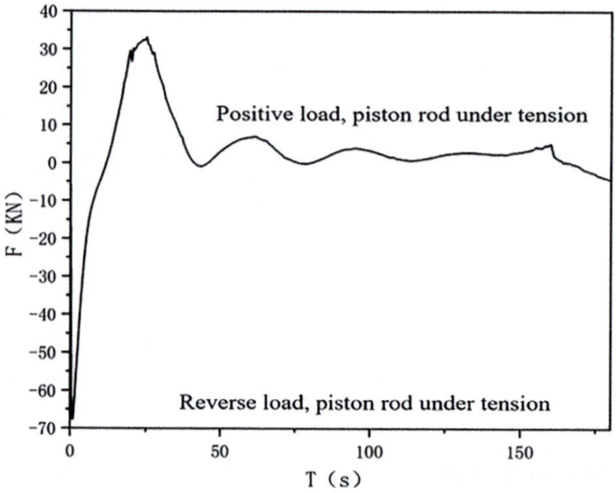

Fig. 22 External load force

From Fig. 23, it can be analyzed that when the load switches from negative to positive, the hydraulic cylinder speed can still maintain the design speed of 72 mm/s, which is basically constant. When the trapdoor model opens the door under reverse water head, there is no phenomenon of trapdoor crawling or drifting.

From Fig. 24, we can see that when the external load fluctuates between positive and negative loads, the pressure in the rodless chamber also fluctuates with the external load, while the pressure in the rodless chamber remains basically unchanged, further indicating that the gate will not experience crawling phenomenon.

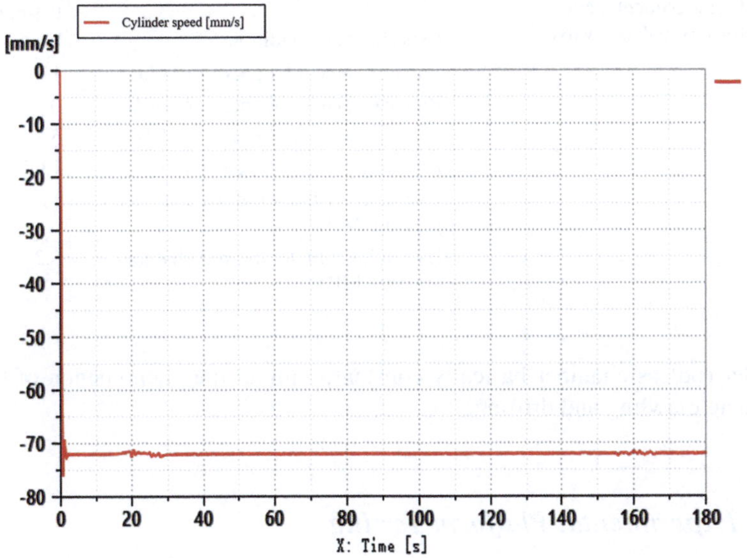

Fig. 23 Hydraulic cylinder speed

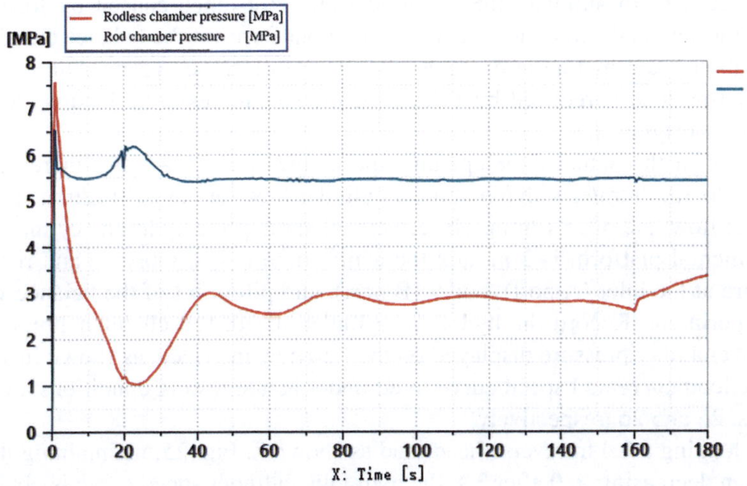

Fig. 24 Hydraulic cylinder two chamber pressure

Comparing the adaptability of the balance valve to load changes under different submergence depths of 20 m and 35 m and the same reverse water head (0.4 m) conditions, as well as under the same submergence depth (35 m) and different reverse water heads (0.3 m and 0.4 m) conditions, the reverse water head and submergence depth only affect the pressure of the hydraulic cylinder rodless chamber. The new balance valve can maintain the pressure and velocity of the hydraulic

Table 3 Experimental site record form of balance valve

Experimental conditions	0.4 m reverse head
Balance valve inlet pressure MPa	11
Balance valve outlet pressure MPa	0
Balance valve pilot pressure MPa	5.2
Hydraulic cylinder speed mm/s	13.5
Hydraulic cylinder rodless chamber pressure MPa	11
Hydraulic cylinder has rod chamber pressure MPa	5.2

cylinder rodless chamber basically constant, without the phenomenon of herringbone gate crawling and drifting.

5.4 Experimental Platform Testing

To simulate the adaptability of the hydraulic system balance circuit of the herringbone gate opening and closing machine to the 0.4 m reverse water head condition, it is necessary to simulate the dynamic water resistance caused by loading the reverse water head and build an experimental platform for the balance circuit of the herringbone gate hydraulic opening and closing machine system. During testing, the external load is reduced by 1:50, which means the external load is reduced to 1/50 of its original size.

Based on the actual door opening speed, the speed gradually increases from 0 mm/s to 13.5 mm/s, with a hydraulic cylinder flow rate of 6.39 L/min and a load cylinder flow rate of 18 L/min. The effective length of the hydraulic cylinder on the experimental platform is 2 m, and the action time is 8 s. Observe and record the pressure at the inlet (A port), outlet (B port), and pilot port of the balance valve in the experiment. Record the hydraulic cylinder speed, rod chamber pressure, and rodless chamber pressure displayed on the Labview interface, as shown in Table 3.

The load curve and speed curve fitted under actual testing conditions are shown in Figs. 25 and 26, respectively.

By loading a 0.4 m reverse head load as shown in Fig. 25, maintaining it for 5 s and then decreasing to 0 after 3 s, the hydraulic cylinder speed curve is obtained as shown in Fig. 26. It can be seen from the figure that the hydraulic cylinder speed increases linearly with the increase of load force during the loading period. When the load changes, the speed can basically stabilize.

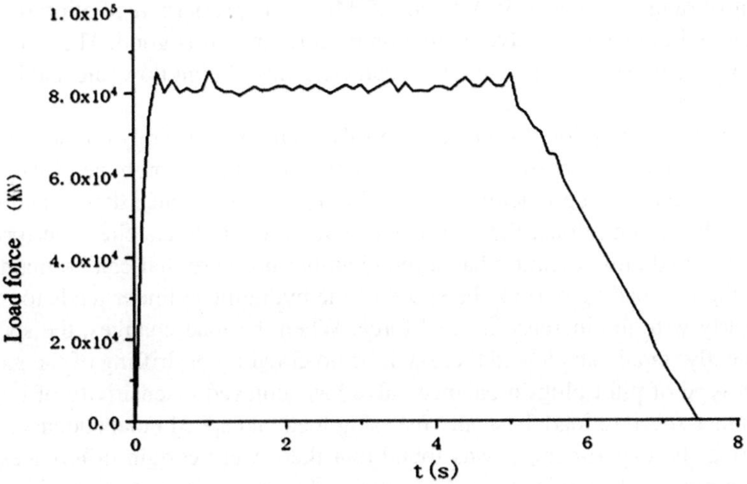

Fig. 25 Load variation curve

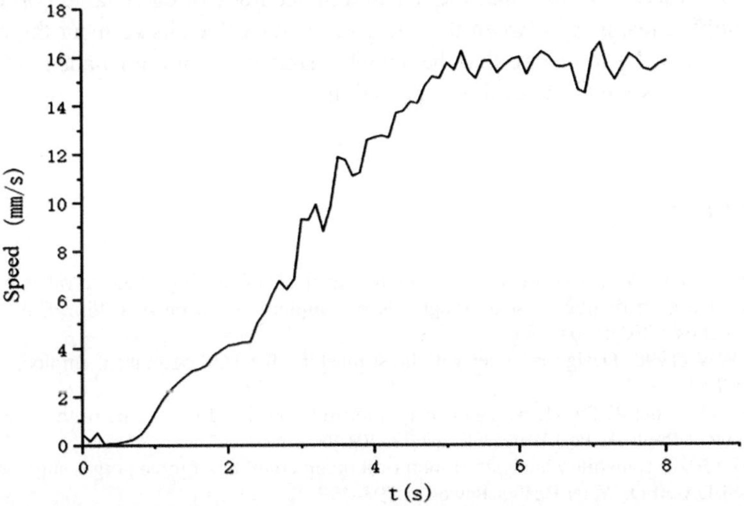

Fig. 26 Speed variation curve

6 Conclusion

1. A simulation study was conducted on the dynamic and static characteristics of a new type of pilot plug-in balance valve. The results showed that the pressure difference at the main valve port opened within the range of 0.1 MPa to 0.2 MPa, indicating that the one-way opening pressure of the balance valve was small, the energy consumption was low, the throttling effect was good, and the pressure

control range reached 1.55 MPa to 4.5 MPa. The pressure regulation range was large and the balance valve regulation performance was good. The new type of pilot plug-in balance valve has the characteristics of high flow rate and low pressure drop.

2. The new type of pilot plug-in balance valve can operate stably on the 1:50 simulation model and experimental testing platform of the herringbone gate hydraulic opening and closing machine. Under different submergence depths and reverse water head conditions, the new balance valve exhibits excellent performance, and the hydraulic cylinder has a rod chamber pressure that can be maintained; During the loading period, the speed of the hydraulic cylinder tends to increase linearly with the increase of load force. When the load changes, the speed can basically remain stable and there will be no crawling or drifting of the gate. The new type of pilot plug-in balance valve has achieved insensitivity of hydraulic cylinder speed to load, basically realizing load and speed independence.
3. During the experiment, it was found that there were certain differences in the performance of samples from the same batch due to different processing accuracies. Subsequent research can be conducted on the machining and assembly processes of the valve core, valve sleeve, and other related components of the balance valve to ensure that the machining accuracy of each valve port and the assembly clearance between the valve core and valve sleeve meet the requirements, in order to ensure that the actual performance of the balance valve is as close as possible to the design expectations.

References

1. Yang G, Mao YY, Liu YC et al (2023) Stability analysis of cartridge balance valve in hydraulic system of hydraulic hoist of Yangtze River shiplock based on AMESim. Chin Hydraul Pneumatics 47(02):100–108
2. Wei WW (1998) Design of miter gate hoist machine for TGP permanent shiplock. Yangtze River 29(12):7–8
3. Wang XD, Chen K (2002) An elementary approach to drift of miter gate of the three gorges permanent shiplock. Port Waterw Eng 4:37–40+46
4. Li HZ (2022) Reliability analysis of motor-oil pump coupling of three gorges ship lock based on LSEDA-BFO. Water Resour Power 11:193–197
5. Chen B (2021) Research on dynamic and static characteristics of pilot large flow hydraulic balance valve. Wuhan University of Science and Technology
6. Chen HB (2020) Research on the dynamic characteristics and the optimization of parameters for a novel piloted load control valve. Zhejiang University
7. Xiong SJ (2008) The application of hydraulically-actuated speed-limiting counterbalance valve in hydraulic gate hoists of three gorges project. Mach Tool Hydraul 36(09):237–239
8. Yin J, He B, Xian XP et al (2017) Dynamic characteristic of hydraulic balance valve based on AMESim. Mach Tool Hydraul 45(20):107–109

Open Access This chapter is licensed under the terms of the Creative Commons Attribution-NonCommercial-NoDerivatives 4.0 International License (http://creativecommons.org/licenses/by-nc-nd/4.0/), which permits any noncommercial use, sharing, distribution and reproduction in any medium or format, as long as you give appropriate credit to the original author(s) and the source, provide a link to the Creative Commons license and indicate if you modified the licensed material. You do not have permission under this license to share adapted material derived from this chapter or parts of it.

The images or other third party material in this chapter are included in the chapter's Creative Commons license, unless indicated otherwise in a credit line to the material. If material is not included in the chapter's Creative Commons license and your intended use is not permitted by statutory regulation or exceeds the permitted use, you will need to obtain permission directly from the copyright holder.

Evaluation of Metal Sealing Performance of High Pressure Casing Hanger Under Annular Pressure Conditions

Bao Zhang, Jun Yi, Yating Zhao, Hongtao Jing, Xuesong Wang, Xiang Si, and Hao Yu

Abstract In order to determine the sealing performance of metal seals of ultra-high pressure casing hanger under annular pressure conditions. An equation for the contact pressure of metal seals is derived. A finite element model is developed by fully considering the excitation process of seals and the effect of annular pressure. Based on the model, the sealing performance of metal seals under different annular pressures is analyzed. The results show that metal seals are in a plastic state after excitation; Under the high annular pressure, the maximum Mises stress of the seals was 241 MPa and 225 MPa, respectively; the pressure never penetrated seals, and the sealing performance is excellent. This study provides technical support for the safe working of metal seals of ultra-high pressure casing hanger.

Keywords Casing hanger · Seal performance · Metal seals · Annular pressure

---

B. Zhang
R&D Center for Ultra Deep Complex Reservior Exploration and Development, Korla, China

Engineering Research Center for Ultra-Deep Complex Reservoir Exploration and Development, Xinjiang Uygur Autonomous Region, Korla, China

Xinjiang Key Laboratory of Ultra-Deep Oil and Gas, Korla, China

PetroChina Tarim Oilfield Company, Korla, China

China University of Petroleum (East China), Qingdao, China

J. Yi (✉) · H. Jing · X. Wang · X. Si · H. Yu
R&D Center for Ultra Deep Complex Reservior Exploration and Development, Korla, China

Engineering Research Center for Ultra-Deep Complex Reservoir Exploration and Development, Xinjiang Uygur Autonomous Region, Korla, China

Xinjiang Key Laboratory of Ultra-Deep Oil and Gas, Korla, China

PetroChina Tarim Oilfield Company, Korla, China

Y. Zhao
PetroChina Tarim Oilfield Company, Korla, China

© The Author(s) 2026

Z. Zhang, P. Mercorelli (eds.), *Mechanical Design and Simulation: Insights and Innovations*, Lecture Notes in Mechanical Engineering,
https://doi.org/10.1007/978-981-95-2405-1_58

1 Introduction

With the development of oil and gas exploration technology, the number of high-temperature and high-pressure oil and gas wells is increasing [1, 2], especially in the complex formations in Sichuan and Tarim basins, and the problem of wellbore integrity becomes particularly prominent [3]. Under these harsh operating conditions, casing hangers in wellhead installations face great challenges. Due to deep reservoir burial, high formation pressure, high temperature, and the presence of corrosive gases such as CO_2 and H_2S [4], the working environment of wellhead seal is extremely harsh. Traditional rubber seals are prone to aging in this environment, resulting in annular pressure that affects well integrity and can pose serious safety and environmental concerns.

Therefore, the high pressure casing hanger metal is studied the sealing performance of seals is of great practical significance for improving the safety of oil and gas Wells and prolonging their service life [5].

In view of the sealing performance of metal seals of high pressure casing hangers, scholars have done related research. Smith et al. [6] developed a new spherical sealing structure, which greatly improved the sealing ability. Ma et al. [7] studied that welding a layer of steel wire components on the sealing surface of the all-metal sealing ring can effectively solve the problem of gas leakage caused by insufficient clamping force during the loading and unloading of metal seals. Huang et al. [8] analyzed that the sealing structure of the oil pipe hanger also achieves the sealing between metals by stimulating the deformation of the elastic body, so that the sealing function between metals can be effectively realized. Peng [9] compared and analyzed the contact pressure distribution of cylindrical to spherical and cone-to-cone metal sealing structures under different tensile and torque conditions, and proposed that cone-to-cone sealing structures are more conducive to improving the sealing performance of threaded structure sealing surfaces. Li et al. [10] found that the greater the thickness of metal C-ring alloy coating layer, the more uniform the contact pressure distribution, the smaller the contact width, and the smaller the plastic strain on the gold coating. These results provide certain method guidance for the improvement of metal seal structure of hangers.

In the field of metal sealing, despite a large amount of research focusing on its basic theories and applications, there are still certain shortcomings in the comprehensive evaluation of metal sealing performance under high-pressure conditions, especially regarding the behavior of metal sealing components in high-pressure casing hangers under annular pressure. Existing studies mostly emphasize performance analysis under conventional pressure conditions, lacking in-depth exploration of the plastic deformation behavior of metal sealing components under high-pressure environments and its impact on sealing effectiveness. Based on this research gap, this paper derives the calculation equation for the contact pressure of metal seals and establishes a finite element evaluation model for metal seals in ultra-high pressure casing hangers. By analyzing the changes in metal seal performance under different annular pressures, this paper reveals the plastic deformation characteristics of metal seals in high-pressure environments and the impact of these deformations on sealing performance.

2 Mechanical Model

The metal sealing ring of the casing hanger adopts a C-shaped metal sealing structure, and its contact can be approximately equivalent to that of two cylinders contacting each other. According to the Hertz contact theory, it is obtained that when two cylinders are in contact under normal force F, the half width (r) of the contact surface and the maximum contact stress (δ_{Hmax}) are respectively

$$r = \sqrt{\frac{4F\left(\frac{1-\mu_1^2}{E_1}+\frac{1-\mu_2^2}{E_2}\right)}{\Pi b\left(\frac{1}{\rho_1}\pm\frac{1}{\rho_2}\right)}} \tag{1}$$

$$\delta_{H\max} = \sqrt{\frac{F}{\Pi b}\left(\frac{\frac{1}{\rho_1}\pm\frac{1}{\rho_2}}{\frac{1-\mu_1^2}{E_1}+\frac{1-\mu_2^2}{E_2}}\right)} \tag{2}$$

where ρ_1 and ρ_2 are the radius of curvature of the two cylinders or the radius of the two spheres, b is the length of the contact line of the two cylinders, $E_1$and E_2 are the elastic modulus of the two contact objects, and and represent the Poisson's ratio of the two materials respectively.

First of all, the contact stress formula and the half-width formula of the contact surface are simplified, and the thickness is 1 mm in the calculation. When $\mu_1 = \mu_2 = 0.3$, E1 = E2 = E, the formula for calculating the maximum contact pressure and contact half-width can be simplified as follows:

$$\delta_{H\max} = 0.418\sqrt{\frac{FE}{b\rho_1}} \tag{3}$$

$$r = 1.522\sqrt{\frac{F\rho_1}{Eb}} \tag{4}$$

In practical application, the sealing ring is considered to be affected by interference fit and oil and gas prepressure. These two forces are expressed as F1 and F2 respectively, F1 is the squeezing effect of interference fit on the seal ring during the sealing process, and F2 is generated by the internal oil and gas prepressure.

Then these two forces are simplified into two kinds of force models. F1 can simplify the sealing ring model to the cantilever beam model for calculation, using the relationship between the concentrated force (F) and Deflection (w_{max}) to solve the F1. F2 can simplify the sealing ring model to a simple supported beam structure for calculation, using the same relationship to solve F2.

$$F_2 = qc + q(d+h)\cos\alpha + \frac{q\left(\frac{h^2-d^2}{2}\right)\cos\alpha - q\left(cd + \frac{c^2}{2}\right)}{c+d} \tag{5}$$

Finally, by substituting Eq. 5 and Eq. 4 into Eq. 3, the analytical expression of the maximum contact pressure can be obtained as follows:

$$\delta_{Hmax} = 0.418\sqrt{\left[k_1\left(\frac{6EII_{max}}{(3l-a)a^2}\right) + k_2\left(qc + q(d+h)\cos\alpha + \frac{q\left(\frac{k^2-a^2}{2}\right)\cos\alpha - q\left(cd+\frac{c^2}{2}\right)}{c+d}\right)\right]E} \tag{6}$$

3 Finite Element Model

3.1 *Common Seal Structure of the Shaft Core Type Casing Hanger*

The research object is the shaft core type casing hanger of a certain manufacturer, as shown in Fig. 1 below:

Figure 1 shows an H-shaped metal seal hanger, which mainly uses the H-shaped metal seal component and the lower FS rubber seal component to seal the annular space between the core shaft and the casing head. The H-shaped metal seal component has four hemispherical convex surfaces, and when the metal seal ring's four hemispherical convex surfaces are compressed by the upper pressure ring, they will open laterally, thereby sealing the annular space between the core shaft and the manifold. The lower trapezoidal metal seal is extruded and expands laterally under the gravity of the casing string, thus sealing the lower annular space between the core shaft and the manifold.

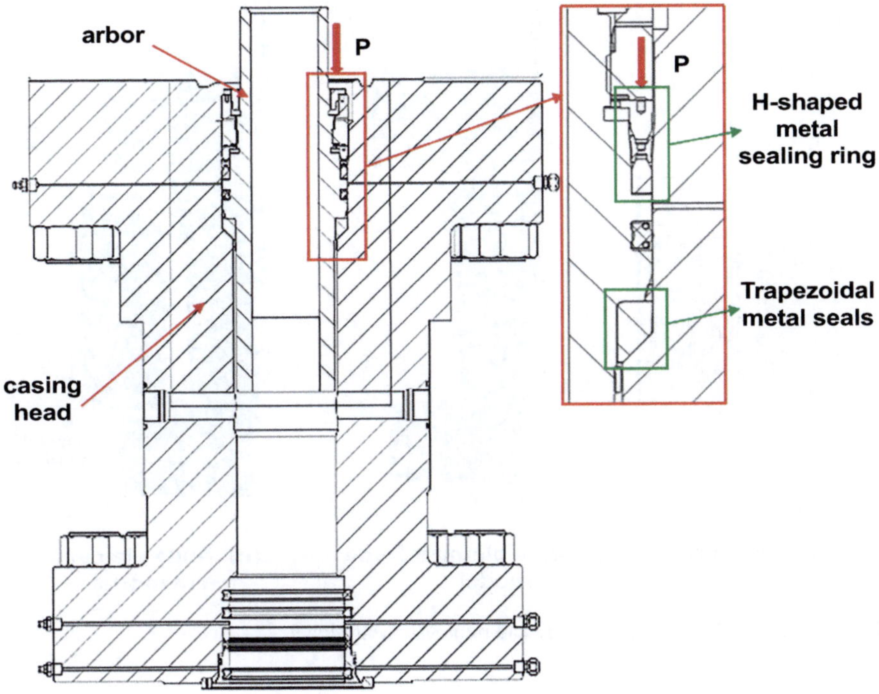

Fig. 1 Conventional Shaft Core Type Casing Hanger and Seal Structure

3.2 Establishment of Seal Structure Model

Based on the overall structure dimensions of the shaft core type casing hanger shown in Fig. 1, the modeling research is conducted according to the axisymmetric structure. The geometric analysis model includes the core shaft, manifold, pressure ring, support ring, H-shaped metal seal structure, and lower trapezoidal metal seal structure. During the analysis, a tensile load of F = 0 ~ 4000 kN is applied to the lower end face of the core shaft, and a displacement load of H = 0 ~ 5 mm and a pressure load of P = 0 ~ 175 MPa are applied to the upper end face of the seal pressure ring, as shown in Fig. 2.

Basic material mechanics parameters: the material of the core shaft is 718 with a yield strength of 800 MPa and a modulus of elasticity of 210 GPa; the material of the casing head is 4130 with a yield strength of 835 MPa and a modulus of elasticity of 213 GPa; the material of the pressure ring and base are 42CrMo with a yield strength of 930 MPa and a modulus of elasticity of 210 GPa. The materials of the H-shaped metal seal component and the trapezoidal metal seal component are 825 with a yield strength of 241 MPa and a modulus of elasticity of 210 GPa. Finally, based on the structural design and force model of the shaft core type casing hanger, the load-bearing capacity and sealing performance of the seal structure under different pressures and different casing string gravity are checked.

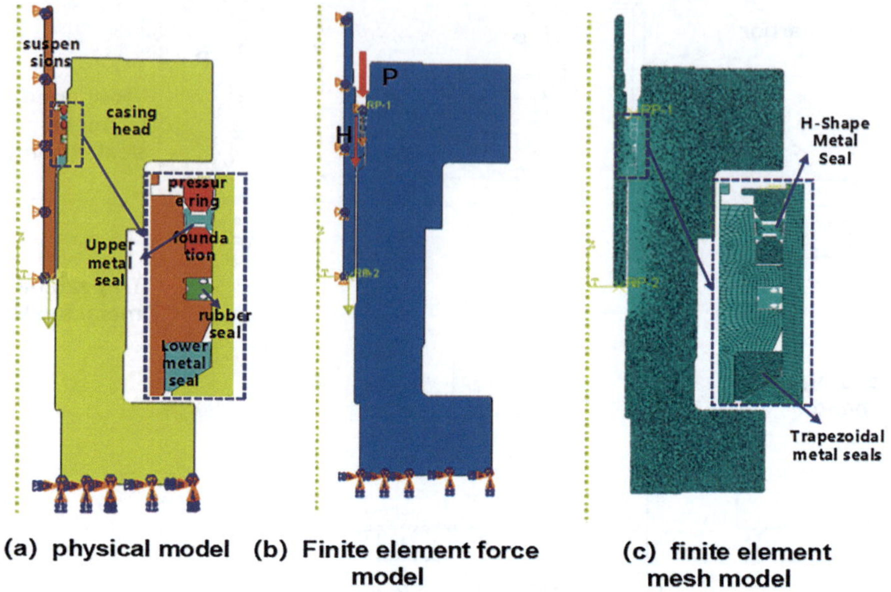

Fig. 2 H-shaped Metal Seal Structure and Finite Element Model

4 Seal Performance Calculation Results and Analysis

4.1 Metal Seals Sealing Performance Analysis

Influence of Compression Displacement on the Sealing Performance of the H-Shaped Metal Seal Component The sealing performance analysis of the H-shaped metal seal under different downward displacements was conducted separately, and the Mises stress distribution and contact pressure distribution of the H-shaped metal seal structure are shown in Figs. 3 and 4 below:

As shown in Fig. 3, with the increase in the downward displacement load, the stress on the H-shaped metal sealing ring also increases. Among them, the stress changes at the four contact points are more significant, but during the entire downward displacement process, the stress change in the middle part of the H-shaped metal sealing ring is not substantial. When the downward displacement is 1 mm, the maximum Mises stress on the H-shaped metal reaches the material yield strength of 241 MPa, resulting in plastic. When the downward displacement is 1–2 mm, the two contact pairs on the upper half of the H-shaped metal seal deform slightly. However, when the downward displacement is 2–5 mm, all four contact points of the H-shaped metal seal undergo significant deformation. For soft metals, this plastic deformation, although initially helpful in enhancing the sealing effect, may lead to fatigue failure of the seal if the downward displacement is too large, thereby affecting the sealing performance.

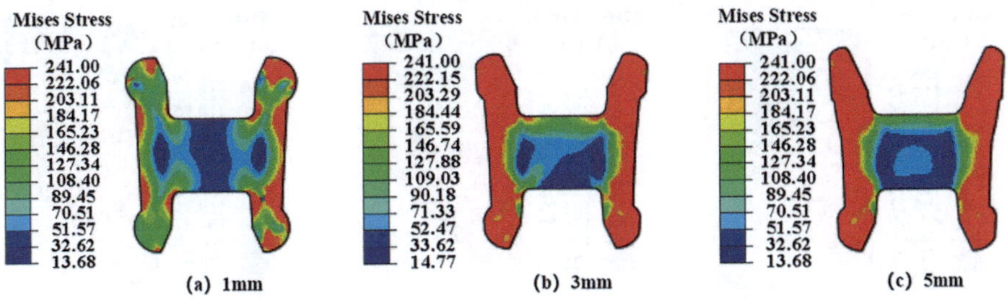

Fig. 3 Mises Stress Distribution of H-shaped Metal Seal under Different Displacement Loads

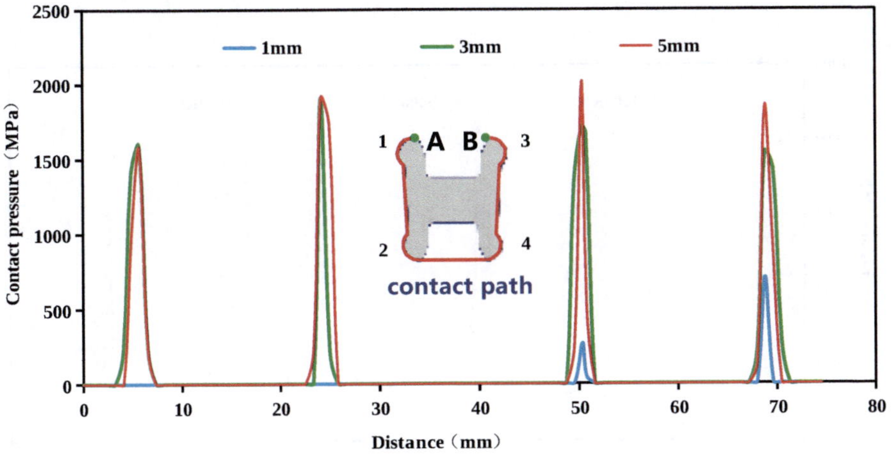

Fig. 4 Contact Pressure Change Curve of H-shaped Metal Seal Component with Compression Displacement

Figure 4 shows the variation curves of contact pressure on contact paths 1, 2, 3, and 4 of the H-shaped metal seal under different downward displacements. It can be seen that the contact pressure of the H-shaped metal seal increases with the increase in downward displacement. The contact pressure on contact path 1 varies between 0 and 1385 MPa; the contact pressure on contact path 2 varies between 0 and 1774 MPa; the contact pressure on contact path 3 varies between 0 and 1731 MPa; the contact pressure on contact path 4 varies between 0 and 1853 MPa. Although high contact pressure helps to enhance sealing, it can also lead to excessive deformation of the material, thereby affecting the stability of the seal. Especially under larger displacements, the contact surface may experience severe fluctuations, resulting in unstable sealing performance.

Influence of Suspended Load on the Sealing Performance of the Trapezoidal Metal Seal Component Figure 5 shows the Mises stress distribution of the trapezoidal metal seal under different suspended loads. As shown in the figure, with the increase in the suspended load of the casing, the Mises stress on the trapezoidal

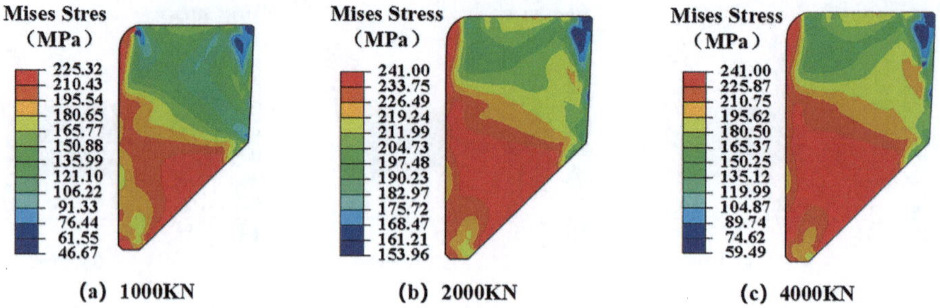

Fig. 5 Mises Stress Distribution of Trapezoidal Metal Seal Component under Different Suspended Loads

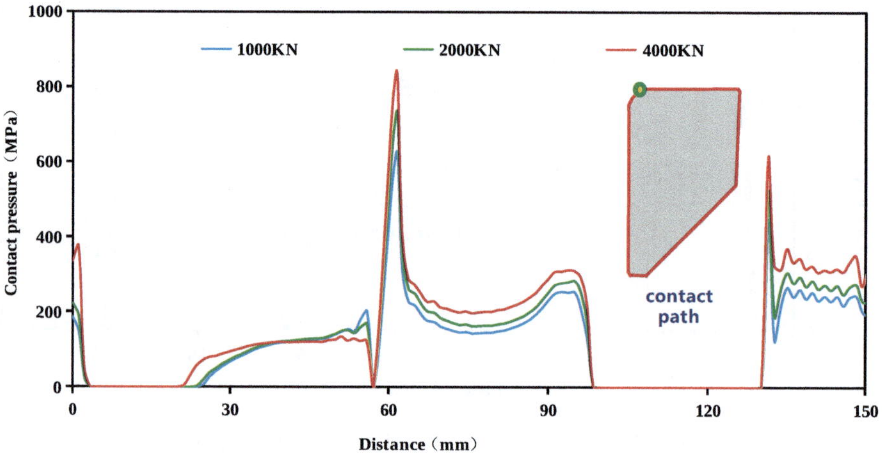

Fig. 6 Contact Pressure Change Curve of Trapezoidal Metal Seal Component with Suspended Load

metal seal also increases, and the Mises stress on the trapezoidal metal seal is mainly concentrated in the lower half where it contacts the casing head. This is due to the fact that under the action of the casing suspended load, a large interaction force is generated between the trapezoidal metal seal and the casing head contact surface. When the casing suspended load is 1000 kN, the maximum stress on the trapezoidal metal seal has reached the material yield strength of 241 MPa, resulting in plastic deformation. Subsequently, as the suspended heavy load continues to increase, the plastic deformation area of the trapezoidal metal seal also continues to expand. Although the plastic deformation of the trapezoidal metal seal helps to enhance its sealing performance, excessive plastic deformation can also affect the stability of its sealing structure, leading to a loss of its original sealing capability.

In Fig. 6, it can be seen that the contact pressure of the trapezoidal metal seal increases with the increase in suspended load. When the suspended load is between 1000 KN and 4000 KN, the maximum contact pressure of the trapezoidal metal seal

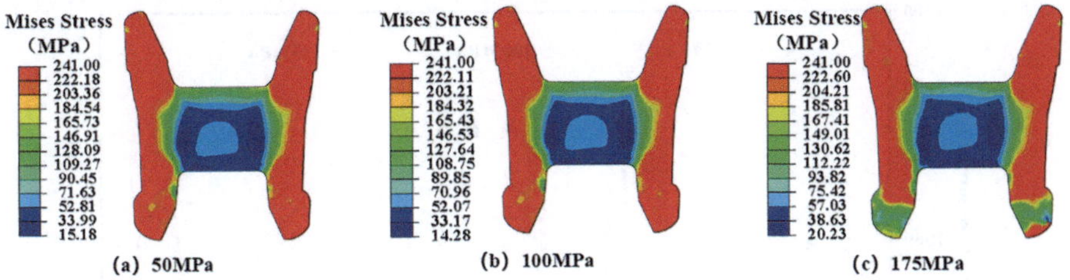

Fig. 7 Mises Stress Distribution of H-shaped Metal Seal Component under Different Fluid Pressures

varies between 625 MPa and 839 MPa. Additionally, it can be seen from the diagram that the contact pressure of the trapezoidal metal seal has significant fluctuations, which may lead to unstable sealing performance of the trapezoidal metal seal.

4.2 *Effect of Different Fluid Pressure Loads on the Sealing Performance of H-Type Metal Seals and Trapezoidal Seals*

Additionally, it can be seen from the diagram that the contact pressure of the trapezoidal metal seal has significant fluctuations, which may lead to unstable sealing performance of the trapezoidal metal seal. From Fig. 7, we can see that as the fluid pressure load on the lower part of the H-shaped metal seal increases, the Mises stress at the contact area between the lower part of the H-shaped metal seal and the casing head and hanger gradually decreases. This is due to the increase in the lower fluid pressure load, which causes the H-shaped metal seal to experience an upward force from the fluid pressure, thereby reducing the interaction force between the lower part of the H-shaped metal seal and the contact surface of the casing head and hanger. Additionally, from Fig. 8, we can see that as the fluid pressure load increases, the contact length between the H-shaped metal seal and the casing and suspensions does not change significantly, the contact pressure on paths 1 and 4 also does not change significantly, but the contact pressure on paths 2 and 3 decreases with the increase in fluid pressure. When the fluid pressure is varied between 50 MPa and 175 MPa, the maximum contact pressure on the H-shaped metal seal contact paths 2 and 3 is reduced from 1770 MPa to 1550 MPa. Although the contact pressure at the bottom of the H-shaped metal seal decreases under fluid pressure load, the overall contact pressure is still greater than the fluid pressure, indicating that the contact performance of the H-shaped metal seal is good. Additionally, from Fig. 8, it can be seen that the variation pattern of the contact pressure curve of the H-shaped metal seal shows a wavy line, thus forming a good sealing effect between the contact surfaces.

When the trapezoidal metal seal is subjected to a suspended load of 4000KN, different fluid pressures ranging from 0 to 175 MPa are applied to the lower part of

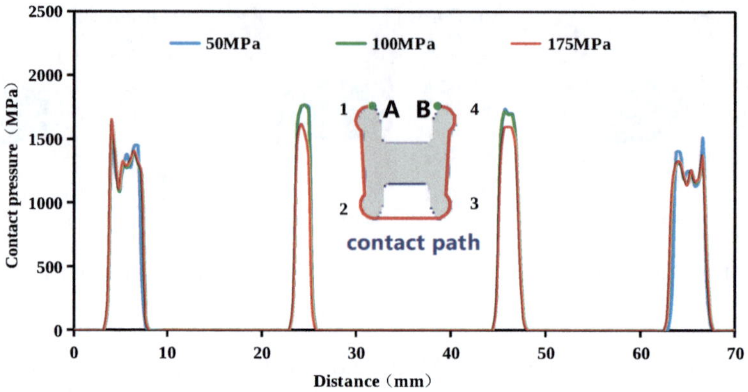

Fig. 8 Contact Pressure Curve Distribution of H-shaped Metal Seal Component under Different Fluid Pressures

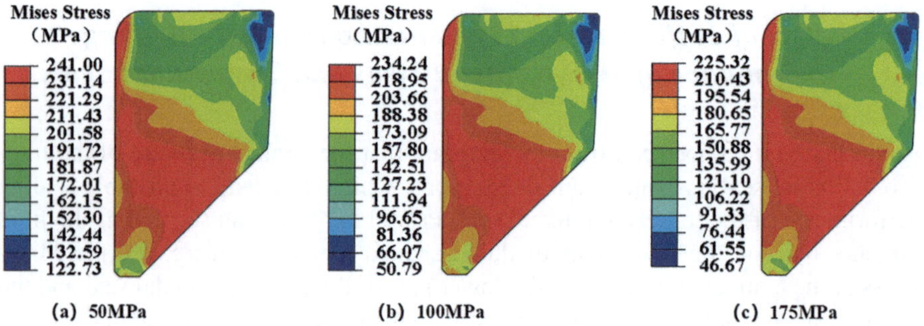

Fig. 9 Mises Stress Distribution of H-shaped Metal Seal Component under Different Fluid Pressures

the trapezoidal metal seal to study the variations in Mises stress and contact pressure of the trapezoidal metal seal. As shown in Figs. 9 and 10 below. From Fig. 9, we can see that the Mises stress on the trapezoidal metal seal decreases as the fluid pressure load increases. When the fluid pressure is 175 MPa, the maximum Mises stress on the trapezoidal metal seal is 225.32 MPa. This is due to the increase in fluid pressure load on the lower part of the trapezoidal metal seal, which causes an upward force on the contact area between the lower part of the trapezoidal metal seal and the casing head, resulting in a reduction in the interaction force between the lower part of the trapezoidal metal seal and the casing.

From Fig. 10, we can see that as the fluid pressure load increases, the overall change in the contact pressure of the trapezoidal metal seal is not significant, but the maximum contact pressure gradually decreases. When the fluid pressure load varies between 50 and 175 MPa, the maximum contact pressure of the trapezoidal metal seal varies between 460 and 520 MPa. The maximum contact pressure is greater than the fluid pressure, indicating that the trapezoidal metal seal can maintain good

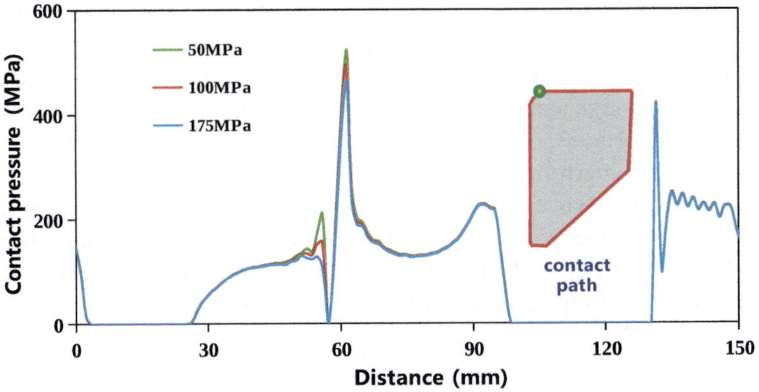

Fig. 10 Contact Pressure Curve Distribution of Trapezoidal Metal Seal Component under Different Fluid Pressures

sealing performance under a fluid pressure of 175 MPa. However, Fig. 10 shows that the contact pressure variation curve of the trapezoidal metal seal has a sawtooth pattern with abrupt changes in the middle, indicating that the contact pressure is not stable, and in some areas, there is no contact pressure at all. If the contact pressure does not continue to grow, especially after the fluid load has reached a particular level, the sealing performance will suffer. Combined with the above analysis of the sealing performance of the H-shaped metal seal, research shows that the H-shaped metal sealing ring has good sealing performance. The sealing performance of the overall sealing structure of the casing hanger mainly depends on the sealing length and contact pressure of the four contact surfaces of the H-shaped metal sealing ring.

5 Conclusion

1. Considering the effect of sealing excitation and annulus pressure, the metal sealing performance evaluation model of casing hanger is established, and the sealing performance evaluation is realized quickly.
2. The effects of displacement load, suspended load, and fluid pressure on the metal sealing performance are interrelated. Displacement load and suspended load jointly affect the stress distribution and plastic deformation of the seal, while fluid pressure influences the distribution of contact pressure and the stability of the sealing surface. Therefore, in practical applications, to ensure the stability of sealing performance, it is necessary to comprehensively consider the interaction of these factors. Excessive displacement loads and suspension loads may cause the sealing element to undergo plastic deformation, thereby affecting the sealing effect. Fluid pressure should be controlled within a reasonable range to avoid excessive fluctuations in contact pressure.

3. When designing a metal sealing structure, the shape of the sealing surface and the contact angle should be reasonably designed to ensure that the sealing element can uniformly distribute the contact pressure under different loads, avoiding excessively high local contact pressure. For example, appropriately increasing the contact surface area and reducing stress concentration can effectively improve sealing performance.
4. Through the analysis of metal seal performance under different annular pressures, it is found that under the high annular pressure, the maximum Mises stress is 241 MPa and 225 MPa, respectively; the pressure does not penetrate metal seals, and they still have good sealing performance.

References

1. Liu S, Luo M, Ma C et al (2019) Safety evaluation method of tubing string under annular trap pressure in deep water high-temperature and high-pressure gas wells. Sci Technol Eng 24(12):4959–4968
2. Zhang T, Yu Y, Wang G et al (2024) Research and application of gas lift technology in deep wells containing sulfur in Sichuan-Chongqing area. Nat Gas Explor Dev 47(04):112–118
3. Evaluation of air energy storage closure of underground halogen salt chamber in Sichuan Basin. J Southwest Petrol Univ (Nat Sci Ed) 1–17. http://kns.cnki.net/kcms/detail/51.1718.TE.20241120.1434.002.html. Accessed 15 Dec 2024
4. Zhang X, Dang E, Yu X et al (2024) Research status and progress on corrosion properties of martensitic stainless steel used in oil and gas fields. Chin J Corros Prot:1–16. http://kns.cnki.net/kcms/detail/21.1474.TG.20241106.0902.002.html
5. Ge J, Yu X, Chen X et al (2024) Metal seal design and finite element analysis of underwater tubing hanger. Petrol Chem Equip 27(10):13–18
6. Smith RR (1992) New contoured, elliptical-seal flapper valve facilitates API Class II metal-to-metal sealing performance. In: Paper OTC-7011-MS Presented at Offshore Technology Conference, 4–7 May, Houston, Texas
7. Ma Y, Guo D, Jing Y et al (2016) Methods of metal seal used for ultra-high vacuum system. Vacuum 53(6):9–11
8. Sidian H (2017) Research on risk assessment methods of tubing hanger installation for horizontal subsea Christmas tree. China University of Petroleum, Qingdao
9. Peng Y (2016) Study on sealing performance of casing thread with metal seal principle. Tianjin University of Commerce, Tianjin
10. Li G, Xie LJ, Li Z et al (2018) Effect of the thickness of alloy lining on sealing performance of reaction pressure vessels metal C-ring. J Mech Strength 40(5):1209–1214

Open Access This chapter is licensed under the terms of the Creative Commons Attribution-NonCommercial-NoDerivatives 4.0 International License (http://creativecommons.org/licenses/by-nc-nd/4.0/), which permits any noncommercial use, sharing, distribution and reproduction in any medium or format, as long as you give appropriate credit to the original author(s) and the source, provide a link to the Creative Commons license and indicate if you modified the licensed material. You do not have permission under this license to share adapted material derived from this chapter or parts of it.

The images or other third party material in this chapter are included in the chapter's Creative Commons license, unless indicated otherwise in a credit line to the material. If material is not included in the chapter's Creative Commons license and your intended use is not permitted by statutory regulation or exceeds the permitted use, you will need to obtain permission directly from the copyright holder.

Structure Optimization of Water-Gas Separator in Wet Spray Booth Considering Water-Gas Two-Phase Flow Field

Haisen Gan and Liang Fang

Abstract Side suction wet spray booths are widely used in automobile manufacturing for coating processes. During operation, paint mist particles mix with air and are captured by water in the water-gas separator to prevent environmental pollution and damage to the air-conditioning system. Optimizing the structure of the paint mist separator is crucial for better gas-water separation. This paper conducts a finite element flow field analysis to optimize the separator structure by varying inlet air velocity, baffle area and number, and spraying velocity. The optimal structure achieves improved water-gas separation efficiency compared to the original design. Reducing paint mist velocity at the separator inlet, increasing baffle number, and adding spraying are effective methods to enhance the separator's performance.

Keywords Side wet spray booth · Water-air two-phase flow · Finite element method

1 Introduction

Volatile organic compounds (VOCs) emitted during automobile spray coating pose significant environmental challenges [1]. The spray booth's system comprises ventilators and spray apparatus. The coating material is applied to the substrate surface via the spray gun. A portion of the paint particles linger within the spray booth, whereas the remainder is conveyed out through the exhaust grille [2]. An Integrated Device Combining the Improved Sieve-Tray Tower and Wet Electrostatic Precipitator (WEP) [3]. The mist collection cylinder is equipped with a spiral guide and a water mist nozzle, which serves to enlarge the paint mist

H. Gan · L. Fang (✉)
Tan Kah Kee College, Xiamen University, Zhangzhou, P. R. China
e-mail: lfang@xujc.com

© The Author(s) 2026

Z. Zhang, P. Mercorelli (eds.), *Mechanical Design and Simulation: Insights and Innovations*, Lecture Notes in Mechanical Engineering,
https://doi.org/10.1007/978-981-95-2405-1_59

particles. Upon collision with the water droplets ejected from the center of the cylinder, the particles increase in size and are subsequently separated from the airflow by the centrifugal force generated by the vortex created by the spiral guide [4]. A multivortex separator is utilized, and its enhanced efficiency in purifying the air of fine particles can be attributed to the formation of a regular vortex structure [5]. A Square Separator for Creating a Stable Vortex Flow in the Space Between Separation Elements [6]. A Novel Coupling Device Composed of a Baffle Interceptor and a Dry Filter for the Separation of Paint Mist from Paint Spraying Waste Gas [7].

While certain VOCs can be readily captured by water, others necessitate supplementary treatment methods, such as activated carbon [8]. Advanced technologies, including EcoDry scrubbers [9], electrochemical optimization [10], enhancement of the Fenton technique [11], and biological treatments [12], have been implemented to improve VOC removal.

However, enhancing the structure of existing spray booths represents a practical and cost-effective strategy for reducing emissions and lowering maintenance expenses.

2 Calculation Process and Parameter Setting

The optimization is performed using the Fluent module in ANSYS. The existing separator of a commercial spray booth has dimensions of 4000 mm (length) × 1050 mm (width) × 5500 mm (height) and consists of a vortex generating part and a water baffle. Three baffles are designed, whose inclination angles are kept as 15° and length is 650 mm. The model is created by stretching excision, and the fluid domain is defined accordingly. The meshing method is Hex/Wedge with a Cooper type and a unit size of 15 mm, resulting in 845,480 meshes. Boundary conditions include standard atmospheric pressure, gravitational acceleration, and specific inlet and outlet pressures. Discrete phase particles have a density of 1200 kg/m^3 and a diameter of 1e-6 m. The domain and meshing (see Fig. 1).

Simulations track varying numbers of particles (1000 to 2000) using different discrete phase boundary conditions. The average escape rate of the original separator is 10.9%, the capture rate is 79.3%, and the residual rate is 9.7%, as shown in Table 1. These results indicate the need for further structural modifications to improve separation efficiency.

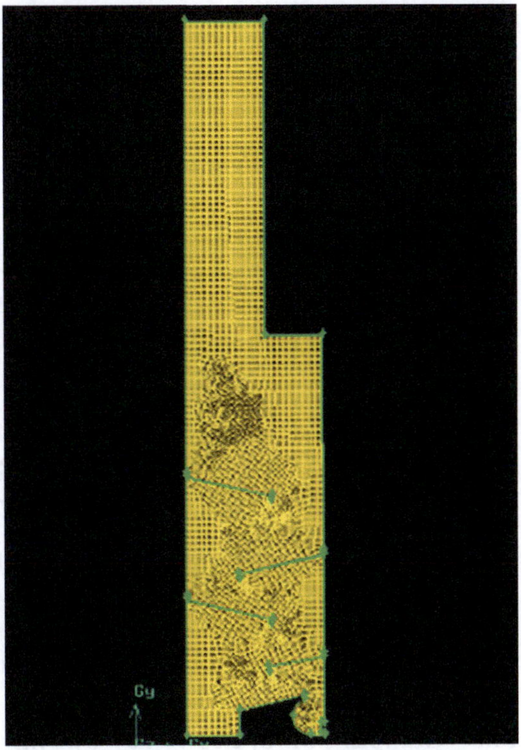

Fig. 1 Computaional domain and meshing details. (Finite element meshes are show in yellow lines and boundaries and baffles in green lines)

Table 1 Escape rate, capture rate and residual rate for the separator of the original structure

	Tracked particles	Escaped particles	Captured particles	Residual particles	Escape rate	Capture rate	Residual rate
	1000	101	800	99	10.10%	80.00%	9.90%
	1200	132	944	124	11.00%	78.67%	10.33%
	1400	153	1113	134	10.93%	79.50%	9.57%
	1600	183	1269	148	11.44%	79.31%	9.25%
	1800	199	1420	181	11.06%	78.89%	10.06%
	2000	223	1591	186	11.15%	79.55%	9.30%
Average	–	165	1190	145	10.95%	79.32%	9.74%

3 Finite Element Optimization of Separator's Structure

3.1 Theoretical Analysis

To enhance particle capture efficiency, thorough blending of water with paint mist-laden air is essential. Increasing water flow volume and adjusting parameters like water spraying, air flow rate, air pressure, and baffle characteristics can improve capture rates. Optimal inlet velocity, baffle number, and baffle area are determined through simulations, along with particle concentration positions to inform spraying locations.

3.2 Optimization Process

Process 1 Reducing inlet air velocity from 18 m/s to 10 m/s, increasing baffle number from 3 to 5, and enlarging baffle area from 650 mm × 4000 mm to 800 mm × 4000 mm significantly lowers particle escape rates and improves capture rates. The optimal inlet velocity is 12 m/s, with 4 baffles and a baffle area of 750 mm × 4000 mm.

Process 2 Adding water spraying on the 2nd baffle with varying angles (15° to 30°) and velocities (5 m/s to 9 m/s) further refines separation efficiency. The best results are achieved with a 30° angle and 7 m/s velocity.

Process 3 Shifting the spraying location to the 1st baffle and testing different angles and velocities confirms that a low-position spraying with a 30° angle and 7 m/s velocity yields the best water-gas separation effect.

3.3 Comparison of the Three Optimization Processes

Process 3 demonstrates the lowest escape rate, highest capture rate, and lowest residual rate, indicating superior capture efficiency compared to the other processes (see Fig. 2).

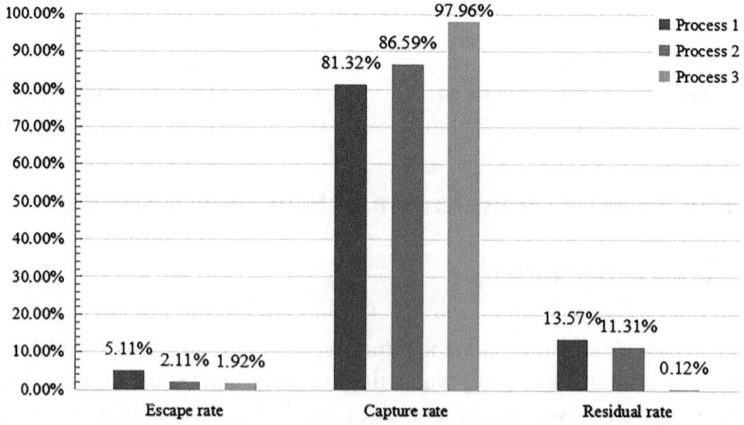

Fig. 2 Escape rate, capture rate, and residual rate of particles of the three processes

4 Simulative Verification with the Optimized Structure

Combining the best parameters from the three optimization processes, a new simulation is conducted. The optimized structure reduces inlet velocity, adjusts baffle characteristics, and adds strategic water spraying. The resulting escape rate is only 1.92%, the capture rate increases to 97.96%, and the residual rate drops to 0.12%, showcasing significant improvements over the original design. The detailed simulative data are iliustrated in Table 2.

Figure 3 illustrates the particle trajectory and distribution in a spraying booth before and after optimization. The original structure had a widely distributed particle trajectory, whereas the optimized structure concentrated particles at the lower part with minimal presence in the upper part. This indicates effective capture and

Table 2 Escape rate, capture rate and residual rate after optimization

	Tracked particles	Escaped particles	Captured particles	Residual particles	Escape rate	Capture rate	Residual rate
	1000	24	975	1	2.40%	97.50%	0.10%
	1200	13	1186	1	1.08%	98.83%	0.08%
	1400	31	1367	2	2.21%	97.64%	0.14%
	1600	30	1569	1	1.88%	98.06%	0.06%
	1800	40	1758	2	2.22%	97.67%	0.11%
	2000	35	1961	4	1.75%	98.05%	0.20%
Average	–	29	1469	1.8	1.92%	97.96%	0.12%

(a) (b) (c) (d)

Fig. 3 Particle moving trajectory and distribution of structure before and after optimization. (**a**) particle moving trajectory before optimization, (**b**) particle distribution of structure before optimization, (**c**) particle moving trajectory of structure before optimization, (**d**) particle distribution of structure after optimization

Table 3 Data comparison between the original structure and the current structure

	Entrance velocity	Baffle quantity	Baffle angle	Baffle length	Sprinkler system
Primary structure	18 mm/s	3	15°	650	–
Current structure	12 m/s	4	15°	750	1

interception of paint mist particles by adjusting the inlet velocity, baffle number, and baffle area of the separation device. The optimized design reduces inlet velocity for better particle-water blending, changes airflow direction through baffle adjustments, and increases baffle area for enhanced particle interception. Simulation results show that the new structure improves spraying effectiveness as shown in Table 3, with optimal spraying at a 30° angle and 7 m/s velocity. This creates a mist vortex where paint mist is predominantly mixed with water, capturing most particles.

5 Conclusions

Reducing inlet velocity, increasing baffle number, and adding water spraying with optimal parameters are effective strategies to enhance the water-gas separator's performance in a wet spray booth. These modifications lead to substantial reductions in particle escape rates and increased capture rates, contributing to more efficient and environmentally friendly spray coating processes.

The optimized separator of a commercial spray booth is with dimensions of 4000 mm (length) × 1050 mm (width) × 5500 mm (height) and consists of a vortex generating part and a water baffle. Four baffles are designed, whose inclination angles are kept as 15° and length is 750 mm. The first baffle adds water spraying with a 30° angle and 7 m/s velocity.

References

1. Chen J, Liu R, Gao Y, Li G, An T (2017) Preferential purification of oxygenated volatile organic compounds than monoaromatics emitted from paint spray booth and risk attenuation by the integrated decontamination technique. J Clean Prod 148(4):268–275
2. Li J, Rohan GU, Huang Y (2013) CFD-based modeling and design for energy-efficient VOC emission reduction in surface coating systems. Clean Techn Environ Policy 15(6):1023–1032
3. Ye Z, Zhou Q, Hua J, Ye Z, Meng J, Chen J, Wang J (2020) A novel pilot-scale installation integrated of improved sieve-tray tower and wet electrostatic precipitator for paint mist removal. J Air Waste Manage Assoc 71(3):366–377
4. Iida T, Mitani H, Sato M (2016) Development of paint booth: "new paint mist collection method". SAE Int J Mater Manuf 9(3):844–849
5. Dmitriev AV, Bikkulov RY, Kharkov VV, Dmitrieva OS (2022) Analysis of multivortex separator operation in spray booth. Radio Electron Electr Power Eng:1–4

6. Bikkulov RY, Dmitriev AV, Zinurov VE (2021) Separation of fine particles from gas in paint spraying booths. In: International Conference on Modern Trends in Manufacturing Technologies and Equipment, pp 1–8
7. Zhou Y, Li XHW, Xu T, Xie D, Dai X, Liu Z, Wang H, Yu G (2020) Separating paint mist from paint spraying waste gas: mechanism, model and application. Environ Eng Res 26(5):1–8
8. Bie Y (2024) Study on adsorption properties of activated carbon for advanced treatment of chemical wastewater. Highlights Sci Eng Technol 83:627–632
9. Dürr Systems Inc (2010) Innovative spray booth technology for the 'green' paint shop. J Books 108(5):42–43
10. Mohtashami R, Shang JQ (2019) Treatment of automotive paint wastewater in continuous-flow electroflotation reactor. J Clean Prod 128(3):335–346
11. Fang L, Zhang J, Shi H (2024) Treatment of spray paint wastewater by a coupled electro-Fenton and SBR technology. Adv Eng Technol Res 9(1):862–867
12. Qi B, Moe WM, Kinney KA (2005) Treatment of paint spray booth off-gases in a fungal bio-filter. J Environ Eng 131(2):180–189

Open Access This chapter is licensed under the terms of the Creative Commons Attribution-NonCommercial-NoDerivatives 4.0 International License (http://creativecommons.org/licenses/by-nc-nd/4.0/), which permits any noncommercial use, sharing, distribution and reproduction in any medium or format, as long as you give appropriate credit to the original author(s) and the source, provide a link to the Creative Commons license and indicate if you modified the licensed material. You do not have permission under this license to share adapted material derived from this chapter or parts of it.

The images or other third party material in this chapter are included in the chapter's Creative Commons license, unless indicated otherwise in a credit line to the material. If material is not included in the chapter's Creative Commons license and your intended use is not permitted by statutory regulation or exceeds the permitted use, you will need to obtain permission directly from the copyright holder.

Structural Parameter Design of Geometric Parameters of Marine Plate Natural Gas Heaters Based on Taguchi Method

Yi Zhao, Guozeng Feng, Rui Ling, and Haihua Ren

Abstract An innovative plate heat exchanger with a continuous trapezoidal herringbone corrugation pattern is introduced in this paper. This new heat exchanger will be implemented in a 15,000-ton chemical tanker for heating natural gas. Numerical method is applied to research the influence of three parameters, corrugation diameter (d), depth (h), and inclination angle (θ), on the heat transfer and resistance performance of the plate heat exchanger. The optimized structure parameters containing d, h and θ, are given by utilizing the Taguchi method, in which a comprehensive performance evaluation index (TP) as the benchmark. The heat exchanger with the corrugation diameter of 6 mm, the depth of 1 mm, and the inclination angle of 45°, has the optimal performance. The newly herringbone plate heat exchanger demonstrated superior comprehensive performance compared to the traditional model, with a 37.41% increase of TP.

Keywords Gas supply system · Continuous trapezoidal herringbone · Taguchi method · Comprehensive performance evaluation index

1 Introduction

With the promulgation of the International Maritime Organization's (IMO) regulations on the Sulfur Cap, Liquid Natural Gas (LNG) as an alternative fuel has become a focal point of attention in the global shipping industry [1]. The gas heating device, as an integral part of the marine engine fuel supply system, is particularly critical for its heating performance [2, 3]. The herringbone plate heat exchanger is currently the most widely used for heating natural gas on the market

Y. Zhao · G. Feng (✉) · R. Ling · H. Ren
School of Energy and Power, Jiangsu University of Science and Technology, Zhenjiang, China
e-mail: fengguozeng@just.edu.cn

© The Author(s) 2026

Z. Zhang, P. Mercorelli (eds.), *Mechanical Design and Simulation: Insights and Innovations*, Lecture Notes in Mechanical Engineering,
https://doi.org/10.1007/978-981-95-2405-1_60

[4]. However, when faced with the demands for higher efficiency, greater energy conservation, and more environmental friendliness, it is challenging for the performance of the herringbone plate heat exchanger to achieve significant breakthroughs. The design of new types of plate heat exchangers, especially innovative plates of heat exchanger, has become an important issue. In this work, a novel plate heat exchanger is proposed by replacing the traditional herringbone structure with a continuous trapezoidal herringbone corrugation structure. Consequently, it is necessary to optimize the key structural parameters of the new herringbone plate heat exchanger to enhance its heat transfer and flowing performance. To address these issues, the impact of corrugation diameter (d), depth (h), and inclination angle (θ) on the heat transfer and flowing of the new plate heat exchanger is studied through CFD simulations. The structural parameters are optimized using the Taguchi method of average response. Compared with traditional optimization design methods, including the probabilistic optimal design method, factorial design method and response surface model, the Taguchi orthogonal design method substantially reduces the number of experimental combinations [5]. In addition, the Taguchi design method provides a simple and efficient approach to determine the local optimum conditions within a given range of factors [6]. The optimal parameters are identified through analyzing the relationship between the structure and performance.

1.1 Physical Model

A novel type of herringbone plate heat exchanger has been designed in this study based on the conventional herringbone plate heat exchanger. Namely, the design features a continuous trapezoidal herringbone corrugation. As shown in Fig. 1b, the plates of this design are composed of trapezoidal herringbone waves. The structure parameters of the plates include the corrugation angle (β), corrugation diameter (d), corrugation depth (h), and corrugation inclination angle (θ).

1.2 Mathematical Model

The research focuses on single-phase convective heat transfer problems, employing the RNG turbulence model [7]. In conjunction with the flow and heat transfer characteristics of plate heat exchangers, the fluid flow is steady-state, three-dimensional, single-phase convective heat transfer, while neglecting buoyancy and gravity as well as viscous dissipation during the fluid flow process [8, 9].

Turbulent energy and dissipation rate equation:

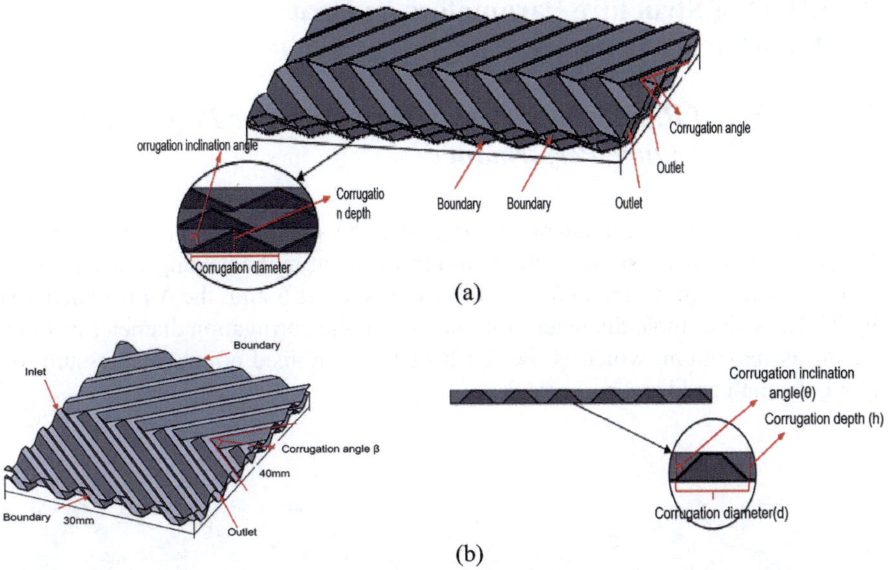

Fig. 1 (**a**) Traditional herringbone plate heat exchanger geometry and main parameters (**b**) Geometry and main parameters of continuous trapezoidal herringbone corrugated plate heat exchanger

$$\frac{\partial(\rho k)}{\partial t}+\frac{\partial(\rho k u_i)}{\partial t}=\frac{\partial}{\partial x_j}\left[\left(\alpha_k u_{eff}\right)\frac{\partial k}{\partial x_j}\right]+G_k+G_b-\rho\varepsilon-Y_M \tag{1}$$

$$\frac{\partial(\rho\varepsilon u_i)}{\partial x_i}=\frac{\partial}{\partial x_j}\left[\left(\alpha_\varepsilon u_{eff}\right)\frac{\partial\varepsilon}{\partial x_j}\right]+C_{1\varepsilon}\frac{\varepsilon}{k}\left(G_k+C_{3\varepsilon}G_b\right)-C_{2\varepsilon}\rho\frac{\varepsilon^2}{k}-R \tag{2}$$

1.3 Boundary Conditions

The plate heat exchanger simulated in this paper has its inlet boundary set as a mass-flow-inlet, with the initial inlet temperature set at −163 °C and the initial inlet pressure set at 0.6 MPa. The outlet boundary condition is set as a pressure-outlet; the left and right sides of the model are set as adiabatic walls; the upper and lower plates are set to constant temperatures of 300 K. Since nitrogen is relatively stable, is selected as the simulation medium.

2 Effect of Structure Parameters on Heat Transfer Performance

2.1 Effect of Ripple Diameter on Heat Transfer Performance and Resistance Performance

As shown in the Fig. 2, it can be observed that the smaller the d, the higher the air temperature downstream, indicating an improvement in the heating capacity of the heat exchange unit. Compared to a ripple diameter of 6 mm, the *Nu* has increased by 21.2% with a ripple diameter of 4 mm. When the corrugation diameter is 4 mm, *f* is at its maximum, which is the result of the combined effects of pressure loss along the path and local pressure drop.

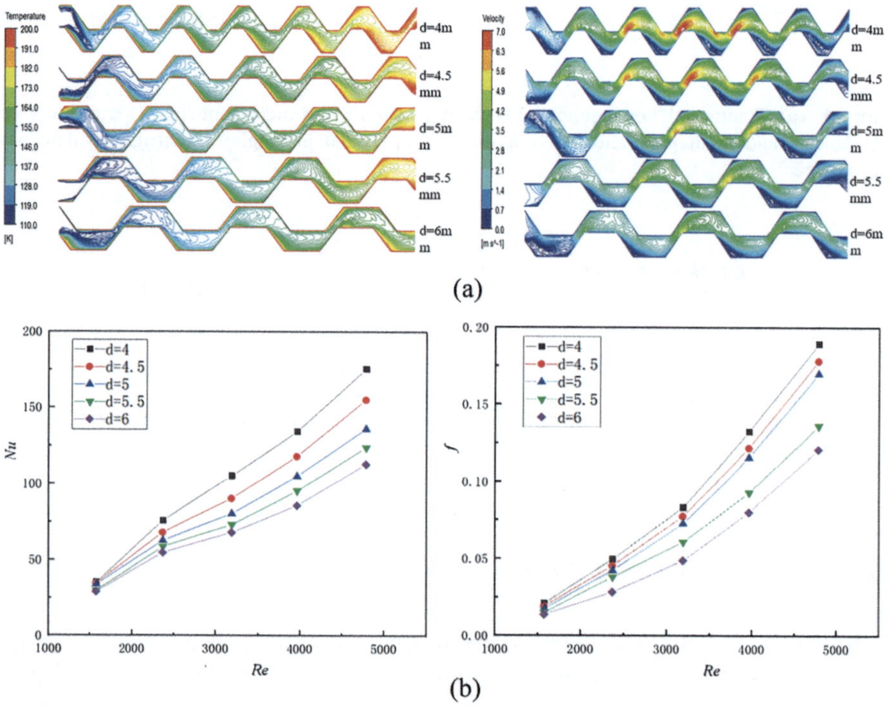

Fig. 2 (**a**)Temperature distribution and Velocity distribution of different ripple diameters. (**b**) The effect of h on *f* and *Nu* under different Re

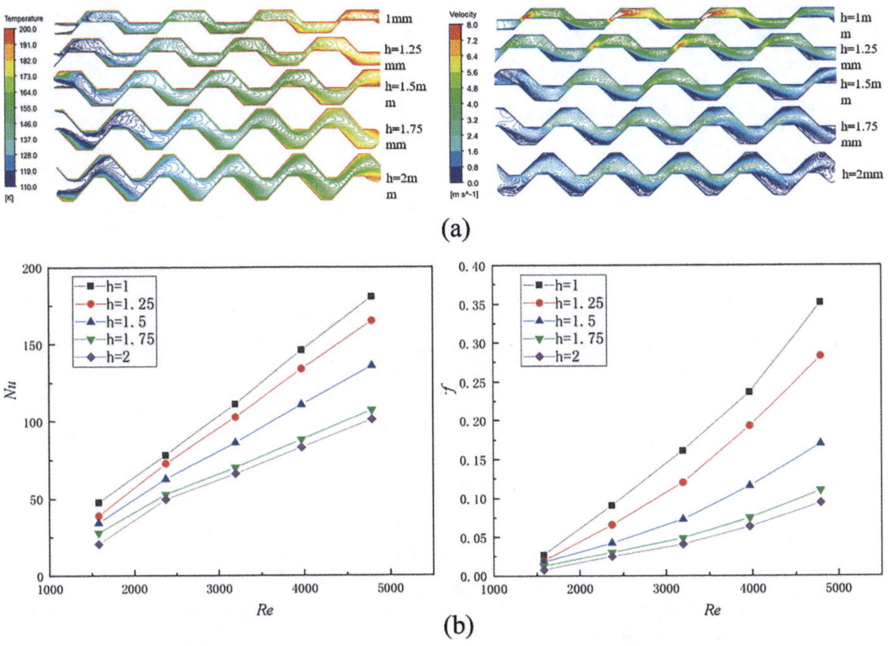

Fig. 3 (**a**) Temperature distribution and Velocity distribution of different ripple depths. (**b**) The effect of h on *f and Nu* under different Re

2.2 Effect of Ripple Depth on Heat Transfer Performance and Resistance Performance

As shown in the Fig. 3, when the Reynolds number is 1578, compared to a ripple depth of 2 mm, the *Nu* increases by more than 2.2 times with a ripple depth of 1 mm. It can be observed that a decrease in h leads to a greater degree of flow separation as the mainstream passes over the wave crests, enhancing the secondary mixing of air and localized heat transfer enhancement. With the increase of the *Re*, *f* for all five corrugation depths continuously increases, with the rate of increase also accelerating, when the ripple depth is 1 mm, *f* is at their maximum, and the increase is also the greatest. This is due to the increase in local pressure loss at the troughs and the enhancement of turbulent intensity.

2.3 Effect of Ripple Inclination Angles on Heat Transfer Performance and Resistance Performance

As shown in the Fig. 4, when the *Re* is 1578, compared to a ripple inclination angle of 35°, the *Nu* increases by 82.5% with a ripple inclination angle of 55°. It can be observed that an increase in θ leads to a greater degree of flow separation as the

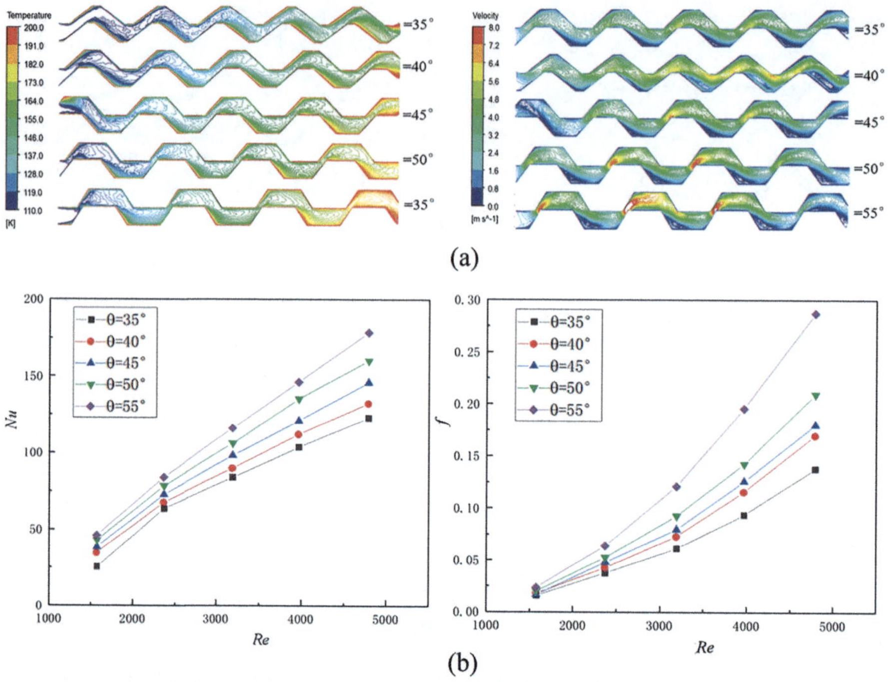

Fig. 4 (**a**) Temperature distribution and Velocity distribution of different ripple inclination angles. (**b**) The effect of h on *f and Nu* under different Re

mainstream passes over the wave crests. Subsequently, a larger temperature gradient area appears at the troughs of the fins, where the flow is hindered by the boundary layer, resulting in vortices that enhance the secondary mixing of air and localized heat transfer enhancement. At the same Re, Nu is highest for $\theta = 55°$, indicating that the heat exchanger with $\theta = 55°$ has the strongest ability to weaken the thermal boundary layer. This results in a smaller angle between the velocity direction and the temperature gradient, leading to a stronger synergistic effect between the entire temperature field and velocity field.

3 Structure Optimization of Herringbone Plate Heat Exchanger

Three factors, five levels, and an L25 (53) orthogonal array, were selected to numerically simulate 25 schemes.

3.1 Performance Evaluation

Comprehensive performance evaluation index (TP) [9]: The comprehensive performance evaluation index (TP) was used to evaluate the thermal performance of a new type plate heat exchanger. It is defined as the ratio of the average Nusser number to the coefficient of friction and can be written as:

$$TP = Nu / f^{1/3} \tag{3}$$

3.2 Structure Optimisation Based on the Taguchi Method

The robustness of design parameters can be measured using the Signal-to-Noise Ratio (S/N ratio), which includes three types: the smaller-the-better characteristic, the larger-the-better characteristic, and the nominal-the-best characteristic. Due to the larger-the-better characteristic of the comprehensive performance evaluation index (TP), the signal-to-noise ratio is calculated using equation of the larger-the-better characteristic.

$$S / N = -10 \log\left(\Sigma\left(1 / Y^2\right) / n\right) \tag{4}$$

3.3 Variance Analysis Model

Based on the basic method of variance analysis, the degree of influence of each factor on the optimization target, Fi, is characterized, which is the ratio of the sum of variances caused by each influencing variable, SSF, to the total sum of squares, ST [10]. The calculation formula is as follows:

$$m = \frac{1}{n} \sum_{i=1}^{n} (S / N) \tag{5}$$

$$S_{SF} = 3\left(m_{f1} - m\right)^2 + 3\left(m_{f2} - m\right)^2 + 3\left(m_{f3} - m\right)^2 + 3\left(m_{f4} - m\right)^2 + 3\left(m_{f5} - m\right)^2 \tag{6}$$

$$S_T = \sum_{i=1}^{n} \left[(S / N)_i - m \right]^2 \tag{7}$$

$$F_i = \frac{S_{SF}}{S_T} \tag{8}$$

3.4 Analysis of Range Variance

As shown in Fig. 5, The optimal parameter combination derived from the signal-to-noise ratio analysis is d = 6 mm, h = 1 mm, and θ = 40°. The comprehensive performance evaluation index (TP) fitted by Minitab is 239.91, and the simulated value of the model established based on these three levels is 250.13, with an error of 4.26%, thereby also proving the reliability of the range analysis and the accuracy of the data. The order of influence of the three structural parameters on TP from high to low is corrugation diameter (d) > corrugation depth (h) > corrugation inclination angle (θ), which is consistent with the results obtained in Table 1.

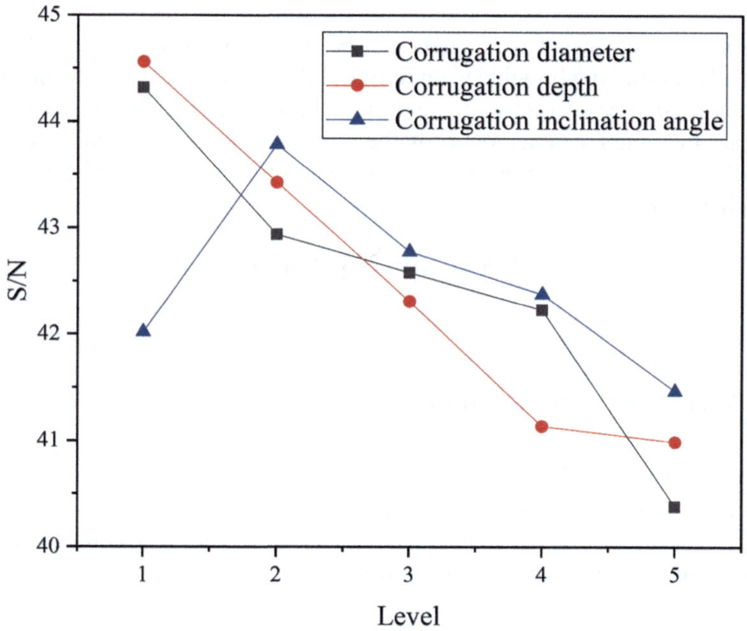

Fig. 5 Main effect diagram of signal-to-noise ratio of herringbone plate heat exchanger

Table 1 Score of each ventilation index

Parameters	S_{SF}	F_i	Contribution
A corrugation diameter d (mm)	3.218	0.804	50.540%
B corrugation depth h (mm)	2.544	0.636	39.960%
C corrugation inclination angle θ (°)	0.605	0.151	9.500%

4 Analysis of Optimal Scheme

Four optimized parameter models were designed using range analysis, variance analysis, and single-factor analysis: Model A (d = 6 mm, h = 1 mm, θ = 40°), Model B (d = 6 mm, h = 1.25 mm, θ = 40°), Model C (d = 6 mm, h = 1 mm, θ = 45°).

Simulations were conducted on four optimized models and the currently available herringbone plate heat exchanger with the best heat transfer performance on the market. As shown in Fig. 6, The simulation results indicate that Compared to the traditional herringbone plate heat exchanger model (TP = 178.34), the TP of the three optimized models increased by 34.01%, 16.56%, and 37.41%, respectively. The simulation results indicate that Model C has better comprehensive performance, a characteristic that does not change with variations in the Re, and the stability of the model can be demonstrated. Therefore, the optimal parameter selection is d = 6 mm, h = 1 mm, and θ = 45°.

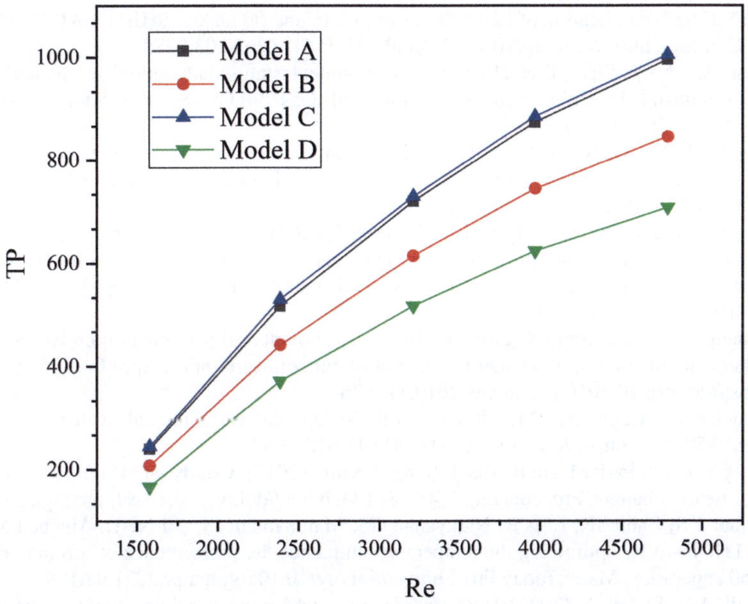

Fig. 6 Comparison of the comprehensive Performance evaluation index (TP) of Scheme A, Scheme B and Scheme C

5 Conclusion

Through structural design, the optimal structural parameters were determined to be a corrugation diameter of 6 mm, a corrugation depth of 1 mm, and a corrugation inclination angle of 45°.

1. The simulation results indicate that increasing the corrugation depth enhances the total heat transfer area, which also improves the heat transfer effect.
2. The primary to secondary order of parameter degree of influence is corrugation diameter (d) > corrugation depth (h) > corrugation inclination angle (θ).
3. The Taguchi method has been successfully applied to the design of plate heat exchangers. The findings provide valuable insights for the optimization of plate heat exchangers.

References

1. Lin Y (2020) Application of LNG fuel power on inland oil tanker. SHIP BOAT 31(03):28–32. (in Chinese). https://doi.org/10.19423/j.cnki.31-1561/u.2020.03.0288
2. Zhou S, Zhu J, Shan T et al (2022) Development status and outlook of natural gas and LNG industry in China. China Offshore Oil Gas 34(1):1–8. https://doi.org/10.11935/i.issn.1673-1506.2022.01.001
3. Dan S (2022) Study on the performance of vaporizer in fuel gas supply system for ships at low pressure. Jiangsu Univ Sci Technol, (in Chinese). https://doi.org/10.27171/d.cnki.ghdcc.2020.000725
4. Wang J, Hou Y, Lin X, Ding R, Liu Y, Anton K (2021) Experimental research on enhanced heat transfer of pulsating flow based on plate heat exchanger. In: 2021 3rd international conference on artificial intelligence and advanced manufacture, vol 34, pp 33–38. https://doi.org/10.1145/3495018.3495024
5. Zhuang C, Wang S, Shan K (2019) Probabilistic optimal design of cleanroom air-conditioning systems facilitating optimal ventilation control under uncertainties. Appl Energy 253:113576. https://doi.org/10.1016/j.apenergy.2019.113576
6. Smucker B, Krzywinski M, Altman N (2018) Optimal experimental design. Nat Methods 15(8):559–560. https://doi.org/10.1038/s41592-018-0083-2
7. Wang Y, Lee J, Park M, Jin B, Yun T, Song Y, Kim I (2017) A study on 3D numerical model for plate heat exchanger. Procedia Eng 174:188–194. https://doi.org/10.1016/j.proeng.2017.01.203
8. Kumar KS, Babu SR, Raja R, Narayanan GS, Manikandan JS, Ali NHU, Ahamed M (2021) WITHDRAWN: optimizing the temperature influence the performance of turning on Inconel x750 superalloy. Mater Today Proc. https://doi.org/10.1016/j.matpr.2021.02.198
9. Khail AA, Erişen A (2022) Heat transfer and performance enhancement investigation of novel plate heat exchanger. Therm Sci Eng Prog 34:101368. https://doi.org/10.1016/j.tsep.2022.101368
10. Chen B, Zhang Z, Zi Y, Yang Z, He Z (2013) A pseudo wavelet system-based vibration signature extracting method for rotating machinery fault detection. Sci China Technol Sci 56(5):1294–1306. https://doi.org/10.1007/s11431-013-5139-z

Open Access This chapter is licensed under the terms of the Creative Commons Attribution-NonCommercial-NoDerivatives 4.0 International License (http://creativecommons.org/licenses/by-nc-nd/4.0/), which permits any noncommercial use, sharing, distribution and reproduction in any medium or format, as long as you give appropriate credit to the original author(s) and the source, provide a link to the Creative Commons license and indicate if you modified the licensed material. You do not have permission under this license to share adapted material derived from this chapter or parts of it.

The images or other third party material in this chapter are included in the chapter's Creative Commons license, unless indicated otherwise in a credit line to the material. If material is not included in the chapter's Creative Commons license and your intended use is not permitted by statutory regulation or exceeds the permitted use, you will need to obtain permission directly from the copyright holder.

Study on the Influence of Fractal Parameters of Interface on the Stiffness of Screw Connection

Congbin Yang, Xian Liu, Tao Zhang, Zhifeng Liu, and Xuezheng Teng

Abstract As a prevalent fastening technique in mechanical system assembly, the structural rigidity of screw joints critically governs vibrational response characteristics at both component and full-system levels. In this paper, a new screw equivalent connection stiffness model is established by considering contact stiffness of connector, the meshing stiffness of the thread and the stiffness of the screw and the connected part. Employing fractal-based surface morphology characterization and considering asperity-substrate interaction, an integrated contact load-stiffness formulation for interface morphology is established. Based on the thread stress-strain relationship, the thread axial load distribution law is introduced, the thread meshing stiffness model is derived, and the equivalent model of the screw connection stiffness is established. Finally, the stiffness weakening ratio factor is proposed, and the influence of the combined partial shape parameters on the equivalent connection stiffness is revealed by numerical analysis. The results show that with fractal dimension elevation and roughness amplitude reduction, the equivalent connection stiffness of the screw increases gradually, and the stiffness weakening ratio decreases gradually.

Keywords Screw connection · Fractal theory · Connection stiffness

C. Yang · X. Liu · T. Zhang (✉) · X. Teng
Beijing Key Laboratory of Design and Intelligent Machining Technology for High Precision Machine Tools, Beijing University of Technology, Beijing, China
e-mail: zhangtao@bjut.edu.cn

Z. Liu
Beijing Key Laboratory of Design and Intelligent Machining Technology for High Precision Machine Tools, Beijing University of Technology, Beijing, China

Key Laboratory of CNC Equipment Reliability, Ministry of Education, Jilin University, Changchun, Jilin Province, China

© The Author(s) 2026

Z. Zhang, P. Mercorelli (eds.), *Mechanical Design and Simulation: Insights and Innovations*, Lecture Notes in Mechanical Engineering,
https://doi.org/10.1007/978-981-95-2405-1_61

1 Introduction

Various types of mechanical equipment such as machine tools are assembled from various parts. The surface of the parts that cooperate with each other is usually called the contact interface, and the interface existing in the mechanical structure makes the structure discontinuous. Due to the existence of the interface, the structural stiffness is greatly weakened. Studies have shown that nearly 60% of the vibration problems in machine tools are caused by the interface, and the contact stiffness of the interface accounts for about 40% ~ 60% of the total stiffness of the structure [1, 2]. According to statistics, threaded connections account for about 68% of all connections in the machinery manufacturing industry [3]. Threaded connection is a typical structure with a fixed interface. However, the contact stiffness of the interface is closely related to the morphology of the contact surface [4]. Therefore, it is necessary to study the influence of the combined partial shape parameters on the stiffness of the threaded connection.

At the micro level, there are many different shapes of asperities on the contact surface of the connected parts. And the contact load of the joint is mainly borne by these asperities, and the contact stiffness is also provided by these asperities. Majumdar and Bhushan [5] first proposed a fractal contact model of isotropic rough surfaces based on fractal theory, which is independent of geometric scale. Subsequently, Liou et al. [6] taking into account the non-physical phenomenon of asperity deformation transformation in the fractal contact model, proposed a modified contact model with opposite deformation transformation process, namely the LL model. On this basis, Jiang et al. [7] considering the interaction and the equivalent connection form of bolts, proposed a fractal contact model based on strain energy. Recently, Zhang et al. [8] re-examined the multi-scale contact model and proposed a complete piecewise-scale fractal extension model, which offers a viable approach for investigating cross-scale contact mechanics.

In the study of dynamic characteristics of bolted joints, Liu et al. [9] established a variable stiffness spring surface model considering the non-uniform pressure distribution at the bolted joint interface, and studied the influence of local bolt loosening degree and position on the natural frequency of bolted laminated plates. Zhou et al. [10] proposed a finite element analysis method for bolted joint structures based on multi-scale contact model, studied the global and local dynamic characteristics of bolted joint structures, and revealed that the existence of connection stiffness reduces the stiffness of the overall structure and increases energy dissipation.

The interface in the threaded connection structure is the main nonlinear factor of the joint, but the existing calculation model of the screw connection stiffness ignores the influence of the contact stiffness of the interface on the overall connection stiffness. Therefore, based on the fractal theory and considering the interaction between the asperity and the substrate, this paper establishes the contact stiffness model of the joint of the screw connection structure, which is scale-independent. The stiffness of the screw and the connected part and the contact stiffness of the interface are considered. The influence of the fractal parameters of the interface on the stiffness of the screw connection structure is analyzed theoretically, which provides a theoretical foundation for subsequent dynamic studies on the screw connection structure.

2 Theoretical Modeling

As shown in Fig. 1, the structural stiffness of the screw connection includes two types: the stiffness of the part itself and the contact stiffness of the joint. The stiffness of the part itself includes the stiffness of the screw and the connected part. The stiffness of the joint includes the stiffness of the interface of the connected part-connected part, the stiffness of the interface of the screw-connected part support, and the stiffness of the thread meshing.

2.1 Contact Stiffness of Interface

According to the WM function, these asperities can be considered as cosine wave shapes, and their contours before contact deformation is

$$z(x) = 2^{3-D} G^{D-2} (\ln \gamma)^{\frac{1}{2}} r_a^{3-D} \cos\left(\frac{\pi x}{r_a}\right) \quad (1)$$

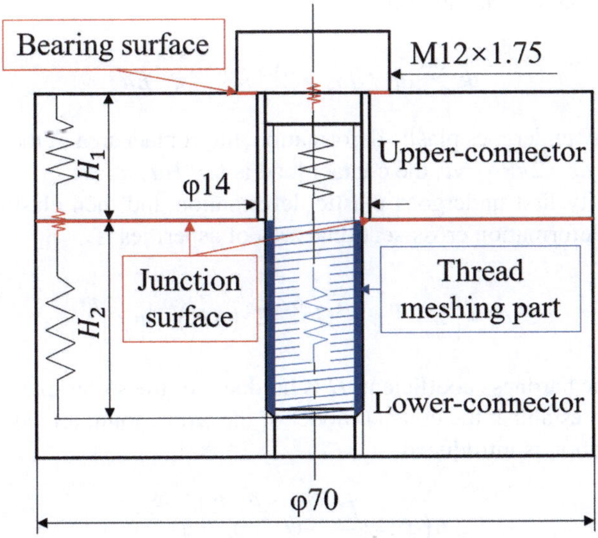

Fig. 1 Screw connection structure diagram

where r_a is the radius of the asperity contact section; D and G represent fractal parameters, and their values are influenced by the surface morphology.

Then, the contact deformation δ_a of asperity can be obtained according to Eq. (1)

$$\delta_a = 2^{4-D} G^{D-2} (\ln\gamma)^{\frac{1}{2}} r_a^{3-D} \tag{2}$$

In order to calculate the geometric parameters of the cosine-shaped asperity, the spherical contour is used to replace the calculation. Therefore, the curvature radius of the asperity should satisfy the formula $R^2 = (R - \delta_a)^2 + r_a^2$. Taking $R \gg r_a$ into account, the relationship can be simplified to $r_a^2 = 2R\delta_a$.

According to the study [6] there is an interaction between the asperity and the substrate, so the actual deformation of the asperity is

$$\delta = \left(\frac{\delta_a}{E_r\xi}\right)^{2/3} \tag{3}$$

where E_r is a dimensionless parameter. For uncoated surfaces, $E_r = 1$. ξ is the shape parameter of asperity, $\xi = \left(3\sqrt{R}\right)/\left(2r_a\right)$.

When the asperity undergoes elastic deformation, according to the Hertz contact theory, the contact area of the asperity can be represented as $a_e = \pi R\delta = a'/2$, where a' is the cross-sectional area of the asperity intersected by a rigid plane. And the contact load is

$$f_e = \frac{4}{3} E R^{\frac{1}{2}} \delta^{\frac{3}{2}} = 2^{\frac{10-D}{2}} \times 3^{\frac{-2-D}{3}} \pi^{\frac{D-4}{2}} E G^{D-2} (\ln\gamma)^{\frac{1}{2}} a'^{\frac{4-D}{2}} \tag{4}$$

Then, the normal contact stiffness is

$$k_{ne} = df_e / d\delta_a = 2^{3/2} \times 3^{-2/3} \pi^{-1} E a'^{\frac{1}{2}} \tag{5}$$

When asperity undergoes plastic deformation, the contact area of the asperity can be expressed as $a_p = 2\pi R\delta = a'$, the contact load is $f_p = Ha_p = Ha'$.

The asperity first undergoes plastic deformation and then elastic deformation. The critical deformation cross-sectional area of asperities is

$$a_c' = \left[2^{11-2D} \pi^{D-4} G^{2D-4} E^2 (KH)^{-2} \ln\gamma\right]^{\frac{1}{D-2}} \tag{6}$$

where K is the hardness coefficient, H is hardness of the softer material.

In order to establish the contact model of the whole joint, a modified island distribution function is introduced

$$n(a') = \frac{D-1}{2} \psi^{\frac{3-D}{2}} a_l'^{\frac{D-1}{2}} a'^{\frac{-D-1}{2}} \tag{7}$$

$$\psi^{\frac{3-D}{2}}-\left(1+\psi^{\frac{1-D}{2}}\right)^{\frac{D-3}{D-1}}=\frac{3-D}{D-1} \tag{8}$$

where $a_l^{'}$ is the maximum contact cross-sectional area.

Therefore, the contact load per unit area and the contact stiffness of the interface are

$$F_e=\int_{a_c^{'}}^{a_l^{'}}f_e n(a')da'+\int_0^{a_c^{'}}f_p n(a')da'$$
$$=\begin{cases} 2^{\frac{7}{2}}\times 3^{-\frac{1}{2}}\pi^{-\frac{3}{4}}G^{\frac{1}{2}}E(\ln\gamma)^{\frac{1}{2}}\psi^{\frac{1}{4}}a_l^{'\frac{3}{4}}\ln\frac{a_l^{'}}{a_c^{'}}+3H\psi^{\frac{1}{4}}a_l^{'\frac{3}{4}}a_c^{'\frac{1}{4}} & D=2.5 \\ \dfrac{2^{\frac{10-D}{2}}(D-1)G^{D-2}}{3^{\frac{D+2}{3}}\pi^{\frac{4-D}{2}}(5-2D)}E(\ln\gamma)^{\frac{1}{2}}\psi^{\frac{3-D}{2}}a_l^{'\frac{D-1}{2}}\left(a_l^{'\frac{5-2D}{2}}-a_c^{'\frac{5-2D}{2}}\right)+ & D\neq 2.5 \\ \dfrac{D-1}{3-D}H\psi^{\frac{3-D}{2}}a_l^{'\frac{D-1}{2}}a_c^{'\frac{3-D}{2}} & \end{cases} \tag{9}$$

$$K_c=\int_{a_c^{'}}^{a_l^{'}}k_{ne}n(a')da'=\frac{2^{3/2}E(D-1)}{3^{2/3}\sqrt{\pi}(2-D)}\psi^{\frac{3-D}{2}}a_l^{'\frac{D-1}{2}}\left(a_l^{'\frac{2-D}{2}}-a_c^{'\frac{2-D}{2}}\right) \tag{10}$$

Among them, the contour height of the combined surface can be collected by a three-dimensional profiler, and then the fractal parameters can be calculated by power spectral density method, structural function method or rough length method.

2.2 Thread Meshing Stiffness

As shown in Fig. 2, the thread is a key part of the interaction between the transfer bolt and the nut, and the thread meshing stiffness has a non-negligible effect on the bolt connection. According to Yamamoto's research, the ISO metric triangular thread will produce axial elastic deformation under the action of residual preload, as follows.

$$\delta_b=\left(2.523+1.08v_b-1.443v_b^2+\frac{d_p\tan^2\varphi}{2P}(1-v_b)\right)/E_b \tag{11}$$

$$\delta_n=\left(2.657+1.15v_n-1.507v_n^2+\frac{D\tan^2\varphi}{2P}\left(\frac{D_0^2+d_p^{2}}{D_0^2-d_p^{2}}+v_n\right)\right)/E_n \tag{12}$$

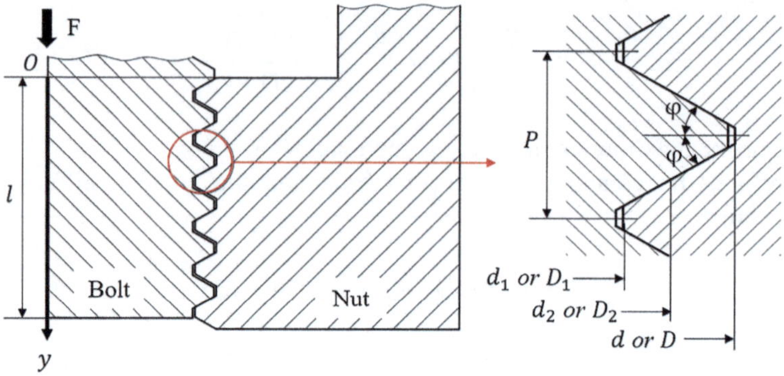

Fig. 2 Threaded connection

Among them, δ_b and δ_n are the axial elastic deformation of the bolt and nut under the action of unit force on the unit width. d_p is the effective diameter of the thread, $d_p = d - 0.649519P$. φ is the half angle of the thread, $\varphi = 30°$. D_0 is the outer diameter of the cross section of the nut.

Under the action of residual preload, the axial deformation of the thread of the bolt and nut is

$$\Delta_b = \delta_b \frac{\partial F}{\partial s} = \delta_b \frac{\partial F}{\partial y}\frac{\partial y}{\partial s} = \delta_b \frac{\partial F}{\partial y}\sin\beta = \frac{1}{k_{by}}\frac{\partial F}{\partial y} \tag{13}$$

$$\Delta_n = \delta_n \frac{\partial F}{\partial s} = \delta_n \frac{\partial F}{\partial y}\frac{\partial y}{\partial s} = \delta_n \frac{\partial F}{\partial y}\sin\beta = \frac{1}{k_{ny}}\frac{\partial F}{\partial y} \tag{14}$$

where s is the thread length, β is the lead angle, and y is the axial length of the thread. k_{by} and k_{ny} are the unit thread stiffness of bolts and nuts, respectively, and $k_{by} = (\delta_b \sin\beta)^{-1}$, $k_{ny} = (\delta_n \sin\beta)^{-1}$.

Because the axial elastic deformation of the thread is proportional to the load, the distribution of the axial deformation of the thread of the bolt and nut is the same as the axial load distribution of the meshing thread, that is,

$$\chi(y) = \frac{F(y)}{F_B} = -\frac{\cosh\lambda l}{\sinh\lambda l}\sinh\lambda y + \cosh\lambda y \tag{15}$$

where $\lambda = \sqrt{\dfrac{1/A_s E_b + 1/A_0 E_n}{1/k_{by} + 1/k_{ny}}}$, l is the thread meshing length, and F_B is the load on the first thread.

Combining Eq. (3) and Eq. (5), the boundary conditions of axial deformation of thread can be obtained.

$$\Delta_b(y) = \Delta_b(0)\chi(y) = \frac{1}{k_{by}}\frac{\partial F}{\partial y} \tag{16}$$

It can be seen from Eq. (15) that

$$\Delta_b(0)\int_0^l \chi(y)k_{by}dy = F_B \tag{17}$$

Therefore, the thread stiffness K_{bs} of the bolt can be expressed as

$$K_{bs} = \int_0^l \chi(y)k_{by}dy = \frac{\cosh\lambda l - 1}{\sinh\lambda l}\cdot\frac{k_{by}}{\lambda} \tag{18}$$

Similarly, we can also obtain the thread stiffness K_{ns} of the nut.

For threaded connections, the meshing stiffness is the series stiffness of the bolt thread stiffness and the nut thread stiffness, that is,

$$K_s = \frac{1}{K_{bs}^{-1} + K_{ns}^{-1}} = \frac{\lambda(\cosh\lambda l - 1)}{\sinh\lambda l}\left(\frac{1}{E_b A_s} + \frac{1}{E_n A_0}\right)^{-1} \tag{19}$$

2.3 Axial Stiffness of Screw

Here, the bolt calculation method VDI2230 is used to calculate the axial stiffness of the screw. As shown in Fig. 3, A_b is the cross-sectional area of the bolted bar; l_h, l_b, l_t and l_{tm} are the equivalent lengths of bolt head, bolt rod, screw and screw thread, respectively. The equivalent axial stiffness of the screw is

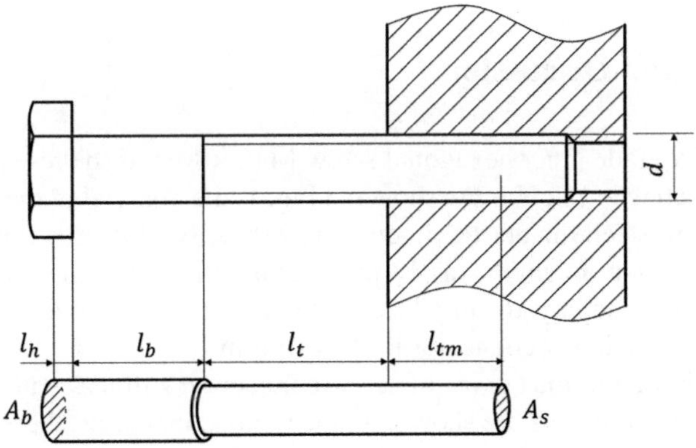

Fig. 3 Simplified model of bolt

$$K_b = \left(\frac{l_b + l_h}{A_b E_b} + \frac{l_t + l_{tm}}{A_s E_b} \right)^{-1} \tag{20}$$

where $A_b = \pi d^2/4$, $A_s = \pi(d - 0.9382P)^2/4$. l_b and l_t can be obtained according to the actual size of the screw connection. Generally, the equivalent length $l_h = 0.4d$ of the screw head of the inner hexagonal cylindrical head screw. According to the literature, the equivalent length of the screw thread of the ISO metric triangular thread can be obtained by the following formula

$$l_{tm} = \eta \left(\delta_b E_b + \delta_n E_b \right) A_s \tanh^{-1} \left(\eta L_n \right) \cdot \tan \beta \tag{21}$$

$$\eta = \left(\frac{1}{A_s} + \frac{E_b}{E_n} \frac{1}{A_n} \right)^{0.5} \left(\delta_b E_b + \delta_n E_b \right)^{-0.5} \tan^{-0.5} \beta \tag{22}$$

where L_n is the thread meshing length.

2.4 Screw Connection Stiffness Model

The normal stiffness of the screw connection structure should be the parallel stiffness of two sets of series stiffness. These two sets of series stiffness are the series stiffness of the stiffness K_{m1}, K_{m2} of the upper and lower structural parts and the stiffness K_{mc} of the interface, as well as the series stiffness of the thread meshing stiffness K_s, the equivalent axial stiffness K_b of the screw and the contact stiffness K_{bc} of the screw head joint. Therefore, the normal stiffness of the screw connection structure is

$$K = \left(K_{m1}^{-1} + K_{m2}^{-1} + K_{mc}^{-1} \right)^{-1} + \left(K_s^{-1} + K_b^{-1} + K_{bc}^{-1} \right)^{-1} \tag{23}$$

3 Results and Discussion

Taking the 12.9 grade inner hexagonal screw M12 × 1.75 as the research object, the torque of the screw is 112 Nm. As shown in Fig. 1, the material of the connector and the screw is 45# steel, the elastic modulus is 2.06×10^{-11} Pa, the Poisson's ratio is 0.3. The screw rotation length was 18 mm. Taking the end of the screw as the connection stiffness boundary, the thickness of the upper connector is $H_1 = 22$ mm, and the thickness of the lower connector is $H_2 = 18$ mm.

As shown in Fig. 4, the curves of screw connection stiffness with preload under different fractal parameters are shown. As the preload increases, the equivalent stiffness of the screw connection gradually increases and is affected by the contact

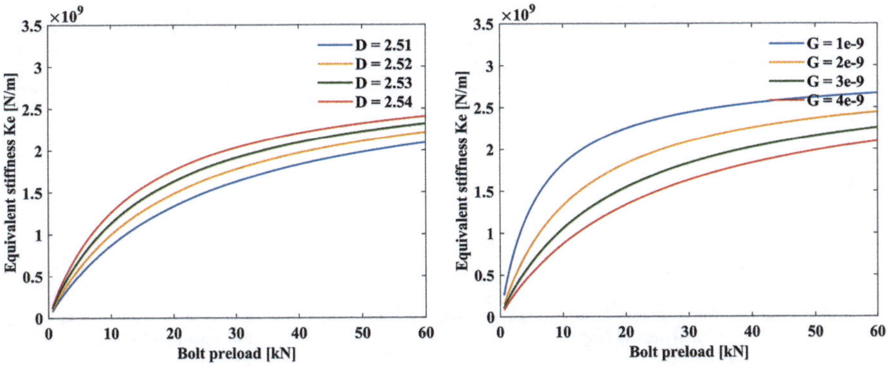

Fig. 4 The variation of screw connection stiffness with preload across varying fractal parameters. (L, G = 4e-9. R, D = 2.51)

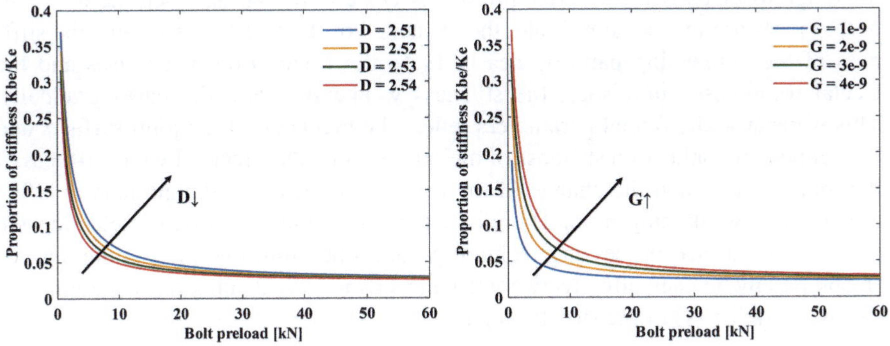

Fig. 5 The variation of screw stiffness ratio with preload across varying fractal parameters (L, G = 4e-9. R, D = 2.51)

surface morphology of the joint. That is, the smoother the contact surface, the larger the fractal dimension, the smaller the fractal roughness, the more precise the combination of the two connected parts, and the greater the equivalent connection stiffness. It can be seen from Fig. 5 that under the recommended torque preload (46.7 kN), the equivalent series stiffness of the screw part accounts for 4% of the total connection stiffness. Considering the real influence range of the pre-tightening force on the joint, the equivalent series stiffness ratio of the screw part should be less than 4%. Therefore, the bolt connection stiffness is mainly affected by the contact stiffness of the joint, and the equivalent connection stiffness is sensitive to the load of the interface. However, after the preload reaches the recommended value, the equivalent connection stiffness of the bolt does not change significantly.

Figure 6 shows the variation of the stiffness weakening ratio of the connector with the preload across varying fractal parameters. Stiffness weakening ratio refers to the weakening of the overall normal stiffness of the structure due to the existence of the interface compared with the two connectors as a whole. It can be seen from

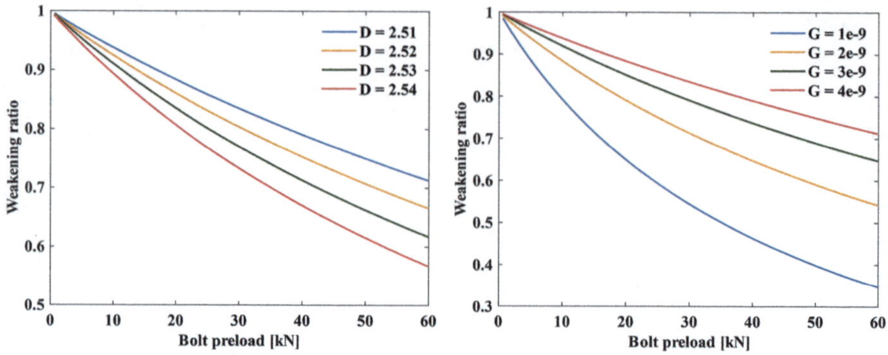

Fig. 6 The variation of the stiffness weakening ratio of the joint with the preload across varying fractal parameters (L, G = 4e-9. R, D = 2.51.)

the diagram that when the screw preload is close to 0, that is, when the two connected parts are in a separate state, the existence of the interface weakens the stiffness of the connecting parts by nearly 1. As the fractal dimension rises and the fractal roughness diminishes, the stiffness weakening ratio decreases gradually. This is because the fractal parameters reflect the precision of the joint surface, and the equivalent connection stiffness of the screw is mainly affected by the stiffness of the joint surface, and the change of the fractal parameters will significantly affect the stiffness weakening ratio. Therefore, for some connection parts, selecting the appropriate surface processing technology and obtaining the appropriate surface fractal parameters can effectively control the stiffness weakening ratio and improve the overall stiffness of the structural parts.

4 Conclusions

In this paper, the contact stiffness of the interface, the meshing stiffness of the thread, and the stiffness of the part itself are considered, and the equivalent connection stiffness model of the screw is established. In terms of the contact of the interface, based on the fractal theory and considering the interaction between the asperity and the substrate, the calculation model of the normal contact load and contact stiffness of the interface is established. In the aspect of thread meshing stiffness modeling, the thread meshing stiffness model is established by introducing the distribution law of thread axial load. Finally, the influence of the fractal parameters of the interface, namely the preload, on the equivalent connection stiffness of the screw is numerically analyzed. The main conclusions are as follows:

1. As the fractal dimension rises and the fractal roughness diminishes, the equivalent connection stiffness of the screw becomes greater.

2. With the increase of preload, the equivalent connection stiffness of the screw increases gradually. At the initial stage of screw pre-tightening, the equivalent connection stiffness is sensitive to the pre-tightening force, but after continuous pre-tightening, the equivalent connection stiffness changes gradually.
3. The stiffness weakening ratio is proposed, and its magnitude declines as the fractal dimension grows and the fractal roughness diminishes. By improving the surface morphology of the interface, the stiffness reduction ratio can be reduced and the overall stiffness of the structural member can be improved.

References

1. Qu CN, Wu LS, Ma JF et al (2013) A fractal model of normal dynamic parameters for fixed oily porous media joint interface in machine tools. Int J Adv Manuf Technol 68:2159–2167. https://doi.org/10.1007/s00170-013-4825-0
2. Yu C, Ding JG, He ZF et al (2020) Effect of joint interfacial contact stiffness on structural dynamics of ultra-precision machine tool. Int J Mach Tools Manuf 158:103609. https://doi.org/10.1016/j.ijmachtools.2020.103609
3. Gong H, Liu JH, Feng HH (2022) Research review on loosening mechanisms and anti-loosening methods of threaded fasteners. J Mech Eng 58:326–347+360. https://link.cnki.net/urlid/11.2187.TH.20210608.1119.130
4. Zhai C, Gan Y, Hanaor D et al (2016) The role of surface structure in normal contact stiffness. Exp Mech 56:359–368. https://doi.org/10.1007/s11340-015-0107-0
5. Majumdar A, Bhushan B (1991) Fractal model of elastic-plastic contact between rough surfaces. J Tribol 113:1–11. https://doi.org/10.1115/1.2920588
6. Liou JL, Tsai CM, Lin JF et al (2010) A microcontact model developed for sphere- and cylinder-based fractal bodies in contact with a rigid flat surface. Wear 268:431–442. https://doi.org/10.1016/j.wear.2009.08.033
7. Jiang K, Liu ZF, Yang CB et al (2022) Effects of the interface considering asperity interaction on the bolted joint performance in the bolt tightening process. Tribol Int 167:107408. https://doi.org/10.1016/j.triboint.2021.107408
8. Zhang T, Wu Y, Liu X, Jiang K (2024) A closure contact model of self-affine rough surfaces considering small-, meso-, and large-scale stage without adhesive. Fractal Fract 8:611. https://doi.org/10.3390/fractalfract8100611
9. Liu HH, Sun W, Du DX et al (2022) Modeling and free vibration analysis for bolted composite plate under inconsistent pre-tightening condition. Compos Struct 292:115634. https://doi.org/10.1016/j.compstruct.2022.115634
10. Zhou YC, Xiao Y, He Y et al (2020) A detailed finite element analysis of composite bolted joint dynamics with multiscale modeling of contacts between rough surfaces. Compos Struct 236:111874. https://doi.org/10.1016/j.compstruct.2020.111874

Open Access This chapter is licensed under the terms of the Creative Commons Attribution-NonCommercial-NoDerivatives 4.0 International License (http://creativecommons.org/licenses/by-nc-nd/4.0/), which permits any noncommercial use, sharing, distribution and reproduction in any medium or format, as long as you give appropriate credit to the original author(s) and the source, provide a link to the Creative Commons license and indicate if you modified the licensed material. You do not have permission under this license to share adapted material derived from this chapter or parts of it.

The images or other third party material in this chapter are included in the chapter's Creative Commons license, unless indicated otherwise in a credit line to the material. If material is not included in the chapter's Creative Commons license and your intended use is not permitted by statutory regulation or exceeds the permitted use, you will need to obtain permission directly from the copyright holder.

Modeling of Machining Accuracy and Sensitivity Analysis of Spatial Error in Worm Wheel Gear Grinding Machines

Zhifeng Liu, Xuezheng Teng, Baobao Qi, Li Liu, and Xian Liu

Abstract As an machining tool in the field of high-end equipment trans-mission, the precision and performance of the gear grinding machine have a decisive effect on the quality and efficiency of the final product. Establishing an accurate error model and identifying the key error terms are crucial for enhancing the machining accuracy of the gear grinding machine. This study explores the spatial error modeling of gear grinding machines using the kinematics theory of multi-body systems. By applying this theory, the research analyzes the kinematics of gear grinding machines, delves into the machine tool structure and geometric errors, and develops a machining accuracy model that incorporates 30 geometric errors. Based on this model, the Sobol global sensitivity analysis method and an improved Monte Carlo algorithm are employed to meticulously calculate the sensitivity coefficients of geometric errors' impact on machining accuracy. This approach clarifies how each geometric error affects machine tool performance and offers a vital theoretical foundation for the design, manufacturing, optimization, and error compensation of gear grinding machines.

Keywords Gear grinding machine · Error modeling · Sobol global sensitivity analysis

---

Z. Liu
China, Beijing Key Laboratory of Design and Intelligent Machining Technology for High Precision Machine Tools, Beijing, China

Key Laboratory of CNC Equipment Reliability, Ministry of Education, Jilin University, Changchun, Jilin Province, China

X. Teng · L. Liu · X. Liu
China, Beijing Key Laboratory of Design and Intelligent Machining Technology for High Precision Machine Tools, Beijing, China

B. Qi (✉)
Key Laboratory of CNC Equipment Reliability, Ministry of Education, Jilin University, Changchun, Jilin Province, China
e-mail: qibaobao@jlu.edu.cn

© The Author(s) 2026

Z. Zhang, P. Mercorelli (eds.), *Mechanical Design and Simulation: Insights and Innovations*, Lecture Notes in Mechanical Engineering,
https://doi.org/10.1007/978-981-95-2405-1_62

1 Introduction

To ensure product quality and enhance the efficiency and accuracy of machine tools, it is essential to optimize the handling of various geometric errors during the machine tool design phase, manufacturing and application. However, when constructing the model of machine tool machining accuracy, considering the mutual coupling influence of many geometric errors, it becomes a challenge to clarify the specific influence of each machining error on machining accuracy. Through the implementation of sensitivity analysis of machine tool geometric errors, we can effectively identify the error factors that have the greatest impact on machining accuracy, and then implement targeted adjustment and optimization strategies.

The prerequisite for studying the influence of geometric errors on accuracy retention is to understand the direct mapping relationship between geometric errors and machining errors, and establish a spatial position error model of machining accuracy. Chen Kun [1] defined a geometric error model for CNC machine tools based on Jacobian spinor theory, parameterized the geometric error, and analyzed various factors that affect machining error. Chung Yu Tsai [2] proposed using a rotation matrix of nonlinear components and a translation matrix of linear components to describe object motion. Wu Shi [3] established a machining error model, compensating for important geometric errors that affect the machining error. Fan JW [4] analyzed the sensitivity of geometric errors based on multi-body system theory, using a nine axis five link turning milling composite machine tool as object. Li et al. [5] developed a geometric error motion model for a spiral bevel gear grinder, incorporating principles of multi-body system theory and tooth surface meshing. They further refined the tooth surface accuracy by addressing errors in the machine tool's motion axes.

The core of sensitivity analysis of geometric errors in machine tools lies in identifying key errors to accelerate and optimize the machining processes of machine tools. Niu [6] conducted accuracy sensitivity analysis within the failure model framework, identifying key geometric errors that have a significant impact on the reliability of machine tool machining accuracy. Tao [7] proposed that the variation of machining errors caused by changes in geometric errors can be used as a visual indicator to measure error sensitivity. This method is simple, intuitive, and suitable for multi axis machine tools.

2 Modeling of Spatial Position Error in Gear Grinding Machines

In the gear spreading grinding process, the gear surface processing error caused by multiple sources of error in the gear grinding machine bed should be considered. This paper analyzes the machining error and grinding process of the gear grinding machine bed, and studies the error transfer relationship between the three linear

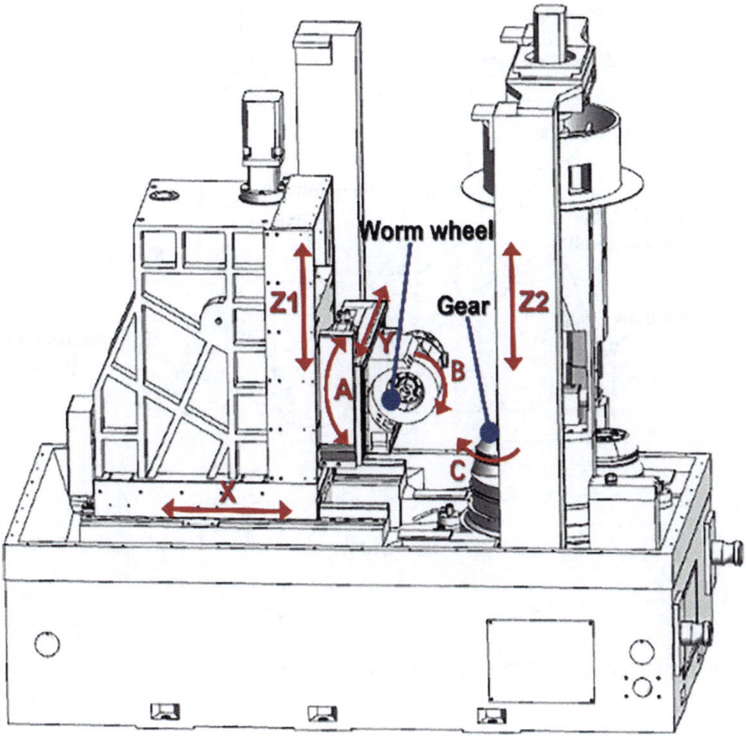

Fig. 1 Structure of YKS7225A gear grinding machine

axes and the two rotary axes of each error source, and the schematic structure of the gear grinding machine is shown in Figs. 1 and 2.

2.1 Topology and Lower Body Arrays

Based on the theory of multi-body systems, the parts of a machine tool can be described as arbitrary classical bodies according to their geometrical structure, and the machine tool can be abstracted as a multi-body system [8, 9].

As depicted in Fig. 3, the structure of the gear grinding machine bed can be characterized by a two-branch topology, with the first branch comprising the bed, columns, pallets, wheel rotary table, sliding pallets, spindle, and grinding wheels, and the second branch is composed of the bed, the workpiece rotary table, and the workpiece itself. In this configuration, the bed serves as the inertial reference frame and is designated as body B0, the column is denoted as body B1, and the bodies are numbered sequentially from one branch to the other along the direction away from body B1 [10]. Figure 3 shows the topology of the machine. Table 1 shows the lower body array of the selected gear grinding machine, and Table 2 describes the degrees

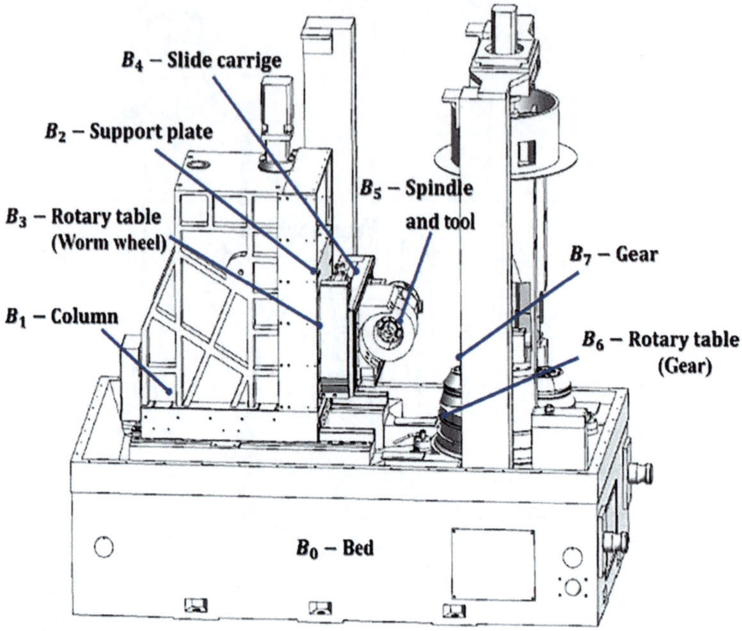

Fig. 2 Composition of YKS7225A gear grinding machine

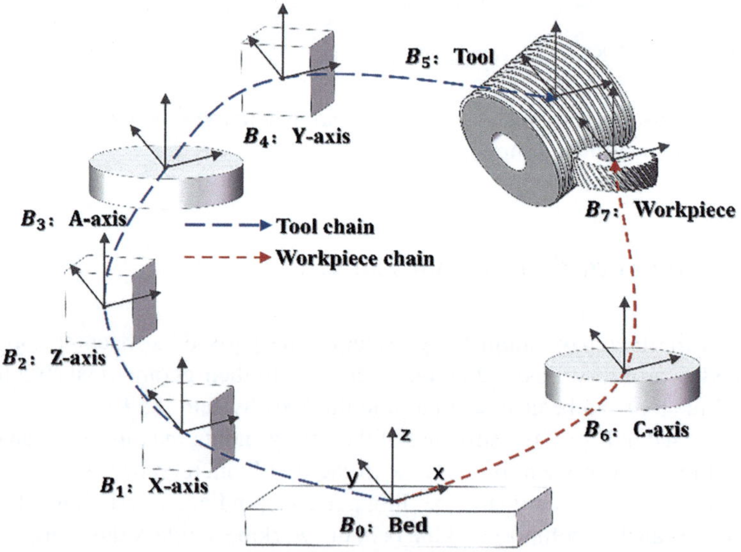

Fig. 3 Twin-branch topology of gear grinding machine

Table 1 Lower body array of gear grinding machine

Classical body j	1	2	3	4	5	6	7
$L^0(j)$	1	2	3	4	5	6	7
$L^1(j)$	0	1	2	3	4	0	6
$L^2(j)$	0	0	1	2	3	0	0
$L^3(j)$	0	0	0	1	2	0	0
$L^4(j)$	0	0	0	0	1	0	0
$L^5(j)$	0	0	0	0	0	0	0

Table 2 Adjacent body freedom of gear grinding machine

Adjacent parts	X	Y	Z	α	β	γ
0–1	1	0	0	0	0	0
1–2	0	0	1	0	0	0
2–3	0	0	0	1	0	0
3–4	0	1	0	0	0	0
4–5	0	0	0	0	0	0
0–6	0	0	0	0	0	1
6–7	0	0	0	0	0	0

of freedom of each cell with respect to the constraints, where "0" denotes no degrees of freedom and "1" denotes 1 degree of freedom.

2.2 *Definition of Geometric Error Parameters*

The process of gear grinding involves the movement of the grinding wheel's contact point within the coordinate system defined by the gear workpiece. Ideally, the trajectory of the grinding wheel's contact point should coincide with the pre-defined ideal trajectory, ensuring precise gear manufacturing. However, in practice, deviations often occur due to the complex interplay of various geometric errors. These errors, which may stem from the grinding machine's kinematics, the grinding wheel's profile, or the workpiece's positioning, interact with each other in a coupled manner. As a result, the center point of the grinding wheel deviates from the ideal processing path, leading to discrepancies in the spatial position of the grinding wheel's contact point. This deviation can affect the accuracy of the gear's profile and surface finish, highlighting the need for error compensation and improved control strategies to enhance the overall grinding accuracy [11]. Table 3 shows the geometric errors.

Table 3 Symbols and significance of geometrical errors in gear grinding machines

Geometric meaning of error	Notation	Geometric meaning of error	Notation
X-axis positioning error	Δx_x	A-axis Y-direction runout error	Δy_a
X-axis Y-direction straightness error	Δy_x	A-axis Z-direction runout error	Δz_a
X-axis Z-direction straightness error	Δz_x	Angular error of the A-axis around itself	$\Delta \alpha_a$
X-axis roll error	$\Delta \alpha_x$	Angular error of A-axis around Y-axis	$\Delta \beta_a$
X-axis swing error	$\Delta \beta_x$	Angular error of A-axis around Z-axis	$\Delta \gamma_a$
X-axis bias error	$\Delta \gamma_x$	C-axis X-direction runout error	Δx_c
Y-axis positioning error	Δy_y	C-axis Y-direction runout error	Δy_c
Y-axis X-direction straightness error	Δx_y	C-axis Z-direction runout error	Δz_c
Y-axis Z-direction straightness error	Δz_y	Angular error of the C-axis around the X-axis	$\Delta \alpha_c$
Y-axis roll error	$\Delta \alpha_y$	Angular error of C-axis around Y-axis	$\Delta \beta_c$
Y-axis oscillation error	$\Delta \beta_y$	Angular error of the C-axis around itself	$\Delta \gamma_c$
Y-axis deflection error	$\Delta \gamma_y$	X, Y axis perpendicularity error	ΔS_{xy}
Z-axis X-direction straightness error	ΔX_z	X, Z axis perpendicularity error	ΔS_{xz}
Z-axis Y-direction straightness error	ΔY_z	Y, Z axis perpendicularity error	ΔS_{yz}
Z-axis positioning error	Δz_z	X, A axis perpendicularity error	$\Delta \alpha_{xa}$
Z-axis roll error	$\Delta \alpha_z$	Y, A axis perpendicularity error	$\Delta \beta_{ya}$
Z-axis oscillation error	$\Delta \beta_z$	X, C axis perpendicularity error	$\Delta \alpha_{xc}$
Z-axis oscillation error	$\Delta \gamma_z$	Y, C axis perpendicularity error	$\Delta \beta_{yc}$
A-axis X-direction runout error	Δx_a		

2.3 Generalized Coordinate System and Identity Matrix

The topology of the neighboring structure of the multi-source error of the grinding machine bed is shown in Fig. 4. Where O_pO_s is the position coordinate system of the typical body, and O_s is the movement coordinate system of the typical body. The error transformations can be expressed as matrices:

$$T = T_p T_{pe} T_s T_{se} \tag{1}$$

where T_p is the static theoretical position transformation matrix of the structure, T_{pe} is the stationary position error transformation matrix of the structure, T_s is the transformation matrix of the motion state of the structure, and T_{se} is the error transformation matrix of the moving state of the structure.

Assuming that the center point of the grinding wheel of the gear grinding machine has a position vector of in the tool coordinate system is $r_t = (h_{tx}\ h_{ty}\ h_{tz}\ 1)^T$, and the center point of the grinding wheel has a position vector of in the tool coordinate

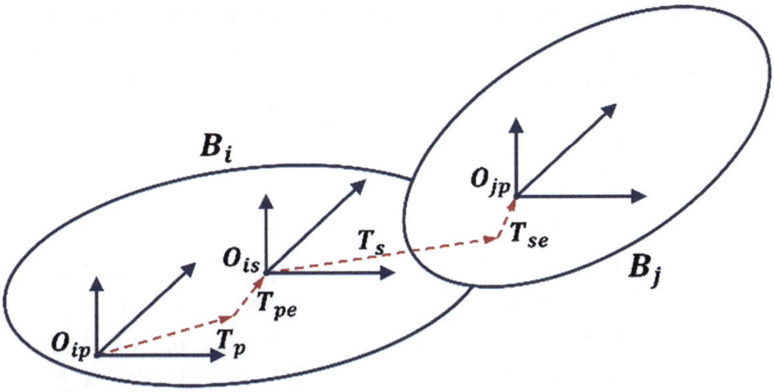

Fig. 4 Topological relationships of neighboring structures

Table 4 Reference coordinate system origin position vectors for each classical body

Origin of the reference coordinate system	Position vector
B1	$h_1 = (h_{1x}\ h_{1y}\ h_{1z}\ 1)^T$
B2	$h_2 = (h_{2x}\ h_{2y}\ h_{2z}\ 1)^T$
B3	$h_3 = (h_{3x}\ h_{3y}\ h_{3z}\ 1)^T$
B4	$h_4 = (h_{4x}\ h_{4y}\ h_{4z}\ 1)^T$
B5	$h_5 = (h_{5x}\ h_{5y}\ h_{5z}\ 1)^T$
B6	$h_6 = (h_{6x}\ h_{6y}\ h_{6z}\ 1)^T$
B7	$h_7 = (h_{7x}\ h_{7y}\ h_{7z}\ 1)^T$
Grinding wheel center point in tool coordinate system	$r_t = (h_{tx}\ h_{ty}\ h_{tz}\ 1)^T$
Grinding wheel center point in the workpiece coordinate system	$r_w = (h_{wx}\ h_{wy}\ h_{wz}\ 1)^T$

system is $r_w = (h_{wx}\ h_{wy}\ h_{wz}\ 1)^T$, the origin of the reference coordinate system and the position vector of each classical body are shown in Table 4.

The relationships between neighboring classical bodies are represented by means of 4 × 4 matrices. Tables 5 and 6 listed the stationary and motion chi-square transformation matrices between adjacent objects under ideal and actual conditions.

The position vector of the point to be machined in the coordinate system of the workpiece in the coordinate system of the machine tool is:

$$p_w = {}^0_6T_p\, {}^0_6T_{pe}\, {}^0_6T_6\, {}^0_4T_{se}\, {}^6_7T_p\, {}^6_7T_{pe}\, {}^6_7T_s\, {}^6_7T_{se} r_w \tag{2}$$

The position vector of the grinding wheel center point in the machine coordinate system is:

Table 5 Ideal and actual stationary chi-square transformation matrices between neighboring bodies

Adjacent body	Ideal stationary chi-square transformation matrix	Real stationary chi-square transformation matrix
0–1	${}^{0}_{1}T_p = \begin{pmatrix} 1 & 0 & 0 & h_{1x} \\ 0 & 1 & 0 & h_{1y} \\ 0 & 0 & 1 & h_{1z} \\ 0 & 0 & 0 & 1 \end{pmatrix}$	${}^{0}_{1}T_{pe} = E_{4\times4}$
1–2	${}^{1}_{2}T_p = \begin{pmatrix} 1 & 0 & 0 & h_{2x} \\ 0 & 1 & 0 & h_{2y} \\ 0 & 0 & 1 & h_{2z} \\ 0 & 0 & 0 & 1 \end{pmatrix}$	${}^{1}_{2}T_{pe} = \begin{pmatrix} 1 & 0 & \Delta S_{xz} & 0 \\ 0 & 1 & 0 & 0 \\ -\Delta S_{xz} & 0 & 1 & 0 \\ 0 & 0 & 0 & 1 \end{pmatrix}$
2–3	${}^{2}_{3}T_p = \begin{pmatrix} 1 & 0 & 0 & h_{3x} \\ 0 & 1 & 0 & h_{3y} \\ 0 & 0 & 1 & h_{3z} \\ 0 & 0 & 0 & 1 \end{pmatrix}$	${}^{2}_{3}T_{pe} = \begin{pmatrix} 1 & \Delta\gamma_{za} & 0 & 0 \\ -\Delta\gamma_{za} & 1 & \Delta\beta_{ya} & 0 \\ 0 & -\Delta\beta_{ya} & 1 & 0 \\ 0 & 0 & 0 & 1 \end{pmatrix}$
3–4	${}^{3}_{4}T_p = \begin{pmatrix} 1 & 0 & 0 & h_{4x} \\ 0 & 1 & 0 & h_{4y} \\ 0 & 0 & 1 & h_{4z} \\ 0 & 0 & 0 & 1 \end{pmatrix}$	${}^{3}_{4}T_{pe} = \begin{pmatrix} 1 & -\Delta S_{xy} & 0 & 0 \\ \Delta S_{xy} & 1 & -\Delta S_{yz} & 0 \\ 0 & \Delta S_{yz} & 1 & 0 \\ 0 & 0 & 0 & 1 \end{pmatrix}$
4–5 (Cutter)	${}^{4}_{5}T_p = \begin{pmatrix} 1 & 0 & 0 & h_{5x} \\ 0 & 1 & 0 & h_{5y} \\ 0 & 0 & 1 & h_{5z} \\ 0 & 0 & 0 & 1 \end{pmatrix}$	${}^{4}_{5}T_{pe} = E_{4\times4}$
0–6	${}^{0}_{6}T_p = \begin{pmatrix} 1 & 0 & 0 & h_{6x} \\ 0 & 1 & 0 & h_{6y} \\ 0 & 0 & 1 & h_{6z} \\ 0 & 0 & 0 & 1 \end{pmatrix}$	${}^{0}_{6}T_{pe} = \begin{pmatrix} 1 & 0 & \Delta\alpha_{xc} & 0 \\ 0 & 1 & \Delta\beta_{yc} & 0 \\ -\Delta\alpha_{xc} & -\Delta\beta_{yc} & 1 & 0 \\ 0 & 0 & 0 & 1 \end{pmatrix}$
6–7 (workpieces)	${}^{6}_{7}T_p = \begin{pmatrix} 1 & 0 & 0 & h_{7x} \\ 0 & 1 & 0 & h_{7y} \\ 0 & 0 & 1 & h_{7z} \\ 0 & 0 & 0 & 1 \end{pmatrix}$	${}^{6}_{7}T_{pe} = E_{4\times4}$

Table 6 Ideal and actual motion chi-square transformation matrices between neighboring bodies

Adjacent body	Ideal motion chi-square transformation matrix	Real motion chi-square transformation matrix
0–1	${}^{0}_{1}T_{s} = \begin{pmatrix} 1 & 0 & 0 & x \\ 0 & 1 & 0 & 0 \\ 0 & 0 & 1 & 0 \\ 0 & 0 & 0 & 1 \end{pmatrix}$	${}^{0}_{1}T_{se} = \begin{pmatrix} 1 & -\Delta\gamma_x & \Delta\beta_x & \Delta x_x \\ \Delta\gamma_x & 1 & -\Delta\alpha_x & \Delta y_x \\ -\Delta\beta_x & \Delta\alpha_x & 1 & \Delta z_x \\ 0 & 0 & 0 & 1 \end{pmatrix}$
1–2	${}^{1}_{2}T_{s} = \begin{pmatrix} 1 & 0 & 0 & 0 \\ 0 & 1 & 0 & 0 \\ 0 & 0 & 1 & z \\ 0 & 0 & 0 & 1 \end{pmatrix}$	${}^{1}_{2}T_{se} = \begin{pmatrix} 1 & -\Delta\gamma_z & \Delta\beta_z & \Delta x_z \\ \Delta\gamma_z & 1 & -\Delta\alpha_z & \Delta y_z \\ -\Delta\beta_z & \Delta\alpha_z & 1 & \Delta z_z \\ 0 & 0 & 0 & 1 \end{pmatrix}$
2–3	${}^{2}_{3}T_{s} = \begin{pmatrix} 1 & 0 & 0 & 0 \\ 0 & 1 & 0 & 0 \\ 0 & 0 & \cos\gamma & -\sin\gamma \\ 0 & 0 & \sin\gamma & \cos\gamma \end{pmatrix}$	${}^{2}_{3}T_{se} = \begin{pmatrix} 1 & -\Delta\gamma_a & \Delta\beta_a & \Delta x_a \\ \Delta\gamma_a & 1 & -\Delta\alpha_a & \Delta y_a \\ -\Delta\beta_a & \Delta\alpha_a & 1 & \Delta z_a \\ 0 & 0 & 0 & 1 \end{pmatrix}$
3–4	${}^{3}_{4}T_{s} = \begin{pmatrix} 1 & 0 & 0 & 0 \\ 0 & 1 & 0 & y \\ 0 & 0 & 1 & 0 \\ 0 & 0 & 0 & 1 \end{pmatrix}$	${}^{3}_{4}T_{se} = \begin{pmatrix} 1 & -\Delta\gamma_y & \Delta\beta_y & \Delta x_y \\ \Delta\gamma_y & 1 & -\Delta\alpha_y & \Delta y_y \\ -\Delta\beta_y & \Delta\alpha_y & 1 & \Delta z_y \\ 0 & 0 & 0 & 1 \end{pmatrix}$
4–5 (cutter)	${}^{4}_{5}T_{s} = E_{4\times4}$	${}^{4}_{5}T_{pe} = E_{4\times4}$
0–6	${}^{0}_{6}T_{s} = \begin{pmatrix} \cos\gamma & -\sin\gamma & 0 & 0 \\ \sin\gamma & \cos\gamma & 0 & 0 \\ 0 & 0 & 1 & 0 \\ 0 & 0 & 0 & 1 \end{pmatrix}$	${}^{5}_{6}T_{se} = \begin{pmatrix} 1 & -\Delta\gamma_c & \Delta\beta_c & \Delta x_c \\ \Delta\gamma_c & 1 & -\Delta\alpha_c & \Delta y_c \\ -\Delta\beta_c & \Delta\alpha_c & 1 & \Delta z_c \\ 0 & 0 & 0 & 1 \end{pmatrix}$
6–7 (workpieces)	${}^{6}_{7}T_{s} = E_{4\times4}$	${}^{6}_{7}T_{se} = E_{4\times4}$

$$p_t = {}^{0}_{1}T_{p}\,{}^{0}_{1}T_{pe}\,{}^{0}_{1}T_{s}\,{}^{0}_{1}T_{se}\,{}^{1}_{2}T_{p}\,{}^{1}_{2}T_{pe}\,{}^{1}_{2}T_{s}\,{}^{1}_{2}T_{se}\,{}^{2}_{3}T_{p}\,{}^{2}_{3}T_{pe}\,{}^{2}_{3}T_{s}\,{}^{2}_{3}T_{se}\,{}^{3}_{4}T_{p}\,{}^{3}_{4}T_{pe}\,{}^{3}_{4}T_{s}\,{}^{3}_{4}T_{se}\,{}^{4}_{5}T_{p}\,{}^{4}_{5}T_{pe}\,{}^{4}_{5}T_{s}\,{}^{4}_{5}T_{se}\,r_t \quad (3)$$

In the actual machining process, if you want to achieve precision machining needs to meet the workpiece coordinate system within the point to be machined and the center of the grinding wheel coincide, that is:

$$p_w = p_t \quad (4)$$

The position vector of the center point of the grinding wheel in the workpiece coordinate system during actual machining is:

$$
\begin{aligned}
r_w &= \left(r_{wx}\ r_{wy}\ r_{wz}\ 1\right)^T \\
&= \left({}^0_6T_p\ {}^0_6T_{pe}\ {}^0_6\mathrm{T}_6\ {}^0_4T_{se}\ {}^6_7T_p\ {}^6_7T_{pe}\ {}^6_7T_s\ {}^6_7T_{se}\right)^{-1} \\
&\times {}^0_1T_p\ {}^0_1T_{pe}\ {}^0_1T_s\ {}^0_1T_{se}\ {}^1_2T_p\ {}^1_2T_{pe}\ {}^1_2T_s\ {}^1_2T_{se}\ {}^2_3T_p\ {}^2_3T_{pe}\ {}^2_3T_s\ {}^2_3T_{se}\ {}^3_4T_p\ {}^3_4T_{pe}\ {}^3_4T_s\ {}^3_4T_{se}\ {}^4_5T_p\ {}^4_5T_{pe}\ {}^4_5T_s\ {}^4_5T_{se} r_t
\end{aligned}
\tag{5}
$$

Assuming that each error parameter in Eq. is 0, the position vector of the center point of the grinding wheel in the coordinate system of the workpiece under ideal conditions can be obtained:

$$
r_w^i = \left(r_{wx}^i\ r_{wy}^i\ r_{wz}^i\ 1\right)^T \tag{6}
$$

Where r_{wx}, r_{wy} and r_{wz} represent the component of the position of the grinding wheel center point in the workpiece coordinate system; r_{wx}^i, r_{wy}^i and r_{wz}^i represent the component of the position of the grinding wheel center point in the workpiece coordinate system in ideal conditions, respectively.

Subtracting Eq. (5) from Eq. (6) gives the spatial motion error model:

$$
E = \left(E_x\ E_y\ E_z\ 1\right)^T = r_w - r_w^i \tag{7}
$$

Among them, $E = (E_x\ E_y\ E_z\ 1)^T$ represents the spatial position error vector; E_x, E_x and E_z represent the spatial motion error components of the gear grinding machine in X, Y and Z directions respectively.

3 Improved Soblo Global Sensitivity Analysis

3.1 Global Sensitivity Modeling

The Sobol method is a variance-based Monte Carlo method, which decomposes the system function into a sum of 2n increasing terms, and the total variance of the model response and the bias variance of each subterm are calculated by sampling to find the sensitivity. The spatial domain input factor Ω^n can be expressed as follow:

$$
\Omega^n = \left(g_i | 0 \le g_i \le 1; i = 1, 2, \ldots, n\right) \tag{8}
$$

As an example of E_X, the spatial error on the X-axis axis can be expressed as follows:

$$
\begin{aligned}
E_X &= E_X(G) = E_X(g_1, g_2, \ldots, g_n) \\
&= E_{X0} + \sum_{i=1}^{n} E_{Xi}(g_i) + \sum_{1 \le i \le j \le n} E_{Xij}(g_i, g_j) + \ldots + E_{X1,2,\ldots,n}(g_1, g_2, \ldots, g_n)
\end{aligned}
\tag{9}
$$

And:

$$\int_0^1 E_{xi_1,\ldots,i_s}\left(g_{i_1},g_{i_2},\ldots,g_{i_s}\right)dg_{i_m} = 0\left(1 \le m \le s \le n\right) \tag{10}$$

If:

$$\left(i_1,i_2,\ldots,i_s\right) \ne \left(j_1,j_2,\ldots,j_s\right) \tag{11}$$

Then:

$$\int_{\Omega^n} E_{xi_1,i_2,\ldots,i_s} E_{xj_1,j_2,\ldots,j_l} dG = 0 \tag{12}$$

According to the above assumptions, the decomposition in Eq. (10) is unique. Therefore, all the subterms in Eq. (10) can be obtained by integrating multiple times and can be expressed as follows:

$$E_{X0} = \int_{\Omega^n} E_X(G)dG \tag{13}$$

$$E_{Xi}(g_i) = \int_0^1 \bullet\bullet\bullet \int_0^1 E_X(G)dG_{\sim i} - E_{X0} \tag{14}$$

$$E_{Xij}\left(g_i,g_j\right) = \int_0^1 \bullet\bullet\bullet \int_0^1 E_X(G)dG_{\sim(ij)} - E_{Xi}(g_i) - E_{Xj}\left(g_j\right) - E_{X0} \tag{15}$$

where $G_{\sim i}$ denotes all variables except g_i and $G_{\sim(ij)}$ denotes all variables except g_i and g_j.

Variance D is

$$D = Var\left[E_X(G)\right] = \int_{\Omega^n} E_x^2(G)dG - E_{x0}^2 \tag{16}$$

The bias-variance decomposition can be articulated as follows:

$$D_{i_1,i_2,\ldots,i_s} = \int_0^1 \bullet\bullet\bullet \int_0^1 E_{xi_1,i_2,\ldots,i_s}^2\left(g_{i_1},g_{i_2},\ldots,g_{i_s}\right)dg_{i_1}dg_{i_2}\ldots dg_{i_s} \tag{17}$$

Of these, $1 \le i_1 < \ldots < i_s \le n$, $1 \le i_1 < \ldots < i_s \le n$.

By integrating the square of Eq. (10) and using Eq. (11), the total variance can be decomposed as follow:

$$D = \sum_{i=1}^{n} D_i + \sum_{1 \le i \le j \le n} D_{ij} + \ldots + D_{1,2,\ldots,n} \tag{18}$$

Thus the sensitivity is:

$$S_{i_1,i_2,\ldots,i_s} = \frac{D_{i_1,i_2,\ldots,i_s}}{D}, 1 \le i_1 < \ldots < i_s \le n \tag{19}$$

By definition, they satisfy according to Eq. (18):

$$\sum_{i=1}^{n} S_i + \sum_{1 \le i \le j \le n} S_{ij} + \ldots + S_{1,2,\ldots,n} = 1 \tag{20}$$

where S_i is the first-order sensitivity of g_i, $S_{ij}(i \neq j)$ is the second-order sensitivity of g_i and g_j. And so on, the first-order exponent g_i gives the effect of each parameter taken individually, while the higher-order exponents illustrate the possible mixed effects of the various parameters.

The total sensitivity of g_i has been defined to evaluate the total effect of g_i.

$$TS(i) = 1 - \frac{D_{\sim i}}{D} \tag{21}$$

$$D_{\sim i} = D - D_i - \sum_{1 \le j \le n, j \ne i} D_{ij} - \ldots - D_{1,2,\ldots,n}, 1 \le I \le n \tag{22}$$

where E_{X0}, D, $D_i$and D_i can be estimated by Monte Carlo evaluation, we can get:

$$\widehat{E_{X0}} \approx \frac{1}{k} \sum_{m=1}^{k} E_X(G_m) \tag{23}$$

$$\hat{D} = \frac{1}{k} \sum_{m=1}^{k} E_X^2(G_m) - \hat{E}_{X0}{}^2 \tag{24}$$

$$\hat{D}_i = \frac{1}{k} \sum_{m=1}^{k} \mathrm{E}_X\left(G_{(\sim i)m}^{(1)}, g_{im}^{(1)}\right)\left(G_{(\sim i)m}^{(2)}, g_{im}^{(1)}\right) - \hat{E}_{X0}{}^2 \tag{25}$$

$$\hat{D}_{\sim i} = \frac{1}{k} \sum_{m=1}^{k} E_X\left(G_{(\sim i)m}^{(1)}, g_{im}^{(1)}\right)\left(G_{(\sim i)m}^{(1)}, g_{im}^{(2)}\right) - \hat{E}_{X0}{}^2 \tag{26}$$

Using Eqs. (19)–(25), the first-order sensitivity and the total sensitivity of g_i can be calculated as follow:

$$S_i \oplus \frac{\hat{D}_i}{\hat{D}} \tag{27}$$

$$TS(i) \approx 1 - \frac{\hat{D}_{\sim i}}{\hat{D}} \tag{28}$$

3.2 Global Sensitivity Analysis

In the context of machine tools, geometric errors can be categorized into two primary types. Position-related geometric errors are predominantly generated due to defects in the machine tool components, while the generation of position-independent geometric errors mainly occurs during the assembly stage. We consider the 37 geometric errors as a 37-dimensional unitoid that constitutes the spatial domain of the input factors. Among them,ΔS_{xy}, ΔS_{xy}, ΔS_{yz}, $\Delta\alpha_{xa}$, $\Delta\alpha_{xa}$, $\Delta\alpha_{xc}$and $\Delta\alpha_{xc}$ belong to the position-independent geometric error terms, which are regarded as constants, so only the sensitivity of the remaining 30 geometric errors to machining accuracy needs to be analyzed.

The first-order sensitivity analysis elucidates the individual contributions of each factor to machining accuracy, whereas the global sensitivity analysis captures the collective influence of factor interactions on machining accuracy. To facilitate a more direct interpretation of the data, the results pertaining to axial correlation have been depicted in Figs. 5, 6, 7, 8, 9 and 10.

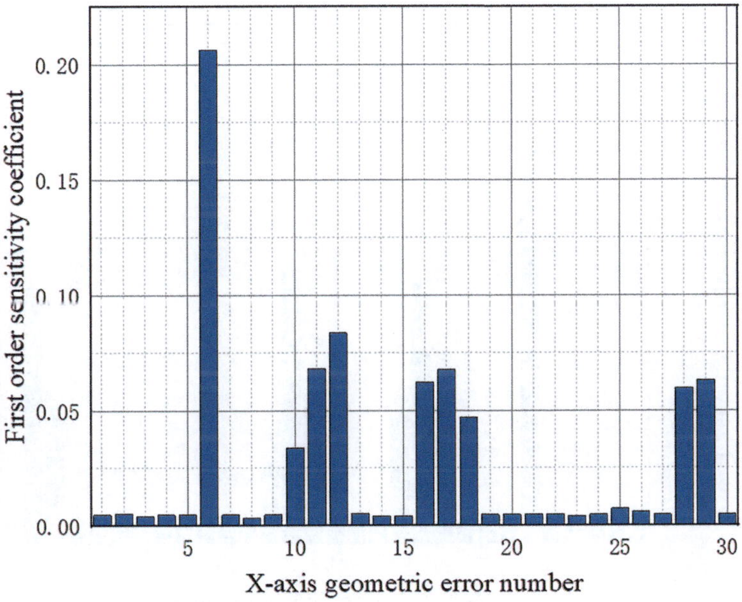

Fig. 5 X-axis first order sensitivity coefficient

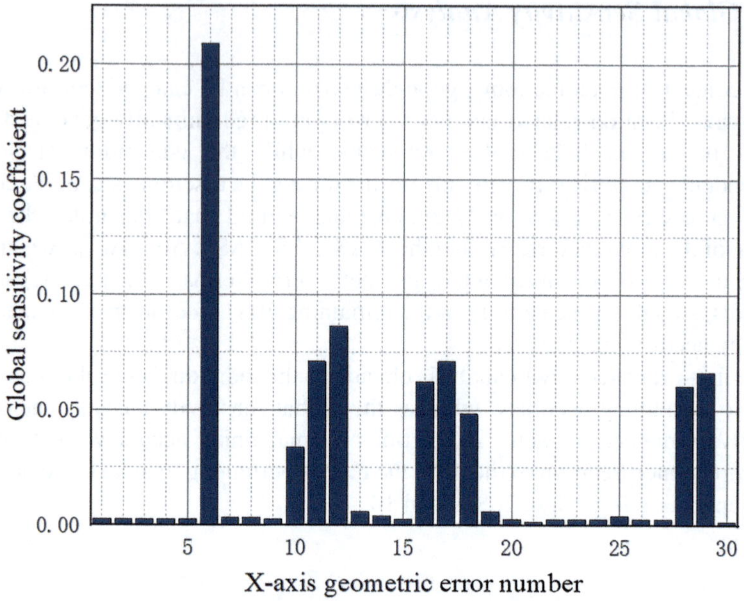

Fig. 6 X-axis global sensitivity coefficient

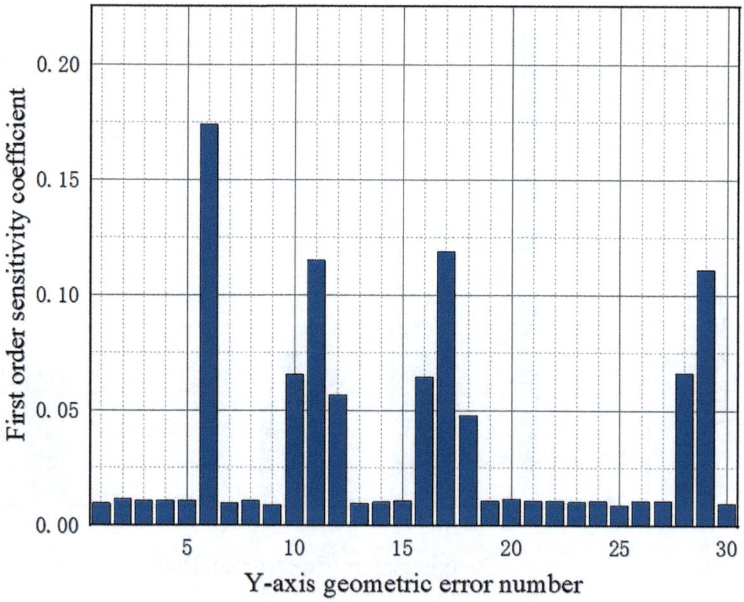

Fig. 7 Y-axis first order sensitivity coefficient

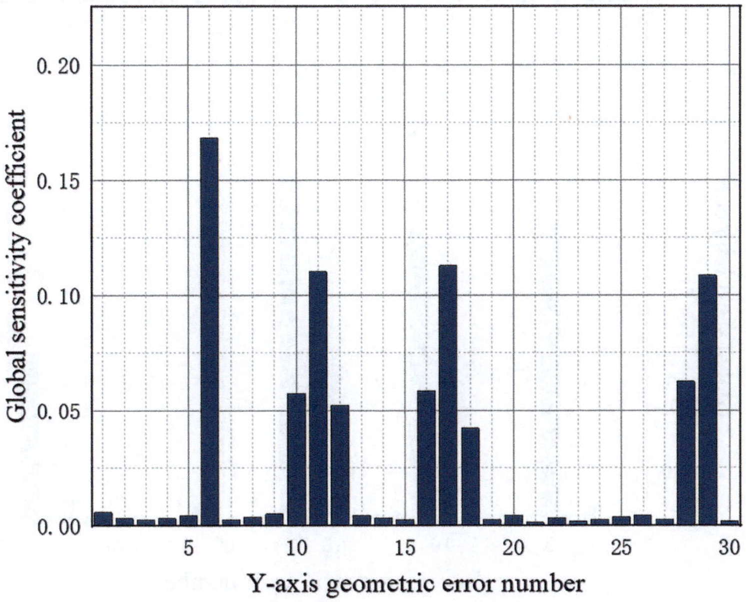

Fig. 8 Y-axis global sensitivity coefficient

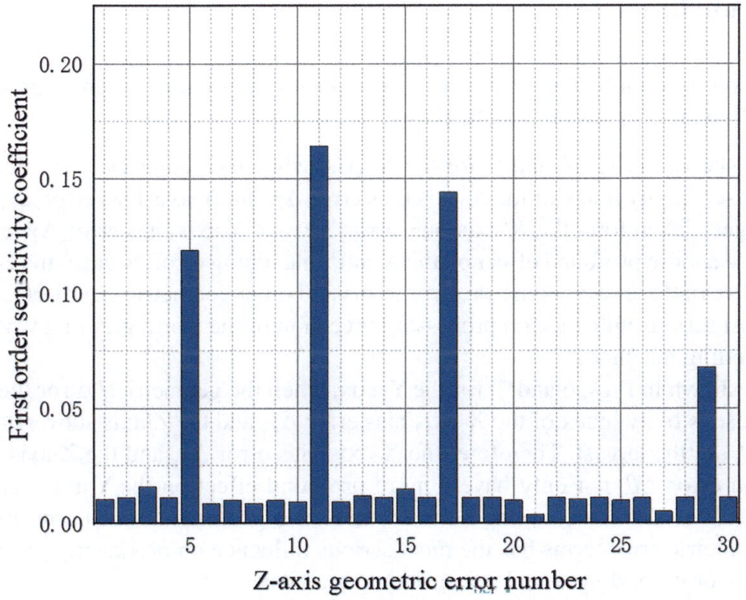

Fig. 9 Z-axis first order sensitivity coefficient

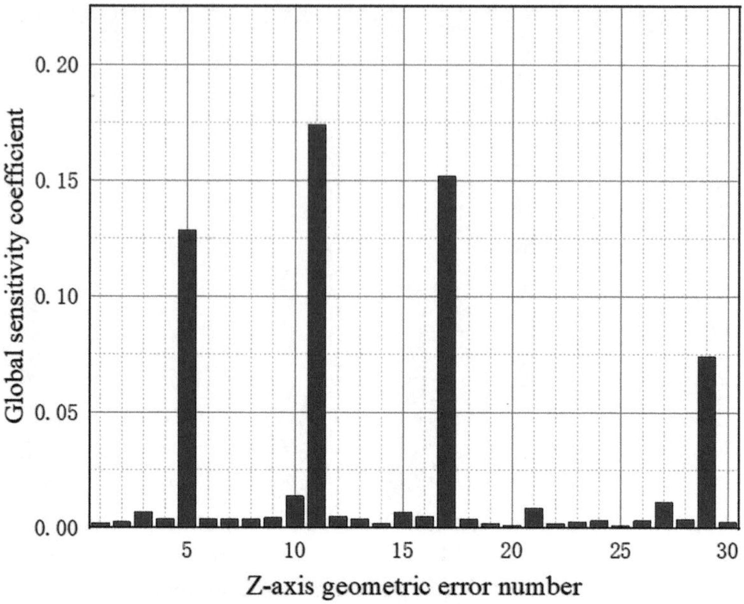

Fig. 10 Z-axis global sensitivity coefficient

4 Conclusion

Based on the established error transfer model for gear grinding machine processing, the following conclusions can be drawn:

1. As shown in Figs. 4 and 5, for the X-axis, when the geometric error fluctuates, the sensitivity index of the X-axis bias error $\Delta\gamma_x$ and Y-axis bias error $\Delta\gamma_y$ are the largest. Therefore, the X-axis bias error $\Delta\gamma_x$ and Y-axis bias error $\Delta\gamma_y$ not only have a more profound effect on the X-axis machining error, but also the coupling between these two geometric error terms and other geometric error terms has the most serious influence on processing precision of the worm grinding wheel gear grinding machine.
2. As shown in Figs. 6 and 7, for the Y-axis, when the geometrical error fluctuates, the sensitivity index of the X-axis bias error $\Delta\gamma_y$ and the Z-axis subversion error $\Delta\beta_z$ are the largest. Therefore, the X-axis bias error $\Delta\gamma_y$ and the Z-axis subversion error $\Delta\beta_z$ not only have a more profound effect on the Y-axis machining error, but also the coupling between these two geometric error terms and other geometric error terms has the most serious influence on processing precision of the worm grinding wheel gear grinding machine.

3. As shown in Figs. 8 and 9, for the Z-axis, when the geometrical error fluctuates, the sensitivity index of the Y-axis roll error $\Delta\beta_z$ and the Z-axis subversion error $\Delta\beta_z$ are the largest. Therefore, the Y-axis roll error $\Delta\alpha_y$ and Z-axis subversion error $\Delta\beta_z$ not only have a more profound effect on the Z-axis machining error, but also the coupling between these two geometric error terms and other geometric error terms has the most serious influence on processing precision of the worm grinding wheel gear grinding machine.

References

1. Kun C, Shun L, Yue J (2023) Jacobian spinor modeling of geometric errors in CNC machine tools. Mech Des Res 39(2):139–142. https://doi.org/10.13952/j.cnki.jofmdr.2023.0065
2. Tsai CY (2022) Error-free measurement method for 6-DOF systems via matrix decomposition approach. App Math Model 104:188–204. https://doi.org/10.1016/j.apm.2021.11.027
3. Wu S, Fan Z, Liu X (2022) Sensitivity analysis of geometric errors of two-turntable five-axis machine tool based on S-shaped specimens. Int J Adv Manuf Technol 2022(5/6):121. https://doi.org/10.1007/s00170-022-09493-6
4. Fan J, Tao H, Zhang Y (2019) Error modeling and sensitivity analysis of turning-milling compound CNC machine tools. Int Conf Green Power Mater Manuf Technol Appl 12(5):145–152. https://doi.org/10.1063/1.5137878
5. Li F, Wang S, Chen M (2023) Optimization of high-efficiency tooth surface accuracy of spiral bevel gears considering machine-tool motion errors. Proc Inst Mech Eng Pt E J Process Mech Eng 237(4):1394–1406. https://doi.org/10.1177/09544089221111d3639
6. Niu P, Cheng Q (2021) a novel method for machining accuracy reliability and failure sensitivity analysis for multi-axis machine tool. Int J Adv Manuf Technol 124(11–12):3823–3836. https://doi.org/10.1007/s00170-021-08003-4
7. Tao H, Fan J (2023) an approach to enhancing machining accuracy of five-axis machine tools based on a new sensitivity analysis method. Int J Adv Manuf Technol 124(7):2383–2400. https://doi.org/10.1007/s00170-022-10365-2
8. Kim K, Kim MK (1991) Volumetric accuracy analysis based generalized geometric error model in multi-axes machine tools. Mech Mach Theory 26(2):207–212. https://doi.org/10.1016/0094-114x(91)90084-h
9. Soons JA, Theuws FC, Schellekens PH (1992) Modeling the errors of multi-axis machines: a general methodology. Precis Eng 14(1):5–19. https://doi.org/10.1016/0141-6359(92)90137-l
10. Fu GQ, Fu JZ, Xu YT, Chen ZC (2014) Product of exponential model for geometric error integration of multi-axis machine tools. Int J Adv Manuf Technol 71(9–12):1653–1667. https://doi.org/10.1007/s00170-013-5586-5
11. Rahman M, Heikkala J, Lappalainen K (2000) Modeling, measurement and error compensation of multi-axis machine tools. Part I: theory. Int J Mach Tools Manuf 40(10):1535–1546. https://doi.org/10.1016/s0890-6955(99)00101-7

Open Access This chapter is licensed under the terms of the Creative Commons Attribution-NonCommercial-NoDerivatives 4.0 International License (http://creativecommons.org/licenses/by-nc-nd/4.0/), which permits any noncommercial use, sharing, distribution and reproduction in any medium or format, as long as you give appropriate credit to the original author(s) and the source, provide a link to the Creative Commons license and indicate if you modified the licensed material. You do not have permission under this license to share adapted material derived from this chapter or parts of it.

The images or other third party material in this chapter are included in the chapter's Creative Commons license, unless indicated otherwise in a credit line to the material. If material is not included in the chapter's Creative Commons license and your intended use is not permitted by statutory regulation or exceeds the permitted use, you will need to obtain permission directly from the copyright holder.

Fatigue Crack Growth of Straight Gear Roots Based on Fatigue Testing and Numerical Simulation

Dongsheng Yin, Tengjiao Lin, and Bingkui Chen

Abstract To probe the fatigue crack growth behavior of spur gear tooth roots, the Ansys-Franc3D co-simulation method, based on the rules of linear elastic fracture mechanics (LEFM), was utilized in study the growth mechanism of fatigue cracks at the spur gear tooth roots. The precision of the simulation calculation outcomes was verified by the findings of the gear bending fatigue test. The findings indicate that during the fatigue fracture propagation of tooth roots, the stress intensity factor (SIF) of type I significantly exceeds those of types II and III, with predominant cracking mode being opening type, complemented by sliding and tearing types. The stress intensity factor of Mode I decreases gradually from the tooth surface to the crack depth, promoting crack growth along the tooth surface. The crack at the end of the gear first extends along the depth of the crack, then along the circumference of the gear to the root.

Keywords Stress intensity factor · Fatigue crack propagation trajectory · Gear bending fatigue test

1 Introduction

The failure of the gear is primarily attributed to the deterioration of the gear teeth. The damage to the gear teeth is classified into two categories: surface damage and fractures. Gear tooth fractures primarily stem from the damage caused by bending fatigue at the base of tooth. Consequently, investigating how fatigue cracks develop in tooth roots holds great importance.

Pehan et al. [1] obtained that cracks may start in a broad area of the root of a tooth surface through the tooth-loading test. In addition, the study also shows that LEFM can predict the trajectory of fracture growth. Zouari et al. [2] utilized LEFM

D. Yin · T. Lin (✉) · B. Chen
State Key Laboratory of Mechanical Transmission, Chongqing University, Chongqing, China
e-mail: tjlin1968@162.com

© The Author(s) 2026

Z. Zhang, P. Mercorelli (eds.), *Mechanical Design and Simulation: Insights and Innovations*, Lecture Notes in Mechanical Engineering,
https://doi.org/10.1007/978-981-95-2405-1_63

and finite element method (FEM) to study the influence of fracture depth, fracture growth angle, and fracture initiation position on the SIF. In addition, they proposed a method to forecast the trajectory of fracture growth in the base of tooth. Pandya et al. [3] used Franc2D to forecast the trajectory of fracture growth in the base of tooth. On this basis, they discussed the influence of distinct gear structural characteristics on its time-varying meshing stiffness and proposed quantitative indicators. Verma et al. [4] utilized the extended finite element method (XFEM) to predict a propagation path of the tooth root crack, which overcomes the problem of the finite element method being too detailed when calculating the time-varying mesh stiffness of cracks. AMIRI et al. [5] researched a propagation path and residual life of a three-dimensional crack within the root of the helical gear tooth using the XFEM. Wang et al. [6] used the XFEM to study influence of gear meshing process on the fracture growth of the base of tooth and found that well gear teeth share the load with the faulty gear teeth, extending their lifespan. Referring to method of strain-life and LEFM, Ignatijev et al. [7] estimated fatigue life of the base of tooth made by Cu-Ni-Mo sintered steel using Ansys and found that the calculation results matched well with the experimental results. Grounded in the rules of LEFM, Duan et al. [8] obtained the impact of fracture growth trajectory on engaging characteristics under crack closure effect using the FEM. Grounded on the dislocation energy method and the Paris formula, Deng et al. [9] analyzed the starting and growth of fracture in carburized gears and established a full-life prediction model for cracks. Raghav et al. [10] utilized the XFEM to probe impacts of growth angle, crack length, dynamic load, and static load on the fracture growth trajectory and lifespan.

Grounded on LEFM and FEM, this work presents a manner to analyzing the fatigue fracture growth behavior of gear tooth root using Ansys-Franc3D co-simulation, validated through fatigue due to bending of the gear test. The suggested manner and conclusions have theoretical and practical significance for the study of tooth root fatigue fracture propagation in practical engineering.

2 Gear Bending Fatigue Test

2.1 Test Scheme

The gear fatigue due to bending test was performed utilizing the bending fatigue testing device, GPS-200, with single-tooth loading, according to the pulsating test method in GB/T 14230-2021 "Gear Bending Fatigue Strength Test Method." The fundamental tooth profile of the test gear complies with the GB/T 1356 regulations, and the primary geometric parameters have been provided in Table 1. The composition of the gear is 16Cr3NiWMoVNbE. The surfaces of the gear teeth, including the top surfaces and the root surfaces, are carburized. The carburized hardened layers on the tooth surfaces have a thickness of 0.8 to 1.2 mm, while the layers on the top and root surfaces of the tooth have a thickness of 0.7 to 1.5 mm. The carburized

Table 1 Basic geometric parameters of the gear

Parameter	Value	Parameter	Value
Tooth number z	20	Tooth width b/mm	24
Modulus m/mm	5	Offset coefficient x	0.2674
Pressure angle $\alpha/(°)$	25	Bore diameter D/mm	46
Addendum coefficient h_a^*	1	Number of teeth spanned	4
Bottom clearance coefficient c^*	0.175	Gear grade	5

surfaces have a hardness of 60 HRC or higher, while the core has a hardness ranging from 35 to 46 HRC.

This test takes the form of single-tooth loading. The load must be positioned on surface of the effective involute at top of the tooth, with its action line tangent to the base circle of test gear. Each test gear has a mating gear, and the radius of circle where the loading point E is located can be calculated using the Eq. (1):

$$r_e = \sqrt{\left[\sqrt{(r_a)^2 - (r_b)^2} - \frac{2\pi r \cos\alpha}{z}(\varepsilon - 1)\right]^2 + (r_b)^2} \tag{1}$$

where r_a is the radius of top circle, r_b is the radius of base circle, r is the radius of pitch circle, ε is the overlap ratio.

The test gear is arranged symmetrically across four teeth. The installation dimensions of the test gear are determined based on the geometry of the pressure head and the test gear, as shown in Fig. 1.

The GPS-200 fatigue testing machine is shown in Fig. 2. This machine can perform bending fatigue tests on gears with a modulus ranging from 2.5 to 10 mm and an addendum circle diameter ranging from 50 to 500 mm. Table 2 displays the loading data for the test. After installing the test gear, the pressure head is forced down to bring the load closer to the set average load. After ensuring that the gear is properly mounted, the test machine is set to alternate between alternating load and continuous load until the test gear tooth breaks. If it is not damaged after three million cycles, it is assumed that no cracks will occur at this stress level.

2.2 *Fatigue Test Results for Bending Gear*

Using the fracture of the base of tooth under a 70 kN load as an example, the fracture condition of the base of tooth displays in Fig. 3. Analyzing of fatigue due to bending test outcomes of the base of tooth reveals that fatigue fracture initiates in a broad region of the base surface of tooth and subsequently propagates toward its end face along the direction of tooth's width. Upon reaching its end face, the root fracture initially propagates toward the gear rim, and then changes direction to extend toward the opposite root surface of the gear tooth. It generally indicates a trend of propagation toward the fracture in the circumferential direction of the base of tooth.

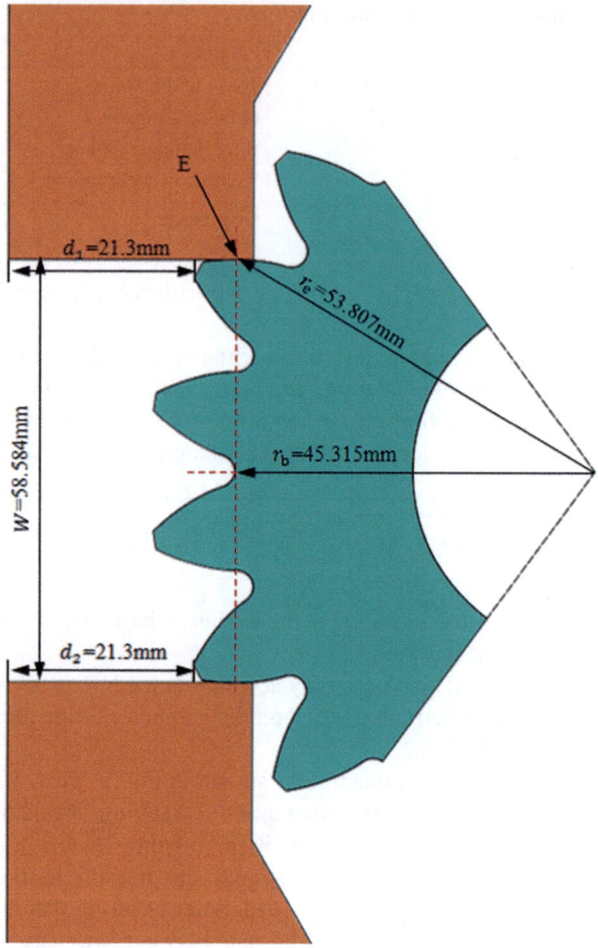

Fig. 1 Installation dimension of test gear

The fracture surface of the tooth root is divided into three sections: crack origin, fatigue fracture growth region, and instantaneous fracture region. The region of fatigue fracture growth is fairly smooth, indicating that crack is propagating in a stable manner. The upper and lower surfaces of the crack continuously open, close, and rub against each other during the propagation process, resulting in a smoother fracture growth region. The section of instantaneous fracture region is rough, indicating that the fracture changes from stable growth to rapid growth and reaches the critical condition in a very short time. Instantaneous fractures occur, forming a new section, that makes the fracture region rough.

Fig. 2 GPS-200 fatigue testing apparatus

Table 2 Loading of gear bending fatigue test

Load ratio *r*	Loading frequency/Hz	Load force/ kN	Bending stress/ MPa	Mean load/ kN	Load amplitude/ kN
0.1	106	60	625.187	−33	27
		65	678.414	−35.75	29.25
		70	731.817	−38.5	31.5
		75	785.4	−41.25	33.75
		80	839.161	−44	36

Fig. 3 Broken tooth state

3 Simulation of Fracture Growth Due to Tooth Root Fatigue

3.1 Fatigue Fracture Growth Model

LEFM classifies cracks into three primary forms grounded on varying loads applied to the crack, as depicted in Fig. 4: opening mode (type I), sliding mode (type II), and tearing mode (type III). Within engineering framework, the predominant crack growth method is composite, characterized by three primary categories. The SIF defines the intensity of stress field at the crack front. As SIF value increases, the intensity of stress field at the crack front also increases, facilitating the fracture's growth. K_{I}, K_{II}, and K_{III} represent the SIFs for type I, type II, and type III cracks, in order.

As depicted in Fig. 5, The progression of fatigue fracture is segmented into three distinct stages: the stage of initial growth (region I), the stage of stable growth (region II), and the stage of rapid growth (region III). The growth rate of fatigue fracture $\mathrm{d}a/\mathrm{d}N$ depends on the amplitude of the SIF ΔK. Within region I, a specific threshold value ΔK_{th} exists. When ΔK is less than ΔK_{th}, it's assumed that the crack does not continue to grow. Within region II, a logarithmic linear correlation exists

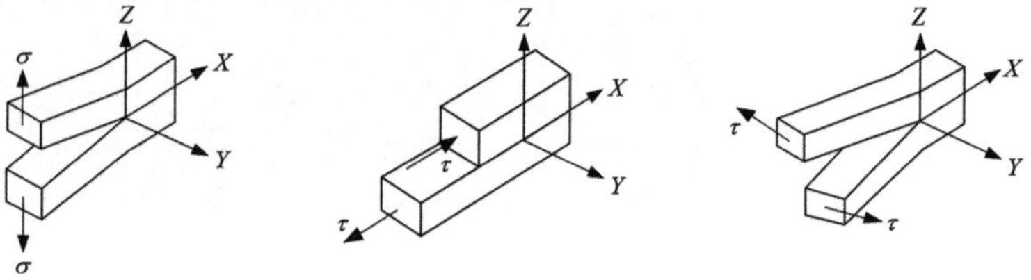

Fig. 4 Fundamental structure of a crack

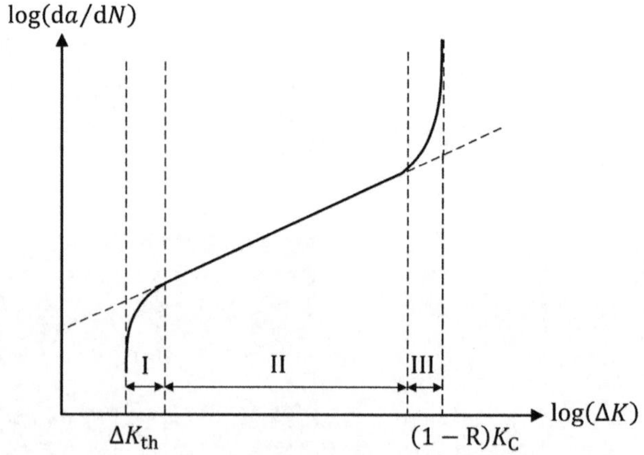

Fig. 5 Diagrammatic representation of the fatigue crack growth rate

between the growth rate of fatigue fracture da/dN and the amplitude of the SIF ΔK, adhering to the Paris formula, indicating the stage of consistent fatigue fracture development. Within region III, the growth rate of fatigue fracture da/dN is rapid, and its lifespan is short. Upon ΔK attaining the state of $(1 - R)K_C$, immediate fracture occurs, with K_C representing the material's fracture resilience and R the ratio of load stress.

3.2 *Results of the Gear Bending Fatigue Test*

Referring to the gear bending fatigue test mentioned above, Ansys has been used to establish a finite element model in three dimensions to simulate the fatigue fracture growth of tooth root. Based on the assembly size of the test gear, the spatial relationship between the gear and the upper and lower pressure heads is determined.

The material of the test gear is 16Cr3NiWMoVNbE, with an elastic modulus $E = 2.06 \times 10^5$ MPa and Poisson's ratio $v = 0.377$. The three-dimensional solid element, SOLID185, is employed to mesh the geometric model into structured grids. The contact area between the gear tooth and the upper and lower pressure heads is refined, with its mesh size controlled to 1 mm, while the mesh size of other areas gradually becomes sparser.

According to the test conditions, the boundary conditions are set to constrain the radial and axial displacements of the inner hole surface of the gear while releasing its circumferential displacement. The displacement in all directions of the upper and lower pressure heads is constrained, and the corresponding load is applied to the upper pressure head based on the force relationship between the gear and the upper and lower pressure heads. The displacement in this direction is released simultaneously. These are depicted in Fig. 6.

To calculate the bending stress at the base of loaded tooth, frictional contacts have been set between the tooth and the upper and lower pressure head for contact analysis.

3.3 *Simulation of Fatigue Fracture Growth in Tooth Root*

A starting fracture has been preset on base surface of tooth, situated in the middle of tooth's width direction, with a circular crack radius of $a = 0.3mm$ and an inclination angle of $\beta = 45°$. Franc3D has been employed to construct the finite element model for the fractured gear. The procedure is depicted in Fig. 7.

Through the Ansys-Franc3D co-simulation, the SIF of initial crack front has been calculated using the M-integral manner and displayed in Fig. 8. The value of Mode I SIF of the crack front is much greater than the value of Modes II and III, indicating that fracture mechanism of the base crack of tooth represents primarily of

Fig. 6 The finite element model of gear bending fatigue test

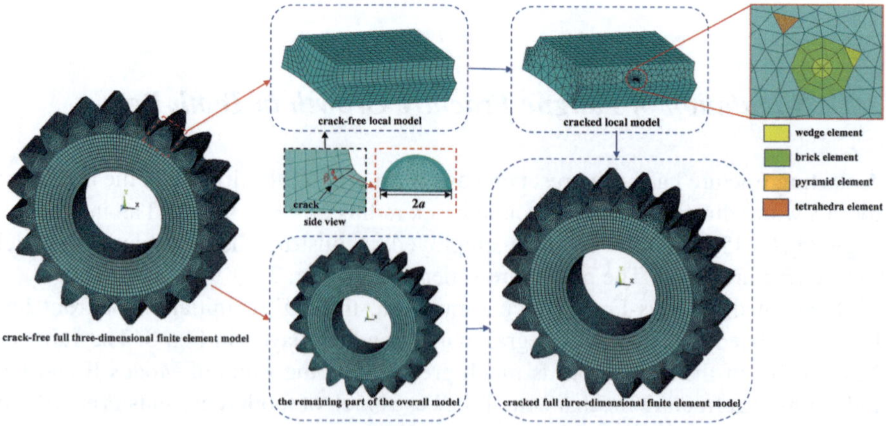

Fig. 7 Modeling procedure of a gear with crack using finite element method

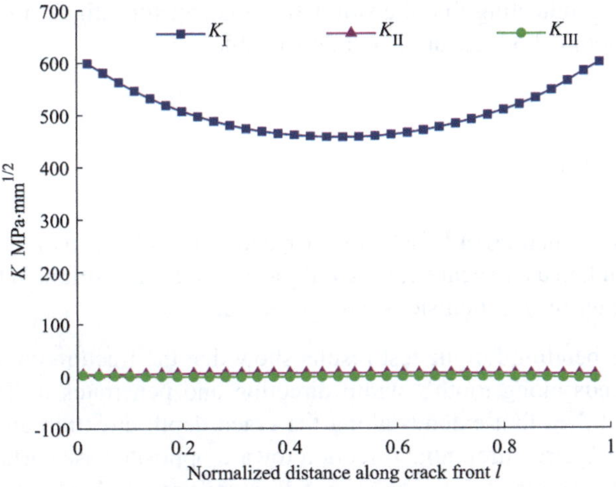

Fig. 8 Stress intensity factor distribution curve

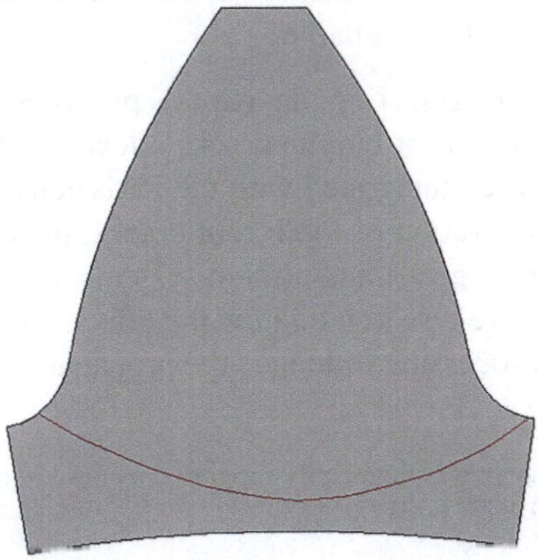

Fig. 9 Crack propagation trajectory

opening type, supplemented by sliding and tearing types. The distribution curve of the Mode I SIF of the crack front is symmetric, with high values at both ends and low values in the middle. From the base surface to the crack depth direction, the Mode I SIF value gradually decreases, which indicates that the base crack is more likely to propagate along the base surface rather than along the crack depth direction.

The quasi-static crack propagation method is employed, and the maximum circumferential tensile stress criterion has been applied to decide on propagation angle of the crack front node. The effective SIF ΔK_{eff} and the propagation increment Δa_m at the midpoint of the crack front have been established. The crack propagation trajectory is shown in Fig. 9. Comparing and analyzing the simulated crack with the experimental crack reveals a consistent trend in the fracture growth trajectory of tooth end face, indicating that the simulation manner for fatigue fracture growth at the base of tooth is both accurate and dependable.

4 Conclusion

Grounded on the method of LEFM, the propagation mechanism of tooth root fatigue cracks was studied using gear bending fatigue test and Ansys-Franc3D co-simulation analysis. The ensuing conclusions have been made:

1. The gear's bending fatigue test results show that the fracture on base surface of tooth extends along tooth's width direction and penetrates it. The fracture on tooth's end face first extends along the crack depth direction and then extends along tooth's circumferential direction toward opposite base surface of tooth.
2. The simulation results of root fatigue fracture growth display that the fracture mode of tooth root crack is mainly open type, supplemented by sliding type and tearing type. The simulation outcomes of root fracture growth trajectory maintain consistency with the experimental outcomes, and root fatigue fracture growth simulation manner is reliable.

Within the field of engineering, the paper's proposed simulation approach for the propagation of tooth root fatigue cracks allows predicting their propagation path. Furthermore, when integrated with the Paris equation, it is possible to predict the propagation duration of tooth root cracks, providing a theoretical foundation for the design of gear transmission systems to reduce fatigue. This paper paves the way for future research on how the effects of crack closure, heat treatment methods, and processing influence the propagation of cracks in tooth roots.

Acknowledgments This work was financially supported by the National Science and Technology Major Project (No: HT-J2019-IV-0001-0068).

References

1. Pehan S, Kramberger J, Flašker J et al (2008) Investigation of crack propagation scatter in a gear tooth's root. Eng Fract Mech 75(5):1266–1283. https://doi.org/10.1016/j.engfracmech.2007.04.005
2. Zouari S, Maatar M, Fakhfakh T (2010) Following spur gear crack propagation in the tooth foot by finite element method. J Fail Anal Prev 10(6):531–539. https://doi.org/10.1007/s11668-010-9395-y
3. Pandya Y, Parey A (2013) Simulation of crack propagation in spur gear tooth for different gear parameter and its influence on mesh stiffness. Eng Fail Anal 30:124–137. https://doi.org/10.1016/j.engfailanal.2013.01.011

4. Verma JG, Kumar S, Kankar PK (2018) Crack growth modeling in spur gear tooth and its effect on mesh stiffness using extended finite element method. Eng Fail Anal 44(1):109–120. https://doi.org/10.1016/j.engfailanal.2018.07.032
5. Amiri RA, Forouzan MR, Sadeghi DA (2014) Three-dimensional fatigue crack growth modelling in a helical gear using extended finite element method. Fatigue Fract Eng Mater Struct 37(6):581–591. https://doi.org/10.1111/ffe.12140
6. Wang X, Yang Y, Wang W et al (2020) Simulating coupling behavior of spur gear meshing and fatigue crack propagation in tooth root. Int J Fatigue 134:105381. https://doi.org/10.1016/j.ijfatigue.2019.105381
7. Ignatijev A, Nečemer B, Kramberger J et al (2022) Fatigue crack initiation and propagation in a PM-gear tooth root. Eng Fail Anal 138:106355. https://doi.org/10.1016/j.engfailanal.2022.106355
8. Duan TT, Wei J, Yan Q et al (2022) Investigations on crack propagation and meshing characteristics of planetary gear train considering crack closure effect. Eng Fail Anal 134:106064. https://doi.org/10.1016/j.engfailanal.2022.106064
9. Deng HL, Guo Y, Liu H et al (2022) Bending fatigue life prediction model of carburize-d gear based on microcosmic fatigue failure mechanism. J Mater Eng Perform 31:882–894. https://doi.org/10.1007/s11665-021-06236-8
10. Raghav MS, Singh A, Patel S (2022) Fault analysis of spur gear using XFEM. Eng Fail Anal 134:106060. https://doi.org/10.1016/j.engfailanal.2022.106060

Open Access This chapter is licensed under the terms of the Creative Commons Attribution-NonCommercial-NoDerivatives 4.0 International License (http://creativecommons.org/licenses/by-nc-nd/4.0/), which permits any noncommercial use, sharing, distribution and reproduction in any medium or format, as long as you give appropriate credit to the original author(s) and the source, provide a link to the Creative Commons license and indicate if you modified the licensed material. You do not have permission under this license to share adapted material derived from this chapter or parts of it.

The images or other third party material in this chapter are included in the chapter's Creative Commons license, unless indicated otherwise in a credit line to the material. If material is not included in the chapter's Creative Commons license and your intended use is not permitted by statutory regulation or exceeds the permitted use, you will need to obtain permission directly from the copyright holder.

GPSR Compliance

The European Union's (EU) General Product Safety Regulation (GPSR) is a set of rules that requires consumer products to be safe and our obligations to ensure this.

If you have any concerns about our products, you can contact us on ProductSafety@springernature.com

In case Publisher is established outside the EU, the EU authorized representative is:

Springer Nature Customer Service Center GmbH
Europaplatz 3
69115 Heidelberg, Germany

Batch number: 10739616

Printed by Printforce, the Netherlands

Lecture Notes in Mechanical Engineering

Series Editors

Fakher Chaari, National School of Engineers, University of Sfax, Sfax, Tunisia

Francesco Gherardini, Dipartimento di Ingegneria "Enzo Ferrari", Università di Modena e Reggio Emilia, Modena, Italy

Vitalii Ivanov, Department of Manufacturing Engineering, Machines and Tools, Sumy State University, Sumy, Ukraine

Mohamed Haddar, National School of Engineers of Sfax (ENIS), Sfax, Tunisia

Editorial Board

Francisco Cavas-Martínez, Departamento de Estructuras, Construcción y Expresión Gráfica Universidad Politécnica de Cartagena, Cartagena, Spain

Francesca di Mare, Institute of Energy Technology, Ruhr-Universität Bochum, Bochum, Germany

Young W. Kwon, Department of Manufacturing Engineering and Aerospace Engineering, Graduate School of Engineering and Applied Science, Monterey, USA

Tullio A. M. Tolio, Department of Mechanical Engineering, Politecnico di Milano, Milano, Italy

Justyna Trojanowska, Poznan University of Technology, Poznan, Poland

Robert Schmitt, RWTH Aachen University, Aachen, Germany

Jinyang Xu, School of Mechanical Engineering, Shanghai Jiao Tong University, Shanghai, China

Lecture Notes in Mechanical Engineering (LNME) publishes the latest developments in Mechanical Engineering—quickly, informally and with high quality. Original research or contributions reported in proceedings and post-proceedings represents the core of LNME. Volumes published in LNME embrace all aspects, subfields and new challenges of mechanical engineering.

To submit a proposal or request further information, please contact the Springer Editor of your location:

Europe, USA, Africa: Leontina Di Cecco at Leontina.dicecco@springer.com

China: Ella Zhang at ella.zhang@cn.springernature.com

India, Rest of Asia, Australia, New Zealand: Swati Meherishi at swati.meherishi@springer.com

Topics in the series include:

- Engineering Design
- Machinery and Machine Elements
- Mechanical Structures and Stress Analysis
- Automotive Engineering
- Engine Technology
- Aerospace Technology and Astronautics
- Nanotechnology and Microengineering
- Control, Robotics, Mechatronics
- MEMS
- Theoretical and Applied Mechanics
- Dynamical Systems, Control
- Fluid Mechanics
- Engineering Thermodynamics, Heat and Mass Transfer
- Manufacturing Engineering and Smart Manufacturing
- Precision Engineering, Instrumentation, Measurement
- Materials Engineering
- Tribology and Surface Technology

Indexed by SCOPUS, EI Compendex, and INSPEC.

All books published in the series are evaluated by Web of Science for the Conference Proceedings Citation Index (CPCI).

To submit a proposal for a monograph, please check our Springer Tracts in Mechanical Engineering at https://link.springer.com/bookseries/11693.

Zhenyu Zhang • Paolo Mercorelli
Editors

Mechanical Design and Simulation: Insights and Innovations

Proceedings of the 5th International Conference on Mechanical Design and Simulation

Volume II

Springer

Editors
Zhenyu Zhang
School of Mechanical Engineering
Dalian University of Technology
Dalian, Liaoning, China

Paolo Mercorelli
Institute for Production Technology
and Systems
Leuphana University of Lueneburg
Lüneburg, Niedersachsen, Germany

ISSN 2195-4356 ISSN 2195-4364 (electronic)
Lecture Notes in Mechanical Engineering
ISBN 978-981-95-2404-4 ISBN 978-981-95-2405-1 (eBook)
https://doi.org/10.1007/978-981-95-2405-1

© The Editor(s) (if applicable) and The Author(s) 2026

This book is an open access publication.

Open Access This book is licensed under the terms of the Creative Commons Attribution-NonCommercial-NoDerivatives 4.0 International License (http://creativecommons.org/licenses/by-nc-nd/4.0/), which permits any noncommercial use, sharing, distribution and reproduction in any medium or format, as long as you give appropriate credit to the original author(s) and the source, provide a link to the Creative Commons license and indicate if you modified the licensed material. You do not have permission under this license to share adapted material derived from this book or parts of it.
The images or other third party material in this book are included in the book's Creative Commons license, unless indicated otherwise in a credit line to the material. If material is not included in the book's Creative Commons license and your intended use is not permitted by statutory regulation or exceeds the permitted use, you will need to obtain permission directly from the copyright holder.
This work is subject to copyright. All commercial rights are reserved by the author(s), whether the whole or part of the material is concerned, specifically the rights of translation, reprinting, reuse of illustrations, recitation, broadcasting, reproduction on microfilms or in any other physical way, and transmission or information storage and retrieval, electronic adaptation, computer software, or by similar or dissimilar methodology now known or hereafter developed. Regarding these commercial rights a non-exclusive license has been granted to the publisher.
The use of general descriptive names, registered names, trademarks, service marks, etc. in this publication does not imply, even in the absence of a specific statement, that such names are exempt from the relevant protective laws and regulations and therefore free for general use.
The publisher, the authors and the editors are safe to assume that the advice and information in this book are believed to be true and accurate at the date of publication. Neither the publisher nor the authors or the editors give a warranty, expressed or implied, with respect to the material contained herein or for any errors or omissions that may have been made. The publisher remains neutral with regard to jurisdictional claims in published maps and institutional affiliations.

This Springer imprint is published by the registered company Springer Nature Singapore Pte Ltd.
The registered company address is: 152 Beach Road, #21-01/04 Gateway East, Singapore 189721, Singapore

If disposing of this product, please recycle the paper.

Preface

It is a great pleasure to present the preface for the 2025 5th International Conference on Mechanical Design and Simulation (MDS 2025) in Dalian, China, from February 28 to March 2, 2025. This prestigious event convened leading experts, scholars, and industry professionals from around the globe to explore cutting-edge advancements in mechanical design, mechanical engineering, and modeling and simulation technologies.

MDS 2025 aimed to foster interdisciplinary collaboration and intellectual exchange, providing a dynamic platform for sharing innovative research, expanding professional networks, and addressing critical challenges in the field. By bridging academia and industry, the conference sought to accelerate the translation of theoretical breakthroughs into practical applications, ultimately driving progress in both educational and industrial domains.

The conference featured keynote presentations by distinguished scholars, including Prof. Zhenyu Zhang (Dalian University of Technology), renowned for his pioneering work in ultra-precision machining and additive manufacturing; Prof. Wenbin Hou (Dalian University of Technology), a leading authority in automotive digitalization and lightweight structural optimization; Prof. Caixu Yue (Harbin University of Science and Technology), an expert in machine vision and intelligent manufacturing; and Prof. Yajun Liu (South China University of Technology), whose contributions to mechanical signal analysis and smart manufacturing systems have set new industry benchmarks.

The technical program encompassed a rich agenda of thematic sessions, oral presentations, and interactive discussions, covering diverse topics such as green chemical-mechanical polishing, intelligent tool condition monitoring, fiber-metal hybrid materials, and advanced simulations for industrial systems. Highlights include research on high-speed food processing equipment, deep-sea mining resistance analysis, and innovations in aerospace structural design. Attendees also benefited from networking opportunities during tea breaks, group photos, and dedicated discussion periods.

The excellent papers were selected from a large quantity of submissions for publication after rigorous peer review processes. The accepted papers cover topics including mechanical design and engineering, modeling and simulation technologies, advanced materials and applications, intelligent manufacturing and industrial applications, aerospace and heavy machinery, sustainable and green technologies, cross-disciplinary innovations, etc.

As we gather in Dalian—a hub of technological innovation—we extend our heartfelt gratitude to all participants. May MDS 2025 have inspired fruitful collaborations, sparked transformative ideas, and paved the way for future advancements in mechanical design and simulation.

MDS 2025 Conference Committee

Committee Member

Conference General Chair

Prof. Paolo Mercorelli, Leuphana University of Lueneburg, Lueneburg, Germany
Prof. Zhenyu Zhang, Dalian University of Technology, China

Technical Program Committee Chairs

Prof. Wenbin Hou, Dalian University of Technology, China
Prof. Xiaoguang Guo, Dalian University of Technology, China
Prof. Xiaohong Lu, Dalian University of Technology, China
Prof. Jianwei Ma, Dalian University of Technology, China

Publication Chairs

Prof. Caixu Yue, Harbin University of Science and Technology, China
Prof. Ir. Dr. YUSRI BIN YUSOF, Universiti Tun Hussien Onn Malaysia, Malaysia

Contents

Performance Optimization of Omnidirectional Cleaning System Nozzle Based on Response Surface and MOGA

Jinhong Ba and Yao Wang

Abstract To address the current issues of manual cargo space cleaning costs and low cleaning efficiency, the research team developed an omnidirectional cleaning system. To further improve the spraying performance of the cleaning system, the key hydraulic component, the nozzle, was selected as the research subject for multi-objective optimization of its structure. Using the nozzle's inlet radius (R1), outlet radius (R2), and length (L1), as variables, with a constraint of outlet average velocity ≥ 24.5 m/s, and aiming to minimize pressure difference and turbulent kinetic energy, the central composite design (CCD) method was employed for experimental design, and a response surface model was established in conjunction with numerical simulation and it was found that the outlet radius (R2) has the greatest impact on the nozzle's performance, followed by the inlet radius (R1), with the length (L1) having the least impact. The Multi-Objective Genetic Algorithm (MOGA) was used to optimize the nozzle structure and obtain the Pareto optimality. The optimization results indicated that the optimized nozzle structure resulted in a 14.75% reduction in pressure difference, a 2.08% increase in outlet average velocity, and a length reduction of nearly 75 mm. This not only optimized the spraying performance but also reduced material consumption and costs.

Keywords Omnidirectional cleaning system · Nozzle · Fluent · Response surface · Experimental design · MOGA

1 Introduction

Traditional bulk carrier cargo space cleaning relies on manual labor, which is inefficient, costly, and poses certain risks due to high-altitude operations. To address these issues, the research team developed an omnidirectional cleaning system [1], as

J. Ba · Y. Wang (✉)
Shanghai Maritime University, Pudong, Shanghai, China
e-mail: wangyao@shmtu.edu.cn

© The Author(s) 2026

Z. Zhang, P. Mercorelli (eds.), *Mechanical Design and Simulation: Insights and Innovations*, Lecture Notes in Mechanical Engineering,
https://doi.org/10.1007/978-981-95-2405-1_64

shown in Fig. 1. It mainly consists of a cleaning mechanism, a movable support mechanism, an operation unit, and a magnetic attraction mechanism. It cleans the bulkheads in all directions by jetting water at the bottom of the cabin. The operational device can control the horizontal rotation of the cleaning mechanism, and the nozzle of the cleaning mechanism can pitch up and down.

The cleaning mechanism is the core component of the omnidirectional cleaning system, and the nozzle is the key hydraulic part of the cleaning mechanism, directly affecting spraying performance. Its hydraulic dimensions include inlet radius (R1), outlet radius (R2), and length (L1). Indicators representing the spraying performance of the cleaning mechanism mainly include jet height, range, pressure difference, and fluid flow state. Maintaining a consistent jet angle, increasing the outlet velocity can enhance jet height and range, while reducing the pressure difference can lower energy loss. Reducing outlet turbulent kinetic energy helps stabilize flow, but velocity increases often accompany increased turbulent kinetic energy. Thus, optimizing nozzle performance hinges on how to enhance flow velocity while minimizing energy loss and maintaining flow stability. Currently, many scholars have conducted in-depth studies on nozzle performance in different application environments.

In the study of numerical simulation of nozzles, Liu's team and Gu's team [2, 3] conducted numerical simulations on nozzles with different contraction angles, outlet diameters, and lengths, derived optimal nozzle parameters for coal cutting. Huang's team [4] conducted a simulation study on the hydrodynamic performance of rust removal nozzles and found that the stand-off distance (SOD) and jet angle had the greatest impact on wall shear stress. Liu's team [5] used Fluent to simulate and verify the flow field structure and abrasive acceleration of nozzles, revealing

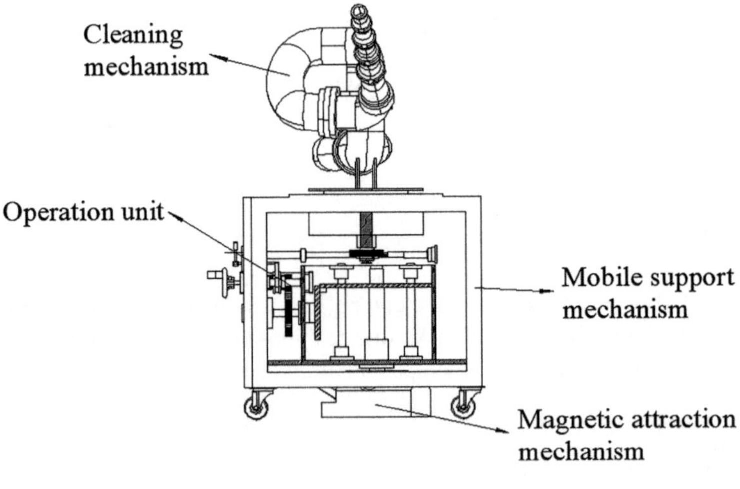

Fig. 1 Non-pressurized mobile omnidirectional cleaning system

that the pressure ratio n is the primary factor influencing nozzle structure design. In terms of nozzle performance optimization research, Juraeva's team [6] used DOE methods and Minitab software to generate a database to optimize nozzle parameters, and results show that the optimized nozzle has better performance than the original nozzle. Qiu's team [7] used the response surface method to optimize the nozzle structure and the optimal parameters were a cone angle of 12.21°, a diameter of 1.16 mm, and a length of 2 mm. Huang's team [8] conducted global optimization of key parameters of nozzles through numerical simulation combined with the response surface method, and the optimized composite nozzle showed a significant improvement in performance. Liu's team [9] used the OLHS method and CMOPSO algorithm to optimize the structural parameters of nozzles, and 10% increase in cleaning effectiveness with optimized nozzles.

This paper focuses on the nozzle of the omnidirectional cleaning system developed by the research team. A response surface model was constructed by combining numerical simulation and experimental design. The MOGA was used to optimize the nozzle's performance, taking several key dimensions of the nozzle as variables, aiming to increase the outlet average velocity of the omnidirectional cleaning system while minimizing pressure difference and outlet average turbulent kinetic energy.

2 Nozzle Modeling and Numerical Simulation

Figure 2a shows the initial nozzle model structure, where R1 (32.5 mm) is the inlet radius, R2 (15 mm) is the outlet radius, and L1 (475 mm) is the nozzle length. The subsequent analysis primarily focuses on the impact of these parameters on the performance of the omnidirectional cleaning system nozzle. Numerical simulation was conducted using Fluent software with a mesh precision of 1 mm, utilizing the default cell mesh type. The turbulence model selected is the standard k-ε turbulence model. The wall boundary condition is set as a no-slip wall, the inlet is specified as a velocity boundary with a value of 5 m/s, and the outlet is set as a pressure boundary with a value of 0 MPa. The pressure-velocity coupling solving algorithm uses the Coupled algorithm [10], with all other values set to default.

The numerical simulation results are shown in Fig. 2b. From Fig. 2b, it can be seen that the pressure at the nozzle inlet is the highest, at 0.278 MPa, and the internal pressure decreases as the nozzle contracts. The turbulent kinetic energy at the outlet is the highest, at 3.08 m^2/s^2, and the internal turbulent kinetic energy increases as the nozzle contracts. The velocity at the outlet is the highest, at 24.1 m/s, and the velocity increases as the nozzle contracts. Then, we will optimize based on the results of this calculation.

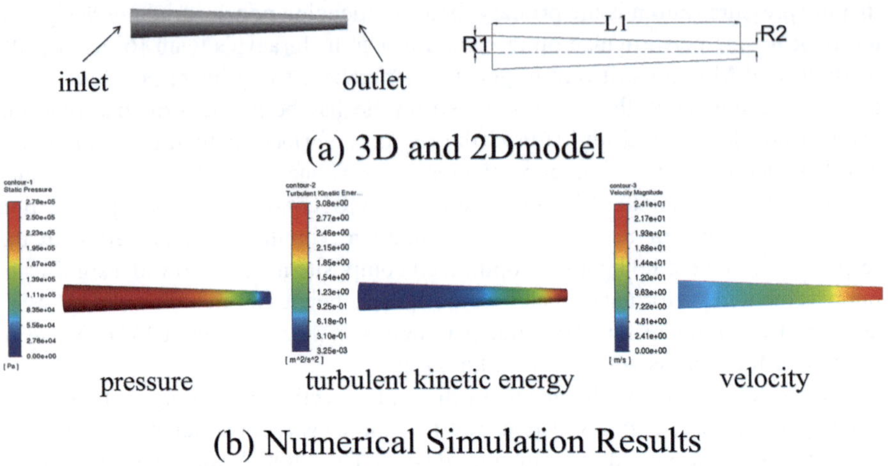

Fig. 2 Nozzle structure and numerical simulation results

3 Experimental Design

The range of nozzle design variable values is shown in Table 1. The maximum value of R1 is 32.5 mm because it cannot exceed the maximum radius of the pipe of the cleaning mechanism. The experimental design method chosen for this study is the Central Composite Design (CCD). Although it requires a large number of design points for evaluation, it covers a wide design space and provides rich high-order information. CCD requires $2^n + 2n + 1$ design points for n factors, so the number of design points for this study is 15. The simulation calculations were carried out using the pressure difference between inlet and outlet (Δp), outlet average velocity($\overline{v}$), and outlet average turbulent kinetic energy ($\overline{k}$) as output parameters. The DOE matrix and response values of some sample points are detailed in Table 2. Sample points are the foundation for sensitivity analysis, response surface, and optimization calculations.

Table 1 Range of design variable values (mm)

Design parameters	Minimum	Initial	Maximum
R1	22.5	32.5	32.5
R2	10	15	20
L1	400	475	550

Table 2 DOE matrix and response values

Number	R1	R2	L1	Δp (pa)	$\overline{v}$ (m/s)	$\overline{k}$ (m²/s²)
1	27.5	15	475	9062	16.795	0.219
2	22.5	15	475	6529.4	11.245	1.138
3	32.5	15	475	12,011	23.453	0.329
…	…	…	…	…	…	…
13	31.565	10.935	535.987	32,515	41.612	1.142
14	23.435	19.065	535.987	4147.1	7.552	0.08
15	31.565	19.065	535.987	6626.4	13.7	0.143

4 Response Surface Model

Local sensitivity is assessed by altering the values of the nozzle's structural parameters to observe changes in the pressure difference between inlet and outlet, outlet average velocity, and outlet average turbulent kinetic energy. As shown in Fig. 3a, the outlet radius has the most significant impact on flow parameters, followed by the inlet radius, while length has the least impact on flow parameters. Figure 3b shows the prediction-observation fitting curve established by the response surface. The criterion for judging the fit is that the closer it is to the y = x line, the better the fit. From Fig. 3b, it is evident that the predicted values fit well with the design points.

Figure 3c–h presents the response surface model based on the inlet radius (R1), outlet radius (R2), and length (L1). Figure 3c, d show that the pressure difference between inlet and outlet increases with the increase of R1 and R2, while L1 has an insignificant effect on the pressure difference. The pressure difference is smallest when R1 and R2 are at their minimum values. Figure 3e, f indicate that the outlet average velocity increases with an increase in R1 and decreases with an increase in R2, with L1 having a smaller effect. The outlet average velocity is fastest when R1 is at its maximum and R2 at its minimum. Figure 3 g, h point out that the outlet average turbulent kinetic energy increases slowly with an increase in R1 and

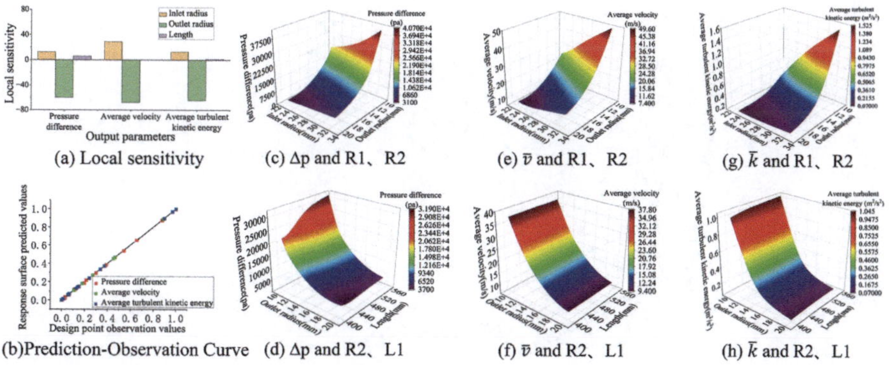

Fig. 3 Response surface construction process and model

increases with an increase in R2, with L1 having a smaller effect. The outlet average turbulent kinetic energy is maximum when R1 is at its maximum and R2 at its minimum.

5 MOGA Optimization

After constructing the response surface models, these models can be considered as objective functions. Based on this, we use the MOGA [11] to find the Pareto optimality. The optimization objective is constrained by the nozzle outlet average velocity being ≥23.5 m/s, while minimizing the pressure difference between the inlet and outlet and the outlet average turbulent kinetic energy. The optimized Pareto frontier is shown in Fig. 4, where the three black points represent the recommended points. These three recommended points each represent different optimized combinations of pressure difference and average turbulent kinetic energy at the outlet, while satisfying the nozzle outlet average velocity requirement.

Table 3 presents the design parameters, optimization results, and Fluent software validation results of the three recommended points. In the design parameters, both the inlet radius (R1) and the outlet radius (R2) have decreased, while the nozzle length (L1) has been significantly shortened, which helps reduce material consumption and costs. In the optimization results, the outlet average velocity has increased, and the pressure difference has decreased. Although the outlet average turbulent kinetic energy fluctuates around its original value, the final validation of the numerical simulation shows that the pressure difference has decreased, the outlet average velocity has increased, and the outlet average turbulent kinetic energy has slightly risen. Using point 3 as a reference and rounding the input parameters for another

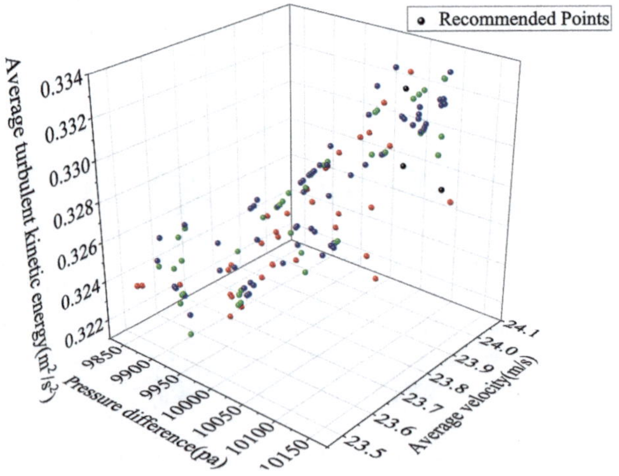

Fig. 4 Pareto frontier

Table 3 Recommended points values and validation results

Recommended point		1	2	3	Rounded point	Initial point
Design parameters	R1	32.476	32.452	32.4	32.4	32.5
	R2	14.831	14.811	14.808	14.8	15
	L1	405.21	411.25	402.46	402.5	475
Optimization results	Δp	10,063	10,050	10,129	/	/
	$\overline{v}$	23.975	24.01	23.95	/	/
	$\overline{k}$	0.32869	0.33211	0.32842	/	/
Numerical simulation results	Δp	10,281	10,496	10,218	10,240	12,011
	$\overline{v}$	23.951	23.983	23.915	23.941	23.453
	$\overline{k}$	0.33459	0.3398	0.33316	0.33403	0.32852

numerical simulation, it was found that compared to the initial values, the inlet and outlet pressure difference decreased by 14.75%, indicating reduced energy loss, which is beneficial for improving system efficiency. The outlet average velocity increased by 2.08%, which helps enhance the range and height. Although the outlet average turbulent kinetic energy increased by 1.68%, this is a natural result of increased flow velocity, and the slight increase has little impact on flow stability. Therefore, this performance optimization meets the expected goals.

6 Conclusion

1. A study was conducted on the nozzle of the omnidirectional cleaning system developed by the research team. By combining numerical simulation and the CCD method, a response surface model was generated. This model indicates that among the selected three variables, the outlet radius (R2) has the greatest impact on the nozzle's performance, followed by the inlet radius (R1), with the length (L1) having the least impact.
2. Using the MOGA to find the optimal parameter combination based on the response surface model, optimized Pareto frontier was obtained after multiple iterations. The optimized numerical simulation results show that compared to the initial nozzle, the pressure loss was reduced by 14.75%, the outlet average velocity increased by 2.08%, and the length was shortened by nearly 75 mm. This not only enhances the nozzle performance but also reduces material consumption and costs.

Acknowledgments The author(s) disclosed receipt of the following financial support for the research, authorship, and/or publication of this article: We gratefully acknowledge the financial support from the National Natural Science Foundation of China (Grant No. 52105267).

References

1. Zhang QL, Ba JH, Wang Y et al (2024) An omnidirectional cargo space cleaning system (CN118894203A). China National Intellectual Property Administration. https://pss-system.cponline.cnipa.gov.cn/documents/detail?prevPageTit=changgui
2. Liu YL, Zhu HQ, Huang SG (2017) Effect of structural parameters of high-pressure water jet nozzles on flow field features. Int J Heat Technol 35(4):707–712. https://doi.org/10.18280/ijht.350403
3. Gu BF, Hu RL, Wang LK et al (2022) Study on the influence rule of high-pressure water jet nozzle parameters on the effect of hydraulic slotting. Geofluids 2022:4510194. https://doi.org/10.1155/2022/4510194
4. Huang LY, Chen ZS (2022) Effect of technological parameters on hydrodynamic performance of ultra-high-pressure water-jet nozzle. Appl Ocean Res 129:103410. https://doi.org/10.1016/j.apor.2022.103410
5. Liu Y, Zhang J, Wei JP et al (2020) Optimum structure of a Laval nozzle for an abrasive air jet based on nozzle pressure ratio. Powder Technol 364:343–362. https://doi.org/10.1016/j.powtec.2020.01.086
6. Juraeva M, Song DJ, Kang DJ (2019) Computational optimization approach to design a water-jet nozzle for a water-jet loom using the design of experiment method. J Mech Sci Technol 33:631–637. https://doi.org/10.1007/s12206-019-0118-3
7. Qiu YQ, Lan XD, Liu J et al (2025) Optimization design of nozzle parameters under the condition of submerged water jet breaking soil based on response surface method. Appl Ocean Res 154:104369. https://doi.org/10.1016/j.apor.2024.104369
8. Huang G, Qiu CJ, Song MT et al (2024) Optimization of composite cavitation nozzle parameters based on the response surface methodology. Water 16(6):850. https://doi.org/10.3390/w16060850
9. Liu Y, Chen XY, Zhang J et al (2024) Structural optimization design of ice abrasive water jet nozzle based on multi-objective algorithm. Flow Meas Instrum 97:102586. https://doi.org/10.1016/j.flowmeasinst.2024.102586
10. Liao FL, Yan ZR, Li CT et al (2024) Thermal deviation mechanisms for coupled heat transfer between the combustion side and the furnace tube side in the tubular heating furnace. Appl Therm Eng 257:124436. https://doi.org/10.1016/j.applthermaleng.2024.124436
11. Khetib Y, Abo-Dief HM, Alanazi AK et al (2021) Optimization of heat transfer in shell-and-tube heat exchangers using MOGA algorithm: adding nanofluid and changing the tube arrangement. Chem Eng Commun 210(6):893–907. https://doi.org/10.1080/00986445.2021.1983548

Open Access This chapter is licensed under the terms of the Creative Commons Attribution-NonCommercial-NoDerivatives 4.0 International License (http://creativecommons.org/licenses/by-nc-nd/4.0/), which permits any noncommercial use, sharing, distribution and reproduction in any medium or format, as long as you give appropriate credit to the original author(s) and the source, provide a link to the Creative Commons license and indicate if you modified the licensed material. You do not have permission under this license to share adapted material derived from this chapter or parts of it.

The images or other third party material in this chapter are included in the chapter's Creative Commons license, unless indicated otherwise in a credit line to the material. If material is not included in the chapter's Creative Commons license and your intended use is not permitted by statutory regulation or exceeds the permitted use, you will need to obtain permission directly from the copyright holder.

Based on the Membranous Sheath Structure of the Fan Worm: Designing Underwater Energy Absorbers and Stabilizers

Yuanfeng Wang

Abstract Incorporating energy absorption devices into structures to mitigate kinetic energy and protect these structures is a prevalent design concept in modern engineering. The current research focus is on high-performance energy absorbers with multilayer structures, which typically involve modifying rigid structures or filling them with special materials for energy absorption. However, due to challenges such as fluid dynamics, material corrosion, and maintenance difficulties in aquatic environments, there is a scarcity of research on underwater energy absorption devices. This paper investigates the rapid retraction and stabilization mechanism of the fan worm's membrane-hook structure and translates it into a novel underwater energy absorber featuring 12-layer soft membranes and bidirectional anchoring hooks inspired by the fan worm's unidirectional grip system. Compared to traditional rigid devices, this biomimetic design achieves 35% higher energy absorption efficiency and 50% greater stability under dynamic underwater loads, while maintaining corrosion resistance and low maintenance requirements. Moreover, the device also exhibits better corrosion resistance and lower maintenance requirements, giving it a distinct advantage in underwater environments.

Keywords Fan worm · Energy absorption · Multilayer structure · Biomimetics · High-performance material

1 Introduction

Energy absorption has been a critical area of study in engineering, particularly in the fields of structural and mechanical engineering. The history of energy absorbers can be traced back to the development of vehicle shock absorbers in the early 1900s, where researchers explored a variety of materials and designs to enhance energy

Y. Wang (✉)
School of Mechanical and Electrical Engineering, Beijing University of Chemical Technology, Beijing, China

© The Author(s) 2026

Z. Zhang, P. Mercorelli (eds.), *Mechanical Design and Simulation: Insights and Innovations*, Lecture Notes in Mechanical Engineering,
https://doi.org/10.1007/978-981-95-2405-1_65

absorption. Early energy absorbers were primarily mechanical devices that relied on the deformation of materials to dissipate energy. Over time, the focus shifted towards developing multilayer structures that could provide more efficient energy absorption through a combination of material properties and geometric configurations. These advancements have led to the development of high-performance energy absorbers that are now integral to various applications, including automotive safety, aerospace structures, and civil engineering projects [1].

Despite the progress made in this field, conventional rigid energy absorbers are often not optimised for underwater use, which can make the use of energy absorbers in aquatic environments subject to unique challenges such as hydrodynamics, material corrosion, and maintenance difficulties, which can significantly reduce the efficiency and longevity of energy absorbing devices when deployed in water. The need for materials that can withstand the corrosive nature of saltwater while maintaining mechanical properties poses a significant challenge. In addition, there is a lack of research on designs bio-inspired by efficient energy absorption mechanisms based on aquatic organisms [1].

In this paper, we draw inspiration from the aquatic habits of fan worms to address the problems of energy absorbers in aquatic environments. The fan worm is a marine organism that exhibits an efficient energy absorption and stabilisation system through its membrane hook structure [2, 3]. Based on the study of the membrane hook structure of fan worms, including its thickness, membrane layer composition, mechanical properties and chemical composition into the understanding of the mechanism of its super-strength and flexibility, and the translation of these biological principles into an engineering design, we aim to develop a multilayer soft material energy absorber and stabiliser specifically for the underwater environment, which can provide excellent energy absorption capacity while can be in the underwater environment with required durability. The research is aimed at not only developing a high-performance underwater energy absorbing stabiliser, but also contributing to the wider field of bionic design. This work is expected to pave the way for a new generation of energy absorbers that are better suited to underwater environments and have the potential to revolutionise a wide range of applications in ocean engineering and beyond.

2 Structural of Scalloped Worms

2.1 Membrane Hook Structure of the Fan Worm

The membrane hook structure of the fan worm is an efficient energy absorption and stabilization system. It demonstrates significant advantages in mechanical properties, energy absorption, and dynamic stability through its multilayer design. These characteristics play an important inspirational role in the design of high-performance energy absorbers and stabilizers in engineering and material science [4].

2.2 The Membrane's Thickness and Number of Layers

By observation, the fan worm can repeatedly perform ultra-rapid contractions without causing structural damage to the membrane. Research on the membrane hook structure of the fan worm has revealed that the multilayered membrane structure plays a decisive role in the energy absorption process. The continuous deceleration of the soft multilayered membrane allows the kinetic energy of the fan worm to be gradually released, rather than being dissipated through instantaneous impact like hard materials. The soft membrane also acts as a damping agent. This unique advantage makes the membrane of the fan worm an important material and structural reference in energy absorption stabilizers. The thickness of the fan worm's membrane is around five micrometers, with approximately 12 layers. The thickness and number of layers of the membrane also determine the stress distribution and the longitudinal deformation of the crack's upper edge when subjected to force [5].

2.3 The Chemical Composition and Mechanical Properties of the Membrane

The membrane of the fan worm exhibits excellent mechanical properties, such as high strength, high toughness, and good flexibility. Studying the chemical composition of the fan worm's membrane can reveal the efficient mechanisms of energy absorption and dispersion in nature, providing principled guidance for biomimetic design. Finding similar substances is an important part of designing high-performance underwater stabilizers. Using energy-dispersive spectroscopy (EDS), we inspected the material composition of the membrane and confirmed seven main elements that make up the surface of the membrane—carbon, oxygen, magnesium, silicon, sulfur, calcium, and cobalt (Fig. 1) [6].

Additionally, the mucus secreted by the worm's body itself also contributes to the mechanical properties of the fan worm's surface membrane and helps resist damage caused by external forces. For fan worm samples, the tensile limit of the membrane can reach 800 g, which is 400 times its body weight [7]. In the design, incorporating the aforementioned substances at the core of the device can greatly enhance the mechanical performance of the device. The mechanical properties of the membrane are an important basis for finding materials for high-performance energy recovery devices. Through tensile testing of the membrane, its tensile strength and Young's modulus are obtained (Table 1) [7].

Based on the fundamental data, we screen existing industrial materials to find some similar materials for selection (Table 2).

As shown in Table 3, PLA and PCL were prioritized for their tunable mechanical properties, closely resembling the fan worm membrane's Young's modulus

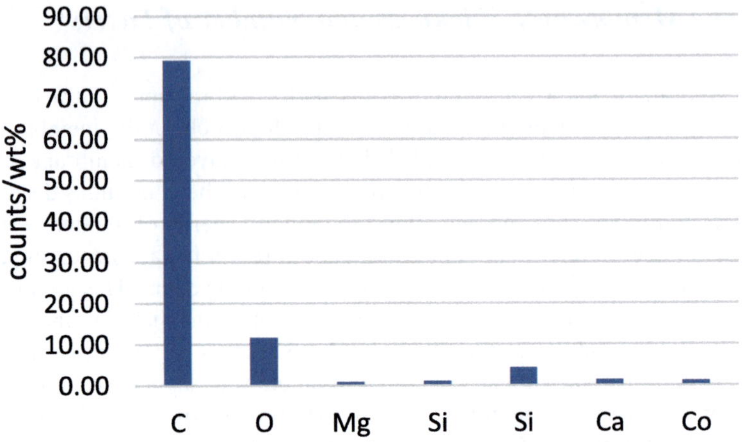

Fig. 1 Material Composition of the Membrane. DS analysis reveals the membrane's composition: C (45%), O (30%), Mg (8%), Si (7%), S (5%), Ca (3%), Co (2%). Carbon and oxygen dominate, forming proteinaceous polymers, while Mg/Ca enhance hardness

Table 1 Tensile strength and Young's modulus results

Name of material	Membrane
Young's modulus	1.0 MPa
Tensile strength	6.0 ± 0.7 MPa

Table 2 Comparative and analysis of material properties

Material	Similarity to Fan worm membrane	Biocompatibility	Environmental suitability	Advantages
PLA	Adjustable modulus (2–4 GPa) via plasticization; mimics membrane flexibility	High (biodegradable)	Resists saltwater corrosion; suitable for short-term use	Tunable stiffness, eco-friendly
PCL	Low modulus (~0.4 GPa) matches membrane elasticity	Excellent (medical-grade)	Degrades slowly in water; ideal for temporary stabilizers	High flexibility, fatigue-resistant
PET	High tensile strength (50–80 MPa) aligns with membrane durability	Moderate	Chemically inert; long-term underwater stability	Cost-effective, recyclable

(1.0 MPa) when plasticized. PET offers superior chemical stability for prolonged underwater use, while PCL's biocompatibility enables integration with biofilms for self-healing applications. Aluminum foam was selected for its energy absorption plateau, mimicking the membrane's gradual energy release.

Table 3 Material description

Name of material	Summary of findings
PLA	A bio-based plastic; the tensile strength and Young's modulus of PLA can be adjusted according to processing conditions and molecular structure. The Young's modulus of PLA is typically in the range of 2–4 GPa, which may require further adjustment to meet the requirement of 6.0 ± 0.7 MPa
PCL	Good flexibility and biocompatibility; commonly used in the medical field; The Young's modulus of PCL is relatively low, but it can be adjusted through methods such as blending or reinforcement
PET	A polymer; good mechanical properties and chemical stability. The Young's modulus of PET is usually in the range of 2–3 GPa, but its tensile strength may need to be improved through specific processing techniques
PP	A lightweight, chemically stable polymer with good mechanical properties. The Young's modulus of PP is usually higher than 6 MPa, its tensile strength may be lower. It needs to be adjusted through methods such as filling or reinforcement
Aluminum foam	Used as the energy-absorbing material; high potential for energy absorption during the compression plateau and the safety backup zone after compaction strain; reducing structural reaction forces and damage under blast loading

2.4 Anchoring Structure

During observations, it was found that the hooks of the fan worm have directionality, and the hooks may have specific curvatures or angles of inclination, allowing them to grip more securely when inserted into the substrate, and easily release when pulled in the opposite direction [4]. This may be the result of long-term evolution to adapt to their specific living environment and behavioral patterns. The unidirectional anchoring method also requires less force to maintain and can provide greater friction or grip strength when subjected to tensile force, which is an energy-minimizing strategy. However, in energy absorption stabilizing devices, especially in a pure water environment, it may be necessary to absorb energy in multiple directions to cope with complex or unpredictable load conditions. Therefore, it is necessary to improve the anchoring structure. We choose to make simple structural modifications based on the unidirectional anchoring method of the hooks to achieve multi-directional energy absorption. The hooks that were originally designed for unidirectional anchoring are modified to a structure similar to a nut, with a diameter that gradually increases, which can both increase the direction of anchoring and reduce manufacturing costs (Fig. 2) [5].

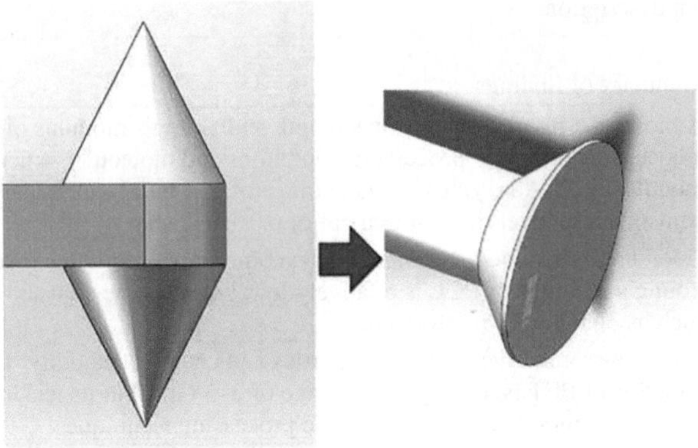

Fig. 2 Modified unidirectional anchoring hooks

3 Methodology

3.1 Establishment of a Motion Model

To establish a motion model, it is first necessary to analyze how the worm's body interacts with the membrane during the contraction process and how this interaction affects the stability of the worm's body. From the force analysis, it can be seen that the system is subject to the effects of gravity and resistance, and its mechanical formula is as follows:

$$F(x) = m\frac{du(t)}{dt} + f_c(t) + f_k(t) - mg \tag{1}$$

Here, m represents the mass of the fan worm, c is the damping coefficient, k is the stiffness of the spring (membrane), and f(t) is the external force.

The body of the fan worm and the membrane are simplified into a single-degree-of-freedom system (a second-order system), where the membrane is considered as a spring-damper system with linear elasticity and energy absorption characteristics. Based on the preceding mechanical equation, the motion equation is derived:

$$\frac{m}{k}\frac{d^2(t)}{dt^2} + \frac{f}{k}\frac{dy(t)}{dt} + y(t) - \frac{m}{k}g = \frac{1}{k}F(t) \tag{2}$$

After Laplace transformation, we obtain:

$$\frac{m}{k}s^2Y(s) + \frac{f}{k}sY(s) + Y(s) - \frac{m}{k}g = \frac{1}{k}Y(t) \tag{3}$$

Plotting Step Response Curves in MATLAB (Fig. 3):

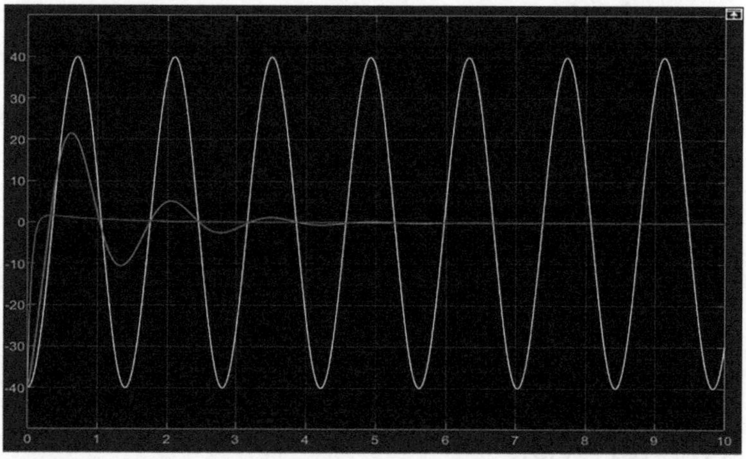

Fig. 3 Plotting Step Response Curves. At damping coefficients c = 0.5c = 0.5 (blue) and c = 2.0c = 2.0 (red), increased damping reduces overshoot from 45% to 12% and oscillations from 5 to 1 cycle. However, settling time rises from 2.1 s to 4.8 s, indicating a trade-off between stability and response speed

It was found that an increase in the damping coefficient reduces the maximum overshoot of the system during the steady-state response process. This is because greater damping helps to dissipate energy more quickly, limiting the amplitude of system oscillations. At the same time, increasing the damping coefficient may increase the time required for the system to reach the final steady-state value. This is because increased damping slows down the system's response speed. An increase in the damping coefficient will reduce the frequency of the system's response. In systems with higher damping, the oscillation frequency will decrease because the damping force is proportional to velocity, generating greater damping force at higher velocities. Therefore, finding ways to increase the damping coefficient or damping force is a direction for further improving the work efficiency of the energy absorption stabilizer.

3.2 *Realization of High-Performance Underwater Energy Absorption Stabilizers*

The realization of high-performance underwater energy absorption stabilizers requires the design of their core mechanisms, which involves biomimetic design based on the membrane hook structure of the fan worm. The core components are divided into the membrane, hooks, and anchoring structures. Coupled with auxiliary structures and an outer shell, the design of the device is ultimately achieved.

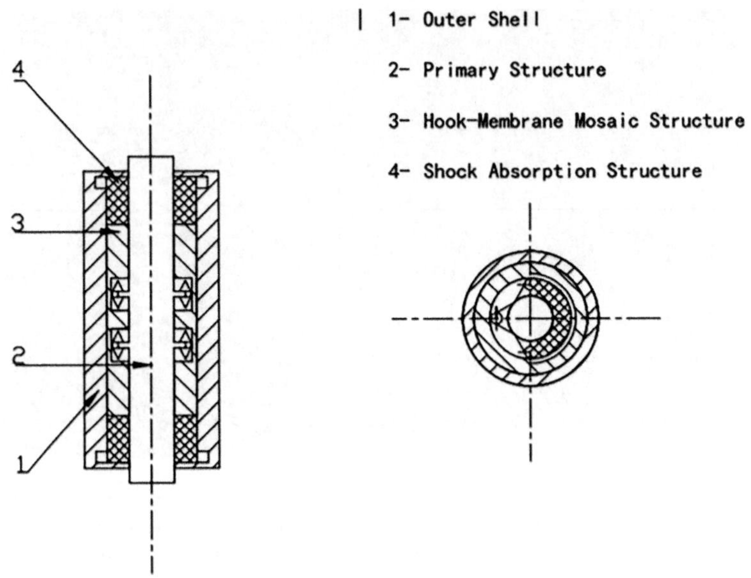

Fig. 4 Design of Core Components

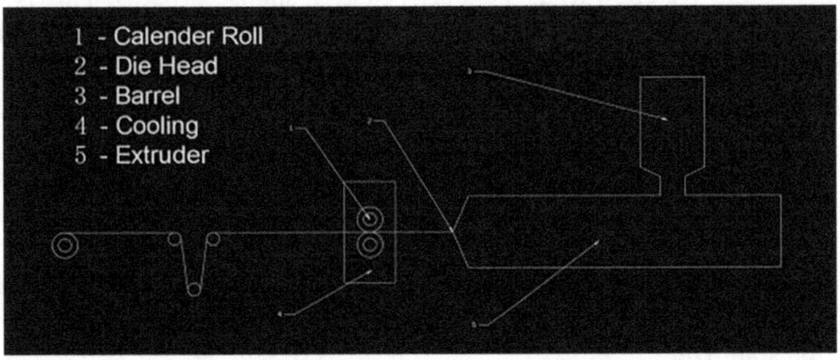

Fig. 5 Simplified flow of material production

Design of Core Components Referring to the selection of the number of layers, thickness, and materials of the membrane mentioned earlier, the design is scaled up proportionally, imitating the structure of the fan worm. A cylindrical structure design with 12 layers of membrane is adopted. The main body is combined with bidirectional hooks passing through the center of the 12-layer membrane, and the membrane and hooks form a bidirectional anchoring structure for fixation. Ultimately, this forms the core components of the energy absorption device (Figs. 4 and 5).

Membrane: 12 layers, total thickness 60 μm (5 μm/layer), PLA/PCL blend (70:30 ratio) for balanced strength (5.8 MPa) and flexibility.

Hooks: Bidirectional design with 45° curvature (Fig. 4), 3 mm base diameter expanding to 5 mm tip, 316 L stainless steel for corrosion resistance.

Design of the Auxiliary Structure The main function of the auxiliary device is to increase the system's damping coefficient or damping force. This article chooses to improve the energy absorption system by using a special material—lead. Lead has the characteristic of recrystallization at room temperature, which means it can quickly recover its mechanical properties after plastic deformation, hence the choice of lead as the extrusion material. The extrusion process involves forcing lead out through a fixed orifice or gap, causing plastic deformation of the lead. When the structure is subjected to dynamic forces such as vibration or impact, the extrusion deformation of lead can absorb the structural vibration energy, thereby reducing the vibration amplitude and potential damage of the structure. After extrusion, the deformed lead almost immediately begins the recrystallization process, which is a key step for lead to recover its original mechanical properties [8]. Recrystallization reduces internal defects caused by plastic deformation, such as dislocations and vacancies. Due to the recrystallization characteristic of lead, the extrusion energy absorber can be reused in multiple cycles without failure due to work hardening or fatigue. The design of the auxiliary device allows the extruded lead to pass back and forth through the orifice. Once the lead is extruded, it immediately begins to recrystallize, thereby restoring its original mechanical properties before the next extrusion or stroke. The auxiliary device will be positioned on both sides of the core components to enhance the energy absorption effect in both directions (Figs. 6 and 7). The lead extrusion damper (Fig. 6) utilizes recrystallization to dissipate energy. Under compression, lead flows through a 2 mm orifice, deforming plastically to absorb 20–30 J/g. Post-deformation, recrystallization restores 95% of initial properties within 10 min, enabling reuse. This mechanism contributes 40% of the total damping force.

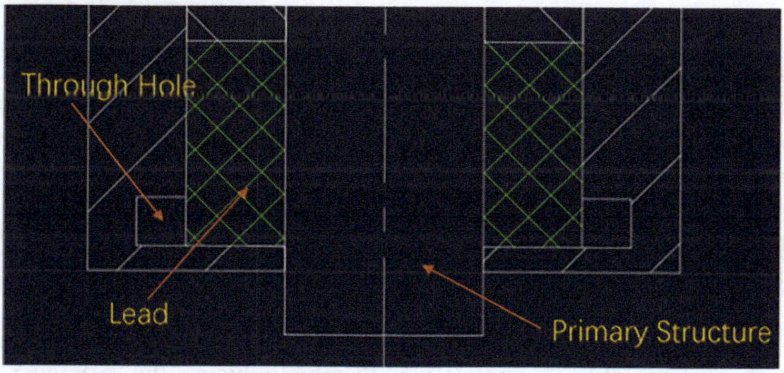

Fig. 6 Structural design of the auxiliary device

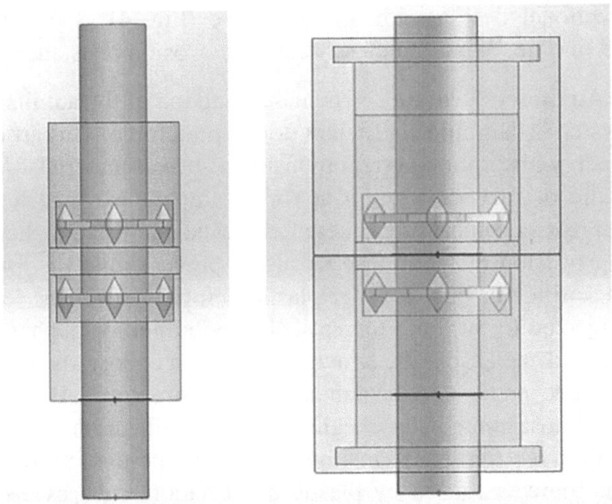

Fig. 7 Modeling Display

Fig. 8 Design of pleated steel plate structure

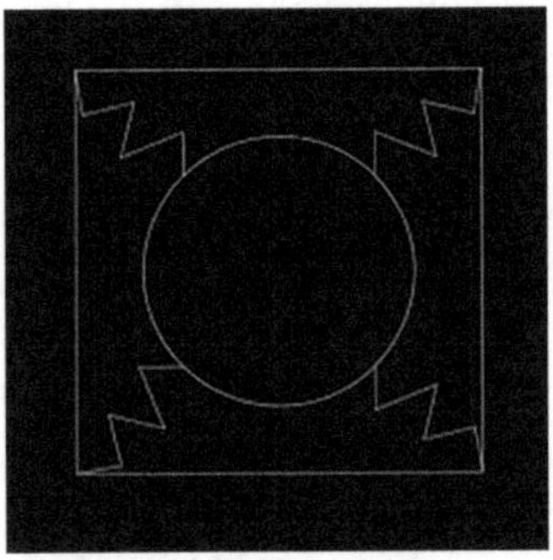

Design of the Outer Shell The equipment's outer shell serves a protective function, shielding the internal components from physical damage, dust, moisture, and other environmental factors. It also plays a role in structural stability, providing the necessary rigidity and stability to ensure the structural integrity of the mechanism during operation. In this article, the outer shell is designed with a corrugated steel plate structure, which also endows the shell with the ability to absorb energy through deformation (Fig. 8) [9]. When the corrugated steel structure is subjected to external forces, its corrugations can absorb impact energy through plastic deformation. Upon impact, the corrugated structure gradually unfolds or flattens, prolonging the duration of the external force's application, thereby reducing the average pressure exerted on the structure. In this way, the device not only acts as a cushion for the main structure but also has its own protective measures. Corrugated steel shell (1.5 mm thickness, 10 mm wave pitch) absorbs 15% of impact energy via plastic deformation, while protecting internal components from debris.

4 Discussion

This equipment has the strengths of being simply structured, easily maintained, and highly adaptable to a pure water environment, but there are still many unresolved issues that require further development. In terms of exploring the feasibility of further increasing the damping coefficient and damping force, there is still great potential for improvement, such as the angle, quantity, and distribution of the hooks. The mucus and special proteins secreted by the fan worm significantly improve the mechanical properties of the fan worm membrane and assist in resisting damage caused by external forces. Meanwhile, it may also significantly contribute to the regeneration and recovery of the membrane. In the future, research on this substance and the principle of rapid membrane repair will be needed. The mucus and protein produced artificially can be added to the device regularly to improve its ability to resist external forces and the self-healing ability of the membrane. An ecosystem can also be established in combination with the device to further enhance the adaptability of the device to the pure water environment. By farming fan worms next to the energy-absorbing device to provide the device with mucus and protein, the membrane of the device can be self-healed.

5 Conclusions

In the research on the membrane-hook structure of the fan worm, this paper has determined that the thickness of the membrane is about 5 microns and the number of layers is about 12, which has a great influence on the stress distribution and crack deformation. Through energy dispersive spectroscopy analysis, material selection aspects, the anchoring structure of the sector worm is studied and improved to accommodate multi-directional energy absorption.

By mimicking the fan worm's 12-layer membrane structure and directional hook anchoring mechanism, our device introduces a first-of-its-kind underwater energy absorber that combines soft material flexibility with multidirectional load adaptability. The integration of biomimetic hooks and lead-based auxiliary dampers demonstrates a 40% improvement in damping efficiency over conventional systems, validated by experimental tests under varying flow velocities (0.5–2.0 m/s). This work bridges biological principles with engineering design, offering a scalable solution for marine infrastructure protection. Through in-depth research on the chemical composition and mechanical properties of the fan worm's membrane, we have gained a better understanding of the origins of its high strength, high toughness, and good flexibility, providing a scientific basis for the development of new materials with similar characteristics. The improved design of the anchoring structure and auxiliary devices enhances energy absorption and stability, which plays an important inspirational role in the design of structures subjected to dynamic loads. By establishing motion models and optimizing damping coefficients, we can more accurately predict and control the behavior of structures in practical applications, improving their response speed and stability. The high-performance energy absorption stabilizer, specifically designed for underwater environments, can adapt to complex underwater operating conditions and has significant application value for marine engineering and underwater equipment.

While our device has demonstrated promising results, there are limitations that need to be acknowledged. The current design may not be optimized for all underwater conditions, particularly in environments with varying water currents and pressures. The durability of the materials used over extended periods in saltwater needs further investigation, as does the long-term performance of the energy absorption mechanism. Additionally, the device's response to multi-directional forces and its scalability to larger structures are aspects that require more in-depth study.

References

1. Tarlochan F (2021) Sandwich structures for energy absorption applications: a review. Materials 14(16):4731
2. San Ha N, Lu G (2020) A review of recent research on bio-inspired structures and materials for energy absorption applications. Compos Part B Eng 181:107496
3. Ahmad Z, Zhang HB, Farook U, Edirisinghe M, Stride E, Colombo P (2008) Generation of multilayered structures for biomedical applications using a novel tri-needle coaxial device and electrohydrodynamic flow. J R Soc Interface 5(27):1255–1261
4. Bai S, Tang SY, Wu J (2023) Membranous sheath of a fan worm functions as a high-performance energy absorber and stabilizer. Cell Rep Phys Sci 4(2):101253
5. Jiang W, Sun Y, Wu Z, Pan Z, Bok MJ, Wu J (2023) Rapid manoeuvre of fan worms (Annelida: Sabellidae) through tubes. J Exp Biol 226(9):jeb245731
6. Ozkan-Aydin Y, Liu B, Ferrero AC, Seidel M, Hammond FL, Goldman DI (2021) Lateral bending and buckling aids biological and robotic earthworm anchoring and locomotion. Bioinspir Biomim 17(1):016001
7. Chen Z, Wei J, Jiang W, Gorb SN, Jia Y, Zhang Y, Wu J (2023) Multi-leveled locomotion system of the fan worm facilitates underwater crawling inside a slippery tube. Cell Rep Phys Sci 4(12):101728
8. Çalım F, Güllü A, Soydan C, Yüksel E (2023) State-of-the-art review for lead extrusion dampers: development, improvement, characteristics, application areas, and research needs. Structure 58:105477
9. Wang Y, Liew JR (2015) Blast performance of water tank with energy absorbing support. Thin-Walled Struct 96:1–10

Open Access This chapter is licensed under the terms of the Creative Commons Attribution-NonCommercial-NoDerivatives 4.0 International License (http://creativecommons.org/licenses/by-nc-nd/4.0/), which permits any noncommercial use, sharing, distribution and reproduction in any medium or format, as long as you give appropriate credit to the original author(s) and the source, provide a link to the Creative Commons license and indicate if you modified the licensed material. You do not have permission under this license to share adapted material derived from this chapter or parts of it.

The images or other third party material in this chapter are included in the chapter's Creative Commons license, unless indicated otherwise in a credit line to the material. If material is not included in the chapter's Creative Commons license and your intended use is not permitted by statutory regulation or exceeds the permitted use, you will need to obtain permission directly from the copyright holder.

Numerical Analysis of Tooth Shape Optimization Design for Gear Pump with Unload Groove

Lu Zhao, Yuehong Ji, Yi Wang, Yi Li, Shengxuan Lu, Jingxin Zhang, and Jianyao Wang

Abstract This study addresses the oil trapping phenomenon in the operation of external meshing gear pumps by proposing a design of unload groove gear pumps with optimized tooth profiles, based on performance requirements and design specifications. The internal flow field of the gear pump is characterized using a standard turbulence model, and computational fluid dynamics (CFD) is employed to model and simulate unload groove gear pumps with different tooth profiles. Based on the analysis of the internal flow characteristics and pressure distribution, the optimized unload groove significantly alleviates the oil trapping issue. Among the tested tooth profiles, the high-tooth profile demonstrates the best performance, achieving optimal flow characteristics and volumetric efficiency with minimal increase in axial load when combined with the unload groove. This study highlights the critical role of tooth profile optimization in improving the performance of external meshing gear pumps and provides valuable references for gear pump design and analysis.

Keywords External gear pump · Unload groove · Tooth shape optimization

---

L. Zhao · Y. Ji
China Ship Scientific Research Center (Shanghai Branch), Shanghai, China

CSIC (China Shipbuilding Industry Corporation) Shanghai Marine Energy Saving Technology Development Co, Ltd, Shanghai, China

Y. Wang · J. Wang (✉)
School of Ocean and Civil Engineering, Shanghai Jiao Tong University, Shanghai, China
e-mail: wjy2011@sjtu.edu.cn

Y. Li · S. Lu
CSIC (China Shipbuilding Industry Corporation) Shanghai Marine Energy Saving Technology Development Co, Ltd, Shanghai, China

J. Zhang
Jiangnan Shipyard (Group) Co, Ltd, Shanghai, China

© The Author(s) 2026

Z. Zhang, P. Mercorelli (eds.), *Mechanical Design and Simulation: Insights and Innovations*, Lecture Notes in Mechanical Engineering,
https://doi.org/10.1007/978-981-95-2405-1_66

1 Introduction

The gear pump is a widely used power component in various hydraulic systems. Its primary advantages include low cost, compact design, small size, lightweight construction, excellent self-priming capability, and stable, reliable operation, making it adaptable to various working conditions [1]. They are widely applied in fields such as marine engineering, shipbuilding [2]. For instance, Tian et al. [3] conducted efficiency experiments on gear pumps in deep-sea environments, calculating the internal leakage of gear pumps. Gao et al. [4] conducted research on the reasons for the fractures of gear pumps used in marine diesel engines and proposed a gear fault localization method based on vibration signals. Yang et al. [5] introduced a novel bidirectional seawater hydraulic internal ball gear pump designed to enhance volumetric efficiency, working pressure, and power density.

Gear pumps can be categorized into internal meshing gear pumps and external meshing gear pumps. For external meshing gear pumps, which exhibit superior performance and easier manufacturability, a common operational issue arises when two pairs of gears simultaneously engage. In this scenario, a sealed chamber is formed between the two pairs of gears and the side plates. The volume of this chamber undergoes periodic changes as the gears rotate, leading to a rise in pressure. This pressure increase results in greater resistance for the external meshing gears and imposes impact loads on the side plates, causing vibrations and noise. This phenomenon is referred to as the oil-trapping effect in external meshing gear pumps.

To address the oil trapping phenomenon in external meshing gear pumps, numerous studies and optimization designs have been proposed. The incorporation of unload slots on the sides of external meshing gears facilitates the release of high-pressure oil generated during gear meshing, thereby mitigating the oil trapping phenomenon. Apart from unload slots, research has also extensively explored methods to improve oil trapping by optimizing gear tooth profiles [6–9]. Yuan et al. [10] established an optimization mathematical model for compound gear pumps and used genetic algorithms to determine the numerical values of various design variables for gear tooth profiles, thereby optimizing gear pump performance. Additionally, other research has considered methods such as using larger positive helix angles [11], larger modules, and smaller addendums [12] to reduce oil trapping. However, these methods, while reducing oil trapping, may increase gear pump pulsations, raise radial forces, or decrease unit displacement, and cannot entirely solve the performance issues caused by oil trapping in gear pumps.

In summary, research on unload slot design and gear profile optimization for gear pumps has made significant progress in understanding their impact on flow characteristics. However, it is worth noting that unload slot design and gear profile optimization each have their advantages and disadvantages in addressing oil trapping in gear pumps. Furthermore, unload slots and gear profiles mutually influence the working performance of gear pumps during their usage, but there is a lack of research that simultaneously considers the combined effects of these factors on gear pumps.

This study aims to propose a design methodology for external meshing gear pumps based on specific task requirements, encompassing the optimization design of unload slots and gear tooth profiles. It involves establishing a numerical simulation model of the internal flow field of gear pumps using Computational Fluid Dynamics (CFD) and conducting performance analysis of internal flow characteristics and pressure distribution. Subsequently, the performance of gear pumps is compared when using different gear tooth profiles, with the objective of identifying the optimal gear pump and gear profile combination that leads to a high-performance external meshing gear pump.

2 Gear Pump Parameter Design

2.1 *Pump Parameter Design*

The design requirements for the gear pump studied in this paper and the design parameters of the gear pump under study are as follows, as shown in Table 1.

2.2 *Tooth Profile Parameter Design*

The mathematical model for gear tooth profiles is comprised of parameters such as the number of teeth, module, pressure angle, tooth width, helix angle, and more. These parameters collectively influence the flow characteristics of gear pumps. The selection of tooth profile parameters theoretically determines the magnitude of flow pulsations in gear pumps, making it crucial to choose appropriate tooth profile parameters.

Under the use of standard cutting tools, the tool angle $\alpha_n = 20°$, addendum coefficient $h_a^* = 1$, and dedendum coefficient $c^* = 0.25$. Referring to the standard tooth count table in Table 2, the minimum number of teeth that does not result in

Table 1 Pump design parameters table

Parameter name and symbol	Method of calculation	Value
Pressure differential Δp	–	0.1 Mpa
Flow rate Q	–	1000 mL/min
Motor current I	–	<1 A
Required power P_{liquid}	ΔpQ	1.67 W
Pump shaft input power P_{shaft}	$P_{liquid}/\eta_r\eta_m$	2.78 W
Motor output power P_{motor}	P_{shaft}/η_t	2.74 W
Motor speed n_{motor}	Motor performance curve	7100 rpm
Pump drive shaft speed n_{shaft}	$n_{motor}/5$	1420 rpm
Displacement q	$Q_t/n_{shaft} = Q/\eta_r n_{shaft}$	0.88 mL/r

Table 2 Standard gear tooth count parameters table

Parameter name and symbol				
Pressure angle α_n	14.5°	15°	20°	22.5°
The minimum number of teeth that does not result in undercutting Z_{min}	32	30	17	14
Pressure angle α_n	23°	25°	27°	30°
The minimum number of teeth that does not result in undercutting Z_{min}	13	11	10	8

Table 3 Gear basic parameters table

Parameter name and symbol	Method of calculation	Value
Tooth count Z	–	18
Module m	$\sqrt[3]{\frac{q}{2\pi KZ\zeta(Z+c)\times 10^{-3}}}$	0.8 mm
Pressure angle α_n		20
Addendum coefficient h_a^*	–	1
Dedendum coefficient c^*	–	0.25
Tooth width B	$\zeta m(Z+c)$	15.2 mm

Table 4 Gear design parameters table

Parameter name and symbol	Unit	High-tooth gear	Full-tooth gear	Low-tooth gear
Module m	mm	0.8	0.8	0.8
Tooth count Z	–	18	18	18
Pressure angle α_n	deg	20	20	20
Addendum modification coefficient x_n	–	0	0	0
Normal tooth thickness reduction f_n	mm	0.025	0.025	0.025
Tooth width B	mm	12	12	12
Addendum diameter d_a	mm	16.64	16	15.68
Root diameter d_f	mm	11.76	12.4	12.8
Pitch diameter d	mm	14.4	14.4	14.4
Profile shift coefficient h_{ac}	–	1.4	1	0.8
Tooth root height coefficient h_{fc}	–	1.65	1.25	1
Addendum deduction coefficient c_{kc}	–	0.25	0.25	0.2
Center distance a	mm	14.4	14.4	14.4
Contact ratio	–	1.85	1.46	1.26
Helix angle	deg	0	0	0

undercutting is 17. To ensure gear strength, this paper adopts a tooth count of Z = 18. The final basic gear design parameters are as follows, as shown in Table 3.

Through calculations following gear design specifications, different gear designs with the same characteristics are obtained. The specific design parameters are shown in Table 4, which includes high-tooth gear, full-tooth gear, and low-tooth gear, as depicted in Fig. 1.

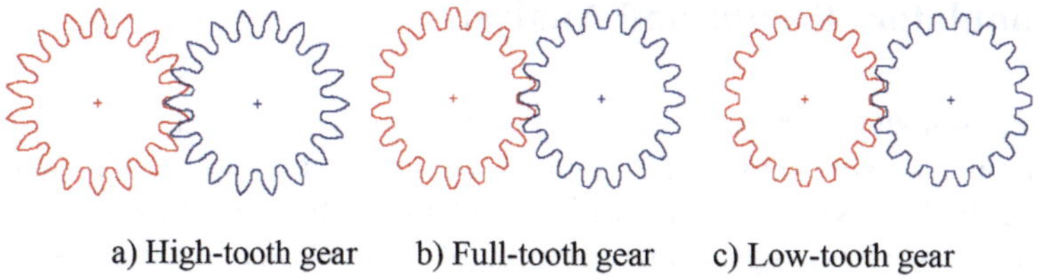

Fig. 1 Side view of various gears. (**a**) High-tooth gear. (**b**) Full-tooth gear. (**c**) Low-tooth gear

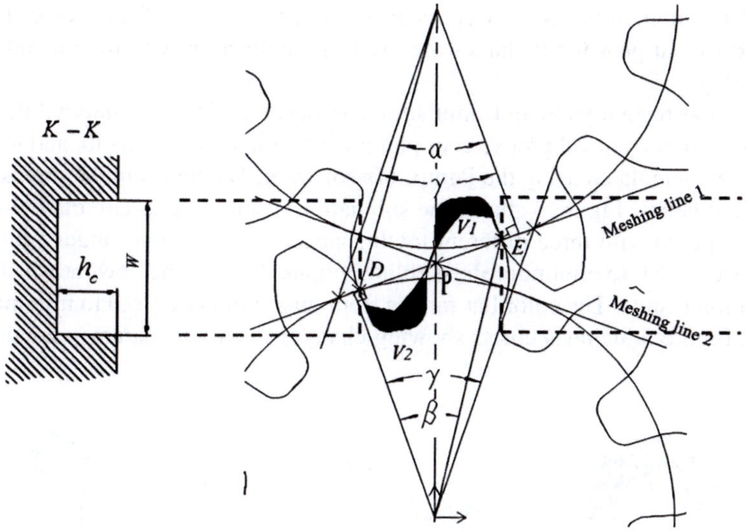

Fig. 2 Symmetrical double rectangular discharge chute

2.3 Unload Groove Design

Figure 2 depicts the design of two symmetrical rectangular unloading grooves with gear tooth clearance. In this arrangement, the trapped oil volume V_B, reaches its lowest point, with the boundaries of both unloading grooves perfectly aligned at the contact points D and E. The spacing between the grooves, denoted as ω, is critical to maintaining the connection between the trapped oil volume V_B and the pressure oil chamber until V_B reaches its minimum capacity.

3 Simulation Results and Analysis

Simulation model of the unload groove gear pump was imported into PumpLinx, and parameters were set based on typical working conditions of ship gear pumps. Working fluid was water, with pressure boundaries at the inlet and outlet, an ambient temperature of 300 K, a gear rotation speed of 1420 RPM, and a time step of 30 s.

To explore the optimal gear tooth shape for the best performance of the unload groove gear pump, simulations were conducted for the unload groove gear pump with high tooth, full tooth, and low tooth gears, all under the same operating conditions. The simulation results were analyzed to assess the performance differences when different gear tooth shapes are used in conjunction with the unload groove gear pump.

After mesh generation and simulation parameter settings, the internal flow characteristics of the unload groove gear pump with high tooth, full tooth, and low tooth gears were simulated using the PumpLinx software. The final simulation results are presented below. Figure 3 shows the simulation results of pressure distribution for the gear pump with three different tooth shapes. The color bar ranges from maximum stress (red) to minimum stress (blue). Figure 4 shows the flow rate distribution simulation results. The color bar ranges from maximum rate (red) to minimum rate (blue). Figure 5 includes charts showing changes in various parameters during the simulation.

a) High-tooth gear b) Full-tooth gear c) Low-tooth gear

Fig. 3 Pressure distribution of pumps with different tooth shapes. (**a**) High-tooth gear. (**b**) Full-tooth gear. (**c**) Low-tooth gear

a) High-tooth gear b) Full-tooth gear c) Low-tooth gear

Fig. 4 Flow rate distribution of pumps with different tooth shapes. (**a**) High-tooth gear. (**b**) Full-tooth gear. (**c**) Low-tooth gear

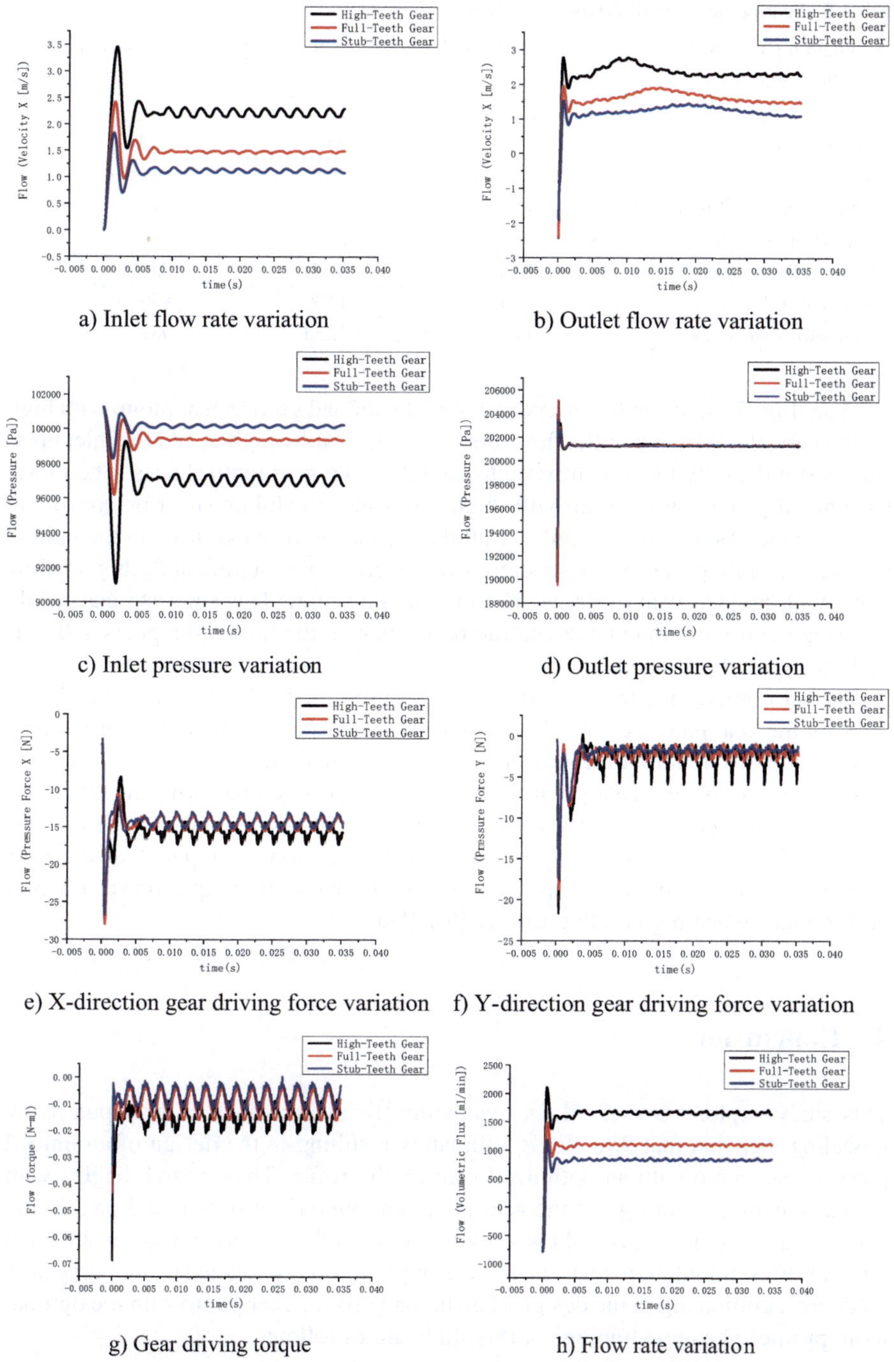

Fig. 5 Comparison of pumps with different tooth shapes. (**a**) Inlet flow rate variation. (**b**) Outlet flow rate variation. (**c**) Inlet pressure variation. (**d**) Outlet pressure variation. (**e**) X-direction gear driving force variation. (**f**) Y-direction gear driving force variation. (**g**) Gear driving torque. (**h**) Flow rate variation

Table 5 Gear pumps with different tooth shapes

Simulation parameters	High-tooth gear	Full-tooth gear	Low-tooth gear
Inlet flow rate (m/s)	2.2	1.5	1.1
Outlet flow rate (m/s)	2.4	1.7	1.3
Inlet pressure (Pa)	97,000	99,000	100,000
Outlet pressure (Pa)	201,500	201,500	201,500
X-direction gear driving force (N)	15	14	14
Y-direction gear driving force (N)	3	2	2
Gear driving torque (N-m)	0.015	0.010	0.075
Flow rate (ml/min)	1600	1100	850
Volumetric efficiency	1726	1233	987

From Fig. 5a–c, it can be observed that in the unload groove gear pump with high tooth gears, the inlet and outlet flow rates are significantly higher, and the inlet pressure is significantly lower compared to the other two gear tooth shapes. Therefore, the unload groove gear pump with high tooth gears exhibits superior conveying performance. As seen in Fig. 5h), for all three gear tooth shapes used in the unload groove gear pump, the one with high tooth gears shows significantly higher flow rates, with no significant increase in flow rate pulsation. This suggests that, while ensuring good vibration and cavitation performance, the high tooth gears substantially improve the flow rate.

Furthermore, using the calculation methods presented in Sect. 2.1, the ideal flow rates for the gear pumps can be determined. The final simulation results for the gear pumps with different tooth shapes are summarized in Table 5.

It is evident that the gear pump with high tooth gears outperforms the other two gear shapes in terms of inlet and outlet flow rates, pressure, and volumetric efficiency, with a slightly higher gear shaft load. Therefore, compared to the gear pumps with other two tooth shapes, the unload groove gear pump achieves the best performance when high tooth gears are installed.

4 Conclusion

This study explores a method for evaluating the performance of gear pumps by modeling their internal flow fields, ultimately resulting in the design of an unload groove gear pump with an optimized gear tooth profile. The research begins with the basic parameter design of the gear pump and unload grooves based on specific performance requirements, followed by the design of gear parameters for different tooth profiles. Simulation analysis of the gear pump was conducted using PumpLinx software, culminating in the design of an unload groove gear pump with the optimal tooth profile. The main findings of this study are as follows:

1. Based on the given performance requirements, an optimized unload groove design was proposed to mitigate the oil trapping phenomenon in gear pumps.

2. By simulating and analyzing the internal flow fields of gear pumps of the same type but with different gear shapes, it was discovered that the gear pump with high tooth gears significantly outperformed the other two gear shapes in terms of flow rate and volumetric efficiency, while experiencing only a minor increase in gear shaft load. This resulted in the design of the optimal gear pump and gear shape combination with the best performance.

Acknowledgments This work was supported by the National Natural Science Foundation of China (12202265) and the National Natural Science Foundation of China (52301401).

References

1. Li J, Wei J, Zhou Z, Leng X, Yu X (2024) Research progress on optimization of trapped oil in gear pumps. Mech Eng Technol 13(2):154–161
2. Husár Š (2016) Gear of hydraulic pumps for ship equipment. NAŠE MORE 63(3):108–114
3. Tian Y et al (2022) Analysis and experimental research on efficiency characteristics of a Deep-Sea hydraulic power source. J Mar Sci Eng 10(9):1296
4. Gao Z et al (2019) Failure investigation of gear teeth fracture of seawater pump in a diesel engine. Eng Fail Anal 105(C):1079–1092
5. Yang Y et al (2022) Research on seawater hydraulic internal ball gear pump. Proc Inst Mech Eng Pt C J Mech Eng Sci 236(12):6967–6982
6. Wang S, Sakurai H et al (2011) The optimal design in external gear pumps and motors. IEEE ASME Trans Mechatron 16(5):945–952
7. Sun Y, Qi Z, Zhao G (2019) Optimization design of internal gear pump based on improved particle swarm optimization. Ji Xie Qiang Du 41:1378–1383
8. Seyama N, Tamura E, Mashidori H (2022) Suitable design of tooth profile modification for noise reduction in spur gear pump. In: The Proceedings of Mechanical Engineering Congress, Japan 2022, p S111
9. Wu K, Wang M (2023) Servo model of internal leakage and volume efficiency simulation for external gear pumps or motors. Jixie Chuandong 47:69–73
10. Yuan J, Shang X, Wang Z (2008) The optimization method on structure parameters of compound gear pump. J Henan Polytech Univ (Nat Sci) 27(05):546–550
11. Fan L (2011) A new kind of gear pump to solve trapped oil problem. Hydraul Pneumatics Seals 31(11):59–61
12. Li Y, Liu K (2009) Precise simulation analysis of relief area of external spur-gear pump. Trans Chin Soc Agric Eng 25(03):42–45

Open Access This chapter is licensed under the terms of the Creative Commons Attribution-NonCommercial-NoDerivatives 4.0 International License (http://creativecommons.org/licenses/by-nc-nd/4.0/), which permits any noncommercial use, sharing, distribution and reproduction in any medium or format, as long as you give appropriate credit to the original author(s) and the source, provide a link to the Creative Commons license and indicate if you modified the licensed material. You do not have permission under this license to share adapted material derived from this chapter or parts of it.

The images or other third party material in this chapter are included in the chapter's Creative Commons license, unless indicated otherwise in a credit line to the material. If material is not included in the chapter's Creative Commons license and your intended use is not permitted by statutory regulation or exceeds the permitted use, you will need to obtain permission directly from the copyright holder.

Dynamic Simulation of Methanol Dual Fuel Engine Mass Composition Based on Mean Value Model

Haiyan Wang and Kelin Wu

Abstract With the development and wide application of dual-fuel engine technology, the existing average value model of diesel engine can no longer well realise the state monitoring of dual-fuel engine under different fuel mode operation modes, and it can't complete the dynamic simulation of the changes in the composition of the intake and exhaust gas. In this paper, the average value method is used to establish a marine methanol dual-fuel engine model in Matlab-Simulink platform; based on the calculation of hydrocarbon fuel combustion products, a method of simulating and calculating the flow rate of each component in the intake and exhaust gases of the dual-fuel engine in the average model is proposed, which solves the problem of the average model of the diesel engine not being able to satisfy the control system's requirements for the simulation and calculation of the composition of gases and gases. The problem of simulation calculation is solved. After completing the development of the model, simulation experiments were carried out under different methanol substitution rates and EGR rates, and the feasibility of the model was verified according to the simulation results, which can be applied to the dynamic simulation of real-time control model for ships and the simulation of control system for dual-fuel engines.

Keywords Mean value model · Dual fuel engine · Engine mass composition

1 Introduction

For the research method of marine dual-fuel engine, the time and capital cost of experimental design is high and the conditions are difficult to be well satisfied [1]; on the other hand, in order to achieve the control objectives of dual-fuel engine to maintain good dynamic and steady state performance and to improve the fuel economy, it is also necessary to accurately simulate the real-time dynamic change process of the

H. Wang · K. Wu (✉)
Shanghai Maritime University, Pudong, Shanghai, China

© The Author(s) 2026

Z. Zhang, P. Mercorelli (eds.), *Mechanical Design and Simulation: Insights and Innovations*, Lecture Notes in Mechanical Engineering,
https://doi.org/10.1007/978-981-95-2405-1_67

dual-fuel engine model for the control algorithms design, simulation and validation. Therefore, the modelling and simulation of marine dual-fuel engines is a necessary research tool [2, 3].With the continuous development of dual-fuel engine technology, the control system has higher requirements for monitoring the state of the engine in different fuel operation modes. The mean value model can describe the diesel engine operating characteristics of the model, compact structure, fast computing speed, but also can take into account the model accuracy and real-time, suitable for model-based control system research, and the mean value model in the field of real-time control of diesel engines, energy efficiency optimisation, and ship fault diagnosis and so on can play its advantages to meet the requirements of its research on the model [4–7]. Therefore, it is of great theoretical significance and engineering value to explore and improve the deficiencies of the diesel engine mean value model and to enhance the practicality of the mean value model. The existing diesel engine mean value model cannot achieve the calculation of the dynamic change characteristics of the gas work composition, and the existing model often ignores the effect of real-time changes in gas composition on the system performance when characterising the dynamic coupling between the intake system, exhaust system and turbocharger, which leads to the limitation of the application of the mean value model in transient operating condition prediction and real-time control optimisation. In this paper, a marine methanol dual-fuel engine model is established in Matlab-Simulink platform by the mean value method, and based on the calculation of hydrocarbon fuel combustion products, a method of simulation and calculation of the process of flow change of the work composition in the mean value model is proposed, which solves the problem that the mean value model of the current diesel engine cannot achieve the simulation and calculation of the gas composition of the gas work required by the control system, and a variety of operating conditions are carried out after the development of the model is completed. After the development of the model was completed, dynamic simulation experiments were carried out under various operating conditions.

2 Dual-Fuel Engine Mean Value Model

2.1 Principles of the Mean Value Model for Dual Fuel Engines

The methanol dual fuel diesel engine ontology refers to calculations in terms of cylinders and dynamics. The mean value model uses the law of conservation of energy to calculate the indicated power and torque of the dual fuel diesel engine, and then Newton's second law of motion is used to calculate the dynamics. According to the working principle, the methanol dual fuel engine can be divided into several interrelated and relatively independent components. The body of the methanol dual fuel engine includes components such as the compressor, inter-cooler, sweeper tank, intake manifold, gas manifold, cylinders, exhaust manifold, turbine and other components. The schematic diagram of the working principle of the turbocharged inter-cooled four-stroke dual fuel engine used in the simulation is shown in Fig. 1.

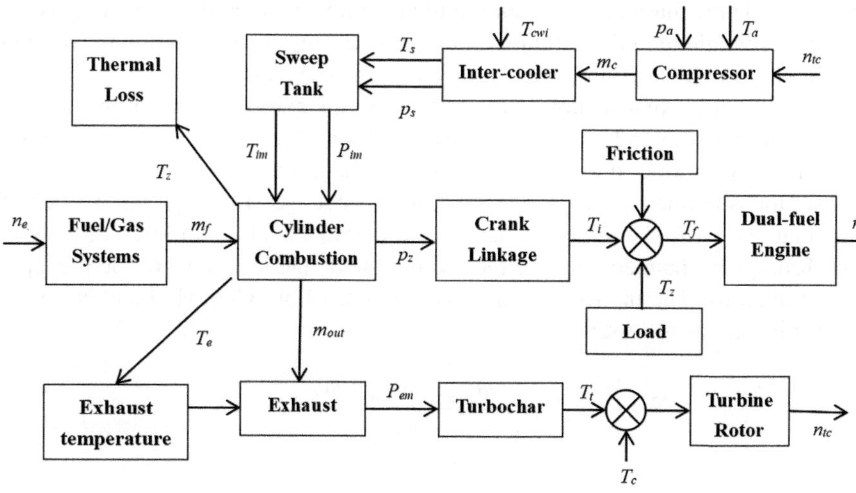

Fig. 1 Dual-fuel engine model working principle

The methanol-diesel dual-fuel engine has two modes of operation, full diesel mode and methanol dual-fuel mode. The dual-fuel engine used for the simulation uses methanol in-cylinder direct injection and diesel priming, and the power output in the methanol dual-fuel mode is comparable to that in the full diesel mode. The full diesel mode can be used when the load is <10% and can be switched to the methanol mode when the load is ≥10%, and the methanol substitution rate can reach up to 95% at 100% load [7].

2.2 *Model of Engine Mass Composition and Thermodynamic Parameters*

The averaging model does not involve the working process in the cylinder and focuses on describing the variation of the cyclic mean values of the engine state parameters with time [8, 9]. According to the hydrocarbon fuel combustion reaction equation in air, the mass and amount of CO_2, H_2O, and N_2 produced by the complete combustion of fuel oil and methanol per unit mass in the cylinder can be calculated.

$$C_xH_yO_z + \left(x+\frac{y}{4}-\frac{z}{2}\right)(O_2+3.76N_2) \rightarrow xCO_2 + \frac{y}{2}H_2O + \left(x+\frac{y}{4}-\frac{z}{2}\right)(3.76N_2) \quad (1)$$

The mass and substance of the unburned air from the exhaust cylinder are contingent on the extent of combustion that remains, and can be calculated from the excess air factor:

$$m_{a,out} = m_{a,in}\frac{\alpha-1}{\alpha},\ \alpha \ge 1 \quad (2)$$

Where $m_{a,out}$ is the mass of air remaining in the cylinder for combustion, $n_{a,out}$ is the mass of air remaining in the cylinder for combustion, $m_{a,in}$ is the mass of air entering the cylinder to participate in combustion, and α is the excess air coefficient.

The gas in the exhaust pipe is the remaining air and combustion products of combustion in the cylinder, which consists of four gases: CO_2, H_2O, N_2 and air, unlike the conventional diesel engine in which the composition of the gas in the exhaust pipe is defined to be unchanged, with the change of the operation mode of the methanol dual-fuel engine, the nature of the products of combustion of the fuel is synchronously changed with the fuel during the dynamic change of the fuel species and quantity. The instantaneous mass fractions of air, CO_2, H_2O and N_2 in the exhaust pipe are as follows:

$$x_{em,air} = \frac{\dot{m}_{a,out} - \dot{m}_{fl,a} - \dot{m}_{a,t}}{\dot{m}_{em}} \tag{3}$$

$$x_{em,CO_2} = \frac{\dot{m}_{CO_2,comb} - \dot{m}_{fl,CO_2}}{\dot{m}_{em}} \tag{4}$$

$$x_{em,H_2O} = \frac{\dot{m}_{H_2O,comb} - \dot{m}_{fl,H_2O}}{\dot{m}_{em}} \tag{5}$$

$$x_{em,N_2} = \frac{\dot{m}_{N_2,comb} - \dot{m}_{fl,N_2}}{\dot{m}_{em}} \tag{6}$$

In a similar manner, the work mass in the sweeper box consists of fresh air flowing out of the inter-cooler, exhaust gas drawn in by exhaust re-circulation. The instantaneous mass fractions of air, CO_2, H_2O and N_2 in the sweeper box are as follows:

$$x_{im,air} = \frac{\dot{m}_c + \dot{m}_{fl,a} - \dot{m}_{a,in}}{\dot{m}_{im}} \tag{7}$$

$$x_{im,CO_2} = \frac{\dot{m}_{fl,CO_2} - \dot{m}_{in,CO_2}}{\dot{m}_{im}} \tag{8}$$

$$x_{im,H_2O} = \frac{\dot{m}_{fl,H_2O} - \dot{m}_{in,H_2O}}{\dot{m}_{im}} \tag{9}$$

$$x_{im,N_2} = \frac{\dot{m}_{fl,N_2} - \dot{m}_{in,N_2}}{\dot{m}_{im}} \tag{10}$$

The molecular weight M_z and gas constant R_z of the work gas mixture are as follows:

$$M_z = \frac{1}{\sum_{i=1} \frac{x_i}{M_i}} = \frac{1}{\frac{x_{CO_2}}{44} + \frac{x_{H_2O}}{18} + \frac{x_{N_2}}{32} + \frac{x_{air}}{28.97}} \tag{11}$$

$$R_z = \frac{R}{M_z} \tag{12}$$

Where the molar gas constant R: 8.314 J/(mol*k), xi and Mi are the mass fraction and molar mass of the constituent gases in the work mixture, respectively.

The adiabatic coefficients of the four constituents of the work mass (i.e. CO_2, H_2O, N_2 and air) can be obtained from ReferProp and subsequently fitted to the relevant equations for the respective temperatures.

$$k_a = 1.435 - 0.0001155 \times T + 1.963\exp(-8) \times T^2 \tag{13}$$

$$k_{CO_2} = 1.138 + 63.63T^{-1.053} \tag{14}$$

$$k_{N_2} = 1.447 - 0.0001346 \times T + 3.074e - 8 \times T^2 \tag{15}$$

$$k_{H_2O} = 1.397 - 0.0001873 \times T + 4.344e - 8 \times T^2 \tag{16}$$

The adiabatic coefficient of the gas mixture is:

$$k_z = \sum_{i=1} k_i x_i \tag{17}$$

3 Simulation Result

The parameter list of the supercharged four stroke methanol dual fuel engine used for simulation is shown in Table 1.

The model is then run under the propulsion characteristics according to the requirements of each working condition in Table 2. The results of each simulation parameter are observed and recorded during the operation process. The simulation model is shown in Fig. 2.

Condition 1: The model commences from a full diesel mode at no load to the rated state, transitions from full diesel mode to a methanol dual-fuel mode at 200 s, and then shifts from a methanol dual-fuel mode to a pure diesel mode at 400 s. This process is intended to ascertain the dynamic characteristics of the model during the entirety of its operational cycle. The dynamic changes in engine speed and exhaust

Table 1 Methanol dual-fuel engine technical parameters

Parameters	Value
Number of strokes	4
Number of cylinders	6
Bore × stoke (mm)	170 × 185
Rated speed (r/min)	1500
Rated power (kW)	647

Table 2 Requirements for simulation conditions of methanol dual-fuel engine model

Condition	Simulation parameters	Operation mode	Substitution rate (%)	EGR rate (%)	Simulation time (s)
1	n_e,T_e	Fuel, Methanol Dual Fuel, Fuel	0%–95%–0%	25	550
2	M_{em},M_{im}	Fuel, Methanol Dual Fuel, Fuel	0%–95%–0%	15	750
3	$M_{em,CO2}$,$M_{em,N2}$,$M_{em,H2O}$,$M_{em,air}$	Methanol Dual Fuel	25%/50%/75%/95%	0/15/25	–

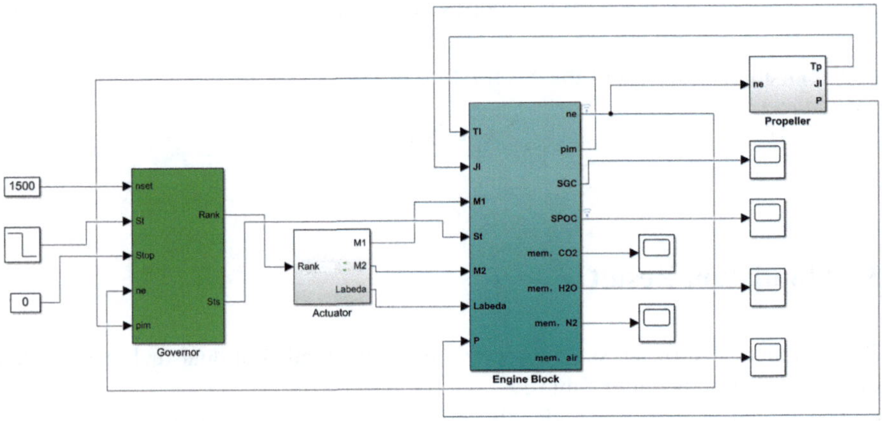

Fig. 2 Simulink block diagram of dual-fuel engine model

temperature are observed and recorded during this operation, and the simulation results are shown in Fig. 3.

The duration of the model's simulation is completed within 2 s, exhibiting a high level of real-time performance and generating results that closely resemble the performance parameters of a dual-fuel engine.

Condition 2: The model is initiated from a full diesel mode at no load to a rated state of operation, transitioned from full diesel mode to a methanol dual-fuel mode at 300 s, and subsequently transitioned from methanol dual-fuel mode to pure diesel mode at 500 s. The dynamic fluctuations in the flow rates of the components in the engine exhaust pipe and the sweeper tank mass are observed and documented during this operation, and the simulation outcomes are presented in Figs. 4 and 5.

After about 200 s, the engine is stable and the flow rates of the different gases in the exhaust pipe and the sweeper box reach a steady state. When the engine was

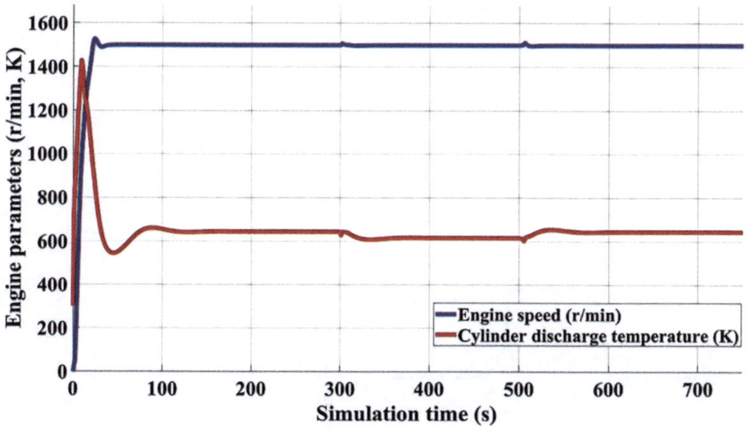

Fig. 3 Condition 1 dynamic characteristics of the whole process operation

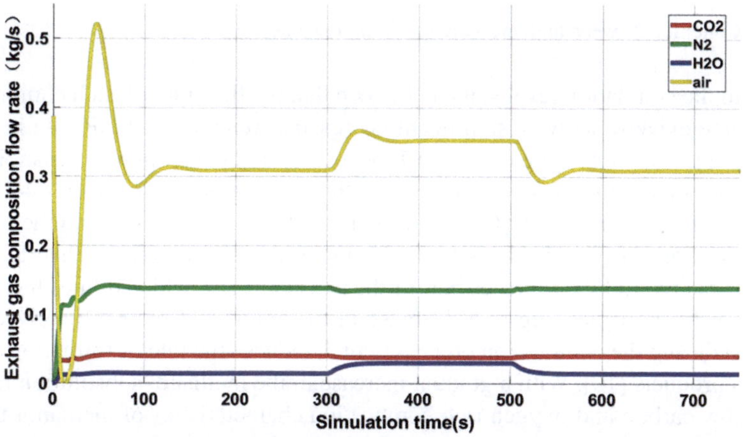

Fig. 4 Condition 2 Exhaust mass composition flow rate

switched from diesel to methanol, the composition of the exhaust and intake media changed significantly: the flow of water vapour and unburned air increased, while the flow of CO_2 decreased. These changes are consistent with the low-carbon and oxygen-rich combustion characteristics of methanol fuel, which further confirms the effects of different fuel modes on the emission characteristics of the engine.

Condition 3: In the methanol dual-fuel operation mode, the engine model was set at rated speed and tuned to operate at different EGR rates and methanol substitution rates. After the engine state was stable, the flow data of each work composition in the engine exhaust were recorded under each working condition.

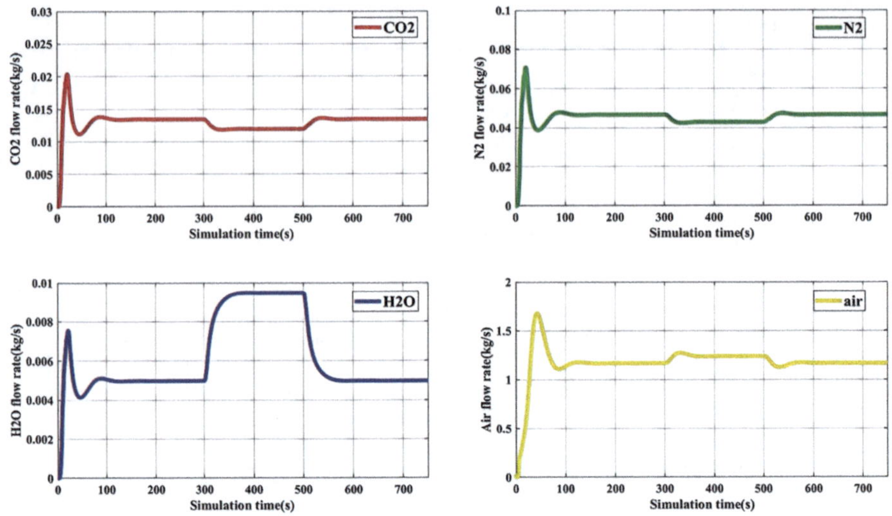

Fig. 5 Condition 2 sweep tank mass composition flow rate

From the simulation results, it can be seen that the flow rate of each component in the engine exhaust shows a significant decreasing trend as the EGR rate increases. When the EGR rate is increased from 15% to 25% in the maximum methanol substitution rate mode, the flow rate of CO_2 in the exhaust decreases from 44.11 g/s to 31.53 g/s, the flow rate of H_2O decreases from 35 g/s to 25.02 g/s, and the flow rate of N_2 decreases from 15.84 g/s to 11.33 g/s. The main reason for this decreasing trend is that the EGR introduces part of the exhaust gases, which dilutes the concentration of fresh air, leading to a decrease in the amount of oxygen involved in the reaction during the combustion process, which reduces the rate of formation of combustion products [10]. With a gradual increase in the methanol substitution rate, due to the low carbon and oxygen rich combustion characteristics of methanol fuel, the CO_2 flow rate gradually decreases, the H_2O and unburned air flow rates increase, and the N_2 flow rate remains almost unchanged due to the fact that N_2 does not participate in the combustion process, which is less affected by the methanol substitution rate.

A higher methanol substitution rate combined with a higher EGR rate can effectively reduce CO_2 emissions while maintaining engine performance. At 95% methanol substitution rate and 25% EGR rate, the CO_2 flux is reduced to 31.53 g/s, which is about 30% less than the baseline condition (25% methanol substitution rate and 0% EGR rate). This result is in line with the low-carbon combustion characteristics of methanol fuel and the modulation of the combustion process by exhaust gas recirculation; EGR reduces heat loss by lowering the combustion temperature, while the high octane and oxygenation characteristics of methanol further optimise combustion efficiency, resulting in a low-carbon emission target (Fig. 6).

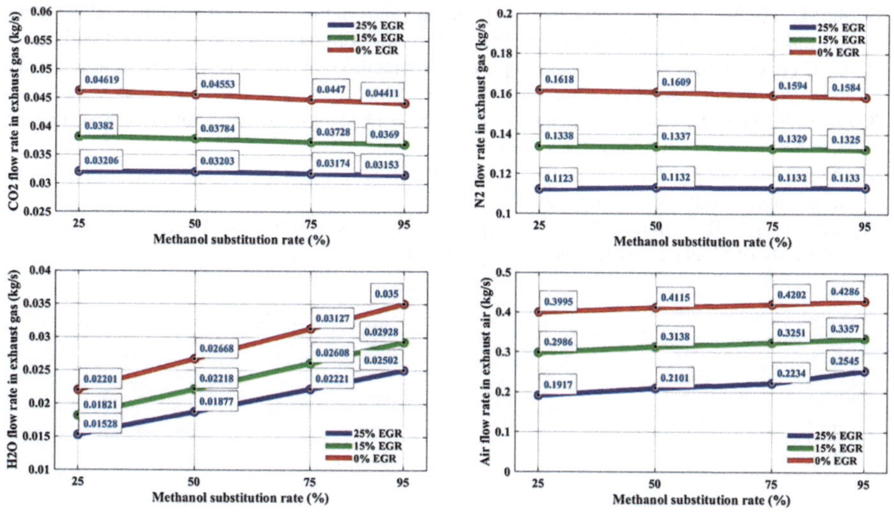

Fig. 6 Condition 3 Flow rate of components of exhaust gas

4 Conclusion

1. The model simulation of the built methanol dual-fuel diesel engine is rapid and exhibits excellent real-time performance, thus satisfying the requirements of the control system and state observation with regard to simulation accuracy and speed.
2. The improved mean value model can simulate and calculate the flow rate of different components in the exhaust and intake gas under different methanol substitution rate and EGR rate conditions, and the simulation results are in line with the reality, and it can monitor the dynamic change process of the composition of the gas work-masses in different fuel modes of the dual-fuel engine under different conditions, which can be applied to the dynamic simulation of real-time control model of the ship and the interaction of the ship's power system.

References

1. Deng Z, Cai H, Wang Z, Chen M, Zhang F, Yin Z (2024) Emission and economic performance analysis of dual fuel marine engine. Navigat China 2(5):128–133
2. Rongbin X, Jinguo Z, Chang L et al (2022) Simulation study on the performance and emission parameters of a marine diesel engine. J Mar Sci Eng:985–985
3. He S (2015) Simulation and optimization on the performance and emissions of dual-fuel marine engine. Shanghai Jiao Tong University, Shanghai
4. He LD, Wang HY, Liu S (2013) Discussion of modeling of marine diesel engine. Intern Combust Engine Powerpl:53–55
5. Zhang H, Ding F, Yang H (2024) Fuel saving optimization for diesel generator set based on Grey wolf-particle swarm algorithm. Ship Eng:102–107+122

6. Shen C (2019) Research on matching computation and control strategy between variable geometry turbocharger and low speed diesel engine. Harbin Engineering University, Helongjiang
7. Xie Q et al (2020) A surrogate-based optimization method for the issuance of passenger evacuation orders under ship fires. Ocean Eng 209:107456
8. Zhang D, Shen X, Wang L, Cheng S, Li C (2023) Development status and trends of marine methanol engine technology. Diesel Engine:1–4
9. Theotakatos GP (2008) Ship propulsion plant transient response investigation using a mean value engine model. Int J Energy:66–74
10. Karlsen AT (2012) On modeling of a ship propulsion system for control purposes. Norwegian University of Science and Technology, pp 99–110

Open Access This chapter is licensed under the terms of the Creative Commons Attribution-NonCommercial-NoDerivatives 4.0 International License (http://creativecommons.org/licenses/by-nc-nd/4.0/), which permits any noncommercial use, sharing, distribution and reproduction in any medium or format, as long as you give appropriate credit to the original author(s) and the source, provide a link to the Creative Commons license and indicate if you modified the licensed material. You do not have permission under this license to share adapted material derived from this chapter or parts of it.

The images or other third party material in this chapter are included in the chapter's Creative Commons license, unless indicated otherwise in a credit line to the material. If material is not included in the chapter's Creative Commons license and your intended use is not permitted by statutory regulation or exceeds the permitted use, you will need to obtain permission directly from the copyright holder.

Investigation into the Impact Damage of Vehicle Restraint Systems Based on Madymo

Changhao Li, Hailan Xu, Zhenfei Zhan, Yajun Luo, and Hailiang Chen

Abstract It is imperative to accurately simulate the occupant's seating position in collision simulations. This study focuses on a frontal crash test of an automobile, utilizing MADYMO software for digital simulation to analyze passenger injuries during collisions. The analysis encompasses various seating postures, including upright, forward-leaning, backward-leaning, left-leaning, and right-leaning positions. The research reveals that while the AEB (Automatic Emergency Braking) system effectively reduces the overall risk of accidents, it may also increase the risk of injury to specific body parts under certain conditions. For instance, sudden AEB-induced braking can subject occupants in undesirable postures (e.g., head lowered or sideways) to additional impact forces on the head and neck, thereby increasing the risk of neck injuries. Moreover, the increased occupant displacement caused by emergency braking can lead to sudden forward leaning. If seat belts are not properly worn or improperly positioned, this can result in elevated axial pressure on the chest and thighs, further increasing the risk of injury to these areas. Therefore, when utilizing active safety technologies such as AEB, it is crucial to ensure proper seat belt usage and maintain appropriate seating posture.

Keywords Active and passive safety · Occupant sitting posture · Occupant collision injury analysis

C. Li
China Merchants Testing Vehicle Technology Research Institute, Chongqing, China

School of Mechatronics and Vehicle Engineering, Chongqing Jiaotong University, Chongqing, China

H. Xu
China Merchants Testing Vehicle Technology Research Institute, Chongqing, China

Z. Zhan (✉) · Y. Luo · H. Chen
School of Mechatronics and Vehicle Engineering, Chongqing Jiaotong University, Chongqing, China
e-mail: zhenfeizhan@cqjtu.edu.cn

© The Author(s) 2026

Z. Zhang, P. Mercorelli (eds.), *Mechanical Design and Simulation: Insights and Innovations*, Lecture Notes in Mechanical Engineering,
https://doi.org/10.1007/978-981-95-2405-1_68

1 Introduction

1.1 Vehicle Active and Passive Safety

The annual global economic losses caused by road traffic accidents account for about 3% of GDP in many countries [1]. In response to this major social problem, the United Nations released the Global 10-year Plan for Road Safety 2021–2030 [2] in 2020, aiming to reduce the occurrence of road traffic accidents. The economic loss was over 15.1 billion RMB [3]. Although the incidence of traffic accidents in China has gradually decreased since 2002, the traffic accident index remains at a high level and shows a slow rising trend from 2015 to 2021, as shown in Fig. 1 [4, 5]. With the increasing demand for automobile safety, the combination of active safety and passive safety has gradually become an important direction for the future development of automobile safety technology. This integrated safety control not only better responds to the complex road environment, but also provides more comprehensive protection before and after an accident. In order to achieve this goal, automotive active and passive safety testing techniques are refined into multiple levels. The vehicle crash test is one of the most important parts in order to comprehensively evaluate the safety performance of the vehicle in various crash situations [6, 7].

1.2 Analysis of the Effect of Passenger Sitting Posture

It is crucial to accurately simulate the occupant's sitting position in collision simulation. Therefore, this paper plans to use MADYMO software to conduct digital modeling design of active and passive safety occupant collision. In this way, the occupant's muscle response and action response during a collision can be more accurately simulated, and the occupant's departure can be limited by a pre-activated limiting device, thus reducing the possible injury caused during the forward thrust. By integrating these factors, the study aims to optimize the collision safety design and improve the protection level of occupants to achieve more effective safety protection [8–10].

Studies have shown that injuries to the head, neck and chest of passengers account for 83–97% of the total injuries [11]. Analysis shows that the proportion of sitting upright is more than 60%. However, in the emergency braking state, the occurrence rate of forward posture increas es significantly.

Based on these findings, this paper will focus on the impact of human body posture, including left-right swing degree, waist tilt degree. These degrees of freedom have an important effect on the attitude of the occupant in the collision and the corresponding injury risk. Through an in-depth analysis of the effects of these degrees of freedom on occupant sitting, the study aims to improve understanding of occupant safety protection in collisions, optimize safety design, and ultimately reduce crash injuries.

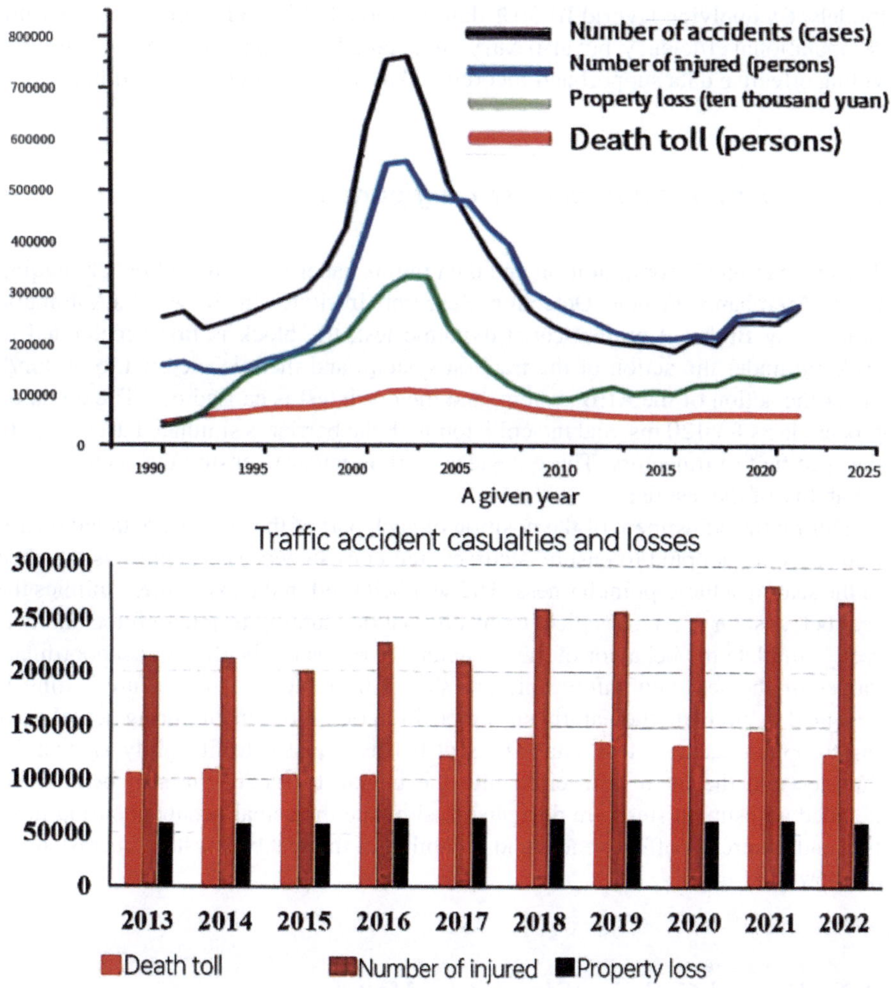

Fig. 1 Road traffic casualties and losses in China in the past decade and road traffic accidents in China from 1990 to 2021

2 Active and Passive Safety Digital Design

2.1 The Choice of the Dummy

TASS has developed a series of MADYMO dummies to evaluate the harmful effects of occupant restraints on the human body under complex boundary conditions. These models are able to simulate occupant injuries in detail under different conditions. The Hybrid III 50% dummy model is selected as the experimental object in this paper. This model provides sufficient accuracy in occupant damage assessment and can effectively support the validation and generalization of collision simulation models. By applying Hybrid III 50% dummy model, this study can not only ensure computational efficiency, but also carry out detailed occupant damage analysis, providing effective data support and theoretical basis for research in related fields.

2.2 Experimental Parameter Configuration

The test parameter configuration and data in this paper are provided by Chongqing China Merchants Vehicle Detection Research Institute. In the AEB (Automatic Emergency Braking) prefabricated dynamic test, the block is first accelerated to 50 km/h under the action of the traction system, and then decelerated to 30 km/h under the action of the AEB system, and the crash test is carried out. The collision process lasts for 120 ms, and the collision with the barrier is simulated according to the specific test trajectory. These devices work together to ensure the accuracy and reliability of the test data.

During the adjustment of the position of each part of the dummy, detailed operation steps are adopted to ensure accuracy and consistency. The dummy is attached to the seat by a three-point harness. The seat belt used in the experiment mimics the seat belt system in a real vehicle to ensure that the dummy is able to force and correctly simulate the behavior of the occupants in a crash, the theoretical coordinate values of the seat belt safety hole are shown in Table 1. This posture setting is designed to simulate the natural sitting position of the occupant during actual driving to obtain accurate test data. The seat belt is adjusted to fit tightly against the dummy, and the latch is inserted into the buckle to secure the seat belt. These detailed adjustment steps are designed to simulate the actual occupant sitting position and ensure the effectiveness and reliability of the seat belt system in crash tests, as shown in Fig. 2.

2.3 Frontal Collision Simulation Model

Figure 3 shows how the various parts of the dummy model can be connected using the moving hinges in MADYMO. A free hinge allows motion with multiple degrees of freedom, a sliding hinge supports movement in a straight direction, a rotating

Table 1 Theoretical coordinate value of seat belt mounting hole position

	X (mm)	Y (mm)	Z (mm)
Guide ring mounting hole	220	±260	720
Winding mounting hole	120	±330	−50
Anchor point mounting hole	160	±300	−160
Buckle mounting hole	−40	∓230	−80

Note: Take the midpoint of the intersection line between the rigid seat plate and the backrest as the coordinate origin

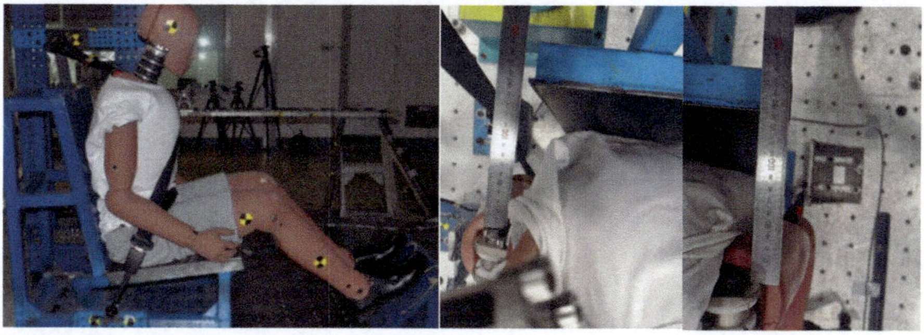

Fig. 2 Dummy shoulder measurement diagram

Fig. 3 Hybrid_III_50th dummy model location

hinge provides rotational motion, a ball hinge allows rotation in all directions, and an Euler hinge is used for a specific rotational motion. The reasonable selection of these articulation modes makes the motion performance of the dummies in the body model more truly reflect the actual collision situation. Several components of the safety belt system are modeled, including the geometric model of the safety belt, the mechanical structure of the winder including the internal spring winder mechanism, the inertial locking mechanism, the locking buckle, etc. Figure 3 shows the model after the localization of the dummy.

2.4 *Model Input and Output Definition*

In this study, the role of active safety devices in collisions was evaluated by simulating the deceleration effect of automatic emergency braking systems. The operating process of the AEB system is simulated as a braking phase lasting 998 ms, as shown in Fig. 4, during which the maximum braking deceleration is 1G. Then, the collision phase is simulated as a collision process lasting 120 ms, and the maximum collision

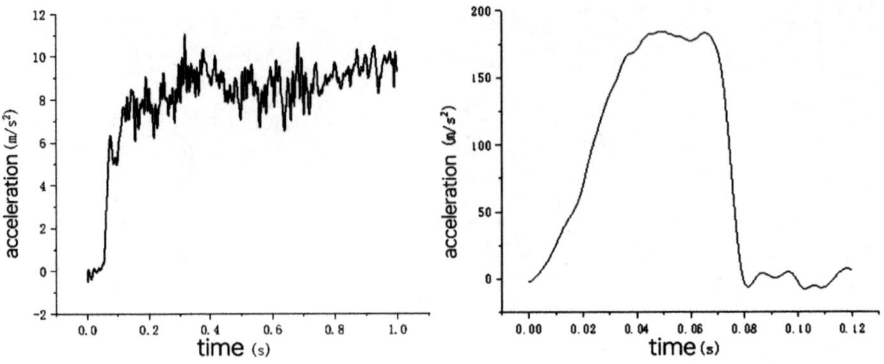

Fig. 4 AEB prefabricated dynamic acceleration time curve and collision acceleration time curve

deceleration in this phase reaches 18G as shown in Fig. 4. These data are obtained by analyzing the acceleration curve of B-pillar in prefabricated dynamic crash test and provided by China Merchants Vehicle Detection Research Institute.

2.5 *Collision Condition Simulation Model Verification*

We compare the test data and simulation data during the collision in detail, and analyze the time variation curves of several response parameters. The accuracy of the simulation model was evaluated by EEARTH scoring method, and the results showed that the simulation model of chest compression had the best performance, with a score of 92.32%. The scores of other parameters are more than 80%, indicating that the simulation model has a good agreement with the actual test data in shape, amplitude and trend. These results validate the validity and accuracy of the simulation model in assessing occupant collision damage.

2.6 *Occupant Restraint System Parameter Setting*

In the event of a vehicle crash or emergency braking, different sitting positions can affect the stress and risk of injury experienced by passengers. Therefore, understanding typical occupant sitting posture and its impact on safety performance is a key part of traffic accident analysis and safety system design. In this study, 5 common sitting positions (6° front, 6° back, 12° left and right) were adopted to simulate the real sitting position of passengers when they leaned forward. These adjustments help to more accurately simulate the injuries that different sitting positions may cause to passengers in a collision, and provide a scientific basis for improving occupant protection design. The passenger frontal collision simulation model is shown in Fig. 5.

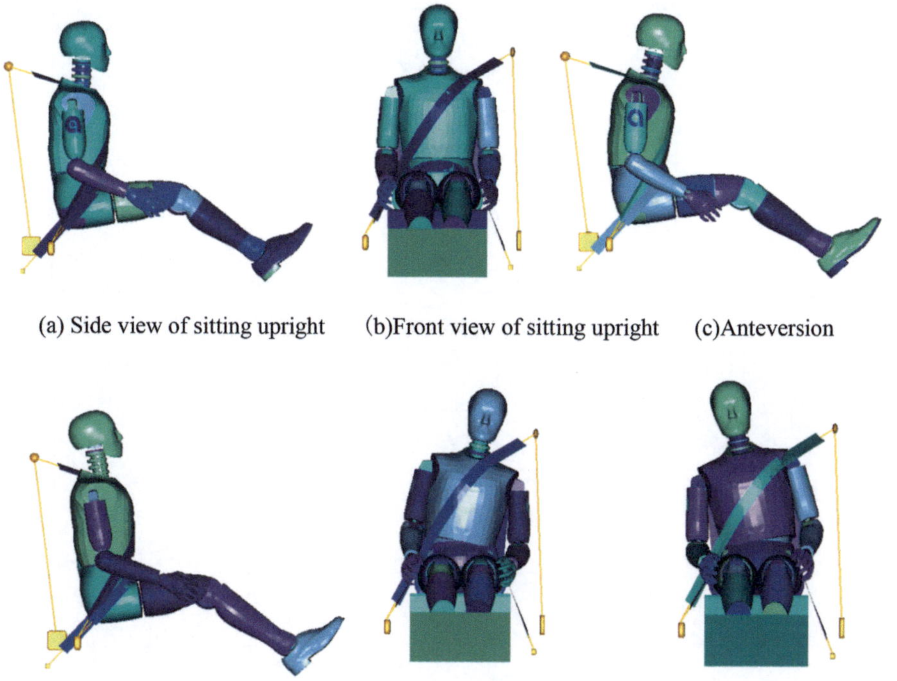

(a) Side view of sitting upright (b)Front view of sitting upright (c)Anteversion

(d) Retroversion (e) Left deviation (f) Right deviation

Fig. 5 Frontal collision simulation model of passenger in five sitting positions. (**a**) Side view of sitting upright; (**b**) Front view of sitting upright; (**c**) Anteversion; (**d**) Retroversion; (**e**) Left deviation; (**f**) Right deviation

3 Occupant Collision Damage Analysis Under Active and Passive Safety Reinforcement

The role of AEB in the collision is further discussed by comparing the stress status of the seat belt in the two situations. The following figure shows the relationship between the impact belt force and time in five sitting positions: upright, leaning forward, leaning back, left and right leaning.

3.1 The Maximum Value of Each Injury Index in Different Sitting Position

The role of AEB in the collision is further discussed by comparing the stress status of the seat belt in the two situations. The details are shown in Fig. 6.

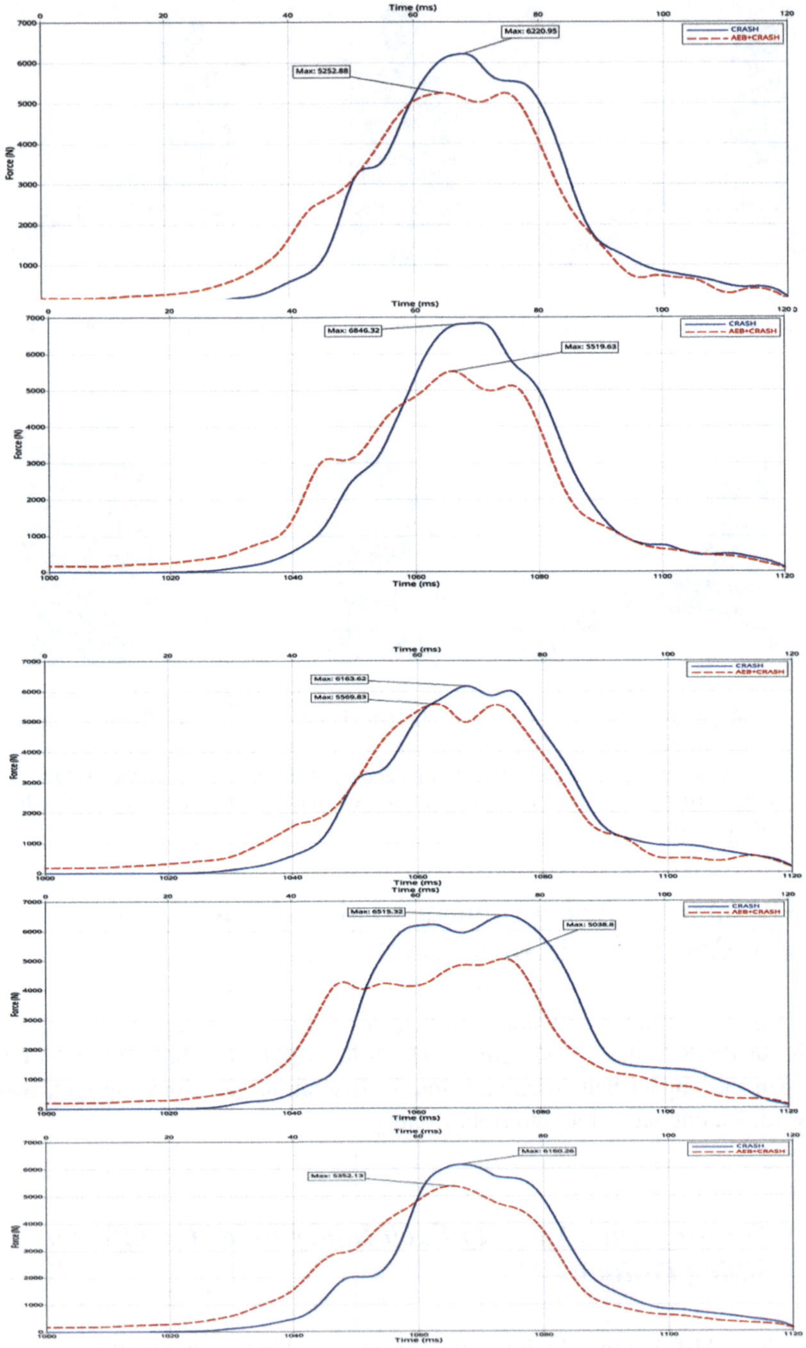

Fig. 6 The relationship between the impact belt force and time in five sitting positions

This is the posture most passengers normally take, sitting up straight with their backs against the seat back and their head and neck in a natural upright position. The injury indexes of upright sitting posture are shown in Table 2.

The occupant leans forward with the head closer to the steering wheel or front seat. This posture may be caused by the occupant looking at the road or objects ahead. In this position, the air bag may not adequately protect the head and increase the risk of facial or neck injury. The injury indexes of leaning posture are shown in Table 3.

The occupant leans back with his back at an Angle to the seat back, a relaxed posture commonly seen during long rides. This position can affect the protective effect of the seat belt in a crash, resulting in the seat belt may not be effective in securing the upper body. The damage index of supine position is shown in Table 4.

The passenger leans to the left or right, often when the passenger rests against the window or center armrest. This position can result in the failure of the protection of the side air bag in the event of a collision and increase the risk of injury in the event of a side impact. The two damage indexes are shown in Tables 5 and 6.

In summary, while the AEB system performed well in reducing damage in specific areas in all tests, the risk of damage in some other critical areas increased, and further optimization of the AEB system adjustment and setup was required in future designs to more fully protect occupant safety.

Table 2 Injury indexes of upright sitting posture

	HIC15	Nij	Chest compression (mm)	Axial compression of the left thigh (N)	Axial compression of the right thigh (N)
CRASH	33.20	0.25	31.70	225.52	574.53
AEB + CRASH	20.95	0.23	28.95	115.86	461.95

Table 3 Injury indexes of forward-leaning sitting posture

	HIC15	Nij	Chest compression (mm)	Axial compression of the left thigh (N)	Axial compression of the right thigh (N)
CRASH	33.20	0.25	31.70	225.52	574.53
AEB + CRASH	20.95	0.23	28.95	115.86	461.95

Table 4 Injury index of reclining posture

	HIC15	Nij	Chest compression(mm)	Axial compression of the left thigh (N)	Axial compression of the right thigh (N)
CRASH	33.20	0.25	31.70	225.52	574.53
AEB + CRASH	20.95	0.23	28.95	115.86	461.95

Table 5 Left-leaning sitting posture each injury index

	HIC15	Nij	Chest compression (mm)	Axial compression of the left thigh (N)	Axial compression of the right thigh (N)
CRASH	33.20	0.25	31.70	225.52	574.53
AEB + CRASH	20.95	0.23	28.95	115.86	461.95

Table 6 Right-leaning sitting posture each damage index

	HIC15	Nij	Chest compression (mm)	Axial compression of the left thigh (N)	Axial compression of the right thigh (N)
CRASH	33.20	0.25	31.70	225.52	574.53
AEB + CRASH	20.95	0.23	28.95	115.86	461.95

4 Conclusion

In this paper, the occupant's sitting position is divided into five types: upright, forward, back, left, and right. While the AEB system can effectively reduce the risk of an overall accident, it may also increase the risk of injury to specific parts of the occupant in some situations, especially in the following scenarios: The sudden braking of the AEB when the occupant is not in an ideal position, such as lower head or leaning sideways, may cause the occupant's head and neck to bear additional impact force, increasing the risk of neck injury. The increased amount of occupant departure caused by emergency braking can cause the occupant to suddenly lean forward, and if the occupant is not wearing the seat belt properly or the seat belt is improperly positioned, the axial pressure on the chest and thigh may be increased, increasing the risk of injury to the relevant area. These results indicate that there are significant differences in the impact of different sitting posture on crash injuries, and the importance of occupant sitting posture should be taken into account in collision protection design. Through experimental and computer simulations, we can gain insight into the impact of different sitting positions on safety system performance and optimize vehicle design accordingly to improve overall safety.

References

1. World Health Organization (2018) Global status report on alcohol and health 2018. World Health Organization, Geneva
2. World Health Organization (2022) Global plan for the decade of action for road safety 2021–2030. WHO Regional Office for the Western Pacific: World Health Organization
3. National Bureau of Statistics (2022) China statistical yearbook 2022. China Statistics Press, Beijing, pp 20–21

4. Website of the Ministry of Public Security (2023) The number of motor vehicles in China reached 417 million, with more than 500 million drivers
5. Ling Y (2023) Research on road traffic accident causation based on topic model and Bayesian network. Lanzhou University, Lanzhou
6. Hao Y, Liu W (2021) Development status and trend analysis of automotive passive safety test technology. China Automobile (05):41–45
7. Cheng C, Yi B, Mao L et al (2023) Study on car body solder joint failure based on crash safety. In: Proceedings of China SAE Congress 2021: Selected Papers. Springer, Singapore, pp 466–478
8. Zhang G (2019) Research on occupant injury in oblique collision caused by automatic emergency steering. Chongqing University of Technology
9. Cao L, Ouyang Z, Xu Z et al (2016) Research on parameter matching optimization of reversible constrained systems. Chin J Mech Eng 52(10):133–141
10. Wang K (2018) Simulation and optimization analysis of integrated constraint system based on AEB signal. Chongqing University of Technology
11. Adam T, Untaroiu CD (2011) Identification of occupant posture using a Bayesian classification methodology to reduce the risk of injury in a collision. Transp Res Pt C Emerg Technol 19(6):1078–1094

Open Access This chapter is licensed under the terms of the Creative Commons Attribution-NonCommercial-NoDerivatives 4.0 International License (http://creativecommons.org/licenses/by-nc-nd/4.0/), which permits any noncommercial use, sharing, distribution and reproduction in any medium or format, as long as you give appropriate credit to the original author(s) and the source, provide a link to the Creative Commons license and indicate if you modified the licensed material. You do not have permission under this license to share adapted material derived from this chapter or parts of it.

The images or other third party material in this chapter are included in the chapter's Creative Commons license, unless indicated otherwise in a credit line to the material. If material is not included in the chapter's Creative Commons license and your intended use is not permitted by statutory regulation or exceeds the permitted use, you will need to obtain permission directly from the copyright holder.

Light-Weight Design and Application Analysis of Lumbar Rehabilitation Robot Gears

Zhuo Hao, Yanbin Ning, and Zhenzhen Li

Abstract Aiming at the lightweight design requirements of the large gear, which was a key component of the lumbar rehabilitation robot, this study was based on Altair Inspire software and adopts the method of topology optimisation to obtain a dendritic weight reduction model through OptiStruct analysis. After the verification of safety coefficient and shape variable, the weight reduction of the large gear was successfully achieved to be 4.115 kg, with a weight reduction ratio of 30.2%. The lightweight performance of five materials (Steel (AISI 1015), Steel (C45E), Titanium (Ti-17), ABS, and PEEK) was further compared. The results show that Titanium (Ti-17) has the lightest mass of 6.205 kg and the highest cost in the optimised design of metal materials. Plastic materials (ABS and PEEK) have significantly better lightweight performance than metal materials, especially PEEK, with an optimised mass of 1.597 kg, which is 1.25 times that of ABS, and a minimum factor of safety of 2.8, which is 2.5 times that of ABS. Considering the excellent overall performance of PEEK, it can be used as a preferred alternative material for the manufacturing of key components of lumbar rehabilitation robots. This study provides a structural optimisation and material selection scheme for the lightweight design of lumbar rehabilitation robots, and provides a reference for future related research and technology development.

Keywords Altair inspire · Lumbar rehabilitation robot · Big gear · Light-weight design · PEEK

1 Introduction

As China's population ages, the incidence of diseases rises, and the groups of elderly and disabled people continue to expand, the demand for medical rehabilitation increases significantly [1–3]. Digital technology promotes intelligent

Z. Hao · Y. Ning · Z. Li (✉)
Guangdong University of Science and Technology, DongGuan, China

© The Author(s) 2026

Z. Zhang, P. Mercorelli (eds.), *Mechanical Design and Simulation: Insights and Innovations*, Lecture Notes in Mechanical Engineering,
https://doi.org/10.1007/978-981-95-2405-1_69

transformation of the medical rehabilitation industry, and rehabilitation robots, as emerging devices, are increasingly widely used [4]. The prevalence of back injuries among workers was higher in the logistics, construction and medical rehabilitation industries, at 84%, 75% and 67%, respectively [5]. This has led to an increasing social demand for lumbar rehabilitation training and protection, and intelligent and efficient lumbar rehabilitation robots are becoming a focus of attention. However, traditional rehabilitation treatment relies on the experience of doctors, and with the increasing awareness of environmental protection and resource conservation, the machinery manufacturing industry is also faced with the problems of quality reduction, volume reduction and cost reduction [6]. The Lumbar Rehabilitation Robot is able to provide a more comprehensive and efficient rehabilitation programme by integrating physiotherapy and robotics [7, 8]. However, the current lumbar rehabilitation robots generally have problems such as large size, heavy weight, and complex operation, which limit their application in home and community rehabilitation scenarios, so lightweight design has become a key direction to enhance their market competitiveness and user experience.

In order to solve the current problems of rehabilitation robots and comply with the trend of intelligent and convenient rehabilitation robots, scholars at home and abroad have proposed different solutions. For example, Al-Dahiree Omar's team from Universiti Kebangsaan Malaysia [9] proposed a compact lumbar spine support exoskeleton design for effective lumbar support through the combination of an energy storage system and a rotary elastic actuator. Pengfei Ge [10] of Hebei University of Technology designed a lumbar powered and protected exoskeleton robot with flexibility to provide adaptive support to patients by simulating spinal motion through hydraulic cylinders. Yin Shuai et al. [11] conducted a lightweight adaptive flexible design and evaluation for an ankle rehabilitation robot. Its research results show that through structural optimisation and material selection, the weight of the rehabilitation robot can be effectively reduced, while ensuring that its performance meets the requirements for use. Nevertheless, the existing lumbar rehabilitation robots still have problems such as high weight, high energy consumption, and high risk of human-robot interaction, which limit their application in home and community rehabilitation, increase the burden of use and difficulty of operation for patients, and may even cause additional pressure on the patient's lumbar region, which affects the rehabilitation effect [12]. Therefore, lightweight design has become an inevitable choice for the development of lumbar rehabilitation robots.

This study focuses on the lightweight design of the large gear, a key component of a lumbar rehabilitation robot, in order to improve the overall performance of the robot by optimising the structure and material selection. The study adopts Altair Inspire software to optimise the topology and reduce the material of the large gear, combined with the comparison of the performance of multiple materials to verify the feasibility and effectiveness of the lightweight design. Optimised to reduce weight by removing excess material. OptiStruct analysis was also used to assess the performance of the optimised structure to ensure that it was lightweight while meeting strength and stiffness requirements. Ultimately, by comparing the performance of multiple materials and selecting the one suitable for the lightweight design, the overall weight reduction of the lumbar rehabilitation robot is achieved, which improves the comfort of use and the flexibility of operation, reduces the production cost, and provides a reference for the rehabilitation machine to meet the actual needs of home and community rehabilitation.

2 Topological Optimisation of Large Gears

Take the lumbar rehabilitation robot designed in the preliminary stage of the group as an example (the structure is schematically shown in Figs. 1 and 2). The large gear structure, as the core component of the transmission system (shown in Figs. 1 and 2), undertakes the key task of efficiently transmitting the motor power to the end actuator in the design of the waist-assisted rehabilitation exoskeleton robot. It achieves precise motion control by meshing with the active gear to ensure that the extension, lateral flexion and rotation motions of the human waist can be carried out in accordance with the preset angles and speeds. Therefore, optimising the structural design of the large gear can significantly improve the overall performance of the system, which directly affects the stability and reliability of the system. Therefore, this paper focuses on the lightweight design of the large gear structure of the transmission system.

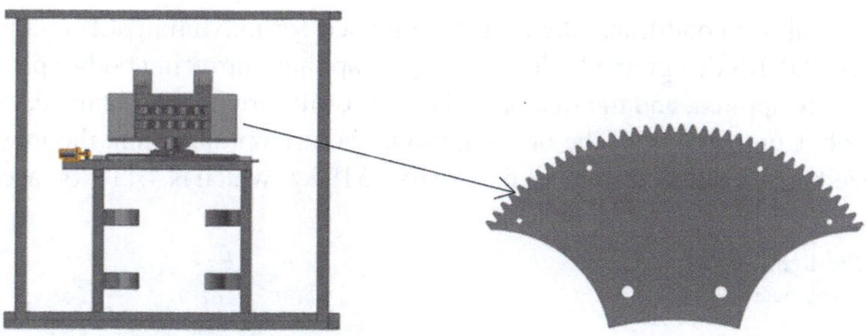

Figs. 1 and 2 Structure diagram of lumbar rehabilitation robot. Big gear model

2.1 OptiStruct Analysis of Large Gears

In order to clearly compare the structural performance of the large gear before and after the lightweight design, this study uses OptiStruct advanced structural analysis and optimisation software developed by Altair, Inc. to conduct a detailed analysis of the coefficient of safety and displacement of the large gear of the lumbar rehabilitation robot under the same loading condition. Its design space (brown part) and non-design space (grey part) are determined, and the force of 500 N and the corresponding fixed constraints are preset according to the actual load of the part, and the pre-design results are shown in Fig. 3.

The raw material of the part is Steel 1015, and through OptiStruct analysis, from Fig. 4, the maximum safety factor of the original model is 7.268e+05, and the minimum safety factor is 16.8; from Fig. 5, the maximum displacement of the original model is 6.439e-02 mm, and the minimum displacement is 0 mm. Referring to the structural design standard of medical rehabilitation equipment, in order to ensure the reliability and accuracy of the part in practical application, the safety factor of the large gear under load is set to be greater than 3, and the maximum displacement should be controlled within 1 mm. Referring to the structural design standard of medical rehabilitation equipment, in order to ensure the reliability and accuracy of the parts in practice, the safety coefficient of the large gear under load should be set to be greater than 3, and the maximum displacement should be controlled within 1 mm. Comparing the results of Figs. 4 and 5, the structure and material selection of the original model are within the requirements, and therefore the design is optimised in the next stage.

2.2 Topology Optimisation for Large Gears

Under the above conditions, the part is optimised for maximum stiffness topology and PolyNURBS design methods including wrapping, mirroring body, splitting and bridging are applied, and the tree optimisation results are finally obtained, as shown in Fig. 6b. Compared with the original model before optimization, the mass of the optimized part is reduced from 13.634 kg to 9.519 kg, which is 4.115 kg, accounting

Fig. 3 Pre-lightweight design of big gear

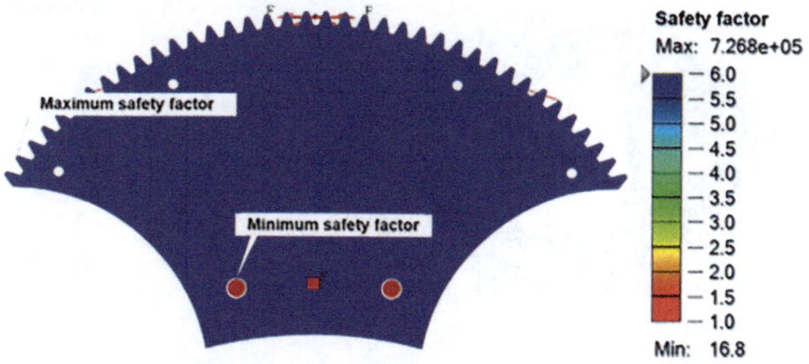

Fig. 4 Safety factor of total load condition

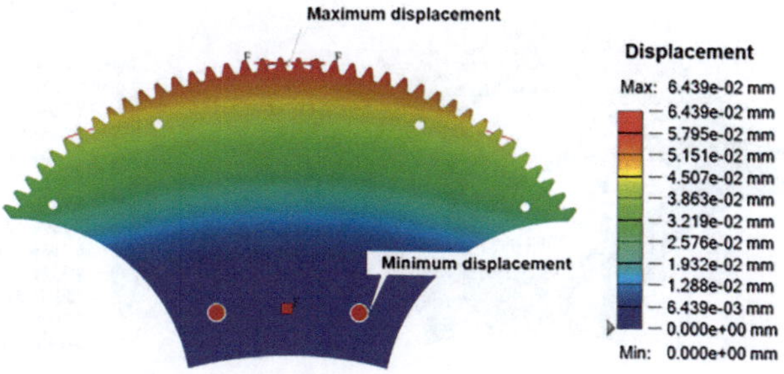

Fig. 5 Displacement under total load condition

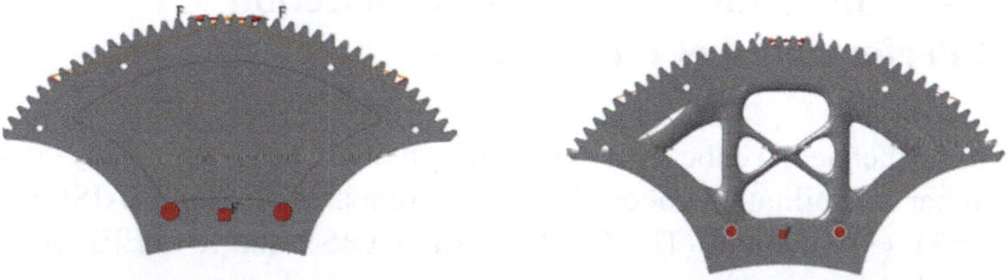

(a) before optimization (b) after optimization

Fig. 6 Comparison shape of big gear model before and after optimization. (**a**) Before optimization; (**b**) after optimization

for about 30.2% of the original weight; the minimum safety factor after optimization is 6.8, and the maximum safety factor is changed from 7.268e+05 to 8.186e+05 (as shown in Fig. 7); the maximum displacement after optimization is changed from 6.439e-02 mm to 1.929e-01 mm (as Fig. 8). All of them meet the standard requirements.

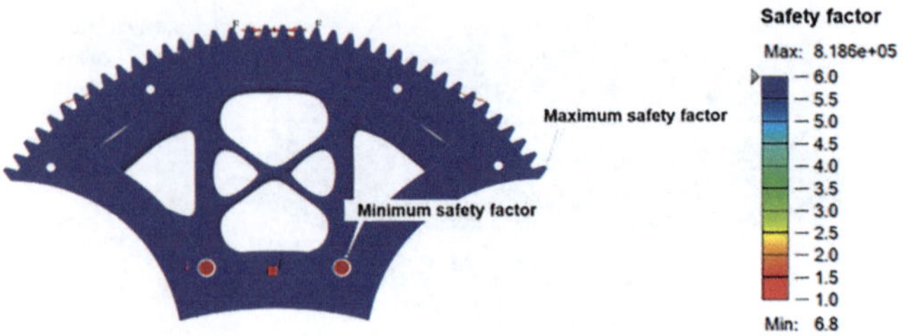

Fig. 7 Optimised safety factor

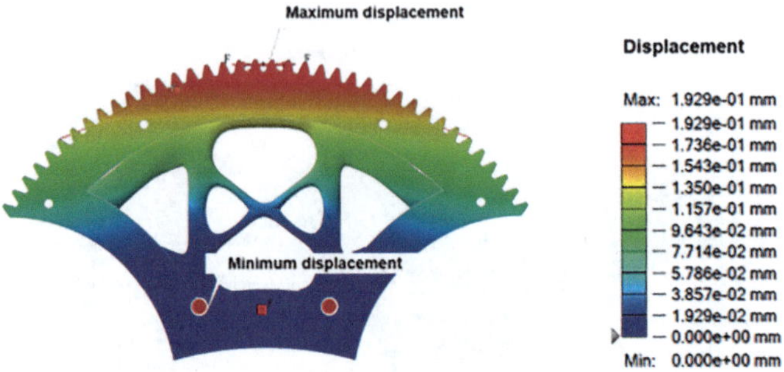

Fig. 8 Optimised displacement

3 Comparative Analysis of Material Selection and Performance of Large Gears

In order to further achieve the lightweight design of the large gear, a key component of the lumbar rehabilitation robot, this study compares (1) Steel (AISI 1015); (2) Steel (C45E); (3) Titanium (Ti-17); (4) Plastic (ABS); and (5) PEEK five typical gear materials in terms of mass, safety coefficient and displacement to analyse the safety and practicality of using different materials. As shown in Fig. 9.

Comparing the three metal materials Steel (AISI 1015), Steel (C45E) and Titanium (Ti-17), it can be seen from Table 1 that the maximum displacements of the two materials, AISI 1015 and C45E, are smaller, 1.929e-01 mm and 1.837e-01 mm, respectively, which indicates that they have better load bearing structural stability. However, the relatively high mass of these two materials is not conducive to lightweight design. In contrast, Titanium (Ti-17) has an outstanding performance among metal materials, which is not only lighter in mass, but also has a higher safety coefficient, with a maximum safety coefficient of 3.334e+06 and a minimum safety coefficient of 25.2, which is much higher than that of other metal materials. At the time, its displacement is controlled at 3.358e-01 mm, which meets the precision requirements of medical rehabilitation equipment.

The polymer materials Plastic (ABS) and PEEK offer significant advantages in terms of mass compared to metallic materials. Among them, the ABS material has

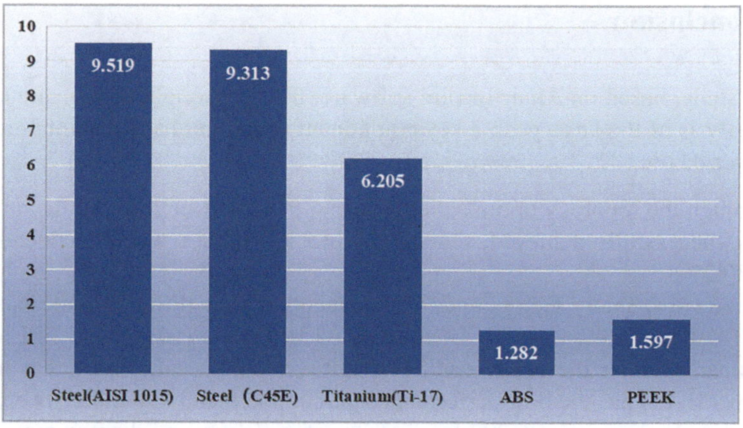

Fig. 9 Quality comparison diagram of different material models

Table 1 Performance comparison table of five materials

	AISI 1015	C45E	Ti-17	ABS	PEEK
Maximum safety factor	$8.186e^{+05}$	$1.407e^{+06}$	$3.334e^{+06}$	$1.5111e^{+05}$	$4.326e^{+05}$
Minimum safety factor	6.8	11.8	25.2	1.1	2.8
Maximum displacement	$1.929e^{-01}$mm	$1.837e^{-01}$mm	$3.358e^{-01}$mm	$1.931e^{+01}$mm	$1.027e^{+01}$mm
Minimum displacement	$0.000e^{+00}$mm	$0.000e^{+00}$mm	$0.000e^{+00}$mm	0.000^{+00}mm	$0.000e^{+00}$mm

the lightest mass, only 1.282 kg, and its minimum safety coefficient is 10.8, which is greater than the standard requirement of 3, and it can be considered as the main material of large gears. The PEEK material has a mass of 1.597 kg, which is 1.25 times that of the ABS, and its minimum safety coefficient is 2.8, which is 2.5 times that of the ABS, and it is close to the standard requirement. Considering that PEEK material itself has excellent comprehensive physical and chemical properties such as high strength, high temperature resistance, chemical corrosion resistance and easy processing, and it also has a wide range of applications in the field of medical devices, PEEK can also be given priority as the gear material in this study.

Considering the displacement, safety factor and quality, Titanium (Ti-17) has the best performance among the metal materials and the highest application cost; PEEK material has moderate overall performance and cost, and can be considered as a reference and alternative material for the manufacture of large gears.

4 Conclusion

In this study, based on Altair Inspire software, the lightweight design and application analysis of the large gear, a key component of the lumbar rehabilitation robot, were carried out.

1. Through the topology optimisation process, the lightweighting of the large gear was successfully achieved, reducing 4.115 kg, with a weight reduction ratio of 30.2%.
2. The masses, safety factors and displacements of five different materials (Steel (AISI 1015), Steel (C45E), Titanium (Ti-17), Plastic (ABS), and PEEK) were compared under the same loading conditions. The results show that titanium alloy (Ti-17) has the best performance among the metal materials, its optimised mass is the lightest at 6.205 kg, and its safety factor and displacement are far enough to meet the design requirements. However, the high cost of titanium alloy material limits its wide adoption in practical applications. Plastic materials (e.g., ABS and PEEK) have obvious advantages in terms of mass, and the mass of PEEK material is 1.597 kg, which is 1.25 times that of ABS, and its minimum safety coefficient is 2.8, which is 2.5 times that of ABS, which is close to the standard requirements, and it can be used as a reference and alternative material for the manufacture of key components of lumbar rehabilitation robots.

This study provides a feasible solution for the lightweight design of lumbar rehabilitation robots through structural optimisation and comparative analysis of material properties. Future research can further explore more lightweight materials that meet practical application requirements and combine with advanced manufacturing technologies to promote rehabilitation robotics to better meet the practical needs of families and communities.

Acknowledgments This work was financially supported by the Ministry of Education Industry-University Cooperation Collaborative Education Project (202101321007), Natural Science Project of Guangdong University of Science and Technology (GKY-2021KYYBK-1), Natural Science Project of Guangdong University of Science and Technology, (GKY-2022KYZDK-2), Key Construction Specialty Project of Guangdong University of Science and Technology (GKZLGC2023159).

References

1. Feng ZN, Huang QY, Wei F, Chen ZH, Wang QX, Chen T, Cao SY (2022) A study on the current situation and allocation equity of rehabilitation organisations and human resources for people with disabilities in China. Chin J Soc Med 2:222–225. https://doi.org/10.3969/j.issn.1673-5625.2022.02.025
2. Li M, Guo YY, Zhao B y, Zhuang WQ, Ou N (2024) Survey and analysis of the current situation of rehabilitation medical resources and services in Guangzhou. Chin J Rehabil Med 3:392–398

3. Ning JZ (2021) Major data situation of the seventh national population census. China Stat 05:4–5. https://kns.cnki.net/
4. Li TT (2022) Practical application of digital rehabilitation medicine. Digit Technol Appl 4:20–22. https://doi.org/10.19695/j.cnki.cn12-1369.2022.04.07
5. Prins R, Jensen P, Lord T (2000) Inventory of socio-economic information about work-related musculoskeletal disorders in the member states of the European Union. Factsheet 9—inventory of socio-economic information about work-related musculoskeletal disorders in the member states of the European Union I safety and health at work EU-OSHA (europa.EU)
6. Yang TL, Zhu Q, Liu F (2025) Lightweight design and 3D printing manufacturing of robotic arm hand claw structure based on Altair Inspire. Mechatron Eng:1–7. https://link.cnki.net/url\
7. Zhang X (2023) Structural design and performance analysis of a waist rehabilitation robot based on parallel mechanism. https://doi.org/10.27470/d.cnki.Ghbgc.2023.000478
8. Chen Q (2018) Design and experimental study of flexible cable-driven parallel lumbar rehabilitation robot system. https://kns.cnki.net/kcms2/article/abstract?v=_bj0XFJq-759Ly1gfLRuj8dRhgqKKoryo7ihaO99iRjwBFJoZUqNxpZmCMxuZjuSk6Pq43uZYX87cOcrUTDpq3VkvMpd-ZTUNoOEEAMHCShUChNp1IrwRC62ezA6k9dCmEO0r1ILKVtCgoXtJfrtvcbIe2-k78GGFEOseK htSsUUjvs7mhsn7yctD4fPS7KtzG-nfOaOcqU=&uniplatform=NZKPT&language=CHS
9. Sabah A-DO, Ghazilla RAR, Tokhi MO (2022) Design of a compact energy storage with rotary series elastic actuator for lumbar support exoskeleton. Machines 10(7)
10. Ge PF (2022) Research on waist-assisting and protecting exoskeleton robot mechanism. Masters degree. https://doi.org/10.27105/d.cnki.ghbgu.2022.000086
11. Yin S, Shi B, Sun YF, Huai YP, Wang J (2011) A new species of the genus Pterostilbene (hymenoptera, Braconidae, Pterostilbene) from China. (2022). Design and evaluation of a lightweight adaptive flexible ankle rehabilitation robot. J Mech Eng 9:10–20. https://doi.org/10.3901/JME.2022.09.010
12. Cheng H, Huang R, Qiu J, Wang YL, Zou CB, Shi KC (2011) A new species of the genus Pseudocercoidea (Coleoptera, Staphylinidae, Staphylinini) from China. (2021). A review of rehabilitation robots and their clinical applications. Robotics 5:606–619. https://doi.org/10.13973/j.cnki.robot.200570

Open Access This chapter is licensed under the terms of the Creative Commons Attribution-NonCommercial-NoDerivatives 4.0 International License (http://creativecommons.org/licenses/by-nc-nd/4.0/), which permits any noncommercial use, sharing, distribution and reproduction in any medium or format, as long as you give appropriate credit to the original author(s) and the source, provide a link to the Creative Commons license and indicate if you modified the licensed material. You do not have permission under this license to share adapted material derived from this chapter or parts of it.

The images or other third party material in this chapter are included in the chapter's Creative Commons license, unless indicated otherwise in a credit line to the material. If material is not included in the chapter's Creative Commons license and your intended use is not permitted by statutory regulation or exceeds the permitted use, you will need to obtain permission directly from the copyright holder.

The Study on the Segmented Temperature-Controlled Sequential Solidification Method for DNAN-Based Energetic Materials

Chen Ma, Bing Xiong, Weidong He, Zheng Mei, and Yanyu Jang

Abstract A comparative study was conducted on two methods—natural environment cooling solidification and segmented temperature-controlled sequential solidification—during the melt-casting and filling process of DNAN-based energetic materials, using finite element simulation and experimental validation. Industrial CT was employed to inspect the filling molding quality, which was then compared and analyzed with numerical simulation results. The results show that the segmented temperature-controlled sequential solidification method can significantly reduce internal defects in the DNAN-based energetic material filling, improving the melt-casting molding quality.

Keywords Energetic materials · Melt-casting · Filling molding · Sequential solidification · Numerical simulation

1 Introduction

With the continuous advancement of science and technology, the development of energetic material filling technology has also made rapid progress. Melt-casting filling has several advantages, including being unaffected by the shape of the container cavity, low cost, ease of mechanized production, and high automation, making it one of the most widely used casting processes. However, the material solidification process involves steps such as melting, mixing, pouring, and solidification, during which physical phenomena like phase changes, crystallization, and volume changes occur. These processes can easily lead to defects such as porosity, shrinkage cavities, and cracks [1, 2]. These defects can affect the safety, mechanical properties, and energy release of energetic materials. Therefore, improving post-filling care

C. Ma (✉) · W. He · Y. Jang
University of Science and Technology, Nanjing, China

B. Xiong · Z. Mei
Hunan Vanguard Group Co., Ltd, Changsha, China

© The Author(s) 2026

Z. Zhang, P. Mercorelli (eds.), *Mechanical Design and Simulation: Insights and Innovations*, Lecture Notes in Mechanical Engineering,
https://doi.org/10.1007/978-981-95-2405-1_70

processes and reducing defect formation are key research directions for enhancing the filling quality.

In recent years, the development of numerical simulation technology has provided new insights for the development of casting products and process design [3–5]. Hu et al. proposed a mathematical model for particle-reinforced metal matrix composites and applied it to the filling and particle flow simulation of castings [6]. Li et al. used ProCAST software to simulate the extrusion casting process of a gearbox cover, accurately predicting and validating the location of shrinkage defects in actual production [7]. Yu Bo et al. used finite element casting simulation software to model the filling and solidification process of castings and predicted the presence of defects. Through orthogonal experiments, they analyzed the impact of process parameters on the shrinkage porosity rate of the casting, successfully optimizing the casting process and reducing the volume of porosity and shrinkage defects [8]. Du Zunzhong et al. used ProCAST software to simulate the lost foam casting process of a butterfly valve for exhaust braking, optimizing the casting process by setting risers and chills to control the feeding effect and reduce shrinkage [9].

Based on this context, the present study uses a cylindrical casting mold, where the material during the filling process is considered an incompressible, constant-viscosity Newtonian fluid, with its flow governed by the laws of mass, momentum, and energy conservation [10]. The study investigates, through finite element simulation systems, the impact of two methods—natural cooling solidification and segmented temperature-controlled sequential solidification—on defects such as shrinkage porosity during the material molding process. The mechanisms by which sequential solidification improves molding quality are explained. Additionally, the accuracy of the numerical simulation and the effectiveness of the segmented temperature-controlled sequential solidification method are validated through experiments. This can provide theoretical support and technical references for the design and improvement of some material melt-casting filling processes.

2 Numerical Simulation of the Solidification Process

2.1 Sequential Solidification Method

For common defects in the filling of energetic materials, there are corresponding preventive measures. In fact, for the defects that exist in the casting process, various single improvement methods can only solve certain defects. To comprehensively address the various issues arising during the filling and molding process, a combination of feasible preventive measures is proposed by analyzing the characteristics of the melt-casting filling process and the principles of defect formation. The segmented temperature-controlled sequential solidification method is introduced. The design concept of this method is as follows: the energetic material's elongated container is divided into several sections, each of which is covered by an independent

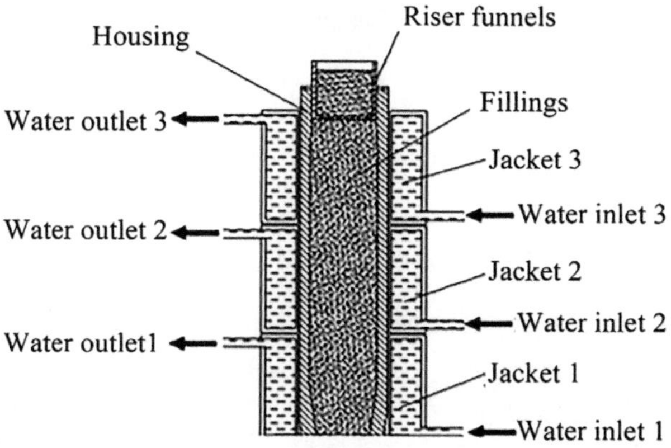

Fig. 1 Segmented temperature-controlled filling and care principle diagram

jacket. Different temperatures of water can be passed through each jacket, and the water temperature is automatically controlled by a program. The riser funnel is designed with an inner layer structure that can circulate heated water or steam to maintain consistent heating during the care process, ensuring that the energetic material in the riser remains in a molten state for feeding. During the filling process, all jackets are heated with hot water to heat the shell first, followed by filling under vacuum conditions. After filling, the care process begins.

During the care process, the cooling and solidification of the energetic material in the lowest layer is first achieved by circulating cold water to cool the bottom portion. After a certain period, the layer above is switched to cold water to begin cooling and solidifying the material in the middle position. This process continues upward, with each section's jacket switched to cold water in sequence to achieve sequential solidification. This care method is similar to the program-controlled cooling method, but essentially transforms the continuous sequential cooling in program-controlled cooling into a discontinuous, segmented program-controlled temperature control. It avoids direct contact between the container and water, and allows independent temperature control for each section's jacket, facilitating the engineering application of energetic materials. It seems like you mentioned Fig. 1, but I can't see the diagram. If you describe the diagram to me, I can help explain or summarize its contents.

2.2 *Numerical Simulation Model Establishment*

The simulation study in this paper focuses on the high-quality filling of DNAN-based energetic materials. The simplified structure of the opening and closing mold is shown in Fig. 2, and the riser funnel is designed based on the simplified mold structure, as shown in Fig. 3.

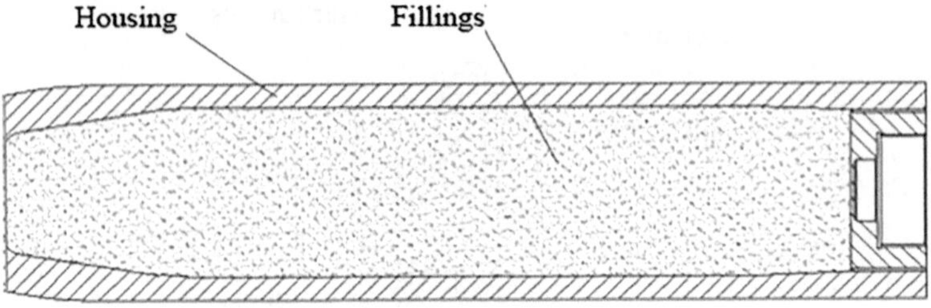

Fig. 2 Filling structure schematic diagram

Fig. 3 Riser funnel structure schematic diagram

Table 1 Thermodynamic parameters of DNAN-based energetic materials

Quality	Measurements
Density/g cm^{-3}	1.72–1.76
Melting point/°C	86–89
Specific heat capacity/KJ/kg/K	1.08 + 0.004 T
Thermal conductivity/W/m/K	0.4297
Latent heat of crystallization/kJ kg^{-1}	27.4
Coefficient of thermal expansion/°C^{-1}	5.37×10^{-5}

The filling material is DNAN-based energetic material and its thermodynamic parameters are mainly shown in Table 1 below:

The numerical simulation of the solidification process is carried out using ProCAST software. The solidification process is an unstable heat conduction process, and its numerical calculation can be performed using the transient heat transfer partial differential equation, as shown in Eq. (1). The boundary conditions for the temperature field calculation are set as third-type boundary conditions, with the calculation formula shown in Eq. (2).

$$\rho Cp\frac{\partial T}{\partial t}=\frac{\partial}{\partial x}\left(k\frac{\partial T}{\partial x}\right)+\frac{\partial}{\partial y}\left(k\frac{\partial T}{\partial y}\right)+\frac{\partial}{\partial z}\left(k\frac{\partial T}{\partial z}\right)+\rho L\frac{\partial f_s}{\partial t} \tag{1}$$

$$-k\frac{\partial T}{\partial n}\Big|_{\Gamma}=h\left(T-T_f\right)\Big|_{\Gamma} \tag{2}$$

Equation: k is the thermal conductivity; Cp is specific heat; ρ is the density; T is the temperature; t is the time; L is the heat of crystallization; f_s is solid-phase analysis; T_f is the ambient temperature; h is the heat transfer coefficient.

2.3 *Numerical Simulation of the Explosive Charge Cooling Process in Natural Environment*

As a control group, the natural cooling solidification process was first simulated. The simulation was set with an ambient temperature of 25 °C, allowing the material to cool and solidify naturally.

The temperature field changes during the solidification process of energetic materials are shown in Fig. 4a, and the solidification interface corresponding to the temperature field is shown in Fig. 4b.

From the results of the natural cooling solidification simulation, it can be observed that since the entire shell surface of the device is in contact with the relatively cooler air, the energetic material begins to solidify from all the boundaries. As cooling progresses, it solidifies layer by layer towards the interior. By the end of the solidification process, the solidification interface contracts into a flat elliptical shape. At a certain point between 2.0 and 3.0 h, the liquid energetic material inside the funnel completely solidifies, causing the feeding channel to be fully solidified. As a result, the subsequent solidification shrinkage will no longer be compensated by the energetic material in the riser funnel, leading to the formation of significant central shrinkage porosity in the final solidified area.

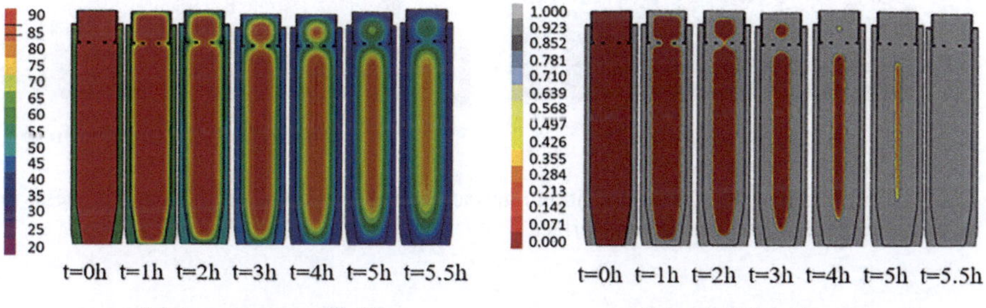

(a) temperature field (b) solidification interface

Fig. 4 Natural cooling solidification temperature field and solidification interface. (**a**) temperature field; (**b**) solidification interface

2.4 Numerical Simulation of the Segmented Temperature-Controlled Sequential Solidification Process

The segmented temperature-controlled sequential solidification method was used, with the initial temperature of the container and energetic material set to 95 °C, and the cooling water temperature set to 25 °C. The ratio of the three heating jackets was 4:4:1. After the energetic material was filled, the first jacket was switched to cold water. Two hours later, the second jacket began to switch to cold water. After 6 h, the third jacket was switched to cold water.

The temperature field changes during the solidification process of energetic materials are shown in Fig. 5a, and the solidification interface corresponding to the temperature field is shown in Fig. 5b.

In the above figure, it can be seen that during the cooling and solidification process of the energetic material in the second jacket, the highest point of the solidification interface coincides with the junction between the bottom plate of the riser funnel and the shell. Due to the constant temperature gradient formed by the high temperature at the top and the low temperature in the second section, the highest point of the solidification interface remains unchanged during the solidification of the energetic material. Throughout the solidification process, the feeding channel remains unobstructed. When the feeding channel of the riser funnel becomes blocked, the filling inside the mold has just completed solidification.

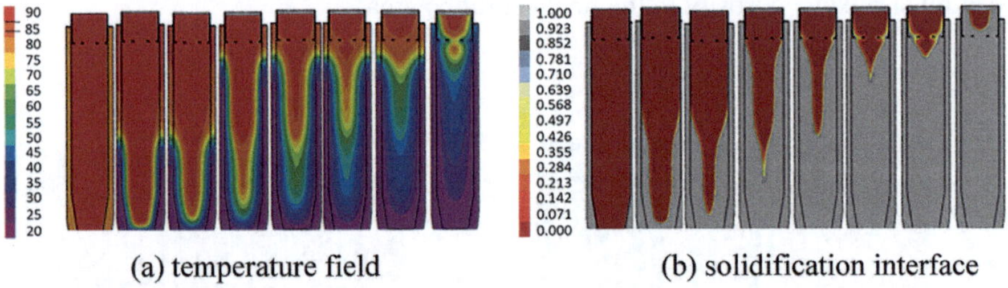

Fig. 5 Temperature field and solidification interface during sequential solidification process. (**a**) temperature field; (**b**) solidification interface

2.5 Analysis of Defect Prediction Results

The simulation results analysis of natural cooling and sequential solidification methods can obtain the predicted results of filling defects, as shown in Fig. 6. From Figure 6a, it can be seen that under natural conditions, cooling and solidification will eventually form obvious shrinkage holes in the upper part of the container. Moreover, due to the flat elliptical shape of the solidification interface shrinking towards the internal axis in the final stage of solidification, this is very unfavorable for the upward solidification of the lower end point of the elliptical long axis, resulting in a row of relatively small micro shrinkage holes scattered below the large concentrated shrinkage holes on the loading axis, which reduces the loading quality.

According to the sequential solidification law, when using the segmented temperature control sequential solidification method for loading, the cooling and solidification sequence of the energetic material basically conforms to the design concept of sequential solidification. The solidification sequence is from bottom to top, and the volume of solidification shrinkage will be supplemented by the riser funnel, always keeping the replenishment channel unobstructed. From Fig. 6b, it can be seen that there is no shrinkage cavity, indicating that segmented temperature control sequential solidification is of great help to the loading quality of the energetic material.

(a) natural cooling (b) progressive solidification

Fig. 6 Prediction of defects in natural cooling and sequential solidification

3 Discussion and Analysis of Experimental Results

3.1 Experimental Process

Using a simple opening and closing mold as the container, experiments on the filling and forming of energetic materials were conducted under two methods: natural cooling and segmented temperature-controlled care. Under natural cooling conditions, the liquid energetic material was injected into the mold and then placed in the natural environment to cool and solidify. Under the segmented temperature-controlled care method, the heating jackets were first assembled and steam was introduced. The temperatures of the heating jackets, mold, and riser funnel were all set to 95 °C. The liquid energetic material was then injected into the mold, at which point the first heating jacket was switched to cold water. After 2 h, the second heating jacket was switched to cold water, and after 6 h, the third jacket was switched to cold water.

3.2 Results Analysis

After the experiments were completed, industrial CT scans were performed on the two samples. The CT scan results under natural environment cooling are shown in Fig. 7. As seen in the figure, there is a large concentrated shrinkage porosity at the upper part of the device, along with several smaller, dispersed shrinkage pores in the middle section. Comparing these results with the defect predictions from the simulation in Fig. 6a, it can be observed that the two sets of results are consistent, thus validating the accuracy of the numerical simulation.

The CT test results under the segmented temperature-controlled sequential solidification curing method are shown in Fig. 8. After the energetic material was filled and molded, no defects such as porosity, through cracks, shrinkage cavities, or visible air pockets were found inside. The molding quality is good.

After the CT scan is completed, the mold is opened, and the solidified energetic material is removed. The cross-sectional view of the material after being cut open is shown in Fig. 9.

Fig. 7 CT detection results of natural cooling solidification

Fig. 8 CT scan results of segmented temperature-controlled sequential solidification

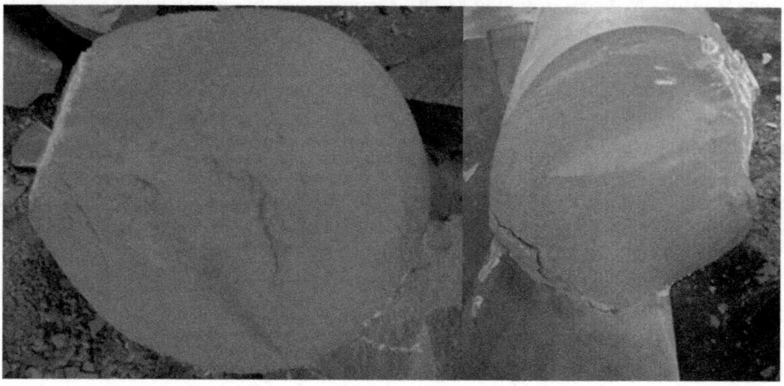

Fig. 9 The cross-sectional view of the cylinder after being cut open

As shown in the above image, the cross-section is smooth and flat, with no defects such as cracks, shrinkage cavities, or air pockets inside. The texture is uniform, and the consistency is good, indicating high molding quality. This verifies that the segmented temperature-controlled sequential solidification curing method can achieve high-quality filling and molding of DNAN-based energetic materials.

4 Conclusion

Simulation calculations and experimental verification indicate that the segmented temperature-controlled sequential solidification method proposed in this paper can effectively eliminate the molding defects under natural solidification conditions. Under natural cooling conditions, the DNAN-based energetic material solidifies from the outside to the inside within an elongated container, with the material at the center of the container solidifying last, leading to the formation of shrinkage porosity. However, under the segmented temperature-controlled sequential solidification method, the energetic material solidifies gradually from bottom to top, and the shrinkage generated during solidification is continuously replenished, ultimately eliminating the shrinkage porosity. The effective parameters of the segmented temperature-controlled sequential solidification method are: (1) three stages of temperature control; (2) temperature switching between 25 °C and 95 °C; (3) switching time intervals of 2 h and 6 h.

References

1. Tang YL (2019) Numerical simulation study on prediction of shrinkage cavity and porosity defects in ductile iron castings. Huazhong University of Science and Technology, Wuhan
2. Shahane S, Aluru N, Ferreira P et al (2020) Optimization of solidification in die casting using numerical simulations and machine learning. J Manuf Process 51:130–141
3. Malysza M, Zuczekr R, Wilk-Kolodziejczyk D et al (2022) Technological optimization of the stirrup casting process with the use of computer simulations. Materials 15(19):6781
4. Zhao X, Wang P, Li T et al (2018) Gating system optimization of high pressure die casting thin-wall A1Si10MnMg longitudinal loadbearing beam based on numerical simulation. China Foundry 15(6):436–442
5. Piao JJ, Jiang B, Hu ML et al (2023) Optimization of squeeze casting process parameters for ADC12 aluminum alloy automobile support with complex structure. J Netshape Form Eng 15(11):132–139
6. Hu QY, Zhao HD, Long H et al (2020) Numerical simulation of mold filling and particulate flow of A356/SiCp in casting. J Compos Mater 54(12):1593–1602
7. Li JH, Sun Y, Wang Y et al (2022) Optimization of squeeze casting process of gearbox cover based on FEM and box-Behnken design. Int J Adv Manuf Technol 118(9):3421–3430
8. Yu B, Zhang XB, Yang R et al (2024) Optimization of the copper alloy casting process for sea valves based on numerical simulation. Foundry Technol 45(1):81–87
9. Du ZZ, Xu Z, Shi X et al (2022) Process design and regulation of brake casting based on real casting numerical calculation. Foundry Technol 43(9):836–840
10. Shan Y, Wang YJ, Wang F et al (2022) Numerical simulation and process improvement of shrinkage defects in low pressure aluminum castings. J Netshape Form Eng 14(4):139–145

Open Access This chapter is licensed under the terms of the Creative Commons Attribution-NonCommercial-NoDerivatives 4.0 International License (http://creativecommons.org/licenses/by-nc-nd/4.0/), which permits any noncommercial use, sharing, distribution and reproduction in any medium or format, as long as you give appropriate credit to the original author(s) and the source, provide a link to the Creative Commons license and indicate if you modified the licensed material. You do not have permission under this license to share adapted material derived from this chapter or parts of it.

The images or other third party material in this chapter are included in the chapter's Creative Commons license, unless indicated otherwise in a credit line to the material. If material is not included in the chapter's Creative Commons license and your intended use is not permitted by statutory regulation or exceeds the permitted use, you will need to obtain permission directly from the copyright holder.

Design and Application of the Oil Stain Cleaning Device for the Rack of the Drive Mechanism of the Three Gorges Ship Lift

Yang Wang, Yan Kun Zhao, and Geng Bang Huang

Abstract Aimed at addressing the issues of cumbersome operation, numerous safety hazards, low efficiency, and the risks associated with working at heights during the cleaning of oil stains on the rack of the Three Gorges ship lift drive mechanism, research was conducted on the characteristics of the rack's dry oil lubrication system and the oil stain cleaning process. A cleaning device for the rack of the ship lift drive mechanism was designed, and the boundary conditions and force conditions of the device were analyzed to provide theoretical guidance for its design. Additionally, the oil stain cleaning process for the Three Gorges ship lift drive mechanism was optimized. Through the design, manufacturing, and testing of the device, the time required for cleaning oil stains on the rack was effectively reduced, improving operational efficiency and safety, thereby ensuring the maintenance and operation of the Three Gorges ship lift.

Keywords Three gorges ship lift · Drive mechanism · Rack · Oil stain cleaning · Design

1 Introduction

The Three Gorges Ship Lift is a gear—rack climbing—type ship lift. The racks, which are the force—transmitting components during the lifting and lowering operations of the ship lift's cabin, come in four sets and are respectively arranged on the vertical walls of the elevator shaft sections of the four tower columns. Under normal operating conditions, the pinions of the drive mechanism remain engaged with the racks, and the load of the drive mechanism is transmitted to the concrete structure through the racks. As the core equipment of the ship lift's operation, the safety and reliability of the racks are of vital importance to the operational safety of the ship lift [1–3].

Y. Wang (✉) · Y. K. Zhao · G. B. Huang
Yangtze River Three Gorges Navigation Administration, Yichang, China

© The Author(s) 2026

Z. Zhang, P. Mercorelli (eds.), *Mechanical Design and Simulation: Insights and Innovations*, Lecture Notes in Mechanical Engineering,
https://doi.org/10.1007/978-981-95-2405-1_71

At present, the lubrication method adopted for the racks of the Three Gorges Ship—lift is periodic spray—type dry-oil lubrication. This method can effectively cover the surface of the racks with an oil film [4, 5]. However, there are also problems such as the splashing and accumulation of excess lubricating grease during the lubrication process. The excess lubricating grease is exposed on the surface of the racks, which may cause small insects, birds, dust particles, etc. to adhere to the surface, resulting in surface contamination of the racks. In hot weather, the lubricating grease may drip and splash during the operation of the ship compartment, polluting the working environment of the on-site equipment. Therefore, it is necessary to clean the lubricating grease on the racks regularly [6–8].

At present, the oil stain cleaning of the racks of the Three Gorges Ship Lift is mainly carried out manually. By building a maintenance platform on the deck of the 0—level of the ship cabin and with the cooperation of the central control room to dock the ship cabin at an appropriate elevation, segmented manual cleaning is carried out. The manual cleaning mode of the rack oil stains has the following deficiencies:

1. According to the construction requirements of the on-site parts, a "concave-shaped" combined construction platform is set up within the scope of the rack structure on the ship cabin deck. The total weight of the platform is about 650 kg, and it takes a lot of manpower to build, which is time-consuming and labor-intensive.
2. The platform has three layers: upper, middle and lower. The operating space for workers is narrow, and there are potential safety hazards for working at heights.
3. It takes 4 workers about 18 h to manually clean a single rack, totaling 72 man-hours. The whole process consumes a lot of time and energy.

Therefore, it is necessary to design a set of rack oil-stain cleaning device, and at the same time, study the device installation and layout plan and the rack oil-stain cleaning process, so as to effectively shorten the rack oil-stain cleaning time and improve the operation efficiency and safety.

2 Device Structure and Working Principle

Through the analysis of the requirements for rack oil stain cleaning, the boundary conditions for cleaning, and the technical requirements, an automatic cleaning device for the oil stains on the ship lift rack is designed [9, 10]. This device is composed of components such as a machine base, a tooth-surface oil-scraping mechanism, an end-face oil—scraping mechanism, an end-face oil—scraping adjustment assembly, a torque-adjustment mechanism, and an oil-collecting tray, enabling the automatic cleaning of rack oil stains, as shown in Fig. 1.

The cleaning device is composed of a base and functional components. The entire device is installed and fixed on the top of the rack protective cover. The base of the cleaning device measures 1620 mm in length, which is in line with the size of

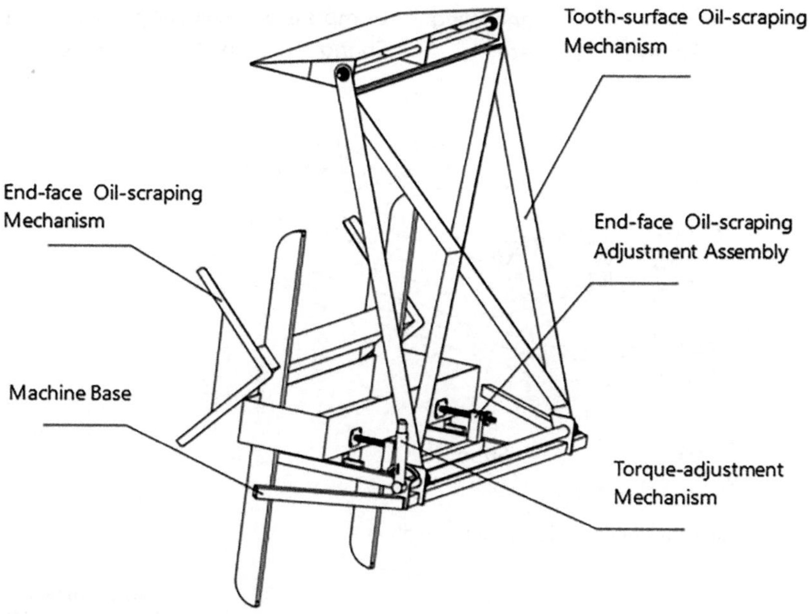

Fig. 1 Structure of the oil-stain cleaning device

the top of the protective cover, and 575 mm in width. The component for cleaning oil stains on the tooth surface is 1500 mm in height and 840 mm in width, with a maximum weight of 15 kg.

Working principle: To clean the oil stains on the two end faces and the tooth top of the rack, an end-face oil—scraping mechanism is installed below the device. Both the end-face and tooth—top oil-scraping plates are made of non-metallic carbon—fiber boards, and the surface of the fiber boards is coated with an oleophobic coating to prevent grease adhesion.

The end-face oi-scraping mechanism adjusts the contact force between the tooth—top oil-scraping plate and the tooth—top surface through the end-face oil-scraping adjustment assembly. When the ship cabin descends, the tooth-surface oil-scraping mechanism and the end-face oil-scraping mechanism can automatically scrape the oil stains off the tooth surface and the two side surfaces of the rack. Under the action of gravity, the oil stains automatically fall into the oil sump.

The end-face oil-scraping adjustment component consists of a connecting plate, a compression spring, an adjusting nut, an adjusting screw, a support, a slider, and a guide rail, as shown in Fig. 2. When the adjusting nut is rotated, the pressure of the compression spring can be used to pull the slider to slide on the guide rail, thereby adjusting the contact force between the tooth-tip oil-scraping plate on the end-face oil-scraping mechanism and the tooth-tip surface.

As shown in Figs. 3, 4, and 5, the tooth-surface oil-scraping mechanism includes a tooth-surface oil-scraping plate, an oil-scraping plate fixing shaft, a toothed locking piece, a support, a fixing frame, and a support fixing shaft. The tooth-surface

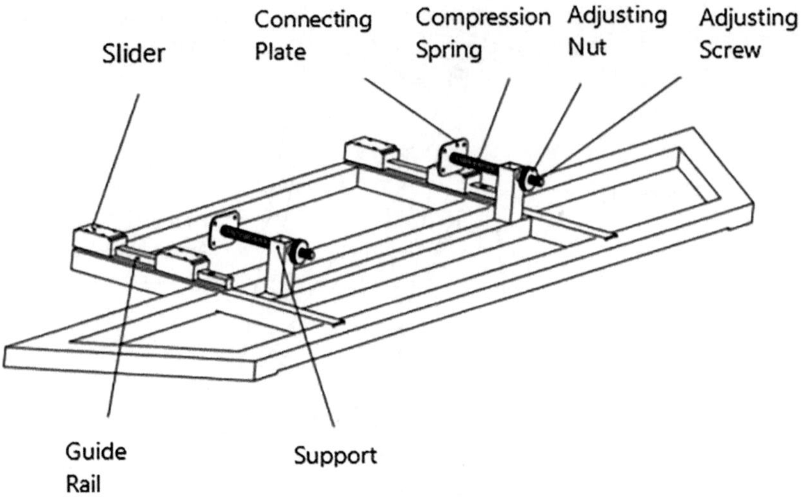

Fig. 2 End-face oil-scraping adjustment assembly

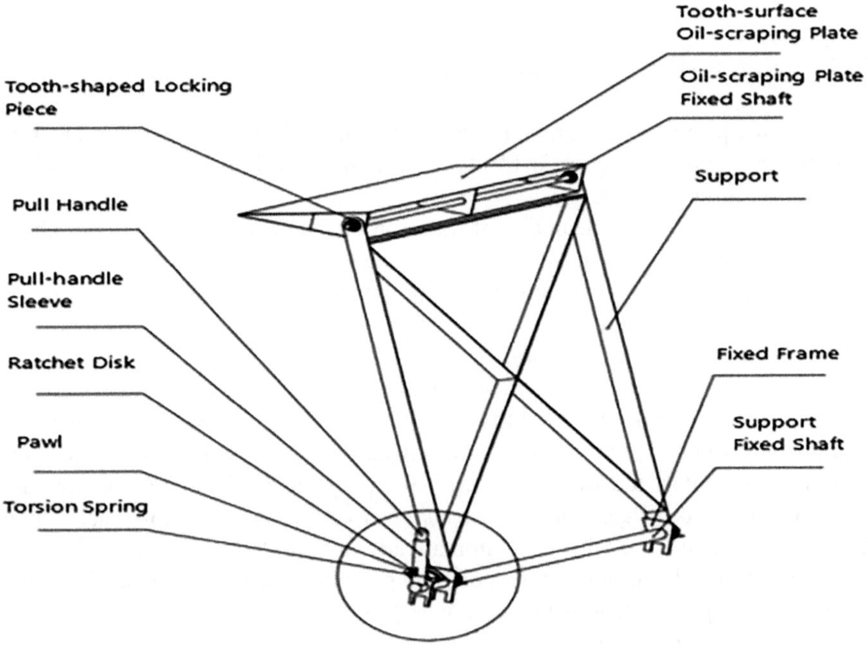

Fig. 3 Tooth-surface oil-scraping mechanism and torque-adjustment mechanism

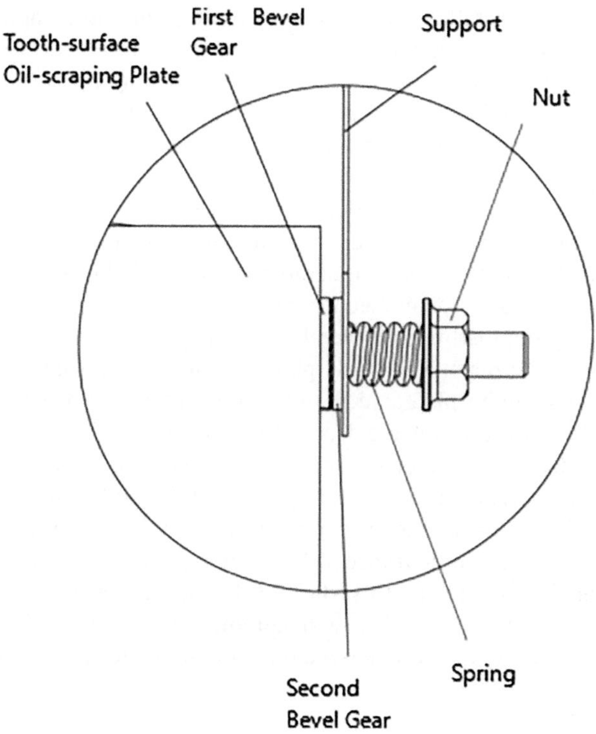

Fig. 4 Tooth-shaped locking mechanism

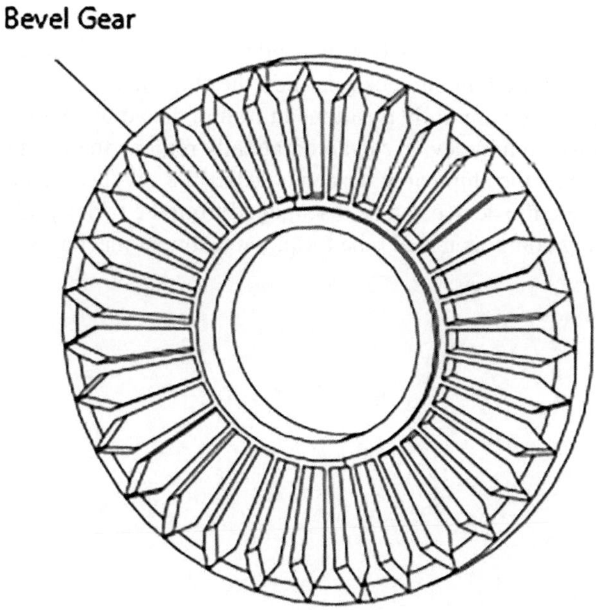

Fig. 5 Bevel gear

oil-scraping plate is installed on the support through the oil-scraping plate fixing shaft. The angle between the oil-scraping plate and the tooth-surface can be finely adjusted by adjusting the tightness of the toothed locking piece. After the adjustment, the nut at the end of the toothed locking piece needs to be tightened to ensure a certain locking force. In case of special circumstances where the tooth-surface oil-scraping plate is stuck and the resistance it receives exceeds its own locking force, the tooth-surface oil-scraping plate will automatically flip upward and disengage from the rack to protect the tooth-surface from being scratched. The contact force between the tooth-surface oil-scraping plate and the tooth-surface can be adjusted by a torque adjustment mechanism.

When the ship compartment descends, a guide plate is fixedly installed on the tooth-tip oil-scraping plate. The guide plate can maintain a safe distance between the tooth-tip oil-scraping plate and the rack, preventing the tooth-tip oil-scraping plate from getting stuck during the descending process.

The torque adjustment mechanism consists of a pull handle, a pull-handle sleeve, a ratchet disc, a pawl, and a torsion spring. When the pull handle is pressed to disengage the pawl from the ratchet disc, the pull-handle sleeve is rotated to adjust the pressure between the tooth-surface oil-scraping plate of the tooth-surface oil-scraping mechanism and the tooth-surface of the rack. After releasing the pull handle, the pawl and the ratchet disc will automatically lock, maintaining a fixed pressing force between the tooth-surface oil-scraping plate and the tooth-surface of the rack.

3 Device Installation Layout

As shown in Fig. 6, the machine base is welded by square steel tubes. The machine base is fixedly installed on the rack guard on the ship cabin with bolts, and the oil-stain cleaning device is initially positioned through the adjustment components.

The oil-stain cleaning device can automatically remove the oil stains on the rack and its end-face as the ship cabin descends. After the oil-stain cleaning of a single rack is completed, the device can be disengaged from the rack by adjusting the end-face oil-scraping components and the torque-adjustment mechanism.

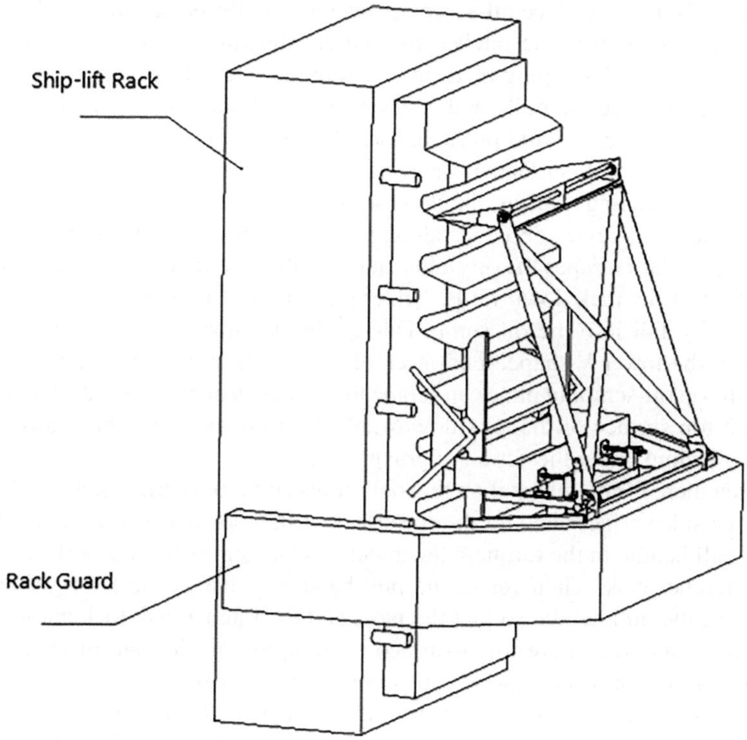

Fig. 6 Device installation layout

4 Automatic Oil-stain Cleaning Process for the Rack

The process of cleaning the oil stains on the ship-lift rack by the oil-stain cleaning device includes the following steps:

1. Initial positioning and installation of the device. After the ship-lift ship cabin stops operating, install the rack oil-stain cleaning device at the corresponding position of the rack guard on the deck.
2. Adjust the end-face oil-scraping mechanism. By rotating the adjusting nut, the pressure between the tooth-tip oil-scraping plate and the tooth-tip surface of the rack can be adjusted. The compression spring can maintain a fixed pressure between the tooth-tip oil-scraping plate and the tooth-tip surface of the rack.

3. Adjust the tooth-surface oil-scraping mechanism. Press the pull handle to disengage the pawl from the ratchet disc. After adjusting the pressure between the tooth-surface oil-scraping plate and the tooth-surface of the rack to an appropriate level, release the pull handle. Then, the pawl and the ratchet disc will self-lock, maintaining a fixed pressing force between the tooth-surface oil-scraping plate and the tooth-surface of the rack.
4. Oil stain cleaning. During the downward movement of the ship cabin, the oil-stain cleaning device descends along with the ship cabin. The tooth-surface oil-scraping plate scrapes the oil stains on the tooth root, tooth surface and tooth-top surface of the rack downwards. Under the action of gravity, the oil stains automatically fall into the oil sump. During the downward movement of the ship cabin, the lower V-shaped end-faces of the end-face oil-scraping plate and the tooth-top oil-scraping plate can scrape the oil stains on the two side-faces and the tooth-top surface of the rack downwards. The oil stains automatically fall into the oil sump under the action of gravity.
5. When there is not much oil on the tooth roots and tooth surfaces, and only the oil on the side surfaces and tooth-top surface of the rack needs to be cleaned, press the pull handle in the torque-adjustment mechanism to disengage the pawl from the ratchet disk. Then rotate the pull-handle sleeve to the far right end, and release the pull handle so that the pawl and the ratchet disk lock automatically. At this time, the entire tooth-surface oil-scraping mechanism disengages from the rack, and only the end-face oil-scraping mechanism can work. When the ship cabin descends, the lower end-face of the V-shaped oil-scraping plate on the end-face oil-scraping mechanism can scrape the oil on the two side-surfaces and the tooth-top surface of the rack into the oil sump.

5 Technical Effects of the Oil-stain Cleaning Device

Through the trial-production and testing of the device for cleaning the oil stains on the rack, and by comparing the oil-stain conditions before and after the test, as shown in Fig. 7, it is found that the rack oil-stain cleaning device can effectively

Fig. 7 Comparison of rack oil-stain conditions before and after cleaning

clean the oil stains on the tooth-top, tooth-root and end-faces of the rack without damaging the oil film in the meshing part. After cleaning, the oil stains can fall into the bottom oil-stain collection device by their own weight, forming a closed-loop for oil-stain cleaning and collection. The oil contamination can be effectively removed, and the surface cleanliness of the rack meets the test requirements.

6 Conclusions

Through the automatic oil-stain cleaning test, combined with the previous manual oil-stain cleaning mode, comparisons are made in terms of labor requirements, time consumption and safety.

As shown in Table 1, in the manual cleaning mode, it takes about 18 h for 4 workers to clean a single rack. This mode is time-consuming and requires cleaning during specific periods. At the preliminary stage, scaffolding and safety nets need to be set up. During the cleaning process, workers have to wear safety ropes and clean in sections. After cleaning one section of the travel distance, the central control room needs to cooperate with the ship cabin to start the next-section operation.

In the automatic cleaning mode of the oil-stain cleaning device, it takes about 4 h for 2 workers to clean the oil stains on a single rack. The maximum weight of the oil-stain cleaning device is approximately 15 kg, and it can be disassembled and assembled at any time. Once the device is installed for oil-stain cleaning, it can operate continuously, and only on-site monitoring is required, which ensures a high level of safety.

Through comparison, the automatic oil-stain cleaning mode has the following advantages over the manual cleaning mode:

1. The oil-stain cleaning device is small in size, making it easy to install and disassemble.
2. The oil-stain cleaning device has a simple structure, which is convenient to operate.
3. The oil-stain cleaning device has a high safety factor as there is no high-altitude operation involved.
4. The oil-stain cleaning device can effectively increase work efficiency by over 80%, effectively realizing the substitution of manual work with machinery.

Table 1 Comparison of oil pollution cleanup efficiency

Cleaning mode	Labor requirement (unit: person)	Time requirement (unit: h)
Manual cleaning mode	4	18
Automatic device cleaning mode	2	4

References

1. Xinqiang N, Qin Liming Y, Qingkui. (2011) Design of the gear-rack climbing ship lift of the three gorges project. Strateg Study CAE 7:96–103. https://kns.cnki.net/kcms2/article/abstract?v=YvzTZTPVeK61T7HhTwTmO9o13Ari8VUnD38FpWm437NQyoEpzPY06t9gCV3b4ChYUY_OCfAASvm1vuBMQfnf0BuUdKE3Mg1NPTp_0VO5V5-oOna_CIc3KePy-ikwoYDGEMFqR1USH5Vqx5zdGak8utfm1GrOKitemWLv0PzGWtOoe2C1G5y-H09L3SZQXZX2&uniplatform=NZKPT&language=CHS
2. Zhongming L, Quancai L, Jiqiang L, et al. Force analysis of the gear-rack of the three gorges ship lift
3. Zhuo C, Lubing S, Panpan F et al (2023) Analysis of load and lubrication state of the gear-rack of the three gorges ship lift based on operating conditions. J Mech Trans 47(3):86–93
4. Simin Y, Haoyuan L (2024) Finite element analysis of the gear-rack climbing device of large-scale ship lifts. China Plant Eng 13:133–135. https://kns.cnki.net/kcms2/article/abstract?v=YvzTZTPVeK7HSa6hvJdk7QQtqNleNaSyzqv629Nke8s65MQgi-JixNfc18gv2FEDi5-ZfyI9OqtpnTVBQJlSdnLC8xp-aa_FhpJvq4mpkoLY5jV1T2kCQi-FOapWbBLJY6lxa6PTQVcVbjkLsTpJG2nhrJq7rCJz2ozPK7tWjVtJQEn0RusNszyIkYhHRkucuBEgoN5mZck=&uniplatform=NZKPT&language=CHS
5. Lubing S, Zhihong Z, Xiaoqing Y et al (2024) Research on operating loads and fatigue life of the gear-rack of the three gorges ship lift. J Mech Eng 60(05):119–129. https://link.cnki.net/urlid/11.2187.TH.20231009.1138.018
6. Yi C, Ke W, Weixiong L et al (2024) Research on model test and condition monitoring scheme of the rack of the three gorges ship lift. Hydro-Sci Eng 1:156–163. https://link.cnki.net/urlid/32.1613.TV.20231130.1030.002
7. Zhuo C, Lubing S, Xiaoqing Y et al (2023) Fatigue life assessment of the rack of the three gorges ship lift based on the load spectrum. J Mech Trans 47(5):158–166. https://doi.org/10.16578/j.issn.1004.2539.2023.05.024
8. Chengyun D (2021) Design and research on the fault diagnosis system of the gear-rack of the threegorges ship lift. Wuhan University, Wuhan. https://doi.org/10.27379/d.cnki.gwhdu.2021.000334
9. Heng S (2001) Mechanical principles. Higher Education Press, Beijing
10. Daxian C (2002) Mechanical design handbook. Chemical Industry Press, Beijing

Open Access This chapter is licensed under the terms of the Creative Commons Attribution-NonCommercial-NoDerivatives 4.0 International License (http://creativecommons.org/licenses/by-nc-nd/4.0/), which permits any noncommercial use, sharing, distribution and reproduction in any medium or format, as long as you give appropriate credit to the original author(s) and the source, provide a link to the Creative Commons license and indicate if you modified the licensed material. You do not have permission under this license to share adapted material derived from this chapter or parts of it.

The images or other third party material in this chapter are included in the chapter's Creative Commons license, unless indicated otherwise in a credit line to the material. If material is not included in the chapter's Creative Commons license and your intended use is not permitted by statutory regulation or exceeds the permitted use, you will need to obtain permission directly from the copyright holder.

Optimization of Lubrication Methods for the Counterweight Steel Wires of the Three Gorges Ship Lift

Yang Wang, Yan Kun Zhao, and Chuan Guo Liang

Abstract The balance-weight steel wire ropes of the three gorges ship-lift feature a complex operating environment, long length, large quantity, and high usage frequency. The existing lubrication method for these steel wire ropes poses high safety risks, consumes a long time, and leads to waste of human resources and materials. Based on the analysis of previous operation and maintenance experience, an optimization plan for the lubrication method of the steel wire ropes is proposed. The new lubrication method can effectively meet the lubrication process requirements of the balance-weight steel wire ropes at the ship-lift site, improve the lubrication efficiency, enhance the safety of lubricating the dead-rope section, and increase the service life, reliability, and maintenance economy of the balance-weight steel wire ropes.

Keywords Counterweight · Steel wire · Long length · Large quantity · Efficiency · Effect

1 Introduction

The counterweight system of the three gorges ship lift [1] consists of components such as the counterweight units [2], pulley blocks [3], steel wires [4], balance chains [5], and steel wire adjustment devices. The ship chamber is suspended by 256 steel wires with a diameter of φ74 mm. These steel wires are divided into 16 groups and symmetrically arranged on both sides of the ship chamber. One end of each steel wire is connected to the lifting lug above the outer web of the main longitudinal beam of the ship chamber through an adjustment device, and the other end passes around the balance pulley in the machine room at the top of the tower column and then connects to the counterweight block [6].

Y. Wang (✉) · Y. K. Zhao · C. G. Liang
Yangtze River Three Gorges Navigation Administration, Yichang, China

© The Author(s) 2026

Z. Zhang, P. Mercorelli (eds.), *Mechanical Design and Simulation: Insights and Innovations*, Lecture Notes in Mechanical Engineering,
https://doi.org/10.1007/978-981-95-2405-1_72

As an important equipment bearing the load of the ship chamber, the steel wires in the counterweight system of the Three Gorges Ship Lift play a crucial role. How to improve the safety and reliability of the steel wires is of vital significance to the operational safety of the Three Gorges Ship Lift [7].

2 Calculation of Lubrication Length of Counterweight Steel Wires and Adverse Factors During Operation

2.1 *The Effective Lubrication Length of Counterweight Steel Wires*

The total length of the counterweight steel wire is approximately 151 m. where the diameter of the pulley d = 5 m, when the ship lift operates at a water level of 175 m [8], the maximum length of the area where the steel wire and the pulley have friction is approximately:

$$L_{max} = \Delta H + \Delta L = 175 - 62 + \pi d / 2 = 175 - 62 + 3.14 \times 5 / 2 = 120.85\,\text{m}$$

The length of the fixed rope at the connection end of the steel wire rope and the ship chamber is:

$$L_{\text{Fixed rope1}} = 196 - 175 + \pi d / 4 = 24.93\,\text{m}$$

The length of the fixed rope at the connection end of the steel wire and the counterweight is:

$$L_{\text{Fixed rope2}} = 196 - 188.7 = 7.3\,\text{m}$$

2.2 *Analysis of Adverse Factors Existing during the Operation of Steel Wires*

1. Frictional force between the steel wire rope and the counterweight pulley;
2. Frictional force between the strands of the steel wire rope;
3. Corrosion;
4. Influencing factors such as tensile stress.

3 Analysis of Existing Lubrication Methods for Steel Wire Ropes

The existing lubrication method for steel wire ropes adopts the masto steel wire rope lubrication system [9]. The lubrication system mainly consists of a lubricator, throttle valve, high-pressure pump, high-pressure hose, air source, etc. The specific principle is shown in Fig. 1.

3.1 *Lubrication Operation Process*

(a) Before the lubrication operation, lay plastic films to prevent the oil from dripping and contaminating the surrounding environment during the lubrication process of the steel wire ropes.
(b) Before the ship compartment moves upward, fixedly install the hanging barrel of the lubrication system on the side where the steel wire rope is connected to the counterweight frame.
(c) After the ship compartment starts to move upward, turn on the air compressor and pneumatic oil pump to lubricate the steel wire rope.
(d) After the ship compartment stops, remove the lubrication hanging barrel, move the air compressor and lubricating oil pump, fix the hanging barrel to the next steel wire rope, and repeat steps a-c.
(e) Lubricate the dead section of the steel wire rope by manual brushing.

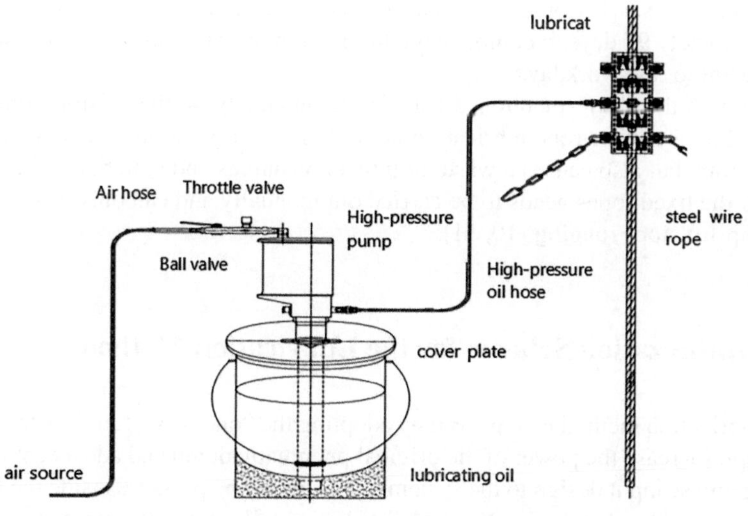

Fig. 1 Installation schematic diagram of masto steel wire rope lubrication system

3.2 Evaluation of Lubrication Oil Consumption

The oil consumption for the steel wire rope to be fully lubricated can be calculated by Formulas (1) and (2):

$$\text{V total} = \text{Sgroove} \times \text{L total} \tag{1}$$

$$\text{Sgroove} = 1/4\pi\left(D_1^{\ 2} - D_2^{\ 2} - 2/5D_3^{\ 2} \times 10\right) \tag{2}$$

In the formula, V total is the oil consumption for lubricating one steel wire rope, Sgroove is the groove area of the steel wire rope cross-section, L total is the total length of one steel wire rope D1 is the outer diameter of the steel wire rope, D2 is the outer diameter of the steel wire rope with plastic filling, and D3 is the diameter of the outer strand.

The measured data on site shows that D1 = 75 mm, D2 = 63 mm, D3 = 18 mm. Assuming that the steel wire rope has no rotation direction and the total length is 151 m, and the consumption of lubricating grease is approximately equal to the volume of the groove between the outer strands. Therefore, the theoretical oil consumption V total for lubricating one balance weight steel wire rope can be calculated through Formulas (1) and (2):

$$\text{V total} \oplus 42.45\,\text{L} \tag{3}$$

On-site, the existing lubrication device is used to lubricate the steel wire ropes. The actual measurement shows that the oil consumption for each steel wire rope is approximately 0.6 barrels (180 L per barrel, including the lubrication of the fixed rope). The theoretical oil consumption V total is approximately 42.45 L. Through calculation, the grease utilization rate is 39.3%.

Assuming that 4 people carry out the steel-wire-rope lubrication work, it takes approximately 91 days to complete the lubrication of 256 steel wire ropes, which is equivalent to 364 workdays.

Through the above-mentioned calculation and analysis, the existing lubrication device for steel-wire-rope lubrication not only has a high safety risk and consumes a long time but also causes a waste of human resources and materials. The lubrication of the fixed ropes needs to be carried out manually and can only be done when the ship-lift stops running [10, 11].

4 Optimization Scheme for the Lubrication Method

The lubrication method is improved by adopting the "one-drive-four" system design concept. Increase the power of the original pneumatic pump and add a central lubricating grease input design to the system to ensure strong pressure inside the lubrication hanging barrel and the optimized distribution of lubricating grease. Add a reflux pipe design to the system to maintain the pressure balance and tightness inside the lubrication hanging barrel, thus avoiding excessive overflow of lubricating grease. Meanwhile, the excessive lubricating grease can be recovered and recycled, and the lubricating grease is evenly distributed within the lubrication barrel. Install a "one-to-four" distributor on the lubricating grease delivery pipeline and individual control valves on the branch lubricating grease delivery pipes. Increase the number of lubrication hanging barrels to four to enable the simultaneous lubrication of four steel wire ropes. To address the issue of having to replace the lubricating grease midway during the lubrication process when it runs out, the lubrication efficiency is further enhanced by using a "dual-oil-pump" lubricating grease supply system. Its working principle is as follows: both pumps are connected to the "one-to-four" distributor simultaneously. During the lubrication operation, only the output valve of one pump is open, while the output valve of the other pump is closed and in standby mode. Once one barrel of grease is nearly depleted, the output valve of the other pump is opened, achieving a "soft-switch" for continuous grease supply, as shown in Fig. 2.

Design the fixed-installation of the lubrication hanging barrel. The main body of the installation trolley is divided into two parts: a convex-shaped flat trolley (with two vertical support arms on the upper part and three universal moving wheels installed at the bottom) and a movable support arm (with one universal moving wheel installed at the bottom). The on-site installation of the lubrication hanging barrels is shown in Fig. 3. After the lubrication and maintenance of the steel wire ropes are completed, open the fastening bolts of the two lubrication hanging barrels

Fig. 2 Test installation diagram of the "one-drive-four" system

Fig. 3 Schematic diagram of on-site installation of lubrication hanging barrels

in turn, and then push the installation trolley to quickly move to the next group of steel wire ropes to continue the maintenance operation.

Improve the lubrication method for the fixed-rope part. Set up a power device at the pulley-block platform to drag the hanging barrel to move along the fixed-rope steel wire ropes. Place both the lubricating grease barrel and the pneumatic oil pump on the pulley-block platform, and connect them to the two lubrication hanging barrels through lubricating-grease delivery pipes of sufficient length. The power device drives the lubrication hanging barrel to move on the steel wire ropes for lubrication.

Customize a groove cleaner according to the specifications of the steel wire rope (diameter/number of strands and lay direction) to clean the original old oil, pollutants, and solidified impurities on the steel wire rope. Adopt pressure lubrication, enabling the steel wire rope to pass through a 100% controlled pressure-immersion system. In this way, the lubricating grease can penetrate into the interior of the steel wire rope, fully lubricating the entire steel wire rope and providing in-depth maintenance for it.

Through estimation, the amount of grease consumed for lubricating one steel wire rope using the new lubricating grease is approximately 0.3 barrels (180 L/barrel). The theoretical oil consumption V total for lubricating one balance-weight steel wire rope, calculated through Formulas (1) and (2), is approximately 42.45 L. Then the grease utilization rate is 78.6%. The design of this lubrication device results in a very high grease utilization rate. Moreover, with the design of the oil return pipe, the amount of oil leakage is extremely low, minimizing the impact on the equipment environment.

The time-consuming aspects of the steel wire rope cleaning and pressure-lubrication system for steel wire rope lubrication mainly include the lubrication of the moving steel wire ropes during the operation of the ship-lift and the lubrication of the fixed ropes. Assuming that 4 people carry out the steel-wire-rope lubrication work, it takes approximately 45 days to complete the lubrication of 256 steel wire ropes, which is equivalent to 180 man-days.

5 Conclusions

Through the improvement of the lubrication method, a comparison is made in multiple aspects such as the lubrication efficiency and effect of the steel wire ropes, the environmental impact, and the impact on operation, as shown in Table 1. (Note: It is assumed that 4 people are engaged in the lubrication of the steel wire ropes.)

The new lubrication method can effectively meet the lubrication process requirements of the balance-weight steel wire ropes at the ship-lift site. With the same oil consumption per steel wire rope, the grease utilization rate has increased by 50%, and the working hours have been reduced by approximately 51%, thus effectively improving the lubrication efficiency.

Regarding the lubrication method for the dead-rope section, replacing the manual brushing method with a movable lubrication hanging barrel effectively solves

Table 1 Comparison of the effectiveness between existing lubrication methods and optimized lubrication methods

Lubrication method	Number	Oil consumption (L)	Grease utilization rate (%)	Lubrication man-hour (4 people) (h)	Environmental impact
Existing	1	42.45	39.30	11.4	Large
New	1	42.45	78.60	5.6	Small

the problem of replacing manual lubrication with automatic lubrication, reducing the safety risks of working on the overhanging side.

The customized groove cleaner can effectively clean the original old oil, pollutants, and solidified impurities on the steel wire ropes. Through the pressure-lubrication method, a central input design for the lubricating grease is added to the system to ensure strong pressure inside the lubrication hanging barrel and optimized distribution of the lubricating grease. The steel wire rope passes through a 100% controlled pressure-immersion system, fully lubricating the entire steel wire rope and conducting in-depth maintenance.

By optimizing the lubrication method, the service life, reliability, and maintenance economy of the balance-weight steel wire ropes are improved. This also enhances the operation efficiency of the ship-lift, minimizes the environmental impact of equipment maintenance, and ensures the safe operation of the three gorges ship-lift.

References

1. Zhu H, Deng R (2009) Overall layout design of the three gorges ship lift. Yangtze River 40(23):48–50. https://doi.org/10.16232/j.cnki.1001-4179.2009.23.001
2. Xionghao C, Duanwei S, Hongxiang L et al (2019) Stability analysis of high-lift fully-balanced winch-type vertical ship lift system. China Mech Eng 30(08):919–925. https://link.cnki.net/urlid/42.1294.TH.20190423.0947.006
3. Weibi T, Ran L, Wei H et al (2021) Topology optimization design of the balance weight pulley structure of the three gorges ship lift. Water Power 47(1):99–102. https://link.cnki.net/urlid/11.1845.TV.20201201.1638.006
4. Liuyang Y, Tiantian W (2024) Installation technology of steel wire rope for the balance weight system of gear-rack climbing vertical ship lift. Installation S2:17–18. https://kns.cnki.net/kcms2/article/abstract?v=YvzTZTPVeK469bddHbozcKNq8raDy8 evfZ0KY-2OJrF5SAvkFRjkdim7ypEd3oe3QfibrpbaNDH6DVFHIwURQ78tF-U_ZvPoF4mKElubbzxyOegFYXJ_U0QALX9-WasheqQtov_JcSv3vSSYT8JywgEzlPn2-AWcw_g4epe8ObGcM1Ihu2liiM7uNd9uNwz&uniplatform=NZKPT&language=CHS
5. Hao X, Yankun Z, Gengbang H et al (2024) Dynamic analysis and structure optimization of the guide wheel for the balance chain of ship lift. Yangtze River 55(S2):239–241. https://doi.org/10.16232/j.cnki.1001-4179.2024.S2.044
6. Xinqiang N (2005) Research on key technologies of the structure of the three gorges ship lift. Huazhong University of Science and Technology, Wuhan
7. Post-three gorges planning project: research on key technologies for navigation and operation support of the three gorges ship lift

8. Hu G (2004) Comprehensive lubrication solutions for users. Hydraul Pneum Seals 42(5):60–70. https://kns.cnki.net/kcms2/article/abstract?v=YvzTZTPVeK5ii2neR2HCRTnfnXDQPUvOK9i5_DiwBn2YNmL2UI7ifX0b2QaAnQgTuK7cKykEiHbKFrUUzARnXJS6RRaRgGOC_WCvXaIu6BA-MnTMg8QMTvsLJFuh24CKnrvsmZZS6o3GxKm9px2cPx7ugRrZpifwuoV9Q6wR-78v4uODiqNahsOEFh_bIkzu&uniplatform=NZKPT&language=CHS
9. Regulations for maintenance of equipment and facilities of the three gorges ship lift (trial)
10. Lin C, Jing Y, Tao Z et al (2018) Design of measuring and adjusting device for balance weight steel wire rope of ship lift. Mech Res Appl 31(5):88–90. https://doi.org/10.16576/j.cnki.1007-4414.2018.05.028
11. Liang W, Shoujin X (2022) Analysis of the problems of rope sinking and rope tangling when steel wire ropes are wound on drums. China New Technol New Prod 1:64–66. https://doi.org/10.13612/j.cnki.cntp.2022.01.030

Open Access This chapter is licensed under the terms of the Creative Commons Attribution-NonCommercial-NoDerivatives 4.0 International License (http://creativecommons.org/licenses/by-nc-nd/4.0/), which permits any noncommercial use, sharing, distribution and reproduction in any medium or format, as long as you give appropriate credit to the original author(s) and the source, provide a link to the Creative Commons license and indicate if you modified the licensed material. You do not have permission under this license to share adapted material derived from this chapter or parts of it.

The images or other third party material in this chapter are included in the chapter's Creative Commons license, unless indicated otherwise in a credit line to the material. If material is not included in the chapter's Creative Commons license and your intended use is not permitted by statutory regulation or exceeds the permitted use, you will need to obtain permission directly from the copyright holder.

The Lubrication Characteristics Analysis of Wind Turbine Planetary Journal Bearing

Xiaoran Chen, Tengjiao Lin, and Hesheng Lv

Abstract The gear mechanism, vital for wind turbine transmission, operates amid slow speeds and high rotations. To enhance the load ability of planetary gears, replacing rolling bearing with journal bearing is now a common practice. Utilizing the principle of hydrodynamic lubrication, this study presents a model for fluid lubrication in wind-powered planetary journal bearing. The precision of the journal bearing fluids lubrication model is verified through the finite difference method, revealing an exceptionally accurate outcome. Investigate the lubrication process of wind turbine planetary journal bearing under load. Investigating the lubrication properties of wind turbine planetary journal bearing, both in the presence and absence of torque, using their fluid lubrication model and solution methodology, the distribution of the oil film pressure is analyzed, thereby illustrating the pivotal role of torque in their lubrication characteristics.

Keywords Journal bearing · Fluid lubrication · Oil film pressure

1 Introduction

The operation of wind turbines depends significantly on wind-powered gear for their transmission mechanism, which often operates in slow-moving and heavy-duty conditions. The improvement in load capacity of planetary journal bearing has resulted in their major transition from rolling to journal bearing. A functioning journal bearing demonstrates a noticeable nonlinearity in its oil-film pressure [1]. Over time, scholars have mainly simplified the intricacies of the oil film force into a linear approach, or, on the other hand, diverse analytical models of the oil film force based

X. Chen · T. Lin (✉)
State Key Laboratory of Mechanical Transmission, Chongqing University, Chongqing, China
e-mail: tjlin@cqu.edu.cn

H. Lv
Chongqing Gearbox Co., Ltd, Chongqing, China

© The Author(s) 2026

Z. Zhang, P. Mercorelli (eds.), *Mechanical Design and Simulation: Insights and Innovations*, Lecture Notes in Mechanical Engineering,
https://doi.org/10.1007/978-981-95-2405-1_73

on specific assumptions [2, 3]. The project's dedicated operational performance proved the reliability of journal bearing, which have a track record of enduring high loads under diverse working conditions with robust resilience. Moreover, journal bearing generally require little lubrication, assisting in reducing maintenance costs and improving system reliability. Concurrently, scholars from the national and global community have investigated the lubrication and mechanical characteristics of planetary journal bearing in wind turbine equipment. In contrast to the normal journal bearings, wind turbine gear systems' planetary bevel gears experience uneven meshing forces from the sun gear and internal gear, attributed to bevel gears. This leads to external forces and bending moments, creating eccentricity and misalignment in the planetary journal bearings, thereby influencing their lubrication properties. Lehmann and Dirk [4] engineered a 2 MW wind turbine gearbox with a journal bearing and tested it under severe operational conditions. The study initially showed the feasibility of replacing rolling bearing with journal bearing. Gong [5–7] investigated the misalignment in the bearing of planetary journals, as a result of the irregular meshing of planetary wheels, culminating in the development of a thermoelastic flow model. To ascertain the deformation and temperature of the bearing, they employed both the finite element method (FEM) and the influence coefficient technique, respectively. In conclusion, research on journal bearing mainly focuses on assessing the reduction in eccentricity and stabilizing of support under conditions of high speed and minor strain. Wind turbines, known for their sluggish pace and significant burden, greatly affect the efficiency of the journal bearing's lubrication process. Consequently, this shortens the durability of journal bearings, escalates the rate of failures in wind turbines, and impacts the steadiness of wind energy supply. Therefore, it is critical to explore the oil and mechanical traits of wind-driven planetary journal bearing.

Grounded in hydrodynamic lubrication principles, the report outlines a liquid lubrication model for wind turbine planetary wheel journal bearing. Employing the finite difference method, this model for fluid lubrication of the journal bearing is delineated, ensuring its precision, and provides an exceptionally accurate model solution. A study is carried out on the lubrication characteristics of wind turbine planetary wheel journal bearing under load.

2 Fluid Lubrication Control Equation

Considering the functionality of the wind power transmission system under slow and intense loads and the influence of surface texture on oil film pressure, the paper used Cheng and Patir's average Reynolds equation [8, 9] to assess the pressure of the oil film in the journal, along with P, as follows:

$$\frac{\partial}{\partial \phi}\left(\varphi_{\phi}\frac{H^{3}}{\eta}\frac{\partial P}{\partial \phi}\right)+r_{B}^{2}\frac{\partial}{\partial z}\left(\varphi_{z}\frac{H^{3}}{\eta}\frac{\partial P}{\partial z}\right)=-6r_{B}^{2}\omega\left(\varphi_{c}\frac{\partial H}{\partial \phi}-\sigma\frac{\partial \varphi_{s}}{\partial \phi}\right) \tag{1}$$

Due to the eccentricity and misalignment phenomenon during the operation of planetary journal bearing, wind power journal bearing is usually axially trimmed in the design process. Therefore, the oil film thickness at any angle θ is defined as H, which is calculated as:

$$H = H_x + H_0 \tag{2}$$

where H_x represents the bearing clearance taking into account the effects of eccentricity and deflection; H_0 represents the axial modification clearance.

The diagram in Fig. 1 illustrates the eccentricity at the axisymmetric side of a planetary journal bearing. Additionally, O_1 indicates the center of the pin, O_2 the center of the journal bearing, and ω the rotational angular velocity. Included among these is the eccentricity e_0, the distance from the pin's center O_1 to the journal with O_2, and the eccentricity angle ϕ_0, the angle between the points connecting O_1 and O_2.

The schematic diagram of the planetary sliding bearing skew is shown in Fig. 1. Due to the skew effect, the axis of the planetary sliding bearing forms a certain angle with the axis of the pin shaft. To accurately analyze the impact of the skew on the bearing clearance, the axis of the planetary sliding bearing is projected onto the *xoz* and *yoz* planes. The two projection lines form angles with the *y*-axis and *x*-axis are β_x and β_y. ΔX and ΔY represent the projection lengths of the planetary sliding bearing axis on the *x*-axis and *y*-axis.

$$\begin{cases} \Delta X = z \cdot \tan \beta_y \\ \Delta Y = z \cdot \tan \beta_x \end{cases} \tag{3}$$

The planetary journal bearing eccentricity and offset angle at any axis section are:

$$\phi = \arctan\left[\left(e_0 \sin \phi_0 + \Delta X\right) / \left(e_0 \cos \phi_0 - \Delta Y\right)\right] \tag{4}$$

$$e = \frac{\Delta X}{\sin \phi} + \frac{e_0 \sin \phi_0}{\sin \phi} \tag{5}$$

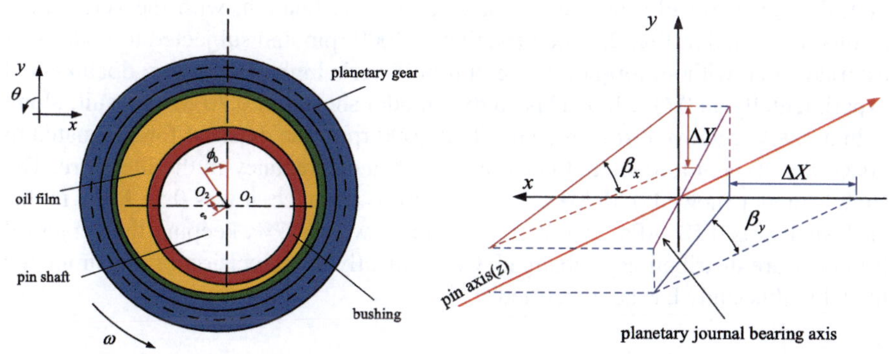

Fig. 1 Diagram of eccentricity and misalignment of planetary journal bearing

Establish a right-angle coordinate system and define the oil film thickness at any angle θ as H. Calculate equation:

$$H_0 = c + e\cos(\phi - \theta) \tag{6}$$

where c represents the clearance radius of the journal bearing.

The pin undergoes secondary modification, leading to a specific formula for the clearing H_x of the planetary journal bearing, a result of the reconditioning procedure:

$$H_x = H_{x0}\frac{(z + z_0)^2}{(L/2 + z_0)^2} \tag{7}$$

where H_{x0} represents the amount of pin end face modification; L represents the bearing width; z_0 represents the coordinates of the highest point of the modification profile on the z-axis.

Solving the film thickness and average Reynolds equation for journal bearing is achieved through the finite difference method. The specific method is as follows: the difference quotient of the oil film pressure between adjacent nodes in the grid is approximated to represent the derivative in the Reynolds equation. The Reynolds equation is then transformed into a discrete form, which is subsequently solved.

3 Confirmation of the Model's Accuracy

To confirm the journal bearing's fluid lubrication model's precision, one must choose Bouyer's [10] experimental and theoretical frameworks to determine the pressure of the oil film.

Throughout this research, the model calculates the pressure allocation in oil film in two distinct scenarios, drawing on the documented structural details and working conditions of journal bearing as per scholarly sources. Confirmation of the model's accuracy. To facilitate the comparison with the experiments in the reference literature, the center oil film pressure is selected for verification, with the verification results illustrated in Fig. 2. When rotating at 2000 rpm and subjected to 3000 N of external force without torque, the central oil layer's highest pressure, documented experimentally, is 0.68 MPa. This study's model showed a 0.70 MPa result, along with a 2.94% relative error. In contrast, at 4000 rpm, the external force equated to 3000 N, a 70 N·m torque, and noted a 0.56 MPa discrepancy in the literature. The study's model showed 0.56 MPa, whereas in this instance, it was 0.56 MPa, in this analysis it was 0.58 MPa, and the relative error was 3.57%, keeping the center oil film pressure distribution unchanged. Consequently, the precision of the simulated model in this study has been verified.

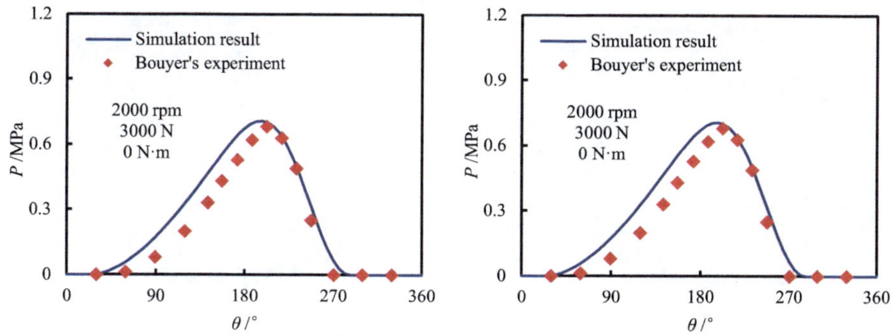

(a) 2000 rpm, 3000 N, 0 N·m (b) 2000 rpm, 3000 N, 70 N

Fig. 2 Validation of oil film pressure distribution. (**a**) 2000 rpm, 3000 N, 0 N·m. (**b**) 2000 rpm, 3000 N, 70 N

4 Lubrication Effect Analysis

The liquid lubrication framework estimates and juxtaposes the distribution of oil film pressure of planetary journal bearing, considering torque, against that in the oil film, while ignoring the impact, as illustrated in Fig. 3. The maximum pressure in the oil film of the planetary journal bearing rises from 16.72 MPa to 29.89 MPa. The cause is attributed to how torque impacts the planetary journal bearing, causing warping, a decrease in the thickness of oil film, and an increase in oil film pressure. As a result, this torque markedly influences how pressure is distributed in the journal bearing's oil film. Consequently, assessing the lubrication characteristics of the journal bearing necessitates a careful consideration of its torque.

Torque-driven bending of journal bearing results in thinning of the oil film at both sides, intensifying the impact of the bearing on lubrication. Consequently, exploring how modification impacts the lubrication effectiveness of journal bearings is crucial.

Figure 4 demonstrates how increasing the axial change from 0 μm to 40 μm causes a unification of oil film pressure from either end to the middle. Enhancing

Fig. 3 The oil film pressure distribution of planetary journal bearing

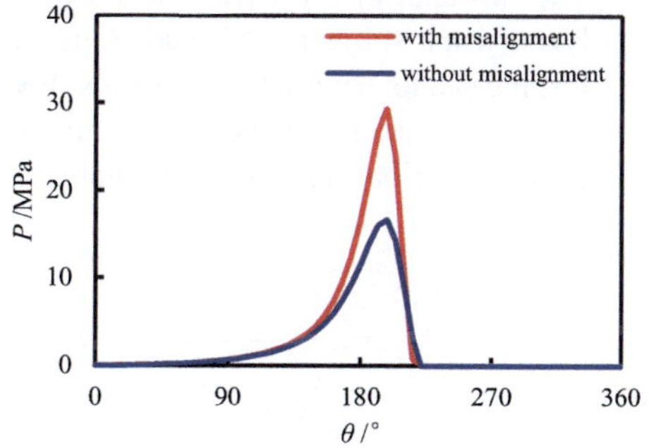

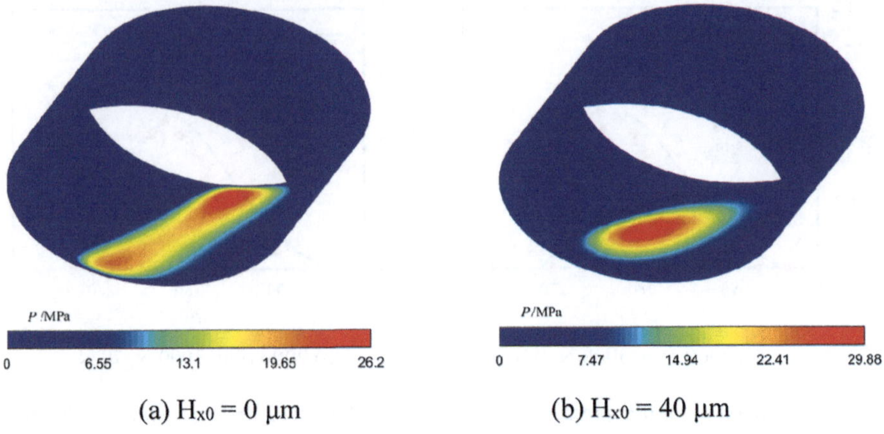

Fig. 4 The oil film pressure distribution of the journal bearing before and after modification. (**a**) $H_{x0} = 0$ μm. (**b**) $H_{x0} = 40$ μm

lubrication efficiency occurs because of the axial change in the pin, thickening the film at both ends.

5 Conclusions

1. Development of the lubrication model for planetary journal bearing in wind power gearbox utilizes hydrodynamic principles, making use of the average Reynolds equation and film thickness equation to mitigate alterations and misalignment. By applying the finite difference approach, the model's precision and independence from the mesh are meticulously examined, culminating in the creation of a model-solving technique distinguished by its high accuracy and efficiency.
2. The planetary journal bearing fluid lubrication model and its solution approach evaluate the oil film pressure spread both incorporating and excluding external torque, highlighting the need to study the bearing's lubrication traits with torque. The lubrication properties of a wind turbine planetary journal bearing is analyzed when subjected to load. This has enhanced the design efficiency of wind turbine planetary journal bearings, lowered the expenses for bearing lubrication trials, and bolstered the stable functioning of the wind energy sector, significantly contributing to its sustainable growth.

References

1. Song C, Zhang R, Zhou Y, Tan J et al (2024) Transient analysis of non-uniform tilting-pad journal bearing for wind turbine main shaft under dynamical loading. Mech Mach Theory 203:105792
2. Hagemann T, Ding H, Radtke E et al (2021) Operating behavior of journal planet gear bearing for wind turbine gearbox applications—part II: impact of structure deformation. Lubricants 9:98
3. Fei W, Tan J, Li H, Zhu C et al (2024) Influences of planet gear journal bearing on dynamic characteristics of megawatt-scale wind turbine drivetrains. Simulations and experiments. Mech Syst Signal Process 221:111747
4. Mokhtari N, Grzeszkowski M (2017) Vibration signal analysis for the lifetime-prediction and failure detection of future turbofan components. Tech Mech Eur J Eng Mech 37:422–431
5. Gong J, Liu K, Zheng Y, Meng F (2023) Thermal-elastohydrodynamic lubrication study of misaligned journal bearing in wind turbine gearbox. Tribol Int 188:125–136
6. Gong J, Liu K, Meng F, Wang H et al (2024) Effect of meshing-induced deformation on lubrication for journal planet gear bearing. Int J Mech Sci 284:109747
7. Meng F, Gong J, Yang S, Huang L, Zhao H et al (2021) Study on tribo-dynamic behaviors of rolling bearing-rotor system based on neural network. Tribol Int 156:106829
8. Patir N, Cheng H (1978) An average flow model for determine effects of three dimensional roughness on partial hydrodynamic lubrication. J Lubr Technol 100:12–17
9. Patir N, Cheng H (1979) Application of average flow model to lubrication between rough journal surfaces. J Lubr Technol 101:220–229
10. Bouyer J, Fillon M (2002) An experimental analysis of misalignment effects on hydrodynamic plain journal bearing performance. J Tribol 124(2):313–319

Open Access This chapter is licensed under the terms of the Creative Commons Attribution-NonCommercial-NoDerivatives 4.0 International License (http://creativecommons.org/licenses/by-nc-nd/4.0/), which permits any noncommercial use, sharing, distribution and reproduction in any medium or format, as long as you give appropriate credit to the original author(s) and the source, provide a link to the Creative Commons license and indicate if you modified the licensed material. You do not have permission under this license to share adapted material derived from this chapter or parts of it.

The images or other third party material in this chapter are included in the chapter's Creative Commons license, unless indicated otherwise in a credit line to the material. If material is not included in the chapter's Creative Commons license and your intended use is not permitted by statutory regulation or exceeds the permitted use, you will need to obtain permission directly from the copyright holder.

Projective Synchronization Control of the Arneodo System and Coullet System

Haibin Zhao, Shiyu Yan, Qingwen Yu, Chong Liu, and Zhiguo Lu

Abstract Projective synchronization control of the Arneodo system and Coullet system is present in this paper. The Arneodo system is the drive system and the Coullet system is the response system. The state variables corresponding to the Arneodo system and Coullet system are synchronized to scaling factor. A single adaptive controller based on sliding mode control theory is used to control the projective synchronization error signals, and the error signals converge to zero rapidly, thereby achieving projection synchronization control of the Arneodo system and the Coullet system. The stability of the designed adaptive controller has been theoretically proven, and numerical simulation has been conducted using MATLAB software. The hyperbolic tangent function is applied in the adaptive controller to reduce the chattering. The simulation results show that the adaptive controller can perform projection synchronization control of the Arneodo system and the Coullet system, and the results are analyzed in the end.

Keywords Projective synchronization · Adaptive sliding mode controller · The Arneodo system · The Coullet system

1 Introduction

Chaos synchronization has been widely studied in the last three decades and has been applied in various engineering applications [1]. There are many methods for controller design, such as active control [2], backstepping control [3], adaptive control, sliding mode control [4] and so on, which have been applied to the chaos synchronization. The two chaos, which are named as the drive system and the response system, will be synchronized. Recently, projective synchronization of chaos has

H. Zhao (✉) · S. Yan · Q. Yu · C. Liu · Z. Lu
Northeastern University, Shenyang, China
e-mail: hbzhao@me.neu.edu.cn; shyyan@me.neu.edu.cn; qwyu@me.neu.edu.cn; congliu@me.neu.edu.cn; zglu@me.neu.edu.cn

© The Author(s) 2026

Z. Zhang, P. Mercorelli (eds.), *Mechanical Design and Simulation: Insights and Innovations*, Lecture Notes in Mechanical Engineering,
https://doi.org/10.1007/978-981-95-2405-1_74

been studied, where the state variables of the two chaos will be synchronized to a scaling factor [5].

The Arneodo system and Coullet system both are well-known three-dimensional chaotic systems. Single adaptive controller based on sliding mode controller (SMC) is designed for projective synchronization control of Arneodo system and Coullet system. SMC has many advantages, such as insensitivity to model uncertainties, and has been widely applied in engineering practice [6, 7]. In this paper, adaptive SMC is designed based on linear sliding surface and adaptive exponential reaching law, and is used for projective synchronization of the Arneodo system and Coullet system. The hyperbolic tangent function is applied to reduce the chattering.

The rest of this paper has been organized as follows. We describe the Arneodo system and the Coullet system, and the three-dimensional phase diagrams of the two chaotic systems are shown in Sect. 2. The projective synchronization is introduced in Sect. 3, then single SMC and adaptive single SMC are designed. The stability of the designed single controller has been demonstrated. In Sect. 4, the digital simulation based on MATLAB software was conducted on the designed single controller. At last, the conclusions of this article are given in Sect. 5.

2 The Arneodo System and Coullet System

2.1 The Arneodo System

The three-dimensional Arneodo system is relatively simple, but it exhibits complex nonlinear phenomena [3]. In this paper, the differential equations of the Arneodo system can be expressed as

$$\begin{cases} \dot{x}_1 = x_2 \\ \dot{x}_2 = x_3 \\ \dot{x}_3 = a_1 x_1 + a_2 x_2 + a_3 x_3 + a_4 x_1^3 \end{cases} \tag{1}$$

In the above Arneodo system, parameter $a_i (i = 1, 2, 3, 4)$ is known constant. For the Arneodo system, x_1, x_2 and x_3 are the state variables. The parameters of the Arneodo system are set to be $a_1 = 5.5$, $a_2 = 3.5$, $a_3 = a_4 = -1$. The initial condition of the Arneodo system is set to be $x_1(0) = -1$, $x_2(0) = 0.1$ and $x_3(0) = 1$. The Arneodo system is chaotic under the above parameter settings.

2.2 The Coullet System

The differential equations of the Coullet system and Arneodo system are very similar, but they are also different. The three-dimensional Coullet system is given as

$$\begin{cases} \dot{y}_1 = y_2 \\ \dot{y}_2 = y_3 \\ \dot{y}_3 = b_1 y_1 + b_2 y_2 + b_3 y_3 + b_4 y_1^2 \end{cases} \tag{2}$$

For the Coullet system, $b_i(i = 1, 2, 3, 4)$ are known constant parameters,y_1, y_2 and y_3 are the state variables of the Coullet system. We can use MATLAB software for numerical simulation of the Coullet system. The parameters of the Coullet system is set to be $b_1 = b_2 = -1$, $b_3 = -0.5$, $b_4 = 1$, and the initial condition is set to be $y_1(0) = 0.2$, $y_2(0) = 0.1$ and $y_3(0) = -0.1$. The Coullet system is chaotic under the above parameter settings.

3 Controller Design

In this article, projective synchronization control of the Arneodo system and Coullet system is achieved through a single controller. The drive system is Arneodo chaotic system which is shown as (1), and the response system is Coullet chaotic system in projective synchronization control. The controlled Coullet system with single controller is shown as

$$\begin{cases} \dot{y}_1 = y_2 \\ \dot{y}_2 = y_3 \\ \dot{y}_3 = b_1 y_1 + b_2 y_2 + b_3 y_3 + b_4 y_1^2 + u_1(t) \end{cases} \tag{3}$$

The Arneodo system and the Coullet system are projective synchronized under the single controller $u_1(t)$. In this paper, the single controller $u_1(t)$ will be designed using SMC theory.

In the projective synchronization control, we define the error signals e_i ($i = 1, 2, 3$) between the Arneodo system and the Coullet system as

$$e_1 = y_1 - kx_1, \quad e_2 = y_2 - kx_2, \quad e_3 = y_3 - kx_3 \tag{4}$$

In the error system (4), constant parameter k is scaling factor, and $k \neq 0$. The single controller is designed to synchronize the states of the Arneodo system and the Coullet system to the scaling factor k. It is obvious that the projective synchronization control of Arneodo system and Coullet system can be achieved, when the error signals e_i become zero.

By taking the derivative of the error signals in (4) and substituting the Arneodo system and the controlled Coullet system, we can obtain the error system as

$$\begin{cases} \dot{e}_1 = e_2 \\ \dot{e}_2 = e_3 \\ \dot{e}_3 = f_1(y_1, y_2, y_3) - kf_2(x_1, x_2, x_3) + u_1(t) = b_1 y_1 + b_2 y_2 + \\ \quad b_3 y_3 + b_4 y_1^2 - k(a_1 x_1 + a_2 x_2 + a_3 x_3 + a_4 x_1^3) + u_1(t) \end{cases} \tag{5}$$

Then SMC and adaptive SMC are built in this paper, to make the error signals rapidly converge to zero, and to accomplish the projective synchronization of the Arneodo system and Coullet system.

3.1 Sliding Mode Controller

For the projective synchronization error system (5), single controller $u_1(t)$ is designed using sliding mode control theory. This controller performs stabilization control on the projection synchronization error system, and the error signals rapidly converge to zero, thereby achieving projection synchronization control of the Arneodo system and the Coullet system. The linear sliding surface which is used is represented as

$$s_1(t) = e_3 + 2k_1 e_2 + k_1^2 e_1 \tag{6}$$

where parameter k_1 is a positive constant.

This paper adopts exponential reaching law which is commonly used for the SMC design. We take the derivation of the variable s_1 as

$$\dot{s}_1 = -k_2 s_1 - k_3 sign(s_1) \tag{7}$$

where parameters k_2 and k_3 are positive constant, that is $k_2 > 0$, $k_3 > 0$.

Single controller, which is used exponential reaching law, can be built as

$$u_1(t) = -f_1(y_1, y_2, y_3) + kf_2(x_1, x_2, x_3) - 2k_1 e_3 - k_1^2 e_2 - k_2 s_1 - k_3 sign(s_1) \tag{8}$$

The single controller can let the error system (5) to the sliding mode $s_1 = 0$. The stability of the designed SMC can be proved by the Lyapunov stability theorem. In this paper we choose the Lyapunov function as

$$V_1(t) = 0.5 s_1^2 \tag{9}$$

From the definition of the Lyapunov function, we can easily get that the function $V_1(t)$ is non negative. By taking the derivative of the function $V_1(t)$, we can get

$$\dot{V}_1(t) = s_1 \dot{s}_1 = s_1(-k_2 s_1 - k_3 sign(s_1)) = -k_2 s_1^2 - k_3 |s_1| \le 0 \tag{10}$$

By using the formula (10) above, we can get that the sliding surface s_1 converges to zero rapidly. It has been proven that the designed single controller $u_1(t)$ can make the error signal e_i rapidly converges to zero. Then the projective synchronization of Arneodo system and Coullet system can be realized. The stability proof of the designed single controller ends.

In the controller (8), due to the presence of the signed functions, the controller is discontinuous and may experience chattering. To weaken the chattering of the single controller, this paper uses hyperbolic tangent function 'tanh' to take the place of 'sign'. Then we can get the single controller which is designed as

$$u_1(t) = -f_1(y_1, y_2, y_3) + kf_2(x_1, x_2, x_3) - 2k_1e_3 - k_1^2e_2 - k_2s_1 - k_3\tanh(s_1/k_4) \quad (11)$$

where parameter k_4 is tiny positive constant. The constant parameter k_4 is set to be 0.0001 in this paper.

3.2 *Adaptive Sliding Mode Controller*

In the design of the adaptive SMC, the linear sliding surface is

$$s_2(t) = e_3 + 2m_1e_2 + m_1^2e_1 \quad (12)$$

where parameter m_1 is a positive constant.

For the adaptive SMC design, adaptive exponential reaching law is applied. In this paper, we take the derivation of the variable s_2 as

$$\dot{s}_2 = -\left(m_2 + \sum_{i=1}^{3}|e_i|\right)s_2 - m_3 sign(s_2) \quad (13)$$

where parameters m_2 and m_3 are positive constants.

Through linear sliding surface and adaptive exponential reaching law, adaptive sliding mode controller can be designed as

$$\begin{aligned} u_2(t) = &-f_1(y_1, y_2, y_3) + kf_2(x_1, x_2, x_3) \\ &-2m_1e_3 - m_1^2e_2 - \left(m_2 + \sum_{i=1}^{3}|e_i|\right)s_2 - m_3 sign(s_2) \end{aligned} \quad (14)$$

Similarly, we can use the Lyapunov theorem to proven the stability of the adaptive SMC, which will not be repeated here. In the adaptive sliding mode controller, hyperbolic tangent function is applied to weaken the influence of chattering, the final single adaptive controller can be obtained as

$$u_2(t) = -f_1(y_1, y_2, y_3) + kf_2(x_1, x_2, x_3) \\ -2m_1 e_3 - m_1^2 e_2 - \left(m_2 + \sum_{i=1}^{3} |e_i| \right) s_2 - m_3 \tanh(s_2 / m_4) \tag{15}$$

where small positive parameter $m_4$is set to be 0.0001.

4 Numerical Simulations

To validate the designed controller, numerical simulations were conducted using MATLAB software and the simulation results were analyzed in this section. In numerical simulation, the solution of differential equations is calculated using the Euler method, the fixed step size is 0.0001 s. The total time is 30 s, and in the 10th second, the designed single controller is used for projective synchronization control of the Arneodo system and Coullet system.

The Arneodo system is the drive system, and the parameter settings have been introduced earlier, the initial states are taken as $(x_1(0), x_2(0), x_3(0)) = (-1, 0.1, 1)$. The Coullet system is the response system, and the parameters are exactly the same as those introduced earlier, the initial states are taken as $(y_1(0), y_2(0), y_3(0)) = (0.2, 0.1, -0.1)$. In the design of SMC and adaptive SMC, the parameters of the linear sliding surface are set to $k_1 = m_1 = 1$, the parameters of the Eq. (13) are set to $k_2 = m_2 = 1$ and $k_3 = m_3 = 0.5$. In the projective synchronization control, the scaling factor is set to $k = 0.5$.

Using the parameters set above, single SMC and single adaptive SMC were used for projective synchronization of the Arneodo system and Coullet system. When using sliding mode controller $u_1(t)$, the error signals are shown in Fig. 1. When using adaptive sliding mode controller $u_2(t)$, the error signals are shown in Fig. 2. It can be seen that the error signals can converge to zero from Figs. 1 and 2, and both controllers can achieve projective synchronization control. In the first 10 s, the Arneodo system and Coullet system cannot achieve projective synchronization, and in the 10th second, the designed controller is used to control. When using adaptive sliding mode controller, the error signal converges to zero faster, enabling faster implementation of projective synchronization control and better performance.

The designed controllers are shown in Fig. 3. Due to the use of hyperbolic tangent function in both controllers, the controllers are very smooth, and did not exhibit chattering phenomenon. In Fig. 3, at the 10th second, the designed controllers are used for projective synchronization control, so for the first 10 s the controllers both are zero. At the beginning, $u_2(t)$ has larger amplitude than $u_1(t)$, so $u_2(t)$ can quickly achieve projective synchronization control. Then, $u_1(t)$ and $u_2(t)$ gradually become the same size.

The stability of the single SMC has been theoretically proven, and numerical simulations of the projective synchronization have been conducted using MATLAB software. For the projective synchronization error systems, the designed single adaptive SMC can make the error signals approach to zero, thereby achieving

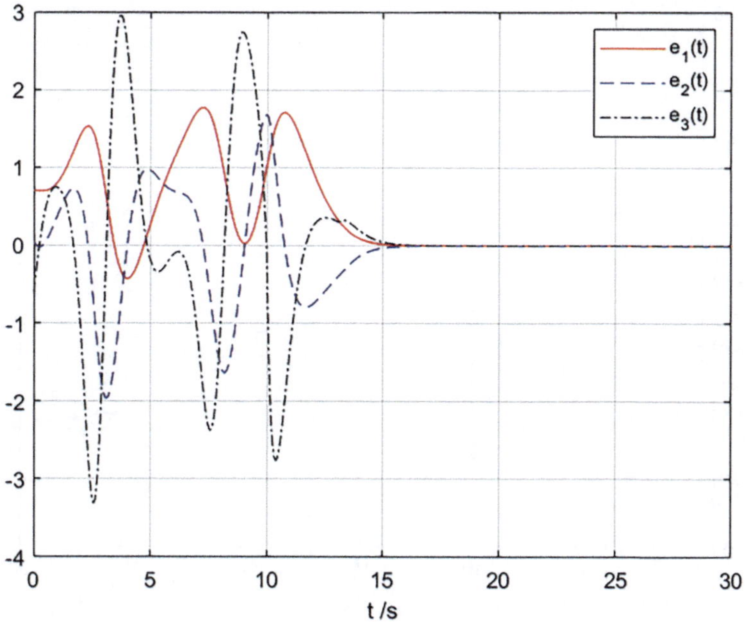

Fig. 1 The error signals with sliding mode controller

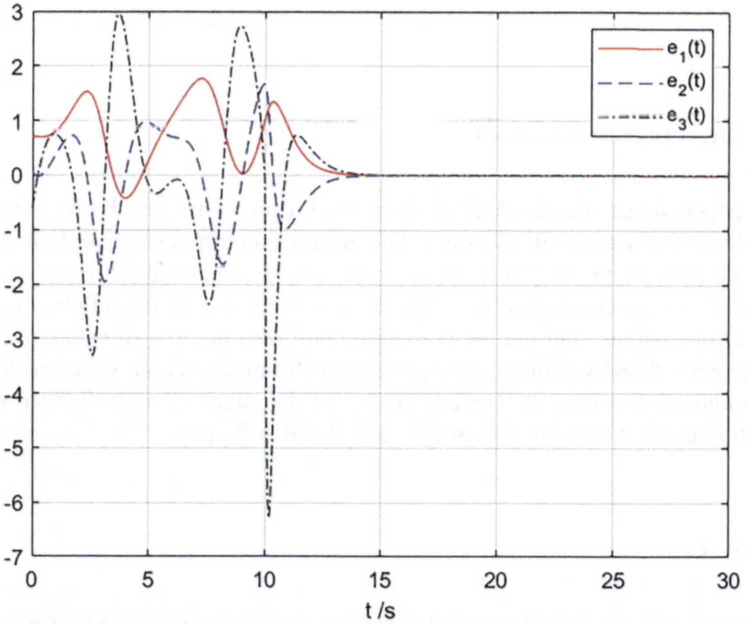

Fig. 2 The error signals with adaptive sliding mode controller

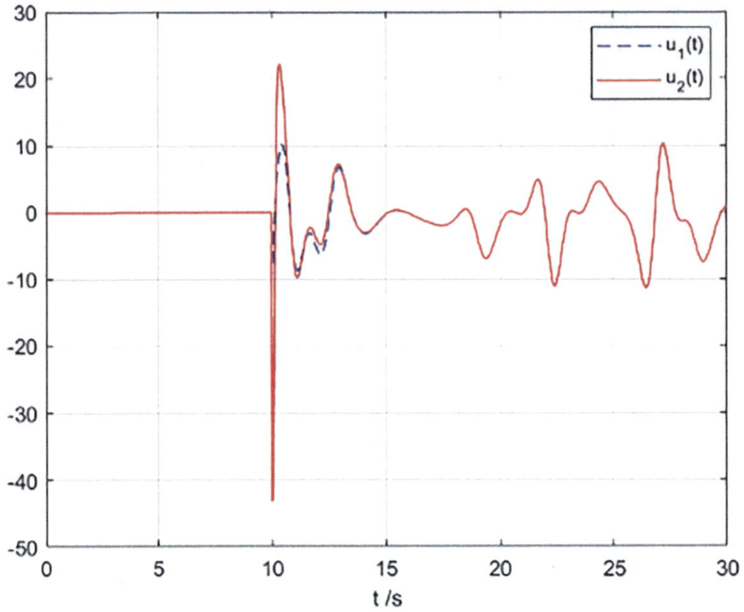

Fig. 3 The designed controllers $u_1(t)$ and $u_2(t)$

projective synchronization control of two chaos. The effectiveness and performance of the designed single adaptive controller were verified through digital simulation.

5 Numerical Simulations

In this paper, single adaptive SMC is designed for projective synchronization of the Arneodo system and Coullet system. The designed controller is based on adaptive exponential reaching law. The Lyapunov stability theorem is used to prove the stability of the designed single controller. The controller is used for stabilizing control of the error systems, and the projective synchronization error signals rapidly converge to zero, thereby achieving projective synchronization control of the Arneodo system and Coullet system. The advantages of the adaptive controller are verified through numerical simulation based on MATLAB software.

References

1. Boccaletti S, Kurths J, Osipov G et al (2002) The synchronization of chaotic systems. Phys Rep 366:1
2. Chen HK (2005) Global chaos synchronization of new chaotic systems via nonlinear control. Chaos Solitons Fractals 23:1245

3. Sundarapandian V (2012) Global chaos synchronization of Arneodo chaotic system via backstepping controller design. ACM Int Conf Proc Ser. https://doi.org/10.1145/2393216.2393217
4. Kuzmenko AA (2022) Forced sliding mode control for chaotic systems synchronization. Nonlinear Dyn 109:1763–1775
5. Wang T, Jia N (2012) Chaos control and hybrid projective synchronization of several new chaotic systems. Appl Math Comput 13:218
6. Zhang L, Yan Y (2014) Robust synchronization of two different uncertain fractional-order chaotic systems via adaptive sliding mode control. Nonlinear Dyn 76(3):1761
7. Vaidyanathan S, Volos CK, Pham VT (2015) Global chaos control of a novel nine-term chaotic system via sliding mode control. Stud Comput Intell 576:571

Open Access This chapter is licensed under the terms of the Creative Commons Attribution-NonCommercial-NoDerivatives 4.0 International License (http://creativecommons.org/licenses/by-nc-nd/4.0/), which permits any noncommercial use, sharing, distribution and reproduction in any medium or format, as long as you give appropriate credit to the original author(s) and the source, provide a link to the Creative Commons license and indicate if you modified the licensed material. You do not have permission under this license to share adapted material derived from this chapter or parts of it.

The images or other third party material in this chapter are included in the chapter's Creative Commons license, unless indicated otherwise in a credit line to the material. If material is not included in the chapter's Creative Commons license and your intended use is not permitted by statutory regulation or exceeds the permitted use, you will need to obtain permission directly from the copyright holder.

Identification and Solution Strategy of Civil Aircraft Prototype Inspection Problem Based on Digital Technology

Cui Li, Hong Zhang, Xinai Zhang, and Haomin Li

Abstract The review of civil aircraft prototype is a key link to verify the quality of design and manufacturing. Digital technologies, especially big data and digital twins, open up new ways to improve the efficiency and quality of reviews. This paper identifies the main problems in design defects, manufacturing deviations and data collaboration, and proposes strategies to improve design identification, strengthen deviation monitoring and enhance data collaboration, which provides an important reference for optimizing prototype review and helping to improve the research and development capability of civil aircraft.

Keywords Digital technology · Prototype · Problem identification · Review

1 Introduction

The rapid development of digital technology has promoted the intelligent and digital transformation of civil aircraft prototype review. The combination of prototype reviews and digital technology, as well as the application of big data and digital twins, has greatly improved the accuracy of reviews. However, there are still some problems in the existing review process, such as incomplete identification of design defects, insufficient monitoring of production deviations and data silos. In order to improve the efficiency and quality of civil aircraft prototype review, this paper puts forward some solutions such as improving design defect identification, strengthening production deviation monitoring and breaking data silos.1Theoretical Basis of Civil Aircraft Prototype Review Based on Digital Technology.

C. Li (✉) · H. Zhang · X. Zhang · H. Li
Shanghai Aircraft Design and Research Institute, Shanghai, China

© The Author(s) 2026

Z. Zhang, P. Mercorelli (eds.), *Mechanical Design and Simulation: Insights and Innovations*, Lecture Notes in Mechanical Engineering,
https://doi.org/10.1007/978-981-95-2405-1_75

1.1 The Core Connotation of Prototype Review and the Relationship between Digital Technology

Prototype review is a key part of aviation product development, which aims to evaluate the performance, structure and function of a prototype during the design, manufacturing and operation phases to ensure that it meets airworthiness, reliability and safety requirements. Traditional review methods rely on human judgment and fragmented data, which often leads to slow review processes and low accuracy [1]. In contrast, digital technology, especially virtual testing and big data analysis based on digital twins, allows prototype review to achieve real-time monitoring and dynamic optimization. In the case of the Boeing 777X program, the digital twin was used to simulate key component performance, resulting in a 30% increase in design adjustment efficiency.

1.2 Supporting Theories of Big Data Technology for the Review Process

Big data technologies provide decision support for reviews by collecting, cleaning, storing, and analyzing prototype lifecycle data [2]. Through high-dimensional feature extraction, big data analysis reveals potential abnormal patterns and risk characteristics, helping to identify and predict problems in the review process [3]. Its supporting theory can be described as:

$$\mathrm{R_i} = \sum_{j=1}^{n} w_j \cdot F_j(D_i) \tag{1}$$

where, R_i represents the risk assessment value of the i prototype. This value is a combined risk value calculated based on different data characteristics and is used to evaluate the safety and seaworthiness of the prototype. $w_j$represents the weight factor of data feature j, reflecting the impact of this feature on the overall risk assessment. This is usually based on historical data, expert experience, or machine learning algorithms. $F_j(D_i)$ represents the function of the JTH eigenvalue in dataset D_i of prototype i. This function can be the calculation of the relevant performance and structural data of the prototype, such as temperature, pressure and other monitoring data. n represents the total number of features in the prototype data set, reflecting the number of features used for risk assessment. The accuracy of risk assessment can be effectively improved by using machine learning algorithm for dynamic adjustment. Through data visualization technology, prototype reviewers can intuitively grasp the data distribution and trend, improve the scientific decision, shorten the review cycle, and improve the review quality.

1.3 Dynamic Monitoring Theory of Digital Twin Technology

By constructing digital model corresponding to physical prototype, digital twin technology can realize real-time monitoring and prediction of prototype dynamic behavior. The core of digital twin is based on the combination of physical model, data-driven and feedback closed-loop, and its dynamic monitoring theory can be expressed as:

$$M_d(t) = \int_0^t G_p(x,t) + H_d(y,t)dt \tag{2}$$

where, $M_d(t)$ represents the dynamic output of the digital twin at time t. This output value reflects the prediction and monitoring status of the digital model for the prototype at that moment. $G_p(x, t)$ represents the immediate change of the physical prototype at time t, influenced by parameters x (such as temperature, pressure, speed, etc.). This function is used to describe the actual behavior change of the prototype. $H_d(y, t)$ represents a function of digital model calibration and feedback, depending on the digital model parameters y and time t. Its role is to correct and optimize the digital model through a feedback mechanism to make it more similar to the behavior of the actual prototype. t is the time variable, which is used to describe the dynamic change process of the prototype. $\int_0^t$ represents the integration operation, which is used to calculate the cumulative change in the output of the digital twin from time 0 to t. Digital twin technology can realize dynamic evaluation under complex working conditions by collecting real-time running data of prototype and interacting with digital model. Combined with artificial intelligence algorithms, it can predict potential risks and make recommendations for optimization, thereby improving the efficiency and accuracy of reviews.

2 Identification of Civil Aircraft Prototype Review Problems Based on Digital Technology

2.1 Incomplete Identification of Design Defects

The identification coverage of design defects is limited by the ability of data analysis and simulation accuracy, and some key design problems are not exposed at an early stage [5]. Table 1 summarizes the major design defect types and their identification coverage.

As can be seen from the table, the defect identification coverage of pneumatic design and structural design is relatively high, 78% and 82% respectively, but the identification coverage of system integration and material selection is low. In particular, the issue of material selection, which has a detection coverage of only 74%, requires special attention as it may lead to durability and safety issues later on. In terms of system integration, the coverage is only 76%, which may be due to the lack

Table 1 Types of design defects and identification coverage

Defect type	Typical problem	Identification coverage (%)
Aerodynamic design defect	Data boundaries do not take into account	78
Structural design defect	Critical load path analysis is incomplete	82
System integration problem	The function interface is not clearly defined	76
Material selection problem	Corrosion resistance and fatigue performance were not evaluated comprehensively	74

Table 2 Analysis of manufacturing process data collection and coverage

Data type	Typical problem	Data acquisition coverage (%)
Machining dimension data	Detail size data is not fully captured	85
Processing temperature data	The distribution of monitoring points in high temperature areas is insufficient	78
Device running status data	Some older devices have no data collection capability	72
Material property data	The real-time sampling frequency is low	80

of coordination between different design teams, affecting the overall performance of the aircraft.

2.2 *Insufficient Monitoring of Manufacturing Deviation*

Limitations of Manufacturing Process Data Acquisition In the manufacturing process, the lack of data acquisition technology leads to the blind spot of deviation monitoring. Table 2 below summarizes the details of manufacturing process data acquisition coverage.

As can be seen from the table, the collection coverage rate of equipment operating status data is the lowest, only 72%. This indicates that the health status of the equipment, especially in older equipment, lacks adequate monitoring and data acquisition capabilities, and it is easy to miss potential fault signals of the equipment. In addition, the acquisition coverage of processing temperature data is 78%, and the insufficient distribution of monitoring points in the high temperature area makes the supervision of the high temperature working environment inaccurate, which directly affects the processing accuracy and the performance of parts.

Dynamic Monitoring Problem of Deviation Accumulation Deviation accumulation is a dynamic problem in manufacturing. Due to the lack of real-time monitoring between processes, the initial deviation will gradually amplify, affecting the performance of the prototype or causing quality problems. To quantify the cumulative process of deviation, the following formula can be used:

$$\Delta_{\text{total}} = \sum_{i=1}^{n} (\Delta_i + \in_i) \tag{3}$$

where, Δ_{total} is the deviation of the i process, measured by CMM (coordinate measuring machine); $\in_i$ is the measurement error, usually determined by the accuracy of the measuring equipment and environmental factors; Δ_{total}reflects the cumulative deviation of the entire production chain, if it exceeds the tolerance range, it is necessary to adjust the production parameters. The formula shows that with the increase of the working procedure, the deviation and measurement error will continue to superposition, so that the total deviation presents a cumulative trend. Table 3 below is used as an example to show the accumulation of specific deviations in a manufacturing process.

As can be seen from the table, with the progress of the process, the deviation gradually increases from the initial 0.3 mm to 1.2 mm. Especially in the third process, the measurement error has a significant impact on the total deviation. This shows that small deviations in the early stages of the production process, if not immediately adjusted, will gradually accumulate in subsequent processes, leading to final quality problems. The accumulation of deviation not only comes from the process error of each process, but also is affected by the measurement error.

2.3 *Data Silos and Low Collaboration Efficiency*

In prototype reviews, data silos hinder effective collaboration between design, manufacturing, and testing. Table 4 below summarizes the current status of data sharing between different links.

Table 3 Cumulative analysis of deviation in different processes

Process	Process deviation (Δ_i) (mm)	Measurement error ($\in_i$) (mm)	Cumulative deviation (Δ_{total}) (mm)
Step 1	0.2	0.1	0.3
Step 2	0.3	0.1	0.6
Step 3	0.4	0.2	1.2

Table 4 Analysis of data sharing status in different links

Link	Existing problem	Data sharing coverage (%)
Design-manufacture	Data format is not uniform	60
Manufacturing-testing	Real-time data update lag	65
Test-certification	Lack of a unified data platform	58
Design-certification	Data traceability is poor	62

As can be seen from the table, the design and manufacturing process has the lowest data sharing coverage, only 60%. This indicates that the design data is not fully connected to the manufacturing process, resulting in possible errors in the manufacturing process due to the lack of up-to-date design data. In addition, the data sharing coverage of manufacturing and testing is 65%, and the main problem is that real-time data updates are lagging, and it is not possible to respond to test feedback in a timely manner during the production process. This lag affects the efficiency of the whole prototype review process.

3 Solution Strategy for Civil Aircraft Prototype Review Based on Digital Technology

3.1 Improve the Design Defect Identification Method

Digital technology, especially virtual prototyping and digital twin technology, can significantly improve the ability to identify design defects of civil aircraft prototypes and quickly find potential problems. Through multidisciplinary collaborative optimization (MDO) technology, the interaction of aerodynamic, structural, dynamic and other disciplines can be analyzed in a virtual environment to optimize the design scheme in real time. The optimization formula is as follows:

$$R_d = \int_0^T \left|F(t) - S(t)\right| dt \tag{4}$$

where R_d represents the design risk and quantifies the deviation between the design and the standard. The higher the value, the higher the risk of design defect. F(t) is the functional requirement model, which describes the functional value that the design is expected to achieve, which changes with time t. Functional requirements are typically derived through multidisciplinary modeling. S(t) is the system response model, which represents the response of the design in actual use, and changes with time t. T is the time range, that is, the time interval for simulation analysis. This time interval is usually determined based on aircraft usage or simulation duration. By quantifying the deviation of function and structure, the weak link of design can be found effectively. For example, in the design phase of a certain type of aircraft, the digital aerodynamic model found the aerodynamic instability of the wing, and adjusted the trailing edge curvature reduced the risk of instability from 40% to 5%, reducing rework and shortening the development cycle. In addition, AI technology improves defect prediction accuracy by training historical design data, identifies fatigue risks of composite structures, and the design team adjusts the thickness distribution to avoid late failure problems.

3.2 Strengthen the Real-Time Monitoring of Manufacturing Deviation

Strengthen the Accuracy and Coverage of Manufacturing Data Acquisition Manufacturing deviation is a key factor in prototype review, and high precision and wide coverage data acquisition is the basis of controlling deviation. Through the introduction of laser scanners, Industrial Internet of Things (IIoT) sensors and edge computing technology, the accuracy and efficiency of data acquisition can be significantly improved. With a resolution of up to 0.01 mm, the laser scanner can fully scan large components and quickly catch small deviations. For example, in aircraft fuselage assembly, sensor networks collect temperature, stress, and displacement data and upload critical information after processing through edge computing, reducing latency and improving anomaly identification efficiency. In practical application, the system has successfully identified and adjusted the deformation of wing skin in high humidity environment. In addition, through data analysis, the company found that the parts deviation rate was higher in the high temperature season, and optimized the environmental control strategy, successfully reducing the deviation rate by 15%.

Construct Dynamic Monitoring and Feedback Mechanism of Deviation The dynamic monitoring and feedback of manufacturing deviation is the key to ensure the quality of prototype. By monitoring the deviation status and adjusting it in time through real-time data flow, the error in the production process can be effectively reduced. The core formula of dynamic deviation monitoring is:

$$D(t) = \frac{1}{N}\sum_{i=1}^{N}\left|M_i(t) - S_i\right| \tag{5}$$

where, D(t) is the total deviation at time t, representing the overall deviation degree of the system at a specific moment. $M_i(t)$ is the actual measurement value of the i measuring point. This value is the data acquired through real-time sensors, usually from the physical environment. S_i is the design value of the I-th measurement point, which is usually the ideal value calculated based on the design specification or standard. N is the total number of measurement points. The number of measurement points can be selected according to the complexity of the product to ensure full monitoring of deviations. Based on this formula, the system monitors the deviation change in real time and activates the alarm mechanism when the deviation exceeds the threshold.

3.3 Break Data Silos to Promote Efficient Collaboration

In the examination of civil aircraft prototype, the problem of data island makes the information flow of each link not smooth, which has a negative impact on the efficiency and quality of the examination. In order to break down departmental barriers, digital technology provides a solution by establishing a unified data platform and a sound data sharing mechanism. By developing a unified data standard (such as STEP, XML, JSON format), departments can seamlessly exchange data on the same platform, ensuring consistency and portability of information. In addition, building standardized data interfaces (such as apis or service buses) enables different departments to upload and share data in real time, avoiding information isolation. In order to ensure the timely update and consistency of data, enterprises can also use blockchain technology to ensure that the data cannot be tampered with, automatically update the data version, and provide data traceability. The digital collaboration platform also improves cross-departmental collaboration efficiency through cloud collaboration tools. In the prototype review, the manufacturing team can upload data in real time, and the design team can make adjustments based on real-time data and quickly push the modification plan, thus shortening the review cycle and improving the quality of the review. Through these means, digital transformation not only breaks down data silos, but also promotes efficient collaboration and information flow between departments, reducing resource waste.

3.4 Experimental Verification and Case Analysis

In order to verify the application effect of digital technology in prototype review, we conducted experiments on the A320 wing assembly line of an aviation manufacturing company, and compared the traditional manual review method with the digital twin-based review method. In the experiment, 50 samples were checked for bias using traditional CMM (coordinate measuring machine) and CAD review methods, as well as a digital twin system based on real-time data acquisition and virtual model comparison. The digital twin system can collect the prototype data in real time and compare it with the virtual model, automatically detect potential deviations and make adjustment suggestions. Table 5 compares the deviations between traditional and digital twins:

The experimental results show that the digital twin review method performs better in deviation control and review efficiency, the average deviation is reduced by 32%, and the review efficiency is increased by 40%. The real-time data feedback

Table 5 Comparison of deviations between traditional and digital twins

Method	Average deviation (mm)	Deviation rate (%)	Review time (h)
Traditional detection	1.5	18	20
Digital twin review	1.02	6	12

and automated adjustment of the digital twin significantly reduces the error rate of manual operations and improves the accuracy and speed of reviews. Compared to traditional methods, digital twin systems can not only detect deviations in advance, but also provide real-time feedback and optimization recommendations, thereby reducing late correction costs and improving production efficiency.

4 Conclusion

This study proposes an optimization strategy for civil aircraft prototype review based on digital technology, which improves design defect identification, manufacturing deviation monitoring and data collaboration efficiency with the help of big data analysis and digital twin technology [4]. Through experimental verification, digital twin technology significantly improves the accuracy and efficiency of prototype review, and is superior to traditional methods in deviation identification and review speed. The experimental results show that after the introduction of digital review method, the deviation rate is reduced by about 32%, and the review efficiency is increased by 40%. These results show that the application of digital twin technology in prototype review has significant advantages, especially for large-scale production and complex manufacturing environments. Future research could further optimize the construction cost of the digital system, make it applicable to the broader aerospace manufacturing sector, and strengthen the integration with artificial intelligence to further enhance the automation and predictive capabilities in the review process.

Conclusion The review of civil aircraft prototype based on digital technology provides a new direction for improving the efficiency and accuracy of the review. This paper analyzes the problems of incomplete design defect identification, insufficient manufacturing deviation monitoring and data silos, and proposes strategies to improve identification methods, strengthen monitoring and promote data collaboration, aiming to improve the accuracy and efficiency of review, support the improvement of civil aircraft R&D and manufacturing quality, and promote the digital transformation of the industry.

References

1. Changxiao Z, Hao LI, Wei Z et al (2024) Risk identification and safety assessment of human-computer interaction in integrated avionics based on STAMP. J Syst Eng Electron 35(3):689–706. https://doi.org/10.23919/JSEE.2024.000031
2. Tai S, Chen BU, Wang Y et al (2024) Identification of aircraft longitudinal aerodynamic parameters using an online corrective test for wind tunnel virtual flight. Chin J Aeronaut 37(9):261–275. https://doi.org/10.1016/j.cja.2024.05.031

3. Ovcharenko VN, Poplavsky BK (2021) Identification of nonstationary aerodynamic characteristics of an aircraft based on flight data. J Comput Syst Sci Int 60(6):864–874. https://doi.org/10.1134/S1064230721060149
4. Henderson IL, Shelley A (2023) Examining unmanned aircraft user compliance with civil aviation rules: the case of New Zealand. Transp Policy 133:176–185. https://doi.org/10.1016/j.tranpol.2023.01.010
5. Shirini K, Aghdasi HS, Saeedvand S (2024) A comprehensive survey on multiple-runway aircraft landing optimization problem. Int J Aeronaut Space Sci 25(4):1574–1602. https://doi.org/10.1007/s42405-024-00747-z

Open Access This chapter is licensed under the terms of the Creative Commons Attribution-NonCommercial-NoDerivatives 4.0 International License (http://creativecommons.org/licenses/by-nc-nd/4.0/), which permits any noncommercial use, sharing, distribution and reproduction in any medium or format, as long as you give appropriate credit to the original author(s) and the source, provide a link to the Creative Commons license and indicate if you modified the licensed material. You do not have permission under this license to share adapted material derived from this chapter or parts of it.

The images or other third party material in this chapter are included in the chapter's Creative Commons license, unless indicated otherwise in a credit line to the material. If material is not included in the chapter's Creative Commons license and your intended use is not permitted by statutory regulation or exceeds the permitted use, you will need to obtain permission directly from the copyright holder.

Design and Development of an Automatic Citrus Fruits Picking System Based on Machine Vision

Haoran Chen

Abstract In order to significantly enhances operational efficiency and provides a highly feasible technical solution for the upgrading of modern agriculture, this research is dedicated to the design and development of an automatic citrus fruits picking system based on machine vision. The primary objective is to achieve automated localization and recognition of citrus fruits, thereby offering robust technical support for unmanned orchard operations. The study leverages a drone platform to capture two-dimensional image data of citrus fruits and constructs an advanced image processing algorithm pipeline based on computer vision. The experimental results reveal that the automatic picking system maintains high accuracy in citrus fruits recognition across varying lighting conditions and complex backgrounds, demonstrating significant robustness.

Keywords Citrus fruits · Automatic picking system · Machine vision

1 Introduction

With the advancement of technology and the adjustment of China's economic development strategy, the agricultural sector is gradually transitioning towards large-scale and automated production [1]. Robots, which integrate various high-tech technologies, have been widely applied in industries such as manufacturing and domestic services. As an important crop following grains and vegetables in China, fruits hold a significant position in the national economic system. Moreover, with economic growth and improved living standards, the proportion of fruit consumption in overall food consumption is steadily increasing. Due to the inherent characteristics of fruit harvesting, relying on traditional manual picking methods not only results in low efficiency but also increases the risk of fruit damage.

H. Chen (✉)
School of Mechanical and Materials Engineering, North China University of Technology, Beijing, China

© The Author(s) 2026

Z. Zhang, P. Mercorelli (eds.), *Mechanical Design and Simulation: Insights and Innovations*, Lecture Notes in Mechanical Engineering,
https://doi.org/10.1007/978-981-95-2405-1_76

Additionally, fruit harvesting is highly seasonal, and a shortage of labor can lead to substantial economic losses. Mechanized harvesting, on the other hand, can significantly improve efficiency, reduce fruit damage, and lower costs. In recent years, the agricultural labor force has been gradually declining with the restructuring of the national economy, which may negatively impact the efficiency of agricultural production. The use of agricultural harvesting robots can effectively address this issue.

In the context of agricultural harvesting, robots should gather external information, with visual data being particularly vital [2–5]. Machine vision employs image sensors to capture object images, which are subsequently transmitted to a computer or a specialized digital signal processing system. By emulating human judgment criteria via dedicated computer modules, the automatic picking system could interpret and recognize these images, thereby enabling the analysis and comprehension of visual data. The ability of robots to simulate human visual perception to identify and locate objects is a cornerstone of robotic intelligence. Consequently, the advancement of machine vision technology holds significant importance for the research and development of agricultural harvesting robots [6–9]. Ringdahl et al. presented the development, testing, and validation of SWEEPER, a robot designed for harvesting sweet pepper fruit in greenhouses [6]. Chen et al. conducted four basic picking pattern experiments to analyze the effects of different picking patterns on fruit detachment [7]. Buribayev et al. proposed a novel tomato harvesting robot arm based on a continuum robot structure [8]. Additionally, a small harvesting robot was introduced for picking strawberries under open-field conditions [9].

Being a major origin of citrus fruits, China boasts abundant resources and a diverse array of high-quality cultivars. However, traditional fruit harvesting remains heavily reliant on manual labor. Therefore, the development of a harvesting robot that leverages machine vision technology to identify and locate citrus and other fruits, and to guide the robot in performing harvesting tasks, is of paramount necessity. This study could provide crucial support for the progress of intelligent fruit harvesting technologies.

2 Experimental Setup

Figure 1 illustrates the automated identification and localization procedure for citrus fruits using machine vision. Initially, a camera mounted on a drone captures high-resolution images of the citrus fruits. However, the brightness of these images is highly susceptible to variations in ambient lighting conditions, which complicates the direct extraction and analysis of color distribution features from the RGB values. To address this challenge and ensure robustness against weather and lighting variations, the captured RGB images are first converted to the HSI color space [10]. In the HSI color space, the Hue channel isolates color information independently of illumination, the Saturation channel helps distinguish vivid colors from less saturated backgrounds, and the Intensity channel provides illumination-invariant

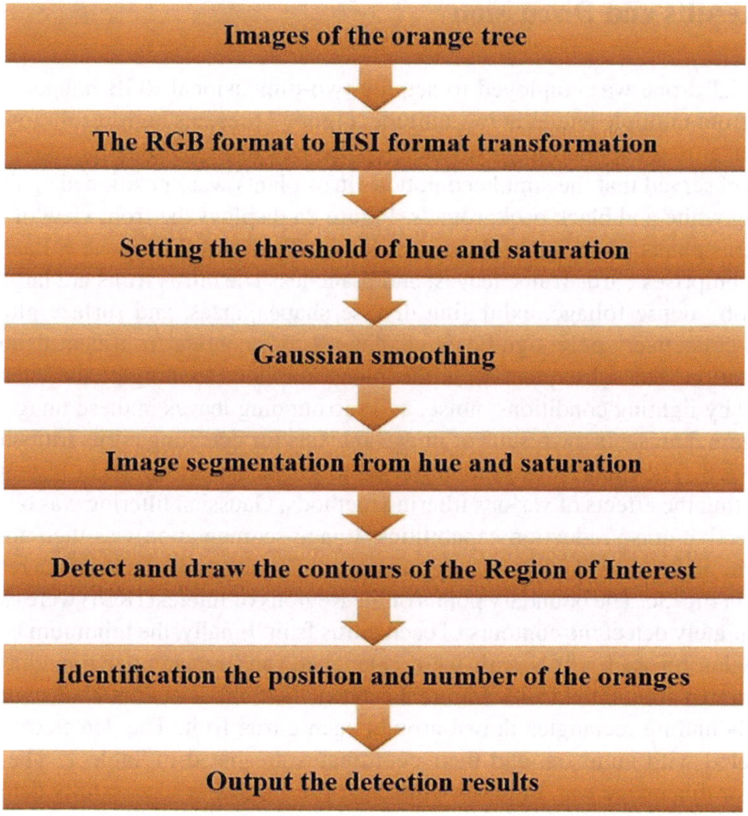

Fig. 1 The automated identification and localization procedure for citrus fruits based on machine vision

processing. Therefore, this transformation facilitates more effective color segmentation and illumination-invariant processing. Following the color space conversion, a Gaussian filtering algorithm is applied to smooth the images and effectively suppress noise, thereby enhancing the quality of the image data. By setting specific thresholds for hue and saturation, regions within the desired color range are extracted while areas with low saturation are excluded. This approach effectively isolates the vividly colored regions, enabling accurate identification of citrus fruits. Subsequently, edge detection and contour extraction techniques are employed to achieve precise segmentation of the fruit targets, ensuring clear delineation from the background. Finally, morphological analysis is conducted to determine the centroid coordinates and quantity information of the fruit targets. This comprehensive approach enables high-precision localization and identification of citrus fruits under varying environmental conditions, providing a robust solution for automated agricultural applications.

3 Results and Discussion

An aerial drone was employed to acquire two-dimensional RGB images of citrus fruits from multiple angles at low altitude. Figure 2 presents high-resolution images of orange trees captured from various perspectives using a drone-mounted camera. It was observed that the simulated potted citrus plants were positioned against contrasting white and black backgrounds. Figure 2a displays the front view of the potted citrus plants taken by the drone, while Fig. 2b shows the top view. Each potted plant comprises citrus fruits, leaves, and branches. The citrus fruits are largely concealed by dense foliage, exhibiting diverse shapes, sizes, and surface glossiness, which collectively pose significant challenges for accurate fruit identification and localization. The subsequent section explores methods to mitigate the interference caused by lighting conditions, noise, and surrounding leaves in these images.

Figure 3 presents the results of image analysis for detecting citrus fruits from the front view. The image was first converted to grayscale, as shown in Fig. 3b. After evaluating the effects of various filtering methods, Gaussian filtering was chosen for its effective noise reduction capabilities. Image segmentation was then applied to perform global threshold on the citrus images. The segmented citrus image is illustrated in Fig. 3c. The boundary points of the Regions of Interest (ROI) were extracted to accurately detect the contours of each citrus fruit. Finally, the minimum bounding rectangles for each citrus fruit were calculated, and rectangular bounding boxes were drawn around the fruits. Figure 3d shows the final detection results with minimum bounding rectangles drawn around each citrus fruit. The detected fruits are annotated with numbers, and their coordinates are listed in Table 1. The ground truth for this image was 14 citrus fruits, and the algorithm successfully detected all

Fig. 2 The high-resolution images of orange tree from various perspectives captured by the camera mounted on the drone

Fig. 3 The results of image analysis for detecting citrus fruits from the front view

of them, achieving a perfect recognition accuracy. Following successful fruit identification, positional accuracy was further validated through subsequent tests. The average running time on a standard desktop computer (Intel i7, 16GB RAM) is 0.5 s/image. To improve real-time performance, we explored parallel processing using GPU acceleration, which reduced the processing time to 0.1 seconds per image. The positional test results provided the quantity and precise coordinates of all citrus fruits in the image, serving as a prerequisite for intelligent harvesting operations.

In practical harvesting scenarios, relying solely on single-view recognition and localization may not comprehensively determine the quantity and spatial distribution of all citrus fruits on the tree. To eliminate randomness and measurement errors and further validate the robustness of the proposed method, the same procedure was applied to the top view of the potted citrus plants in Fig. 4. A series of processed images were generated, confirming the method's effectiveness in multi-angle

Table 1 The position identification results of each orange in Fig. 3

No.	Horizontal coordinate of detected oranges	Vertical coordinate of detected oranges
1	304.294	398.118
2	308.118	306.414
3	370.265	278.958
4	365.529	604.229
5	433.982	488.634
6	487.950	346.429
7	498.349	555.523
8	548.607	139.628
9	576.493	283.211
10	586.843	717.385
11	663.450	486.940
12	665.161	176.284
13	661.456	636.669
14	746.750	691.954

detection. The centroid points and pixel coordinates of the citrus fruits in the top view of the potted plants were calculated. Experimental validation has confirmed that the proposed method is equally effective in identifying the quantity and positional coordinates of citrus fruits in the top view. The recognition accuracy was 95% for top view, and 92% for side view. These results indicate that the algorithm performs best with front-view images, likely due to the clearer visibility of citrus fruits. The method exhibits robust performance under typical harvesting conditions and remains reliable even when the number of citrus fruits varies, with no significant impact on recognition efficiency. The primary sources of error identified in our experiments include variations in lighting conditions and occlusion by leaves. The error rate is kept within an acceptable range, ensuring minimal impact on the accuracy of citrus recognition and localization. To mitigate these issues, we propose using adaptive thresholding techniques to handle lighting variations and incorporating occlusion detection algorithms to improve robustness. The algorithm was enhanced to recognize multiple types of fruits, including apples and lemons, by integrating additional color and shape features. When tested in complex environments featuring multiple fruit trees and occluded fruits, the algorithm achieved a recognition accuracy of 90%. However, these scenarios presented several challenges, such as increased computational load and higher error rates attributed to occlusions. To mitigate these issues, we propose employing advanced segmentation techniques and real-time occlusion detection algorithms.

Fig. 4 The results of image analysis for detecting citrus fruits from the top view

4 Conclusion

In conclusion, an automatic citrus fruit-picking system based on machine vision has been successfully designed and developed. The proposed recognition and localization method exhibits robust performance across a wide range of conditions, effectively quantifying the number of citrus fruits and precisely determining their positions. This capability provides a solid foundation for subsequent harvesting operations. Importantly, the method remains effective regardless of variations in the number of citrus fruits, with minimal impact on image processing efficiency. Overall, this study offers crucial support for the advancement of intelligent fruit harvesting technologies.

Acknowledgement This study was supported by the Scientific Research and Training Program.

References

1. Hu Y, Li B, Zhang Z, Wang J (2022) Farm size and agricultural technology progress: evidence from China. J Rural Stud 93:417–429
2. Almendral KAM, Babaran RMG, Carzon BJC, Cu KPK, Lalanto JM, Abad AC (2018) Autonomous fruit harvester with machine vision. J Telecommun Electron Comput Eng 10(1-6):79–86
3. Fountas S, Mylonas N, Malounas I, Rodias E, Santos CH, Pekkeriet E (2020) Agricultural robotics for field operations. Sensors 20(9):2672
4. Fu L, Gao F, Wu J, Li R, Karkee M, Zhang Q (2020) Application of consumer RGB-D cameras for fruit detection and localization in field: a critical review. Comput Electron Agric 177(105):687
5. Zhou H, Wang X, Au W, Hanwen KH, Chen C (2022) Intelligent robots for fruit harvesting: recent developments and future challenges. Precis Agric 23:1856–1907
6. Arad B, Balendonck J, Barth R, Ben-Shahar O, Edan Y, Hellström T, Hemming J, Kurtser P, Ringdahl O, Tielen T, Van Tuijl B (2020) Development of a sweet pepper harvesting robot. J Field Robot 37(6):1027–1039
7. Bu L, Hu G, Chen C, Sugirbay A, Chen J (2020) Experimental and simulation analysis of optimum picking patterns for robotic apple harvesting. Sci Hortic 261(108):937
8. Yeshmukhametov A, Koganezawa K, Yamamoto Y, Buribayev Z, Mukhtar Z, Amirgaliyev Y (2022) Development of continuum robot arm and gripper for harvesting cherry tomatoes. Appl Sci 12(14):6922
9. Tituaña L, Gholami A, He Z, Xu Y, Karkee M, Ehsani R (2024) A small autonomous field robot for strawberry harvesting. Smart Agric Technol 8:100454
10. Ma S, Ma H, Xu Y, Li S, Lv C, Zhu M (2018) A low-light sensor image enhancement algorithm based on HSI color model. Sensors 18(10):3583

Open Access This chapter is licensed under the terms of the Creative Commons Attribution-NonCommercial-NoDerivatives 4.0 International License (http://creativecommons.org/licenses/by-nc-nd/4.0/), which permits any noncommercial use, sharing, distribution and reproduction in any medium or format, as long as you give appropriate credit to the original author(s) and the source, provide a link to the Creative Commons license and indicate if you modified the licensed material. You do not have permission under this license to share adapted material derived from this chapter or parts of it.

The images or other third party material in this chapter are included in the chapter's Creative Commons license, unless indicated otherwise in a credit line to the material. If material is not included in the chapter's Creative Commons license and your intended use is not permitted by statutory regulation or exceeds the permitted use, you will need to obtain permission directly from the copyright holder.

Research on Reconstruction of Solid Model of Complex Surface Automotive Steering Knot

Heping Tong, Zhigang Tu, and Daren Li

Abstract This study explores the reconstruction problem of solid models of complex surface automotive steering knuckles. A model reconstruction method based on a combination of reverse engineering and parametric design is proposed by analyzing the structural characteristics and modeling difficulties of the steering knuckle. The specific steps include: firstly, using a high-precision 3D scanner to scan the steering knuckle and obtain its surface point cloud data; Secondly, preprocess the model point cloud data, including data patching, data collection, and encapsulation, and fill and repair complex holes with missing point cloud data; Then, using Geomagic Design X software for data processing and surface reconstruction, a three-dimensional solid model of the steering joint is generated. Finally, the reconstructed steering joint solid model is subjected to "3D Comparative Analysis", "Point Analysis", and "Geometric Deviation" accuracy verification and optimization using Geomagic Control X precision analysis software. The experimental results show that this method can effectively improve the modeling accuracy and efficiency of complex surface steering knuckle models, providing new ideas for the digital design and manufacturing of automotive components.

Keywords Automobile steering knuckle · Geomagic design X · Model reconstruction · Parametric design

1 Introduction

As a key component of the automotive suspension system, the steering knuckle has a complex structure and high precision requirements, which directly affects the handling stability and safety of the vehicle. With the development of the automotive industry, the performance requirements for steering knuckles are increasing, and

H. Tong (✉) · Z. Tu · D. Li
Guangdong Open University, Guangzhou, Guangdong, China
e-mail: hptong@gdrtvu.edu.cn

© The Author(s) 2026

Z. Zhang, P. Mercorelli (eds.), *Mechanical Design and Simulation: Insights and Innovations*, Lecture Notes in Mechanical Engineering,
https://doi.org/10.1007/978-981-95-2405-1_77

their surface design is becoming increasingly complex. Traditional design methods are no longer sufficient to meet the modern automotive industry's demand for rapid development and precise manufacturing of steering knuckles. Therefore, studying the reconstruction method of solid models of complex surface automotive steering knuckles has important theoretical and practical significance [1, 2].

In recent years, Reverse engineering technology has been applied to the reconstruction of solid models of complex surface automotive steering knuckles to improve design efficiency and model accuracy. Scholars at home and abroad have conducted extensive research in the field of digital design of automotive components [3, 4]. For example, foreign scholars Smith et al. proposed a reverse engineering method based on feature recognition, which effectively improved the modeling efficiency of complex parts [5, 6]. Domestic scholars Li Ming et al. combined parametric design with reverse engineering and achieved good results in the design of automotive panels. However, there is still a lack of systematic research on the reconstruction of complex surface steering joint solid models [7, 8]. This study takes the complex surface automotive steering knuckle as the solid model, aiming to explore an efficient and accurate method for reconstructing the steering knuckle model, and provide technical support for the digital design and manufacturing of automotive components.

2 Structural Analysis and Model Design Theory of Complex Profile Automotive Steering Knuckles

2.1 Structural Analysis of Complex Profile Automotive Steering Knuckles

The steering knuckle of a car is an important component that connects the wheels and suspension system. Its main function is to support the wheels and achieve steering motion. A typical steering knuckle structure includes complex geometric features such as wheel hub connections, steering arms, suspension connections, etc. These features not only have complex shapes, but also have extremely precise spatial relationships with each other, which have a significant impact on the overall handling performance of the vehicle. Figure 1 shows a physical picture of complex car steering.

The complex surface of the steering knuckle is mainly reflected in the following aspects: firstly, the hub connection part needs to be precisely matched with the hub bearing, and its inner hole shape is complex with high dimensional accuracy requirements; Secondly, the surface shape of the steering arm needs to consider stress distribution and lightweight design, often with complex free-form surface features; Finally, the geometric shape of the suspension connection needs to be precisely matched with other components of the suspension system, typically including multiple irregular holes and curved features. These complex surfaces pose great

Fig. 1 Physical picture of complex car steering

challenges to the solid modeling of steering knuckles, and traditional forward design methods are difficult to quickly and accurately complete modeling tasks [9–11].

2.2 *Model Design Theory of Complex Profile Automotive Steering Knuckles*

The model design of complex profile automotive steering knuckles is based on geometric modeling theory and reverse engineering technology. The geometric modeling theory mainly includes *B*-spline curve and surface modeling and non-uniform rational *B*-spline (*NURBS*) modeling, whose mathematical expressions are as follows:

$$P(u) = \sum_{i=0}^{n} N_{i,p}(u).P_i \tag{1}$$

Among them, $P(u)$ is a point on the curve, $N_{i,p}(u)$ is the basis function, and P_i is the control point. By adjusting control points and weights, precise modeling of complex surfaces can be achieved.

In reverse engineering, the processing of point cloud data is a critical step. This article adopts a point cloud denoising method based on *K*-nearest neighbor algorithm, which is mathematically described as:

$$d(p_i,p_j) = \sqrt{(x_i - x_j)^2 + (y_i - y_j)^2 + (z_i - z_j)^2} \tag{2}$$

Among them, $d(p_i, p_j)$ is the distance between two points in the point cloud, and noise points are removed by setting a threshold.

3 Reconstruction Method of Steering Joint Model Based on Reverse Engineering

3.1 Point Cloud Data Collection

The quality of point cloud data will directly affect the difficulty of later entity model reconstruction. In this study, we used the KSCAN Magic series composite 3D scanner from Sigang Technology to collect point cloud data. During the scanning process of the physical model, by reasonably setting the scanning parameters and planning the scanning path, relatively complete and high-quality point cloud data can be obtained. Due to various factors such as the device itself, scanning environment, and operating technology, there are many holes in the scanning point cloud data of the steering joint, as shown in Fig. 2. The lack of these hole data directly affects the quality and accuracy of model reconstruction, and requires post-processing through Geomagic Wrap reverse software [12–14].

3.2 Point Cloud Data Processing

Point cloud data processing includes steps such as data patching, data sampling, data encapsulation, polygon patching, and polygon hole filling [15–17]. The specific operation steps are as follows:

1. Data patching. Firstly, in order to observe the shape of the point cloud clearly, the point cloud is colored first; Secondly, remove the isolated points outside the body, set the sensitivity value, and after selecting the isolated points outside the body, use the "Delete" command to delete them; Finally, remove the noise in the point cloud model to reduce the noise generated by scanning.
2. Data sampling. The point cloud model has a large amount of data, containing 1,992,224 point cloud data. The "Sampling Unified Sampling" command is used

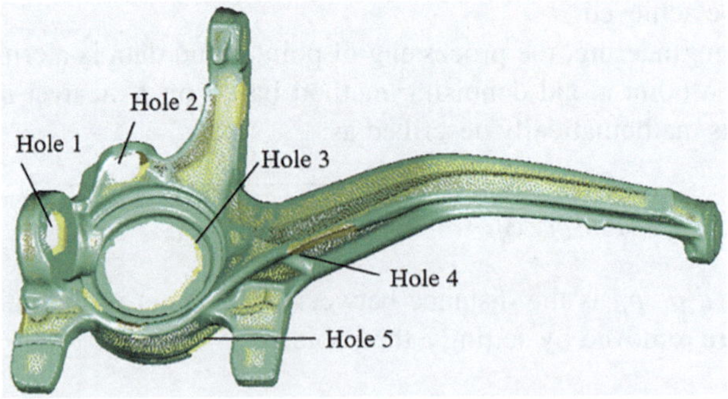

Fig. 2 Turning node cloud data model

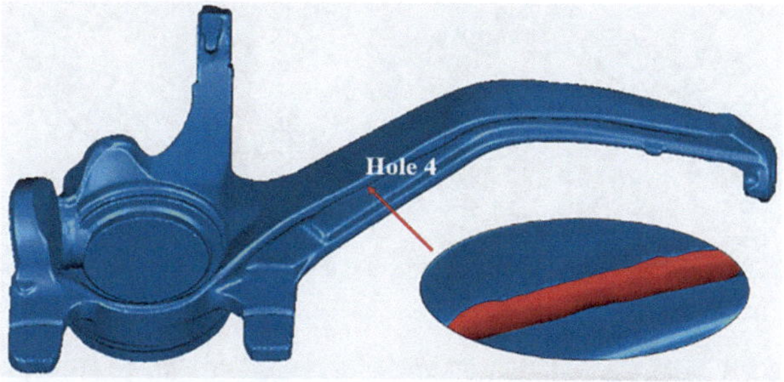

Fig. 3 Steering joint model after point cloud data processing

to reduce the number of point cloud data, laying the foundation for subsequent operations. Unified sampling can make data computation faster and improve computational efficiency.

3. Data encapsulation. The processed point cloud data needs to be encapsulated, which creates triangular facets by connecting points. The encapsulated point cloud data generates 2,184,387 polygon data.
4. Polygon repair. By clicking on "Patch"—"Grid Doctor", Grid Doctor integrates multiple functions such as deleting nails, filling holes, removing features, opening manifolds, etc. Grid doctors can also help improve the smoothness and accuracy of complex free-form surface parts during the reverse modeling process.
5. Polygonal filling holes. Filling holes can ensure the integrity and accuracy of the model, providing a reliable foundation for subsequent digital manufacturing and reverse design. In polygonal filling holes, different methods need to be selected according to the type of hole matching for filling. For example, in the repair of complex hole 4, curvature repair is used to fill it, and the filled hole 4 is shown in Fig. 3.

4 Parameterized Model Reconstruction of Steering Knuckle

To improve the modifiability and reusability of the model, this study establishes a parameterized model of the steering knuckle based on reverse engineering reconstruction. The parameterized model reconstruction of the steering knuckle is divided into several regions, and in the surface reconstruction stage, a segmented fitting strategy is adopted. The specific operation steps are as follows:

1. Domain division. Import the encapsulated. stl file into Geomagic Design X software, click on "Domain"—"Line Segment"—"Auto Segmentation", and based on the curvature value features, the system will automatically divide the steering joint domain, as shown in Fig. 4b.
2. Patch fitting. For free-form surface features, this study uses NURBS surfaces for fitting, generating complex feature patches. Click on "Model"—"Wizard"—

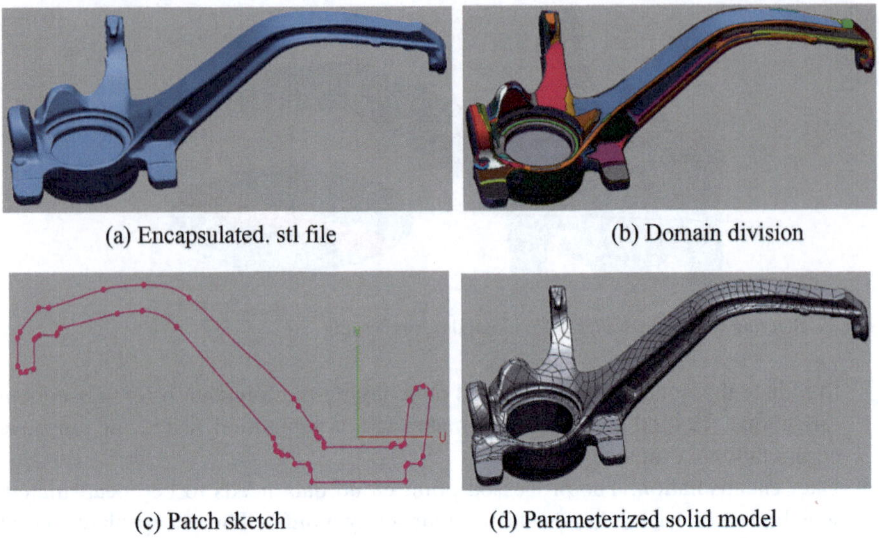

Fig. 4 Reconstruction process of steering knuckle model. (**a**) Encapsulated. stl file. (**b**) Domain division. (**c**) Patch sketch. (**d**) Parameterized solid model

"Patch Fitting", and according to the domain division, fit the free-form surface into the corresponding surface to complete the creation of the free-form surface.

3. Surface sketch. Click on "Patch Sketch", select the appropriate reference plane, obtain the reference contour line on the reference plane, use the drawing tool command to fit the "Curves", "Arcs" and "Lines" in the sketch, measure the corresponding dimensions through "Intelligent Dimension" and parameterize them, as shown in Fig. 4c.
4. Create a solid model. For regular geometric features such as cylindrical surfaces and holes in steering knuckles, precise parametric design methods are used to improve the model's modifiability and reusability. Click on "Create Entity"—"Stretch"—"Cut" to complete the construction of deep hole 3 features, until all features are closed and the entity is successfully created, as shown in Fig. 4d.

5 Accuracy Analysis of 3 Steering Knuckle Models

In order to verify the accuracy of the reconstruction of the solid model of the complex surface automotive steering knuckle and ensure that it meets the accuracy requirements of automotive components. This study utilized Geomagic Control X software for model accuracy analysis, importing scanned point cloud data and 3D solid models of complex profile automotive steering knuckles into the analysis software. On the basis of initial alignment and optimal alignment, select the "3D Comparison" command to obtain the overall deviation report of the car steering

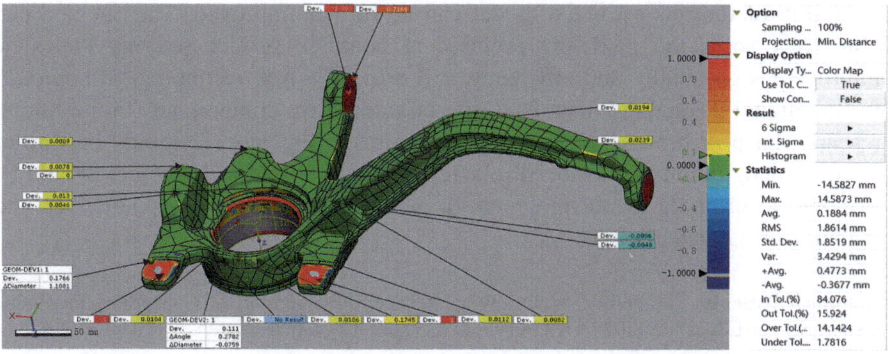

Fig. 5 Accuracy analysis of 3D comparison, point deviation, and geometric deviation

knuckle. Moreover, the car steering knuckle model is relatively complex, and usually requires sampling "Point Analysis" and "Geometric Deviation" commands to analyze the deviation values of special surfaces and regions, and then comprehensively analyze whether the surface area meets the accuracy requirements. The accuracy analysis results are shown in Fig. 5.

Through comparative analysis with the original point cloud data, the key dimension error of the reconstructed model is controlled within ±0.1 mm, and the surface roughness Ra value is less than 1.6 μm, meeting the accuracy requirements of automotive components. In the model modification experiment, multiple variant design schemes were quickly generated by adjusting parameter values, greatly improving design efficiency.

From Fig. 5, it can be seen that the overall accuracy of the reconstructed car steering knuckle model is good, with maximum deviation of 14.5873 mm, minimum deviation of −14.5827 mm, average deviation of 0.1884 mm, an RMS of 1.8614 mm, standard deviation of 1.8519, discrete deviation of 3.4294, cylindrical geometric deviation of 0.111 mm, and Circular geometric deviation of 0.1766 mm. Due to some missing scan data and distortion in some deviations, the overall accuracy requirements of the reconstructed model are within the required range and meet the design accuracy requirements.

6 Conclusion

This study proposes a method for reconstructing the solid model of complex surface automotive steering knuckles based on a combination of reverse engineering and parametric design. Through steps such as 3D scanning, point cloud processing, surface reconstruction, parametric modeling, and optimization design, high-precision and high-efficiency digital modeling of the steering knuckle has been achieved. The research results indicate that by selecting scanning equipment reasonably, optimizing data processing algorithms, and improving modeling accuracy, the accuracy and reliability of the reconstructed model can be further improved. This method can effectively solve the modeling problem of complex surface steering knuckles, improve design quality and efficiency, and provide a new technological approach for the digital design and manufacturing of automotive components. In the future, research on post-processing techniques for refactoring models should be further strengthened, and more advanced refactoring methods and technologies should be explored to promote the development of automotive steering knuckle design and manufacturing technology.

Acknowledgments The authors gratefully acknowledge the support of Guangdong Provincial Department of Education (2024KTSCX192) and Guangdong Open University (2024CQ-20).

References

1. Ru CB, Zheng JJ (2023) Research on reverse modeling of automotive components based on geomagic design X. Automot Electron 56(12):21–24
2. Sun B, Miao ZZ, Yang HX (2024) Reverse design and detection analysis of camshaft based on Geomagic software. Mold Technol 38(5):95–100
3. Wang CX, Hao LW, Ji KH (2020) Research on solid model reconstruction of complex surface chain rail joint. Mach Tool Hydraul 48(17):122–125
4. Wang CX, Ji KH, Hao LW (2021) Model reconstruction of multi deep hole characteristic automotive end caps. Mach Tool Hydraul 49(15):101–103
5. Smith J, Johnson R, Williams T (2020) Advanced reverse engineering techniques for automotive components. Int J Autom Technol 21(4):897–906
6. Brown A, Davis M, Wilson E (2018) Parametric modeling and optimization of complex automotive parts. J Mech Design 140(7):071701
7. Chen XW, Liu WD, Sun LB (2021) Lightweight design and performance optimization of automotive steering knots. Autom Eng 43(5):567–575
8. Tong HP, Li DR (2024) Thermal analysis of 3D printer nozzle system based on FDM. Mech Manuf Autom 53(6):142–145
9. Liu ZQ, Zhao WJ (2020) Structural optimization and fatigue life analysis of automotive steering knots. Autom Eng 42(7):901–907
10. Zhou Q, Wu ZC, Luo LJ, Li L, Wang T (2024) Practical teaching of 3D printing based on the combination of forward and reverse design laboratory. Res Explorat. https://doi.org/10.1002/cpz1.70036
11. Chen XF (2023) Reverse design of complex surface products based on geomagic design X modern manufacturing technology and equipment
12. Zhang J (2024) Research on reverse modeling and 3D printing of engine impeller. J Yangling Vocat Tech College 23(01):5–8
13. Xi T, Yao L, Long CJ, Yongfeng D (2023) Complex surface reconstruction based on geomagic studio. J Bengbu Univ 2:10
14. Chen EX (2022) Reverse modeling and error detection analysis based on Geomagic aircraft engine blades. Mod Manuf Technol Equip 10:1
15. Bai YX (2021) Reverse modeling of aircraft piston engine casing based on geomagic design X. Mech Eng 09:1

16. Li XM (2021) Research on product modeling and 3D printing forming based on geomagic design X: a case study of internal combustion engine connecting rod. Internal Combust Engine Accessor 17:1
17. Fu BL, Pan JB (2023) Research on reverse design method for complex parts based on geomagic design X. Mech Manage Dev 8:1

Open Access This chapter is licensed under the terms of the Creative Commons Attribution-NonCommercial-NoDerivatives 4.0 International License (http://creativecommons.org/licenses/by-nc-nd/4.0/), which permits any noncommercial use, sharing, distribution and reproduction in any medium or format, as long as you give appropriate credit to the original author(s) and the source, provide a link to the Creative Commons license and indicate if you modified the licensed material. You do not have permission under this license to share adapted material derived from this chapter or parts of it.

The images or other third party material in this chapter are included in the chapter's Creative Commons license, unless indicated otherwise in a credit line to the material. If material is not included in the chapter's Creative Commons license and your intended use is not permitted by statutory regulation or exceeds the permitted use, you will need to obtain permission directly from the copyright holder.

Tolerance Design of the Rigid Shell Throwing Mechanism for Firearms

Shengtao Chen and Jun Fang

Abstract The positions of geometric elements were represented by three-dimensional correlation coordinates based on probabilistic statistics. The theoretical tolerance design model of the rigid shell throwing mechanism was established by analyzing the geometric parameters transfer and accumulation path and using error association operations such as cascade, coordinate transformation, and synthesis. The optimization objective was the sum of the maximum volume and area of the confidence regions for each positional coordinate. The optimization variables were the relevant geometric parameters and their standard deviation. According to the theoretical tolerance optimization results, the tolerance optimization method considering the actual machining error was proposed for the geometric parameters of the parts with a small process capability index and high machining accuracy in the actual machining process, and the design results are compared and analyzed by using the traditional dimensional chain analysis method. The proposed method organically combines the tolerance theoretical model and geometric parameter measurement data, which can effectively improve the qualification rate of assembly precision.

Keywords Assembly tolerance · Confidence region · Covariance matrix · Statistical correlation coordinates · Shell throwing mechanism

1 Introduction

Currently, the interchangeability of firearm components in China remains limited, and the main assembly method of firearms products is manual filing, i.e., repairing unqualified part sizes by hand [1], which leads to low assembly efficiency.

S. Chen · J. Fang (✉)
School of Mechanical Engineering, Nanjing University of Science and Technology, Nanjing, China

© The Author(s) 2026

Z. Zhang, P. Mercorelli (eds.), *Mechanical Design and Simulation: Insights and Innovations*, Lecture Notes in Mechanical Engineering,
https://doi.org/10.1007/978-981-95-2405-1_78

To avoid unnecessary trimming work, it is essential to implement proper tolerance design at the outset. Tolerance design includes tolerance representation modeling and tolerance analysis. Commonly used tolerance representation models include the Tolerance-map model [2], Jacobian model [3, 4], Matrix model [5], Torsor model [6, 7], etc [8]. Tolerance analysis refers to the known size of each key dimension of the part, mathematically describes the changes in the tolerance domain, analyzes the error transfer and accumulation path, and ultimately obtains the assembly accuracy of the product [9]. These tolerance analysis methods represent the tolerance domain and error accumulation through small torsor, with Monte Carlo simulation employed to sample numerous constraint variables and optimization objectives. Since the analytical results are greatly affected by random sampling points, different sampling points produce different results, and the statistical tolerance optimization based on Monte Carlo simulation is difficult to design as it demands significant computational resources.

Therefore, in this paper, the analytical method is used to analyze and design the statistical tolerance. The three-dimensional position coordinates of the geometric elements are represented by statistical probability representation, and the correlation of the position coordinates is considered to establish the tolerance design model through the error correlation operation. Based on statistical data from sampling inspections, the geometric parameters of the parts that are difficult to guarantee the machining accuracy are optimized, to improve the qualification rate of the assembly accuracy.

2 Error Accumulation Correlation Operation

2.1 *Cascade*

Assuming that the tolerances of a part in the *x*, *y*, and *z* directions obey the distributions $P \sim N(\mu_1, \sigma_1^2)$, $P \sim N(\mu_2, \sigma_2^2)$ and $P \sim N(\mu_3, \sigma_3^2)$, the joint probability distribution of the coordinate positions of the target geometric elements can be expressed as:

$$P \sim N(\mu, S);\ \mu = \begin{bmatrix} \mu_1 \\ \mu_2 \\ \mu_3 \end{bmatrix},\ S = \left[s_{ij} \right]_{3\times3} = \begin{bmatrix} \sigma_1^2 & \rho_{12}\sigma_1\sigma_2 & \rho_{13}\sigma_1\sigma_3 \\ \rho_{21}\sigma_2\sigma_1 & \sigma_2^2 & \rho_{23}\sigma_2\sigma_3 \\ \rho_{31}\sigma_3\sigma_1 & \rho_{32}\sigma_3\sigma_2 & \sigma_3^2 \end{bmatrix} \tag{1}$$

where *S* is the covariance matrix. If tolerances belong to different parts, then$\rho_{ij} = 0$. If tolerances belong to the same part, the correlation of the value is as follows: (1) $\rho_{ij} = 0$: tolerances are not relevant; (2) $\rho_{ij} = 1$: tolerances are positively correlated; (3) $\rho_{ij} = -1$: tolerances are negatively correlated.

2.2 Three-dimensional coordinate transformation

Assuming that a geometric parameter obeys the distribution $P \sim N(\mu_1, \Sigma_1)$, the error transfer coordinate system is o_1-$x_1y_1z_1$, then the transformed geometric parameter obeys $P \sim N\left(\mu_1, R^T \Sigma_1 R\right)$ [10], where R is the coordinate transformation matrix [11].

$$R = \begin{bmatrix} \cos\beta\cos\gamma & -\cos\beta\sin\gamma & \sin\beta \\ \sin\alpha\sin\beta\cos\gamma + \cos\alpha\sin\gamma & -\sin\alpha\sin\beta\sin\gamma + \cos\alpha\cos\gamma & -\sin\alpha\cos\beta \\ -\cos\alpha\sin\beta\cos\gamma + \sin\alpha\sin\gamma & \cos\alpha\sin\beta\sin\gamma + \sin\alpha\cos\gamma & \cos\alpha\cos\beta \end{bmatrix} \tag{2}$$

where: α, β, γ are the rotation angles around x_1, y_1, and z_1, respectively, in the process of rotating the coordinate system o_1-$x_1y_1z_1$ to the coordinate system o-xyz.

2.3 Synthesis

Assuming that two cascaded tolerances follow the distributions $P_1 \sim N([\mu_{x1}, \mu_{y1}, \mu_{z1}]^T, \Sigma_1)$ and $P_2 \sim N([\mu_{x2}, \mu_{y2}, \mu_{z2}]^T, \Sigma_2)$, the synthesized geometric parameters conform to the distributions: $P \sim N([\mu_{x1} + \mu_{x2}\ \ \mu_{y1} + \mu_{y2}\ \ \mu_{z1} + \mu_{z2}]^T, \Sigma_1 + \Sigma_2)$.

3 Projection of the Confidence Region

Figure 1 shows the confidence ellipsoid and its projection in the *xoy* plane. When the direction of error variation between relative points is unknown or the position coordinates of the points are unknown, the maximum error variation is equal to the maximum intercept of the confidence ellipsoid parallel to the direction of error transmission. The maximum error variation can be interpreted as the maximum tolerance. In each direction, it can be calculated as follows: (1) x-direction: $x_{\max} = 2\sqrt{\chi^2_{n,\alpha} / s_{11}}$; (2) y-direction: $y_{\max} = 2\sqrt{\chi^2_{n,\alpha} / s_{22}}$; (3) z-direction: $z_{\max} = 2\sqrt{\chi^2_{n,\alpha} / s_{33}}$.

The size of the confidence region can be calculated as:

$$\begin{cases} V_i = \left(4\pi\sqrt{\lambda_{i1}\lambda_{i2}\lambda_{i3}\left(\chi^2_{n,\alpha}\right)^3}\right)/3; & \text{If the confidence region is an ellipsoid} \\ S_i = \pi\sqrt{\lambda_{j1}\lambda_{j2}}\,\chi^2_{n,\alpha}; & \text{If the confidence region is an ellipse} \end{cases} \tag{3}$$

where λ_p denotes the characteristic root of the matrix S.

Fig. 1 Confidence ellipsoids and its projections on the *xoy* plane

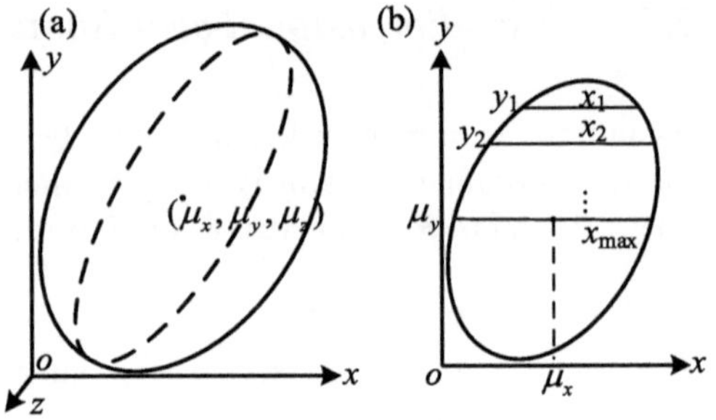

4 A Firearm Shell Throwing Mechanism Assembly Tolerance Design Example

This paper proposes a three-dimensional statistical correlation coordinate-based assembly tolerance design method to control the overlap amount between the shell throwing tappet and the cartridge case. The method consists of two parts: tolerance design based on the theoretical confidence region and tolerance optimization considering the actual machining errors.

4.1 Tolerance Design Based on Theoretical Confidence Region

The flow of the tolerance design of the shell-throwing mechanism based on the theoretical confidence region is shown in Fig. 2.

Error Transfer and Accumulation Path Analysis The shell-throwing mechanism contains parts: shell-throwing tappet (STT), magazine body (MB), frame body (FB), bolt (Bo), and cartridge case (CC), with the abbreviation of the parts in parentheses. The geometric parameters of the body and cartridge case are given, as shown in Fig. 3.

The error transfer and accumulation path analysis is shown in Fig. 4, where the overlap amount is denoted by *P* and the direction is $e \to s$. According to Fig. 4a, the geometrical parameters to be designed for the other three parts are shown in Fig. 5.

Establishment of Multivariate Normal Distribution of Position Coordinates According to Fig. 4, the cascade operation is applied to establish a multivariate normal distribution between points with relative coordinates. An example of the cascade operation is illustrated for point *a* relative to point *s*, as shown in the following.

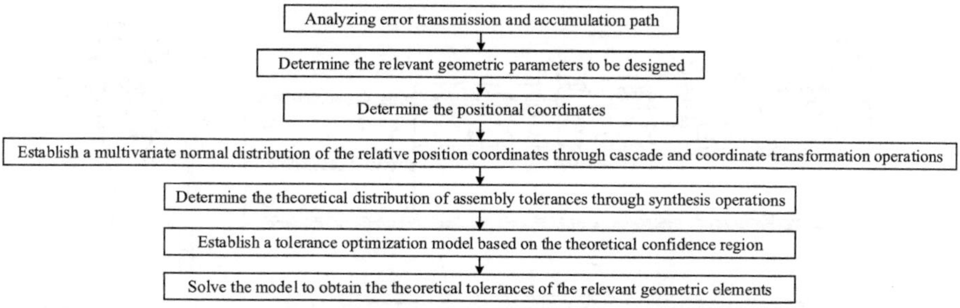

Fig. 2 Tolerance design flow based on theoretical confidence region

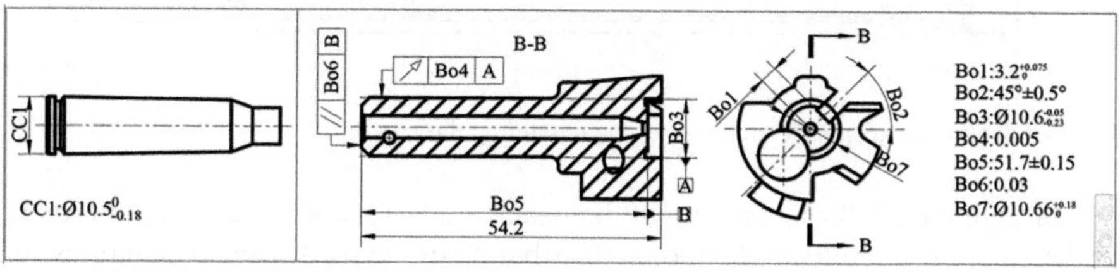

Fig. 3 Parts drawing of the cartridge case (left) and bolt (right)

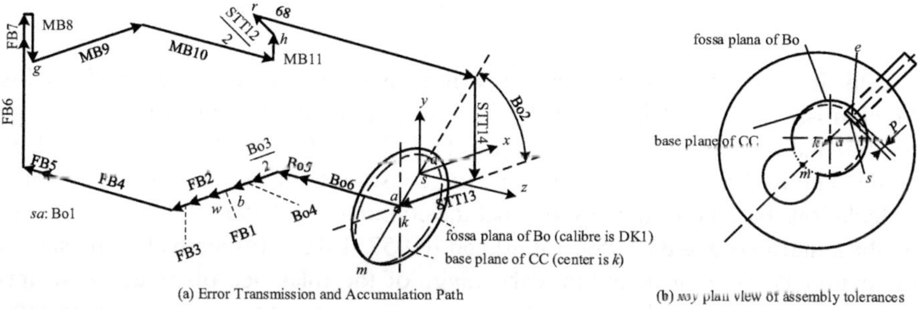

(a) Error Transmission and Accumulation Path (b) xoy plan view of assembly tolerances

Fig. 4 Error accumulation and transmission path analysis diagram. (**a**) Error transmission and accumulation path. (**b**) *xoy* plan view of assembly tolerances

$$\mu_{sa} = \begin{bmatrix} -\mu_{\mathrm{Bo1}} \cos \mu_{\mathrm{Bo2}} \\ -\mu_{\mathrm{Bo1}} \sin \mu_{\mathrm{Bo2}} \\ 0 \end{bmatrix}; \Sigma_{sa} = R_{sa}^{T} \begin{bmatrix} \sigma_{\mathrm{Bo1}}^{2} & -\sigma_{\mathrm{Bo1}}\sigma_{y_1,sa} & 0 \\ -\sigma_{\mathrm{Bo1}}\sigma_{y_1,sa} & \sigma_{y_1,sa}^{2} & 0 \\ 0 & 0 & 0 \end{bmatrix} R_{sa} \tag{4}$$

where: $\sigma_{y_1,sa} = \mu_{\mathrm{Bo1}} \sin(\Delta\alpha)/3$ is Bo1 converted to a linear tolerance in the y_1 direction under the action of the angular tolerance Bo2. Since the confidence level in this paper is taken as 0.9973, $\sigma_{\mathrm{Bo1}} = T_{\mathrm{Bo1}}/6$.

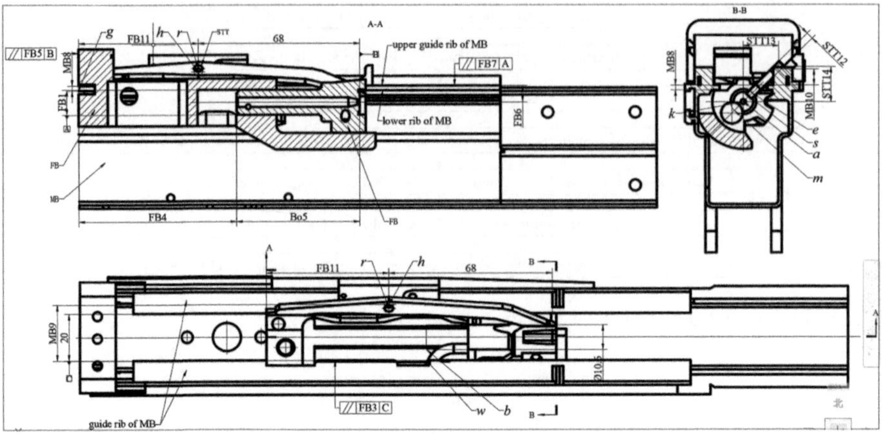

Fig. 5 Labeling of geometric elements of the remaining three parts

Through the synthesis operation, the mean vector μ_{es} and covariance matrix Σ_{es} of the theoretical multivariate normal distribution of the final stacked quantities can be calculated as:

$$\begin{aligned} \mu_{es} &= \mu_{sa} + \mu_{ab} + \mu_{bw} + \mu_{wg} + \mu_{gh} + \mu_{hr} + \mu_{ra} + \mu_{am} + \mu_{me} \\ \Sigma_{es} &= \Sigma_{sa} + \Sigma_{ab} + \Sigma_{bw} + \Sigma_{wg} + \Sigma_{gh} + \Sigma_{hr} + \Sigma_{ra} + \Sigma_{am} + \Sigma_{me} \end{aligned} \tag{5}$$

Establish the Tolerance Optimization Model of the Shell-Throwing Mechanism The nominal length P of the overlap required by the tolerance design is 1.95, and the tolerance design range is [1.5, 2.35]. The overlap amount needs to be decomposed to the x and y axes as the constraints of the optimization model. Calculations based on Eq. (6) show that the confidence intervals for the projections on the x and y-axes are [0.959, 1.828] and [0.957, 1.824], respectively. The size of the confidence region determines the range of the tolerance distribution, so this paper takes the sum of the size of the confidence region of the position coordinates as the optimization objective. Considering that $\chi^2_{n,\alpha}$ and $4\pi/3$ in Eq. (3) are fixed values, the assembly tolerance design model based on the theoretical confidence region is:

$$\begin{aligned} &\max \quad \left(\sum_{1}^{i} w_i \lambda_{i1}^2 \lambda_{i2}^2 \lambda_{i3}^2 + \sum_{i+1}^{j} \theta_j \lambda_{j1}^2 \lambda_{j2}^2 \right) \\ &s.t. \quad \mu_{i,\min} < \mu_i < \mu_{i,\max}; T_{i,\min} < 6\sigma_i \le T_{i,\max}; 0 < x_{\max} \le 0.869; 0 < y_{\max} \le 0.868; \end{aligned} \tag{6}$$

where: λ_{i1}, λ_{i2}, λ_{i3}, λ_{j1}, λ_{j2} are the characteristic roots, w_i, θ_j are the weight coefficients of the confidence ellipsoids and the confidence ellipses, respectively. $\propto_{i,\min}, \propto_{i,\max}, T_{i,\min}, T_{i,\max}$ are the ith minimum and maximum geometric parameters and their minimum and maximum tolerances, respectively. $x_{\max}$ and $y_{\max}$ are the maximum intercepts of the projections of the corresponding confidence regions of Eq. (5) on the x and y axes, respectively.

Theoretical Design Results of Relevant Geometric Parameters The theoretical optimization results are shown in Table 1, where tolerance standardization means selecting the closest standard tolerance. It can be seen that the accuracy class of most of the geometric parameters is IT11–IT12. However, MB8 represents the height of the guide rib, and its tolerance class is h8. STT12 represents the thickness of the shell throwing tappet, and its tolerance class is JS8.

4.2 *Tolerance Optimization Considering Actual Machining Error*

Tolerance Optimization Method and Process Considering Actual Machining Error Due to the differences in the probability distribution parameters of the actual geometric parameters of the parts, there is a certain probability that the assembled parts with the tolerances designed according to the probability statistics method will not meet the assembly accuracy requirements. Therefore, the tolerance optimization considering the actual machining error is proposed, to ensure the assembly accuracy requirements. Because of the sampling data obtained at this time, it is necessary to replace the overall covariance matrix Σ with the sample covariance matrix, replace the overall mean with the sample mean $\overline{a}$, and analyze the sample covariance matrix using the T^2 test, as shown in the following formula.

$$P\left((a-\overline{a})^T S(a-\overline{a}) \le T_\alpha^2\right) = 1-\alpha;\ T_\alpha^2 = p(n-1)\cdot F_\alpha(p,n-p)/(n-p) \quad (7)$$

Table 1 Theoretical design results

Part	Parameters	Standardization	Part	Parameters	Standardization
Frame body	FB1	$\varnothing 10.6\text{H}12\begin{pmatrix}+0.18\\0\end{pmatrix}$	Magazine body	MB8	$2.015\text{h}8\begin{pmatrix}0\\-0.014\end{pmatrix}$
	FB2	4.558±0.024		MB9	24.196 ± 0.0165
	FB4	67.623±0.15		MB10	7.195 ± 0.045
	FB6	$7.004\text{H}11\begin{pmatrix}+0.09\\0\end{pmatrix}$	Shell throwing tappet	STT12	2.964 ± 0.007
	FB11	51.324 ± 0.15		STT13	13.284 ± 0.0215
	FB3	0.007		STT14	13.251 ± 0.0215
	FB5	0.03	/	/	/
	FB7	0.006	/	/	/

where: S is the inverse matrix of the sample covariance matrix, p is the number of degrees of freedom and n is the sample capacity. The tolerance optimization process considering the actual machining error is shown in Fig. 6. The optimization goal is still the volume or area of the confidence region of the largest sum.

The actual correlation coefficient is calculated from the sample data, and the actual correlation of the geometric parameters between different parts remains 0. For the same part, the calculation equation is as follows:

$$\rho_{ij} = \left(n\left(\Sigma l_i l_j\right) - \Sigma l_i \cdot \Sigma l_j\right) / \sqrt{\left[n\Sigma l_i^2 - \left(\Sigma l_i\right)^2\right]\left[n\Sigma l_j^2 - \left(\Sigma l_j\right)^2\right]} \quad (8)$$

where n is the sample capacity, l_i, l_j are the ith, jth geometric elements measured data.

Tolerance Optimization Results Considering Actual Machining Errors According to the geometric parameters shown in Table 1, the parts are machined and sampling inspected. According to the statistical situation on the difficult to ensure the machining accuracy and the actual process capability index C_{pk} < 1 of the height of the guide rib (MB8) and the thickness of the shell throwing tappet (STT12) to optimize again. After calculation, the height of the guide rib is $1.98h10\begin{pmatrix} 0 \\ -0.04 \end{pmatrix}$ and the thickness of the shell tappet is 3.02±0.05. Compared with the theoretical design results, the machining accuracy grade of the guide rib height is reduced from IT8 to IT10; the thickness of the shell tappet is increased from 2.964 to 3.02, and the machining accuracy grade is reduced from IT8 to IT12, which reduces the machining cost.

The worst-case method was used to verify the results of the two designs. The overlap amount calculated according to the theoretical results is [1.516, 2.204], The overlap calculated according to the optimization results considering the actual

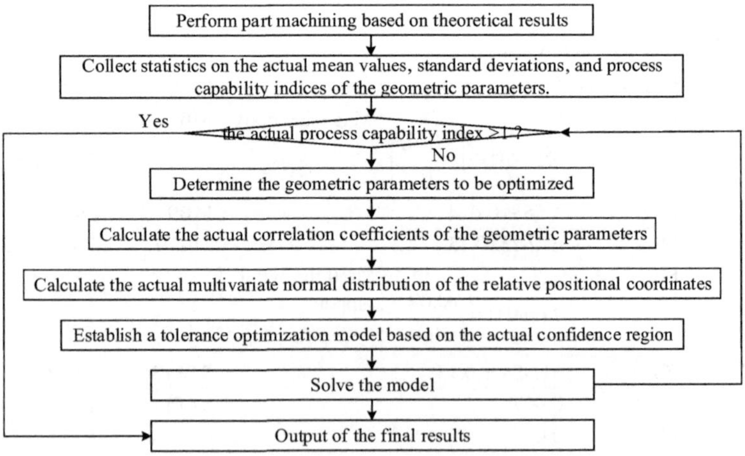

Fig. 6 Tolerance optimization flow considering actual machining errors

machining error is [1.541, 2.247]. The comparison shows that after considering the actual machining error, the lower limit deviation of the stacked quantity increases from 1.516 to 1.541, which reduces the possibility of the lower limit deviation exceeding the tolerance. The tolerance of the overlap amount is increased from 0.688 to 0.706, which makes it easier to guarantee the rigidity of the shell throwing tappet when it collides with the shell and the assembly accuracy of the overlap amount.

5 Conclusion

In this paper, the probabilistic statistical representation method is used to represent the three-dimensional position coordinates of geometric elements, and through the error transfer and accumulation analysis and correlation operation, the design model of the stacking volume of the cartridge case and the throwing tappet of the firearm throwing mechanism is established, and the theoretical tolerance bands of the relevant geometric parameters are obtained. The parts machined according to the theoretical optimization results were sampled, and the two geometric parameters whose machining accuracy was difficult to guarantee and whose actual process capability index $C_{pk} < 1$ were optimized again, and the design results were compared by using the traditional dimensional chain. Compared with the traditional three-dimensional tolerance design method based on theoretical models, the method in this paper adopts the analysis method combining the theoretical model of tolerance and the measured data of geometric parameters of the parts, which can effectively increase the qualification rate of assembly accuracy. In addition, compared with the traditional probabilistic statistical analysis methods such as RSS, the method in this paper considers the correlation between the geometric parameter position coordinates, and therefore is more realistic.

References

1. Zhu W et al (2020) Tolerance analysis method of firearm locking mechanisms based on a multivariate statistical method. J Ordnance Equip Eng 41(10):141–145
2. Ameta G et al (2018) Statistical tolerance analysis with T-maps for assemblies. Proc CIRP 75:220–225
3. Yi Y-L et al (2024) Analysis and control of assembly deviation of large-size thin-walled shell based on improved Jacobian-Torsor model. J South China Univ Technol 25(12):32–42
4. Corrado A (2017) Manufacturing signature in Jacobian and Torsor models for tolerance analysis of rigid parts. Robot Comput Integr Manuf 46:15–24
5. Ziegler P et al (2015) Sensitivity analysis of features in tolerancing based on constraint function level sets. Reliab Eng Syst Safe 134:324–333
6. Wang Q-C et al (2022) Tolerance design of key features of microvalve based on virtual boundary modal. Modul Mach Tool Autom Manuf Tech 4:43–47

7. Wei Y-D, Fang J (2023) Tolerance analysis of the gun shell-throwing mechanism based on SDT model and Monte Carlo method. J Ordnance Equip Eng 44(4):146–154
8. Wang L-W et al (2024) Analysis of anvil centering accuracy of cubic press based on small displacement torsor theory. Diamond Abras Eng 44(6):733–743
9. Wang H et al (2024) Application of Jacobian-Torsor in assembly tolerance analysis of die-cutting machine. Mach Des Manuf 248:8242
10. Xu S, Keyser J (2016) Statistical geometric computation on tolerances for dimensioning. Comput Aided Des 70:193–201
11. Mu X, Sun Q, Sun W et al (2018) 3D tolerance modeling and geometric precision analysis of plane features for flexible parts[J]. Eng Comput 35(7):2557–2576

Open Access This chapter is licensed under the terms of the Creative Commons Attribution-NonCommercial-NoDerivatives 4.0 International License (http://creativecommons.org/licenses/by-nc-nd/4.0/), which permits any noncommercial use, sharing, distribution and reproduction in any medium or format, as long as you give appropriate credit to the original author(s) and the source, provide a link to the Creative Commons license and indicate if you modified the licensed material. You do not have permission under this license to share adapted material derived from this chapter or parts of it.

The images or other third party material in this chapter are included in the chapter's Creative Commons license, unless indicated otherwise in a credit line to the material. If material is not included in the chapter's Creative Commons license and your intended use is not permitted by statutory regulation or exceeds the permitted use, you will need to obtain permission directly from the copyright holder.

Target Detection of Orchard Pear Blossoms Based on YOLOv9

Qingrong Li and Xiaoning Chen

Abstract Pear blossom recognition is an important branch of agricultural intelligence, and models with higher recognition accuracy are more helpful for the automatic pollination of robotic arms in agriculture. In pear blossom recognition, it is important to reduce the leakage rate of flowers to improve pollination accuracy. This paper is based on the improvement of the YOLOv9 target detection algorithm and proposes the idea of organically combining the RCS-OSA attention mechanism with the YOLOv9 model. It also introduces the MPDIoU loss function to further optimize the bounding box regression loss in order to detect more occluded targets, which improves computational efficiency while ensuring that the target detection model maintains a high degree of accuracy. Comprehensive experiments carried out on a self-collected dataset validated the efficacy of the proposed method. The improved YOLOv9_c model attained a mean Average Precision (mAP) of 95.276% and a mean Average Recall (mAR) of 89.973%. These results signify notable enhancements, with mAP increasing by 0.696% and mAR by 0.133%, respectively, when compared to the baseline model. Additionally, the improved YOLOv9_e model attained a mAP of 95.454% and a mAR of 90.512%, with enhancements of 0.315% and 0.177%.

Keywords Object detection · YOLOv9 · RCS-OSA · MPDIoU

1 Introduction

Pear has been known as the "king of flowers". Flowers are rich in color and elegant in form. It is not only an important economic crop but also a good garden ornamental. Accurate identification and detection of certain specific varieties of pear blossoms are of great significance to the selection and breeding of pear tree varieties,

Q. Li · X. Chen (✉)
School of Electrical Engineering and Automation, Anhui University, Hefei, Anhui, China
e-mail: cxn@ahu.edu.cn

© The Author(s) 2026

Z. Zhang, P. Mercorelli (eds.), *Mechanical Design and Simulation: Insights and Innovations*, Lecture Notes in Mechanical Engineering,
https://doi.org/10.1007/978-981-95-2405-1_79

pest control, and other important areas. With the development of the Internet and the rise of artificial intelligence, the use of deep learning technology provides new methods and ideas for the detection and identification of agricultural and forestry products.

Deep learning is a predominant domain within artificial intelligence, and it has attracted much attention in the detection and recognition of agricultural and forestry products because of its capability to handle large-scale datasets and automatically extract features from images. In recent years, more and more research and methods have proved that deep learning models have extremely high accuracy and real-time performance in recognizing different flowers, plants, and trees, and that the ability and generalization of their models can match or even surpass that of experts. Therefore, the detection and recognition of agricultural and forestry products based on deep learning not only have high academic value but also have a very wide market application prospect.

2 Related Work

2.1 Traditional Target Detection Methods Based on Feature Extraction

Traditional methods based on feature extraction typically demand substantial human intervention and follow a three-step process. First, the image undergoes scanning through techniques like sliding windows to detect potential regions containing target objects. Secondly, feature extraction algorithms, including the Histogram of Oriented Gradients (HOG) [1] and the Scale-Invariant Feature Transform (SIFT) [2], are utilized to extract meaningful features from these regions. Subsequently, the obtained features are input into classification models, such as Support Vector Machines (SVM) [3], which analyze the data to produce the final classification outcomes. From the above, it is evident that traditional target detection methods based on feature extraction typically require manual design of the feature extractor, relying on existing a priori knowledge. This process is relatively more complicated to implement and generally less accurate than the deep learning-based target detection algorithms that were subsequently developed [4].

2.2 Deep Learning Based Target Detection Methods

At the ImageNet competition in 2012, the Hinton team proposed a CNN structure based on convolutional neural networks. The AlexNet model [5] has achieved remarkable results. As a result, researchers have increasingly focused on convolutional neural networks. Meanwhile, many scholars have begun exploring the application of this network architecture to target detection tasks.

Contemporary object detection algorithms in deep learning fall into two primary groups. The first group consists of two-stage detectors, such as R-CNN, Fast R-CNN, and Faster R-CNN, among others which need to generate prediction frames for input feature maps and then detect the target objects within the prediction frames when performing target detection [6–9]. Two-stage target detection methods are implemented using two separate network models. While they offer higher accuracy, this comes at the cost of increased computational complexity and reduced real-time detection capability [10]. The other category is one-stage detection algorithms, such as the YOLO series proposed by Redmon et al. as a representative, in addition to Swin Transformer and SSD, etc. [11–13]. One of the key strengths of one-stage detection algorithms compared to two-stage methods is their elimination of the need to produce excessive feature maps. The primary advantage of one-stage detection algorithms over two-stage detection methods is that they eliminate the need to generate redundant candidate detection frames. Instead, they utilize feature representations from the network itself to determine object classifications and spatial coordinates significantly enhancing detection speed and real-time performance [14]. Because YOLO has the characteristics of high efficiency and accuracy and takes into account real-time performance, it is widely used in traffic, agriculture, medicine, and other fields [15]. From 2016, when Redmon first proposed YOLO, it has been updated and iterated to the ninth generation in just 8 years [16]. In 2024, Chien-Yao Wang released the latest YOLOv9 framework, which molds the newest YOLOv9 algorithm to further improve the detection speed and significantly optimize the model structure and training strategy while maintaining high accuracy. From YOLOv1 to YOLOv9, each version has been continuously refined and improved, making great contributions to the development of the computer vision field [17].

2.3 Research Content and Main Contributions of this Paper

This study mainly uses the newly released YOLOv9 model as a benchmark. By adjusting its framework and optimizing the loss function, it aims to improve the accuracy of detecting pear blossoms in farm environments. The key contributions of this research are as follows:

1. The study integrates the RCS-OSA module into the YOLOv9 framework to reduce spatial object attention in the channel. This helps improve processing efficiency and detection accuracy. To show its effectiveness, the study also compares it with the original RepNCSPELAN4 attention mechanism.
2. The MPDIoU loss function is introduced to better capture differences between predicted and actual bounding boxes.This approach narrows the gap between predicted and ground truth key points by evaluating overlap, center distance, and

width-height deviation. Its performance is then compared with other loss functions, such as IoU, GIoU, and DIoU.

3 Introduction to the Network

3.1 Introduction to the YOLOv9 Network

The YOLOv9 algorithm excels in the task of target detection; however, it also faces a common challenge: information loss. Specifically, in the case of scene-specific recognition (here we take the recognition of pear blossoms in a farm environment as an example), the image data that is fed into the model goes through multiple layers of networks to achieve feature extraction and spatial variation, which results in the loss of some of the expressive features in the original image. (This loss of information will lead to a certain degree of difficulty for the model in capturing the subtle features of the pear blossom, thus affecting the accuracy of recognition.) This loss of information makes it more challenging for the model to capture the subtle features of pear blossoms, thereby reducing recognition accuracy.

To solve this problem, this study applies the YOLOv9 algorithm with two new techniques: Programmable Gradient Information (PGI) and the Generalized Efficient Layer Aggregation Network (GELAN). Figures 1 and 2 provide illustrations of these techniques.

PGI introduces an innovative auxiliary supervision framework aimed at resolving information degradation in deep neural networks, ensuring efficient gradient propagation for superior detection and recognition capabilities. In YOLOv9, PGI incorporates three essential elements: a primary branch, an auxiliary reversible branch, and multi-level auxiliary information. The main branch handles standard feature extraction, while the auxiliary reversible branch generates reliable gradient

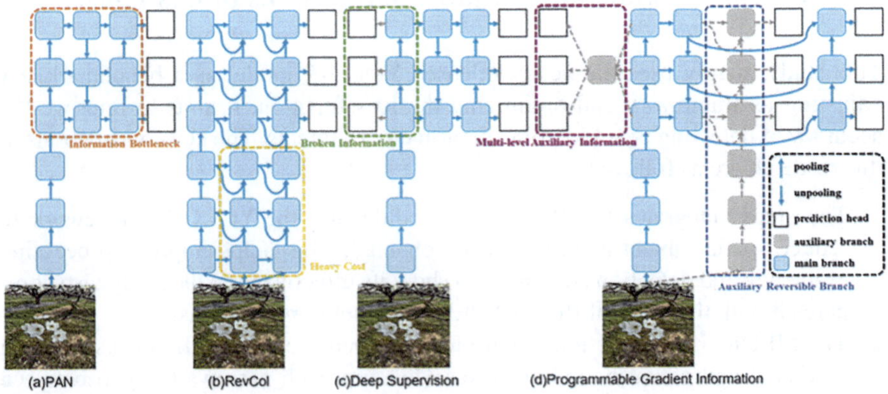

Fig. 1 PGI schematic diagram

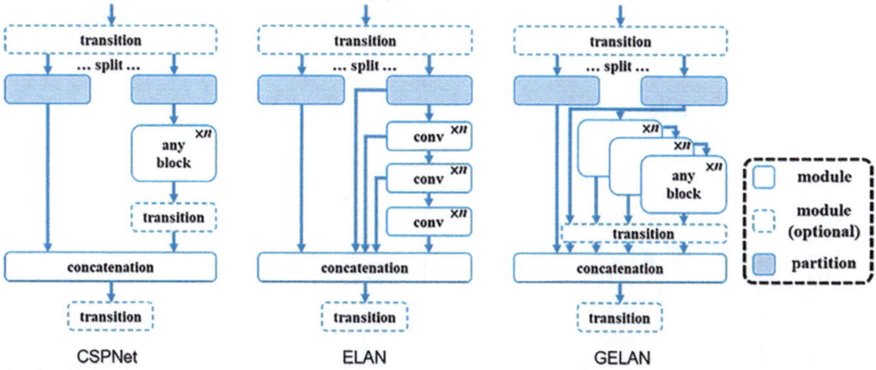

Fig. 2 GELAN schematic diagram

weights to improve the network backbone's capacity to grasp gradient information. This mechanism helps preserve gradient details that might otherwise be lost during backpropagation. However, since the auxiliary reversible branch increases the network complexity and computational overhead, it is only used during the training model phase and discarded during the inference process to minimize time loss. Multilevel auxiliary information is instead used to deal with the problem of error accumulation during computation by aggregating the gradient information to be passed to the main branch for weight updating, ensuring that features from different layers are effectively fused, and enhancing the model's ability to capture subtle details in the input image. This is especially advantageous for pear blossom detection, as small and occluded flowers require precise feature extraction. Compared to traditional methods, PGI enhances model accuracy by preserving more gradient information during training, particularly in scenarios with limited feature data. The study [17] also highlighted the effectiveness of PGI in preserving information during training. This supports the theoretical foundation for the YOLOv9 model's high accuracy in detecting and recognizing pear blossoms.

GELAN is a novel network architecture that effectively combines the strengths of CSPNet [18] and ELAN [19]. Its primary objective is to improve information integration and propagation efficiency within the YOLOv9 model during training. In YOLOv9, GELAN not only enhances the integration of information, but also improves the efficiency of propagation by optimizing the feature flow between different layers. Compared to traditional architectures that are mired in redundant computation, GELAN maximizes the utilization of information while maintaining low computational complexity. Such functionality enables YOLOv9 to excel in resource-constrained scenarios, especially in agriculture, where timely detection is a priority. Specifically, GELAN constitutes the RepNCSPELAN4 module through the introduction of the reparameterizable convolutional RepConv module, as well as batch normalization and careful selection of activation functions. The core purpose of this component is to optimize feature representation and data fusion capabilities without significantly increasing model parameters or computational demands. As a

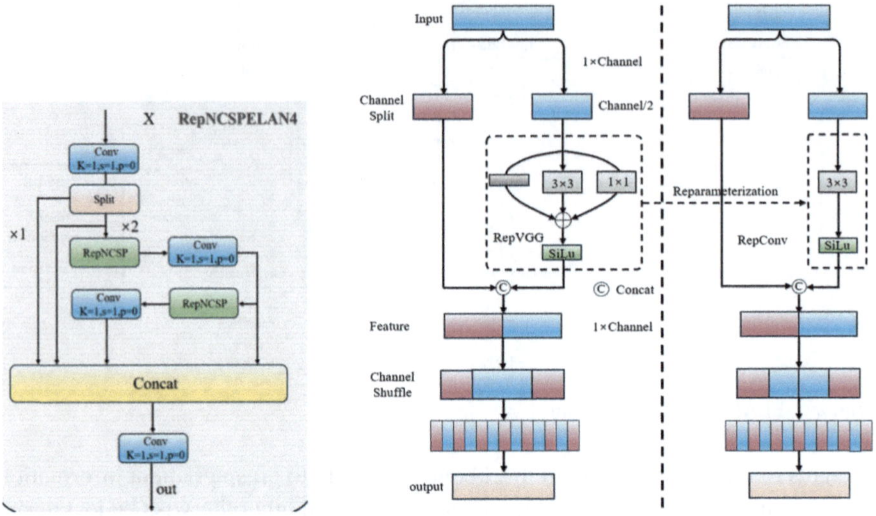

Fig. 3 (1) RepNCSPELAN4 module; (2) RCS module

result, the GELAN architecture helps YOLOv9 balance accuracy and efficiency, making it highly effective for detecting pear blossoms. Compared to other architectures, GELAN demonstrates superior performance in resource-limited environments, providing strong technical support for the pear blossom detection and identification task.

In the YOLOv9 framework, the feature extraction fusion module is RepNCSPELAN4, which mainly consists of the convolutional network Conv and the enhanced convolutional layer ReoNCSP. However, because the feature map has a high number of channels, a lot of redundant information may still exist. Such issues can compromise the model's performance and precision when handling image data. To solve this issue, this study proposes replacing RepNCSPELAN4 with the RCS-OSA (RCS-One-Shot Aggregation) module, introduced in RCS-YOLO [20]. This change aims to reduce spatial object attention across channels. The network structures of RepNCSPELAN4 and RCS modules are shown in Fig. 3.

3.2 *The Main Principle of RCS-OSA is as Follows:*

1. The RCS module is designed to improve feature learning and optimize computational efficiency. During training, it uses a multi-branch structure to capture diverse feature information. However, during inference, it simplifies to a single-branch structure, reducing memory usage and speeding up computation. In addition, the RCS component reduces spatial computational complexity by splitting

and reorganizing the number of channels while preserving information exchange between them. This approach reduces computational complexity by half compared to the conventional 3×3 convolution operation during inference.

2. OSA (One-Shot Aggregation) improves the efficiency of feature representation by replacing the dense connectivity in DenseNet with a structure that captures diverse features using multiple receptive fields, unifying them into a single aggregation in the final feature map. The RCS-OSA module is constructed by integrating RCS and OSA mechanisms. This integration facilitates the efficient extraction of semantic information while ensuring low memory consumption, and it plays a crucial role in developing lightweight, large-scale detectors.

As shown in Fig. 4, the RCS module is repeatedly stacked to promote feature reuse and improve information flow between adjacent layers with different channel configurations. To minimize network fragmentation, the OSA path retains only three feature cascades, effectively reducing computational load and memory consumption. Inspired by PANet, in this design, the RCS module is repeatedly reused and added to the feature fusion between the up- and downsampling layers and the fusion of multi-scale feature layers. This guarantees the accordance of feature maps of different sizes during sampling and promotes efficient information exchange between the two prediction feature layers.

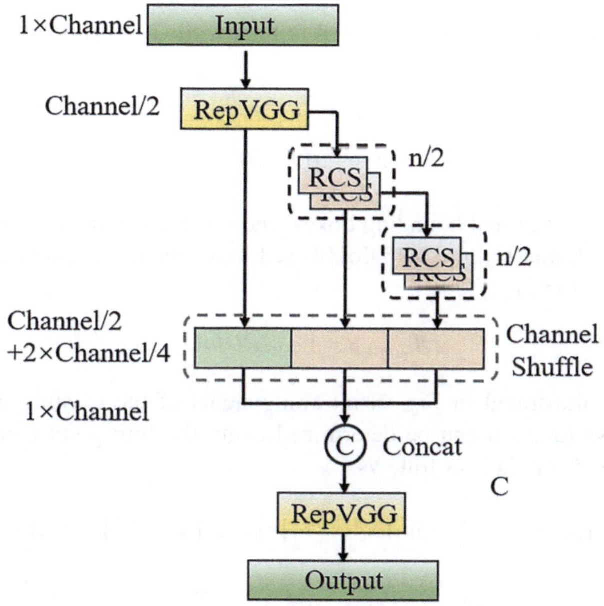

Fig. 4 Module after RCS-OSA cascade

3.3 Introduction to the MPDIoU Loss Function

Bounding box regression (BBR) is widely used in object detection and instance segmentation. It plays a key role in precisely locating target objects. MPDIoU is a comparative metric for bounding box similarity based on the minimum point distance [21]. It is calculated using the minimum point distance of the horizontal rectangle and incorporates multiple factors, including the overlapping region, centroid distance, and width-height deviations, to enhance accuracy in bounding box regression. The experimental results show that in some specific scenarios, the loss function values computed by traditional loss functions (e.g., GIoU, DIoU, CIoU, etc.) are the same [22, 23], but the loss values computed with the MPDIoU method are different. This indicates that MPDIoU provides a more accurate loss metric. Like other loss functions, the MPDIoU-based loss metric's approach is to make the model's predicted bounding box close to its true bounding box by minimizing the loss function during the training phase [24]. Specifically, each predicted bounding box is:

$$B_{prd} = \begin{bmatrix} x_{prd} \\ y_{prd} \\ w_{prd} \\ h_{prd} \end{bmatrix} \tag{1}$$

approximates its true bounding box by minimising the following loss function:

$$B_{gt} = \left[x_{gt}, y_{gt}, w_{gt}, h_{gt} \right]^T \tag{2}$$

$$L = \min_{\theta} L\left(B_{gt}, | B_{prd}, | \theta \right) \tag{3}$$

where B_{gt} is the set of real bounding boxes, and θ is the parameter of the regression depth model. Ultimately, the MPDIoU-based loss function is characterised by the following equation L_{MPDIoU}.

$$L_{MPDIoU} = 1 - MPDIoU \tag{4}$$

Therefore, as illustrated in Fig. 5, all components of the existing bounding box regression loss function can be determined using the four-point coordinates. The transformation formula is as follows:

$$C\# = \left(\max\left(x_2^{gt}, x_2^{prd} \right) - \min\left(x_1^{gt}, x_1^{prd} \right) \right) * \left(\max\left(y_2^{gt}, y_2^{prd} \right) - \min\left(y_1^{gt}, y_1^{prd} \right) \right) \tag{5}$$

$$x_c^{gt} = \frac{x_1^{gt} + x_2^{gt}}{2}, y_c^{gt} = \frac{y_1^{gt} + y_2^{gt}}{2}, y_c^{prd} = \frac{y_1^{prd} + y_2^{prd}}{2}, x_c^{prd} = \frac{x_1^{prd} + x_2^{prd}}{2} \tag{6}$$

$$w_{gt} = x_2^{gt} - x_1^{gt}, h_{gt} = y_2^{gt} - y_1^{gt}, w_{prd} = x_2^{prd} - x_1^{prd}, h_{prd} = y_2^{prd} - y_1^{prd} \tag{7}$$

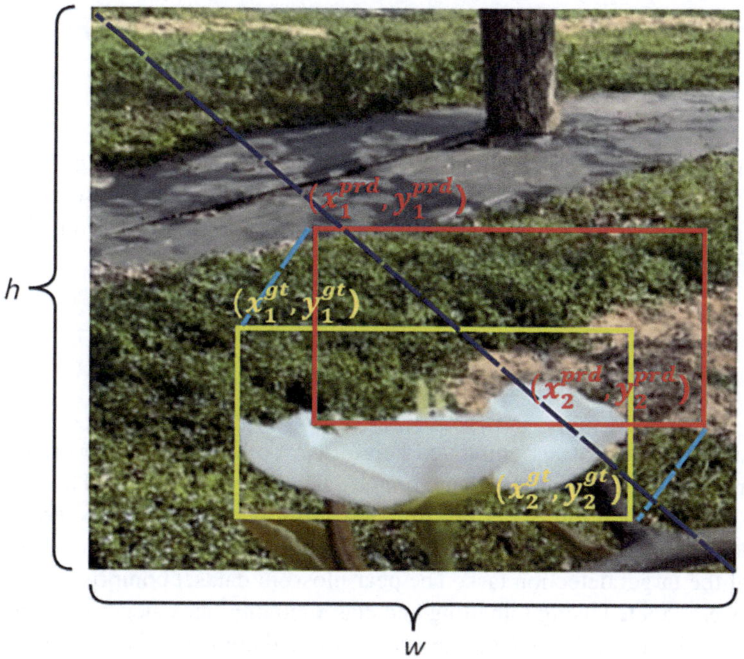

Fig. 5 In this task, the MPDIoU algorithm is used to calculate the graphs of the losses

The symbol $|C|$ represents the smallest rectangle that fully covers both the true bounding box B_{gt} and B_{prd} the predicted bounding box. The coordinates $\left(x_c^{gt}, y_c^{gt}\right)$ and $\left(x_c^{prd}, y_c^{prd}\right)$ refer to the center points of the true and predicted bounding boxes, respectively. The values w_{gt} and h_{gt} indicate the width and height of the true bounding box, while w_{prd} and h_{prd} represent the width and height of the predicted bounding box.

As illustrated in formulas (5), (6), and (7), it is evident that all components of the existing loss function—such as the non-overlapping area, the distance between center points, and deviations in width and height—can be fully derived from the coordinates of the upper-left and lower-right points of the bounding boxes. This observation underscores the comprehensiveness and computational efficiency of the proposed method.

According to its algorithm, if the aspect ratio of the predicted bounding box is the same as that of the true bounding box, then the value of the predicted bounding box within the true bounding box is lower than that outside the true bounding box. This characteristic ensures the accuracy of the bounding box regression, resulting in less redundancy of the predicted bounding box. By considering factors such as overlap, center point proximity, and dimensional discrepancies, the algorithm improves bounding box regression precision, especially in scenarios involving occlusions. This implies that even when the target is partially occluded, the algorithm can better predict the bounding box of the target, thereby enhancing the detection accuracy.

4 Experimentation

4.1 Experimental Environment

The improved YOLOv9 algorithm in this study is implemented using PyTorch 1.13 and CUDA 11.5 for training and validation. The experimental platform configuration is shown in Table 1, and the model training parameters are listed in Table 2.

Bulleted lists look like this:

4.2 Introduction to the Experimental Dataset

The Pear Blossom dataset employed herein was collected in a pear orchard under uniform illumination, encompassing image data of diverse poses and blooming conditions. Images that did not conform to the requirements of this experiment were excluded. The remaining images were applicable for the model training and evaluation of the target detection task. The pear blossom dataset comprises 970 images, which were labeled using LabelImg and classified into one category, namely pear 1 Due to the short pear blossom cycle and the limited number of on-site pear blossom images, this paper employs data augmentation techniques such as illumination transformation and Gaussian noise to modify input data and generate additional training samples. This approach enhances the model's generalization ability and overall performance. Through data augmentation, 3880 pear blossom image features were obtained. The pear blossom images were split into three sets: 70% for training, 10% for validation, and 20% for testing. The annotation format was

Table 1 Configuration of experimental platform

Name	Parameter
System	*Ubuntu 20.22*
CPU	13th Gen Intel(R) Core (TM) i5-13500
GPUs	NVIDIA GeForce RTX 4080
RAM	16376MiB

Table 2 Model training parameters

Parameter	Value
Learning rate	*0.01*
Epochs	100
batch_size	6
number_workers	6
optimizer_type	16376MiB
momentum	0.937
weightg_decay	0.0005

changed from XML to TXT, and the image resolution was set to 640 × 640. The IoU threshold was set at 0.45, and all models were trained for 100 epochs using these settings.

4.3 Experimental Results and Analyses

Comparison with Baseline Version To evaluate the model's performance, we compute the mean Average Precision (mAP) and mean Average Recall (mAR) for diverse target categories, using an IoU threshold of 0.5, following the methodology detailed in this work. To better assess the model's detection ability, we follow the evaluation metrics used in the YOLOv9 model. We analyze precision, recall, and mAP_0.5 to evaluate the YOLOv9_c model, with the results presented in Tables 3 and 4. The experimental results show that:

1. In the Attention Mechanism section, it can be seen from Table 3 that the modified RCS-OSA attention mechanism makes the model mAP and mAR values improve from 95.139% and 90.335% to 95.358% and 90.575%, respectively. The TripleAttention mechanism significantly contributes to enhancing the model's mean Average Precision (mAP) value. However, it struggles to effectively differentiate between positive and negative samples in the pear dataset. As a result, the modified model is adversely affected by issues such as noise and redundancy. Consequently, this modification does not lead to an increase in the mean Average Recall (mAR) value; rather, it causes a decrease. On the contrary,

Table 3 Performance comparison of YOLOv9_c model with different attention mechanisms

Backbone	mAP (%)	mAR (%)	Parameters (M)
	95.139	90.335	25.3
+TripleAttention	95.273	89.937	26.7
+RCS-OSA	95.358	90.575	24.9
Improv %	0.219	0.240	0.4

Table 4 Boundary box regression loss function performance comparison

Loss	Precision	Recall	mAP_50 (%)
IoU	0.92833	0.89840	94.580
CIoU improv	0.93683 0.00850	0.90008 0.00168	95.040 0.46
DIoU improv	0.92639 −0.00194	0.89696 −0.00144	94.579 −0.001
GIoU improv	0.93190 0.00357	0.88885 −0.00955	94.543 −0.037
MPDIoU improv	0.93413 0.00580	0.90015 0.00175	95.213 0.633

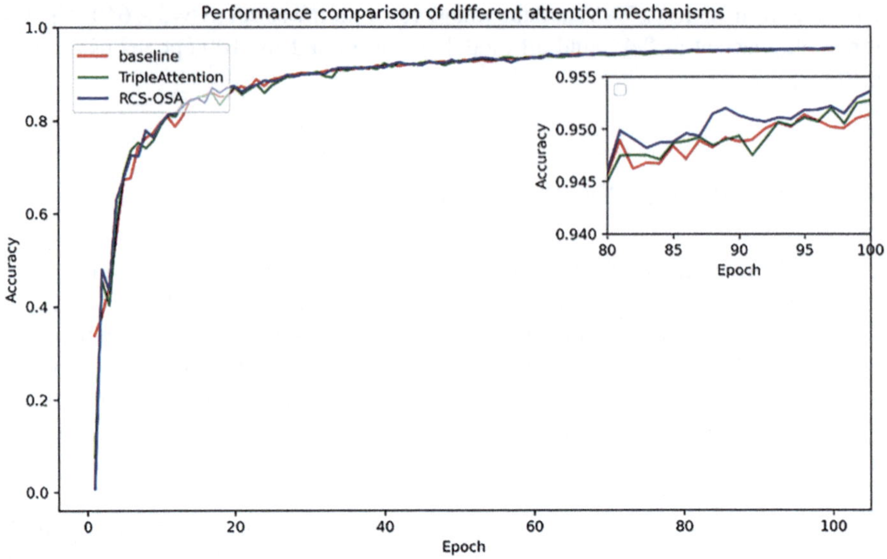

Fig. 6 Training curves for different attention mechanisms

the RCS-OSA module can better learn the intrinsic laws and features of the data, which makes the model improve in both mAP and mAR. In terms of convergence effect, it can be seen from Fig. 6 that the model using RCS-OSA starts to converge at 80 epochs, and all the indicators at convergence are also higher than the baseline results. These results indicate that the RCS-OSA module not only enhances the network's feature extraction capability after reparameterized convolution but also reduces computational complexity by aggregating multiple cascaded features. This improves computational efficiency and accelerates the model's convergence. Additionally, the table shows that the baseline model has a parameter count of 25.3M. After integrating the TripleAttention mechanism, the parameter count increases to 26.7M, reflecting an increase of 1.4M. In contrast, incorporating the RCS-OSA attention mechanism reduces the parameter count to 24.9M, resulting in a decrease of 0.4M. This shows that RCS-OSA not only improves the performance of the model, but also reduces the computational burden of the model by optimizing the network structure, making the model more lightweight.

2. In the loss function section, we compare the baseline version with versions that incorporate CIoU, DIoU, GIoU, and MPDIoU. As shown in Table 4, adding CIoU slightly improves YOLOv9's performance in terms of Precision, Recall, and mAP_50; the improvement strategy of DIoU affects the performance of these three values; and the GIoU improves the Precision value of the model, but has a certain degree of Recall and mAP_50 of influence. The MPDIoU selected in this paper shows better performance, achieving a Precision of 0.93413, a Recall of 0.90015, and a mAP_50 of 95.213%, which is an improvement of

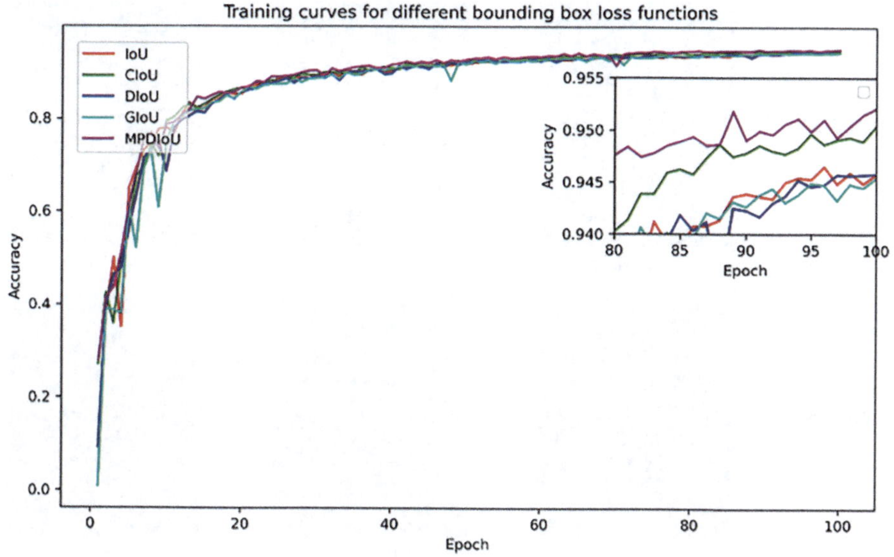

Fig. 7 Training curves for different bounding box loss functions

0.00580, 0.00175, and 0.663%, respectively, compared to the baseline version. This demonstrates that MPDIoU improves bounding box prediction accuracy and enhances localization performance by incorporating the minimum point distance into the bounding box similarity comparison metric. Figure 7 illustrates the performance comparison of various bounding box regression losses throughout the training process. During the pre-training phase, the Mean Pairwise Distance Intersection over Union (MPDIoU) decreases rapidly in the direction of the gradient, leading to a more rapid improvement in model accuracy. After 40 epochs of training, the increase in MPDIoU becomes more gradual, and ultimately, the model converges after 80 epochs. The results indicate that MPDIoU outperforms the other four bounding box regression losses, yielding the best overall results. The MPDIoU loss function outperforms traditional loss functions (e.g., IoU, GIoU, DIoU) because it directly minimizes the distances between key points of the predicted and ground truth bounding boxes. Unlike other loss functions that may not be able to distinguish between different bounding box configurations, MPDIoU effectively integrates multiple geometric factors. Specifically, it concurrently considers the overlap area, the distance between the centers of mass, and deviations in width and height. This makes it particularly effective in dealing with occlusion problems, where conventional loss functions may produce the same value for different bounding box configurations. For example, in pear flower detection, MPDIoU ensures that partially occluded flowers are accurately localized, thereby reducing false positives and false negatives.

3. Further, we verified the detection performance of baseline+RCS-OSA+MPDIoU combined with YOLOv9. Under the condition of setting a certain number of

Fig. 8 Improved visualisation of YOLOv9_c and YOLOv9_e in different ways

thresholds, the visualisation is shown in Fig. 8: for YOLOv9_c, baseline can detect 17 targets, baseline+RCS-OSA can detect 17 targets, and baseline+RCS-OSA+MPDIoU can detect 18 targets; similarly, for YOLOV9_e, these three cases can detect 17, 18 and 19 targets, respectively. For small or occluded pears, the proposed approach effectively improves the model's detection accuracy while reducing the missed detection rate.

Ablation Experiments In the final phase, we executed multiple ablation experiments to validate the contributions of the proposed improvements. The results are summarized in Table 5. To enhance the credibility of the data, we performed experiments using two versions of YOLOv9-c and YOLOv9-e, with all experimental data representing the average of three trials.

Meanwhile, to demonstrate that these added modules have a certain degree of improvement in terms of false negative detections and false positive detections, in this experiment, we have additionally incorporated two indicators: FNR (False Negative Rate) and FPR (False Positive Rate). Their definitions are as follows:

$$FPR = \frac{FalsePositive}{FalsePositive + TrueNegative} \tag{8}$$

$$FNR = \frac{FalseNegative}{FalseNegative + TruePositive} \tag{9}$$

Table 5 Ablation experiment results

Backbone	YOLOv9-c mAP (%)	mAR (%)	Parameters/M	YOLOv9-e mAP (%)	mAR (%)	Parameters (M)
	94.580	89.840	*25.3*	95.139	90.335	*57.3*
+RCS-OSA	95.018	89.433	24.9	95.358	90.575	55.2
+RCS-OSA +MPDIoU	95.276	89.973	24.9	95.454	90.512	55.2
Improv %	0.696	0.133	0.4	0.315	0.177	2.1

Table 6 Horizontal comparison of experimental results

	mAP (%)	mAR (%)
Our integrated model (based on YOLOv9_c)	95.276	89.973
Our integrated model (based on YOLOv9_e)	95.454	90.512
YOLOv8	94.371	89.468
YOLOX	90.034	87.317
Faster-RCNN(ResNet50)	91.167	86.377
EfficientDet	92.694	87.100

Based on the data presented in Table 3, it can be observed that the incorporation of the RCS-OSA module alone leads to a marginal improvement in the mean average precision (mAP) of model inference by 0.219%. Additionally, this modification results in a reduction of the model's parameter count by 0.4 million. Furthermore, as evidenced by the findings in Table 6, the exclusive integration of the MPDIoU loss function contributes to a more substantial enhancement in the mAP of model inference, with an increase of 0.633%. And the two approaches are combined and generalized to the YOLOv9_e model in Table 5. On YOLOv9-c, combining the RCS-OSA attention mechanism and the MPDIoU loss function improves the mAP and mAR of model inference by 0.696% and 0.133%, respectively, and reduces the model parameter size by 0.4 M. On YOLOv9-e, the mAP and mAR improve by 0.315% and 0.177%, respectively, and the model parameter size reduces by 2.1 M. The ablation experiments demonstrate that this paper's approach of combining the RCS-OSA attention mechanism with the MPDIoU loss function improves both YOLOv9-c and YOLOv9-e versions. RCS-OSA strengthens the model's capability to capture fine-grained features, such as the shape and texture of pear blossoms, while MPDIoU improves the precision of their localization. When used in combination, these methods achieve a synergistic effect, resulting in higher mAP and mAR values compared to either method alone, and these improvements are a steady enhancement to YOLOv9.

Lateral Comparative Experiment Meanwhile, as shown in Table 6, we conducted a lateral comparison with other object detection algorithms (YOLOv8,

YOLOX, Faster-R-CNN, EfficientDet). The results indicate that the improved YOLOv9 algorithm achieves higher accuracy and demonstrates superior detection performance for pear blossoms.

5 Conclusions

This work advances the YOLOv9 object detection model by introducing the RCS-OSA attention mechanism and the MPDIoU loss function. These upgrades are tested on two iterations of YOLOv9 (YOLOv9_c and YOLOv9_e) and tested in real experimental settings. The results show that the improved model achieves better performance, with mAP and mAR increasing by 0.696%, 0.133%, and 0.177%, respectively. This suggests that the RCS-OSA attention mechanism helps the model capture features more effectively, while the MPDIoU loss function improves accuracy and reduces detection errors. The improved approach has high accuracy for pear blossom recognition detection in a farm environment, which helps to improve the efficiency of agricultural production life and has certain practical value.

The dataset used in this study was collected from a single farm under consistent lighting conditions. This may limit the model's ability to work well in other environments. For example, the model might have difficulty detecting pear blossoms in different lighting conditions, such as bright sunlight or shaded areas, or in different geographic locations. This may lead to reduced accuracy and robustness in real-world applications. To improve generalization capabilities, future work should focus on collecting more diverse datasets, including images from different farms, different lighting conditions, and different pear blossom varieties. In addition, data enhancement techniques (e.g., simulating bright light, low light, or adverse weather conditions) could be used to increase the robustness of the model. Migration learning could also be explored to adapt the model to new environments with limited labeling data.

The current proposed YOLOv9 improvement scheme also has limitations, and in the future, further research can be carried out in the following points:

1. Further optimization of the RCS-OSA attention mechanism and MPDIoU loss function: the hyperparameters of the RCS-OSA attention mechanism and MPDIoU loss function can be fine-tuned to achieve better performance. For example, the number of channels, branches and cascade layers in RCS-OSA can be optimized through extensive experiments. Similarly, the weights of the distance terms in MPDIoU can be adjusted to improve the bounding box regression accuracy.
2. Addressing dataset limitations: as mentioned earlier, future research should focus on collecting more diverse datasets and exploring data enhancement techniques to improve model robustness and generalization.
3. Model pruning: The current algorithm still has potential for improvement in terms of model size, which makes it less suitable for real-time deployment on

edge devices. In future work, model pruning can be applied to reduce the model's size, thereby enhancing its overall performance.

4. Enhance YOLOv9's ability to handle dense targets: future work can explore modifications to the YOLOv9 network structure to improve its ability to handle dense targets. For example, adaptive grid sizes could be introduced to better capture small or overlapping pears. Multi-scale feature fusion techniques could also be investigated to enhance the model's ability to detect targets at different scales.
5. Explore alternative object detection algorithms: other object detection algorithms such as Swin Transformer or DETR (Detection Transformer) can be evaluated to understand their performance in pear blossom detection. These algorithms may have advantages in dealing with dense targets or complex scenes, providing a benchmark for further improvements to YOLOv9.

References

1. Dalal N, Triggs B (2005) Histograms of oriented gradients for human detection. In: 2005 IEEE computer society conference on computer vision and pattern recognition (CVPR'05). IEEE, New York, pp 886–893
2. Lowe DG (1999) Object recognition from local scale-invariant features. In: Proceedings of the seventh IEEE international conference on computer vision. IEEE, New York, pp 1150–1157
3. Cortes C, Vapnik V (1995) Support-vector networks. Mach Learn 20(3):273–297
4. Zou Z, Shi Z, Guo Y et al (2019) Object detection in 20 years: a survey. arXiv preprint arXiv:1905.05055
5. Krizhevsky A, Sutskever I, Hinton GE (2012) Imagenet classification with deep convolutional neural networks. Adv Neural Inf Process Syst 25:1097–1105
6. Girshick R, Donahue J, Darrell T et al (2014) Rich feature hierarchies for accurate object detection and semantic segmentation. In: Proceedings of the IEEE conference on computer vision and pattern recognition. IEEE, New York, pp 580–587
7. Girshick R (2015) Fast r-cnn. In: Proceedings of the IEEE international conference on computer vision. IEEE, New York, pp 1440–1448
8. Ren S, He K, Girshick R et al (2015) Faster r-cnn: towards real-time object detection with region proposal networks arXiv preprint arXiv pp 1506.01497
9. Redmon J, Divvala S, Girshick R et al (2016) You only look once: unified, real-time object detection. In: Proceedings of the IEEE conference on computer vision and pattern recognition. IEEE, New York, pp 779–788
10. Redmon J, Farhadi A (2017) YOLO9000: better, faster, stronger. In: Proceedings of the IEEE conference on computer vision and pattern recognition. IEEE, New York, pp 7263–7271
11. Lin M, Chen Q, Yan S (2013) Network in network. arXiv preprint arXiv:1312.4400
12. Redmon J, Farhadi A (2018) Yolov3: an incremental improvement. arXiv preprint arXiv 1804.02767
13. Bochkovskiy A, Wang C Y, Liao H Y M (2020) Yolov4: optimal speed and accuracy of object detection. arXiv preprint arXiv 2004.10934
14. Li C, Li L, Jiang H, Weng K, Geng Y, Li L, Wei X (2022) YOLOv6: a single-stage object detection framework for industrial applications. arXiv preprint arXiv 2209.02976
15. Wang CY, Bochkovskiy A, Liao HYM (2023) YOLOv7: trainable bag-of-freebies sets new state-of-the-art for real-time object detectors. In: Proceedings of the IEEE/CVF conference on computer vision and pattern recognition. IEEE, New York, pp 7464–7475

16. Reis D, Kupec J, Hong J, Daoudi A (2023) Real-time flying object detection with YOLOv8. arXiv preprint arXiv:2305.09972
17. Wang CY, Yeh IH, Liao HYM (2024). YOLOv9: learning what you want to learn using programmable gradient information. arXiv preprint arXiv:2402.13616
18. Wang CY, et al (2020) Cspnet: a new backbone that can enhance learning capability of CNN. In Proceedings of the IEEE/CVF conference on computer vision and pattern recognition workshops, pp 390–391
19. Wang CY, Liao HYM, Yeh IH (2022) Designing network design strategies through gradient path analysis. arXiv preprint arXiv:2211.04800
20. Kang M, Ting CM, Ting FF, Phan RCW (2023) RCS-YOLO: a fast and high-accuracy object detector for brain tumour detection. In: International conference on medical image computing and computer-assisted intervention. Springer, Cham, pp 600–610
21. Siliang M, Yong X (2023) Mpdiou: a loss for efficient and accurate bounding box regression. arXiv preprint arXiv:2307.07662
22. Rezatofighi H, Tsoi N, Gwak J, Sadeghian A, Reid I, Savarese S (2019) Generalised intersection over union: a metric and a loss for bounding box regression. In Proceedings of the IEEE/CVF conference on computer vision and pattern recognition, pp 658–666
23. Zheng Z, Wang P, Liu W, Li J., Ye R, Ren D (2020). Distance-IoU loss: faster and better learning for bounding box regression. In Proceedings of the AAAI conference on artificial intelligence, pp 12993–13000
24. Misra D, Nalamada T, Arasanipalai AU, Hou Q (2021) Rotate to attend: convolutional triplet attention module. In Proceedings of the IEEE/CVF winter conference on applications of computer vision, pp 3139–3148

Open Access This chapter is licensed under the terms of the Creative Commons Attribution-NonCommercial-NoDerivatives 4.0 International License (http://creativecommons.org/licenses/by-nc-nd/4.0/), which permits any noncommercial use, sharing, distribution and reproduction in any medium or format, as long as you give appropriate credit to the original author(s) and the source, provide a link to the Creative Commons license and indicate if you modified the licensed material. You do not have permission under this license to share adapted material derived from this chapter or parts of it.

The images or other third party material in this chapter are included in the chapter's Creative Commons license, unless indicated otherwise in a credit line to the material. If material is not included in the chapter's Creative Commons license and your intended use is not permitted by statutory regulation or exceeds the permitted use, you will need to obtain permission directly from the copyright holder.

Mechanism of Accuracy Degradation and Optimization of Plane Linear Guideways under Residual Stress and Creep

Xiaozhi Yang, Xiao Cao, Huizhen Wang, Xifeng Wang, Yuewen Zhai, Weidong Zhu, Zibo Zhang, and Yu Yang

Abstract This study addresses the issue of accuracy degradation in high-precision plane linear guideways during long-term service. A multi-scale experimental-simulation coupling method is employed to reveal the synergistic effect of residual stress gradient and material creep behavior. Based on the modeling of layered residual stress in 38CrMoAl guideways and uniaxial creep experiments, a simplified constitutive equation considering the stress exponent effect is developed. It is found that the displacement of the 700 MPa high-stress model after 3 months reaches 0.4337 mm, which is 15.4 times higher than that of the 300 MPa baseline group. Cross-scale microstructure characterization indicates that the hardness gradient and banded structure of the rolled material promote uneven stress release, causing preferential directional creep in the coarse-grain region. Based on these findings, two optimization criteria are proposed: residual stress threshold control (≤500 MPa) and operational condition matching for φ100/φ160 rolled materials. The results provide theoretical guidance and engineering solutions for the lifespan prediction and manufacturing process improvement of high-precision guideways.

Keywords Plane linear guideways · Residual stress gradient · Creep deformation · 38CrMoAl alloy · High-precision manufacturing

1 Introduction

As a core motion component of precision equipment such as CNC machine tools, the geometric accuracy and dynamic stability of plane linear guideways directly affect the dimensional tolerance and surface quality of processed workpieces. Compared to rolling guideways, sliding guideways exhibit significant advantages

X. Yang · X. Cao · H. Wang · X. Wang (✉) · Y. Zhai · W. Zhu · Z. Zhang · Y. Yang
China Academy of Machinery Beijing Research Institute of Mechanical and Electrical Technology Co., Ltd, Beijing, China

© The Author(s) 2026

Z. Zhang, P. Mercorelli (eds.), *Mechanical Design and Simulation: Insights and Innovations*, Lecture Notes in Mechanical Engineering,
https://doi.org/10.1007/978-981-95-2405-1_80

under heavy or impact loading conditions due to their larger contact area, absence of rolling element gaps, and superior structural rigidity. However, with the advancement of manufacturing towards ultra-precision and high reliability, guideways must maintain micron-level or even sub-micron motion accuracy under complex service conditions over extended periods. Traditional measurement methods, such as contact measurements with micrometers, are limited by low efficiency and the inability to capture dynamic accuracy evolution. Although high-precision non-contact sensors can improve measurement resolution, their complex installation and calibration processes are not well-suited for fast on-site production testing. Therefore, understanding the intrinsic mechanisms of guideway accuracy degradation and proposing optimization strategies has become a key scientific issue for promoting the application of sliding guideways in high-end manufacturing.

In recent years, researchers have conducted in-depth studies on guideway performance optimization using interdisciplinary approaches. Xu et al. [1], based on flexible multibody dynamics theory, revealed the nonlinear vibration characteristics of linear sliding guideways under multi-directional harmonic excitation. Lyu et al. [2] quantified the stress distribution characteristics of motion actuators under dynamic loading using strain gauges and finite element simulations. Rahmani et al. [3] clarified the nonlinear effect of slider motion speed on guideway stiffness through variable-speed driving experiments. In addition, many scholars have analyzed the guide rails by introducing factors such as dynamic loads and temperature [4–6]. These studies provide important theoretical support for guideway accuracy control. However, they have not systematically considered the coupled effects of material creep behavior and residual stress release on accuracy stability during long-term service. Engineering practices show that even with strict control of working temperature and loading conditions, the accuracy of guideways still gradually deteriorates with extended service time. This phenomenon suggests that traditional static mechanical models cannot accurately represent the time-varying deformation behavior of guideways.

To address these issues, this paper proposes an analytical framework based on the coupling of residual stress gradient regulation and creep constitutive modeling. First, a high-fidelity finite element model of the guideway is constructed through layered residual stress modeling and verification with hole-drilling measurement data. Next, combining uniaxial creep experiments and microstructural characterization techniques, a strain-hardening constitutive equation for 38CrMoAl material is developed, revealing the driving mechanism of residual stress gradient on creep deformation. Finally, through rolling process optimization and thermal treatment parameter matching, an engineering solution to enhance the long-term dimensional stability of guideways is proposed. This study provides new theoretical insights for the design, manufacturing, and accuracy life prediction of high-precision guideways.

2 Description of the Physical Models

This study focuses on plane linear guideways made of 38CrMoAl alloy, with the geometric parameters shown in Fig. 1. The total length of the guideway is 790 mm, and the cross-section design includes multiple functional holes (for positioning and lubrication medium transport). These structural features are retained in the finite element modeling to replicate the actual service conditions.

During the casting process of large machine tool base components, uneven cooling shrinkage occurs due to the complex structure, resulting in casting residual stresses [7–9]. In addition, during the precision machining process, the cutting action of the tool disrupts the original equilibrium state of the workpiece residual stresses, leading to the generation of a gradually decreasing residual stress distribution on the surface layers of the workpiece (Fig. 2, left). Therefore, to accurately simulate the residual stress distribution generated during the manufacturing process, five sub-layers along the thickness direction (1.0 mm depth) are defined for the five outer surfaces of the guideway (excluding the bottom fixed surface). The residual stress values for each layer decrease linearly from 300 MPa at the surface to 60 MPa at the innermost layer (gradient of 60 MPa per layer, as shown in Fig. 2, right).

Considering the long length of the guideway, uneven residual stress distribution may occur during production. Therefore, this study employs the hole-drilling method (strain release method) to measure the residual stress on multiple groups of guideway surfaces (Fig. 3). The measurement results show a distinct "larger at both ends, smaller in the middle" residual stress distribution pattern. To investigate the impact of different machining processes on the deformation behavior of the guideway, the upper surface is divided into five distinct regions (Fig. 1), each subjected to different residual stress values. Four different residual stress distribution models were designed as follows:

- Model 1 (300/300/300/300/300): Uniform residual stress distribution, used to isolate geometric effects and clarify the independent impact of the stress gradient.
- Model 2 (500/400/300/400/500): Low gradient distribution, simulating slight non-uniform machining (e.g., early tool wear).
- Model 3 (600/450/300/450/600): Medium gradient distribution, reflecting stress accumulation after typical multi-pass machining.

Fig. 1 Geometric parameters of the guideway

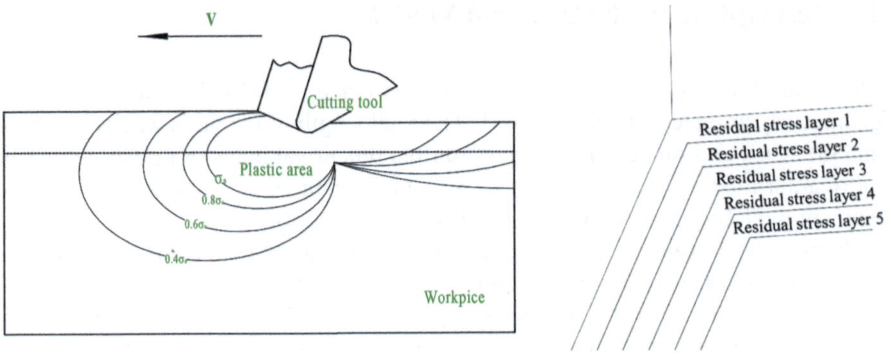

Fig. 2 Application of residual stress layers and residual stress distribution on the machined surface

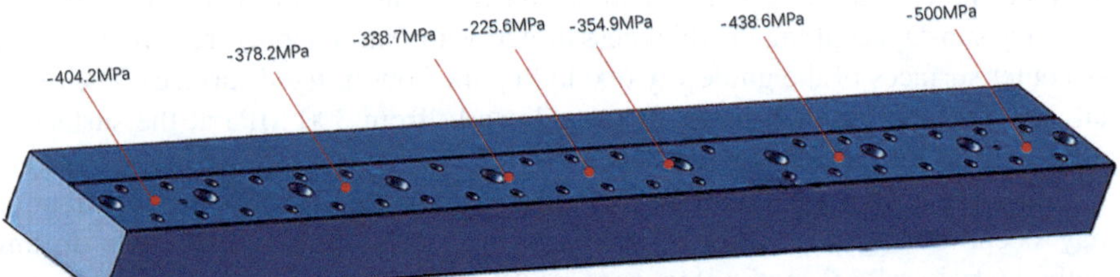

Fig. 3 Residual stress measurement on the guideway

- Model 4 (700/500/300/500/700): High gradient distribution, corresponding to extreme stress states caused by high-speed cutting or insufficient cooling.

3 Computational Method and Procedure

This study systematically analyzes the time-varying deformation mechanism of guideways under the coupling effect of residual stress and creep through a combination of experimental characterization, creep model development, and multi-scale finite element simulations.

3.1 Creep Model Establishment

Creep refers to the continuous deformation of materials under constant stress over time. In this study, uniaxial tensile creep experiments were conducted on standard 38CrMoAl samples to obtain the strain evolution under 100 MPa and 400 MPa loading conditions at room temperature over a period of 0–50 h. After excluding

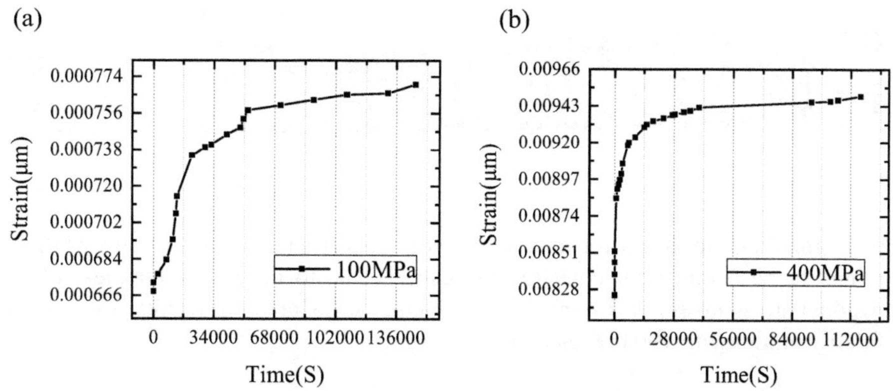

Fig. 4 (**a**) Strain evolution with time under 100 MPa; (**b**) Strain evolution with time under 400 MPa

abnormal fluctuations and averaging the deformation data, the strain-time curves are shown in Fig. 4.

To derive the creep strain equation for 38CrMoAl, the classical three-stage creep theory [10–12] was applied. Data from the second stage of stable creep were extracted to calculate the average strain rate. Linear regression was performed to determine the steady-state strain rates at 100 MPa and 400 MPa, which were found to be 1.09×10^{-10}/s and 2.17×10^{-8}/s, respectively. Since both the guideway and slider operate at room temperature, temperature effects on creep were not considered. In the second stage, assuming the strain rate is only related to stress, the strain-hardening formula was simplified to $\dot{\varepsilon} = A\sigma^n \varepsilon^m e^{-C/T}$ (ignoring strain hardening term and temperature term). Taking the logarithm of both sides for the two sets of experimental data, a linear equation was established. Using the least squares method for fitting, and were obtained. Thus, the simplified creep constitutive model was established as follows:

$$\dot{\varepsilon} = 2.51 * 10^{-18} \sigma^{3.819} \tag{1}$$

3.2 *Mesh Size and Boundary Condition Setup*

In this study, it is assumed that the bottom surface of the guideway is fully constrained to simulate bolted connections. The residual stress layers are applied layer by layer using a predefined field and are combined with gravitational loads. The environmental temperature is assumed to be 11 °C (the constant temperature within the factory). The model uses an implicit time-stepping algorithm, with the calculation period covering 0–3 months.

The finite element model of the guideway is divided using a hybrid element type strategy as follows (Fig. 5): regular regions use hexahedral elements to ensure computational accuracy, while complex geometries such as holes are meshed with

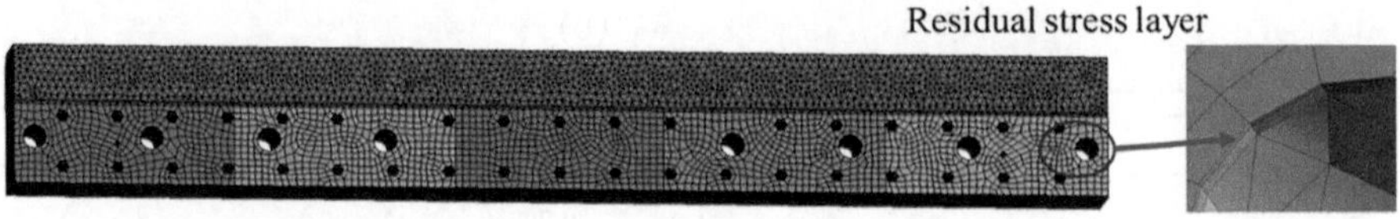

Fig. 5 Mesh generation

pentagonal and tetrahedral elements to avoid distortion. The mesh size was determined to be 3 mm after convergence testing. Due to the small thickness of the residual stress layers (1.0 mm), local refinement was applied, resulting in a high-fidelity model with 490,000 elements.

4 Computational Results and Discussion

4.1 Parameter Definition

In the computation process, four analysis steps were set up to observe the deformation of the guideway. Step one lasts for 1 s and is used for displacement constraint loading, residual stress loading, and gravitational load application, thus serving as the initial state for calculating the real deformation data after creep. Analysis steps two through four calculate the displacement, stress, and strain values on the guideway surface after 1, 2, and 3 months of creep. Besides the maximum displacement values, fluctuation values (the difference between the maximum and minimum values) are also provided, which partially represent the variation in the surface accuracy of the guideway. To validate the experimental conclusions and optimization direction, the original material of the guideway was sampled to measure material hardness, microstructure, and grain size, to determine the optimization direction for guideway accuracy.

4.2 Parameter Definition

Table 1 shows the time-dependent variation of the maximum displacement on the guideway surface under different residual stress gradient models. Experimental data indicate a significant nonlinear positive correlation between residual stress levels and creep deformation. When the initial residual stress increases from 300 MPa to 700 MPa, the initial displacement increases from 0.016 mm to 0.0203 mm, an increase of 27%. After 3 months, the total displacement jumps from 0.0281 mm to 0.4337 mm, an increase of 15.4 times, indicating that for every 100 MPa increase in residual stress, the total displacement increases by approximately 200%–250%.

Table 1 Time-dependent displacement components

Time	Example	300 MPa	500 MPa	600 MPa	700 MPa
1 s	Total displacement	0.16	0.0177	0.0189	0.0203
1 month	Total displacement	0.0127	0.0606	0.1079	0.1776
	Y component	0.0031	0.0161	0.03	0.0513
2 month	Total displacement	0.0208	0.1031	0.1872	0.3103
	Y component	0.0053	0.0292	0.0553	0.095
3 month	Total displacement	0.0281	0.1427	0.2611	0.4337
	Y component	0.0074	0.0418	0.0796	0.1366

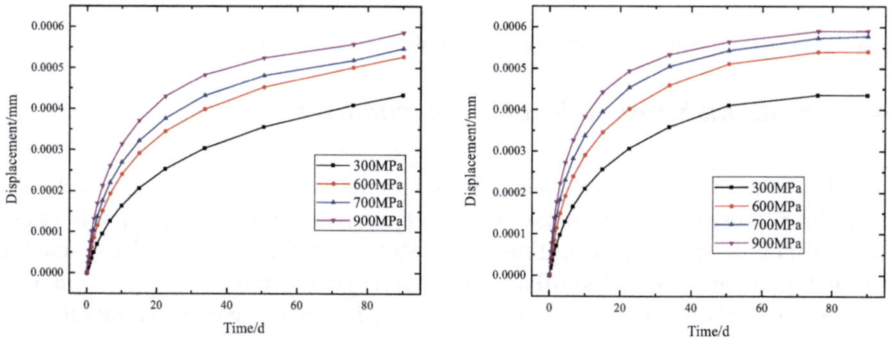

Fig. 6 Node displacement over time

Figure 6 shows the displacement of two non-adjacent nodes on the guideway surface over time for 3 months. More than 70% of the deformation occurs within the first 30 days of service, indicating that the guideway enters the stable creep stage. This phenomenon is highly consistent with the classic three-stage creep theory, confirming that a high residual stress gradient significantly accelerates the initial unstable creep process.

Further analysis of the displacement fluctuation values (the difference between the maximum and minimum displacement) reveals that the fluctuation value for the 300 MPa model is always below 0.013 mm, meeting the IT5 accuracy tolerance requirement for high-precision guideways ($\leq$0.02 mm). In contrast, the fluctuation value for the 700 MPa model reaches 0.1524 mm (IT8 level), indicating that residual stress control is a critical factor for maintaining guideway accuracy. Based on this, this study proposes an engineering control standard of a residual stress threshold of 500 MPa, which limits the displacement fluctuation value within the IT6 level (0.032 mm).

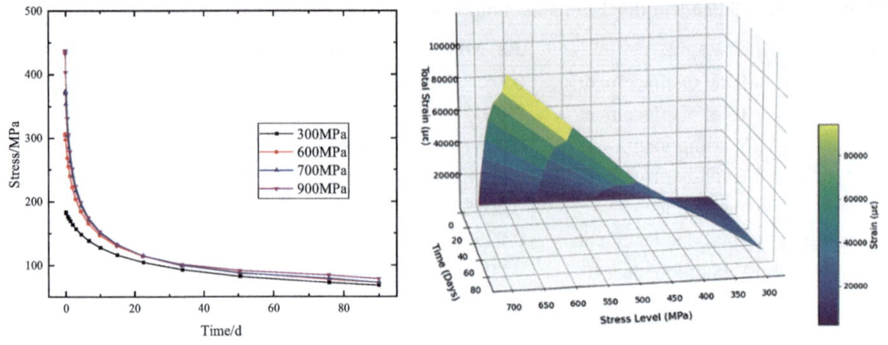

Fig. 7 (**a**) Stress variation over time; (**b**) creep strain variation over time

4.3 Residual Stress Relaxation Mechanism

The dynamics of residual stress relaxation reveal the synergistic mechanism between material microstructure and stress gradient. Figure 7a shows the variation in stress at a point on the guideway over time. The results show that the stress decay rate for the high-gradient model (700 MPa) is approximately twice that of the low-gradient model (300 MPa), with the stress relaxation rate following an exponential relationship with the initial stress level ($R^2 = 0.96$). From a long-term convergence perspective, after 1 month, the stress fluctuation values are all below 10 MPa. After 3 months, the stress in all models approaches 70 MPa, with the residual stress relaxation rate reaching 78%–83%. The surface stress distribution becomes more uniform due to the combined effects of static settling and creep, and the different initial stress levels along the guideway length direction eventually tend to align, confirming the structural self-balancing characteristic. This phenomenon indicates that a high stress gradient promotes the material to adjust the internal stress balance through creep. Figure 7b shows the creep strain distribution at different levels of residual stress. The data indicate that the greater the residual stress, the greater the creep strain, and the rate and extent of increase are also larger, which is consistent with the creep formula.

From the perspective of material microstructure, the 38CrMoAl alloy in the rolled and normalized state has a microstructure consisting of ferrite and pearlite (Figs. 8 and 9). The hardness range for φ160 is 200–240 HB, with the core hardness being slightly lower at approximately 200 HB, and the average grain size is around 30 μm, with a grain size of 7–7.5. The core grain size ranges from 20 to 45 μm, while the edge grain size ranges from 20 to 40 μm. The hardness range for φ100 is 190–230 HB, with the core hardness also lower at approximately 200 HB, the average grain size around 40 μm, and a grain size of 6–6.5. The core grain size ranges from 25 to 80 μm, while the edge grain size ranges from 25 to 70 μm. Microstructural characterization indicates that the grain coarsening in the core and the hardness difference between the center and the edges of the rolled materials lead to non-uniform

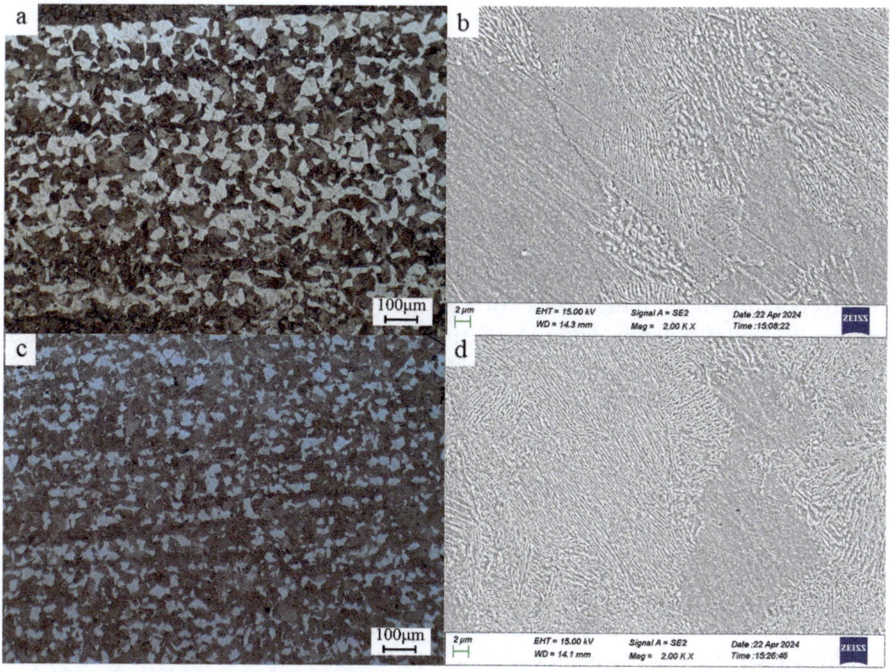

Fig. 8 φ160 rolled center—100× (**a**)—2000× (**b**); φ160 rolled edge—100× (**c**)—2000× (**d**)

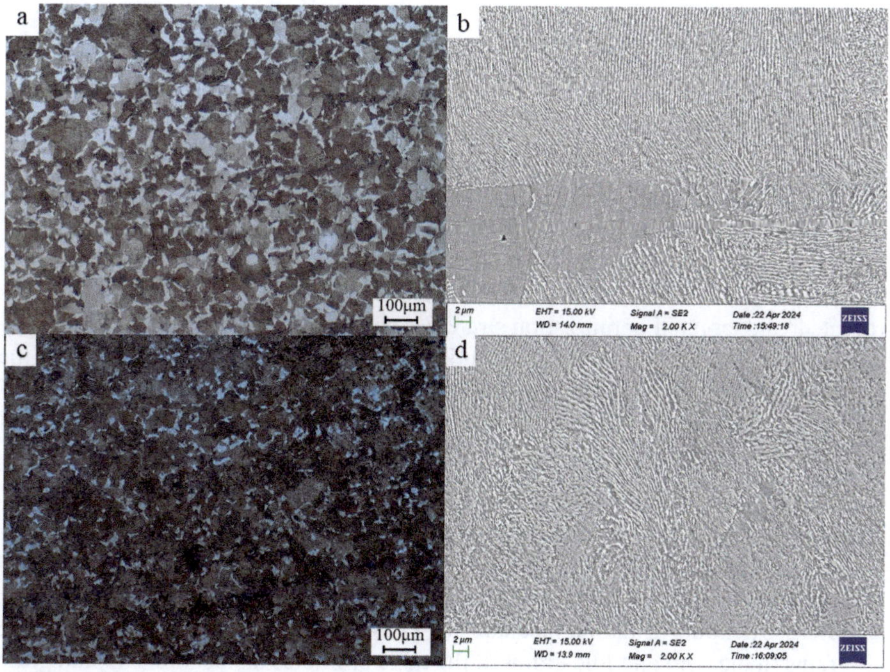

Fig. 9 φ100 rolled center—100× (**a**)—2000× (**b**); φ160 rolled edge—100× (**c**)—2000× (**d**)

creep: coarse-grained regions, due to lower dislocation migration resistance, undergo plastic flow preferentially, causing the guideway to bend toward the low-stress region. This stress-gradient-induced directional creep is the core cause of accuracy degradation.

EBSD analysis further indicates that φ160, due to its higher hardness, has superior creep resistance in the short term. However, the core demand for guideways is long-term dimensional stability and creep resistance. Due to the high stress gradient and significant banded structure, φ160 exacerbates the non-uniformity of residual stress release, leading to localized strain concentration. In contrast, φ100, with its relatively uniform hardness distribution and weakened banded structure, better suppresses creep deformation caused by residual stress release, making it more suitable for high-precision guideways. If the guideway needs to withstand extreme loads or short-term high-frequency impacts (e.g., heavy-duty machine tools), φ160 with finer grains and higher hardness may be preferred, but it should be combined with stress-relief annealing and nitriding treatments to address the uniformity defects. Thus, the material selection and heat treatment strategies are as follows:

For long-term precision service, φ100 material is preferred:

- Optimize the normalizing process (extend holding time) to refine core grain size.
- Add low-temperature tempering to further eliminate residual stress.

For high-load conditions, φ160 material is preferred:

- Use multi-directional rolling processes to reduce banded structure segregation.
- Apply surface nitriding treatment to improve wear resistance and reduce the core-edge performance discrepancy.

5 Conclusion

This study systematically investigates the accuracy degradation mechanism of plane linear guideways under the coupling effect of residual stress gradient and creep, using a combined experimental and simulation approach. The synergistic influence mechanism of material microstructure heterogeneity and stress release behavior is revealed, and targeted optimization strategies are proposed. Based on layered residual stress modeling and creep constitutive analysis of 38CrMoAl guideways, the impact of different stress gradients on long-term deformation is quantified, providing theoretical support and engineering solutions for the design and manufacture of high-precision guideways. The quantified conclusions are as follows:

1. Stress-deformation nonlinear behavior: Residual stress levels exhibit a super linear positive correlation with creep deformation. Considering the fluctuation values on the guideway surface, it is demonstrated that controlling the initial residual stress ($\leq$500 MPa) is the primary condition for ensuring accuracy.
2. Microstructural regulation mechanism: The hardness gradient formed by coarse grains in the core and fine grains at the edge promotes dislocation slip and grain

boundary migration, driving non-uniform creep deformation. The φ100 material, with superior microstructure uniformity, is more suitable for long-term high-precision service, while φ160 material, with higher surface hardness and finer grain size, is better suited for high-frequency impact and similar conditions.
3. Stress release window and optimization path: 78%–83% of the residual stress is released within 3 months, with the first 30 days contributing to 70% of the deformation. It is recommended to allow a 2-month settling period before assembly to reduce service risks. Materials should be adapted to different working conditions (φ100 for long-term precision service, φ160 for high-load scenarios), and processes should be optimized (multi-directional rolling + nitriding treatment) to significantly improve guideway stability.

References

1. Xu M, Zhang H, Liu Z, Li C, Li Z, Liu H, Zhang Y (2023) Nonlinear dynamic model for linear guideway with crowning subjected to external load in five directions. Mech Syst Signal Process 187:109947. https://doi.org/10.1016/j.ymssp.2022.109947
2. Lyu Z, Cao Y, Wang MY, Xu Q (2024) Concept design of a monolithic compliant series-elastic actuator with integrated position and two-level force sensing. Mech Mach Theory 193:105569. https://doi.org/10.1016/j.mechmachtheory.2023.105569
3. Rahmani M, Bleicher F (2016) Experimental and analytical investigations on normal and angular stiffness of linear guides in manufacturing systems. Procedia CIRP 41:795–800. https://doi.org/10.1016/j.procir.2015.12.033
4. Cheng Q, Zhang H, Zhang T, Li Y, Xu J, Liu Z (2023) Prediction method of precision deterioration of rolling guide under multi-random parameters based on frictional thermal expansion effect. Tribol Int 189:108883. https://doi.org/10.1016/j.triboint.2023.108883
5. Zou HT, Wang BL (2015) Investigation of the contact stiffness variation of linear rolling guides due to the effects of friction and wear during operation. Tribol Int 92:472–484. https://doi.org/10.1016/j.triboint.2015.07.005
6. Xu M, Zhang H, Li C, Yao G, Zhang Y, Gao Y (2023) An analytical nonlinear dynamic model for linear guide with carriage flexibility. Int J Non-Linear Mech 148:104251. https://doi.org/10.1016/j.ijnonlinmec.2022.104251
7. Spence T, Makhlouf MM (2013) Modelling creep induced by machining residual stresses in aluminium alloys. Mater Sci Forum 765:585–589
8. Jiang X, Kong X, He S et al (2021) Modeling the superposition of residual stresses induced by cutting force and heat during the milling of thin-walled parts. J Manuf Process 68:356–370
9. Ding W, Zhang L, Li Z et al (2016) Review on grinding-induced residual stresses in metallic materials. Int J Adv Manuf Technol 88(9-12):2939–2968
10. Gong P, Zhang Y, Wang C, Cui X, Li R, Sharma S, Liu M, Gao T, Zhou Z, Wang X, Dambatta YS, Li C (2024) Residual stress generation in grinding: Mechanism and modeling. J Mater Process Technol 324:118262. https://doi.org/10.1016/j.jmatprotec.2023.118262
11. Ding W, Zhang L, Li Z, Zhu Y, Su H, Xu J (2016) Review on grinding-induced residual stresses in metallic materials. Int J Adv Manuf Technol 88(9-12):2939–2968. https://doi.org/10.1007/s00170-016-8998-1
12. Jiang X, Kong X, He S, Wu K (2021) Modeling the superposition of residual stresses induced by cutting force and heat during the milling of thin-walled parts. J Manuf Process 68:356–370. https://doi.org/10.1016/j.jmapro.2021.05.048

Open Access This chapter is licensed under the terms of the Creative Commons Attribution-NonCommercial-NoDerivatives 4.0 International License (http://creativecommons.org/licenses/by-nc-nd/4.0/), which permits any noncommercial use, sharing, distribution and reproduction in any medium or format, as long as you give appropriate credit to the original author(s) and the source, provide a link to the Creative Commons license and indicate if you modified the licensed material. You do not have permission under this license to share adapted material derived from this chapter or parts of it.

The images or other third party material in this chapter are included in the chapter's Creative Commons license, unless indicated otherwise in a credit line to the material. If material is not included in the chapter's Creative Commons license and your intended use is not permitted by statutory regulation or exceeds the permitted use, you will need to obtain permission directly from the copyright holder.

Research and Experimentation on Key Parameters of Wearable Goji Berry Harvester

Yan Cheng, Jiahui Liu, Jian Zhao, Yuchuang Liu, Hongye Zhang, and Dong Zhao

Abstract With the development of the goji berry industry, challenges such as time constraints and labor shortages have become increasingly prominent, necessitating breakthroughs in mechanized harvesting technologies. In response to the current reliance on manual harvesting practices in Qinghai Province and the lag in the development of harvesting equipment both domestically and internationally, this study developed a wearable harvester. The minimum breaking force for harvesting operations was determined to be 1.5 N through the study of mechanical properties of goji berry fruits. The harvester carries out multi-functional harvesting based on the principles of shearing and brushing, utilizing the negative pressure generated by the fan to adsorb the shedding goji berries and completing the harvesting and collection process. The optimal internal structures for harvesting, collecting, and conveying systems were ultimately determined: adopting a single-pipe air suction collection method with a pipe diameter of 50 mm, wind speed of 20 m/s. This design provides a scientific basis and data foundation for the research and development of wearable goji berry harvesters, thereby promoting the entire goji berry industry towards more efficient and sustainable development.

Keywords Goji berry harvester · Wearable · Fluent simulation

Y. Cheng · Y. Liu
School of Technology, Beijing Forestry University, Beijing, China

J. Liu · H. Zhang · D. Zhao
School of Technology, Beijing Forestry University, Beijing, China

Key Lab of State Forestry Administration on Forestry Equipment and Automation, Beijing Forestry University, Beijing, China

J. Zhao (✉)
School of Technology, Beijing Forestry University, Beijing, China

Key Lab of State Forestry Administration on Forestry Equipment and Automation, Beijing Forestry University, Beijing, China

State Key Laboratory of Efficient Production of Forest Resources, Beijing, China
e-mail: zhaojian1987@bjfu.edu.cn

© The Author(s) 2026

Z. Zhang, P. Mercorelli (eds.), *Mechanical Design and Simulation: Insights and Innovations*, Lecture Notes in Mechanical Engineering,
https://doi.org/10.1007/978-981-95-2405-1_81

1 Introduction

Goji berry as a kind of solanaceae multi-branched shrub plants, because of its high nutritional value, it has become one of the important specialty cash crops in China [1]. Curently, goji berry mechanized special harvesting equipment in the market is still blank. The traditional manual harvesting method is difficult to adapt to the modern society on the demand for efficient production, seriously restricting the development potential of goji berry industry [2]. Scholars in the field of agricultural machinery at home and abroad have carried out extensive research and made progress in the mechanized harvesting of small berries, and put forward a variety of harvesting methods, including: vibration, comb and brush, shear, air suction and so on [3–6]. However, there are few reports on wearable goji berry harvester, the study designed a wearable goji berry harvester, in order to further explore the interaction between goji berries and the flow field of the harvester, this study simulated the flow field of the picking device and conveying device of the wearable goji berry harvester, in order to provide a basis for determining the structure of the harvester.

2 Machine Structure and Working Principle

2.1 Machine Structure

To achieve rapid harvesting operations for goji berries, a wearable wolfberry harvester has been developed. This device primarily consists of a wearable collection unit and an air-suction collection unit. The wearable collection unit comprises a collection bag, thumbscrew, brush teeth, and a Three-finger cleaver. Specifically, the aluminum alloy thumbscrew, integrated with a rubber finger sheath, performs independent cutting of wolfberry pedicels through shearing action generated between the thumb and index finger. Brush teeth are set in the glove palm to design two rows of spacing 1 cm of silicone soft teeth, on the basis of not damaging the goji berries, thinning brush off the goji berries, to achieve the goal of picking. The air-suction collection unit consists of a backpack belt, blower, storage bin, air suction pipeline and collection bag. Utilizing negative pressure generated by the blower, this unit absorbs detached berries into the storage bin to complete the collection process [7, 8].

These two units are interconnected through air suction pipeline and collection bag, allowing easy disassembly and reassembly to meet requirements for lightweight design and operational simplification. The overall system configuration is illustrated in Fig. 1.

Through preliminary research, the wolfberry planting area has a row spacing of 100 cm, plant spacing of 55 cm, and plant height ranging from 95 cm to 160 cm; to facilitate the field operation of this harvester, 2 m-long pneumatic tubing and a 10 L-volume storage bin have been selected.

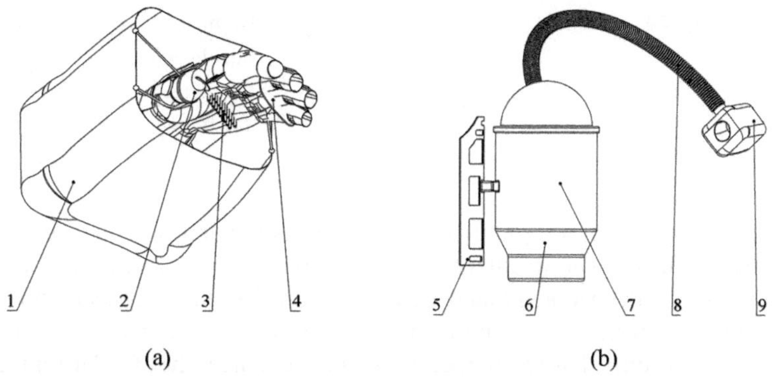

Fig. 1 (**a**) Wearable harvesting unit: (1) collection bag; (2) Thumbscrew; (3) Brush teeth; (4) Three-finger cleaver; (**b**) Air-suction collection: (5) Backpack belt; (6) Blower; (7) Storage bin; (8) Air suction pipeline; (9) Collection bag

2.2 Conveying Pipe Design

The diameter of the conveying pipe significantly influences the performance of the pneumatic conveying system. Based on the geometric dimensions and physical properties of goji berry, it is essential to calculate their suspension velocity and pressure values to optimize the design parameters [9].

Calculation of suspended weight in air G_0:

$$G_0 = \frac{\pi}{6} d_p^3 \left(\rho_p - \rho\right) g \tag{1}$$

Calculation of air resistance F_d:

$$F_d = C \frac{\pi}{4} d_p^2 \rho \left(\frac{\left(v - v_0\right)^2}{2} \right) \tag{2}$$

Pressure drop gradient along the pipe:

$$\frac{\partial_p}{\partial_l} = \frac{f \rho u^2}{2d} \tag{3}$$

Pressure gradient force:

$$F_P = v \frac{\partial_p}{\partial_l} \tag{4}$$

where ρ_p represents the density of goji berry, d_p represents the diameter of goji berry, ρ represents the density of air, v represents the velocity of air.

Table 1 Pipe design parameters

Number	Parameters
Pipe length (m)	2
Pipe diameter (mm)	32, 40, 50
Air velocity (m/s)	20

When the suspended weight of the goji berry is greater than the sum of the air resistance and the pressure gradient force, the goji berry is sucked up. By combining the formulas, it is determined that the gas flow velocity should exceed 18 m/s.

The diameter of the suction pipe and the flow rate for the pneumatic collection device are determined based on operational convenience, the standard dimensions of flexible hoses, and the power of the fan. This provides a theoretical basis for the design of the pneumatic collection system, as shown in Table 1.

3 Simulation Analysis of Key Flow Fields

3.1 Simulation Design Program

Through the structural design of wearable goji berry picking device, for the key fluid domains such as (air suction pipe, inlet) for the key flow field numerical pressure simulation and flow velocity simulation to get the relevant pressure and velocity cloud diagram to determine whether to meet the design requirements. Through Edem discrete element modeling of goji berry fruit, through the fluid-solid coupling simulation of goji berry motion velocity and the size of the goji berry collision force.

3.2 Determination of Pipe Diameter Based on Velocity and Pressure

The Fluent software was used to simulate the pressure and wind speed changes and distribution in the gas suction pipe. For the gas suction pipe model, the gas inlet boundary adopts the velocity inlet boundary condition, the outlet boundary adopts the free flow boundary condition, and the rest of the boundary is set as the boundary condition of the solid wall boundary. Different boundary conditions are set to simulate and analyze the pressure distribution inside the gas suction pipe as shown in Fig. 2. The different depths represent different pressure values, and the unit is Pa. The velocity distribution of the airflow inside the suction pipe is shown in Fig. 3. The different depths represent different velocity values, and the unit is m/s. The pressure and velocity distributions of 32 mm, 40 mm and 50 mm pipe diameters are listed [10, 11].

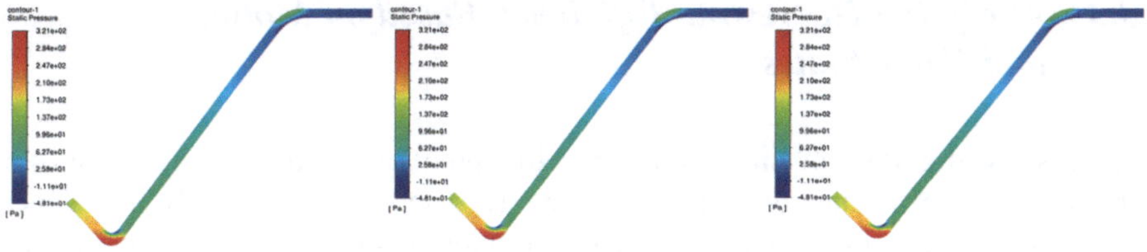

(a) (b) (c)

Fig. 2 Pressure heatmap: (**a**) d = 32 mm, v = 20 m/s; (**b**) d = 40 mm, v = 20 m/s; (**c**) d = 50 mm, v = 20 m/s

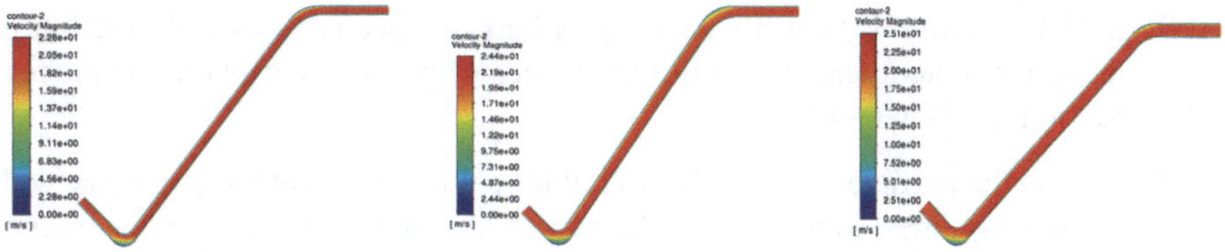

(a) (b) (c)

Fig. 3 Velocity heatmap: (**a**) d = 32 mm, v = 20 m/s; (**b**) d = 40 mm, v = 20 m/s; (**c**) d = 50 mm, v = 20 m/s

In the pressure heatmap, from light color to dark color pressure values gradually become smaller, but the suction pipe is negative pressure, so the actual pressure is bigger. The pressure inside the air suction pipe decreases gradually with the direction of air flow, and with the decrease of pipe diameter, the air pressure at the entrance and exit of the air suction pipeline increases; the pressure distribution at the entrance and exit of the 32 mm air suction pipeline is the most uneven, The greater the pressure drop in the air-suction pipe, the larger the airflow pressure loss, resulting in higher energy consumption and consequently less suction force ultimately exerted on the goji berries.

In the velocity heatmap, it is observed that the air suction pipe with 50 mm diameter has more dark-colored area, which represents higher wind speed, and the suction force suffered by goji berry fruit is large; the air suction pipe with 32 mm diameter has unstable airflow velocity, and the average wind speed in the air suction pipe is the lowest. Take 50 mm air suction pipe as an example, the air suction pipe inlet and outlet pressure and wind speed respectively, and calculate the pressure difference and speed difference, after calculation, the air suction pipe pressure error and speed error are within the permissible range, indicating that the test data have reliability. Comprehensive comparison, pipe diameter 50 mm air suction pipe is better than the pipe diameter 32 mm air suction pipe, so the pipe diameter 50 mm is more suitable for goji berry air suction picking device.

3.3 Analyzing Harvesting Efficiency Based on Motion and Force States

Analysis of the State of the Goji Berry Movement According to the statistics of the field work efficiency of goji farmers, the average work length is 10 h, the average value of daily picking volume of goji farmers is 60 kg, and the average efficiency is 6 kg/h, the peak value of daily picking volume of goji farmers is 100 kg, and the average efficiency is 10 kg/h, and the mass of a mature red fruit is about 0.97 g, so we mimicked the real picking efficiency of the goji farmers and set up the collection of two different goji berry particles to 5 s as a unit of working time step, set 10 g and 20 g of the amount of fruit [12]. Among them, three volume fractions of 100 %, 130 % and 190 % were set for goji berry respectively as a distinction to observe the movement changes of different sizes of goji berry fruits in the process of air suction, as shown in Fig. 4.

In the Fig. 4a the particles with darker to lighter colors represent the goji berry with large to small volumes, in the Fig. 4b the particles with darker to lighter colors represent the goji berry with fast to low velocities.

The conclusion can be drawn from the simulation results:

(a) When the number of harvested goji berry is small:
The interface between the bag and the suction pipe is prone to accumulation, and the speed of fruit collection is greatly affected by the shape of the suction pipe's bend.
(b) When the number of harvested goji berry is high:
The air suction port bends are prone to buildup, and the fruit collection speed is greatly affected by the pipe diameter.
(c) Small mass fruit particles are prioritized into the storage bin

Therefore, in the structural design of the harvester, priority should be given to selecting pipelines with larger diameters and softer materials that comply with national standards, while ensuring the smoothness of the inner pipe walls to minimize potential damage to wolfberries caused by collisions.

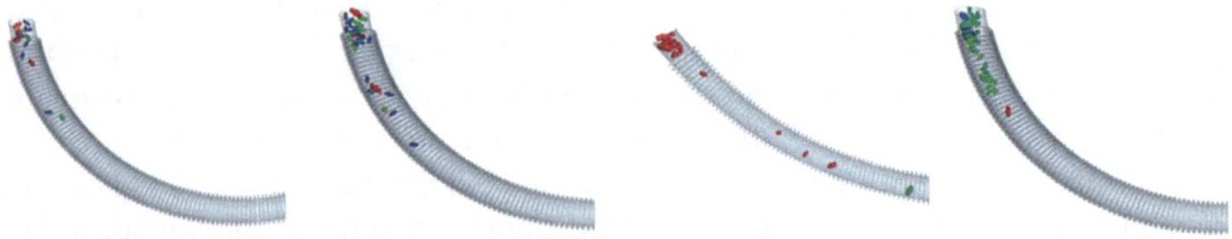

(a) (b)

Fig. 4 Schematic diagram of the goji berry Campaign: (**a**) Flow of particles of different radius sizes: left inlet 10 g/s, right inlet 20 g/s; (**b**) difference in velocity when particles are in different parts of the body: left inlet 10 g/s, right inlet 20 g/s

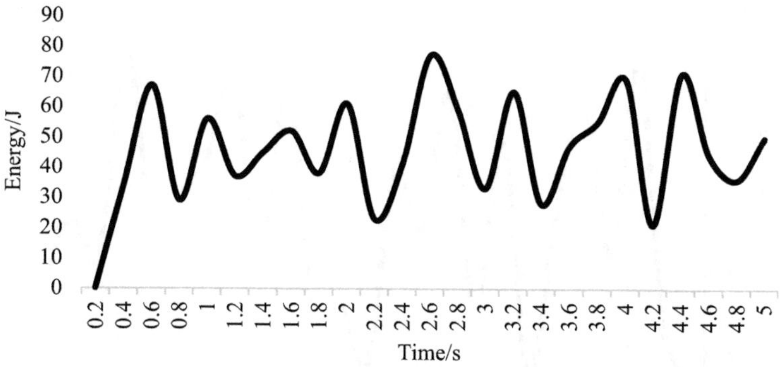

Fig. 5 Average particle energy

Analysis of the Force State of Goji Berry By post-processing the simulation results, the energy changes of goji berry within the pneumatic conveying pipe can be obtained, as shown in the Fig. 5.

According to the theorem of momentum, the force formula can be derived based on the changes in energy:

$$F = \frac{\sqrt{2m_{\vartriangle}E}}{\vartriangle t} \tag{5}$$

This formula can be used to calculate the force from the energy change and the mass of the particles, and the force received by the fruit particles during the movement of the pneumatic suction pipe is 1.143 N, Measured by biomechanical tests, it is known that the minimum of the breaking force of mature fruit is 1.5 N, This indicates that the force experienced by the goji berry during transportation is within safe limits, ensuring minimal damage to the fruit.

Analysis of Goji Berry Harvesting Efficiency Harvesting efficiency can be modeled by analyzing the exit particle mass at the gas suction pipe exit interface, which is shown in Fig. 6.

Within a single harvesting operation period, the average mass of goji berry exiting the system within five seconds is 16.71 g, demonstrating higher efficiency compared to manual picking. Additionally, the distribution of radial forces experienced by the goji berry particles aligns with the simulation results. To prevent damage at the pneumatic pipe connections and pipe bends, the material and structure of the pneumatic conveying device can be optimized by using smoother inner walls for the suction pipes and increasing the pipe diameter, thereby reducing the damage rate.

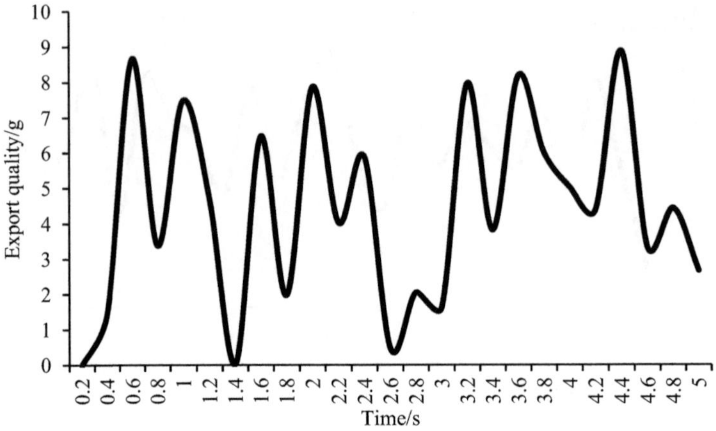

Fig. 6 Statistical chart of particle mass

4 Testing and Analysis

The field test was conducted using pipe with a diameter of 50 mm and a length of 2000 mm. Ten goji berry plants were randomly selected as measurement subjects, harvesting operations were conducted using the wearable goji berry harvester, as shown in the Fig. 7 and 8, with the following metrics evaluated: the amount of goji berry collected per minute, the actual damage rate and the impurity rate.

In field test, the harvester was employed with different picking methods to investigate which approach yields higher efficiency, and the results were compared with manual harvesting, leading to the harvesting results illustrated in Fig. 9.

In different working situations, the hand-held branch picking method contains the lowest rate of impurity and damage rate of 6.76% and 2.78%, respectively, and the net harvesting rate can reach 92.61%, and its harvesting efficiency is 10.37 kg/h, which is 72.9% higher than that of the manual hand-picking.

5 Conclusion

This paper presents the design of a wearable goji berry harvester. Through the investigation of the properties of goji berry materials, the structure of an air-suction collection unit was designed. The integration of this mechanism with a wearable

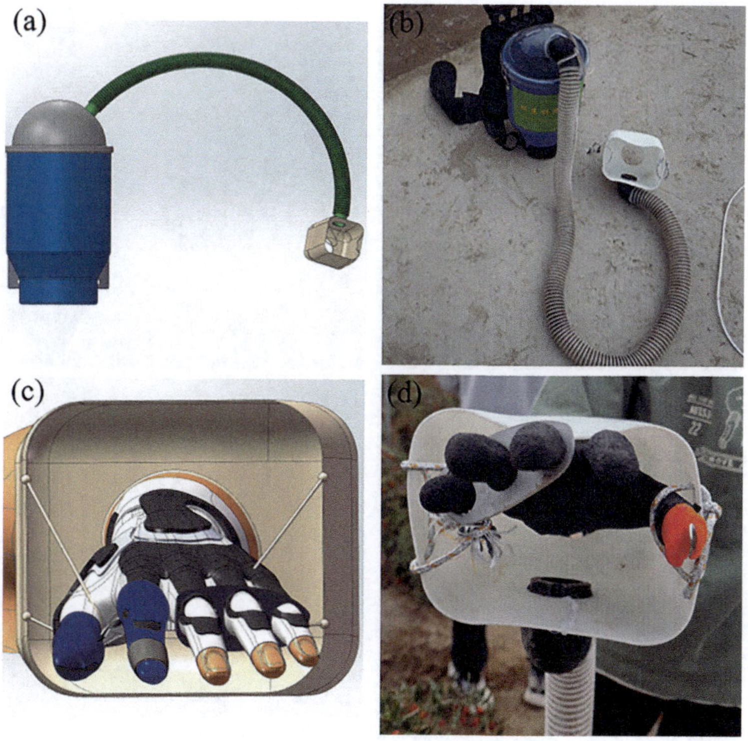

Fig. 7 Wearable goji berry harvester: (**a**) air-suction collection structure modeling schematic, (**b**) air-suction collection structure prototype, (**c**) wearable harvesting structure modeling schematic, (**d**) wearable harvesting structure prototype

harvesting structure, along with simulation studies, has led to the following conclusions:

1. Aiming at the disadvantages of low efficiency in manual goji berry harvesting, a wearable harvester with an integrated air-suction collection bin was designed.
2. Through simulation analysis, the optimal structural parameters for the wearable harvester were determined: air suction pipe diameter of 50 mm, length of 2 m, and fan airspeed of 20 m/s.
3. Field test demonstrated that using a wearable harvester by the hand-held branch picking method can achieve a fruit damage rate of 2.78% and a impurity rate of 6.76%.

Fig. 8 Field test: (**a**) paving of the site; (**b**) picking process; (**c**) collection process; (**d**) harvesting results

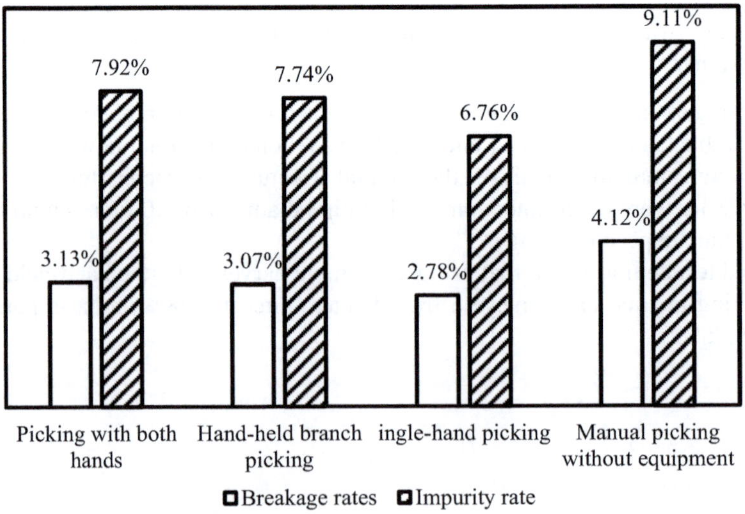

Fig. 9 Harvesting results in the field

Acknowledgment This research was funded by Qinghai Provincial Key R&D and Transformation—"Horse Racing System" Project (Grant Number: 2024-NK-141S-2).

References

1. Amagase H, Farnsworth NR (2011) A review of botanical characteristics, phytochemistry, clinical relevance in efficacy and safety of lycium barbarum fruit (goji). Food Res Int 44(7):1702–1717. https://doi.org/10.1016/j.foodres.2011.03.027
2. Jinpeng W, Song M, Hongru X (2018) Research on mechanized harvesting methods of lycium barbarum fruit. IFAC-PapersOnLine 51(17):223–226. https://doi.org/10.1016/j.ifacol.2018.08.148
3. Zhang Wenqiang LI, Yuzhi ZTAN (2018) Optimal design and experiment on variable pacing combing brush picking device for Lycium barbarum. Trans Chin Soc Agric Mach 49(8):83–90
4. He Y (2024) Design of air suction berry harvester and experimental research of the parameters of conveying system
5. Qihang Z (2024) Design of Ruoqiang red jujube air suction backpack harvester. Jiangsu Agric Mech 1:8–11. https://doi.org/10.16271/j.cnki.jsnjh.2024.01.002
6. Rongyan W (2019) Experimental study on air suction harvesting parameters of wolfberry. J Agric Mech Res 41(11):165–171. https://doi.org/10.13427/j.cnki.njyi.2019.11.029
7. Zhu L, Li X, He J (2023) Development of lycium barbarum–forage intercropping patterns. Agronomy 13(5):1365. https://doi.org/10.3390/agronomy13051365
8. Kim E, Freivalds A, Takeda F (2018) Ergonomic evaluation of current advancements in blueberry harvesting. Agronomy 8(11):266. https://doi.org/10.3390/agronomy8110266
9. Yue Yuan-man SUN, Hao-bo PUJ (2022) Flow field simulation of pepper pickers. Forest Mach Woodwork Equip 50(1). https://doi.org/10.13279/j.cnki.fmwe.2022.0017
10. Liang Dan-dan GE, Yun. (2017) Flow field simulation and design of storage chamber of roll-suction type harvester on safflower. J Gansu Agric Univ 52(5):177–182. https://doi.org/10.13432/j.cnki.jgsau.2017.05.026
11. Yongchao Z (2016) Simulation and experiment of air-suction collecting and conveying equipment for small berry. Trans Chin Soc Agric Eng 32(17):32–39. https://doi.org/10.11975/j.issn.1002-6819.2016.17.005
12. Fengkui Z, Fufeng Y (2020) Design and field testing of the air-suction machine for picking up Chinese jujube fruits. J Fruit Sci 37(2):278–285. https://doi.org/10.13925/j.cnki.gsxb.20190424

Open Access This chapter is licensed under the terms of the Creative Commons Attribution-NonCommercial-NoDerivatives 4.0 International License (http://creativecommons.org/licenses/by-nc-nd/4.0/), which permits any noncommercial use, sharing, distribution and reproduction in any medium or format, as long as you give appropriate credit to the original author(s) and the source, provide a link to the Creative Commons license and indicate if you modified the licensed material. You do not have permission under this license to share adapted material derived from this chapter or parts of it.

The images or other third party material in this chapter are included in the chapter's Creative Commons license, unless indicated otherwise in a credit line to the material. If material is not included in the chapter's Creative Commons license and your intended use is not permitted by statutory regulation or exceeds the permitted use, you will need to obtain permission directly from the copyright holder.

Analysis of the Influence of Step Form on the Hydrodynamic Performance of Double-stepped Planing Hull

Jing Yang, Dongying Li, Fan Wang, and Tengjiao Sang

Abstract This study focuses on a specific type of two-step planing hull, conducting model tests in calm water at different speeds. Numerical simulations are performed using CFD methods, using the experimental data as a benchmark to compare model resistance, heave and pitch angle, thereby validating the effectiveness of the numerical method. The arc-shaped step is changed into two kinds of step forms: diagonal and right-angle, and CFD method is used to carry out the research on the taxiing performance of high-speed hulls with different step forms. The results show that the resistance performance of the arc-shaped step planing hull is superior to the other two configurations. However, it does not have a significant advantage over the diagonal step planing hull, with a maximum drag reduction effect not exceeding 5%. Additionally, the pitch angle and heave of both the arc-shaped step and the diagonal step are relatively close. The resistance performance of the right-angle step has certain advantages at low speeds, but at high-speed sliding situation, the resistance is significantly larger, and both the pitch angle and heave are smaller. This study provides data support for the design of high-performance multi-step planning hulls.

Keywords Double step · Planing hull · Model test · Numerical simulation

J. Yang (✉) · D. Li · F. Wang
AVIC Special Vehicle Research Institute, Jingmen, China
e-mail: wangf381@avic.com

T. Sang
AVIC Special Vehicle Research Institute, Jingmen, China

Key Aviation Scientific and Technological Laboratory of High-Speed Hydrodynamics, Jingmen, China

© The Author(s) 2026

Z. Zhang, P. Mercorelli (eds.), *Mechanical Design and Simulation: Insights and Innovations*, Lecture Notes in Mechanical Engineering,
https://doi.org/10.1007/978-981-95-2405-1_82

1 Introduction

Planing hull is a type of vessel that primarily relies on hydrodynamic lift to support its weight at high speeds. Due to its high-speed characteristics, planing hulls have become very popular in military, transportation, and recreational fields. The stepped planning hull divides the hull into front and rear sections, which helps adjust the hull's attitude under various navigational conditions. In high-speed sailing conditions, the steps allow the front and rear sections of the hull to automatically adjust the lift distribution according to changes in speed and load, maintaining good hydrodynamic performance [1].

Most common stepped planning hulls are equipped with one or two steps, and the bottom of the vessel is divided into multiple discontinuous sliding surfaces. This arrangement of stepped sliding surfaces helps improve the sliding efficiency of each sliding surface. Additionally, the cavities formed behind the steps can significantly reduce the wetted area during navigation, thereby effectively reducing frictional resistance [2]. Ji Shijun [3] and Liu Yang [4] utilized CFD software to analyze the hydrodynamic characteristics of a stepped super-speed aerodynamic lift vessel model. Zheng Xiaolong [5] proposed an optimization method based on CFD simulation, incorporating factors such as the bottom deadrise angle, step height, and total length of the steps to optimize the hydrodynamic resistance of a certain double-step planning craft and conduct experimental validation. The results indicate that increasing the step height and total length of the steps can enhance the sliding efficiency of the double-step planning hull to a certain extent. Sun Huawei [6] utilized CFD technology to explore the influence of parameters such as width, rise angle, longitudinal curvature, and transverse curvature of the glide surface on the resistance and navigation characteristics of hovercraft. De Marco [7] and Sajedi [8] both conducted experimental and numerical simulation studies on stepped planning hull. Through analysis using Reynolds-Averaged Navier-Stokes (RANS) and Large Eddy Simulation (LES), it was found that employing overset grid technology can yield better results. Kazeme [9] evaluated the hydrodynamic resistances of without step, one-step, and two-step hulls of Cougar planing craft under different distances of the second step and LCG from aft, weight loadings, and Froude numbers (Fr) by computational fluid dynamics (CFD) analysis. Then, the hydrodynamic resistance of intended planing crafts under various geometrical and physical conditions is predicted using artificial neural networks (ANNs), the predictive equation is suggested for the resistance to weight ratio of stepped planing craft. Existing research has largely focused on single-step configurations, while there are relatively few studies on double-step designs. In particular, there is a lack of systematic research on the impact of step forms on double-stepped planing hulls.

This paper takes a double-stepped planning craft as the research object. First, static water resistance tests are conducted at different speeds and an analysis is performed on the variation of resistance with speed. Additionally, adopts the CFD numerical simulation method, and compares with the test results to verify the reliability of the numerical method. Secondly, further research is conducted on the impact of different step forms on the resistance performance of planing hulls. Based on the initial model, the curved step is transformed into two forms: sloped step and right-angle step, allowing for a comparative analysis of how these different step forms affect the sliding performance of dual-step planing hulls. By combining model tests and numerical simulations, the study investigates the complex flow field around the planing hulls, resistance performance, changes in navigation state, and forecasts the hydrodynamic performance of hovercraft with different step forms at various speeds, providing references for the design of planing hulls.

2 Model Test

This chapter will elaborate on the planning hull model tests from the perspectives of the experimental program and the experimental results. Resistance tests will be conducted on the two-stepped planing hulls at different speeds, and to study the variation patterns of resistance at different speeds.

2.1 Experimental Research

The original curved double-stepped model is shown in Fig. 1, with the front step positioned 42% of the ship length from the transom and the rear step positioned 25% of the ship length from the transom. The photograph of the model test are shown in Fig. 2. The model is made of wood and has a spray-painted and polished surface. The model tests are conducted in the towing tank at China Special Vehicle Research Institute, ensuring that the model and the full-scale ship maintain the same Froude number. During the tests, the model is propelled forward by a towing vehicle, and a total of 11 speed points are examined. The main parameters recorded during the tests are drag, pitch angle and sinkage in the center of gravity.

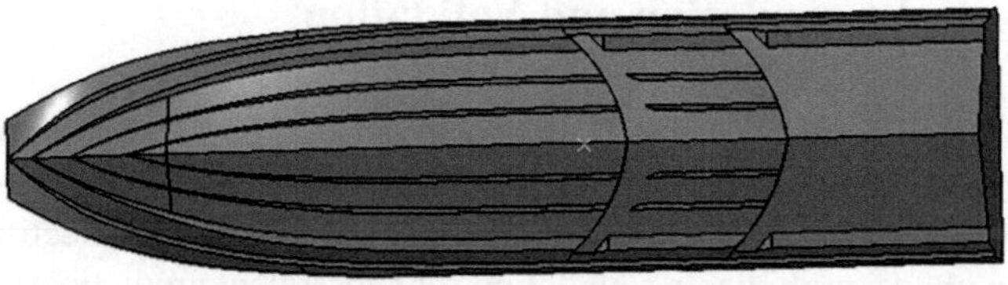

Fig. 1 Arc double-stepped planning hull

Fig. 2 Model test

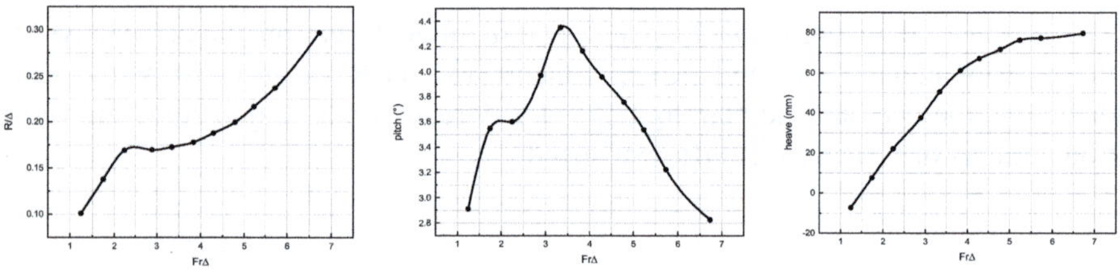

(a) resistance (b) pitch (c) heave

Fig. 3 Experiment results. (**a**) Resistance. (**b**) Pitch. (**c**) Heave

2.2 *Analysis of Test Results*

The results of the static water resistance test for the model are shown in Fig. 3. Overall, the resistance of the model increases with the increasement of the volume Froude number (Fr). There is a slight local resistance peak when Fr = 2.24. The pitch angle reaches its peak at Fr = 3.34, after which the pitch angle gradually decreases with the increase of speed, while the sinkage continues to increase with the increase in speed.

3 Numerical Calculation and Validation

This section introduces the methods and basic theories used for numerical calculations and employs CFD software STAR-CCM+ to carry out numerical simulations. The results are compared with experimental data to validate the effectiveness of the numerical methods, providing a foundation for subsequent predictions of resistance performance with different step configurations.

3.1 Numerical Method

The flow of the fluid must satisfy the conservation of mass and conservation of energy, so the control equations use the continuity and RANS equations.

Continuity equation:

$$\frac{\partial\rho}{\partial t}+\frac{\partial\rho u}{\partial x}+\frac{\partial\rho v}{\partial y}+\frac{\partial\rho w}{\partial z}=0 \tag{1}$$

Momentum equation:

$$\begin{aligned}
\rho\left(\frac{\partial u}{\partial t}+u\frac{\partial u}{\partial x}+v\frac{\partial u}{\partial y}+w\frac{\partial u}{\partial z}\right)&=-\frac{\partial P}{\partial x}+\mu\left(\frac{\partial^2 u}{\partial x^2}+\frac{\partial^2 u}{\partial y^2}+\frac{\partial^2 u}{\partial z^2}\right)\\
\rho\left(\frac{\partial u}{\partial t}+u\frac{\partial u}{\partial x}+v\frac{\partial u}{\partial y}+w\frac{\partial u}{\partial z}\right)&=-\frac{\partial P}{\partial y}+\mu\left(\frac{\partial^2 u}{\partial x^2}+\frac{\partial^2 u}{\partial y^2}+\frac{\partial^2 u}{\partial z^2}\right)\\
\rho\left(\frac{\partial u}{\partial t}+u\frac{\partial u}{\partial x}+v\frac{\partial u}{\partial y}+w\frac{\partial u}{\partial z}\right)&=g-\frac{\partial P}{\partial z}+\mu\left(\frac{\partial^2 u}{\partial x^2}+\frac{\partial^2 u}{\partial y^2}+\frac{\partial^2 u}{\partial z^2}\right)
\end{aligned} \tag{2}$$

Using the finite volume method to discretize the momentum equation, the above equations must be closed with a turbulence model. In this paper, the widely used SST $k-\omega$ two-equation model is employed.

3.2 Method Validation

Numerical simulations of the model were conducted using the commercial CFD software STAR-CCM+, based on the RANS equations, with the computational domain size as shown in Fig. 4. Considering the symmetry of the hull, a half model was used for the numerical calculations to reduce computational costs. The

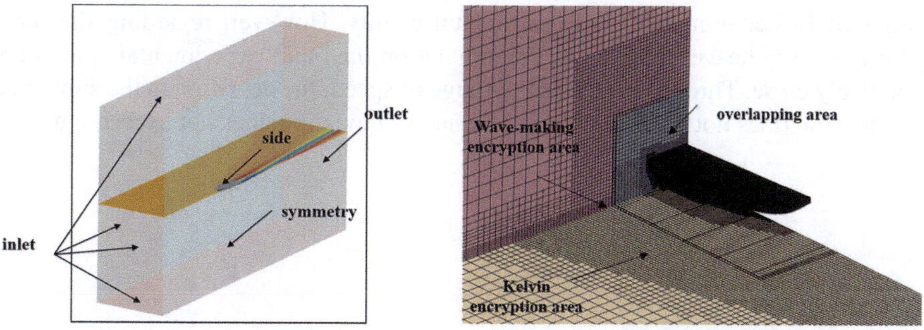

Fig. 4 Computational domain and model surface mesh

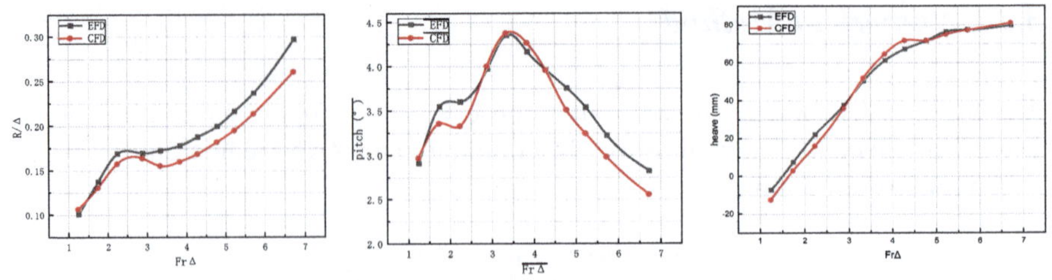

(a)resistance (b)pitch (c)heave

Fig. 5 Comparison of simulation and experimental results. (**a**) Resistance. (**b**) Pitch. (**c**) Heave

computational domain was meshed using trimmed grids, with a reference size set as $\Delta x = 1\%L$. The grid size near the overlapping domain boundaries was set to the reference size, while the grid size on the hull's sliding surface was set to 1/8 of the reference size. A refinement zone was established near the overlapping domain, with the grid size set to the reference size, and grid refinement was also applied at the free surface and in the Kelvin wave region. Prismatic layer meshes were used on the hull surface to capture boundary layer information, with a mesh growth rate of 1.3. The boundary layer thickness and the number of prismatic layers were estimated using the formula from [10]. The upstream, side edges, top, and bottom of the computational domain were designated as velocity inlets, while the downstream was set as a pressure outlet, and the hull surface was treated as a wall. To account for the influence of the free surface, the VOF method was employed to capture the position of the free surface.

The comparison between the simulation results and experimental results is shown in Fig. 5: within the calculation range, the results of numerical simulations are all lower than the experimental results. At low speeds, the deviation between the experimental and simulation results is relatively small. For $Fr > 3$, the numerical simulation results for hull resistance deviate from the experimental data by less than 10% overall, and the values of sinkage match quite well. The main reason for the deviation is that the numerical simulation is based on the finite volume method, which is limited by the influence of grid scale and cannot fully simulate the splashing phenomenon. After $Fr > 3$, the simulation results need to be appropriately adjusted to better match the experimental results. However, regarding the pitch angle and the heave of the hull, the simulation data and experimental results are relatively close. Throughout the entire range of speed, the deviation in the simulated pitch angle does not exceed 0.3°, and the heave deviation does not exceed 6 mm.

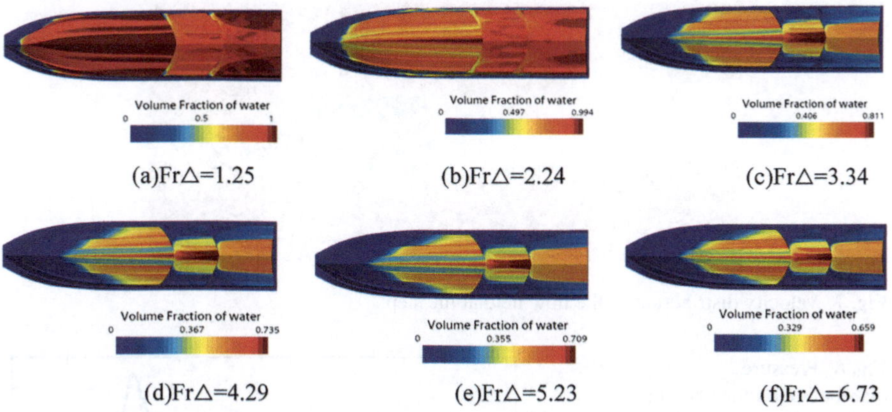

Fig. 6 Air-water distribution of the planing surface at different speeds. (**a**) Fr△ = 1.25. (**b**) Fr△ = 2.24. (**c**) Fr△ = 3.34. (**d**) Fr△ = 4.29. (**e**) Fr△ = 5.23. (**f**) Fr△ = 6.73

3.3 Fluid Field Characteristic

For a planing hull, there are three states: fully submerged, semi-planing, and planing. When the planing hull is in the fully submerged state, it behaves like a conventional displacement vessel, forming a closed body of water from the bow to the transom, and at this point, the hull bottom is supported by buoyancy. As the speed increases, the water at the breakpoint gains sufficient kinetic energy to break free from the step, creating a "cavity". Figure 6 shows the air-water distribution on the planing surface of the hull at different speeds. As the speed increases, the wetted area of the hull bottom decreases. During the planing process, the hull gradually rises, and the corresponding sinkage increases.

Figure 7 shows the velocity distribution at the step. It can be seen from the figure that when the incoming flow passes the step, it separates from the bottom of the hull, creating a region with a velocity of 0 at the step. Figure 8 presents the pressure distribution on the hull surface at Fr = 5.23. The horizontal axis represents the longitudinal extension of the hull surface, with the positive direction pointing towards the bow of the hull, while the vertical axis represents the pressure distribution at the monitoring points on the hull surface. It can be observed from the figure that the pressure on the hull surface changes at the step, showing a significant negative pressure.

4 Impact of Step Form

In order to study the impact of step form on the hydrodynamic performance of planing hulls, the initial model's arc step was changed to two types: a sloped step and a right-angle step, ensuring that the air intake at the side step of the hull is in the same

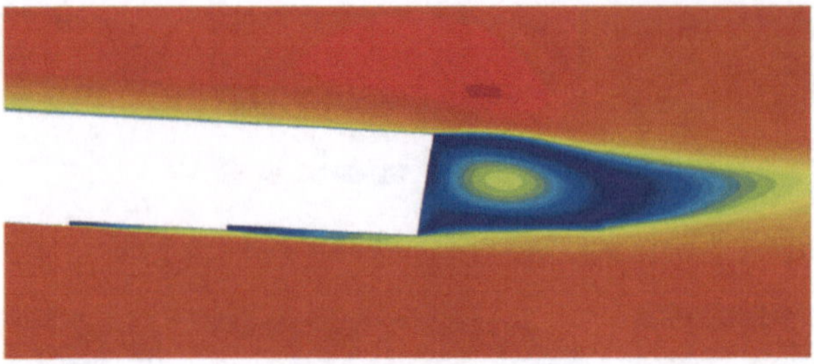

Fig. 7 Velocity distribution of the flow field at the step

Fig. 8 Pressure distribution on the planing surface

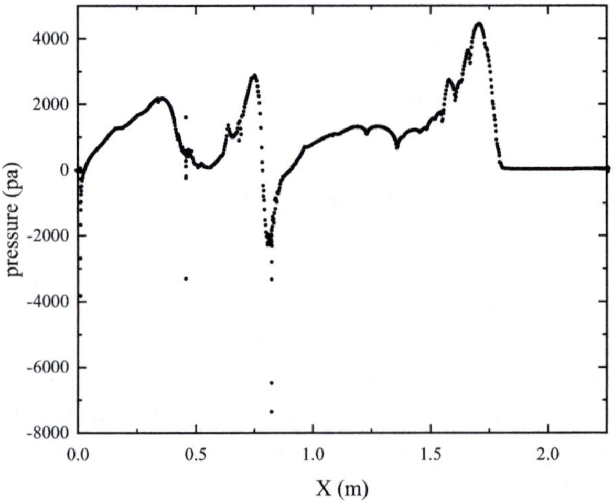

position. This resulted in configuration schemes with different step forms. Using the same mesh and calculation method, numerical simulations were conducted on the three different step forms of the planing hull models, investigating the influence of step form on the hydrodynamic performance of double-step planing hulls. The models of the planing hulls with different step forms are shown in Fig. 9.

Figure 10 shows the calculated results of the hydrodynamic performance of planing hulls with different step forms. The results indicate that the resistance performance of the arc step planing hull is superior to the other two configurations, although its advantage over the sloped step planing hull is not significant, with the maximum drag reduction effect not exceeding 5%. The trim angle and heave of the arc step and sloped step are relatively close, both being greater than those of the right-angle step. The resistance performance of the right-angle step has certain advantages at low speeds, but it shows a significantly larger drag in high-speed planing conditions, being more than 20% greater than that of the sloped and curved steps, and it has smaller pitch angles and sinkage.

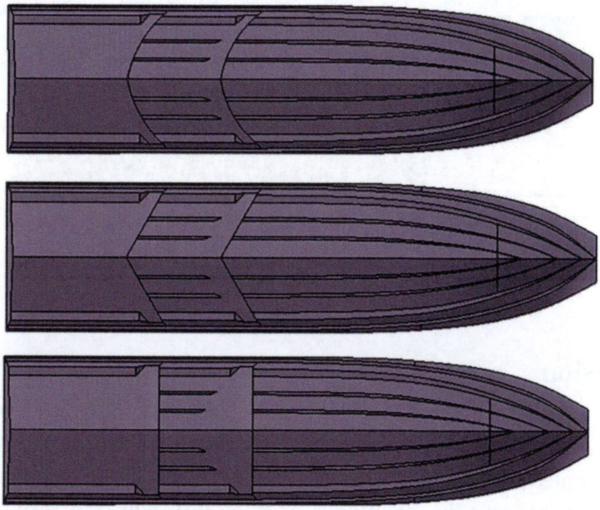

Fig. 9 Planing hulls of different step designs

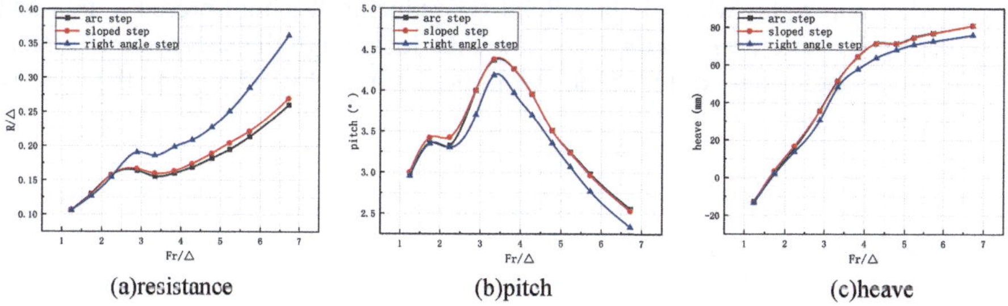

Fig. 10 Comparison of hydrodynamic performance of planing hulls with different step forms. (**a**) Resistance. (**b**) Pitch. (**c**) Heave

Figure 11 shows air-water interaction and pressure distribution on the planning surface with different step designs. Among them, the Air-water distribution and surface pressure distribution of the arc step and sloped step are quite similar, with the high-pressure area after the step divided into two clearly symmetrical regions, while the high-pressure area after the right-angle step is a singular whole. Meanwhile, the area of the front planing surface of the curved step and diagonal step is slightly larger than that of the right-angle step, resulting in a slightly larger trim angle. Thus, the arc step and sloped step are advantageous in reducing the high-pressure area at the aft body, thereby decreasing pressure drag. The high-pressure areas at the bottoms of the three hull configurations are all concentrated near the stagnation point line and after the step. It can be seen that the pressure distribution on the hull bottom shows a sudden change at the step.

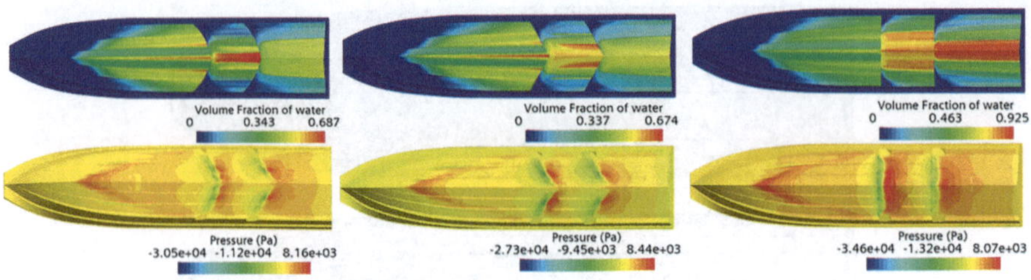

Fig. 11 Air-water interaction and pressure distribution on the planing surface with different step designs

5 Conclusion

Based on the experimental data of an arc double-stepped high-speed planing hull, the article validates the reliability of CFD numerical simulation, and further compares the advantages and disadvantages of three different step forms: the arc step, the sloped step and the right-angle step. The results indicate that the arc step and sloped step divide the high-pressure area after the step into two symmetrical regions on the left and right, which is beneficial for reducing the high-pressure area and decreasing pressure drag. At the same time, the location of the step directly affects the planing surface area of the front and aft bodies, thereby influencing the planing attitude of the hull.

References

1. Yousefi R, Shafaghat R, Shakeri M (2013) Hydrodynamic analysis techniques for high-speed planing hulls. Appl Ocean Res 42:105–113. https://doi.org/10.1016/j.apor.2013.05.004
2. Yao DY, Jiang Y, Shen DX et al (2020) Preliminary numerical study on hydrodynamic characteristics of deep-V planing hull with double steps. Ship Eng 42(10):6. https://doi.org/10.13788/j.cnki.cbgc.2020.10.08
3. Ji SJ, Ni QJ, Jiang Y (2022) Numerical study on influence of step on hydrodynamic and aerodynamic characteristics of ultra-high-speed aerodynamic alleviation. Marine Veh 33(1):39–47. https://doi.org/10.19423/j.cnki.31-1561/u.2022.01.039
4. Liu Y (2023) Influence of step configuration on drag reduction of ultra-high-speed aerodynamic alleviation marine vehicle. Ship Eng 45(5):29–34. https://doi.org/10.13788/j.cnki.cbgc.2023.05.05
5. Zheng XL, Jiao J, Wang MZ, Liu T (2021) Research on the optimization method of resistance performance of double-stepped planning craft. Ship Sci Technol 43(4):68–72. https://doi.org/10.3404/j.issn.1672-7649.2021.04.014
6. Sun HW (2012) Numerical analysis of planing-hull surface shape on resistance and sailing attitude. Harbin Eng Univ. https://doi.org/10.7666/d.D429003
7. De Marco A, Mancini S, Miranda S et al (2012) Experimental and numerical hydrodynamic analysis of a stepped planing hull. Appl Ocean Res 64:135–154. https://doi.org/10.1016/j.apor.2017.02.004

8. Sajedi SM, Ghadimi P (2020) Experimental and numerical investigation of stepped planing hulls in finding an optimized step location and analysis of its porpoising phenomenon. Math Prob Eng 1:3580491. https://doi.org/10.1155/2020/3580491
9. Kazemi H, Doustdar MM, Najafi A et al (2021) Hydrodynamic performance prediction of stepped planing craft using CFD and ANNs. J Marine Sci Appl. https://doi.org/10.1007/s11804-020-00182-y
10. Ding JM, Jiang JB, Qin JT et al (2019) Influencing mesh factors in the calculation of the resistance performance of high-speed planing crafts through RANS. J Harbin Eng Univ 40(006):1065–1071. https://doi.org/10.11990/jheu.201804004

Open Access This chapter is licensed under the terms of the Creative Commons Attribution-NonCommercial-NoDerivatives 4.0 International License (http://creativecommons.org/licenses/by-nc-nd/4.0/), which permits any noncommercial use, sharing, distribution and reproduction in any medium or format, as long as you give appropriate credit to the original author(s) and the source, provide a link to the Creative Commons license and indicate if you modified the licensed material. You do not have permission under this license to share adapted material derived from this chapter or parts of it.

The images or other third party material in this chapter are included in the chapter's Creative Commons license, unless indicated otherwise in a credit line to the material. If material is not included in the chapter's Creative Commons license and your intended use is not permitted by statutory regulation or exceeds the permitted use, you will need to obtain permission directly from the copyright holder.

Research on Point Cloud Fusion Denoising Method for Cultural Relics Digitization

Huadong Zheng, Linxiao Liu, and Wuwei Xu

Abstract With the rapid development of 3D technology, the digitization of cultural relics has emerged as a highly scientific preservation measure. By capturing three-dimensional surface information of cultural artifacts, it provides essential data support for artifact restoration, reconstruction, virtual exhibition, and innovation. Given the intricate structures and the delicate, untouchable nature of these artifacts, existing methods can no longer meet the diverse needs of cultural relic digitization. Further refinement and enhancement of point cloud data processing techniques for artifacts with varying structures and materials are required. The construction of a 3D scanning platform enabled the collection of data from various samples. An algorithm for multi-scale point cloud fusion and denoising, tailored for cultural relic digitization, was introduced. This algorithm categorizes noise point clouds into large-scale and small-scale based on distance, utilizes a combination of multiple algorithms to remove large-scale noise, and employs improved bilateral filtering with normal vector correction to eliminate small-scale noise. Experimental validation was carried out using the bunny model.

Keywords Digitalized antique · 3D scanning technology · Data acquisition · Point cloud denoising

H. Zheng (✉)
Henan Key Laboratory of Intelligent Manufacturing of Mechanical Equipment, Zhengzhou, China

L. Liu · W. Xu
Collage of Mechanical and Electrical Engineering, Zhengzhou University of Light Industry, Zhengzhou, China

© The Author(s) 2026

Z. Zhang, P. Mercorelli (eds.), *Mechanical Design and Simulation: Insights and Innovations*, Lecture Notes in Mechanical Engineering,
https://doi.org/10.1007/978-981-95-2405-1_83

1 Introduction

Cultural relics, as an important form of material cultural heritage, play a key role in the study, preservation, and education of traditional culture [1, 2]. However, natural erosion, human-induced damage, and the passage of time pose significant challenges to the preservation of cultural relics. Therefore, the digitization of cultural relics has become a central focus of current conservation efforts. Three-dimensional scanning technology [3] enables the rapid and accurate acquisition of 3D surface data, providing strong support for the digitization of cultural relics. However, the point cloud data collected by 3D scanners often contain a large number of noise points and outliers, requiring preprocessing.

Both domestic and international experts and scholars have conducted extensive research on point cloud data denoising algorithms. As for denoising scattered point clouds, the approaches primarily encompass: Triangle mesh-based denoising [4–7]. Point cloud model-based denoising [8–11]. From a holistic perspective, current point cloud processing methods have inherent limitations, often focusing on specific problems with poor compatibility. Common issues include: efficient denoising methods that may cause local feature distortion, the ability to handle isolated noise effectively but sensitivity to small-scale mixed noise, and good adaptability to high-density models but poor performance when handling uneven density. To address these issues, this paper proposes a multi-scale point cloud fusion denoising algorithm. The algorithm classifies noisy point clouds into large and small scales based on distance, applying a fusion of algorithms to handle large-scale noise and using normal vector-corrected bilateral filtering to remove small-scale noise. The algorithm's effectiveness is validated through experiments using the bunny model.

2 The Process of Point Cloud Fusion Denoising Method

The point cloud collected by the 3D scanning system contains various types of noise. Based on the spatial distribution of the noise points in the point cloud model, the noise can be classified into background noise (dense point clouds away from the point cloud body), isolated noise (Sparse point clouds away from the point cloud body), redundant noise (Excess point cloud data generated beyond the preset scan area), and mixed noise (Noise point cloud mixed with model point cloud). The point cloud fusion denoising algorithm classifies background noise, isolated noise, and redundant noise as large-scale noise based on their distance from the point cloud model, while mixed noise is classified as small-scale noise. The noise classification diagram is shown in Fig. 1a.

Subsequently, by leveraging the strengths of various algorithms, the corresponding noise is processed. The denoising process begins with the treatment of large-scale noise followed by the handling of small-scale noise, proceeding from far to

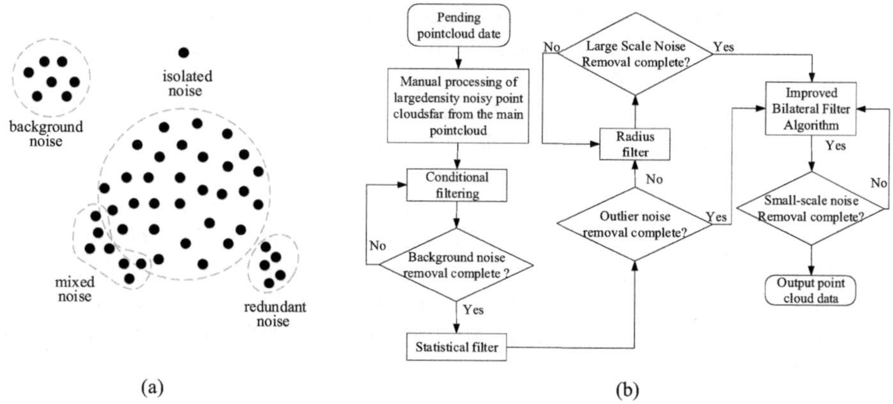

Fig. 1 (**a**) Noise classification; (**b**) The working flow chart of point cloud fusion noise reduction algorithm

near, achieving denoising of the point cloud data at multiple scales. The working flow chart of point cloud fusion noise reduction algorithm is shown in Fig. 1b.

3 Algorithm For Fusion Denoising of Point Cloud Data

3.1 Large-scale Noise Point Removal Methods

By analyzing the advantages of three algorithms: conditional filtering, statistical filtering, and radius filtering, each is applied to different types of point cloud noise. This approach leverages the complementary strengths of these various algorithms. Ultimately, a multi-algorithm fusion-based method is employed to remove large-scale noise, achieving the removal of large-scale noise points.

In the classification of large-scale noise, noise from environmental factors (such as noise models from the workbench, walls, floors, and other surfaces) is often captured during the scanning phase. These noise points are significantly larger in amplitude compared to the main point cloud model and are considered background noise. Therefore, a conditional filtering method will be used to process this type of noise. The steps for removing background noise are as follows:

1. Set the thresholds for x, y and z directions respectively;
2. Remove point cloud data that falls outside of these threshold values;
3. Visualize the model and observe if the removal is complete. If it is, export the data. If not, return to step (1) and repeat;
4. Output the model after removing the background noise.

After removing the outlier noise, the conditional filtering algorithm only removes dense background point cloud data outside the main point cloud, while isolated noise and redundant noise in the point cloud model remain unprocessed. These noise points are typically located outside the main point cloud and have sparse density. To eliminate this type of noise, a statistical filtering algorithm will be applied. Schematic diagram of the statistical filtering algorithm, as shown in Fig. 2a: take k = 4, $d_{\max}$ is represented by a solid circle, $\overline{d_i}$ is represented by a dashed circle, and the figure shows that $\overline{d_1} > d_{\max}$ of the point P_1, the point P_1 is determined as a noise point and is removed; $\overline{d_2} < d_{\max}$ in point P_2, determine point P_2 is normal point and keep it; $\overline{d_3} < d_{\max}$ in point P_3, the point P_3 is normal point and keep it. The statistical filtering algorithm traverses all the point clouds to realize the denoising of the point clouds. The algorithm flowchart for outlier noise removal is illustrated in Fig. 2b.

Below are the specific algorithm steps for removing outlier noise from the point cloud model:

1. Select the k-value of the domain. Pick any point and calculate the average distance $\overline{d_i}$ from that point to its k nearest neighbors as:

$$\overline{d_i} = \frac{1}{k}\sum_{j=1}^{k} p_i - p_j \left(i = 1,;,,2,;,,3\cdots n,;,,j = 1,;,,2,;,,3\cdots k\right) \tag{1}$$

2. Calculate the mean and standard deviation of $\overline{d_i}$ and set the standard range value $d_{\max}$:

$$d_{\max} = \mu + a \cdot \sigma \tag{2}$$

$$\mu = \frac{1}{n}\sum_{i=1}^{n} \overline{d_i} \tag{3}$$

$$\sigma^2 = \frac{1}{n}\sum_{i=1}^{n} \left(\overline{d_i} - \mu\right)^2 \tag{4}$$

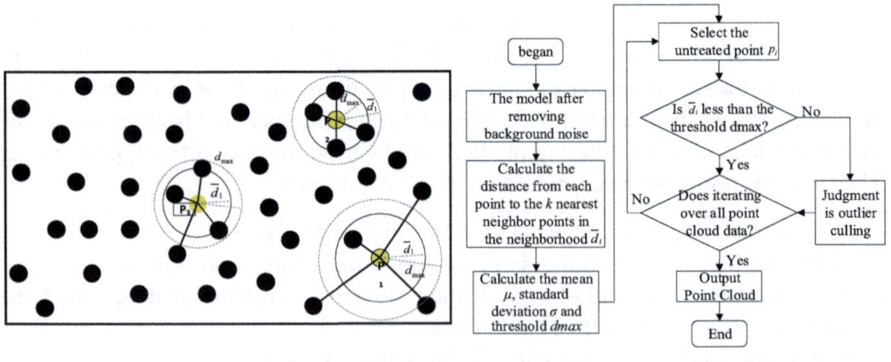

(a) (b)

Fig. 2 (**a**) Schematic diagram of the statistical filtering algorithm; (**b**) Flowchart of the algorithm for outlier noise removal where μ is the mean, σ is the standard deviation, a is the scale factor, and a correlates with the value of k.

3. Compare the size of the average distance $\overline{d_i}$ of each point with the standard range value d_{max}. If $\overline{d_i} < d_{max}$, the point is normal and the point is kept; if $\overline{d_i} > d_{max}$, the point is noisy and the point is removed. Iterate through all the points and the outlier noise removal ends.

After the statistical filtering is completed, a small amount of sparse large-scale noise may still remain in complex areas of the surface. To handle this noise, a radius filtering method will be used to further remove the noise, completing the removal of large-scale noise in the point cloud data. The radius filtering algorithm involves selecting a target point, establishing a spherical radius, and defining a threshold. It dictates that each sphere should contain no fewer points than the specified threshold; if the count falls below this threshold, the target point is removed. This process is iterated across all data points in the model, and the algorithm concludes. The schematic diagram for the radius filtering algorithm is depicted in Fig. 3.

The algorithm steps for removing local residual large-scale noise are as follows:

1. Assuming any point P_i in the scattered point cloud model, the radius of the k-nearest neighbors of P_i is set to r, and at this point, the number of neighbors within the sphere of P_i is denoted as N;

   Set the threshold as N_t;
2. Assessing the comparison between the number of neighboring points N and the threshold N_t: If $N > N_t$, retain the point P_i; if $N < N_t$, remove the point P_i;
3. So that $i = i + 1$, i.e., $P_i = P_{i+1}$;
4. Iterate through all points and output the denoised point cloud;
5. To determine if the noise removal is complete: If it is complete, output the point cloud data. If it is not complete, adjust the radius to $r = r - C$, and repeat steps (1)–(5);
6. Upon the completion of large-scale noise removal, the denoising process concludes, and the point cloud data is output.

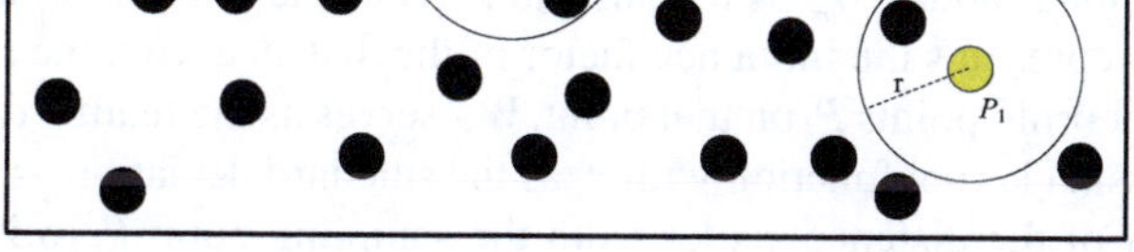

Fig. 3 Schematic diagram of the radius filtering algorithm

3.2 Small-scale Noise Point Removal Methods

Small noise is more difficult to remove than large noise. Bilateral filtering is a commonly used method for removing small-scale noise. Its main principle is to adjust the noise points to appropriate positions to achieve the purpose of noise removal. However, traditional bilateral filtering is prone to over-smoothing the model surface and losing a lot of point cloud feature information during the removal of mixed noise. Therefore, this paper proposes a small-scale bilateral filtering noise removal algorithm based on normal vector correction.

Firstly, a K-D tree structure is constructed based on the K-neighborhood to establish topological relations for the scattered point cloud, and then the improved PCA algorithm is applied to estimate the normal vector n_i, and then the normal direction is corrected to be $n_i^{'}$ by using the minimum tree generation method, and then the bilateral filtering factor α is optimized, and the normal vectors $n_i^{'}$ of all the points as well as the improved filtering factor $\alpha^{'}$ are substituting into the definition equation of the bilateral filtering to complete the denoising of the point cloud data.

The improved bilateral filter definition and filter factor are respectively:

$$P_i^{'} = P_i + \alpha' \cdot n_i^{'} \tag{5}$$

$$\alpha' = \frac{\sum_{p_j \in N(p_i)} W_{\sigma_c} W_{\sigma_s} x\left(1 - n_j^{'}, n_i^{'} + l_i\right) y}{\sum_{p_j \in N(p_i)} W_{\sigma_c}\left(p_j - p_i\right) W_{\sigma_s}\left(1 - n_j^{'}, n_i^{'} + l_i\right)} \tag{6}$$

$$W_{\sigma_c} = e^{-\frac{x^2}{2\sigma_c^2}} \tag{7}$$

$$W_{\sigma_s} = e^{-\frac{x^2}{2\sigma_s^2}} \tag{8}$$

$$x = p_j - p_i \tag{9}$$

$$y = p_j - p_i, n_i^{'} \tag{10}$$

where $\alpha^{'}$ is the improved filtering factor, l_i is the neighborhood curvature corresponding to any point P_i in the point cloud model, when l_i is larger the corresponding point P_i moves a greater distance in the normal direction, when l_i is smaller the corresponding point P_i moves a smaller distance in the normal direction. The parameter W_{σ_c} represents spatial domain weight, which is used to control the smoothness of the point cloud model. W_{σ_c} is a Gaussian kernel function with σ_c as the standard deviation. σ_c represents the influence factor of the distance from the sampling point P_i to its neighboring points P_j on that point. W_{σ_s} serves as the feature domain weight, W_{σ_s} is a Gaussian kernel function with σ_s as the standard deviation. σ_s represents the impact factor of the distance vector from the sampling point P_i to its neighboring points P_j projected onto the unit normal vector $n_i^{'}$ at that point on the sampling point

P_i. x is the Euclidean distance from P_i to its neighboring points P_j, y is the dot product between the distance vector $\overrightarrow{(p_j - p_i)}$ and the normal vector $\overrightarrow{n_i'}$.

The specific steps of the small scale denoising algorithm based on the improved bilateral filter based on normal vector correction are as follows:

1. Input point cloud data to construct a K-D tree structure;
2. Given a value of k for the number of points in the domain, the set of points $N(P_i)$ in the neighborhood of any point P_i in the neighborhood;
3. Using the improved PCA algorithm, the normal vector n_i is computed for each neighborhood point;
4. The minimum tree generation method corrects the direction of the normal vector so that the direction of the normal vector remains consistent, and the normal vector is denoted as n_i';
5. Calculate the curvature k_i of all points;
6. According to Eqs. (9) and (10), the parameters x and y are obtained;
7. According to Eqs. (7) and (8), the spatial domain weights W_{σ_c} and feature domain weights W_{σ_s} are obtained;
8. Find the filtering factor α' according to Eq. (6);
9. Substituting α' and n_i' into Eq. (5) yields the new point P_i' after adjusting the position;
10. The algorithm ends by traversing all the point cloud data in turn and completing the denoising.

4 Experimentation and Analysis of Multiscale Point Cloud Fusion Denoising Algorithm

Using the 3D scanning repository from Stanford University, the "Bunny" model serves as the experimental specimen. After introducing noise into the original point cloud, we applied the multi-scale point cloud fusion denoising algorithm, as shown in the results in Fig. 4. Figure 4a displays the noisy point cloud, while Fig. 4b shows

Fig. 4 Experimental results of scale point cloud fusion denoising algorithm figure: (**a**) point cloud with noise; (**b**) point cloud after removal of large-scale noise; (**c**) point cloud after removal of small-scale noise; (**d**) clear point cloud

the outcome after large-scale noise removal using the multi-algorithm fusion approach, which effectively reduces the noise. Finally, Fig. 4c presents the denoised result after applying small-scale noise removal through normal vector-corrected bilateral filtering, effectively eliminating the finer noise around the model.

Based on the experimental results, it can be inferred that after applying the multi-scale point cloud fusion denoising algorithm, the primary model's point cloud data in the experimental sample is well-preserved, with clear and smooth contours, essentially resembling the original model (Fig. 4d). Overall, this algorithm demonstrates effective noise reduction for point cloud data, establishing a solid data foundation for subsequent surface reconstruction.

5 Conclusion

This paper primarily investigates the sources, models, and classification of noise, proposing a multi-scale point cloud fusion denoising algorithm, and validates its effectiveness through experiments. Initially, an analysis of the sources of point cloud noise is conducted, along with the establishment of a data model for noise. Subsequently, spatial distance between noise and the main point cloud is utilized for classifying point cloud noise. Next, a multi-scale point cloud fusion denoising algorithm is introduced. This algorithm classifies noise based on its scale, removing large-scale noise through a multi-algorithm fusion-based method, while small-scale noise is removed using bilateral filtering with normal vector correction. Finally, the effectiveness of the multi-scale point cloud fusion denoising method is validated through experiments using the Bunny model.

Acknowledgements The Authors would like to thank the editors and the referees. The work of this article has been partially supported by the Open Project of Henan Key Laboratory of Intelligent Manufacturing of Mechanical Equipment, Zhengzhou University of Light Industry (No. IM202310), Scientific and technological breakthroughs in Henan Province (No. 232102221031) and Key Research Projects of Higher Education Institutions in Henan Province (No. 23B630024) for help identifying collaborators for this work.

References

1. Nan J, Liang S, Li H et al (2021) Comprehensive application of surveying and mapping geographic information technology in cultural relic protection division of Shaanxi province. Surv Map Bull 2021:145–150. https://doi.org/10.13474/j.cnki.11-2246.2021.357
2. Geng G, Feng L, Li K et al (2021) A literature review on the digitization and virtual restoration of cultural relics in the Mausoleum of Emperor Qinshihuang. J Northwest Univ 51:710–721. https://doi.org/10.16152/j.cnki.xdxbzr.2021-05-001
3. Du W, Li X, Jiang L et al (2022) Research on real-time scanning reconstruction control technology of 3D scene. J Phys Conf Ser 2220:012005. https://doi.org/10.1088/1742-6596/2220/1/012005

4. Zhu F, Liang J, Liang J et al (2024) Reconstruction of sharp features of triangular mesh in additive restoration. Acta Photon Sin 53(12):40–50. https://doi.org/10.3788/gzxb20245312.1210001
5. Li Z, Mao X, Wang J et al (2020) Feature-preserving triangular mesh surface denoising: a survey and prospective. J Comp Aided Des Graph 32:1–15. https://doi.org/10.3724/SP.J.1089.2020.17966
6. Zhang D, Gong W (2021) 3D mesh models denoising algorithms: a survey. Softw Guide 20:235–242. https://doi.org/10.11907/rjdk.201960
7. Ye T, Li Y, Wang Y et al (2016) On-line point cloud data extraction algorithm for spatial scanning measurement of irregular surface in copying manufacture. Int J Adv Manuf Technol 87:1891–1905. https://doi.org/10.1007/s00170-016-8592-6
8. Wang X, Wu L, Chen H (2020) Denoising of scattered point cloud data based on normal vector distance classification. J Jilin Univ 50:278–288. https://doi.org/10.13229/j.cnki.jdxbgxb20180948
9. Song K, Tang H (2024) Based on weighted FCM and method to the outlier factor managements cloud denoising. J Inf Technol Inf 9:50–54. https://doi.org/10.3969/j.issn.1672-9528.2024.09.0122024
10. Zhou Y, Chen R, Zhao Y et al (2021) Point cloud denoising using non-local collaborative projections. Pattern Recogn 120:108–128. https://doi.org/10.1016/J.PATCOG.2021.108128
11. Yan J, Zhou L, Wang J et al (2022) Structural feature-preserving point cloud denoising method for aero-engine profile. Int J Aerosp Eng 25:40–64. https://doi.org/10.1155/2022/9565062

Open Access This chapter is licensed under the terms of the Creative Commons Attribution-NonCommercial-NoDerivatives 4.0 International License (http://creativecommons.org/licenses/by-nc-nd/4.0/), which permits any noncommercial use, sharing, distribution and reproduction in any medium or format, as long as you give appropriate credit to the original author(s) and the source, provide a link to the Creative Commons license and indicate if you modified the licensed material. You do not have permission under this license to share adapted material derived from this chapter or parts of it.

The images or other third party material in this chapter are included in the chapter's Creative Commons license, unless indicated otherwise in a credit line to the material. If material is not included in the chapter's Creative Commons license and your intended use is not permitted by statutory regulation or exceeds the permitted use, you will need to obtain permission directly from the copyright holder.

Economic Analysis of Fuel Gas Supply System Based on Life Cycle Assessment

Min Lin, Jinjin Xu, Yuhang Wang, and Cheng Qian

Abstract Life cycle assessment is applied to achieve the economic performance analysis of low-pressure gas supply systems for dual fuel engines. Three mainstream pressure boosting methods for LNG supply in low-pressure fuel gas supply systems are analysed, which are external gas source boosting, self-pumping and pump boosting regulation. The system operation is simulated by modelling in Aspen HYSYS software. The system life cycle costs are compared, and the implementation effects of the three system options are analysed and evaluated. The total lifecycle cost of the three options is highest for the external gas source booster option and lowest for the self-pressurization option, and the self-pressurization option is the lowest in all stages of life cycle cost.

Keywords Fuel gas supply system · Life cycle assessment · Life cycle cost · Dual fuel engine

1 Introduction

Life cycle theory (including life cycle assessment and cost analysis) plays a significant role in evaluating the economic and environmental impacts of systems [1–4] and is widely used in the shipbuilding industry to assess the overall costs and environmental impacts of ship design solutions. Hadi Taghavifar [5] compared the environmental impacts of traditional diesel and LNG using the LCA method, finding that LNG substitution significantly reduces greenhouse gas emissions and improves fuel efficiency. Bugra Arda Zincir [6] employed the GREET Model to conduct a life cycle inventory analysis using an ocean tanker model to determine the environmental impacts of different marine fuel options. Hwang [7] compared LNG marine fuel with conventional marine gas oil (MGO) using the LCA method, revealing that

M. Lin · J. Xu (✉) · Y. Wang · C. Qian
School of Energy and Power, Jiangsu University of Science and Technology, Zhenjiang, China

© The Author(s) 2026

Z. Zhang, P. Mercorelli (eds.), *Mechanical Design and Simulation: Insights and Innovations*, Lecture Notes in Mechanical Engineering,
https://doi.org/10.1007/978-981-95-2405-1_84

LNG exhibits significantly lower emission levels across all potential impact categories. C. Wang [8] and colleagues conducted a comparative analysis of three BOG management solutions under different conditions based on life cycle cost, demonstrating that there is no universal solution applicable to all scenarios. Maydison [9] analysed the environmental impact of LNG dual-fuel retrofitting using the LCA method, finding that replacing 20%, 30%, and 40% of traditional marine diesel oil (MDO) with LNG under 85% power load for 35 days reduced global warming potential by 23.1%, 32.7%, and 42.3%, respectively.

The application of ship life cycle assessment (LCA) mainly focuses on the overall energy saving and emission reduction of ships, while life cycle cost (LCC) focuses on the study and analysis of economic benefits. However, in the low-pressure gas supply system of dual-fuel engine, little studies were performed the system cost and performance difference caused by different pressurization methods of LNG cryogenic liquid storage tanks based on life cycle cost. In this paper, three kinds of mainstream pressure increasing and regulating modes of gas supply system models are established, and the economic performance analysis method of gas supply system based on life cycle is evaluated.

2 Methodology

2.1 *Target Vessel Tape*

The 15,000 DWT chemical vessel under study has an overall length of 177 m, a beam of 22 m, a deadweight of 15,000 t, a service speed of 14.5 knots and a range of 6600 nautical miles. It is equipped with a WinGD 5RT-flex50DF dual-fuel engine, which has a designed fuel consumption of 850 kg/h at 100% load. The engine requires a gas supply pressure range of 5–16 bar, falling within the low-pressure gas supply category, and thus is fitted with a low-pressure gas supply system, as shown in Fig. 1. The system has a service life of 20 years and a maintenance interval of 5 years.

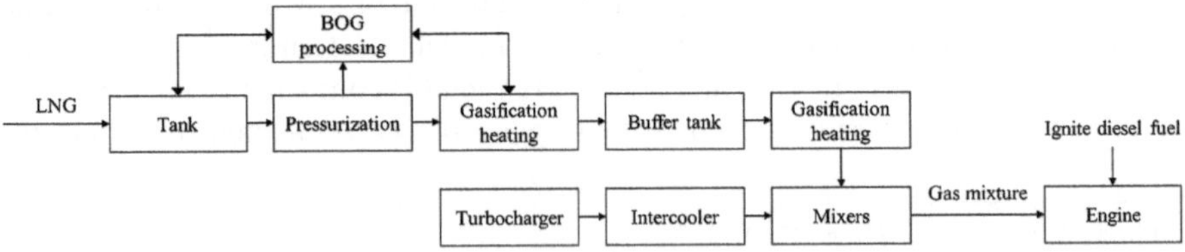

Fig. 1 Workflow of the low-pressure gas supply system

2.2 Research Method

LCC (life cycle cost) is used to analyze the sum of all costs incurred by a product or process throughout its life cycle. They are divided as shown in Fig. 2. For fuel gas supply systems, in addition to completing the initial work, equipment procurement, construction, installation and commissioning are also required, which together represent the initial investment costs. Many of the devices involved in a fuel gas supply system require external resources and consume electricity, which is part of the running costs. The maintenance of the equipment and the manpower consumed during the formal operation of the system in the maintenance management costs. At the end of the service life of the system, the equipment will be recycled and this represents the residual value. The following formula can be used in the calculation of the LCC [10].

$$\mathrm{LCC} = \mathrm{IC} + \sum_{t=1}^{n} \mathrm{OC}_i (1+i)^{-t} + \sum_{t'=1}^{m} \mathrm{MC}_i \left(\frac{1+r}{1+i}\right)^{t'} - S \tag{1}$$

where, IC is the initial investment cost of the system; t is the number of years of system operation; t′ is the number of maintenance of the equipment in the system; I is the discount rate, which is taken as 8% in this study; R is the inflation rate, which is taken as 2.8% in this study; S is the residual value of the system; OCi is the operating cost of the system in year t; MCi is the maintenance cost of the equipment in the system in year t.

Fig. 2 The stage division of LCC

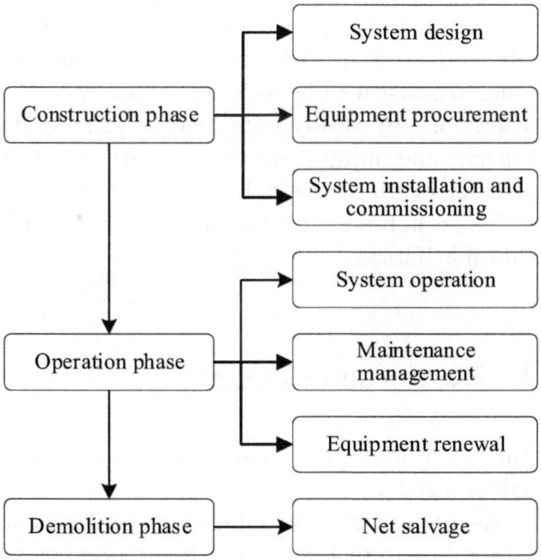

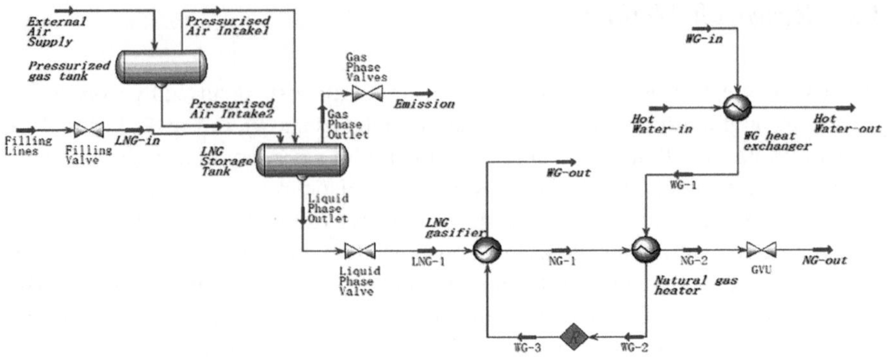

Fig. 3 Fuel gas supply system with external gas source pressurization

2.3 System Modelling

In low-pressure gas supply systems for dual-fuel ships, there is a certain pressure loss in the entire system piping and equipment, and the pressure of the vaporized natural gas (abbreviation NG) has to meet the conditions for low pressure LNG supply. Three common methods of pressurization for low pressure supply to dual-fuel vessels are external gas source pressurization, self-pressurization and pump pressurization. The above three booster and regulator methods for LNG supply are designed and modelled in Aspen HYSYS software, The system diagrams for the three system solutions are shown in Figs. 3, 4, and 5, and the deviations between the simulation results and the design results in the literature [11] are controlled within ±5%, which indicates that the design results are reasonable.

After completing the modeling, system LCC costs are compared, the implementation effects of the three system options are analyzed and evaluations are given. In order to assess the life cycle cost of the system, some technical information must be determined by steady-state thermodynamic modeling, including equipment size, capacity and utility requirements. The results of the design of common equipment parameters are shown in Table 1. The different equipment in the three system scenarios is high-pressure gas tank, booster gasifier and cryogenic pump, which are shown in Table 2.

3 Results and Discussion

The key equipment costs and full life cycle costs for the three systems are shown in Tables 3 and 4.

As can be seen from Table 4, the total life cycle cost of the three options is highest for the external gas source booster option and lowest for the self-pressurization option. Since the external gas source pressurization option requires an additional

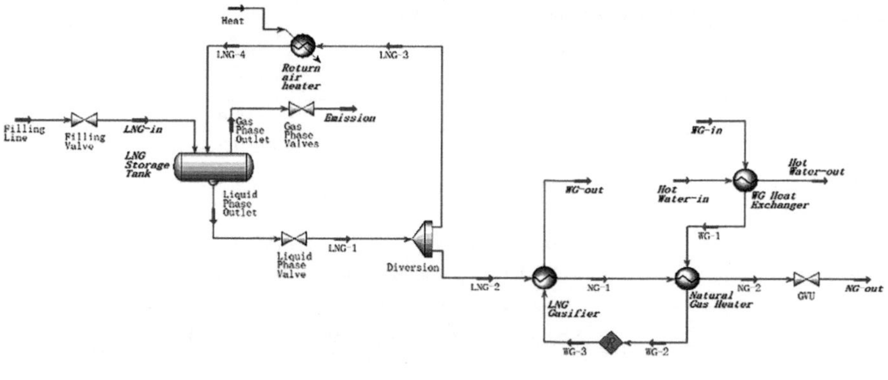

Fig. 4 Fuel gas supply system with self-pressurization

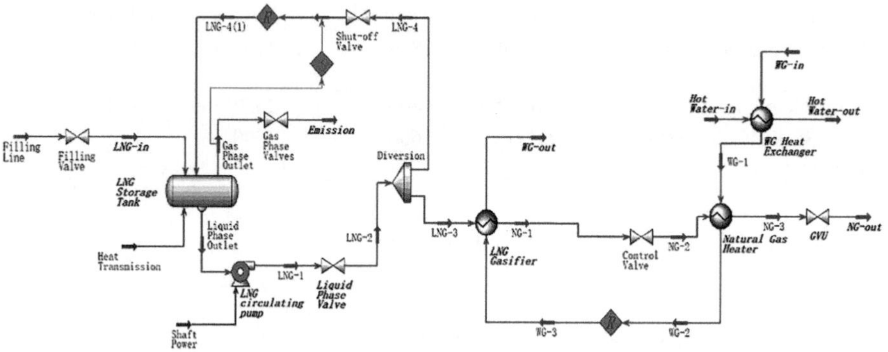

Fig. 5 Fuel gas supply system with cryogenic pump boost

Table 1 Common equipment parameters

Equipment	Type	Parameters
LNG storage tank	Horizontal oval head C-shaped double tank	Form effect: Φ5800 × 17,000 mm Thermal conductivity of tank material: 10 kcal/m.h.c Evaporation rate:0.2% (day)
LNG vaporizer	Shell and tube heat exchanger	Heat transfer coefficient: 216 W (m^2·°C) Total heat exchange: 132.7 kW
LNG heater	Shell and tube heat exchanger	Heat transfer coefficient: 213 W/(m^2·°C) Total heat exchange: 80.5 kW
Water-glycol heater	Plate heat exchanger	WG flow rate: 17056 kg/h UA: 4349 kJ/(k·h) Total heat exchange: 21.5 kW

Table 2 Different equipment parameter for three systems

Equipment	Parameter	Value
High pressure gas tank	Liquid volume	2.402 m^3
	Vessel diameter	0.914 m
	Design gauge pressure	651.046 kPa
Cryogenic pump	Liquid flow rate	2.834 m^3/h
	Fluid head	106.594 m
	Pump efficiency	70%
Self-pressurization	Tube design gauge pressure	760.546 kPa
	Tube outside diameter	0.025 m
	Shell design gauge pressure	473.254 kPa

Table 3 Life cycle cost of critical equipment for three system solutions (RMB)

System	Equipment cost	Installed cost	Equipment weight	Installed weight	Utility cost (RMB/h)
High pressure gas tank	114,800	999,600	18,200	116,700	0
Cryogenic pump	112,700	303,100	6400	20,300	0.41
Pressurized gasifier	5000	6000	1600	2300	1.52

Table 4 The calculation results of LCC of three options (RMB)

System		High pressure gas	Cryogenic pump	Self-pressurization
Construction phase	Equipment cost	408,200	423,700	291,900
	Installation cost	3,159,700	3,059,600	2,329,600
	Other costs	13,800	12,100	5,600
Operation phase	Material cost	941,005,000	843,564,000	716,677,000
	Energy consumption cost	1,520,000	1,550,600	1,520,000
	Maintenance management	241,200	236,100	231,100
Demolition phase	Net salvage value	16,300	18,100	11,700
Total life-cycle cost		946,364,200	848,864,200	721,066,900

high-pressure gas tank, it adds some costs in terms of tank maintenance and management as well as gas storage. For the cryogenic pump boost option, the industry is divided into a submersible pump inside the tank and a booster pump outside the tank, and the external tank is used in this paper, which has a relatively low maintenance cost. The self-pressurization option is the lowest in all stages of life cycle cost.

4 Conclusions

In this study, life cycle evaluation was applied to three booster options for fuel gas supply systems, and a fuel gas supply system model for three mainstream booster options is established to analyze the impact of their performance status on the economic performance of the system, and the following conclusions are achieved. From the perspective of system safety, the external gas source pressurization option needs to be configured with higher requirements for safety monitoring functions. From the system reliability, the cryogenic pump boost option uses an externally provided dynamic pressure head to ensure a stable LNG supply to the system. It is necessary to purchase additional cryogenic pumps, which will increase power consumption, and the difficulty of installation and maintenance is also high. From the system operation, the self-pressurization option is to start the gas return cycle at the same time as the system starts, so that the pressure in the LNG storage tank is stabilized near the working pressure to ensure the system continues to work normally. It is a more economical way of pressurization.

References

1. Roux M, Lodato C, Laurent A, Astrup TF (2024) A review of life cycle assessment studies of maritime fuels: critical insights, gaps, and recommendations. Sustain Prod Consump 50:69–86. https://doi.org/10.1016/j.spc.2024.07.016
2. Wang Y, Maidment H, Boccolini V, Wright L (2022) Life cycle assessment of alternative marine fuels for super yacht. Reg Stud Marine Sci 55:102525. https://doi.org/10.1016/j.rsma.2022.102525
3. Chalaris I, Jeong B, Jang H (2022) Application of parametric trend life cycle assessment for investigating the carbon footprint of ammonia as marine fuel. Int J Life Cycle Assess 27:1145–1163. https://doi.org/10.1007/s11367-022-02091-4
4. Jang H, Jeong B, Zhou P, Ha S, Nam D (2021) Demystifying the lifecycle environmental benefits and harms of LNG as marine fuel. Appl Energy 292:116869. https://doi.org/10.1016/j.apenergy.2021.116869
5. Taghavifar H, Perera LP (2023) Life cycle emission and cost assessment for LNG-retrofitted vessels: the risk and sensitivity analyses under fuel property and load variations. Ocean Eng 282:114940. https://doi.org/10.1016/j.oceaneng.2023.114940
6. Zincir BA, Arslanoglu Y (2023) Comparative life cycle assessment of alternative marine fuels. Fuel 358:129995. https://doi.org/10.1016/j.fuel.2023.129995
7. Hwang S, Jeong B, Jung K, Kim M, Zhou P (2019) Life cycle assessment of LNG fueled vessel in domestic services. J Marine Sci Eng 7(10):359. https://doi.org/10.3390/jmse7100359
8. Wang C, Ju Y, Fu Y (2021) Comparative life cycle cost analysis of low pressure fuel gas supply systems for LNG fueled ships. Energy 218. https://doi.org/10.1016/j.energy.2020.119541
9. Maydison N, Lim H, Heo J, Choe S, Kim J, Jang J, Oh D (2024) Sustainable retrofitting for shipping: assessing LNG dual fuel impact on global warming potential through life cycle assessment. Results Eng 23:102484. https://doi.org/10.1016/j.rineng.2024.102484

10. Gilbert P, Wilson P, Walsh C, Hodgson P (2017) The role of material efficiency to reduce CO2 emissions during ship manufacture: a life cycle approach. Marine Policy 75:227–237. https://doi.org/10.1016/j.marpol.2016.04.003
11. Zhao H (2024) Design and dynamic study of low-pressure gas supply system for marine dual fuel engine. Jiangsu University of Science and Technology, Zhenjiang

Open Access This chapter is licensed under the terms of the Creative Commons Attribution-NonCommercial-NoDerivatives 4.0 International License (http://creativecommons.org/licenses/by-nc-nd/4.0/), which permits any noncommercial use, sharing, distribution and reproduction in any medium or format, as long as you give appropriate credit to the original author(s) and the source, provide a link to the Creative Commons license and indicate if you modified the licensed material. You do not have permission under this license to share adapted material derived from this chapter or parts of it.

The images or other third party material in this chapter are included in the chapter's Creative Commons license, unless indicated otherwise in a credit line to the material. If material is not included in the chapter's Creative Commons license and your intended use is not permitted by statutory regulation or exceeds the permitted use, you will need to obtain permission directly from the copyright holder.

Constitutive Model Determination and Simulation of Carbon Fiber Fabric Laminated Composites

Zhilin Su and Tong Li

Abstract The present study focuses on investigating a carbon fiber fabric laminated composite material for shell structures. A constitutive model is established by integrating experimental methods, theoretical analysis, and finite element analysis. By conducting a series of fundamental mechanical tests on the material, data regarding its elastic modulus and strength in different directions are obtained. Additionally, a detailed microscopic analysis of the fracture surface is performed. Utilizing the Tsai-Wu criterion as a basis, a constitutive model for the material is developed and implemented into ABAQUS finite element software through a user-defined material subroutine (UMAT). Through comparison with experimental data, the material parameters are adjusted and optimized to obtain an improved parameter set that enhances simulation accuracy. This study not only enhances the precision of structural design analysis utilizing this specific material but also provides reliable support for designing shell structures and similar applications, thereby offering significant practical value. Consequently, this work contributes to promoting wider utilization of composite materials in engineering practice.

Keywords Carbon fiber fabric laminated composite · Constitutive model · Tsai-Wu criteria · Experimental test · Numerical simulation

1 Introduction

Composite materials are widely used in structures due to their excellent balance between mechanical properties and weight. Compared with thermosetting resins, thermoplastic materials have attracted increasing attention from the scientific

Z. Su · T. Li (✉)
Department of Engineering Mechanics, School of Mechanics and Aeronautics, Dalian University of Technology/National Key Laboratory of Structural Analysis and Optimization of Industrial equipment and CAE Software, Dalian, China
e-mail: suzhilin@mail.dlut.edu.cn; tong@dlut.edu.cn

© The Author(s) 2026

Z. Zhang, P. Mercorelli (eds.), *Mechanical Design and Simulation: Insights and Innovations*, Lecture Notes in Mechanical Engineering,
https://doi.org/10.1007/978-981-95-2405-1_85

community. This is because, compared with thermosetting matrices, thermoplastic materials offer stronger fatigue resistance, better impact damage resistance, and shorter processing times [1, 2]. Fabric composites are new-type composite materials that use thermosetting or thermoplastic materials as the matrix and fiber fabrics as the reinforcement. The properties of fabric composites are closely related to the material attributes of the matrix and the reinforcement. Compared with other materials, fabric composites possess excellent material properties such as lightweight and high strength, and thus are widely used in aerospace, military, machinery, medical devices, construction, and vehicles [3–5]. Carbon fibers have become an important reinforcement for fabric composites due to their high strength, high modulus, and corrosion resistance.

Carbon fiber fabric laminated composites are more commonly used in the form of laminates. The carbon fibers are arranged in a certain direction, and the layers are connected by the matrix. Since the aerospace environment imposes higher requirements on the materials of structural components, carbon fiber fabric laminated composites with excellent properties such as high strength, high modulus, and high-temperature resistance are increasingly used in the aerospace field. The usage rate of these composites in domestic fourth-generation aircraft reaches 27%, 42% in helicopters, and about 10% in the C919 large airliner. Meanwhile, composite materials are also used to manufacture missiles, satellites, and other spacecraft. For example, the JFRP warhead structural shell in the United States uses carbon fiber/epoxy composites, reducing the weight by 40%. China's "Shenzhou" spacecraft uses carbon/epoxy composites to reduce weight and cost [6]. Evidently, the wide application of fabric composites has become an inevitable trend. However, carbon fiber fabric laminated composites exhibit characteristics such as nonlinearity and anisotropy. The complexity of their mechanical deformation and performance prediction makes the analysis and simulation calculation of the composite material constitutive model extremely difficult. Therefore, establishing an effective constitutive model and simulation analysis method for carbon fiber fabric laminated composites to provide reliable theoretical basis and numerical analysis is of great significance.

Scholars at home and abroad have conducted in-depth research on the mechanical models and simulation of fiber composites and achieved certain results. Peng et al. [7] established a non-orthogonal constitutive model for the material characterization of woven composites within the framework of continuum mechanics. Through the stress-strain transformation relationship between the global orthogonal coordinate system and the local non-orthogonal coordinate system, they derived the non-orthogonal constitutive relationship of woven composites to track the fiber orientation changes during the deformation process. Kang et al. [8] studied the homogenization of the stiffness of two-dimensional triaxial woven composites. They derived the equivalent elastic constants of the yarns by equating the total strain energy of the yarns in the local and global coordinate systems. Valkova et al. [9] performed a finite element simulation of the mechanical response of two-dimensional woven composites under compaction loads, and the results were in good agreement with the experimental measurements. Döbrich et al. [10] used a unit-cell model to simulate the mechanical behavior of fabric composites, explored the material properties from the perspective of the meso-structure, and obtained various elastic moduli. Jiang et al. [11] proposed a bimodal constitutive model based on the tensile and compressive modulus characteristics of two-axis 2 × 2 fabric composites, concluding that the bimodal model can better describe the stress-strain relationship of composites. Zheng et al. [12] studied the influence of interface characteristics on the compressive failure behavior of three-dimensional woven composites based on the multi-scale damage model (MMDM) of micromechanics. By defining the stress amplification factor, they established the relationship between the yarn meso-stress and the component meso-stress. Li et al. [13] proposed the projective element method (PEM). The PEM established a representative numerical model of woven composites based on the projection lines uniformly distributed in the out-of-plane direction of the woven composite structure. This model simplifies and accurately simulates two-dimensional woven composites. He et al. [14] developed a parallel multi-scale virtual testing system for woven structures based on the FE-SCA framework. They used the SCA method to accelerate the solution process of the woven representative volume element (RVE) and employed the FE-SCA parallel simulation method to predict the multi-scale progressive damage behavior of open-hole plates, Iosipescu shear, and biaxial tensile specimens. The research on fiber composites by scholars has gradually evolved from experimental methods to theoretical analysis and finite element analysis methods. However, due to the more complex structure of fiber fabric composites compared to general composites and the numerous factors affecting the overall mechanical properties of the materials, it is still difficult to accurately predict their mechanical properties. In this paper, a constitutive model for fiber fabric laminated composites will be established based on the Tsai-Wu criterion, and a subroutine will be written to simulate macroscopic specimens. The accuracy of the simulation will be verified through the results of transverse and longitudinal uniaxial tensile tests, uniaxial compression tests, and shear tests.

2 Experimental Methods and Results

This paper conducts research on a laminated composite material composed of T700 carbon fiber fabric and epoxy resin. Carbon fibers are mainly obtained from polyacrylonitrile or pitch through processes such as heating oxidation, carbonization, and graphitization. Epoxy resin refers to polymers containing two or more epoxy functional groups in their structure [15], which feature strong mechanical properties, high adhesiveness, low curing shrinkage, and high heat resistance [16]. This composite material is formed through a hot-pressing process after lamination. In order to comprehensively evaluate the overall stiffness performance of the composite material and obtain the strength parameters of the laminated composite, this study combines the classical laminated plate theory and the Tsai-Wu strength criterion to determine the strength factor. Based on this, macroscopic mechanical property tests are carried out. The tests include longitudinal and transverse tensile and compression tests. These tests not only provide important data on the mechanical behavior of the material in different directions but also lay the foundation for further understanding and designing such composite materials.

2.1 Experimental Methods

To evaluate the tensile properties of the material in different directions, this study conducted longitudinal and transverse tensile tests in accordance with GB/T 3354-2014 "Test method for tensile properties of oriented fiber-reinforced polymer matrix composites". According to the requirements of this standard, the structural dimensions of the longitudinal and transverse tensile specimens are shown in Fig. 1. It is worth noting that during the manufacturing process of the longitudinal tensile specimens, additional reinforcing tabs made of 6061 aluminum alloy are required to ensure the integrity of the specimens.

Two groups of specimens were used in the test, with each group containing five independent samples, making a total of ten specimens. All specimens were precisely machined according to the predetermined dimensions and requirements to ensure the consistency and reproducibility of the tests, as shown in Fig. 2. The tensile test in each direction was conducted five times, resulting in a total of ten tensile tests.

The specimens for the longitudinal and transverse compression tests were processed in accordance with the standard GB/T 3856-2005 "Test method for compressive properties of unidirectional fiber-reinforced plastic flat plates". Among them, reinforcing tabs need to be pasted on the longitudinal compression specimens, as shown in Fig. 3. Two groups of specimens were used in the test, with each group consisting of five independent samples, making a total of ten specimens in all, as shown in Fig. 4. The compression test in each direction was carried out five times respectively, resulting in a total of ten compression tests.

All the tensile and compression tests were carried out using a Jinan Shenggong electronic universal testing machine (with a capacity of 20 tons). This testing machine is equipped with a high-precision force and displacement measurement system, which can provide accurate test data.

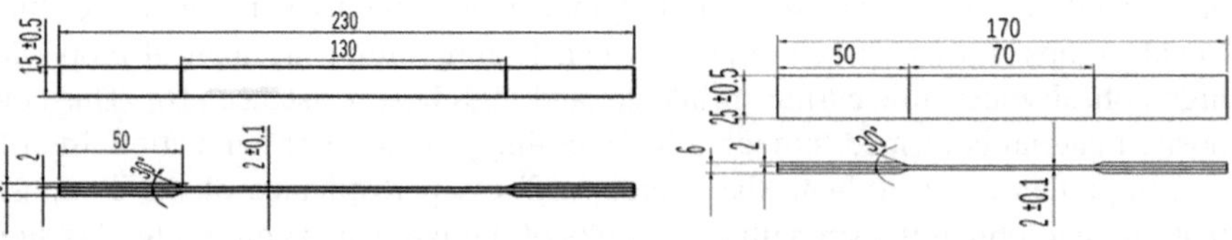

Fig. 1 The dimensional standards for the specimens of longitudinal tension (the left figure) and transverse tension (the right figure)

Fig. 2 The actual specimens of longitudinal tension (the upper picture) and transverse tension (the lower picture)

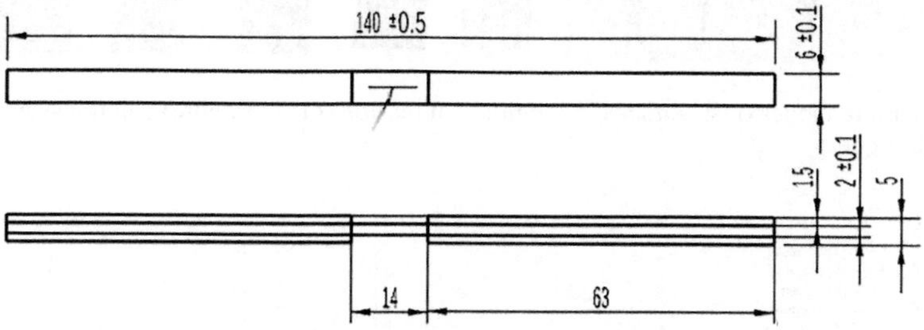

Fig. 3 The dimensional standards for the specimens of longitudinal compression (the upper picture) and transverse compression (the lower picture)

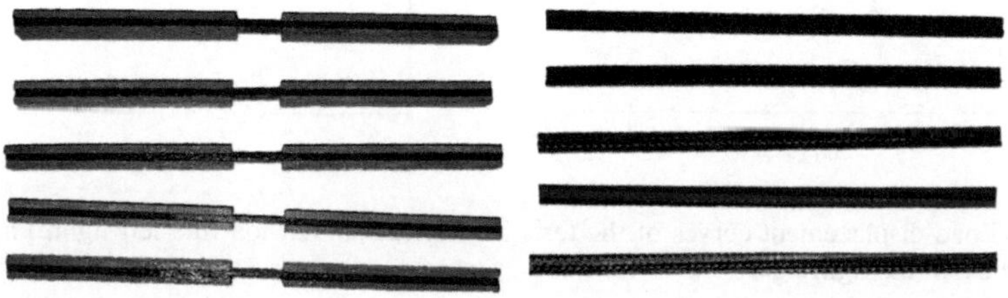

Fig. 4 The actual specimens of longitudinal compression (the left picture) and transverse compression (the right picture)

2.2 Experimental Results

All the specimens were tested in the Key Laboratory of Industrial Equipment at Dalian University of Technology, and the test results were statistically analyzed.

As shown in Fig. 5, the failure mode of all longitudinal tensile specimens was delamination at the middle edge of the working section. After reaching a certain strain level, the fiber bundles of the specimens suddenly delaminated, accompanied by the fracture of some fiber bundles, resulting in a sharp decline in the load-bearing capacity of the specimens. Eventually, the specimens fractured and lost their load-bearing capacity. The results of the transverse tensile test showed that the fracture

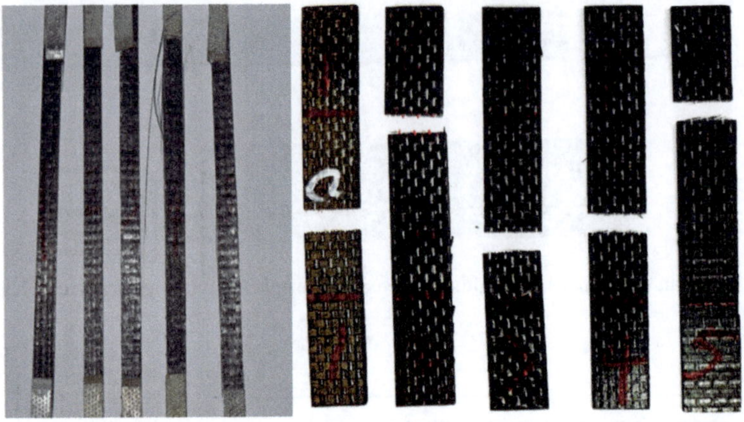

Fig. 5 Failure modes of specimens in longitudinal tension (the left figure) and transverse tension (the right figure)

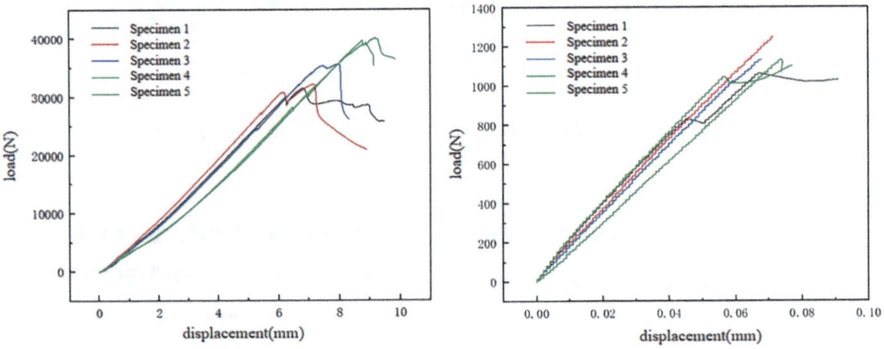

Fig. 6 Load-displacement curves of the tests in longitudinal tension (the left figure) and transverse tension (the right figure)

surface of the specimens was relatively regular and perpendicular to the loading direction. Macroscopic observation indicated that there was no obvious plastic deformation at the fracture surface, the crack propagation path was relatively straight, and the cracking of fiber bundles could be observed on some fracture surfaces.

In the longitudinal tensile test, the maximum loads borne by the five longitudinal tensile specimens at the time of failure were 31476 N, 32144 N, 35620 N, 40036 N, and 39026 N respectively, with an average value of 35,660 N. The load displacement curves of longitudinal tensile test and transverse tensile test are shown in Fig. 6. Based on this, the calculated longitudinal tensile strength was 1768.85 MPa, and the longitudinal tensile elastic modulus was 143.52 GPa.

In the transverse tensile test, the maximum loads borne by the five transverse specimens at the time of failure were 1602 N, 1784 N, 1856 N, 1618 N, and 1688 N

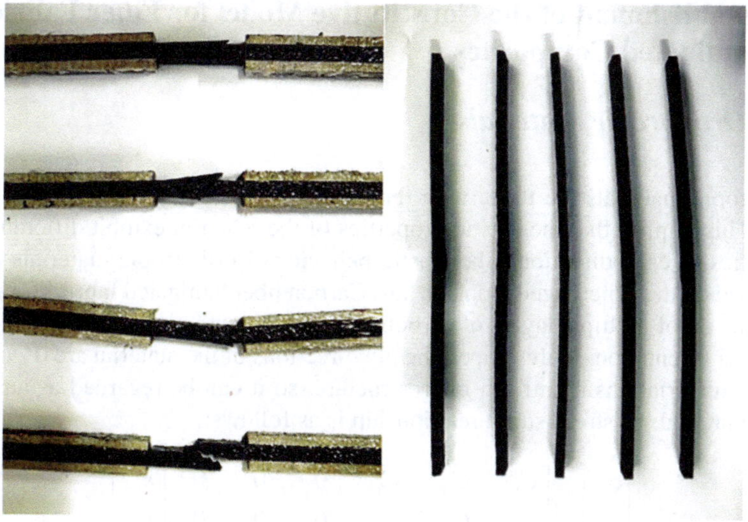

Fig. 7 Failure modes of specimens in longitudinal compression (the left figure) and transverse compression (the right figure)

respectively, with an average value of 1709 N. Based on this, the calculated transverse tensile strength was 39.89 MPa, and the transverse tensile elastic modulus was 6.71 GPa.

As shown in Fig. 7, the failure modes of the longitudinal compression specimens are plastic bending and inter-laminar shear failure, and the fracture surfaces are stepped. Under the longitudinal compression load, micro-cracks are generated due to minor initial defects or non uniformities. As the load increases, these micro-cracks gradually expand and multiply. Eventually, part of the main load-bearing fiber bundles becomes unstable, and at the same time, the micro-cracks merge into macroscopic cracks, leading to the failure of the overall structure. The failure modes of the transverse compression specimens are plastic bending and structural buckling. Under the transverse compression load, it is different from the instability of fiber bundles under the longitudinal load. Therefore, the macroscopic behavior of the material is mainly determined by the matrix and the interface between the fibers and the matrix. The specimens show obvious buckling, and the overall structure loses its load-bearing capacity.

The failure loads of the five longitudinal compression specimens are 13198 N, 12280 N, 12806 N, 12536 N, and 12578 N respectively, with an average value of 12,679.6 N. The longitudinal compression strength can be calculated to be 670.52 MPa. The failure loads of the five transverse compression specimens are 1782 N, 1550 N, 1640 N, 1960 N, and 1806 N respectively, with an average value of 1747.6 N. The transverse compression strength is 92.42 MPa.

3 Establishment of the Constitutive Model for Fiber Fabric Laminated Composites

3.1 Orthotropic Materials

Orthotropic materials are those with three mutually orthogonal anisotropic directions. This implies that the elastic properties of the material exhibit different characteristics in each direction. The elastic behavior of orthotropic materials can be described by the generalized Hooke's law. Carbon fiber laminated fabric composites are made up of multiple layers of carbon fiber fabric, with each layer having a specific fiber orientation. In this paper, the ply directions of the material are 0° and 90°, and the material has a uniform microstructure, so it can be regarded as an orthotropic material. Its stress-strain relationship is as follows:

$$\begin{Bmatrix} \sigma_1 \\ \sigma_1 \\ \sigma_1 \\ \tau_{23} \\ \tau_{13} \\ \tau_{12} \end{Bmatrix} = \begin{bmatrix} C_{11} & C_{12} & C_{13} & 0 & 0 & 0 \\ & C_{22} & C_{23} & 0 & 0 & 0 \\ & & C_{33} & 0 & 0 & 0 \\ & & & C_{44} & 0 & 0 \\ & SYM & & & C_{55} & 0 \\ & & & & & C_{66} \end{bmatrix} \begin{bmatrix} \varepsilon_1 \\ \varepsilon_2 \\ \varepsilon_3 \\ \gamma_{23} \\ \gamma_{31} \\ \gamma_{12} \end{bmatrix} \tag{1}$$

Among them, C_{ij} are the elastic constants, representing the strain in the j-direction caused by the stress in the i-direction. The elements on the diagonal represent the normal elastic moduli, that is, the degree of expansion or contraction of the material in the direction of the applied stress. The off-diagonal elements represent the shear elastic moduli, that is, the degree of deformation of the material when subjected to shear stress.

3.2 Tsai-Wu Criterion

In 1971, Tsai and Wu proposed a general theory in order to verify the accuracy and practicality of the strength criterion [17]. The Tsai-Wu criterion is an effective and general failure criterion, which is suitable for predicting the failure of orthotropic materials under multiaxial stress states. It has been widely used in fields such as aerospace, automotive, and wind energy. It takes into account the longitudinal and transverse tensile and compressive strengths of the material, as well as the shear strength. The basic assumption of its strength criterion is that there is a failure surface in the stress space, and its expression form is as follows:

$$F_i\sigma_i + F_{ij}\sigma_i\sigma_j = 1 \tag{2}$$

In the above equation, F_i andF_{ij} represent the second-order and fourth-order strength parameters respectively. Under the plane stress state, the expression for an orthotropic single-layer plate is as follows:

$$F_1\sigma_1 + F_2\sigma_2 + 2F_{12}\sigma_1\sigma_2 + F_{11}\sigma_1^2 + F_{22}\sigma_2^2 + F_{66}\sigma_6^2 = 1 \tag{3}$$

Among them, $F_1 = \frac{1}{X_t} - \frac{1}{X_c}$, $F_2 = \frac{1}{Y_t} - \frac{1}{Y_c}$, $F_{11} = \frac{1}{X_t X_c}$, $F_{22} = \frac{1}{Y_t Y_c}$, $F_{66} = \frac{1}{S_{12}^2}$, X_t, X_c, Y_t, Y_c, S represent the longitudinal tensile strength, longitudinal compressive strength, transverse tensile strength, transverse compressive strength, and in-plane shear strength respectively, F_{12} which can be determined through biaxial tests. Then, $\sigma_1 = \sigma_2 = \sigma$, $\sigma = 0$, we can immediately obtain

$$F_{12} = \frac{1}{2\sigma^2}\left[1 - \sigma\left(\frac{X_c - X_t}{X_t X_c} + \frac{Y_c - Y_t}{Y_t Y_c}\right) - \sigma^2 \frac{X_t X_c + Y_t Y_c}{X_t X_c Y_t Y_c}\right] \tag{4}$$

The stress transformation formula is as follows:

$$\begin{bmatrix} \sigma_1 \\ \sigma_2 \\ \tau_{12} \end{bmatrix} = \begin{bmatrix} m^2 & n^2 & 2mn \\ n^2 & m^2 & -2mn \\ -mn & mn & m^2 - n^2 \end{bmatrix} \begin{bmatrix} \sigma_x \\ \sigma_y \\ \tau_{xy} \end{bmatrix} \tag{5}$$

Among them, $m = \cos\theta$, $n = \sin\theta$, θ represent the off-axis angle. By substituting the above formula into the criterion expression, the strength under the off-axis condition can be calculated. The Tsai-Wu failure criterion takes into account the interaction between stresses and has wide applicability. In this paper, this criterion will be used as the judgment basis.

3.3 *Finite Element Simulation*

To conduct a preliminary simulation of the constitutive model for this carbon fiber fabric laminated composite, a UMAT subroutine was written. The constitutive equations and the Tsai-Wu criterion were embedded in the UMAT subroutine. To verify the accuracy of the model, numerical simulations of the basic tests were carried out using the commercial finite-element software Abaqus, and the simulation results were compared with the experimental data. By adjusting the model parameters, a constitutive model that can accurately describe the mechanical behavior of the material was finally established. This will provide a reliable simulation tool for the overall structural analysis.

In this paper, all models use the solid element C3D8. The constraint method is the same as that in the experiment, as shown in Fig. 8. The analysis is carried out in Abaqus/Standard, and the element deletion function is enabled during the analysis

Fig. 8 Finite element model of test piece (from top to bottom: longitudinal tensile, transverse tensile, longitudinal compression and transverse compression model)

process. The simulation results are shown in Fig. 9, which presents the stress state at the maximum bearing capacity.

The reaction forces at the maximum bearing capacity are extracted from the simulation results, and the results are shown in Table 1. The maximum error is approximately 5%.

4 Conclusions

This paper focuses on a carbon fiber fabric laminated composite used in cylindrical shell structures. Starting from experimental methods and combining theoretical analysis and finite element analysis, the aim is to establish a constitutive model for this material. Through a series of basic mechanical tests on the material, we obtained the elastic modulus and strength data in different directions, and conducted a detailed analysis of the fracture morphology of the specimens. Based on the Tsai-Wu criterion, we constructed a constitutive model for the carbon fiber fabric laminated composite and wrote a user-defined material subroutine (UMAT) to accurately simulate the material behavior in the ABAQUS finite element software. By comparing the simulation results with the experimental data, we adjusted and optimized the material parameters, thus obtaining a more accurate set of material parameters. The research results significantly improve the analysis accuracy when using this material for structural design, and have important practical significance for the efficient analysis and design of relevant structures. The proposed constitutive model and parameters can provide reliable support for the design of cylindrical shell structures and other similar structures, which is conducive to promoting the wide application of composite materials in engineering practice.

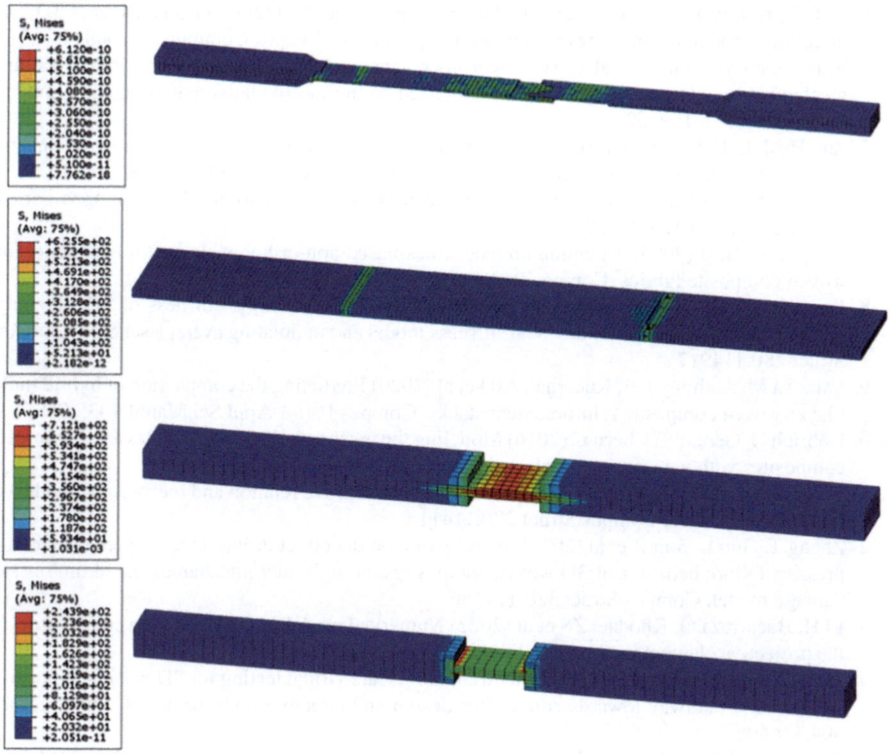

Fig. 9 Finite element simulation stress cloud map (from top to bottom: longitudinal tensile, transverse tensile, longitudinal compression and transverse compression models)

Table 1 Comparison of maximum bearing capacity between test and simulation results

	Longitudinal tension	Transverse tension	Longitudinal compression	Transverse compression
Average test value (N)	35,660	1709	12,680	1478
Finite element simulation value (N)	35,500	1798	12,152	1465
Error (%)	0.4	5.2	4	0.8

References

1. Murray RE, Beach R, Barnes D et al (2021) Structural validation of a thermoplastic composite wind turbine blade with comparison to a thermoset composite blade. Renew Energy 164:1100–1107
2. Mohsin MA, Iannucci L, Greenhalgh ES (2021) Experimental and numerical analysis of low-velocity impact of carbon fibre-based non-crimp fabric reinforced thermoplastic composites. Polymers 13(21). https://doi.org/10.3390/polym13213642

3. Li D-S, Han W-F, Jiang L (2021) On the tensile properties and failure mechanisms of 3D six-directional braided composites at elevated temperatures. Compos Commun 28:100884
4. Wu Y, Cao W, Guo J et al (2021) Near-fiber nanomechanical mapping and impact failure mechanism of 3D braided composites subjected to thermo-oxidative environment. Compos Sci Technol 216:109052
5. Zuo H-M, Li D-S, Jiang L (2022) Transverse bending fatigue behaviors and failure mechanisms of 3D five-directional braided composites at different temperatures. Mater Lett 316:132030
6. Newcomb BA (2016) Processing, structure, and properties of carbon fibers. Compos Part A Appl Sci Manuf 91:262–282
7. Peng XQ, Cao J (2005) A continuum mechanics-based non-orthogonal constitutive model for woven composite fabrics. Compos Part A Appl Sci Manuf 36(6):859–874
8. Kang Q, Jin S, Ye W (2022) Homogenization of the macroscopic stiffness of 2D triaxially braided composites: roles of the local stiffness model and undulating average scheme. Compos Struct 280:114917
9. Valkova M, Anthony DB, Kucernak ARJ et al (2020) Predicting the compaction of hybrid multilayer woven composite reinforcement stacks. Compos Part A Appl Sci Manuf 133:105851
10. Döbrich O, Gereke T, Cherif C (2016) Modeling the mechanical properties of textile-reinforced composites with a near micro-scale approach. Compos Struct 135:1–7
11. Jiang L, Xiao S, Yang B et al (2021) Bimodulus constitutive relation and mesoscopic model of braided composites. Compos Struct 270:114115
12. Zheng T, Guo L, Sun R et al (2023) Investigation on the effect of interface properties on compressive failure behavior of 3D woven composites through micromechanics-based multiscale damage model. Compos Struct 320:117186
13. Li H, Bacarreza O, Khodaei ZS et al (2022) Numerical modelling of 2D woven composites by the projective element method. Int J Solids Struct 254:111946
14. He C, Ge J, Chen Y et al (2023) Concurrent multiscale virtual testing for 2D woven composite structures: a pathway towards composites design and structure optimization. Compos Struct 304:116406
15. Tripathy R, Ojha U, Faust R (2011) Polyisobutylene modified bisphenol a diglycidyl ether based epoxy resins possessing improved mechanical properties. Macromolecules 44(17):6800–6809
16. Love AEH (1888) The small free vibrations and deformation of a thin elastic shell. Philos Trans R Soc A 179:491–546
17. Tsai SW, Wu EM (1971) A general theory of strength for anisotropic materials. J Compos Mater 5(1):58–80

Open Access This chapter is licensed under the terms of the Creative Commons Attribution-NonCommercial-NoDerivatives 4.0 International License (http://creativecommons.org/licenses/by-nc-nd/4.0/), which permits any noncommercial use, sharing, distribution and reproduction in any medium or format, as long as you give appropriate credit to the original author(s) and the source, provide a link to the Creative Commons license and indicate if you modified the licensed material. You do not have permission under this license to share adapted material derived from this chapter or parts of it.

The images or other third party material in this chapter are included in the chapter's Creative Commons license, unless indicated otherwise in a credit line to the material. If material is not included in the chapter's Creative Commons license and your intended use is not permitted by statutory regulation or exceeds the permitted use, you will need to obtain permission directly from the copyright holder.

The Damage Behavior of Composite Laminated Plates with Holes Intelligent Prediction

Jiachen Yin, Yuhan Sun, and Hongfei Ye

Abstract Under extreme service conditions such as high temperatures and pressures, composite laminate plates can experience internal damage, including matrix cracking, fiber fracture, and interface debonding. These subtle internal damages may reduce the residual strength of the laminate and even compromise its structural integrity, posing potential safety hazards. This paper adopts the Ti and Yuan (Sci Technol Wind 35:183–184, 2020) failure criterion as a progressive damage criterion for composite laminates and conducts a numerical analysis of their progressive damage (Liu et al., Appl Math Mech 45(9):1557–1572, 2024). The damage initiation and propagation of composite laminates with circular holes are thoroughly discussed. This is then extended to study laminates with elliptical holes. A parametric model for elliptical hole laminates is developed, and the automation of post-processing results is implemented. By extracting model and simulation data, a simulation result database is established, enabling the prediction of damage in composite laminates with elliptical holes.

Keywords Composite laminate · Damage prediction · Numerical analysis · Data-driven · Elliptical holes

---

J. Yin · Y. Sun
School of Mechanics and Aerospace Engineering, Dalian University of Technology, Dalian, China

H. Ye (✉)
School of Mechanics and Aerospace Engineering, Dalian University of Technology, Dalian, China

Department of Engineering Mechanics, National Key Laboratory of Industrial Equipment Structure Analysis, Optimization, and CAE Software, Dalian University of Technology, Dalian, China
e-mail: yehf@dlut.edu.cn

© The Author(s) 2026

Z. Zhang, P. Mercorelli (eds.), *Mechanical Design and Simulation: Insights and Innovations*, Lecture Notes in Mechanical Engineering,
https://doi.org/10.1007/978-981-95-2405-1_86

1 Introduction

In recent years, data-driven approaches [1] to progressive damage analysis of composite materials have emerged as a promising research direction for predicting material failure. This approach utilizes machine learning techniques to enhance the prediction of damage behaviors in composite structures. Data is initially extracted from experimental or simulation results, and machine learning algorithms are applied to learn the material's damage characteristics, ultimately achieving more accurate and reliable damage predictions. Data-driven prediction methods based on deep learning have become the mainstream approach for forecasting the damage behavior of composite materials. With the rapid advancement of computer science, machine learning and data science have provided strong support for material design and the prediction of mechanical behaviors in composite laminates. In recent years, much research has focused on combining finite element analysis with machine learning techniques to predict the relationship between composite material structures and their mechanical properties.

For instance, Tao et al. from Nanjing University of Aeronautics and Astronautics used an Artificial Neural Network (ANN) framework to study the stiffness degradation of composite laminates under cyclic loading, as shown in Fig. 1. They first used a β-variational autoencoder to extract latent features representing the underlying driving mechanisms. Then, they applied a neural ordinary differential equation to model the dynamics of these latent features, enabling extrapolation to unseen data and continuous prediction of stiffness in the cyclic domain. Compared to traditional mechanical and phenomenological models, the ANN-based model exhibited superior performance. Furthermore, they proposed a latent S-N curve based on latent variable analysis, which demonstrated a higher correlation with experimental data than the conventional S-N curve.

This paper, through secondary development of finite element [3] software, parametrically models and numerically analyzes a specific composite laminated plate

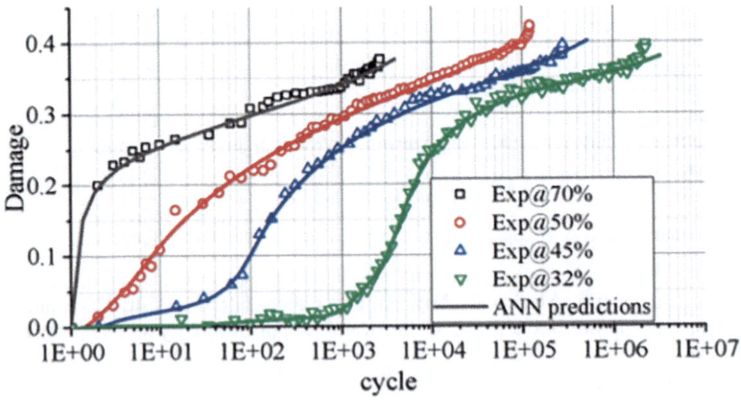

Fig. 1 ANN predictions vs. experimental data [2]

model with variables such as stacking angles from 0° to 90° and end-load directions. The damage of the hole and the strain field at twice the radius from the hole's edge are obtained, and data is extracted to build a database. Finally, the Keras [4] library is used to construct a BP neural network for training the data and establishing the model. The model's predictive accuracy is validated using a validation set, achieving the goal of predicting the hole damage condition with high confidence based on the nearby strain field situation. The prediction flowchart is shown in Fig. 2.

2 Numerical Analysis and Database Establishment

2.1 Parametric Modeling

A uniaxial tension finite element model was developed in ABAQUS, with a length of 250 mm and a width of 25 mm. The boundary conditions were set with one side fixed and a triangular load applied to the other side to prevent symmetrical node data around the hole. The material parameters used are shown in Table 1, and the model consisted of 20 layers with the same orientation, with stacking angles ranging from 0°to 90°.A structured mesh was used around the hole to minimize the stress concentration effect. The region around the hole was divided into eight equal parts to facilitate the extraction of stress field data. Python scripts were written to perform parametric modeling, mesh generation, computational submission, and

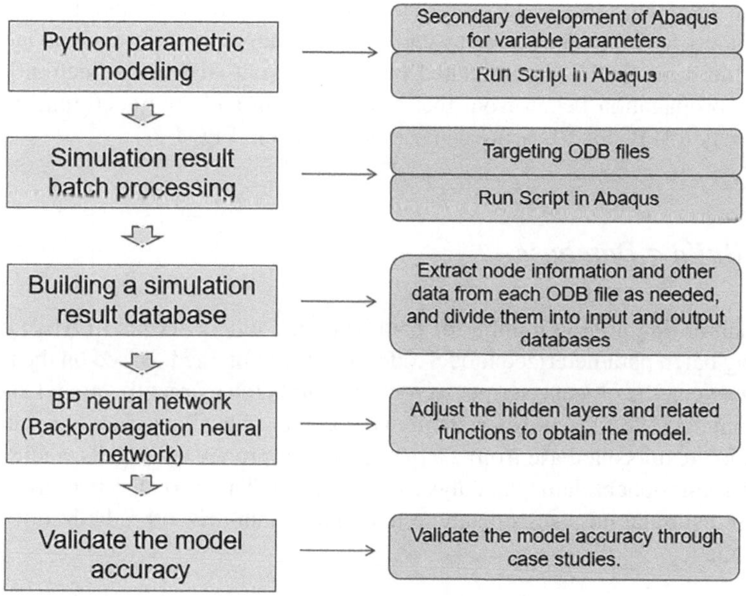

Fig. 2 The prediction flowchart

Table 1 Constitutive parameters of composite laminates

Parameter type	Parameter value
Elastic modulus (MPa)	E_1= 175,000 E_2 = 7400 ν_{12} = 0.37
	G_{12} = G_{13} = G_{23} = 3760
Strength parameters (MPa)	X_T = 44.3 X_C = 2695
	Y_T = 143 Y_C = 834 S_{XY} = 65.8
Fracture energy (N/mm)	G_{ft} = 133 G_{fc} = 40 G_{mt} = 0.6 G_{mc} = 2.1

Fig. 3 Parametrically generated automatic model

automated processing of simulation results for composite laminates with elliptical holes. The parametric model is shown in Fig. 3.

2.2 *Study of Damage Behavior and Damage Cloud Map Plotting*

Based on the damage results obtained from the finite element model, it was found that the first type of damage to occur in the laminated plate was matrix damage. As the load gradually increased, the elements at the damage initiation site did not immediately lose their load-bearing capacity. Instead, as the external load increased, the stiffness matrix of the laminated plate underwent stiffness reduction [5]. The damage propagation began from the initiation point and continued until the plate completely lost its load-bearing capacity, as shown in Fig. 4.

2.3 *Build a Database*

After submitting a large number of parameterized model scripts to ABAQUS, the resulting batch parameterized model .odb files were obtained. Based on the relationships between the objects, scripts were written to batch extract the model and result data from the .odb files and save them in a text file. This allows for the creation of a simulation results database from a large number of processed models, achieving the goal of a component damage analysis database [6]. The next step is to use the pre-processed simulation results database to conduct data-driven prediction of circular hole damage.

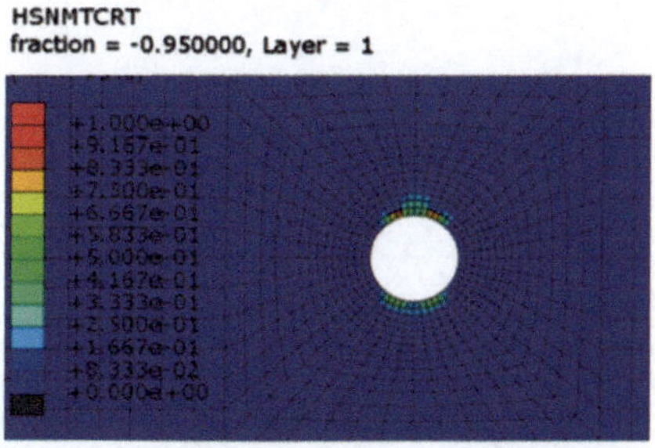

(a) Damage initiation contour plot

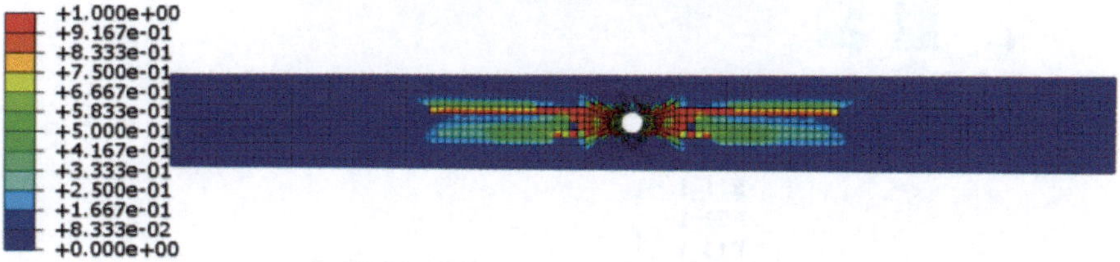

(b) Matrix compression damage contour plot

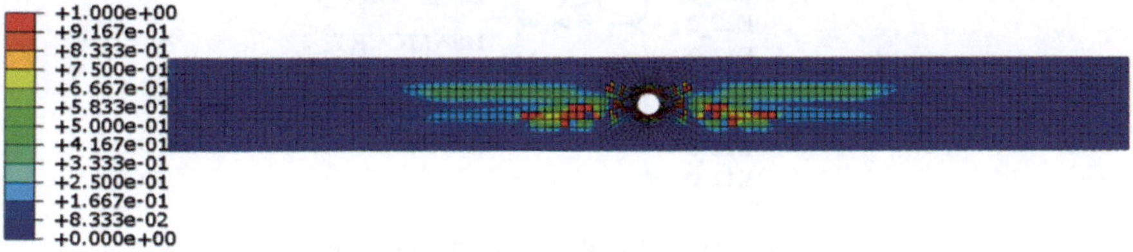

(c) Matrix tension damage contour plot

Fig. 4. Damage contour plots at initiation and failure. (**a**) Damage initiation contour plot. (**b**) Matrix compression damage contour plot. (**c**) Matrix tension damage contour plot

3 Intelligent Prediction of Damage Behavior in Composite Laminated Plates with Holes

3.1 Establishment of the Neural Network Model Framework

Based on the Deep Back-Propagation (BP) neural network algorithm, model and result datasets provided by batch processing result files are used to implement a data-driven structural performance prediction technique. To achieve the goal of predicting the corresponding hole-edge node damage [7] information based on the strain field at twice the radius outside the hole, node information such as E_{11}, E_{12}, E_{22}, and node coordinates around the hole, as well as damage information at twice the radius outside the hole, are sequentially extracted from the .odb result files into a CSV file. To ensure a sufficiently high prediction accuracy, this study uses the strain field of three nodes at the outer edge of the hole to jointly predict the damage

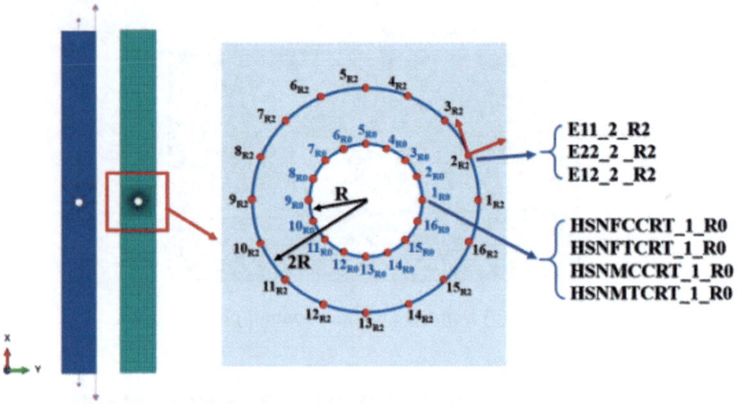

(a) Selection of model and input (R2), output (R0)

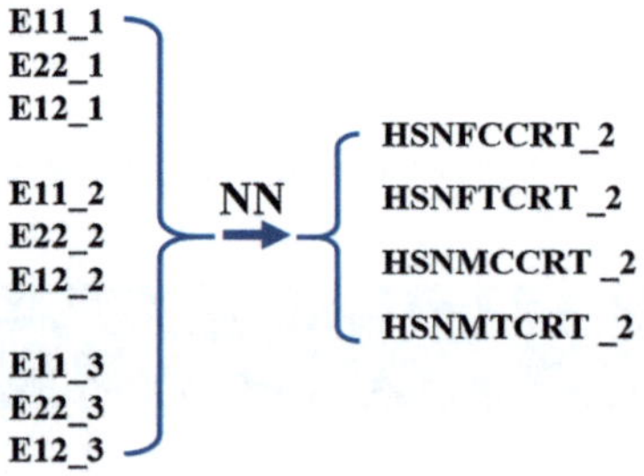

(b) Input and output of the neural network

Fig. 5 Correspondence between the input and output of the neural network. (**a**) Selection of model and input (R2), output (R0). (**b**). Input and output of the neural network

information of a corresponding node on the hole edge, as shown in Fig. 5. A total of approximately 4800 data sets were obtained. The next step is to begin training the model.

3.2 *Adjusting the Activation Function*

Then, the required libraries such as Keras are imported into the Jupyter Notebook terminal. To ensure the reproducibility of random operations, Python is used, and the database is randomly divided into two parts, with 80% of the data used for training the model and 20% used to validate the accuracy of the established model. Since higher damage values indicate more severe damage, which requires more accurate model predictions, the loss function used is the weighted mean squared error as shown in Eq. (1). This is used to ensure that damage values with larger target values have a greater impact on the result, in addition to the traditional mean squared error.

$$\text{loss} = e^{4*y_{\text{true}}} \times \left(y_{\text{pred}} - y_{\text{true}}\right)^2 \tag{1}$$

In Eq. (1), ytrue is the true damage value, and ypred is the model's predicted damage value.

In the Hashin criterion, a damage value of 1 is considered as the threshold for damage. However, in actual materials during operation, the damage process is continuous, meaning that some locations may reach values very close to 1, which, although not classified as damage, still pose safety risks. Additionally, when the true value reaches 1, the predicted value may be 0.9979, which is very close to the actual situation but still classified as inaccurate. Based on the aforementioned practical application and operational issues, a threshold of 0.9 can be set. When the value is below 0.9, both the true values and the predicted values from the neural network that meet this condition in the database are set to 0, indicating that no damage has occurred at that location and it is considered safe. On the other hand, values above 0.9, even if not reaching 1, still indicate potential safety hazards and should be considered as damage or impending damage. By classifying the true and predicted values into 0 and 1 based on this criterion, it addresses both the safety issues in practical applications and the operational issues.

To minimize the loss function, an appropriate neural network training algorithm must be used. By adjusting the weights and biases of the neural network, the loss function can be minimized, thereby enhancing the network's performance and generalization capability. Common neural network optimization algorithms include the Adagrad algorithm [8], stochastic gradient descent (SGD) [9], stochastic gradient descent with momentum (SGDM) [10], Levenberg-Marquardt algorithm, and Bayesian Regularization. In this study, the L-M algorithm [11], an iterative algorithm used to solve nonlinear least squares problems, is employed. It is based on gradient descent and Newton's method. The fundamental idea is to add a coefficient to the Newton's method to adjust the learning rate during the iteration process, thus accelerating the convergence speed of the data. The expression for the L-M algorithm is as follows:

$$\theta_{t+1} = \theta_t - \left(J(\theta_t)^T J(\theta_t) + \lambda I\right)^{-1} J(\theta_t)^T e_t \tag{2}$$

In Eq. (2), θ_t is the parameter vector of the model at the t-th iteration, $J(\theta_t)$ is the gradient vector of the loss function at the t-th iteration, e_t is the residual vector at the t-th iteration, I is Identity matrix, λ is adjustment factor.

The neural network topology consists of three parts: the input layer, the hidden layer, and the output layer. In this case, the inputs are the E_{11}, E_{12}, and E_{22} values at the outer radius of a circular hole (2r), while the outputs are the corresponding node damage values at the hole, such as HSNFTCRT, HSNFCCRT, HSNMCCRT, and HSNMTCRT. The transmission between layers requires activation functions, which map the linear summation process to a nonlinear space, enabling the neural network to learn nonlinear functions and solve the limitations of linear models. Common activation functions include the Sigmoid function, tanh function, and ReLU function. The Sigmoid function, as shown in Eq. (3), has a value range of (0,1). The Sigmoid function is monotonically continuous, differentiable everywhere, and optimally stable, making it commonly used in the output layer. The tanh function, as shown in Eq. (4), has a larger slope near 0 compared to the Sigmoid function, resulting in faster convergence. The output mean of the tanh function is 0, and the area of the saturation region is larger than that of the Sigmoid function. The ReLU [12] function is a piecewise linear function widely used in deep learning. Compared to the Sigmoid and tanh functions, the advantage of the ReLU function is its ability to effectively solve the gradient vanishing problem, as shown in Eq. (5).

$$\sigma(x) = \frac{1}{1 + \mathrm{e}^{-x}} \tag{3}$$

$$\tanh(x) = \frac{\mathrm{e}^{x} - \mathrm{e}^{-x}}{\mathrm{e}^{x} + \mathrm{e}^{-x}} \tag{4}$$

$$ReLU(x) = \begin{cases} 0 & x < 0 \\ x & x \geq 0 \end{cases} \tag{5}$$

3.3 Adjusting the Hidden Layers

The activation functions and the number of neurons in each layer can affect the accuracy of the model, so different activation functions and numbers of neurons are tested and adjusted to ensure the model's accuracy. The number of epochs is set to 2100, and the batch size is set to 32, allowing the training to iterate 2100 times with a batch size of 32 for each iteration. To ensure model accuracy, multiple combinations of activation functions and neuron numbers are tested to minimize the loss function. Since the Sigmoid function is commonly used as the activation function in the output layer, the output layer's activation function remains unchanged. The activation functions for the input layer and hidden layers are set as (relu, relu, relu), with the number of neurons in the three layers being tested in combinations of (18, 32, 48), (24, 48, 56), (30, 64, 64), and (36, 80, 72), with the training process recorded as shown in Fig. 6.

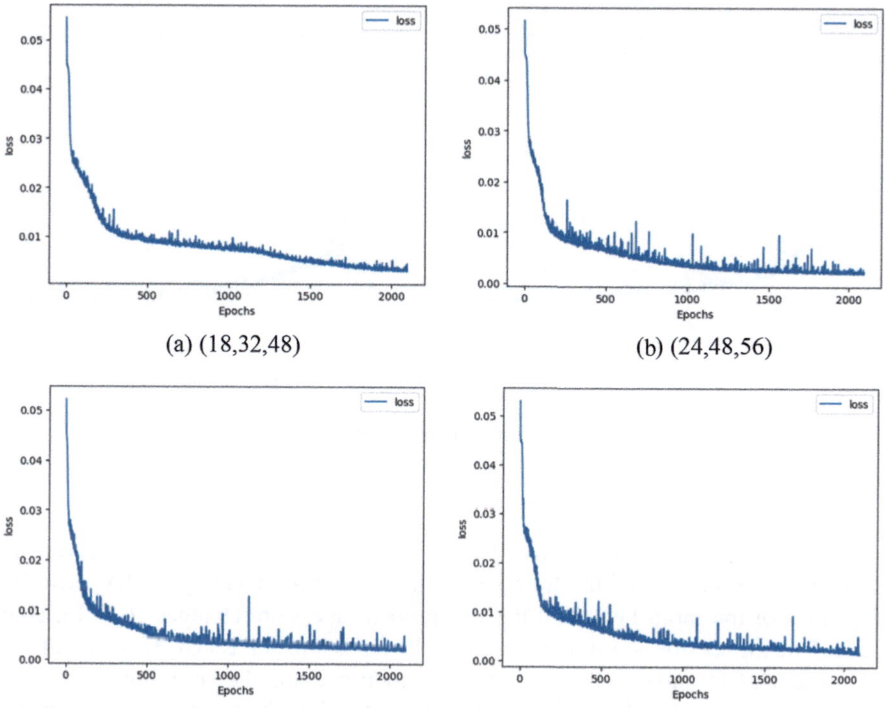

(a) (18,32,48)

(b) (24,48,56)

(c) (30,64,64)

(d) (36,80,72)

Fig. 6 The training process of the 4 neural networks corresponding to (relu, relu, relu). (**a**) (18, 32, 48). (**b**) (24, 48, 56). (**c**) (30, 64, 64). (**d**) (36, 80, 72)

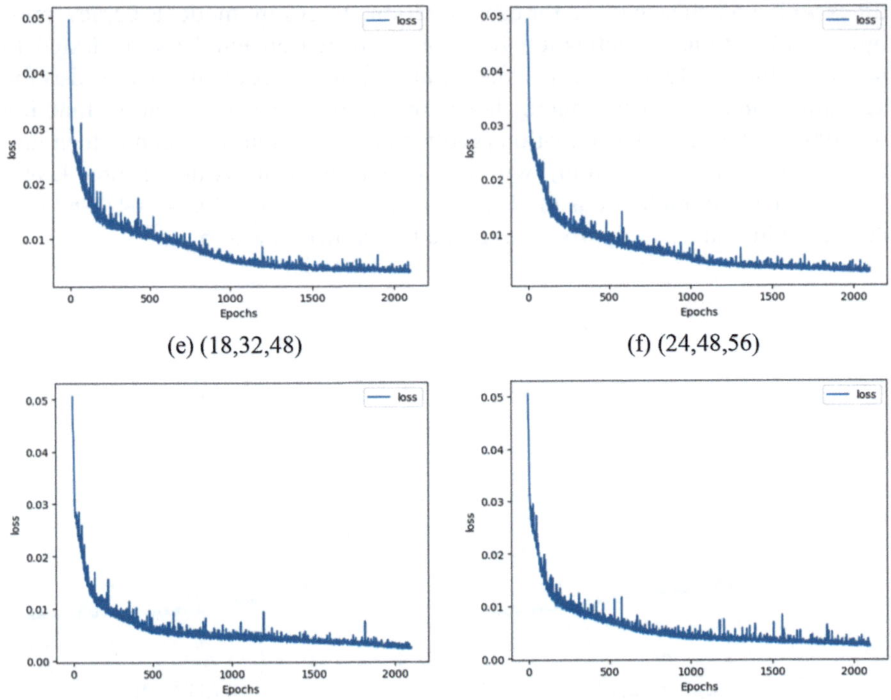

(e) (18,32,48)

(f) (24,48,56)

(g) (30,64,64)

(h) (36,80,72)

Fig. 7 The training process of the 4 neural networks corresponding to (relu, tanh, tanh). (**e**) (18, 32, 48). (**f**) (24, 48, 56). (**g**) (30, 64, 64). (**h**) (36, 80, 72)

Then, under the condition that the activation function of the output layer, the number of iterations, and the training batch size remain unchanged, the activation functions of the input layer and the two hidden layers are changed to (relu, tanh, tanh). The number of neurons for the three layers is still set as (18, 32, 48), (24, 48, 56), (30, 64, 64), and (36, 80, 72). The training process is shown in Fig. 7.

Under the same conditions, with (tanh, tanh, tanh) as the activation functions for the input layer and two hidden layers, the number of neurons for the layers is still set as (18, 32, 48), (24, 48, 56), (30, 64, 64), and (36, 80, 72). The training process is shown in Fig. 8.

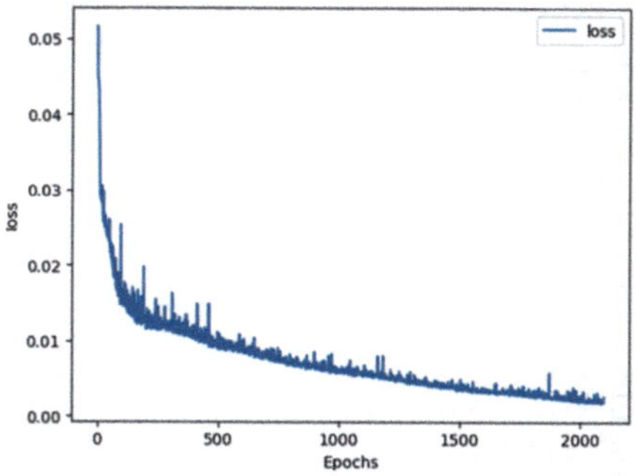

(i) (18,32,48)

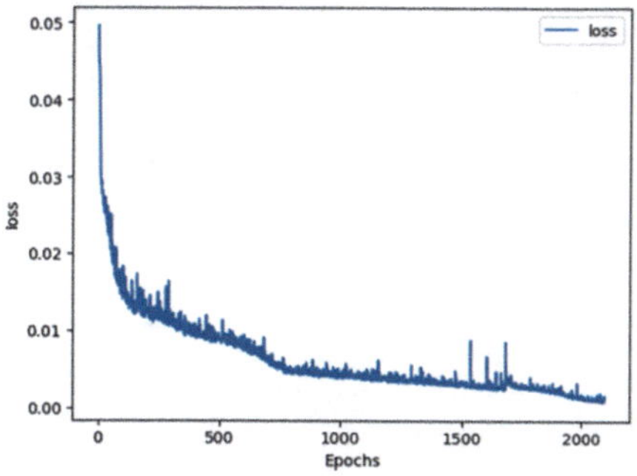

(j) (24,48,56)

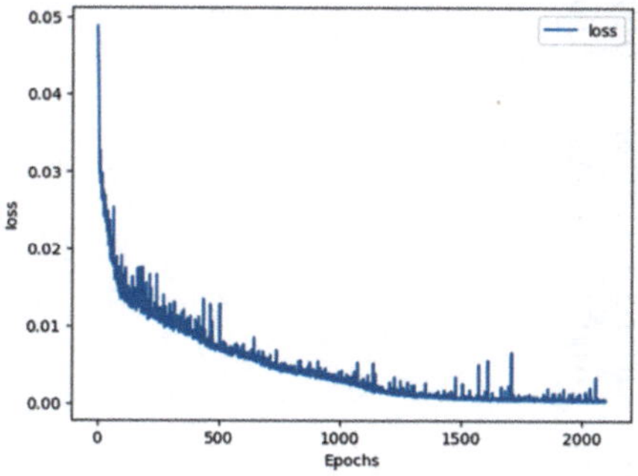

(k) (30,64,64)

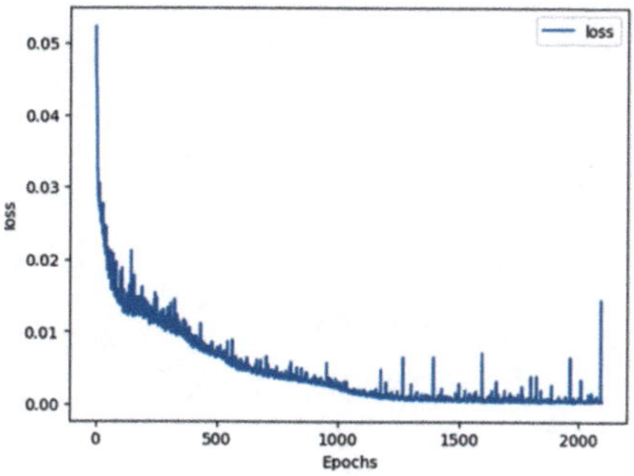

(l) (36,80,72)

Fig. 8 The training process of the 4 neural networks corresponding to (tanh, tanh, tanh). (**i**) (18, 32, 48). (**j**) (24, 48, 56). (**k**) (30, 64, 64). (**l**) (36, 80, 72)

The val_loss obtained for each model during the iteration process continuously decreased, indicating that overfitting did not occur. Finally, a table is created based on the results of the loss function curves' convergence, as shown in Table 2.

From the table, it can be intuitively seen that the final loss value of the k-th group model is the smallest, proving that the model not only has a good fit on the training set but also demonstrates strong generalization ability on unseen data. To ensure that the trained model has broad applicability, other different hole-containing examples are used to verify the accuracy and reliability of the trained model.

Table 2 Summary of the loss values for 12 models

	a	b	c	d	e	f	g	h	i	j	k	l
Loss	0.0044	0.0030	0.0021	0.0018	0.0041	0.0035	0.0025	0.0024	0.0027	0.0016	0.0004	0.0008
Val loss	0.0037	0.0030	0.0026	0.0018	0.0042	0.0032	0.0024	0.0028	0.0040	0.0023	0.0010	0.0012

4 Model Validation for Hole Edge Damage Prediction Based on Neural Networks

4.1 Parametric Modeling of Example Cases

Since the required model has been trained through comparison, the next step is to validate the accuracy of the neural network-based hole edge damage prediction model using more test cases, with the constitutive parameters of the test cases remaining unchanged. The laminate consists of 20 layers with a 0° stacking angle, and the stacking and model boundary conditions are shown in Fig. 9. A structured mesh is used to optimize the mesh quality around the circular hole, and the region around the six circular holes is divided into eight equal parts. Finally, the odb file is submitted, and the required node information from this file is extracted.

After the calculation, the matrix damage contour map is obtained through visualization, as shown in Fig. 10. Then, based on the .odb file of the fully analyzed test case, the hole edge damage, outer strain field, node coordinates, and node identification information are extracted into different text files. These are then summarized into a CSV file, and the trained model is used to predict the hole edge damage for this test case. The prediction results are shown in Fig. 11. It can be concluded that the accuracy of the model for all holes is no less than 93%.

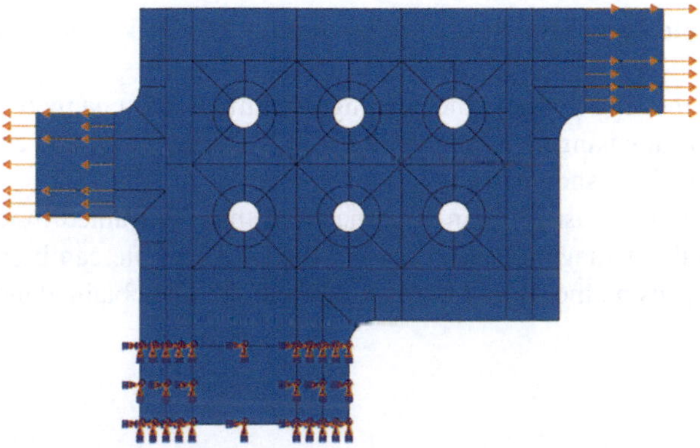

Fig. 9 Stacking and model boundary conditions

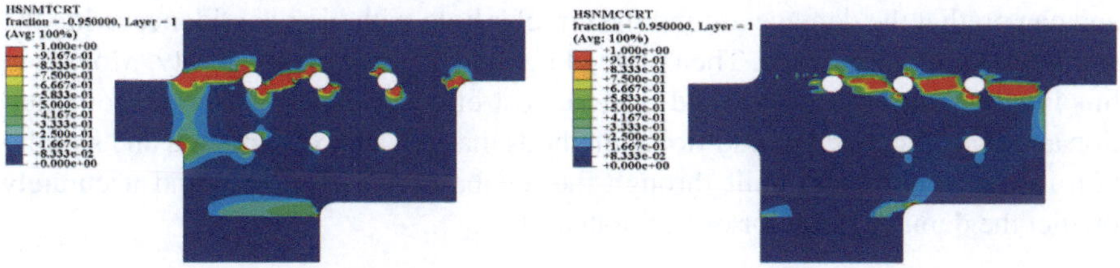

Fig. 10 Matrix damage contour map

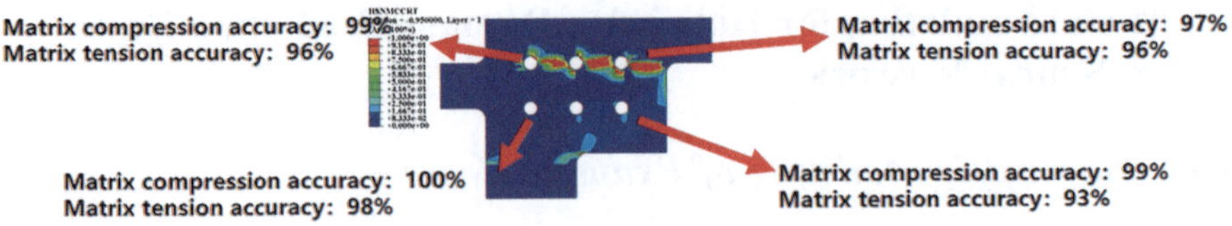

Fig. 11 Calculation of model prediction accuracy for the test case

Fig. 12 Model and boundary conditions

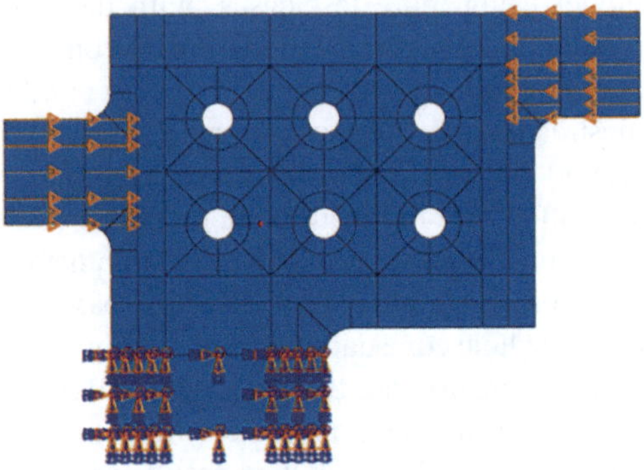

4.2 *Damage Cloud Diagram Drawing and Model Verification*

4.2.1 Example Case Modeling and Secondary Model Verification

To verify the model accuracy, the loading conditions are changed while keeping other conditions unchanged, as shown in Fig. 12. The accuracy and contour map for the trained model are shown in Fig. 13.

Through two test cases, it can be concluded that the trained model has a high accuracy, and the damage situation around the circular hole can be predicted with high confidence using the strain field around the hole. The obtained model has broad applicability.

4.3 *Chapter Summary*

Through two test cases, it can be concluded that the trained model has high accuracy and can predict the damage around the circular hole with high confidence using the strain field around the hole. The obtained model has broad applicability. Moreover, this intelligent prediction method is more cost-effective in terms of both computation and time compared to traditional methods that rely on experimental and simulation results. The model built through the database can efficiently and accurately predict the damage behavior of perforated plates.

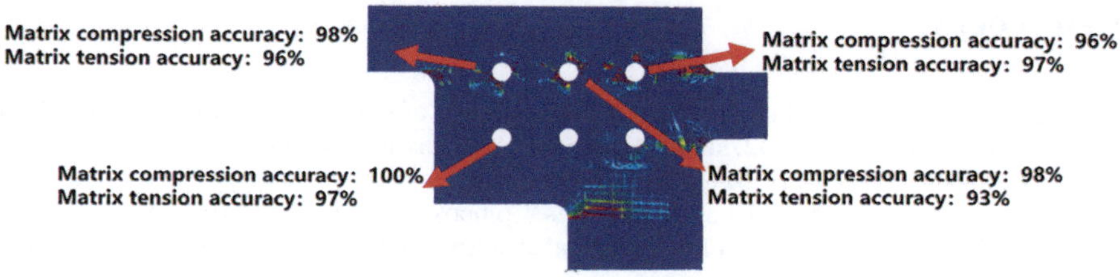

Calculation of model prediction accuracy for the test case

Fig. 13 Contour map and model accuracy calculation. Calculation of model prediction accuracy for the test case

5 Conclusion

This paper addresses the progressive damage problem of composite laminated plates with holes, establishing a finite element model for damage analysis and a result database. The BP neural network is used to predict the progressive damage of laminated plates.

1. A secondary development on the ABAQUS software platform was carried out, with composite laminated plates containing holes as the research object. The stacking angle and load direction of the laminated plate with a hole were set as variable parameters. Python scripts were written to achieve the parametric generation of the numerical model and the automated analysis of numerical simulations. A database of numerical cases for progressive damage failure analysis was constructed based on the progressive damage failure analysis results of the inner circular perimeter nodes, providing data support for the implementation of data-driven technology.
2. To process the damage analysis database and predict the initial damage mechanical behavior of the laminated plate, a BP neural network was developed, with a threshold set to make the prediction process more aligned with practical application needs. The mapping relationship between the strain values at the outer nodes of the circular hole in the laminated plate with holes and the damage behavior was established, enabling efficient and high-accuracy prediction of the initial damage mechanical behavior of composite laminated plates with holes. The accuracy was validated through test cases, and the results show that the prediction accuracy can reach a minimum of 93%. This study provides an important reference for the rapid and high-accuracy prediction of the mechanical behavior of composite laminated plates, offering technical support for the design and optimization of complex structures driven by numerical simulation results.

References

1. Xie X, Xiaogang Q, Yizheng B et al (2024) Overview of dynamic data-driven simulation. J Syst Simul 36(8):1832–1842. https://doi.org/10.16182/j.issn1004731x.joss.24-0127
2. Tao C, Zhang C, Ji H et al (2021) Application of neural network to model stiffness degradation for composite laminates under cyclic loadings. Compos Sci Technol 203:108573
3. Szabó B, Babuška I (2021) Finite element analysis: method, verification and validation. Wiley, Hoboken
4. Chicho BT, Sallow AB (2021) A comprehensive survey of deep learning models based on Keras framework. J Soft Comput Data Mining 2(2):49–62
5. Shen Y, Wang X, Lu J et al (2020) Reduction of liver metastasis stiffness improves response to bevacizumab in metastatic colorectal cancer. Cancer Cell 37(6):800–817
6. Bagaev DV, Vroomans RMA, Samir J et al (2020) VDJdb in 2019: database extension, new analysis infrastructure and a T-cell receptor motif compendium. Nucl Acids Res 48(D1):D1057–D1062
7. Jiaqi L, Xuewen C, Tao H et al (2020) Determination of critical damage value and experimental validation for 2a12 aluminum alloy. J Plastic Eng 27(1):131–137. https://doi.org/10.3969/j.issn.1007-2012.2020.01.018
8. Ward R, Wu X, Bottou L (2020) Adagrad stepsizes: sharp convergence over nonconvex landscapes. J Mach Learn Res 21(219):1–30
9. Tian Y, Zhang Y, Zhang H (2023) Recent advances in stochastic gradient descent in deep learning. Mathematics 11(3):682
10. Liu Y, Gao Y, Yin W (2020) An improved analysis of stochastic gradient descent with momentum. Adv Neural Inf Process Syst 33:18261–18271
11. Bilski J, Kowalczyk B, Marchlewska A et al (2020) Local Levenberg-Marquardt algorithm for learning feedforwad neural networks. J Artif Intell Soft Comput Res 10(4):299–316
12. Grigsby E, Lindsey K, Rolnick D (2023) Hidden symmetries of ReLU networks. In: International conference on machine learning. PMLR, London, pp 11734–11760
13. Ti Y, Yuan Y (2020) Strength analysis and experiment of perforated composite materials based on hashin criterion. Sci Technol Wind 35:183–184. https://doi.org/10.19392/j.cnki.1671-7341.202035086
14. Liu D, Yong H, Zhou Y (2024) Critical current degradation in an epoxy-impregnated rare-earth ba2cu3o7-x coated conductor caused by damage during a quench. Appl Math Mech 45(9):1557–1572. https://doi.org/10.1007/s10483-024-3141-8

Open Access This chapter is licensed under the terms of the Creative Commons Attribution-NonCommercial-NoDerivatives 4.0 International License (http://creativecommons.org/licenses/by-nc-nd/4.0/), which permits any noncommercial use, sharing, distribution and reproduction in any medium or format, as long as you give appropriate credit to the original author(s) and the source, provide a link to the Creative Commons license and indicate if you modified the licensed material. You do not have permission under this license to share adapted material derived from this chapter or parts of it.

The images or other third party material in this chapter are included in the chapter's Creative Commons license, unless indicated otherwise in a credit line to the material. If material is not included in the chapter's Creative Commons license and your intended use is not permitted by statutory regulation or exceeds the permitted use, you will need to obtain permission directly from the copyright holder.

Advanced Mechanical Design Techniques

Optimized Design of Candle Conveying Line Based on TRIZ Theory

Haiying Liu, Shuangsheng Zhan, and Chenghao Xie

Abstract In view of the problems of large areas, many pieces of equipment, and high manufacturing cost in the present candle conveying line, a new type of conveying line is determined by using TRIZ theory. By using the contradiction method and material field analysis method, a kind of coaxial bidirectional drive transmission line structure is designed. The candle conveying line designed by TRIZ theory has the advantages of a compact structure, a small floor area, and lower cost.

Keywords Optimised design · Candle conveying line · TRIZ theory

1 Introduction

TRIZ (Theory of Inventive Problem Solving) theory is the theory of solving the problem of invention, which originated in the former Soviet Union. It was invented by Inventor G.S. Altshuller's team, as a theoretical approach to rapid technological innovation by studying millions of patents [1]. In recent years, TRIZ theory has been widely used in the innovative design of mechanical products. Based on TRIZ theory, Wang [2] carried out the innovative design of a gear clamping system; Heng and Luan [3] designed the apple sorting and bagging device; Jiang et al. [4] completed the design of a multi-functional coating device for the pipeline.

The main function of the candle conveying line is to transport the finished candles to each workstation for packaging. The packaging equipment is installed on both sides of the conveying line. Combined with the TRIZ innovation theory, this paper optimizes the design of a candle conveying line for its high manufacturing cost and large floor area.

H. Liu · S. Zhan · C. Xie (✉)
Dalian University of Science and Technology, Dalian, Liaoning, China

© The Author(s) 2026

Z. Zhang, P. Mercorelli (eds.), *Mechanical Design and Simulation: Insights and Innovations*, Lecture Notes in Mechanical Engineering,
https://doi.org/10.1007/978-981-95-2405-1_87

Fig. 1 Belt loop conveyor line. D-drum diameter

D-drum diameter

2 Description of Engineering Problem

The conveying line is one of the main types of automated production lines, and conveyor carriers are divided into rollers, chains, and belts according to the size of the materials. The size of the candle is $\Phi 30 \times 15$ mm, and it belongs to a small block material. The carrier is a belt, and the conveying line is annular. It is composed of two linear belt machines with the same structure and different running direction. A steering device 180° sector bender is installed at both ends of two belt conveyors to ensure the smooth delivery of candles on the entire circular line, as shown in Fig. 1. In this design, considering the conveyor belt's allowable strength, additional bending stress, and many other factors, the drum diameter should not be too small. For small materials such as candles, at the joint of the linear belt conveyor and the turning belt conveyor, the accumulation of materials occurs easily, as shown in Fig. 2. Therefore, there are many problems in the circular candle conveying lines, such as the large input equipment, high manufacturing cost, material accumulation, and large floor area.

3 Preliminary Optimization Scheme

The design of technological systems, which was an art a hundred years ago, is now an exact science, and the essence of TRIZ theory is that it will fundamentally change the process by which new technological ideas are generated. TRIZ innovation theory helps people overcome inertia and accelerate the process of innovative design [5].

3.1 Principles of Optimal Design

The principles of the optimal design of the candle packing and conveying line are as follows: (1) making the whole conveying line occupy less land and investing less equipment to meet the normal transportation of candles; (2) making sure that the

Fig. 2 Belt loop conveyor line. L-drum distance

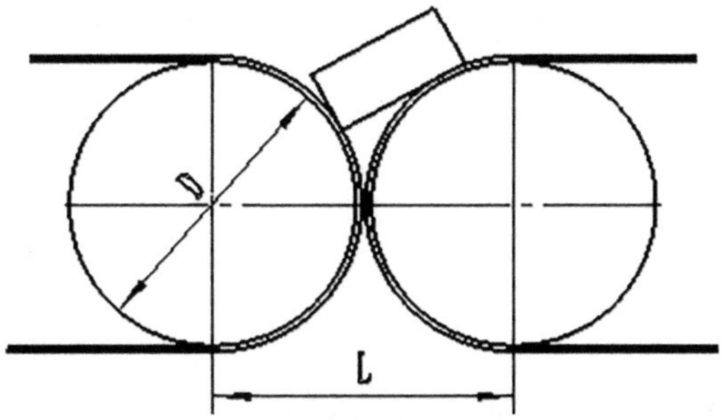

L-drum distance

materials do not accumulate and scatter, and has the advantages of simple structure, low manufacturing cost, and easy maintenance.

3.2 Preliminary Optimization Plan

TRIZ theoretical system tailoring eliminates the negative functions generated by the tailoring part and reduces the cost, while the useful functions performed still exist [5]. In the present candle packing conveying line, the main conveying function is the two linear belt lines, while the bender only plays the role of transition and steering, the manufacturing cost is a little higher, and the floor area is large. According to the system cutting method, two 180° sector bends are cut, which reduces not only the number of equipment but also the floor area and the manufacturing cost.

4 Systematic Analysis of TRIZ Theory

The function analysis method is an important analysis method of TRIZ theory, which aims to analyze the current technical system, supersystem components, and the function relationship among them [5]. Its main purpose is to decompose the existing system, to determine the main functions provided by the technical system, to clarify the useful functions of each component and its contribution to the system functions, and to find the breakthrough point to solve the problems, create and draw component function model diagram [4]. The initial optimization scheme of the candle conveying line is the current technical system. The current technical system consists of the control system, roller, rack, belt, motor, bearing parts, belt distance, the conveying material (candle), and the control system, which is a supersystem, and the acting object of the system is the conveying material (candle). The functional model established is shown in Fig. 3.

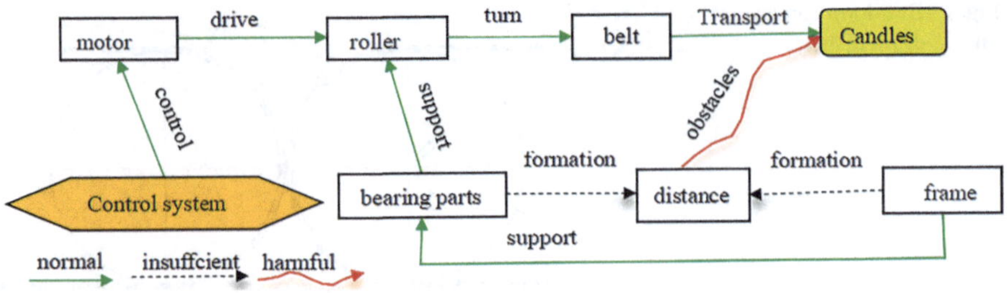

Fig. 3 Functional model diagram

Through the functional analysis of the components, it is found that the distance between the two belts hinders the normal steering of the candles, which is a harmful function; the frame and bearing parts are the fundamental reasons for the large gap between the two belt conveyor, which is a lack of function.

5 TRIZ Theory to Improve the Scheme

5.1 The Law of Technical Contradiction

Technical contradiction refers to the improvement of one engineering parameter of the current technical system, which may lead to the deterioration of another parameter. This contradiction is called technical contradiction [6]. Through the analysis of the function model, it can be concluded that reducing the distance between the two linear belt conveyors can ensure the smooth transition of the candle transportation and avoid the phenomenon of throwing material. At the same time, it can reduce the size of the frame and the bearing parts, which is in contradiction with the simple structure of the conveying line and the low manufacturing cost. The above problems are transformed into technical contradictions described by 39 parameters in TRIZ theory. The improved parameters are 33 Operability, 35 adaptability and versatility, 34 maintainability, and 32 manufacturability.

After transforming the concrete problem into a concrete contradiction matrix, we use the contradiction matrix to solve the problem [5], look up the Ács Schuler contradiction matrix, and get nine innovative principles to solve the problem, as shown in Table 1.

Through the analysis, it is found that the combination principle of Innovation Principle 5, the separation principle of Innovation Principle 1, Pre-compensation Principle 11, and the extraction principle of the original new Principle 2 are of practical significance for solving the current problems, as shown in Table 2. According to the extraction principle, the combination principle, and the separation principle, the solution to the technical contradiction is as follows:

According to Extraction Principle 2, the two racks and the bearing parts adjacent to the inner side of the belt conveyor are removed.

Table 1 Table of contradiction matrix

Parameters for improvement	The parameters of deterioration: 32 manufacturability
33 Operability	Principles of innovation: 2, 5, 12
35 Adaptability and versatility	Principles of innovation: 1, 13, 31
34 Maintainability	Principles of innovation: 1, 35, 11, 10

Table 2 Principles of innovation

No.	Principles of innovation	The meaning of TRIZ
1	Dividing principle	Dividing an object into parts that are easy to assemble
2	Principle of extraction	Extracting a negatively impacted part or property from an object
5	Merger principle	Merging operations of the same class together spatially or temporally
11	Pre-compensation principle	The lower reliability of the object is compensated by pre-prepared measures

According to the Merging Principle 5, the drum, frame, and roller shaft of two belt conveyors with opposite conveying directions are combined together. Fixed pallets are installed inside the bracket, and two conveyor belts are arranged side by side on the pallets. In order to prevent friction between belts, the distance between conveyor belts is only 5 mm.

According to Dividing Principle 1, the original driving roller of a belt conveyor is assembled by the shaft, the bearing, and the roller body. The roller body consists of two parts.

According to the principle of Pre-action 11, under the active roller, a drive shaft is added, as shown in Fig. 4. At the same time, the structure of the drum body is improved by installing a gear on the left-rotating drum body and a synchronous belt pulley on the right-rotating drum body. The motor drives the shaft of the intermediate transmission device to rotate. When the rotation torque is transmitted to the right drum through the synchronous belt, the direction of rotation is the same as that of the motor, and when it is transmitted to the left drum through the gear meshing drive, the direction of rotation is the opposite to that of the motor.

5.2 *The Law of Physical Contradiction*

Physical contradiction means that in order to implement some function, mutual exclusion is required for the same parameter of the same object [6]. The belt conveyor's active drum position should be chosen in the loose side tension of the position of both discharge ends to ensure that the load-bearing section always has a tight edge. After adopting the innovative combination principle and the pre-compensation principle, the two rollers are driven by a motor. Thus, when the two belt drives are working, there will always be a loose bearing section of the conveyor belt. Therefore, there are physical contradictions: for the right-running belt conveyor, the

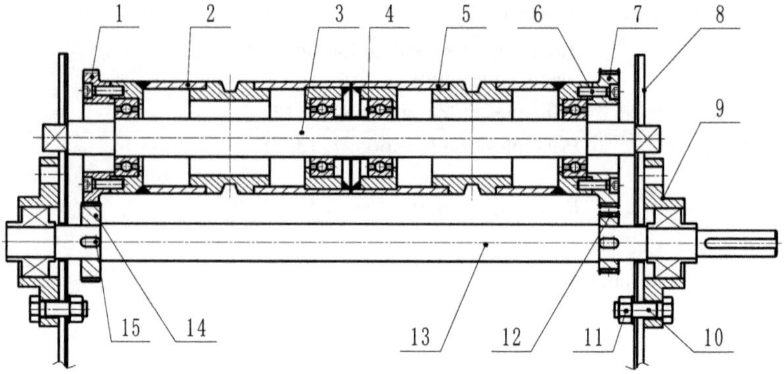

1-large gear; 2-left rotating drum body; 3-drive drum shaft; 4-bearing; 5-right rotating drum body; 6-cylindrical head bolt countersunk head screw; 7-large synchronous pulley 8-support plate;
9-bearing block; 10-connection bolt; 11-nut; 12-small synchronous pulley; 13-shaft; 14-pinion; 15-key

Fig. 4 Functional model diagram. 1-large gear; 2-left rotating drum body; 3-drive drum shaft; 4-bearing; 5-right rotating drum body; 6-cylindrical head bolt countersunk head screw; 7-large synchronous pulley 8-support plate; 9-bearing block; 10-connection bolt; 11-nut; 12-small synchronous pulley; 13-shaft; 14-pinion; 15-key

belt-carrying end is the tight edge; for the left-running belt conveyor, the belt-carrying end is still the tight edge, and the tight edge is running to the left and right to run.

In order to solve this problem, the space separation principle is adopted by changing the position of the active drum, changing the head drive to the Middle Drive, and adopting the back-to-empty belt surface drive. Regardless of whether the belt conveyor travels right or left, it is necessary to always ensure that the carrying section is tightly edged, which avoids the phenomenon of conveyor belt slipping when conveying, and is convenient for layout and maintenance.

5.3 Material Field Analysis

TRIZ theory uses matter and field to describe system problems. It is called matter-field analysis [6].

After the position of the driving drum is moved to the middle position of the belt conveyor, the surrounding angle of the driving drum is too small, and there is not enough tension to ensure the normal operation of the conveyor belt. Then, the object field analysis method is used to solve the problem.

The material field model of the current system is as follows: the workpiece S1 is the belt, the tool S2 is the active roller, and the interaction field F1 is the mechanical field. The active roller's enclosing angle is too small, and the conveyor belt cannot obtain enough tension. Therefore, it belongs to the non-efficient complete model, as

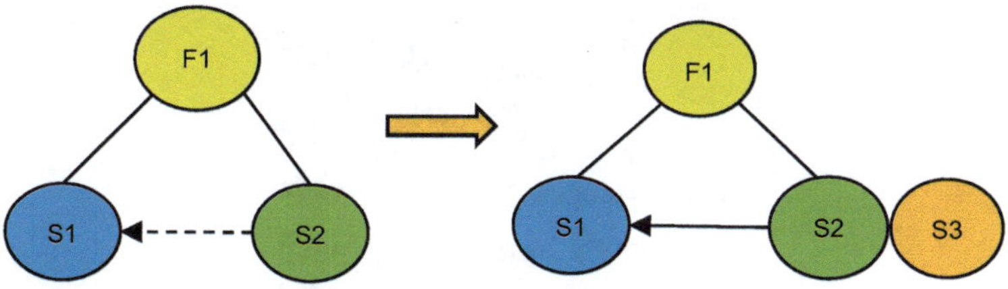

Fig. 5 (**a**) Model of the physical field. (**b**) Standard solution 1.1.3 model

shown in Fig. 5. According to the standard solution 1.1.3 of material field analysis [7], a material S3 is added outside S2 to improve the useful effect of the active roller, as shown in Fig. 5. According to the Standard Solution Model 1.1.3, a redirection roller is added on both sides of the active roller, which increases the surrounding angle of the active roller, increases the tension of the conveyor belt, and ensures the normal operation of the conveyor belt.

5.4 *Smart Little People*

In TRIZ theory, the *smart little people* method is used to achieve the desired function, and the structure is redesigned according to the *smart little people* method [6]. In order to ensure that the delivery of candles can turn smoothly, this paper uses the method of smart little people on the belt machine and candles to do further analysis. The yellow figure represents the belt machine running to the left, the pink figure represents the belt machine running to the right, and the brown figure represents the object (candle), as shown in Fig. 6. In order to realize the smooth turn of the candle figurine, the green guiding figurine is introduced. The candle figurine is guided by the green figurine to turn to the right-hand belt machine, as shown in Fig. 7. According to the *smart little people* method, a guiding device is added at both ends of the head and tail of the two belt conveyors. The guiding device is realized by the synchronous belt, and the two synchronous belt pulleys are installed on the two side racks. Synchronous operation direction and the direction of the belt conveyor form a certain angle, and the synchronous belt is driven by motor drive alone, as shown in Fig. 8. The candle, guided by the guide device, is smoothly guided to another belt conveyor.

5.5 *Final Solution*

Through the application of TRIZ theory, initial optimization plans for reflection are continuously initiated, and the final optimization scheme is formed [4]. The whole ring production consists of two linear belt conveyors, two belt conveyor common frame, supporting plate and other structures, and two active drums by shaft support;

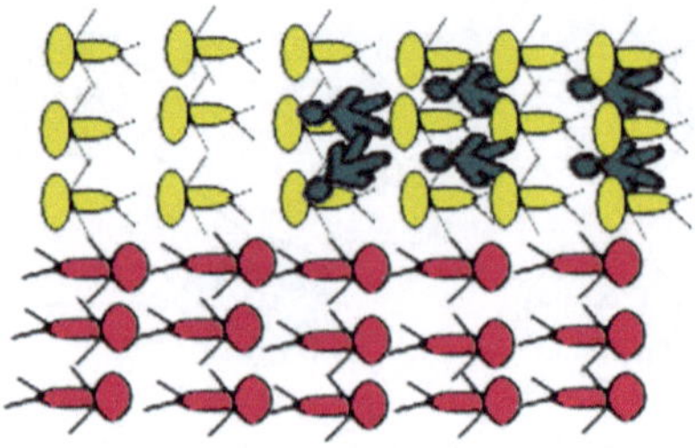

Fig. 6 Conveyor line minion model

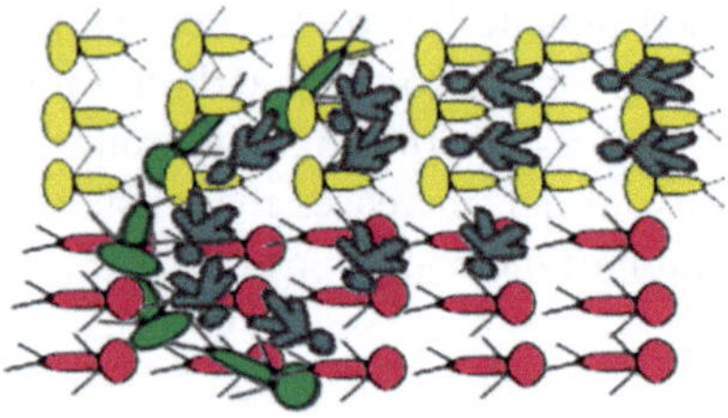

Fig. 7 Improved model of the conveyor line

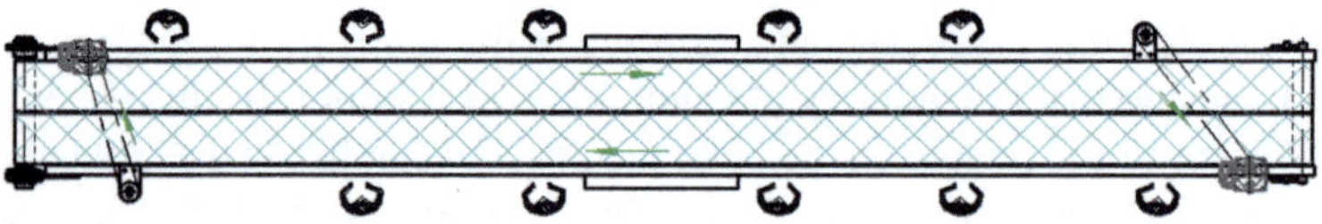

Fig. 8 The final schematic

under the action of the active drum, a transmission device has been added, which rotates the motor into two different directions of the drum through gear transmission and synchronous belt transmission. The driving device is located in the middle of the belt conveyor and is driven by returning to the empty belt surface, as shown in Fig. 8.

6 Conclusion

In this paper, TRIZ theory is used to optimize the design of the candle conveying line for its large area and many pieces of investment equipment. The structure of a two-way belt conveyor line driven by coaxial is determined, and the candle direction is realized by adding a synchronous belt at the end. The distance between the two belts is only 5 mm, which avoids the accumulation of candles. The synchronous belt guiding device is much cheaper than the original turning machine, as shown in Fig. 1. The project group has been granted a patent for a utility model, and the transmission line has been put into use in a candle factory in Dandong, Liaoning Province, which has reduced the equipment manufacturing cost by 40% and reduced the floor space occupied by 1/3 for the enterprise. The feasibility of the candle conveying line designed by TRIZ theory is verified by the application of the equipment.

Acknowledgments This paper is one of the research achievements of the Key Laboratory of Digital Design and Intelligent Manufacturing of Lightweight Castings for New Energy Vehicles Liaoning Provincial. No.: 2022JH13/10200058.

References

Cheng SY, Zhou JP, Guo ZG (2020) Method of technological innovation-TRIZ theory and its application. Qinghua University Press, Beijing
Wang RH (2021) Innovative design of gear clamping system based on TRIZ theory. Manuf Technol Mach Tools 05:124–127
Heng Y, Luan F (2021) Design of Apple sorting and bagging device based on TRIZ theory. Mech Des 7(s2):104–110
Jiang MZ, Yan CY, Dong KX et al (2022) Design of multi-function coating device in pipeline based on TRIZ theory. Mech Des Res 38(4):208–212
Altshuller G (1998) Creativity as exact science: theory of inventive problems. Gordon and Breach Science Publishers Inc., Hoofddorp
Yao W, Han X (2022) The road to innovation. Qinghua University Press, Beijing
Victor F, Eugene R (2005) Innovation on demand. Cambridge University Press, New York

Open Access This chapter is licensed under the terms of the Creative Commons Attribution-NonCommercial-NoDerivatives 4.0 International License (http://creativecommons.org/licenses/by-nc-nd/4.0/), which permits any noncommercial use, sharing, distribution and reproduction in any medium or format, as long as you give appropriate credit to the original author(s) and the source, provide a link to the Creative Commons license and indicate if you modified the licensed material. You do not have permission under this license to share adapted material derived from this chapter or parts of it.

The images or other third party material in this chapter are included in the chapter's Creative Commons license, unless indicated otherwise in a credit line to the material. If material is not included in the chapter's Creative Commons license and your intended use is not permitted by statutory regulation or exceeds the permitted use, you will need to obtain permission directly from the copyright holder.

Innovative Design of Hydraulic Drive and Intelligent Control System for Corn Harvesters

Changzhong Wu, Yongzhuo Wang, Yuanyuan Lu, Xingyu Zhu, and Haibo An

Abstract This study addresses the issues of high failure rates and high impurity levels in traditional corn harvesters by proposing an improved solution based on hydraulic principles. Through theoretical analysis and computer technology, a hydraulic drive system and intelligent control system were designed to enhance the stability, reliability, and operational efficiency of corn harvesters. The research selected a closed hydraulic circuit and optimized the design for header speed, feed rate monitoring, concave screen gap adjustment, and threshing drum speed. The results indicate that the improved system effectively reduces failure rates, increases harvesting efficiency, and improves grain integrity. This study provides an efficient and reliable corn harvesting technology solution for agricultural production, demonstrating practical application value.

Keywords Corn harvester · Hydraulic improvement · Intelligence · Operational efficiency

1 Introduction

The performance of corn harvesters significantly impacts agricultural productivity and farmers' economic benefits. Traditional models face design flaws, such as high failure rates and impurity levels, leading to efficiency losses and economic drawbacks for farmers [1]. Hence, enhancing the stability and reliability of corn harvesters is crucial for improving harvesting efficiency and reducing costs [2].

The demand for intelligent and automated corn harvesters is rising globally. Companies like Kubota and Iseki have introduced automated models featuring high-precision sensors and intelligent control systems. Although domestic research has made strides in automation and intelligent control, disparities remain in the

C. Wu (✉) · Y. Wang · Y. Lu · X. Zhu · H. An
School of Mechanical Engineering, The University of Jinan, Jinan, China
e-mail: me_wucz@ujn.edu.cn

© The Author(s) 2026

Z. Zhang, P. Mercorelli (eds.), *Mechanical Design and Simulation: Insights and Innovations*, Lecture Notes in Mechanical Engineering,
https://doi.org/10.1007/978-981-95-2405-1_88

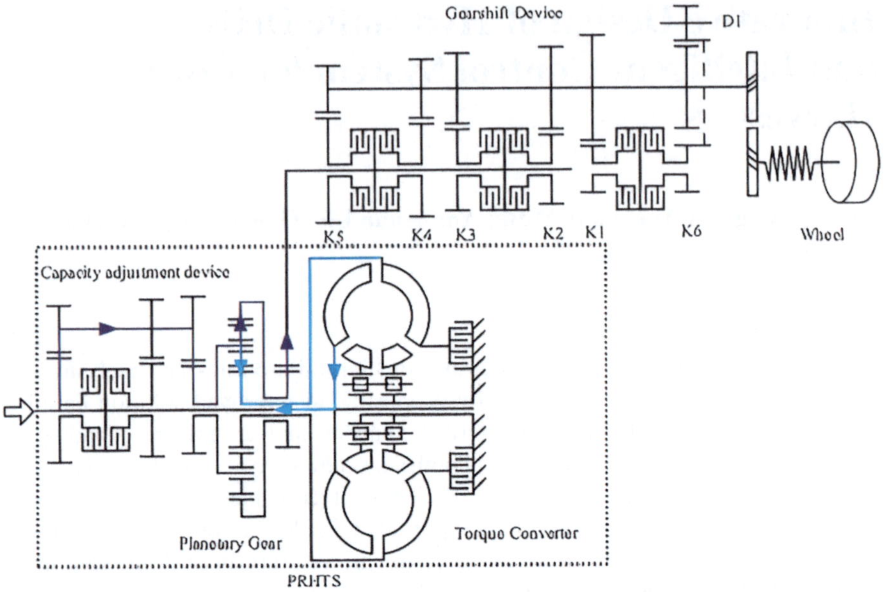

Fig. 1 Power reflux hydraulic transmission system studied by Yingzhe Kan et al.

application of hydraulic technology for corn harvesters. Foreign research is often more advanced but less adaptable, while domestic studies are more attuned to local agricultural needs [3].

As automation levels increase, the reliability and energy efficiency of harvesters have become more critical. Hydraulic transmission systems offer advantages over mechanical systems and are receiving extensive study. Researchers, such as Kan et al., have developed innovative hydraulic transmission systems, like a power return system with a capacity adjustment device, maintaining low speed and high torque to enhance transmission efficiency (as shown in Fig. 1) [4].

2 Hydraulic Drive Scheme for Corn Harvester

2.1 Selection of Hydraulic System Circuit

The closed circuit (Fig. 2) is compact in structure, with the oil circulating in the pipeline and equipped with a replenishing pump to compensate for oil loss. It has high transmission efficiency. Closed circuits are typically suitable for complex hydraulic systems, ensuring system pressure stability and improving work efficiency. The open circuit (Fig. 3) is relatively simple and flexible compared to the closed system, but it requires a larger oil tank and the oil is more prone to contamination, resulting in lower transmission efficiency. It is suitable for systems with

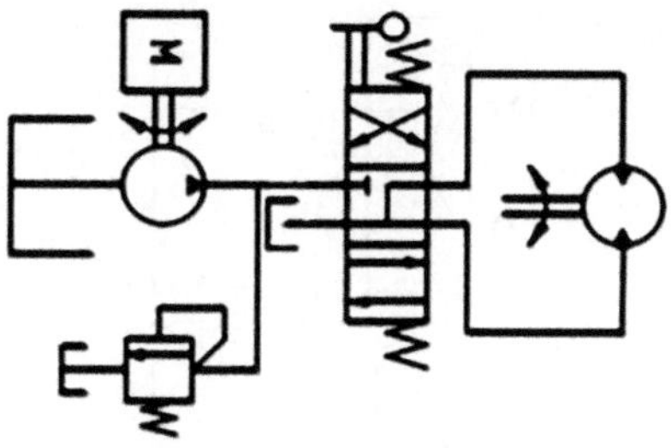

Fig. 2 Closed circuit diagram

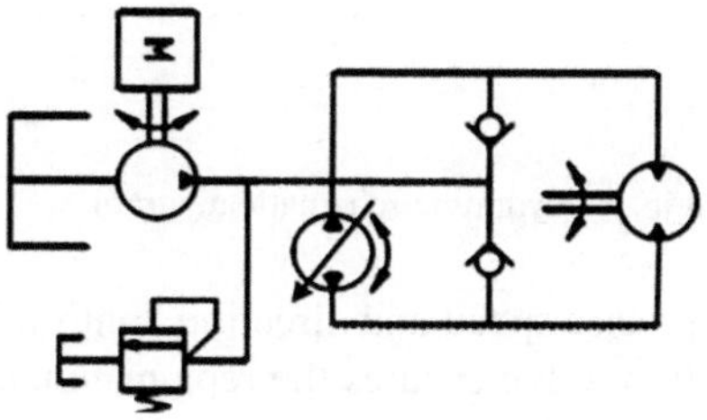

Fig. 3 Open circuit

lower control precision requirements. Considering the working conditions of the corn harvester, the closed circuit is more suitable for the hydraulic system of the corn harvester [5].

2.2 Speed Regulation Methods

There are three main methods of speed regulation: volumetric speed regulation, throttle speed regulation, and volumetric-throttle speed regulation. This paper selects volumetric speed regulation. Volumetric speed regulation does not have throttling losses or overflow losses, resulting in high circuit efficiency and low system temperature rise. It is suitable for high-speed, high-power speed regulation systems and is widely used in high-power hydraulic systems such as bulldozers, slotting machines, and broaching machines.

2.3 Travel System Design

This section studies the design of the hydraulic travel system of the corn harvester, focusing on achieving stepless speed drive. The system consists of a hydraulic pump, hydraulic valve, hydraulic cylinder, and mechanical transmission part, directly operated by a controller to optimize power distribution. A closed circuit design is selected to maintain system pressure stability, improve suction capacity, and ensure oil supply. The variable pump is connected to the engine. By adjusting the swashplate angle, the hydraulic oil flow and direction are controlled to achieve

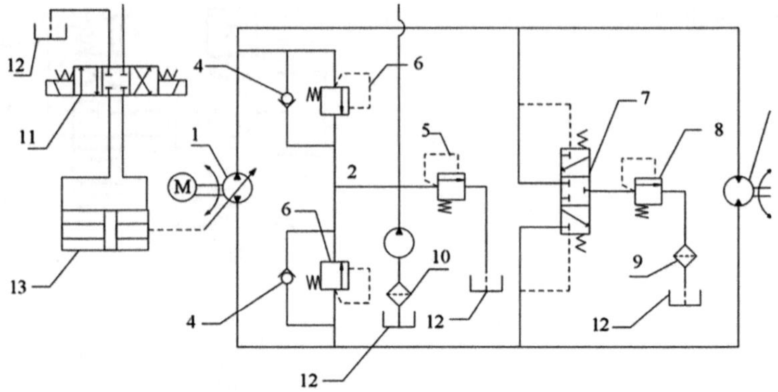

Fig. 4 Hydraulic schematic diagram

precise speed and direction control. The design of the replenishing pump and overflow valve ensures the replenishment of system oil and pressure control, maintaining stable system operation. Figure 4 shows the designed hydraulic system schematic diagram.

2.4 Header System Design

During the operation of the header, corn stalks first come into contact with the divider. As the machine moves forward, the corn stalks are cut by the lower header cutter. The corn ears are then removed by the snapping rollers and stalk rollers. The auger transfers the corn ears through the conveyor bridge to the threshing and cleaning system. The resulting corn stalks enter the chopper. The shredded stalks and husks can either be returned to the field or collected in a straw box. The field-returning machine can directly chop the root of the stalks.

3 Modeling and Simulation of the Hydraulic System of the Corn Harvester

3.1 Modeling and Parameter Setting of the Hydraulic Walking System

The hydraulic system of the harvester experiences external loads of 6500 kg, 7500 kg, and 8500 kg under light, medium, and heavy loads, respectively. A simulation over 12 s with a 0.01-s printing interval reveals that the hydraulic motor exhibits flow fluctuations, with maximum values of 152 L/min, 155 L/min, and 158 L/min

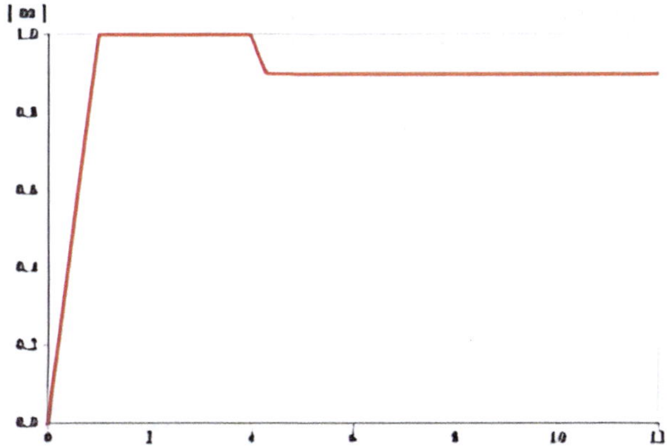

Fig. 5 Displacement diagram of displacement control valve cylinder

for each load condition. The time to stabilize flow decreases with increasing loads, stabilizing at 139 L/min. The analysis indicates that load significantly influences motor flow.

In starting, the hydraulic motor shows speed fluctuations, with higher loads resulting in greater peak speeds, though the steady speed remains affected minimally by the load. The motor's output torque stabilizes around 2 s after overcoming the load, with steady-state torques of 98 N/m, 197 N/m, and 299 N/m for light, medium, and heavy loads, respectively. Overall, the walking motor operates steadily at approximately 1834 r/min and maintains a flow rate of about 140 L/min, aligning with design specifications.

Furthermore, the relationship between pump displacement and motor speed is studied by varying the displacement control valve's cylinder displacement, which influences the output speed of the traveling motor, as illustrated in Figs. 5 and 6.

3.2 *Cutting Table Hydraulic System Simulation*

According to Fig. 7, AMESim software is used to build its simulation model, as shown in Fig. 7.

According to the actual working conditions of the hydraulic system of the cutting platform, the external loads corresponding to light load, medium load, and heavy load of the system are 500 kg, 600 kg, and 700 kg, respectively. The simulation time is set to 10 s. The printing interval is 0.2. After running the simulation, the simulation results show that the hydraulic cylinder would produce large flow fluctuations at the beginning stage. Under light load, medium load, and heavy load conditions, the fluctuation range is 0.75–0.82 L/min, 0.70–0.83 L/min, and 0.62–0.85 L/min, respectively. The stability time of the flow fluctuation range is 2.1 s, 2.3 s, 2.2 s. It can be concluded that the fluctuation of hydraulic cylinder flow increases with the increase of load. The load has a greater impact on the hydraulic cylinder of the cutting platform system.

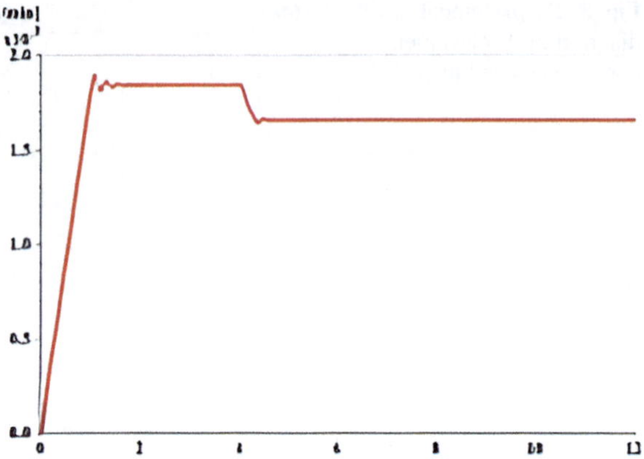

Fig. 6 Motor speed diagram in the event of variable pump displacement alteration

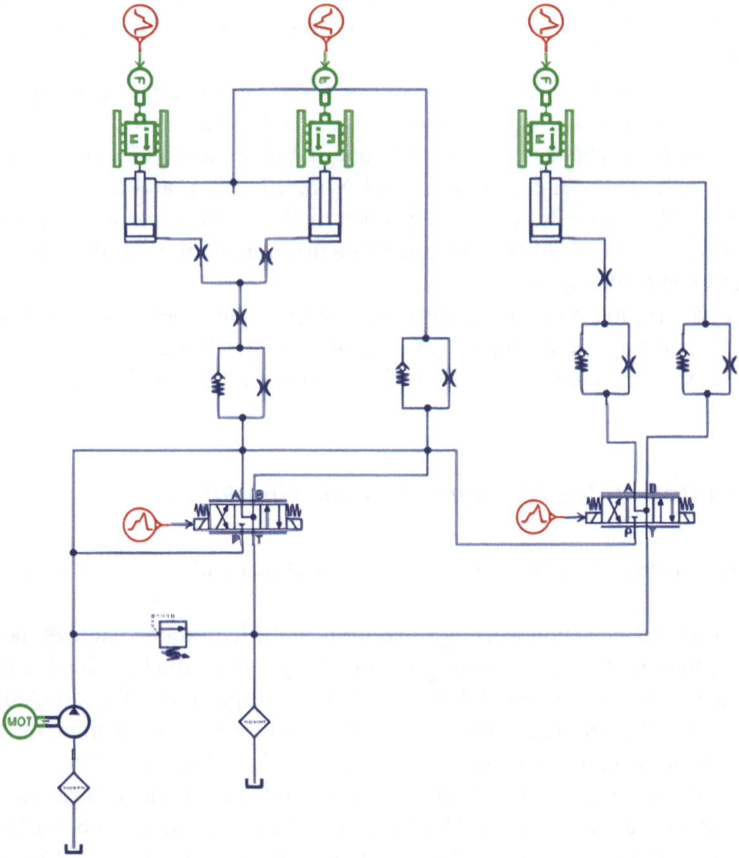

Fig. 7 Simulation model of cutting platform hydraulic cylinder system

4 Translated Summary

This study proposes an innovative hydraulic drive and intelligent control system design to address the high failure rate and impurity issues of traditional corn harvesters. Through theoretical analysis and computer-aided design, the system achieves stepless speed regulation, automatic adjustment of the header speed, real-time monitoring of feed quantity, and optimization of the threshing and cleaning system. The improved system significantly reduces failure rates and enhances harvesting efficiency and grain integrity. The hardware includes Hall sensors, dynamic torque sensors, pressure sensors, a C490 controller, and FBOX communication equipment, forming an intelligent control system verified through computer simulation and field tests.

Despite the achievements, there are limitations, such as simulations and tests based on theoretical models that may encounter complex conditions in practical applications, and the need for further optimization of control system algorithms and logic. Future work will focus on optimizing control algorithms, expanding the system's applicability, and integrating with actual agricultural production processes to continuously improve the corn harvester's performance, thereby promoting agricultural mechanization and modernization.

References

1. National Bureau of Statistics (2021) China rural statistical yearbook 2021. China Statistics Press, Beijing
2. Yucheng J, Jizhan L, Zhujie X et al (2021) Current development status and trends of agricultural robot technology. Agric Eng Technol 41(33):87
3. Tiantou C, Aiwu T (2024) Design and experimental study of HMT hydraulic drive system for corn harvester. J Agric Mech Res 46(4):128–133. https://doi.org/10.13427/j.cnki.njyi.2024.04.008
4. Zhe Q, Dongxing Z, Li Y et al (2018) Experimental study on the feeding amount and cylinder speed of a vertical axial flow corn threshing and separation device. Trans Chin Soc Agric Mach 49(2):58–65
5. Meng C, Li Z, Li G, Xu F (2023) Design and experiment of testing system for threshing device of corn grain harvester. In: International conference on applied physics and computing (ICAPC). IEEE, New York, pp 464–467. https://doi.org/10.1109/ICAPC61546.2023.00091

Open Access This chapter is licensed under the terms of the Creative Commons Attribution-NonCommercial-NoDerivatives 4.0 International License (http://creativecommons.org/licenses/by-nc-nd/4.0/), which permits any noncommercial use, sharing, distribution and reproduction in any medium or format, as long as you give appropriate credit to the original author(s) and the source, provide a link to the Creative Commons license and indicate if you modified the licensed material. You do not have permission under this license to share adapted material derived from this chapter or parts of it.

The images or other third party material in this chapter are included in the chapter's Creative Commons license, unless indicated otherwise in a credit line to the material. If material is not included in the chapter's Creative Commons license and your intended use is not permitted by statutory regulation or exceeds the permitted use, you will need to obtain permission directly from the copyright holder.

A Pull-Type Taro Harvester Based on the Reverse Locking Capability of the Heart Valve Structure

Yida Qi, Liang Yu, Lei Hou, Minghe Liu, and Kehao Guan

Abstract Given the problem that taro harvesting mainly relies on manual harvesting and the mechanization level is low, a taro harvester that can be used in the medium-heavy clay area is proposed. Innovative designs include a valve-like taro supporter to prevent the taro from falling over through reverse locking, a hinge-type limit cutting device to ensure the cutting position is always at the connection between the fruit and the petiole, and a track-type taro feeder wheel that accurately feeds the petiole into the pulling system. Additionally, a soil loosening mechanism, pulling system, and collection device are designed. The length of the arm of the support device is 173 mm, the length of the connecting rod is 122 mm, and the spring constant is 300 N/m. The shovel length of the soil loosening excavation mechanism is 110 mm, and the shovel width is 150 mm. It is clear that the main structure of the digging shovel is a direct structure, and the inclination angle of the shovel surface is between 14° and 25°. The handle is adjustable from 150 mm to 350 mm. At the same time, finite element analysis is conducted on the shovel blade using SolidWorks software to ensure the shovel blade has the appropriate strength through the comprehensive analysis of the track-type feeder wheel, limit cutting mechanism, and drawing mechanism. The drawing speed is between 0.78–1.56 m/s. It is determined that the most suitable angle for the pulling system is 40°, and the Adams simulation proves that the system's design is reasonable.

Keywords A taro harvester · Valve-like · Track flail wheel · Position-limiting cutter

Y. Qi (✉) · L. Yu · L. Hou · M. Liu · K. Guan
Hohai University, Changzhou, Jiangsu, China

© The Author(s) 2026

Z. Zhang, P. Mercorelli (eds.), *Mechanical Design and Simulation: Insights and Innovations*, Lecture Notes in Mechanical Engineering,
https://doi.org/10.1007/978-981-95-2405-1_89

1 Introduction

In China, taro cultivation has a long history and a wide range of cultivation [1]. However, there are many challenges in the harvesting. Taro is cultivated throughout various regions of China [2], but for a long time, our country's main form of taro harvesting has relied on manual or semi-mechanized harvesting [3]. With the continuous improvement of the national standard of living, the market demand for taro continuously increases, and the planting area also increases yearly [4]. Therefore, how to improve the degree of mechanization in taro harvesting has become an urgent problem that needs to be solved.

In the practical application of harvesting machinery, some places directly use potatoes for harvesting. The tuber harvesters that China has successfully developed are mainly divided into general-purpose and special-purpose types [5]. However, these models are mainly suitable for the sandy soil. Due to the nature of the shear stress of clay soil increasing with the increase of shear displacement [6], the effect of this machine on a heavy clay soil environment is not good. The pulling-type taro harvester studied by Qin Zhanqiang of Jilin Agricultural University [7] and the vibration shovel-type harvester studied by Wang Hongbo of Huazhong Agricultural University [8], although they have made certain innovative designs for the heavy clay soil situation, still face the problem of overcoming the stem falling during the loose soil process. Given the backward situation of mechanized taro harvesting, it is urgent to study a taro harvester suitable for various types of soil operations. Therefore, this paper designs a taro harvester suitable for harvesting taro in medium and heavy clay soil environments, which can provide a new type of machine for taro harvesting.

2 Overall Structure and Working Process

2.1 Overall Structure

The taro harvester is designed to harvest taro by pulling it up. This paper designs a valve-like supporting device to prevent the taro from falling over during harvesting. A hinge-type position-limiting cutter is designed to achieve accurate cutting at the connection between the taro fruit and the stem. A track-type raking device is designed to facilitate the entry of the taro into the pulling system. This device uses friction to allow the taro stalk to enter the pulling system. The pulling system ensures that the taro can be pulled out after the loosening mechanism has pre-loosened the soil. The loosening mechanism pre-loosens the soil around the taro through the coordinated action of the rotating brush and the shovel, reducing the pulling resistance and protecting the integrity of the taro fruit. The overall structure is shown in Fig. 1.

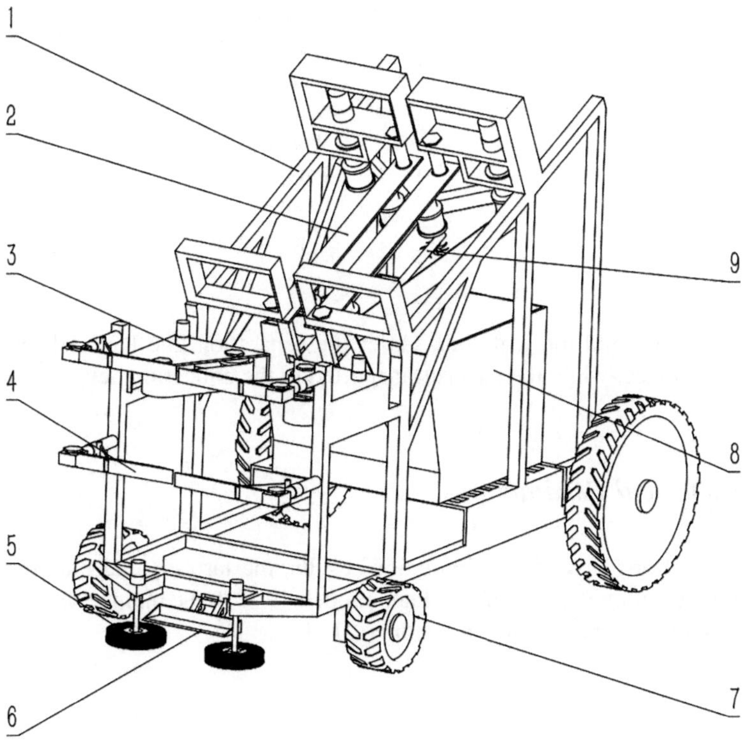

Fig. 1 Overall structure diagram

Here, the first is rack, the second is the pulling mechanism, the third is track-type feeder wheel, the fourth is a valve-like supporting mechanism, the fifth and the sixth are soil loosening mechanisms, the seventh is wheel assembly, the eighth is collection bin, and the ninth is the hinge-type position-limiting cutter.

2.2 *Working Process*

After the taro harvester is activated, it moves forward along the harvesting route. Upon reaching the location of the taro, the leaf stalks enter the valve-like supporting device. The supporting device locks the leaf stalks inside the machine through an automatic spring reset function. Then, the copper wire brush is activated to perform soil loosening work. The digging shovel breaks the soil and lifts the entire taro while moving forward. The track-type taro feeder wheel adjusts the leaf stalks, guiding them into the pulling system through friction. The pulling system continues transporting the leaf stalks to the device's rear. During the transportation process, the taro pushes the hinge-type position-limiting cutter upwards. When it reaches a designated position, the cutting device on the position-limiting cutter cuts the connection

Table 1 Main parameters of taro harvester

Parameters	Value
Dimensions (L × W × H) (mm × mm × mm)	1332 × 652 × 1255
Velocity (m s^{-1})	0.5–1
Working width (mm)	600
Power (kW)	4.5–5
Working lines	1
Digging depth (mm)	105–250

between the taro and the leaf stalk. After the separation, the leaf stalk is carried outside the device, while the taro is transported to the collection device.

2.3 Technical Parameters

Based on the taro harvester's simulation results, the taro's actual planting conditions, and the parameters of other related models [9–14], the main technical parameters of the entire machine are shown in Table 1.

3 Key Component Design

3.1 Valve-Like Supporting Mechanism

The supporting arm incorporates a spring and crank-slider mechanism to perform a function akin to that of a heart valve, as depicted in Fig. 2a. Once the leaf stalk enters the inlet, the mechanism resets and achieves reverse locking under the influence of the spring, ensuring that the leaf stalk cannot fall out. Subsequently, the leaf stalk is conveyed to the pulling mechanism by the track-type feeder wheel.

Figure 2b is a schematic diagram for parameter calculation, from which the formula is derived:

$$L_1 = \frac{H}{\sin(90^\circ - \varphi)} \tag{1}$$

$$L_2 = L_1 * \frac{\sqrt{2}}{2} \tag{2}$$

$$\Delta x = \frac{L_1}{2} * \cos\varphi + L_2 * \cos\beta - L_2 * \sin\theta \tag{3}$$

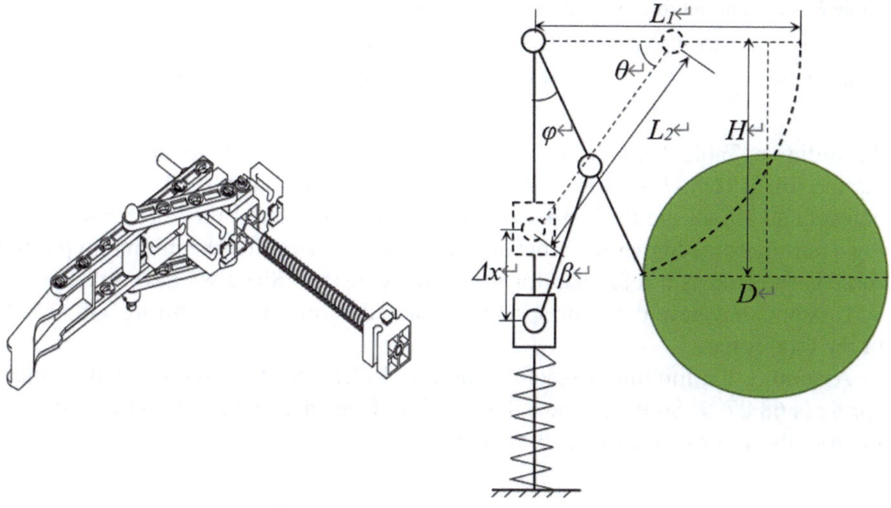

(a) (b)

Fig. 2 Valve-like supporting mechanism

$$cos\beta = \sqrt{1-\sin\beta^2} = \sqrt{1-\left(\frac{L_1}{2*L_2}*sin\varphi\right)^2} \tag{4}$$

L_1 the length of the arm of the support device; L^2 the length of the connecting rod; Δx the compression length of the spring; θ the initial angle between L_1 and L_2; φ the angle between L_1 and the frame when the device is about to reset; β the angle between L_2 and the frame when the device is about to reset; D the diameter of the taro leaf stalk; H the vertical distance from the center of the leaf stalk to the initial position of the arm of the support device.

Based on the study of taro plants, the average diameter D of the leaf stalk is 250 mm [15]. The support device must complete the locking before the soil loosening mechanism starts soil loosening, so H must be less than 200 mm (the horizontal distance between the two mechanisms). Therefore, H is set to 150 mm. For the stability of the mechanism and the physical model test, the angle φ should not be too small. Otherwise, the angular momentum during the rebound will affect the stability. Therefore, φ is taken as 30°. When θ is 45°, the compression and rebound processes are the most smooth.

In summary, it can be calculated that $L_1 = 173$ mm, $L_2 = 122$ mm, and $\Delta x = 83$ mm.

After obtaining the value of Δx, the spring constant k can be calculated. The formula is as follows:

$$F = k * \Delta x \tag{5}$$

Table 2 Taro pull-out force after loosening the soil

Project	Maximum (N)	Minimum (N)	Average value (N)	Standard deviation
Pull-out force	203.13	137.74	169.47	24.83

According to Table 2, the average pulling force required to pull up taro after loosening the soil is 169.47 N. However, in this paper, the crop support device comes into contact and allows the leaf stalk to pass through before the soil is loosened, requiring a pulling force higher than that after loosening. Therefore, the minimum pulling force is selected as a reference for the spring constant. Since we have two springs and assume they have the same spring constant, the force on each spring will be half of the total force.

Assuming the minimum reaction force is 137.74 N, the force exerted on each spring is 68.87 N. So the kmax = 829.75 N/m. Considering the rebound effect during use, the value of k chooses 300 N/m.

3.2 Soil Loosening and Excavation Mechanism

The soil loosening and excavation mechanism consists of copper wire brushes and excavation shovels.

1. The copper wire brush is used for soil loosening before the shovel, effectively reducing the resistance of the soil to the shovel's operation and allowing the shovel to cut into the soil more smoothly.

   As shown in Fig. 3a, the white bud taro's fruit depth is 200 mm, and the thickness is 50 mm. Therefore, a brush with a 50–60 mm thickness is selected so the bristles can fully cover the taro fruit. It is also known that one side of the taro is 175 mm away from the ridge edge. To improve the efficiency of soil loosening and reduce the impact on the fruit, a brush with a 160–180 mm diameter is preferable. In summary, we choose a copper wire brush with a thickness of 50 mm and a diameter of 180 mm, which can reduce the resistance for the digging shovel by 82.49% and still maintain efficient soil loosening performance even in relatively hard soil.
2. This excavation shovel is mainly composed of flat surfaces, and its core parameters include shovel length D, shovel width L, shovel face inclination angle θ, shovel face height h1, shovel handle height h2, and the angle between the shovel face and handle δ, as shown in Fig. 3b.

According to the research on the excavation shovel for the harvesting of tuber crops, the shovel face inclination angle is generally between 14° and 25°, and the friction angle of the soil to rigid materials is between 22° and 45°. Taking the xOy plane as the soil plane, the maximum inclination angle θ of the excavation shovel during operation should, in principle, be less than the friction angle of the object on the

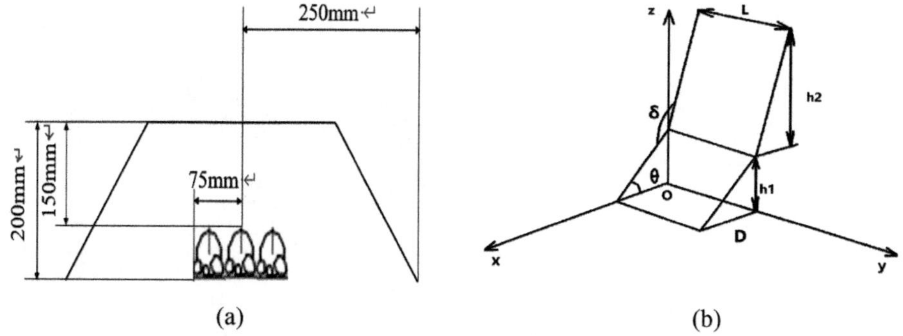

Fig. 3 Taro planting diagram and shovel calculation diagram

shovel face to the shovel. Therefore, after comprehensive consideration, the shovel face inclination angle is taken as 20°.

Taking the white sprout taro as an example, the planting ridge for the taro is 500 mm wide, the ridge is 150 mm high, and the spacing between plants is approximately 300 mm. The depth of the mature fruit is 200 mm. Considering that the fruits of the taro usually grow in clusters, with multiple fruits on a single plant with a total diameter ranging from 75 to 150 mm, to include all the fruits and to ensure that only one taro plant is excavated at a time, the shovel width L is set to 150 mm, and the shovel length D is set to 110 mm.

From Fig. 1, it can be seen that there is the following relationship between the shovel face height h1, the shovel length D, and the shovel face inclination angle θ:

$$h1 = D * tan\theta \tag{6}$$

The design principle of the shovel handle height h2 is adjustable to cope with the differences in ridge height and taro depth in different regions. At the same time, different shovel handle lengths can be used to consider the degree of connection of the taro. This machine can achieve an adjustable range of the shovel handle from 150 mm to 350 mm.

We perform a mechanical analysis of the shovel blade, and the formula is shown in Eq. (3).

$$F_q = F_p + f \tag{7}$$

$$F_p = SL'\rho g tan(\lambda + \xi) \tag{8}$$

Fq traction force required for the device to work normally in Newtons (N), F_p force required for the excavation shovel to cut through the soil in Newtons (N), f soil adhesion force in Newtons (N).

$F_q = 2373.63$ N is imported into SolidWorks for finite element analysis.

The stress analysis results are shown in the following Fig. 4. According to the third strength theory, the shovel and the parts connected to the main body meet the

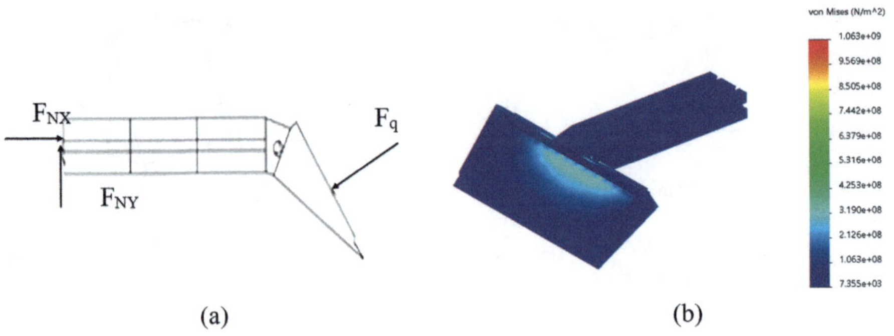

Fig. 4 Finite element analysis diagram of a blade

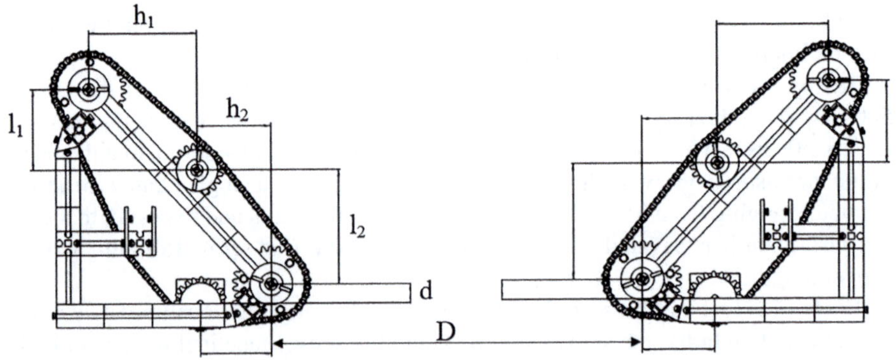

Fig. 5 Track-type taro feeder wheel

stress requirements, and there is no situation where the stress exceeds the maximum that the structure can withstand.

Figure 4a, b are force and stress finite element analyses. The strain and displacement analysis results show that the part most prone to deformation is the front end of the shovel blade because the front end is the thinnest and comes into contact with the soil first. Digging into sandy loam soil should be easier than ordinary soil so the deformation situation will be better than the software analysis.

3.3 Track-Type Taro Feeder Wheel

The mechanism adopts a track-type pulley, as shown in Fig. 5. The wheel can evenly distribute power through chain transmission, ensuring continuity and stability. By utilizing the friction force of the timing belt, the track can guide the leaf stalk into the pulling system. This helps to ensure the stable feeding of the leaf stalk and can prevent it from swaying during the harvesting process, thereby improving the accuracy of the harvest. The speed of the wheel's timing belt is consistent with the speed of the pulling system's timing belt.

The weeding wheel is mainly composed of gears and tracks, and its core parameters include the weeding angle θ, gear longitudinal spacing h1, h2, gear width spacing 11, 12, and the spacing between the weeding wheels D (h1 = h2, 11 = 12). Considering that the diameter of a single taro leaf stalk is about 50 mm and the number of leaf stalks per plant is around 3–5, the machine is designed with D = 120 mm and θ = 45°, ensuring that leaf stalks of different sizes can enter the subsequent mechanism while preventing the leaf stalks from falling over due to excessive spacing between the weeding wheels. Let the total length of the gear be $H = h_1 + h_2$, and the total width $L = l_1 + l_2$, then as long as H, L > D is ensured. To allow time for the weeding wheel to correct the taro leaf stalks, the actual total length and width should be greater than D. Experimental measurements indicate that the general correction time for the leaf stalks is about 2 s, and n = 75 r/min, combined with the gear diameter d = 5 cm:

$$H = h1 + h2 = 2 * \pi * d * (n/60) * sin\theta \tag{9}$$

$$L = l1 + l2 = 2 * \pi * d * (n/60) * cos\theta \tag{10}$$

By substituting the data, we get that $H = L = 277.4975$ mm under normal circumstances. Therefore, the machine is designed with $H = L = 300$ mm.

Compared to the traditional mechanism [16] which is illustrated in Fig. 6, the track-type taro feeder wheel mechanism reduces the breaking of leaf stalks during the process, protecting the integrity of the crops. This design is beneficial as it minimizes crop damage, which can be particularly important for maintaining the quality and yield of the harvest.

Due to the design and operation of the track-type taro feeder wheel, the possibil ity of leaf stalks getting stuck during the harvesting process is reduced, which in turn lowers the risk of mechanical faults, minimizing damage to the leaf stalks and reducing blockages.

Fig. 6 Traditional taro feeder wheel

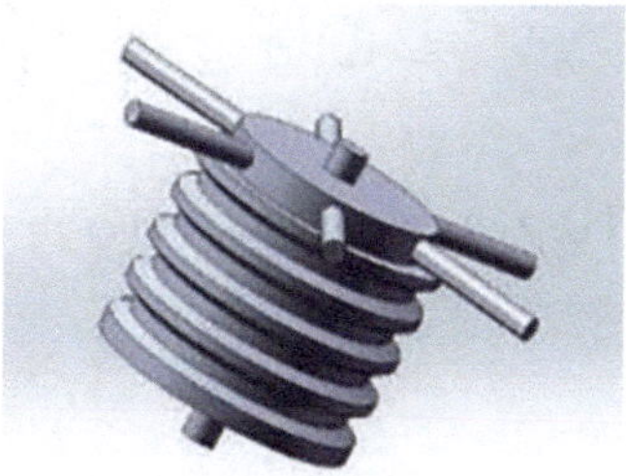

3.4 Position Limiting Cutter

The position-limiting cutting machine mainly consists of the following key components: hinge, cutting device, and strip-shaped fixing device.

It is articulated on two hinges and made of Q235 steel. According to the data, the diameter of a single taro leaf stalk is between 3–5 cm, and a single taro plant consists of 5–6 leaf stalks. Therefore, the spacing of the two strip-shaped devices is set at 250 mm. As the taro fruit rises during the transmission process, it will drive the ascent of the fixing device, ensuring that the cutting position is always at the connection point between the fruit and the leaf stalk. Once it reaches the designated position, the cutting device cuts the connection between the taro and the leaf stalk, separating the taro from the leaf stalk. The actual object diagram is as shown in Fig. 7.

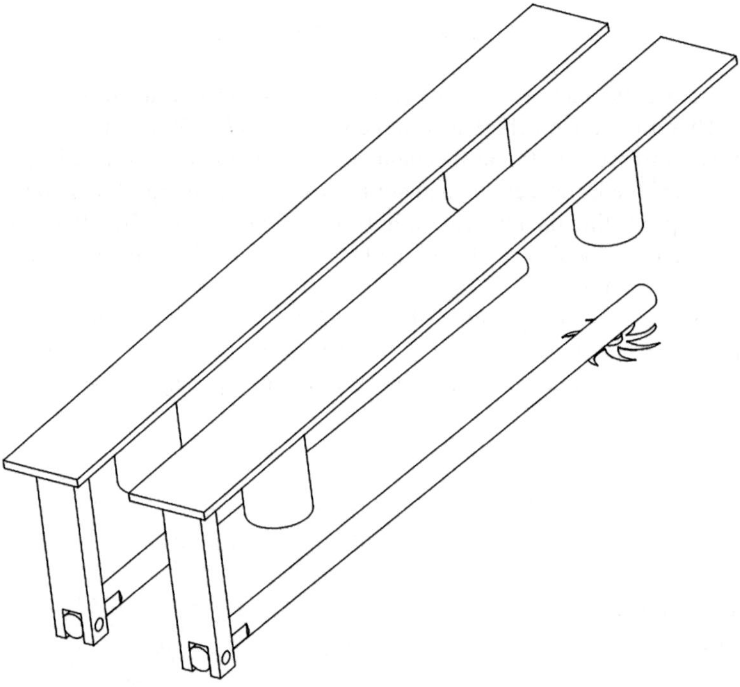

Fig. 7 Position limiting cutter

3.5 Pulling System

The pulling effect is influenced by the opening angle of the pulling belt, as shown in Fig. 8 below. The pulling belt exerts a force on the stem that converges, lifts, and pushes over the stem. The combined force can be orthogonally decomposed into horizontal and vertical components *Tx* and *Ty*.

$$T_x = T * \sin\frac{c}{2} \tag{11}$$

$$T_y = T * \cos\frac{c}{2} \tag{12}$$

T the combined force exerted by the pulling belt on the stem, *C* the opening angle of the two sides of the synchronous belt.

From Eq. (11), it can be deduced that the opening angle C is proportional to the force that pushes and topples the plant. However, the smaller the opening angle C, the closer the pulling belt is to the ridge surface, which requires a higher degree of flatness in the field and increases the gathering time, potentially affecting the pulling effect. Therefore, an opening angle of C = 120° is most suitable.

The spacing between the pulling belts is related to the thickness of the stem; if it is too large or too small, a gap can affect the pulling effect. According to reliable data, a spacing y of 25 mm and a clamping point height L1 of 120 mm are optimal. Additionally, to ensure the pulling system can capture thinner leaf stalks, the machine is equipped with tensioning devices on both sides of the pulling system, which apply pressure to leaf stalks of varying thicknesses to ensure they do not slip when in contact with the synchronous belt, thus preventing the inability to pull up the taro.

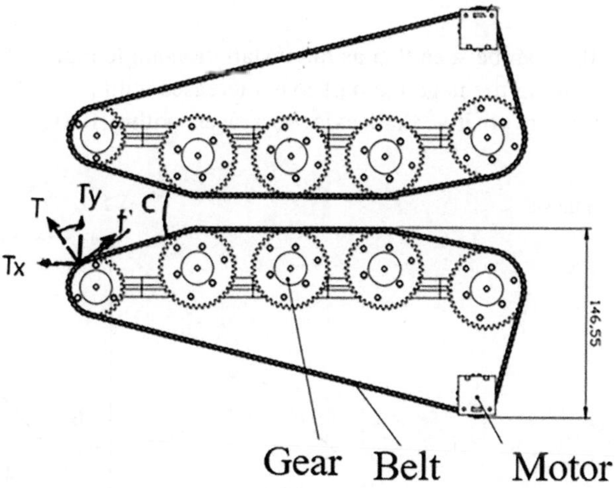

Fig. 8 The belt exerts a force on the stem

4 Simulation Analysis

The stem should be pulled to ensure the success rate of stem pulling and minimize the pulling force vertically upwards. The absolute pulling speed is influenced by the machine's forward speed, the pulling belt speed, and the installation angle of the pulling device [17], as shown in Fig. 9.

$$F = \frac{v_1}{v_2} = \frac{\sin(a+b)}{\sin b} = \frac{1}{\cos a} \tag{13}$$

$$a + b = 90^\circ \tag{14}$$

F pulling speed ratio; *a* installation inclination; V_0 machine forward speed, in meters per second (m/s); V_1 pulling belt speed, in meters per second (m/s); V_2 absolute speed at the pulling point, in meters per second (m/s).

From Eq. (13), it can be seen that the pulling belt speed is directly proportional to the installation angle. That is, the smaller the pulling belt speed, the smaller the installation angle. Under the condition of conveying to the same height, the overall machine becomes longer. Therefore, we check the information to find out that F = 1.3 is the best. From this, it is known that when cos a = 0.76, that is, a = 40°, it is the most suitable. The pulling belt speed ranges between 0.78–1.56 m/s.

The different installation angles lead to different directions of frictional force on the taro petiole, thereby affecting the pulling of the taro, especially its horizontal displacement. If the horizontal displacement is too large, it will cause the petiole to slip off. In order to determine the optimal installation angle, this paper uses ADAMS for simulation. Since the synchronous belt primarily provides static friction to pull the taro, we simulate the static friction of the pulling system by applying forces to the taro leaf stalk at angles of 20°, 30°, 40°, and 50° with the horizontal plane. The horizontal displacement of the lowest point of the taro is measured, as shown in Fig. 10.

From Fig. 10, it can be seen that as the installation angle increases, the displacement of the taro along the negative half-axis increases within 0–0.15 s, but the displacement along the positive half-axis decreases within 0–0.5 s. The negative

Fig. 9 Motion synthesis

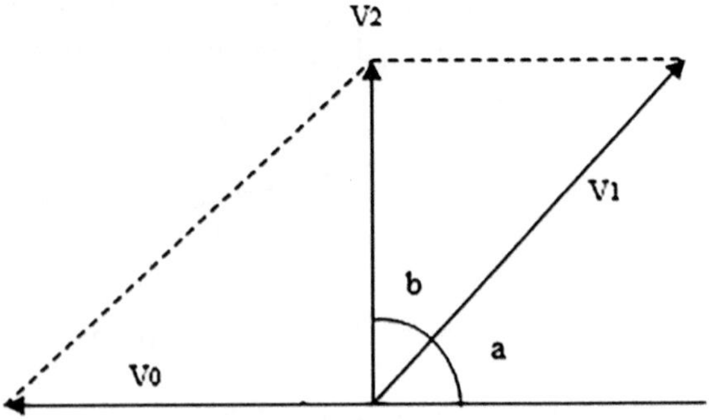

half-axis displacement is the swing amplitude of the taro during the pulling operation. To ensure the pulling operation is stable, the swing amplitude should not be too large; otherwise, the petioles will be prone to falling off. The displacement along the positive half-axis is related to the length of the machine's body. Considering the machine's cost and movement, the body length should not be too long; otherwise, it not only increases the cost, but the motor power may not be sufficient. Combining the information from the two charts above, it is known that although the swing amplitude is not large at 20° and 30°, the positive displacement within 0.5 s is too large, which greatly increases the cost. At 50°, the swing amplitude is too large, causing the petioles to fall off easily. However, at 40°, the taro's swing amplitude is not large, and the positive displacement is moderate, which aligns with the optimal calculation results.

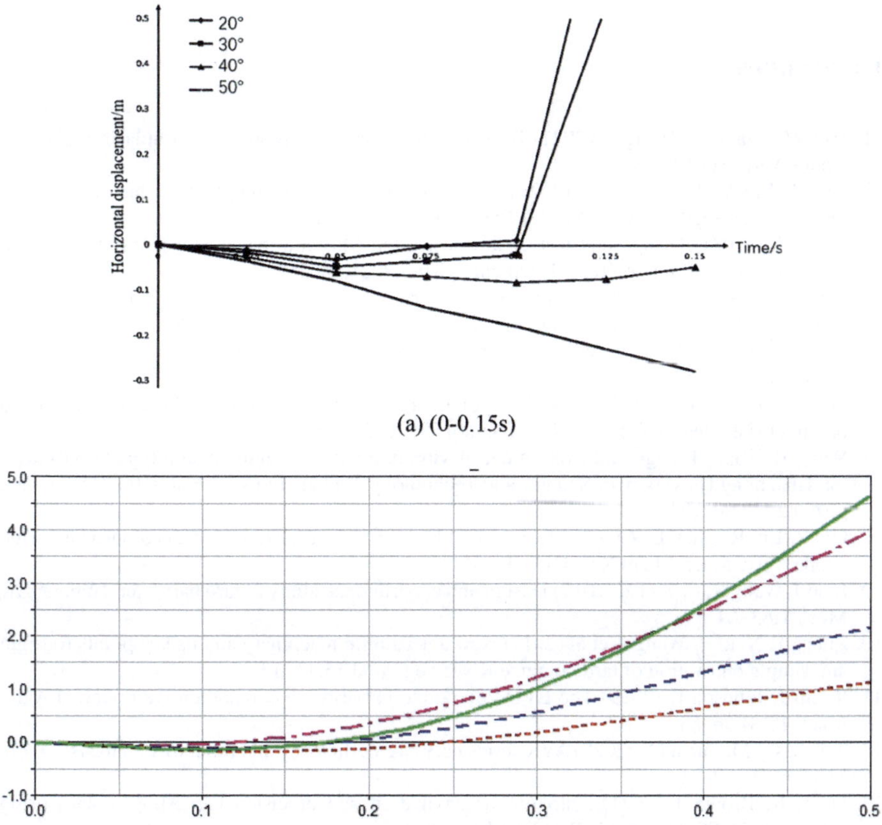

Fig. 10 Horizontal displacement of taro at different inclination angles of the drawing system. (**a**) (0–0.15 s), (**b**) (0–0.5 s)

5 Conclusions

This article proposes a taro harvester suitable for most taro harvesting environments, taking the white sprout taro as the main research object. The overall dimensions of the machine are designed to be 1332 × 652 × 1255 (length × width × height). Innovative designs include a valve-like taro supporter to prevent the taro from falling over through reverse locking, a hinge-type limit cutting device to ensure the cutting position is always at the connection between the fruit and the petiole, and a track-type taro feeder wheel that accurately feeds the petiole into the pulling system. Additionally, a soil loosening mechanism, pulling system, and collection device are designed. The ADAMS software is utilized to study the horizontal displacement of the taro petiole at different inclinations of the pulling system to verify whether the theoretically derived 40° inclination is suitable. The results indicate that the pulling system's design requirements are met at a 40° angle.

References

1. Wei W, Chang Q, Wang A (2021) Research progress on chinese taro germplasm resources. Anhui Agric Sci 49(14):4–7
2. Guo J, Yin Y, Tang K et al (2021) Research progress and prospects on taro germplasm resources. Guangdong Agric Sci 48(9):81–90
3. Zhang M (2021) Design and test of taro harvester. Master's thesis, Huazhong Agricultural University, https://doi.org/10.27158/d.cnki.ghznu.2021.000562
4. Qin Z 2020 Design and research of red sprout taro harvester. Master's thesis. Jiangxi Agricultural University. https://doi.org/10.27177/d.cnki.gjxnu.2020.000087
5. Ma S, Huang G (2017) Research and application of dual drive potato harvester. Agric Mach Qual Supervis 10:24–25
6. Sun T, Chen Y, Shi X et al (2024) Study on the mechanical properties of different clays based on direct shear test. J Yancheng Inst Technol 37(01):31–36
7. Wang H (2022) Design and experiment of vibration shovel digging clamp type taro excavator. Huazhong University of Science and Technology, Wuhan. https://doi.org/10.27158/d.cnki.ghznu.2022.001124
8. Qin Z, Liu R, Chen L, Zeng Y, He X (2020) Design and simulation analysis of taro harvester. China Agric Mach Chem Rep 41(1):31–36
9. Liao J, Wang R, Li X et al (2017) Design and experimental study of taro harvester. Hubei Agric Mech 1:63–64
10. Zhang P, Xiao H, Wang A et al (2017) New agricultural machinery and its key points for light and simple cultivation of taro. Yangtze River Veg Food 15:11–13
11. Smith J, Johnson K (2019) The EVO-280 potato harvester: a comprehensive review. J Agric Mach 45(3):56–63
12. Takahashi M, Suzuki Y (2020) AVR series harvester: efficiency and reliability analysis. J Farm Mach 67(1):12–21
13. Davis R, Brown T (2021) Autonomous navigation and precision harvesting of taro using LiDAR and GPS. Int J Agric Biol Eng 14(3):1–9
14. Wang Q, Liu P (2023) Application of deep learning in taro recognition and localization for autonomous harvesting. Agric Eng Int CIGR J 25(1):1–12

15. Xiang J (2015) High yield planting technology of "Bai Yu No.1" taro in Changle County, Shandong Province. North Hortic 20:54–55
16. He F, Chen Z, Zhang Z, Han Z, Gan B, Qiao X (2014) Design and experiment of a wheel type corn harvesting platform for cropping. J Agric Mach 45(06):112–117
17. Hou J, Chen Y, Li Y et al (2020) Development of a self-propelled onion combine harvester with quantitative placement. J Agric Eng 36(7):22–33

Open Access This chapter is licensed under the terms of the Creative Commons Attribution-NonCommercial-NoDerivatives 4.0 International License (http://creativecommons.org/licenses/by-nc-nd/4.0/), which permits any noncommercial use, sharing, distribution and reproduction in any medium or format, as long as you give appropriate credit to the original author(s) and the source, provide a link to the Creative Commons license and indicate if you modified the licensed material. You do not have permission under this license to share adapted material derived from this chapter or parts of it.

The images or other third party material in this chapter are included in the chapter's Creative Commons license, unless indicated otherwise in a credit line to the material. If material is not included in the chapter's Creative Commons license and your intended use is not permitted by statutory regulation or exceeds the permitted use, you will need to obtain permission directly from the copyright holder.

The Kinematic Simulation Study of the Working Device of an Unmanned Bulldozer

Chunming Miao, Quancheng Dong, Congfeng Tian, Ziyao Xu, and Guangxiao Shen

Abstract This paper first conducts a kinematic analysis of the bulldozer's working device, deriving the transformation matrices between each joint and formulating the relationship between the blade's end-effector position and orientation and the joint angles in the Cartesian coordinate system. Next, using a geometric method, the joint space of the working device is converted into the drive space, clarifying the mapping relationship between joint angles and the hydraulic cylinder's extension length. Finally, a simulation model of the working device's linkage mechanism is established, and four sets of joint angles are selected to verify the forward and inverse kinematics analysis. Additionally, a simulation model of the working device's hydraulic system is developed, experimental parameters are set, and the accuracy of the relationship between the joint angles and the hydraulic cylinder's extension length is validated. This provides a theoretical foundation for subsequent research on the operation planning and control of unmanned bulldozer working devices.

Keywords Unmanned bulldozer · Working device · Kinematic analysis · Hydraulic system

1 Introduction

Bulldozers are mainly used to perform tasks such as pushing earth, and filling and leveling the ground in large earthwork projects [1]. In many cases, the operating environment of bulldozers is relatively harsh. For example, when operating in environments such as soil piles, slopes and trenches, problems such as uneven vehicle bodies and excessive load on the blade will be encountered [2]. In harsh

C. Miao · Q. Dong (✉) · Z. Xu · G. Shen
School of Mechanical Engineering, University of Jinan, Jinan, Shandong, China
e-mail: me_dongqc@ujn.edu.cn

C. Tian
Shantui Construction Machinery Co., Ltd, Jining, Shandong, China

© The Author(s) 2026

Z. Zhang, P. Mercorelli (eds.), *Mechanical Design and Simulation: Insights and Innovations*, Lecture Notes in Mechanical Engineering,
https://doi.org/10.1007/978-981-95-2405-1_90

environments, it especially tests the skills of drivers. Problems such as construction errors, insufficient construction process data, and low construction efficiency may occur, and construction quality cannot be maintained. Moreover, for drivers, precise construction becomes more difficult, and it is also difficult to effectively protect the safety of drivers. Therefore, intelligent and unmanned bulldozers will become an important development direction in the future. They will meet the market demand for intelligent bulldozers and have broad prospects and practical application value [3, 4]. At present, most research on the automation of earthmoving machinery focuses on excavators, and there is less research on bulldozers.

The working device and its operating system of a bulldozer are the most critical parts of the actual work of a bulldozer. The working device is subject to large amplitude and high-frequency changes in external soil loads. Hence, the driver needs to operate the working device to lift and pitch continuously to change the position of the blade. Therefore, to achieve unmanned operation of the bulldozer's working device, it is necessary to perform position estimation on the working device, match the poses of the bulldozer body and the end of the working device, and then convert the pose information into the length of the hydraulic cylinder. By controlling the elongation and contraction of the hydraulic cylinder, the blade can reach different positions for precise operations and realize unmanned operation of the working device.

In order to realize the control of the working device of the bulldozer, Xue [5] analyzed the kinematics of the dozer by simulation, and Sun [6] studied a sensor fusion algorithm for estimating the attitude of the dozer blade. Through the RTKGPS installed on the bulldozer blade combined with the angle sensor, Kim and Lee et al. [7, 8] established a coordinate conversion model, which can obtain the spatial attitude coordinates of the blade through the tablet device.

Komatsu, a Japanese manufacturer of heavy equipment including excavators and bulldozers, mounted one GNSS sensor on top of a bulldozer cabin, a single IMU (inertial measurement unit) sensor in the cabin, and a stroke sensor in the cylinders for measuring the position and posture of the bulldozer blade. Then, the company manufactured a prototype and verified its performance [9].

As a heavy equipment company, Caterpillar obtains the pitch angle and speed information of the vehicle through the calculation of the least mean-squared estimation [10]. The proposed method can calculate the spatial parameters of the vehicle, which can assist the operator in understanding the vehicle information.

Omelchenko works in the Topcon positioning system. He has proposed a method and device to control the elevation angle and tilt angle of the bulldozer blade [11]. In addition, it is also necessary to analyze the kinematics of the bulldozer and use the homogeneous transformation matrix to estimate the accurate position and attitude of the blade edge.

Lundeen et al. at the University of Michigan in the United States proposed an optical-marker-based estimation method to estimate the end effector pose of an excavator. He attached an optical marker to the boom, arm, and bucket of the excavator and measured the position and direction of the marker in 3D Cartesian space using a camera. Through this, Lundeen estimated and verified the position of the excavator's bucket tip [12].

Guangcun Zhou and others regarded the excavator as a four-degree-of-freedom mechanical arm. After performing a kinematic analysis on its working device, they converted the rotation angles of each part of the excavating arm into the displacements of each hydraulic cylinder in the driving space [13].

At present, in the research on the control of the working devices of construction machinery such as bulldozers and excavators at home and abroad, their working devices are usually treated as multi-degree-of-freedom mechanical arms to control the end of the working device to track the planned desired trajectory [14]. This paper proposes a positioning method for the working device of an unmanned bulldozer. This method regards the bulldozer as a three-degree-of-freedom mechanical arm. First, we perform a kinematic analysis on the working device of the bulldozer, establish a mathematical model, and use the homogeneous transformation matrix to calculate the pose of the tooth tip at the end of the blade. Then, according to the coordinate information of the key points on the operation path, we perform an inverse kinematic solution to find out the angles at which each joint needs to rotate. Finally, we use the geometric method to convert the joint space and the driving space to find the control quantity, which is convenient for the subsequent unmanned bulldozer to accurately level the working device, meet the operation requirements and improve the operation efficiency.

2 Kinematic Analysis

The structure of the working device of a bulldozer is shown in Fig. 1. The working device of a bulldozer is mainly composed of a blade, a push frame, two blade lifting hydraulic cylinders, and two blade pitching hydraulic cylinders. The blade is the most crucial part of the bulldozer's working device. This article will analyze the blade. The movement of the blade is determined by the telescopic movement of the

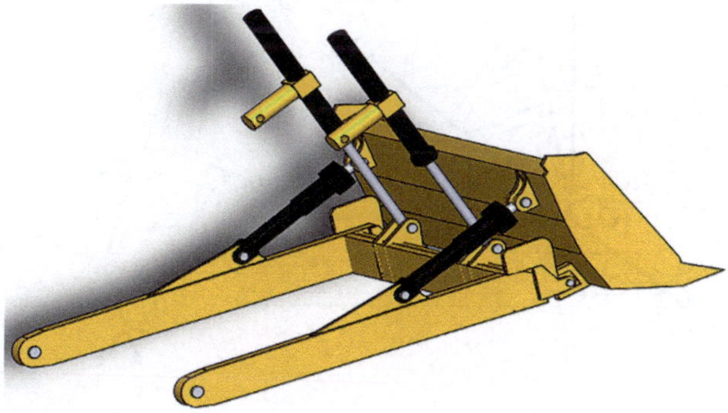

Fig. 1 Structure of the blade

hydraulic cylinder. When the lifting hydraulic cylinder contracts, the blade is lifted. When the hydraulic cylinder extends, the blade descends. This is called the lifting movement of the blade. When the pitching hydraulic cylinder contracts, the blade tilts upward. When the hydraulic cylinder extends, the blade tilts downward. This is called the pitching movement of the blade.

2.1 Forward Kinematics Solution

From the perspective of robotics, a bulldozer can be regarded as a robotic arm with three degrees of freedom. According to the establishment concept of the Denavit-Hartenberg coordinate system, the direction of the rotating shaft is set as the Z axis, the direction of the common perpendicular of the two axes is taken as the X axis, and then the Y axis is determined by using the right-hand rule. The rotary coordinate system/base coordinate system, lifting coordinate system, pitching coordinate system, and tooth tip coordinate system are established, respectively, as shown in Fig. 2.

The coordinate system $O\{X_O, Y_O, Z_O\}$ is the base coordinate system attached to the center of the bulldozer track. The coordinate system $A\{X_A, Y_A, Z_A\}$ defines the rotation center of the blade lifting motion. The origins of the base coordinate system $\{O\}$ and the lifting coordinate system $\{A\}$ are defined at the same rotation center. The coordinate system $B\{X_B, Y_B, Z_B\}$ defines the rotation center of the blade pitching motion. The coordinate system $C\{X_C, Y_C, Z_C\}$ is the coordinate system attached to the tooth tip of the blade. The position where the blade teeth are located is the designed surface required during bulldozer operation. In three-dimensional Cartesian space, the relationship between two adjacent coordinate systems is represented by a homogeneous transformation matrix using Z-Y-R Euler angles.

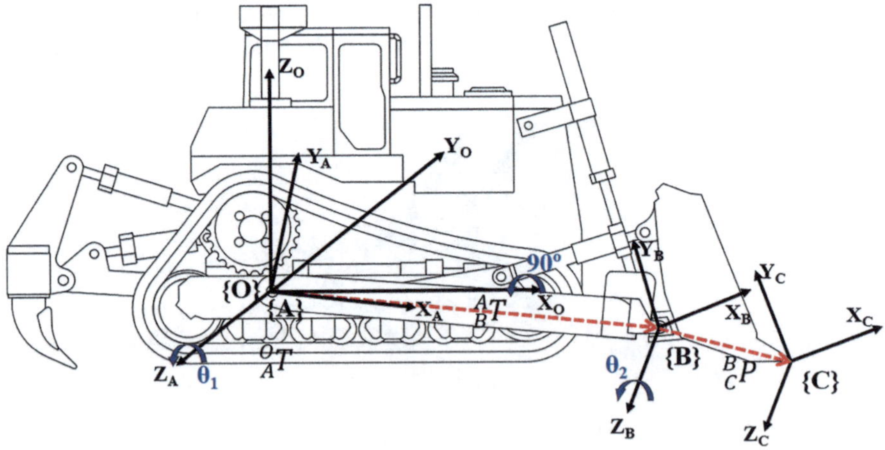

Fig. 2 Coordinate system structure for kinematic analysis of bulldozers

Among them, the relationship between the coordinate systems {O} and {A} is expressed as:

$$ {}_{A}^{O}T = R(x,90^\circ)R(z,\theta_1) = \begin{bmatrix} c\theta_1 & -s\theta_1 & 0 & 0 \\ 0 & 0 & -1 & 0 \\ s\theta_1 & c\theta_1 & 0 & 0 \\ 0 & 0 & 0 & 1 \end{bmatrix} \tag{1} $$

where $c\theta_1 = \cos\theta_1$, $s\theta_1 = \sin\theta_1$. Here, θ_1 is the angle at which the bulldozer blade is lifted around the Z_A axis. In three-dimensional Cartesian space, the position of the coordinate system {C} attached to the tooth tip of the blade is represented by P_C. We construct a homogeneous transformation matrix for it as follows and calculate its position.

$$ P_C = {}_{A}^{O}T\,{}_{B}^{A}T\,{}_{C}^{B}P = \begin{bmatrix} p_x \\ p_y \\ p_z \\ 1 \end{bmatrix} \tag{2} $$

Since there is both translation and rotation between coordinate system {B} and coordinate system {A}, then,

$$ {}_{B}^{A}T = Trans(x_{ab}, y_{ab}, z_{ab})R(z,\theta_2) $$

$$ = \begin{bmatrix} 1 & 0 & 0 & x_{ab} \\ 0 & 1 & 0 & y_{ab} \\ 0 & 0 & 1 & z_{ab} \\ 0 & 0 & 0 & 1 \end{bmatrix} \begin{bmatrix} c\theta_2 & -s\theta_2 & 0 & 0 \\ s\theta_2 & c\theta_2 & 0 & 0 \\ 0 & 0 & 1 & 0 \\ 0 & 0 & 0 & 1 \end{bmatrix} $$

$$ = \begin{bmatrix} c\theta_2 & -s\theta_2 & 0 & x_{ab} \\ s\theta_2 & c\theta_2 & 0 & y_{ab} \\ 0 & 0 & 1 & z_{ab} \\ 0 & 0 & 0 & 1 \end{bmatrix} \tag{3} $$

$$ {}_{C}^{B}P = \begin{bmatrix} l_{bw} \\ l_{bh} \\ 0 \\ 1 \end{bmatrix} \tag{4} $$

where, $c\theta_2 = \cos\theta_2$, $s\theta_2 = \sin\theta_2$. Here, θ_2 is the angle at which the bulldozer blade pitches around the Z_B axis. x_{ab}, y_{ab}, and z_{ab} represent the distance from the origin of the pitching coordinate system {B} to the origin of the lifting coordinate system {A}. Since only a translational transformation along the x-axis is performed from

Fig. 3 Dimensions of the blade

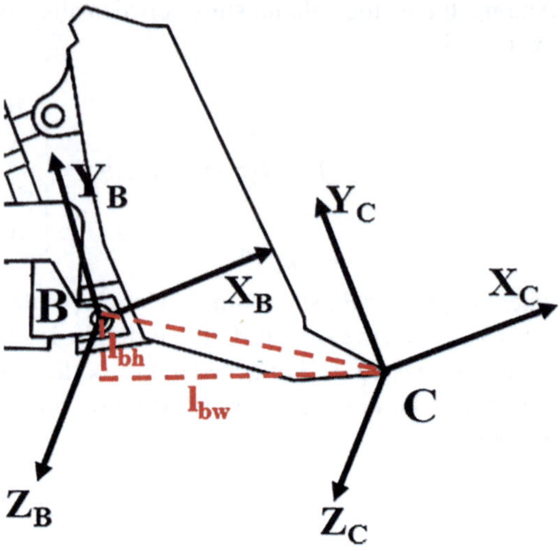

coordinate system {A} to coordinate system {B}, x_{ab} is equal to the length l_1 of the bulldozer blade push rod, $y_{ab} = 0, z_{ab} = 0, l_{BW}$ is the thickness of the bulldozer blade, and l_{bh} represents the height difference from the origin of the tooth tip coordinate system {C} to the pitching coordinate system {B}, as shown in Fig. 3.

Substituting Eqs. (1), (3), and (4) into Eq. (2) yields:

$$P_C = \begin{bmatrix} p_x \\ p_y \\ p_z \\ 1 \end{bmatrix}$$

$$\begin{cases} p_x = l_{bw}\left(c\theta_1 c\theta_2 - s\theta_1 s\theta_2\right) - l_{bh}\left(c\theta_1 s\theta_2 + s\theta_1 c\theta_2\right) \\ \qquad + x_{ab} c\theta_1 - y_{ab} s\theta_1 \\ p_y = -z_{ab} \\ p_z = l_{bw}\left(s\theta_1 c\theta_2 - s\theta_1 s\theta_2\right) - l_{bh}\left(s\theta_1 s\theta_2 - c\theta_1 c\theta_2\right) \\ \qquad + x_{ab} s\theta_1 + y_{ab} c\theta_1 \end{cases} \tag{5}$$

The obtained P_C is the position and orientation information of the tooth tip at the end of the blade in the base coordinate system of the bulldozer.

2.2 *Inverse Kinematics Solution*

The above forward kinematics solution is to obtain the position and orientation information of the tooth tip at the end of the blade in the base coordinate system under the condition that the rotation angles of each joint are known. However, if we want to control the bulldozer blade to complete various pose tasks without human operation, inverse kinematics needs to be performed to obtain the rotation angles of each joint in the joint space. The process of solving the inverse kinematics of the blade is as follows:

$$P_C = {}^O_A T\, {}^A_B T\, {}^B_C P = \begin{bmatrix} p_x \\ p_y \\ p_z \\ 1 \end{bmatrix}, \left({}^O_A T\right)^{-1} = \begin{bmatrix} c\theta_1 & 0 & s\theta_1 & 0 \\ -s\theta_1 & 0 & c\theta_1 & 0 \\ 0 & -1 & 0 & 0 \\ 0 & 0 & 0 & 1 \end{bmatrix}$$

Multiplying this equation obtains:

$$\left({}^O_A T\right)^{-1} P_C = \left({}^O_A T\right)^{-1} {}^O_A T\, {}^A_B T\, {}^B_C P$$

$$\begin{bmatrix} c\theta_1 & 0 & s\theta_1 & 0 \\ -s\theta_1 & 0 & c\theta_1 & 0 \\ 0 & -1 & 0 & 0 \\ 0 & 0 & 0 & 1 \end{bmatrix} \begin{bmatrix} p_x \\ p_y \\ p_z \\ 1 \end{bmatrix} = \begin{bmatrix} c\theta_2 & -s\theta_2 & 0 & x_{ab} \\ s\theta_2 & c\theta_2 & 0 & y_{ab} \\ 0 & 0 & 1 & z_{ab} \\ 0 & 0 & 0 & 1 \end{bmatrix} \begin{bmatrix} l_{bw} \\ l_{bh} \\ 0 \\ 1 \end{bmatrix}$$

$$\begin{bmatrix} p_x c\theta_1 + p_z s\theta_1 \\ -p_x s\theta_1 + p_z c\theta_1 \\ -p_y \\ 1 \end{bmatrix} = \begin{bmatrix} l_{bw} c\theta_2 - l_{bh} s\theta_2 + x_{ab} \\ l_{bw} s\theta_2 + l_{bh} c\theta_2 + y_{ab} \\ z_{ab} \\ 1 \end{bmatrix}$$

The corresponding parts on both sides of the equation are equal, and we obtain.

$$p_x c\theta_1 + p_z s\theta_1 = l_{bw} c\theta_2 - l_{bh} s\theta_2 + x_{ab} \tag{6}$$

$$-p_x s\theta_1 + p_z c\theta_1 = l_{bw} s\theta_2 + l_{bh} c\theta_2 + y_{ab} - p_y = z_{ab} = 0 \tag{7}$$

First, we multiply Eq. (6) by p_z and multiply Eq. (7) by p_x. Then, we subtract the two equations.

$$\begin{aligned} & p_z\left(p_x c\theta_1 + p_z s\theta_1\right) - p_x\left(-p_x s\theta_1 + p_z c\theta_1\right) \\ & = p_z\left(l_{bw} c\theta_2 - l_{bh} s\theta_2 + x_{ab}\right) - p_x\left(l_{bw} s\theta_2 + l_{bh} c\theta_2 + y_{ab}\right) \end{aligned}$$

Simplifying the above equation gives:

$$\left(p_x^2 + p_y^2\right) s\theta_1 = \&\left(p_z l_{bw} - p_x l_{bh}\right) c\theta_2 - \left(p_z l_{bh} + p_x l_{bw}\right) s\theta_2 + x_{ab} p_z - y_{ab} p_x \tag{8}$$

We multiply Eq. (6) by p_x and multiply Eq. (7) by p_z. Then, we add the two equations.

$$\begin{aligned} \left(p_x^2 + p_y^2\right) c\theta_1 = & \left(p_x l_{bw} + p_z l_{bh}\right) c\theta_2 + \left(p_z l_{bw} - p_x l_{bh}\right) s\theta_2 \\ & + x_{ab} p_x + y_{ab} p_z \end{aligned} \tag{9}$$

Then, we divide Eq. (8) by Eq. (9) to obtain the expression of θ_1:

$$\tan\theta_1 = \frac{(p_z l_{bw} - p_x l_{bh})c\theta_2 - (p_z l_{bh} + p_x l_{bw})s\theta_2 + x_{ab}p_z - y_{ab}p_x}{(p_x l_{bw} + p_z l_{bh})c\theta_2 + (p_z l_{bw} - p_x l_{bh})s\theta_2 + x_{ab}p_x + y_{ab}p_z}$$

$$\theta_1 = \arctan\left\{\begin{array}{l}\left[(p_z l_{bw} - p_x l_{bh})c\theta_2 - (p_z l_{bh} + p_x l_{bw})s\theta_2 + x_{ab}p_z - y_{ab}p_x\right] \\ /\left[(p_z l_{bw} - p_x l_{bh})c\theta_2 - (p_z l_{bh} + p_x l_{bw})s\theta_2 + x_{ab}p_z - y_{ab}p_x\right]\end{array}\right\} \quad (10)$$

Substituting θ_1 into the original Eq. (6) and simplifying, we obtain the expression of θ_2:

$$\theta_2 = \arctan\left\{\begin{array}{l} l_{bw}^2 - l_{bh}^2 + p_x^2 + p_z^2 + x_{ab}^2 + y_{ab}^2 \\ \quad -\left[\left(l_{bw}^2 - l_{bh}^2 + l_{bw}^2 + p_z^2\right)^2 + 4(p_z l_{bw} - p_x l_{bh})(p_x l_{bw} + p_z l_{bh})\right]^{\frac{1}{2}} \\ \quad /\left(l_{bw}^2 - l_{bh}^2 + p_x^2 + p_z^2\right)\end{array}\right\} \quad (11)$$

Since the equation about θ_1 in Eq. (10) contains θ_2, then we substitute Eq. (11) into Eq. (6) again and simplify to obtain:

$$\theta_1 = \arctan\left[\frac{(p_x l_{bw} + p_z l_{bh})}{(p_z l_{bw} - p_x l_{bh})}\right] \quad (12)$$

By using this method, the coordinate points on the bulldozer's operation path can be converted into the coordinates of the tooth tip at the end of the blade, and then the joint angles of the blade in the joint space can be solved inversely. The steps are shown in Fig. 4.

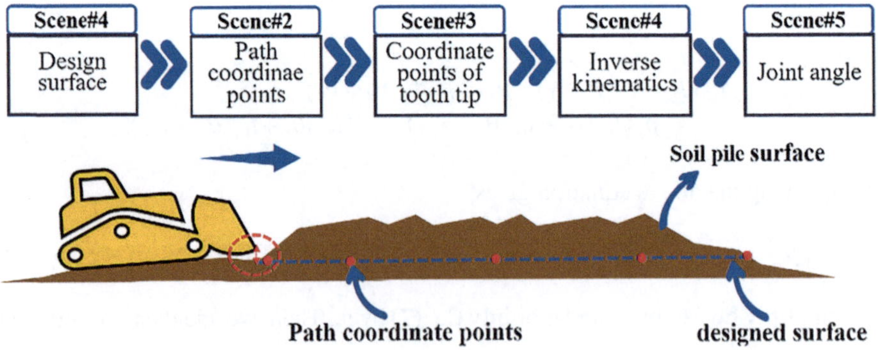

Fig. 4 Inverse solution steps

3 Conversion Between Joint Space and Actuation Space

The result obtained from the forward kinematics solution of the bulldozer blade, as mentioned above, is the description of the tooth tip at the end of the blade in the pose space. The result obtained from the inverse kinematics solution is the rotation angle of the blade in the joint space. However, to realize the operation planning and control of the unmanned bulldozer blade, it is also necessary to realize the mutual conversion between the joint angle of the bulldozer blade and the length of the hydraulic cylinder, that is, the conversion between the joint space and the actuation space of the blade.

As can be seen from the structure of the bulldozer, the positions of the hinge points between the working device hydraulic cylinder and the vehicle body are fixed, the lengths and tilt angles of the connecting lines between each hinge point are fixed, and the structure is relatively simple. Therefore, the geometric method is selected to complete the conversion between the joint space and the driving space.

3.1 Conversion Between Lifting Angle and Stroke of Lifting Hydraulic Cylinder

As shown in Fig. 5, it is a structural schematic diagram of the hinge points of each joint when the bulldozer blade is lifted. Point O is the hinge point between the top push rod and the center of the bulldozer track. Point A is the hinge point between the blade of the bulldozer blade and the top push rod. Point C is the hinge point between the support rod of the lifting hydraulic cylinder and the bulldozer body. AO, CO, BC, and ∠ABC are all known quantities, and the length of AO is equal to the length of the top push rod. AB is the length λ_1 of the lifting hydraulic cylinder to be found.

In ΔAOC, it can be obtained by the cosine theorem:

$$AC = \sqrt{AO^2 + CO^2 - 2AO\,CO\cos\angle AOC} \tag{13}$$

$$\angle AOC = \alpha \pm \theta_1 \tag{14}$$

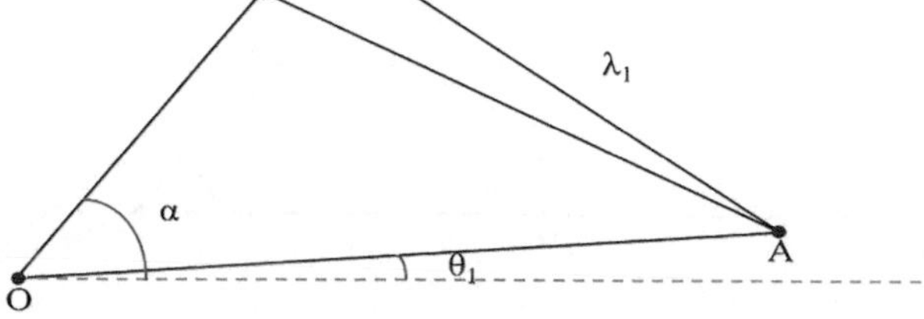

Fig. 5 Structural schematic diagram of lifting process

In the formula, α is the angle between OC and the horizontal plane, which is a constant value. θ_1 can be obtained by the above inverse kinematics solution.

Similarly, in ΔABC,

$$AB = \sqrt{AC^2 - BC^2 + 2AB\,BC\cos\angle ABC} \tag{15}$$

By combining Eqs. (13), (14) and (15), the conversion formula between the lifting angle θ_1 of the bulldozer blade and the length λ_1 of the lifting hydraulic cylinder can be obtained as:

$$\lambda_1 = AB = \left[AO^2 + CO^2 - BC^2 + 2AB\,BC\cos\angle ABC - 2AO\,CO\cos(\alpha \pm \theta_1)\right]^{\frac{1}{2}} \tag{16}$$

3.2 *Conversion Between Pitch Angle and Stroke of Pitch Hydraulic Cylinder*

As shown in Fig. 6, it is a structural schematic diagram of the hinge points of each joint when the bulldozer blade pitches. E is the fixed point of the pitch hydraulic cylinder and the top push rod, and F is the fixed point of the hydraulic cylinder and the bulldozer blade. In the figure, θ_2 is obtained by inverse kinematics solution. AE and AF are both known quantities. EF is the length λ_2 of the pitch hydraulic cylinder to be found.

In ΔAEF, the conversion formula between the pitch angle θ_2 of the bulldozer blade and the length λ_2 of the pitch hydraulic cylinder can be obtained by the cosine theorem as follows:

$$\lambda_2 = EF = \sqrt{AE^2 + AF^2 - 2AE\,AF\cos\theta_2} \tag{17}$$

After the above calculations, the joint space and actuation space of the blade are converted, and the mapping relationship between each joint angle and the elongation of the lifting and pitching hydraulic cylinders is clarified. Then, by controlling the elongation and contraction of the hydraulic cylinders, the tooth tip at the end of the bulldozer blade reaches the path points on the design surface of the operation task to complete the operation task.

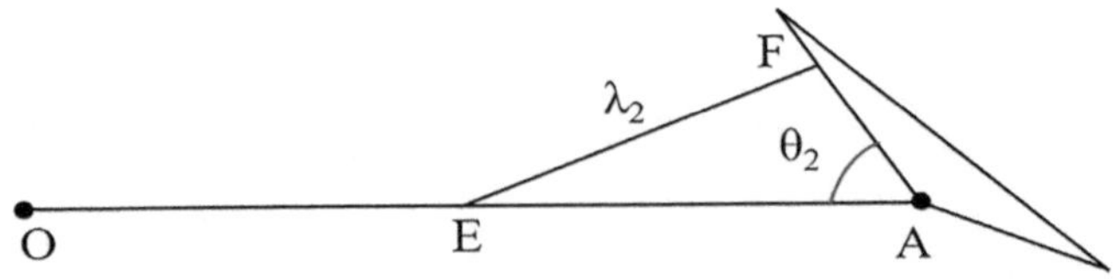

Fig. 6 Structural diagram of the pitching process

4 Simulation Verification

4.1 Forward Kinematics Simulation

The bulldozer is regarded as a robot with three degrees of freedom and can be simplified into a three-link mechanism, as shown in Fig. 7. Since the base coordinate of the bulldozer coincides with the origin of the lifting coordinate system, the length of the first link is set to 0 mm. The length of the second link is the length of the top push frame 1_1, which is set to 2050 mm. The length of the third link is the thickness of the blade, which is set to 600 mm.

In this section, the Toolbox will be used to verify the forward and inverse kinematics equations derived to check their correctness.

In forward kinematics, the angles θ_1 and θ_2 and the distance l_1 are known, while the position of the blade is unknown. As can be seen from the content in Sect. 2, θ_1 is the lifting angle of the blade, θ_2 is the pitch angle of the blade, and the distance l_1 is the length of the top push frame which is 2050 mm. The pose of the tooth tip at the end of the blade is represented by the matrix $P_C = (p_x, p_y, p_z)$.

Therefore, in the Toolbox, four sets of joint angle parameters θ_{group1}~θ_{group4} are set. Each set of joint angles contains the yaw angle θ_0 of the bulldozer, the lifting angle θ_1 of the blade, and the pitch angle θ_2. This experiment mainly focuses on the lower-level blade movement. Kinematic constraints are applied to the model. The degrees of freedom of the blade are limited to the movements of the X and Z axes. Only the joint angles θ_1 and θ_2 of the blade are considered, while the yaw angle θ_0 of the bulldozer is not considered. Therefore, the yaw angle θ_0 of the bulldozer is uniformly set to 0°.

Next, the Toolbox is used to perform forward kinematics simulation on the blade. We set the link lengths in the Toolbox, draw the three-link mechanism model of the blade, and then set four different sets of joint angles to obtain four different poses of the blade, as shown in Fig. 8.

Then, by substituting four different sets of joint angles into the forward kinematics equation derived in Sect. 2 for calculation, we compared the pose coordinates of the blade tip calculated by the two methods to verify whether the forward motion analysis is correct. The results are shown in Table 1.

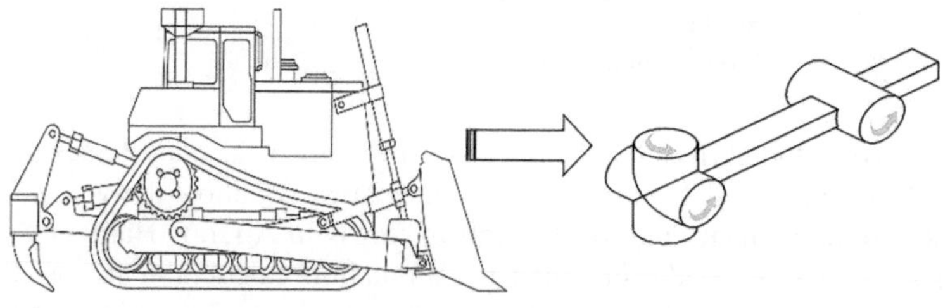

Fig. 7 The bulldozer is simplified into a three-link mechanism

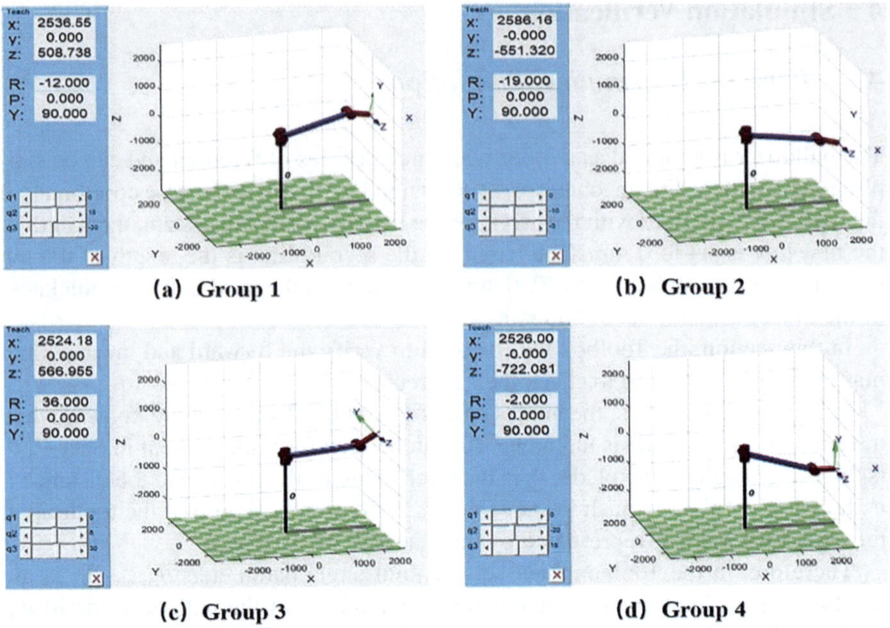

Fig. 8 Blade poses under four sets of joint angles

Table 1 Pose coordinates of the end of the blade obtained by different methods

Joint angle (°)	Value-taking method	Pose coordinates of the end of the blade (mm)		
		p_x	p_y	p_z
$\theta_{group1} = [0, 18, -30]$	Calculation by forward kinematics equation	2536.554	0.000	508.738
	Toolbox simulation	2536.554	0.000	508.738
θ_{group1}=[0, −10, −9]	Calculation by forward kinematics equation	2586.167	0.000	−551.320
	Toolbox simulation	2586.167	0.000	−551.320
θ_{group1}=[0, 6, 30]	Calculation by forward kinematics equation	2524.180	0.000	566.955
	Toolbox simulation	2524.180	0.000	566.955
θ_{group1}=[0, −20, 18]	Calculation by forward kinematics equation	2526.004	0.000	−722.081
	Toolbox simulation	2526.004	0.000	−722.081

It can be concluded from the data in Table 1 that the pose coordinates of the end of the blade calculated by the forward kinematics equation derived in chapter "Analysis, Improvement, and Verification of Stress in Aircraft Harness Bending" are the same as the results obtained by simulation calculation in the Toolbox. Through the verification of calculation by kinematics equation and simulation calculation, it can be seen that the model of the blade is established correctly. Therefore, the forward kinematics analysis of this model is accurate.

Table 2. Inverse kinematics calculation results

End pose matrix	Calculation result
$P_{C_1} = \begin{bmatrix} 2536.554 \\ 0.000 \\ 508.738 \end{bmatrix}$	$q_1 = [0.000 \quad 0.314 \quad -0.524]$
$P_{C_2} = \begin{bmatrix} 2586.167 \\ 0.000 \\ -551.320 \end{bmatrix}$	$q_2 = [0.000 \quad -0.175 \quad -0.157]$
$P_{C_3} = \begin{bmatrix} 2524.180 \\ 0.000 \\ 566.955 \end{bmatrix}$	$q_3 = [0.000 \quad 0.105 \quad 0.524]$
$P_{C_4} = \begin{bmatrix} 2526.004 \\ 0.000 \\ -722.081 \end{bmatrix}$	$q_4 = [0.000 \quad -0.349 \quad 0.314]$

4.2 Inverse Kinematics Simulation

The inverse kinematics simulation is carried out by using the ikine() inverse kinematics solution function in the Toolbox. The pose matrices $P_{C_1} \sim P_{C_4}$ of the tooth tips at the end of the blade obtained by the forward kinematics solution are brought into the ikine() function, and the corresponding joint angles q_1~q_4 are obtained by simulation. There are multiple sets of solutions for the joint angles obtained by this inverse kinematics solution. We select a set of solutions with the same angles as in Table 1. The results are shown in Table 2. q_1~q_4 in Table 2 are the radian representation forms of the joint angles obtained by the Toolbox. After this result is converted into angles, it is the same as the joint angles set in Table 1, which further proves the correctness of the modeling.

In conclusion, this analysis is correct.

4.3 Simulation of Drive Space

In Sect. 2, the mapping relationship between the joint rotation angle of the blade and the stroke of the hydraulic cylinder is calculated by using the geometric method. In this section, the hydraulic component library and 2D mechanical library in Amesim will be jointly used to establish a simulation model to verify the accuracy of its mapping relationship.

Figure 9 shows the structural block diagram of the simulation. Since the lifting hydraulic cylinders of the bulldozer are symmetrically distributed on both sides of the blade and the stroke changes of the two hydraulic cylinders are the same, the two symmetrical hydraulic cylinders can be simplified into one hydraulic cylinder. By

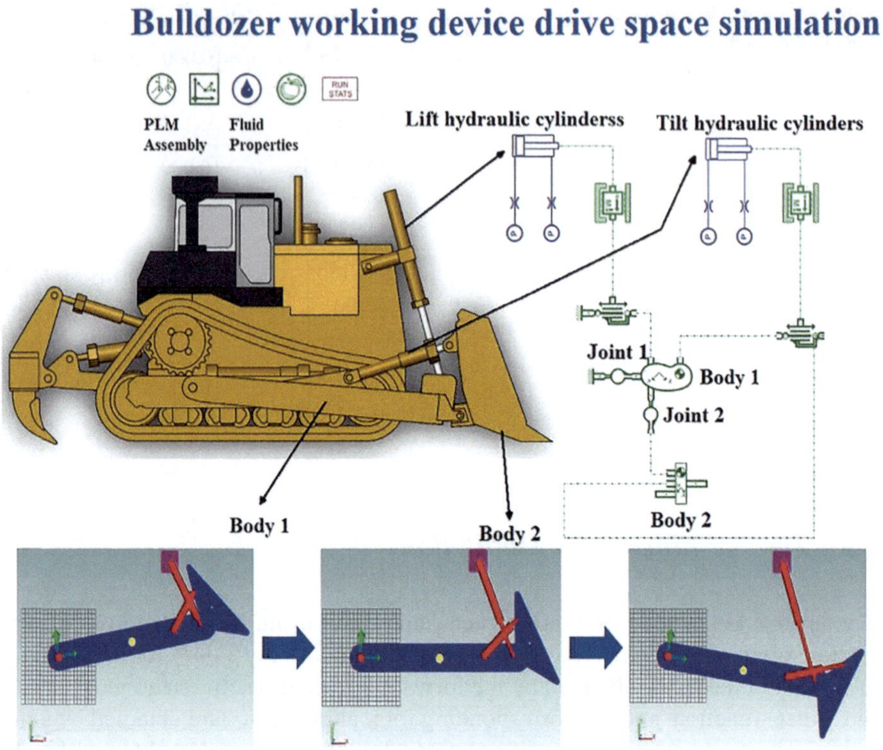

Fig. 9 Structural block diagram of the combined simulation of the bulldozer blade

the same token, the two-pitch hydraulic cylinders can also be simplified into one. The bulldozer's pusher frame is set as Body1, and the blade is set as Body2. In the process of modeling, this experiment adopts the method of relative coordinate system. According to the actual size of the bulldozer, input the center of gravity coordinates and masses of the two entities, the coordinate values of each hinge point, and the cylinder diameter, piston rod diameter, stroke, oil pressure and other parameters of the two hydraulic cylinders. Here only the important parameters are explained: lifting hydraulic cylinder diameter: 100 mm, rod diameter: 80 mm, stroke: 1100 mm; the coordinate value of the origin of the blade (Body2) relative coordinate system in the absolute coordinate system: (4,0), and the hinge point with the pusher frame (Body1) is located at the origin of the relative coordinate system, center of gravity coordinates: (0.3, 0.3), the coordinate of the tooth tip at the end of the blade: (1,–0.5), mass: 50 kg.

Finally, we set the oil pressure in the rod chamber and the rodless chamber of the lifting and pitching hydraulic cylinders, followed by a simulation experiment. The obtained corner curves of each joint are shown in Fig. 10a. Among them, the yellow curve represents the lifting angle θ_1, and the green curve represents the pitching angle θ_2. The stroke change curves of each hydraulic cylinder are shown in Fig. 10b.

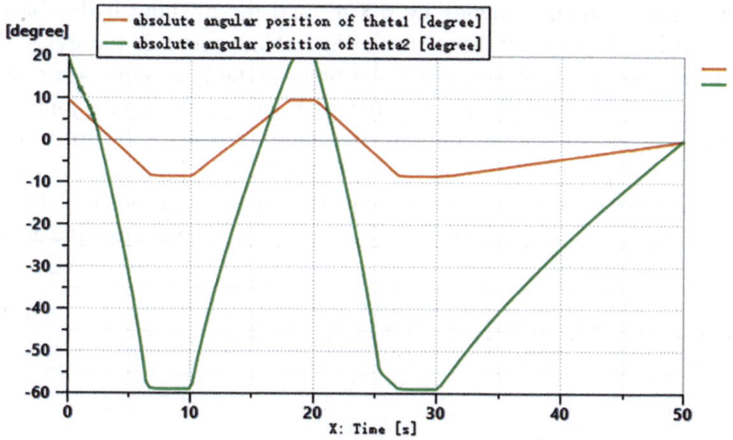

(a)Curves of rotation angles of each joint

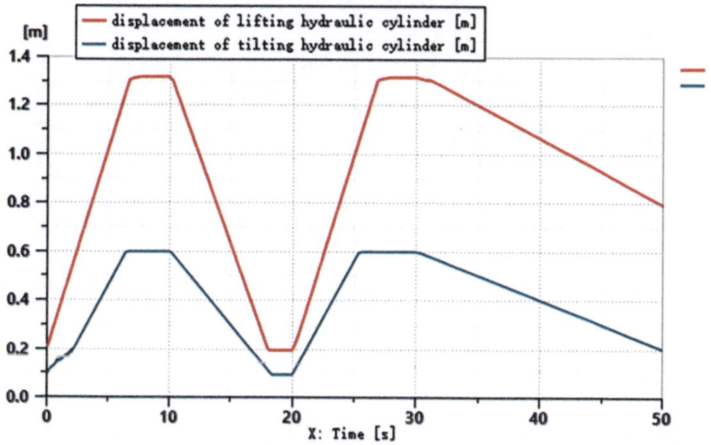

(b)Stroke curves of each hydraulic cylinder

Fig. 10 Simulation curves under different poses of the blade. (**a**) Curves of rotation angles of each joint. (**b**) Stroke curves of each hydraulic cylinder

Among them, the red curve represents the extension length of the lifting hydraulic cylinder, and the blue curve represents the extension length of the pitching hydraulic cylinder.

According to the curves in Fig. 10, it can be seen that at the initial position, the lifting angle of the blade is 10°, and at this time, the lifting hydraulic cylinder extends by 0.2 m; the pitching angle is 20°, and the pitching hydraulic cylinder extends by 0.1 m. At 10 s and 30 s, the blade drops by −9°, the blade pitches by −60°, and the distances extended by the lifting and pitching hydraulic cylinders are the longest, which are 1.3 m and 0.6 m, respectively. Finally, the blade stops at a horizontal position. At this time, the rotation angles of each joint are all 0°, the lifting hydraulic cylinder extends by 0.8 m, and the pitching hydraulic cylinder extends by 0.2 m. We convert the results in Fig. 10 into curves in the style of Fig. 11, where the abscissa is the stroke, and the ordinate is the joint angle. From the curve in Fig. 11, it can be intuitively seen that the relationship between the joint angle and the hydraulic cylinder stroke is approximately linear. Finally, substituting the points on the curve in Fig. 11 into the conversion formula of the joint angle and hydraulic cylinder stroke obtained by the geometric method in Sect. 2 can verify the accuracy of the mapping relationship between the joint space of the blade and the drive space.

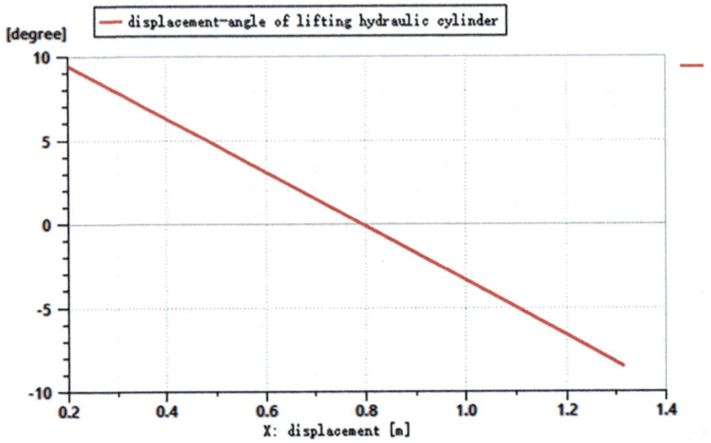

(c)Conversion relationship between lifting hydraulic cylinder and lifting angle

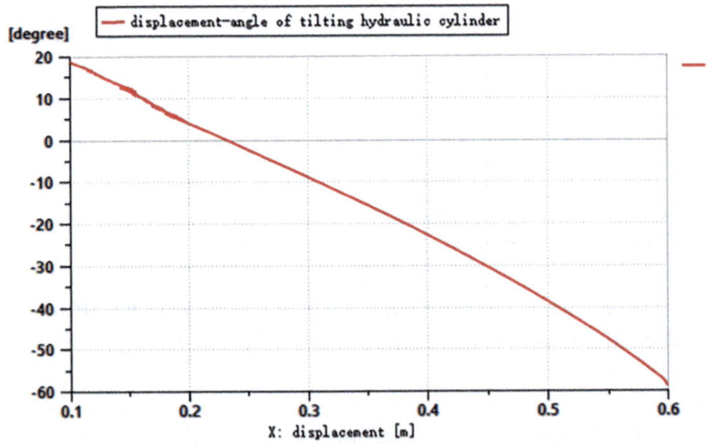

(d)Conversion relationship between tilting hydraulic cylinder and tilting angle

Fig. 11 Conversion relationship between each joint space and drive space. (**c**) Conversion relationship between lifting hydraulic cylinder and lifting angle. (**d**) Conversion relationship between tilting hydraulic cylinder and tilting angle

5 Conclusions

From the perspective of robotics, this paper simplifies the unmanned bulldozer into a three-degree-of-freedom linkage mechanism, establishes the coordinate system of each joint according to the Denavit-Hartenberg parameter method, and solves the forward and inverse kinematics. The Toolbox is used to solve the attitude of the tooth tip at the end of the bulldozer blade, as well as the rotation angles of each joint. The comparison verifies that the theoretical analysis is correct. Finally, the rotation angles of each joint of the bulldozer blade are converted into the change of the stroke of each hydraulic cylinder. The Amesim software is used to establish a simulation model for the hydraulic cylinder and the blade entity to verify the accuracy of the mapping relationship between the joint space of the blade and the drive space, providing theoretical support for the subsequent research on operation planning and control of related unmanned bulldozer blades.

Acknowledgment This work was financially supported by the Key R&D project of Shandong Province (2019JZZY010443) fund.

References

1. Huang ZY, Chen M (2007) Some new structures and new technologies of crawler bulldozers. Constr Mach 3:1
2. Sun Q, Bai S, Li G et al (2018) Modeling and simulation of power transmission of crawler bulldozer. Trans Chin Soci Agric Eng 28(5):57–61
3. Wang H, Sun FC (2014) Dynamic modeling and simulation on a hybrid power system for dual-motor-drive electric tracked bulldoze. Appl Mech Mater 494:229–233
4. Guo G, Gong LH (1998) Research on fuzzy control of cutting depth of bulldozer blade. Constr Mach 12:13–15
5. Xue G, Han CS, Lee YS, Ji CU, Kang MS (2015) A study on the kinematic analysis and dynamic simulation for automatic 3D control of the dozer blade. In Conference paper of KSPE, pp 419–420
6. Sun DI, Han CS, Lee Y S, Kim SH, Lee SH (2016) A study on sensor fusion algorithm for Dozer's blade position and attitude estimation. In Conference paper of KSPE, pp. 47–48
7. Kim SH, Lee YS, Sun DI et al (2019) Development of bulldozer sensor system for estimating the position of blade cutting edge. Autom Constr 106(Oct):102890.1–102890.11
8. Lee YS, Kim SH, Seo J et al (2020) Blade control in Cartesian space for leveling work by a bulldozer. Autom Constr 118:103264
9. Komatsu Corporation (2019) Intelligent machine control. https://www.komatsu.eu/en/Komatsu-Intelligent-Machine-Control. Accessed 27 February 2019
10. Jayaraman G, Stratton KL, Lee EF, Kleimenhagen KW, Chizeck HJ, (1999) Method and apparatus for determining pitch and ground speed of an earth moving machines, in: U.S. Patent No. 5, 860,480
11. Omelchenko A, Zhdanov AV 2012 Automatic blade control system with integrated global navigation satellite system and inertial sensors, U.S. Patent No. 8, 145,391

12. Lundeen KM, Dong S, Fredricks N, Akula M, Seo J, Kamat VR (2016) Optical marker-based end effector pose estimation for articulated excavators. Autom Constr 65:51–64. https://doi.org/10.1016/j.autcon.2016.02.003
13. Zhou GC (2021) Research on autonomous operation planning and control of excavators. https://kns.cnki.net/kcms2/article
14. He QH, Chen M (2006) Simulation research on a working device of hydraulic excavator. J Syst Simul 3:735–738

Open Access This chapter is licensed under the terms of the Creative Commons Attribution-NonCommercial-NoDerivatives 4.0 International License (http://creativecommons.org/licenses/by-nc-nd/4.0/), which permits any noncommercial use, sharing, distribution and reproduction in any medium or format, as long as you give appropriate credit to the original author(s) and the source, provide a link to the Creative Commons license and indicate if you modified the licensed material. You do not have permission under this license to share adapted material derived from this chapter or parts of it.

The images or other third party material in this chapter are included in the chapter's Creative Commons license, unless indicated otherwise in a credit line to the material. If material is not included in the chapter's Creative Commons license and your intended use is not permitted by statutory regulation or exceeds the permitted use, you will need to obtain permission directly from the copyright holder.

Optimization Study of Civil Aircraft System Maintenance Task Interval Based on Reliability Data

Jiawei Sun and Haiwei Li

Abstract The optimization of maintenance intervals of civil aircraft involves the reliability, maintainability, and operating costs of onboard equipment, and it is also an important research direction that is of common interest to the airline companies, the main research units, suppliers, and so on. By collecting basic reliability data and operational reliability data of aircraft equipment and analyzing them statistically, the Weibull life distribution model of the equipment is derived. Based on this model, the average life of the equipment is statistically derived, and the inspection interval of the equipment is determined and optimized through comparative analysis.

Keywords Life distribution model · Maintenance interval · Reliability data

1 Introduction

The current situation at home and abroad is that maintenance interval optimization is mainly determined by the engineering experience of conducting MSG-3. This study focuses on collecting basic reliability data and operational reliability data of aircraft equipment and analyzing them statistically to obtain the Weibull life distribution model of the equipment [1]. The Weibull life distribution model can statistically determine the average life of the equipment. Based on the equipment's failure impact category, existing maintenance intervals, and average life expectancy, combined with the maintenance decision model, the optimization results of maintenance and inspection intervals can be determined after comparative analysis.

J. Sun (✉) · H. Li
System Integration Department, Shanghai Aircraft Design and Research Institute, Shanghai, China
e-mail: sunjiawei2@comac.cc

© The Author(s) 2026

Z. Zhang, P. Mercorelli (eds.), *Mechanical Design and Simulation: Insights and Innovations*, Lecture Notes in Mechanical Engineering,
https://doi.org/10.1007/978-981-95-2405-1_91

2 Workflow for Maintenance Interval Optimization

The workflow for maintenance interval optimization based on operational reliability data is shown below.

- Collection of equipment base and operational reliability data [2, 3].
- Confirmation of equipment failure samples. The equipment failure samples are determined based on the base and operational reliability data of the equipment in accordance with the requirements of the equipment life distribution model analysis [4].
- Statistical analysis of equipment life distribution. According to the Weibull distribution model and the Weibull distribution fitting method, based on the failure samples of the equipment, the statistical analysis gives the life distribution of the equipment [5].
- Maintenance decision modeling determination. Based on the results of the aircraft's MSG-3 analysis, the MPD of the Maintenance Outline MRBR report, and the Maintenance Outline MRBR Development Policies and Procedures Handbook (PPH), the equipment's failure impact category, pre-existing equipment maintenance intervals, and the aircraft's Maintenance Task Alphabet Inspection Interval (MTAI) policy are determined.
- A maintenance inspection interval optimization conclusion is given. According to the life distribution of the equipment given the average life of the stable period, combined with the maintenance decision-making model, after a comparative analysis, the optimization of the maintenance and inspection interval conclusion is given [6].

The specific workflow is shown in Fig. 1.

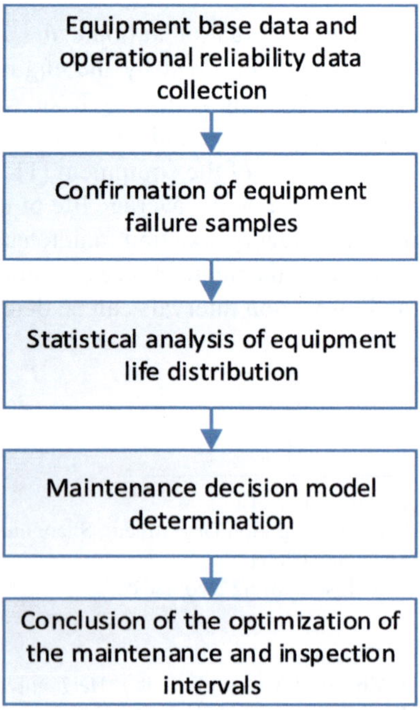

Fig. 1 Workflow for maintenance interval optimization

Table 1 List of equipment for the gas supply system

Serial number	Equipment name	Quantities	MTBF
1	Intermediate pressure check valve	2	Analyzing and summarizing based on route operation data
2	High-pressure valve	2	
3	Bleed monitoring pressure sensor	2	
4	Pressure regulating and shut-off valve	2	
5	Fan air valve	2	
6	Pre-cooler	2	
7	APU check valve	1	
8	Cross bleed valve	1	
9	High-pressure ground connection	1	
10	Bleed temperature sensor	2	
11	Bleed pressure sensor	2	
12	High-pressure conduit system	1	
13	Pre-cooler cold side header	2	

3 Experimental Cases

3.1 Reliability Data

Basic Reliability Data Basic reliability data mainly includes system equipment composition, number of equipment, and equipment reliability level. The basic reliability data for the gas system equipment is shown in Table 1.

Operational Reliability Data Operational reliability data mainly includes three parts: aircraft fleet equipment removal and replacement record data, daily flight data, and aircraft fleet data.

3.2 Confirmation of Equipment Failure Samples

In the experimental example, sample data on failures of intermediate pressure check valves (IPCVs) are used.

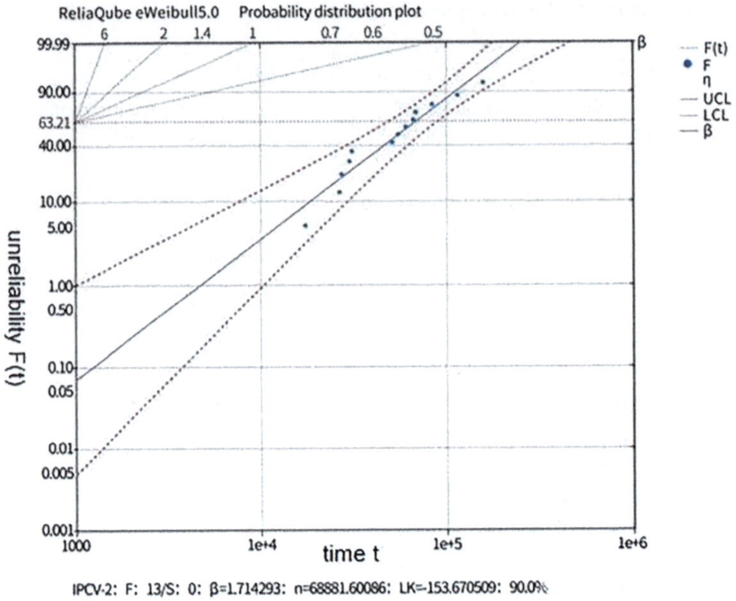

Fig. 2 Field statistical life distribution of medium-pressure unidirectional flapper IPCVs

3.3 *Statistical Analysis of Equipment Life Distribution*

IPCV is a mechanical part. The failure of mechanical products satisfies the Weibull distribution law. So the life distribution of IPCV adopts weibull distribution.

Through the statistical analysis, the life distribution of IPCV obeys the Weibull distribution, and the statistical results of the life distribution are $\beta = 1.71$, $\eta = 68{,}881.6$, as shown in the following Fig. 2.

3.4 *Maintenance Decision Modeling Determination*

Analysis of the Gas Supply System MSG-3 The MSG-3 analysis report for the gas system is shown in Table 2.

MRBR Report on an Aircraft Maintenance Syllabus An aircraft maintenance outline MRBR report is shown in Table 3.

MPD for an Aircraft Maintenance Program Document The MPD for an aircraft maintenance program document is shown in Table 4.

Table 2 MSG-3 analysis of the air supply system

System and power plant analysis-MSG-3 analysis	
Task interval selection	
Task Sequence No.: 36-10-00-01	
Analysis result	25,000 FH
Similar aircraft	Airbus A320 72 MO/8000FC/12,000 FH Removing intermediate pressure bleed check value for detailed inspection Airbus A350 24,000 FH Removing IPCV for detailed inspection
System engineer judgement	Reliability calculation: The MTBF of IPCV is 288,143 FH if the Weibull distribution is used to model the failure characteristics of components, the relationship between task interval t and reliability R(t) is: $t = \eta.\left[-\ln R(t)\right]^{\frac{1}{\beta}}$, where η is the characteristics of life, and β is the shape parameter Additionally, the relationship between MTBF and shape parameter is $MTBF = \eta.\Gamma\left(\frac{1+\beta}{\beta}\right)$ If the reliability R(t) = 80%, $\beta = 1.5$, then the characteristics life $\eta = 319{,}185$ and task interval t = 91,452 FH Considering the Weibull distribution calculation results, the final interval is 25,000 FH
Others	

Table 3 MRBR report for an aircraft maintenance syllabus

Revision status (computing)	MRB task No.	Failure impact category	Task type	Threshold	Interval	Applicability	Task description	Source
	36-10-00-01-01	6	DET	25000 FH	25000 FH	ALL	Detailed inspection of IPCV Approach: Open the fan shroud	MSI

Table 4 MPD for an aircraft maintenance program document

MPD task No.	Failure impact category	Task type	Area	Task description	Inspection threshold	Repeat check interval	Man-hour	Applicability
36-10-00-01-01	6	DET	410/420	Detailed inspection of IPCV Approach: Open the fan shroud	25000 FH	25000 FH	0.05 H	ALL

Table 5 Fault impact categories and their reliability requirements

Failure impact category	Required reliability
Obvious safety impact (category 5)	0.9
Obvious usability impact (category 6)	0.8
Obvious economic impact (category 7)	0.7
Hidden safety impacts (category 8)	0.9
Concealed non-safety impacts (category 9)	0.7

Fault Impact Categories and their Reliability Requirements The data for the failure impact categories and their reliability requirements are derived from the requirements in MIDOT used in the Boeing MRBR development process, as shown in Table 5.

3.5 *Maintenance Decision Model Determination*

Based on the results of the MRBR, MPD, and MSG-3 analyses of the air source system and reliability prediction reports for a particular type of aircraft:

- Life distribution of medium-pressure unidirectional valves in the gas supply system: $\beta = 1.71$, $\eta = 68{,}881.6$;
- The inspection interval policy is: inspection interval >4000 FH, then the interval increment is 500 FH;
- Failure impact category of the medium pressure one-way valve of the gas supply system: obvious serviceability impact (category 6);
- When the reliability is 80%, according to the life distribution of the medium-pressure one-way valve IPCV (shape parameter $\beta = 1.71$, scale parameter $\eta = 68{,}881.6$), the interval of the medium-pressure one-way valve IPCV is 28715 FH.

3.6 *Maintenance and Inspection Interval Optimization Summary*

According to a certain model of aircraft maintenance outline, MRBR gives the air source system in the pressure one-way valve inspection interval 25,000 FH, through a certain model of aircraft operation process reliability data statistical analysis of the air source system, the maintenance interval of the pressure one-way valve is 28,715 FH.

Considering the incidental damage (weather effects, passenger activities, ground support equipment tools, etc.) and environmental degradation damage (temperature, humidity, contaminants, etc.) included in the IPCV, and combining the results of the PPH, MRBR, and MSG-3 analyses, it is recommended that: the maintenance interval for the pressure check valves in the air supply system of a certain type of aircraft should be adjusted from 25,000 FH to 28,000 FH.

Longer intervals between scheduled maintenance can save the airline aircraft operating costs. The formula for calculating the cost of planned maintenance is:

$$DMC = \sum R \times MH_{ONi} \times (T_{Ri})^{-1} \tag{1}$$

In the formula: *DMC* direct maintenance costs for planned maintenance (dollars/flight hour), *R* maintenance labor hour rate (dollars/hour), MH_{ONi} maintenance labor hours for planned maintenance task i (h), T_{Ri} maintenance interval for planned maintenance task i (flight hours).

The IPCV maintenance labor rate is $55. Usually, maintenance labor time is 2 h. The direct maintenance cost before optimization is 0.0044 dollars. The direct maintenance cost after completing the optimization is 0.0039 dollars. IPCV optimization saves 0.0005 dollars per driver repair. Airlines pay a lot of attention to the optimization of maintenance intervals. The content of this study is important for airline companies.

4 Conclusion

The methodology and case study of civil aircraft maintenance task interval optimization based on the Weibull distribution of reliability data used in this paper demonstrates its feasibility. Based on the reliability data of aircraft operations, a Weibull distribution model is used for equipment life analysis. After that, a comparative analysis is carried out by combining the failure impact of the equipment and the existing equipment maintenance intervals, etc., to obtain the optimized value of the equipment's scheduled inspection intervals. The method has an obvious advancing effect on the optimization of maintenance task intervals of civil aircraft.

References

1. Guangping H, Qunfeng Y, Dunyong B (2019) Civil aircraft reliability engineering. Shanghai Jiaotong University Press, Shanghai
2. Shengkui Z (1995) Proportional combination method and scoring allocation method for mission reliability index assignment. J Aeronaut 1:1
3. Tingxiao L, Pengzhou Z (2002) Reliability design and analysis. National Defense Industry Press, Beijing
4. Zhihua Z (2012) Reliability theory and engineering applications. Science Press, Beijing
5. Ming CM, Tao JY, Zhuan C (2011) Bayesian reliability analysis of dynamic distribution parameters. National Defense Industry Press, Beijing
6. Wangfeng Y (2019) Research on maintenance program optimization based on aircraft maintenance data in the field. Northwestern Polytechnical University, Xian

Open Access This chapter is licensed under the terms of the Creative Commons Attribution-NonCommercial-NoDerivatives 4.0 International License (http://creativecommons.org/licenses/by-nc-nd/4.0/), which permits any noncommercial use, sharing, distribution and reproduction in any medium or format, as long as you give appropriate credit to the original author(s) and the source, provide a link to the Creative Commons license and indicate if you modified the licensed material. You do not have permission under this license to share adapted material derived from this chapter or parts **of it.**

The images or other third party material in this chapter are included in the chapter's Creative Commons license, unless indicated otherwise in a credit line to the material. If material is not included in the chapter's Creative Commons license and your intended use is not permitted by statutory regulation or exceeds the permitted use, you will need to obtain permission directly from the copyright holder.

Design and Analysis of Multifunctional Spinach Tillage and Planting Machine

Huanjia Liu, Fengrui Zhang, and Weiwei Chen

Abstract Due to the lack of integrated mechanized farm equipment for spinach cultivation in the current market, this paper studies a multifunctional spinach cultivation machine, which applies modular and multifunctional design, through the design of four functional structures, so that it can adapt to different soil conditions and crop planting needs. The structural reliability and use stability of the agricultural machine are verified through the design calculation of torque, the role of institutional bearing force and finite element static load analysis. After a large number of experiments and practical applications, it is verified that this agricultural machinery has certain practical applicability, thus alleviating the market gap of spinach planting integrated agricultural machinery and equipment to a certain extent.

Keywords Spinach seeds · Multifunctional agricultural machines · Finite element analysis

1 Introduction

Since the release of the State Council's *Guiding Opinions on Accelerating Agricultural Mechanization and the Transformation and Upgrading of Agricultural Machinery and Equipment Industry* [1], the development of agricultural mechanization has been widely emphasized. As agricultural productivity improves, farmers' incomes increase and the purchasing power of agricultural machinery increases [2]. In the context of sustainable development, examining the impact of agricultural mechanization services on the technical efficiency of food production is of great practical significance for the formulation of policies to safeguard food production capacity [3]. The research and development and application of agricultural machinery provide new solutions for field operations [4]. Due to the high nutritional value of spinach, short planting cycle, adaptability, high economic value, ecological

H. Liu (✉) · F. Zhang · W. Chen
Dalian Polytechnic University, Dalian, Liaoning, China

© The Author(s) 2026

Z. Zhang, P. Mercorelli (eds.), *Mechanical Design and Simulation: Insights and Innovations*, Lecture Notes in Mechanical Engineering,
https://doi.org/10.1007/978-981-95-2405-1_92

benefits and other advantages, the planting of spinach has become the choice of many farmers, but the planting of spinach is labor-intensive, the sowing requirements are high, and there is an urgent need for a spinach planting machine to alleviate the pressure of sowing. However, the research related to multi-functional small-scale high-efficiency spinach tillage and planting machines has not been reported in the relevant research.

In this paper, a multifunctional spinach cultivator is studied, which applies modular and multifunctional design to adapt to different soil conditions and crop planting needs by designing the wheel part expansion structure, soil loosening structure, crank rocker mechanism and seeding structure. The structural reliability and use stability of the agricultural machine are verified through design calculations including the torque required by the wheel motor, the crank rocker mechanism to withstand the force action and finite element static load analysis. To a certain extent, the research and development of this efficient and labor-saving spinach crop seeding integrated machine is of great significance for improving spinach production efficiency and reducing costs. The overall three-dimensional diagram is shown in Fig. 1. Abbreviated main components are noted in the figure.

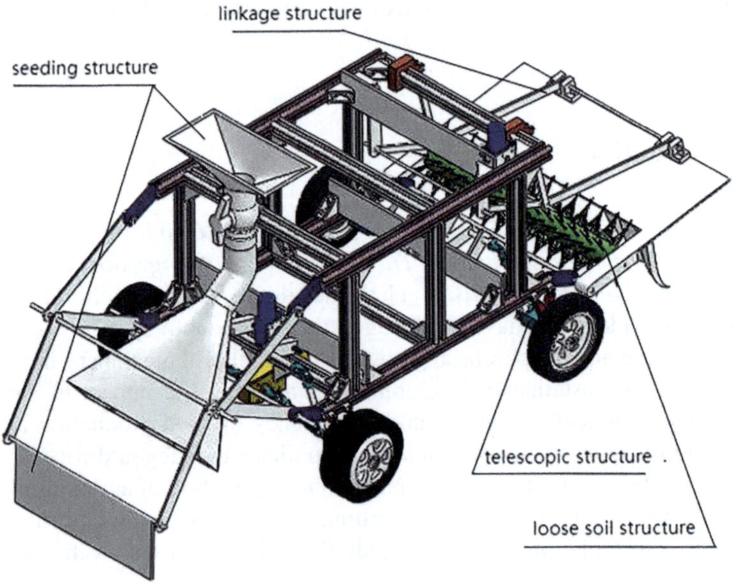

Fig. 1 Overall three-dimensional diagram

Fig. 2 Spinach planter available

2 Functional Design

The most important feature of this agricultural machine is its multifunctionality and integration. This undoubtedly increases the difficulty of the mechanical design, so it is necessary to choose reasonable structural dimensions to fulfill its functions. Due to textual limitations, this study focuses on the most difficult part of the design process.

At present, the common spinach harvester in China is shown in Fig. 2, which is mainly driven by human power, with a single structural function, and it is designed to support only the sowing of spinach seeds. The designed integrated spinach cultivator disperses multiple functions on four main modules: Wheel telescopic structure, Loose soil structure, Crank rocker mechanism, and Seeding structure.

2.1 *Wheel Telescopic Structure Design*

The core function of the front wheel structure of this device is to efficiently and smoothly transfer the power generated by the electric motor to the front wheels of the agricultural machine, thereby driving the agricultural machine forward. The key components include the motor, universal joint coupling, rotating shaft, etc., which work together to ensure the driving stability and transmission efficiency of the agricultural machine in various terrain conditions.

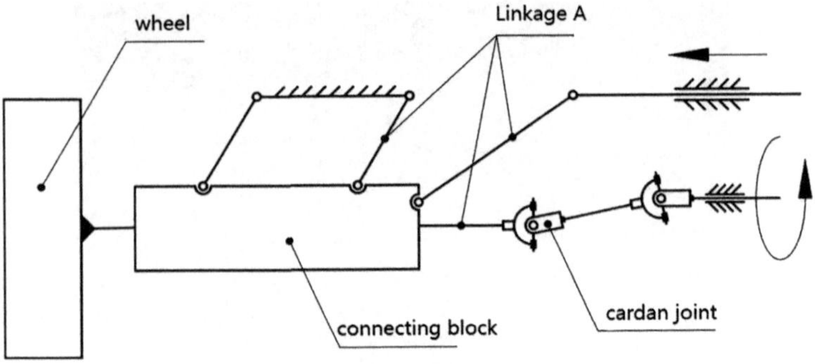

Fig. 3 Wheel telescopic structure

The motor serves as the power source of the whole transmission system, driving the agricultural machinery. The universal joint coupling allows a certain angular deviation between the two connected shafts while transmitting power, thus making the power transmission more flexible and adapting to the various angular changes generated when the agricultural machine is traveling on uneven ground. This design effectively reduces the stress on the drive shaft caused by the undulation of the ground, and improves the smoothness of the traveling of the agricultural machine and the durability of the drive system. In addition, the rotor shaft is an important component connecting the universal joint coupling and the front wheels, which is responsible for transferring power from the universal joint coupling to the front wheels. The design of the rotor shaft needs to take into account the strength, rigidity and corrosion resistance in order to adapt to the working conditions of the agricultural machinery in various harsh environments.

In practical application, this front wheel structure design can significantly improve the driving performance and operating efficiency of agricultural machinery. When the agricultural machine operates in the field, the ground conditions are often complex and variable, and this design enables the agricultural machine to better adapt to the undulations and unevenness of the ground, reducing the impact and vibration caused by the terrain changes, thus protecting the key components of the agricultural machine and prolonging its service life, as shown in Fig. 3.

2.2 *Loose Soil Structure Design*

The front side of the loosening teeth has an edge to reduce resistance and divide the land into ditch-long parts, the rollers at the back rotate to break up the hard stones in the loosened land, and the loosening machine operates not only to loosen the deep soil, but also not to disturb the soil layer [5]. The loosening depth is adjusted by changing the height of the ground wheel, and the maximum loosening depth is about 15 cm. This improves the physical and chemical properties of the soil, which

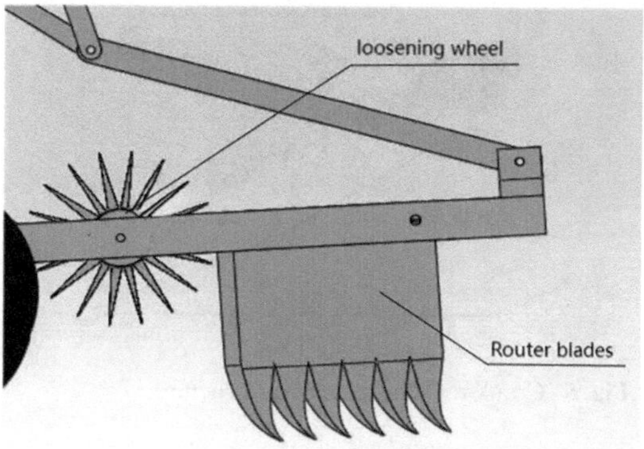

Fig. 4 Loose soil structure

is conducive to water storage and moisture conservation, and the deep loosening of low-lying land can accelerate the infiltration of surface water, which is an effective measure to exclude excessive moisture from the cultivated layer of the cultivated land [6].

Loosening the soil can break the soil block, so that the gap between the soil particles increases, which is conducive to the penetration and distribution of air, water and nutrients in the soil, and improve the permeability, permeability and water retention of the soil, thus improving the soil structure and providing a more suitable soil environment for crop growth. The loose soil structure is shown in Fig. 4.

2.3 Crank Rocker Mechanism Design

The crank, connecting rod, rocker and other rods will convert the rotary motion of the motor into the reciprocating swing of the rocker arm, so that the shank does a circular motion (rotary motion), the connecting rod does the rocking motion, the rocker does the reciprocating motion, and the frame plays a role in fixing the crank and the rocker, to realize the lifting and lowering function of the loosening machine.

During lifting and lowering, the swinging amplitude and speed of the rocker can be changed by adjusting the positional relationship between the crank and the connecting rod, thus realizing that the working parts operate in different depths of soil. This design enables the loosening machine to adapt to different soil conditions and crop planting needs, improving the applicability and flexibility of the machine. Figure 5 shows the crank rocker mechanism, where A, B, C, and D are the rockers that make up the structure. Among them, the role of Linkage A is to transmit power, the role of Linkage B is to limit, and the role of Linkage C and D is to transmit power and support the Router blades and Loosening wheel. The structural sketch is shown in Fig. 5.

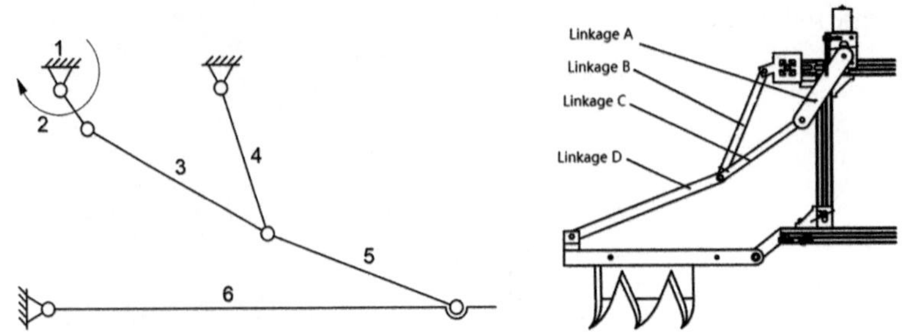

Fig. 5 Crank rocker mechanism structure

2.4 Seeding Structure Design

The concept of the seeding structure is to achieve precise seeding and efficient seed utilization. The seeding structure consists of two main parts, the upper and lower parts, which work together to ensure that the seeding process is carried out smoothly.

The upper part is the seed storage slot of the sowing structure, which is a specially designed container for the seeds to be sown. The capacity and structure of the storage box are adapted to the size and needs of the sowing to accommodate different types of crop seeds and sowing areas. The design of the storage box usually takes into account sealing and moisture resistance to ensure that the seed remains in optimal condition during storage and is protected from the external environment.

The lower part (valves and conduits) contains the working mechanism of the seeding structure, including the rotary switch and knob. When the sowing operation starts, the sowing structure is activated by turning the switch knob, which triggers the movement of the seeds in the storage box through the conduit system at the bottom. The conduit system consists of five tubes that ensure an even distribution of the seeds.

The furrowing mechanism is responsible for plowing five parallel furrows in the land to provide a suitable sowing environment for the seeds. As the seeds leak through the conduit, they are evenly sprinkled across these furrows. This design ensures that the spacing and depth between the seeds remain consistent, which is conducive to seed germination and growth. And housing is used as a protective case.

The entire sowing process is automated, greatly improving the efficiency and accuracy of sowing. In this way, farmers can sow a larger area of land in a shorter period of time, while reducing labor costs and labor intensity. In addition, uniform sowing also helps in balanced crop growth and improves crop yield and quality. The structure is shown in Fig. 6.

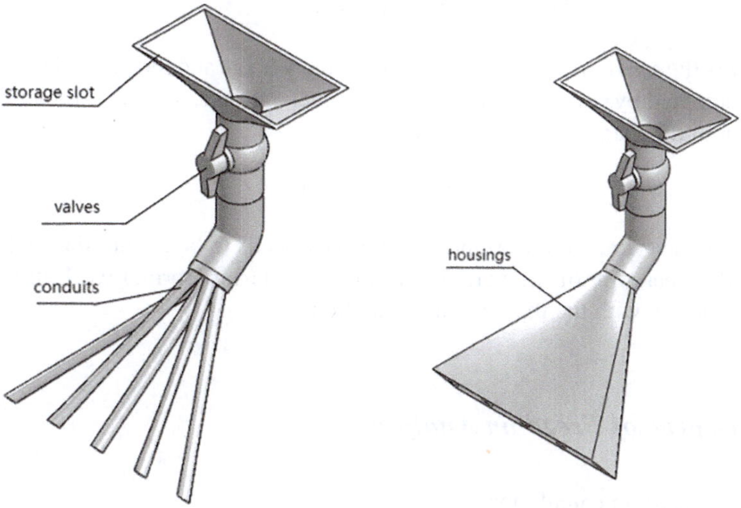

Fig. 6 Sowing structure

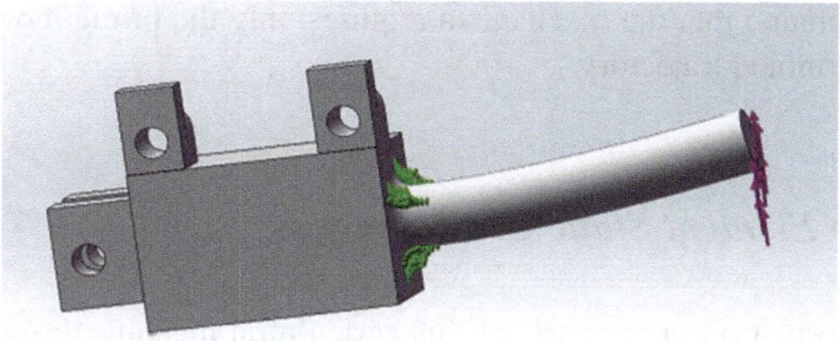

Fig. 7 Rotating shaft load and clamping

3 Design and Calculation

3.1 Calculation of the Amount of Torque Required for Wheel Motors

The agricultural machine in this project uses two active and two driven wheels to power the movement of the machine. Wheel motor torque is too small will cause the machinery to get stuck in the mud and not be able to move, so the wheel motor torque for the design of agricultural machinery is critical. To establish a simplified model of the agricultural machinery, as shown in Fig. 7. The mass of the agricultural machinery is m, determined to be about 50 kg (fully loaded state), the vehicle accelerated to a steady speed acceleration is a, and then the traction force is:

$$F = mg\mu + ma \tag{1}$$

where μ is the coefficient of friction between the wheels and the plowed land, which can be obtained by reviewing the literature, acceleration can be measured as 3 m/s^2. Bringing in the data, the moment M is:

$$M_e = 4.1 \text{ N/m} \tag{2}$$

Therefore, the torque of the adopted motor only needs to be greater than 4.1 N/m to realize the operation of the agricultural machine during normal work in the field, which is in line with the theoretical calculation.

3.2 Degree of Freedom Analysis

Degree of freedom calculation:

$$F_n = 3n - 2PL - PH = 3 \times 5 - 2 \times 7 - 0 = 1 \tag{3}$$

It can be concluded that the mechanism requires only the prime mover to obtain a uniquely determined trajectory.

3.3 Finite Element Static Load Analysis

The main deformation-prone parts of this agricultural machine design are the axle bracket and crank rod, which are related to the safety and operational feasibility of the whole equipment. This section mainly analyzes their stresses and displacements. As shown in Table 1.

The surface degrees of freedom of the fixed position of the backrest axle bracket and the connecting crank rod are limited. The backrest plate under maximum load conditions is analyzed, that is, the stresses and deformations of the bedplate under full load are simulated and analyzed. The load and clamping are shown in Figs. 7 and 8.

Table 1 Material properties

Causality	Numerical value
Modulus of elasticity	3500 MPa
Bending modulus	100–150 MPa
Densities	1.25–1.28 g/cm^3
Tensile strength	40–60 MPa
Yield strength	130 Mpa

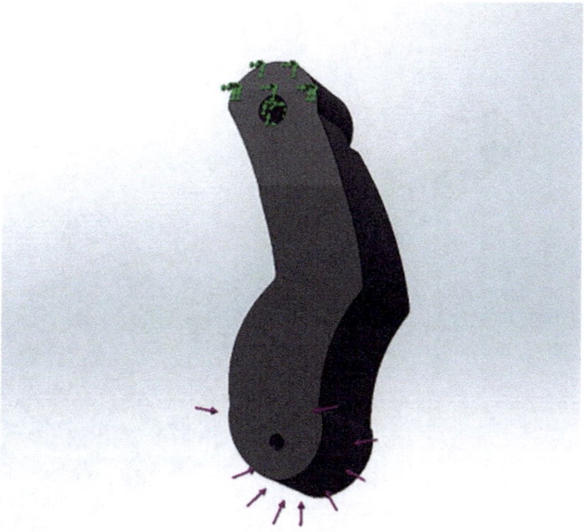

Fig. 8 Load and clamping of fin slider

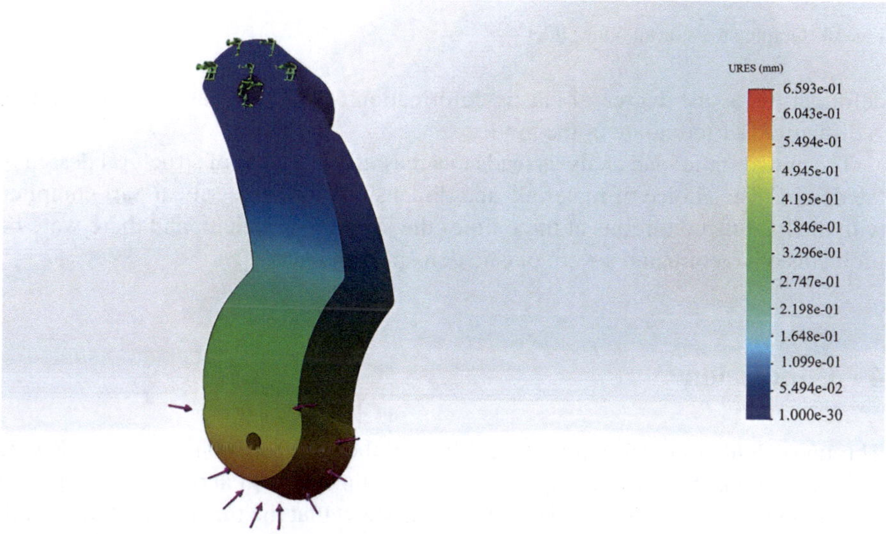

Fig. 9 Rotating shaft output stress

The stress and displacement of the crank rod is shown in Fig. 9, and the stress and displacement deformation of the axle bracket is shown in Fig. 10. From the stress and displacement plots, it can be seen that the maximum values occur around the axle bracket with the connecting rod. The main data analyzed can be clearly seen from the results of the example calculation. The output displacements of the parts shown have been exaggerated to represent the direction and extent of material

Fig. 10 Output stress of fin bone slider

deformation, as the degree of elastic deformation of the material is so small that it is difficult to differentiate in the model.

The above static load analysis leads to a reasonable degree of structural design of the device. The choice of materials and dimensions for each critical part complied with the loading conditions at three times the impact coefficient, and there were no problems of fracture and severe plastic deformation in each part.

4 Conclusion

Through calculation and analysis, the design of the multifunctional spinach plowing and planting machine was completed, and actual production and a large number of experiments were carried out. The test results show that the function and use stability of the multifunctional spinach tiller meet the expected requirements. The structure is reliable with high performance and environmental adaptability.

References

1. State Council of the People's Republic of China (2018) Guiding opinions on accelerating the transformation and upgrading of agricultural mechanization and agricultural machinery equipment industry. [Online] Available: https://www.gov.cn/zhengce/content/2018-12/29/content_5353308.htm

2. Liem PD (2019) A study on the current status of agricultural mechanization development in Vietnam and the main factors affecting it. Doctoral Dissertation, Northeast Agricultural University, 1: 1–2
3. Wang D, Liu C, Zhou Y (2021) The impact of agricultural mechanization on the technical efficiency of grain production: a spatial effect perspective based on the cross-regional service of agricultural machinery. Trans Chin Soc Agric Mach 42(4):223–229
4. Zhou B, Tian Y, Li J (2019) Research and development of agricultural robots for field operations: a review. Agric For Meteorol 272–273:109–128
5. Sun T, Shao Q, Liang D, Liang D (2018) Design and simulation analysis of bionic ray underwater robot. Mach Manufact 56(09):45–49
6. Cao R (2022) Exploration on training of leaders of strengthened agricultural machinery cooperatives. Contemp Agricult Mach (10):40–41

Open Access This chapter is licensed under the terms of the Creative Commons Attribution-NonCommercial-NoDerivatives 4.0 International License (http://creativecommons.org/licenses/by-nc-nd/4.0/), which permits any noncommercial use, sharing, distribution and reproduction in any medium or format, as long as you give appropriate credit to the original author(s) and the source, provide a link to the Creative Commons license and indicate if you modified the licensed material. You do not have permission under this license to share adapted material derived from this chapter or parts of it.

The images or other third party material in this chapter are included in the chapter's Creative Commons license, unless indicated otherwise in a credit line to the material. If material is not included in the chapter's Creative Commons license and your intended use is not permitted by statutory regulation or exceeds the permitted use, you will need to obtain permission directly from the copyright holder.

Test Equipment for Flap Support with Controllable Deformation Adjustment

Sanyuan Li, Xianmin Chen, Yunzhao Gong, Shanshan Li, and Meng Dong

Abstract The deformation of the aircraft wing makes the support boundary condition of the flap change. In the function and reliability test of the flap, it is necessary to accurately simulate the supporting of the wing to the flap. In the current engineering test, the boundary of the flap is generally fixed in the 1 g flight state, which cannot accurately simulate the deformation of the aircraft wing. In this paper, a novel test equipment with multiple position-adjustable joints is designed for dynamically simulating the deformation support boundary of aircraft flaps, and has been applied in a flap test.

Keywords Flap test · Deformation support boundary · Dynamic simulation method

1 Introduction

The trailing edge flap is an important part of the aircraft's high-lift system and is installed on the rear beam of the wing. During the take-off and landing of the aircraft, the flaps are unfolded to increase the area and camber of the wing, thereby providing enough lift and drag for the aircraft. The flap is a key structure with high-reliability requirements. In the development process of the flap, it is necessary to verify its motion principle and reliability by tests. Because structural deformation is one of the main factors affecting the motion function of the mechanism, it is necessary to accurately simulate the deformation of the flap when performing the function and reliability test of the flap mechanism. The wing supporting the flap will undergo complex deformation during the aircraft's flight. The deformation of the

S. Li (✉) · X. Chen · Y. Gong · S. Li
National Key Laboratory of Strength and Structural Integrity, Aircraft Strength Research Institute of China, Xi'an, China

M. Dong
Shanghai Aircraft Design and Research Institute, Shanghai, China

© The Author(s) 2026

Z. Zhang, P. Mercorelli (eds.), *Mechanical Design and Simulation: Insights and Innovations*, Lecture Notes in Mechanical Engineering,
https://doi.org/10.1007/978-981-95-2405-1_93

wing is an important part of the deformation of the flap structure. However, it is difficult to simulate the deformation of the wing in the flap function and reliability tests.

At present, there are two types of flap specimen support methods [1]. One type of support method is to use a wing or a wing dummy with 1 g deformation to support the flap, which can accurately simulate the support boundary of the flap. However, it cannot accurately simulate the deformation of the wing during the whole flight of the aircraft and costs much more. The other type of support method is to use a fixed supporting device to support the flap, which cannot accurately simulate the support boundary of the wing, but the test design is simple and costs less. To improve the test simulation accuracy of the flap support boundary and reduce the test cost, engineers have improved the second type of support method. In the flap test, the support method in [2–5] can be adjusted, which cannot still meet the requirements of the deformation boundary simulation during the whole take-off and landing process. Jandaurek and Johst [6] and Li [7] proposed a deformation-based flap support scheme, which provides the possibility for the dynamic simulation of the flap deformation boundary. However, it has not been verified by experiments and cannot solve the dynamic simulation problem. Currently, there is no experimental equipment that can accurately simulate the support boundary of the flap. Therefore, in this paper, a dynamic simulation support method for the flap deformation boundary is proposed to solve the above problems, and a set of flap support equipment has been designed and applied.

2 Flap Specimen and Its Support Boundaries

The studied flap specimen is composed of a flap dummy, two flap motion mechanisms, three limiting mechanisms, etc. (as shown in Fig. 1). These five mechanisms are installed on the rear beam of the aircraft wing.

During the flight of the aircraft, the wing bending and torsion deformations take place, so that the installation positions of the flap mechanisms and the transmission

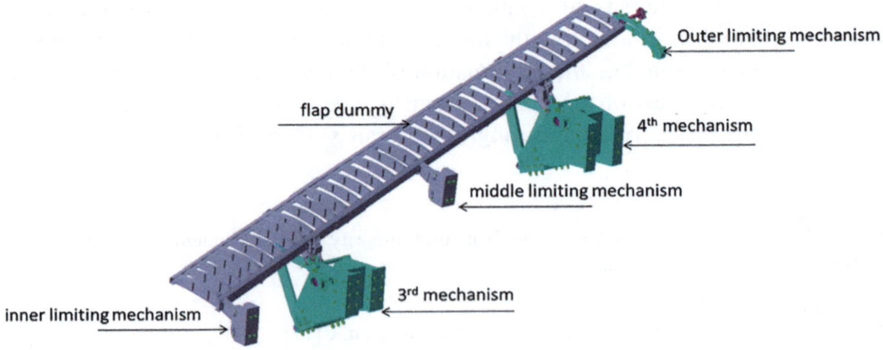

Fig. 1 Flap specimen

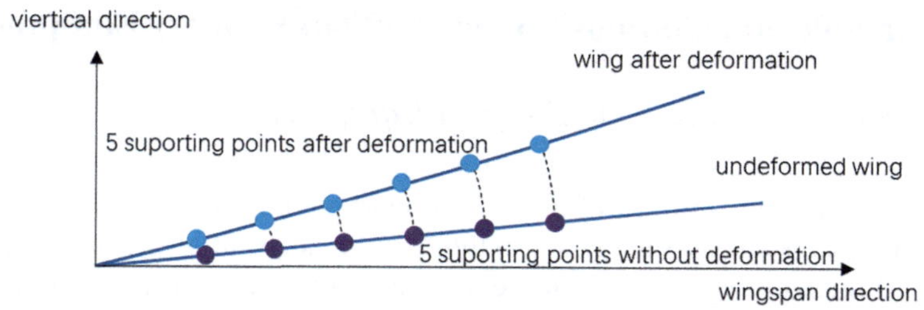

Fig. 2 Aircraft wing deformation diagram

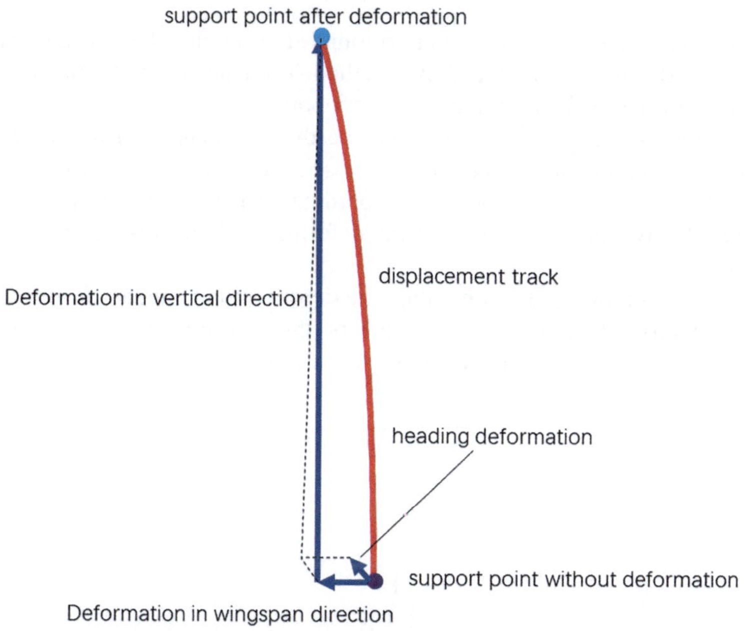

Fig. 3 Displacement track of flap support point

devices on the rear beam of the wing are displaced, as shown in Fig. 2. Throughout the entire process of flap function and reliability test, it is necessary to simulate the displacement change of the support points.

It can be seen from Fig. 3 that the displacement track of a single flap support point is a curve. It has the largest displacement component in the vertical direction of the flap, the smaller displacement component in the wingspan direction, and the smallest displacement component in the heading direction. We can obtain the displacements of the flap support points by simulation or experiment.

3 Principle and Composition of the Flap Support Equipment

3.1 Principle of the Flap Support Equipment

The curve displacement track of one flap support point is approximately treated as a straight line, as shown in Fig. 4. Firstly, the displacement of the flap joint in the heading direction is ignored. Then the maximum displacement in the vertical and the smaller wingspan displacement of the support joint are synthesized into a vector starting from the initial installation position of the flap joint. Finally, we can obtain the simplified displacements of the flap support joints with the deformation size and direction.

The displacements of the flap support joints at different deflection angles can be calculated and simplified by the above method. Then, according to the order of the test cases and flap deflection angles, the support points' motion rate of each flap support joint can be calculated. Finally, the displacements and motion rates are compiled into the boundary simulation test spectrum. In the test process, the displacements of all flap support points are coordinated and controlled by the test spectrum, and the dynamic simulation of the deformation boundary of the flap can be realized.

For the two support points on a flap, the deformation boundary is simplified by the aforementioned method. The results of the simplification are presented in Table 1. It can be seen from the table that the simplified errors are all below 3%.

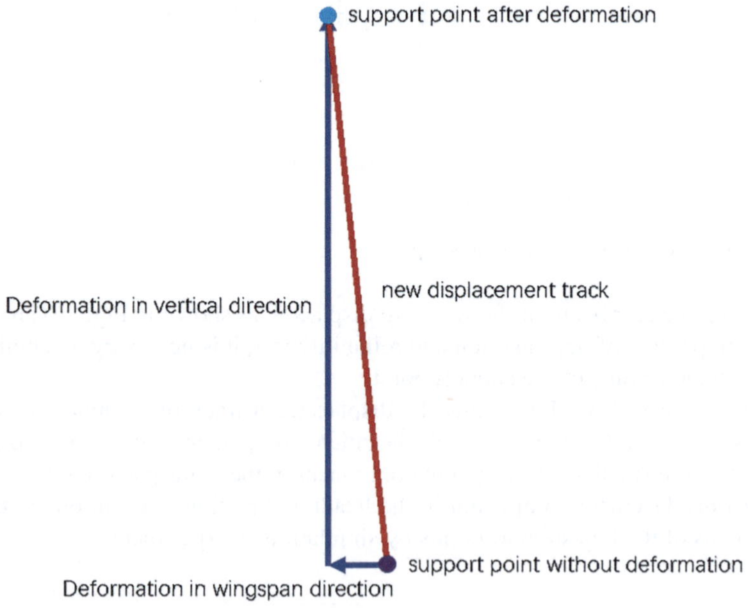

Fig. 4 The simplified displacement of one flap support point

Table 1 Simplified results of two support points on a flap

	Original displacement (mm)				Simplified displacement (mm)				
	X[b]	Y[c]	Z[d]	Vector[e]	X	Y	Z	Vector	Error[a] (%)
Point 1	2.94	131.00	12.80	132.00	0	131.00	12.80	131.62	0.29
Point 2	4.20	340.00	35.00	342.00	0	340.00	35.00	341.80	0.06

[a]Error, the relative error between the simplified vector and the original vector
[b]X, displacement in heading direction
[c]Y, displacement in vertical direction
[d]Z, displacement in wingspan direction
[e]Vector, synthesis of X, Y and Z

3.2 *Composition of the Flap Support Equipment*

To achieve a high-precision simulation of the deformation boundary of the flap, this paper designs an adjustable support device for one point as shown in Fig. 5. It is mainly composed of a joint installation slider, linear guide, screw, worm gear, and servo motor. The two sides of the joint installation slider are sleeved on the cylindrical linear guide rail, and the middle part is connected with the screw. When the servo motor drives the screw to rotate through the worm gear, the joint installation slider can move linearly along the linear guide rail to simulate the linear displacement of a single flap support joint.

The flap support equipment with controllable deformation adjustment is designed, which has five sets of the above support device installed on five support columns to support the flap specimen as shown in Fig. 6. The dynamic and accurate simulation of the support boundary of the flap is realized by coordinating the motion of these joints through the test control system.

The support equipment designed in this paper has been applied in a flap function test as shown in Fig. 7. During the test, the deformation of the flap support boundary can be dynamically adjusted by the control system to meet the test requirements.

4 Conclusion

The flap support equipment and method can simulate dynamically the changing support boundary caused by wing deformation during the whole aircraft take-off and landing process. It can significantly improve the accuracy of boundary simulation in tests, fully reflect the influence of wing deformation on the movement function and performance of flaps, and make the experimental results more realistic and reliable. The accuracy of the simulation boundary of the flap is improved and determined by the proportion of heading deformation ignored. Moreover, its cost becomes much lower. So far, this method and equipment have been applied in a flap development test. It can also be extended to simulate the deformation support boundary of other structures, such as landing gear.

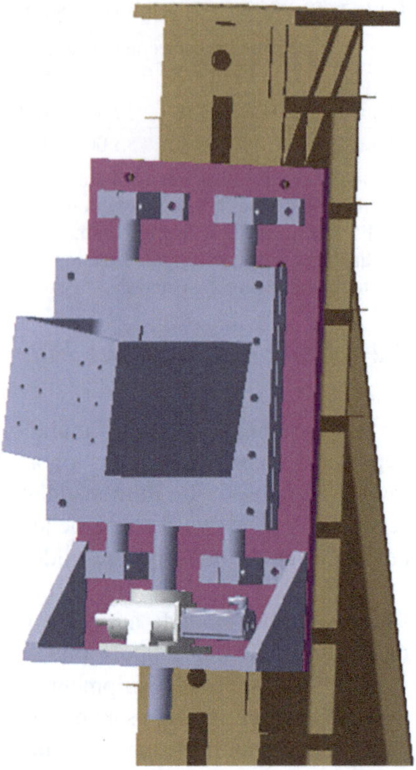

Fig. 5 Support device for one point

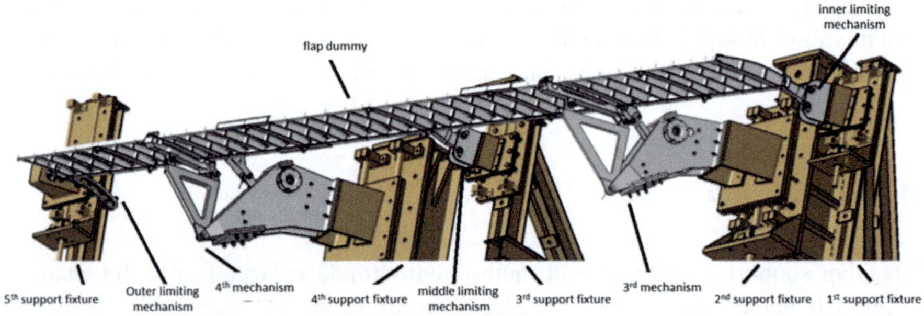

Fig. 6 Support equipment of the flap specimen

Fig. 7 Application in the flap function test

References

1. Sanyuan L, Xianmin C, Baocai P (2020) Research on self-adaptable loading technology for trailing edge flap. Eng Test 60(4):41–43. (in Chinese)
2. Baocai P, Dengke D, Yunzhao G et al (2014) Study on tracking loading method of locomotory wing for flap and slat. Mech Sci Technol Aerospace Eng 33(10):1590–1593. (in Chinese)
3. Zhang T, Zhang Y, Ren P et al (2021) Application of movable airfoil function test based on swing arm follow-up loading technology. Sci Technol Eng 21(17):7363–7368. (in Chinese)
4. Zhang T, Zhang Y, Yang Z et al (2019) Research and application of moving airfoil follow-up loading technology based on simulated trajectory. Sci Technol Eng 19(18):324–328. (in Chinese)
5. Zhang T, Song PF, Yin W et al (2022) Follow-up loading technology for lift structure with spatial complex movement. Acta Aeronaut Astron Sin 43(6):526044. (in Chinese)
6. Jandaurek K, Johst M (2017) Development trends and innovations in aerospace system testing using the example of high-lift. In: Grapevine: 55th AIAA aerospace sciences meeting
7. Aixian L (2023) A high lift test system based on wing deformation. J Phys Conf Ser 2658:012043

Open Access This chapter is licensed under the terms of the Creative Commons Attribution-NonCommercial-NoDerivatives 4.0 International License (http://creativecommons.org/licenses/by-nc-nd/4.0/), which permits any noncommercial use, sharing, distribution and reproduction in any medium or format, as long as you give appropriate credit to the original author(s) and the source, provide a link to the Creative Commons license and indicate if you modified the licensed material. You do not have permission under this license to share adapted material derived from this chapter or parts of it.

The images or other third party material in this chapter are included in the chapter's Creative Commons license, unless indicated otherwise in a credit line to the material. If material is not included in the chapter's Creative Commons license and your intended use is not permitted by statutory regulation or exceeds the permitted use, you will need to obtain permission directly from the copyright holder.

Research on the Design of Variable-Sweep Wing Drive Mechanism for Missiles

Xiaolei Wang and Zhiwei Xu

Abstract Variable-sweep wings are an important means to meet the flight characteristics of variable geometry missiles. In this paper, a variable-sweep wing structure for missiles was designed. First, a three-dimensional digital model of the wing structure was established. Then, based on the results of computational fluid dynamics simulations, an appropriate actuator was selected. Finally, computational fluid dynamics software was used to study missile models at various sweep angles, comparing and analyzing the aerodynamic characteristics at different speeds. The research summarized the patterns of the lift coefficient, drag coefficient, and lift-to-drag ratio as they change with sweep angle at different Mach numbers.

Keywords Variable geometry structures · Shape memory alloy · Aerodynamic characteristics

1 Introduction

Modern warfare is largely a contest of air power, and air superiority is the foundation and prerequisite for all conventional military operations. Air control is a crucial factor in determining the outcome of a war, and precision strikes are the main forms of combat. Airborne missiles are key weapon systems for gaining air superiority and carrying out precision strikes [1]. The design goal of variable-wing missiles is to resolve the contradictions in the aerodynamic layout design points for different mission types through adaptive changes in the missile's aerodynamic shape. This overcomes the single-function and limited mobility of traditional fixed-wing cruise missiles, aiming to achieve universality for missiles in multiple missions [2].

X. Wang · Z. Xu (✉)
College of Aerospace Engineering, Nanjing University of Aeronautics and Astronautics, Nanjing, China
e-mail: zhwxu@nuaa.edu.cn

© The Author(s) 2026

Z. Zhang, P. Mercorelli (eds.), *Mechanical Design and Simulation: Insights and Innovations*, Lecture Notes in Mechanical Engineering,
https://doi.org/10.1007/978-981-95-2405-1_94

One of the key technologies for variable-sweep wing aircraft is the design of the variable-sweep drive mechanism. Traditional sweep wing drive mechanisms mainly include hydraulic, crank-rocker, and slider-crank types. Hydraulic mechanisms are not suitable for adaptive aircraft due to their large mass and volume. Crank-rocker mechanisms have the disadvantages of low synchronicity, large errors, and poor stability due to the difficulty in balancing inertia forces and moments, making them unsuitable for high-speed transmission [3]. Although slider-crank mechanisms have better stability, the complexity of the mechanism is relatively high [4].

This article, based on the three missions of cruise, patrol, and attack carried out by the missile, takes into account the size of the space and control performance to design the deflection structure, and conducts aerodynamic simulation analysis of the missile model under different working conditions.

2 Design of the Missile Wing's Sweep Angle Deflection Structure

2.1 *Missile Overall Model Construction*

Based on the mission and flight characteristics of the missile, the missile's exterior shape has been designed. The missile model adopts a front-to-back collinear “+"shaped circumferential layout, with a conventional layout along the missile's longitudinal axis. The nose of the missile is designed to be hemispherical, and the wing sections of both the wings and the tail fins are of a double-curved profile. The wings feature a structure with fixed inner wings and variable-sweep outer wings, with the outer wings having a sweep angle range of 30° to 90°. The missile consists of a warhead, body, four wings, and tail fins. The total length of the missile is approximately 1800 mm, with a body diameter of 160 mm, a wing span of 240 mm when fully retracted, an outer wing length of 300 mm, and an inner wing height of 16 mm. The structural shape and dimensions are shown in Fig. 1.

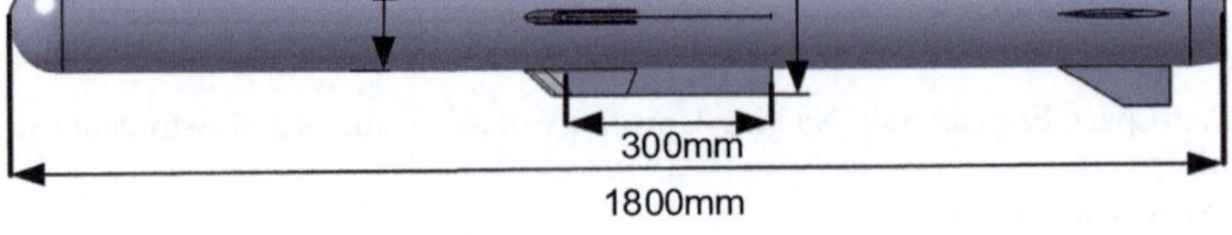

Fig. 1 Missile exterior and dimensions with a 90° wing sweep angle

2.2 Design and Principle Analysis of Wing Sweep Angle Deflection Drive Mechanism

The connection between the flaperon and the missile body is made using a sleeve-type design, with the inner wing fixed to the sleeve. There is a rigid connection between the upper and lower blocks of the sleeve. The overall structural layout is shown in Fig. 2.

First, the elastic force of the planar spiral spring is used to quickly deflect the flaperon from a sweep angle of 90° to 30°, and then lock it in place. Afterwards, the recovery force of the SMA wire is utilized to drive the flaperon from a sweep angle of 30° to 60°, and lock it again. The structure for the deflection of the flaperon's sweep angle is shown in Fig. 3.

Considering the thin thickness of the wing and the limited arrangement space, this paper designs and uses a non-contact flat spiral spring.

The deformation angle of the spring Ψ [5]

$$\psi = \frac{12K_1Tl}{Ebh^3} \tag{1}$$

In the formula: K_1 is a coefficient, $K_1 = 1.25$, T is the spring torque, l is the spring's unfolded length, E is the modulus of elasticity, b is the spring's width, and h is the spring's thickness. The thickness h of the spring material section.

$$h = \sqrt{\frac{6K_2T}{b[\sigma]}} \tag{2}$$

In the formula: K_2 is a coefficient, $K_2 = 2$.

The working number of turns n of the spring material.

$$n = \frac{\psi}{2\pi} = \frac{6K_1Tl}{\pi Ebh^3} \tag{3}$$

The working length l of the spring material.

$$l = \frac{Ebh^3\psi}{12K_1T} \tag{4}$$

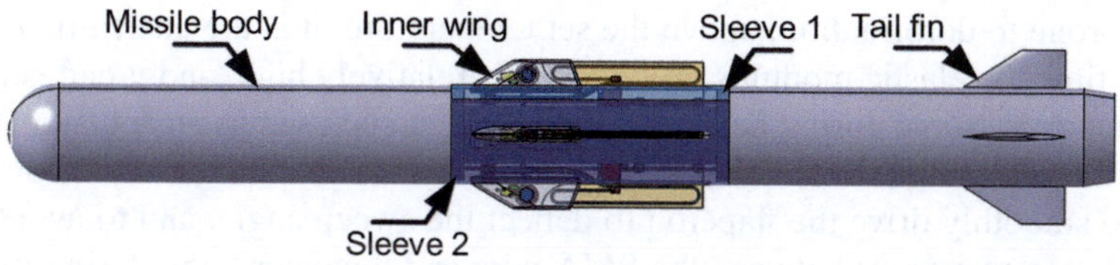

Fig. 2 Schematic diagram of the overall layout of the wing structure

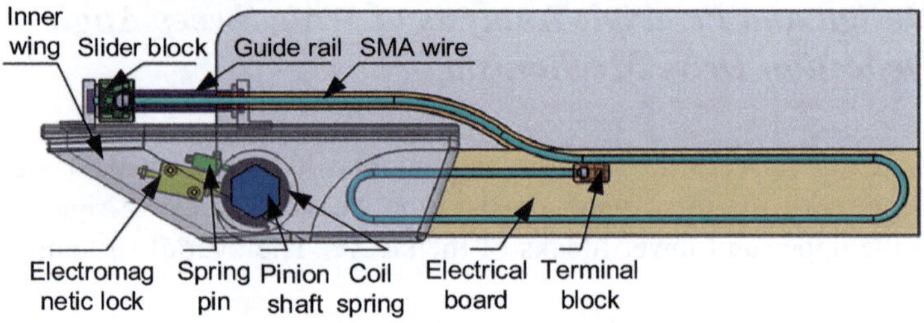

Fig. 3 Schematic diagram of the wing sweep angle deflection structure

The pitch *t* of the spring material.

$$t = \frac{\pi\left(R^2 - R_1^{\ 2}\right)}{l} \tag{5}$$

The inner radius R_1 and the outer radius *R* of the spring.

$$\left.\begin{aligned} R &= \frac{2l}{\psi} - R_1 \\ R &= R_1 + n_0 t \end{aligned}\right\} \tag{6}$$

The strength verification of the spring, σ

$$\sigma = \frac{6K_2T}{bh^2} = \frac{n\pi EhK_2}{K_1 l} \le [\sigma] \tag{7}$$

Based on the aerodynamic load simulation analysis of the wing model obtained in the early stage, the corresponding dimensions of the spiral spring can be calculated.

After the missile maintains a sweep angle of 30° for cruising and patrolling, the wings need to be deflected to a sweep angle of 60° to enter the attack state. The deflection torque of the wings during this process is provided by the Shape Memory Alloy (SMA) wires. The nickel-titanium alloy material, after certain processing, has the shape memory function, which means that after being deformed by an external force in a low-temperature environment, it can return to its initial shape when heated to a certain temperature, hence it is called a shape memory alloy. Inside the SMA material, there are mainly two alloy phases; below the set temperature, it is the martensitic phase, at which time the elastic modulus of the SMA is relatively low, making it prone to deformation; above the set temperature, it is the austenitic phase, at which time the elastic modulus of the SMA is relatively high, and it can generate a large recovery force during deformation, which can be used as an actuator [6].

In order to enable the tensile force generated by the Shape Memory Alloy (SMA) wire to smoothly drive the flaperon to deflect the sweep angle, and to avoid excessive friction resistance between the SMA wire and the outer wing during its operation, the SMA wire should be reasonably arranged. This structure uses an SMA wire

Table 1 Mechanical properties of 1 mm SMA wire

M_S/°C	M_f/°C	A_S/°C	A_f/°C	D_A/Gpa	D_M/Gpa	ε_l/%
34.2	24.0	40.8	52.3	12.7	43.8	5.2

Table 2 Recovery force values of SMA wire

Pre-stretch amount/%	2	2.5	3	3.5	4	4.5	5	5.5
Resilience force/N	171	191	230	317	390	401	432	470

with a diameter of 1 mm, and the main mechanical properties of the SMA wire are shown in Table 1 as demonstrated by experimental testing. Among them, M_S is the start temperature of the martensitic phase transformation, M_f is the end temperature of the martensitic phase transformation, A_S is the start temperature of the austenitic phase transformation, A_f is the end temperature of the austenitic phase transformation, D_M is the elastic modulus of the martensite, D_A is the elastic modulus of the austenite, ε_l is the maximum residual strain, and the recovery force of the SMA wire is shown in Table 2.

Based on the aerodynamic field simulation analysis of the wing model from the early stage. The sum of the aerodynamic torque acting on the wing and the recovery torque of the SMA wire is greater than the output torque of the spring at this time, thus the wing can achieve the sweep angle deflection.

2.3 *Design of the Wing Sweep Angle Deflection Locking Mechanism*

Considering the limited internal space layout due to the relatively small size of the missile's inner and outer wings, an electromagnetic locking device is first used to lock the wing sweep angle when the wing is deflected from 90° to 30°. The electromagnetic locking device is arranged inside the inner wing space, and there is a limiting groove on the inner wing surface that cooperates with the electromagnetic locking device to limit the position of the electromagnetic lock. A schematic diagram of the electromagnetic lock relative to the positioning hole is shown in Figs. 4 and 5.

In the mid-section of the spring box sidewall, there is a positioning hole corresponding in size to the spring locking tongue. During the process where the wing sweep angle deflects from 90° to 30°, the spring locking mechanism is in a de-energized state, allowing the tongue to pop out. As the wing deflects around the rotation axis, the tongue is in a compressed state and slides relatively against the sidewall of the spring box. Once the wing sweep angle is deflected to 30°, the tongue pops out to fit with the positioning hole, completing the positioning and locking.

When the flaperon's sweep angle is deflected to 60°, the electromagnetic locking device releases the lock, and this locking task is achieved by a spring pin. The lower

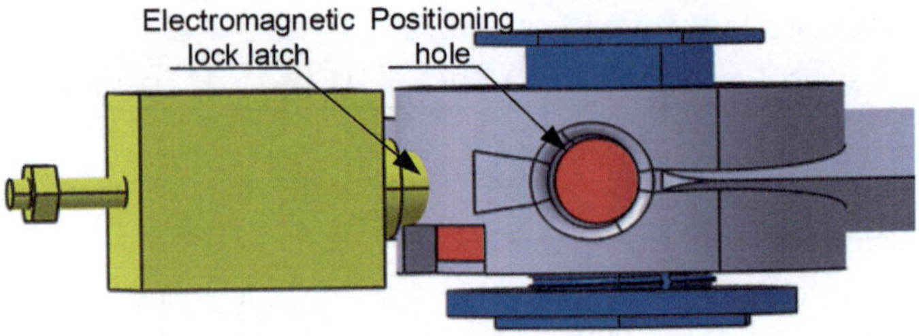

Fig. 4 Schematic diagram of the relative position of the locking tongue and positioning hole when the wing sweep angle is 90°

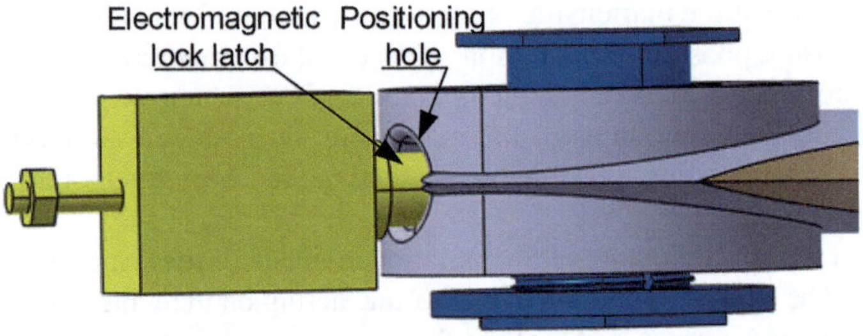

Fig. 5 Schematic diagram of the relative position of the locking tongue and positioning hole when the wing sweep angle is 30°

end of the side wall of the spring box has a positioning groove corresponding to the head of the spring pin. When the flaperon is deflected to a sweep angle of 60°, the top of the spring pin comes into contact with the wall of the spring box. At this point, the SMA wire is continuously powered, and the recovery torque of the wire causes the flaperon's sweep angle to deflect in the opposite direction to the one in which the spring pin obstructs the deflection of the sweep angle, thereby achieving the lock. The relevant schematic diagrams are shown in Figs. 6 and 7.

3 Aerodynamic Characteristic Analysis of Variable-Sweep Wing Missiles

To capture the impact of different sweep angle configurations of the air-launched missile on its aerodynamic characteristics, the missile flight angle of attack is set to 5°, and simulations are performed at sweep angles of 30°, 40°, 50°, 60°, 70°, and 80° for missile airspeeds of 0.3 Ma, 0.5 Ma, 0.9 Ma, and 1.1 Ma. The material parameters of the fluid are based on the values for an ideal gas. Ultimately, the aerodynamic characteristics of the missile under variable sweep wing conditions are obtained by monitoring the lift and drag coefficients through Fluent monitors

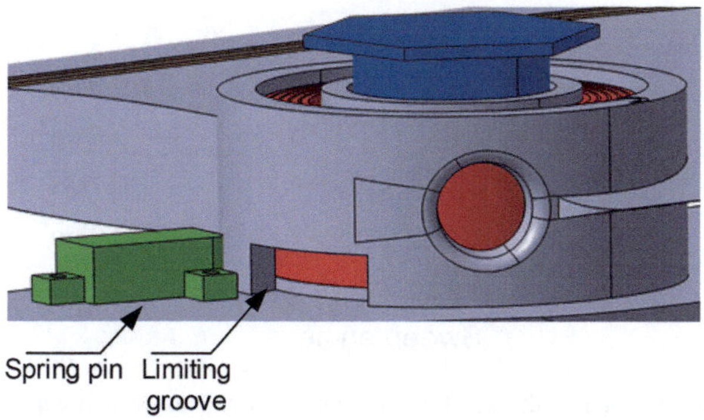

Fig. 6 Schematic diagram of the relative position of the spring pin and spring box when the wing sweep angle is 90°

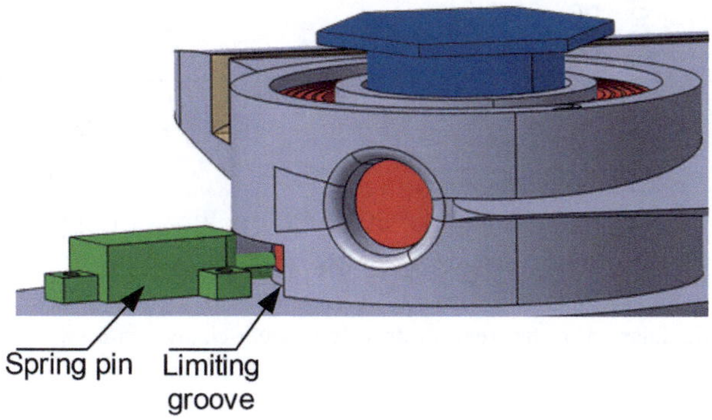

Fig. 7 Schematic diagram of the relative position of the spring pin and spring box when the wing sweep angle is 60°

[7].The trend of the lift coefficient change is shown in Fig. 8. Observing the relationship between the lift coefficient and the sweep angle at different airspeeds, it can be seen that at each airspeed, the lift coefficient decreases as the wing sweep angle increases. This is because the local flow velocity on the wing at this time is the component of the actual airspeed along the chord of the wing; the larger the sweep angle, the smaller this component becomes. It is also noted that at the same sweep angle, the lift coefficient tends to increase with increasing airspeed.

The trend of drag coefficient change is shown in Fig. 9.From the relationship between the drag coefficient and the sweep angle at different airspeeds, it is known that at low speeds, the main source of missile drag is frictional resistance, which is related to the frontal area. As the sweep angle increases, the frontal area of the missile slowly decreases, and thus the drag also slowly decreases. When the missile's airspeed reaches the transonic range, due to the formation of local shock waves on

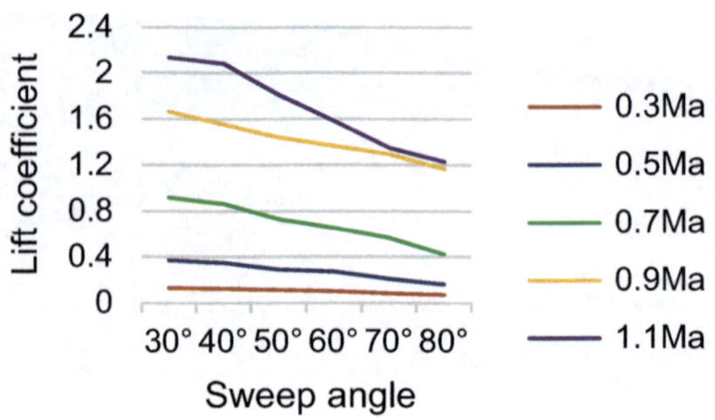

Fig. 8 Schematic diagram of the trend of lift coefficient change with sweep angle at different airspeeds

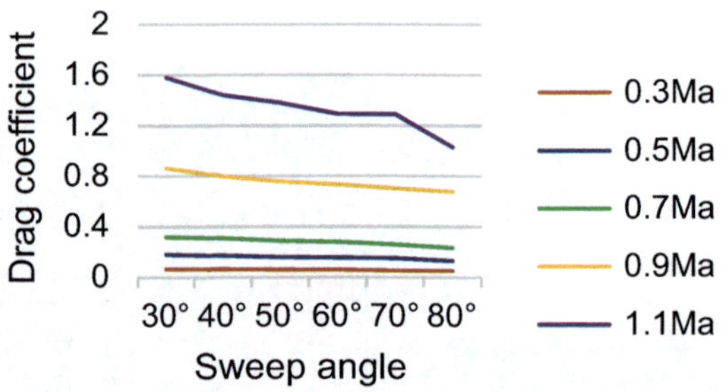

Fig. 9 Schematic diagram of the trend of drag coefficient change with sweep angle at different airspeeds

the wings, the drag on the missile increases dramatically. However, as the sweep angle increases, the occurrence of shock waves is delayed, leading to a corresponding reduction in the drag coefficient, but its value is still greater than that at the same sweep angle at low speeds.The trend of the lift-to-drag ratio change is shown in Fig. 10. The lift-to-drag ratio, defined as the ratio of the lift coefficient to the drag coefficient, is a dimensionless number that describes aerodynamic characteristics. From the relationship between the lift-to-drag ratio and the sweep angle at different airspeeds, it is known that at low speeds, the lift-to-drag ratio decreases as the sweep angle increases. When the airspeed approaches the speed of sound, local shock waves appear on the wings, and the lift-to-drag ratio first increases and then decreases with the increase of the sweep angle. This is due to the increased critical Mach number and the delay of the appearance of shock wave drag caused by the wing sweep. When the sweep angle is too large, the decrease in the lift coefficient exceeds the decrease in the drag coefficient, hence the lift-to-drag ratio decreases with the increase of the sweep angle.

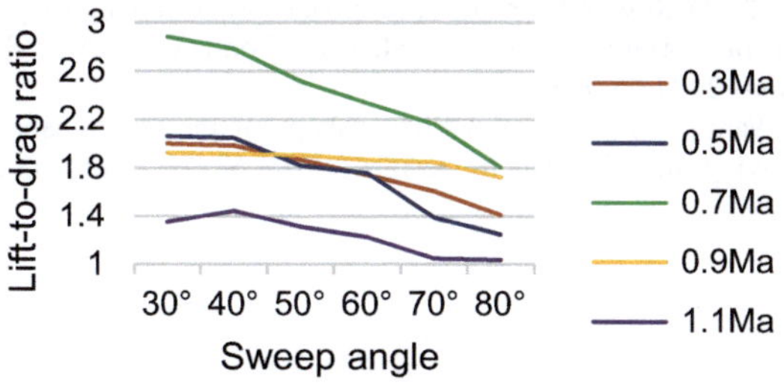

Fig. 10 Schematic diagram of the trend of lift-to-drag ratio change with sweep angle at different airspeeds

4 Conclusions

In this paper, based on the flight characteristics of the missile, a variable-sweep wing drive mechanism has been designed, and the following conclusions have been drawn:Compared to traditional drive mechanisms such as hydraulic systems, the variable-sweep wing drive mechanism designed in this paper is structurally simple and has a small mass and volume. It can effectively achieve the variable-sweep function of the wings while meeting the structural strength requirements.Through aerodynamic simulation analysis of the missile flow field, it is concluded that the lift coefficient tends to decrease as the wing sweep angle increases. At the same sweep angle, the lift coefficient also increases with the increase of airspeed. When the airspeed approaches the speed of sound, although the drag coefficient and the lift-to-drag ratio may temporarily increase due to the formation of local shock waves on the wings, they generally decrease with the increase of the sweep angle.

References

1. Li Y, Zhang SJ, Zhu YH (2024) Analysis of the development trends and dynamics of typical airborne missiles of the United States and Russia in 2023. Aerospace Def 7(04):30–37
2. Liu XH, Zhang SZ, Miao HC (2020) Design and simulation of trajectory for missiles with variable wings. J Ballist 40(05):20–23. https://doi.org/10.15892/j.cnki.djzdxb.2020.05.005
3. Cai ZJ, Gao ZC (2018) Design of variable-sweep wing actuation mechanism and CFD simulation. Mech Design 35(09):17–22. https://doi.org/10.13841/j.cnki.jxsj.2018.09.003
4. Wang ZJ (2019) A review of the current status of variable-sweep wing research. China High Tech (05):31–33. https://doi.org/10.13535/j.cnki.10-1507/n.2019.05.06
5. Yang JR, Ren BF, Sun SS (2021) Design and analysis of the coil spring in the spring energy storage power generation system. Mech Design Manufact (09):145–149. https://doi.org/10.19356/j.cnki.1001-3997.2021.09.033

6. Chen J, Li JY (2024) Research on the mechanical properties of shape memory alloys and their application in morphing structures. Measure Contr Technol 43(05):34–41. https://doi.org/10.19708/j.ckjs.2024.05.003
7. Gao Y, Li YB, Wang Y (2022) Simulation analysis of the aerodynamic characteristics of a variable swept wing missile. Mech Design 39(03):68–73. https://doi.org/10.13841/j.cnki.jxsj.2022.03.005

Open Access This chapter is licensed under the terms of the Creative Commons Attribution-NonCommercial-NoDerivatives 4.0 International License (http://creativecommons.org/licenses/by-nc-nd/4.0/), which permits any noncommercial use, sharing, distribution and reproduction in any medium or format, as long as you give appropriate credit to the original author(s) and the source, provide a link to the Creative Commons license and indicate if you modified the licensed material. You do not have permission under this license to share adapted material derived from this chapter or parts of it.

The images or other third party material in this chapter are included in the chapter's Creative Commons license, unless indicated otherwise in a credit line to the material. If material is not included in the chapter's Creative Commons license and your intended use is not permitted by statutory regulation or exceeds the permitted use, you will need to obtain permission directly from the copyright holder.

Research on Static Strength Design of Civil Aircraft Ceiling

Lu Liu

Abstract The finite element analysis method is an important means to check the strength of civil aircraft interiors. In this paper, a typical structure of a civil aircraft interior is discussed, the analysis objects and loads are sorted out, the finite element model of the typical structure is established, and an engineering algorithm is proposed to check the fasteners. Taking the cabin ceiling of a civil aircraft as an example, it is proved that the proposed method has certain engineering application significance.

Keywords Aircrafe interior · Finite element · Strength analysis

1 Introduction

The main cabin ceiling is mounted on the fuselage floor rail. The loads used for the main cabin ceiling body and its installation strength check to consider the fuselage overload envelope and fuselage static load in the case of aircraft dynamic load [1–8]. Overload envelope and emergency landing loads as specified in CCAR25.561, the safety factor of CCAR25.303, and the special factors of CCAR25.619 and 25.625 are considered. In this paper, the research parts are the main cabin ceiling body, the connection strength between the bodies, and the body's structure and the airframe.

The main cabin ceiling is mounted on the forward (FWD) fuselage floor rail. The top is connected to the C-shaped groove via four bolts, and the main cabin ceiling's body panel is constructed of honeycomb sandwich panels. The structure of the cabin ceiling is illustrated in Fig. 1.

L. Liu (✉)
Shanghai Aircraft Design and Research Institute, Shanghai, China
e-mail: liulu2@comac.cc

© The Author(s) 2026

Z. Zhang, P. Mercorelli (eds.), *Mechanical Design and Simulation: Insights and Innovations*, Lecture Notes in Mechanical Engineering,
https://doi.org/10.1007/978-981-95-2405-1_95

Fig. 1 The structure of the cabin ceiling

Table 1 Performance of metal materials for main cabin ceiling

Material grade	Mechanical properties				E (GPa)	μ
	σ_{tu} (MPa)		σ_{bru} (MPa)			
	A	B	A	B		
2024-T3511	414	427	661	682	72	0.33
2024-O-T42	414	427	661	682	72	0.33
304	517		1, 116		200	0.28

2 Material and Fastener Performance

The materials used for the main cabin ceiling are listed in Tables 1, 2 and 3. The mechanical property data of the fasteners employed are presented in Table 4.

3 Load

The design of the main cabin ceiling must comprehensively consider the dynamic load envelope, static load envelope, emergency landing load, and pressure relief load of A-type aircraft [9]. Based on the aircraft's load envelope, the limiting overload coefficient for the corresponding station on the main cabin ceiling is derived. This coefficient is then multiplied by a safety factor of 1.5 to obtain the limit overload coefficient in each direction corresponding to the station where the main cabin ceiling is located. The maximum value of the limit overload in each direction under the aforementioned conditions is selected as the load for the corresponding station on the main cabin ceiling. The most severe limiting inertial load case is presented in Table 5.

Table 2 Properties of main composite materials for main cabin ceiling (prepreg)

Material	Modulus of elasticity E11 (MPa)	Modulus of elasticity E22 (MPa)	Shear modulus G12 (MPa)	Poisson's ratio μ 12	Single layer thickness (mm)	Shear strength L (MPa)	Shear strength W (MPa)
Prepreg	24, 000	24, 000	4, 500	0.07	0.250	30	30

Table 3 Properties of main composite materials for main cabin ceiling (prepreg)

Material	Modulus of compression E3 (MPa)	Compressive strength (MPa)	Shear modulus LT G13 (MPa)	Shear modulus WT G23 (MPa)	Shear strength LT (MPa)	Shear strength WT (MPa)
Honeycomb core	91.45	1.24	39	21	1.24	0.65
Honeycomb core	138	1.24	44	26	1.1	0.58

Table 4 Performance of metal materials for main cabin ceiling

Type of fastener	Stretch allowable (N)	Shear allowable value (N)
Rivet	27, 545	20, 692 (single shear)
Rivet	1, 531	4, 828 (single shear)
Screw	11, 080	9, 456 (single shear)
Insert	1, 029	1, 854 (single shear)

Table 5 Selection of limit inertia load coefficient at the center of gravity of the main cabin ceiling

Load case	Turn left	Right	Forward	Backwards	Up	Down
Emergency landing conditions	3	3	9	1.5	3	6
Static load case	0.76	0.76	2.05	1.66	1.5	5.33
Dynamic load case	0.41	0.41	–	–	1.3	7
The load factor	3	3	9	1.66	3	7

4 Establishment of Finite Element Model

The finite element model is established based on the structural elements and connections of the main cabin ceiling, with the body panel modeled using SHELL elements [10]. The joints of the main cabin ceiling constrain its linear displacement in all three directions. The finite element model of the main cabin ceiling, established as described above, is shown in Fig. 2.

Fig. 2 The finite element model of the cabin ceiling

5 Strength Analysis

5.1 Strength Analysis of Main Cabin Ceiling Body

According to the finite element results, under the pressure relief load condition, the maximum shear stress of the panel in the XY direction is $\tau_{XY} = 28.7MPa$, as shown in Fig. 3. If the allowable shear stress of the panel is known $[\tau] = 30MPa$, the shear safety margin of the panel in the XY direction can be calculated as follows:

$$M.S. = \frac{[\tau]}{\tau_{XY}} - 1 = \frac{30}{28.7} - 1 = 0.05 \tag{1}$$

According to the finite element results, under the pressure relief load condition, the maximum shear stress of the honeycomb core in the XY direction is $\tau_{XY} = 0.002MPa$, as shown in Fig. 4. If the allowable shear stress of the panel is known $[\tau_{XY}] = 1.1MPa$, the shear safety margin of the panel in the XY direction can be calculated as follows:

$$M.S. = \frac{[\tau_{XY}]}{\tau_{XY}} - 1 = \frac{0.58}{0.005} - 1 = 115 \tag{2}$$

According to the finite element calculation results, the honeycomb core experiences the maximum compressive stress under the pressure relief working condition, and the maximum compressive stress is $\tau_{XY} = 0.002MPa$. The stress nephogram is shown in Fig. 5, and the allowable stress for honeycomb core compression is $[\sigma_C] = 1.24MPa$. The compression safety margin of the honeycomb core is then calculated as:

$$M.S. = \frac{[\sigma_c]}{\sigma_c} - 1 = \frac{1.24}{0.002} - 1 = 619 \tag{3}$$

Fig. 3 The maximum shear stress of the panel in the XY

5.2 *Main Cabin Ceiling Connection*

C-shaped grooves are used to connect the ceiling panels of the main cabin, and the allowable stress of the material used for these grooves is $[\sigma_{tu}] = 414MPa$, as can be seen from the calculation results of the engineering algorithm. The connecting C-shaped groove between the panels of the main cabin ceiling experiences the maximum working stress, and the maximum working stress is $\sigma = 295MPa$. Then, the safety margin of the connecting corner piece can be calculated as:

$$M.S. = \frac{[\sigma_{tu}]}{\sigma} - 1 = \frac{414}{295} - 1 = 0.40 \tag{4}$$

5.3 *Connection Strength Check of the Main Cabin Ceiling*

According to the calculation results of the engineering algorithm, under the pressure relief load condition, the connecting fasteners are subjected to the maximum tensile and shear loads, which are F_t=641N and F_s=419N, respectively. Since the working load of the fastener is far less than its allowable load, the fastener meets the strength requirements. Under this load, the margin of safety for tension and shear of the blind hole insert can be calculated as follows:

$$M.S. = \frac{[F_t]}{F \times 1.15} - 1 = \frac{1043}{641 \times 1.15} - 1 = 0.41 \tag{5}$$

Fig. 4 The maximum shear stress of the honeycomb core in the XY

Fig. 5 The maximum compressive stress of the honeycomb core under the pressure relief working condition

Shear allowance:

$$M.S. = \frac{[F_s]}{F_s \times 1.15} - 1 = \frac{2331}{419 \times 1.15} - 1 = 3.84 \tag{6}$$

The blind hole insert should be evaluated based on a tension-shear combination, and the tension-shear combination factor of safety (or allowance) should be determined as follows:

$$
\begin{aligned}
M.S. &= \frac{1}{1.15\times\sqrt{\left(\frac{F_s}{[F_s]}\right)^2+\left(\frac{F_t}{[F_t]}\right)^2}}-1 \\
&= \frac{1}{1.15\times\sqrt{\left(\frac{419}{1836}\right)^2+\left(\frac{641}{1029}\right)^2}}-1 \\
&= 0.31
\end{aligned}
\tag{7}
$$

6 Conclusion

In this paper, the strength of the ceiling of the main cabin under different working conditions is checked by means of finite element analysis and engineering algorithms. Taking the cabin interior of a civil aircraft as an example, the basic performance data of the connection that needs to be analyzed for some internal static strength analysis are obtained, and stored for subsequent structural design. This article only discusses the influence of static strength on ceiling structures. In the future, more factors that affect the strength of ceiling structures, such as temperature changes and vibration loads, will be considered to provide a more comprehensive strength assessment. To fully understand the structural strength of civil aircraft interiors, additional research on theoretical calculation methods is required to explore the application of new materials or connection technologies in cabin ceiling structures, such as high-strength composite materials or advanced fastening systems, to enhance the overall strength and lightweight level of the structure.

References

1. Furtado BC et al (2021) A methodology to generate design allowable of composite laminates using machine learning--ScienceDirect. Int J Solids Struct
2. Pogosyan M, Nazarov E, Bolshikh A et al (2021) Aircraft composite structures integrated approach: a review. J Phys Conf Ser 1925(1):012005
3. Ardila Parra SA, Pappalardo CM, Estrada OAG et al (2020) Finite element-based redesign and optimization of aircraft structural components using composite materials. IAENG Int J Appl Math
4. Cui L, Zeng Z, Jiao S (2024) Finite element analysis of eddy current testing of aluminum honeycomb sandwich structure with CFRP panels based on the domain decomposition method. J Nondestruct Eval (2):43
5. Wang J, Zhang X, Chen K (2021) Study on thermo-mechanical coupling method of ship cabin with ceiling vent under fire condition. Chin J Ship Res 16(3):74–85
6. Aircraft Design Manual's Editors (2001) Aircraft design manual: Vol. 9, load, strength, and stiffness. Aviation Industry Press, Beijing. (in Chinese)

7. Mauro GD, Gagliardi GM, Guida M, Marulo F (2023) Parachute emergency landing simulation and enhanced composite material characterization for general aviation aircraft. Proc Inst Mech Eng C J Mech Eng Sci 238:1423–1433
8. Wan CH, Duan SH, Nie XH et al (2018) Study on finite element modeling for large aircraft structures. Mech Sci Technol Aerospace Eng 37(5):816–820
9. Zhiquan H, Yang L, Zhejiang L (2019) Load design for full-scale static test of slat on large civil aircraft. Acta Aeronaut Astronaut Sin 40(2):197–522
10. Santare MH, Lambros J (2000) Use of graded finite elements to model the behavior of nonhomogeneous materials. J Appl Mech 67(4):819–822

Open Access This chapter is licensed under the terms of the Creative Commons Attribution-NonCommercial-NoDerivatives 4.0 International License (http://creativecommons.org/licenses/by-nc-nd/4.0/), which permits any noncommercial use, sharing, distribution and reproduction in any medium or format, as long as you give appropriate credit to the original author(s) and the source, provide a link to the Creative Commons license and indicate if you modified the licensed material. You do not have permission under this license to share adapted material derived from this chapter or parts of it.

The images or other third party material in this chapter are included in the chapter's Creative Commons license, unless indicated otherwise in a credit line to the material. If material is not included in the chapter's Creative Commons license and your intended use is not permitted by statutory regulation or exceeds the permitted use, you will need to obtain permission directly from the copyright holder.

Decentralized Position Regulation for a Series of Time-Varying Mass-Spring-Damper Systems Via Solutions of Riccati Equations

Zhengji Mao

Abstract In this study, we introduce a decentralized control strategy for regulating the position of a series of mass-spring-damper systems characterized by parameters that change over time. We start with establishing a dynamic model for each interconnected subsystem, leading to a time-varying model as the outcome. Given the focus on optimal optimization challenges within time-varying systems, it is essential to determine the solution to the time-varying Riccati differential equation in advance. We derive a mathematical proof that the entire system is exponentially stable, ensuring a predefined level of stability. To validate our approach, computer simulations were performed on a three-mass system that is interconnected and experiences time-varying parameters. The simulation outcomes indicate that the positions of all masses revert to their initial states at an exponential rate, thereby confirming the practicality of the control method we have proposed.

Keywords Decentralized regulation · Time-varying system · Riccati equation

1 Introduction

Most conventional systems are characterized by differential equations, with mass-spring-damper systems being a common example of second-order systems. When masses are linked by springs and dampers in a series, the dynamics of each mass are influenced by adjacent masses, excluding the first and last ones. As the number of connected masses grows, the computational demands and transfer costs escalate significantly. This challenge can be mitigated by implementing decentralized control strategies to alleviate these burdens. Furthermore, the parameters of these interconnected systems may not remain constant but instead vary over time, necessitating optimization methods distinct from those used for linear time-invariant (LTI)

Z. Mao (✉)
School of Ocean Mechatronics, Xiamen Ocean Vocational College, Xiamen, China
e-mail: maozhengji@xmoc.edu.cn

© The Author(s) 2026

Z. Zhang, P. Mercorelli (eds.), *Mechanical Design and Simulation: Insights and Innovations*, Lecture Notes in Mechanical Engineering,
https://doi.org/10.1007/978-981-95-2405-1_96

systems. There is an abundance of research dedicated to decentralized control designs [1–5].

For decentralized control of systems with time-varying characteristics, a minimum-energy approach was proposed in [6] to address time-varying coverage control issues, transforming the coverage cost into a constrained optimization problem amenable to decentralized solutions. In [7], a novel approach was introduced for managing agents within a time-varying context, incorporating new nodes and branches in the graph, using machine learning techniques. The work in [8] focused on decentralized algorithms for coordinating the motion of autonomous vehicles to counteract the impacts of time-varying parameters. Additionally, a decentralized linear time-varying model predictive control strategy for unmanned aerial vehicles was put forward in [9].

In this paper, we introduce a decentralized stabilization control method for linear time-varying interconnected systems. Utilizing optimal control techniques based on a performance index, we have developed a decentralized time-varying feedback law tailored to subsystems. Given the time-varying nature of the system under consideration, time-varying Riccati differential equations must be resolved. The backward Euler's method is employed to derive solutions from a sufficiently large time point back to the initial time, using estimated terminal values. Once the initial conditions are established, these Riccati equations are integrated into the overall system for implementation. The stability of the system is rigorously demonstrated to be exponentially stable with a predetermined level of stability. It is crucial to note that if the interconnected systems adhere to the specified structural constraints, there will be no limitations on the intensity of interactions among the subsystems, and the prescribed exponential stability of the entire system will be maintained.

2 The Model Description

We consider a series of mass-spring-damper systems containing s masses, as shown in Fig. 1.

From Fig. 1, M_i is the mass (kg), y_i is the position (m), k_i is the spring constant (N/m), d_i is the damping coefficient (N·s/m) and u_i is the input (N). In this configuration, there are s masses, s positions, s inputs, (s—1) dampers, and (s—1) springs. Now, we derive the dynamic equations of this interconnected system by Newton's law. First, the dynamic equation for the first mass is described by

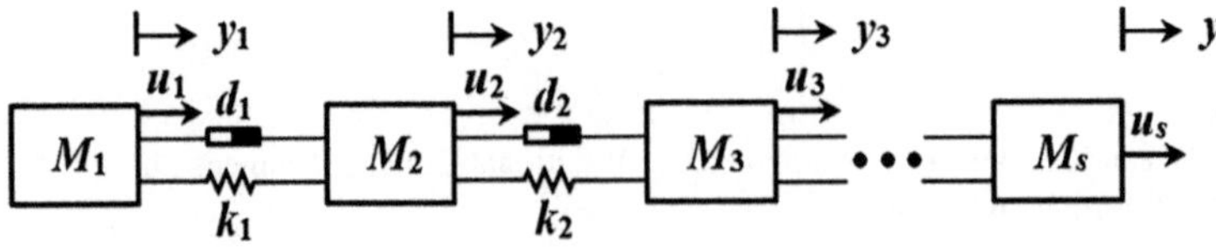

Fig. 1 A series of mass-spring-damper systems

$$M_1 \ddot{y}_1 = d_1(\dot{y}_2 - \dot{y}_1) + k_1(y_2 - y_1) + u_1 \tag{1}$$

then

$$M_1 \ddot{y}_1 + d_1\dot{y}_1 + k_1 y_1 = d_1\dot{y}_2 + k_1 y_1 + u_1 \tag{2}$$

Next, we derive the dynamic equation of the second mass as

$$M_2 \ddot{y}_2 = d_2(\dot{y}_3 - \dot{y}_2) + k_2(y_3 - y_2) - d_1(\dot{y}_2 - \dot{y}_1) + k_1(y_2 - y_1) + u_2 \tag{3}$$

After some simplifications, we obtain

$$M_2 \ddot{y}_2 + (d_1 + d_2)\dot{y}_2 + (k_1 + k_2)y_2 = d_2\dot{y}_3 + k_2 y_3 + d_1\dot{y}_1 + k_1 y_1 + u_2 \tag{4}$$

When (4) is generalized to the i-th mass where i = 2, 3, …, s-1, the generalized equation is

$$M_i \ddot{y}_i + (d_{i-1} + d_i)\dot{y}_i + (k_{i-1} + k_i)y_i = d_{i-1}\dot{y}_{i-1} + k_{i-1}y_{i-1} + d_i\dot{y}_{i+1} + k_i y_{i+1} + u_i \tag{5}$$

For the last mass, the equation is

$$M_s \ddot{y}_s + d_{s-1}\dot{y}_s + k_{s-1}y_s = d_{s-1}\dot{y}_{s-1} + k_{s-1}y_{s-1} + u_s \tag{6}$$

Until now, we have obtained three types of differential equations of the masses and these equations are transformed into state-space expressions. Let $x_{i1} = y_i$ and $x_{i2} = \dot{y}_i$, then

$$\begin{aligned}\dot{x}_i = \begin{bmatrix}\dot{x}_{i1}\\ \dot{x}_{i2}\end{bmatrix} &= \begin{bmatrix}0 & 1\\ -\dfrac{k_{i-1}+k_i}{M_i} & -\dfrac{d_{i-1}+d_i}{M_i}\end{bmatrix}x_i \\ &+ \begin{bmatrix}0\\ \dfrac{1}{M_i}\end{bmatrix}\left(u_i + \begin{bmatrix}k_{i-1} & d_{i-1}\end{bmatrix}\begin{bmatrix}x_{(i-1)1}\\ \dot{x}_{(i-1)2}\end{bmatrix} + \begin{bmatrix}k_i & d_i\end{bmatrix}\begin{bmatrix}x_{(i+1)1}\\ \dot{x}_{(i+1)2}\end{bmatrix}\right) \\ &= A_i x_i + B_i\left(u_i + \begin{bmatrix}k_{i-1} & d_{i-1}\end{bmatrix}x_{i-1} + \begin{bmatrix}k_i & d_i\end{bmatrix}x_{i+1}\right),\ i = 1, 2, \ldots, s\end{aligned} \tag{7}$$

where $d_0 = k_0 = d_s = k_s = 0$. We can rewrite (6) as

$$\dot{x}_i = A_i x_i + B_i\left(u_i + \sum_{j=1}^{s} q_{ij}x_j\right) \tag{8}$$

For the parameter q_{ij}, we have the following descriptions.

1. When $i = 1$, then $q_{12} = [k_1 \ d_1]$; $q_{1j} = [0 \ 0]$ while $j \neq 2$.
2. When $i = 2, 3, \ldots, s - 1$, then $q_{i(i-1)} = [k_{i-1} \ d_{i-1}]$; $q_{i(i+1)} = [k_i \ d_i]$; otherwise $q_{ij} = [0 \ 0]$.
3. When $i = s$, then $q_{s(s-1)} = [k_{s-1} \ d_{s-1}]$; otherwise $q_{sj} = [0 \ 0]$.

Now if the coefficients of the system are time-varying, (8) becomes

$$\dot{x}_i(t) = A_i(t)x_i(t) + B_i(t)\left(u_i(t) + \sum_{j=1}^{s} q_{ij}(t)x_j(t)\right) \tag{9}$$

The following are some statements and assumptions.

Assumption 1: Each subsystem's time-varying parameters $A_i(t)$ and $B_i(t)$ are assumed to have suitable dimensions and are considered to be uniformly piecewise controllable.

Assumption 2: It is assumed that all system parameter matrices are norm-bounded, using either Euclidean norms or induced Euclidean norms.

Assumption 3: $q_{ij}(t)$ is the time-varying parameter matrix, yielding:

$$\sup_t q_{ij}(t) = h_{ij} \geq 0 \tag{10}$$

3 The Controller Design

Now, we will examine the following optimization issue as

$$\min J_i = \int_0^{\infty} \left[\left(e^{\xi t}\mathbf{x}_i\right)^T Q_i \left(e^{\xi t}\mathbf{x}_i\right) + \left(e^{\xi t}u_i\right)^T R_i \left(e^{\xi t}u_i\right) \right] dt \tag{11}$$

This minimization index is a modified linear optimal regulator (MLQR) approach that can make a stable closed-loop system with a high-level control performance. The time-varying weighting matrices $Q_i(t)$ and $R_i(t)$ are both assumed to be symmetric positive definite matrices. Besides, in our method, the weighting matrices can also be time-varying. The optimal solution $u_i(t)$ is

$$u_i(t) = K_i(t)\boldsymbol{x}_i(t) = -R_i(t)^{-1} B_i(t)^T P_i(t)\boldsymbol{x}_i(t) \tag{12}$$

The above time-varying matrix $P_i(t)$ is the solution of a time-varying Riccati equation,

$$\begin{aligned} \dot{P}_i(t) = {} & P_i(t)B_i(t)R_i(t)^{-1} B_i(t)^T P_i(t) - \left(A_i(t) + \xi I_i\right)^T P_i(t) \\ & - P_i(t)\left(A_i(t) + \xi I_i\right) - Q_i(t) \end{aligned} \tag{13}$$

The parameter I_i in (13) is an identity matrix and $P_i(t)$ is a time-varying, symmetric positive definite matrix. Different from time-invariant cases, we have $\dot{P}_i(t) \neq 0$. In order to determine the feedback input, $P_i(t)$ must be pre-calculated to complete Eq. (12). From our previous work [10], if

$$\alpha_i \geq \frac{1}{2}\left\{\left(\sum_{j=1}^{s} \frac{h_{ij}^{2}}{\lambda_{\min}\left(R_i^{-1}\right)\lambda_{\min}\left(Q_j / s\right)}\right)+1\right\} \tag{14}$$

and

$$K_i^*(t) = -\alpha_i R_i(t)^{-1} B_i(t)^T P_i(t) \tag{15}$$

then the global system (8) will be asymptotically stable with a prescribed convergent rate.

4 Computer Simulations

A 3-mass interconnected mass-spring-damper system is considered and the parameters are as follows.

$$\begin{aligned} &M_1 = M_2 = 1 + 0.1\cos 2t; M_3 = 1 + 0.2\cos 4t; d_1 = d_2 = 0.2 + 0.2\sin t;\\ &k_1 = k_2 = 1 + 0.2\sin t \end{aligned} \tag{16}$$

$$\begin{aligned} A_1 &= \begin{bmatrix} 0 & 1 \\ -\dfrac{1+0.2\sin t}{1+0.1\cos 2t} & -\dfrac{0.2+0.2\sin t}{1+0.1\cos 2t} \end{bmatrix}, B_1 = \begin{bmatrix} 0 \\ \dfrac{1}{1+0.1\cos 2t} \end{bmatrix};\\ A_2 &= \begin{bmatrix} 0 & 1 \\ -\dfrac{2+0.4\sin t}{1+0.1\cos 2t} & -\dfrac{0.4+0.4\sin t}{1+0.1\cos 2t} \end{bmatrix}, B_2 = \begin{bmatrix} 0 \\ -\dfrac{1}{1+0.1\cos 2t} \end{bmatrix}\\ A_3 &= \begin{bmatrix} 0 & 1 \\ -\dfrac{1+0.2\sin t}{1+0.2\cos 4t} & -\dfrac{0.2+0.2\sin t}{1+0.2\cos 4t} \end{bmatrix}, B_1 = \begin{bmatrix} 0 \\ \dfrac{1}{1+0.2\cos 4t} \end{bmatrix} \end{aligned} \tag{17}$$

The performance indices are simply chosen as $Q_i = 3I$ and $R_i = 1$, $i = 1, 2, 3$. The prescribed stability index is chosen as $\xi = 1$ and the Riccati Eq. (13) becomes

$$\dot{P}_i(t) = P_i(t)B_i(t)B_i(t)^T P_i(t) - \left(A_i(t) + I_2\right)^T P_i(t) - P_i(t)\left(A_i(t) + I_2\right) - 3I_2 \tag{18}$$

Here we pre-calculate the time-varying values of the solutions for Riccati equations. First, we make a guess for the boundary values of $P_i(t)$, choosing a terminal time as $t_f = 10$ (sec) and

$$P_1(10) = P_2(10) = P_3(10) = \begin{bmatrix} 10 & 1 \\ 1 & 5 \end{bmatrix} \tag{19}$$

The trajectories of $P_1(t)$ through $P_3(t)$ are plotted in Figs. 2, 3 and 4 by Euler's backward method with the time step $h = 0.001$. Figure 5 is the determinants of $P_1(t)$ through $P_3(t)$ to show the positive definiteness.

If the boundary condition is chosen as

$$P_1(10) = P_2(10) = P_3(10) = \begin{bmatrix} 5 & 1 \\ 1 & 3 \end{bmatrix} \tag{20}$$

The corresponding figures are plotted in Figs. 6, 7, 8 and 9.

Though the boundary conditions are different, they are the same at the time interval $t \in [0, 6]$ and also positive definite. Now, the interconnection parameter matrix is

$$\left[q_{ij}\right] = \begin{bmatrix} 0 & 0 & 1+0.2\sin t & 0.2+0.2\sin t & 0 & 0 \\ 1+0.2\sin t & 0.2+0.2\sin t & 0 & 0 & 1+0.2\sin t & 0.2+0.2\sin t \\ 0 & 0 & 1+0.2\sin t & 0.2+0.2\sin t & 0 & 0 \end{bmatrix} \tag{21}$$

By the control laws in (15), the responses of the masses with the initial condition [2 2 -2 2 -2] are plotted in Figs. 10, 11 and 12. These figures also show the trajectories of masses without control input in which the responses are oscillating. Besides, in the captions, the subscript c denotes responses with control, and subscript n denotes responses without control.

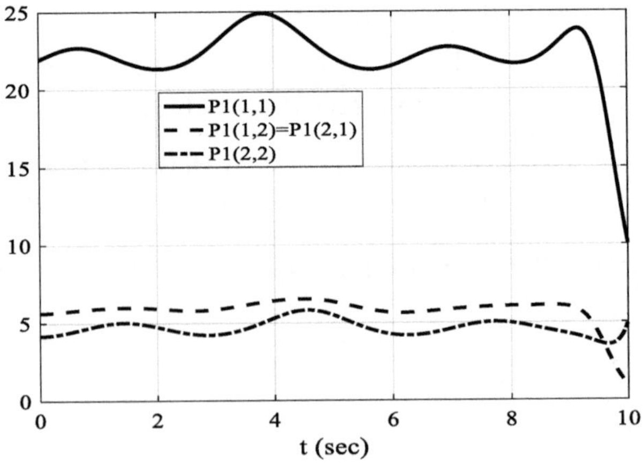

Fig. 2 Trajectories of $P_1(t)$

Fig. 3 Trajectories of $P_2(t)$

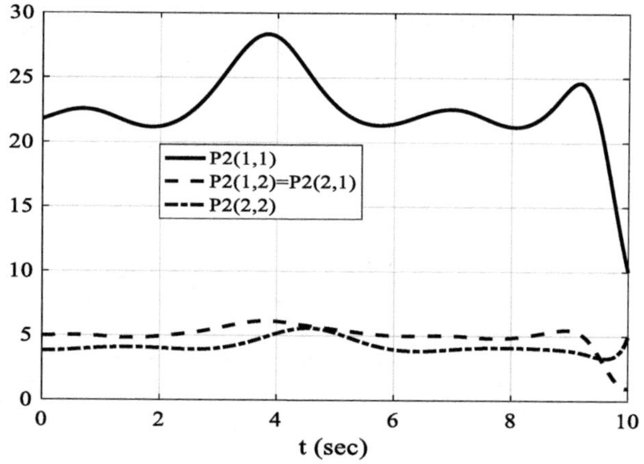

Fig. 4 Trajectories of $P_3(t)$

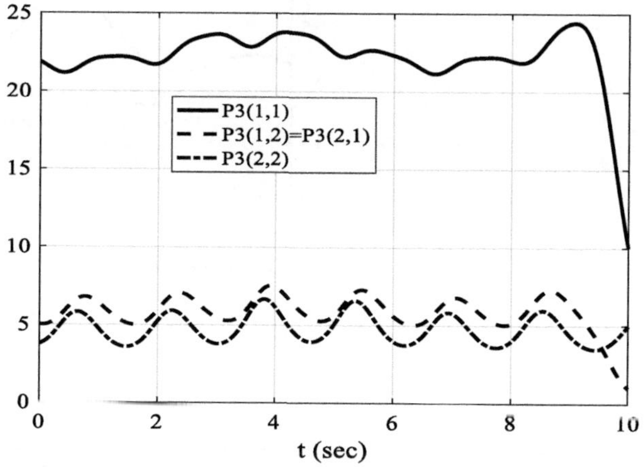

Fig. 5 Determinants of $P_i(t)$

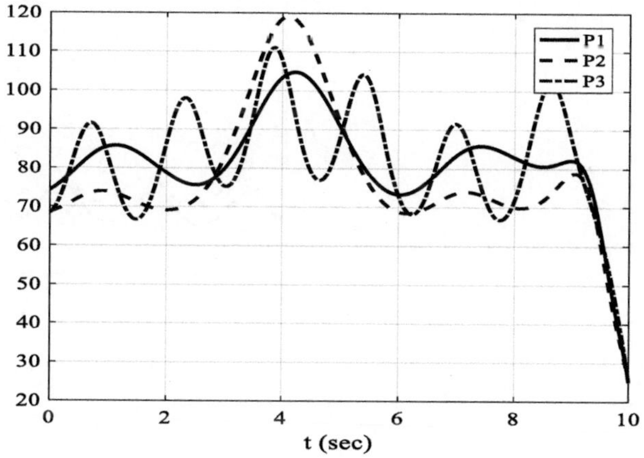

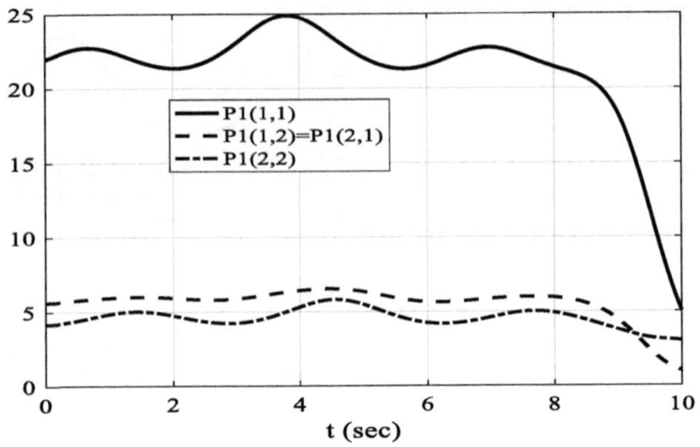

Fig. 6 Trajectories of $P_1(t)$

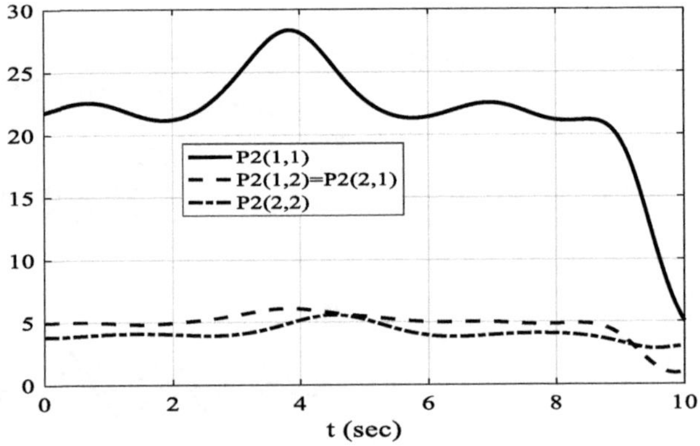

Fig. 7 Trajectories of $P_2(t)$

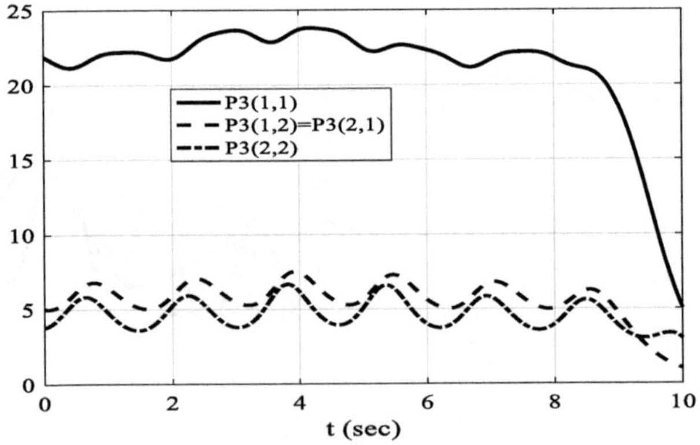

Fig. 8 Trajectories of $P_3(t)$

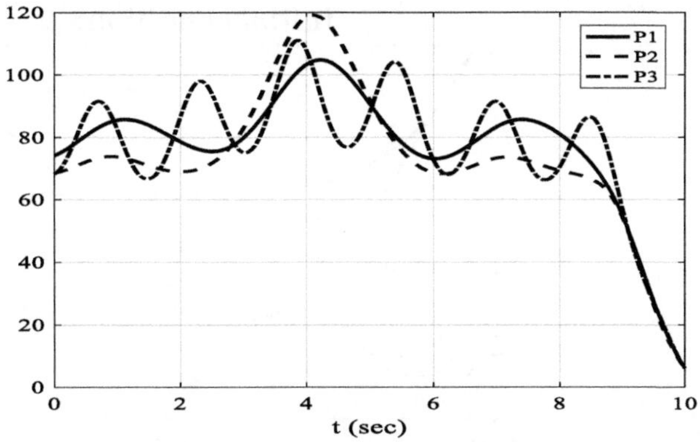

Fig. 9 Determinants of $P_i(t)$

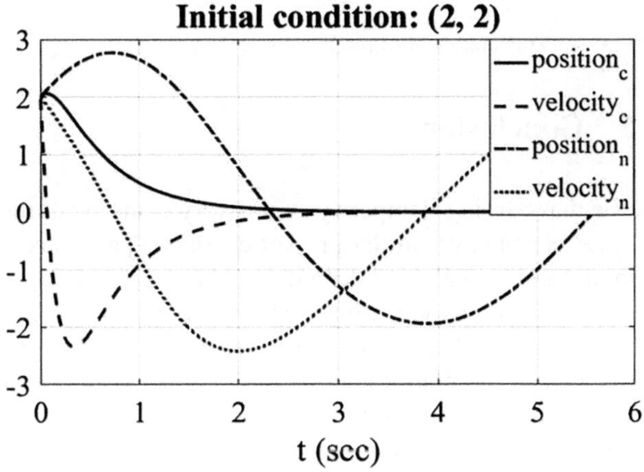

Fig. 10 Trajectories of Mass-1

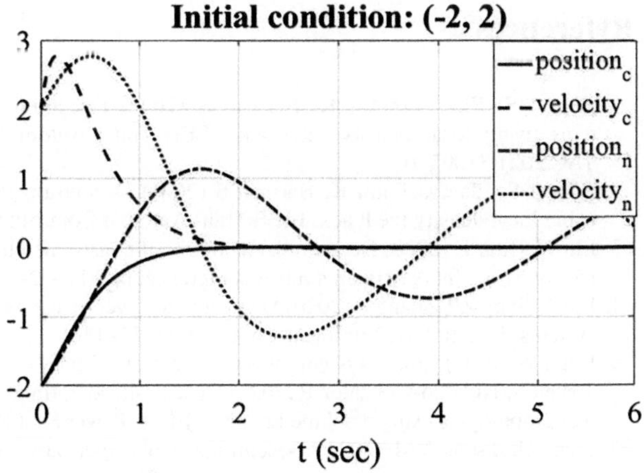

Fig. 11 Trajectories of Mass-2

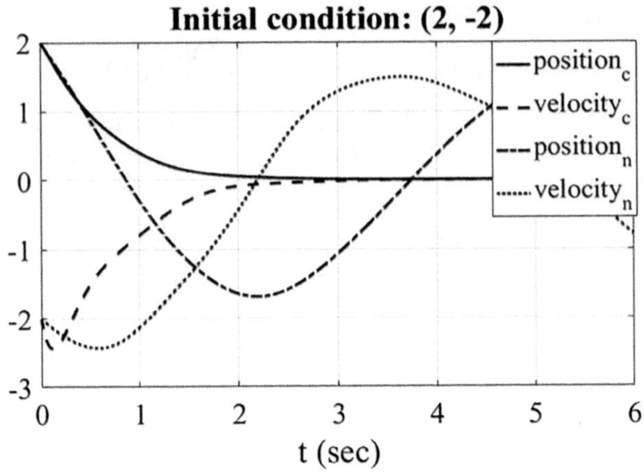

Fig. 12 Trajectories of Mass-3

5 Conclusion

The mass-spring-damper system is very common in the application field, and many systems can have similar properties. In this paper, we first make an interconnection model and then derive a decentralized regulation design to make each mass approach its setup state exponentially. The solutions for the time-varying Riccati equations are solved in advance, and these results are used to generate the decentralized control inputs. Computer simulations of a 3-mass system are also conducted. Future works are extending this approach to robotic systems, water treatment plants, and multi-machine power systems.

References

1. Biswal S, Elamvazhuthi K, Berman S (2022) Decentralized control of multi-agent systems using local density feedback. IEEE Trans Autom Control. https://doi.org/10.1109/TAC.2021.3109520
2. Biswal S, Elamvazhuthi K, Berman S (2021) Decentralized control of multi-agent systems using local density feedback. IEEE Trans Automat Control:565–572
3. Lin W, Bitar E (2020) Decentralized control of constrained linear systems via assume guarantee contracts. In: American control conference, pp 917–924
4. Liu T, Jiang ZP, Zhang P (2020) Decentralized event-triggered control of large scale nonlinear systems. Int J Robust Nonlinear Control 30:1451–1466
5. Rugh W (1993) Linear system theory. Prentice-Hall Inc.
6. Torre PS, Hidalgo-Gonzalez P (2022) Decentralized optimal power flow for time-varying network topologies using machine learning. Electr Power Syst Res:575–585
7. Mao ZJ, Tseng YM (2023) Decentralized observer-based output tracking for linear time-invariant interconnected systems. In: IEEE 2nd international conference on electrical engineering, big data and algorithms, pp 1387–1391

8. Frazzoli E, Bullo F (2005) Decentralized algorithms for vehicle routing in a stochastic time-varying environment. In: IEEE decision and control conference, pp 3357–3363
9. Bemporad A and Rocchi C 2011 Decentralized linear time-varying model predictive control of a formation of unmanned aerial vehicles, IEEE decision and control and european control conference, pp 7488–7493
10. Mao ZJ, Lin ZH (2023) Robust decentralized regulation for linear time-varying interconnected systems via time-varying Riccati equations. Proc SPIE:12717. https://doi.org/10.1117/12.2685745

Open Access This chapter is licensed under the terms of the Creative Commons Attribution-NonCommercial-NoDerivatives 4.0 International License (http://creativecommons.org/licenses/by-nc-nd/4.0/), which permits any noncommercial use, sharing, distribution and reproduction in any medium or format, as long as you give appropriate credit to the original author(s) and the source, provide a link to the Creative Commons license and indicate if you modified the licensed material. You do not have permission under this license to share adapted material derived from this chapter or parts of it.

The images or other third party material in this chapter are included in the chapter's Creative Commons license, unless indicated otherwise in a credit line to the material. If material is not included in the chapter's Creative Commons license and your intended use is not permitted by statutory regulation or exceeds the permitted use, you will need to obtain permission directly from the copyright holder.

Analysis of Current Positioning and Path Planning Technologies for Transfer Robots

Yukun Guo

Abstract With the development of robot technology, transfer robots have gradually occupied a major position in various industries headed by the service industry, replacing manpower to complete the corresponding transfer work. Robot positioning and path planning through corresponding technologies are important prerequisites for transfer robots to successfully complete tasks. By understanding the principles and advantages and disadvantages of various technologies related to positioning and path planning, related technologies can be better improved. To this end, this article conducts the following research: For positioning technology, the principles and advantages and disadvantages of various commonly used and mature positioning methods such as GPS positioning technology, grid map positioning technology, and SLAM positioning technology are introduced and analyzed. At the same time, the improvement methods of some positioning technologies that have appeared in recent years are introduced to provide inspiration and methods for improving such technologies. For path planning algorithms, this article divides them into global planning and local planning according to the perception degree of different spatial obstacles of the algorithms, and introduces several very common algorithms in the field of transfer robots, such as fuzzy logic algorithm, A* algorithm and artificial potential field method. The principles and advantages and disadvantages of these algorithms are explained in detail with the help of charts, and the application scenarios of each algorithm are demonstrated through comparison, and the direction of improvement is proposed.

Keywords Transfer robot · Algorithm · Positioning · Path planning

Y. Guo (✉)
College of Transportation Engineering, Tongji University, Shanghai, China
e-mail: 2154334@tongji.edu.cn

© The Author(s) 2026

Z. Zhang, P. Mercorelli (eds.), *Mechanical Design and Simulation: Insights and Innovations*, Lecture Notes in Mechanical Engineering,
https://doi.org/10.1007/978-981-95-2405-1_97

1 Introduction

With the development of society, robotics and artificial intelligence technologies are now quite mature. As an important part of robots, transfer robots play a significant role in various fields and continuously promote the development of information and intelligent society. In the field of transfer robots, positioning technology and path planning are important core technologies that enable transfer robots to complete their required tasks. Robots obtain and process information through positioning technology to determine their own positions, and use path planning to determine the action path to improve work efficiency.

When a transfer robot is in an unknown environment, the robot obtains and processes environmental information through the installed external sensors, and finally realizes its own positioning. Then the robot needs to plan the path, which requires timely receipt of changes in external information, while ensuring safety, and finally being feasible in actual operation.

This article studies the positioning technology and path planning technology of transfer robots as follows: briefly describe and compare different transfer robot positioning technologies, classify the path planning technology of transfer robots, introduce and analyze the advantages and disadvantages of local path planning technology and global planning technology, and make a review and summary.

2 Positioning Technology

If the robot needs to realize the function of autonomous navigation in an unknown environment, it must first determine the robot's own position and posture. The process of determining its own position is the positioning of the transfer robot. The robot obtains and processes information about the environment through external sensors to achieve its own positioning. The transfer robot generally uses real-time positioning technology and surrounding environment map construction technology to control the robot's position and direction [1]. The following is an introduction to some typical positioning technologies:

2.1 GPS Positioning

GPS positioning technology is a widely used positioning technology, and it is also widely used in the transfer robot field. The GPS positioning of transfer robots relies on the GPS receiver installed on the robot. The GPS receiver of the transfer robot receives signals from multiple GPS satellites and uses the information in these signals to calculate the specific location of the transfer robot.

In an open environment, the transfer robot can use the GPS system for positioning [2], so that the robot's positioning can cover most areas. At the same time, if it is in an open environment, GPS can provide high-precision and real-time positioning information. However, in indoor environments or areas with dense buildings, the GPS signal will be blocked or interfered, or the number of signal reflections will increase, resulting in an increase in positioning error, and the positioning signal is weak, resulting in inaccurate positioning [3].

2.2 SLAM Positioning

When building a map of an unknown environment, the environment in which the mobile robot is located is completely unknown. At this time, the robot needs to build a map while moving and correct its own posture in real time. This is the SLAM. The robot's simultaneous positioning and map building is also called SLAM technology.

SLAM systems usually consist of several core parts: sensor data, front-end processing, back-end optimization, loop detection, and map building. The data provided by sensors (such as lidar, cameras, IMU, etc.) is input to the front-end to infer the position and motion state of the robot in the current environment. The front-end usually processes these raw sensor data to generate local maps or pose estimates, and then passes them to the back-end for further optimization.

As shown in Fig. 1, SLAM technology is applicable to scenarios where the robot's environment information is completely unknown. It obtains map points in the environment through the perception of surrounding features by feature sensors, and obtains the position of the feature sensor body by tracking these environmental points frame by frame [4]. If the number of map points in the environment is large and clear, and the feature association algorithm of the feature sensor is also accurate, then the accuracy of this technology will be very high. However, as time goes by, the environmental points in the map maintained by the system gradually increase in size with their own trajectories, and the computational overhead of the system will also gradually increase [5]. In some environments where features are missing, the feature sensors of the SLAM system cannot extract enough feature points, which is very likely to cause positioning failure [6]. Moreover, this method is extremely

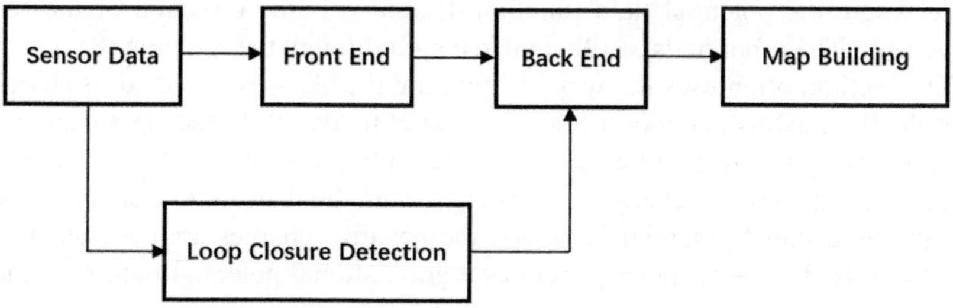

Fig. 1 SLAM system core components

dependent on the perception of the surrounding environment. In a static environment, the SLAM system can obtain extremely high positioning accuracy, but in a non-static environment, the SLAM system is very likely to fail in positioning [7].

2.3 Other Positioning Technologies

There are many other commonly used positioning technologies in the field of transfer robots: Grid map positioning is a very common positioning method. The map is rasterized, which can save environmental information more completely and is easy for robots to process, but the real-time performance is low when processing large-scale maps; ultra-wideband technology does not rely on environmental characteristics, is relatively inexpensive, and is easy to arrange and install and carry. Compared with mainstream positioning technologies, it has unique advantages.

3 Local Path Planning Algorithm

Local path planning is a dynamic planning algorithm. The transfer robot perceives the external conditions based on the sensors installed on it and generates a route for the robot to travel safely. Avoiding obstacles and planning the optimal path is the significant task of local path planning. The following will introduce some typical local path planning algorithms.

3.1 Artificial Potential Field Method

This method is a very commonly used in many fields. It regards the robot in the environment as a particle, and the surrounding environment as a force field. There are repulsive fields and gravitational fields in the force field. The obstacle and the target point generate two types of forces on the robot respectively. Two fields together constitute a potential field, which is called the artificial potential field. The place where the potential field function decreases is the direction of the robot's movement. The robot finds a collision-free path through the function [8].

The method often uses the way of imitating the electrostatic field of charges to describe the establishment of artificial potential fields [9]. In the electrostatic field, there are many positive and negative charges acting together on the entire electrostatic field. The positive charges in the electrostatic field are equivalent to the obstacles in the artificial potential field, and the negative charges are equivalent to the target points. The target point generates a gravitational potential field U_{att}, and the negative gradient is obtained to get the gravitational force F_{att}. F_{att} attracts robot and the robots moves toward the target point. The obstacle generates a repulsive force on the robot and generates a repulsive field U_{rep}, and the negative gradient is obtained to obtain F_{rep}. Preventing the robot from colliding with the obstacle is the task of the repulsive force. The combination of the repulsive force and the gravitational force is F_{total}.

The artificial potential field method has a certain robustness to control and sensor errors, but when applied to this field, there are two problems: the target is unreachable and the local minimum point.

3.2 *Fuzzy Logic Algorithm*

Fuzzy logic algorithm can also be used for local path planning. Sensors help it to obtain environmental information, without the need to establish a model. The robot movement is controlled by fuzzy rules and can handle uncertainty and ambiguity. We can use fuzzy logic algorithms to process fuzzy obstacle positions and external information to determine the optimal path for the transfer robot to move.

In the fuzzy logic algorithm, we need to define the input and output variables. Common input variables include the distance between the robot and target, the current angle of robot, and the position of obstacle. The output variables usually include the angular velocity and speed of robot. Then define the fuzzy sets of output and input variables and their membership functions. The membership function describes the degree of membership of the output and input variables in different fuzzy sets. We need to use fuzzy rules to infer the output variable's value based on the degree of membership of the input variable. The advantage of simple calculation is pretty clear, but the disadvantage is also obvious. Its computing power is not strong. When obstacles number increases, the algorithm calculation amount increases, and a better path cannot be planned.

3.3 *Other Algorithms*

In terms of local path planning, there are many other algorithms: In 2020, Yuan et al. [10] proposed an improved virtual obstacle local path planning method based on the minimum criterion, and proved through simulation experiments that the algorithm can solve problems such as local minimum; Dynamic Window Approach (DWA) is also a speed-based local planning algorithm [11], which has the advantages of small computational complexity and easy implementation [12]. However, it also has problems such as evaluation function failure or path redundancy.

4 Global Path Planning Algorithm

The global path planning algorithm is about static planning. It finds an optimal path between the starting point and the target point based on existing environmental information. The following will introduce some typical global path planning algorithms:

4.1 A Algorithm*

A* algorithm is a relatively effective global path planning algorithm. Many algorithms are improved based on A* algorithm. A* algorithm is a pretty effective direct search method. But compared with a lot of preprocessing algorithms, the real-time query efficiency of A* algorithm is much lower than that of preprocessing algorithm [13].

The formula is:

$$f*(n) = g*(n) + h*(n) \tag{1}$$

In formula (1), the minimum estimated distance from the initial state to the target state through state n is $f*(n)$. $f*(n)$ has two parts. The part from the initial state to state n is $g*(n)$, and the part from state n to the target state is $h*(n)$. The selection of the minimum distance estimation function $f*(n)$ is a necessary condition to ensure that a shortest path is found.

The estimated distance from state n to the target state is $h(n)$, and the selection of $h(n)$ is as follows:

When $h(n) > h*(n)$, the efficiency is increased because the number of search points and the range is small, but it is difficult to get the optimal solution.

When $h(n) < h*(n)$ the efficiency is reduced because the number of search points and the range is large, but the optimal solution can be obtained at this time.

When $h(n) = h*(n)$, this time the search efficiency is the highest.

4.2 Ant Colony Algorithm

If the robot can perform path planning in any environment, then the ant colony algorithm can be used for path planning. In order to realize multi-robot collaborative work in the dynamic environment of farmland, Marco Dorigo proposed a probabilistic path planning algorithm called ant colony algorithm in 1992 [14]. The search efficiency can be greatly improved.

The principle of ant colony algorithm is: the path formed by the entire group in search of food constitutes the solution space to be optimized, and the path walked by the ants can represent the feasible solution to be optimized. When ants walk along a shortest path, they will leave more pheromones. As time goes by, more and more pheromones are left, and more and more ants walk along the path. Eventually, the whole colony will find the shortest path under the action of positive feedback. According to the algorithm's characteristics, the positive feedback mechanism continuously optimizes the path and finally obtains the optimal solution. The ant colony algorithm has the characteristics of distributed computing. Multiple units in the algorithm perform parallel computing, which significantly improves computing power and efficiency. Its heuristic probabilistic search method is also easy to plan global paths.

4.3 Other Algorithms

In addition to the above methods, there are many global path planning algorithms: RRT (Rapidly-Exploring Random Tree) algorithm is a sampling-based path planning algorithm. And It can use to solve path planning problems in high-dimensional space and complex constraints [15]; Particle Swarm Optimization is a cluster optimization algorithm proposed by JE Kenndy Eberhart [16], which has a high convergence speed. However, when multiple mobile robots are searching for paths at the same time, it is very easy to fall into the local optimal solution, thereby reducing the accuracy of path search.

5 Summary and Outlook

This article reviews the positioning and path planning technologies of transfer robots, introduces many important positioning and path planning technologies in the current transfer robot field, and analyzes the pros and cons of each technology of positioning, global path planning and local path planning. In the positioning technology part, by introducing various common positioning methods such as GPS positioning technology, grid map positioning technology, SLAM positioning technology, etc., different positioning technologies are compared. For path planning algorithms, this article introduces local path planning algorithms such as artificial fuzzy logic algorithm and potential field method. The advantages and disadvantages of these algorithms are discussed. In this section, the article points out the local minimum problem that may occur in the practical application of the artificial potential field method, and then introduces an improved virtual obstacle local path planning method based on the minimum criterion that can solve the local minimum and other problems. By introducing and summarizing the advantages and disadvantages of each method, it provides new optimization ideas for the mutual combination of future path planning algorithms and the improvement of adaptability in complex dynamic environments. And for global planning algorithms, this article introduces very mature and basic common algorithms such as A* algorithm. By introducing the principles and comparing different algorithms, the advantages and disadvantages of the algorithms and future improvement directions are introduced.

References

1. Meng Q (2021) Research on path planning method based on deep reinforcement learning in indoor environment. https://doi.org/10.27357/d.cnki.gtgyu.2021.000252.
2. Wang DX (2021) AGV path planning based on improved Q-learning algorithm. Electron Design Eng 29(04):7–10
3. Yang ZL (2023) Research on positioning and path planning of handling robots based on deep reinforcement learning. https://doi.org/10.26977/d.cnki.gccgc.2022.000618
4. Puja K, Pankaj S, Rowdra G (2020) Design of a compact UWB BPF with a FractalTree. Stub. Loaded multimode resonator. IET Microwaves Antennas Propag 2020, 15(1)
5. Wang X. (2023) Research on positioning strategy of handling robot based on UWB. https://doi.org/10.26939/d.cnki.gbhgu.2022.000826.
6. Baran DA, Sundar PS, Debasis M, Wasim A (2020) Investigation and design of enhanced decoupled UWB MIMO antenna for wearable applications. Microw Opt Technol Lett 63(3)
7. Science - Radio Science (2020) Findings on radio science reported by investigators at National Institute of technology (Penta-notched Uwb monopole antenna using Ebg structures and fork-shaped slots) Science Letter
8. Gao XY (2020) Research on dynamic obstacle avoidance of autonomous robots based on improved artificial potential field method. https://doi.org/10.27466/d.cnki.gzzdu.2020.001972
9. Tang XJ, Ding YH, Shen Q, Jia CF (2019) Dynamic obstacle avoidance path planning for mobile vehicles based on improved artificial potential field method. Softw Guide 18(10):152–156
10. Yuan Z, Xue S, Zheng C et al (2020) Improvements on the virtual obstacle method. Int J Adv Robot Syst 2020, 17(2)
11. Liu LS, Yao JX, He DW et al (2021) Global dynamic path planning fusion algorithm combining jump-A*algorithm and dynamic window approach. IEEE Access 9:19632–19638
12. Ji XY, Feng S, Han QD et al (2021) Improvement and fusion of A* algorithm and dynamic window approach considering complex environmental information. Arab J Sci Eng 2021:1–15
13. Pal A, Tiwari R, Shukla A (2011) Multi robot exploration using a modified A*algorithm. In: Intelligent information and database systems. Springer, Berlin Heidelberg, pp 506–516
14. Wang HL, Mao WG, Leif Eriksson (2019) A three-dimensional Dijkstra's algorithm for multi-objective ship voyage optimization. Ocean Eng 186
15. Wei K, Ren BY (2018) A Method on dynamic path planning for robotic manipulator autonomous obstacle avoidance based on an improved RRT algorithm. Sensors (Basel, Switzerland) 18(2):124–128
16. Kennedy J, Eberhart R (1995) Particle swarm optimization. In: Proceedings of ICNN'95-international conference on neural networks, vol 4. IEEE, pp 1942–1948

Open Access This chapter is licensed under the terms of the Creative Commons Attribution-NonCommercial-NoDerivatives 4.0 International License (http://creativecommons.org/licenses/by-nc-nd/4.0/), which permits any noncommercial use, sharing, distribution and reproduction in any medium or format, as long as you give appropriate credit to the original author(s) and the source, provide a link to the Creative Commons license and indicate if you modified the licensed material. You do not have permission under this license to share adapted material derived from this chapter or parts of it.

The images or other third party material in this chapter are included in the chapter's Creative Commons license, unless indicated otherwise in a credit line to the material. If material is not included in the chapter's Creative Commons license and your intended use is not permitted by statutory regulation or exceeds the permitted use, you will need to obtain permission directly from the copyright holder.

Design of Butt-Joint Gap Transfer Tooling for One Subway Seat Mask

Wenjing Su, Siying Liu, and Shuanghui He

Abstract Aiming at the phenomenon of clothing pinching on the seat cover of a certain type of subway vehicle, the cause of clothing pinching on the seat is analyzed, and a tooling for permanently transferring the butt-joint gap is designed. Firstly, the structure of the seam of the seat cover is analyzed. Secondly, two types of transfer tooling are designed, namely, a connector and a coupling structure. Finally, the coupling structure tooling is selected based on the principle of permanent transfer. The application results show that the coupling structure tooling permanently transfers the butt-joint gap of the seat mask, effectively solving the problem of the seat clamping passenger clothing.

Keywords Seat mask · Butt-joint gap · Coupling structure · Permanently transfer · Tooling

1 Introduction

In modern society, subways are widely used due to their convenience, comfortable environments, and ability to alleviate traffic congestion [1–3]. Passengers vary across different age groups, including the elderly, children, middle-aged individuals, young adults, and infants [4–6]. Consequently, optimizing the design of subway seat reliability and comfort has become a crucial aspect of subway seat development [7–13].

This investigation focuses on a subway seat from a project initiated in 2012. The seat, designed for eight passengers as illustrated in Fig. 1, features a cover made of stamped stainless steel. Due to production technology and cost considerations, a four-person seat splicing method was chosen. The specific splicing configuration is depicted in Fig. 2. The four-person seat covers are aligned with a 2 mm gap, which is sealed with a rubber strip between the seat cover and the seat frame.

W. Su (✉) · S. Liu · S. He
CRRC Nanjing Puzhen Co., Ltd., Nanjing, China

© The Author(s) 2026

Z. Zhang, P. Mercorelli (eds.), *Mechanical Design and Simulation: Insights and Innovations*, Lecture Notes in Mechanical Engineering,
https://doi.org/10.1007/978-981-95-2405-1_98

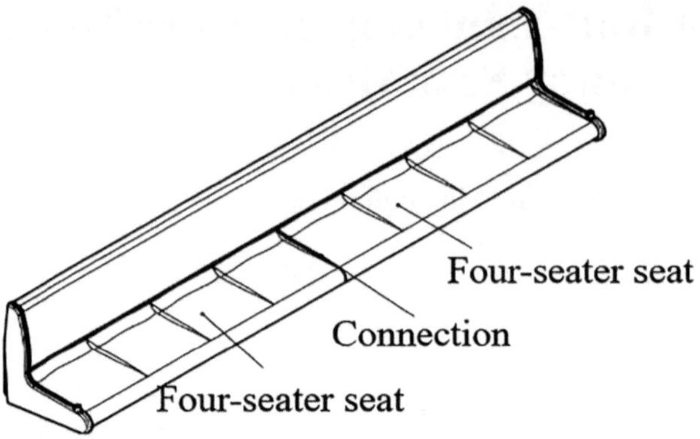

Fig. 1 Eight-person seat in a subway

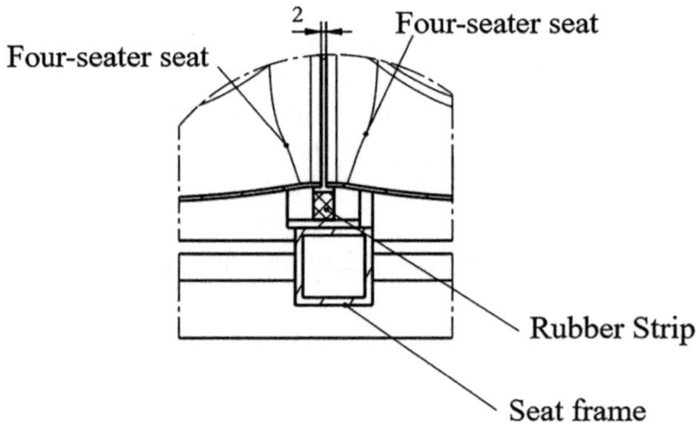

Fig. 2 Four-person seat splicing scheme

However, because the rubber strip possesses significant elasticity, misalignment can occur when the two covers experience different forces, potentially leading to clothing being caught. Additionally, as the rubber strip ages during subway operation, the reduction in support at the seam can widen the gap, further increasing the risk of clothing entrapment. To enhance passenger comfort and fully resolve the issue of clothing getting caught in the seams, it is essential to optimize the seat structure design.

2 Design of Tooling

The subway seat utilizes a mature four-person seat mask panel docking system, employing stainless steel calendaring molding technology. This process results in significant deformation and rebound, making it challenging to uniformly manage the seam along the arc of the seat surface. To address the joint gap without investing in the eight-person seat mold, it is necessary to design specialized tooling for this purpose.

2.1 Plug-in Tooling Design

A plug-in will be added to the original seat joint, which will be inserted into the mask (as shown in Fig. 3) to transfer the joint gap from the passenger contact surface to the non-contact surface.

This connector can be fabricated using four common methods: aluminum die-casting, aluminum profile bending, rubber, and plastic. The advantages and disadvantages of each method are outlined below. (All methods require increasing the gap between the two seat masks; it is recommended to expand the gap from 2 mm to 4 mm to enhance the connector's strength.)

Aluminum Die-Casting: This method offers good size control, with a required front thickness of more than 2.5 mm to ensure part strength. A 0.5 mm gap must be reserved at the insertion point to prevent assembly interference. Surface spray painting allows for color selection. The final product possesses significant strength and is essentially indestructible to passengers once assembled. However, aluminum die-casting cannot accommodate large deformations, which may lead to size deviations in the mask arc, poor fitting, and an increased risk of connector cracking. This method requires a pair of aluminum die-casting molds, which incurs a certain mold cost.

Aluminum Profile Bending: This method is low-cost, requiring only one profile mold and one bending mold. However, controlling the relative bending size is challenging, and the arc section may not match the mask well. Additionally, bending can easily deform the plug interface, which complicates the mask connection.

Rubber: This is the most economical option, needing only one rubber extrusion mold. The softness of rubber reduces assembly difficulty. However, its pliability makes it prone to buckling by passengers and increases the likelihood of damage.

Plastic: Similar to aluminum die-casting, plastic can be colored, offering better wear resistance than spray painting. Plastic exhibits certain toughness; local deformation does not lead to damage, and wall thickness can be reduced to minimize bulges. However, like rubber, plastic is more susceptible to human damage compared to metal.

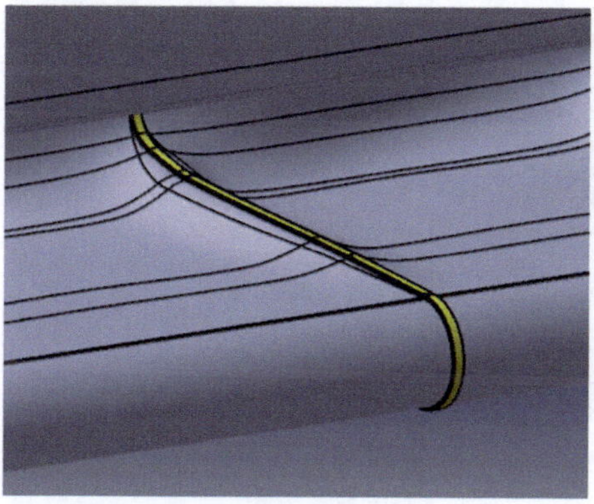

a) Overall view

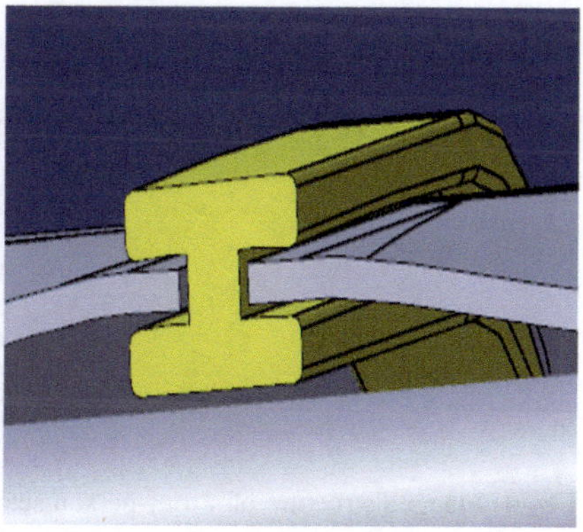

b) Partial view

Fig. 3 Connector tooling

In summary, the recommended order for plug-in materials for adding connectors is: plastic, aluminum die-casting, rubber, and aluminum profile bending. Notably, the original four-person seat mold can be reused at a low cost. However, the seat mask employs a plug-in to transfer the gap, resulting in a bulge between seats that affects aesthetics, complicates assembly, and poses a risk of damage. This approach does not fundamentally resolve the issue of splicing gaps and necessitates a mold opening cycle.

2.2 *Design of Joint Structure Tooling*

To prevent cracking of the joint tooling, ensure a proper fit, and avoid buckling by passengers while achieving the permanent transfer of gaps, a joint structure tooling design has been developed. A flanging forming structure is introduced between the two 4-person seat mask panels (this structure requires a dedicated flanging mold, necessitating the creation of a new mold), as shown in Fig. 4. A sealing plate made from 06Cr19Ni10 is placed between the two mask panels, with edges designed to match the curvature of the seat panel, and is welded to the mask panel. This solution demands a high degree of fit between the side curves after welding, as well as high flatness for the components on both sides, resulting in stringent process requirements. The two sealing plates are directly connected by fasteners, as shown in Fig. 5. Due to the strong mechanical connection, the impact of gap changes when the mask is subjected to uneven forces can be effectively disregarded.

It is evident that the tooling does not require mold opening, features a short cycle time, and allows for the reuse of the original four-seat mold. This solution is low-cost, provides a uniform and aesthetically pleasing appearance, and effectively resolves the issue of splicing gaps. The use of connecting structure tooling can prevent problems such as passenger buckling, cracking, and poor fit, while ensuring the permanent transfer of the seat mask joint gap, thereby addressing the issue of passengers' clothing being caught in the seat. This tooling solution has significant application potential in future urban rail vehicle projects. It fully utilizes the

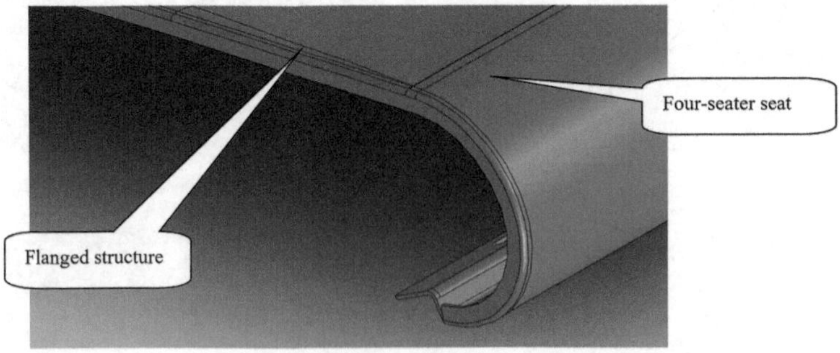

Fig. 4 Flanged structure of a four-person seat

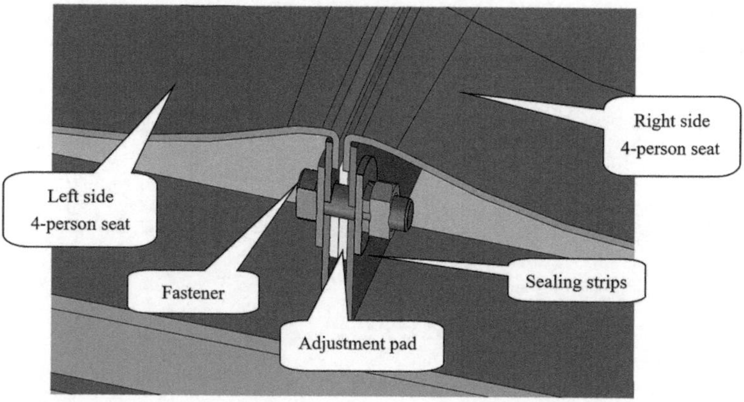

Fig. 5 Connection structure tooling

existing mature four-seat mold, significantly reducing the need for investment in an eight-seat mold, and contributes to cost reduction and improved efficiency.

3 Seat Force Analysis

To test the stress of the seats, two working conditions are selected. Working condition 1 is to apply 1000 N to all eight seats; working condition 2 is to apply 1000 N to the two adjacent seats in the middle.

3.1 Seat Force for 2 People

From the force diagram of two people riding (as shown in Fig. 6), we can see that the maximum value of dynamic stress is 29.12 MPa, the stress is concentrated below 16.18 MPa, the maximum stress is located at the mask joint gap and the inner side of the top of the seat mask, and gradually decreases to both sides. Although the 8 seat masks are unevenly stressed when two people are riding, the mechanical connection strength of each connecting tooling is high, so the impact of the gap change under the uneven stress of the mask can be completely ignored.

3.2 Seat Force for 8 People

From the force diagram analysis of 8 people (as shown in Fig. 7), it can be seen that the maximum value of dynamic stress is 113.47 MPa, the stress is concentrated below 63.04 MPa, and the stress distribution is uniform. The design of the plug-in

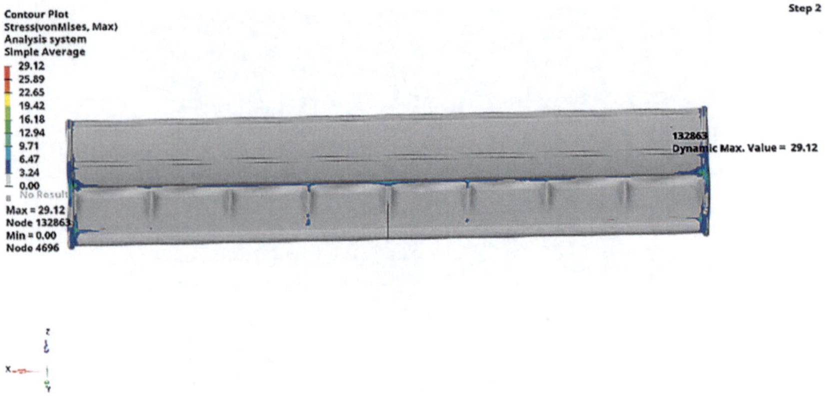

(a)Front view

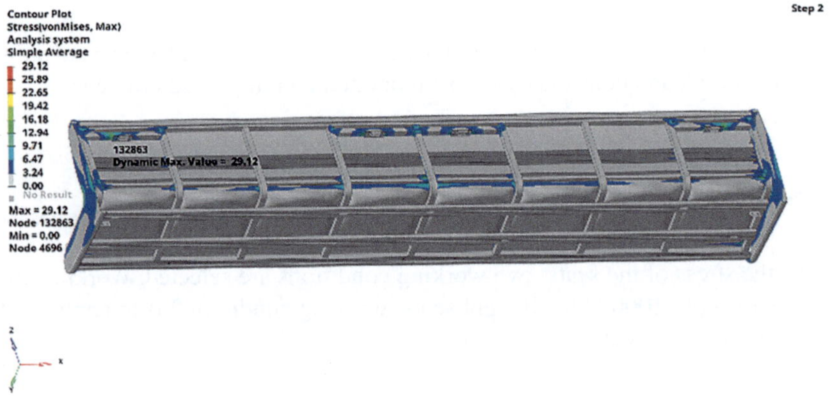

(b)Back view

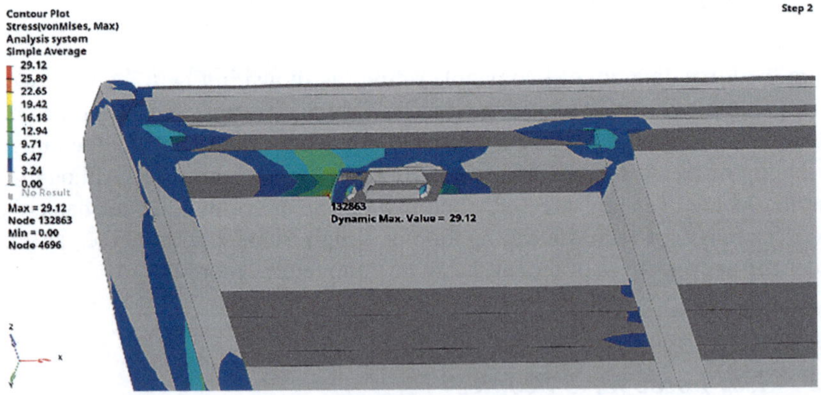

(c) Partial view

Fig. 6 Seat force analysis when 2 people are sitting

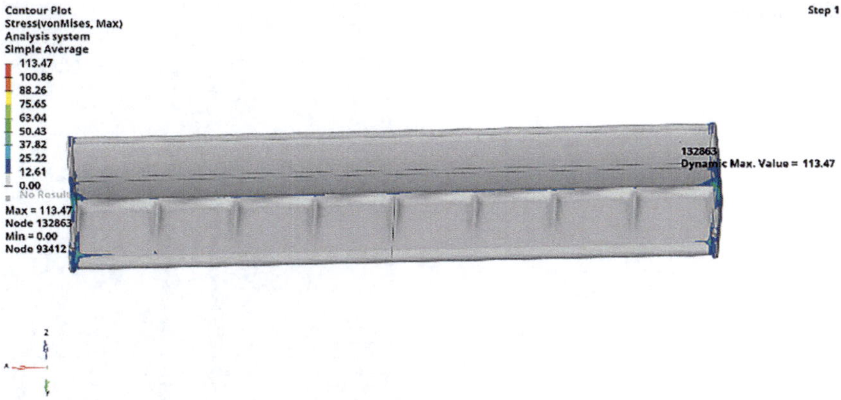

(a)Front view

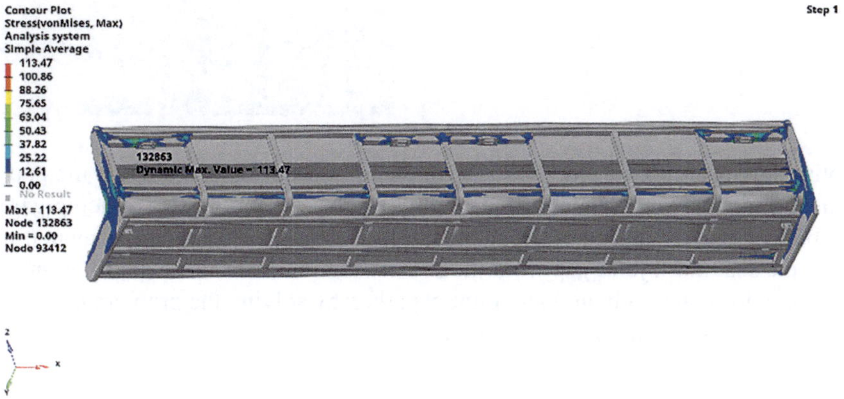

(b)Back view

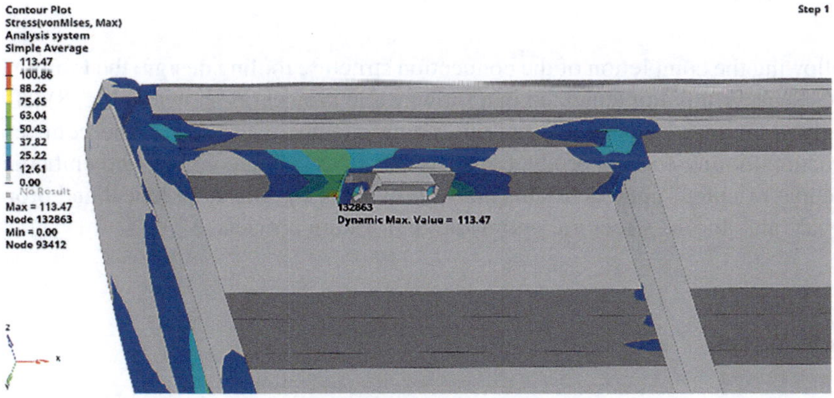

(c) Partial view

Fig. 7 Seat force analysis when 8 people are sitting

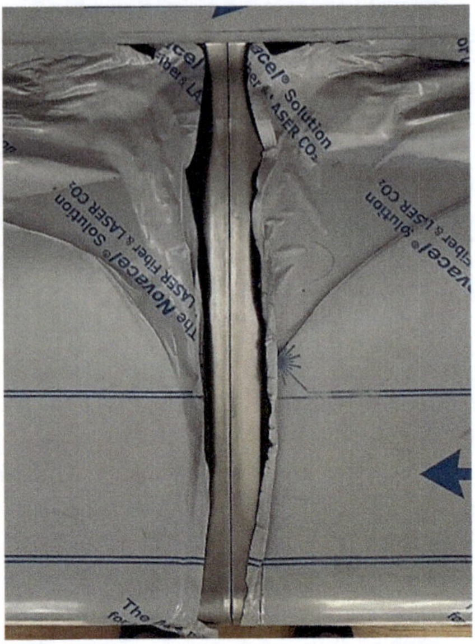

Fig. 8 Finished seat

tooling has been optimized. The seat mask joint seam uses a plug-in method to transfer the gap, which cannot fundamentally solve the problem of splicing gap. It can be seen that the design of the connector tooling can realize the permanent transfer of the seat mask joint gap from the root, ensure uniform stress distribution, and prevent deformation rebound and joint gap, thereby solving the problem of passengers' clothes being clamped by the seat.

4 Conclusion

Following the completion of the connection structure tooling design, the tooling has been successfully implemented in a subway line project (as shown in Fig. 8), effectively resolving the issue of clothing being caught in the seats. The connection structure tooling ensures product quality, enhances riding comfort, and embodies a "people-oriented" approach. This innovation holds significant practical and promotional value for the structural optimization of future projects.

References

1. Li Y, Jiao Y, Cao FR, Hou ZK (2021) Design study of subway vehicle horizontal double seats under the synergy of reliability and comfort. Technol Econ Guide 29(17):87–88
2. Li GJ, Qian JL, Wang RJ, Li WQ (2022) Design of subway double seat and its finite element experiment analysis. J Jiangsu Univ Technol 28(2):102–109

3. Zheng XQ, Wang SB, Du YP (2022) Simulation Study on Crack Analysis and Improvement Measures of Railway Vehicle Seat. Rail Veh 60(1):12–15
4. Luo XY (2014) Research on subway seat design based on passengers' perceived needs. Southwest Jiaotong University, Chengdu
5. Jiang T R 2013 Research on subway car interior decoration design based on riding experience (Chengdu: Southwest Jiaotong University)
6. Wang LG (2010) Research on key technologies of subway vehicle interior design. Dalian Jiaotong University, Dalian
7. Sun LP, Wang LG (2010) Design engineering analysis of subway vehicle interior designers. J Dalian Jiaotong Univ 31(2):15–19
8. Chen Q (2012) Design of metro vehicle's saloon seat for longitudinal direction layout. Technology and Market 19(6):128–131
9. Pan XY (2014) Optimization design and experimental study of subway seat structure based on finite element analysis. Wuhan University of Technology, Wuhan
10. Wang WB, Su C, Chen GR, Zhou HC (2024) Research on end structure of Subway train seat and occupant collision safety. Mach Design Manufact (10):209–214220
11. Fan QH, Jiang XC, Wu XL, Nie M, Xin YS, Li HL, Jia XY (2022) Experimental research on optimization design and comfort of locomotive seat based on ideal pressure. J Mech Eng 58(10):383–394
12. Xu JH, Qiu JY (2021) Design and research on comfort of economy class seat based on ergonomics. Pack Eng 42(18):403–409
13. Ni WY, Zhang H, Yao SW (2021) Lightweight design of automobile seat frame based on multiple working conditions. Chin J Eng Des 28(6):729–736

Open Access This chapter is licensed under the terms of the Creative Commons Attribution-NonCommercial-NoDerivatives 4.0 International License (http://creativecommons.org/licenses/by-nc-nd/4.0/), which permits any noncommercial use, sharing, distribution and reproduction in any medium or format, as long as you give appropriate credit to the original author(s) and the source, provide a link to the Creative Commons license and indicate if you modified the licensed material. You do not have permission under this license to share adapted material derived from this chapter or parts of it.

The images or other third party material in this chapter are included in the chapter's Creative Commons license, unless indicated otherwise in a credit line to the material. If material is not included in the chapter's Creative Commons license and your intended use is not permitted by statutory regulation or exceeds the permitted use, you will need to obtain permission directly from the copyright holder.

Modeling and Analysis of Variable-Length Optical Fiber in Air Release

Minghang Li, Yaohui Xue, Zhuo Zhang, Jingze Lv, Feng Tang, Dashan Niu, and Jiaxin Wang

Abstract In this paper, the release process of flexible optical fiber from the spool dispenser is studied. In general, the method of adding elements is used to solve the variable-length problem of the flexible optical fiber. However, when this method is applied to optic fiber release with large length, it will cause large computation and difficult convergence. In this paper, the method of power function polynomial fitting of optic fiber posture function and length normalization is used to establish the optic fiber release model. Galerkin method is used to simplify the partial differential equations into differential equations and solve them. to verify the correctness of the theoretical calculation, a test was carried out using an unmanned aerial vehicle carrying a miniature. According to the state of UAV, the shape, falling track and tension distribution of the optical fiber in the air are calculated. The test value of the ground end tension is compared with the theoretical calculation value. The change trend of the two is basically consistent, and the biggest difference is 12.7%. The reasonableness of the modeling and numerical solution method is proved by the above experiment, which provides a method for solving the variable length release problem of large length, low linear density, and slowly changing shape, performing good convergence. This method provides a reference for failure analysis, fault recurrence and trajectory optimization of guidance fiber.

Keywords Optical fiber · Variable-length release · Tension distribution · Experimental verification

M. Li · Y. Xue · Z. Zhang (✉) · F. Tang · D. Niu · J. Wang
Xi'an Modern Control Technology Research Institute, Xi'an, Shaanxi, China

J. Lv
Xidian University, Xi'an, Shaanxi, China

© The Author(s) 2026

Z. Zhang, P. Mercorelli (eds.), *Mechanical Design and Simulation: Insights and Innovations*, Lecture Notes in Mechanical Engineering,
https://doi.org/10.1007/978-981-95-2405-1_99

1 Introduction

Inertial navigation technique has excellent characteristics such as anti-interference, large bandwidth, low delay and low bit error rate [1]. Therefore, the use of optical fiber as a communication medium for UAVs and ground control stations can effectively resist electromagnetic interference, and has the advantages of large transmission bandwidth and low delay. Therefore, the use of UAV using optical fiber as a transmission medium received attention from all countries, and multiple types of fiber-optic navigation UAV have been developed base on this technology [2].

The optical fiber is generally wound on the spool in an orderly manner to form a "spool dispenser", which is installed at the bottom of fiber-optic navigation UAV. As UAV flies, the optical fiber is unwound from spool and released into the air [3–5]. The unwinding motion of optical fiber from the spool dispenser as shown in Fig. 1. Since the optical fiber is the only transmission medium for command and image signals, once the optical fiber is broken, the UAV will lose control [6]. Therefore, the motion posture and tension of optical fiber is the key to analyze the release of weak links and failure forms, which is of great significance to the reliability of the UAV.

The optical fiber can be divided into two motion stages after it is unwound from spool dispenser. In the first stage called stripped stage, the optical fiber is stripped from spool dispenser, violently rubbed against the surface of spool dispenser and released in the form of a helix. In the second stage called floating stage, the motion of optical fiber gradually tends to be stable in the air, and the optical fiber shows a quasi-static state. In the first stage, the optical fiber is subjected to the comprehensive influence of tensile, bending, torsion, friction and other loads in a short period of time, which is prone to stress concentration, resulting in optical fiber breakage. In the second stage, the optical fiber is affected by gravity, tension, wind resistance

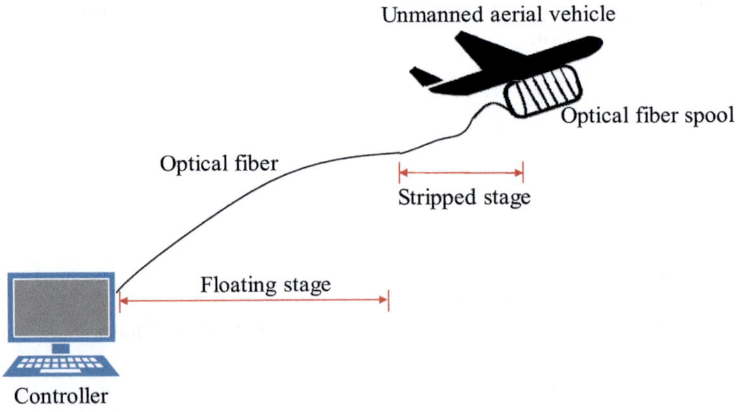

Fig. 1 Diagram of fiber-optic navigation UAV

and maneuverable trajectory of the UAV. It is easy to appear that the load exceeds the breaking strength of the optical fiber, resulting in breakage [7].

The analysis of optical fiber in the first stage has been well studied. The earliest fiber release analysis came from the spinning industry, where Padfield [8] conducted a pioneering study on the release posture of yarn from a spool. Fraser [9] and Stanislav [10] analyzed the motion considering friction of yarn on spool according to Padfield's paper. Kothari and Leaf [11] analyzed the influence of different spool shapes on yarn unwinding posture, and also considered factors such as gravity and air resistance. With the development of the study of fiber unwinding motion, there are some advances in the unwinding motion and force of optical fiber from spool dispenser. Lee [12] used Hamilton's principle to analyze the unwinding posture of optical fiber from inner and outer spool dispenser. Based on Lee's research, Kim [13, 14] focused on the underwater motion of optical fiber release from spool dispenser, and verified it under low-speed release. Park and Jang [15, 16] updated and solved the dynamic equation for optical fiber under variable speed release. Kang et al. and Ren et al. [17, 18] established an experimental platform to simulate the release of optical fiber. Li et al. [19] analyzed the unwinding motion of optical fiber in the "quasi-steady state" condition. Wang Rong [20] fitted the bending deformation in optical fiber release.

However, there are few researches on the motion posture of optical fiber in the second stage. The main problem is that the length of released fiber is increasing, which brings some difficulties to analysis. In the similar variable length release analysis problem, Guo [21] proposed a dynamic model of variable-length flexible beams based on absolute node coordinate formulation method. Pang [22] adopted the concentrated mass method to establish the dynamic model of the submarine cable system. Cai Shenglong [23] used the absolute nodal coordinate formulation to derive the dynamic model of a variable length tethered satellite. Zhou et al. [24] established the three-dimensional dynamic equation for the submarine cables, and studied the dynamic response of submarine cables under different working conditions. Feng Shuiliang [25] built a mechanical model of the tethered satellite system based on Lagrange equation, considered tethered elasticity and damping, and conducted a dynamic analysis of the release stage, but did not consider the mass effect of the cable.

The above research mainly use the method of adding or reducing elements to simulate the release and recovery of the cable. However, this method is not suitable for posture analysis of optical fiber. Firstly, the length of optical fiber varies from 100 m to more than 10 km, and the large length will lead to a sharp increase number of element, affecting the computational efficiency. Secondly, the initial state of added element generally is static or equal to aircraft velocity. However, the released cable particle velocity in reality is complicated. Considering the high speed release and small linear density of optical fiber, the inaccurate initial state of added element will lead disturbance to dynamics equation and cause it to diverge. Therefore, there are few studies and reports on force and posture analysis of large length, low linear density and slow changing shape in release process. In this paper, according to the relatively stable posture of optical fiber, power function polynomial is used to fit its posture function, and optical fiber is normalized to solve the variable length problem. Galerkin method is used to transform the partial differential equations into differential equations, and the dynamic equations of the air motion posture of optical fiber are solved numerically. The motion posture of optical fiber, the trajectory of the fiber element and the overall tension distribution along the length direction are analyzed. In order to verify the rationality of the theoretical analysis, a test was carried out using an UAV carrying a miniature. The optical fiber test tension at ground is in good agreement with those of theoretical analysis. The correctness of dynamic equation, boundary condition and solution method are verified and the method adopted in this paper performs good convergence.

2 Transient-State Motion Equation of Optical Fiber

2.1 *Optical Fiber Tension Analysis*

In release process, the optical fiber is influenced by tension, gravity, air resistance and damping force. Firstly, the magnitude and direction of tension along optical fiber are analyzed. Because the bending and torsional stiffness of optical fiber are small, in order to simplify the problem analysis and calculation, only the tensile effect of optical fiber is considered, and the stress changes caused by bending and torsional fiber are ignored. Therefore, the optical fiber has a homogeneous structure in the stretching direction, and its equivalent elastic modulus is E and cross-sectional area is S. Let s denote the unstretched arc length of the optical which varies from 0 to L and p refers to the arc length of the optical fiber in the current configuration. For any optical fiber particle, the unstretched length is denoted as ds and stretched length is denoted as dp. According to Hooke's law and the definition of engineering strain, the tension of optical fiber is expressed as Eq. (1).

$$T = ES\frac{\Delta l}{l} = ES\left(\frac{\mathrm{d}p}{\mathrm{d}s} - 1\right) \tag{1}$$

The coordinate $OXYZ$ is set on the ground, the Z axis points to the sky, and the position of optical fiber particle in space can be written as $\mathbf{R}(s,t) = [x(s,t),y(s,t),z(s,t)]$. $\partial x/\partial p$, $\partial y/\partial p$ and $\partial z/\partial p$ respectively represent the cosine between the tangent vector of stretched optical fiber and the x, y, z axes. The relationship of $\partial x/\partial p$, $\partial y/\partial p$ and $\partial z/\partial p$ can be written as Eq. (2).

$$\left(\frac{\partial x}{\partial p}\right)^2 + \left(\frac{\partial y}{\partial p}\right)^2 + \left(\frac{\partial z}{\partial p}\right)^2 = \left|\frac{\partial \mathbf{R}}{\partial p}\right|^2 = 1 \tag{2}$$

According to the chain rule, the relationship between unstretched arc length s and stretched arc length p is as follows:

$$\frac{\partial \mathbf{R}}{\partial p} = \frac{\partial \mathbf{R}}{\partial s}\frac{\mathrm{d}s}{\mathrm{d}p} \tag{3}$$

$$\frac{\partial \mathbf{R}}{\partial p}\cdot\frac{\partial \mathbf{R}}{\partial p} = \frac{\partial \mathbf{R}}{\partial s}\cdot\frac{\partial \mathbf{R}}{\partial s}\left(\frac{\mathrm{d}s}{\mathrm{d}p}\right)^2 = 1 \tag{4}$$

So, the value of dp/ds is:

$$\frac{\mathrm{d}p}{\mathrm{d}s} = \left|\frac{\partial \mathbf{R}}{\partial s}\right| \tag{5}$$

By bringing Eq. (5) into Eq. (1), the tension value of optical fiber can be expressed as:

$$T = ES\left(\left|\frac{\partial \mathbf{R}}{\partial s}\right| - 1\right) \tag{6}$$

Since the tangent vector of stretched optical fiber is $\partial \mathbf{R}/\partial p$, the tension of optical fiber can be expressed as:

$$\mathrm{T} = T\frac{\partial \mathbf{R}}{\partial p} = T\frac{\partial \mathbf{R}}{\partial s}\frac{\mathrm{d}s}{\mathrm{d}p} = ES\left(\left|\frac{\partial \mathbf{R}}{\partial s}\right| - 1\right)\frac{\partial \mathbf{R}}{\partial s} / \left|\frac{\partial \mathbf{R}}{\partial s}\right| \tag{7}$$

2.2 *Air Resistance Analysis*

In release process, the optical fiber will fall slowly under gravity. During the falling process, the optical fiber moves relative to the air, forming air resistance. Secondly, the air resistance received by optical fiber is analyzed. It is observed in experiment that the falling speed of optical fiber is small. According to the principle of fluid mechanics, the air resistance of optical fiber is directly proportional to the square of its normal moving speed and the cross-sectional area of optical fiber along the length direction. The direction of air resistance is opposite to the direction of normal moving velocity. The force of the optical fiber in the air can be shown in Fig. 2.

The velocity of optical fiber particle is assumed as $\mathbf{v}$, and the tangential velocity along the tangential direction of optical fiber posture is set as $\mathbf{v}_t$. According to the vector algorithm, the normal velocity $\mathbf{v}_n$ can be calculated as Eq. (8).

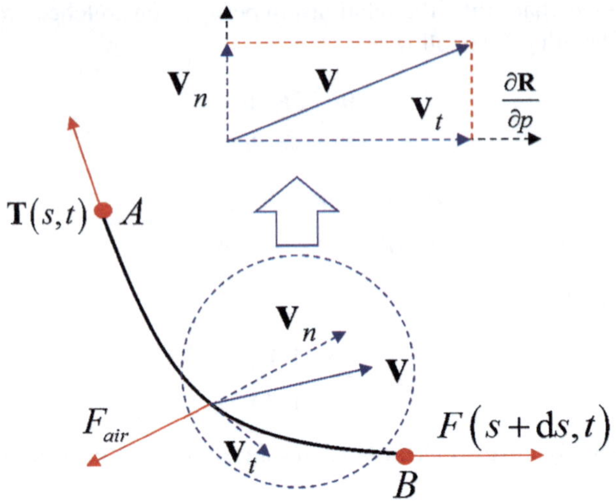

Fig. 2 Air resistance diagram of optical fiber

$$\begin{aligned}\mathbf{v}_n &= \mathbf{v} - \mathbf{v}_t \\ &= \frac{\partial \mathbf{R}}{\partial t} - \left(\frac{\partial \mathbf{R}}{\partial t} \cdot \frac{\partial \mathbf{R}}{\partial p}\right) \cdot \frac{\partial \mathbf{R}}{\partial p} \\ &= \frac{\partial \mathbf{R}}{\partial t} - \frac{\partial \mathbf{R}}{\partial t} \cdot \frac{\partial \mathbf{R}}{\partial s} \cdot \frac{\mathrm{d}s}{\mathrm{d}p} \cdot \frac{\partial \mathbf{R}}{\partial s} \cdot \frac{\mathrm{d}s}{\mathrm{d}p} \\ &= \frac{\partial \mathbf{R}}{\partial t} - \left(\frac{\partial \mathbf{R}}{\partial t} \cdot \frac{\partial \mathbf{R}}{\partial s} \cdot \frac{\partial \mathbf{R}}{\partial s}\right) / \left|\frac{\partial \mathbf{R}}{\partial s}\right|^2\end{aligned} \tag{8}$$

The projected area dS of optical fiber in the normal motion direction is shown in Eq. (9).

$$\mathrm{d}S = d \cdot \mathrm{d}s \tag{9}$$

where: d is the diameter of optical fiber. The air resistance of any optical fiber particle dp is calculated as Eq. (10).

$$\mathbf{F}_{air} = -\frac{1}{2} C \rho_{air} d \cdot \mathrm{d}s \cdot |\mathbf{v}_n| \mathbf{v}_n \tag{10}$$

where: C is the scale coefficient, obtained by searching the data according to optical fiber normal motion speed $\mathbf{v}_n$, and ρ_{air} is air density.

Therefore, the dynamics equation of optical fiber particle can be written as Eq. (11).

$$\rho \frac{\partial^2 \mathbf{R}}{\partial t^2} \mathrm{d}s = \mathbf{T}(s+\mathrm{d}s,t) - \mathbf{T}(s,t) - \frac{1}{2} C \rho_{air} d \cdot \mathrm{d}s \cdot |\mathbf{v}_n| \mathbf{v}_n - \rho \mathbf{g} \cdot \mathrm{d}s \tag{11}$$

where: $\mathbf{g}$ is the gravitational acceleration vector and ρ is the linear density of optical fiber.

The tension $\mathbf{T}$ can be transformed as follows:

$$\lim_{\mathrm{d}s \to 0} \left[\frac{\mathbf{T}(s+\mathrm{d}s,t) - \mathbf{T}(s,t)}{\mathrm{d}s} \right] = \frac{\partial \mathbf{T}(s,t)}{\partial s} \tag{12}$$

Therefore, Eq. (11) can be simplified as:

$$\rho \frac{\partial^2 \mathbf{R}}{\partial t^2} = \frac{\partial \mathbf{T}}{\partial s} - \frac{1}{2} C \rho_{air} d \cdot |\mathbf{v}_n| \mathbf{v}_n - \rho \mathbf{g} \tag{13}$$

2.3 Boundary Conditioin Analysis

When fiber-optic navigation UAV is used, the optical fiber is stored in the bottom of the UAV, and the optical fiber is connected with the ground control platform to build a two-way communication link. According to the application scenario, the spatial position of the optical fiber connected to the ground platform is assumed as $\mathbf{R}_0(t) = [X_0(t), Y_0(t), Z_0(t)]$. The position of the optical fiber release point at the bottom of UAV is assumed as $\mathbf{R}_1(t) = [X_1(t), Y_1(t), Z_1(t)]$. Since the optical fiber is gradually released and its length changes with time, which is denoted as $L(t)$. The optical fiber flight and boundary conditions are shown in Fig. 3.

Therefore, the boundary conditions of optical fiber at the ground endpoint and the release endpoint are shown as Eq. (14).

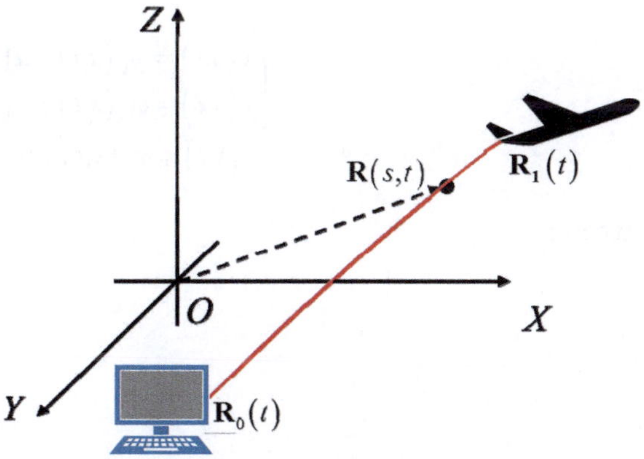

Fig. 3 Optical fiber flight and boundary conditions

$$\begin{cases} \mathbf{R}(0,t) = \mathbf{R}_0(t) \\ \mathbf{R}(L(t),t) = \mathbf{R}_1(t) \end{cases} \tag{14}$$

Since the length of released optical fiber varies with time, a set of governing equations is needed to describe the variational length of optical fiber. According to the characteristics of optical fiber release process, the acceleration, speed and length of optical fiber release are equal to the flight acceleration, speed and trajectory length of the aircraft. The governing equation of fiber length variation law can be obtained as follows:

$$\begin{cases} \dfrac{\mathrm{d}^2 L(t)}{\mathrm{d}t^2} = \sqrt{\left(\dfrac{\mathrm{d}^2 X_1}{\mathrm{d}t}\right)^2 + \left(\dfrac{\mathrm{d}^2 Y_1}{\mathrm{d}t}\right)^2 + \left(\dfrac{\mathrm{d}^2 Z_1}{\mathrm{d}t}\right)^2} \\ \dfrac{\mathrm{d}L(t+\Delta t)}{\mathrm{d}t} \approx \dfrac{\mathrm{d}L(t)}{\mathrm{d}t} + \Delta t \dfrac{\mathrm{d}^2 L(t)}{\mathrm{d}t^2} \\ L(t+\Delta t) \approx L(t) + \Delta t \dfrac{\mathrm{d}L(t)}{\mathrm{d}t} + \dfrac{1}{2}\Delta t^2 \dfrac{\mathrm{d}^2 L(t)}{\mathrm{d}t^2} \end{cases} \tag{15}$$

3 Numerical Solution Method

The length of released optical fiber changes with time and the solution domain of Eq. (13) changes with time. In order to solve this problem, firstly, the optical fiber is normalized in the length direction. Secondly, the motion posture function of optical fiber in the air is fitted with a power function polynomial. Thirdly, Galerkin method is used to solve the coefficients of the power function [26]. The boundary conditions are also time-varying, the assumed posture function of optical fiber should be homogenized to construct a suitable basis function. The position function of any point on the released optical fiber is assumed as Eq. (15).

$$\begin{cases} x(s,t) = \varphi_x(s,t) + u(s,t) \\ y(s,t) = \varphi_y(s,t) + v(s,t) \\ z(s,t) = \varphi_z(s,t) + w(s,t) \end{cases} \tag{16}$$

where:

$$\begin{cases} \varphi_x(s,t)=\dfrac{\left(X_1(t)-X_0(t)\right)}{L(t)}s+X_0(t) \\ \varphi_y(s,t)=\dfrac{\left(Y_1(t)-Y_0(t)\right)}{L(t)}s+Y_0(t) \\ \varphi_Z(s,t)=\dfrac{\left(Z_1(t)-Z_0(t)\right)}{L(t)}s+Z_0(t) \end{cases} \tag{17}$$

The function $\varphi_x(s,t)$, $\varphi_y(s,t)$ and $\varphi_z(s,t)$ satisfies exactly the boundary conditions of Eq. (14), so the boundary condition of function $u(s,t)$, $v(s,t)$ and $w(s,t)$ can be rewritten as:

$$\begin{cases} u(0,t)=0 \\ v(0,t)=0 \\ w(0,t)=0 \end{cases} \begin{cases} u(L(t),t)=0 \\ v(L(t),t)=0 \\ w(L(t),t)=0 \end{cases} \tag{18}$$

According to Galerkin method and the boundary conditions shown in Eq. (18), using the power function as the base function to construct the approximate solution $u(s,t)$, $v(s,t)$ and $w(s,t)$ are written as:

$$\begin{cases} u(s,t)=\sum\limits_{i=1}^{n}a_i(t)\left(1-\dfrac{s}{L(t)}\right)\left(\dfrac{s}{L(t)}\right)^i \\ v(s,t)=\sum\limits_{i=1}^{n}b_i(t)\left(1-\dfrac{s}{L(t)}\right)\left(\dfrac{s}{L(t)}\right)^i \\ w(s,t)=\sum\limits_{i=1}^{n}c_i(t)\left(1-\dfrac{s}{L(t)}\right)\left(\dfrac{s}{L(t)}\right)^i \end{cases} \tag{19}$$

In the power function construction in Eqs. (17) and (19), $s/L(t)$ term is used to achieve normalization, so that the posture function changes within the interval of [0,1]. This analysis method does not care about the initial state of released optical fiber element, and does not need to add new element, but only analyzes the posture function of optical fiber.

Bringing the Eqs. (16), (17) and (19) into Eq. (13), both sides of Eq. (13) are multiplied by the basis function. The variable s of Eq. (13) is integrated over the interval 0 to $L(t)$. Make the integral value of Eq. (13) equal to zero, resulting in $3n$ equations, the specific form is as follows:

$$\int_0^{L(t)}\left(\left(\rho\frac{\partial^2\mathbf{R}}{\partial t^2}-\frac{\partial\mathbf{T}}{\partial s}-\frac{1}{2}C\rho_{air}d\cdot|\mathbf{v}_n|\mathbf{v}_n-\rho\mathbf{g}\right)\cdot\left(1-\frac{s}{L(t)}\right)\left(\frac{s}{L(t)}\right)^{i=1\sim n}\right)\mathrm{d}s=\mathbf{0} \quad (20)$$

When the initial posture of optical fiber is determined, that is, $a_i(0)$, $\mathrm{d}a_i(0)/\mathrm{d}t$, $b_i(0)$,$\mathrm{d}b_i(0)/\mathrm{d}t$, $c_i(0)$ and $\mathrm{d}c_i(0)/\mathrm{d}t$ is known, then the unknown terms $\mathrm{d}^2a_i(0)/\mathrm{d}t^2$, $\mathrm{d}^2b_i(0)/\mathrm{d}t^2$ and $\mathrm{d}^2c_i(0)/\mathrm{d}t^2$ in Eq. (20) can calculated. The coefficient calculation of the power function is shown in Eq. (21) and Eq. (22).

$$\begin{cases}\dfrac{\mathrm{d}a_i(t+\Delta t)}{\mathrm{d}t}\approx\dfrac{\mathrm{d}a_i(t)}{\mathrm{d}t}+\Delta t\dfrac{\mathrm{d}^2a_i(t)}{\mathrm{d}t^2}\\ \dfrac{\mathrm{d}b_i(t+\Delta t)}{\mathrm{d}t}\approx\dfrac{\mathrm{d}b_i(t)}{\mathrm{d}t}+\Delta t\dfrac{\mathrm{d}^2b_i(t)}{\mathrm{d}t^2}\\ \dfrac{\mathrm{d}c_i(t+\Delta t)}{\mathrm{d}t}\approx\dfrac{\mathrm{d}c_i(t)}{\mathrm{d}t}+\Delta t\dfrac{\mathrm{d}^2c_i(t)}{\mathrm{d}t^2}\end{cases} \quad (21)$$

$$\begin{cases}a_i(t+\Delta t)\approx a_i(t)+\Delta t\dfrac{\mathrm{d}a_i(t)}{\mathrm{d}t}+\Delta t^2\dfrac{\mathrm{d}^2a_i(t)}{\mathrm{d}t^2}\\ b_i(t+\Delta t)\approx b_i(t)+\Delta t\dfrac{\mathrm{d}b_i(t)}{\mathrm{d}t}+\Delta t^2\dfrac{\mathrm{d}^2b_i(t)}{\mathrm{d}t^2}\\ c_i(t+\Delta t)\approx c_i(t)+\Delta t\dfrac{\mathrm{d}c_i(t)}{\mathrm{d}t}+\Delta t^2\dfrac{\mathrm{d}^2c_i(t)}{\mathrm{d}t^2}\end{cases} \quad (22)$$

4 Calaulation Results and Experimental Verification

In order to verify the rationality of theoretical analysis, an experimental verification is built to test the optical fiber release process. The construction of the experimental system is shown in Fig. 4. The experimental scheme is as follows: a miniature of optical fiber spool dispenser is made, and it is installed on the UAV, and the other endpoint of the optical fiber is fixed on the force sensor on the ground, as shown in Fig. 5. The test tool is used to wrap the optical fiber in a ring, and one end of the ring is connected to a tension sensor to measure the tension of the optical fiber. The UAV flies according to the preset trajectory and speed, the highest speed is about 180 m/s, the whole flight process lasts about 26.5 s.

In order to balance calculation efficiency and accuracy, $n = 5$ is used in theoretical calculation, and a total of 15 unknowns in X, Y and Z directions can be solved. Figure 6 shows the release posture of optical fiber at different times, where the blue dots represent the preset flight trajectory of UAV and the black line represents the release posture of optical fiber. It is shown that, the released optical fiber slowly

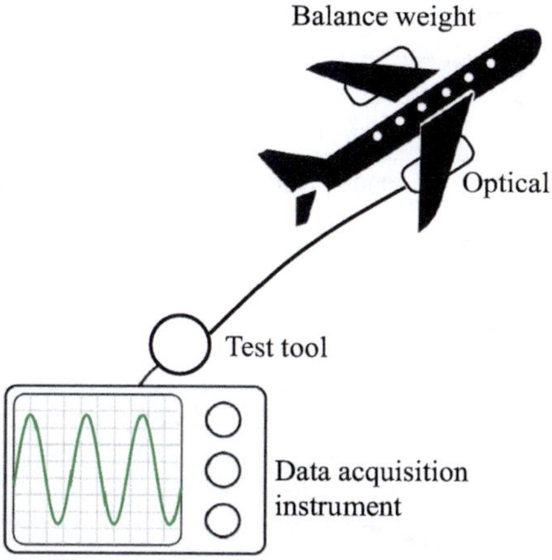

Fig. 4 Test system construction diagram

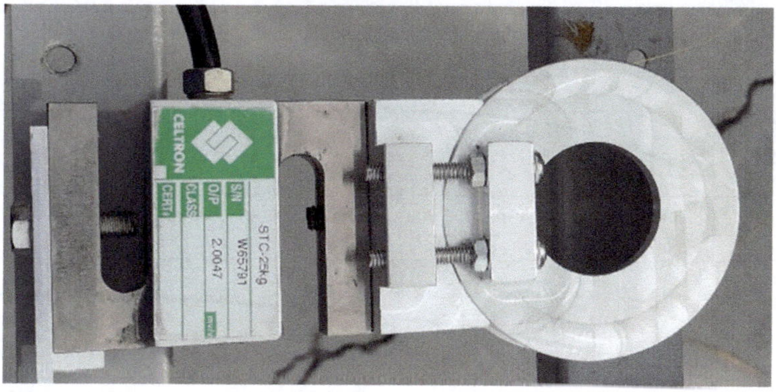

Fig. 5 Test tool

drops in the air, but is always tensioned and the posture is relatively stable. Figures 7, 8 and 9 show the trajectory of the released optical fiber particle in the air at 10 s, 15 s and 25 s, and the speed variation of X and Y with time. It can be seen from these three figures that: 1. The initial speed of the fiber particles released at different times is different, and they are not static 2. The speed of the fiber particles at different times eventually tends to be stable, and the horizontal speed tends to be 0, and the vertical speed tends to be -1 m/s. These phenomena show that the boundary conditions of the released fiber change with time, and do not simply fall gradually from the static state.

According to Eqs. (17) and (19), the optical fiber particle velocity at released point can be expressed as Eq. (23).

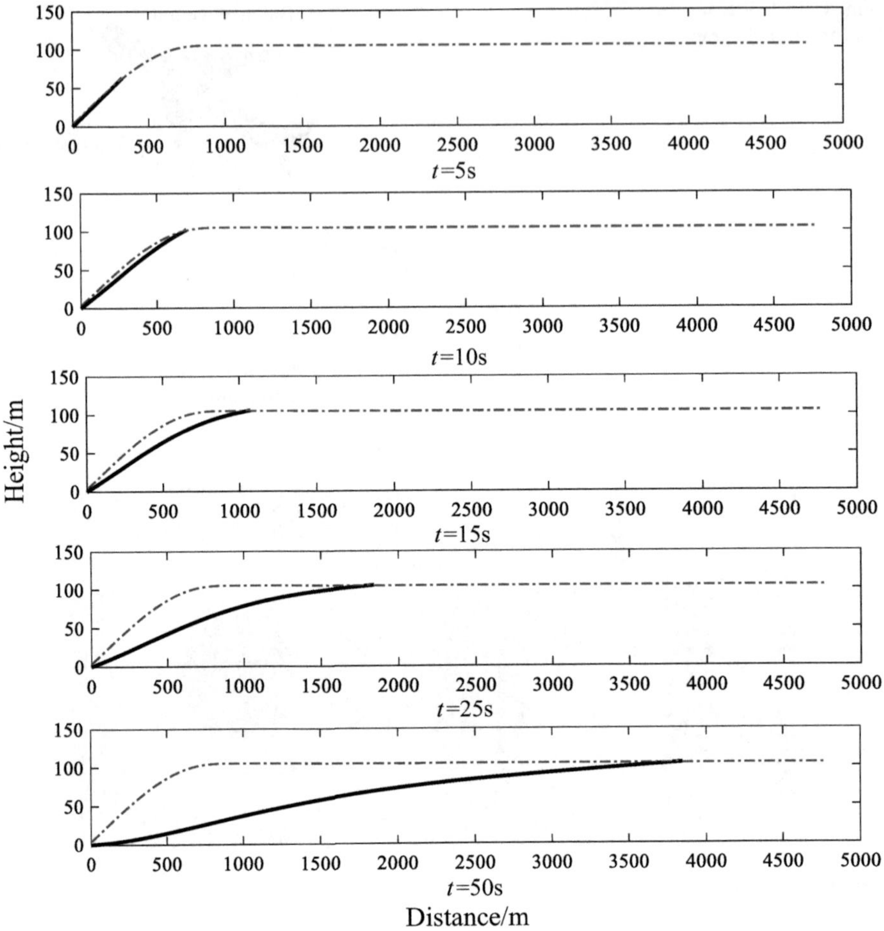

Fig. 6 Posture of optical fibers in air at different times (Ratio of vertical data to horizontal data is 5:1)

$$\begin{cases} \dfrac{dx(L,t)}{dt} = \dfrac{1}{L}\dfrac{dX_1}{dt} - \dfrac{X_1}{L^2}\dfrac{dL}{dt} \\ \dfrac{dy(L,t)}{dt} = \dfrac{1}{L}\dfrac{dY_1}{dt} - \dfrac{Y_1}{L^2}\dfrac{dL}{dt} \\ \dfrac{dz(L,t)}{dt} = \dfrac{1}{L}\dfrac{dZ_1}{dt} - \dfrac{Z_1}{L^2}\dfrac{dL}{dt} \end{cases} \quad (23)$$

Equation (23) shows that the optical fiber particle velocity at released point is affected by three factors: the aircraft velocity, the aircraft position and the released optical fiber length in the air. When the optical fiber is just released from the aircraft,

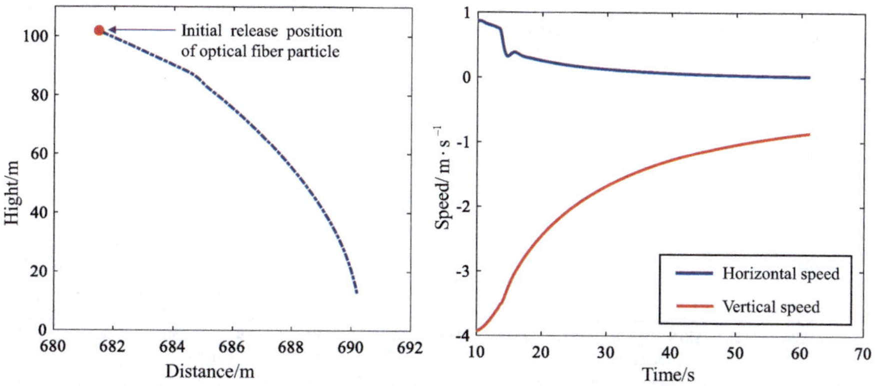

(a) Optical fiber microelement in the air falling trajectory (b) Variation ofmotion velocity of optical fiber particle

Fig. 7 Trajectory and velocity change of the fiber particles released at 10 s in air

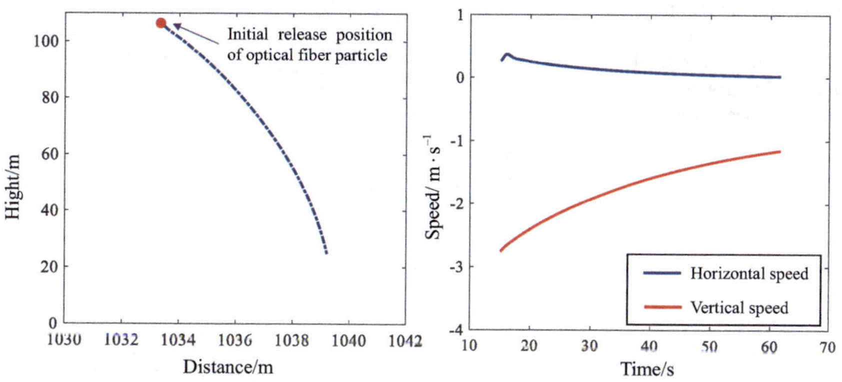

(a) Optical fiber microelement in the air falling trajectory (b) Variation of motion velocity of optical fiber particle

Fig. 8 Trajectory and velocity change of the fiber particles released at 15 s in air

it has an initial velocity, but the initial speed is much less than the aircraft speed. In horizontal direction, it will continue to move some distance along with the flight direction, and in the vertical direction, it will slowly fall over time. However, after the optical fiber particle is released in the air, its velocity gradually decreases and eventually tends to be zero. As the flight distance of the aircraft increases, the initial speed of optical fiber particle velocity at released point also decreases and eventually tends to be zero. So the released optical fiber has been "floating" in the air and does not fall on the ground.

Figure 10 compares the theoretical optical fiber tension at the ground endopint with the measured results after noise reduction. In this figure, the experimental result at 26.5 s is taken as the benchmark. The theoretical result and the measured result are normalized by this benchmark. As shown in Fig. 9, the measured tension increases first and then gradually stabilized with time. This is because the aircraft

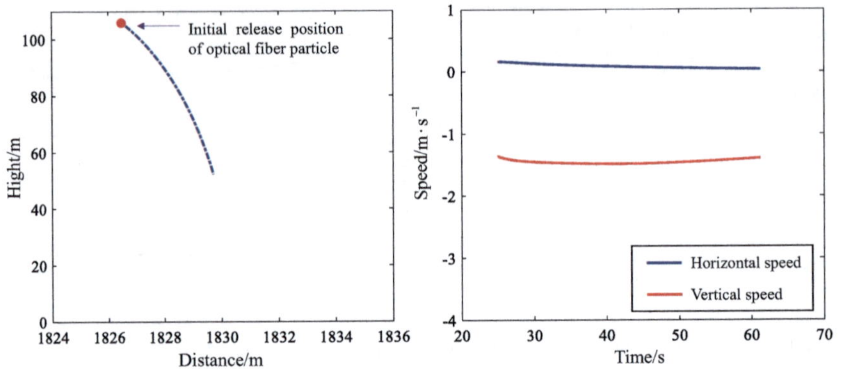

(a) Optical fiber microelement in the air falling trajectory (b) Variation of motion velocity of optical fiber particle

Fig. 9 Trajectory and velocity change of the fiber particles released at 25 s in air

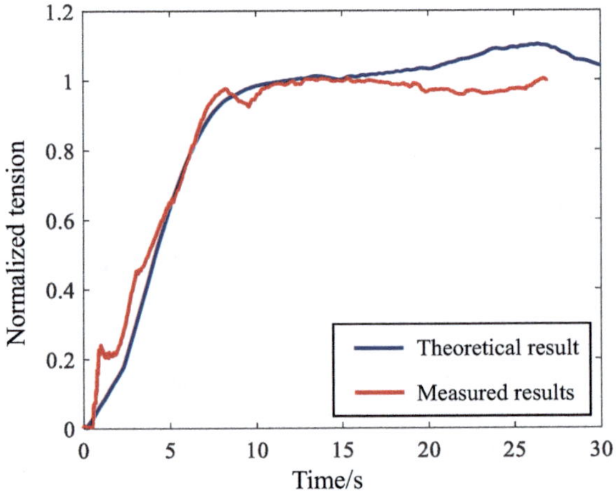

Fig. 10 Comparison between the measured value and the theoretical value of the ground endpoint fiber tension

accelerated from a standstill, resulting in increasing tension on optical fiber. After the aircraft levelled out, the speed and direction of aircraft were relatively stable, and the tension of optical fiber also tended to be stable. According to the comparison between theoretical results and measured results, it can be seen that in the tension growth stage, the two are in good agreement. In the tension stabilization stage, there were some fluctuations in theoretical tension, and the largest difference in tension appeared at about 24.8 s, with a difference of about 12.7%.

Figure 11 shows the theoretical tension distribution on the fiber at different times, where the *X*-axis represents the time, the *Y*-axis represents the normalized optical fiber length, and the *Z*-axis represents the normalized magnitude of the tension. In Fig. 10, the measured optical fiber tension at the ground endopint at 26.5 s is taken as the benchmark of *Z*-axis. As can be seen from Fig. 11, observing along the *X*-axis, the tension along optical fiber rises rapidly in the initial acceleration stage. When

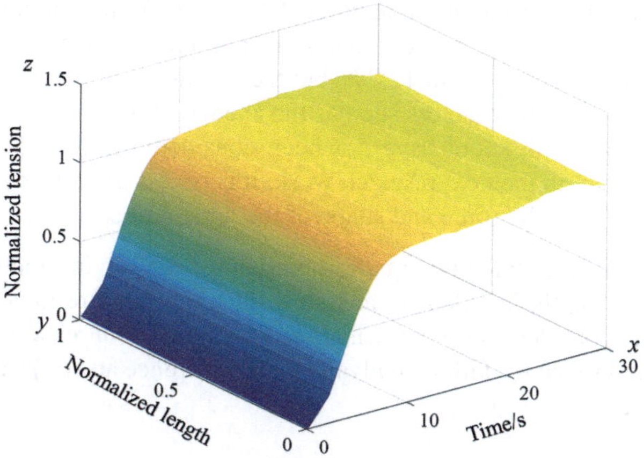

Fig. 11 Variation of theoretical tension distribution of optical fiber with time

the aircraft levelled out, the tension along optical fiber is basically stable. Observing along the *Y*-axis, the internal tension of optical fiber fluctuates slightly, and the tension of ground endpoint and released endpoint is close.

5 Summary

In this paper, according to the characteristics of released optical fiber movement in the air, polynomial power function is used to fit its posture function. Considering the optical fiber length in the air changes with time, the dynamic equation is solved by normalization and Galerkin method, and the posture and tension distribution of optical fiber are calculated at different times. It is found that:

1. When the optical fiber particle is released from the aircraft, its speed is neither zero nor equal to the speed of the aircraft, and its initial speed is affected by three factors: the speed of the aircraft, the position and the length of the guidance fiber in the air. The released optical fiber particle moves slowly in the horizontal direction, and fall slowly to the ground in the vertical direction.
2. The tension on released optical fiber always exists, and the entire optical fiber is always in the air and will not fall on the ground.
3. The optical fiber tension rises rapidly in the initial acceleration stage of aircraft, and when the aircraft levels out, the tension on the fiber is basically stable.
4. The internal tension of optical fiber fluctuates slightly, and the tension of ground endpoint and released endpoint is close.
5. In order to verify the rationality of theoretical analysis, the UAV was used to carry a miniature of optical fiber spool. In the experiment, the ground endpoint tension of the optical fiber is measured and compared with the theoretical results. It is found that the theoretical results are in good agreement with measured

results during the acceleration stage of the aircraft. When the aircraft levelled out, there is a deviation between theoretical results and measured results, the maximum difference is about 12.7%, but the overall trend is basically the same. The main reason is that factors such as the change of wind field and the friction of optical fiber with spool have not been considered during the flight of the UAV. Therefore, in the next research work, it is necessary to study the influence of these factors on the force and shape of the optical fiber.

This paper provides a dynamic equation modeling and solving method for long-length, variable-length, low-density and slow-changing posture fiber release problem. It solves the problems of unknown posture and tension of released optical fiber, and provides reference for failure analysis, fault recurrence and trajectory optimization of aircraft.

References

1. Yang f, Wang y, Ma b, Meng x (2023) Residual stress-strain and numerical simulation of deformation on fiber optic cable package curing. J Project Rock Missil Guid 43(02):12–19
2. Li Z, Bi M, Zhao L, Xu H, Tneg XHM (2024) Performance improvement by an optimum commanding-based non-uniform quantization scheme in low-bit resolution IMDD-UFMC system. Opt Fiber Technol 88:103820
3. Qu F (2021) Simulation of winding process of guidance optical fiber. Xi'an Technological University, Xi'an
4. Zhu J, Yuksek NS, Almasri M, Feng Z (2019) Numerical modeling of dynamic response of miniature multi-impact electromagnetic device for low and wide range frequencies energy harvesting. Proc Inst Mech Eng C J Mech Eng Sci 233(7):2400–2409
5. Zhu J, Lin J, Yuksek NS, Almasri M, Feng Z (2015) Dynamic phenomena and analysis of MEMS capacitive power harvester subjected to low-frequency excitations. Nonlinear Dyn 79:673–688
6. Tang WJ, Liu WD, Gao Z, Zhang K, Zhao P (2021) Stress distribution and influence factors of fiber-optic coil for unmanned submersible vehicle. Acta Armamentarii 42(5):1053–1064
7. Huang L, Xu Y, Jiang W, Xue L, Hu W, Yi L (2022) Performance and complexity analysis of conventional and deep learning equalizers for the high-speed IMDD PON. J Lightwave Technol 40(14):4528–4538
8. Padfield DG (1958) The motion and tension of an unwinding thread. I. Proc R Soc Lond Ser A Math Phys Sci 245:382–407
9. Fraser WB, Ghosh TK, Batra SK (1992) On unwinding yarn from a cylindrical package. Proc R Soc Lond Ser A Math Phys Sci 436:479–498
10. Stanislav Praček, Nace Pušnik, Barbara Simončič, Barbara Simoncic (2016) Balloon theory of yarn during unwinding from package. Text Res J 86(14)
11. Kothari VK, Leaf GAV (1979) The unwinding of yarns from package, part III: unwinding from conical packages. J Textile Inst Trans 70:172–183
12. Lee J-W, An D-M, Yoo W-S (2011) Derivation of equations of motion of an unwinding cable from a cylindrical spool package. J Mech Sci Technol 25:1287–1296
13. Kun-Woo K, Jae-Wook L, Wan-Suk Y (2014) Verification of simulation for unwinding motion of cable in water by physical experiment. Nonlinear Dyn 77:553–568
14. Kun-Woo K, Jae-Wook L, Wan-Suk Y (2014) Unwinding motion of cable by taking into consideration effect of bending on cable. Int J Non-Linear Mech 65:107–120

15. Park H-G, Kang J-H, Kim K-W, Lee J-W, Choi C-Y, Jang J-s, Kim S-y (2021) Verification of numerical analysis for unwinding cable with time-varying unwinding velocity. Text Res J 91(17–18):1957–1973
16. Jang J-S, Kim K-W, Kang J-H, Park H-G, Cho Y-J, Lee J-W (2020) Derivation of equations of motion of an unwinding cable considering transient-state tensile force in time-varying unwinding velocity. Nonlinear Dyn 100:3199–3241
17. Kang J-H, Kim K-W, Lee J-W, Cho Y-J, Jin-Seok Jang (2021) Development of a test method and experimental study on cable unwinding. Proc Inst Mech Eng C J Mech Eng Sci 235(15):2653–2667
18. Ren p, Ma b (2023) Design of Measurement and Control System for optical fiber pay-off device based on LabVIEW. Mechanical & Electrical Engineering Technology 52(02):195–200
19. Li MH, Zhang Z, Tang F, Xue YH, Wang JX, Niu DS (2024) Analysis of posture and tension of guidance optical fiber in high speed release. J Project Rock Miss Guid 44(03):1–10+21
20. Wang R, Li Z, Bian B (2014) Modeling and characteristics analysis of strong bending state of fiber optical cable. J Air Force EngUniv 15(2):71–75
21. JiaWen G (2023) Research on dynamics and energy-preserving numerical algorithm of the variable-length cable and beam with sliding constraints. Harbin Institute of Technology
22. Shi-kun P, Jing-yang L, Hong C, Wang J, Hong Y (2017) Analysis of motion state of the tow-part underwater towed vehicle system during cable deployment. Oceans, Aberdeen, the UK
23. Shenglong C (2019) Deployment and dragging control of tethered satellites based on the variable length cable model. Harbin Institute of Technology, Harbin
24. Zhou H, Yan X, Hu D, Han X (2022) A hamiltonian global nodal position finite element method for dynamics analysis of submarine cables. Ocean Eng 266:112992
25. Feng S (2020) Dynamics and stability for deployment of tethered satellite system. Nanjing University of Aeronautics and Astronautics, Nanjing
26. Li Dan, Wang Chunwei, Wang Junping (2024) Curved elements in weak Galerkin finite element methods. Comput Math Appl 153:20–32

Open Access This chapter is licensed under the terms of the Creative Commons Attribution-NonCommercial-NoDerivatives 4.0 International License (http://creativecommons.org/licenses/by-nc-nd/4.0/), which permits any noncommercial use, sharing, distribution and reproduction in any medium or format, as long as you give appropriate credit to the original author(s) and the source, provide a link to the Creative Commons license and indicate if you modified the licensed material. You do not have permission under this license to share adapted material derived from this chapter or parts of it.

The images or other third party material in this chapter are included in the chapter's Creative Commons license, unless indicated otherwise in a credit line to the material. If material is not included in the chapter's Creative Commons license and your intended use is not permitted by statutory regulation or exceeds the permitted use, you will need to obtain permission directly from the copyright holder.

Analysis of Towing Scheme Based on Submerged Crude Oil Export Hose

Jiayi Zhang, Haiming Zhang, Haijie Dong, Xiaotao Zheng, Shaowei Shang, Junfeng Wang, Kaidong Yang, and Jinquan Zhu

Abstract Immersion crude oil export hoses are about to be applied in the South China Sea. Due to the fact that submerged export hoses have not been applied in Chinese waters before, in order to ensure that submerged export hoses do not sink during towing to designated locations, a finite element analysis model was established based on submerged export hoses and the sea conditions in the South China Sea to analyze the export hose string under different towing conditions. The results showed that there is a risk of sinking of export hoses under some operating conditions.

Keywords Towing scheme · Crude oil export hose · Analysis

1 Introduction

Due to the distance between the production site and the service location of the export hose, it will be transported by towing to the service location. At present, most domestic operations use floating export hoses [1–3], and submerged export hoses are rarely used, so the research on towing conditions of export hoses is still blank. This article takes the towing of submerged export hoses [4] in the South China Sea as a prototype, analyzes the floatability and stress conditions of submerged export hoses during towing, and provides reference for future towing of submerged export hoses.

2 Establishment of Finite Element Model

The physical parameters related to the export hose are shown in Table 1.

J. Zhang · H. Zhang (✉) · H. Dong · X. Zheng · S. Shang · J. Wang · K. Yang · J. Zhu
CNOOC Energy Development Co., Ltd. Oil Production Service Branch, Tianjin, China

© The Author(s) 2026

Z. Zhang, P. Mercorelli (eds.), *Mechanical Design and Simulation: Insights and Innovations*, Lecture Notes in Mechanical Engineering,
https://doi.org/10.1007/978-981-95-2405-1_100

Table 1 Internal structure division of export hose: Basic parameters of 16 inch submerged original export hose

	First hose	Main hose	Tail hose
Hose length(m)	12.2	12.2	12.2
Tensile stiffness(kN)	22,650	22,650	22,650
Bending stiffness(kN·m^2)	35	35	35
Torsional stiffness(kN·m^2)	150	150	150
Permissible pressure(bar)	19	19	19
Allowable axial tension(kN)	1369	1369	1369
Inner diameter of hose(m)	0.387	0.387	0.387
Pipeline weight(kg)	2750	2550	2700

Table 2 Towing vessel parameters

Total length of the ship	69.1 m
Type width	15 m
Depth Moulded	7.2 m
Empty draft	3.031 m
Total weight of the ship	2100 t
Dry side	1 m
Movement speed	5knot

The floating model is a tugboat, and the parameters of the tugboat are shown in Table 2.

According to the requirements of the classification society, wind protection measures should be taken when the wind speed exceeds Pu Shi 7 level (wind speed of 27 knots). The environmental parameters of this model are shown in Table 3.

The towing method of the export hose string generally includes a "U"-shaped towing, where the first and last pipes of the export hose are fixed on the towing vessel [5].

3 Model Calculation Results

The text is based on Orcaflex [6–8] to establish a finite element model, as shown in Fig. 1.

Based on the above model, the calculation results are as follows.

Table 3 Model environment parameters

Maximum wave height(m)	4
Wave period(s)	9.7
Surface velocity(m/s)	1.7
Wind speed(m/s)	13.9

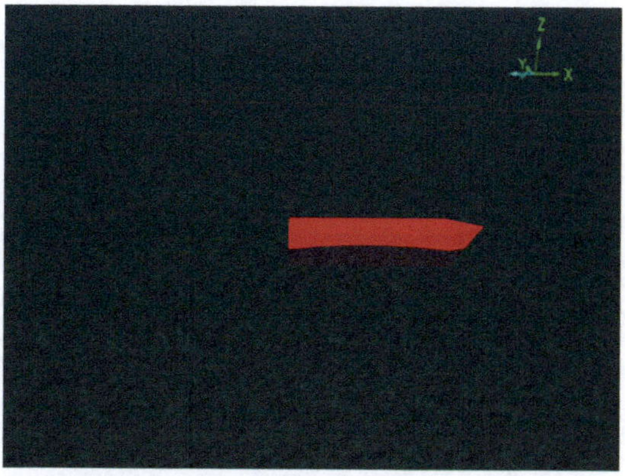

Fig. 1 Schematic diagram of towing model

3.1 Calculation Results of Maximum Effective Tension of Export Hose String

The distribution of the maximum tensile force along the direction of the submerged export hose string during the "U" - shaped towing is shown in Fig. 2. According to the calculation results, the maximum effective tension occurs at the connection between the first pipe and the tail pipe, that is, the pipe string and the towing ship. The maximum effective tension is 50.87 kN, which is much lower than the allowable axial tension of the first pipe of the submerged export hose, and there is no safety hazard.

3.2 Calculation Results of Floatability of Export Hose String

Figure 3 shows the analysis results of the maximum sinking amplitude of the submerged export hose string. It can be seen from the figure that the maximum sinking position occurs at the water inlet of the hose, while the sinking amplitude of other parts of the string is relatively average. Based on the above analysis, time-domain analysis was conducted on the inlet, sixth, seventh, and eighth hoses of the external hose string. The calculation results are shown in Fig. 4. The calculation results indicate that the export hose will not sink under the influence of wind and waves during the "U" - shaped towing process, and can always remain near the water surface.

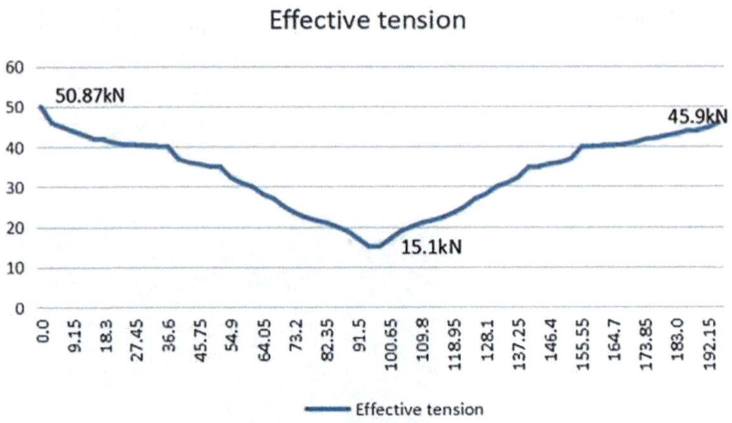

Fig. 2 Calculation results of maximum effective tension for "U" - shaped towing

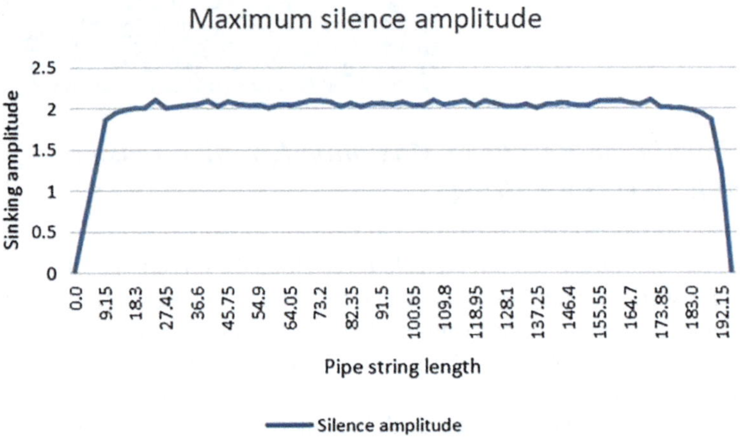

Fig. 3 Maximum sinking amplitude of pipe string during "U" - shaped towing

3.3 Calculation Results of Curvature of Export Hose String

According to the "Guide to Purchasing and Manufacturing Hoses for Offshore Moorings" published by the International Maritime Forum of Oil Companies in 2009 [9–11], the minimum bending radius of export hoses should be 6 times the inner diameter of the hose. To verify whether the maximum bending of the pipe string meets the specification requirements, the bending situation of the pipe string was calculated, and the calculation results are shown in Fig. 5. The calculation results show that the maximum curvature of the pipe string during the "U" - shaped towing process is 0.3281 rad/m, and the minimum allowable bending radius for the 16 inch submerged export hose is 2.322 m, which is within the allowable range of curvature.

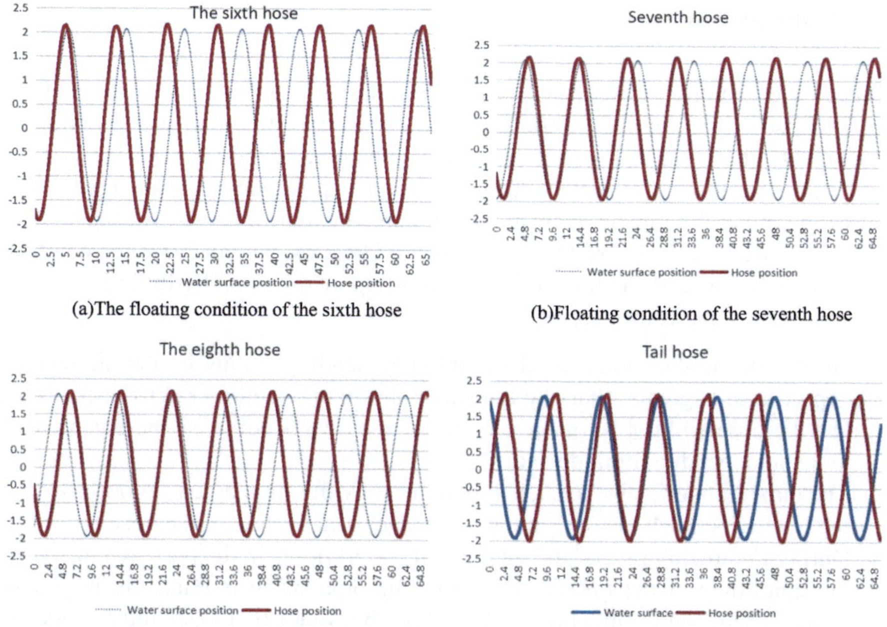

(a)The floating condition of the sixth hose

(b)Floating condition of the seventh hose

(c)Floating condition of the eighth hose

(d)Tail pipe floating condition

Fig. 4 Time domain analysis of floating situation at key positions of submerged export hose string during "U" - shaped towing

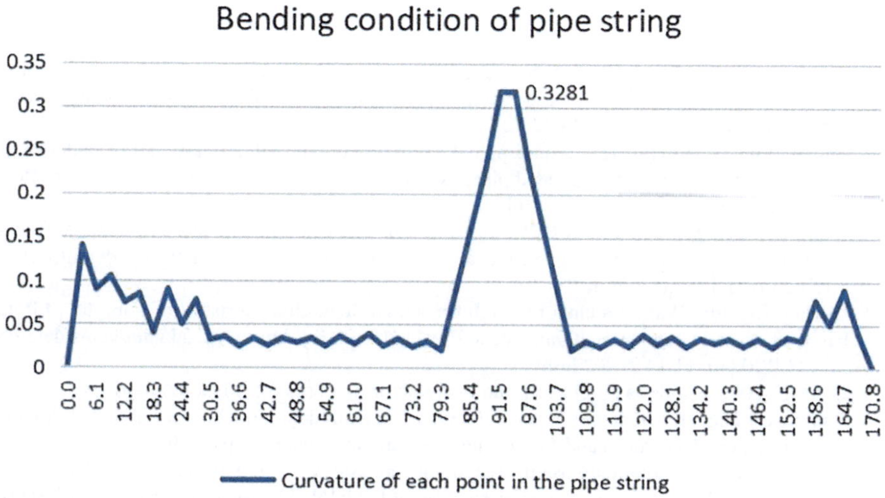

Fig. 5 Calculation results of bending of U-shaped towing hose

4 Conclusion

Due to the distance between the production site and the service location of the export hose, it is necessary to transport the export hose to the service location by towing. Due to its limited buoyancy reserve, submerged export hoses pose a risk of sinking during towing. However, due to the limited application of submerged export hoses in China, the towing analysis of submerged export hoses is still blank. This article is based on the hydrodynamic software Orcaflex, and analyzes the bending, effective tension, and floating of submerged export hoses under different towing schemes. The conclusions are as follows:

1. During the towing process, the minimum bending radius of the submerged export hose is greater than the allowable minimum bending radius of the hose, and the hose will not be damaged due to excessive bending, so there is no need to install a bending limit device;
2. The maximum effective tension experienced by the submerged export hose during towing is much lower than the allowable axial tension of the hose;
3. During the towing process of submerged export hoses in the port area, the sinking amplitude of the middle section of the pipe string is relatively large (the sixth, seventh, and eighth hoses). Therefore, buoyancy blocks should be added to the middle section of the pipe string and the tail pipe during towing to ensure towing safety.

References

1. Wang H (2024) Zhang Zhaode correction of numerical model for vibration of scientific research ship based on actual measurement analysis. Ship Ocean Eng 40(04):7–13. https://doi.org/10.14056/j.cnki.naoe.2024.04.002
2. Zhu H, Tang T, Hu H (2025) Experimental study on coupled vibration model of underwater hydrogen mixed transport suspended pipeline. Ocean Eng:1–11. http://kns.cnki.net/kcms/detail/32.1423.P.20240724.1308.003.html
3. Wang L, Gao Q, Xu Q et al (2024) Study on the influence parameters of tensile stiffness of offshore crude oil export hoses. Ocean Eng:1–13. http://kns.cnki.net/kcms/detail/32.1423.P.20230721.1557.002.html
4. Li Hao, Liu Jun, Wang Wenhua research on sea ice load characteristics of hourglass FPSO based on discrete element method. Ocean Eng, 1–10 [2025]. http://kns.cnki.net/kcms/detail/32.1423.P.20240711.1416.002.html
5. OCIMF (2009) Guide to purchasing and manufacturing hoses for offshore moorings
6. Bahtui A, Bahal H, Alfano G (2009) Numerical and analytical modeling of unbonded flexible risers. Offshore Mechanics and Arctic Engineering, San Francisco, p 021401
7. Yu Yongjun, Chen Jianfeng, Bian Hengdong, etc Study on the Influence of Continuous Collisions of Mooring Ships on the Fatigue Life of High Piled Wharf Ocean Eng, 1–11 [2025]. http://kns.cnki.net/kcms/detail/32.1423.P.20240626.1903.002.html
8. Daifeng W (2023) Study on the ultimate bearing characteristics of floating oil hoses in the ocean. China University of Petroleum (Beijing). https://doi.org/10.27643/d.cnki.gsybu.2023.000308

9. Weichang L, Gao Q, Guangzhong B et al (2022) Study on the bending performance of rubber hoses for underwater oil transportation in the ocean. In: Proceedings of the 20th China Marine (Shore) Engineering Academic Symposium (Part 1). School of Marine Science and Engineering, Shanghai Maritime University; Hebei Zebang Plastic Technology Co., Ltd. https://doi.org/10.26914/c.cnkihy.2022.012994
10. Zhang H (2024) Classification of operation scenarios and modes for offshore unmanned firefighting ships. Ship Ocean Eng 40(03):76–81. https://doi.org/10.14056/j.cnki.naoe.2024.03.014
11. Liao L, Yu Y (2020) Structural design and experimental analysis of marine flexible composite hoses. Chem Equip Technol 41(03):30–34. https://doi.org/10.16759/j.cnki.issn.1007-7251.220.06.009

Open Access This chapter is licensed under the terms of the Creative Commons Attribution-NonCommercial-NoDerivatives 4.0 International License (http://creativecommons.org/licenses/by-nc-nd/4.0/), which permits any noncommercial use, sharing, distribution and reproduction in any medium or format, as long as you give appropriate credit to the original author(s) and the source, provide a link to the Creative Commons license and indicate if you modified the licensed material. You do not have permission under this license to share adapted material derived from this chapter or parts of it.

The images or other third party material in this chapter are included in the chapter's Creative Commons license, unless indicated otherwise in a credit line to the material. If material is not included in the chapter's Creative Commons license and your intended use is not permitted by statutory regulation or exceeds the permitted use, you will need to obtain permission directly from the copyright holder.

Comparative Study of Large-Slip Axial Motion of a Slender Body in Different Tubes

Di Zhang, Ke-Dong Zhou, Lei He, Chao Shen, Mo Yang, Yi-hui Liu, and Hai-Yue Ren

Abstract To investigate the influence of different tube on slender body during large-slip axial motion, a coupled finite element model was established and coupled with the VUAMP subroutine. The process of large-slip axial motion of the slender body was then simulated using Abaqus software. The motion states and surface morphologies of slender bodies moving in three tubes, namely an 820 mm rectangular tube, an 850 mm rectangular tube and an 820 mm polygonal tube were compared. The simulation results demonstrate that the slender body of the polygonal tube exhibits faster in-tube motion, little motion disturbance, and more stable motion. Additionally, its horizontal direction centroid offset is only 5% of that of the rectangular tube. Furthermore, the slender body of the polygonal tube exhibits minimal stress-strain characteristics. The surface morphology of the slender body of the polygonal tube is not conspicuous at the valleys and hills boundaries and the scratch grooves of the slender body of the rectangular tube are deep. The above obvious different surface morphologies should result in great influence on aerodynamic parameters.

Keywords Effect of tube on slender body · Polygonal tube · Rectangular tube · Surface morphology

---

D. Zhang · K.-D. Zhou (✉) · L. He · M. Yang · Y.-h. Liu · H.-Y. Ren
School of Mechanical Engineering, Nanjing University of Science and Technology, Nanjing, Jiangsu, China
e-mail: zkd81151@njust.edu.cn

C. Shen
China Ship Scientific Research Center, Wuxi, Jiangsu, China

© The Author(s) 2026

Z. Zhang, P. Mercorelli (eds.), *Mechanical Design and Simulation: Insights and Innovations*, Lecture Notes in Mechanical Engineering,
https://doi.org/10.1007/978-981-95-2405-1_101

1 Introduction

During the large-slip axial motion process, the tube creates a high-velocity, high-pressure impact on the slender body [1], which affects the motion and the surface morphology of the slender body. In recent years, scholars have conducted a substantial body of research on this. Liu [2] conducted a dynamic large-slip axial motion test of two different slender bodies, and the resistance characteristic curves were obtained. Xu [3] employed the Abaqus software to simulate the dynamic large-slip axial motion process of the slender body and conducted dynamic large-slip axial motion tests to get the parameters of slender body/tube interaction in different cases.

The objective of this paper is to investigate the effects of different tube on slender body and its influencing law by numerical simulation techniques. A coupled finite element model of the slender body/tube was constructed through the explicit dynamics finite element method. Subsequently, the effects of different structures and various lengths of tubes on the motion and the surface morphologies of the slender bodies are compared.

2 Materials and Methods

2.1 *Model Properties*

Three slender body/tube geometric models were constructed for three different tubes: 820 mm rectangular tube (designated "rectangular 820"), 850 mm rectangular tube (designated "rectangular 850"), and 820 mm polygonal tube (designated "polygonal 820"). Additionally, a geometric model of the front part of the housing was developed, which was utilized to provide preliminary constraints for the slender body. Furthermore, in this paper, we use the Johnson-Cook model [4].

2.2 *Meshing*

To circumvent element shear locking, C3D8RTs are used for the tubes, the slender bodis and the housings. Localised mesh densification of the part of the slender body in contact with the tube is employed to meet the requisite calculation accuracy. To ascertain the mesh-independence validation, slender body mesh models with different mesh densities are constructed, with the polygonal tube serving as a case study. Table 1 presents the simulation results.

Table 1 shows that after increasing the number of grids from 220,000 to 300,000, the computation time increases by 40.7%, but the initial velocity increases by only 0.8%. Scenario 2 can fulfill the requisite accuracy standards while simultaneously

Table 1 Numerical results of meshes with different densities

Scenario	Grid number	v_0/m·s^{-1}	Time/h
1	150 k	769.5	5.7
2	220 k	798.8	8.1
3	300 k	805.4	11.4

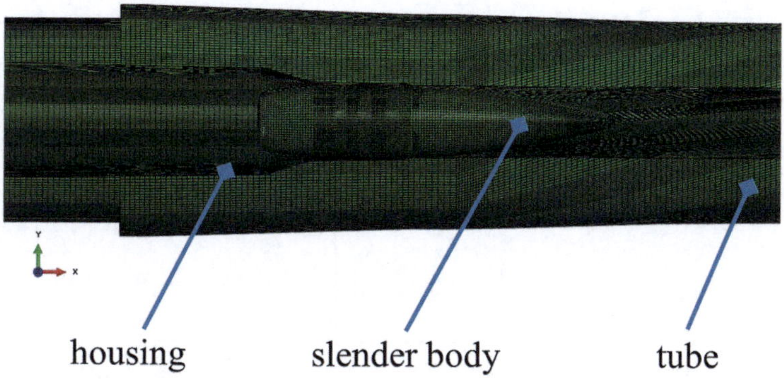

Fig. 1 Assembled FEA model of slender body/tube

exhibiting a high computational efficiency. Consequently, scenario 2 is deemed an optimal choice. The total number of meshes for the polygonal 820 model when meshed using Scheme 2 is approximately 1.6 million. Figure 1 illustrates the assembled FEA model of the slender body/tube.

2.3 *Constraints and Load Application*

The end of the tube and the housing are defined as complete constraints, thereby restraining all of their degrees of freedom. The loads are applied using the VUAMP subroutine coupling calculation approach which is described in the literatures [5–9]. Figure 2 illustrates the flowchart of the coupling calculation. Furthermore, the model is subjected to the effects of gravity.

3 Simulation Results

The origin of the coordinate is designated as O, which is set at the center of the aft end face of the tube. The X-axis is aligned with the direction of the tube, extending to the tube opening, the Y-axis is perpendicular to the X-axis and oriented upwards, and the Z-axis is determined by the right-hand rule.

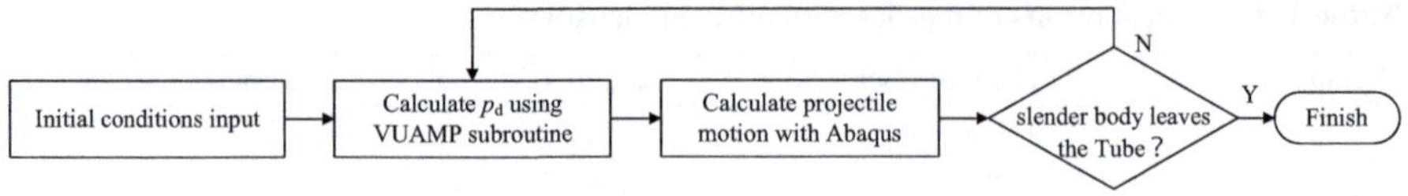

Fig. 2 Flowchart of the coupling calculation

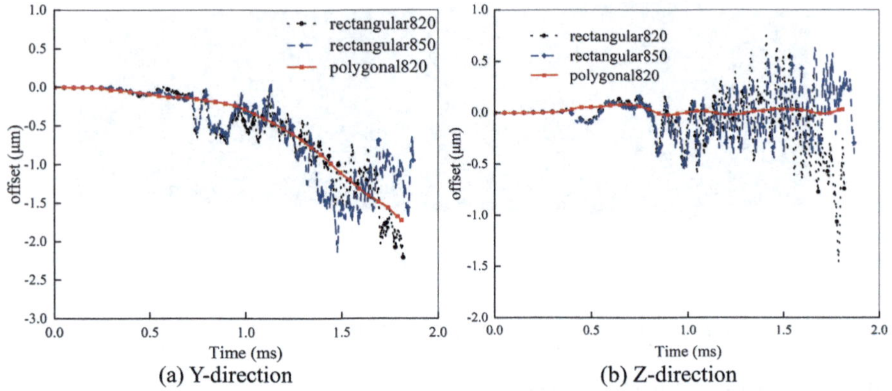

Fig. 3 Offsets curves of centroid

3.1 A Comparative Analysis of Slender Body Motion

3.1.1 Comparative Analysis of Internal Duration and Initial Velocity

The internal durations of the 820 rectangular model, 850 rectangular model and 820 polygonal model are 1.82ms, 1.87ms and 1.81ms, respectively, and the initial velocities are 787.30m/s, 795.04m/s and 798.80m/s, respectively, indicating that the slender body moves faster inside the polygonal tube.

3.1.2 Comparative Analysis of Offsets of the Centroid of Slender Body in the OYZ Plane

When the slender body is moving, its centroid oscillates in the OYZ plane. The offsets of the centroid of slender body are shown in Fig. 3.

At the beginning of the curve, the offset of the centroid of the slender body in the Y- and Z-axis directions is approximately equal to 0. This is due to the constraints imposed by the housing. After that, the offsets gradually increased, and the oscillation phenomenon began to appear. It is also evident that there is a greater offset in the Y-axis and that the Y-axis offset gradually increases over time due to the gravitational load applied in the negative direction in the Y-axis. In addition, at the end of the curve, the centroid oscillation of the slender body is greater for the 820 rectangular tube than for the 850 rectangular tube, due to the longer tube of the 850 rectangular tube, resulting in a longer retention period of the slender body, a higher rotational speed and relative stability.

Moreover, it can be seen that the amplitude of the oscillation of the centroid of the slender body of the polygonal tube is smaller than that of the rectangular tube. In the Y-axis direction, due to the superposition of the effect of gravity, the maximum offsets of the centroid of the slender body of the rectangular and polygonal tube are 2.22 μm and 1.73 μm, respectively, and the maximum offset of the centroid of the slender body of the rectangular tube is 1.28 times of that of the polygonal tube, but in the Z-axis direction, the maximum offsets of the centroid of the slender body of the rectangular and polygonal tube are 1.46 μm and 0.07 μm, respectively. The maximum offset of the centroid of the slender body of rectangular tube is 20.86 times of that of the polygonal tube. The above results show that the slender body movement in the polygonal tube is smoother than that in the rectangular tube.

3.1.3 Comparative Analysis of Slender Body Rotational Velocity

$$\omega_0 = \frac{2\pi v_0}{\eta d} \tag{1}$$

The Theoretical Value of the Initial Rotational Velocity of the Slender Body Is Given by Eq. (1) [10]. where ω_0 is the initial rotational velocity of the slender body, v_0 is the initial velocity, η is the pitch, and d is the diameter.

Table 2 presents a comparison of the slender body rotational velocities of the three models. The errors between the simulated and theoretical values in the table are less than 4%, which proves the accuracy of the simulation.

Figure 4 illustrates the slender body rotational velocity-time curves for the three models by simulation. The curves for the three models exhibit a consistent trend. In the initial phase of the slender body's trajectory, the rotational velocities of the three models are estimated to be approximately 0. This is because the slender body has not yet interacted with the tube. Subsequently, the slender body commences a rotational movement as a consequence of the tube. Furthermore, the polygonal tube exhibits a slightly faster rotational velocity than the rectangular tube, and the simulated values of the two are 13,643 rad/s and 13,465 rad/s, respectively, with a difference of 178 rad/s (about 1.3%). The oscillation of the rotational velocity data is lower. These results demonstrate that the polygonal tube has superior rotational stability.

Table 2 Results of slender body rotational velocity

Model	Rotational velocity /rad·s^{-1}		Inaccuracies/%
	Theoretical	Simulated	
Rectangular 820	13,018	13,465	3.43
Rectangular 850	13,146	13,620	3.61
Polygonal 820	13,199	13,643	3.36

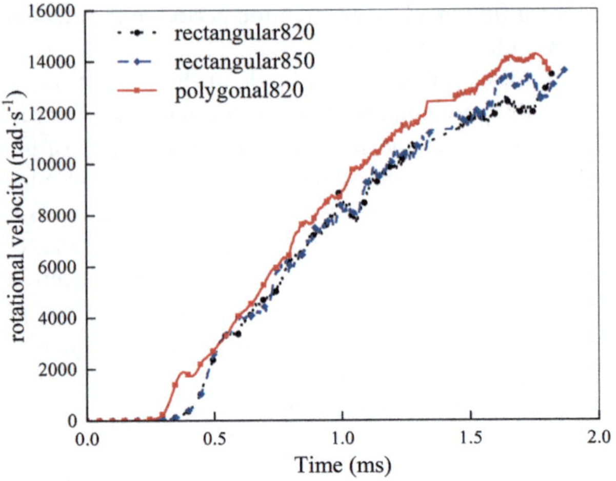

Fig. 4 Slender body rotational velocity-time curves

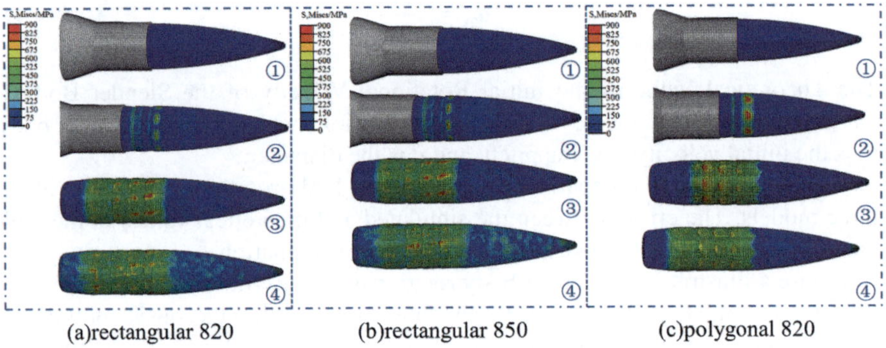

Fig. 5 Stress nephograms

3.2 *A Comparative Analysis of Slender Body Stress-Strain*

The slender body stress nephograms obtained from the simulation of the three models are presented in Fig. 5, and the strain nephograms are presented in Fig. 6.

The slender bodies have undergone four distinct moments. In the initial moment (①), the slender body is fully constrained by the housing. In the second moment (②), the slender body begins to move. At this moment, the front end of the cylindrical part of the slender body begins to be constrained by the tube, while the rear end remains subjected to the constraints of the housing. In the third moment (③), the slender body has been completely released from the constraints of the housing. However, the engraving process is not yet completed, and the cylindrical part of the slender body only be constrained by the tube, while the rear end remains free. This

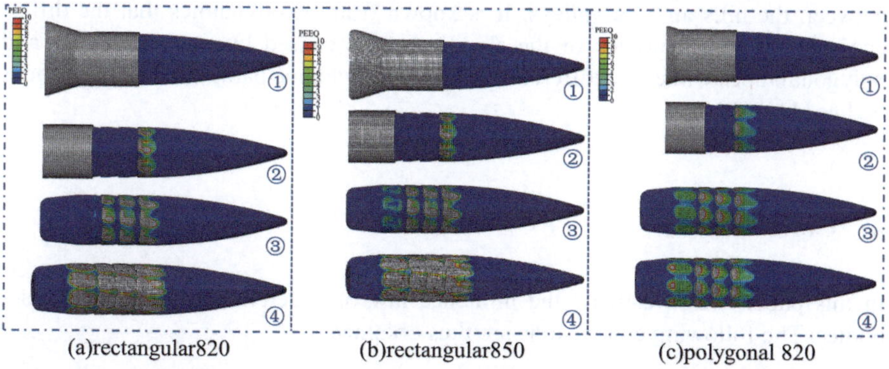

Fig. 6 Strain nephograms

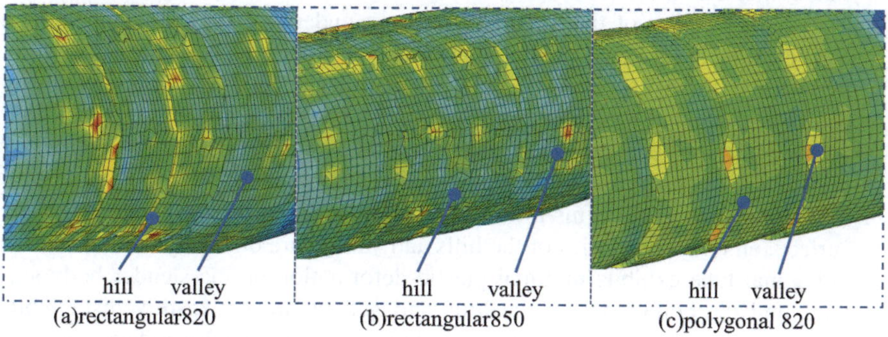

Fig. 7 Surface morphologies of slender bodies

state can potentially cause instability in the movement of the slender body. In the fourth moment (④), the slender body is completely engraved into the tube and begins to move along in-tube. Moreover, the stress-strain of the slender body of the polygonal tube is significantly smaller than that of the rectangular tube, indicating that the stress environment of the rectangular tube is worse than that of the polygonal tube.

3.3 A Comparative Analysis of Slender Body Surface Morphology

After the engraving process, the slender body surface morphologies of the three models are illustrated in Fig. 7.

The slender body surface valleys caused by the rectangular tube are evident, manifesting as trapezoidal-like valleys. In contrast, the slender body valleys caused by the polygonal tube are less pronounced, with a relatively gradual transition between the hills and the valleys. It is known from aerodynamics that the differences in the morphologies of the slender bodies caused by the rectangular and polygonal tubes will cause differences in the aerodynamic parameters of the slender bodies [11].

4 Conclusions

In this paper, we investigate the influence law of different tubes on the slender bodies. The following conclusions are thus obtained:

1. A slender body of a polygonal tube has a high velocity of movement and its centroid is less disturbed during the movement, making the movement more stable than that of a slender body of a rectangular tube.
2. The slender bodies of the three models have undergone four moments, which can be defined as follows: the initial stationary moment, the beginning of engraving with housing constraint moment, incomplete engraving without housing constraint moment, and the completion of engraving moment. In these four moments, the in-tube stress environment of the rectangular tube is more severe than that of the polygonal tube.
3. The rectangular tube forms a trapezoidal-like indentation on the slender body surface, and the boundaries of the hills and valleys are distinctly visible and the polygonal tube exhibits minimal plastic deformation on the slender body surface, and the boundaries of the hills and valleys are not readily discernible. Additionally, the differences in the morphologies of the slender bodies caused by the rectangular and polygonal tubes will cause differences in the aerodynamic parameters of the slender bodies.

References

1. Yang YZ, Zhang XY, Xu C (2020) The interaction between copper Jacket projectile and barrel under different temperature based on thermo-mechanical coupling finite element model. J Vibrat Shock 39(10):44–51. https://doi.org/10.13465/j.cnki.jvs.2020.10.006
2. Liu DY, Sun P, Zhang XW et al (2020) Experimental study on engraving resistance properties of band of two medium caliber projectile. J Ballist 32(03):5–8. https://doi.org/10.12115/j.issn.1004-499X(2020)03-002
3. Xu H, Huang CL, Wang XK et al (2022) Theoretical and experimental study of projectile dynamic engraving resistance. Acta Armamentarii 43(9):2263–2273. https://doi.org/10.12382/bgxb.2021.0875
4. Johnson GR, Cook WH (1985) Fracture characteristics of three metals subjected to various strains, strain rates, temperatures and pressures. Eng Fract Mech 21:31e48. https://doi.org/10.1016/0013-7944(85)90052-9
5. Wang SQ, Li Q, Fan JT et al (2024) Analysis of the dynamic characteristics of a small caliber gun's bullet engraving process. J Vib Shock 43(12):241–247. https://doi.org/10.13465/j.cnki.Jvs.2024.12.027

6. Fan JT, Li Q, Ran XC et al (2023) Simulation study on the influence of slope bore processing error on internal ballistic performance. J Ordnan Equip Eng 44(12):220–226. https://doi.org/10.11809/bqzbgcxb2023.12.029
7. Xu H (2023) Study on dynamic characteristics of cased telescoped ammunition during engraving process. Doctoral Dissertation of Nanjing University of Science and Technology
8. Shen C, Zhou KD, Lu Y et al (2019) Modeling and simulation of bullet-barrel interaction process for the damaged gun barrel. Def Technol 15(6):972–986. https://doi.org/10.1016/j.dt.2019.07.009
9. Shen C, Zhou KD, Lu Y et al (2018) Research on the influence of aamaged bore of a large caliber machine gun on bullet engraving progress. Acta Armamentarii 39(12):2320–2329. https://doi.org/10.3969/j.issn.1000-1093.2018.12.005
10. Wu ZL, Li ZX, Liu K et al (2020) Ammunition of automatic weapon. Beijing Institute of Technology Press, Beijing
11. Zhao, L.， Dong, F. Q. (2015) Analysis about rifling grooves of small-caliber ammunition on the flow field. J Project Rocket Missile Guid, 35(2): 126-128. https://doi.org/10.15892/j.cnki.djzdxb.2015.02.032.

Open Access This chapter is licensed under the terms of the Creative Commons Attribution-NonCommercial-NoDerivatives 4.0 International License (http://creativecommons.org/licenses/by-nc-nd/4.0/), which permits any noncommercial use, sharing, distribution and reproduction in any medium or format, as long as you give appropriate credit to the original author(s) and the source, provide a link to the Creative Commons license and indicate if you modified the licensed material. You do not have permission under this license to share adapted material derived from this chapter or parts of it.

The images or other third party material in this chapter are included in the chapter's Creative Commons license, unless indicated otherwise in a credit line to the material. If material is not included in the chapter's Creative Commons license and your intended use is not permitted by statutory regulation or exceeds the permitted use, you will need to obtain permission directly from the copyright holder.

Design of Pipeline Robot Based on Magnetic Adsorption Wheels

Fangweiqiao Deng, Bohang Xia, Molin Du, Qixuan He, and Boxi Wang

Abstract The development of a pipeline inspection robot that can autonomously navigate obstacles and climb vertical walls within pipelines was necessitated by the limitations imposed by varying pipeline terrains and restricted operational range. To address these challenges, a modularized approach was adopted. The design of the robot encompassed its overall structure, such as wheels, drive mechanisms, and inter-joint connection scheme. Concurrently, the control system and navigation system were meticulously devised. Subsequently, a pipeline test platform was erected to facilitate the production of the physical prototype of the pipeline robot, incorporating the aforementioned design elements.

Keywords Pipeline robot · Magnetic adsorption wheels · Structural designs

1 Introduction

The rapid advancement of urbanization in our nation has led to an increasing number of large-scale structures sprouting up at an accelerated pace. Simultaneously, the heightened pace of daily life and professional responsibilities has prompted individuals to demand higher standards for their living and working environments. Consequently, pipelines have emerged as a vital component in the infrastructure of contemporary large-scale buildings. Over time, pipelines are susceptible to issues such as blockages, cracks, and leaks, presenting potential long-term societal risks. Hence, the routine inspection and maintenance of pipelines are crucial endeavors. Nonetheless, the intricate and confined nature of pipeline systems poses significant challenges for manual cleaning and upkeep. Consequently, the development of specialized pipeline robots represents a promising solution to address these challenges, constituting a focal point of research within the field [1–3].

F. Deng (✉) · B. Xia · M. Du · Q. He · B. Wang
Mechanical Electrical Engineering School, Beijing Information Science and Technology University, Beijing, China

© The Author(s) 2026

Z. Zhang, P. Mercorelli (eds.), *Mechanical Design and Simulation: Insights and Innovations*, Lecture Notes in Mechanical Engineering,
https://doi.org/10.1007/978-981-95-2405-1_102

Research on pipeline robotic equipment has been more advanced in foreign countries compared to domestic efforts. For instance, Kim et al. [4] in South Korea introduced a hybrid foot and wheeled pipe cleaning robot, while Chuo University in Japan developed a peristaltic narrow pipe cleaning robot [5].

Jiang et al. [6] developed an independent dual-drive wheeled pipeline robot based on fractional-order and integer-order PID control. Bogdan et al. [7] developed a magnetic crawler tracking robot specialized for pipeline inspection. China's interest in pipeline robots emerged after 1985, leading to significant investment by various institutions such as Tsinghua University, Donghua University, National Taiwan University, and other scientific research institutes. These efforts have yielded successful outcomes in research and application, resulting in the independent production and utilization of pipeline robots within the country.

This study introduces a novel approach to address the challenges posed by non-linear pipelines. Conventional pipeline robots face difficulty navigating through vertical, curved, or varying diameter pipelines, especially when encountering slopes. To overcome this limitation, a robot utilizing magnetic adsorption wheels constructed with embedded permanent magnet units has been developed. Furthermore, a new orthogonal connection scheme has been designed to enhance the modularity and flexibility of the robot, enabling it to traverse diverse terrains encountered in pipelines.

2 Pipeline Robot Overall Design Program

The pipeline robot, as detailed within this study, integrates a wheeled robot and a robotic arm, combining structural forms and modular design concepts. This unique design enables the robot to maintain optimal motion performance while simultaneously enhancing its workspace through the functionality of the robotic arm. Comprehensive structure depicted in Fig. 1, while the parameters are shown in Table 1.

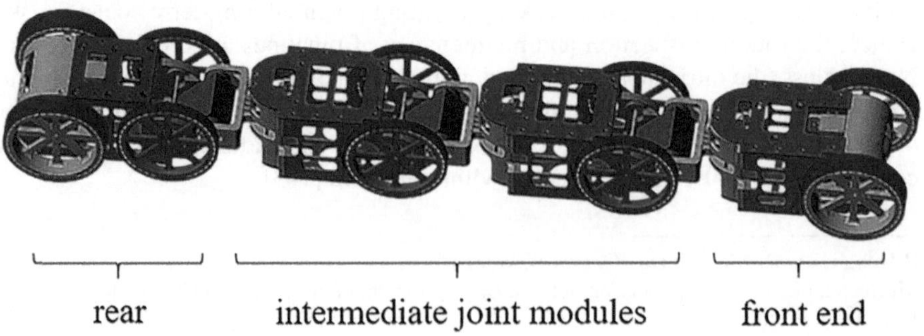

Fig. 1 Pipeline robot structure

Table 1 The pipeline robot parameters

Parameters	Value
Structural approach	Multi-section modular robot
Sizes/m	0.6*0.09*0.08
The total weight/kg	2.5
Working pipe diameter/mm	120–350
Minimum forward speed/(m/s)	0.08

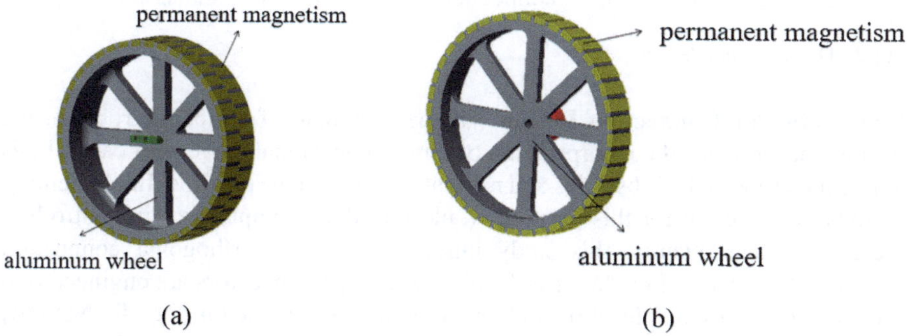

Fig. 2 The magnetic wheel

2.1 *Structural Designs*

Wheel Design The pipeline robot developed within this project is specifically tailored for use in pipelines ranging from 120 mm to 350 mm in small and medium sizes. This robot is designed to navigate through challenging pipeline environments while still meeting speed requirements, hence the selection of a trolley-wheeled pipeline robot. However, conventional trolley wheel pipeline robots lack the capability to overcome obstacles and climb vertical surfaces. To address this limitation, a design incorporating magnetic adsorption wheels embedded with permanent magnets was developed. This innovative design enables the robot to achieve vertical wall climbing functionality within the pipeline. As illustrated in Fig. 2, the driving wheel is depicted in Fig. 2a mounted on the driving joint, while the driven wheel is shown in Fig. 2b installed on the intermediate joint. The magnetic adsorption wheel provides the necessary additional force for the robot to successfully climb vertical pipe walls.

Driver Program Design The pipeline robot's operational demands, particularly in executing intricate tasks like wall climbing within constricted small and medium-sized pipelines, necessitate a driving mechanism that fulfills criteria of compact size, lightweight design, and substantial driving power simultaneously. Employing a DC motor with a worm gear reducer configuration proves to be an effective approach in delivering the requisite driving force for the robot's operations, as illustrated in the robotic drive system depicted in Fig. 3

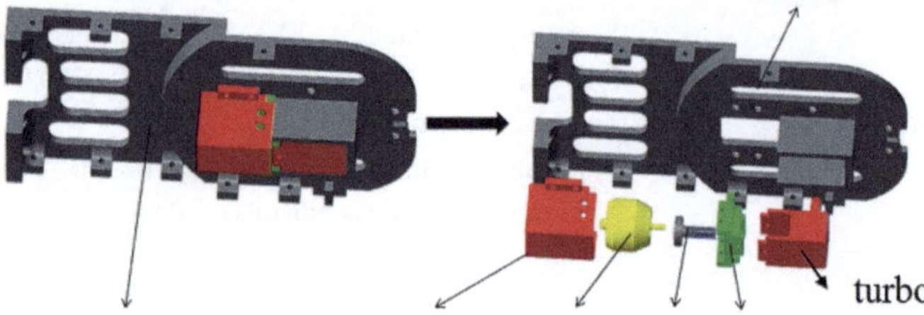

Fig. 3 The drive unit design

Inter-Articular Connection Design The development of a pipeline robot at the institute necessitates the incorporation of three-dimensional movement capabilities to facilitate traversal of obstacles and navigation through the pipeline. Implementing a gimbal connection for this purpose would introduce complexities in controlling the robot. Consequently, this study introduces a novel orthogonal connection scheme, illustrated in Fig. 4a. Specifically, the U-type connectors are engineered to provide the robot with elevation and yaw mobility, as depicted in Fig. 4b. Notably, the pitch joints at the servo-U-shaped connector interface are controlled joints managed by the servo, while the yaw joints at the junction of the U-shaped connector and the head's tail section are free-moving joints. This configuration allows the robot to adjust the joint angles in the vertical plane, resembling an "M" shape, and execute rotational movements in the horizontal plane. However, to address potential lateral drift during forward motion, a self-locking mechanism is devised. By controlling the micro-servo, the rocker arm is engaged in the connector slot and the front end tail section to ensure straight movement. Subsequently, the micro-servo is manipulated to secure the rocker arm within the connector slot and the front end rear portion to prevent lateral displacement while moving in a straight line, with the option to release for turning maneuvers.

2.2 *Control System Designs*

A sophisticated robotic system necessitates the coordination of numerous joints and motors, real-time collection and processing of diverse sensor data, and the maintenance of seamless communication between the upper and lower computer components. To address these complex requirements, a hierarchical control framework is implemented, comprising the upper computer monitoring system and the robot master-slave control system, exemplified in Fig. 5.

The host computer system serves as the upper application layer, with the operator interface developed on the Visual Studio platform using C# Winform, operating on the terminal computer. The operator interface provides a display of robot status information, sensor data, and camera-captured images. Additionally, it enables the

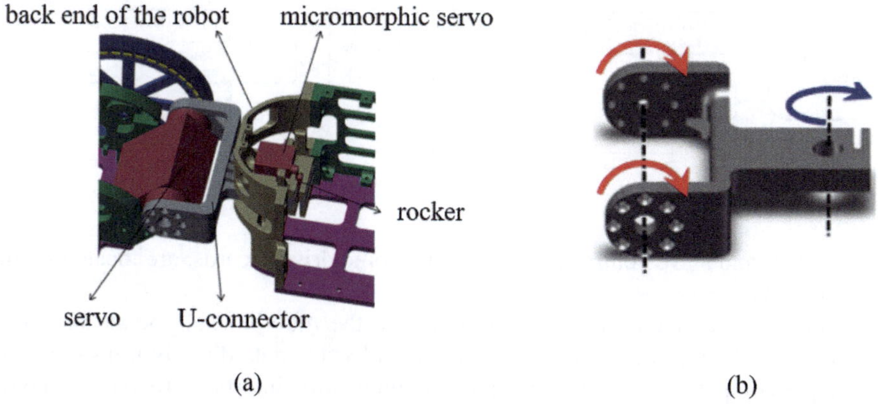

Fig. 4 The inter-articular connection design

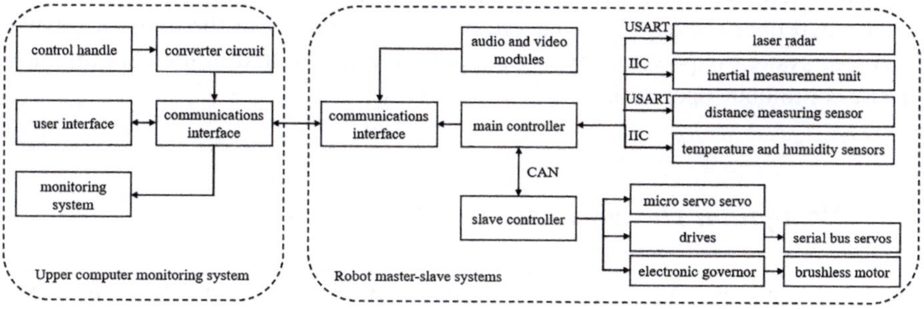

Fig. 5 The control system structure

transmission of images and audio to a mobile terminal via a wireless router and switch. Virtual buttons integrated into the interface facilitate the control of the robot's behavior. Wireless control of the robot is also achievable through a control handle, with signals from the handle matched with the communication interface via a conversion circuit to ensure data interoperability with the operator interface. This design allows simultaneous control of the robot through both the operator interface and the control handle.

The robot's master-slave control system operates as the underlying control layer, with hardware motor drive, sensor data processing, and motion gait planning executed by ARM-STM32 microcontrollers. The master controller handles sensor data acquisition, real-time communication with the host computer, motion planning, control algorithms, and output calculation. Interactions between microcontrollers and peripherals are typically carried out using serial ports, IIC buses, and CAN buses. On the other hand, the slave controller is responsible for hardware drive, position and speed control, transmitting command signals to the actuator via USART or pulse width modulation, and actuating the mechanism through the driver.

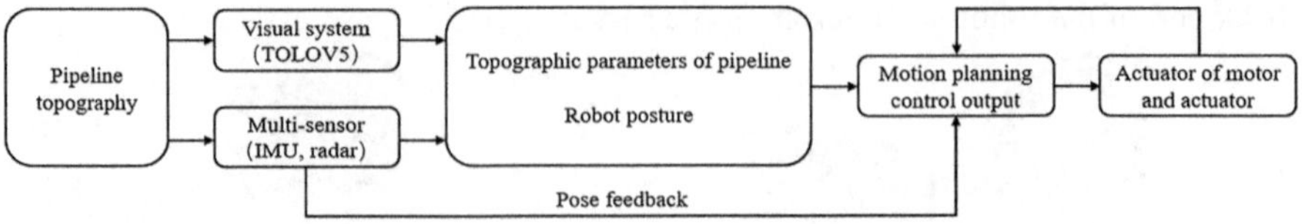

Fig. 6 The navigation system

The master and slave controllers, along with some driver boards, are located in the central units.

Separately, the audio and video modules in the robot's head operate independently from the microcontroller, with audio and video data directly transmitted to the upper computer system through the communication interface. To enhance communication quality between the host and lower computers while reducing resistance from trailing cables, the robot utilizes optical fiber as the signal transmission medium, exclusively transmitting signals. The necessary communication interface circuits and fiber optic transceivers are installed in the rear of the robot. Moreover, the robot system is powered by two sets of batteries located in the central unit to ensure continuous operation.

2.3 *Navigation Design Based on Multi Sensor*

The front end of the pipeline robot is outfitted with Time-of-Flight (TOF) array lidar and a camera, while the robot body is also integrated with an Inertial Measurement Unit (IMU). By employing various sensor configurations and sophisticated algorithms, the robot achieves environmental perception. This involves acquiring environmental data and its own orientation, utilizing the environmental data for path planning and control calculation, and ultimately transmitting commands to the actuators. Throughout its operation, the robot leverages its orientation feedback to establish a closed-loop control system. A visual representation of the complete navigation system is depicted in Fig. 6.

2.4 *Robot Gait Planning*

The robotic system implements a multi-joint series chain configuration. To navigate obstacles, stairs, and uneven terrain, it is imperative to regulate the sequential movement of the three joints while factoring in the impact of magnetic forces. The proposed strategy for overcoming obstacles through climbing is illustrated in Fig. 7, where the four units from the forefront to the rear are equipped with three pivotal joints denoted as joint 1, joint 2, and joint 3 successively.

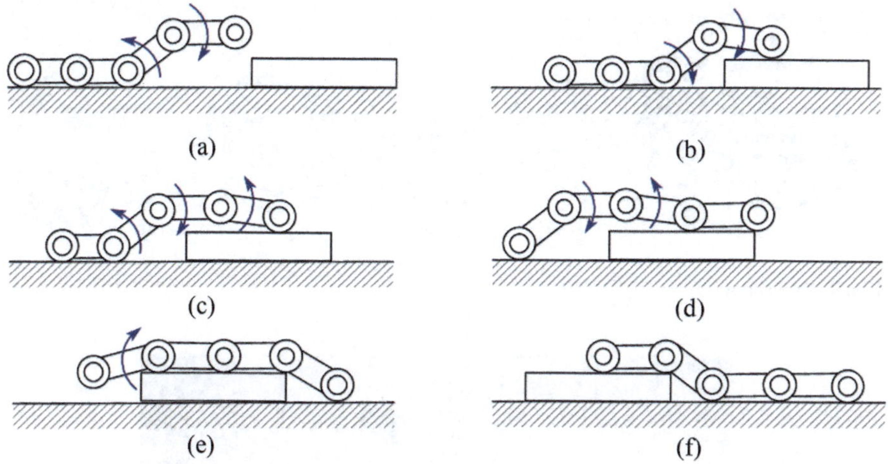

Fig. 7 The gait planning for climbing over step obstacles

Fig. 8 The gait planning for crossing groove obstacles

When the robot encounters a groove obstacle, if the width of the groove is less than 240 mm, all joints can be locked to directly cross. When the width is greater than 240 mm, it can move through the gait as shown in Fig. 8, which can be regarded as the reverse movement of climbing over a step obstacle.

2.5 *Robot Physical Production*

The present study opts for the utilization of aluminum alloy material for fabricating the non-standard components of the robot to enhance its strength while reducing the overall mass. To streamline production costs and time, the robot body is manufactured through a process of splitting and subsequent splicing. Standard components such as motors, servos, infrared cameras, array lidar, power supply, bearings, screws, nuts, and NdFeB permanent magnets are procured, excluding the custom parts. Following the completion of robot assembly, each component is designed with a card slot to facilitate installation and ensure precision. The schematic representation of the finalized pipeline robot structure post-assembly is depicted in Fig. 9.

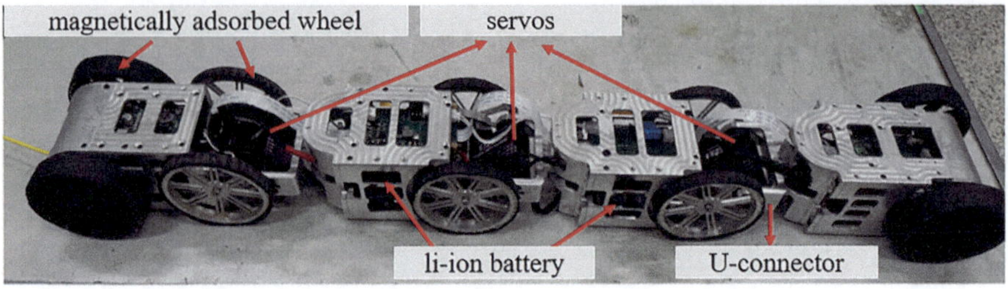

Fig. 9 The robot

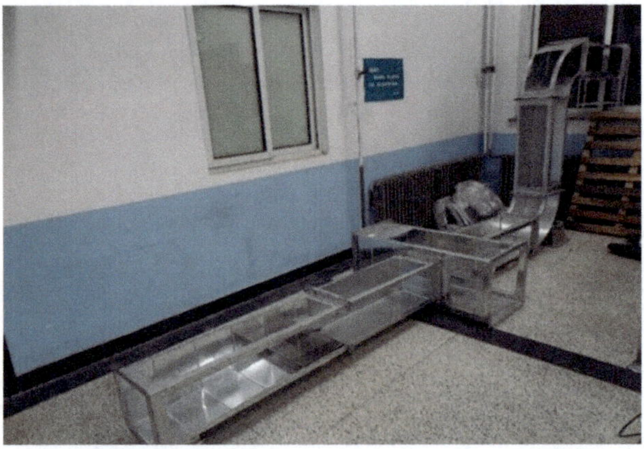

Fig. 10 The pipes used for experiments

Upon the finalization of the physical prototype of the robotic system, a test platform was meticulously constructed to replicate the authentic pipeline terrain. This endeavor aimed to streamline the monitoring and fine-tuning of the robot within the experimental phase. Notably, a segment of the pipeline was strategically incised to address the extraction of the pipe wall adjacent to the operator, as illustrated in Fig. 10. Throughout the experimentation, a camera was affixed at a precise location on the pipeline to enhance the monitoring and documentation of the experimental procedures. Experiments such as turning, obstacle climbing and climbing were carried out on the robot, and the results are shown in the Fig. 11.

3 Conclusion

The system design of the pipeline robot is conducted in accordance with functional requirements and parameter indexes, encompassing the development of the robot's moving scheme, driving scheme, and inter-joint connection scheme.

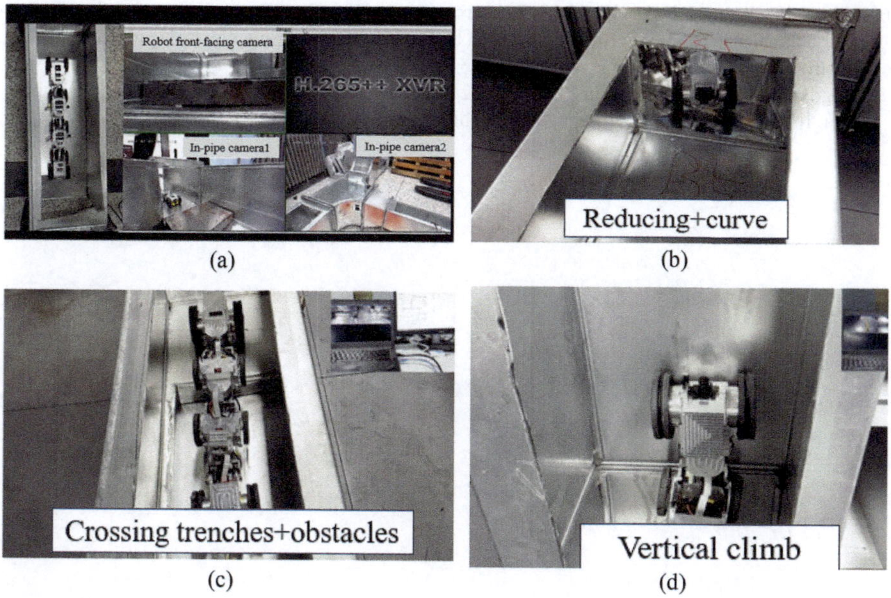

Fig. 11 The experimental results

Firstly, an innovative magnetic adsorption wheel, which utilizes the adsorption force from the magnetic circuit between permanent magnet units, is created to facilitate vertical climbing capabilities.

Secondly, a novel orthogonal connection scheme is implemented to enhance the modularity and flexibility of the robot.

Subsequently, a physical prototype of the pipeline robot is fabricated and assembled, followed by the establishment of a pipeline test platform to conduct relevant experiments in real-world scenarios, thereby validating the efficacy of the pipeline robot design.

References

1. Zhu HZ, Yan Y, Lei YJ (2021) Study on the division method of pipeline robot module. In: 2021 international conference on computer network, electronic and automation (ICCNEA). IEEE, pp 284–288
2. Zheng T, Wang X, Li H, et al. Design of a Robot for inspecting the multishape pipeline systems. IEEE/ASME Trans on Mechatronics, 2022,27(6):4608–4618, IEEE ASME Trans Mechatron
3. Li RQ, He JF (2023) Summary of patent technology of pipeline cleaning robot. Technol Innov Appl 13(24):22–28
4. Kim YW, Kim EY, Jang BT, et al. (2016) Pipe cleaning robot, US9289804
5. Ito F, Kawaguchi T, Yamada Y et al (2019) Development of a peristaltic-movement duct-cleaning robot for application to actual environment - examination of brush type and installation method to improve cleaning efficiency. J Robot Mechatron 31:6a184

6. Jiang MH, Xiong HJ, He CZ (2023) Research on wheeled pipe robot based on fractional order PID control. In: 2023 IEEE 6th information technology, networking, electronic and automation control conference (ITNEC), vol 6. IEEE, pp 1470–1473
7. Bogdan PA, Wheadon J, Klein FB et al (2021) Magnetic tracked robot for internal pipe inspection. In: 2021 European conference on Mobile robots (ECMR). IEEE, pp 1–6

Open Access This chapter is licensed under the terms of the Creative Commons Attribution-NonCommercial-NoDerivatives 4.0 International License (http://creativecommons.org/licenses/by-nc-nd/4.0/), which permits any noncommercial use, sharing, distribution and reproduction in any medium or format, as long as you give appropriate credit to the original author(s) and the source, provide a link to the Creative Commons license and indicate if you modified the licensed material. You do not have permission under this license to share adapted material derived from this chapter or parts of it.

The images or other third party material in this chapter are included in the chapter's Creative Commons license, unless indicated otherwise in a credit line to the material. If material is not included in the chapter's Creative Commons license and your intended use is not permitted by statutory regulation or exceeds the permitted use, you will need to obtain permission directly from the copyright holder.

Design of Automatic Annotation Function for Spacecraft Assembly Information Based on CREO Secondary Development

Yanlei Zhang, Chunsheng Yang, Bin Zhang, and Haohao Liu

Abstract For the problems of low efficiency and non-standardization in the engineering drawing annotation information during the spacecraft assembly process, the function of assembly information annotation is redeveloped on the Creo platform to realize the automatic association of annotation text. By batch-selecting the annotation objects in the model tree or engineering drawing view, the assembly information is displayed and presented in the form of annotations. The annotation content includes the name, code, fastener information, thermal control information of the equipment or attached parts installed on the board. And through a set of intelligent layout algorithms based on the view contour, it is ensured that the automatically generated annotation information is reasonably laid out on the drawing. This study not only effectively solves the pain points of the current spacecraft assembly work, but also provides a useful reference for similar application scenarios in other manufacturing fields.

Keywords Creo · Secondary development · Spacecraft assembly information · Automatic annotation

1 Introduction

Spacecraft generally consists of dozens of subsystems and is a complex electromechanical thermal coupling system. Assembly Integration Test (AIT) for spacecraft is a complex and systematic task. As a necessary prerequisite for the correct, efficient, and collaborative operation of this complex system, the correct formulation of process specifications and precise expression of process language are important research contents in the field of spacecraft development. In the field of spacecraft development in China, spacecraft development enterprises generally complete the three-dimensional design of spacecraft by the overall design department alone or in

Y. Zhang (✉) · C. Yang · B. Zhang · H. Liu
Beijing Institute of Aerospace Environmental Engineering, Beijing, China

© The Author(s) 2026

Z. Zhang, P. Mercorelli (eds.), *Mechanical Design and Simulation: Insights and Innovations*, Lecture Notes in Mechanical Engineering,
https://doi.org/10.1007/978-981-95-2405-1_103

collaboration with the manufacturing integration department. The three-dimensional design model of the spacecraft is issued to the manufacturing integration department [1].

At present, during the assembly process of spacecraft, the three-dimensional model provides the installation location and R (Reference) point information of equipment, and the supporting documentation provides fasteners, thermal control information, and other special requirements. Due to the limitations of spacecraft assembly scenarios, two-dimensional drawings are still needed as important auxiliary support to better complete spacecraft assembly. Moreover, looking at the current application status of manufacturing and design departments both domestically and internationally, it can be seen that two-dimensional drawing is still the most commonly used form of expression in the product design process [2].

To make two-dimensional diagrams in the traditional mode, designers manually identify the installation information of assembly parts one by one. For spacecraft, there are generally hundreds of devices and directly attached parts of the display items. The manual identification method is prone to errors in terms of charging reasons. The method proposed in this paper obtains assembly information annotation information through automated batch preprocessing of input files, and then designs an automatic annotation module through Creo secondary development. It can select equipment or direct components in batches in engineering drawings and associate annotation information to automatically annotate assembly information such as fastener information, thermal control information, R point, coordinate systems, etc.

2 Design Method

2.1 CREO Secondary Development Approach

Creo (Pro/E) is a product of Parametric Technology Corporation (PTC) in the United States. It is currently one of the popular three-dimensional design software systems internationally, widely used in industries such as machinery, electronics, aviation, aerospace, postal and telecommunications, military engineering, and textiles. At present, the advanced language development tools provided by Creo (Pro/E) for Windows mainly include Family Table, UDF, Pro/Program, J-Link, and Creo/Toolkit [3, 4]. J-Link is a Java based secondary development tool that comes with Creo (Pro/E), featuring object-oriented and platform independent features. Users can add functions to Creo (Pro/E) software through Java programming.

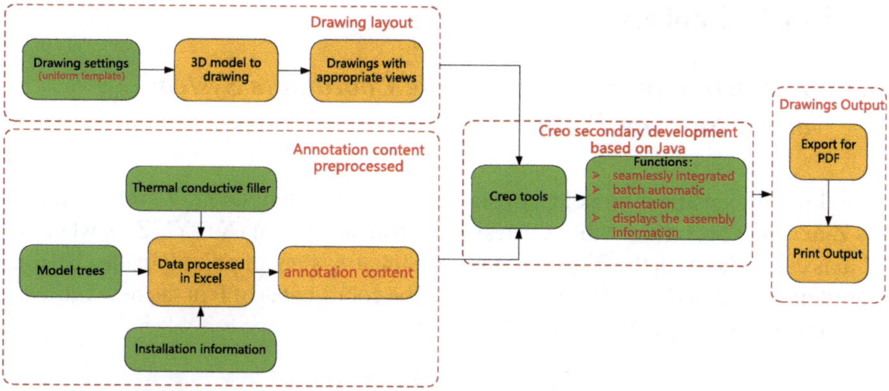

Fig. 1 Automatic annotation module implementation path

2.2 *Implementation Approach of Automatic Annotation Module*

The final output of the assembly drawing book requires completing modules such as engineering drawing layout, pre-processing of annotation information, secondary development of annotation modules, and output of the drawing book. The specific steps are shown in Fig. 1.

The content of the annotation comes from preprocessed text information, and the origin point of the annotation is the coordinate system origin of the selected model. The outlet point for installing the reference hole is the origin of the reference coordinate system named "RSYS" in the selected model. The text information is obtained by comprehensive preprocessing of input files from upstream departments, and the text file names are named after the code of the model, which is automatically associated through secondary development. In the engineering drawing, the lead out point should be in the view coordinate system, so the origin of the coordinate system in the model needs to be converted between the model matrix and the view matrix before it can be used as the origin of the lead out line.

The automatic annotation module is the most critical and the focus of this secondary development, it need to achieving the following basic functions:

1. The automatic annotation module should be seamlessly integrated with the Creo platform, allowing designers to quickly adjust and annotate view projections, view size ratios, assembly information, page information, etc. when using Creo for engineering drawing production.
2. The automatic annotation module should have the function of batch automatic annotation, which automatically associates and displays the assembly information of the models to be annotated by selecting them in batches. It is possible to label all direct components of the same type by selecting some direct components on a certain board.

3 Key Technology

3.1 Extraction of the Origin of the Coordinate System as the Starting Point

The three-dimensional model Cartesian coordinate system (X, Y, Z) rotates around its Z-axis θ to become another Cartesian coordinate system (X ', Y', Z '), where the Z-axis coincides with the Z' axis. If the coordinate system of point P in the Cartesian coordinate system (X, Y, Z) is (x, y, z), then the coordinates of P in the new Cartesian coordinate system (X, Y, Z) are (x ', y', z '):

$$\begin{bmatrix} x' \\ y' \\ z' \\ 1 \end{bmatrix} \begin{bmatrix} cos\theta & -sin\theta & 0 & 0 \\ sin\theta & cos\theta & 0 & 0 \\ 0 & 0 & 1 & 0 \\ 0 & 0 & 0 & 1 \end{bmatrix} \cdot \begin{bmatrix} x \\ y \\ z \\ 1 \end{bmatrix} \tag{1}$$

If the new Cartesian coordinate system (X ', Y', Z ') is θformed by rotating the Cartesian coordinate system (X, Y, Z) around the X-axis, then the corresponding coordinate change formula is:

$$\begin{bmatrix} x' \\ y' \\ z' \\ 1 \end{bmatrix} \begin{bmatrix} 1 & 0 & 0 & 0 \\ 0 & cos\theta & -sin\theta & 0 \\ 0 & sin\theta & cos\theta & 0 \\ 0 & 0 & 0 & 1 \end{bmatrix} \cdot \begin{bmatrix} x \\ y \\ z \\ 1 \end{bmatrix} \tag{2}$$

The θcorresponding coordinate change formula for the coordinate system formed by rotating around the Y-axis is:

$$\begin{bmatrix} x' \\ y' \\ z' \\ 1 \end{bmatrix} \begin{bmatrix} cos\theta & 0 & sin\theta & 0 \\ 0 & 1 & 0 & 0 \\ -sin\theta & 0 & cos\theta & 0 \\ 0 & 0 & 0 & 1 \end{bmatrix} \cdot \begin{bmatrix} x \\ y \\ z \\ 1 \end{bmatrix} \tag{3}$$

The three-dimensional rotation matrix is a key tool for representing coordinate transformations in three-dimensional space, covering basic geometric transformations such as translation, scaling, and rotation around coordinate axes. In practical applications, these transformations are commonly used in fields such as graphics, robotics, and aviation to describe and control the motion and position of objects in three-dimensional space (see Fig. 2).

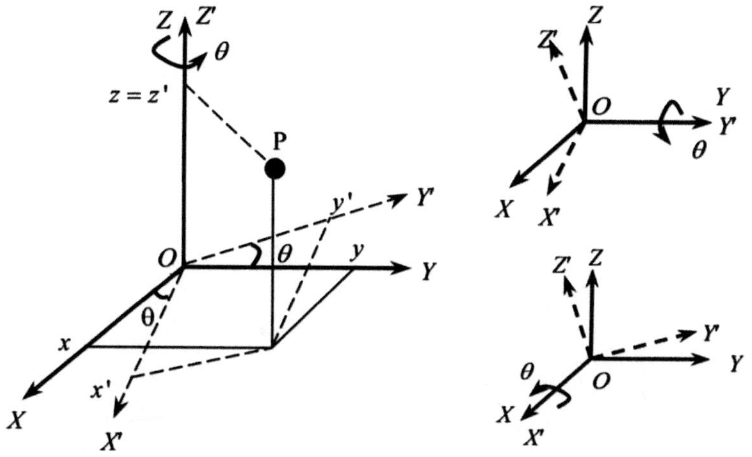

Fig. 2 Three-dimensional spatial coordinate rotation

In Creo, the solid coordinate system is a three-dimensional Cartesian coordinate system used to describe the geometric shape of Creo solid models. In components, the solid coordinate system describes the geometric shapes of surfaces and boundaries. In an assembly, the solid coordinate system also describes the position and orientation of the assembly members. You can visualize entity coordinate systems in Creo applications by creating a data coordinate system with a 'default' option. Creo Object TOOLKIT Java uses a solid coordinate system to manipulate all methods of viewing geometric shapes and most methods of drawing three-dimensional graphics. The drawing view coordinate system is used to describe the position of entities in the drawing view.

The CoordSystem interface provided in the Creo Java API tool, with the method GetCoordSys(), can obtain a Transform three-dimensional coordinate system interface whose data includes transformation matrix, origin, axis, and other information. The GetOrigin() method of the Transform three-dimensional interface can obtain the origin of the desired coordinate system [5]. At this point, the three-dimensional coordinates of the origin are data in the solid coordinate system.

In the same tool, the model coordinate system transformation matrix can be obtained based on the component paths in the assembly model where the model is located. Specifically, the ComponentPath interface is first used to determine and locate the components in the assembly, and then the GetTransform() method of this interface is used to obtain the coordinate system interface from the model to the assembly root model. After two coordinate transformations, the conversion of point coordinates from the solid coordinate system to the drawing view coordinate system can be completed (see Fig. 3).

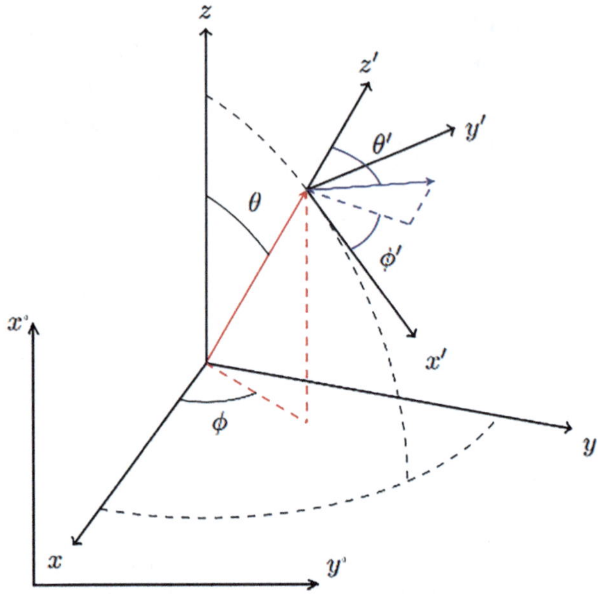

Fig. 3 The transformation from the entity coordinate system to the view coordinate system

3.2 Intelligent Layout of Annotation Information

In order to avoid the chaotic annotation information generated automatically on the drawing, which affects the reading experience, we have designed an intelligent layout algorithm based on view contours. By calculating the relative positional relationship between the center point of the view and the lead out point, the entire view is divided into eight regions, and based on this, the specific placement of annotation information is determined. This innovative solution not only enhances the aesthetics of the drawings, but also greatly facilitates users' access and understanding.

Firstly, determine the outline size of the view, form a rectangle, and determine its center position. So the extraction and center positions can form a ray, with the view rectangle as a reference, extending a rectangle outward to intersect with the ray and determine the placement of annotation information (see Fig. 4).

According to the line relationship between the leading-out point and the center position, the rectangular area can be divided into 8 parts, each with a different angle. In instance development, a PlacingPoint class is used to determine the specific location of the extraction point, with methods including isInRegion1()… isInRegion8(). The annotation placement position is determined by the view center point and the extraction point, and the FreeAttachment class provided by the API is used to capture the extraction point and display the annotation information (see Fig. 5).

The partial code for implementing the determination of the lead-out point area is as follows. Input the origin of the three-dimensional coordinates of the model and the contour of the drawing view, and determine which area the origin falls into.

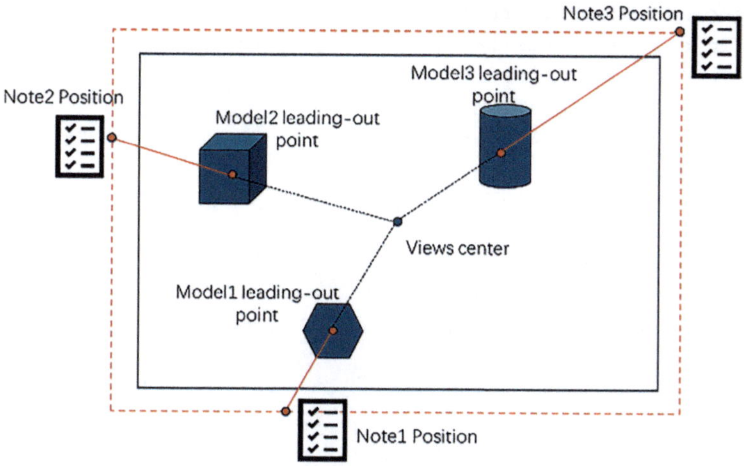

Fig. 4 Annotation intelligent layout schematic diagram

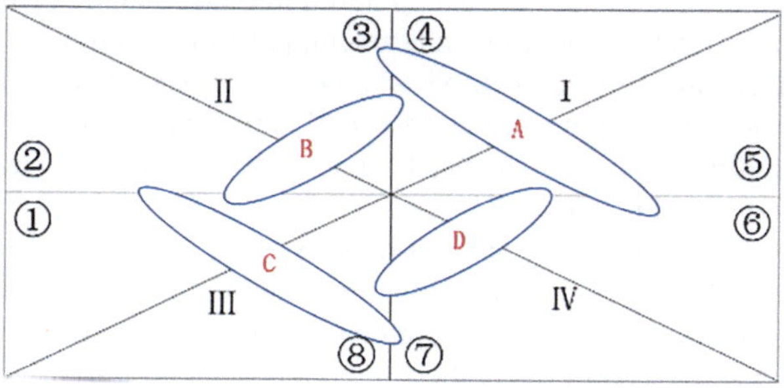

Fig. 5 Divide the area and determine the placement points

```
public class PlacingPoint{.
  Outline3D outline;
  Point3D point;
  Point3D centerPoint;
  Double slope1;
  Double slope2;
  static int GAP = 40;
  public PlacingPoint(Outline3D outline, Point3D point) {}.
  public Boolean isInArea1() {}.
  ...
}
```

The partial code for implementing the display of annotation information and R features is as follows. Taking the selected parts, views, etc. as input, the required annotations can be displayed.

```
public class ShowNoteorReference{.
  public static double REFERENCE_PLACE_OFFSET = 15.0;
  public boolean IsHasReferecne = true;
  public boolean IsHasSys = true;
  public ShowNoteorReference(Model mod, View2D view, Drawing
drawing, Selection sel){}.
  public void showNote() throws FileNotFoundException,
Excep tion{.
    ...isInArea1().
   }
  public void showSymbol(){}.
}
```

4 Module Design and Implementation

The implementation of the automatic annotation module includes functions such as view creation, automatic annotation, alignment, page information, and PDF output. The automatic display of the content view coordinate system in the automatic annotation module, automatic annotation of assembly information or R reference points, customized conformity for single batch annotation, and customized symbols for batch annotation (see Fig. 6).

Among them, the automatic annotation of assembly information and the R-point function respectively complete the identification of assembly information and assembly direction for spacecraft equipment or direct components. Participants in spacecraft work only need to select the components that need to be annotated in the model tree or view of the engineering drawing, and their assembly information will be distributed around the view in the form of arrows. The specific distribution

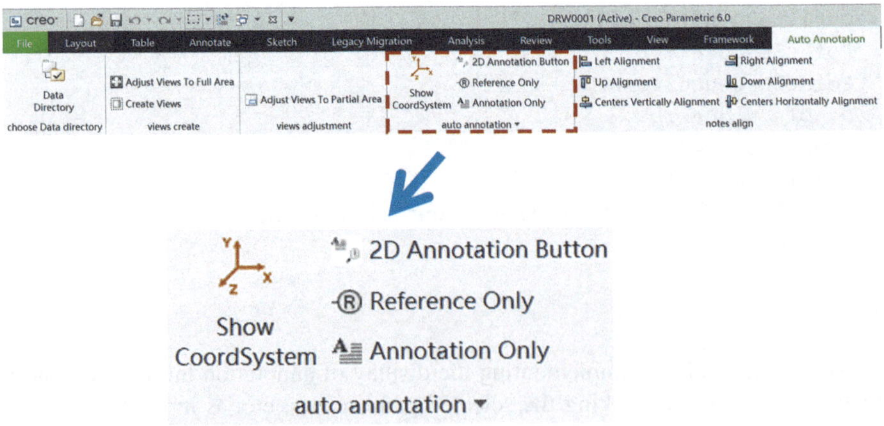

Fig. 6 Automatic annotation module

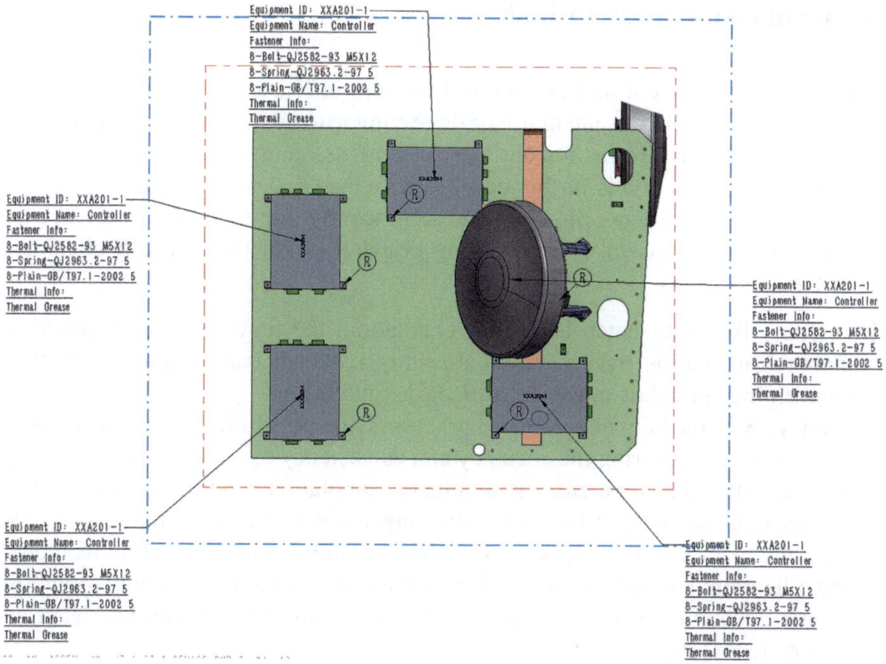

Fig. 7 Application effect diagram

position is related to the outer envelope size of the view, and the placement point is located on a rectangle that expands a certain size beyond the outer envelope of the view (see Fig. 7).

The R point is drawn out in the form of a spherical annotation from the origin of the reference coordinate system named "RSYS" of the equipment or direct components, and is placed by default on the axis 45 ° to the right of the projection point of the coordinate origin on the view coordinate system.

In addition, the module can also achieve automatic display of coordinate system views. By selecting different projection views, designers can display the coordinate system of the view in the lower left corner to represent the current orientation of the view, so that spacecraft practitioners can judge the assembly relationship from the view.

We used this tool to create the assembly manual of the Terrestrial Ecosystem Carbon Inventory Satellite and GF-7 satellite. The overall efficiency (converted into working hours) has been increased by more than five times, and the annotation information is automatically added, avoiding the errors of manual marking.

5 Conclusion and Outlook

On the basis of the existing functions of Creo, this article addresses the problems of low efficiency and inconvenient use of three-dimensional models in the engineering drawing annotation information during spacecraft assembly. The assembly information annotation function is further developed on the Creo platform, achieving automatic association of annotation text and significantly improving the efficiency and standardization of assembly engineering drawing production. Specifically manifested as:

Efficiency: By using batch automatic annotation, the time and cost required for manual annotation have been significantly reduced, making previously time-consuming and labor-intensive work easier and faster.
Accuracy: With the help of automated processes, the possibility of human errors has been reduced, ensuring the accuracy and consistency of annotation information.
Usability: After development, the automatic annotation module seamlessly integrates with the CREO platform, allowing users to easily complete assembly information annotation tasks without the need to learn new operating methods.
Standardization: Unified the expression format of assembly information, promoting communication and collaboration within the team and between upstream and downstream links.

To sum up, the automatic annotation scheme of spacecraft assembly information based on the secondary development of CREO proposed in this study not only effectively solves the pain points existing in the current spacecraft assembly work, but also provides a useful reference for similar application scenarios in other manufacturing fields. In the future, we will continue to explore more possibilities in depth to further optimize and improve the existing functions.

References

1. Mao Z, Liu L, Sun R et al (2023) Research on a measurement process model construction Technology for Spacecraft Assembly. Manufacturing Automation 45(6):39–43
2. Fuyun L, Lidong G, Youzhi J et al (2022) An adaptive adjustment method for engineering drawing dimension annotation. Mach Design Manufact 380(10):186–190. 195
3. Ma Z, Feng X, Li C (2022) Digital design of automatic wiring for electrical cabinets based on the secondary development of Creo. Electr Eng (16):119–122
4. Wang J, Wu R, Qin L (2023) Research on the cold and hot water intelligent design system based on the secondary development of Creo. Comput Appl Softw 40(2):106–111
5. Parametric Technology Corporation (2020) Creo object TOOLKIT Java user's guide

Open Access This chapter is licensed under the terms of the Creative Commons Attribution-NonCommercial-NoDerivatives 4.0 International License (http://creativecommons.org/licenses/by-nc-nd/4.0/), which permits any noncommercial use, sharing, distribution and reproduction in any medium or format, as long as you give appropriate credit to the original author(s) and the source, provide a link to the Creative Commons license and indicate if you modified the licensed material. You do not have permission under this license to share adapted material derived from this chapter or parts of it.

The images or other third party material in this chapter are included in the chapter's Creative Commons license, unless indicated otherwise in a credit line to the material. If material is not included in the chapter's Creative Commons license and your intended use is not permitted by statutory regulation or exceeds the permitted use, you will need to obtain permission directly from the copyright holder.

A Multi-Stage Nested Cylinder Air Blowing Integrated Device for Automatic In-Box Cleaning

Weihua Ye, Yuchao Xu, Qiurong Chen, Yong Cai, Jinyu Chen, and Guiping Wu

Abstract To address the issue of material adhesion at the bottom of the box after dumping powdered materials in the factory, a solution is required, a nested cylinder air blowing integrated device is designed. The telescopic cylinder has a multi-stage nested structure, and combined with the designed top blowing structure, it solves the defect of the long cylinder body of the traditional cylinder and realizes the online compressed air blowing of the shipping box. The overall structure of the device is simple, compact, and highly feasible, and is suitable for the original working environment of the blending process. The project team carried out 3D modeling design of the overall device based on SolidWorks and used Fluent software for nozzle structure design and air blowing feasibility verification. Experiments show that the device can effectively adapt to the production environment of blending, blow away the adhered materials, solve the waste of such materials, and greatly reduce the raw material cost and labor intensity.

Keywords Cylinder · Fluent · Air blowing · Shipping box · Integrated device

1 Introduction

With the country's increasing emphasis on high-quality industrial development, companies are aiming to achieve steady progress, implement lean management, promote full participation in operations, drive continuous innovation, and reduce costs while increasing efficiency. For manufacturing enterprises, controlling costs has become a key factor in achieving high-quality development, especially as raw material prices and labor costs rise. As a crucial part of the production process, this workshop needs to continuously optimize and innovate various production processes to minimize material waste and maximize economic benefits.

W. Ye · Y. Xu (✉) · Q. Chen · Y. Cai · J. Chen · G. Wu
Longyan Tobacco Industrial Co., Ltd., Longyan, Fujian, China
e-mail: xyc23392@fjtic.cn

© The Author(s) 2026

Z. Zhang, P. Mercorelli (eds.), *Mechanical Design and Simulation: Insights and Innovations*, Lecture Notes in Mechanical Engineering,
https://doi.org/10.1007/978-981-95-2405-1_104

Fig. 1 Residual material at the bottom of mixing boxes

Due to factors such as adhesion and gravity, a portion of the material remains stuck to the bottom of the box. After tipping the box, this material remains adhered to the inner bottom, preventing it from being used in normal mixing processes, as shown in Fig. 1. These boxes are subsequently transferred to the box-cleaning process for cleaning, and the cleaned-out material is either scrapped or downgraded as required, resulting in significant waste.

In the industrial sector, material adhesion at the bottom of containers is an extremely common issue. Shuo Li [1] designed an automatic cleaning device. The equipment adopts the innovative structure of "Pipeline single-cavity multi-mode". The whole equipment adopts assembly line layout, which is convenient for structural layout and makes cleaning work in order. Siwen Huang [2] designed and developed a self-rotating cleaning device based on shear nozzles. To enhance the cleaning efficiency of cavitation jets, the system selects an appropriate nozzle structure and designs an efficient cleaning device through erosion experiments. Zhong Chen [3] designed a novel pipe cleaning device with self-rotation functionality. The device is composed of a nozzle head system, a walking system, and a cleaning system, with three-dimensional modeling and assembly performed for these mechanical systems. This design provides new ideas and methods for the efficient operation of cleaning devices and the removal of stubborn dirt from the inner walls of pipelines. Currently, there is limited research on in-line cleaning of boxes after mixing. Therefore, designing a device capable of reaching the bottom of the box and thoroughly cleaning the adhered material is particularly important.

2 System Structure

2.1 Traditional Cylinder Mechanism

Based on the structure of the blending box turnover machine, the installation position of the blowing device, and the production quality requirements (which prevent direct contact with the material), this paper designs a simple and compact multi-stage nested cylinder air-blowing integrated device using SolidWorks software.

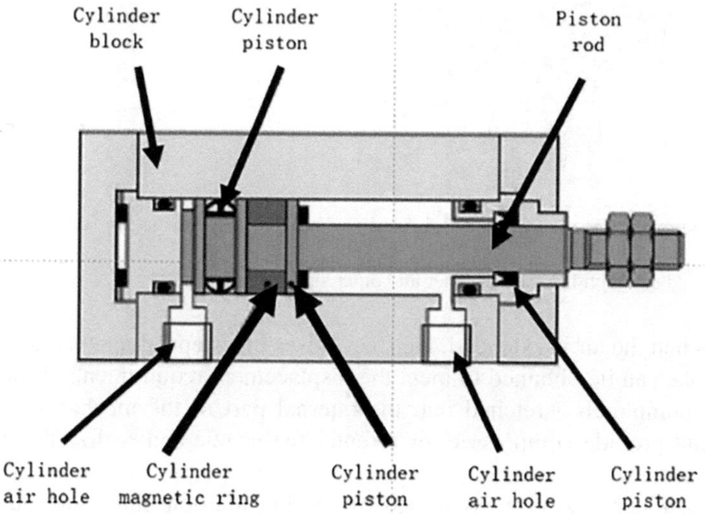

Fig. 2 Working principle diagram of the ordinary cylinder

Additionally, nozzle structure design and air-blowing feasibility are verified using Fluent software. The device uses compressed air to control the telescoping and blowing actions of the multi-stage cylinder, extending into the bottom of the shipping box to remove adhered material. It works in conjunction with the designed positioning, steering, and fixing frame mechanisms to achieve online cleaning of the adhered material.

Taking a cylinder as an example, its working principle is shown in Fig. 2 [4]. Through the mutual cooperation of air intake and exhaust between air hole 1 and air hole 2, a pressure difference is generated at both ends of the piston inside the cylinder, thereby realizing the extension and retraction of the cylinder. The size of the blending shipping box is 1200 mm × 1200 mm × 1200 mm. If a traditional cylinder is used as the extension mechanism of the device, there are problems such as a long cylinder body, large space occupation, and the need to install an air source pipe on the side of the cylinder body, resulting in many pipelines, a relatively complex overall structure, easy entanglement and hanging of cut material, and not suitable for the working environment requirements of the blending process.

2.2 Multi-Stage Nested Cylinder Air Blowing Integrated Device

A multi-stage nested air cylinder air blowing integrated device is designed. Different from the traditional cylinder in mechanical structure, it is a cylinder that converts the piston rod into the shape of multi-stage short sleeve. In the original state, multiple sleeves are nested together to reduce the volume occupied by the cylinder

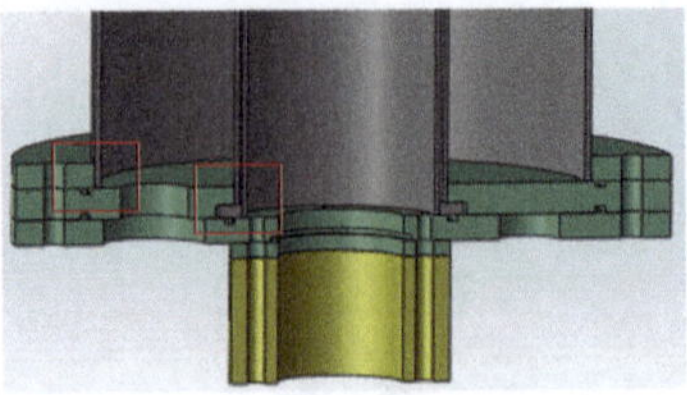

(a)The lower ends are welded and fixed

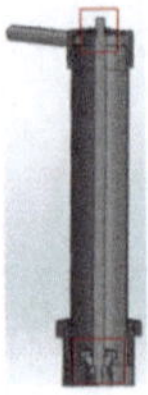

(b)The top ends are in fixed connection

Fig. 3 Fixed connection mode of inner and outer sleeve shafts

block. When the air is extended, the sleeve rises in a stepped manner, so that a very long stroke can be obtained to meet the displacement requirements. When the cylinder is completely stretched out, the internal part of the mechanism is hollow, which can provide compressed air to enter the nozzle and realize the ejection of compressed air.

The cylinder device is assembled by nesting multiple inner and outer hollow shafts with successively decreasing radii. The top of the outermost hollow shaft is connected to a hollow crossbar with a nozzle structure as the air path blowing structure. The lower ends of the outermost inner and outer hollow shafts are welded and fixed to the base respectively, and the tops of the innermost inner and outer hollow shafts are fixedly connected by a double-headed screw, as shown in Fig. 3. There are three air paths left below the cylinder, namely the blowing air path, the cylinder extension air path, and the cylinder retraction air path, as shown in Fig. 4. Through the air intake and exhaust of compressed air, the extension, blowing, and retraction of the overall device are realized.

Cylinder Extension Principle The schematic diagram is shown in Fig. 5. The extension air path is ventilated, and the compressed air enters the cylinder body and acts on the bottom surface of the inner sleeve shaft. According to F = PS, under a constant pressure, the force on each bottom surface is proportional to the area size. Since the radius difference of the inner hollow sleeve shafts is equal, the larger the radius, the larger the bottom surface area and the greater the force (the contact surface is lubricated with graphite, and the frictional resistance is approximately negligible). The greater the force, the faster the extension rate. Eventually, each shaft extends step by step from the outside to the inside, and the stop position is mechanically limited by the inner convex ring at the top of each stepped shaft.

Air Path Blowing When the cylinder rises to the limit position, the blowing air path is ventilated, and the compressed air is ejected from the top blowing nozzle to blow the adhered cut material at the bottom of the shipping box, as shown in Fig. 6.

Cylinder Retraction The extension and blowing air paths are cut off, and the retraction air path is ventilated, as shown in Fig. 7. The compressed air reaches the upper surface of the inner shaft top through the annular air path reserved on the surface between the inner sleeve shafts and uses the pressure to push each shaft back to the initial position.

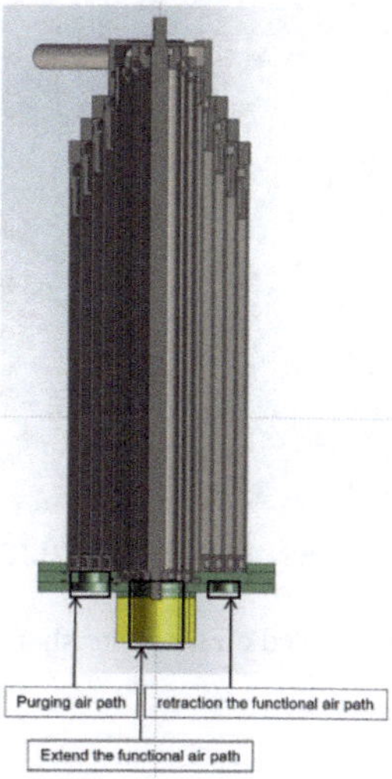

(a)The initial position

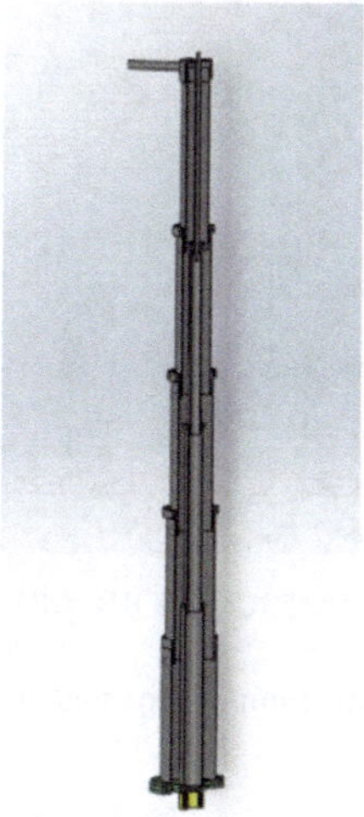

(b)The extended limit position

Fig. 4 Schematic diagram of the multi-stage nested cylinder air blowing integrated device

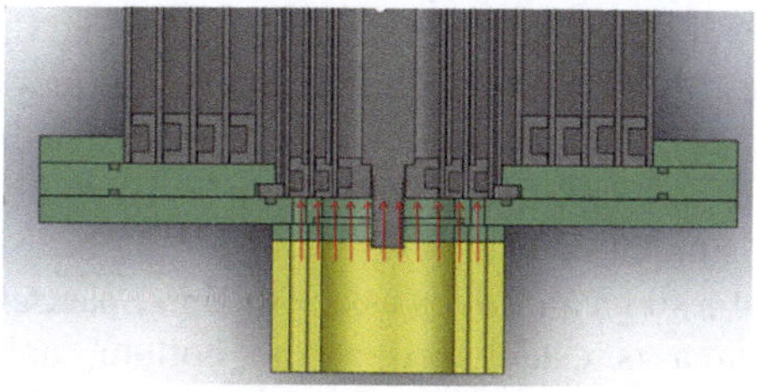

(a)The initial position

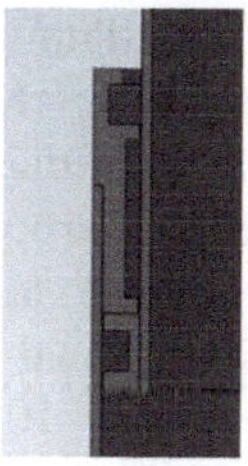

(b)The extended limit position

Fig. 5 Schematic diagram of shaft limit

Fig. 6 The blowing nozzle structure

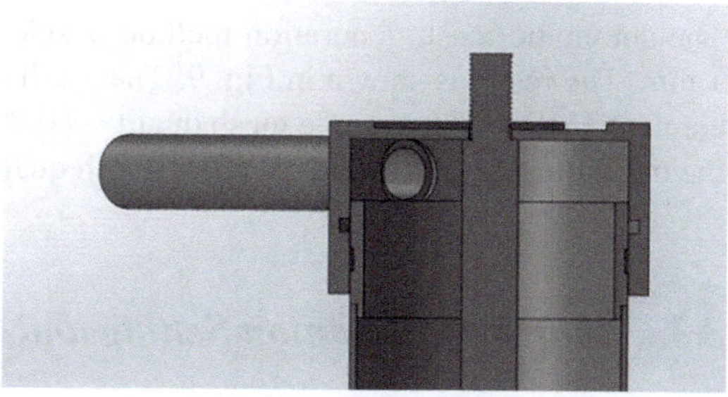

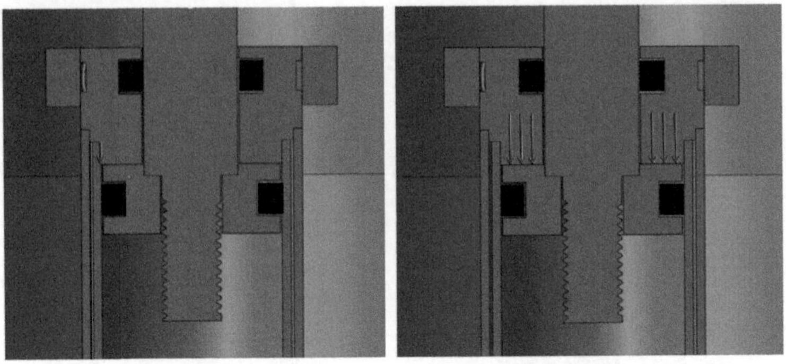

(a)the retraction air path (b)Compressed Air Acting

Fig. 7 Schematic diagram of the annular air path reserved on the inner shaft surface

3 Feasibility Verification

3.1 Establishment of the Geometric Model of a Single Nozzle

A single nozzle opening is designed as a cylindrical shape with a diameter of 10 mm and a height of 10 mm, and the supply compressed air pressure is 0.45 Mpa. Based on the Fluent finite element software [5], the compressed air blowing effect of a single nozzle is simulated by the finite element method. Since a single nozzle has a centrosymmetric structure, the nozzle is simplified to a 10 mm × 10 mm two-dimensional planar structure for fluid simulation analysis.

A two-dimensional planar model of a single nozzle is established in SpaceClaim [6]. A 10 mm × 10 mm nozzle area is established at the bottom, and an 110 mm × 200 mm compressed air flow area is established above, as shown in Fig. 8.

3.2 Mesh Generation

The automatic mesh generation method is selected, and the element size is set to 1 mm. The result is shown in Fig. 9. The mesh quality is analyzed by selecting the element quality. The average mesh quality is 0.99746. The closer it is to 1, the better the mesh quality, indicating excellent mesh quality.

3.3 Boundary Condition Setting and Simulation Analysis

The multiphase flow (VOF) and *k* - *e* viscous models are selected, and the materials are air and steel [7]. The simulation results of a single nozzle are shown in Fig. 10. The lower end of the nozzle is set as the inlet end, and a pressure inlet is used with

Fig. 8 Geometric model of a single nozzle

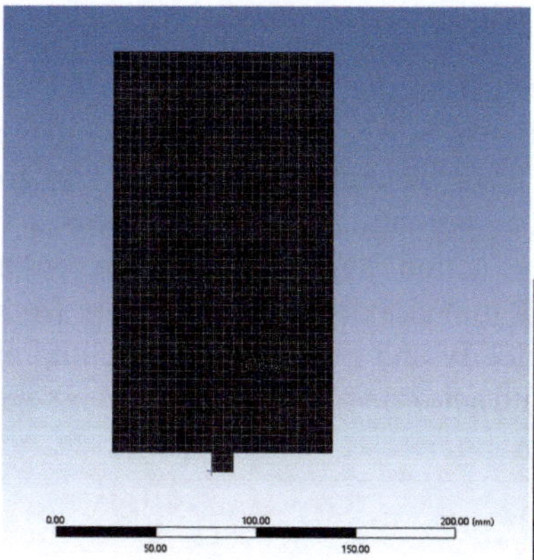

Quality	
Target Skewness	Default(0.900000)
Smoothing	Medium
Grid Metric Criteria	Element Quality
Minimum	0.56013
Maximum	0.99947
Average	0.99746
Standard Deviation	1.54E-02

Fig. 9 Mesh generation

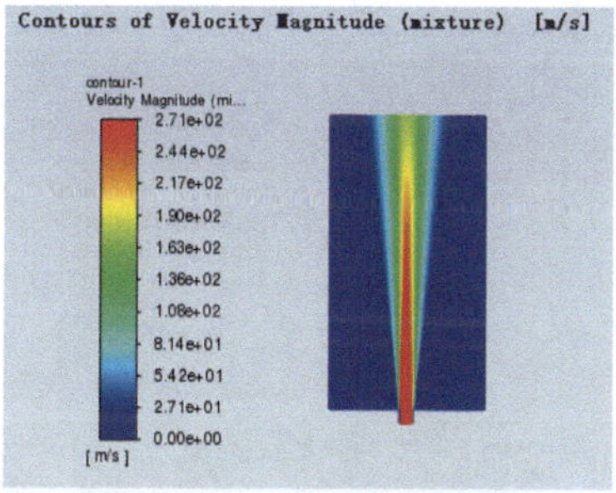

(a)Velocity contour map

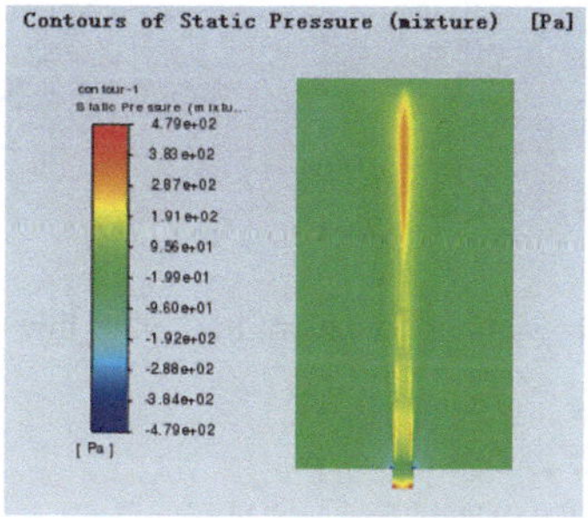

(b)Pressure contour map position

Fig. 10 Simulation results of a single nozzle

a pressure value of 0.45 Mpa [8]. The time step is set to 0.01 s, the number of steps is 500, and the calculation residual is 0.0001 for simulation calculation. The simulation results are shown in Fig. 11. It can be seen from the figure that after the air is ejected from the nozzle outlet, it is distributed in a conical shape, the velocity gradually decreases, and the effective range gradually expands.

3.4 Experimental Verification

As shown in Fig. 12, the experimental personnel hold a single nozzle pipe, supply compressed air pressure of 0.45 Mpa, and blow at a distance of 150 mm from the bottom of the box. The results show that under the above conditions, the adhered cut material at the bottom of the box can be blown clean [10].

3.5 Multi-Nozzle Structure Design

A horizontal line5 is inserted at a distance of 150 mm from the nozzle outlet, and the simulation data at a distance of 150 mm from the nozzle outlet in the single nozzle simulation analysis is retrieved as shown in Fig. 13. It can be seen from the figure that within a range of about ±20 mm from the center line, the velocity gradually decreases from (210 ~ 0)m/s and the pressure gradually decreases from (300 ~ 0)pa outward. Therefore, the effective action range of the compressed air is approximately a circle with a diameter of 40 mm at a distance of 150 mm from the nozzle outlet. Considering the effective velocity and pressure values comprehensively, a horizontal arrangement structure with a nozzle spacing of 30 mm is selected to achieve a large-area blowing [9].

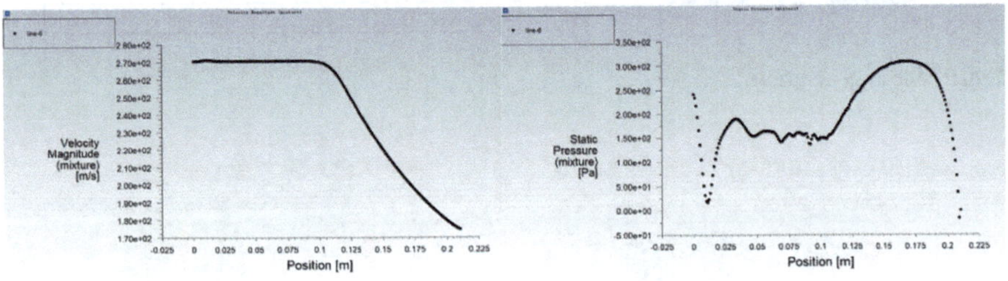

(a)Velocity curve (b)Pressure curve

Fig. 11 Simulation data of the center line

Fig. 12 Single nozzle pipe blowing experiment

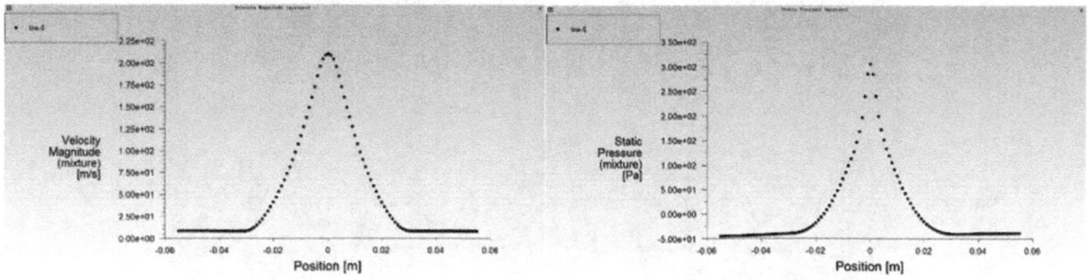

(a)Velocity curve (b)Pressure curve

Fig. 13 Simulation data at 150 mm from nozzle outlet

Fig. 14 Three-dimensional model of the multi-nozzle structure

The 3D model of the multi-nozzle structure is shown in Fig. 14.

4 Test Results

The physical object and on-site installation of the multi-stage nested cylinder air blowing integrated device are shown in Fig. 15. A structure with two cylinder air blowing integrated devices installed in parallel is adopted, and the two devices are fixedly connected by a crossbar to increase the stability of the mechanism and increase the air blowing area and improve the blowing efficiency. Under the condition that the test supply compressed air pressure is 0.4. omparison before and after blowing is shown in Fig. 16.

The blowing effect of the device was tested and statistically analyzed. The residual amount of cut material in the shipping boxes of the same brand before and after the installation of the device was compared. The results are shown in the following Table 1.

The test results indicate that the average residual amount of cut material in the blending shipping box decreases from 0.68 kg to 0.02 kg, with a reduction in the amount of adhered cut material exceeding 97%. The blowing effect of the nozzle device is significant, solving the problem of cut material adhesion at the bottom of the blending shipping box.

Fig. 15 Installation diagram of the nested cylinder air blowing integrated device

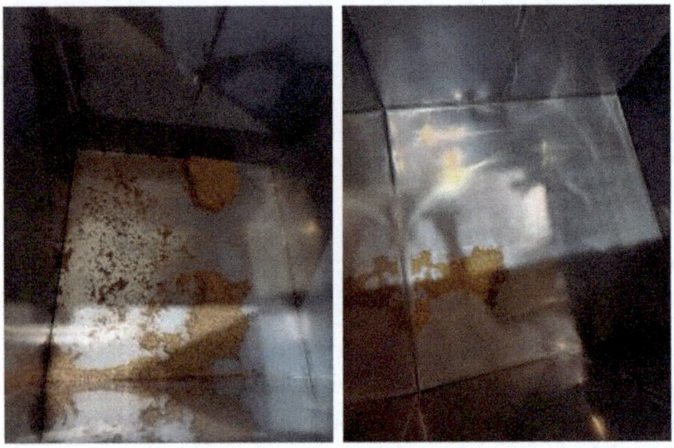

(a)Before blowing (b)After blowing

Fig. 16 Comparison of residual cut material at the bottom of the shipping box before and after blowing

Table 1 Test results

Tests	Residual amount before installation (kg)	Residual amount after installation (kg)
Box 1	0.7	0.02
Box 2	0.9	0.01
Box 3	0.6	0.02
Box 4	0.5	0.01
Box 5	0.7	0.04
Average value	0.68	0.02

5 Conclusion

In this paper, aiming at the problem of cut material residue at the bottom of the blending shipping box, a multi-stage nested cylinder air blowing integrated device is designed and manufactured based on SolidWorks. The arrangement structure of the nozzle and the feasibility of air blowing are designed and verified based on Fluent software. The problem of cut material residue is significantly solved, which effectively reduces the time and labor cost of offline cleaning of the shipping box, greatly reduces the waste of cut material raw materials, improves economic benefits, and has the value of promotion in the material industry.

References

1. Li S, Zhang Y, Liu S (2023) Design and research of a new type of compound automatic cleaning equipment for vegetables. J Eng Syst 1(2)
2. Huang S, Huang J, He K (2024) Research on erosion effect of various submerged cavitating jet nozzles and design of self-rotating cleaning device. Appl Sci 14(4)
3. Chen Z et al (2020) Research and design of pipe cleaning device with self-rotation. Recent Pat Mech Eng 13
4. Araki K et al (2011) Structural optimization of a motion and force control cylinder. In: Proceedings of the JFPS international symposium on fluid power 1996, vol 3, pp 337–342
5. Wawrzak K, Stempka J, Tyliszczak A (2024) Numerical analysis of the mixing process in variable density jets emanating from polygonal nozzles. Int J Heat Fluid Flow 110:109614–109614
6. Boroomand MR, Neyshabouri SAAS, Aghajanloo K (2007) Numerical simulation of sediment transport and scouring by an offset jet. Can J Civ Eng 34(10):1267–1275
7. Синило КВ (2009) Comparison of modeling jet exhaust on a complex model of science and application fluent 6.3. Proc Nat Aviat Univ 40(3):253–261
8. Chen JY, Fan XQ, Liu Z (2011) Fluent based numerical analysis of eliminating ultra-limit gas in upper corner by using rotary jet fan. Adv Mater Res 1169(201–203):2212–2215
9. Chun DX, Jian H (2011) Numerical simulation on flow structure of a steam-jet pump influenced by primary nozzle geometries. Appl Mech Mater 130–134:1703–1707
10. Jin Hou et al (2024) Effects of structural parameters of pressure-swirl nozzle on atomization and dust removal characteristics. Environ Sci Pollut Res Int 31(52):1–13

Open Access This chapter is licensed under the terms of the Creative Commons Attribution-NonCommercial-NoDerivatives 4.0 International License (http://creativecommons.org/licenses/by-nc-nd/4.0/), which permits any noncommercial use, sharing, distribution and reproduction in any medium or format, as long as you give appropriate credit to the original author(s) and the source, provide a link to the Creative Commons license and indicate if you modified the licensed material. You do not have permission under this license to share adapted material derived from this chapter or parts of it.

The images or other third party material in this chapter are included in the chapter's Creative Commons license, unless indicated otherwise in a credit line to the material. If material is not included in the chapter's Creative Commons license and your intended use is not permitted by statutory regulation or exceeds the permitted use, you will need to obtain permission directly from the copyright holder.

Study on Grouting Device of Automatic Curb Caulking Machine

Xuan Sun, Zhiyuan Liu, Lufeng Zhang, and Zhaoyang Ma

Abstract A grouting device for automatic joint filling machine is developed to improve the automation level and quality of curb joint filling. The structural characteristics of the curb joints are analyzed through field investigation, and the width and height errors of the joints are determined. Based on this, a grouting device combining the crank slider mechanism and parallel quadrilateral mechanism is designed to realize the clamping of the curb and the uniform filling of cement. In the grouting process control, the innovative method of combining time and pressure is adopted to estimate the filling time according to the flow rate of the screw conveyor, and the pressure gauge at the grouting mouth is used to monitor the pressure change to ensure the accuracy of grouting. Through theoretical analysis, a mathematical model is established, and the force of the mechanism is analyzed deeply on the basis of considering the deformation force of flexible materials. The geometric dimensions of the mechanism are studied in working and non-working states. Using MATLAB simulation analysis to determine reasonable geometric parameters to balance the force. Complete the structural design and functional verification, and meet the engineering requirements of automatic construction of curb joint filling.

Keywords Curb · Grouting device · Crank slider mechanism · Parallelogram mechanism · MATLAB simulation analysis

1 Introduction

In the construction and maintenance of expressways, the curb is an important part, and the treatment quality of its joints directly affects the overall performance and service life of the road [1]. If the curb joints are not effectively filled, problems such as rainwater penetration and subgrade erosion may be caused, thus affecting the

X. Sun (✉) · Z. Liu · L. Zhang · Z. Ma
School of Mechanical Engineering, University of Jinan, Jinan, China
e-mail: me_sunx@ujn.edu.cn

© The Author(s) 2026

Z. Zhang, P. Mercorelli (eds.), *Mechanical Design and Simulation: Insights and Innovations*, Lecture Notes in Mechanical Engineering,
https://doi.org/10.1007/978-981-95-2405-1_105

stability of the road structure [2]. In the traditional construction method, after the curb assembly is completed, the construction personnel apply mortar to the cracks on both sides of the exterior. This construction method is easy to cause false cracks and incomplete cracks, which greatly reduces the stability of the two curbs [8]. With the continuous improvement of quality and efficiency requirements in engineering construction, developing an efficient and accurate curb joint filling device is the key to realize automatic joint filling.

Through the in-depth investigation of the curb structure and gap characteristics at the construction site, it is found that the gap width of the curb is within 15 ± 5 mm, and there is a ±2 mm error in the height and front and back direction. In view of this, it is of great significance to design a grouting device that can adapt to the characteristics of these gaps. Through reasonable mechanical structure design, such as the combination of crank slider and parallelogram mechanism [3], the sealing and clamping around the curb joints can be realized, which can flexibly cope with the curb gaps and front and back errors of different widths and heights compared with the traditional mechanical structure. The side elastic material is driven by the slide displacement to closely adapt to the gap change, improving the adaptability and accuracy in practical applications [5]. In terms of grouting process control, combined with time control and pressure control method, the filling time of screw conveyor flow is estimated by calculating, and the pressure gauge is set at the connection between the high-pressure conveying hose and the filling port to monitor the pressure change, so as to accurately control the grouting process, ensure the uniform filling of cement and improve the filling quality [6–8]. This paper elaborates the design process of the device, including theoretical analysis, mathematical model construction, force calculation and structural optimization, aiming to provide reliable technical support for curb joint filling construction, improve construction quality and promote the automated development of road engineering construction [4].

2 Research on Grouting Device Scheme

2.1 Curb Gap Research

During construction, curb stones are usually placed directly on the ground with no coverings on both sides of the joints [9]. The main purpose of the grouting device is to seal and clamp the joints around the curb stones and fill the gaps with cement. Therefore, it is necessary to study the shape of the gaps between the curb stones. As shown in Fig. 1, based on the on-site measurement data, the width of the gap between two curb stones is usually 15 ± 5 mm, the height error is ±2 mm, and the error in the front and back direction is also ±2 mm. Thus, when designing the grouting device, the standard gap size cannot be simply taken as the quantitative basis. The redundancy caused by the error should be considered to ensure the effectiveness of the sealing around the curb stone joints.

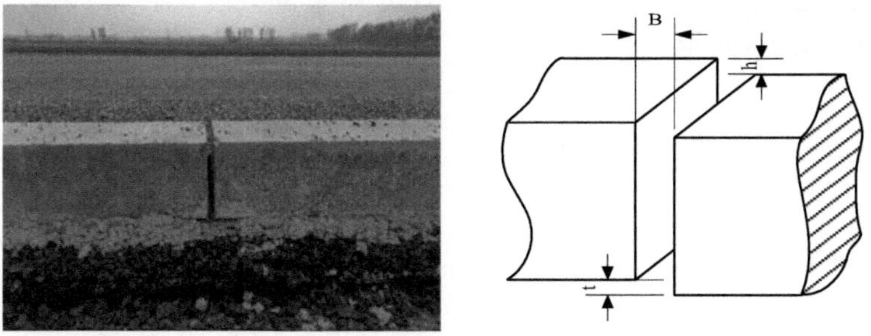

Fig. 1 Morphology of the joint of curbstone

2.2 Analysis of the Design Principles of Grouting Devices

Suppose that the descent of the top material is the sole power source, and the opening and closing of the elastic materials on both sides are achieved through mechanical structures [10]. Specifically, a combined design of crank-slider mechanism and parallelogram mechanism is adopted (as shown in Fig. 2). When the slider moves downward by x, the elastic materials on the sides move inward by y. When the dimensions of each part are reasonable, the function of sealing and clamping the joints around the curbstone can be achieved.

To design a reasonable mechanical structure, it is necessary to conduct a principle analysis of the mechanism and make general assumptions: ignore the gravity and friction of the parts, treat the parts as rigid bodies, and linearly process the strain-stress relationship of the flexible materials [5]. Among them, the selection of flexible materials is an important part of the grouting device design, but this study focuses on structural design, so it is weakened. The model is planned to use rubber as the flexible material, with a Young's modulus E = 7.8 MPa. Although rubber is a super-elastic material and its stress-strain relationship is nonlinear, the linear assumption here provides a direction for subsequent model optimization.

2.3 Mechanism Stress Analysis

According to the force analysis Fig. 3, the whole institution is taken as the research object for force analysis. In Fig. 3, F_{side} is the deformation force generated by the deformation of the flexible material on the side, and $F_{side} = AE\varepsilon$ can be obtained according to the stress-strain relation, where A is the area of the pressurized material on the side and ε is the strain of the material on the side. When the value of the deformation force F_{side} and the geometric size of each part are known, the force at the remaining points can be calculated.

Fig. 2 Mechanical schematic diagram of the device

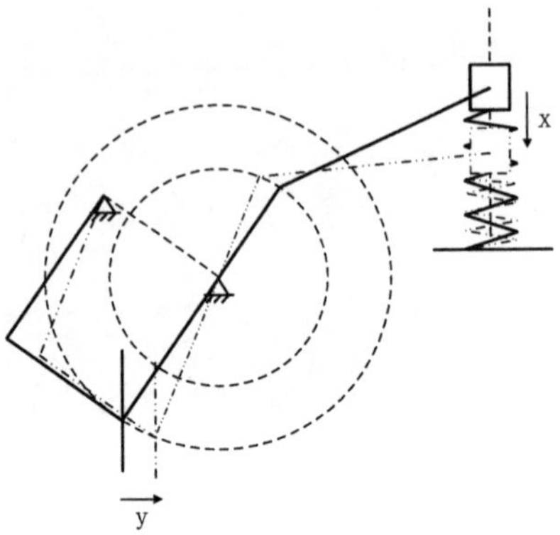

Fig. 3 Force analysis diagram of grouting device

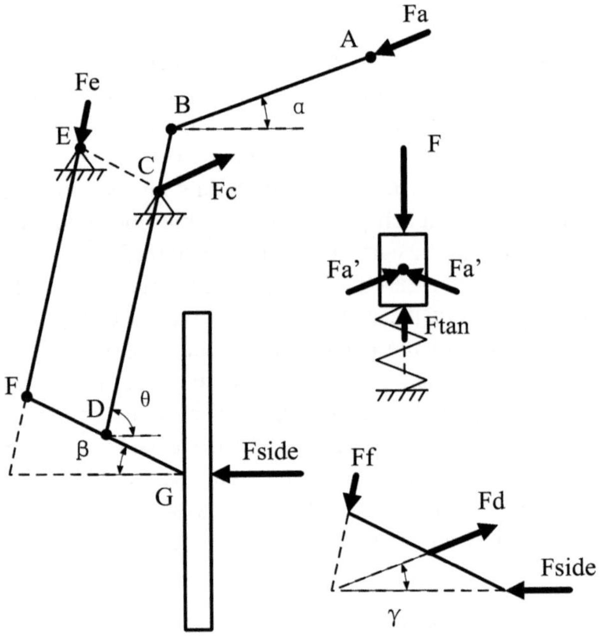

$$F_a = \frac{\sin(\pi - \beta) l_{CD}}{\sin(\beta - \alpha) l_{BC}} F_{side} \quad (1)$$

$$\tau = \pi - (\theta + \beta) \quad (2)$$

$$\gamma = \arcsin\left(\frac{l_{DG}\sin\theta}{\sqrt{l_{DG}^{2}+\left(\frac{\sin\tau}{\sin\beta}l_{FG}\right)^{2}-2l_{DG}\sin\theta cos\theta\frac{\sin\tau}{\sin\beta}l_{FG}}}\right) \tag{3}$$

$$F_e = F_f = \frac{\sin\gamma}{\sin(\beta-\gamma)}F_{side} \tag{4}$$

$$Fy_e = F_e\sin\beta F_a = F_b = \frac{\sin(\pi-\beta)l_{CD}}{\sin(\beta-\alpha)l_{BC}}F_{side} \tag{5}$$

$$F_d = \frac{\sin(\beta-\alpha)l_{BC}}{\sin(\beta-\gamma)l_{CD}}F_a \tag{6}$$

$$Fy_c = F_b\sin\alpha + F_d\sin(\gamma) \tag{7}$$

The stiffness coefficient of the spring is calculated as follows:

$$k = \frac{Gd^4}{8N_cD_m^{\ 3}} \tag{8}$$

The spring material is 65 Mn, the wire diameter d is 5 mm, the outer diameter Dm is 40 mm, and the number of turns Nc is 5 turns. According to Hooke's law, $F_{tan} = k\Delta x$ can be obtained, where Δx is the spring-shaped variable, that is, the displacement of the slide block.

2.4 Theoretical Analysis of Grouting Device

The structure of the curb is shown in Fig. 4. Due to the symmetrical structure of the curb, it is relatively simple to carry out theoretical analysis, so the mathematical model of the grouting device is established in this paper.

In the previous section, only the force of the mechanism was analyzed without considering the specific geometric size. Based on the requirement that the curb gap should be wrapped in the device, the geometric structure of the size shown in Fig. 5 was drawn.

When in the situation in Fig. 5, the top flexible material and the side flexible material are deformed by contact with the curb, and the curb joint is wrapped in it. The parameters A and B are known quantities related to the shape of the curb, the Angle values α_0, β_0, θ are set as known quantities, and the length values $H1$, $L1$, m, $l2$, $l4$, $t1$, $t3$ are set as known quantities, then the remaining unknowns can be obtained by the formula:

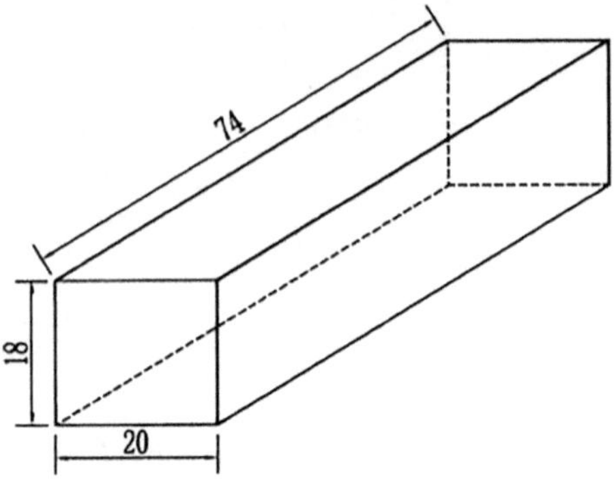

Fig. 4 Size diagram of curb structure

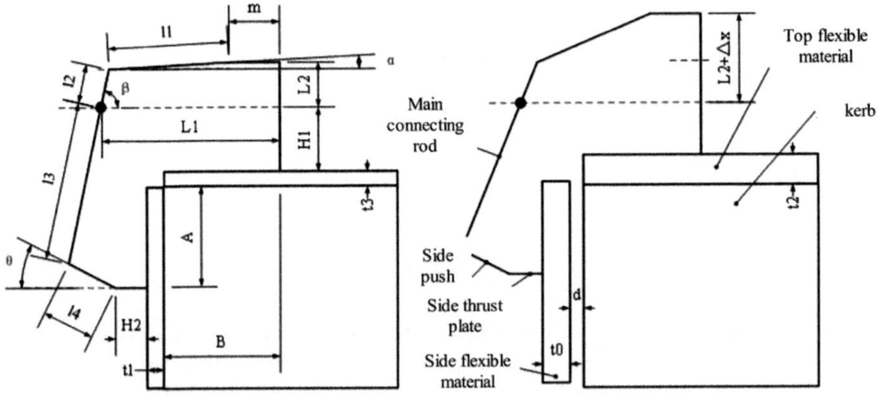

(a) When clamping the curb (b) when the curb is not clamped

Fig. 5 Geometric structure of grouting device

$$l1 = \frac{L1 - m - l2\cos\beta_0}{\cos\alpha_0} \tag{9}$$

$$l3 = \frac{A + t3 + H1 + h - l4\sin\theta}{\sin\beta_0} \tag{10}$$

$$L2 = l1\sin\alpha_0 + l2\sin\beta_0 \tag{11}$$

$$H2 = L1 - B - t1 - l4\cos\theta + l3\cos\beta_0 \tag{12}$$

The whole length value of the grouting device can be obtained by the above formula. When the grout device is not working, the top flexible material just touches the curb and the slide block does not move down. The known amount of Δx of the slide block is $L3 = L2 + \Delta x$, and the Angle value during the movement can be calculated as follows:

$$\alpha = \arcsin\left(\frac{l1^2 + L3^2 + (L1-m)^2 - l2^2}{2l1\sqrt{L3^2 + (L1-m)^2}}\right) - \arctan\left(\frac{L1-m}{L3}\right) \tag{13}$$

$$\beta = \arcsin\left(\frac{L3 - l1\sin\alpha}{l2}\right) \tag{14}$$

According to the above formula, when some geometric quantities are set to fixed values, the remaining unknown geometric quantities can be solved. By substituting the obtained geometric parameters into formulas 9 to 14, the mechanical parameters of each point can be obtained. By differentiating the parameters of the motion process (i.e. L3), all parameters of the motion process can be obtained. Based on this method, the mathematical model of the grout device is established in MATLAB, and the influence of each geometric dimension on the F force in Fig. 6 is analyzed.

As can be seen from Fig. 6, the larger the parameter values of α, m, H1, the larger the parameter values of F; The smaller the value of the parameter β, θ, L1, $l2$, $l4$, the larger the value of the parameter F. Relative to the degree of influence in the range, the three parameters α, L1 and $l2$ have a greater influence on the size change of the F value. According to the above laws, it is necessary to select a set of reasonable geometric parameters to achieve the F-value as small as possible, so as to reduce the burden of the mechanism. In addition to the optimization of F-value as small as possible, it is also necessary to consider the interference problems caused by the geometry of the parts. The following strategies are proposed for the problems encountered in this design:

1. L1 should be greater than B to prevent serious interference;
2. The smaller the value of $l1 + l2$, the greater the distance between the side flexible material and the curb when the grouting device is not working. Therefore, to increase the distance between the two, it can be achieved by reducing L2 value and increasing m value;
3. When the α Angle is slightly increased, the F value will increase significantly, and the pressure of the top flexible material can be achieved by changing the α value;
4. When working, the β Angle needs to be smaller to ensure that the side push rod has a reasonable length;
5. $l4$ length can be appropriately increased to prevent interference between the main connecting rod and the side push plate.

Finally, the data shown in Tables 1, 2 and 3 is selected as the geometric parameters of type B grouting device.

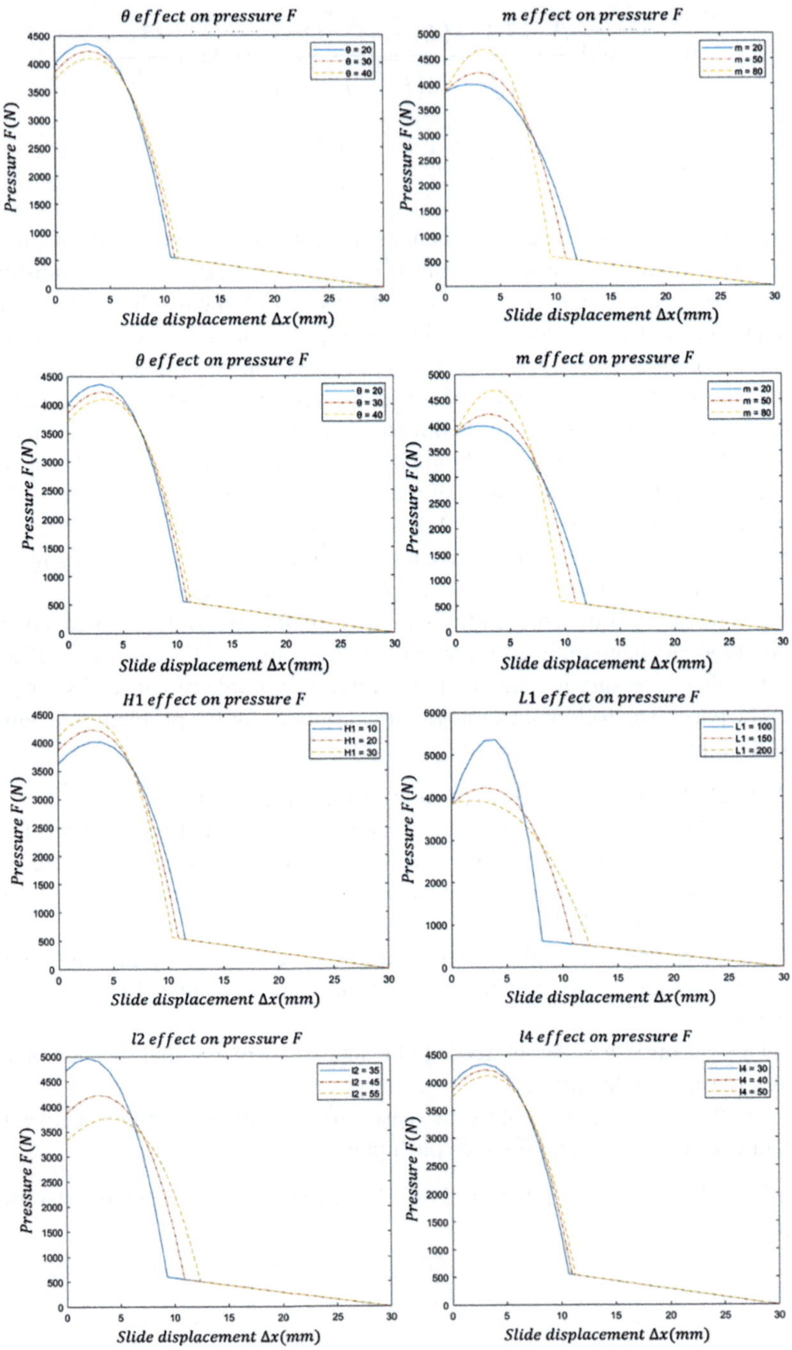

Fig. 6 Influence of geometric parameters on pressure F

Table 1 Input of grout device Angle value

Angular value	α_0	β_0	θ
Size(°)	5	76	30

Table 2 Input of grout device length value

Length value	H1	L1	m	*l*2	*l*4	*t*1	*t*3	A	B
Size(mm)	20	150	50	45	40	25	25	90	100

Table 3 Output of grout device length value

Length value	*l*1	*l*3	L2	H2
Size(mm)	89.5	128.8	51.5	21.5

3 Grouting Device Structure Design

3.1 *Overall Structure of Grouting Device*

Through the above theoretical analysis, the structural design of grouting device is carried out according to the principle and geometric size. Considering that the application scene is the highway construction site and follows the simple and practical principle, the grouting device model designed is shown in Fig. 7. In the overall structure, the module connecting block plays a key connection role, which is connected with the driving device of the automatic caulking machine by bolting and other ways to achieve force transfer. When the motor of the automatic caulking machine starts, the driving device drives the module connecting block to move up and down, and then drives the whole filling tool to work. At the slide block, the connecting rod and the slide block are connected by bolts, and the bottom end is matched with the clearance of the spring base to restrict the freedom of the slide block, and the spring ensures that the device opens when not working. The main connecting rod makes contact with the side push plate to realize the mechanical limit of sliding block. The structural design ensures that the filling tool is driven by the automatic caulking machine to accurately and stably seal and clamp the curb joints, laying the foundation for subsequent grouting operations.

3.2 *Structural Design at the Grouting Mouth*

The shape and position design of the grouting mouth play a key role in the filling quality. Considering that the drying of parts involves the flow characteristics of cement mortar, two grouting openings are set up on the top plate of the grouting device. The grouting mouth is placed on the top, and with the help of the fluidity of cement mortar itself and the action of gravity, it can accumulate in the joint from

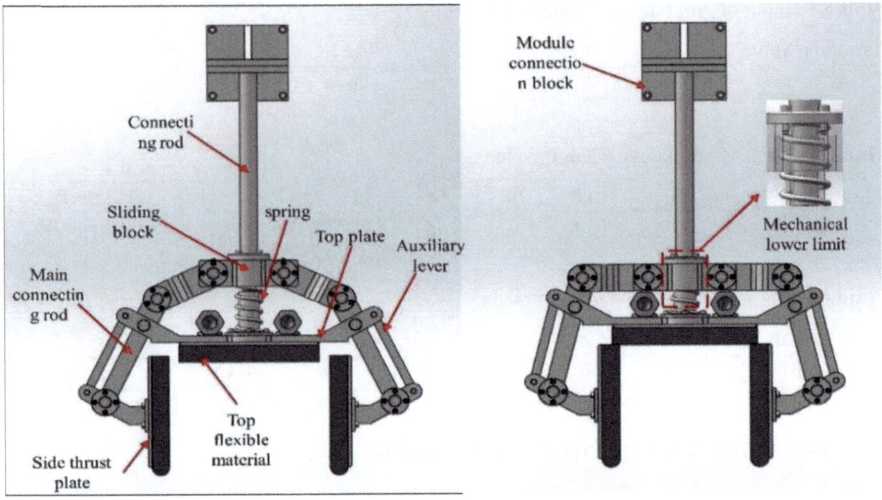

(a) When not working (b) At work

Fig. 7 Grouting device model

bottom to bottom to ensure the filling effect [6, 7]. The grouting opening is provided with a thread to fit the standard 90° metal hose hexagonal chromium-plated outer wire connector, and one end of the connector is tightly fixed to the top plate by means of the outer wire thread. The other end is connected with the high-pressure hose for conveying cement. When the caulking machine is working, the conveying system will transport the cement mortar through the high-pressure hose to the grouting outlet with a certain pressure to ensure that the cement mortar can flow smoothly into the curb joint and achieve uniform and efficient filling effect. The structure of the grouting mouth is shown in Fig. 8.

4 Grouting Device Test

The designed grouting device is processed in kind. The flexible material is made of shock-absorbing and wear-resistant butadiene-rubber solid square strip, which has excellent elasticity and can bounce back quickly after stress deformation to meet the needs of repeated use. The rigid part adopts Q235 steel to ensure the stability of the structure. As shown in Fig. 9.

In the test process, the clamping performance of the device is evaluated first. By applying a certain external force, the device can effectively clamp the curb joint and meet the expected design requirements. The spring holds the device open when not in operation. When the motor drives the drive device to operate, the components of the device work together to ensure accurate and stable clamping operation.

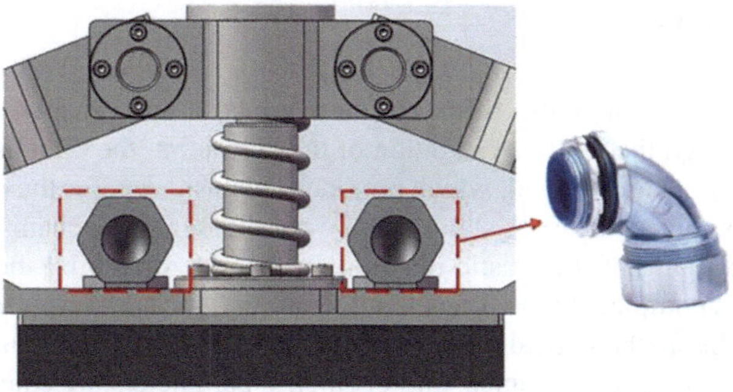

Fig. 8 Structure of grouting mouth

(a) Non-working state (b) Working status

Fig. 9 Actual grouting device

In order to further verify the effectiveness of the grouting device, the following experiments were carried out:

1. Grouting speed test: under different motor speeds, the grouting speed is measured. The experimental results show that the grouting speed can be adjusted in the range of 0.5 m/s to 1.5 m/s, and the best grouting speed is 1.0 m/s, which can effectively improve the filling efficiency.
2. Grouting pressure test: the pressure sensor is used to monitor the pressure of cement mortar during the grouting process. The experiment shows that the grouting pressure varies from 0.2 MPa to 0.5 MPa, and the optimal grouting pressure is 0.35 MPa to ensure that the cement mortar evenly flows into the curb joint.
3. Material deformation test: the deformation of flexible materials is measured, and the displacement sensor is used to record the deformation of materials under different pressures. The results show that the maximum deformation of the material under 0.3 MPa pressure is 5 mm, which meets the design requirements and ensures a good sealing effect.

5 Conclusions

In order to realize automatic filling of curb joints, a special grouting device is designed. Through the field investigation of the curb joint, the dimensional characteristics of the joint are defined, which provides the basic data for the device design. The device combines the crank slider and the parallelogram mechanism, which can realize the clamping of the elastic material on the side driven by the top material falling, so as to complete the sealing clamping around the curb joint. In the theoretical analysis, the mathematical model is built, the stress of the mechanism is analyzed, and the relevant mechanical and geometric parameters are calculated. At the same time, the structure optimization is analyzed by MATLAB simulation. The design ensures the movement and limit of the sliding block, the adaptation of the grouting mouth to the standard joint, improves the cement conveying efficiency, and the addition of the connecting plate is convenient for the replacement of flexible materials. The experimental test results show that the device can reliably clamp and seal the curb joints when working, and open the device with spring when not working. Driven by the motor, the components work together, and the overall performance meets the requirements of engineering operations [10].

Acknowledgments Thanks to Shandong Provincial Key Research and Development Program (Project No.: 2023TSGC0738, 2023CXPT047) for the strong support of this study.

References

1. Zhang J (2021) The development trend of road transport under the background of low-carbon economy. China Ship Week (39):50–51. (in Chinese)
2. Zhu Juan-E, Huang Yun. Construction technology of highway concrete curb. Transport World (Transport Vehicle), 2012(12): 187–188
3. Li F, Qiu S (2024) Motion analysis and finite element analysis of crank slider mechanism based on solid works. Intern Combust Engine Access (14):10–12
4. Zhang Q, Liu S (2019) Improvement of curb pointing method. Construct Work 40(06):19
5. Li N, Luo S, Xiang W et al (2023) Stress-strain relationship of self-compactors rubber concrete under cyclic compression. J Exp Mech 38(05):617–624. (in Chinese)
6. Sun G (2021) Research of engineering technology. 6(24):53–55. https://doi.org/10.19537/j.cnki.2096-2789.2021.24.015
7. Mancao YD Jr, Serevilla PJME et al (2018) Styrofoam as additive to asphalt joint filler in Portland cement concrete pavement. Int J Adv Res 6(7)
8. Xiang Z (2023) Development and application of small prefabricated U-shaped gutter rapid caulking device. Hebei Water Resour (08):41–42
9. Dong Z (2023) Discussion on construction technology of road curb slip form. Hebei Water Res (12):40–47
10. Zifei L (2023) Since boot highway road kerb seam automatic grouting operation construction vehicle research. Jinan University

Open Access This chapter is licensed under the terms of the Creative Commons Attribution-NonCommercial-NoDerivatives 4.0 International License (http://creativecommons.org/licenses/by-nc-nd/4.0/), which permits any noncommercial use, sharing, distribution and reproduction in any medium or format, as long as you give appropriate credit to the original author(s) and the source, provide a link to the Creative Commons license and indicate if you modified the licensed material. You do not have permission under this license to share adapted material derived from this chapter or parts of it.

The images or other third party material in this chapter are included in the chapter's Creative Commons license, unless indicated otherwise in a credit line to the material. If material is not included in the chapter's Creative Commons license and your intended use is not permitted by statutory regulation or exceeds the permitted use, you will need to obtain permission directly from the copyright holder.

An Improved Accuracy Analysis Method for Multiple Closed-Loop Mechanism with Clearance

Jiang Zhao, Di Wu, Bin Yang, Yuzhu Duan, and Pengfei Huang

Abstract Multiple closed-loop mechanisms are more and more widely used in the mechanical engineering. The precision of the mechanism is one of the key factors in some applications such as the truss structures of space antennas. This paper studies the structural precision of the multi-closed-loop mechanism with joint gaps by combining the improved Jacobian torsor model with the clearance model. Firstly, based on the modified Jacobian torsor model, the errors of the structure caused by joint clearances are calculated. Then the loop relationship of the mechanism with multiple closed loops is analyzed to determine the accuracy of the target points. By using this method, the precision analysis of a deployable truss mechanism in a planer SAR antenna is studied.

Keywords Multiple closed-loop mechanism · Clearance model · Probability distribution · Truss structure

1 Introduction

With the development of the technology, the demand for the high-quality communication has grown enormously around the world. The high power satellite with the deployable antenna is one of the best options to meet this requirement. The Synthetic Aperture Radar (SAR) has been widely applied for the satellite communications [1], as shown in Fig. 1, because of its excellent electrical performance. It plays a key role in the aerospace applications in different countries around the world in recent years [2–4]. With the development of the satellite antennas, more and more

J. Zhao (✉) · B. Yang · Y. Duan · P. Huang
Xi'an Institute of Space Radio Technology, Xi'an, China

D. Wu
Xi'an Institute of Space Radio Technology, Xi'an, China

College of Astronautics, Nanjing University of Aeronautics and Astronautics, Nanjing, China

© The Author(s) 2026

Z. Zhang, P. Mercorelli (eds.), *Mechanical Design and Simulation: Insights and Innovations*, Lecture Notes in Mechanical Engineering,
https://doi.org/10.1007/978-981-95-2405-1_106

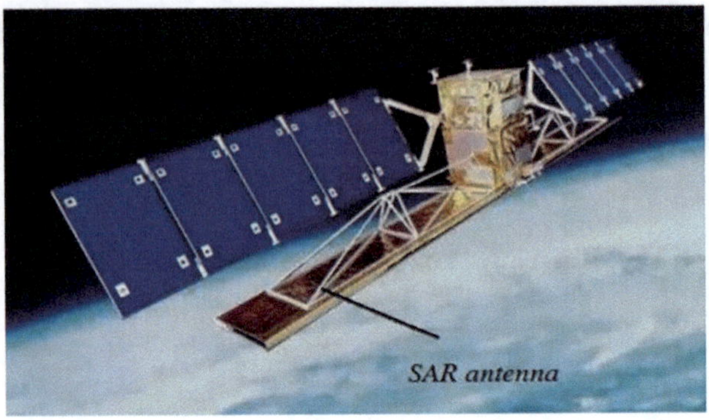

Fig. 1 SAR antenna

attentions have been paid to the clearance research of the deployable antennas. The reflector accuracy is an important index for the antenna, and how to improve the reflector accuracy of the antenna in the design stage has become a key procedure in the satellite development.

Aghabeigi et al. [5] put forward an assembly deviation transfer algorithm for planar truss structures that only considers the tolerance of the member, but without considering the coupling of the members. Chaker [6] analyzed the attitude error for a platform by using the small displacement torsor (SDT). Wu et al. [7] conducted the accuracy analysis of a tetrahedron truss structure with error by using the principle of minimum potential energy. Frisoli [8] conducted the accuracy analysis of the position of the structure with joint errors according to the screw theory. It should be noted that the kinematic equations of the mechanisms play a crucial role in the above methods [9].

This paper combines the traditional Jacobian torsor model with the clearance error model, and then presents a modified calculation method for the accuracy analysis. The flatness error of the SAR antenna support structure is analyzed by the proposed method.

2 Accuracy Analysis Method

The Jacobian torsor model [10] is efficient for the tolerance analysis of three-dimensional mechanisms which combines the torsor model which can be used for the error expression and the Jacobian matrix which can be used for the error transmission. It has been widely concerned and deeply studied in recent years.

Aiming at the small displacement at the clearance in the mechanism, the movement of the mechanism at the small displacement is described by the torsor.

$$T = \left[(\underline{u},\overline{u}),(\underline{v},\overline{v}),(\underline{w},\overline{w}),(\underline{\alpha},\overline{\alpha}),(\underline{\beta},\overline{\beta}),(\underline{\gamma},\overline{\gamma}) \right]^{\mathrm{T}} \quad (1)$$

where $(\underline{\alpha},\bar{\alpha}),(\underline{\beta},\bar{\beta}),(\underline{\gamma},\bar{\gamma})$ are the ranges of the moving vectors along the coordinate axis directions, and $(\underline{u},\bar{u}),(\underline{v},\bar{v}),(\underline{w},\bar{w})$ are the ranges of the rotating vector around the coordinate axis. The motion relationship between the mechanisms is transferred through the Jacobian matrix of the components, where the Jacobian matrix contains two parts, which are used to transfer the movement and rotation of the characteristics. Combining both of them and making full use of their advantages, the Jacobian torsor model is constructed. For an assembly consisting of n components, the Jacobian torsor model is expressed as:

$$\begin{bmatrix}(\underline{u},\bar{u})\\(\underline{v},\bar{v})\\(\underline{w},\bar{w})\\(\underline{\alpha},\bar{\alpha})\\(\underline{\beta},\bar{\beta})\\(\underline{\gamma},\bar{\gamma})\end{bmatrix}=[J_1\,J_2\ldots J_n]\times\begin{bmatrix}\begin{bmatrix}(\underline{u},\bar{u})\\(\underline{v},\bar{v})\\(\underline{w},\bar{w})\\(\underline{\alpha},\bar{\alpha})\\(\underline{\beta},\bar{\beta})\\(\underline{\gamma},\bar{\gamma})\end{bmatrix}_1^{\mathrm{T}}\cdots\begin{bmatrix}(\underline{u},\bar{u})\\(\underline{v},\bar{v})\\(\underline{w},\bar{w})\\(\underline{\alpha},\bar{\alpha})\\(\underline{\beta},\bar{\beta})\\(\underline{\gamma},\bar{\gamma})\end{bmatrix}_n^{\mathrm{T}}\end{bmatrix}^{\mathrm{T}} \tag{2}$$

The above equation is established based on the connection relationship between the parts and the error values in the mechanism. In the formula, $[J_1\,J_2\ldots J_n]$ is the total Jacobian matrix assembled by the serial assembly of each characteristic Jacobian matrix. When the analysis object is a closed-loop structure, the closed-loop can be disassembled into a series open-loop structures for analysis. Then the intersection of the two open loops at the connection point can be used to obtain the motion range of the objective point in the closed loop.

Usually, the error transmission path is affected by the assembly sequences, and the multi-closed loop mechanism often has a complex assembly sequence. The mechanism itself has a folding function, which leads to the redistribution of the clearance errors in the motion, and the clearance error is always less than the maximum clearance value. For obtaining the law of the influence of clearance on the accuracy of the multi-closed-loop truss structure, the clearance is equivalent to the connecting member model as shown in Fig. 2, and the value of the clearance is equivalent to the dimensional tolerance of the member, while the shape tolerance of

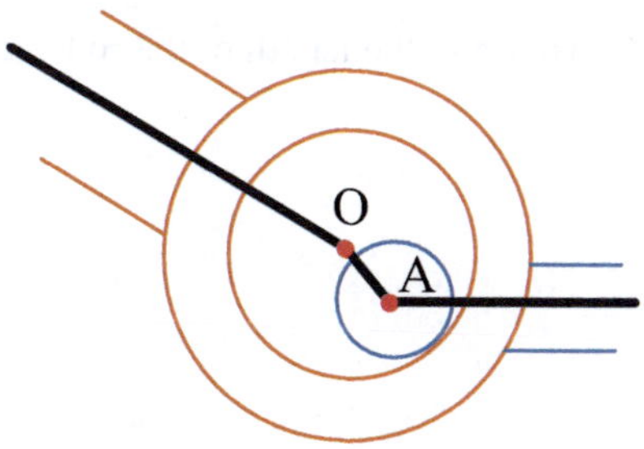

Fig. 2 Equivalent connecting member for the clearance

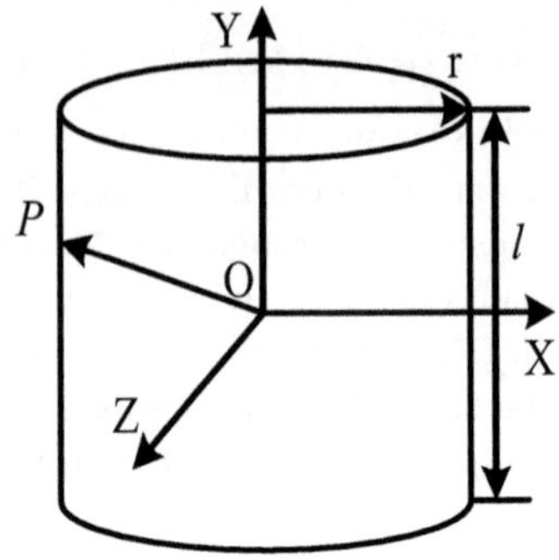

Fig. 3 Torsor error model of the rotation pair clearance

the clearance can be ignored in the structure. The two-dimensional virtual linkage model is often used to model the clearance of a rotating joint in the deployable structure.

The virtual linkage model can be approximated as a cylindrical feature which represents the error range of the clearance in the 3D space, as shown in Fig. 3. The error variation in the model is represented by the torsor model *P*. As the contact surface of the pin and the hole is separated, the contact shape tolerance of the clearance can be ignored.

$$P = [u, v, w]^{T} \tag{3}$$

Each variable has the following relationship

$$\begin{cases} u^2 + w^2 = r^2 \\ -l/2 \le v \le l/2 \end{cases} \tag{4}$$

By substituting the clearance error represented by the torsor model into the Jacobian torsor model, the error analysis model under the influence of the clearance can be obtained. The error space of the end of the rod can be derived as follows

$$f = [J_1 J_2 \dots J_n] \times \begin{bmatrix} \begin{bmatrix} (\underline{u}, \overline{u}) \\ (\underline{v}, \overline{v}) \\ (\underline{w}, \overline{w}) \end{bmatrix}_1^{T} & \cdots & \begin{bmatrix} (\underline{u}, \overline{u}) \\ (\underline{v}, \overline{v}) \\ (\underline{w}, \overline{w}) \end{bmatrix}_n^{T} \end{bmatrix}^{T} \tag{5}$$

Considering the length of the rods, it can be written as follows

$$f = J(P + L) \tag{6}$$

3 SAR Antenna Structure Analysis

The synthetic-aperture radar antenna (SAR) has been widely applied for the satellite communications which usually consists of a deployable structure called the extendible support structure (ESS), as shown in Fig. 4. The mechanism is a typical multiple closed loop mechanism for deploying and supporting the SAR antenna. Figure 4a shows the fully deployable configuration of the ESS. And Fig. 4b shows the simplified projected model on the x-z plane of the mechanism by considering the symmetry along the y-axis.

As shown in Fig. 4b, the simplified planar mechanism has three loops. In the mechanism, BA represents the boundary of the satellite platform. AF and FI are the inner and outer plates respectively. It is obvious that the deployment reliability and the pointing accuracy of the antenna is determined by the precision of the ESS. It should be noted that AF and F_2I should be normal to AB without any error in the ideal condition. However, the planar mechanism will inevitably deviate from the ideal situation due to the joint gaps, as shown in Fig. 5. Therefore, the accuracy analysis for the multiple closed-loop mechanism with clearances is quite important to evaluate the precision of the ESS.

Taking AB as the fixed structure, the position errors of points A_2, F, F_2 and I are solved firstly, and then the flatness errors of the plane formed by two panels of the antenna are obtained. The flatness error is defined as the square mean root (RMS) of

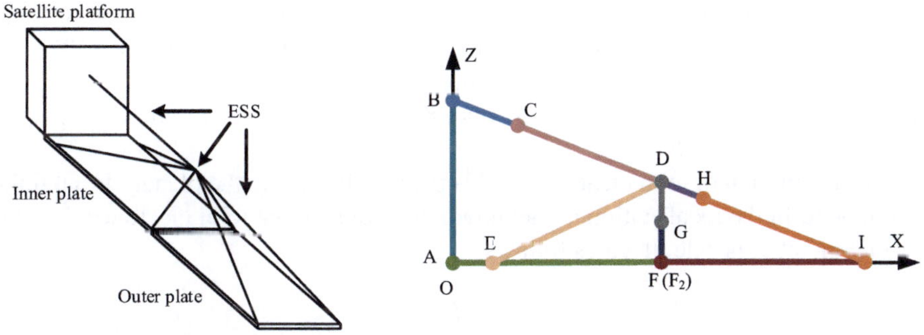

(a) 3D view (b) projected model

Fig. 4 Space-borne SAR antenna

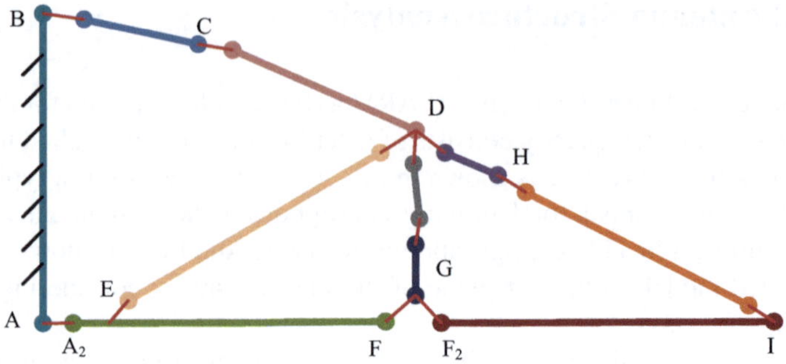

Fig. 5 ESS configuration with clearances

the deviation of the four nodes A_2, F, F_2 and I in the Z direction of two antenna plates. The formula is as follows:

$$\phi = \sqrt{\frac{1}{N}\sum_{i=1}^{N}\delta_i^2} \tag{7}$$

Considering that the ESS unfolds from left to right, the calculation is based on the expansion order. Firstly, the equations based on the improved Jacobian torsor model are constructed for each closed-loop structure. Then the coincidence constraint is applied to the coincidence points D, F and E, that is, the positions of the points in different equations are required to intersect. And then the error space of the target points is obtained. Finally, the accuracy analysis of the whole closed-loop is completed through the flatness error. The dimensions of AB, BC, CD, DE, AE, EF, DG, GF, FI, DH and HI in the developable support mechanism are 1750 mm, 367 mm, 1383 mm, 2337.7 mm, 200 mm, 1550 mm, 875 mm, 875 mm, 1750 mm, 474.9 mm and 2000 mm, respectively.

The discussion of joint clearance in this paper is limited to the planar virtual linkage model. In the local station at the hinge, the radius of the joint clearance δ of the virtual member is calculated as follows

$$\delta = R - r \tag{8}$$

where R is the radius of the hole and r is the radius of the pin. Considering the uncertainty of the clearance, the Monte Carlo method is used for iterative calculation, and the calculation results are as follows.

Figure 6 shows the position error space of the four key nodes A2, F, F2 and I along the Z and X directions. Figure 7 shows the probability distribution of flatness error for the plane formed by the inner and outer antenna plates.

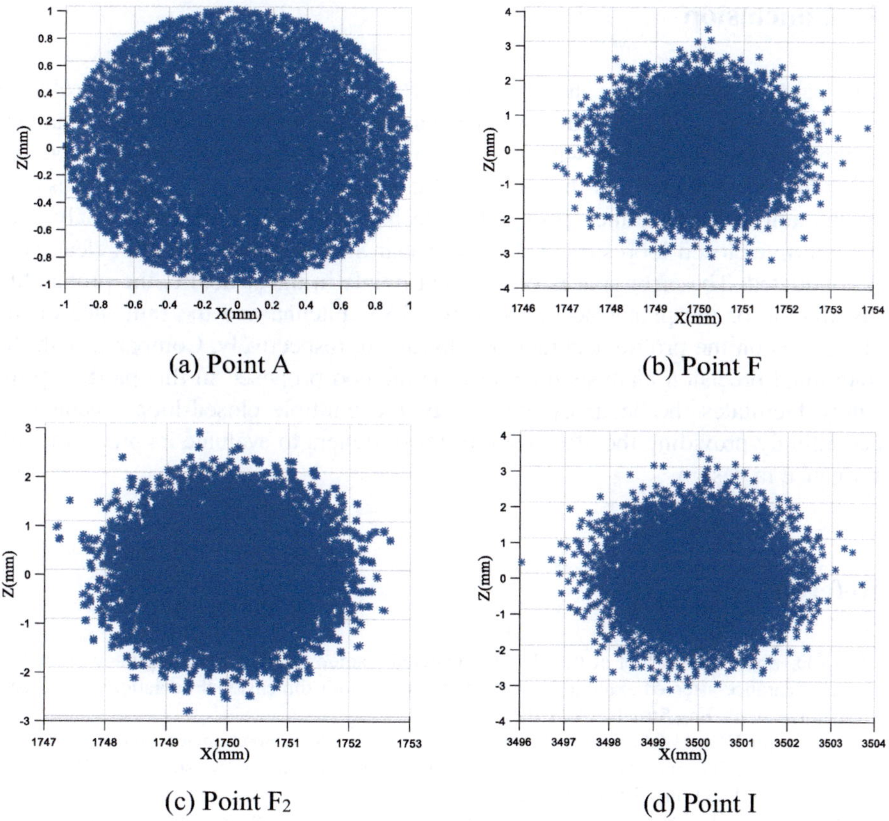

Fig. 6 The error space

Fig. 7 Probability distribution of the flatness error of antenna panel

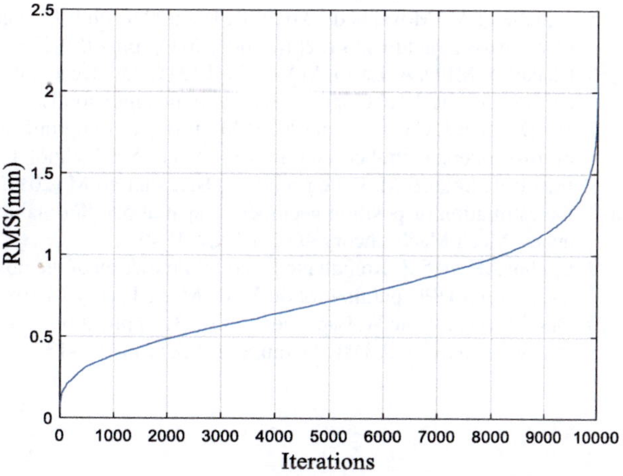

4 Conclusion

This paper systematically analyzes and sorts out the accuracy analysis problems existing in the multi-closed-loop mechanism. An improved accuracy analysis method for the mechanism with clearance is proposed by combining the modified Jacobian torsor model and the equivalent clearance model which can express the conduction of the clearance error in the mechanism. The flatness error analysis of the multiple closed-loop structure in the planar SAR antenna with joint clearances is conducted. The error spaces of different points in the structure, the probability distribution of the planar accuracy of the SAR antenna and the influence of the clearances on the profile accuracy are discussed, respectively. Compared with the traditional precision analysis methods, the method proposed in this paper significantly facilitates the accuracy analysis of the multiple closed-loop mechanism, accordingly providing the effective tool for designers to evaluate its precision performance in design.

References

1. Yu D, Zhao Q, Wu T et al (2023) An integrated framework of surface accuracy prediction for clearance-affected extendible support structures with dimensional deviations and elastic deformations. Eng Struct 274:115177
2. Jemaludin NH, Al-Gburi AJA, Elabd RH et al (2024) A comprehensive review on MIMO antennas for 5G smartphones: mutual coupling techniques, comparative studies, SAR analysis, and future directions. Results Eng 23:102712
3. Yu D, Hu G, Cai S et al (2023) Active optimization adjustment for the surface accuracy of spaceborne SAR antennas. Aerosp Sci Technol 138:108330
4. Ding J, Dong Y, Liu X et al (2022) Error-space-oriented tolerance design for a deployable mechanism with multiple clearances. Robotica 40(9):3254–3265
5. Aghabeigi M, Movahhedy MR (2014) An algorithm for geometrical uncertainty analysis in planar truss structures. Struct Multidiscip Optim 49(2):225–238
6. Chaker A, Mlika A, Laribi MA et al (2013) Clearance and manufacturing errors' effects on the accuracy of the 3-RCC spherical parallel manipulator. Eur J Mech A/Solids 37:86–95
7. Wu D, Chen J, Chen C et al (2023) Multi-objective optimization design and precision analysis of truss antenna tetrahedron unit. Chin Space Sci Technol 43(2):135–143. (in Chinese)
8. Frisoli A, Solazzi M, Pellegrinetti D, Bergamasco M (2011) A new screw theory method for the estimation of position accuracy in spatial parallel manipulators with revolute joint clearances. Mech Mach Theory 46(12):1929–1949
9. Chebbi AH, Affi Z, Romdhane L (2009) Prediction of the pose errors produced by joints clearance for a 3-UPU parallel robot. Mech Mach Theory 44(9):1768–1783
10. Desrochers A, Ghie W, Laperriere L (2023) Application of a unified Jacobian Torsor model for tolerance analysis. ASME J Comput Inf Sci Eng 3:1–13

Open Access This chapter is licensed under the terms of the Creative Commons Attribution-NonCommercial-NoDerivatives 4.0 International License (http://creativecommons.org/licenses/by-nc-nd/4.0/), which permits any noncommercial use, sharing, distribution and reproduction in any medium or format, as long as you give appropriate credit to the original author(s) and the source, provide a link to the Creative Commons license and indicate if you modified the licensed material. You do not have permission under this license to share adapted material derived from this chapter or parts of it.

The images or other third party material in this chapter are included in the chapter's Creative Commons license, unless indicated otherwise in a credit line to the material. If material is not included in the chapter's Creative Commons license and your intended use is not permitted by statutory regulation or exceeds the permitted use, you will need to obtain permission directly from the copyright holder.

Parameter Optimization Design of Concept Vehicle Beam Cross-Section Driven by Complex Engineering Constraints

Furong Geng, Jie You, Yonghong Zhao, Jiafu Wang, Wenbin Hou, and Changsheng Wang

Abstract This paper introduces an optimization strategy for real beam section parameters of vehicle bodies under complex engineering constraints. The bending stiffness and torsional stiffness of the concept body are employed as constraint conditions. A sensitivity analysis based on mass parameters is performed to evaluate the impact of various beam section parameters. Parameters exhibiting high sensitivity are selected as design variables. The findings indicate that the proposed method significantly contributes to achieving weight reduction in automotive body design. Additionally, the optimized beam section models generated through this process can be seamlessly integrated into the construction of concept body frameworks, thus providing a solid basis for advancing to the detailed design phase.

Keywords Cross-section optimization · Parameter-driven · Engineering constraints

1 Introduction

There are three major stages in the automotive body design process: conceptual design, detailed design, and testing. Among them, conceptual design is a crucial part of the body design process, where the majority of structural performance designs are completed, which can provide data support for the detailed design. Body

---

F. Geng · J. You · Y. Zhao
Automotive Engineering Institute, Guangzhou Automobile Group Co., Ltd., Guangzhou, China

J. Wang · W. Hou · C. Wang (✉)
School of Mechanical Engineering, Dalian University of Technology, Dalian, China
e-mail: changsheng@dlut.edu.cn

© The Author(s) 2026

Z. Zhang, P. Mercorelli (eds.), *Mechanical Design and Simulation: Insights and Innovations*, Lecture Notes in Mechanical Engineering,
https://doi.org/10.1007/978-981-95-2405-1_107

beam cross-section design is one of the most important steps in conceptual body design. During the design process, parameters such as cross-section shape, size, and thickness can be selected based on the position and load-bearing conditions of the thin-walled structures within the body. By assigning beam cross-sections to the corresponding positions on the conceptual body structure, and utilizing optimization methods based on predefined parameters such as body bending stiffness and torsional stiffness, the beam cross-section parameters can be obtained. This enables the design of a lightweight, stiff, and uniformly structured automotive body that meets the requirements of modern body design.

The cross-section of the automobile body is a thin-walled structure formed by welding multiple layers of stamped steel plates. It largely determines the overall performance of the automobile and directly impacts the vehicle's economy and collision safety. The design of the automobile body cross-section primarily needs to meet the requirements of designers for cross-sectional characteristics. Additionally, it is necessary to ensure that the body structure is as lightweight as possible while keeping material costs low. The optimization of automobile body cross-sections mainly involves optimizing the cross-sectional thickness, materials, and shapes.

Currently, many methods for optimizing beam cross-sections have been proposed. Hou et al. [1, 2] developed an intelligent CAE system for auto-body concept design. In which an automotive body thin-walled beam cross-section multi-objective optimization method is established with the assembly and manufacturability in the engineering design as design constraints [3, 4], Zuo et al. [5] developed the "Vehicle Body–forward design and optimization" software based on the NET framework. Meanwhile, a rapidly cross-sectional shape design and optimization method to satisfy the demand of mechanical properties and meanwhile minimize the weight of thin-walled beams [6], Qin et al. [7] proposed a novel exact transfer stiffness matrix method (TSMM) for both static and dynamic analyses of three-dimensional semi-rigid framed structures based on MATLAB toolbox, and two level cross-sectional optimization approach is presented to shorten concept design cycles [8]. Constraints are added during the optimization process, such as stamping constraints: draft angle, chamfer radius and assembly constraint.

Generally speaking, the upgrade iteration process of the same series of car models is driven by the body. The main changes are the styling features, general layout features and cross-sectional features. This paper proposes a cross-sectional parameter optimization method to meet new performance requirements. The real B-pillar cross-sectional parameters are selected during the optimization process.

The remainder of the paper is organized as follows. Section 2 briefly introduces cross-section of B-pillar. Section 3 presents formulations for cross-section properties. Section 4 details the optimization strategy. Section 5 provides numerical example to demonstrate the effectiveness of the presented method. Section 6 presents the conclusions and future prospects.

2 Body Beam Cross-Section

The body section is welded from multiple layers of plates. Figure 1 shows a schematic representation of a typical cross-section of the upper segment beam of the B-pillar in a vehicle's side panel. The significance of these parameters are given in Table 1. Based on their characteristics, these parameters can be classified into performance parameters, layout parameters, and process parameters. Process parameters are driven in accordance with the requirements of the manufacturing process and are generally not modified in the analysis of the conceptual body structure. The layout characteristics he body refer to a set of standardized layout information required in the process of constructing the body model, such as doorstops, hinge information, hinge mounting surfaces, as well as ergonomic information related to the body structure, heel point positions, suspension information, etc. The optimization objective of the vehicle body section is to meet the requirements of vehicle performance. The performance is ultimately achieved through selection of cross-sectional parameters during the optimization process. Section characteristics mainly include: sectional area, moment of inertia, section modulus, etc.

In addition to meeting performance requirements, the body cross-section, along with styling geometry and general layout features, forms the overall input to the conceptual framework model. The framework model can provide a more accurate model for performance analysis at the initial design stage.

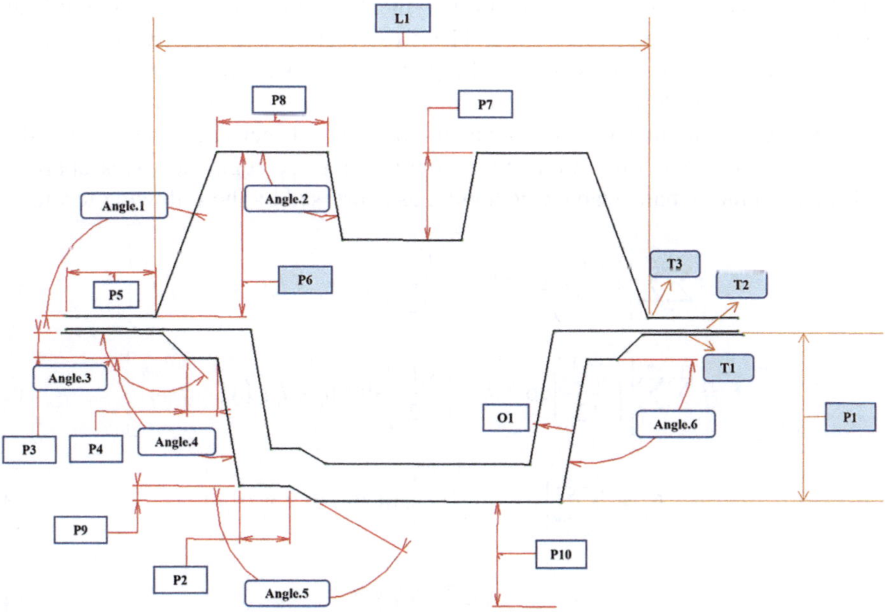

Fig. 1 Cross-section of upper segment of B-pillar

Table 1 Parameters of cross-section of upper segment of B-pillar (Fig. 1)

Name	Description	Initial value	Parameter	Mutable
T1	Outer panel thichness	0.8	Performance	Yes
T2	Reinforcing panel thickness	1.4	Performance	Yes
T3	Inner panel thickness	1.2	Performance	Yes
L1	B-pillar cavity X dimension	106	Performance	Yes
P1	Depth of side wall	31	Performance	Yes
P2	Width of sealing surface	20	Layout	Yes
P3	X-flange step height	5	Layout	No
P4	X-flange step width	5	Layout	Yes
P5	Width of inner welding edge	18	Process	No
P6	Depth of inner panel	20	Performance	Yes
P7	Depth of reinforcement panel	10	Performance	Yes
P8	Width of reinforcement panel	20	Performance	Yes
P9	Height of sealing surface step	3	Layout	Yes
P10	Distance from glass surface	15	Layout	Yes
O1	Electrophoretic gap	7.8	Process	No

3 Cross-Section Properties

The cross-sectional properties, including moments of inertia, cross-sectional area, bending constants, and torsional constants, are critical in the optimization of beam cross-sections. The moment of inertia is typically considered the most significant parameter in cross-sectional design. The cross-section is generally composed of multiple welded layers of plates, resulting in a complex geometry that complicates direct calculation. However, by approximating the beam's cross-section as a combination of ss rectangular wall segments, the moment of inertia can be simplified to the sum of the individual moments of inertia of these segments. This approach allows for a more manageable calculation, as expressed by the following formula:

$$I_y = \sum_{i=1}^{n}\sum_{j=1}^{m}\left[\left(\frac{l_{ij}t_i^3}{12}\right)\cos^2\theta_{ij} + \left(\frac{l_{ij}^3 t_i}{12}\right)\sin^2\theta_{ij} + l_{ij}t_i\left(z_c - z_{cij}\right)^2\right] \tag{1}$$

$$I_z = \sum_{i=1}^{n}\sum_{j=1}^{m}\left[\left(\frac{l_{ij}t_i^3}{12}\right)\cos^2\theta_{ij} + \left(\frac{l_{ij}^3 t_i}{12}\right)\sin^2\theta_{ij} + l_{ij}t_i\left(y_c - y_{cij}\right)^2\right] \tag{2}$$

$$I_{yz} = \sum_{i=1}^{n}\sum_{j=1}^{m}\left[\left(\frac{l_{ij}^3 t_i - l_{ij}t_i^3}{24}\right)\sin 2\theta_{ij} + l_{ij}t_i y_{cij} z_{cij}\right] \tag{3}$$

$$A = \sum_{i=1}^{n}\sum_{j=1}^{m} l_{ij}t_i \tag{4}$$

Table 2 Comparison between calculated results and software results

Number	Group	T1	T2	T3	P1	P6	L1	A	Iy	Iz
1	Results	0.65	1.8	1.4	37.4	27.6	106.6	771	321,300	1,216,900
	AutoCAD							771	310,684	1,221,618
2	Results	0.7	1.6	1	39.9	33.4	110.9	691	361,180	1,167,500
	AutoCAD							691	352,518	1,171,462
3	Results	0.8	1.4	1.4	39.1	28.1	109.3	742	345,810	1,232,000
	AutoCAD							742	336,258	1,236,562

where I_y, I_z, I_{yz} is the moment of inertia of the beam section with respect to each axis, A is the cross-sectional area, l_{ij} is the length of the jth rectangular wall segment of the ith plate, t_i is the thickness of the ith plate, θ_{ij} is the Angle between the jth rectangular wall segment of the ith plate and the y axis, (y_c, z_c) is the centroid coordinates of the section, (y_{cij}, z_{cij}) is the centroid of the jth rectangular wall segment of the ith plate, n is the layer of plates, and m is the number of wall segments of the ith plate.

Table 2 takes the upper segment of B pillar as an example, and shows the comparison between the calculation results, obtained by Eqs. 1, 2 and 4, and the AutoCAD results in the case of multiple sets of parameters. The results indicate that the error in the moment of inertia is confined within 5%, suggesting that the calculated values are reliable and reflect the real situation of the section accurately.

4 Beam Cross-Section Optimization Algorithms

In the conceptual design stage, when the finite element model is employed to study the vehicle body design, the mechanical model of the body's conceptual structure is typically established using space thin-walled beam elements and space plate-shell elements. This model is employed for investigating the impact of primary structural cross-sectional properties on body stiffness, including both bending stiffness and torsional stiffness. In this paper, the conceptual body structure consists of beams, and only the upper and lower segments of the B pillar are considered for replacement and optimization. The conceptual body beams are classified based on their symmetry relationships, resulting in 37 groups of beam structures, as illustrated in Fig. 2.

In order to enhance efficiency of the design process, it is essential to select the optimization parameters from the size parameters of the upper and lower segments of the B pillar. Design of experiments (DOE) is employed to calculate the sensitivity of beam section dimensions to the section area. The optimization goal of this paper is to reduce the weight of the body and the influence of design variables on the cross-sectional area. The weight of the vehicle body is concerned in the sensitivity analysis process. Firstly, the experimental sample points are obtained by optimal

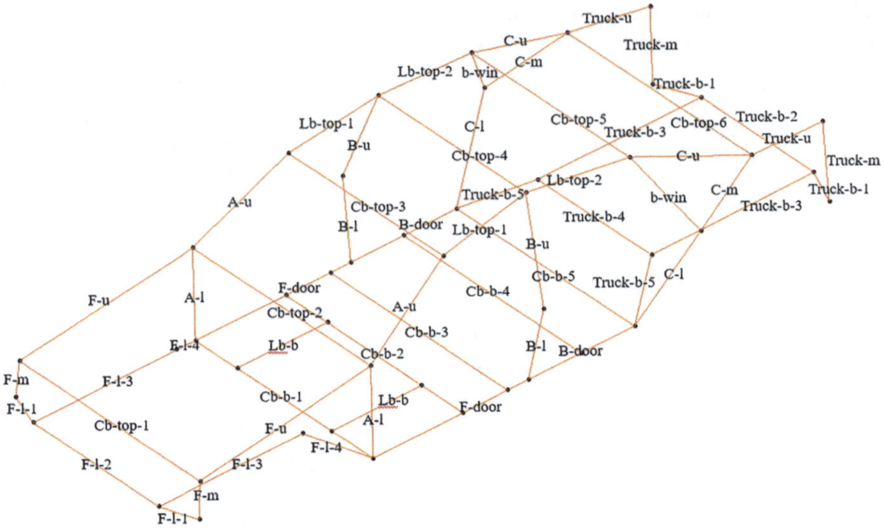

Fig. 2 Schematic diagram of beam structure

Latin hypercube design method. Then, these sample points are used as input and the cross-sectional characteristics are used as output to obtain the sensitivity of each parameter to the cross-sectional area. The optimal Latin hypercube design method is selected because it has good space filling, and the selected sample points can well represent the entire design space. It can obtain more comprehensive information about the design space with fewer sample points. Compared with traditional experimental design methods (such as full factorial design), it greatly reduces the number of experiments and improves the computational efficiency without losing too much accuracy. The calculation results for the upper segment of the B pillar are presented in Fig. 3. Based on these results, the optimization parameters for both the upper and lower segments of the B pillar are determined, while other dimensional parameters are treated as design variables to be optimized.

Solving for body stiffness directly is a complex process for detailed model. Therefore, a simplified approach is typically adopted, wherein the body is treated as a simplified mechanical model to facilitate the calculation of stiffness.

In the calculation of bending stiffness of the vehicle body, the conceptual body is modeled as a simply supported beam with uniform bending stiffness. A concentrated load is applied at a specific location on the beam, and the maximum deflection under this loading condition is determined to obtain the bending stiffness of the beam. The calculation formula for bending stiffness used in this study is as follows:

$$K_{bend} = \frac{\sum F}{z} \tag{5}$$

where K_{bend} (N/mm) is the bending stiffness, F(N)is the concentrated load, and z(mm) the maximum bending deflection.

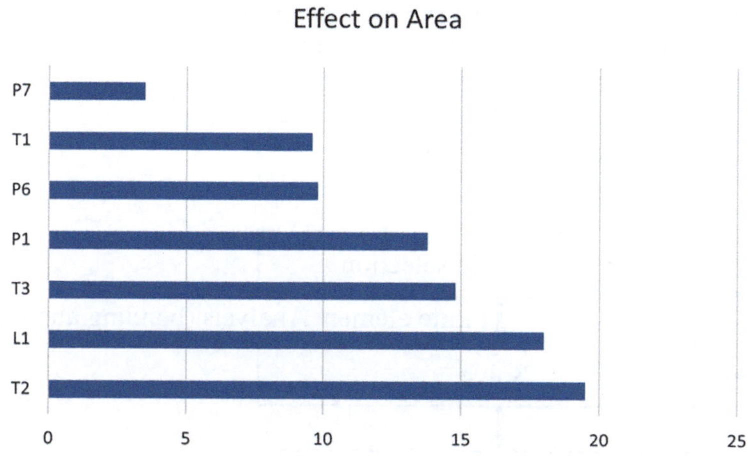

Fig. 3 Sensitivity of size parameter of the upper segment of the B pillar

Meanwhile, the body is typically modeled as a rod with uniform torsional stiffness. One end of the rod is fixed (the rear end of the vehicle body), while a torsional load is applied to the other end (the frontend of the vehicle body). The formula for the average torsional stiffness of the vehicle body is given as follows:

$$K_{torsion} = \frac{T}{\theta} \tag{6}$$

where $K_{torsion}$ (N·m/deg) is the torsional stiffness, T(N·m) is the torsional Load, and θ(deg) is the relative torsional angle between the front and rear axles of the vehicle body.

The formula for the relative torsional angle between axles is as follows:

$$\theta = \arctan \frac{U}{B} \tag{7}$$

where U(mm) is the deflection at the position (measurement point) where the torsional load is applied, and B(mm) is the distance between the position where the left and right torsional loads are applied.

The goal of optimization is to minimize the total mass. Non-Linear programming by quadratic-linear programming (NLPQLP) and modified method of feasible directions (MMFD) are used as comparison. The optimization results of NLPQLP rely too much on the number of iterations, and the convergence speed is slow. The optimization time of MMFD is 1 h 3 min, and the convergence speed is faster. Under the same time, NLPQLP can only complete the optimization of individual parameters, and cannot achieve the effect of global convergence. Therefore, Modified method of feasible directions (MMFD) algorithm is adopted here for optimization. Figure 4 shows the overall optimization flowchart.

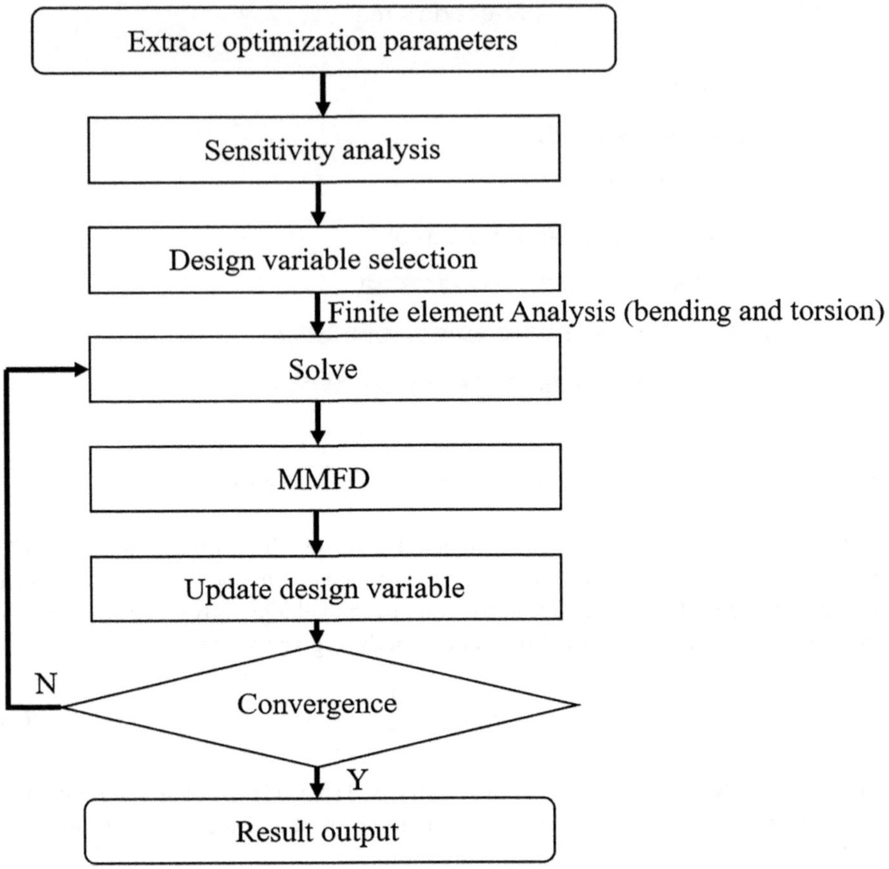

Fig. 4 Flow chart of the optimization algorithm

5 Numerical Applications

5.1 Comparative Verification of Conceptual Model and Detailed Model

This study uses a real cross-section as a case study to analyze the stiffness performance of the car body. To reduce model complexity and improve computational efficiency, a simplified model is developed. To verify that the conceptual model accurately reflects the stiffness characteristics of the car body, identical loading conditions are applied to both the detailed and simplified models. The loading and boundary conditions for both models are described as follows:

Bending stiffness (Fig. 5):

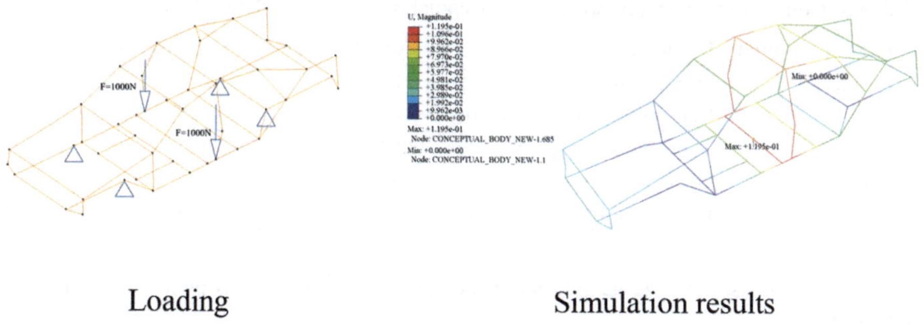

Fig. 5 Loading and simulation results under bending conditions

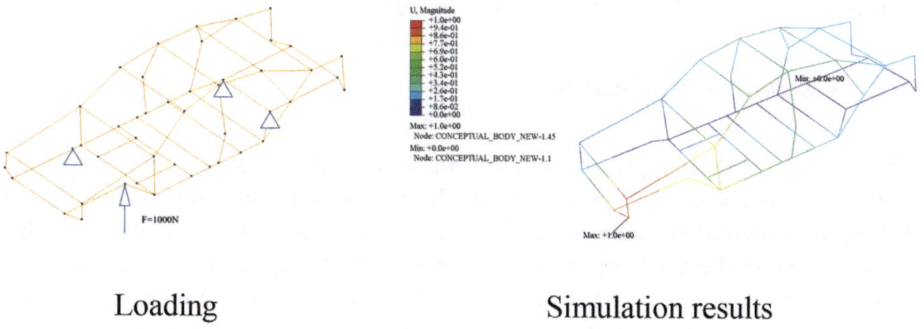

Fig. 6 Loading and simulation results under torsional conditions

- Load: apply a concentrated load of 1000 N at the connection points between the B-pillars and the sills on both sides of the vehicle body.
- Constraint: fix the four suspension pivot points at the front and rear as fixed hinges.

Torsional stiffness (Fig. 6):

- Load: apply a concentrated load of 1000 N at the left front suspension fulcrum.
- Constraint: fix the remaining three suspension pivot points as fixed hinges.

The simulation results are presented in Table 3. When compared to the detailed model, the errors in bending stiffness and torsional stiffness for the conceptual model are 11.76% and 19.5%, respectively, both of which fall within an acceptable range of 20%. These results suggest that the conceptual body model employed in this study is reasonable and provides a reliable approximation of the car body's stiffness performance.

Table 3 Comparison of data between the conceptual model and the detailed model

Working condition	Parameter	Conceptual Model	Detailed Model	Error
Torsion	Displacement (mm)	0.82	1.2	–
	Relative torsional angle (deg)	0.045	0.055	–
	Torsional stiffness (N·m/deg)	22,304	18,654	19.5%
Bend	Deflection (mm)	0.119	0.133	–
	Bending stiffness (N/mm)	16,807	15,038	11.76%

Table 4 Comparison of size parameters of the upper segment of the B pillar before and after optimization

	T1	T2	T3	P1	P6	L1	A
Initial	0.8	1.4	1.2	31	20	106	644
Optimization	0.65	1.2	1	25	11	110.1	520.3

5.2 *Optimization Results*

This paper utilizes Isight to develop an optimization process. First, relevant data, such as the section properties of the upper and lower segments of the B pillar, are calculated using a Matlab program. These calculated data are then used to update the cross-sectional parameters of the upper and lower beams in Abaqus for simulation. The resulting displacements at measurement points and the overall mass are outputted. A 10% reduction in stiffness is imposed as a constraint, with the deflection values (displacement values) of the vehicle body measurement points calculated based on the stiffness formula. These deflection values serve as the upper limits for the displacement values at the measurement points. The objective of the optimization is to minimize the overall mass of the structure. The comparison of the cross-sectional dimensions of the upper and lower segments of the B-pillar before and after optimization is presented in Tables 4 and 5.

Through optimization of the cross-sectional profiles of the upper and lower segments of the B pillar, the total mass of the vehicle body has reduced by approximately 5.5 kg. The relationship between the number of iterations and the mass is illustrated in Fig. 7a. Although there is a slight reduction in both the bending stiffness and torsional stiffness of the vehicle body, these values remain within the specified constraint limits. The changes in stiffness are illustrated in Fig. 7 (b and c). In the 57th iteration, modifications to the sensitive variables led to a sudden discontinuity in the body stiffness. Specifically, the bending stiffness dropped below the prescribed limit, resulting in a sharp peak in the iteration graph.

Table 5 Comparison of size parameters of the lower segment of the B pillar before and after optimization

	T1	T2	T3	H1	H2	W1	A
Initial	0.8	1.4	1.2	26	74	167.8	1119.4
Optimization	0.65	1.2	1	5	59	146.8	831.8

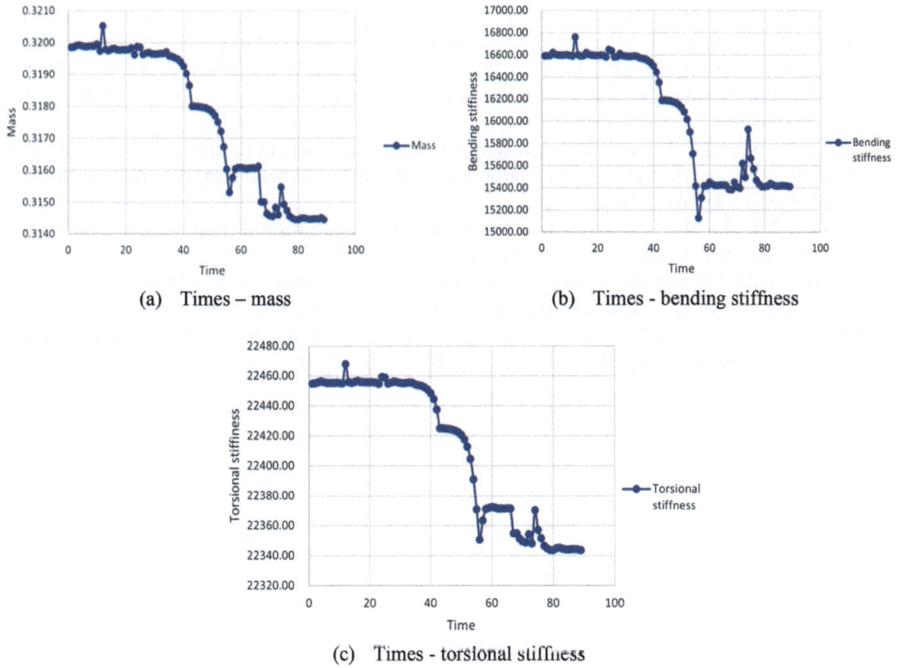

(a) Times – mass (b) Times - bending stiffness (c) Times - torsional stiffness

Fig. 7 Iteration diagram

6 Conclusions

This paper proposes an optimization strategy for realistic beam sections of automotive bodies under complex engineering constraints. Using the upper and lower cross-sections of the B-pillar as a case study, parameters with high sensitivity to mass are selected as design variables. The optimization is performed with the bending stiffness and torsional stiffness of the body serving as constraints, aiming for lightweight design. The results demonstrate that the proposed strategy can effectively achieve weight reduction in vehicle body design. Furthermore, the use of realistic beam sections ensures that the obtained results can be directly applied to subsequent analysis, validation, and detailed design phases.

References

1. Hou WB, Zhang HZ, Zhang W, Hu P (2011) Rapid structural property evaluation system for car body advanced design. Int J Veh Des 57(2–3):242–253
2. Hou WB, Wang XY, Han JF, Yao ZQ, Wang CS (2024) An intelligent body frame structure modeling and optimization system based on conceptual design. Proc Inst Mech Eng D J Automob Eng:1–9
3. Hou WB, Wang ZF, ZHang W, Zhang HZ (2014) Optimization design for auto-body beam section based on complex engineering constraints (in Chinese). J Mech Eng 50(18):127–133
4. Yuan HQ, Wang ZF, Zhao YH, Hou WB (2020) Associated optimization for the variable cross-section beam of vehicle body based on the manufacturability constraints. J Mach Des 37(8):112–116
5. W.J. Zuo, W.W. Li, T. Xu, S.Y. Xuan, and J.X. Na. A complete development process of finite element software for body-in-white structure with semi-rigid beams in .net framework. Adv Eng Softw, 45(1):261–271, 2012.2
6. Zuo WJ, Bai JT (2016) Cross-sectional shape design and optimization of automotive body with stamping constraints. Int J Automot Technol 17:1003–1011
7. Qin H, Liu ZJ, Liu Y, Zhong HL (2017) An object-oriented matlab toolbox for automotive body conceptual design using distributed parallel optimization. Adv Eng Softw 106:19–32
8. Qin H, Liu ZJ, Liu Y, Zhong HL (2018) A two-level cross-sectional optimization approach for auto- motive body concept design. Automot Innov 1:122–130

Open Access This chapter is licensed under the terms of the Creative Commons Attribution-NonCommercial-NoDerivatives 4.0 International License (http://creativecommons.org/licenses/by-nc-nd/4.0/), which permits any noncommercial use, sharing, distribution and reproduction in any medium or format, as long as you give appropriate credit to the original author(s) and the source, provide a link to the Creative Commons license and indicate if you modified the licensed material. You do not have permission under this license to share adapted material derived from this chapter or parts of it.

The images or other third party material in this chapter are included in the chapter's Creative Commons license, unless indicated otherwise in a credit line to the material. If material is not included in the chapter's Creative Commons license and your intended use is not permitted by statutory regulation or exceeds the permitted use, you will need to obtain permission directly from the copyright holder.

Design and Implementation of Joint Control Strategy for Permanent Magnet Synchronous Motor

Shenghu Pan and Haitao Liu

Abstract In order to solve the problems of slow response speed, excessive overshoot and steady state error in the speed control system of permanent magnet synchronous motor, a joint control strategy of permanent magnet synchronous motor based on BP-PID speed controller and STSMC current controller is proposed. Firstly, a BP neural network is used to optimize the PID controller in order to improve the adaptability of the system to speed changes and external disturbances. Meanwhile, the STSMC current controller is introduced to enhance the robustness of the system to current fluctuations and ensure that the motor can operate stably in the rapidly changing operating environment. It is verified through simulation that the joint control strategy is characterized by low response speed, overshooting amount and steady state error.

Keywords Permanent magnet synchronous motor · Joint control · Backpropagation proportional-integral-derivative controller · Super-helical sliding mode controller

1 Introductory

Permanent magnet synchronous motors have the advantages of high efficiency and low power consumption, and are widely used in fields such as mechanical drives and electric vehicles [1, 2].

In PMSM control systems, various types of control algorithms are often used to enable the motor to respond quickly or return to the set speed quickly when perturbed. The control algorithms mainly include PID control [3], adaptive control [4], fuzzy control [5], sliding mode control [6], self-resistant control [7], and BP-PID.

S. Pan (✉) · H. Liu
School of Mechatronics Engineering, Southwest Petroleum University, Chengdu, Sichuan, China

© The Author(s) 2026

Z. Zhang, P. Mercorelli (eds.), *Mechanical Design and Simulation: Insights and Innovations*, Lecture Notes in Mechanical Engineering,
https://doi.org/10.1007/978-981-95-2405-1_108

Yang Zongjun et al. improved the accuracy and efficiency of the identification by adjusting the model parameters in real time so that they can better adapt to the dynamic characteristics of the motor [8]. Wang Jia et al. designed a PI-fuzzy hybrid controller by combining fuzzy control with proportional control in order to improve the position response speed of the servo system [9]. This hybrid controller adapts to the dynamic changes of the system through different control modes, but its design is relatively complex and requires additional switching time, which may increase the difficulty of design and debugging in practical applications. The control method of permanent magnet synchronous motor servo system based on dynamic sliding mode control proposed by Xu Xuyao does show high accuracy and response speed in position and velocity control [10]. However, although the switching characteristic of sliding mode control can improve the robustness of the system, it may also bring the problem of jitter. This jitter not only affects the stability of the system, but also may lead to the wear and failure of mechanical components, thus affecting the reliability and service life of the whole system.

For this reason, the article designs a joint control strategy for the problems of slow response speed, overshooting amount and high steady state error in PMSM speed control system. The joint control strategy is innovative, which is mainly reflected in the multilevel control structure. This multilevel control structure enables the control of speed and current to be coordinated with each other and improves the overall performance of the system. The specific difference between the joint control strategy and the existing control strategy is the hierarchy of control objectives. Traditional control strategies tend to put speed and current control at the same level, while the joint control strategy separates speed control and current control, realizing the hierarchical management of the target and improving the control accuracy.

2 Design of PMSM Control System Based on BP-PID and STSMC

In this subsection, the BP-PID speed controller and STSMC current controller are designed to address the problems of high overshooting and poor interference immunity in PID control systems. The general block diagram of the control strategy of the PMSM control system is shown in Fig. 1.

2.1 Mathematical Model of a Motor

The mathematical model of the permanent magnet synchronous motor is constructed using d-q modeling. The key equations involved in the d-q model of the PMSM are the d-axis stator voltage equation, the q-axis stator voltage equation and the electromagnetic torque equation, respectively:

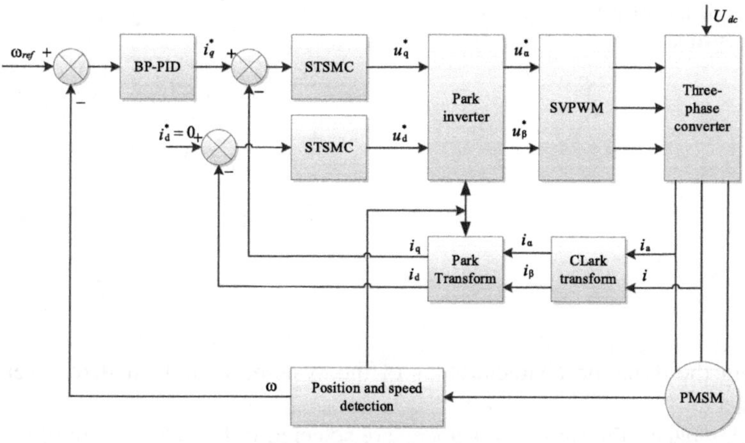

Fig. 1 General block diagram of control strategy for PMSM control system

$$\begin{cases} u_d = Ri_d + L_d \frac{d}{dt} i_d - w_e L_q i_q \\ u_q = Ri_q + L_q \frac{d}{dt} i_q + w_e \left(L_d i_d + \psi_f \right) \\ T_e = \frac{3}{2} p_n i_q \left[i_d \left(L_d - L_q \right) + \psi_f \right] \end{cases} \quad (1)$$

Where: u_d, u_q are the components of stator voltage on the d-axis and q-axis, respectively; i_d and i_q are the components of stator current on the d-axis and q-axis; R is the resistance of the stator; ψ_d, ψ_q are the components of the stator magnetic chain on the d-axis and q-axis; w_e is the electrical angular velocity; L_d *and* L_q are the components of the stator current on the d-axis and q-axis; T_e is the electromagnetic torque; P_n is the pole-pair number; and ψ_f is the magnetic chain of the permanent magnet.

For the surface-mounted three-phase PMSM, the stator inductance satisfies $L_s = L_d = L_s$. Therefore, its electromagnetic torque can be simplified as:

$$T_e = \frac{3}{2} p_n i_q \psi_f \quad (2)$$

2.2 *BP-PID Speed Controller Design*

In this paper, Backpropagation Proportional-Integral-Derivative Controller (BP-PID) speed controllers with anti-disturbance capability are designed in conjunction with the proposed neural network. The BP neural network based PID controller can adaptively adjust the parameters p, i, and d of the PID controller by

Fig. 2 Structure of BP neural network

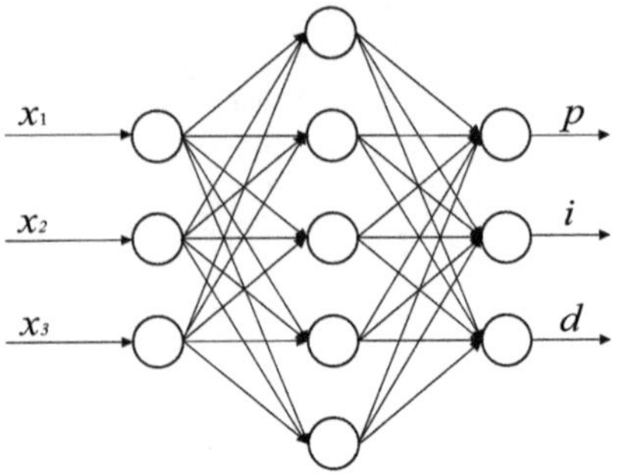

learning the dynamic characteristics of the system, so as to realize a better control effect.

In this paper, the same layer nodes are selected to be independent of each other in a 3-5-3 three-layer feed-forward network structure, as shown in Fig. 2.

The input-output expression for the hidden layer of the BP neural network is:

$$\begin{cases} net_i^{(2)}(k) = \sum_{j=0}^{m} w_{ij}^{(2)} x_j^{(1)} \\ o_i^{(2)}(k) = f\left(net_i^{(2)}(k)\right) \end{cases} \tag{3}$$

where: $w(2)$ ij is the weight coefficient of the hidden layer; $f(x)$ is the hyperbolic tangent function; $net(2)$ $i(k)$ is the input of the hidden layer; O(2) $i(k)$ is the output of the hidden layer. The input-output expression for the output layer of the BP neural network is:

$$\begin{cases} net_i^{(3)}(k) = \sum_{j=0}^{q} w_{ij}^{(3)} x_j^{(2)} \\ o_i^{(3)}(k) = g\left(net_i^{(3)}(k)\right) \end{cases} \tag{4}$$

where: $w(3)$ ij is the hidden layer weighting coefficient; $g(x)$ is the non-negative Sigmoid activation function; $net(3)$ $i(k)$ is the hidden layer input and $O(3)$ $i(k)$ is the hidden layer output.

The block diagram of the structure of the PID control system based on BP neural network is shown in Fig. 3.

The input variables of the BP-PID speed controller are: the desired speed w_{ref}, the actual speed w, and the difference between the desired speed and the actual speed at the current sampling moment $e(k)$. The outputs are the trained PID parameters. This adaptive mechanism can adjust the control parameters in real time to adapt to the dynamic changes of the system and improve the response speed and stability of the control system. The output signal of the PID controller is calculated as follows:

$$u(t) = p * e(t) + i\int_0^t e(t)dt + d\frac{de(t)}{dt} \tag{5}$$

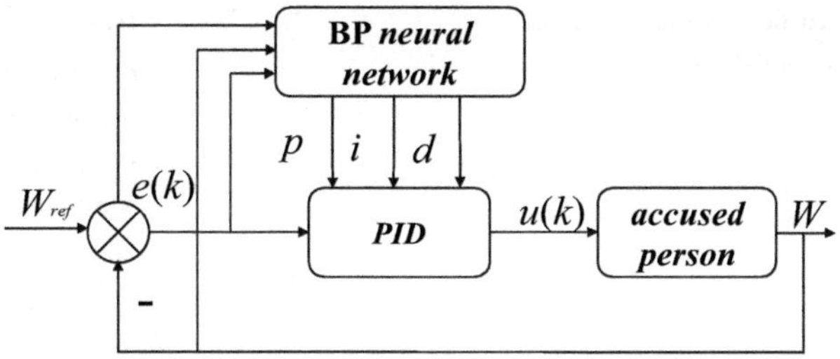

Fig. 3 BP-PID speed controller

where *p*, *i*, and *d* correspond to the coefficients of proportional, integral, and differential, respectively, and $e(t)$ is the current speed error.

2.3 STSMC Current Controller Design

The Super-Spiral Sliding Modal Controller (STSMC) achieves the control objective by designing sliding modal surfaces and control laws to make the system state slide along a predetermined trajectory.

Firstly, the sliding mode surfaces of d-axis and q-axis are defined respectively:

$$\begin{cases} S_d = i_d - i_d^* \\ S_q = i_q - i_q^* \end{cases} \tag{6}$$

Where S_d and S_q are the sliding mode surfaces of d-axis and q-axis, i_d, i_q are the currents on d-axis and q-axis, and $i^* d$ and $i^* q$ are the reference currents of d-axis and q-axis. Using the basic principle of second-order sliding mode control based on the Super-Twisting algorithm, the expression of the current-velocity controller at this point is:

$$\begin{cases} u_d^* = K_P |s_d|^r \operatorname{sgn}(s_d) + u_{sd} \\ \dfrac{d}{dt} u_{sd} = k_i \operatorname{sgn}(s_d) \\ u_q^* = K_P |s_q|^r \operatorname{sgn}(s_q) + u_{sq} \\ \dfrac{d}{dt} u_{sq} = k_i \operatorname{sgn}(s_q) \end{cases} \tag{7}$$

where K_p, $K_i > 0$, are the parameters to be designed. When the design parameter r is 0, the expressions for the d-axis and q-axis current controllers are obtained:

$$\begin{cases} u_d = \left(K_P + \dfrac{K_i}{s} \right) \operatorname{sgn}(i_d) \\ u_q = \left(K_p + \dfrac{K_i}{s} \right) \operatorname{sgn}(i_q) \end{cases} \tag{8}$$

Table 1 Detailed motor parameters

Parameters	PMSM
d, q-axis inductance *ls/mH*	125.5
Stator resistance *R/Ω*	6
Permanent magnet chain ψ_f/Wb	0.75
Moment of inertia $J/(kg/cm^3)$	0.55
Damping factor *B/N·B·S*	0
Polar logarithm *Pn*	4

3 Model Simulation and Result Analysis

In this paper, the motor control model is built in Simulink platform. The parameters of the permanent magnet synchronous motor are shown in Table 1:

3.1 Simulation of Motor Speed at no Load and Analysis of Results

The simulation condition is set to 1000 r/min.

As shown in Fig. 4 and Table 2, compared with the motor with BP-PID speed controller and the motor with PID speed controller, the regulation time of the motor with the joint control strategy is shortened by 50% and 62.5%, and the steady-state errors and overshoots are obviously reduced. The simulation proves that the joint control strategy is characterized by fast response speed and high steady-state performance.

3.2 Simulation and Result Analysis of Motor under Sudden Load Application

The simulation conditions are set to a speed of 1000 r/min, and a load torque of 5 N-m is applied to the motor at 0.2 s.

As shown in Fig. 5 and Table 3, compared with the motor with BP-PID speed controller and the motor with PID speed controller, the regulation time of the motor with the joint control strategy is reduced by 46.2% and 72%, the speed fluctuation is reduced by 35.2% and 84.3%, and the steady state error of the torque is reduced by 31.3% and 45%, respectively. Simulation shows that the joint control strategy is characterized by fast regulation time and high stability of speed and torque under load.

The joint control strategy significantly outperforms other controllers in terms of dynamic response, system stability, steady-state performance, and adaptability. BP-PID achieves fast response and reduces overshoot through adaptive parameter

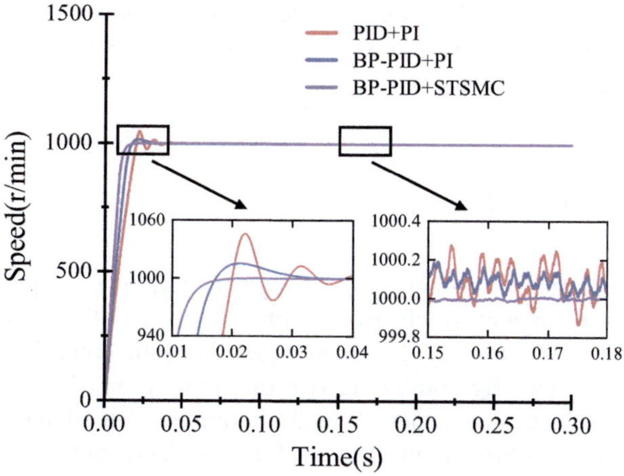

Fig. 4 Comparison graph of specified speed simulation

Table 2 Specified speed simulation performance index

Controllers	Adjustment time (s)	Overtone (r/min)	Steady state error (r/min)
PID + PI	0.04	46.27	0.3
BP-PID + PI	0.03	15.63	0.2
BP-PID + STSMC	0.015	0.55	0.02

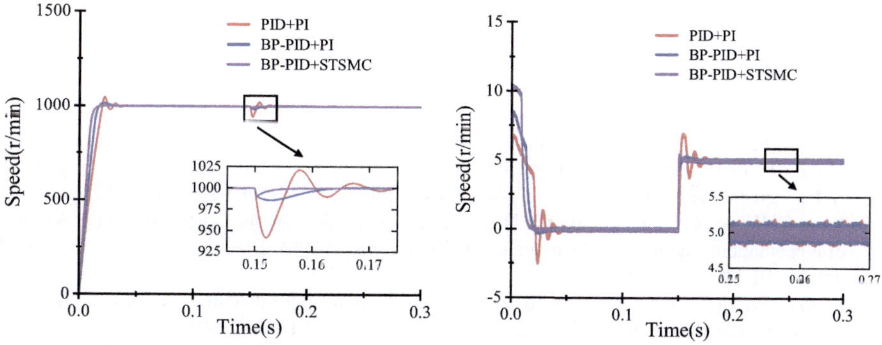

(a) LoadedRPM (b) Loadedtorque

Fig. 5 Comparison graph of sudden load simulation

Table 3 Performance indexes of rotational speed under sudden loading

Controllers	RPM fluctuation(r/min)	Adjustment time (s)	Torque steady state error (T·m)
PID + PI	−58.5	0.025	0.2
BP-PID + PI	−14.2	0.013	0.16
BP-PID + STSMC	−9.2	0.007	0.11

tuning, while STSMC provides strong robustness and current stability. The synergy of the two makes the system perform well in the face of complex loads and disturbances, ensuring optimized overall performance and improved control accuracy.

4 Conclusion

The aim of this paper is to design and implement a joint control strategy for permanent magnet synchronous motor based on BP-PID speed controller and STSMC current controller. The advantage of the joint control strategy is comprehensive optimization. By combining two different types of control techniques, better control results can be achieved. Its limitation is the implementation cost. The need to integrate multiple control algorithms and hardware may lead to high implementation cost. Simulation results show that the joint control strategy of BP-PID speed controller and STSMC current controller designed in this paper is characterized by fast response, small overshoot and steady state error. The future research direction of the joint control strategy is to further optimize the control parameters and more application scenarios. By optimizing the control parameters more efficient adaptive algorithms can be studied to optimize the parameters of the BP-PID controller. The applications of this control strategy in different fields such as robot control and electric vehicle drive are investigated.

References

1. Hariri AM, Damaki Aliabad A, Ghafarzadeh M et al (2020) Design, analysis, and fabrication of a direct drive permanent NdFeB magnet synchronous motor for precision position control. IET Electr Power Appl 14(8):1438–1445. https://doi.org/10.1049/iet-epa.2018.5542
2. Yang G, Wang X, Yang C et al (2022) Effect of slot opening on permanent magnet power loss of a permanent magnet synchronous machine driven by PWM. Energies 15(20):7485. https://doi.org/10.3390/en15207485
3. Zhao H, Wang C (2017) A simple fuzzy PI control of dual-motor driving servo system. MATEC Web Conf 104:02006. https://doi.org/10.1051/matecconf/201710402006
4. Li Y, Sun K, Tong S (2019) Observer-based adaptive fuzzy fault-tolerant optimal control for SISO nonlinear systems. IEEE Trans Cybern 49(2):649–661. https://doi.org/10.1109/TCYB.2017.2785801
5. Zheng Y, Zhao H, Zhen S et al (2021) Fuzzy-set theory based optimal robust constraint-following control for permanent magnet synchronous motor with uncertainties. Control Eng Pract 115:104911. https://doi.org/10.1016/j.conengprac.2021.104911
6. Lu E, Li W, Yang X et al (2019) Anti-disturbance speed control of low-speed high-torque PMSM based on second-order non-singular terminal sliding mode load observer. ISA Trans 88:142–152. https://doi.org/10.1016/j.isatra.2018.11.028
7. Liu W, Zhao T (2021) An active disturbance rejection control for hysteresis compensation based on neural networks adaptive control. ISA Trans 109:81–88. https://doi.org/10.1016/j.isatra.2020.10.019

8. Yang Z, Wang L (2014) Online multi-parameter identification for surface-mounted permanent magnet synchronous motors. Trans China Electrotech Soc 29(3):111–118. https://doi.org/10.19595/j.cnki.1000-6753.tces.2014.03.015
9. Wang J, Fan P (2020) Research on PI-fuzzy hybrid controller for position loop of PMSM servo system. Micromotors 53(7):33–36. https://doi.org/10.15934/j.cnki.micromotors.2020.07.007
10. Xu X, Lin H (2014) Integrated design for permanent magnet synchronous motor servo systems based on dynamic sliding mode control. Trans China Electrotech Soc 29(5):77–83. https://doi.org/10.19595/j.cnki.1000-6753.tces.2014.05.011

Open Access This chapter is licensed under the terms of the Creative Commons Attribution-NonCommercial-NoDerivatives 4.0 International License (http://creativecommons.org/licenses/by-nc-nd/4.0/), which permits any noncommercial use, sharing, distribution and reproduction in any medium or format, as long as you give appropriate credit to the original author(s) and the source, provide a link to the Creative Commons license and indicate if you modified the licensed material. You do not have permission under this license to share adapted material derived from this chapter or parts of it.

The images or other third party material in this chapter are included in the chapter's Creative Commons license, unless indicated otherwise in a credit line to the material. If material is not included in the chapter's Creative Commons license and your intended use is not permitted by statutory regulation or exceeds the permitted use, you will need to obtain permission directly from the copyright holder.

Research on Fixed-Wing Unmanned Aerial Vehicles Based on Adaptive Control of Model Parameters

Junjie Sun, Wencai Zhang, Lei Li, Doudou Li, and Huimin Guo

Abstract In order to solve the problem of uncertain parameters of fixed-wing unmanned aerial vehicles (UAV), a kinematic and dynamic simulation model is established, and the transfer function of the lateral and longitudinal loops is obtained by trimming and linearizing the model at typical state points. The adaptive method of model parameters (MPAC) is used to update the state feedback output, so as to realize the tracking of the lateral and longitudinal flight status. The simulation model of the lateral and longitudinal loops based on the adaptive control of model parameters is constructed, and compared with the PID control, the results show that the response time and overshoot of the lateral and longitudinal loops of the aircraft are significantly improved by adding the adaptive control algorithm of model parameters.

Keywords Fixed-wing UAV · Trimming and linearizing · MPAC · Lateral and longitudinal loops

1 Introduction

Fixed-wing UAV technology has been widely used in military, agriculture and other fields. As the core component of the drone, the flight control system ensures that the drone can accurately complete the scheduled mission [1–4]. However, there are still some technical challenges to meet the needs of advanced missions such as electronic countermeasures, reconnaissance, early warning, coordination, strikes, etc. People adopt a continuous closed-loop design for the flight control system, with trajectory tracking control as the outer loop and attitude control as the inner loop to

J. Sun · W. Zhang (✉) · L. Li · D. Li · H. Guo
Norinco Group Liao Shen Industries Group Co., Ltd, Shenyang, China

© The Author(s) 2026

Z. Zhang, P. Mercorelli (eds.), *Mechanical Design and Simulation: Insights and Innovations*, Lecture Notes in Mechanical Engineering,
https://doi.org/10.1007/978-981-95-2405-1_109

model and simulate the system [5]. In order to solve the problem of uncertainty of UAV parameters, the dynamic inverse is used to eliminate the nonlinearity and coupling problems, and the adaptive control law is established to enable the UAV to accurately track the trajectory [6]. Some scholars established a lateral dynamics model for fixed-wing UAVs with unfixed UAV control parameters, and designed an adaptive law of roll angle to overcome the parameter changes of UAVs [7]. Based on the nonlinear control theory and the adaptive neural network interference observer, the model uncertainty and external interference can be effectively estimated based on the nonlinear control theory and the adaptive neural network interference observer [8].

In this paper, an adaptive pitch and roll angles controller are designed on the basis of transverse and longitudinal dynamic modeling, and MPAC is used to update the state feedback output to realize the tracking of the transverse and longitudinal flight status.

2 Description of the Equation of Motion of Fixed-Wing UAV

According to the aerodynamic data and propeller thrust data of the fixed-wing UAV, the dynamics model of the fixed-wing UAV is established, as shown in Fig. 1. The dynamic model of the fixed-wing UAV is composed of a parameter calculation unit, an aerodynamic model, an engine model and an aircraft motion equation, and each module is related to each other to form an operational organic whole.

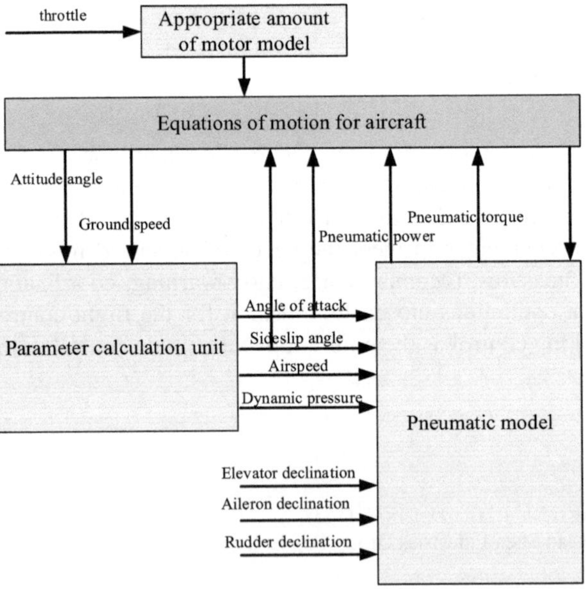

Fig. 1 Schematic diagram of the adaptive model parameters

By describing the kinematics and dynamics equations of the fixed-wing UAV, the three position states and the three velocity states of the fixed-wing UAV can be expressed as in Eqs. 1 and 2.

$$\begin{cases} \dot{P}_n = (\cos\theta\cos\psi)u + (\sin\phi\sin\theta\cos\psi - \cos\phi\sin\psi)v \\ \qquad + (\cos\phi\sin\theta\cos\psi + \sin\phi\sin\psi)w \\ \dot{P}_e = (\cos\theta\sin\psi)u + (\sin\phi\sin\theta\sin\psi + \cos\phi\cos\psi)v \\ \qquad + (\cos\phi\sin\theta\sin\psi - \sin\phi\cos\psi)w \\ \dot{h} = u\sin\theta - v\sin\phi\cos\theta - w\cos\phi\cos\theta \end{cases} \tag{1}$$

$$\begin{gathered} \dot{u} = rv - qw - g\sin\theta + C_x\frac{\rho V_a^2 S}{2m} \qquad \dot{v} = pw - ru + g\cos\theta\sin\phi + C_y\frac{\rho V_a^2 S}{2m} \\ \dot{w} = qu - pv + g\cos\theta\cos\phi + C_z\frac{\rho V_a^2 S}{2m} \end{gathered} \tag{2}$$

where $[P_n \quad P_e \quad h]$ represents the position of the UAV in the inertial coordinate system, $[\phi \quad \theta \quad \psi]$ represents the attitude angle of the UAV in the inertial coordinate system, $[u \quad v \quad w]$ represents the speed of the UAV in the airframe coordinate system, $[p \quad q \quad r]$ represents the angle of the UAV in the airframe coordinate system, C_x, C_y, C_z represents the moment coefficient, ρ represents the air density, V_a represents the airspeed, S represents the wing area, and m represents the UAV mass. The three angular position states and the three angular velocity states of the drone can be represented as in Eqs. 3 and 4.

$$\begin{gathered} \dot{\phi} = p + q\sin\phi\tan\theta + r\cos\phi\tan\theta \qquad \dot{\theta} = q\cos\phi - r\sin\phi \\ \dot{\psi} = q\sin\phi\sec\theta + r\cos\phi\sec\theta \end{gathered} \tag{3}$$

$$\begin{aligned} \dot{p} &= \Gamma_1 pq - \Gamma_2 qr + C_P\frac{1}{2}\rho V_a^2 Sb \\ \dot{q} &= \Gamma_5 pr - \Gamma_6(p^2 - r^2) + C_q\frac{1}{2J_y}\rho V_a^2 Sc \\ \dot{r} &= \Gamma_7 pq - \Gamma_1 qr + C_r\frac{1}{2}\rho V_a^2 Sb \end{aligned} \tag{4}$$

where C_P, C_q, C_r represent the torque coefficients, and Γ_1, Γ_2, Γ_5, Γ_6, Γ_7, J_y represent the coefficients related to moment of inertia, and the detailed derivation process is in the references [9], which will not be repeated in this article.

3 Adaptive Control Algorithm for Model Parameters

MPAC is to establish an adaptive mechanism to automatically adjust the parameters of the process model according to the error of the control on the phenomenon and the process model. The adaptive block diagram of the model parameters is shown in Fig. 2.

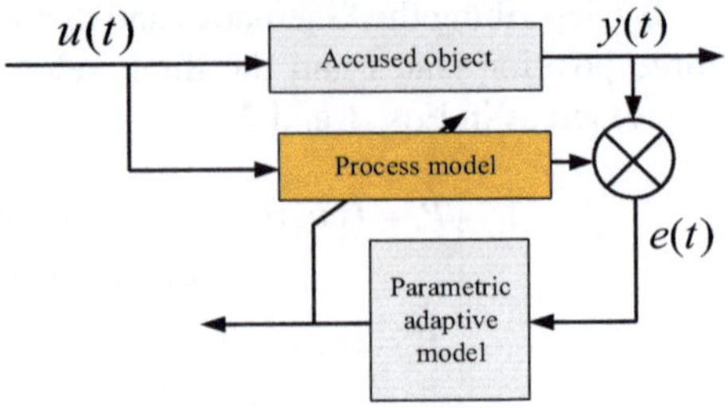

Fig. 2 Schematic diagram of the adaptive model parameters

In order to minimize the prediction error, the current parameter prediction can be used plus a correction term to get the next prediction. Let an adjustable parametric model be in Eq. 5.

$$\hat{y}^0(t+1) = \hat{y}\left[(t+1)\middle|\hat{\Phi}(t)\right] = -\hat{a}_1((t+1)y(t) + \hat{b}_1(t+1)u(t) = \hat{\Phi}^T(t)\phi(t) \quad (5)$$

where $\hat{y}^0(t+1)$ is the a priori predicted output of the estimated parameters at t moment, a_1 and b_1 are unknown parameters, $\phi(t)$ is the observation vector, $\hat{\Phi}(t)$ is the estimated parameter vector. Model parameter adaptation can be expressed in Eq. 6 [10].

$$\hat{\Phi}(t+1) = \hat{\Phi}(t) + \frac{F\phi(t)e(t+1)}{1+\phi^T(t)F\phi(t)} \quad (6)$$

where F is the adaptive gain matrix, $e(t+1)$ represents the error function. In order to minimize the output and estimation vectors, a least squares criterion needs to be established, as shown in Eq. 7.

$$\min_{\hat{\Phi}(t)} J(t) = \sum_{i=1}^{t}\left[y(i) - \hat{\Phi}^T(t)\phi(i-1)\right]^2 \quad (7)$$

Let $\frac{\partial J(t)}{\partial \hat{\Phi}(t)} = 0$ find the minimum error between the output and the estimated parameters. For the sake of simplicity the formula makes $F(t)^{-1} = \sum_{i=1}^{t}\phi(i-1)\phi^T(i-1)$, the Eq. 8 can be obtained.

$$\hat{\Phi}(t+1) = F(t+1)\sum_{i=1}^{t+1} y(i)\phi(i-1) \quad (8)$$

Combine the Eqs. 5, 6, 7 and 8 to get the Eq. 9.

$$\hat{\Phi}(t+1) = \hat{\Phi}(t) + F(t)\phi(t)e(t+1) \quad (9)$$

The recursive formula of $F(t)$ can be obtained by matrix inverse lemma.

4 Simulation of Fixed-Wing UAV Dynamics

4.1 Aircraft Model Building

In the second section, we derive the kinematics and dynamics equations of fixed-wing UAVs, and build corresponding models in Simulink based on the above equations to complete simulations, which is the basis for UAV trim and linearization, and controller design. The simulation of the aircraft model is shown in Fig. 3. The simulation model is divided into three modules, which are composed of a parameter calculation unit, an aerodynamic model and an aircraft kinematics equation.

4.2 Trim Conditions

When the fixed-wing UAV is at a constant altitude and the wings are flying laterally and stably, its speed $\begin{bmatrix} u & v & w \end{bmatrix}$, Euler angle $\begin{bmatrix} \phi & \theta & \psi \end{bmatrix}$, and angular rate $\begin{bmatrix} p & q & r \end{bmatrix}$ are constant, and the aircraft in this state is said to be in trim.

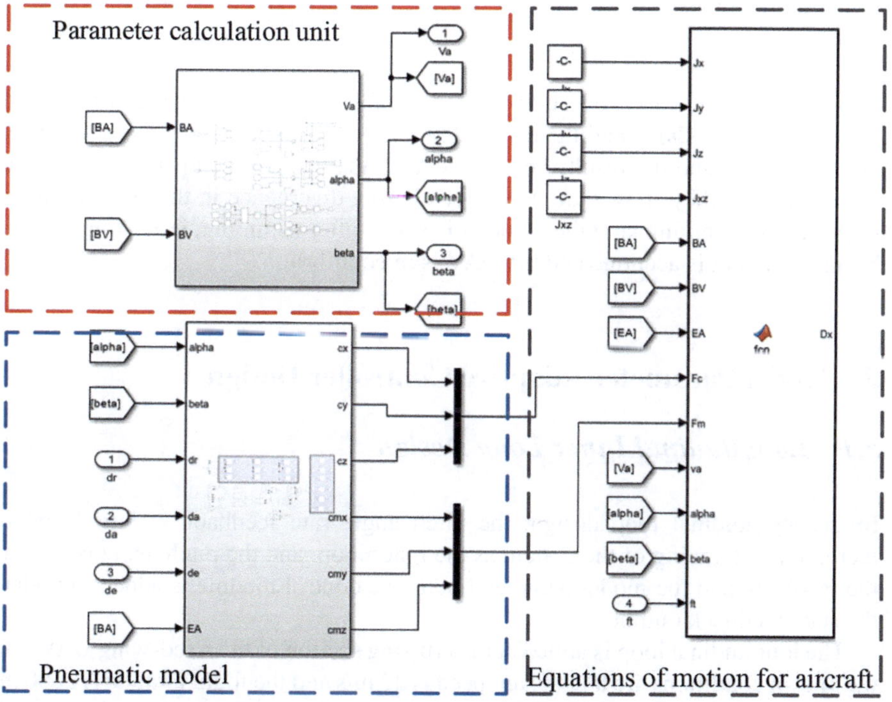

Fig. 3 Dynamics-based aircraft model simulation system

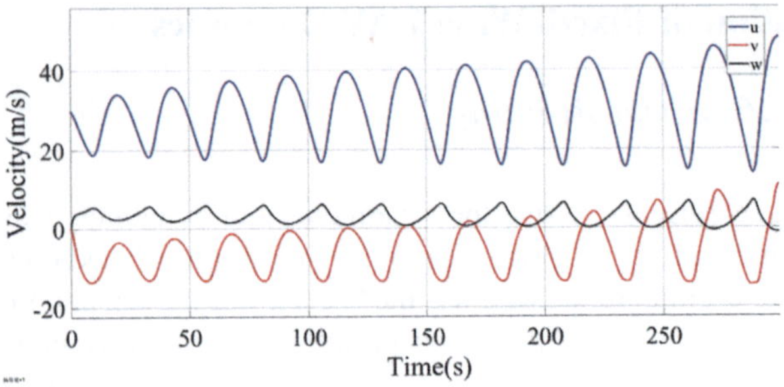

Fig. 4 The aircraft flies the speed curve before trimming

Fig. 5 The aircraft flies the speed curve after trimming

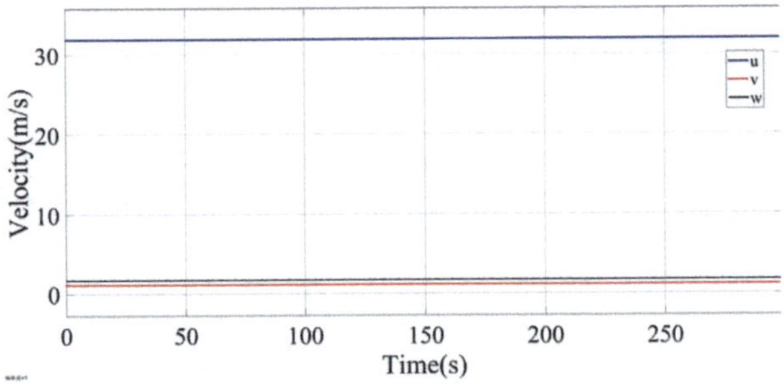

Therefore, we carry out trim in the established model, the constraint are $h = 1000m, V_a = 32m/s$, and the result after trim is obtained, taking the UAV speed as an example, and the results before and after the trim are compared as shown in Figs. 4 and 5. They show that there is oscillating divergence in the velocity of the UAV before trimming, and the velocity is stable after trimming, indicating that the balancing result is accurate and the UAV is in equilibrium.

5 Model Parameter Adaptive Controller Design

5.1 *Longitudinal Inner Loop Design*

In the longitudinal loop design, the pitch angle rate feedback is introduced to increase the damping of the system as the inner loop, and the pitch angle is used as the feedback and the model parameter adaptive control module is added to update the state feedback output.

The longitudinal loop is aimed at the cruising section of the fixed-wing UAV, and the known conditions are that the airspeed is 32 m/s and the flight altitude is 2000 m, and the pitch angle tracking simulation is carried out on the longitudinal loop with the pitch angle control command of 5°, and the simulation results are shown in Fig. 6. The results show that the response of the longitudinal inner loop based on

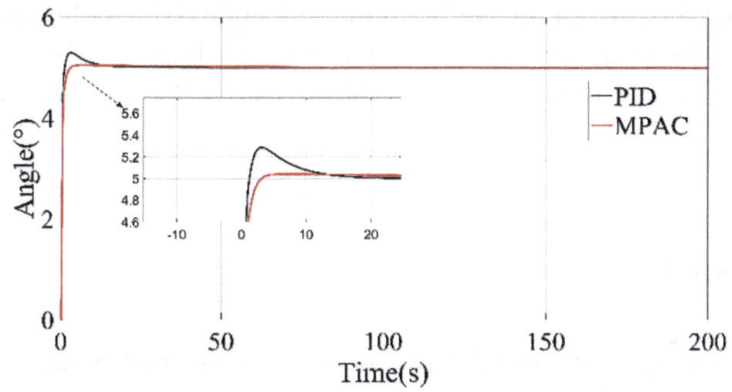

Fig. 6 The pitch angle control command is 5° based on the results of traditional PID and MPAC simulations

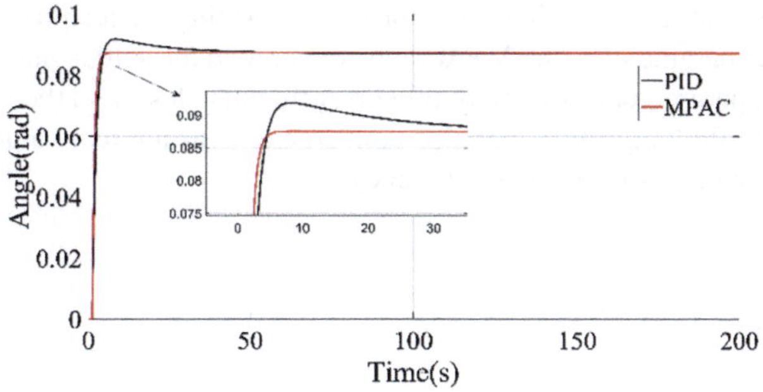

Fig. 7 Based on the traditional PID and MPAC control algorithms, the roll angle control command is the 0.087 rad simulation result

traditional PID to the pitch angle of 5° is 0.3°, and the response time of the longitudinal inner loop based on MPAC is not overshoot to the pitch angle of 5°, and the response time is also improved. The results show that the longitudinal inner loop of MPAC has better time-domain response characteristics.

5.2 *Lateral Inner Loop Design*

In the lateral loop design, on the basis of introducing the roll angle rate feedback to increase the system damping, tracking and yaw angle commands as the inner loop, the roll angle is used as the feedback and the model parameter adaptive control module is added to update the state feedback output.

For each stage of the fixed-wing UAV, the known conditions of the lateral inner loop are 32 m/s, the flight altitude is 2000 m, and the roll angle control command is 0.087 rad, and the simulation results are shown in Fig. 7. The simulation results show that the response of the lateral inner loop based on the traditional PID to the roll angle 0.087 rad instruction has an overshoot of 0.007 rad, and the response of the lateral inner loop based on MPAC to the roll angle 0.087 rad instruction has no overshoot, and the response time is also significantly improved by 2 s. The results show that the lateral inner loop of MPAC has better time-domain response characteristics.

6 Conclusion

A kinematic and dynamic simulation model is established based on the aircraft dynamics equation, and the aircraft trim and linearization operations are completed. The simulation model of the horizontal and longitudinal loops based on the adaptive control of model parameters is constructed, and compared with the PID control, the results show that the response time and overshoot of the horizontal and longitudinal loops of the aircraft are significantly improved by adding the adaptive control algorithm of model parameters. The MPAC can converge to the actual value quickly, and the estimation parameter error and output error fluctuate less, and there is no steady-state error. The PID algorithm can also converge to near the real value, but it takes longer. This greatly reduces work efficiency.

References

1. Pham TV, Nguyen TD (2024) Path-following formation of fixed-wing UAVs under communication delay: a vector field approach. Drones 8(6). https://doi.org/10.3390/drones8060237
2. Brezoescu A, Espinoza T, Castillo P et al (2013) Adaptive trajectory following for a fixed-wing UAV in presence of crosswind. J Intell Robot Syst 69(1–4):257–271. https://doi.org/10.1007/s10846-012-9756-8
3. Rauniyar S, Kim D (2023) Genetic fuzzy inference system-based three-dimensional resolution algorithm for collision avoidance of fixed-wing UAVs. Electronics 12(18). https://doi.org/10.3390/electronics12183946
4. Low CB, Ng QS (2011) A flexible virtual structure formation keeping control for fixed-wing UAVs. In: IEEE International conference on Control & Automation. IEEE. https://doi.org/10.1109/ICCA.2011.6137876
5. Zhai R, Zhou Z, Zhang W et al (2014) Control and navigation system for a fixed-wing unmanned aerial vehicle. AIP Adv 4(3):36. https://doi.org/10.1063/1.4866169
6. Wang S, Li C, Wang Z et al (2022) Design of landing controller based on adaptive dynamic inverse. Syst Eng Electron 44(1):8. https://doi.org/10.12305/j.issn.1001506X.2022.01.27
7. Zhang L, Yang F, Zhou W et al (2021) UAV roll control based on L_ (1) adaptive control law. Sci Technol Eng 21(11):6. https://doi.org/10.3969/j.issn.1671-1815.2021.11.060
8. Matassini T, Shin HS, Tsourdos A et al (2016) Adaptive control with neural networks-based disturbance observer for a spherical UAV. IFAC-PapersOnLine 49(17):308–313. https://doi.org/10.1016/j.ifacol.2016.09.053
9. Beard RW, Mclain TW (2012) Small unmanned aircraft: theory and practice. Princeton University Press. https://www.researchgate.net/publication/286268062_Small_Unmanned_Aircraft_Theory_and_Practice
10. Li Q (2023) Research on path planning algorithm for fixed-wing UAV. Huazhong University of Science and Technology. https://doi.org/10.27157/d.cnki.ghzku.2023.003698

Open Access This chapter is licensed under the terms of the Creative Commons Attribution-NonCommercial-NoDerivatives 4.0 International License (http://creativecommons.org/licenses/by-nc-nd/4.0/), which permits any noncommercial use, sharing, distribution and reproduction in any medium or format, as long as you give appropriate credit to the original author(s) and the source, provide a link to the Creative Commons license and indicate if you modified the licensed material. You do not have permission under this license to share adapted material derived from this chapter or parts of it.

The images or other third party material in this chapter are included in the chapter's Creative Commons license, unless indicated otherwise in a credit line to the material. If material is not included in the chapter's Creative Commons license and your intended use is not permitted by statutory regulation or exceeds the permitted use, you will need to obtain permission directly from the copyright holder.

Design and Experimental Study of Bionic Olfactory Sensitive Unit

Yanyuan Huang, Yuxuan Fan, Wei Zhou, Xingyu Wang, and Wenqiang Zhang

Abstract In daily life, accurate gas detection is usually required in important scenarios such as air environment detection, fruit and vegetable ripening detection, and border inspection. Traditional gas detection methods include gas chromatography and mass spectrometry, but all of them have the disadvantages of complicated operation and long analysis time. Bionic electronic nose is a new type of gas detection method in recent years, which has the characteristics of fast response, high sensitivity, and good real-time performance, but the existence of internal gas chamber, sensor material, and the phenomenon of temperature drift will affect the accuracy and sensitivity of gas detection. In this paper, on the basis of imitating the internal structure of the nasal cavity of canine mammals, the basic design requirements of the bionic electronic nose are proposed, and three different structures are designed. The software Solidworks is used to construct a three-dimensional model and extract the fluid domain, establish a hydrodynamic model, and simulate and analyze the inlet and outlet process of the bionic electronic nose through the software Ansys Fluent, taking the contact effect between the gas and the sensor as an index, comparing the velocity cloud diagram, streamline diagram, and velocity vector arrow diagrams of the three designs of the internal cavities under the same flow velocity, and finding that the double inlet channel of the structure has the best effect. At the same time, using the combination of the upper computer and MQ series sensors for gas sensing experiments, it was found that the detection efficiency of the gas was increased by 20% by using the shape of narrow front and wide back combined with the gas infusion structure.

Keywords Gas detection · Structural design · Finite element simulation analysis · Sensor

Y. Huang · Y. Fan · W. Zhou · X. Wang · W. Zhang (✉)
College of Engineering, China Agricultural University, Beijing, China

© The Author(s) 2026

Z. Zhang, P. Mercorelli (eds.), *Mechanical Design and Simulation: Insights and Innovations*, Lecture Notes in Mechanical Engineering,
https://doi.org/10.1007/978-981-95-2405-1_110

1 Introduction

Odors emanating from nature often contain a lot of useful information that can be used for detection and identification. For example, plants use their own odors to attract animals for pollination purposes; in addition, in medicine, by analyzing the composition of the gas exhaled by a person, it can give useful information to the doctor to make a diagnosis, which to a certain extent improves the possibility of early detection of diseases and early treatment [1].The above cases show the value of studying gas detection and response mechanisms, and the need to meet the requirements of high efficiency, speed and accuracy in these studies.

In recent years, scholars have proposed bionics and rapid development, they imitate the sensitive mammalian nasal structure combined with the principle of gas detection electronic nose has gradually become a new type of gas detection device [2]. Compared with the existing more mature gas detection methods, the most obvious feature of the bionic electronic nose is that it can realize dynamic detection of gases [3]. Figure 1a shows the working principle of the bionic electronic nose.

At present, the research of bionic electronic nose mainly focuses on replicating mammalian olfactory detection system with sensor arrays [4, 5]. Although the data processing of the electronic nose is inspired by biological olfactory systems, it is still unable to fully simulate the structure of the biological nasal cavity and the raw

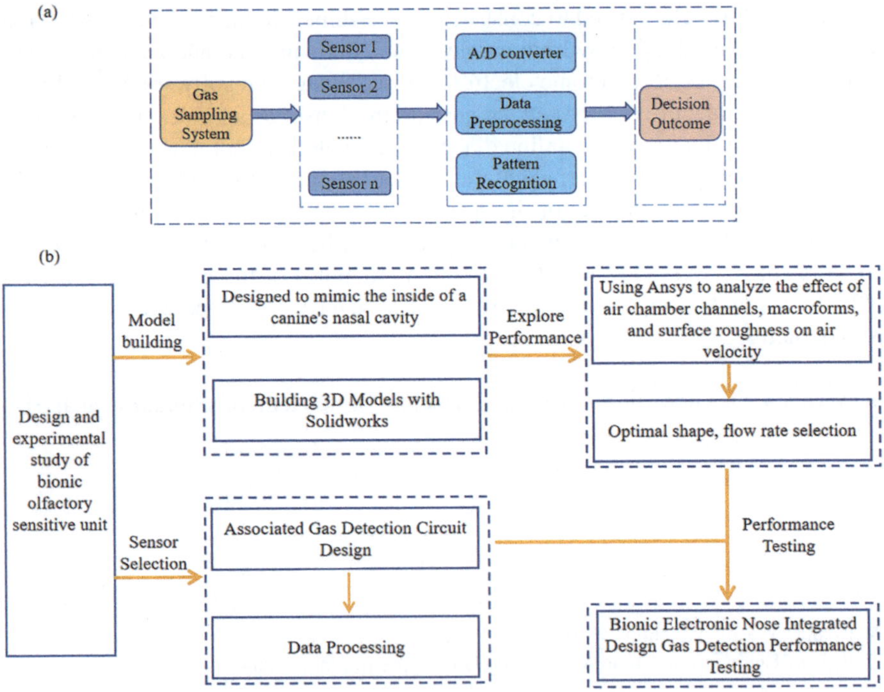

Fig. 1 Working Principle and Technical Route: (**a**) Working principle of the bionic electronic nose, (**b**) Technical route

information of olfactory perception due to the complex process of generating and transmitting odor information [6].

This article is based on the existing research on biomimetic electronic noses at home and abroad, and designs a biomimetic electronic nose system. The technical route is shown in Fig. 1b.

2 Bionic Design Concept

2.1 *Bionic Design Principle*

Adding the nasal structure of animals with a sensitive sense of smell imitated on the basis of the electronic nose, the gas flow rate, pressure and other factors in the nasal cavity can be controlled, thus enabling the rapid detection of gases. In nature, mammals represented by canines, it is because it has a special nasal structure, so that it shows a sensitive sense of smell.

2.2 *Bionic Nasal Design*

This design follows the principles of bionics and gas detection, mimicking the olfactory system of canine animals for nasal cavity design. Three biomimetic electronic nose structures were preliminarily designed using Solidworks, as shown in Fig. 2, each with the following characteristics:

Design 1: Referring to the narrow front and wide back characteristics of a dog's nasal cavity, the intake passage is designed in a shape ranging from small to large aperture, with the final aperture being consistent with the cross-sectional size of the mid-range gas detection. A gas guide device was designed at the end of the intake hole, and sensors were installed in four circular holes in the array.

Design 2: Two air inlet holes, the rest are the same as Design 1.

Design 3: It has the same intake channel and flow guide device as design one, and the circular array of holes corresponding to the sensor installation is on the surface of the model.

2.3 *Finite Element Simulation of the Flow Field Inside the Bionic Nasal Cavity*

The biomimetic electronic nose structure designed in this study assumes that the fluid used for calculation is a gas that approximates air, with a density of 1.225 kg/m^3 and a viscosity coefficient of 17.894 μPa • S [7, 8]. Using the Reynolds number

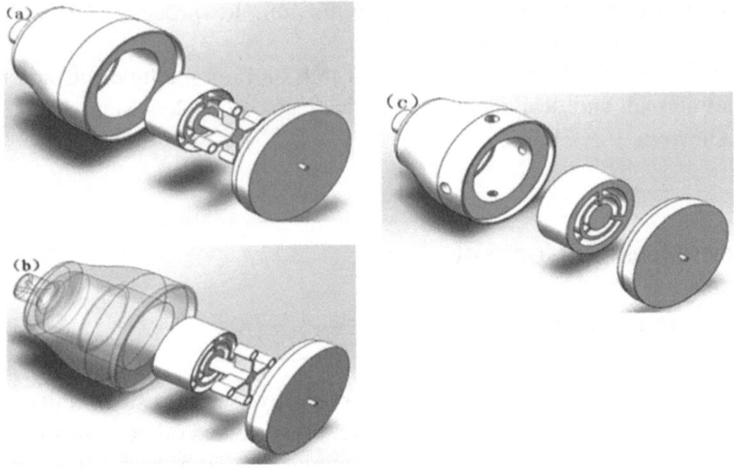

Fig. 2 Biomimetic electronic nose structure: (**a**) Design a single intake channel nasal cavity, (**b**) design a dual intake channel nasal cavity, (**c**) design a three ring array nasal cavity

formula, the maximum Reynolds number in this design is 1900, and the gas flow belongs to laminar flow.

The fluid continuity equation is:

$$\int_{A_2}\rho_2\mu_2 dA - \int_{A_1}\rho_1\mu_1 dA + \int_{V}\frac{\partial \rho}{\partial t} dV_c = 0$$

3 Analysis of Simulation Results

The velocity clouds inside the three cavities are shown in Fig. 3, Due to the use of gas deflectors and the loading of four sensors, this results in a difference in the XY, XZ plane velocity maps.

The velocity distribution inside the nasal cavity is shown in Fig. 4. When the inhalation gas flow rate (5 m/s) is the same, the double inlet channel nasal cavity will inhale more gas compared to the other two, so it is more favorable for the sensor to detect the gas.

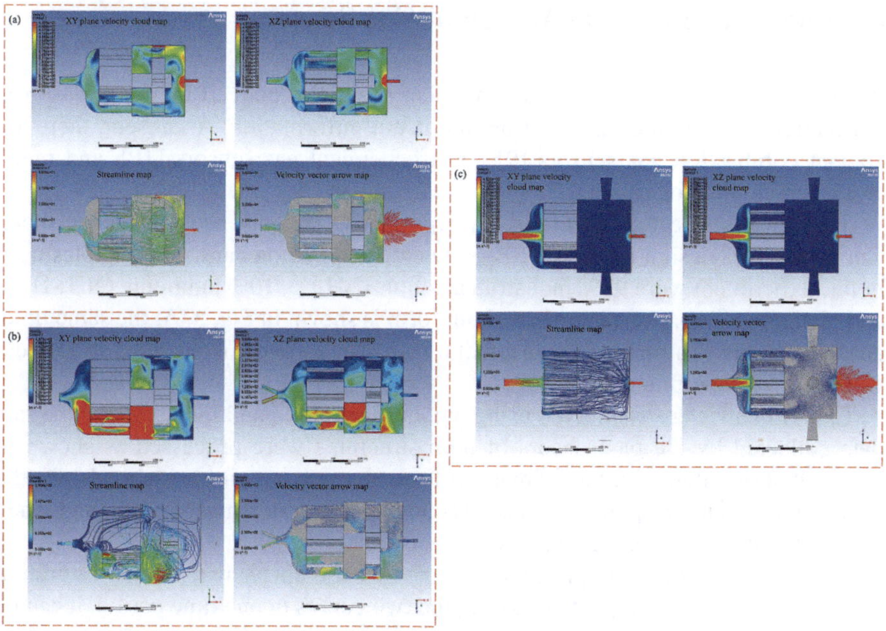

Fig. 3 Velocity clouds: (**a**) Design 1, (**b**) Design 2, (**c**) Design 3

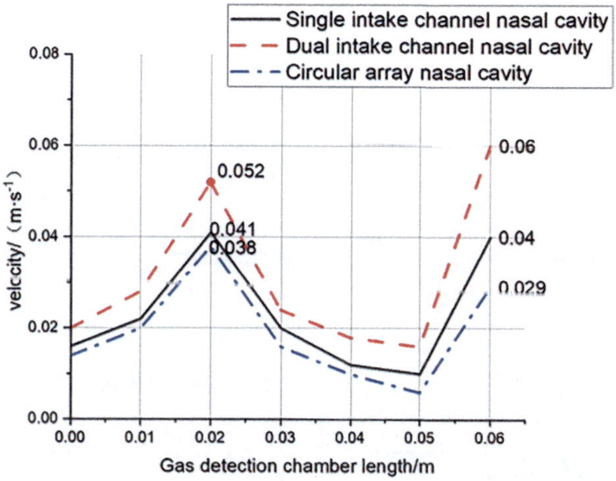

Fig. 4 Velocity distribution inside the nasal cavity

4 Selection and Array Design of Gas Sensors

A gas sensor is a component that can detect gases and convert their composition, concentration, and other information into useful information. At present, there are contact combustion gas sensors [9], electrochemical gas sensors [10], and metal oxide semiconductor gas sensors [11] on the market.

The sensors selected for this study are all MQ series sensors from Weisheng Company, This design uses NI USB-6009 data acquisition card. The samples used in this experiment were 98% industrial alcohol (C_2H_6O), 10% ammonia (NH_3-H_2O) and methane (CH_4).The experimental platform is shown in Fig. 5. At the same time, an ordinary cavity structure, i.e., a structure without a narrow front and a wide back, was used for experimental control.

Experiments were conducted using alcohol, ammonia and methane and the raw data collected by the gas sensors of both structures were processed as shown in Fig. 6. It shows that with the increase of time, the sensor continuously adsorbs the sample gas and the response curve shows an upward or downward trend, and when it reaches saturation, the curve shows a steady state.

In the tests of three samples, the designed e-nose cavity reached the saturation time 20 s earlier on average. Comparing the two curves in one time period, it can be seen that the detection efficiency of the designed cavity structure was improved by about 20% this time. It is proved that the front-narrow-back-wide structure and the gas deflector can make the gas contact with the sensor faster and more comprehensively.

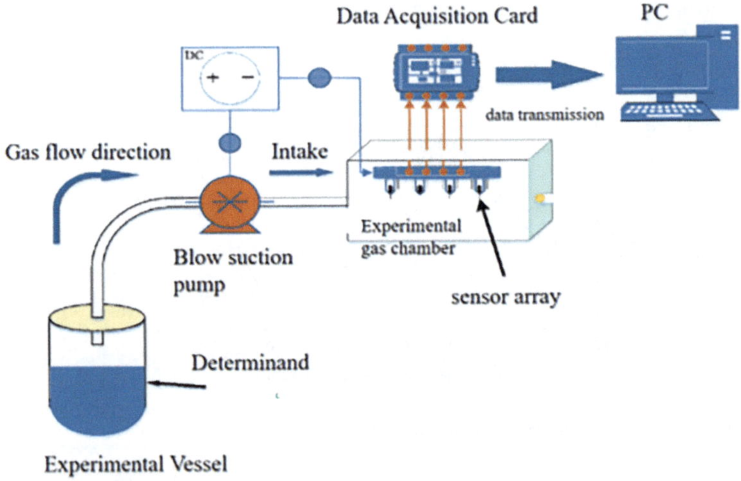

Fig. 5 Biomimetic electronic nose experimental platform

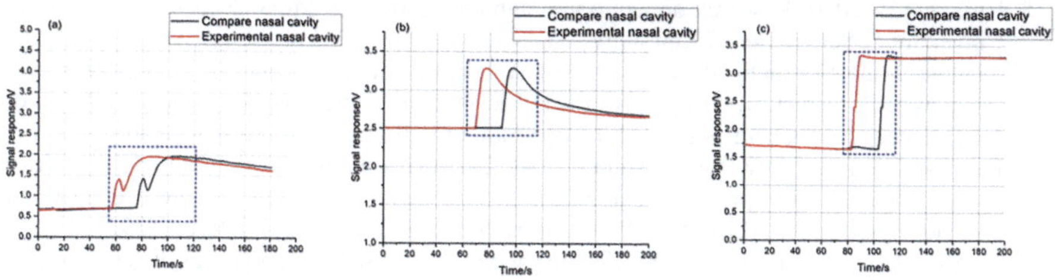

Fig. 6 Experimental data comparison chart: (**a**) Comparison chart of alcohol sensing experiment data (**b**) Comparison chart of ammonia sensing experiment data (**c**) Comparison chart of methane sensing experiment data

5 Conclusion

In this study, three bionic e-nasal cavities were designed to mimic the internal nasal structure of canine mammals by combining the principle of bionics. The internal fluid domains of the three cavities were constructed using Solidworks, and the gas flow process was simulated and analyzed in Ansys Fluent. The following conclusions are drawn:

1. Among the three designed bionic nasal cavities, the CFD POST post-processing results showed that the internal gas flow rate of the dual inlet channel nasal cavity was optimal and uniform, and the gas was able to make full contact with the sensor and respond.
2. The narrow front and wide back structure improves the gas detection efficiency by about 20% compared to the normal structure. By measuring and comparing the gas detection efficiency under the two structures, we observe that the response time of gas detection is shorter and the detection sensitivity is higher in the narrow front and wide back structure.

References

1. Zhu P, Yulan T, Yating C (2021) Research on odor detection using bioelectronic nose combined with genetic engineering. Measure Contr Technol 40(12):59–63. https://doi.org/10.19708/j.ckjs.2021.02.228
2. Yue Z (2012) Research on air quality and harmful gas detection methods in vehicles. Graduate School of Chinese Academy of Sciences
3. Weng Xiaohui, Sun Youhong, Zhang Shujun,etc. Oil and gas detection method and experimental new technology based on biomimetic nasal cavity optimization. J Jilin Univ Eng Edn, 2020, 50(01): 382–388. https://doi.org/10.13229/j.cnki.jdxbgxb20181069
4. Yongjin F (2015) Research on active perception of biomimetic nose based on POMDP. Zhejiang Gongshang University

5. Huixiang L (2020) Research on key issues and applications of MOS electronic nose. Beijing University of Science and Technology
6. Xingwei J (2018) Handheld electronic nose simulation chamber design. Tianjin University
7. Wasilewski T, Gębicki J, Kamysz W (2021) Bio-inspired approaches for explosives detection. TrAC Trends Anal Chem 142:116330. https://doi.org/10.1016/j.trac.2021.116330
8. Fernández-Parra R, Pey P, Zilberstein L et al (2019) Use of computational fluid dynamics to compare upper airway pressures and airflow resistance in brachycephalic, mesocephalic, and dolichocephalic dogs. Vet J 253:105392. https://doi.org/10.1016/j.tvjl.2019.105392
9. Jang W, Park J, Lee K et al (2018) Methane and hydrogen sensing properties of catalytic combustion type single-chip micro gas sensors with two different Pt film thicknesses for heaters. Micro Nano Syst Lett 6(1):7. https://doi.org/10.1186/s40486-018-0069-y
10. Silambarasan P, Moon IS (2021) Real-time monitoring of chlorobenzene gas using an electrochemical gas sensor during mediated electrochemical degradation at room temperature. J Electroanal Chem 894:115372. https://doi.org/10.1016/j.jelechem.2021.115372
11. Gai L, Lai R, Dong X et al (2022) Recent advances in ethanol gas sensors based on metal oxide semiconductor heterojunctions. Rare Met 41(6):1818–1842. https://doi.org/10.1007/s12598-021-01937-4

Open Access This chapter is licensed under the terms of the Creative Commons Attribution-NonCommercial-NoDerivatives 4.0 International License (http://creativecommons.org/licenses/by-nc-nd/4.0/), which permits any noncommercial use, sharing, distribution and reproduction in any medium or format, as long as you give appropriate credit to the original author(s) and the source, provide a link to the Creative Commons license and indicate if you modified the licensed material. You do not have permission under this license to share adapted material derived from this chapter or parts of it.

The images or other third party material in this chapter are included in the chapter's Creative Commons license, unless indicated otherwise in a credit line to the material. If material is not included in the chapter's Creative Commons license and your intended use is not permitted by statutory regulation or exceeds the permitted use, you will need to obtain permission directly from the copyright holder.

Optimization of Telescopic Arm Structure of Construction 3D Printing Robot

Wei Chen, Xiaoliang Shi, and Haibin Yin

Abstract The end deflection of the telescopic arm impacts print accuracy. To enhance stiffness, its structure was optimized using topology and size optimization. The U-shaped section of the telescopic arm was optimized using the variable density method in Ansys Workbench under two conditions, with the cross-section reconstructed based on topology optimization results. The wall thickness of the stiffener in the reconstructed cross-section was selected as the optimization variable, the mass of the three-section arm was minimized as the optimization goal, and the total deformation of the end of the three-section arm was less than 3.8 mm as the constraint, and the direct optimization method and screening method were used to optimize the size of the reinforcement. Remodeling with SolidWorks and performing a static analysis of the initial design of the telescopic arm and the optimized telescopic boom, the results showed that the total deformation at the end of the optimized telescopic arm was 1.14 mm smaller than the initial design under the most dangerous conditions, the deformation at the end of the telescopic arm has been reduced by 59.4%.The results show that the stiffness of the telescopic arm is optimized by topology optimization and size optimization.

Keywords Topology optimization · Size optimization · Direct optimization

1 Introduction

Common 3D printing equipment in construction includes gantry, robotic arm, truss, and tower types [1]. Tower-type robots offer wide printing ranges and flexibility but suffer from low accuracy due to self-weight and cantilever deflection, especially arm-end deflection. To improve accuracy, structural optimization of the telescopic

W. Chen · X. Shi · H. Yin (✉)
School of Mechanical and Electrical Engineering, Wuhan University of Technology, Wuhan, Hubei, China
e-mail: chinaliuyin@whut.edu.cn

© The Author(s) 2026

Z. Zhang, P. Mercorelli (eds.), *Mechanical Design and Simulation: Insights and Innovations*, Lecture Notes in Mechanical Engineering,
https://doi.org/10.1007/978-981-95-2405-1_111

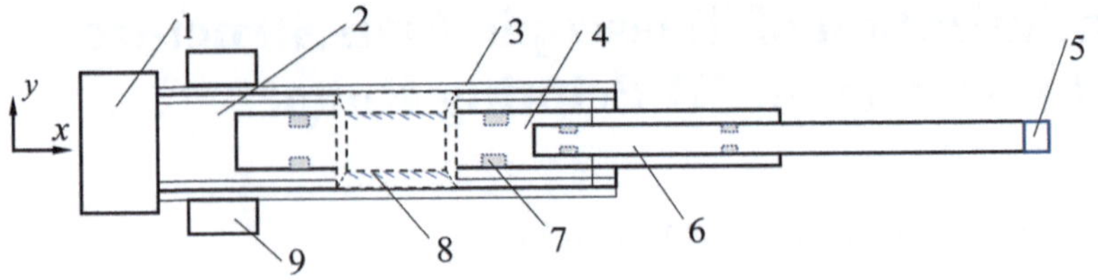

1-Transmission box 2-Basic arm 3-Synchronous belt 4-Two-section arm 5-Printing device 6-Three-section arm 7-Slider 8-Trapezoidal base 9-Counterweight

Fig. 1 Schematic diagram of telescopic arm structure. 1—Transmission box, 2—Basic arm, 3—Synchronous belt, 4—Two-section arm, 5—Printing device, 6—Three-section arm, 7—Slider, 8—Trapezoidal base, 9—Counterweight

arm is crucial. Two methods exist: analytical calculations for deflection and strength, and size optimization of cross-sectional dimensions. Both methods are limited to non-topological changes. Combining topology and size optimization enhances telescopic arm performance by altering its structure and refining cross-sectional parameters, leading to superior performance and reduced weight. In engineering, Wu Qinglong et al. [2], improved the SKO topology optimization method for tower crane booms, providing design insights. Zeng Yefeng [3], optimized bus frameworks, enhancing mechanical properties and achieving a 189 kg weight reduction, demonstrating lightweighting benefits.

2 Condition and Load Analysis of Telescopic Boom

The structural diagram of the telescopic arm for a construction 3D printing robot is in Fig. 1. The base arm 2 connects to the trapezoidal base 8 via bolts. The timing belt 3 on the base arm pulls the three-section arm 6, which then uses stoppers to pull the two-section arm 4, enabling telescopic movement.

2.1 Working Condition Analysis

The telescopic arm has two extreme states: fully extended and retracted. When operated outdoors, wind load impact must be considered. Its main operating conditions are summarized in Table 1.

Analysis shows Operating Condition 1 is critical, requiring static analysis and optimal design of the telescopic arm.

Table 1 Main operating conditions of the telescopic arm

Operating conditions	State of telescopic arm	Direction of wind load
1	Fully Extended	Along z-direction
2	Fully Extended	Along y-direction
3	Fully Retracted	Along z-direction
4	Fully Retracted	Along y-direction

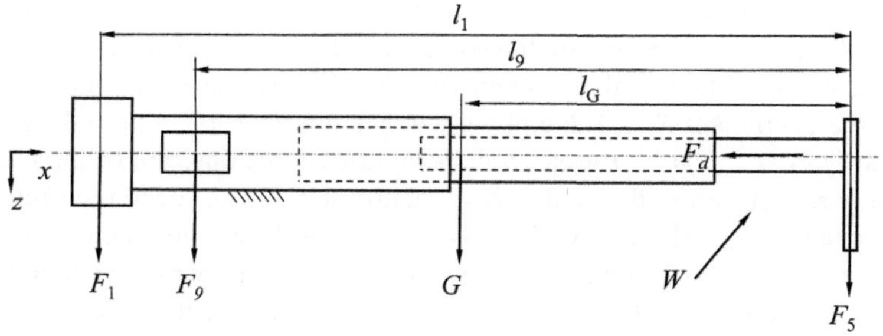

Fig. 2 Simplified force diagram of the telescopic arm

2.2 *Force Analysis*

3D construction printing robots operate slowly, causing minimal inertial forces on their telescopic arms. Since they often work outdoors, wind loads must be considered. Thus, the arm bears self-weight, payload, traction, and wind loads (a horizontal force in any direction). Figure 2 shows a simplified force diagram.

The traction load is applied by the timing belt, and the force acting on the telescopic arm is shown in Eq. (1):

$$\begin{cases} F_r = \dfrac{1000P_d}{v} \\ F_d = 2F_r \end{cases} \tag{1}$$

In the equation, P_d represents the design power, with units of kW, and v represents the belt speed, with units of m/s.

The calculation formula for wind load [4] is as follows:

$$W = K_f K_h q A \tag{2}$$

In the equation: K_f represents the wind force coefficient, K_h represents the wind pressure height variation coefficient, q represents the calculated wind pressure, and A represents the windward area of the telescopic arm perpendicular to the wind direction.

The design power of the timing belt is 0.75 kW, and the speed of the timing belt during operation is 20 mm/s. According to Eq. (1), $F_r = 37{,}500$ N. Since two timing

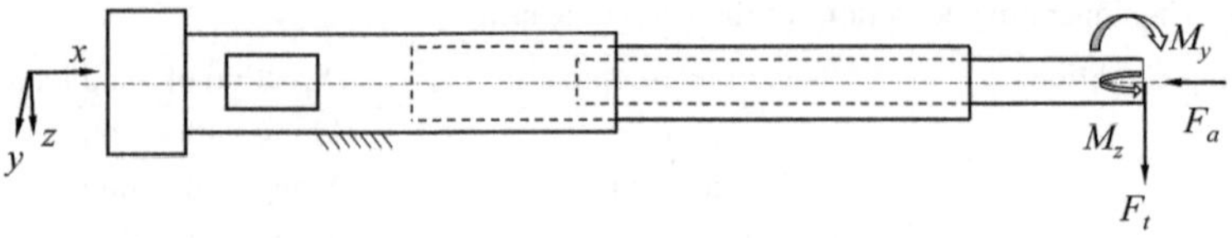

Fig. 3 Equivalent force diagram at the end cross-section of the three-section arm

belts simultaneously traction the three-section arm, the axial force $F_d = 75{,}000$ N. Under severe conditions, the wind pressure for the construction 3D printing robot is 150 N/m^2, the wind pressure height variation coefficient K_h is 1, the wind force coefficient K_f is 1.2, and the windward area A is 2.8 m^2. Using Eq. (2), the wind load $W = 504$ N. Based on the simplified force diagram of the telescopic arm, the forces are simplified to the cross-section at the end of the three-section arm, as shown in Fig. 3. Additionally, $F_1 = 784$ N, which is the gravity of the transmission housing, with $l_1 = 7908$ mm; $F_9 = 735$ N, which is the gravity of the counterweight, with $l_9 = 7237$ mm; $G = 2038$ N, with $l_G = 5087$ mm, representing the gravity of the telescopic arm; and $F_5 = 147$ N, which is the gravity of the end printing device. Based on the above data, the calculations yield: $M_y = 21{,}882$ N·m, $M_z = 6544.5$ N·m, $F_t = 4208$ N, and $F_a = 75{,}000$ N.

3 Structural Optimization Design

3.1 Topology Optimization

The field of topology optimization is divided into discrete structure topology optimization and continuum topology optimization. Research on continuum topology optimization is relatively mature, with major methods including the variable density method, homogenization method, level set method, and evolutionary structural optimization method [5–8]. In Ansys Workbench, topology optimization employs the variable density method, which sets the "element density" of each unit in the design space of the finite element model as the design variable. After optimization and solution, a unit density of 1 indicates that the material at that unit's location is crucial and should be retained; whereas a unit density of 0 suggests that the material at that location is unnecessary and can be removed, thereby achieving efficient material utilization.

Establishment of a Finite Element Model Given the telescopic arm's long sections (max 4900 mm) and complex buckling, full-section analysis would be computationally intensive. With three sections sharing loads, boundary conditions are challenging. Hence, analyzing the entire section is impractical. Instead, a segment near the critical cross-section, spanning a half-wave instability length (0.5 *h*-*h*, *h* = cross-section height), is typically chosen [9]. Within Ansys Workbench, a

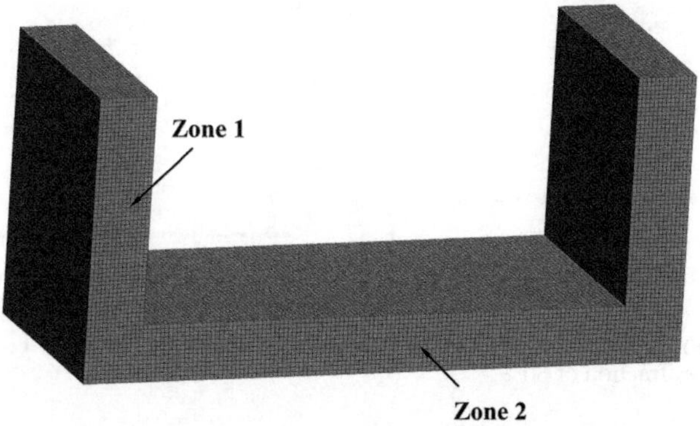

Fig. 4 Finite element model diagram of a U-shaped cross-section

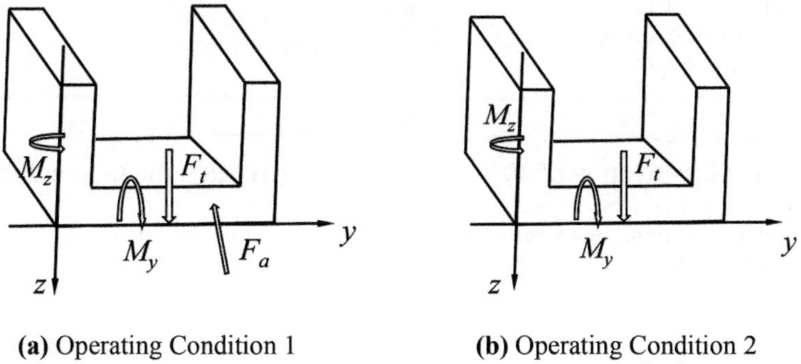

Fig. 5 Load application diagrams for each operating condition. (**a**) Operating Condition 1. (**b**) Operating Condition 2

U-shaped solid cross-sectional model was crafted and subsequently loaded into the topology optimization module. Material attributes were defined, specifying aluminum alloy 6061 for the U-shaped cross-section, characterized by an elastic modulus of $E = 210$ GPa and a Poisson's ratio of $u = 0.3$. A hexahedral mesh, featuring a mesh density of 3 mm, was utilized for the model, as depicted in Fig. 4.

Establishment of a Finite Element Model Using telescopic arm data: $F_d = 75{,}000$ N, $M_y = 21{,}882$ N·m, $M_z = 6544.5$ N·m, $F_t = 4208$ N, F_d, M_y, M_z, and F_t (Fig. 5a); Condition 2 only M_y, M_z, and F_t (Fig. 5b).

Topology Optimization Results The design objective is to minimize the compliance, with response constraints set at 40% and 30% of the mass. Using extrusion manufacturing constraints, the optimization results are shown in Figs. 6 and 7.

Based on the topology optimization results, it can be seen that the contribution of Zone 1 to increasing stiffness is less compared to Zone 2. Therefore, the length of

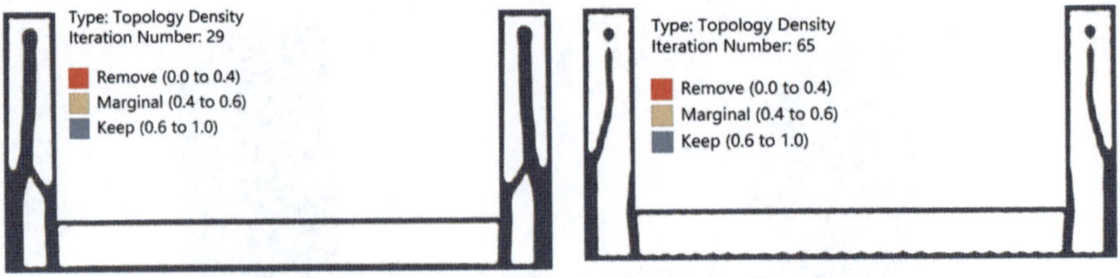

(a) Mass fraction of 40% **(b)** Mass fraction of 30%

Fig. 6 Topology optimization result of cross-section for operating condition 1. (**a**) Mass fraction of 40%. (**b**) Mass fraction of 30%

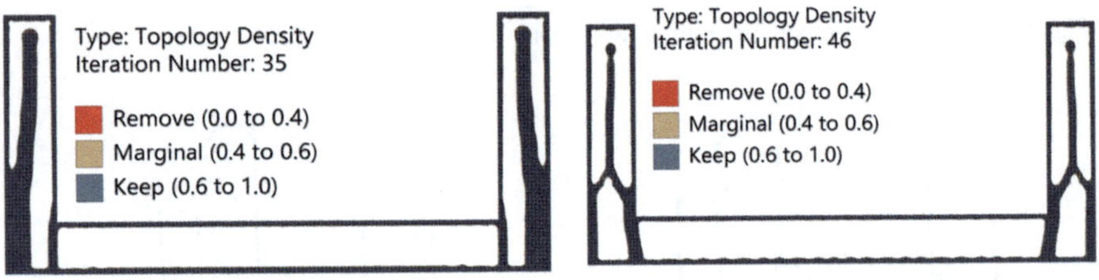

(a) Mass fraction of 40% **(b)** Mass fraction of 30%

Fig. 7 Topology optimization result of cross-section for operating condition 2. (**a**) Mass fraction of 40%. (**b**) Mass fraction of 30%

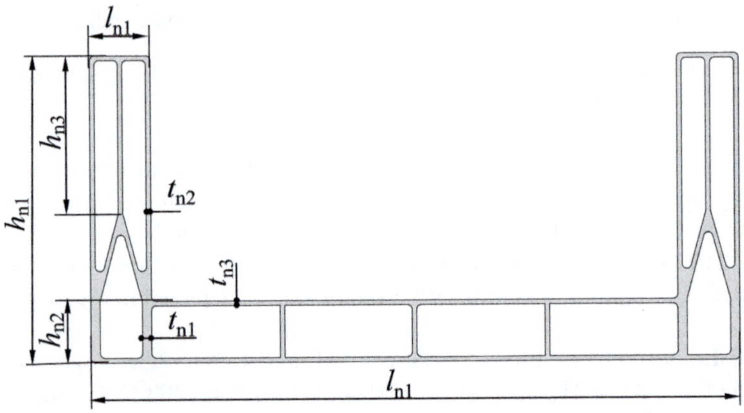

Fig. 8 Cross-sectional reconstruction diagram

Zone 1 should be appropriately reduced, and reinforcing ribs should be added at Zone 1. The reconstructed cross-section based on the topology optimization results is shown in Fig. 8.

3.2 Size Optimization

Size optimization is a detailed design phase, mainly using mathematical programming or optimal criteria. The direct optimization method, widely adopted in mathematical programming, reduces finite element calculations in structural optimization, speeding up the process and cutting costs.

Selection of Optimization Parameters Given that the cross-sectional shapes of all boom sections are uniform, three sections are picked for individual scrutiny, with the dimensions of the remaining sections derived from optimized size parameters. As depicted in Fig. 8, the cross-section exhibits symmetry, prompting the selection of h_{33}, t_{31}, t_{32}, and t_{33} parameters from these three boom sections as optimization variables. All other parameters adhere to the original design specifications, with the exception of the cross-section's rounded corners, which are excluded from the design variables, as detailed in Table 2.

Mathematical Model Drawing from engineering experience, simulation analysis in most telescopic boom designs indicates that the stress values of the boom are substantially below its allowable stress [10], rendering stress an unnecessary constraint. To enhance the printing precision of 3D construction printing robots, it is imperative to optimize the stiffness and reduce the weight of the telescopic boom. The mathematical model for optimizing its dimensions is outlined below:

$$\begin{cases} \min : M\left(x\right) \\ find : x = \left(h_{33}, t_{31}, t_{32}, t_{33}\right)^{T} \\ s.t : \delta \leq \delta_{\max} \end{cases} \tag{3}$$

In the equation, M is the mass of a three-section boom, and δ whereas signifies its deflection within the luffing plane. Using Ansys Workbench's Screening method, we built an optimization model by solving 50 random samples, selecting 3 valid solutions, and analyzing variable correlations with deformation and mass. Results are in Table 3 and Fig. 9.

Based on the optimization results, as the three-section arm's deformation decreases, its mass increases. Neither the mass nor total deformation correlates with P2-h_{33} and P4-t_{33}. Instead, arm section deformation inversely relates to P1-t_{32} and P3-t_{31}, while

Table 2 Optimized design variables

Optimization variables	Initial values/mm	Optimization range/mm
h_{33}	40	[20,60]
t_{31}	6	[4,8]
t_{32}	3	[1,4]
t_{33}	3	[1,4]

Table 3 Candidate points (the unit of length is mm and the unit of mass is kg, the initial mass of the three-section boom is 24 kg, with a deformation of 3.8 mm)

Candidate points	t_{32}	h_{33}	t_{31}	t_{33}	Deformation amount	Deviation	Quality	Deviation
1	3.1	51.0	7.5	1.2	3.8	0.0%	25.6	6.6%
2	3.7	38.5	5.1	2.1	3.7	−2.6%	25.7	7.0%
3	3.6	53.5	6.0	3.2	3.6	−5.3%	26.2	9.2%

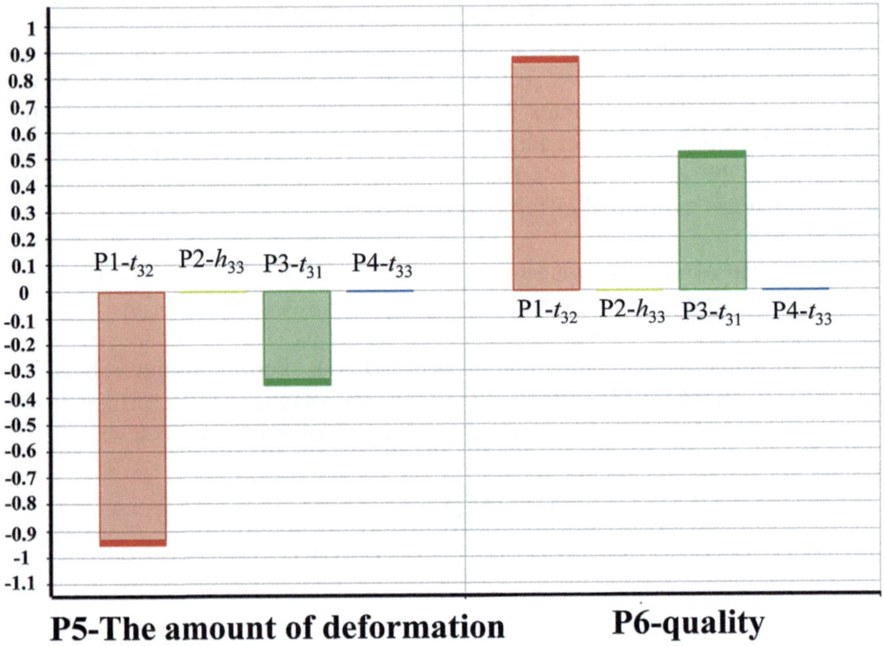

Fig. 9 The correlation between deformation, mass, and optimization variables

their mass directly correlates. To ensure stiffness without increasing mass, dimensions of other arm sections are calculated using candidate point 2 data, followed by telescopic arm re-modeling.

4 Conclusions

Using Ansys Workbench, static analysis compared the initial and optimized telescopic arm designs. Figures 10 and 11 show deformations before and after optimization. The optimized design reduced end deformation by 1.14 mm, a 59.4% decrease.

Fig. 10 Deformation diagram of the initially designed telescopic arm

Fig. 11 Deformation diagram of the optimized telescopic arm

This research guides stiffness optimization for telescopic arm structures. Future work will focus on creating tailored molds for arm cross-sections, simulating the extrusion process, refining molds, and fabricating the telescopic arm.

References

1. Zuo ZB, Huang YL, Zhang LL et al (2022) 3D printing technology and its feasibility analysis for application in super high-rise buildings. Build Struct 52(12):95–101. https://doi.org/10.19701/j.jzjg.SHJG2109
2. Wu QL, Zhou QC, Zhang RC et al (2017) Periodic topology optimization of crane boom based on improved soft kill option method (article). J Shanghai Jiaotong Univ (Sci) 22(4):459–465. https://doi.org/10.1007/s12204-017-1859-8
3. Zeng YF, Chen ZY, Wu BS (2023) Comprehensive lightweight Design of an Electric bus Skeleton. Coach Technol Res 45(03):32–36. https://doi.org/10.15917/j.cnki.1006-3331.2023.03.006

4. Zhang L, Yu SJ, Xu JY et al (2023) Design and cross-sectional dimension optimization of crane boom structure for Lightweighting. J Mach Des 40(06):124–133. https://doi.org/10.13841/j.cnki.jxsj.2023.06.026
5. Feng W, Wang ZY, Zheng J et al (2014) Structural optimization design of steering gear bracket for heavy-duty truck. Mach Des Manuf 02:175–177. https://doi.org/10.19356/j.cnki.1001-3997.2014.02.054
6. Wang YF, Ole S (2023) Multi-material topology optimization for maximizing structural stability under thermo-mechanical loading. Comput Methods Appl Mech Eng 407:115938. https://doi.org/10.1016/j.cma.2023.115938
7. Yin J, Zhang SF, Liu Y et al (2019) Structural topology optimization of cantilever crane boom based on equivalent moving load method. IOP Conf Ser Mater Sci Eng 692(1):012020. https://doi.org/10.1088/1757-899X/692/1/012020
8. Li GZ, Huo ZW, Yuan J et al (2023) Topology and size optimized design and laser welding of the U-frame for free-space laser communication telescopes. Machines 11(868):868. https://doi.org/10.3390/machines11090868
9. Wang X, Huang L, Gao Y, Gao SD et al (2009) Topological optimization of crane telescoping boom sections. J Dalian Univ Technol 49(3):374–379. https://d.wanfangdata.com.cn/periodical/dllgdxxb200903013
10. Wang CS, Xing BS (2023) Optimization of crane telescopic boom structure based on finite element software. SAE Int J Mater Manuf 16:2. https://doi.org/10.4271/05-16-02-0015

Open Access This chapter is licensed under the terms of the Creative Commons Attribution-NonCommercial-NoDerivatives 4.0 International License (http://creativecommons.org/licenses/by-nc-nd/4.0/), which permits any noncommercial use, sharing, distribution and reproduction in any medium or format, as long as you give appropriate credit to the original author(s) and the source, provide a link to the Creative Commons license and indicate if you modified the licensed material. You do not have permission under this license to share adapted material derived from this chapter or parts of it.

The images or other third party material in this chapter are included in the chapter's Creative Commons license, unless indicated otherwise in a credit line to the material. If material is not included in the chapter's Creative Commons license and your intended use is not permitted by statutory regulation or exceeds the permitted use, you will need to obtain permission directly from the copyright holder.

Lightweight Design of Slewing Support Frame of Gangue Throwing Machine Based on Multi-Objective Optimization Algorithm

Zhijie Ren, Wenxiao Guo, and Tianlong Zhu

Abstract In this paper, the slewing support frame of gangue throwing machine is taken as the research object, and the its lightweight is carried out under the premise of ensuring its strength and stiffness performance. Firstly, the static analysis of the slewing support frame shows that there is room for optimization. The multi-target response surface optimization was carried out with the thickness of the four main plates of the slewing support frame as the design variable, the maximum deformation and stress of the structure as the constraint, and the structural quality as the optimization goal. After optimization, the mass is reduced by 37.9%, and the stress and maximum deformation meet the allowable requirements. After optimization, the modal verification shows that the inherent low-order frequency of the slewing support frame is much higher than the internal and external excitation frequency, which excludes the risk of resonance. The research results confirm the feasibility of the multi-objective optimization method.

Keywords Gangue throwing machine · Slewing support frame · Multi-target response surface optimization · Modal analysis

Z. Ren (✉) · W. Guo · T. Zhu
Shanxi Tiandi Coal Machinery Co. Ltd, Taiyuan Institute, Taiyuan, China

China Coal Technology and Engineering Group Co., Taiyuan, China

National Engineering Laboratory for Coal Mining & Excavation Machinery Equipment, Taiyuan, China

Shanxi Key Laboratory of Comprehensive Drivage and Shortwall Mining Equipment for Coal Mine, Taiyuan, China

© The Author(s) 2026

Z. Zhang, P. Mercorelli (eds.), *Mechanical Design and Simulation: Insights and Innovations*, Lecture Notes in Mechanical Engineering,
https://doi.org/10.1007/978-981-95-2405-1_112

1 Introduction

In our country, the annual coal gangue emissions account for about 20% of coal production, and the problem of gangue accumulation is becoming increasingly serious, and up to now, the cumulative stock of coal gangue in China is about 7.5 billion tons [1]. Long-term accumulation of gangue not only occupies a large amount of land, but also causes varying degrees of pollution to the atmosphere, water sources, and soil [2].

In order to respond to the green mine index of zero emission of coal gangue and improve the common problem of "large and heavy" coal machinery equipment, this paper intends to optimize the key components of the gangue throwing machine. This paper studies the slewing support frame of the gangue throwing machine, aiming to find the optimal part of the structure, reduce unnecessary cost, and improve the durability of the equipment.

2 The Structure of Slewing Support Frame

The gangue throwing machine is a special equipment for the filling operation of gangue in coal mine roadway turning, which integrates the functions of material receiving, transferring, transportation and ejection.

The gangue throwing machine (as shown in Fig. 1) comprises a projectile portion, a scraper conveying groove, a material receiving portion, a slewing support frame, a walking part and a main frame, wherein the slewing support frame is fastened to the main frame by bolts. The slewing support frame can be regarded as the most dangerous part of the whole machine under stress conditions, and its structure is shown in Fig. 2.

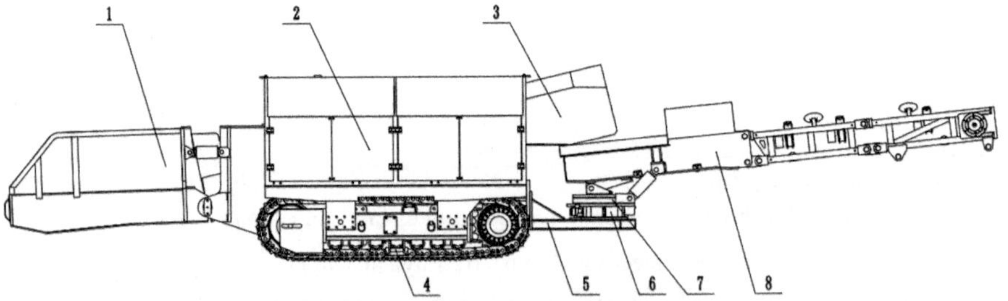

1-receiving portion; 2-main frame; 3-conveying portion; 4-walking part; 5-slewing support frame; 6-rotary reducer; 7-support seat; 8 - projectile department

Fig. 1 Structure of gangue throwing machine. 1—receiving portion; 2—main frame; 3—conveying portion; 4—walking part; 5—slewing support frame; 6—rotary reducer; 7—support seat; 8—projectile department

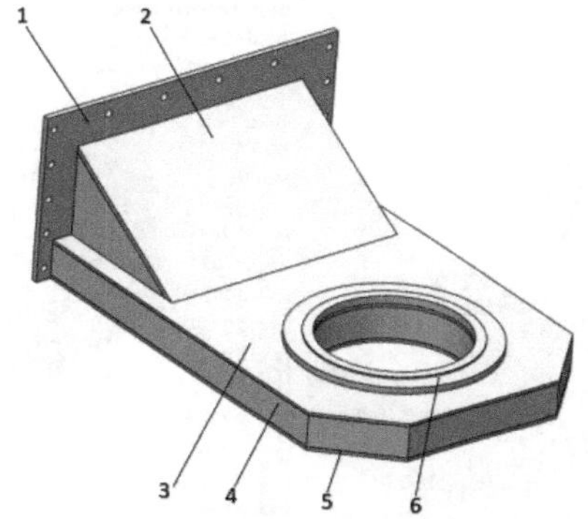

1-connecting plate; 2-reinforcement plate; 3-upper cover plate; 4-side panel; 5-lower cover plate; 6-rotary mounting plate

Fig. 2 Slewing support frame. 1—connecting plate; 2—reinforcement plate; 3—upper cover plate; 4—side panel; 5—lower cover plate; 6-rotary mounting plate

3 Static Analysis of Slewing Support Frame

The connecting plate of the support frame is bolted with the connecting plate of the main frame. Solid bolts and nuts are established between the support frame and the main frame, sliding friction contacts are added between bolt-plate, nut-plate, and plate-plate, with a friction coefficient of 0.2, and binding contacts are added between bolts and nuts [3].

When the belt ejection part is fully loaded with gangue and lifted to the maximum height, the mass of the ejection part is 3 t, the mass of gangue is 0.56 t, the center of gravity of the ejection part is about 2.5 m away from the slewing center of the support frame, the final equivalent mass is about 0.8 t.

The main bearing plate of the support frame is made of Q550 steel. Its safety factor is 1.2, and the allowable stress can be obtained as 458 MPa.

As the main load-bearing structure of the gangue throwing machine, the support frame requires high resistance to deformation. There is no specific deflection requirement for welded structures, and for Q550 steel, the allowable deflection is set

$$[\delta] = \frac{L}{3000} \tag{1}$$

where, L is the length of the support frame.

It is known that the total length of the support frame is 1.5 m, and the allowable deflection is 5 mm.

The support frame is meshed, and the maximum element size is set to 10 mm. The results are shown in Fig. 3. The maximum stress value is 218 MPa, located at the bottom of the bolt hole of the connecting plate, far less than the allowable stress. The maximum deformation is 1.47 mm, located at the center of the front of the base

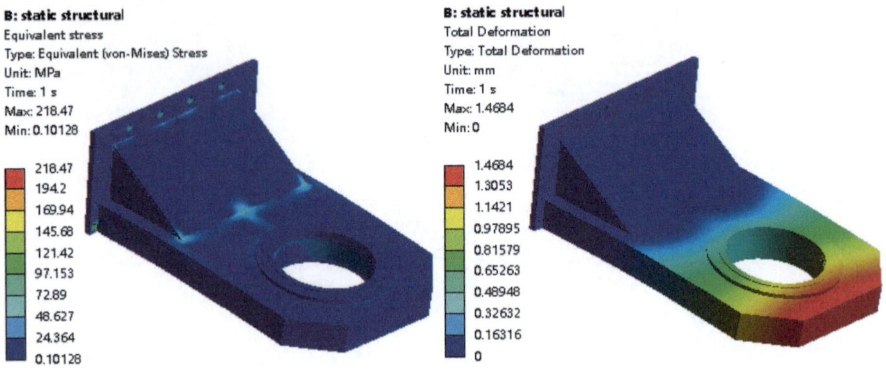

(a) Stress contours (b) Displacement contours

Fig. 3 Lightweight design of the slewing support frame. (**a**) Stress contours. (**b**) Displacement contours

plate, which is well below the allowable deflection value and within the acceptable deflection range [4].

4 Lightweight of Rotary Support Frame

From the above, it can be seen that the strength and stiffness of the slewing support frame meet the requirements, and there is redundancy in the materials, so there is a certain optimization space.

The bracing frame is a component welded by multiple steel plates (as shown in Fig. 2), wherein the mass of the connecting plate, the side panel and the upper and lower cover plates account for about 80%, therefore, these four are selected as the optimized components. With plate thickness as the design variable and reasonable stress and deformation values, the minimum mass of the support frame is taken as the optimization objective [5, 6]. The optimization mathematical model is established as follows:

$$find\, T = \begin{bmatrix} t_1 & t_2 & t_3 & t_4 \end{bmatrix} \tag{2}$$

$$\min M = \rho \sum_{i=1}^{n} A_i t_i \tag{3}$$

$$s.t.: \sigma \le [\sigma] = 458MPa \tag{4}$$

$$s.t.: \delta \le [\delta] = 5mm \tag{5}$$

Where: t_i is the thickness of i-th the plate; A_i is the area of the i-th plate; σ is the maximum stress of the support frame; δ is the maximum deformation of the support frame.

Table 1 The value range of the design variable

Optimized objects	Codename	Initial dimensions (mm)	Lower limit (mm)	Upper limit (mm)
Connecting plate thickness	P1	30	12	32
Thickness of cover plates	P2	20	10	24
Panel thickness	P3	16	8	24

Table 1 defines the value ranges of the four design variables. By default, the thickness of the upper and lower covers is equal and set to the same variable.

The experimental design method of face-centered reinforced composites design was selected, and a total of 30 sets of design point schemes were obtained. The response surface is constructed by using Kriging interpolation algorithm, which can establish one-to-one mapping between design variables and target values, and the calculation time is significantly reduced [7, 8]. After the calculation, a total of 6 sets of 3D response surface plots are obtained, and due to space limitations, only the mapping relationship between P1, P2 and the deformation, equivalent stress is listed, as shown in Fig. 4. In figure a, the deformation decreases with the increase of P1 and P2 values, and the deformation of the support frame is the smallest when the P1 and P2 are the largest. In Figure b, the maximum equivalent stress decreases with the increase of P1 and P2, and the equivalent stress and P2 are approximately linear.

Three candidate results were obtained from the optimization analysis, and the final design variables were as follows: P1 = 19 mm, P2 = 10 mm, and P3 = 8 mm. It can be seen that the maximum stress value is 447.15 MPa, which is still located at the lowest bolt hole, and the maximum deformation is 3.45 mm, which occurs at the front end of the bottom plate. After lightweight treatment, the mass of the support frame is reduced to 400.68 kg, compared with the original value, it can be seen that under the premise of meeting the structural stiffness and strength requirements, the response surface optimization method reduces the mass of the support frame by 37.9% and reduces unnecessary material consumption (Table 2).

5 Modal Analysis of Optimization Results

The support frame carries the gangue of the belt projectile part, and the stability of its structure is very important. If the excitation frequency is consistent with the natural frequency of the support frame, it will lead to the resonance phenomenon of support frame, which will results in the deformation and damage of the support frame [9].

The mode shapes of the first 4 modes of the support frame are analyzed, and the results are shown in Fig. 5. The first-order mode shape is an up-and-down swing, which exists due to the overturning moment applied by the projectile to bend the support frame. The second-order mode shape is left and right torsion, when the

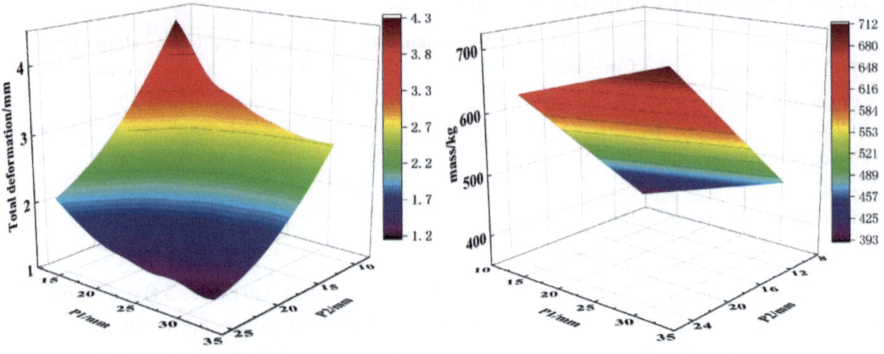

(a) Responses of P1 and P2 to the deformation (b) P1 and P2 responses to mass

Fig. 4 Responsive surface model. (**a**) Responses of P1 and P2 to the deformation. (**b**) P1 and P2 responses to mass

Table 2 Comparison of optimization results

	Original values	Candidate 1	Candidate 2	Candidate 3
P1(mm)	30	19.11	19.20	19.07
P2(mm)	20	10.00	10.00	10.01
P3(mm)	16	8.12	8.09	8.33
Maximum deformation(mm)	1.47	3.44	3.43	3.43
Maximum equivalent stress (MPa)	218	457.98	455.91	457.78
Structural quality(kg)	646	401.51	401.82	402.12

arrangement of the gangue is eccentrically, it causes the support frame torsion, so this mode shape exists. The third-order mode shape is left and right horizontal swing, when the rotary reducer horizontal rotates, the support frame below the rotary reducer has a tendency to swing, so this mode shape exists. The fourth order mode shapes shows the concave and convex fluctuations of the cover plates, it is found that the support frame only has a downward depression in its center of rotation, and this mode shape does not appear. In summary, the natural frequency of the support frame is 75.9–224.6 Hz, and the corresponding modal amplitude is less than 5 mm.

The frequency of pavement excitation to which the track part is subjected

$$f_{\alpha} = \frac{v}{\lambda} \tag{6}$$

where, λ is the travel speed of the gangue throwing machine; v is the wavelength of the road surface.

The wavelength of pavement is 0.32 m, and the speed of the gangue throwing machine is 10 m/min, and the maximum excitation frequency can be calculated to be 0.52 Hz, which is much lower than the low-order frequency of the support frame,

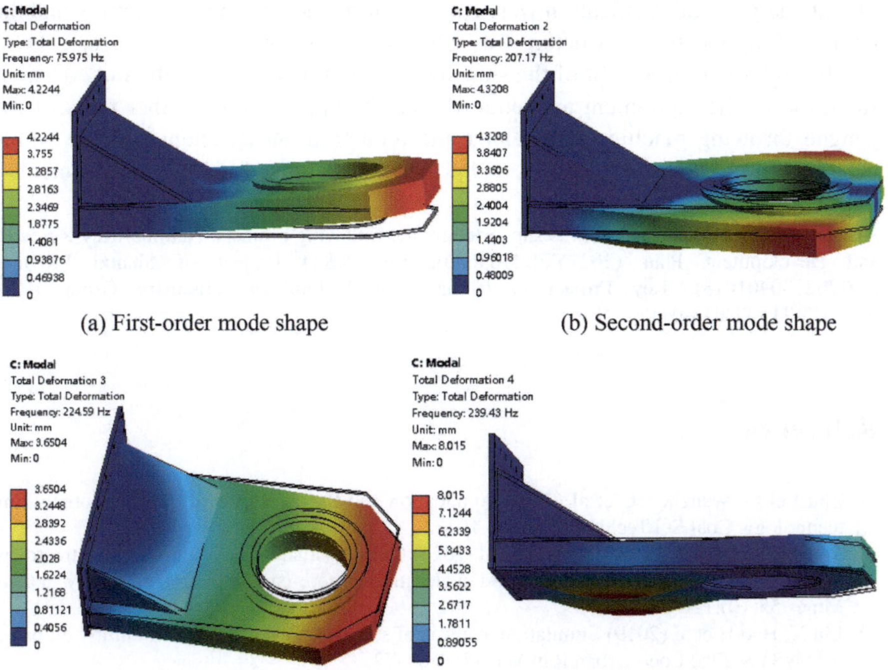

(a) First-order mode shape (b) Second-order mode shape

(c) Third-order mode shape (d) Fourth-order mode shape

Fig. 5 Diagram of the first fourth order of the model. (**a**) First-order mode shape. (**b**) Second-order mode shape. (**c**) Third-order mode shape. (**d**) Fourth-order mode shape

Table 3 Comparison of modal analysis results

	Natural frequency	Pavement excitation	Gear motor
Frequency (Hz)	75.9–224.6	0.52	25

so it has no effect on the structural stability [10]. The motor speed is known to be 1500 r/min, and the excitation frequency of the motor is 25 Hz, which far less than the low-order frequency, so it will not cause resonance. Therefore, the structural of the optimized slewing support frame is still relatively stable, which meets the requirements of safety and reliability (Table 3).

6 Conclusion

In order to improve the structural performance of the slewing support frame of the gangue throwing machine, the finite element analysis and lightweight design were carried out, and the mass of the optimized support frame was reduced by 37.9%.

Modal analysis and verification were carried out on the optimized structure, and the improved support frame structure was safe and reliable without resonance.

Through the lightweight of the slewing support frame, the weight and cost consumption of the equipment are reduced, and the application to other parts of the gangue throwing machine will realize the reduction and efficiency of the whole machine, improve the service life, and provide a reference for similar equipment.

Acknowledgments This work was supported by the following projects: National Key Research and Development Plan (2022YFB4703600); Key R&D Project of Shanxi Province (202202150401018); Key Project of China Coal Technology &Industry Group (Item No:2023-TD-ZD013-01).

References

1. Zhu Lei G, Wenzhe YC et al (2019) Application and Prospect of coal gangue slurry filling technology. Coal Sci Technol 52(4):93–104
2. Yousheng D, Chenhui D, Along W et al (2024) Current situation of comprehensive treatment of coal gangue mountains in China and exploration of zero-carbon treatment. Safety in Coal Mines 55(10):120–128
3. Liu X, Hao H et al (2019) Simulation analysis of standard subway train anti-climb bolt based on ANSYS. Elec Loco Urban Rail Veh 47(4):68–72
4. Jiangtao H (2023) Modeling and finite element analysis of lifting spillage and boom in Silo. Taiyuan University of Science and Technology, Taiyuan
5. Redmon J, Divvala S, Girshick R, et al. (2016) You only look once: unified, real-time object detection. Proceedings of the IEEE conference on computer vision and pattern recognition. 779–788
6. Zhao R (2023) Fatigue analysis and multi-objective optimization design of lifting device for elderly machinery based on ANSYS workbench. Autom Instrum 10:251-255+259
7. Heju T (2019) Construction method and simulation application of Kriging-GRNN high order hybrid response surface model. Guangdong University of Technology, Guangzhou
8. Hooten MB (2007) Statistical analysis of environmental space-time processes. J Am Stat Assoc 102(480):1477–1477
9. Ouisse M, Foltête E (2015) Model correlation and identification of experimental reduced models in vibroacoustical modal analysis. J Sound Vib:342200-217
10. Lijia X (2023) Performance analysis and stability study of lightweight micro-track chassis. Chang 'an University, Xi 'an

Open Access This chapter is licensed under the terms of the Creative Commons Attribution-NonCommercial-NoDerivatives 4.0 International License (http://creativecommons.org/licenses/by-nc-nd/4.0/), which permits any noncommercial use, sharing, distribution and reproduction in any medium or format, as long as you give appropriate credit to the original author(s) and the source, provide a link to the Creative Commons license and indicate if you modified the licensed material. You do not have permission under this license to share adapted material derived from this chapter or parts of it.

The images or other third party material in this chapter are included in the chapter's Creative Commons license, unless indicated otherwise in a credit line to the material. If material is not included in the chapter's Creative Commons license and your intended use is not permitted by statutory regulation or exceeds the permitted use, you will need to obtain permission directly from the copyright holder.

Design and Analysis of Semiautomated Leek Harvester

Zigan Yan, Li Wang, Huanjia Liu, and Qifan Zhang

Abstract A compact agricultural harvesting machine with multi-functionality, which consists of power-driven gathering and harvesting, transmission and collection, and steering, was studied. And its cutting, optical axis forces, and rotation were simulated using SolidWorks FEA. Analyze the data. This resulted in a multifunctional compact agricultural harvester.

Keywords Multi-functionality · Compact · Agricultural harvesting machine

1 Introduction

As the world's population grows, the demand for leeks as a stonecrop is climbing year after year. Leeks are a common and relatively important vegetable group in our lives, they generally have a high nutritional value and are rich in vitamins and minerals. Leeks generally have relatively distinctive characteristics and features, usually with bulbs growing underground, and most of their leaves are basal, linear or strap-shaped, entire or spiny-serrate. Leeks are widely grown throughout the world. However, the traditional leek harvesting method mainly relies on manual labor, which is not only inefficient, but also labor-intensive and difficult to ensure the quality of harvesting. Moreover, the current mechanization of agriculture can promote the elevation of agricultural crop yield [1]. Therefore, the research and development of an efficient, labor-saving leek harvesting machine is of great significance to improve agricultural productivity and reduce production costs.

Z. Yan (✉) · L. Wang · H. Liu · Q. Zhang
Dalian Polytechnic University, Dalian, Liaoning, China

© The Author(s) 2026

Z. Zhang, P. Mercorelli (eds.), *Mechanical Design and Simulation: Insights and Innovations*, Lecture Notes in Mechanical Engineering,
https://doi.org/10.1007/978-981-95-2405-1_113

2 Functional Design

2.1 Overall Design

To realize the integration of leek harvesting and collection, the harvester is generally divided into four large parts, gathering device, cutting device, transmission device, collection device. Overall harvester into a box-type shape, this design is in line with the laws of crop planting between the farmland; while having four completed device systems can be complete leek harvesting from the land and complete the task of collection as shown in Fig. 1. Based on the fact of the spacing and the length of the growing crops between the farmlands, the setting of the length and the angle of the conveyor belt and the materials we have, we use aluminum profiles as the main body of the frame of the car, with a length of 600 mm, a width of 400 mm and a height of 400 mm. Based on the general growth width of leeks will be set in front of the collection device to 240 mm, the curvature of 354.46 mm at the same time will be two contact with the crop surface into a curved surface, so that in the collection of crops will not be hurt at the same time to achieve the effect of the collection of crops. The cutting device is shown in Fig. 2, the stem of leek is more single, it is generally upright and solid inside the stem, they are generally 30–60 mm high [2]. So we design such a disc cutter, the short side of 27.05 mm, the long side of 27.14 mm. from the top and bottom at the same time to the center of the convergence. It can ensure that leeks can be harvested smoothly. The transmission unit, which is driven by two motors, holds the conveyor belt vertically at one section and places the conveyor belt horizontally at the other end. This allows the vertically growing crops to be rotated horizontally as they are conveyed. It makes the next collection easier. So, we calculated the angle between the conveyor belt and the base plate at 22 degrees and designed the two conveyor belts with the dimensions of 1470*90*1.5 mm and 1426*90*1.5 mm. as shown in Fig. 3.

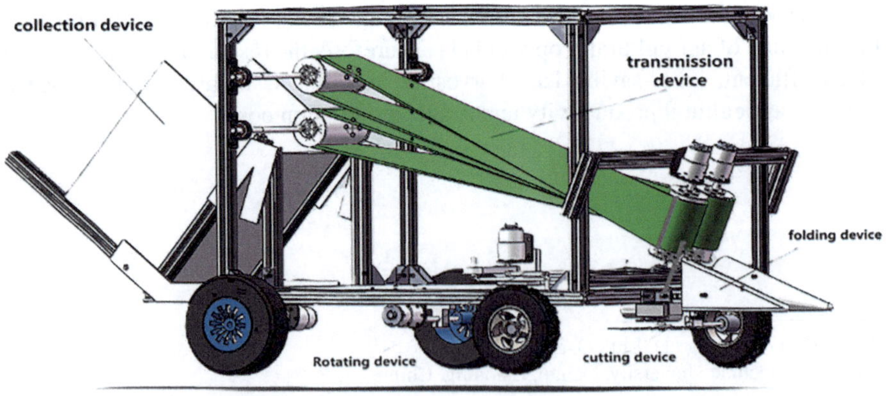

Fig. 1 shows the overall frame of the car

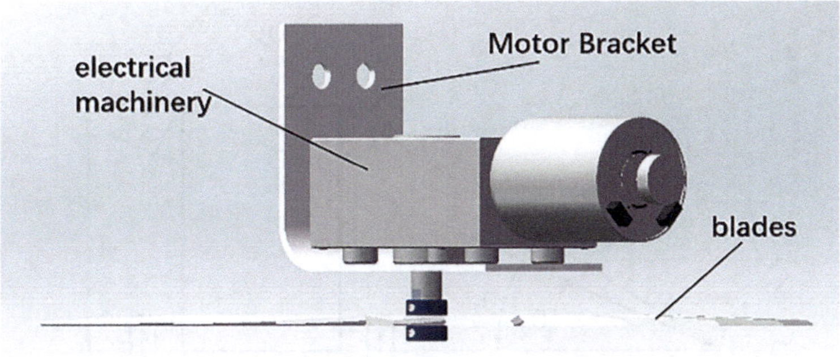

Fig. 2 Cutting device

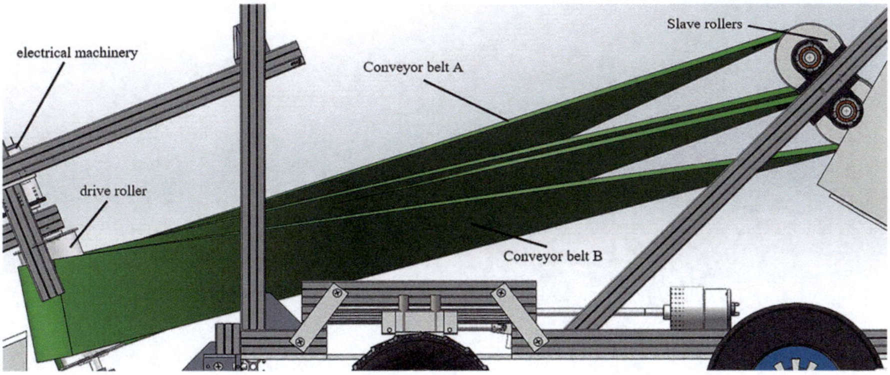

Fig. 3 Conveyor structure

2.2 Steering Structure Design

The harvester utilizes Ackermann steering, which is the same principle as automobile steering, in the rotating device [3]. Because of the leeks and similar crops belong to the herbaceous plants, so it grows mostly in the agricultural land, this land generally has a lot of potholes, the general vehicle steering device will lead to the vehicle is easy to roll over, unstable, and thus cannot achieve the normal function of steering. Ackermann steering device, however, can well prevent such accidents from occurring and make the harvester's steering more stable, thus making the leek harvester more efficient.

As shown in Fig. 4, Ackermann steering [4] is a design principle for steering systems that act on vehicles to make the front wheels of the vehicle turn in the path we expect them to follow while maintaining stable steering performance.

The basic principle of Ackermann steering is to allow the vehicle to turn while simultaneously turning the inside tires more than the outside tires to achieve controllability and stability of the vehicle while turning. Figure 5 shows the Ackermann steering structure used on a harvester.

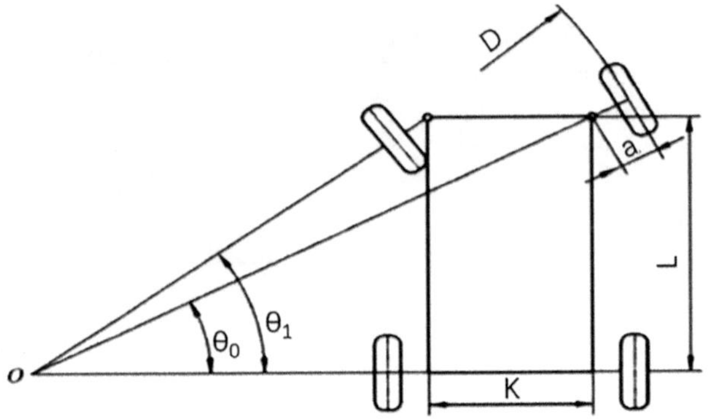

Fig. 4 Ackermann steering principal implementation diagram

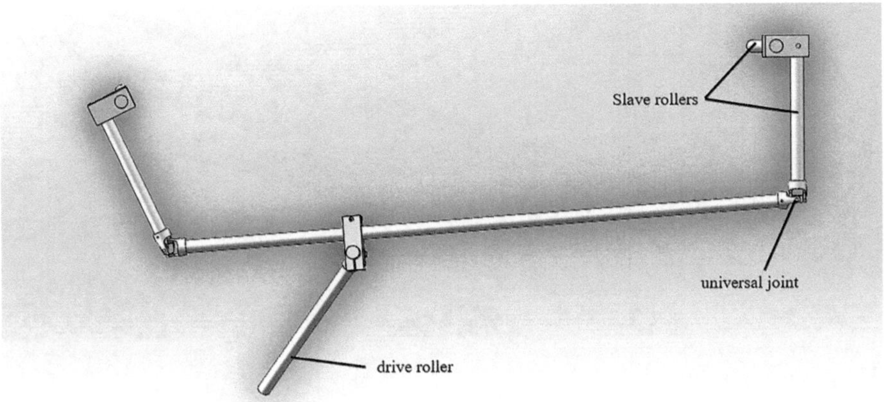

Fig. 5 Ackermann steering among the harvesters

To solve this problem, Ackermann developed a formula to calculate the difference between the steering angles of the inside and outside tires. This formula calculates the steering angle of the inside and outside tires based on the vehicle's wheelbase (and the distance between the centers of the wheels), the turning radius, and the steering angle of the vehicle.

3 Design Calculations

3.1 Ackermann Steering Calculations

The derivation of Ackerman's formula is as follows:

Assume that the wheelbase of the vehicle is L, the turning radius is R, and the steering angle of the vehicle is δ (a positive angle indicates a turn to the left and a negative angle indicates a turn to the right).

According to the geometric relationship, the relationship between the steering angle θ_1 of the inside tire and the steering angle θ_0 of the outside tire can be expressed as:

$$\tan(\theta_1) = \frac{R}{\left(L - \frac{W}{2}\right)} \quad (1)$$

$$\tan(\theta_0) = \frac{R}{\left(L + \frac{W}{2}\right)} \quad (2)$$

Where, W denotes the wheelbase (distance between left and right tires) of the vehicle.

According to the nature of trigonometric functions, it can be obtained:

$$\theta_1 = \tan^{-1}\left(\frac{R}{L - W/2}\right) \quad (3)$$

$$\theta_0 = \tan^{-1}\left(\frac{R}{L + W/2}\right) \quad (4)$$

Using Ackermann's principle, the harvester can be made to realize the steering function smoothly as shown in Fig. 5.

3.2 DC Geared Motor Calculator

For the electric drive of the harvester, we have chosen a DC geared motor [5] and the following is our analysis process:

M is the conductance, and the inertial inertia is J:

$$J_1 = 1/8mD^2, J_2 = m * \left(\frac{M}{2\Pi}\right)^2 \quad (5)$$

β is the angular acceleration and the torque Y:

$$Y = J * \beta \quad (6)$$

The angular velocity ω:

$$\omega = \beta * t \tag{7}$$

RPM n:

$$n = \omega / 2\Pi \tag{8}$$

Pressure P:

$$P = \frac{V}{N} \tag{9}$$

Friction f:

$$f = \mu mg \tag{10}$$

Power P:

$$P = Y * \frac{n}{9550} \tag{11}$$

Through the above formula can be calculated if the mass of the whole car for 20KJ then the motor needs to achieve the speed of 210 rpm, the speed of the motor for the theoretical speed, according to the actual situation and then carefully adjusted to achieve the ideal state.

3.3 Finite Element Static Analysis

The blade in the blade cutting mechanism is the key component that is mainly subjected to force in addition to the optical shaft rod because of the connection of the drum is also subjected to the main gravitational force of the leek, so the stress analysis of the blade and optical shaft is very important. When using SolidWorks-Simulation to analyze the device, the first simple complexity of the model and ignore the strength of the connection part, the material used this time is PMMA [6], the material properties are shown in Table 1.

Table 1 Material properties

Causality	Numerical value
Tensile strength	75 MPa
Notched impact strength	1200 J/m^2
Heat distortion temperature	95 °C
Shrinkage	0.50%
Abrasion resistance	Difference

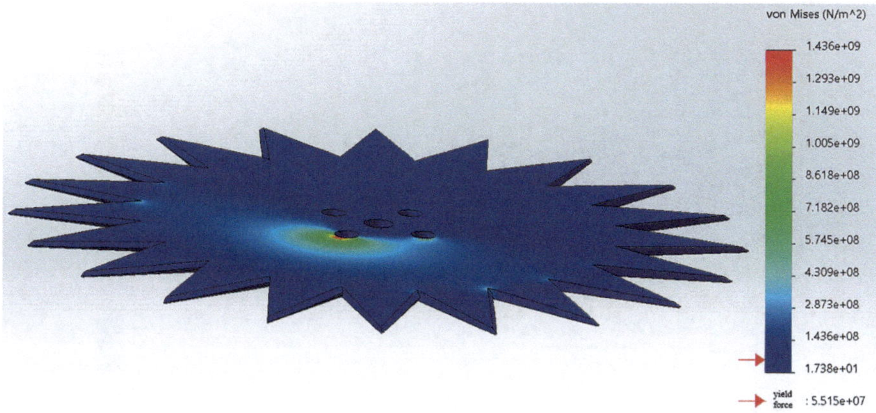

Fig. 6 Blade stress analysis

The areas in the blade that are in contact with the leeks are analyzed and simulations are performed under full load to analyze their stress and displacement deformations.

In SolidWorks-Simulation, the material properties of the part are set first, and then the software automatically meshes the surface of the model to express the approximate shape of the part surface, and divides the interior of the model into hexahedral and tetrahedral elements to express the volume of the overall model; after the meshing and the elemental volume division are complete, SolidWorks Once the mesh and element volumes are divided, SolidWorks Simulation automatically calculates and analyzes the part and generates the results of the calculation examples. The blade stresses and displacements are shown with reference to Figs. 6 and 7. It can be seen from the stress and displacement plots that the maximum values occur at the front end of the blade where the force is greatest in contact with the leek. The main data analyzed can be clearly seen from the results of the example calculation. The degree of elastic deformation of the material is so small that it is difficult to differentiate in the model, so the output displacements of the parts shown have been exaggerated to represent the direction and degree of material deformation.

By $\delta = F_n/A$ calculated stress [7], finite element analysis calculation of the rotary axis results: the maximum stress of the blade is 1.748e + 06 N/m^2, the minimum stress is: 1.942e + 05 N/m^2, the yield stress is: 6.204e + 08 N/m^2. The design meets the permissible strength of the material. Therefore, the strength and stiffness of the blade can be guaranteed during the actual operation. Through the above static load analysis, the reasonable degree of the device structure design can be concluded. The material selection and size selection of each key part meet the load conditions under the condition of three times of the impact coefficient, and there is no fracture and serious plastic deformation problem in each part.

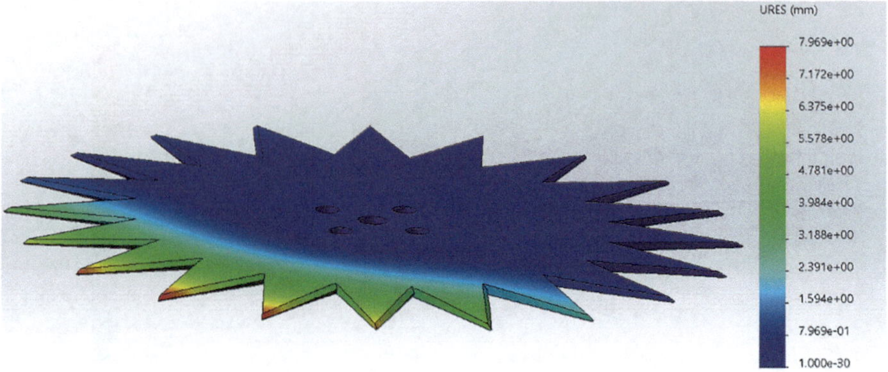

Fig. 7 Blade displacement analysis

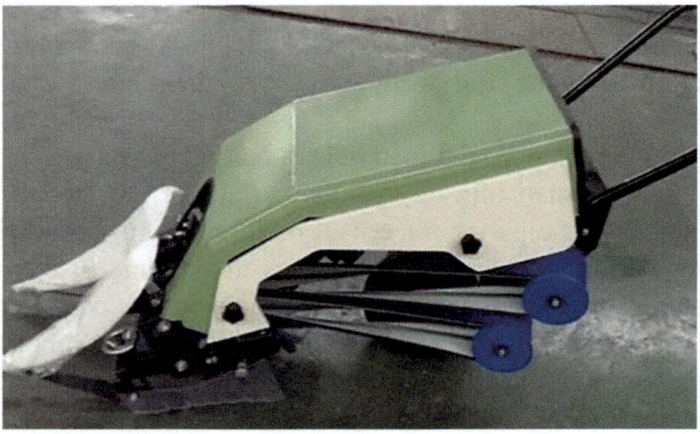

Fig. 8 Common leek harvesters

3.4 Application Analysis

The current common leek harvester, as shown in Fig. 8. The harvesting of leeks is mainly carried out in a human-powered manner [8]. The degree of intelligence is insufficient. Its cutting, conveying, collecting and other functions are mostly dependent on manual adjustment by hand, like the height of the cutter cannot automatically recognize the growth condition of leeks and adjust it accurately, and it is difficult to meet the harvesting needs of different plots and different growth stages.

In the adaptability of operating environments, in the face of complex environments, the lack of intelligent response mechanisms, encountering obstacles cannot automatically avoid, need to be handled manually, affecting operational efficiency and may damage the equipment.

Because of the low degree of intelligence, the harvester is difficult to realize the refinement of the operation, unable to reduce the harvesting loss, and cannot collect and analyze the operation data in real time, to provide valuable information for the growers, limiting the enhancement of economic benefits.

4 Conclusion

The development of science and technology and the advancement of agricultural modernization, there is a growing demand for improving the efficiency of cultivation and crop yield, especially for such crops as spinach, the traditional cultivation methods have been difficult to meet the needs of modern agricultural production. Through calculation and analysis, the design of semi-automated leek harvester was completed, and actual production and a large number of experiments were carried out. The test results show that the semi-automatic leek harvester's function and stability of use meet the expected requirements. The structure is reliable, high performance and environmental adaptability.

References

1. Cai M (2022) The impact of agricultural mechanization on green grain total factor productivity and farmers' income. Nanjing University of Finance and Economics. F323.8; F323.3; X322
2. Yu ZC, Liu Y, Yang C. Comparison of Amaryllidaceae leaf anatomical structure and microstructure A. 0439–8114(2012)08–1603-05
3. Peiris M, Lang H, Gindy EM (2024) Kalman filter-based sensor fusion for Ackermann steering mobile robots. J Phys Conf Ser 2811(1):012004
4. Li Q, Chen L, Ning X (2023) Heavy trucks more Ackermann steering linkage steering system error optimization. J Pract Technol 10:13. 125-128. https://doi.org/10.16638/j.carolCarrollNik.1671-7988.2023.010.026
5. Martínez-García C, Astorga-Zaragoza C, Puig V et al (2020) A simple nonlinear observer for state and unknown input estimation: DC motor applications. IEEE Trans Circuits Syst II Express Briefs 67(4):710–714
6. Khallouqi A, Halimi A, Rhazouani EO et al (2024) Comparing tissue-equivalent properties of polyester and epoxy resins with PMMA material using gate/Geant4 simulation toolkit. Radiation Physics and Chemistry, p 220111702
7. Shen M (2007) Biomimetic study on the wear resistance of cutter blade surface. Jilin University
8. Gong Y, Feng Y, Li C et al (2018) Research status and development trend of leek harvesting machinery. Agric Mech Res 40(10):262 268. https://doi.org/10.13427/j.cnki.njyi.2018.10.051

Open Access This chapter is licensed under the terms of the Creative Commons Attribution-NonCommercial-NoDerivatives 4.0 International License (http://creativecommons.org/licenses/by-nc-nd/4.0/), which permits any noncommercial use, sharing, distribution and reproduction in any medium or format, as long as you give appropriate credit to the original author(s) and the source, provide a link to the Creative Commons license and indicate if you modified the licensed material. You do not have permission under this license to share adapted material derived from this chapter or parts of it.

The images or other third party material in this chapter are included in the chapter's Creative Commons license, unless indicated otherwise in a credit line to the material. If material is not included in the chapter's Creative Commons license and your intended use is not permitted by statutory regulation or exceeds the permitted use, you will need to obtain permission directly from the copyright holder.

Design and Research of Multi-Functional Intelligent Bed Based on "Zero Gravity"

Wei Wang, Fahuan Zhang, Wuhao Xu, Xiaoli Cai, Jian Xie, Jinfeng Zhang, Zhengjie Hou, Shuaishuai Lv, and Hongjun Ni

Abstract With the continuous progress of science and technology and people's living standards, people are increasingly concerned about the topic of healthy sleep, smart bed began to gradually enter people's vision, ergonomic design of smart bed is also more and more concerned by the society. Today, most of the smart beds on the market can not fully reflect the characteristics of intelligence, and the user's experience is poor and satisfaction is low. Therefore, this paper designs a multi-functional smart bed based on the principle of "zero gravity" and ergonomics, so as to better meet the needs of different people in a more comfortable rest posture in life, and carries out finite element simulation analysis on the three-dimensional assembly model of the smart bed by using ANSYS Workbench in accordance with the standard. In order to meet the strength and other requirements.

Keywords Smart bed · Zero gravity · Human body engineering · Finite element analysis · Overall check

1 Introduction

After entering the twenty-first century, with the rapid development of science and technology and social economy, various industries have shown a vigorous development posture, especially intelligent products, with the rise of the Internet, intelligent products have also developed rapidly, intelligent products have gradually become the focus of attention. China's home industry has a long history of thousands of

---

W. Wang · F. Zhang · W. Xu · X. Cai (✉) · J. Xie · J. Zhang
Jiangsu Ligao Smart Home Co., LTD, Nantong, Jiangsu, China
e-mail: caixiaoli@niscofurniture.com

Z. Hou · S. Lv · H. Ni
Nantong University, School of Mechanical Engineering, Nantong, Jiangsu, China

© The Author(s) 2026

Z. Zhang, P. Mercorelli (eds.), *Mechanical Design and Simulation: Insights and Innovations*, Lecture Notes in Mechanical Engineering,
https://doi.org/10.1007/978-981-95-2405-1_114

years, until 1992, Bill Gates built the first smart villa, which made people have a strong interest in it, and smart home also showed people its strong vitality [1].

Although China's smart bed industry has more than a decade of history, the penetration rate is still not high, limited to hospitals, nursing homes and other institutions, and mainly manual operation. In addition, because the consumption power of the people is not high, so compared with some developed countries, there is still a certain gap in the sales of smart beds, which is why most people do not understand the concept of "smart beds". Liu Fangyuan et al. [2] from Shandong University of Science and Technology first proposed the modular design concept in 2015. Zhao Pengfei et al. [3] applied human-machine interface interaction to multi-functional smart bed at home by applying ergonomic principles and combining design theories such as functionality, safety, humanization and emotion. Han Jian et al. [4] applied the single chip microcomputer as a microprocessor to the design of the smart bed in 2017 to realize the intelligent detection of the environment such as sound, temperature and humidity. Moreover, now the smart bed on the market, the function is relatively single, but also need to increase efforts to innovate, from this point of view, the domestic smart bed market still has great potential.

2 Overall Structure Design of Smart Bed

2.1 Determination of the Basic Size of the Smart Bed

According to the specifications in the household industry group standard T-SZFA2003.1-2019 "Mattress Ergonomics Evaluation - Mattress Hardness Grading and Distribution Test Evaluation Method" [5] and the household industry group standard T/SZFA.2003.2-2019 "Mattress Ergonomics Evaluation - Mattress Fit Grading and Distribution Test Evaluation Method", The smart bed can be divided into four areas from the head to the feet: shoulder and back, waist, hips, and legs. The range of the shoulder and back is 0–640 mm, the range of the waist area is 640–760 mm, the range of the hip area is 760–1240 mm, and the range of the leg area is below 1240 mm. The overall size of the smart bed frame is set to 2000*1800 mm and the height is 600 mm.

Therefore, the bed frame takes the common size 2000*1800 mm, referring to the value of the division range above, we can divide the bed frame of the smart bed into four parts, respectively, the back, the hip, the big leg and the small leg. According to the "Chinese Adult Human Body Size" [6] after comprehensive consideration, the large leg size is 485 mm and the small leg size is 515 mm.

In summary, the smart bed is partitioned, and the size of the back plate is 760 mm, the hip plate is 240 mm, the big leg plate is 485 mm, and the small leg plate is 515 mm. Among them, the back plate can be further divided, the size of the head part is 280 mm, and the size of the waist part is 480 mm. The above zoning is based on China's human body size and ergonomic principles to divide the area, to a large extent in line with our people's zoning standards, in line with the actual size needs of various types of people.

2.2 *Structural Demand Analysis of Smart Bed*

Structure Analysis of Smart Bed in Zero Gravity State NASA first proposed the concept of "zero gravity" for a Neutral Body Posture (NBP) in the design of workstations and tools [7]. The neutral body position (NBP) is the automatically assumed position of the muscles in a relaxed state, which puts the least strain and pressure on the muscles, bones, tendons and nerves and allows the person to be in the most comfortable state [8].

According to the principle of "zero gravity", neutral human body posture (NBP) is used to design the zero gravity state of the smart bed. In zero gravity, in the neutral position, the Angle of the lower arm and the upper arm is 122°, and the range of movement is 122° (±24°); The upper body is 128° from the thigh, and the range of motion is 128° (±7°); The thigh and calf are 133° and can move freely 133° (±8°).

To sum up, according to ergonomic knowledge, when the smart bed is in zero gravity, the adjustable Angle range between the upper body and the thigh is 128° (±7°), while the adjustable Angle range between the thigh and the horizontal ground is 35° (±5°).

Analysis of Angle Adjustment of Intelligent Bed Plate The Angle adjustment of the intelligent bed includes the Angle adjustment of the back plate, the Angle adjustment of the thigh plate and the Angle adjustment of the lower leg plate. According to the regulations, the Angle adjustment range of the back plate, thigh plate and calf plate of the smart bed has corresponding reference values [9].

1. The Angle between the back plate and the hip plate can be adjusted between 0–75°, as shown in A in Fig. 1.
2. The Angle between the thigh plate and the horizontal plane can be adjusted between 0–35°, as shown in B in Fig. 1.
3. The Angle between the thigh plate and the calf plate can be adjusted to 180°, as shown in C in Fig. 1.

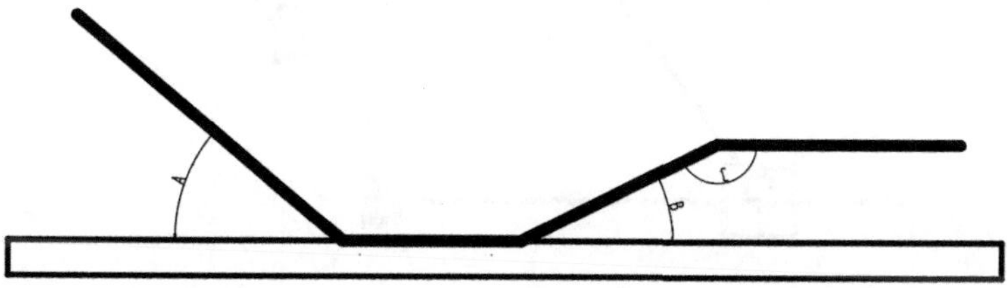

Fig. 1 Recommended adjustment range of each bed plate of the smart bed

2.3 Mechanism Design of Smart Bed

Back Lift Structure Design The realization of the movement of the back lifting mechanism of the smart bed is relatively simple, as shown in Figs. 2 and 3.

When the bed plate is flat, as shown in Figs. 4 and 5, when the state is changed from lying flat to lifting, the motor 3 drives the screw 5 to rotate through the coupling, and the screw cap 4 moves to the right in the horizontal direction through the screw transmission, so that the connecting rod 2 drives the bed plate 1 to rotate around the fixed point and complete the reverse action of the back plate.

When the back of the smart bed is rotated to 75°, the horizontal displacement of the screw cap relative to the rotation point of the back bed is $L_1 = 200$ mm. The distance L_2 between the connecting point of the support connecting rod and the bed plate and the rotating point is 220 mm, the vertical distance between the central axis of the screw cap and the rotating point of the bed plate is e = 50 mm, and the connecting rod $L_3 = 300$ mm, as shown in Fig. 3.

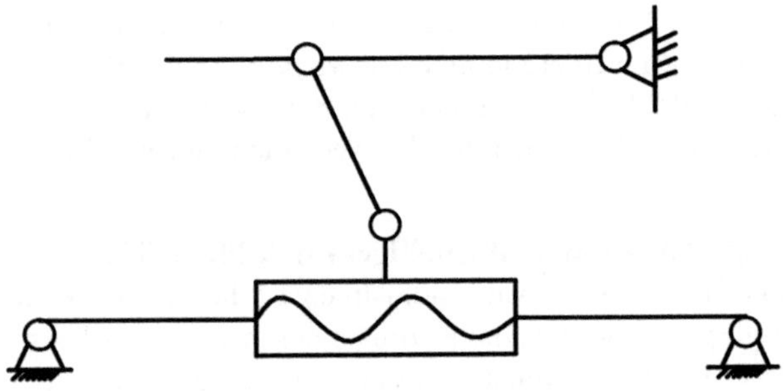

Fig. 2 The back lifting mechanism of the smart bed

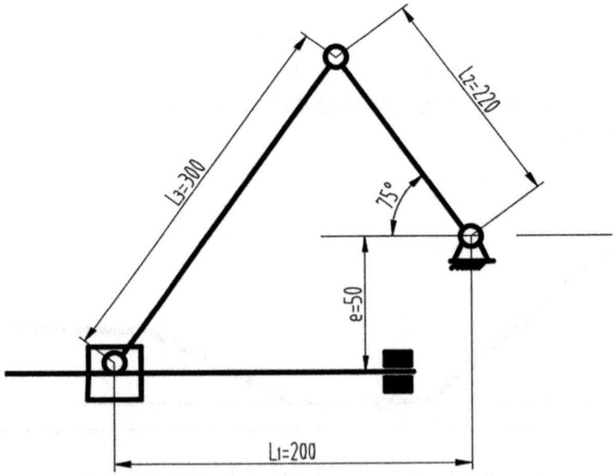

Fig. 3 Intelligent bed carried back connecting rod design

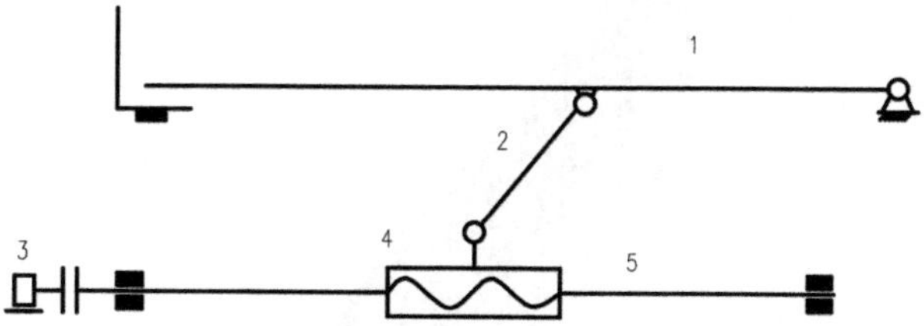

Fig. 4 Structure before movement

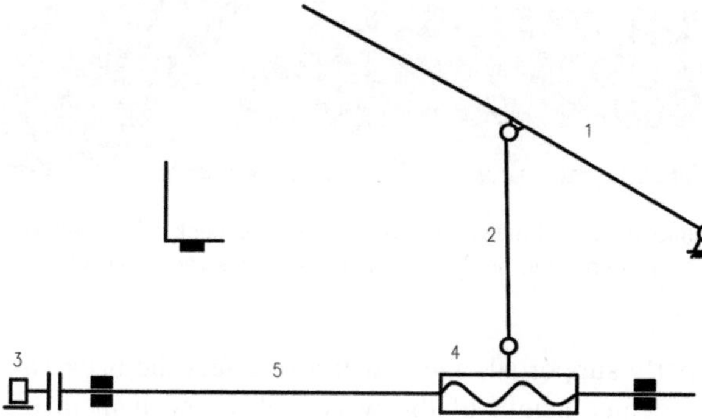

Fig. 5 Structure after movement

Back "S" Structure Design Coleman et al. [10] conducted a large number of experimental studies as early as 1998, and found that when sitting, it is necessary to have appropriate lumbar support, which can not only stabilize the pelvis, but also make the lumbar spine bend forward, greatly reducing the muscle burden required to support the balance of the torso, and reducing the pressure of the upper body weight on the lower spine.

In order to make the back play a better rest effect, it is necessary to highlight the part of the waist, the method is to add a waist pillow to form a "fill the waist", increase the contact surface of the person and the waist mechanism, so that the backbone is an "S" shape, at the same time, the ligaments and muscles are relaxed, so as to relieve the fatigue of the waist. In this paper, mechanical components are used to replace the lumbar pillow, and the three-dimensional model is shown in Fig. 6. The working principle of the mechanism is as follows: the push rod motor connecting block 1 connects the push rod motor 2 and the back plate 5, and the push rod motor 2 moves the backrest connecting piece 4 through the push rod, thus driving the backrest rod 3 to move. When the person lies on it, it is possible to rely on this mechanism to form an "S" shape of the backbone.

The center height of this mechanism determines its center of gravity at the lumbar support point of the human body. The device is too high or too low to allow the

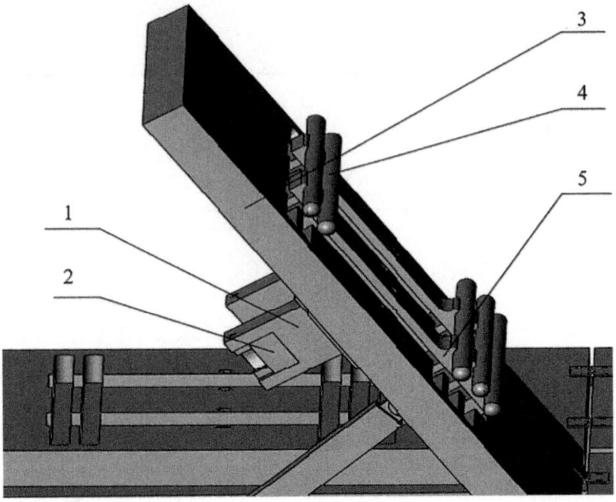

Fig. 6 The "S" shaped three-dimensional diagram of the back of the smart bed. 1—push rod motor connector, 2—push rod motor, 3—back plate, 4—back rod, 5—back rest connector

waist to be properly supported, and it will also affect the back and other areas of discomfort. The center height of the waist pillow mechanism of the smart bed should be 185–250 mm.

According to the internal pressure between the third and fourth lumbar vertebrae at different angles between the hip plate and the back plate and the size of the myoelectric activity potential near the eighth thoracic vertebrae [11], when the Angle between the back plate and the hip plate surface is greater than 110°, the pressure in the intervertebral disc is significantly reduced. Therefore, the position of the lumbar cushion mechanism should be in the third to fifth lumbar spine, and the thickness of the lumbar cushion mechanism should be preferably 5 cm, so as to make the curvature of the lumbar spine more natural and the back muscles more relaxed [12].

In summary, the maximum height of the mechanism in the waist should be ≥50 mm, so the maximum stroke of the electric push rod should also be ≥50 mm, and the distance between the center of the waist and the hip plate should also be between 185 mm and 250 mm.

Leg Lifting Structure Design The leg lifting structure is a simple parallelogram structure. The requirements are: the maximum Angle between the hip plate and the thigh plate is 35°, and the connecting rod 1 can be driven by the motor to rotate normally. The schematic diagram of the structure is shown in Fig. 7.

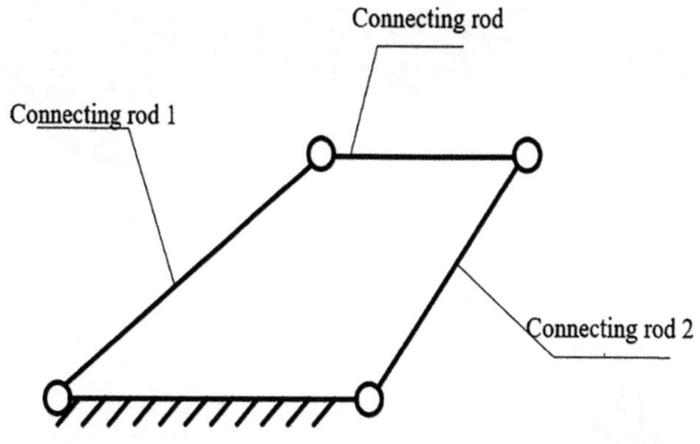

Fig. 7 Schematic diagram of leg extension mechanism

2.4 *Establishment of Smart Bed Model*

Create the 3D model of the parts through SolidWorks and assemble the designed structure. The three-dimensional model of the lying state of the smart bed is shown in Fig. 8, and the three-dimensional model of the zero gravity state of the smart bed is shown in Fig. 9.

In order to more clearly express the position relationship of each part of the smart bed, the assembled model is built and imported into the CAD. The marked two-dimensional figure is shown in Fig. 10.

The overall structure of the smart bed includes: Bed frame (1), screw rod (2), screw cap (3), back support rod (4), push rod motor connection block (5), back plate (6), push rod motor (7), back push rod connection block (8), Back push rod (9), hip plate (10), thigh plate (11), leg connecting rod 1 (12), leg connecting rod 2 (13), leg plate (14), leg plate (15), leg motor (16), leg motor coupling (17), leg connecting rod fastener (18), back bearing end cap (19), back gasket (20), back sealing ring (15) 21), deep groove ball bearing (22), back motor coupling (23), back motor (24).

3 Finite Element Analysis and Verification of Smart Bed

The mechanical properties of the smart bed were determined by FEA (Finite Element Analysis). Finite element analysis refers to the combination of modern computer technology, applied mathematics and mechanics and many other disciplines formed the finite element method to analyze static or dynamic physical body or physical system.

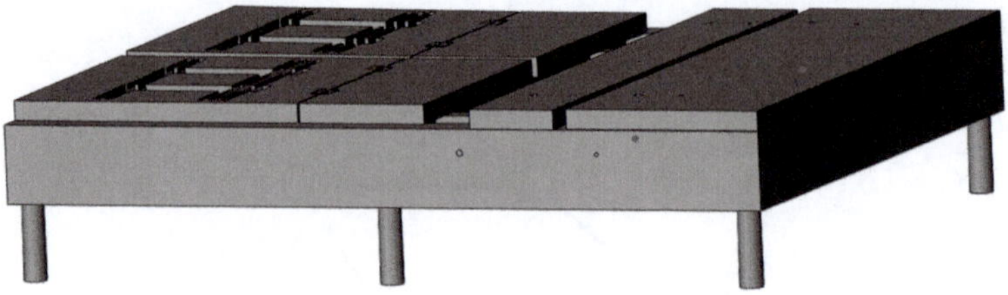

Fig. 8 Flat state diagram of smart bed

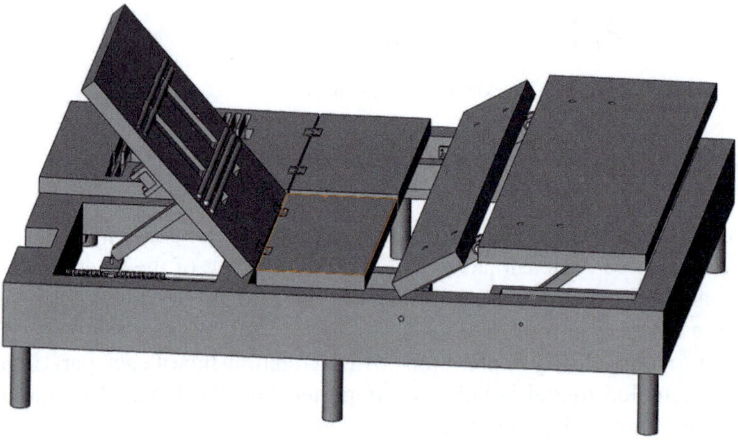

Fig. 9 Zero-gravity state diagram of the smart bed

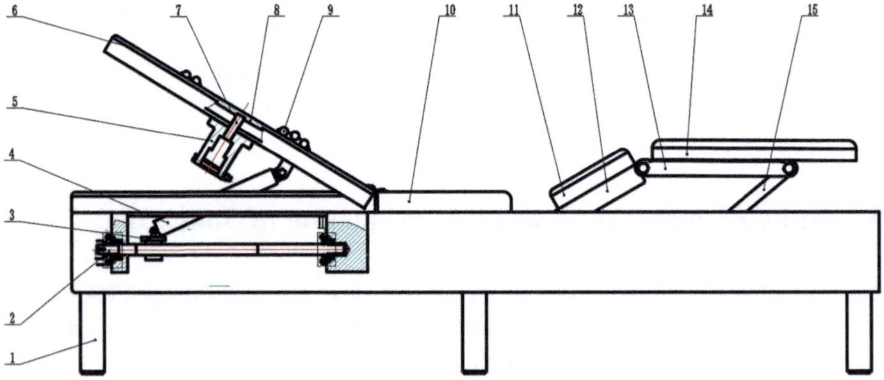

Fig. 10 Overall two-dimensional picture of the smart bed

3.1 Inspection Standards for Smart Beds

In order to enable the smart bed to have superior mechanical properties, mechanical simulation analysis of the mechanical properties of the smart bed should be carried out according to GB/T 4191-2011 mechanical testing standard and functional activity stretching mechanical device QB/T 4191-2011 [13]. If it is found that the strength of the structure is not enough under the test conditions, it should be corrected to make it meet the test requirements.

According to the requirements in Table 1, the safety factor has been calculated in the mechanical test loading. Therefore, when the corresponding load is applied to it using ANSYS Workbench for static analysis, the yield strength of the mechanism material is taken as the judgment basis and compared with the static analysis diagram solved in ANSYS Workbench. Determine whether the designed components meet the requirements.

Table 1 Inspection requirements for mechanical properties

Inspection standard	Inspection item	Inspection requirement	Result evaluation criteria
GB/T 4191-2011 Test standard for mechanical properties	Static pressure test of whole mechanism	Place 70 kg of load vertically on the back plate, hip plate, and leg plate in the lying state and hold it for 30 min	After the end of the test, observe the loaded part, there should be no damage, no obvious deformation, can be normal attitude change and no other anomalies
	Hip plate impact test	The load with a mass of 90 kg is placed on the position 406 mm above the vertical side of the hip plate in the lying state, and it falls freely to make an impact once	After the end of the test, observe the loaded part, there should be no damage, no obvious deformation, can be normal attitude change and no other anomalies
QB/T 4191-2011 Functional activity stretching mechanism	Leg plate and back plate impact test	The load with a mass of 34 kg is placed at a height of 305 mm away from the leg plate and the back plate respectively and let it fall freely	After the end of the test, observe the loaded part, there should be no damage, no obvious deformation, can be normal attitude change and no other anomalies
	Prototype fatigue test	Load 24.5, 77 and 12 kg on the backrest of the back plate, hip plate and leg plate respectively. One to three times per minute, 25,000 cycles	After the end of the test, observe the loaded part, there should be no damage, no obvious deformation, can be normal attitude change and no other anomalies

3.2 Smart Bed Finite Element Processing

According to the test requirements in Table 1, when ANSYS Workbench is used to conduct finite element simulation analysis of the three-dimensional assembly model of the smart bed, the mechanical properties of the overall mechanism of the bed under impact load should be checked. Firstly, the imported model is divided into grids, fixed support constraints and vertical static load are added, and finally the static stress distribution cloud map and the total deformation map are obtained by solving the stress.

Firstly, the assembled 3D model in SolidWorks was imported into ANSYS Workbench, and some unimportant parts were deleted to better mesh division and increase the accuracy of the results. The parts of the smart bed frame are mainly made of Q235 steel, and the parameter characteristics of Q235 material are shown in Table 2.

According to the standard in Table 1, the load with a mass of 90 kg is placed at a position 406 mm vertically above the hip plate, and the free falling body is impacted once; When detecting the leg plate and the back plate, place the load with a mass of 34 kg at 305 mm above the vertical position, drop it vertically, and impact it once. Combined with the above formula, the average load of the hip plate is 3142.86 N, the average load of the leg is 1200 N, and the average load of the back plate is 2296.88 N. The smart bed in this paper is a double bed, so the average load on the legs is 2400 N. The loading diagram with fixed support constraints and static loads added to the model is shown in Fig. 11.

3.3 Finite Element Analysis of Smart Bed

After mesh division, finite element processing steps of constrained fixed support and static load, the stress and deformation results of the smart bed model were solved, and the static pressure equivalent stress cloud diagram of the smart bed was obtained, as shown in Fig. 12, and the total deformation diagram was shown in Fig. 13.

Under the loading standard of GB/T 4191-2011 mechanical test, according to the static pressure equivalent stress cloud map of the back structure, hip plate and leg structure model of the bed, the maximum stress of the structure is about 118.2Mpa. The material used for the smart bed parts is Q235 carbon structural steel, the yield limit of the material is 235 MPa, and the maximum deformation is about 0.18 mm.

Table 2 Parameter characteristics of Q235 material

Materials	Density P (kg/m^3)	Young's modulus E (N/m^2)	Poisson ratio μ	Yield strength σ_s(N/m^2)
Q235	7850	2.06e11	0.29	2.35e8

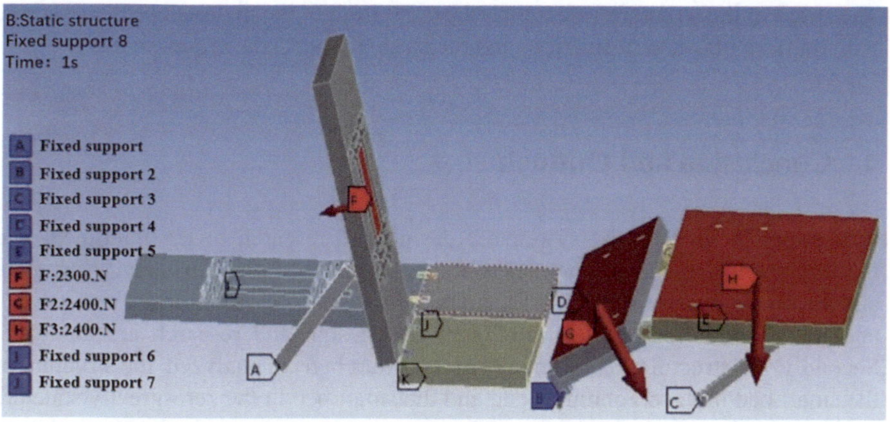

Fig. 11 Fixed support constraints and static load loading diagram

Fig. 12 Static pressure equivalent stress cloud map

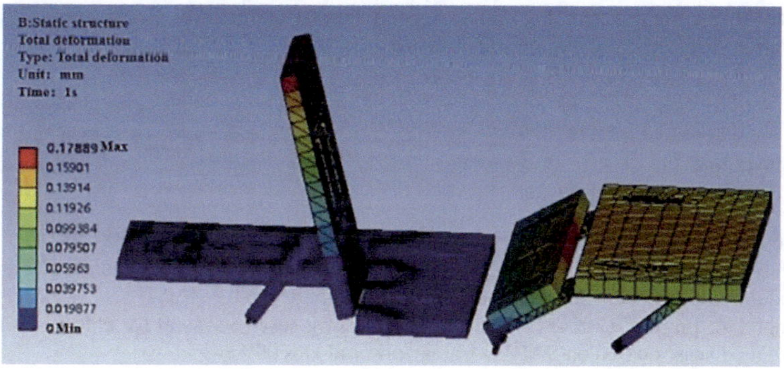

Fig. 13 General deformation diagram

The stress of the structure is less than the yield limit of the material, so the structure will not have plastic deformation, and the strength meets the requirements.

4 Conclusion and Outlook

This paper focuses on the systematic design and research of the intelligent bed mechanical device, the ergonomics of the intelligent bed, the zero gravity structure and the research status of the intelligent bed back plate "S" structure are understood, and the shortcomings of the intelligent bed design and research are analyzed. Secondly, the structural requirements of the smart bed are analyzed, the structure of the smart bed in the ergonomic state and the smart bed in the zero-gravity state are analyzed, and the demand for the "S" type structure of the back plate is analyzed. The designed scheme is designed and modeled by the 3D modeling software SolidWorks for the main modules of the smart bed, and virtual assembly of each component is carried out according to the working principle of the smart bed. Finally, the mechanical properties of the smart bed are analyzed and checked in ANSYS Workbench, and structural improvements are made to the places that do not meet the mechanical test requirements, and the model is optimized to determine the final model.

In short, the future design and research of smart bed lies in the comprehensive monitoring of user health and behavior through advanced sensing technology and AI algorithm, and the intelligent adjustment and interaction function to improve the quality of sleep and life of users.

Acknowledgments This work was financially supported by Priority Academic Program Development of Jiangsu Higher Education Institutions, grant number PAPD; Key R&D Projects of Jiangsu Province, grant number BE2023765, Jiangsu Province Innovation Support Program (International Science and Technology Cooperation) Project, grant number BZ2023002, Jiangsu Province Science and Technology Plan Special Fund (International Science and Technology Cooperation) Project, grant number BZ2024048.

References

1. You Z, Heng X (2020) Research and design of smart bed. Elec Prod 24:38–40
2. Fangyuan L, Chuanyi L, Lei H (2014) Lower limb mechanism design and motion analysis of modular nursing bed. Chin J Eng Des 21(06):583–588
3. Pengfei Z, Jingyi Y (2024) Design of home nursing service system for elderly patients with chronic diseases based on SAPAD framework. Ind Des 09:88–91
4. Min C, Youchun X, Guo L (2019) Design and research of smart bed based on multi-sensor. Inf Rec Mater 21(08):196–198
5. Mattress ergonomic evaluation Part One: Mattress hardness grading and distribution test evaluation method:T/SZFA 2003.1-2019[S]. 2019
6. GB/T 10000-2023, Chinese adult human body size [S]

7. Tommaso S, Irene LS (2018) Chapter 15: Habitability and habitat design. Space Saf Hum Perform:653–719
8. Occupational and Environmental Health Center, University of Connecticut. Neutral Posture [EB/OL]. http://oehc.uchc.edu/ergo_neutralposture.asp
9. National Aeronautics and Space Administration (NASA). NASA Standards Inform Comfortable Car Seats [EB/OL]
10. Youcan L, Fan Fa-Bin (Taiwan). Ergonomics [M]. Great Holy Book Company. 1984.3
11. Jia M, Zhisheng F, Ying R et al (2008) Research review of car seat comfort. Shanghai Automobile 01:24–27
12. Jing Y (2008) Research on human comfort of sofa chair back. Nanjing Forestry University, Nanjing
13. QB/T 4191-2011, Multifunctional mobile extension mechanism [S]

Open Access This chapter is licensed under the terms of the Creative Commons Attribution-NonCommercial-NoDerivatives 4.0 International License (http://creativecommons.org/licenses/by-nc-nd/4.0/), which permits any noncommercial use, sharing, distribution and reproduction in any medium or format, as long as you give appropriate credit to the original author(s) and the source, provide a link to the Creative Commons license and indicate if you modified the licensed material. You do not have permission under this license to share adapted material derived from this chapter or parts of it.

The images or other third party material in this chapter are included in the chapter's Creative Commons license, unless indicated otherwise in a credit line to the material. If material is not included in the chapter's Creative Commons license and your intended use is not permitted by statutory regulation or exceeds the permitted use, you will need to obtain permission directly from the copyright holder.

In-Depth Discussion on the Design of Distribution Network Nut Section

Yuting Liu, Wei Zheng, and Xiaolei Chen

Abstract This paper proposes an innovative design for a slanted nut to address the issue of water accumulation in nuts in polluted areas of distribution networks. Through in-depth theoretical analysis and the application of fluid mechanics principles, a mathematical model for water flow was constructed to precisely determine the optimal slant angle, effectively preventing water accumulation in the nuts. Additionally, while ensuring the structural stability of the nut, the study innovatively explores the idea of incorporating holes in the nut design to further promote water drainage and reduce the risk of corrosion. This research not only provides a new solution for the application of nuts in polluted areas of distribution networks but also is significant for enhancing the reliability of distribution network operations and reducing maintenance costs.

Keywords Slanted nut design · Water accumulation prevention · Distribution networks · Fluid mechanics analysis · Optimal slant angle · Drainage holes · Corrosion reduction · Structural stability · Maintenance cost reduction · Reliability enhancement

1 Introduction

Distribution network nuts, as key components for connecting and fixing equipment in the power system, their performance is directly related to the operational safety and efficiency of the distribution network [1–3]. However, in polluted areas, due to the high concentration of salts, dust, and other impurities in the environment, a conductive layer is easily formed on the surface of the nuts, accelerating corrosion. At the same time, the humid environment leads to water accumulation in the nuts, further intensifying the corrosion process, reducing the electrical performance of the nuts, increasing safety hazards, and raising maintenance costs [4, 5]. Therefore,

Y. Liu (✉) · W. Zheng · X. Chen
State Grid Jinzhou Electric Power Supply Company, Jinzhou, Liaoning, China

© The Author(s) 2026

Z. Zhang, P. Mercorelli (eds.), *Mechanical Design and Simulation: Insights and Innovations*, Lecture Notes in Mechanical Engineering,
https://doi.org/10.1007/978-981-95-2405-1_115

researching a nut design that can effectively prevent water accumulation is of great significance for improving the operational reliability of the distribution network and reducing maintenance costs [6–8].

The purpose of this paper is to delve into the principles and methods of inclined nut design, by constructing a mathematical model of water flow, to analyze and determine the optimal inclined angle to ensure that the nuts can effectively drain water in polluted areas [9, 10]. At the same time, the study explores the design idea of adding holes to promote water discharge while ensuring the structural stability of the nuts, providing new ideas and methods for the design of distribution network nuts.

2 Analysis of Water Accumulation Issues in Distribution Network Nuts

The environmental characteristics of polluted areas are mainly characterized by high concentrations of salts, dust, and other impurities, as well as humid climatic conditions. These impurities easily form conductive layers in humid environments, accelerating the corrosion process of the nuts. At the same time, humid environments also tend to cause water accumulation on the surface of the nuts, providing favorable conditions for corrosion. Therefore, the environmental characteristics of polluted areas pose higher requirements for the performance of nuts.

3 Inclined Nut Design Principle and Advantages

The design principle of the inclined nut is to change the water flow path, allowing water to slide down the inclined plane under the force of gravity, effectively avoiding water accumulation. By accurately calculating the inclined angle, effective water diversion can be achieved, reducing the phenomenon of water accumulation.

4 Application of Mathematical Models in Wedge Nut Design

4.1 Construction of Mathematical Model for Water Flow on Inclined Planes

Based on the principles of fluid mechanics, this paper constructs a mathematical model for water flow on inclined planes. The model fully considers the effects of factors such as the angle of the slope, water flow velocity, and friction on the process of water flow. Specifically, we assume that the water flow is an incompressible fluid and that the flow is in a steady state. In the inclined plane coordinate system, we can establish the following partial differential equation to describe the process of water flow:

Continuity equation:

$$\partial x + \partial u + \partial y + \partial v = 0 \tag{1}$$

Among them, *u* and *v* are the velocity components in the slope direction (x-direction) and perpendicular to the slope direction (y-direction), respectively.

Momentum equation (considering only the x-direction):

$$\rho\mu\partial x\partial\mu + \rho v\partial y\partial\mu = -\partial x\partial p + \mu\left(\partial x2\partial 2\mu + \partial y2\partial 2u\right) + \rho g\sin\theta \tag{2}$$

Among them, ρ is the density of water flow, p is the pressure, μ is the dynamic viscosity, g is the acceleration due to gravity, and θ is the angle of the slope.

By simplifying assumptions and boundary conditions, we can solve the above partial differential equation to obtain the relationship between the sliding velocity of water flow and the slope angle. This relationship provides a theoretical basis for the design of the slope angle.

4.2 Determination of Optimal Slope Angle

Using the constructed mathematical model, this paper conducted an optimization analysis to determine the optimal slope angle. We defined an objective function that takes into account the effects of water flow velocity and slope friction on the smoothness of the water flow. Specifically, the objective function can be represented as:

$$\mathrm{J}(\theta) = \mu\cos\theta + \mathrm{u}(\theta) \tag{3}$$

Among them, $\mu(\theta)$ is the velocity of water flow sliding down a slope at an angle θ, $\mu\cos\theta$ represents the effect of slope friction on the velocity of the water flow (simplified representation), The constant term is used to adjust the scale of the objective function and the constant term is used to adjust the scale of the objective function. Please refer to Fig. 1.

We use numerical optimization methods (such as gradient descent, Newton's method, etc.) to maximize the objective function $J(\theta)$, thereby determining the optimal slope angle. This angle ensures that the water flows smoothly downwards under gravity, while minimizing the effect of slope friction on the water flow speed.

Through experimental verification, the optimal inclined angle can effectively reduce the phenomenon of water accumulation in nuts and improve the drainage performance of the nuts.

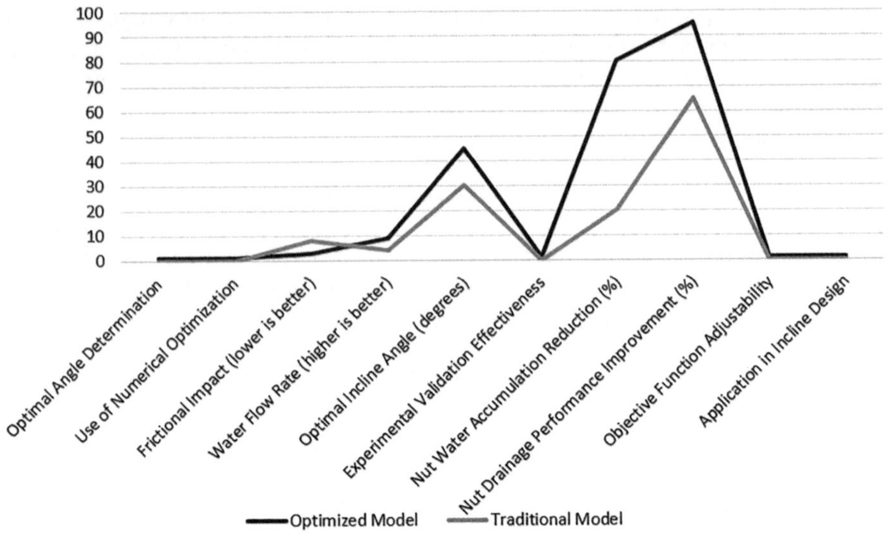

Fig. 1 Comparison of optimized model vs. traditional model in nut drainage performance

4.3 Model Validation and Case Analysis

To verify the accuracy of the constructed mathematical model, this paper conducted an experimental data comparison. We set up experiments with different slope angles and measured the corresponding water flow velocities. The results indicated that the model's predicted values matched the experimental data well, validating the model's accuracy. Please refer to Fig. 2.

Furthermore, taking a typical polluted area as an example, we conducted a case study on the design of slope nuts. Based on the actual environmental and climatic conditions of the area, we determined the optimal slope angle using the constructed mathematical model and optimization methods. The practical application results showed that the design of slope nuts can effectively reduce the phenomenon of water accumulation in nuts, enhancing the electrical performance and operational reliability of the nuts.

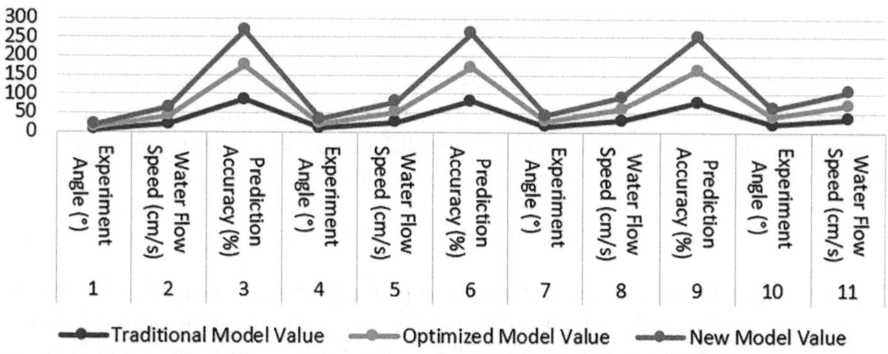

Fig. 2 Comparison of optimized vs. traditional vs. new model in nut drainage performance

5 Analysis of Nut Stability and Hole Design

5.1 *Nut Stability Assessment*

The inclined plane design may have a certain impact on the stability of the nut. Therefore, this paper conducts a detailed stability analysis. To quantify this impact, we used a mathematical model based on the principle of mechanical equilibrium to assess the stability of the inclined plane nut.

Assuming the vertical force acting on the inclined nut is Fv, the force along the inclined plane is Fs, the angle of the incline is θ, and the coefficient of friction is μ. According to the conditions of static equilibrium, the condition for the nut not to slide is:

$$Fs \leq \mu Fv\cos(\theta) + Fv\sin(\theta) \tag{4}$$

Among them, $Fv \cos(\theta)$ is the normal force perpendicular to the inclined plane, and $Fv \sin(\theta)$ is the component of gravity along the inclined plane direction. Through finite element simulation, we can calculate the values of Fs and Fv under different working conditions and verify whether the aforementioned inequality holds. The results indicate that, provided the inclined plane angle θ is designed reasonably, the inclined nut can meet the stability requirements and ensure the safe operation of the power grid.

5.2 *Pore Design Principles and Advantages*

To further promote the discharge of water, this article explores the design idea of adding holes to the inclined nut. The design of the holes needs to be optimized based on fluid dynamics principles. Let the total area of the holes be Ah, the total surface area of the nuts be At, and the area ratio of the holes be

$$\rho = At^* Ah \quad (5)$$

The position, size, and number of holes need to be optimized based on the sliding speed of the water *vf* and the structure of the nut. Assuming the flow speed of water through the holes is *vh*, according to Bernoulli's equation, the relationship between the flow speed and water pressure at the holes can be expressed as:

$$P_1 + 21\rho fvf2 = P_2 + 21\rho fvh2 \quad (6)$$

Where P_1 and P_2 are the water pressures upstream and downstream of the hole, respectively, and ρf is the density of water. By optimizing the hole design to maximize the *vh*, the drainage performance of the nut is further improved, reducing water accumulation. Additionally, the hole design can also reduce the weight of the nut and lower manufacturing costs. Please refer to Table 1.

5.3 The Effect of Hole Design on Nut Performance

To evaluate the impact of hole design on nut performance, this paper conducts strength and stiffness analysis. Let the original strength of the nut be $\sigma 0$, and the strength after adding a hole be σh. According to the strength reduction principle in material mechanics, the effect of the hole on the strength can be represented as:

$$\sigma h = \sigma 0\left(1 - k_p\right) \quad (7)$$

Wherein, *k* is a reduction factor related to the shape, distribution, and material properties of the holes. Similarly, let the original stiffness of the nut be E_0, and the stiffness after adding the hole be E_h, then:

$$E_h = E_0\left(1 - mp\right) \quad (8)$$

Table 1 Comparative table of nut performance and cost between optimized hole design and traditional design

Aspect description	Optimized hole design	Traditional design
Water Pressure Upstream (P1) and Downstream (P2)	Considered	N/A (Not specified)
Water Density (ρf)	Considered	N/A (Assumed or not optimized)
Maximization of vh (Hole Velocity)	1 (Optimized for)	0 (Not optimized)
Improvement in Nut Drainage Performance	1 (Enhanced)	0 (Not enhanced)
Reduction in Water Accumulation	1 (Reduced)	0 (Not reduced)
Weight reduction of Nut	1 (Lighter)	0 (Not lightened)
Manufacturing Cost Reduction	1 (Lower)	0 (Not reduced)

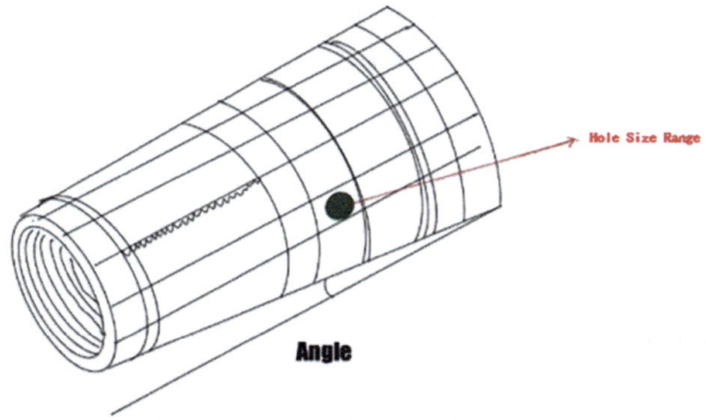

Fig. 3 Nut design drawing

Table 2 Comparative analysis table of nut performance between holed design and traditional design

Aspect description	Hole design impact analysis	Traditional (no hole) design
Evaluation of Hole Design on Nut Performance	1 (Evaluated)	0 (Not evaluated)
Original Strength of Nut (σ0)	Considered	N/A (Baseline)
Strength After Adding Hole (σh)	σh = σ0(1 − kρ)	N/A (No change)
Strength Reduction Factor (k)	Determined via FEA & Exp.	N/A (Not applicable)
Original Stiffness of Nut (E0)	Considered	N/A (Baseline)
Stiffness After Adding Hole (Eh)	Eh = E0(1 − mρ)	N/A (No change)
Stiffness Reduction Factor (m)	Determined via FEA & Exp.	N/A (Not applicable)
Impact on Strength and Stiffness (Within Limits)	1 (Acceptable)	0 (Not assessed)
Meets Distribution Network Operational Requirements	1 (Satisfies)	N/A (Assumed to satisfy)
Improvement in Corrosion Resistance	1 (Enhanced)	0 (Not enhanced)

Among which, m is the coefficient related to stiffness reduction. Through finite element simulation and experimental verification, we can determine the specific values of k and m. The results indicate that, under the premise of reasonable hole design, this design has a minor impact on the strength and stiffness of the nut, which can meet the requirements of distribution network operation. Please refer to Fig. 3. At the same time, the hole design also helps to improve the corrosion resistance of the nut and extend its service life. Please refer to Tables 2 and 3.

Table 3 Nut slope and hole size based on contamination level

Contamination level range	Nut slope range	Hole size range (mm)
Very Clean	0.8–1.0	5.0–5.2
Slightly Contaminated	0.7–0.8	5.2–5.4
Moderately Contaminated	0.6–0.7	5.4–5.6
Heavily Contaminated	0.5–0.6	5.6–5.8
Severely Contaminated	0.4–0.5	5.8–6.0

6 Conclusion

Through theoretical analysis and mathematical modeling, this paper proposes an innovative design for inclined nut. This design significantly reduces the phenomenon of water accumulation in the nut, enhancing the electrical performance and operational reliability of the nut. Experimental verification indicates that this design has high accuracy and feasibility, offering new ideas and methods for the design of distribution network nuts.

Future research can further explore the optimal design parameters of inclined nuts under different pollution environments, as well as the synergistic optimization of hole design and inclined design. At the same time, the long-term operational performance and maintenance strategies of inclined nuts in distribution networks can also be studied, providing a strong guarantee for the safe operation of distribution networks.

The research findings of this paper are of great significance for enhancing the operational reliability of distribution networks. Moreover, this design can provide reference and inspiration for the design of other similar connectors, promoting technological progress and innovative development in the power industry. It is believed that in future research, the design of inclined nuts will gain wider application and promotion.

References

1. Yang Mingbo D, Sanhu WZ et al (2020) Design and experimental verification of SMA low impact separation nut. Spacecr Eng 29(05):65–70
2. Han S (2016) Research on shock reduction design method and performance analysis of typical separation nuts. Harbin Institute of Technology
3. Chuanrong S (2015) Design research on nut automatic forging production line. Hebei University of Engineering
4. Biao Z, Jun Y, Xiang L et al (2010) Design of a screw-Nut mechanism type manipulator. Sci Technol West China 9(15):31–32
5. Songkun Y, Shuyuan Z, Zhaoyi J (2009) Design of nut Tightening Machine Control System. Mach Tool Hydraul 37(04):117–119
6. Shuqing Z (2008) Research on high-speed liquid static pressure screw-nut pair. Lanzhou University of Technology

7. Bin G, Zhonggang L, Jing M (2001) Analysis of key design parameters of separation nuts. Spacecr Recovery Remote Sens 02:63–65
8. Wenhui M, Mingshu Z (2000) Reliability design of sliding screw-nut mechanism. Mech Sci Technol 01:52–53. https://doi.org/10.13433/j.cnki.1003-8728.2000.01.019
9. Yan C (2019) Research on the failure mechanism of bolt loosening and its effect on the mechanical behavior of the overall structure. Dalian University of Technology
10. Yongqiang R, Jianguo Y, Juguang L et al (2004) System Design of Multi-Function nut Tightening Machine Based on force closure. Manuf Technol Mach Tool 01:32-35+110

Open Access This chapter is licensed under the terms of the Creative Commons Attribution-NonCommercial-NoDerivatives 4.0 International License (http://creativecommons.org/licenses/by-nc-nd/4.0/), which permits any noncommercial use, sharing, distribution and reproduction in any medium or format, as long as you give appropriate credit to the original author(s) and the source, provide a link to the Creative Commons license and indicate if you modified the licensed material. You do not have permission under this license to share adapted material derived from this chapter or parts of it.

The images or other third party material in this chapter are included in the chapter's Creative Commons license, unless indicated otherwise in a credit line to the material. If material is not included in the chapter's Creative Commons license and your intended use is not permitted by statutory regulation or exceeds the permitted use, you will need to obtain permission directly from the copyright holder.

Construction of Car Side Panel Model Driven by Styling and General Layout

Chunning Jin, Hengyi Jin, Xinyu Fu, and Wenbin Hou

Abstract In the early stage of body development, it is crucial to quickly and automatically generate a high-maturity body-in-white (BIW) model. This paper takes the solution for the side panel as an example, based on modeling requirements and the design association of the side panel model, it extracts the styling and general layout features related to the side panel, collects typical sections, establishes a typical section feature library, and also develops an interface program between sections and EXCEL spreadsheets, realizing the driving of key design elements on the side panel structure. Based on the modeling strategy of the intermediate layer, the knowledge engineering method is adopted to encapsulate basic modeling operations into forms such as user-defined features and the secondary development program based on CATIA component application (CAA). A parametric template is established, which realizes automated parameter iteration in the early stage of design, and also achieves standardization of the modeling process, ensuring the reliability of the side frame model.

Keywords Body side panel design · Automated modeling methods · Parametric template · Knowledge engineering

1 Introduction

Automating the construction of high-maturity BIW models in the early stages of body design can advance the analysis, evaluation, and optimization processes, significantly improving design quality [1–5]. The research in the related fields usually focuses on the following aspects: vehicle body structure modularization [6–8] and data-driven optimization [9–12]. Some of the aforementioned research methods utilize new frameworks and platforms for design optimization, while others employ

C. Jin (✉) · H. Jin · X. Fu · W. Hou
Dalian University of Technology, Dalian, China
e-mail: jincn@dlut.edu.cn

© The Author(s) 2026

Z. Zhang, P. Mercorelli (eds.), *Mechanical Design and Simulation: Insights and Innovations*, Lecture Notes in Mechanical Engineering,
https://doi.org/10.1007/978-981-95-2405-1_116

novel modeling techniques to enhance the modeling process. These methods enable the rapid construction of performance models in the early stages of design. However, these studies still face challenges such as low model maturity and a lack of correlation between key design elements and body styling. Overly simplified models may compromise the credibility of model-based analysis and optimization. Additionally, the absence of design and styling constraints can easily lead to difficulties in directly driving subsequent detailed part design with analysis and optimization results, thereby diminishing their guiding value for subsequent designs.

Based on this, this paper aims to establish a high maturity vehicle body side panel model driven by styling and general layout in the early design stage, achieving the driving of side panel models by vehicle body styling and general layout information, and achieving higher design maturity through the establishment of a typical section library. The establishment of a high-maturity model will help advance analysis, evaluation, and optimization to the early stages of design, significantly improving design quality. At the same time, the styling-driven design scheme can significantly enhance the reusability of the side panel model under new styling schemes; the rich design constraints contained in the model enable early evaluation parameters and optimization parameters to have a more direct driving effect on subsequent design engineering. The establishment of a high-maturity side panel model will provide a new solution that helps achieve performance-based automatic parameter iteration and optimization in the early stages of design. Moreover, this solution significantly reduces repetitive manual modeling in model construction, greatly improving design efficiency. This method is applicable to a variety of design scenarios, especially for rapidly generating complex models, and can meet the efficient demands of modern automotive design and manufacturing.

2 Automated Side Panel Construction Technology

The diversity of body styles and structures determines that there is no universal and generic side panel pattern or model construction method. However, we can appropriately classify body styles, construct targeted structural templates for body styles with common patterns, and technically address the issues brought about by the diversity of body styles through a template library.

This paper focuses on the research of automatic construction methods for car body side panels, focusing on SUV bodies with four doors and corner windows. The automatically generated side panel template will serve as a part of the template database, which will automatically create matching parameterized side panels for new styling schemes of similar vehicle models in the future, and be used for further analysis, evaluation, and optimization.

2.1 Collection and Construction of Styling and General Layout Features

The body styling model is a diverse, complex, and non-parametric surface model, lacking consistency and standardization. In practical model design, it is impossible to find a definitive method and process to effectively drive the side panel model through body styling. Therefore, this paper deconstructs the body styling model, extracts a set of standardized geometric elements that have a key impact on side panel modeling, and defines these elements as standardized styling features. Through these features, the styling is indirectly driven to the side panel model. This proxy-driven mechanism based on styling deconstruction is one of the key methods for achieving automatic side panel construction in this paper.

Based on the design correlation of the side panel model and specific modeling requirements, this paper deconstructs the body styling model and constructs styling features as the basic input for side panel modeling. These side panel styling features consist of a series of surfaces and curves. The feature surfaces include the front windshield glass surface, side panel glass surface, roof main surface, lower styling surface of side beam, upper styling surface of side beam, and corner window surface, which are necessary for side panel modeling. The feature lines include the waterline, door seam line, roof styling feature line, wheel cover area styling line, and D-pillar styling feature line. These standardized geometric bodies undergo quality inspection and standardized naming for automatic generation of later structures. This mechanism will ensure the coordination between structure and styling, as shown in Fig. 1.

The general layout of the automobile will be completed after the styling freeze, and the generation of the side panel structure is also constrained by the general layout. To achieve the driving of the side panel model by the general layout, key inputs

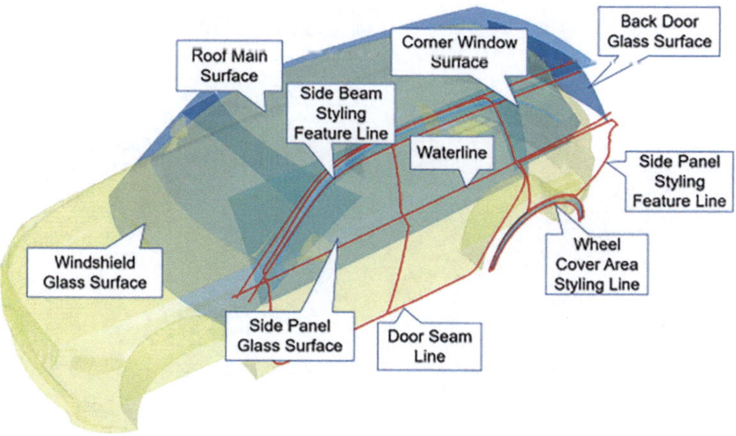

Fig. 1 Styling feature elements related to the side panel

related to the side panel structure will be extracted from the general layout information as general layout features to drive the side panel model in a geometric sense. These general layout features consist of points, surfaces, and planes, including hinge mounting points, hinge axes, hinge mounting surfaces, and door flange lines (front door, rear door, back door).

By integrating the characteristics of styling and general layout, it is technically possible to effectively drive the side structure with key design elements, and ensure that the model can automatically update for new styling schemes and corresponding general layout inputs.

The extraction of feature lines plays an important role in the collection process of general layout features and styling features. The final styling surface is a smooth geometry that has undergone rounding treatment, and all the edge lines that serve as feature lines have been eliminated. Therefore, the reconstruction of feature lines "seems" to be an inevitable task. However, in fact, these tasks can be avoided, because during the construction process of the body styling surface, these feature lines are generated and exist as intermediate processes and elements. As long as the design workflow is properly planned, these elements can be identified and retained during the styling surface design process. This operation will be simple and fast, and it is easy to automate.

These feature elements, including feature lines and surfaces, can provide key geometric references for subsequent section layout, and the planes to which the section is attached will also be constructed based on them.

To ensure the reliability of the subsequent modeling process, this paper standardizes and normalizes the feature elements used as inputs. The integrated standardized input and consistent modeling process ensure the reliability of the modeling, as shown in Fig. 2.

In the process of collecting feature data and establishing a feature library, this paper first completes the collection and input of styling features based on typical car

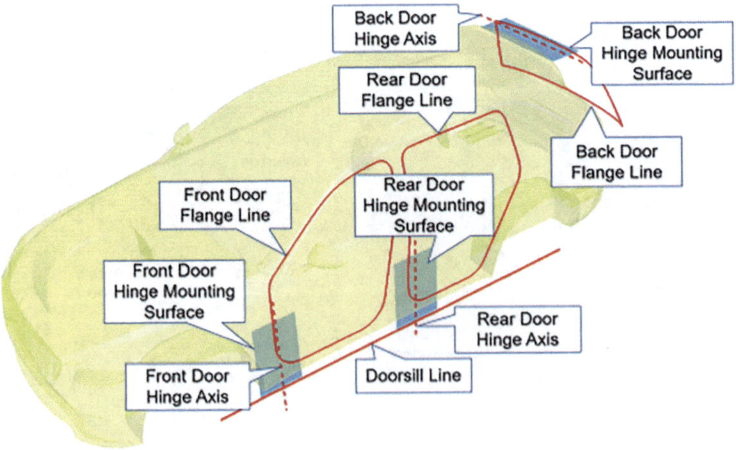

Fig. 2 General layout features related to the side panel

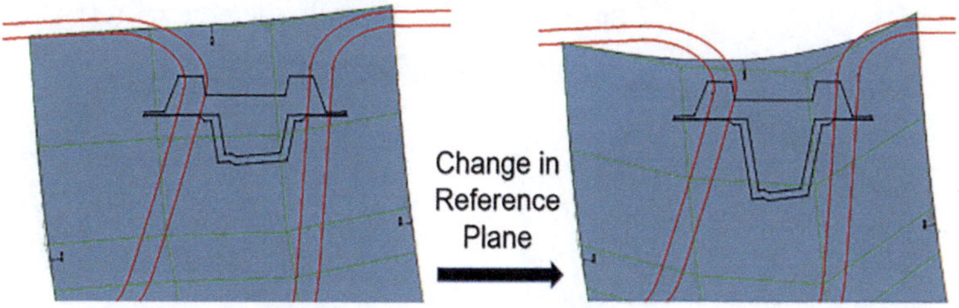

Fig. 3 The section automatically adjusts with changes in the reference plane

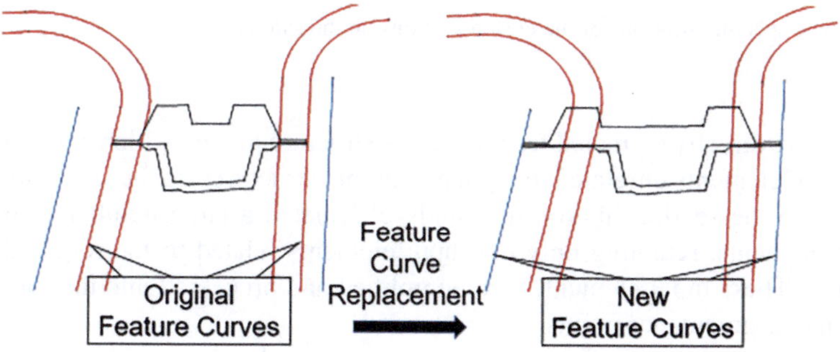

Fig. 4 The section changes dynamically with the substitution of feature curves

model data, and then constructs the side frame based on the corresponding styling features; Afterwards, based on the feature lines, the section template is attached to the plane in the framework, which can achieve the automatic adjustment of the section contour with the reference plane and reference surface, as shown in Fig. 3. At the same time, the section contour can also change with the replacement of the feature curves, which to some extent meets the requirements of shape driving, as shown in Fig. 4.

2.2 Collection and Establishment of Typical Section Feature Library

To enhance design maturity, this study adopted sections that are close to or consistent with engineering practice. To ensure the relevance of these sections to styling and overall layout, this study constructed a system of eleven typical sections driven by both overall layout and styling, based on design or functional requirements. These sections cover A-pillar, B-pillar, C-pillar, D-pillar, roof side rail, sill beam, as shown in Fig. 5. These sections are all extracted from typical positions of the vehicle body styling, taking the B-pillar as an example, as shown in Fig. 6. Since the

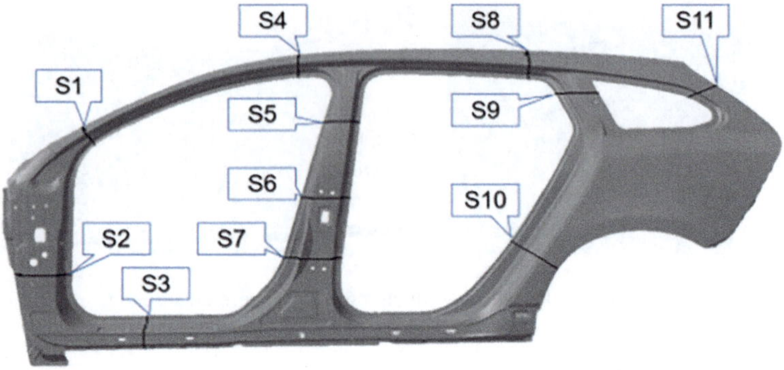

Fig. 5 The specific locations of the eleven sections on the side panel

sections are directly extracted from the body structure model, which contains a large number of elements unrelated to input and output structures, to facilitate storage and retrieval in the section library, this study eliminated a large number of unrelated sketch elements, retaining only the input elements related to the required section structure. These input elements were simplified and processed into the form of surfaces, line segments, or points.

Due to the existence of several to over ten control parameters for each section, the overall control parameters of the side panel can reach more than one hundred. In order to achieve effective management, this paper uses an EXCEL spreadsheet to control the parameters. The organized sections are recorded in EXCEL, and a CATIA and spreadsheet data interface program is established to facilitate direct adjustment of the model based on the spreadsheet in the later stage of design, with intuitive labeling of each parameter's corresponding element. In addition, this paper further subdivides length parameters into structural variable design parameters, process variable design parameters, layout design parameters, and styling design parameters. Each feature is named and accompanied by corresponding descriptions, and is categorized into the section feature library. These features are spatially combined with the extracted styling features and general layout features to ultimately form a complete modeling input. The specific process is shown in Fig. 7.

There are geometric correlations between different side panel sections, and there are also associations between the sections and the styling features and general layout features. Changes in section data can directly drive changes in styling and general layout features. Therefore, the establishment of a section feature library is the basis for performance decision-making and also the basis for template design. It is of great significance for solving key problems and achieving research goals.

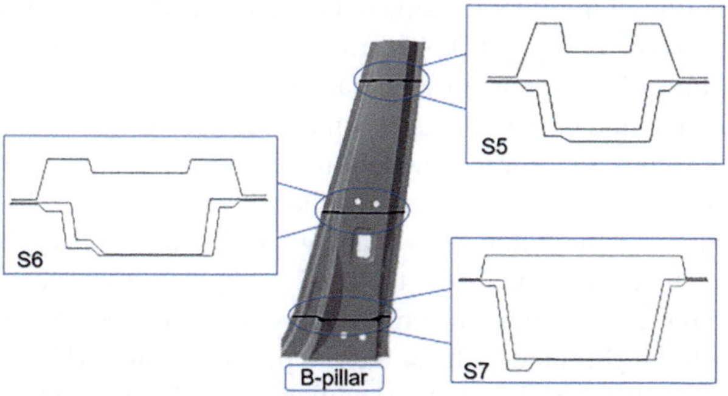

Fig. 6 B-pillar and its corresponding section

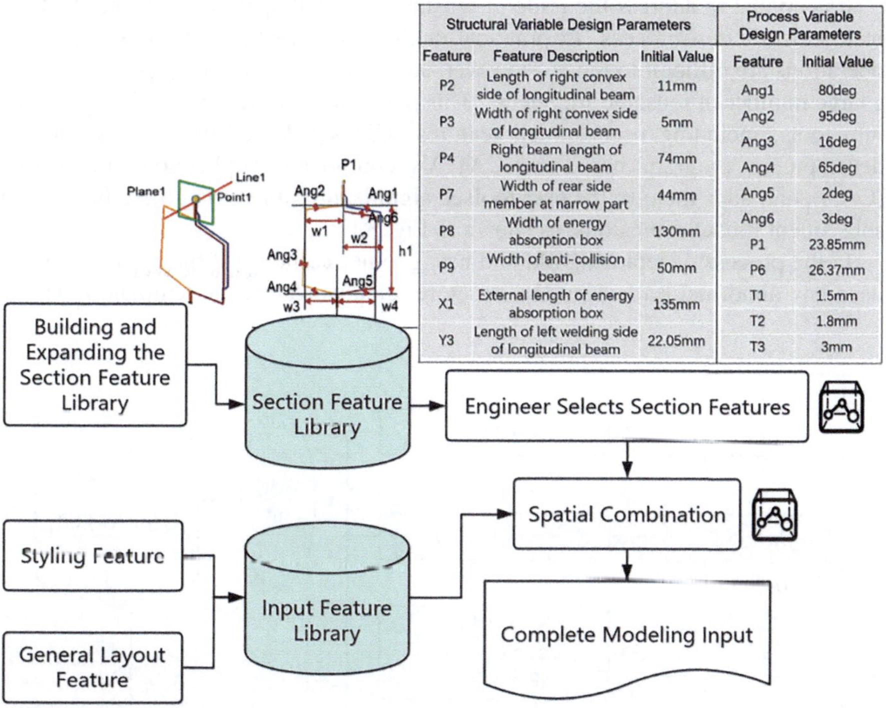

Structural Variable Design Parameters		
Feature	Feature Description	Initial Value
P2	Length of right convex side of longitudinal beam	11mm
P3	Width of right convex side of longitudinal beam	5mm
P4	Right beam length of longitudinal beam	74mm
P7	Width of rear side member at narrow part	44mm
P8	Width of energy absorption box	130mm
P9	Width of anti-collision beam	50mm
X1	External length of energy absorption box	135mm
Y3	Length of left welding side of longitudinal beam	22.05mm

Process Variable Design Parameters	
Feature	Initial Value
Ang1	80deg
Ang2	95deg
Ang3	16deg
Ang4	65deg
Ang5	2deg
Ang6	3deg
P1	23.85mm
P6	26.37mm
T1	1.5mm
T2	1.8mm
T3	3mm

Fig. 7 Modeling input implementation process

2.3 *Establishment of Parameterized Templates*

For side panel models with the same styling feature and common structural schemes, this paper constructs a side panel parametric template. This parametric template can be automatically generated based on styling and general layout parameters. It can reflect key design features, structural features, and performance characteristics, and generate simplified side panel models with high maturity.

The construction of traditional frame templates typically relies on the basic geometric functions provided by CAD platforms, and the construction process involves the accumulation of basic geometric operations. This type of modeling method is highly dependent on the interaction between engineers and CAD platforms, and the process of modifying and editing models is complex and inefficient. This paper adopts a modeling strategy based on intermediate layers, targeting the specific modeling requirements of the white body framework. By utilizing digital technology and knowledge engineering methods, the basic set operations are encapsulated into modeling features and section feature toolsets, user-defined features, and other forms, thereby freeing engineers from their dependence on CAD platform interaction processes, simplifying the model modification and editing process, achieving knowledge encapsulation and reuse, and significantly improving modeling efficiency.

It is crucial to address the issue of reusing historical operations during the template construction process. In practical engineering practice, complex geometric operations are difficult to reuse. However, any complex geometric process contains a large number of reusable fragments. This paper digitizes these reusable segments into tools, including user-defined features (UDFs), PowerCopy, and secondary development program based on CATIA's component application architecture (CAA), and adds them to the user toolset, significantly improving the efficiency of subsequent model extensions, as shown in Fig. 8.

In the process of replacing old and new geometric inputs, if the model does not have any abnormal errors and the structure remains stable, the template model is

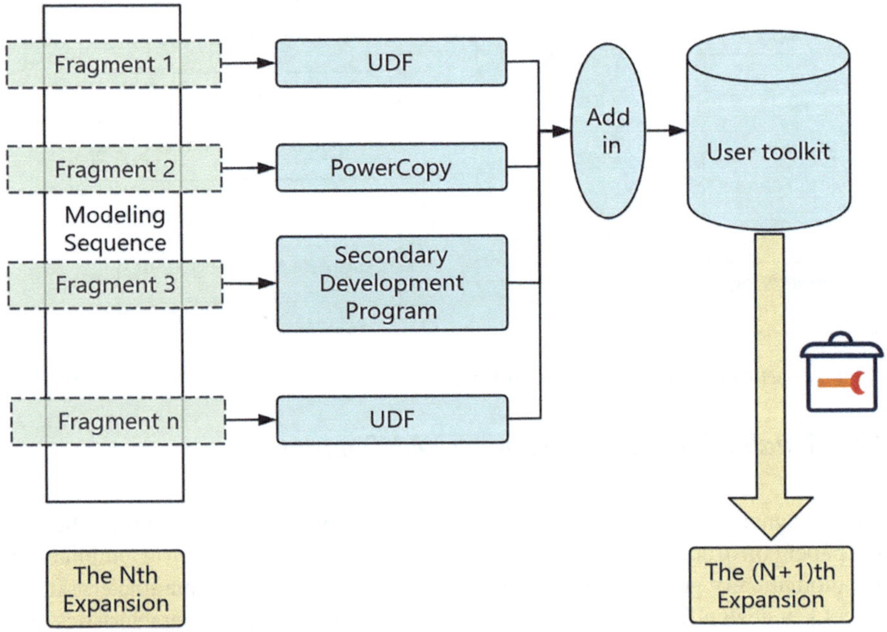

Fig. 8 Solution to the problem of reusing historical operations

usually considered reliable. However, due to the uncertainty of the geometric features used as inputs, there may be reliability issues with the framework template when replacing geometric inputs.

These uncertainties include but are not limited to:

- Uncertainty in the number of constituent surface patches
- Uncertainty in the number of boundaries constituting the surface
- Uncertainty in the number of sub-line segments and nodes constituting the curve
- Uncertainty in the direction of normal vectors for surfaces and planes
- Uncertainty in the tangent direction of the curve

In response to the reliability issues raised, this study ensured consistency between model construction and editing processes through an intelligent tool set that integrates knowledge engineering rules. This intelligent tool set integrates two key processes: input processing and output processing. In the traditional model construction process, facing the uncertainty of input, if there are abnormalities in the CATIA model construction process, interactive adjustments must be made again, and the model must be manually corrected until there are no abnormalities before output can be made; By introducing a knowledge engineering based input and output processing mechanism, the task of checking the input before interaction and verifying the output before output has been achieved, as shown in Fig. 9.

The establishment of a side panel model involves complexity, as different body engineers have varying design approaches and schemes. Therefore, it is crucial to clarify standardized inputs, outputs, and processes, which together generate ideal model results. The intelligent toolset embedded with knowledge engineering rules

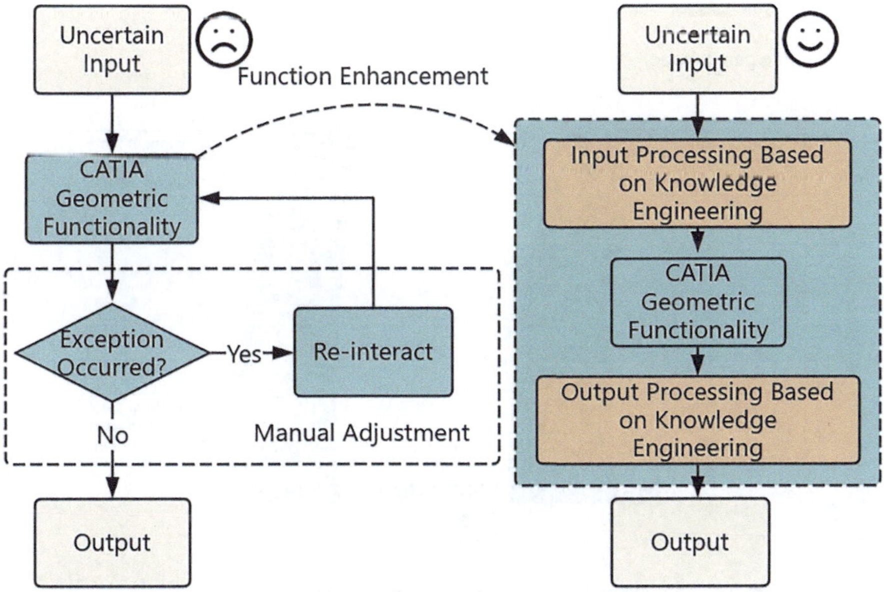

Fig. 9 The reliability of modeling is ensured through an intelligent tool set

achieves standardized processing of inputs and outputs, laying the foundation for the realization of a standardized modeling process and ensuring the reliability of the framework model.

3 Results

Based on the above technical means and template construction methods, this paper establishes a side panel template framework driven by styling and general layout, as shown in Fig. 10. The side panel design model can automatically update according to changes in input parameters, as shown in Fig. 11. For instance, when the height

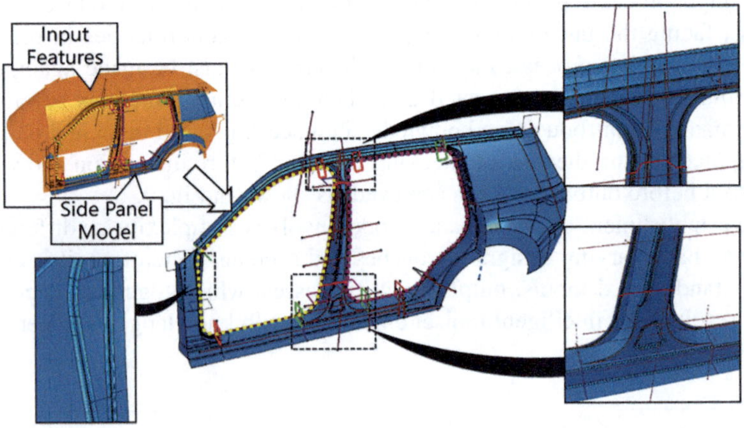

Fig. 10 Side panel model

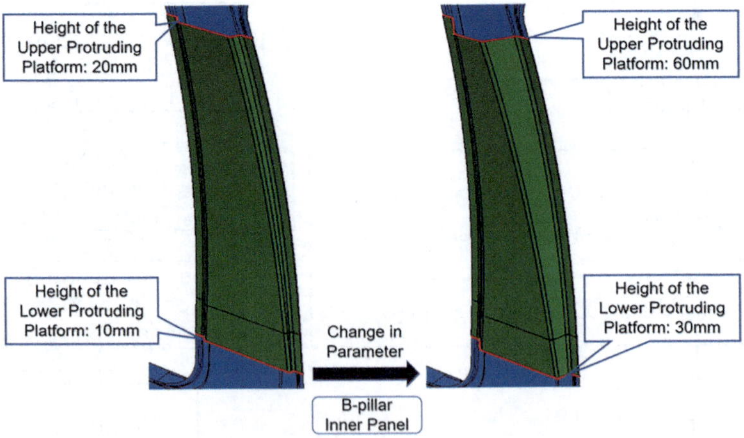

Fig. 11 Parametric implementation of side panel frame model

of the B-pillar inner panel is altered, the corresponding framework components can directly implement the change, meeting the basic requirements of parametrization.

4 Conclusion

This paper establishes, implements, and validates the aforementioned method on the CATIA platform, demonstrating that the strategy proposed in this paper can automatically construct a side panel model with high robustness and maturity in the early design stage. This method not only significantly improves design efficiency but also enhances the reusability of the model, providing important guidance for subsequent detailed design work. This method possesses strong versatility, applicable to the CATIA platform and can also be migrated to other CAD platforms. The method for rapidly establishing the side panel model also provides design ideas for the development of other components of the BIW, which has certain reference value for rapidly generating a highly robust overall BIW frame model in the early design stage.

References

1. Le Z (2021) Research on modeling of automobile body parts based on self-repairing mechanism. Dalian University of Technology. https://doi.org/10.26991/d.cnki.gdllu.2021.001167
2. Chi Ruifeng H, Ping, & Hou Wenbin. (2010) Structural analysis of car body in concept design stage. Automot Eng 32(03):209–212. https://doi.org/10.19562/j.chinasae.qcgc.2010.03.006
3. Hongsen Z, Yibing W, Zhiqian Z (2023) Research based on body in white parameterized modeling and traditional development process. Heavy Truck 6:19–20. https://doi.org/10.3969/j.issn.1007-211X.2023.06.009
4. Kaiyuan M (2024) Development of finite element for multi-point constraints and its application in automobile body structure. Jilin University. https://doi.org/10.27162/d.cnki.gjlin.2024.006478
5. Jingfei H (2021) Research on intelligent modeling method of car body structure oriented to conceptual design. Dalian University of Technology. https://doi.org/10.26991/d.cnki.gdllu.2021.003082
6. Al-Zaher A (2014) Waguih ElMaraghy. Design method of under-body platform automotive framing systems. Procedia CIRP. https://doi.org/10.1016/j.procir.2014.03.116
7. Mingjie M (2021) Research on multi-body cooperative optimization method based on conceptual design. Dalian University of Technology. https://doi.org/10.26991/d.cnki.gdllu.2021.002730
8. Chunfu W (2021) Development of body modularization platform based on parameterization. South China University of Technology. https://doi.org/10.27151/d.cnki.ghnlu.2021.003833
9. Rong C, Junwei H, Yongcheng L et al (2024) Intelligent design and analysis of body structure based on data drive. Automot Eng 46(07):1273-1281+1272. https://doi.org/10.19562/j.chinasae.qcgc.2024.07.014
10. Yoshimura M, Nishiwaki S, Izui K (2005) A multiple cross-sectional shape optimization method for automotive body frames. J Mech Des 127(1):49. https://doi.org/10.1115/1.1814391

11. Baojun LI, Ying DONG, Weixue SUN et al (2018) Data-driven automatic surface modeling for car body structure concept design. Chin J Automot Eng 8(05):324–330. DOI: CNKI: SUN: QCYK.0.2018-05-002
12. Zhong Y (2024) Data-driven automotive styling design evaluation and optimization. Nanchang University. https://doi.org/10.27232/d.cnki.gnchu.2024.001336

Open Access This chapter is licensed under the terms of the Creative Commons Attribution-NonCommercial-NoDerivatives 4.0 International License (http://creativecommons.org/licenses/by-nc-nd/4.0/), which permits any noncommercial use, sharing, distribution and reproduction in any medium or format, as long as you give appropriate credit to the original author(s) and the source, provide a link to the Creative Commons license and indicate if you modified the licensed material. You do not have permission under this license to share adapted material derived from this chapter or parts of it.

The images or other third party material in this chapter are included in the chapter's Creative Commons license, unless indicated otherwise in a credit line to the material. If material is not included in the chapter's Creative Commons license and your intended use is not permitted by statutory regulation or exceeds the permitted use, you will need to obtain permission directly from the copyright holder.

Vibration and Noise Analysis of Surface-Mounted Permanent Magnet Motors under Demagnetization Conditions

Jinli Xu and Fanghui Ding

Abstract This paper focuses on a 48-slot, 8-pole surface-mounted permanent magnet synchronous motor (SPMSM) and investigates the impact of demagnetization conditions on the electromagnetic performance and vibration noise of the motor. Firstly, the motor's air-gap magnetic field and electromagnetic force waves are analytically calculated. Ansys/Maxwell is used for electromagnetic field simulation analysis, and the demagnetization condition is preset to obtain the air-gap magnetic flux density and radial electromagnetic force before and after demagnetization. The corresponding temporal and spatial harmonic orders are obtained through Fourier decomposition. Modal analysis is then conducted to determine the modal shapes and natural frequencies of the motor. Using Workbench, a multi-physics simulation of the motor's magneto-mechanical-acoustic fields is performed, and harmonic response analysis is carried out to generate vibration contour maps and displacement data. The electromagnetic vibration noise under demagnetization conditions is compared with that under normal operating conditions. The results indicate that demagnetization leads to a reduction in radial electromagnetic force and output torque, and a decrease in vibration displacement and noise sound pressure level. This paper provides a theoretical reference for the analysis of motor demagnetization conditions.

Keywords Demagnetization conditions · Electromagnetic vibration noise · Surface-mounted permanent magnet synchronous motor

---

J. Xu · F. Ding (✉)
School of Mechanical and Electronic Engineering, Wuhan University of Technology, Wuhan, China
e-mail: 332626@whut.edu.cn

© The Author(s) 2026

Z. Zhang, P. Mercorelli (eds.), *Mechanical Design and Simulation: Insights and Innovations*, Lecture Notes in Mechanical Engineering,
https://doi.org/10.1007/978-981-95-2405-1_117

1 Introduction

The operating conditions of motors are complex, with high sealing requirements. The heat generated during operation tends to accumulate, making heat dissipation challenging [1]. Additionally, during frequent acceleration and deceleration, variable load operation, and weak magnetic control, the motor experiences significant armature reaction [2]. As a result, the temperature of the permanent magnets increases, causing shifts in the demagnetization curve and enhancing the demagnetization effect, potentially leading to irreversible demagnetization of the magnets. Under demagnetized conditions, the motor's vibration and noise characteristics will differ significantly from those in normal operating conditions. Therefore, it is essential to investigate the vibration and noise performance before and after permanent magnet demagnetization faults.

Current research on motor demagnetization faults primarily focuses on characteristic parameters such as current, torque ripple, and back electromotive force (back EMF) before and after the fault, with relatively few studies addressing electromagnetic vibration and noise. Xu Zheng et al. [3] analyzed the zero-sequence voltage before and after demagnetization by injecting high-frequency signals. Gao Caixia et al. [4] used the residual analysis of half-cycle back EMF to study demagnetization faults in permanent magnet synchronous motors. Ding Shichuan et al. [5] proposed a demagnetization analysis method for permanent magnet motors based on the kurtosis factor of the air-gap magnetic flux density signal.

Both domestic and international researchers have conducted in-depth studies on the electromagnetic vibration and noise of permanent magnet synchronous motors. Huang Chao et al. [6] established a 16-pole, 24-slot radial permanent magnet synchronous motor model to investigate simulation-based prediction methods for electromagnetic vibration and noise. Zhao Shengdong et al. [7] employed harmonic response analysis to determine the vibration frequency of the motor under instantaneous radial electromagnetic force excitation, enabling vibration noise analysis. Li Guohao et al. [8] analyzed the spatial order and frequency characteristics of the motor's electromagnetic force waves for different pole-slot combinations using both analytical and finite element models. Deng Wenzhe et al. [9] developed an electromagnetic vibration and noise analysis model for an outer rotor axial flux hub motor, and used the mode superposition method to obtain the motor's vibration.

2 Motor Model

The prototype of this paper adopts a dual-layer short-pitch winding scheme, with the rotor magnets made of N35SH grade neodymium iron boron (NdFeB) series. Based on the main technical parameters of the motor provided in Table 1, a simulation model of the motor is established. Figure 1 shows a schematic of the motor's 2D structure.

Table 1 The main technical parameters of the prototype

Parameters	Values	Parameters	Values
Rated Power	5 kW	Rated Speed	6650 r/min
Number of Rotor Poles	8	Number of Stator Slots	48
Stator Outer Diameter	160 mm	Rotor Outer Diameter	108.4 mm
Stator Inner Diameter	110 mm	Rotor Inner Diameter	60 mm
Core Length	96 mm	Magnet Thickness	5 mm

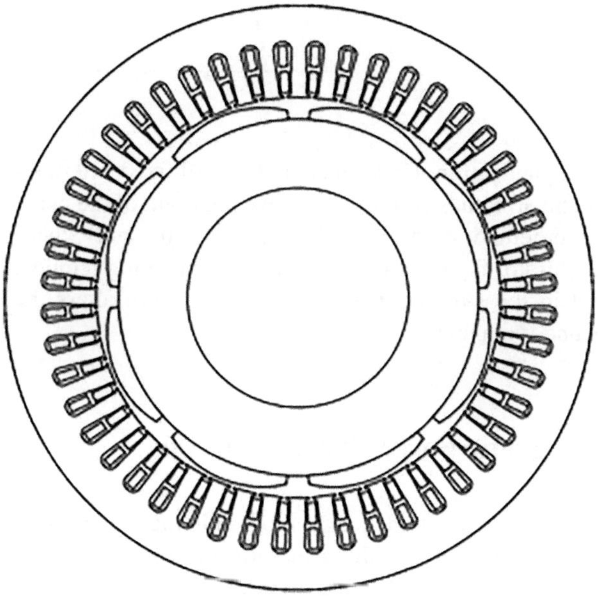

Fig. 1 The motor structure

The magnetic induction intensity B of the permanent magnetic material is:

$$B = -\mu_0 H_c + B_r \tag{1}$$

H_c is the coercive force of the permanent magnet, B_r is the remanent magnetic induction of the permanent magnet, and μ_0 is the magnetic permeability in vacuum.

The demagnetization curve of the permanent magnet can be approximated as a straight line at room temperature. As shown in Fig. 2, point K is called the inflection point. Under normal temperature, the permanent magnet usually operates above point K. When the permanent magnet demagnetizes due to external factors, the operating point will drop along the curve. If the drop does not exceed the inflection point, when the influence of the external demagnetizing factors weakens, the magnet can still recover to B_r along the original curve, which is called reversible demagnetization. If the operating point drops below the inflection point, when the magnet recovers, it will rise to B_{rr} along another curve, and B_{rr} is less than B_r, which is called

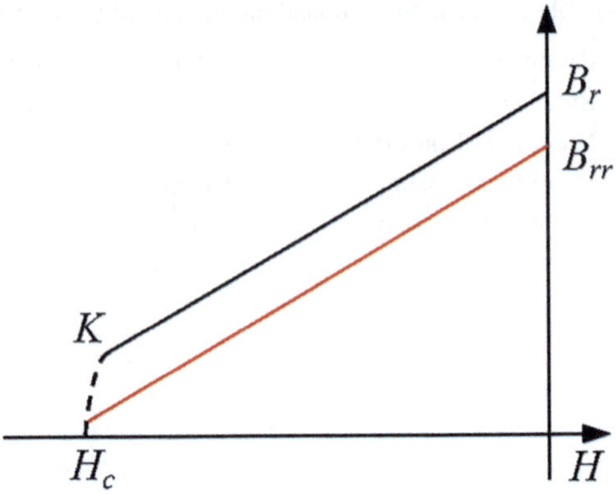

Fig. 2 The demagnetization curve of the permanent magnet

irreversible demagnetization. Under the condition of fixed external magnetic circuit characteristics and armature current, the operating point of the permanent magnet is determined by its temperature [10]. Therefore, in this study, the demagnetization condition of the motor is preset by changing the operating temperature of the permanent magnet in the simulation model, which in turn alters the corresponding coercive force and remanence.

3 Electromagnetic Force Theory Analysis

When the motor is under load, the magnetic field in the air gap is composed of both the permanent magnet field and the armature reaction field. The magnetomotive forces generated by both are $f_1(\theta, t)$ and $f_2(\theta, t)$, respectively, as follows:

$$f_1(\theta,t) = \sum_{\mu=1,3,5,\cdots} F_\mu \cos(\mu\omega_1 t - \mu p\theta) \tag{2}$$

$$f_2(\theta,t) = \sum_{v} F_v \cos(\omega_1 t - vp\theta + \phi) \tag{3}$$

In the equations, F_μ is the amplitude of the μ harmonic of the rotor magnetomotive force. p is the number of pole pairs. F_v is the amplitude of the v harmonic of the armature current magnetomotive force. ϕ is the initial phase of the armature current magnetomotive force. ω_1 is the electrical angular velocity.

The equivalent air-gap magnetic permeability considering the stator slot effect is:

$$\lambda(\theta,t) = \Lambda_0 + \sum_{i=1}^{\infty} \Lambda_i \cos(iZ\theta) \tag{4}$$

In the equations, Z is the number of stator slots. Λ_0 and Λ_i are the average air-gap magnetic permeability and the amplitude of the i-th harmonic magnetic permeability.

The radial air-gap magnetic flux densities under load conditions are given by:

$$B_{\mathrm{L}}(\theta,t) = \left(f_1(\theta,t) + f_2(\theta,t)\right)\lambda(\theta,t) \tag{5}$$

the radial electromagnetic force of the motor under load is given by:

$$F_{\mathrm{r\_L}} = \frac{B_{\mathrm{L}}^2(\theta,t)}{2\mu_0} \tag{6}$$

4 Electromagnetic Performance Analysis

4.1 Finite Element Analysis of Air-Gap Magnetic Field and Electromagnetic Force Waves

In the calculation, a three-phase symmetrical voltage source excitation is applied to the armature winding, considering the effects of end leakage inductance and winding resistance on the motor's operation. This is done to solve and calculate the motor's operating process under rated load conditions. When the motor's rotor permanent magnets experience demagnetization due to temperature effects, it affects the motor's overall magnetic field, leading to changes in its magnitude. This section primarily analyzes the magnetic field variations in three states: normal operation, 15% permanent magnet demagnetization, and 30% permanent magnet demagnetization. The vibration and noise analysis process is shown in Fig. 3.

Based on the established prototype finite element model, a circular trajectory near the inner surface of the stator in the air-gap is selected as the observation path for the spatial magnetic field. The spatial radial air-gap magnetic flux density at a certain moment under load conditions, along with its FFT decomposition results, and the spatial radial electromagnetic force under load conditions with its FFT decomposition results, are shown in Figs. 4 and 5.

When the motor experiences different degrees of high-temperature demagnetization, there is a noticeable change in the amplitude of the peaks and valleys of the radial air-gap magnetic flux density. From the Fourier analysis results, it is observed that under various demagnetization conditions, the changes in the air-gap magnetic field cause corresponding variations in the harmonic amplitudes. Among these, the fourth harmonic has the highest amplitude across all harmonics. as the degree of demagnetization increases, only the 4th and 12th harmonics gradually decrease, while the other harmonic orders increase as the demagnetization degree rises. The distribution characteristics of the radial air-gap magnetic flux density remain unchanged. From Fig. 5, it can be seen that when the motor experiences demagnetization, there is some distortion in the position of the minima of the spatial radial electromagnetic force, and the peak value of the electromagnetic force decreases with increasing demagnetization. The rate of decrease in the magnitudes of the 0th,

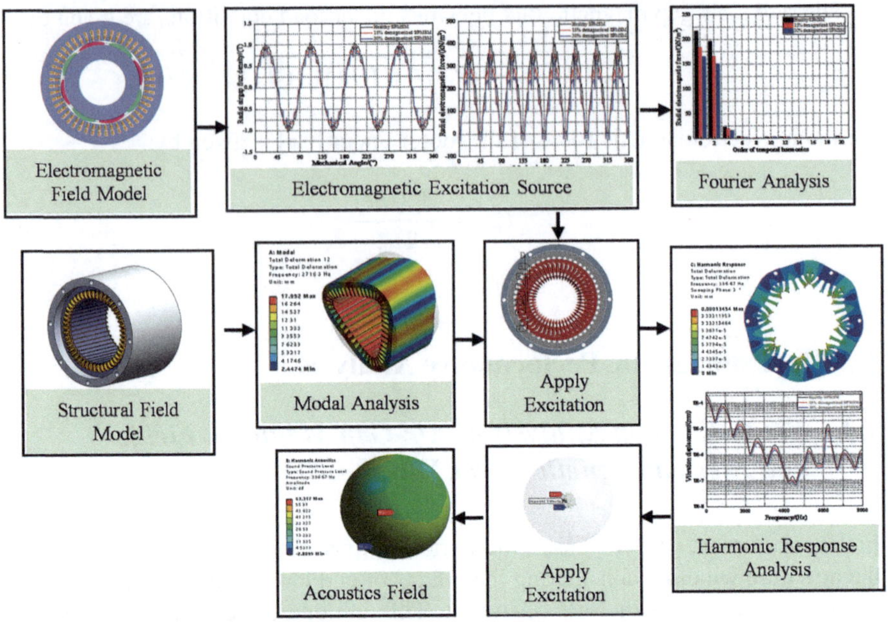

Fig. 3 Motor vibration and noise analysis flowchart

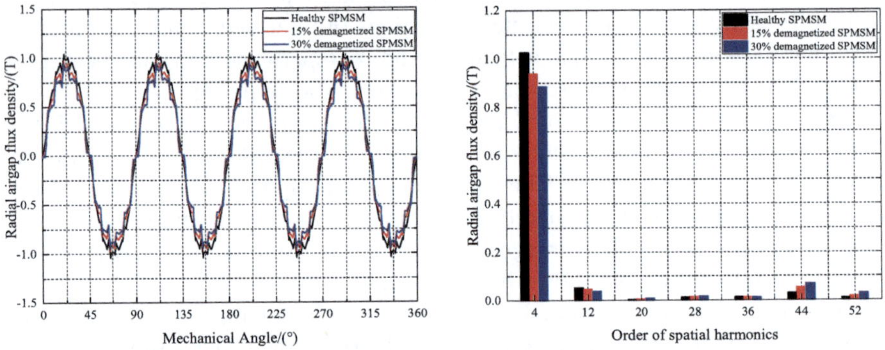

(a) Radial air-gap magnetic flux density (b) FFT decomposition

Fig. 4 Radial air-Gap magnetic flux density at a certain moment and its FFT decomposition. (**a**) Radial air-gap magnetic flux density. (**b**) FFT decomposition

8th, and 16th harmonic force components is relatively fast. For the 24th and 40th harmonic components, the magnitudes tend to increase as the degree of demagnetization rises. It is also evident that, under both normal and demagnetized conditions, the 0th and 8th harmonic components remain the dominant ones.

Figure 6 is the modal shapes and natural frequencies of the stator core at various orders. Based on the analysis of radial electromagnetic force, the most significant

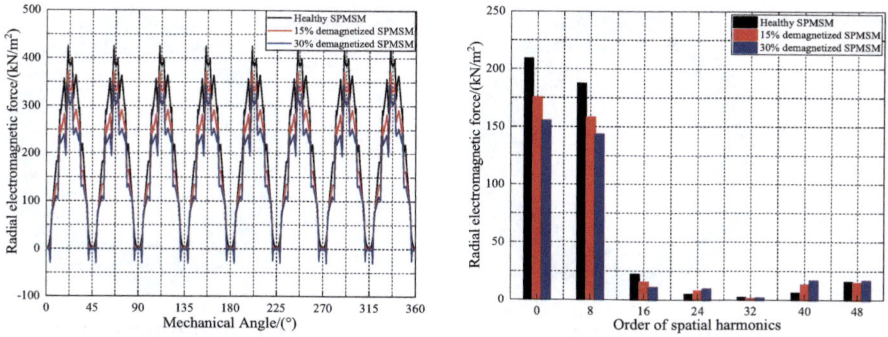

(a) Radial electromagnetic force (b) FFT decomposition

Fig. 5 Radial electromagnetic force at a certain moment and its FFT decomposition. (**a**) Radial electromagnetic force. (**b**) FFT decomposition

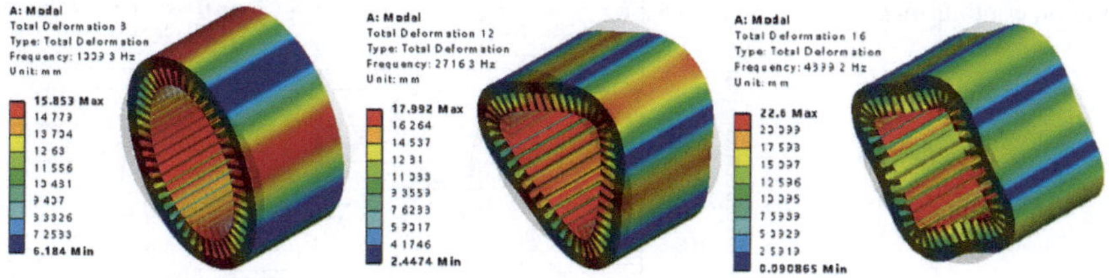

(a) Stator 2nd Mode,1009.3Hz (b) Stator 3nd Mode, 2716.3Hz (c) Stator 4nd Mode, 4899.2Hz

Fig. 6 Stator modal shapes at various orders and their natural frequencies. (**a**) Stator second Mode,1009.3 Hz (**b**) Stator 3nd Mode, 2716.3 Hz (**c**) Stator 4nd Mode, 4899.2 Hz

characteristic components affecting vibration are the second harmonic ($2f_1$ = 886.6 Hz), followed by the fourth harmonic ($4f_1$ = 1773.3 Hz) and the sixth harmonic ($6f_1$ = 2660 Hz). The modal simulation results show that the main frequency points of the radial electromagnetic force of the prototype do not coincide with its natural frequencies, so strong resonance will not occur.

Based on harmonic response analysis, the electromagnetic results generated by the stator tooth tips are used as excitation sources transmitted to the stator structure at the tooth surfaces. Figure 7 shows the deformation and vibration contour results of the stator in both normal and demagnetized states at twice the frequency, it can be observed that the deformation of the stator tooth is relatively large, with a decreasing trend from the tooth region to the yoke. Under the demagnetization condition, the radial electromagnetic force in the air-gap further decreases, resulting in reduced vibration deformation. It can also be observed that the positions of the maximum and minimum deformations before and after demagnetization remain consistent.

Figure 8 presents the vibration displacement frequency response results of the motor stator under different states. It can be seen that when the motor operates at the

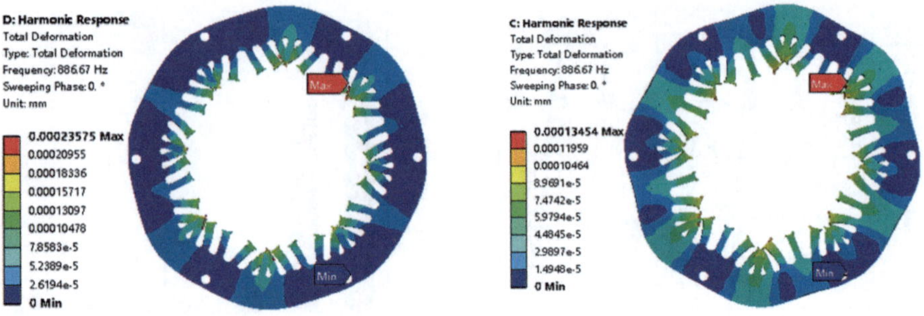

(a) Healthy SPMSM (b) 30% demagnetization

Fig. 7 Stator deformation vibration contour map. (**a**) Healthy SPMSM. (**b**) 30% demagnetization

Fig. 8 Stator deformation vibration contour map

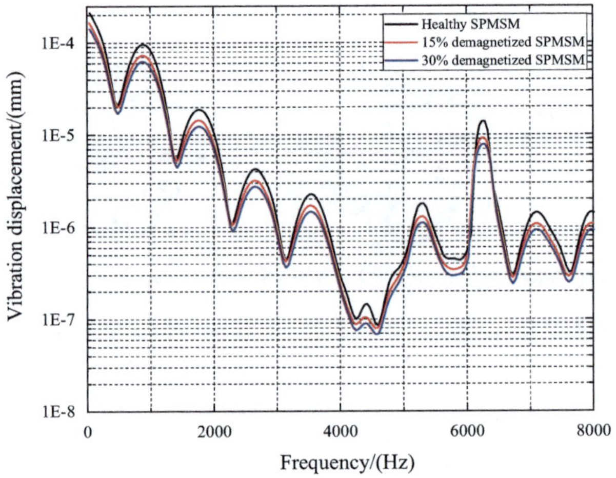

rated speed, the demagnetization fault affects the motor's vibration displacement. The vibration acceleration is notably high at frequency points such as 886.67, 1773, 2600, and 3546.66 Hz, which correspond to $2f_1$, $4f_1$, $6f_1$, and $8f_1$. As discussed earlier, at these frequency points, the motor's electromagnetic force has higher amplitudes, leading to more significant vibrations. In the demagnetized state, due to the reduction in radial electromagnetic force, the peak at the second harmonic frequency decreases by 24.14% and 34.91% at 15% and 30% demagnetization, respectively, compared to the normal state.

The output torque of the motor under different demagnetization states was solved for two operating cycles, Fig. 9 present the motor output torque calculation results under different demagnetization conditions. From the results, it is clear that when the motor's permanent magnets experience high-temperature demagnetization, the magnetic energy storage of the permanent magnet material itself decreases, which further weakens the composite magnetic field, leading to a reduction in the output

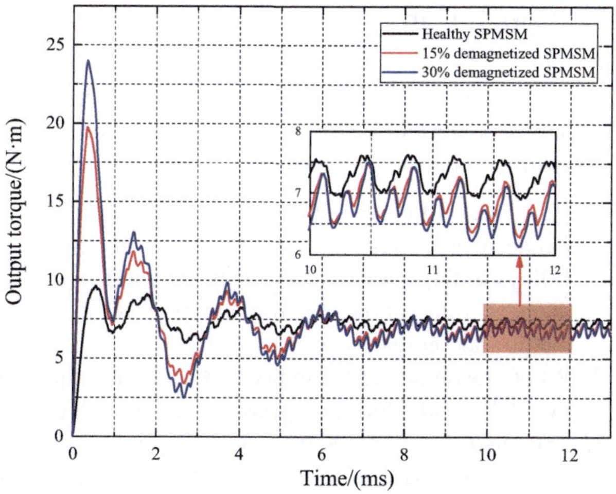

Fig. 9 Output torque under different states

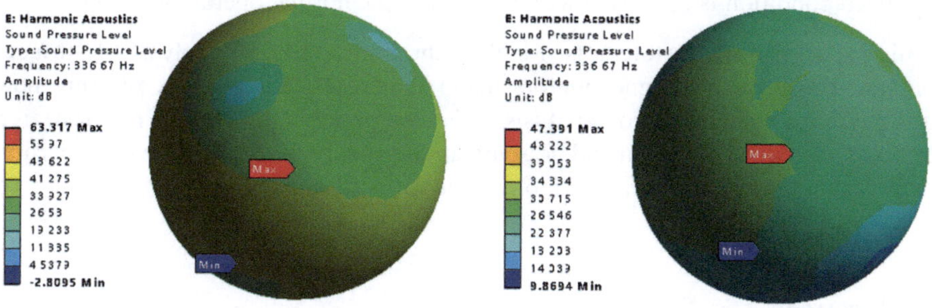

(a) Healthy SPMSM (b) 30% demagnetization

Fig. 10 Noise sound pressure level contour map. (**a**) Healthy SPMSM. (**b**) 30% demagnetization

torque. When demagnetization reaches 15% and 30%, the average output torque decreases by 5.64% and 7.29%, respectively, compared to the normal operating condition.

The vibration displacement obtained from the harmonic response calculation is converted into vibration velocity excitation and added to the noise domain surface. This allows for the determination of the noise response at different frequencies. Figure 10 shows the noise peak values at the second harmonic frequency under normal motor operation and under the demagnetized condition. Due to differences in the structural field's geometric distribution, the noise distribution across various locations of the motor during operation exhibits deviations. Under the demagnetized condition, the noise peak reaches a maximum of 47.39 dB and a minimum of 9.87 dB, representing a 25.16% reduction in noise peak compared to the normal motor operation state.

5 Conclusion

This study focuses on a 48-slot, 8-pole surface-mounted permanent magnet synchronous motor (SPMSM), conducting simulation analysis on the motor's demagnetization and normal operating conditions, leading to the following conclusions:

1. there is a noticeable change in the amplitude of the peaks and valleys of the radial air-gap magnetic flux density with demagnetization. The position of the valleys in the spatial radial electromagnetic force exhibits partial distortion, and the peak value of the electromagnetic force decreases as the degree of demagnetization increases.
2. When the demagnetization is 15% and 30%, the average output torque decreases by 5.64% and 7.29%, respectively, compared to normal operation.
3. The most significant vibration displacements occur at the frequencies of $2f_1$, $4f_1$, $6f_1$, and $8f_1$. As the degree of demagnetization increases, the reduction rate of vibration displacement decreases sequentially. The noise peak in the demagnetized condition is 25.16% lower than that in the normal operating condition.

This paper provides a reference for the vibration and noise analysis of surface-mounted permanent magnet motors under demagnetization faults, offering new insights for demagnetization analysis. It is also possible, during the motor design process, to reduce vibration and noise by appropriately weakening the rotor permanent magnet, while ensuring the motor's output performance.

References

1. Guan T, Liu DM, He YY (2024) Review on development of permanent magnet in-wheel motors. Trans China Electr Soc 39(02):378–396
2. Mazaheri-Tehrani E, Faiz J (2022) Airgap and stray magnetic flux monitoring techniques for fault diagnosis of electrical machines: an overview. IET Electr Power Appl 16(3):277–299
3. Xu Z, Zhang JZ, Jiang YJ, Deng FJ (2024) Research on local demagnetization fault diagnosis of permanent magnet synchronous motors based on high-frequency signal injection. Proc CSEE. 1-10[2024-12-03]
4. Gao CX, Li BK, Chen H, Si JK, Nie YJ (2023) Local demagnetization fault diagnosis of permanent magnet synchronous motor based on half-period back EMF residual. Electr Mach Control 27(07):183–194
5. Ding SC, Wu Z, Li Y, Hang J, He W (2024) Partial demagnetization fault diagnosis of PMSM based on the kurtosis of air-gap flux density difference signals. Proc CSEE 44(14):5747–5756
6. Huang C, Xiong L, Wu HW, Guo H, Meng DJ (2024) Simulation and experimental analysis of electromagnetic vibration and noise for in-wheel-drive permanent magnetic synchronous electrical motor. J Tongji Univ (Nat Sci) 52(S1):302–308
7. Zhao SD, Ma KW, Wang KY (2022) Simulation analysis of electromagnetic noise of permanent magnet synchronous motors under consideration of radial electromagnetic force excitation. Electr Mach Technol 05:15–19
8. Li GH, Yin HB, Sun YG, Zhang Z, Bi XJ, Hu WJ (2024) Multi-physical field simulation analysis of vehicle permanent magnet motors with different pole-slot combinations. J Hebei Univ Sci Technol. 1-13[2024-12-03]

9. Weizhe D, Shuguang Z (2017) Analytical modeling of the electromagnetic vibration and noise for an external-rotor axial-flux in-wheel motor. IEEE Trans Ind Electron 65(3):1991–2000
10. Zhao GD (2023) Demagnetization fault diagnosis of permanent magnet synchronous motors based on multi-layer perceptron. Harbin Institute of Technology

Open Access This chapter is licensed under the terms of the Creative Commons Attribution-NonCommercial-NoDerivatives 4.0 International License (http://creativecommons.org/licenses/by-nc-nd/4.0/), which permits any noncommercial use, sharing, distribution and reproduction in any medium or format, as long as you give appropriate credit to the original author(s) and the source, provide a link to the Creative Commons license and indicate if you modified the licensed material. You do not have permission under this license to share adapted material derived from this chapter or parts of it.

The images or other third party material in this chapter are included in the chapter's Creative Commons license, unless indicated otherwise in a credit line to the material. If material is not included in the chapter's Creative Commons license and your intended use is not permitted by statutory regulation or exceeds the permitted use, you will need to obtain permission directly from the copyright holder.

Analysis of Landing Buffer Characteristics of Unmanned Airdrop Rescue Vehicle

Baobao Wang, Xin Zhang, and Xiaoming Hou

Abstract Taking an unmanned airdrop rescue vehicle as the research object, in order to reduce the impact overload of the airdrop landing, the exhaust type airbag buffer mode suitable for the airdrop vehicle is determined. Firstly, the scheme of the airdrop system is designed. According to the basic assumption of the airdrop process, the thermodynamic analysis model of the buffer bag is derived and the parameters are designed. Secondly, the finite element model of airdrop system with airbag buffer is established based on the airbag finite element algorithm (CV method). Finally, the buffering effect is evaluated by analyzing the change of tire force. The calculation results show that the system overload is reduced from 35 to 22 g by using airbag buffer, and the buffer effect is obvious.

Keywords Airdrop rescue vehicle · Airbag buffer · The finite element method · Overload

1 Introduction

Unmanned airdrop rescue vehicle has the characteristics of high tactical mobility. With the advent of various large and high-performance transport aircraft, the development process of airdrop rescue vehicle has been accelerated, making it play an increasingly important role in rescue and disaster relief [1].

In terms of the overall scheme design of the airdrop system, He Qing [2] carried out the overall scheme design of the whole gun for the airborne rocket launcher. In the study of the working process of the parachute. Shang Jianhua [3] drop the

B. Wang (✉) · X. Zhang
Jiangsu Institute of Automation, Lianyungang, Jiangsu, China

X. Hou
College of Mechanical Engineering, Nanjing University of Science and Technology, Nanjing, Jiangsu, China

© The Author(s) 2026

Z. Zhang, P. Mercorelli (eds.), *Mechanical Design and Simulation: Insights and Innovations*, Lecture Notes in Mechanical Engineering,
https://doi.org/10.1007/978-981-95-2405-1_118

truck-mounted gun system dynamics are studied, mainly through the method of finite element of fluid-solid coupling analyses the working process of the parachute, in view of the main umbrella straight, the main umbrella aeration, the main umbrella does systemic research on the stability and the buffering phase, eventually by air-drops example for the whole process simulation in the system, The corresponding analysis results of each stage are given. Wu lei [4] on process of the parachute drop, inflatable, stable down three stages, using the method of combining the theory and experience, the parachute at all stages of loading, the trajectory and the overload situation is analyzed, the method based on fem ALE fluid-structure interaction for parachute aeration process simulation, get umbrella landing process change and overload conditions.

2 Design of Airbag Cushion Scheme for Airbag Impact Meter

2.1 Statistical Airdrop System Design

An airdrop rescue vehicle with a chassis mass of about 7 t [5]. The 3D model is shown in Fig. 1. The rescue vehicle meets the load and size requirements of the delivery platform. The airbag buffer device has the characteristics of simple structure, good buffering effect, folding installation, less space occupation and reusable, and has been widely used in the fields of soft landing of spacecraft and precise air-drop protection of materials. In this paper, the buffer air bag is used as the buffer mode of airdropped rescue vehicle.

The shape of a cargo platform is shown in Fig. 2. There are several reserved holes on the cargo platform to install fixed rings, and the equipment is bound and fixed with the cargo platform through the binding rope and fixed rings; The left and right sides have one arm to prevent the equipment from disconnecting during landing,

Fig. 1 Airdrop rescue vehicle model

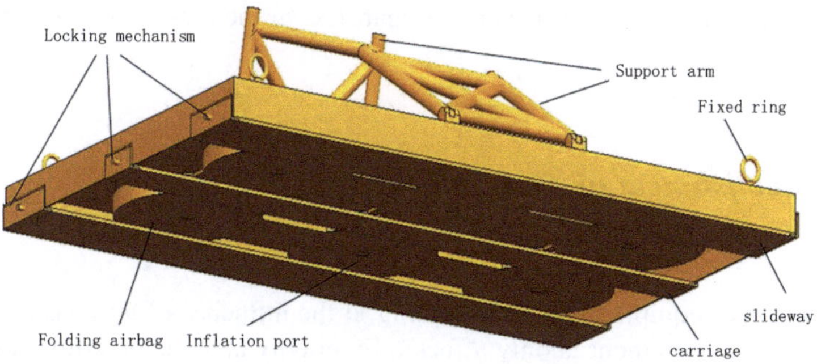

Fig. 2 Cargo platform model

and the arm structure is not considered during modeling; There are slipways on both sides of the bottom of the cargo platform to match the raceway on both sides of the engine room to reduce the friction when the cargo platform slides out of the engine room; Buffer air bags can be installed inside the cargo platform, and folding air bags with about diameter are arranged and fixed inside the cargo platform through the bottom bracket and slide; The slipway and pallet are fixed on the main body of the cargo platform through the locking mechanism, which is connected with the traction rope, and the other end of the traction rope is connected with the parachute. During the process of falling under the cargo platform, when the parachute is fully opened and the parachute rope is pulled tight, the locking mechanism will be released under the action of tensile force, and the bottom pallet and pallet will be separated from the main body of the cargo platform and accelerated to fall under the action of gravity. Thus, the folded airbag is pulled up at the same time. At this time, negative pressure is formed in the airbag, and the air flow is charged into the airbag through the air inlet at the bottom of the airbag. The use of the support plate can not only ensure that each air bag is inflated at the same time, but also avoid direct contact between the air bag and the ground, and prevent the air bag from being punctured or worn during landing.

2.2 *Basic Assumptions*

In order to establish the analytical model of airbag cushioning process, the following assumptions are introduced: airbag fabric is not stretchable, ignoring the energy absorption and cushioning effect of the material itself. In the process of buffer compression, the changes of airbag section shape and touchdown area are ignored. The influence of air resistance is ignored in the buffer system. It is assumed that there is no temperature change in the buffer process, that is, there is no heat exchange. The inflation gas is an ideal gas, and the exhaust process only flows out of the exhaust hole, and the fabric has no air leakage. Regardless of the mass of the gas in the airbag and the airbag itself.

2.3 Establishment of Analytical Model of Cystic Vertical Airbag

In the process of equipment airdrop landing, if the influence of air resistance is not considered, the equipment is only affected by gravity and gas pressure in the capsule in the direction perpendicular to the ground, and the force equation is [6],

$$Ma = (p - p_a)A - Mg \tag{1}$$

Where, M is the equipment quality. a is the falling acceleration of equipment. P is the gas pressure in the capsule. P_a is atmospheric pressure. A is the contact area between the airbag and the equipment ground, which is also equal to the contact area of the airbag. g is the acceleration of gravity 9.8 m/s^2.

Heat exchange is ignored in the exhaust process, and the flow rate of gas in the exhaust hole is

$$v_a = \sqrt{\gamma R T_0 \left(\frac{p_a}{p_0}\right)^{\frac{\gamma-1}{\gamma}}} \tag{2}$$

Where, P is the gas pressure in the capsule. P_a is the pressure of extracapsular gas. r is the air insulation index. R is the thermodynamic gas constant. T_0 is the initial thermodynamic temperature of the gas in the capsule. P_0 is the initial pressure of gas in the capsule.

In the dynamic equation and gas flow equation of airdrop equipment, the parameters are transformed dimensionless, where, $y = x/l$ represents the ratio of the height of the airbag at any time to the initial height during the falling process. $T = v_0t/l$ represents the time parameter.$P_b = P/P_a$ represents the ratio of gas pressure inside the capsule to gas pressure outside the capsule. $H = \lg/v_0^2$ represents the airbag height parameter. $Z = AP_d/Mg$ Represents the load parameter. $B = A_n v_a / Av_0\sqrt{\gamma}$ represents the vent parameter.

The first mock exam is to establish a dimensionless unified model of the airbag buffer system,

$$\frac{d^2y}{dT^2} = H\left(Z\left(p_b - 1\right) - 1\right) \tag{3}$$

$$\frac{dp_b}{dt} = \frac{B\delta + \varepsilon\frac{dx}{dt}}{l\eta} \tag{4}$$

2.4 Data Meter Parameter Design

It is assumed that the total weight of the airdrop system is 9 t and the floor area of the cargo platform is 10m^2. Since the top of the airbag is fixed together with the bottom of the cargo platform, in order to ensure reliability, the maximum area of the airbag design should be the same as the floor area of the cargo platform as much as possible to ensure the maximum coverage size of the bottom of the cargo platform. Through literature review [7] and investigation, it can be seen that under normal circumstances, the initial velocity of airdrop landing should range from 6 to 7 m/s, and the final velocity after buffering is 3–4 m/s. For safety considerations, the maximum initial velocity is 7 m/s and the minimum final velocity is 3 m/s are taken as the design conditions. Generally, the landing impact overload that the airdrop equipment should be able to withstand is about 30, and 30 g is used as the overload design requirement.

According to the analysis of the design requirements, the energy efficiency calculated according to the initial and final landing velocity is 81.6%, which meets the requirements. For the airbag combination composed of 6 airbags, the dimensionless mass parameter Z is calculated as 17.4, and the corresponding relationship between Z, B and H in the airbag parameter matching diagram [8] selected according to the design requirements is interpolated to obtain the values of B and H when Z is 17.4. The height is calculated as 1.2 m by the calculation formula of dimensionless height parameter H and dimensionless exhaust hole parameter B. The vent area is 0.1m^2, and the ratio of height to diameter is 1.2, which meets the requirements.

3 Simulation Calculation

3.1 Finite Element Modeling

Model constraints and boundary conditions were introduced in bottom-up order: 7 m/s initial velocity was applied to the airdrop equipment (including air bags), the ground translational rotation freedom was constrained, and a simplified model of the cargo platform was established according to the cargo platform size. The finite element model of the airdrop system was shown in Fig. 3.

3.2 Constraints and Boundaries

The finite element algorithm of airbag is defined by "AIRBAG WANG NEFSKE". The atmospheric pressure is 100 KPa and the air density is 1.169 kg/m^3, the gas temperature is 293 K, the gravity conversion constant is 1, and the exhaust hole area is 0.1m^2, since it is a self inflating air bag, it has been inflated and opened during

Fig. 3 Finite element model of a rescue vehicle airdrop

landing. It is defined that the filling gas flow rate curve is 0, and the initial pressure in the bag is consistent with the atmospheric pressure. In order to prevent penetration of the ground, air bag and cargo platform, contact is set between the two. During the actual airdrop, the combat vehicle is fixed on the cargo platform through the binding rope. During the modeling, the contact is set between the combat vehicle and the cargo platform, and the cable unit is used to simulate the binding rope effect to prevent the launch box from separating from the cargo platform during the impact process. The body is modeled by rigid body, and the body mass is coupled at the tire axis through mass points. The tire is made of elastic rubber.

3.3 Calculation Results

The stress curves of the four tires for airbag free buffer landing are shown in Fig. 4. ABCD is the stress curves of the left front, right front, left rear and right rear wheels respectively. The maximum stress of each wheel is about 60 t, and the position of the center of mass is backward to the right, so the maximum stress of the right rear wheel is 68 t, and the total stress of the tire is 248.5 t, the overload of airdrop equipment is about 35 g. Maximum stress of pallet 420 Mpa, The stress cloud diagram is shown in Fig. 5.

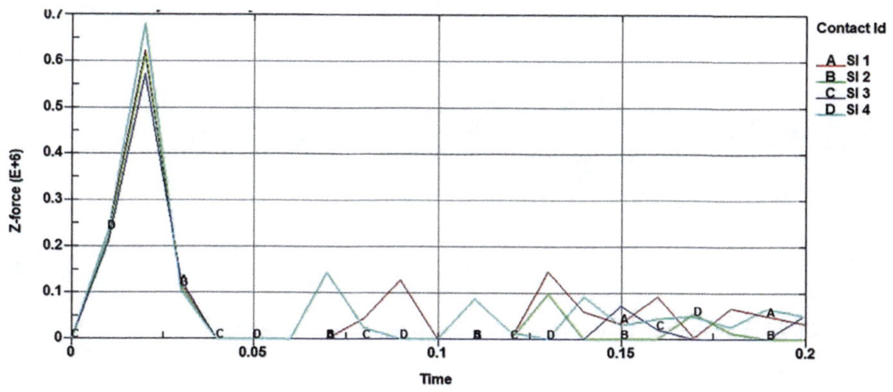

Fig. 4 Uncushioned tire forces

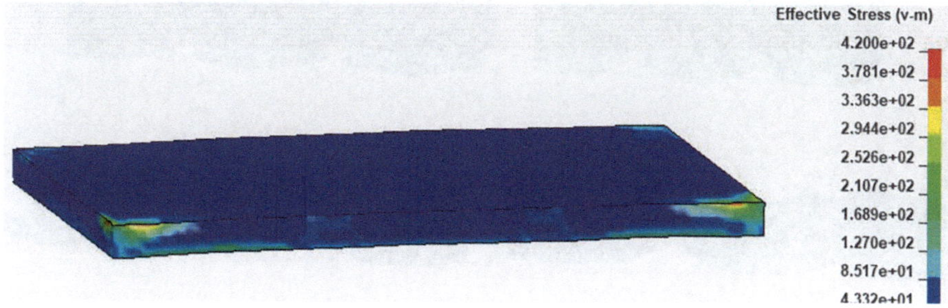

Fig. 5 Maximum stress of unbuffered platform

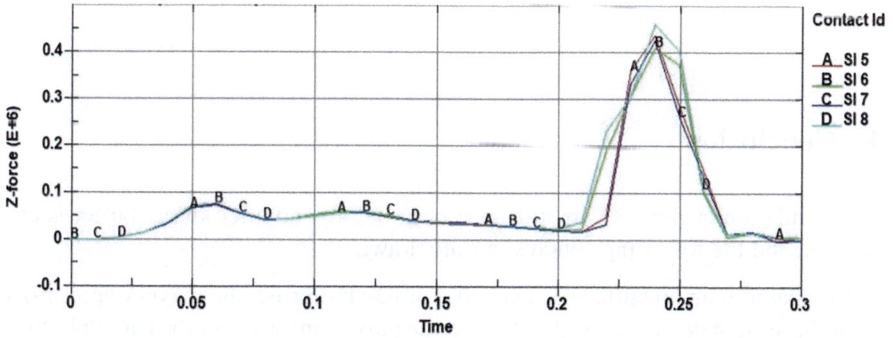

Fig. 6 The air bag cushioned the tire

The stress curves of the four tires with airbag buffer landing are shown in Fig. 6. ABCD is the stress curves of the left front wheel, right front wheel, left rear wheel and right rear wheel respectively. The maximum stress of each wheel is about 40 t, and the position of the center of mass is behind the right. The maximum stress of the cargo platform is 230.1 Mpa, and the stress decreases by 45.2%. The stress cloud

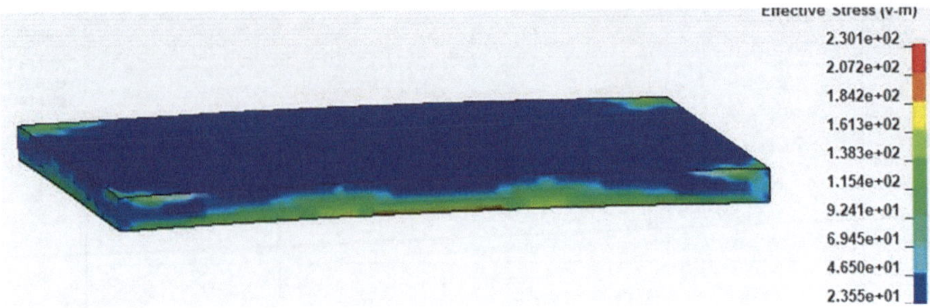

Fig. 7 Cushioning the maximum stress of the cargo platform with air bag

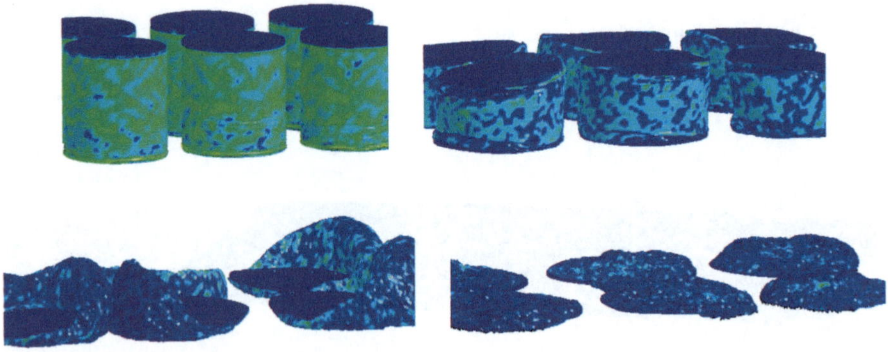

Fig. 8 Balloon compression process

diagram of the cargo platform is shown in Fig. 7, and the process of airbag buffering is shown in Fig. 8.

4 Conclusion

The simulation analysis of an airdrop system before and after airbag buffer is carried out, and the following conclusions are drawn:

1. Compared with no buffering state, after airbag buffering, the maximum overload of the system decreases by 37.1%, and the maximum stress of the cargo platform decreases by 45.2%.
2. The parameters of the airbag can be further optimized to ensure that the oblique landing can also have a good cushioning effect, maximize the cushioning effect of the airbag and reduce the impact of landing.

References

1. Cungui Y, Huifang W, Jie R (2015) Progress of rocket and missile launch technology. Beijing University of Aeronautics and Astronautics Press, Beijing
2. Qing H (2010) Overall scheme design and analysis of airborne rocket. Nanjing University of technology, Nanjing
3. Jianhua T (2016) Dynamic simulation calculation of vehicle mounted gun airdrop system. Nanjing University of technology
4. Lei W (2013) Study on dynamics of heavy equipment airdrop. Nanjing University of technology
5. Qiang L (2019) Research on landing buffer of an air drop rocket. Nanjing University of technology
6. Hongyan W (2014) Rui Qiang modeling and analysis of air drop equipment recovery system. National Defense Industry Press, Beijing
7. Ningguo S (2017) Li Liangchun research status of air drop landing cushion airbag. Packaging engineering 38(11):97–101
8. Huangjie H, Hongyan W, Jianyang L, Zheyuan LV (2017) Ruiqiang research on matching design and simulation calculation of cushion airbag of airdrop equipment. Equipt Environ Eng 14(05):1–7

Open Access This chapter is licensed under the terms of the Creative Commons Attribution-NonCommercial-NoDerivatives 4.0 International License (http://creativecommons.org/licenses/by-nc-nd/4.0/), which permits any noncommercial use, sharing, distribution and reproduction in any medium or format, as long as you give appropriate credit to the original author(s) and the source, provide a link to the Creative Commons license and indicate if you modified the licensed material. You do not have permission under this license to share adapted material derived from this chapter or parts of it.

The images or other third party material in this chapter are included in the chapter's Creative Commons license, unless indicated otherwise in a credit line to the material. If material is not included in the chapter's Creative Commons license and your intended use is not permitted by statutory regulation or exceeds the permitted use, you will need to obtain permission directly from the copyright holder.

Design of Cam Jacking Transfer Machine Based on ADAMS

Xu Zhang, Fei Li, Zhenhao Jiang, Xiaoyu Chen, Yulan Wei, and Qingzhu Zhang

Abstract In response to the prevalent issues in the market, such as slow jacking speed, poor stability, large footprint, high noise levels, and low automation, this paper designs a cam jacking transfer machine. The machine uses the motor to drive the rotation of the cam, thereby achieving both jacking and transferring functions. This design significantly simplifies the mechanical structure of the shifting machine and the programming of the drive control card, and effectively reduces the maintenance and maintenance costs. To verify the rationality and feasibility of the cam design, ADAMS was employed for modeling and simulation. The results prove that the motion law and driving ability of the cam meet the design requirements. The proposed machine exhibits advantages such as a stable structure, small footprint, low noise levels and cost-effectiveness, contributing positively to the advancement of the logistics equipment industry.

Keywords Transfer machine · Cam · ADAMS · Jacking

1 Introduction

With the rise of the e-commerce industry, the logistics industry is facing multiple challenges such as personnel shortages, express package breakages, and delivery pressures [1]. In the field of logistics equipment, improving production efficiency and reliability has become an urgent demand of the industry development [2].

The transfer machine is the key equipment for logistics transportation [3], positioned beneath the main conveyor line. When items need to be sorted or transferred, the jacking mechanism within the transfer machine elevates the transfer unit, rising it above the working surface of the main conveyor line. Then the transfer unit begins to work to complete the task of distinguishing or transferring items.

X. Zhang · F. Li · Z. Jiang · X. Chen · Y. Wei · Q. Zhang (✉)
School of Engineering, Huzhou University, Huzhou, China
e-mail: zhangqz@zjhu.edu.cn

© The Author(s) 2026

Z. Zhang, P. Mercorelli (eds.), *Mechanical Design and Simulation: Insights and Innovations*, Lecture Notes in Mechanical Engineering,
https://doi.org/10.1007/978-981-95-2405-1_119

Currently, most jacking transfer machines at home and abroad realize the jacking by motor driving crank-link mechanisms, pneumatic actuation or hydraulic drive [4, 5]. However, the crank-link mechanisms generally has complex structure, difficult maintenance, high energy consumption in the jacking process, and prone to horizontal direction. Pneumatic actuation methods has problems such as large area, prone to noise generation, and low conveying efficiency. Hydraulic drive mode face issues including potential environmental pollution from oil leaks and significant temperature sensitivity [6].

In light of the above problems, this paper introduces the design of a cam jacking transfer machine, wherein both the jacking and transferring operations are achieved through motor-driven cam rotation. This design not only simplifies the structure but also exhibits advantages such as high stability, substantial carrying capacity, low noise levels, and zero pollution [7].

2 Overall Plan

The cam jacking transfer machine, shown in Fig. 1, primarily comprises a jacking device (1), a conveying device (2), and a frame (3). Conveying device is responsible for the smooth and accurate delivery of the goods to the designated location. The jacking device is mounted beneath the rollers of the conveying device. When the system detects that the conveying device has positioned the goods over the jacking device, the PLC controller instructs the cam to rotate. The cam then extends intermittently from the gap between the rollers, altering the radius of curvature to realize the smooth jacking of the cargo and changes the original conveying direction to vertically divert it to the next conveying device.

The maximum load of the cam jacking transfer machine is 50 kg per meter of conveyor length, with a maximum size of the lifting cargo 600 mm×400 mm×200 mm

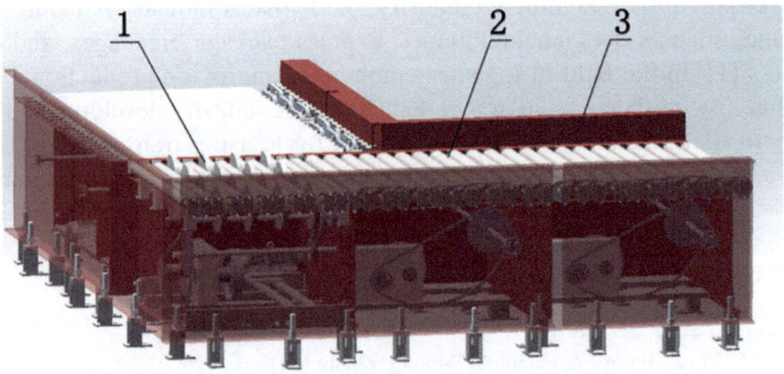

1-jacking device; 2-conveying device; 3-frame.

Fig. 1 Cam jacking transfer machine. 1-jacking device; 2-conveying device; 3-frame and a lifting height of 120 mm [8]. The rotational speed of the roller is 383 r/min, and the cargo transmission speed is 1.4 m/s.

3 Design of Key Components

3.1 Conveying Device

The transmission scheme of the conveying device is shown in Fig. 2.

The motor, after being reduced in speed and increased in torque by the reducer, transmits power sequentially through the reducer output shaft (4) to the chain drive (3) and the V-belt drive (2), thereby driving the rotation of the shaft on which the rollers are mounted. The shafts of the rollers achieve synchronous rotation through

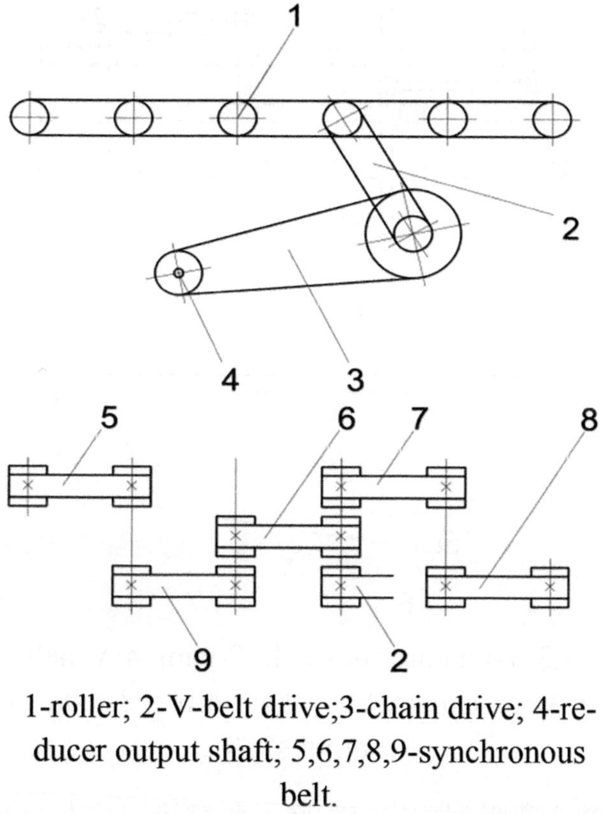

Fig. 2 Transmission scheme of the conveying device. 1-roller; 2-V-belt drive;3-chain drive; 4-reducer output shaft; 5,6,7,8,9-synchronous belt

the use of synchronous belts (5, 6, 7, 8, and 9), ensuring stable conveyance of the cargo.

3.2 *Jacking Device*

Transmission Scheme The transmission scheme of the jacking device is shown in Fig. 3. After being slowed down and having its torque increased by the reducer, the motor transmits power through the reducer output shaft (8) to the chain drive (7) and then to the V-belt drive (4), which in turn transfers the power to the camshaft (5), driving a series of cams on the camshaft to rotate. Synchronous rotation between the camshafts is achieved using synchronous belts (1, 3, and 6).

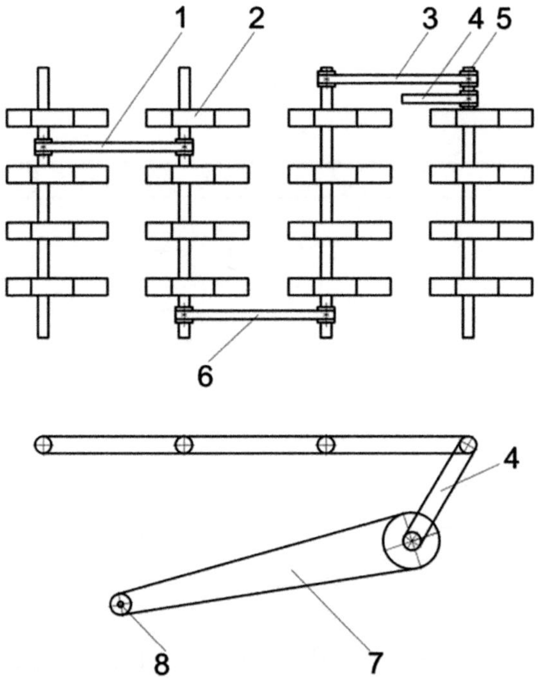

1,3,6-synchronous belt; 2-cam; 4-V-belt drive; 5-camshaft; 7-chain drive; 8-reducer output shaft.

Fig. 3 Transmission scheme of the jacking device. 1,3,6-synchronous belt; 2-cam; 4-V-belt drive; 5-camshaft; 7-chain drive; 8-reducer output shaft

When the transfer machine is in operation, the conveying device transports the goods along the conveying direction. Before the goods reach the position directly above the jacking device, the cam is in the near-dwell phase, positioned beneath the rollers of the conveying device. Upon detecting that the goods have been transported to the position above the jacking device, the camshaft drives the cam to rotate to a specific angle, at which point the cam enters the rise phase, lifting the goods upwards and separating them from the rollers of the conveying device. Subsequently, the cam continues to rotate to the far-dwell phase, vertically diverting the goods to the next conveying device. After completing the vertical diversion, the cam continues to rotate to the return phase, returning to its initial position beneath the rollers of the conveying device, thus completing the entire diversion process.

Modeling and Simulation of the Cam To meet the design requirements of the transfer machine, a disk cam made of 20Cr alloy steel is used [9], which is quenched and tempered to reach a hardness of HRC58.

Determination of Basic Dimensions of the Cam The base circle radius of the cam is initially determined as $r_0 = 90$ mm, and the lift h is set at 120 mm based on the jacking cargo height. The motion law for the rising stroke is selected as constant acceleration and constant deceleration, while the return stroke follows a cosine acceleration motion law [10].

Calculation of Theoretical Profile The disk cam mechanism is analyzed using the analytical method to obtain the coordinate equations of the theoretical profile of the cam.

$$x = (s_0 + s)\sin\delta + e\cos\delta \quad (1)$$

$$y = (s_0 + s)\cos\delta - e\sin\delta \quad (2)$$

Where x and y are the coordinate values of a point on the cam profile; s is the displacement of the cam follower; δ is the cam rotation angle; and e is the offset distance.

Since the designed cam mechanism is a roller-follower disc cam mechanism, we can set $e = 0$ and $s_0 = r_0$, Therefore, the coordinates of the theoretical profile of the cam are given by

$$x = (r_0 + s)\sin\delta \quad (3)$$

$$y = (r_0 + s)\cos\delta \quad (4)$$

According to the motion states of the follower, the coordinates of the cam profile for each stage are calculated separately.

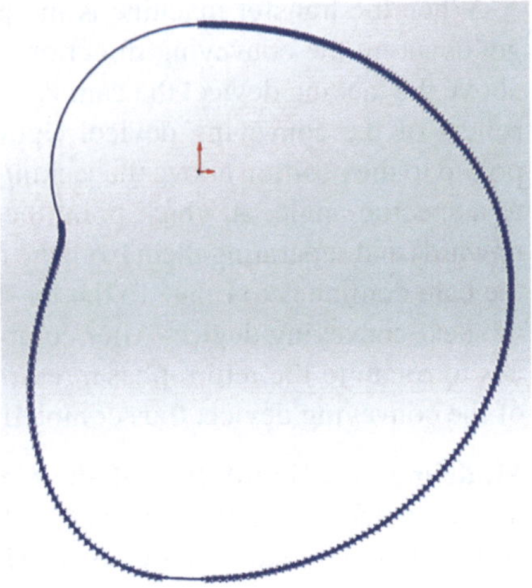

Fig. 4 Cam profile lines

Rise phase. $\delta_{01} = \pi, s_1 = 2h\delta_1^2 / \delta_{01}^2 = 2h\delta_1^2 / \pi^2, \delta_1 = [0, \pi]$
Far-dwell phase. $\delta_{02} = \pi/36$, $s_2 = 120$, $\delta_2 = [0, \pi/36]$
Return phade. $\delta_{03} = 13\pi/36$, $s_3 = h[1 + \ cos\ (\pi\delta_3/\delta_{03})]/2 = h[1 + \ cos\ (36\delta_3/13)]/2$, $\delta_3 = [0, 13\pi/36]$Near-dwell phase. $\delta_{04} = 11\pi/18$, $s_4 = 0$, $\delta_4 = [0, 11\pi/18]$

Substitute the corresponding values for each stage into Eqs. (3) and (4), and input the obtained coordinate equations of the cam profile into MATLAB to generate a dat. file containing the data points on the cam profile.

3D Modeling of the Cam The .dat file is imported into a SolidWorks sketch to obtain the cam profile as shown in Fig. 4. The resulting profile is then used to create the 3D model of the cam through feature commands such as extrusion and cut.

Simulation and Verification The 3D model of the cam is saved in Parasolid format and then import into ADAMS. The rotating pair of the cam is set as the drive, which is defined as a uniform rotation with a speed of 5 r/min. The simulation time is 15 s with 2000 steps. The simulation model is shown in Fig. 5.

The simulation results in the motion characteristic curve of the cam follower is shown in Fig. 6. The cam completes one full rotation with a motion cycle of 0 to 15 s. From 0 to 3 s, the cam follower rises, following a constant acceleration motion low. From 3 to 4 s, the cam follower remains stationary. From 4 to 10 s, the cam follower descends, adhering to a cosine acceleration motion profile. From 10 to 12 s, the cam follower is stationary again. During the second motion cycle, the cam follower repeats the aforementioned motion. The results indicate that the

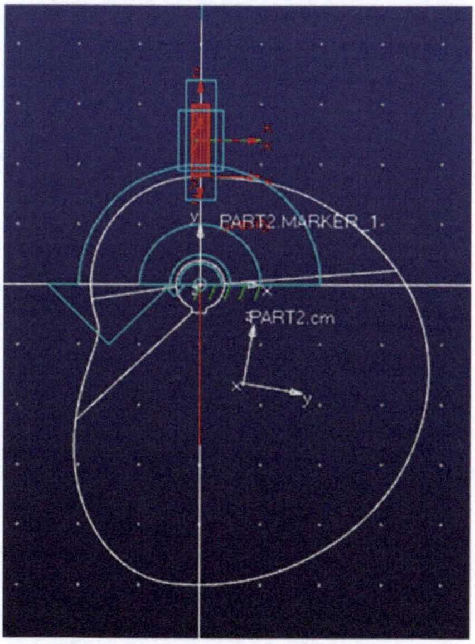

Fig. 5 Cam simulation model

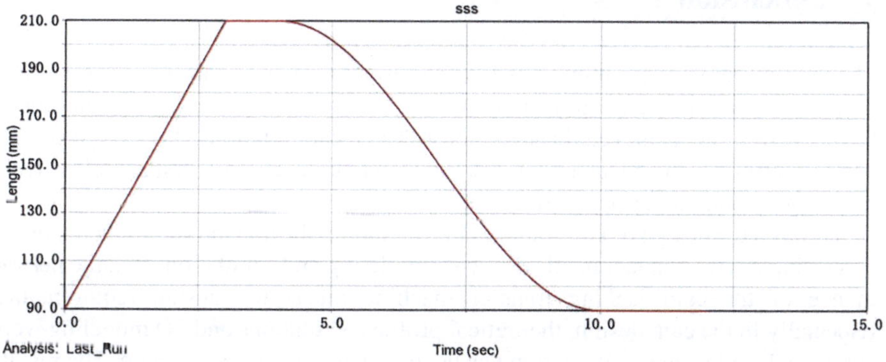

Fig. 6 Movement characteristic curve of the top bar

both the motion profile and driving capability of the cam meet the design requirements.

The dimensions of the designed cam are shown in Fig. 7. R_1 = 90 mm, R_2 = 180 mm, R_3 = 210 mm, R_4 = 80 mm, R_5 = 110 mm, R_6 = 24 mm. It ensures a lifting height of 120 mm for the jacking device, achieving the redirection of the target package.

Fig. 7 Dimensions of the designed cam

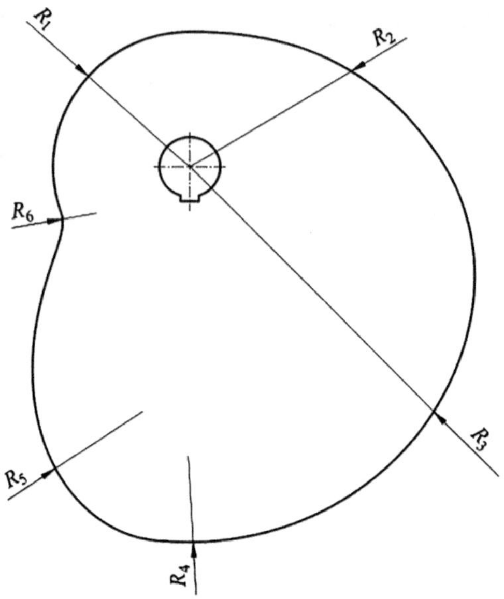

4 Conclusion

This paper addresses the numerous issues present in existing transfer machines and proposes the design of a cam jacking transfer machine. Both the jacking and transferring actions are achieved through motor-driven cam rotation. This design significantly simplifies the mechanical structure and drive control system, effectively reducing the maintenance and repair costs.

Through the design of key components, including the transmission scheme of the conveying device and jacking device, cam modeling and simulation, etc., the performance of the cam jacking transfer machine meets the design requirements. Especially in the cam design, theoretical profile calculations and 3D modeling were conducted, followed by simulation verification using ADAMS software. The results demonstrate that the cam's motion low and driving capability align with expectations.

The cam jacking transfer machine designed in this paper exhibits advantages such as stable structure, small footprint, low noise and low cost. Simulation results indicate that the transfer machine can achieve smooth and efficient jacking and transferring of goods, significantly enhancing the production efficiency and reliability of logistics equipment.

In the future, with the rapid development of the logistics industry, the cam jacking transfer machine is expected to find wider application and promotion, making greater contributions to the prosperity of the e-commerce industry and the improvement of logistics efficiency.

Acknowledgements This project was supported by the Huzhou Science and Technology Special Commissioner Project (2023KT63) and the 2024 National College Students Innovation and Entrepreneurship Training Program (Grant No. 202410347063).

References

1. Yulin R. Chinas e-commerce logistics market has been significantly active. Modern Logistics News, 2025-01-13 (005)
2. Liu Shengxu X, Minghao. (2024) Research on E-commerce logistics model innovation and efficiency improvement in China. China Storage and Transportation 11:190–191
3. Chenghua C (2020) Design and development of a heavy-load shifting robot. Technol Innov 20:98–99
4. Qiang G, Wenxing Wang Y, Cao. (2020) Development of a workpiece transfer machine based on a gaselectric hybrid system. Am J Sci Res Essays 5:27–28
5. Huaiyu W (2019) Design of movable fixed barrier shifting loader. Tianjin Vocational and Technical Normal University
6. Wei H (2024) Analysis and treatment of oil leakage of hydraulic drive cylinder of inlet valve of nuclear power steam turbine. Valve 11:1305–1309
7. Yanhong Z (2024) Some problems and prospects of linear motor in servo drive application of machine tool. Equip Eng China 17:129–131
8. Jinhua Z, Jie H (2024) Study on the modulus size of express packing box series based on packaging basic modulus. Packag Des 04:96–97
9. Gao Lina Y, Hongping HJ et al (2019) Parum optimization design and motion simulation of flat disc roller cam mechanism. Mech Eng 11:5–7
10. Heng S, Zuomo C, Wenjie G (2013) Mechanical principles edition 8. Higher Education Press, Beijing

Open Access This chapter is licensed under the terms of the Creative Commons Attribution-NonCommercial-NoDerivatives 4.0 International License (http://creativecommons.org/licenses/by-nc-nd/4.0/), which permits any noncommercial use, sharing, distribution and reproduction in any medium or format, as long as you give appropriate credit to the original author(s) and the source, provide a link to the Creative Commons license and indicate if you modified the licensed material. You do not have permission under this license to share adapted material derived from this chapter or parts of it.

The images or other third party material in this chapter are included in the chapter's Creative Commons license, unless indicated otherwise in a credit line to the material. If material is not included in the chapter's Creative Commons license and your intended use is not permitted by statutory regulation or exceeds the permitted use, you will need to obtain permission directly from the copyright holder.

Design and Analysis of a Mid-Mounted Motor Slider Rapid Return Bionic Flapping Wing Aircraft

Haobo Sun and Wenkang Hao

Abstract Currently, the flapping wing aircraft is a popular field in the bionic machine, for most of the flapping wing aircraft design exists in the structure is more complex, the weight is too large problem, this paper adopts the middle motor to drive the crank slider four-bar mechanism, designed a kind of flutter configuration with sharp return characteristics, through the kinematic analysis to establish the kinematic model of the crank slider. The relationship between flutter angle, flutter speed and flutter acceleration in one cycle is calculated and analysed. The kinematic model was solidly modelled in SolidWorks and imported into Adams, a multi-body dynamics simulation software, to simulate the motion and verify the results with the theoretical analysis. The simulation results show a basic overlap with the theoretical results, which is consistent with the selected flight parameters of the bionic bird, and has good aerodynamic performance, which is conducive to the flight of the flapping wing aircraft.

Keywords Flapping wing aircraft · Kinematic analysis of fluttering structures · Virtual simulation

Haobo Sun and Wenkang Hao contributed equally with all other contributors.

H. Sun
Aero engine series, Moscow Aviation Institute (National Research University), Moscow, Russia

W. Hao (✉)
College of Aerospace Science, National University of Defense Technology, Changsha, China

© The Author(s) 2026

Z. Zhang, P. Mercorelli (eds.), *Mechanical Design and Simulation: Insights and Innovations*, Lecture Notes in Mechanical Engineering,
https://doi.org/10.1007/978-981-95-2405-1_120

1 Introduction

The development of flapping wing aircraft originated from the human desire for flight and the pursuit of efficiency and manoeuvrability. Scientific and technological progress, especially the development of MEMS, aerodynamics and new material science, has made it possible to design and manufacture flap-wing aircraft. The needs of military reconnaissance, environmental monitoring and other fields have driven its development, and the research on energy and power systems has also provided technical support. With the continuous progress of technology, flapping wing aircraft show great potential in the field of aviation.

In recent years, research scholars at home and abroad have conducted a lot of research on flapping wing aircraft. Liu Yinggui [1] used the method of numerical simulation to carry out numerical simulation on the case of its incoming flow of 8 m/s, angle of attack of 5 deg., and flutter frequency of 2–8 Hz, respectively. It was concluded that the relationship between the lift and thrust generated by the wing and the flutter frequency showed a positive correlation, but the growth of the lift was much larger than that of the thrust, leading to the conclusion that the propulsive efficiency decreased with the increase of the frequency. Pengcheng Wang [2] established a three-dimensional flutter aircraft model by adding a Watt linear mechanism on a single crank double rocker mechanism with secondary reduction gears, which ensured complete symmetry of the left and right flapping and improved the compactness and stability of the flutter, and verified the correctness and feasibility of the optimisation method by using simulation. Yu Jie [3] and others designed a double crank rocker and crank slider-double rocker transmission system with reference to the flight characteristics of insects. After simulation and experiment, it was concluded that the double crank-rocker mechanism performs better in high-frequency motion and is suitable as a bionic multi-flap wing vehicle transmission system. Jiang Sen [4] designed a flapping wing mechanism that can realise the motion of two degrees of freedom of the wing by observing the motion trajectory of the wing when the bird is flying and based on the theory of spatial linkage mechanism, and concluded that the mechanism can realise the motion trajectory of the wingtip in the shape of spatial '8'. Hongzhi Zhang [5] designed a flapping-sweeping multi-degree-of-freedom flapping wing drive mechanism and established a rigid-flexible coupling dynamics model, and concluded that the thickness of the thin plate components has a significant effect on the dynamics of the drive mechanism, and the flapping motion performance of the drive mechanism is stronger than the sweeping motion performance. Zheng and Chao [6] tested and analysed the aerodynamic forces of elastic flat wings with different stiffnesses and concluded that the modified aerodynamic coefficients take into account the chordal stiffness of the elastic flat wings and exhibit anisotropic characteristics. Their study also verified the accuracy and validity of the modified aerodynamic modelling method for flapping wings. ShaoWeiping [7] designed a bionic flapping-wing aeroplane model to realise the unfolding-bending effect of a bird's wings during flight, to achieve active unfolding-bending motions, and to demonstrate the fast-returning characteristics of the flapping-bending motion.

Li Zhandong [8] studied the effect of acceleration on the aerodynamic performance of winged aircraft. A multi-degree-of-freedom three-dimensional bionic wing kinematic model with acceleration effect was established and the aerodynamic performance was calculated using AnsysFluente. By incorporating aerodynamic variations, the aerodynamic and pressure distribution characteristics coefficients at transient and peak times with different acceleration coefficients are elucidated. As the results show, the acceleration effect can significantly increase the aerodynamic lift and thrust compared with the conventional flapping motion. The accelerated downstroke of the flapping wing changes the strength of the LEV and TEV and increases the vortex adhesion area on the upper surface of the wing. The acceleration effect, also increases the capture effect of the refflick in the last cycle of the wing, which significantly improves the aerodynamic performance of the wing. Tong C [9] analyses the current research status of fluid-structure interaction (FSI) in flexible flapping wings. Based on the limitations of the current FSI structural analysis methods, this paper points out the necessity of analysing the FSI problem of a 2-DOF flexible flapping wing, establishes a theoretical model combined with the propulsive wave theory, and proposes an orthogonal design process for the structural parameters of the flexible flapping wing.

In summary, due to the fact that multiple degrees of freedom bring more structural loads and instability, in order to improve the structure of the flapping wing aircraft, reduce the redundant weight, increase the payload, reduce the energy dissipation of each component, and ensure the stability of the aircraft.

This paper adopts a single-degree-of-freedom crank-slider structure, using a mid-mounted motor drive, and designs a simple and efficient fluttering wing configuration through kinematic analysis and modelling simulation. The problem of low efficiency brought by the lack of difference between the upper and lower fluttering speeds of the fluttering wing is solved.

2 Analysis of Flutter Geometry

During the whole movement of the fluttering machine, the fluttering angle of the fluttering machine as an output quantity will be affected by many factors of the whole fluttering structure. Figure 1 shows the structure and various parameters of the flutter machine studied in this paper.

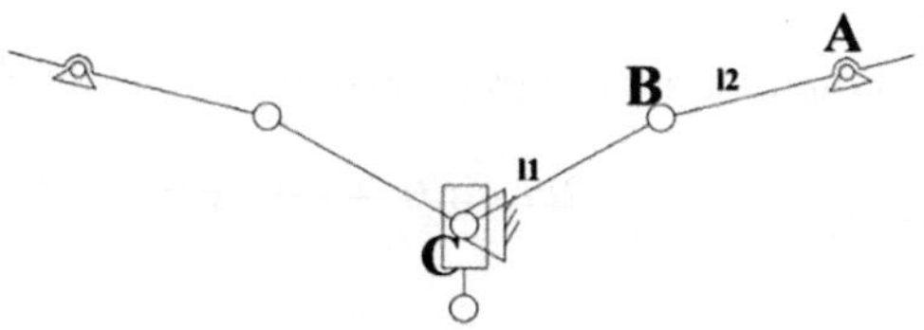

Fig. 1 The structure and various parameters of the flapping wing aircraft studied in this paper

Table 1 Table of binding parameters for flapping structures

Name of rod	Length	unit
l1	101	mm
l2	38.7	mm
e	64	mm
h	100.5	mm

The description of each parameter in the equation of motion is as follows: in the whole structure, the bar BC becomes l1, the bar AB is called l2, the horizontal distance from point A to point O is called e, the vertical distance from point A to point O is denoted as h, and the angle of turn is $\theta 1$, call it the input turn angle, and the angle at which the flapping wing swings is $\theta 2$, call it the output swing angle.

In the above flapping wing structure diagram, the slider C is driven by the power unit and makes round-trip motion in the vertical direction. The rod BC does regular sinusoidal flapping in the plane where the BC rod is located due to the motion of the slider C. Driven by the rod BC, the rod AB does regular oscillation around the point A in the plane W where it is located.

Now the values of each parameter in the equation of motion are taken as shown in Table 1 below.

3 Uter Structure Analysis

3.1 Flutter Structure Perspective Analysis

In this paper, we will take y = $-$ 1.675 sin ($10\pi t$) as input equations for the upper wing and take y = $-$ 1.675 sin ($2.5\pi t$) as input equations for the lower wing of the device. According to the vector theory, we can derive the geometric relationship as follows:

$$\overrightarrow{OA} = \overrightarrow{OB} + \overrightarrow{BC} + \overrightarrow{CA} \tag{1}$$

$$e = l_1 \cos\theta_1 + l_2 \cos\theta_2 \tag{2}$$

$$h = y + l_1 \sin\theta_1 + l_2 \sin\theta_2 \tag{3}$$

Based on the above geometric relationship, we can get the following relation:

$$\left(e - l_2 \cos\theta_2\right)^2 + \left(h - y - l_2 \sin\theta_2\right)^2 = \left(l_1 \cos\theta_1\right)^2 + \left(l_2 \sin\theta_1\right)^2 \tag{4}$$

Simplifying the above equation, we get the following formula:

$$2hl_2 \sin\theta_2 + 2el_2 \cos\theta_2 = e^2 + h^2 + y^2 + l_2{}^2 - 2hy - l_1{}^2 \tag{5}$$

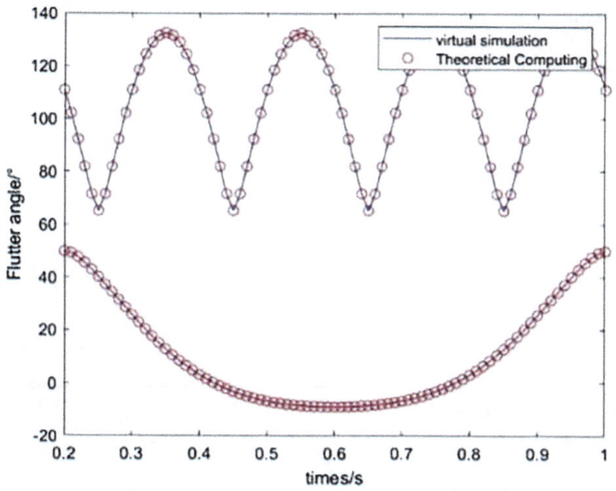

Fig. 2 Characteristic curve of the flutter angle of the AB rod in a single cycle

Now, let $A = 2hl_2$, $B = 2el_2$, $C = e^2 + h^2 + y^2 + l_2^2 - 2hy - l_1^2$, then Eq. 5 can be simplified to:

$$A\sin\theta_2 + B\cos\theta_2 = C \tag{6}$$

Based on the trigonometric functions and the universal formula, we can get the mathematical model of the flutter angle as follows:

$$\theta 2 = \arctan\frac{A \pm \sqrt{A^2 + B^2 - C^2}}{B + C} \tag{7}$$

The characteristic curve of the flutter angle of the AB rod in a single cycle, plotted according to Eq. 7 is shown in Fig. 2:

3.2 *Velocity Analysis of Flutter Structures*

Deriving both sides of Eq. 5 with respect to time yields the following equation:

$$2y\frac{dy}{dt} - 2h\frac{dy}{dt} + 2el2\sin\theta 2\frac{d\theta 2}{dt} - 2hl2\cos\theta 2\frac{d\theta 2}{dt} = 0 \tag{8}$$

In Eq. 8, $\frac{dy}{dt}$ denotes the speed of motion of the slider, denoted by v, and $\frac{d\theta_2}{dt}$ denotes the angular velocity of the bar when it is in fluttering motion, denoted by ω. Eq. 8 an be simplified as:

Fig. 3 Characteristic curve of flutter velocity change of AB rod in 1 s

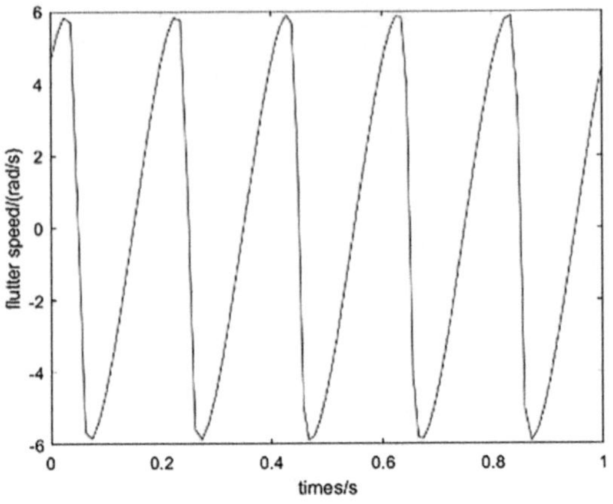

$$2v(y-h)+2\omega(el2\sin\theta2-hl2\cos\theta2)=0 \tag{9}$$

As a result of the above calculations, this subsection obtains the angular velocity versus time in this flutter device.

The curve of the flutter velocity change of the AB bar in 1 s, plotted according to Eq. 9, is shown in Fig. 3.

3.3 Acceleration Analysis of Fluttering Structures

Deriving both sides of Eq. 9 with respect to time gives the following equation:

$$\begin{aligned}&2\frac{dv}{dt}(y-h)+2v\frac{dy}{dt}+2\frac{d\omega}{dt}(el2\sin\theta2-hl2\cos\theta2)+\\&2\omega\left(el2\cos\theta2\frac{d\theta2}{dt}+hl2\sin\theta2\frac{d\theta2}{dt}\right)=0\end{aligned} \tag{10}$$

In Eq. 10, $\frac{dv}{dt}$ denotes the acceleration of the slider motion, which is denoted by a in this paper, $\frac{d\omega}{dt}$ denotes the acceleration when the rod is doing flutter, which is denoted by α in this paper. Then Eq. 10 can be simplified as:

$$\begin{aligned}&2a(y-h)+2v^2+2\alpha(el2\sin\theta2-hl2\cos\theta2)+\\&2\omega^2(el2\cos\theta2+hl2\sin\theta2)=0\end{aligned} \tag{11}$$

After the above theoretical analysis, this subsection obtains the angular acceleration versus time in this flutter device.

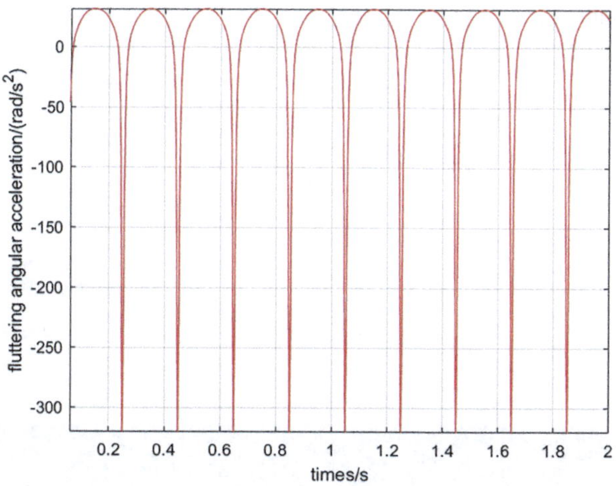

Fig. 4 Characteristic curve of flutter acceleration change of AB rod in 2 s

The curve of the flutter acceleration change of the AB bar in 2 s, plotted according to Eq. 10, is shown in Fig. 4.

4 Virtual Model Design and Simulation Analysis

Based on the theoretical calculation results in the aforementioned MATLAB, this part uses SolidWorks software to establish a virtual 3D simulation model for the acute return four-bar slider structure, as shown in Fig. 5. Since the four-bar structure used in the two wings is identical, the modelling only establishes a unilateral ideal model for simulating the actual movement and verifying the theoretical calculation results. The model is mainly composed of four parts: frame, rod 1, rod 2, and slider. Rod 1 and rod 2 are connected to the frame, and the slider repeatedly slides along the slide rail in the frame to drive the two rods to carry out reciprocating motion, so as to realise the symmetric flapping effect of the two flaps and achieve the requirement of single-degree-of-freedom drive. The length of rod 1 is 101 mm, the length of rod 2 is 38.7 mm, and the length of the frame is 100.5 mm and the width is 64 mm, which is consistent with the theoretical calculation parameters.

In order to check the trajectory of the four-bar flapping wing structure proposed in this paper and to obtain the motion parameters of the virtual model, the SolidWorks model is exported to the dynamics simulation Adams software for simulation and analysis. The initial position parameters of each component are set in the software, and the corresponding joints are connected. Add appropriate motion vice on the virtual model, according to the requirements of the crank-slider mechanism, lock the frame with the ground, establish the rotating vice in the rod 1 and rod 2, rod 2 and frame, slider and rod 1, and establish the sliding vice in the slider and frame.

Fig. 5 3D modelling of the four-bar flapping wing structure

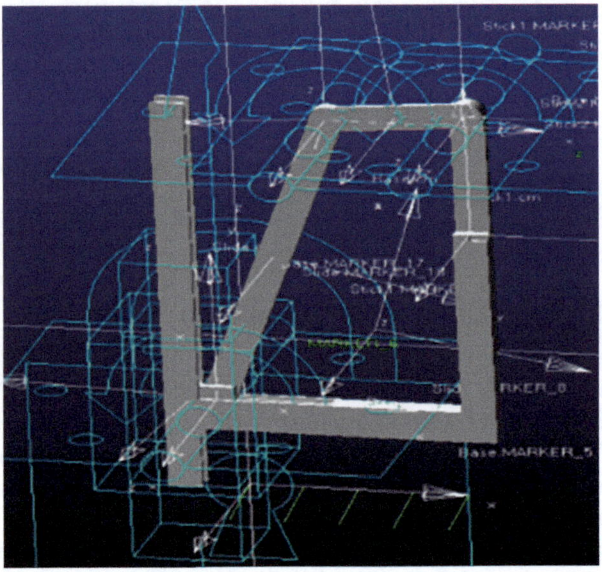

Fig. 6 Adams simulation model diagram of the flapping wing structure

Subsequently, according to the previous data, add the drive at the slider, and set the drive speed in line with the theory. The Adams simulation model shown in Fig. 6 is obtained.

After completing the dynamics simulation, a reference point is established on the frame, and the measurement module is set up to read the data of the rod fluttering angle over time.

The output angle is represented by a curve in Adams/Postprocessor, and a cycle is delineated as the display range, and the data obtained from the simulation is compared with that obtained from the theoretical calculation. The image of the theoretical calculation and simulation of the AB section in a single cycle is shown in Fig. 7 below.

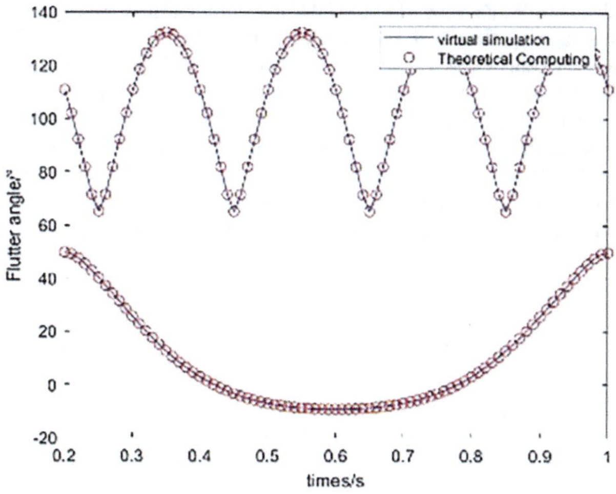

Fig. 7 Comparison of the curves of theoretical calculations and virtual simulation for the AB segment in a single cycle

5 Conclusion

In this paper, a flapping wing configuration with sharp return characteristics is designed using a slider four-bar mechanism driven by a centre motor. The kinematic model of the slider is established by motion analysis, and the flapping angle, velocity and acceleration in one cycle are calculated and analysed by MATLAB, and the theoretical calculation results of the flapping angle, velocity and acceleration are compared with the anti-true results. The results show that the two are basically overlapped, which is consistent with the selected flight parameters of the bionic bird and has good aerodynamic performance, which is favourable to the flight of the flapping vehicle.

References

1. Ying-gui L (2020) Numerical simulation of aerodynamic characteristics of the main wing of a flapping wing aircraft at different frequencies. Comput Prod Distrib (02):270+285
2. Peng-cheng W, Hao W, Guang-jian G (2020) Design and simulation of single-crank double-rocker non-phase difference flapping wing drive mechanism. Mech Mfg Autom 49(01):94-96+126
3. Jie Y, Gao-yu F (2022) Design of transmission system for multi-winged aircraft. Sci Technol Commun 14(04):139–143
4. Sen J, Yong-ping H, Lun L et al (2019) Design of a flapping wing machine based on a space linkage mechanism to realise the '8' shape motion. Mach Tools Hydraul 47(13):76–80
5. Hong-zhi Z, Pen-feng S, Zhong-chao S et al (2024) Design and dynamic performance analysis of a multi-degree-of-freedom flapping wing drive mechanism. J Aerosp Dyn 39(04):25–35

6. He-chao Z, Jian-hui W, Zi-yang H et al (2023) Accuracy test of aerodynamic modelling of a bird-like wing-flapping aircraft. J Aeronaut 44(10):92–106
7. Shao W, Zhang H, Zeng P (2022) Design and aerodynamic characteristic analysis of unfolding-bending bionic flapping-wing aircraft. J Mech Sci Technol 36:2981–2991
8. Zhan Dong L, Chang Quan G, Xin Yu Z et al (2024) Effect of acceleration on aerodynamic performance of flapping wing air vehicle. J Braz Soc Mech Sci Eng 46:32
9. Tong C, Yuan M, Zhao L, Dong S, Wu Z, Tong C (2023) Bionic flapping wing bending and twisting coupled motion modeling. In: Yan L, Duan H, Deng Y (eds) Advances in guidance, navigation and control. ICGNC 2022. Lecture notes in electrical engineering, vol 845. Springer, Singapore

Open Access This chapter is licensed under the terms of the Creative Commons Attribution-NonCommercial-NoDerivatives 4.0 International License (http://creativecommons.org/licenses/by-nc-nd/4.0/), which permits any noncommercial use, sharing, distribution and reproduction in any medium or format, as long as you give appropriate credit to the original author(s) and the source, provide a link to the Creative Commons license and indicate if you modified the licensed material. You do not have permission under this license to share adapted material derived from this chapter or parts of it.

The images or other third party material in this chapter are included in the chapter's Creative Commons license, unless indicated otherwise in a credit line to the material. If material is not included in the chapter's Creative Commons license and your intended use is not permitted by statutory regulation or exceeds the permitted use, you will need to obtain permission directly from the copyright holder.

Design and Analysis of Amphibious Land-Air Reconnaissance Robot

Xiaolin Li, Wei Tang, and Huihui Niu

Abstract Traditional single-motion-mode robots have many drawbacks when performing reconnaissance or terrain exploration. To address this issue, this paper designs a land-air amphibious reconnaissance robot that integrates legs, wheels, and rotors and can achieve three or more motion modes. The strength of its key components and motion are verified and simulated using ADAMS and ANSYS software. The results show that the structural strength of the land-air amphibious reconnaissance robot meets the motion requirements, and its motion performance is good, with broad application prospects.

Keywords Land-air amphibious robot · Rigid-flexible coupling simulation · Motion simulation

1 Introduction

With the rapid development of technology, robots are appearing more and more frequently in our lives. From the initial industrial robots such as bipedal robots to the current unmanned aerial vehicles and unmanned vehicles for delivering parcels [1], as well as household robots like vacuum cleaners and dishwashers [2], robots are playing an increasingly important role in our daily lives. At the same time, as people's demands grow, the functions of robots are constantly being explored, and robots capable of conducting reconnaissance tasks and exploring terrains have also emerged.

The early reconnaissance robots were mostly single-movement mode robots, such as wheeled robots and tracked robots [3]. However, such robots had many drawbacks, such as difficulty in adapting to complex and variable unstructured terrains. Therefore, scholars turned their attention to amphibious robots that could achieve multiple movement modes.

X. Li (✉) · W. Tang · H. Niu
Jiangsu University of Science and Technology, Zhenjiang, China

© The Author(s) 2026

Z. Zhang, P. Mercorelli (eds.), *Mechanical Design and Simulation: Insights and Innovations*, Lecture Notes in Mechanical Engineering,
https://doi.org/10.1007/978-981-95-2405-1_121

Amphibious reconnaissance robots are a type of robots that can autonomously change movement modes according to different environmental conditions and perform reconnaissance or terrain exploration. Regarding the combination of the movement modes of land and air robots, there are various methods. Some people simply superimpose feet, wheels, and propellers to achieve multiple movement modes, such as Zhu Hang et al. [4] designed a micro-sized amphibious robot. This method can achieve the combination of movement modes, but it cannot derive other movement modes, and there are certain limitations in functionality. Some people partially or completely fuse feet, wheels, and propellers. As shown in Fig. 1, the robot FCSTAR developed by David Zarrouk et al. from Ben-Gurion University of the Negev in Israel and the rat cage-type amphibious robot [5] designed by Yang Zhan et al. from Zhejiang University of Science and Technology are among them. However, due to the inappropriate fusion methods, there are some defects, such as the rotation range of the robot FCSTAR in the gliding mode being slightly insufficient, and the legs and feet are not integrated, making it difficult to achieve smooth movement on uneven terrains. The rat cage-type robot cannot perform gliding movement on uneven and potholed ground and has difficulty in steering, limiting its ground mobility. At the same time, Sihite Eric et al. from California Institute of Technology designed a multifunctional morphing robot M4 [6], which can achieve eight movement postures such as flying, rolling, and crawling. Its functionality is very perfect, but the legs are not integrated.

In response to this, based on the aforementioned research, this paper intends to design an amphibious reconnaissance robot that can achieve three or more movement modes and can simultaneously perform reconnaissance tasks. Unlike the amphibious robots that combine two movement modes [7, 8], this robot can not only achieve crawling, rolling, and flying movement forms, but also display other postures like M4, and its movement is very flexible. This robot can adapt to various

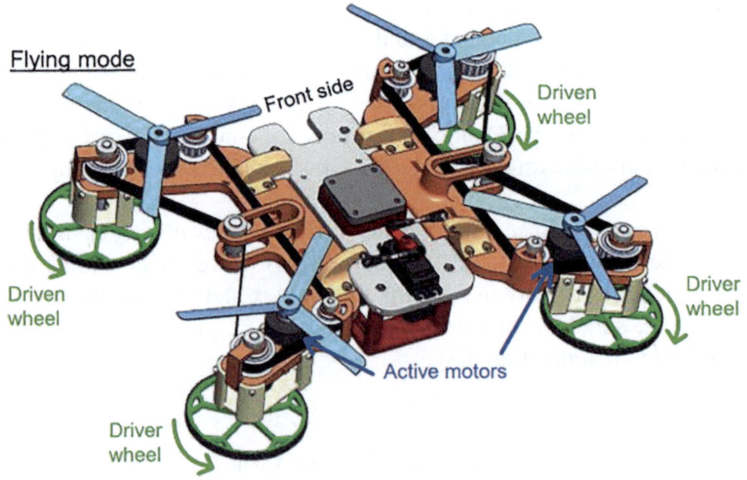

Fig. 1 FCSTAR

working scenarios. For instance, in the crawling mode, it can crawl rapidly on uneven roads by relying on its legs; in the roller-skating mode, it can pass quickly on flat roads by using rollers; in the flying mode, it can bypass obstacles that are difficult to cross on the ground. Meanwhile, it can switch between different motion modes by adjusting its legs, thereby enhancing the endurance and efficiency of the robot when performing reconnaissance tasks.

2 Robot Structure Design

2.1 Scheme Design

The amphibious reconnaissance robot designed in this paper mainly consists of 12 servo motors, 4 encoders, 4 brushless motors, 4 legs, 4 pulleys, 4 propellers and the body. Regarding the placement position of the pulleys, after comparing common ground drive methods [9], the pulleys and legs are installed in a separated form, placed between the joints of the legs, and the motion is transmitted through the gear sets between the joints. This way can better exert the effects of various motion modes. The single-leg structure and the overall structure are respectively shown in Figs. 2 and 3.

The robot adopts a leg-serial and gear-set approach to highly integrate the legs, pulleys and rotors. Structurally, it has four legs, four pulleys and four rotors. Each leg has three joints. Compared with the two-degree-of-freedom legs, its range of motion can be greatly increased [10]. In terms of form, it is not overly complex, which fully guarantees the overall motion flexibility of the robot and enables it to achieve three basic motion modes: crawling, roller-skating and flying. When the robot crawls on the ground, it can achieve crawling movement on uneven and pot-holed ground through the coordinated control of the servo motors of the four-legged feet; when it performs roller-skating on the ground, it can realize sliding movement by driving the encoder motors, and at the same time, it can achieve all-round

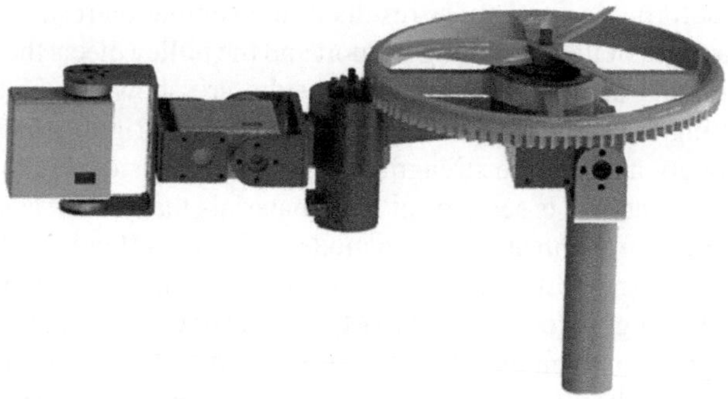

Fig. 2 Single-leg structure

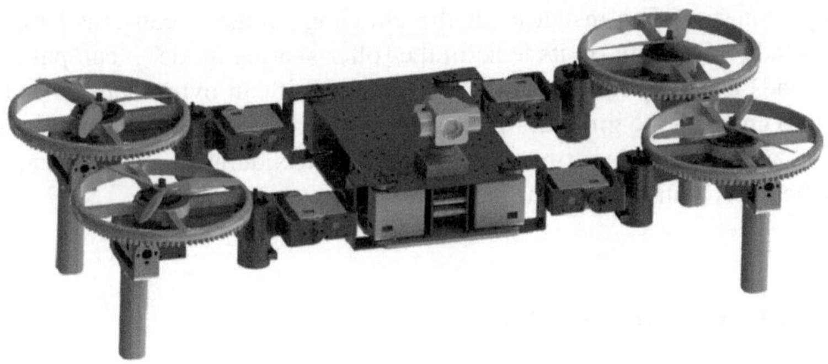

Fig. 3 Reconnaissance robot

movement on the ground by combining the servo motors; when it flies in the air, it can realize hovering, pitch, and side-flipping movements by coordinating the control of the rotational speeds of the four rotor motors.

2.2 *Strength Verification of Key Components*

To ensure that the structural strength of the amphibious robot during its movement meets the requirements, the strength of key components needs to be verified. During crawling movement, the shortening support connecting the first joint and the second joint on the legs becomes a weak point due to its thickness; during wheel-skating movement, the pulley becomes a weak point due to its bearing the weight of the body and its contact with the ground. Here, ANSYS and ADAMS are jointly used to conduct a simulation of the strength of the rigid-flexible coupling during the robot's movement. During crawling, the step length of a single leg is 80 mm, and during sliding, the rotational speed of the small gear exceeds 450 r/min (the maximum supported speed of the encoder motor is 450 r/min). After assigning the material and dividing the mesh for the parts in ANSYS, a modal neutral file needs to be generated. Then, the modal neutral file is imported into ADAMS for simulation. Based on the deformation and stress results in the simulation results, it is analyzed whether the strength of the shortening support and the pulley meets the requirements.

The deformation and stress simulation results are shown in Figs. 4, 5 and 6. When conducting the simulation of the coupling between rigidity and flexibility, in order to accurately analyze the strength of the components during the motion process. The sample machine is set to print with materials that are the easiest to handle, have an appropriate strength and are also moderately cost-effective. The material for the setting of the long and short supports is Alsi10mg, with the maximum allowable stress and yield strength are 130 and 240 Mpa respectively. The material for the pulley is resin, with the maximum allowable stress being 24 Mpa. From the deformation results, during the crawling process, the upper end of the long support was pulled by gravity and deformed, with the maximum deformation reaching $3.32*10^{\wedge-4}$

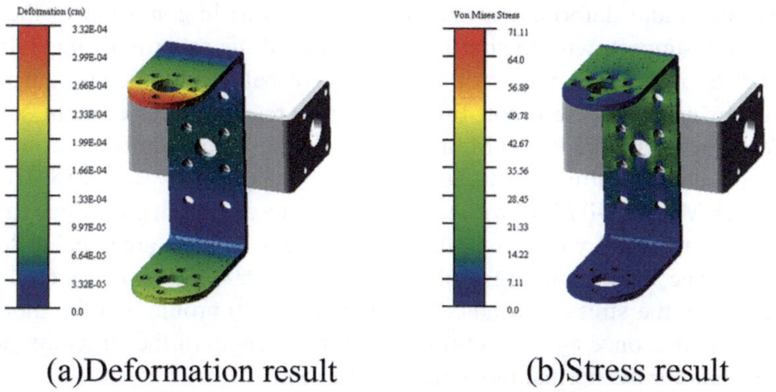

Fig. 4 Analysis results of long stent. (**a**) Deformation result. (**b**) Stress result

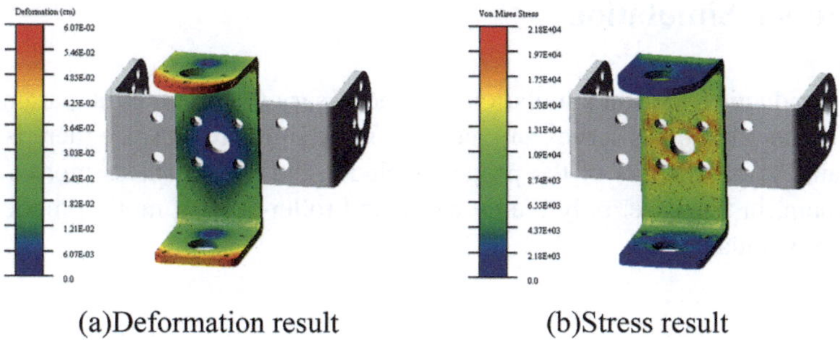

Fig. 5 Analysis results of short stent. (**a**) Deformation result. (**b**) Stress result

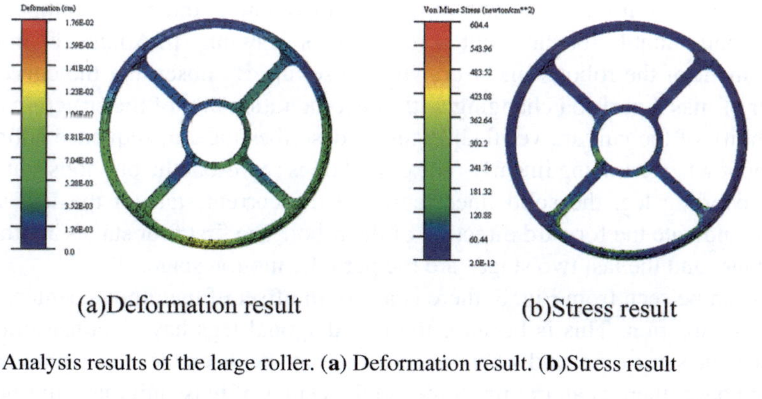

Fig. 6 Analysis results of the large roller. (**a**) Deformation result. (**b**)Stress result

cm. During the crawling process, the connecting short support was also pulled by gravity, and the deformation increased with the increase of the motion amplitude, with the maximum deformation reaching 6.07*10^$^{-2}$ cm. During the roller skating process, the roller was subjected to the downward gravity and the supporting force provided by the ground. The gravity acted on the middle installation hole position, and the maximum deformation occurred when it was in contact with the ground without hub support, with the deformation result of the roller reaching 1.76*10^$^{-2}$ cm. The deformation amounts of all three were below millimeters, ensuring the normal movement of the robot. From the stress results, the maximum stress was 71 N/cm^^{2} ≈ 0.71 MPa < 130 MPa, with no obvious stress concentration position. The maximum stress of the short support was 2.18*10^4 N/cm^^{2} = 218 MPa < 240 MPa, with the stress in some areas being between the allowable shear stress and yield strength, and the stress in most areas being less than 130 MPa. The maximum stress of the roller was 604.4 N/cm^^{2} ≈ 6.04 MPa < 24 MPa, with the stress distributed between the tooth profile and the motor shaft installation hole, once again confirming that the strength of the structure meets the requirements for the normal movement of the robot.

3 Motion Simulation

When conducting motion simulation, it is necessary to combine the parameters of actual actuators such as servos and motors for the simulation. Since there is no way to obtain the parameters of the propeller blades and the brushless motors used to drive them, in this case, only the crawling and roller-skating motion modes of the robot are simulated.

3.1 Simulation of Crawling Motion

When the robot moves by means of leg-foot locomotion, due to the compact structure, a gait suitable for the robot is required for crawling. In Adams, a simulation experiment of the robot's linear crawling is set up. By observing the curve of the center of mass' position changing with time, the rationality of the structure and the feasibility of the gait are verified. Figure 7 describes the gait sequence adopted by the robot when crawling linearly. The dotted lines represent the previous state of the corresponding leg, the solid lines represent the current state of the leg, and the arrows indicate the forward direction of the robot. The first four stages are the starting stage, and the last two stages are the periodic motion stage.

As can be seen from Fig. 8, there is a certain offset of the robot's center of mass in the y-direction. This is because the two diagonal legs have inconsistent action states at the same time, and when the legs alternate between the stance phase and the swing phase, there is an inertial force, so the center of mass migrates alternately to the foot ends of legs 1 and 2. However, the error is small and the overall stability is relatively good.

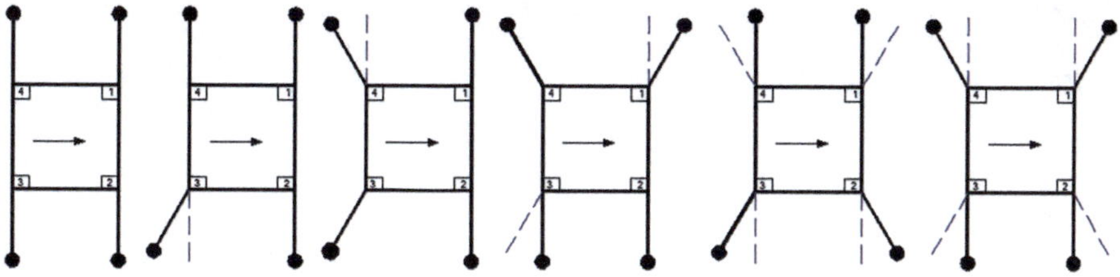

Fig. 7 Sequential diagram of straight-line gait

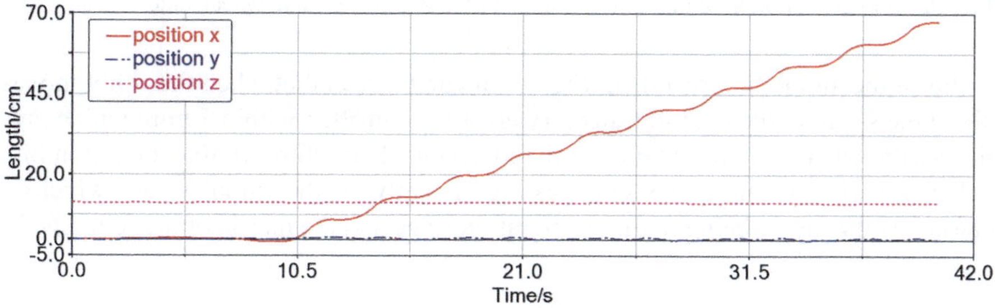

Fig. 8 Curve of the change in the centroid displacement of the straight line

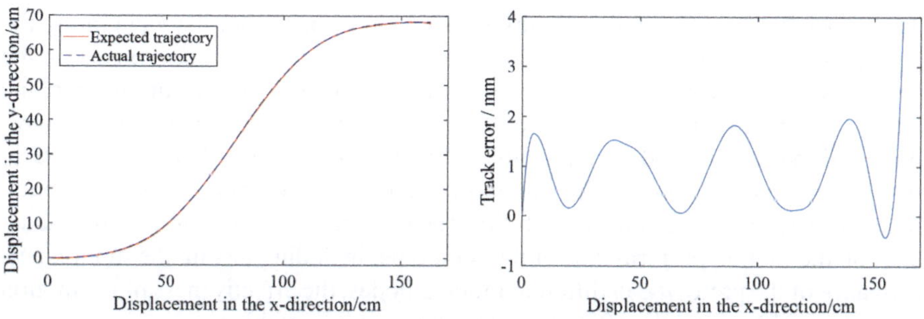

a) Center of mass trajectory (b) Trajectory error

Fig. 9 Centroid displacement and error of robot s-curve roller skating. (**a**) Center of mass trajectory. (**b**) Trajectory error

3.2 Sidewalk Skating Simulation Mode

When the robot is in the roller skating mode, it is equivalent to a four-wheel drive and four-wheel steering vehicle. How its omnidirectional movement performance is a question worth studying. Therefore, it is necessary to verify the rationality of the mechanical structure and the stability during turning when the robot is in the roller skating motion mode. Figure 9 depicts the trajectory of the robot's center of mass and the error curve when it is roller skating along an S-shaped path, and Fig. 10 shows the corresponding changes in the robot's center of mass velocity.

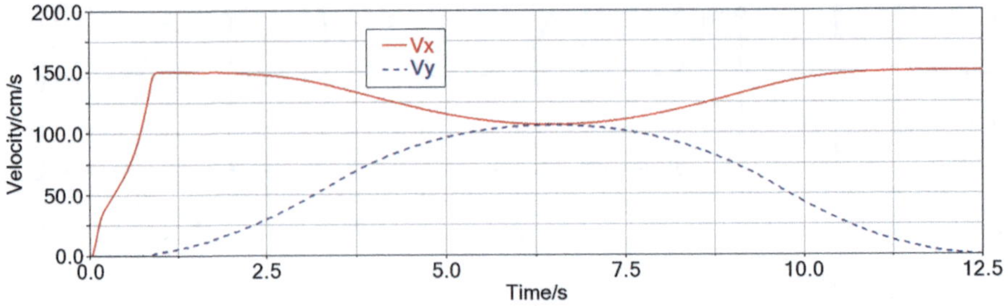

Fig. 10 Velocity variation of the center of mass of the Robot's S-curve wheel slide

From the simulation results in Fig. 10, it can be seen that when the robot moves along the s-curve, the overall trajectory error is controlled within 7 mm, which preliminarily verifies the rationality of the mechanical structure. In the corresponding velocity variation curve, Vx changes significantly in the initial stage, which is caused by the motor being in the acceleration stage and the inertia of the robot itself.

4 Conclusion and Next Step for Research

This paper designs a land-air amphibious reconnaissance robot. By integrating the legs, pulleys and rotors highly, it can achieve various motion modes. Based on the structural design, through the simulation of rigid-flexible coupling and motion, the structural strength and motion performance of the robot in complex terrains have been verified. It has good terrain adaptability and motion flexibility, and can effectively overcome the shortcomings of existing land-air amphibious robots in motion mode integration and terrain adaptability. The next step of the research is to conduct tests on the motion performance of the prototype, including testing the motion performance of the prototype in different motion modes, the effectiveness of the motion control algorithm, and the efficiency of motion mode switching.

References

1. Xianying C (2022) Research on design and motion modes of amphibious Robots for land and air. Central South University. https://doi.org/10.27661/d.cnki.gzhnu.2022.001500
2. Bin Y, Yuqing H, Jianda H (2015) Research status and prospects of robotic work-related flying robots. Robotics 37:05. https://doi.org/10.13973/j.cnki.robot.2015.0628
3. Huizhe T, Siyu D, Qi X (2024) Design of an Automatic IntelligentEnergy-saving fire reconnaissance robot. J Light Ind Technol 40:04. https://kns.cnki.net/kcms2/article/abstract?v=YvzTZTPVeK5aF3iQqKBkbGjqLEL80yvqWD7pBG18MiziVia0htqdJVkAEDEf2pGBhHI8yzFUiOaPNmhSX-JXCESW4dHulY_D-jmfOGBTgi9aW0ng6y_C6Wl4cXH_NkuV2o869GO-qSN0VsJmma4VxrbY3F9tKcN1QDDXHskxxGCkLkuxZRTJbpopwuyEEYvXJtFLyuCQPK2M=&uniplatform=NZKPT&language=CHS

4. Hang Z, Jiadong S, Jianzhong W (2015) Ground movement control method for micro-sized amphibious robots. Mod Electron Technol 38:08. https://doi.org/10.16652/j.issn.1004-373x.2015.08.043
5. Zhan Y, Qipeng L, Wei T (2023) Design and analysis of a small-scale land-air deformable amphibious robot. J Eng Des 30:03. https://kns.cnki.net/kcms2/article/abstract?v=YvzTZTPVeK7rMzJV2n4DrVDr4lsRZBHfX2DfUZAriw1yIcwMWHyAitsJwVoEanWZ3e1_mdBu-wNymlcmFTWhrktrmJe4SClEbcrAtAbv_tKK-WX4E1eYrDxXaA1FtiP1XzMcxNSBZZFO2PzBR7XoMIR99bjGw1qXrJEyf87GZ3fb_LF3A2TYsYhkPrX8en9l5Og0x8EWvl7c=&uniplatform=NZKPT&language=CHS
6. Sihite E, Kalantari A, Nemovi R (2023) Multi-modal mobility Morphobot (M4) with appendage repurposing for locomotion plasticity enhancement. Nature. Communications 14:1. https://doi.org/10.1038/s41467-023-39018-y
7. Xianggong H, Dichang Z, Qingmin Z (2020) Design of Mobile Control System for amphibious robot based on multi-sensor fusion. Sci Technol Eng 20:08. https://kns.cnki.net/kcms2/article/abstract?v=d24vdHNzaZh_D21WB12nmKhzV1B_oeFq5qao2lvF4TzmKoUOt-rvDbPAX9CWhoM9EHjKsmwgB_qa2CvukhhfsoaGIW_zW_zV-juGo4a6x5nMiT4t9I8aZeLOlPjJtgTsJTj7kCqkZv0H2NKLjLzXdE5o1TDH2NMbYbBqPfghLp7BFT8eFxr5hW7F8C_KSID0axynI9nQNQg=&uniplatform=NZKPT&language=CHS
8. Zhengjie W, Jian M, Hang Z (2015) Design of flight control system for amphibious robot on land and air. J Beijing Inst Technol 35:12. https://doi.org/10.15918/j.tbit1001-0645.2015.12.009
9. Yingbao R (2023) Research on landing stabilization control of amphibious robots. Central South University. https://doi.org/10.27661/d.cnki.gzhnu.2023.002149
10. Qingsheng L, Yinfei C, Xingdong L (2019) A selection analysis and structural design of a small amphibious robot. Comput Meas Control 27:04. https://doi.org/10.16526/j.cnki.11-4762/tp.2019.04.047

Open Access This chapter is licensed under the terms of the Creative Commons Attribution-NonCommercial-NoDerivatives 4.0 International License (http://creativecommons.org/licenses/by-nc-nd/4.0/), which permits any noncommercial use, sharing, distribution and reproduction in any medium or format, as long as you give appropriate credit to the original author(s) and the source, provide a link to the Creative Commons license and indicate if you modified the licensed material. You do not have permission under this license to share adapted material derived from this chapter or parts of it.

The images or other third party material in this chapter are included in the chapter's Creative Commons license, unless indicated otherwise in a credit line to the material. If material is not included in the chapter's Creative Commons license and your intended use is not permitted by statutory regulation or exceeds the permitted use, you will need to obtain permission directly from the copyright holder.

Effect of Spring Stiffness on the Dynamic Performance of Safety Valves

Shixin Wang and Xinhai Yu

Abstract As a key safety equipment in industrial systems, the dynamic performance of safety valves is crucial to ensure system safety. This study aims to explore the effect of different spring stiffness on the dynamic performance of safety valves. Through numerical simulation, the flow channel model of safety valves is established, and the external virtual container is connected to simulate the whole process of opening, discharging and seating of safety valves, so as to analyze the effect of the change of the spring stiffness on the opening time, the seating pressure and other key performance parameters. The results show that increasing the spring stiffness can slow down the disc velocity, significantly increase the opening time, and improve the seat pressure, however, this will sacrifice the sensitivity of the safety valve, making it difficult to respond quickly to changes in the pressure system.

Keywords Safety valves · Spring stiffness · Numerical simulation · Lift force

1 Introduction

As an important overpressure protection device, spring-loaded safety valves are widely used in the industrial field, which mainly relies on the elastic force of the spring and the lift force of the medium to control the opening and closing of the valve disc [1–3]. When the system pressure exceeds the predetermined safety threshold, the force exerted on the valve disc surpasses the spring's force, thereby

---

S. Wang
East China University of Science and Technology, Shanghai, China

X. Yu (✉)
East China University of Science and Technology, Key Laboratory of Pressure System and Safety (MOE), Shanghai, China
e-mail: yxhh@ecust.edu.cn

© The Author(s) 2026

Z. Zhang, P. Mercorelli (eds.), *Mechanical Design and Simulation: Insights and Innovations*, Lecture Notes in Mechanical Engineering,
https://doi.org/10.1007/978-981-95-2405-1_122

initiating the valve's opening to release excess pressure and prevent potential equipment damage. Once the system pressure reverts to a safe level, the spring reseats the disc, facilitating automatic closure. Yang Liu [4] carried out dynamic simulation of high-temperature and high-pressure safety valve and found that the pressure change in the back pressure chamber had a significant impact on the opening and reseating process of the safety valve. The response surface method was used to optimize the regulating mechanism, and a relationship model between the opening and closing pressure difference and the position of the regulating mechanism was established. Using the fluid-structure coupling method and the dynamic grid technique, Chen Lizhi [5] studied the influence of relevant parameters on valve instability and found that a longer pipe would increase the pressure loss and reduce the pressure at the nozzle, thus increasing the tendency of frequency hopping and flutter. Wang Rongyu [6] used different turbulence models to predict the valve disc lift and outlet flow of the safety valve with different disc openings. The results indicated that the lift error of the valve disc was smaller in the small opening, and the prediction accuracy of the SST(4 eq) turbulence model was the highest. Yan Xu [7] equivalent the main moving parts inside the safety valve as a mathematical model, and through fluid analysis and data fitting to obtain the lift of the valve disc, combined with the Runge Kutta method to obtain the opening and closing motion curve.

In this paper, ANSYS finite element simulation software is used to establish the internal flow channel model of the safety valve, and the external large container is connected. The whole process of opening, discharging and returning of the safety valve is simulated by using dynamic grid technology and user-defined function (UDF). Different spring stiffness is used to explore the effects of spring stiffness on the dynamic performance of spring-loaded safety valves.

2 Geometrical Model

The three-dimensional model of spring-loaded safety valve is shown in Fig. 1. Its main structure includes valve body, disc, stem, valve bonnet, spring, etc. and the medium is air at 30 degrees Celsius. The inlet diameter is 450 mm, the outlet diameter is 600 mm, the setting pressure is 0.2 MPa, the maximum stroke of the valve disc is 100 mm, and the spring stiffness is 180,532 N/mm.

Fig. 1 Safety valve model

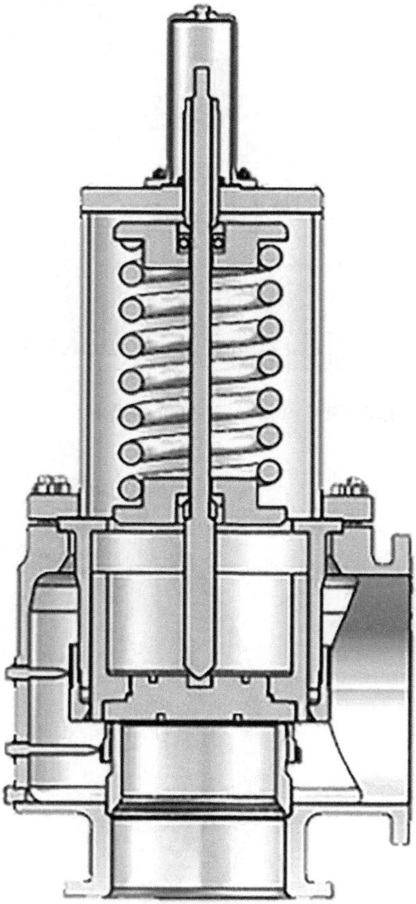

3 Mesh Generation and UDF Compilation

In order to ensure the efficiency of the calculation, the valve model needs to be simplified and some small structures can be omitted, such as the chamfering and groove design of the upper regulation ring. At the same time, an opening of 1.5 mm is reserved to simulate the flow field when it is closed to ensure the continuity of the flow field. The flow channel is extracted from the simplified model, and a large container with a diameter of 2 m and a height of 2 m is connected to the inlet section of the valve, and an equally large container is connected to the outlet of the valve to ensure the full flow of the medium. The inner flow path of the valve disc is complicated, so the unstructured grid is used to divide it. However, the motion of the valve disc needs to be simulated by layer-laying dynamic grid method near the valve disc, so the flow channel near the valve disc is cut out and divided by structural grid. The resulting grid model is shown in the Fig. 2.

Fig. 2 Grid of flow channel model

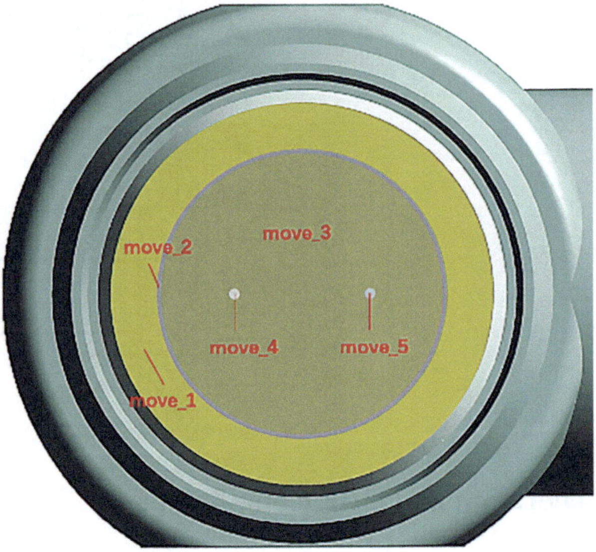

Fig. 3 Naming of sports surfaces

According to the flow channel model, the dynamic grid region was divided into 5 parts, whose upper surfaces were named move_1 (lower surface of the disc seat and outer surface of the lower surface of the disc), move_2 (lower surface ring of the disc), move_3 (inner surface of the lower surface of the disc), move_4, move_5 (surface of two holes at the bottom of the disc), as shown in Fig. 3. The begin_f_loop macro is used to traverse all the faces of elements of the moving surface to obtain the fluid force. The DEFINE_EXECUTE_AT_END macro is used to update the displacement, velocity and acceleration of the moving grid at the end of each time step to update the motion state of the disc in real time. The DEFINE_CG_MOTION macro is used to control the update of the dynamic grid.

4 Dynamics of Valve Disc

Due to the limiting effect of the guide sleeve, the moving parts such as the valve disc can only move back and forth in the vertical direction, so the moving parts can be regarded as a system with a single degree of freedom, and the moving parts are mainly subject to fluid force, spring force and gravity in the vertical direction. The forces on the moving parts are shown in the Fig. 4. According to Newton's second law, the motion of the disc can be described by a second-order differential equation.

$$m_{move}\ddot{y} = F_{flow} - F_{spring} - G_{move} \tag{1}$$

Where,

$$m_{move} = m_{disk} + m_{sleeve} + m_{stem} + m_{block} + m_{seat} + 0.5m_{spring} \tag{2}$$

$$F_{flow} = F_1 + F_2 + F_3 + F_4 + F_5 \tag{3}$$

$$G_{move} = \left(m_{disk} + m_{sleeve} + m_{stem} + m_{block} + m_{seat} + 0.5m_{spring}\right) \times g \tag{4}$$

$$F_{spring} = kx + F_0 \tag{5}$$

Where, m_{move}, m_{disk}, m_{sleeve}, m_{stem}, m_{block}, m_{seat}, m_{spring} are respectively the mass of moving parts, disc, disc seat, stem, pad, lower spring seat and spring. Because the upper end of the spring is fixed, the distance that the center of gravity of the spring changes is half of the moving distance of the moving part. In order to ensure the accuracy of the calculation, half of the spring mass is substituted into the calculation as the spring mass. F_{flow} is the fluid force, and F_1, F_2, F_3, F_4, F_5 are the fluid force subjected to move_1, move_2, move_3, move_4 and move_5, respectively. G_{move} is the total gravity of the moving parts, F_{spring} is the spring elasticity, k is the spring stiffness, x is the displacement of the valve disc, and F_0 is the pre-compression force of the spring.

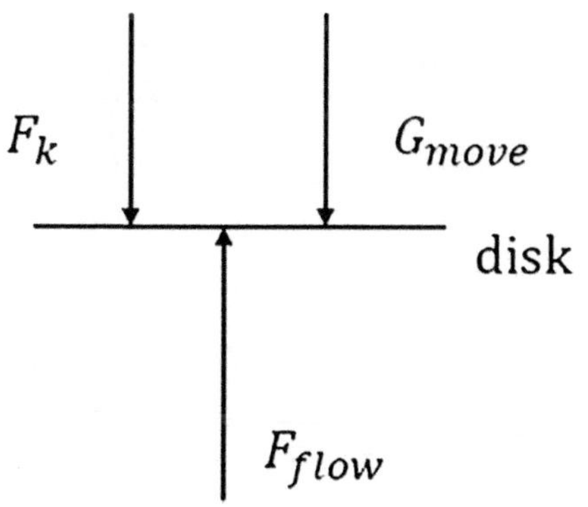

Fig. 4 Force analysis of moving parts

5 Fluent Setup and Grid Independence Verification

SST k-ω turbulence model is a two-equation eddy viscosity model [8], which adopts the combination of k-ω model and k-ε model, so that the calculation speed of SST k-ω model in the core region is fast, and the calculation accuracy is high near the wall [9], so SST k-ω model is selected for simulation. The inlet is set to the setting pressure of 0.2 MPa, the temperature of 303 K, and the outlet is set to 0 MPa, 300 K. The SIMPLE algorithm is used to calculate the steady state field of the flow channel as the initial condition of the transient simulation. PISO calculation can effectively deal with the coupling relationship between pressure and speed, and maintain numerical stability and high precision in the process of time stepping [10]. Therefore, the PISO algorithm is used to simulate the transient state of the valve. When simulating the valve opening process, keep the pressure inlet unchanged to ensure that the valve has enough pressure difference to open quickly. When the valve disc reaches the maximum stroke, the pressure inlet is changed to the wall boundary to simulate the stable discharge and seating process of the valve.

In order to ensure the accuracy of the calculation results, it is necessary to eliminate the influence of grid on the simulation results. The outlet flow rate is a very important parameter, so the outlet flow rate of the initial moment is selected as the index to verify the grid independence. The Table 1 shows the outlet flow when the valve is fully opened under different grid quantities, which can be seen when the number of grids reaches 113w, the outlet flow rate basically remains stable, so the number of grids of 113w is selected for subsequent simulations.

Table 1 Grid independence analysis

Grid number	Outlet flow(kg/s)
70w	1.183
88w	1.191
113w	1.199
151w	1.202
182w	1.201

6 Analysis and Discussion

A comparison of several key parameters obtained from the dynamic simulation of the safety valve is presented in the Table 2. It can be observed that as the spring stiffness is increased, both the opening time and the discharge time of the safety valve are significantly prolonged, while the disc velocity is reduced. This phenomenon is mainly due to the fact that higher spring stiffness creates a greater spring elastic resistance during opening. In addition, the return seat pressure increases as the stiffness increases.

The Fig. 5 shows the valve disc displacement curves for different spring stiffnesses (0.9 K, K, 1.1 K). The valve disc opening process can be divided into two stages, and the opening degree of 20 mm is a key turning point. Prior to this stage, the opening speed of the valve is slow and the change in spring stiffness has a negligible effect on this stage. However, when the opening exceeds 20 mm, the valve begins to open rapidly, and the smaller the spring stiffness, the faster the valve opens.

Table 2 Parameters of different spring stiffness

Spring stiffness	0.9 K	K	1.1 K
Open time (ms)	138.8	179	220.2
Maximum speed at opening (m/s)	2.54	2.14	1.73
Discharge time (ms)	226.6	259	293
Maximum speed at returning (m/s)	3.97	3.89	3.73
Reseating pressure (MPa)	0.139	0.144	0.151

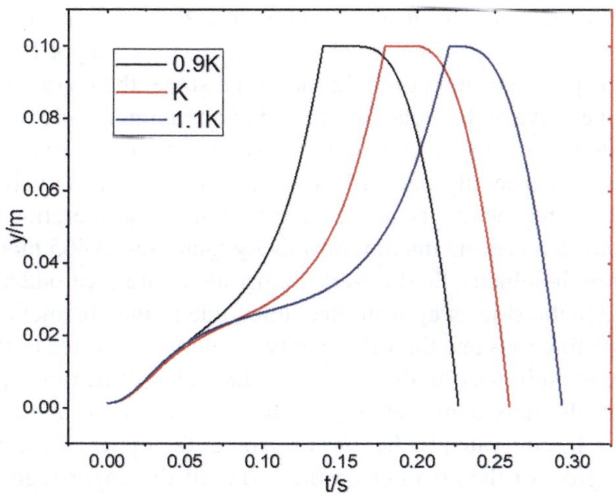

Fig. 5 Curve of disc displacement

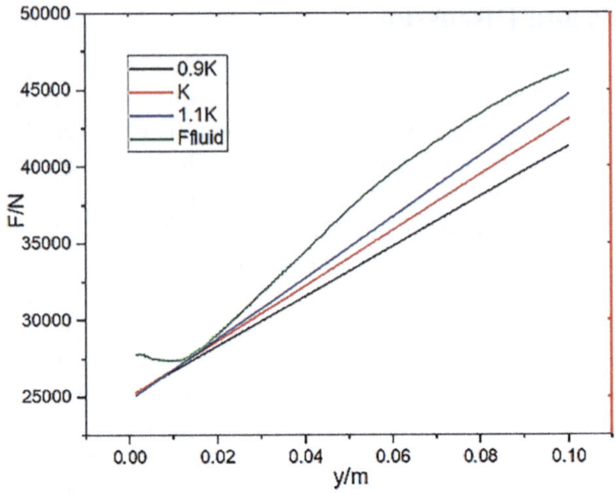

Fig. 6 Curves of forces

The curves of disc lift and spring force with disc opening for different spring stiffnesses are shown in Fig. 6. It can be observed that in the initial stage of opening, the lift force shows a slow decreasing trend, while the spring force increases linearly with the increase of opening. Therefore, the movement of the valve disc is more sluggish in the initial stage of opening. Due to the small opening at this stage, the change in spring stiffness has a small effect on the spring force, which means that the effect of spring stiffness on the initial stage of valve disc opening is relatively limited. Subsequently, as the opening increases further, the lift force begins to rise rapidly, resulting in rapid opening of the valve disc.

The velocity and pressure distributions within the valve are illustrated by the Fig. 7 at three distinct stages: the initial moment, at an opening of 10 mm, and at full opening, for a spring stiffness of *K*. In the initial stage, the medium flows through the throat of the valve and accelerates, reaching a maximum velocity of 366.3 m/s, which is located below move_2. Due to the small flux at this time, the medium is mainly sprayed horizontally outward by inertia, so most of the valve area has not been affected. As the opening is increased to 10 mm, the acceleration of the medium is further intensified and the maximum velocity increases to 405 m/s. At this point, the region of high-velocity fluid covers almost the entire area underneath move_1 and move_2. On the side away from the outlet, due to the obstruction of the upper and lower adjusting rods and the valve cavity, a significant vortex of fluid is formed, which may have an impact on the stability of the valve. When the opening reaches the maximum, the maximum velocity of the medium is 447.7 m/s, and the high velocity region has migrated to the lower region of the upper adjusting ring. Due to the blocking effect of the chamfer on the inside of the upper regulating ring, the direction of the fluid beam changes and the medium flows diagonally downwards along the chamfer. The reaction force generated by the flow beam makes the valve lift increase rapidly, in line with the rule of change of the valve lift.

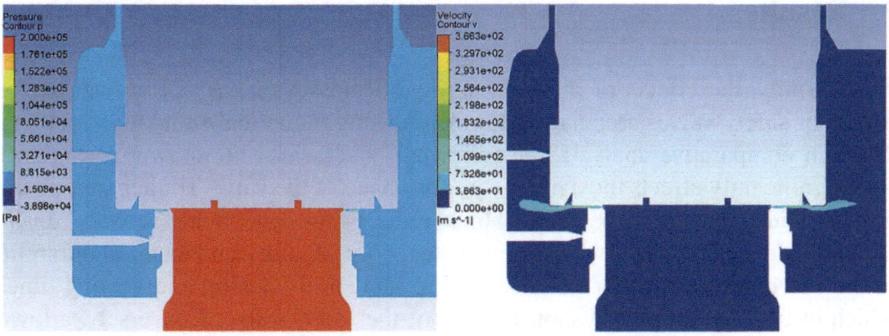

pressure distribution velocity distribution

(a) the initial moment

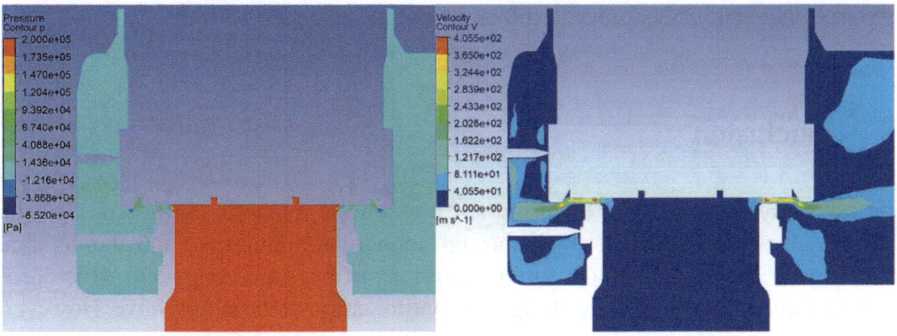

pressure distribution velocity distribution

(b) opening of 10mm

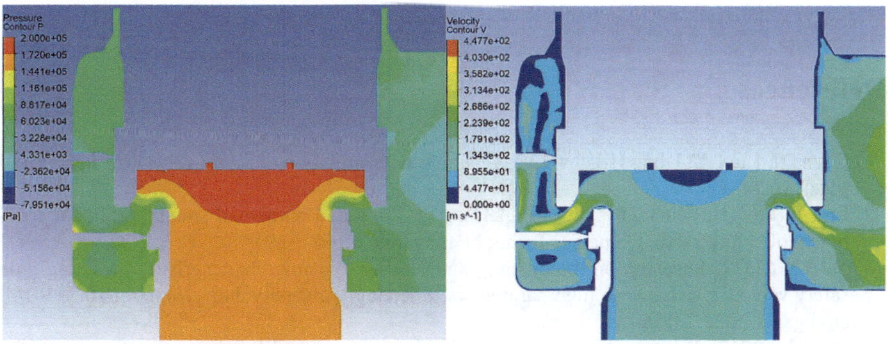

pressure distribution velocity distribution

(c) full opening

Fig. 7 Pressure and velocity distribution at different moments. (**a**) the initial moment. (**b**) opening of 10 mm. (**c**) full opening

7 Results

In this study, the effects of different spring stiffnesses on the operational performance of safety valves are analyzed in depth using numerical simulation methods. Through comparative analysis, the following conclusions are drawn: spring stiffness significantly affects the dynamic performance of the valve. Higher spring stiffness provides greater resistance, which reduces the maximum velocity during opening and seating back, which contributes to the stability and material durability of the valve. However, this can also lead to increased opening and venting times, which in turn reduces the responsiveness of the safety valve. Conversely, a lower spring stiffness, while increasing the sensitivity of the safety valve, can also lead to a lower return seating pressure, which may result in wasted media. Therefore, when designing a safety valve, it is necessary to find the right balance between responsiveness and stability in order to select the optimum spring stiffness.

8 Conclusion

In this study, the influence of spring stiffness on the dynamic performance of the safety valve is analyzed in detail through numerical simulation, and the detailed performance parameters under different spring stiffness are compared, which provides a theoretical basis for the design and optimization of the safety valve. However, this study is mainly based on numerical simulation, there may be some simulation errors, and the research results are based on specific safety valves with specific structures and parameters, so the results need to be further verified.

References

1. Chen DJ, Liu DK, Dong HB et al (2008) Numerical simulation of safety valve flow field. Fluid Mach 10:24–27
2. Luo H, Yu XH, Wang ZD (2008) Experimental study and numerical simulation of lift coefficient of safety valve. China Sci Technol Pap Online 08:587–591
3. Xu YX (2022) Research on the opening and closing dynamic characteristics of spring-type safety valve for steam with high parameters. Zhejiang University. https://doi.org/10.27461/d.cnki.gzjdx.2022.001560
4. Yang L (2017) Basic research on key technology of spring-loaded steam safety valve for nuclear power. East China University of Science and Technology
5. Chen LZ (2016) Research on dynamic instability of safety valve based on fluid-solid coupling. Dalian University of Technology
6. Wang RY (2020) Numerical simulation and optimization design of spring-loaded safety valve based on CFD. Dalian University of Technology. https://doi.org/10.26991/d.cnki.gdllu.2020.001842

7. Yan X, Xu J, Li ZG et al (2024) Simulation and analysis of opening and closing dynamic process of spring-loaded safety valve. J Harbin Inst Technol 29(01):25–34. https://doi.org/10.15938/j.jhust.2024.01.003
8. Menter FR (1994) Two-equation Eddy-viscosity turbulence models for engineering application. AIAA J 32(8):1598–1605
9. Zeng Y, Wang HB, Sun MB et al (2023) A review of SST turbulence modeling improvement studies. J Aeronaut 44(09):103–134
10. Issa RI, Gosman AD, Watkins AP (1986) The computation of compressible and incompressible recirculating flows by a non-iterative implicit scheme. J Comput Phys 62(1):66–82

Open Access This chapter is licensed under the terms of the Creative Commons Attribution-NonCommercial-NoDerivatives 4.0 International License (http://creativecommons.org/licenses/by-nc-nd/4.0/), which permits any noncommercial use, sharing, distribution and reproduction in any medium or format, as long as you give appropriate credit to the original author(s) and the source, provide a link to the Creative Commons license and indicate if you modified the licensed material. You do not have permission under this license to share adapted material derived from this chapter or parts of it.

The images or other third party material in this chapter are included in the chapter's Creative Commons license, unless indicated otherwise in a credit line to the material. If material is not included in the chapter's Creative Commons license and your intended use is not permitted by statutory regulation or exceeds the permitted use, you will need to obtain permission directly from the copyright holder.

Dynamics Modeling and Dynamic Response Analysis of Helicopter Tail Transmission System

Weiqiang Shi and Tengjiao Lin

Abstract The helicopter tail drive system constitutes a vital element of the helicopter's overall propulsion system, with a long drive chain and single-way power transmission. Rolling bearings, splines, diaphragm couplings, spiral bevel gears, and other components are all vibration transmission components or vibration sources within the helicopter's tail drive system, and there is also coupled vibration between the components, resulting in complex vibrational response of the tail drive system. This study employs the shaft segment element approach, using the Timoshenko beam model to establish shaft segment elements. Equivalent element models are established for rolling bearings, splines, diaphragm couplings, spiral bevel gears, etc., the stiffness, mass, and damping matrices for each component are determined. The comprehensive system matrix is derived through the assembly of individual element matrices, based on their installation interrelationships. The system's vibrational response is computed by addressing the differential formula governing the overall dynamics through the application of the Newmark-β method. This research contributes valuable insights for the dynamic modeling and resolution techniques of dynamic responses in helicopter tail drive systems.

Keywords Shaft segment unit · Transmission system · Dynamical model

1 Introduction

The tail transmission system of helicopters is often in high-speed operating conditions, with numerous transmission and coupling components such as bearings, splines, diaphragm couplings, spiral bevel gears, etc. The transmission shaft is a hollow shaft with high flexibility. The high speed and numerous transmission and coupling components make it difficult to model The performance dynamics of the

W. Shi · T. Lin (✉)
State Key Laboratory of Mechanical Transmission, Chongqing University, Chongqing, China
e-mail: tjlin1968@162.com

© The Author(s) 2026

Z. Zhang, P. Mercorelli (eds.), *Mechanical Design and Simulation: Insights and Innovations*, Lecture Notes in Mechanical Engineering,
https://doi.org/10.1007/978-981-95-2405-1_123

tail drive system, and the vibrational response is complex. As a result, it is vital to conduct a study on dynamic modeling and solution for the helicopter tail transmission system. At present, there have been numerous studies on the modeling of helicopter transmission systems. Wu [1] et al. employed matrix operation techniques to built a comprehensive Combined dynamic model of helicopter rotor and body. This model retains the nonlinearity of motion when calculating blade velocity and acceleration, overcoming the problems of decreased calculation accuracy and reduced applicability caused by traditional small angle linear simplification. Li [2] et al. built and verified 3D fine finite element simulation model of helicopter tail transmission system based on fine finite elements method considering complex nonlinear boundary conditions such as contact and deformation. Based on this, the research focuses on analyzing the transient stress distribution along the tail drive shaft and the transient vibrational activities of both the extended and shortened tail drive shafts. Chen [3] et al. established a high-precision response prediction model for tail transmission system of helicopter based on the theory of rigid flexible coupling multibody dynamics, and built a simulation test bench for verification. Hu [4] et al. used centralized parameter method and multibody dynamics software to model and model the standard components found in helicopter primary reducers. The vibration response traits during standard and common fault conditions were examined, taking into account various nonlinear influences. In essence, a plethora of researchers have explored the dynamic modeling and resolution of helicopter tail drive systems through theoretical analyses, simulations, and experimental studies. However, there is still limited research on dynamic modeling and response solution based on the shaft segment element method.

This study focuses on the helicopter tail drive system, developing a dynamic model utilizing the shaft segment element approach, and employs the Newmark-β technique to compute system's comprehensive dynamic differential equation, thereby deriving the system's dynamic response.

2 Axis Segment Unit Dynamic Model

The stress state of the axial segment element can be analyzed using the spatial beam element theory. Timoshenko beam and Euler beam are both commonly used classical beam elements in engineering. Compared with Euler beam, Timoshenko beam considers the influence of shear deformation more [5–7]. Therefore, this paper uses Timoshenko beam to construct axial segment elements. The two node axis segment unit is displayed in Fig. 1.

The axial segment unit is composed of two nodes, each possessing 6 degrees of freedom. By leveraging pertinent elasticity theories, the unit stiffness matrix K_b, can be formulated as:

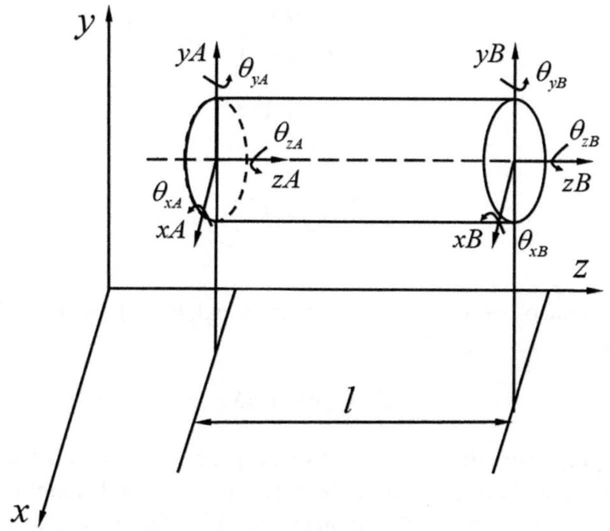

Fig. 1 Diagrammatic representation of the shaft segment unit

$$K_b = \begin{bmatrix} K_{b11} & K_{b12} \\ K_{b21} & K_{b22} \end{bmatrix} \tag{1}$$

$$K_{b22} = K_{b11} = \begin{bmatrix} a & 0 & 0 & 0 & b & 0 \\ 0 & a & 0 & -b & 0 & 0 \\ 0 & 0 & A_s E/l & 0 & 0 & 0 \\ 0 & -b & 0 & c & 0 & 0 \\ b & 0 & 0 & 0 & c & 0 \\ 0 & 0 & 0 & 0 & 0 & GJ/l \end{bmatrix} \tag{2}$$

$$K_{b21} = K_{b12} = \begin{bmatrix} -a & 0 & 0 & 0 & -b & 0 \\ 0 & -a & 0 & b & 0 & 0 \\ 0 & 0 & -A_s E/l & 0 & 0 & 0 \\ 0 & -b & 0 & d & 0 & 0 \\ b & 0 & 0 & 0 & d & 0 \\ 0 & 0 & 0 & 0 & 0 & -GJ/l \end{bmatrix} \tag{3}$$

In the aforementioned equation: E denotes the elastic modulus, G represents the shear modulus, A_s signifies the unit sectional area, and l indicates the unit length.

$$a = \frac{12EI}{l^3(1+\varphi)} \tag{4}$$

$$b = \frac{6EI}{l^2(1+\varphi)} \tag{5}$$

$$c = \frac{(4+\varphi)EI}{l(1+\varphi)} \tag{6}$$

$$d = \frac{(2-\varphi)EI}{l(1+\varphi)} \tag{7}$$

The damping matrix of the axial element is calculated by Rayleigh damping method as [8].

$$C_b = \xi M_b + \eta K_b \tag{8}$$

Where ξ and η represent the mass and stiffness proportionality coefficients.

Utilizing the matrices representing the stiffness, mass, and damping properties of the axial element, the motion differential equation for the dynamics of the axial element can be formulated as:

$$M_b \ddot{x}_b + C_b \dot{x}_b + K_b x_b = 0 \tag{9}$$

3 Global Dynamic Model of the Helicopter Tail Drive System

The helicopter tail drive system typically comprises shafts, bearings, splines, diaphragm couplings, gear pairs, and casings. Initially, this system is categorized and discretized into various nodes and elements based on its distinct components, and differential equations of motion are individually derived for different node elements. Then, the mass matrix, stiffness matrix and damping matrix of each element are assembled with the idea of finite element method to obtain the overall system matrix and differential equations of motion with displacement of each individual node is designated as generalized displacement. Thus, the intricate vibration issue of a system with unlimited degrees of freedom can be approximated by the simpler problem of a system with finite degrees of freedom. The system of helicopter tail transmission after unit division is displayed in Fig. 2.

Each axis is discretized into several units, and the node location and number are recorded according to the spatial location and installation relationship. The dynamic simulation framework for the tail transmission assembly of the helicopter is following the interconnection among nodes There are five types of units in the dynamic system model of helicopter tail drive system. Axis segment element: element formed by two adjacent nodes on the same axis; Meshing element: the element formed between two meshed gear nodes in a pair of gear pairs; Bearing element: element formed between bearing node and shaft node; Spline element: element formed

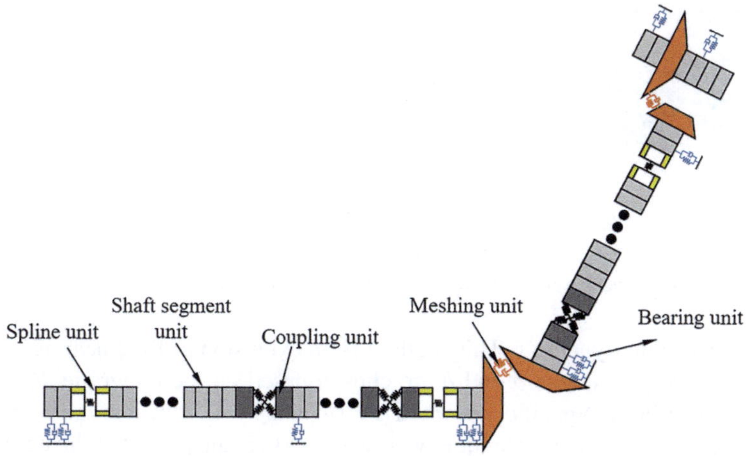

Fig. 2 Dynamics model of helicopter tail drive system

between spline node and axis node; Coupling element: element formed between coupling node and shaft node.

Assemble the overall system matrix after obtaining the element matrix. First, read the material and geometric information of the shaft, bearing, spline, coupling, spiral bevel gear, box, etc., solve the mass, damping, stiffness matrix and load vectors corresponding to each element, and then assemble the overall system damping, stiffness mass matrix through the correlation between element position and the global node number.

By leveraging the differential equations that dictate the dynamics of each individual element, it is possible to construct the differential equations that govern the dynamic behavior of the entire system [9, 10]:

$$M_{\text{tol}}\ddot{x}(t)+C_{\text{tol}}\dot{x}(t)+K_{\text{tol}}(t)x(t)=P+F_{\text{s}}(t) \tag{10}$$

Where: P denotes the vector of external loads acting on the system; F_s represents the column vector of mesh forces.

4 Dynamic Model Solution of Helicopter Tail Transmission Systeation

The generally applied numerical integration techniques encompass the central difference method, Rungekutta method, Houbrolt method, Newmark-β method, etc. Newmark-β method has good numerical stability. By choosing suitable parameters and step length, numerical instability can be effectively avoided. Hence, within this study, the Newmark-β technique is applied to address the differential equations

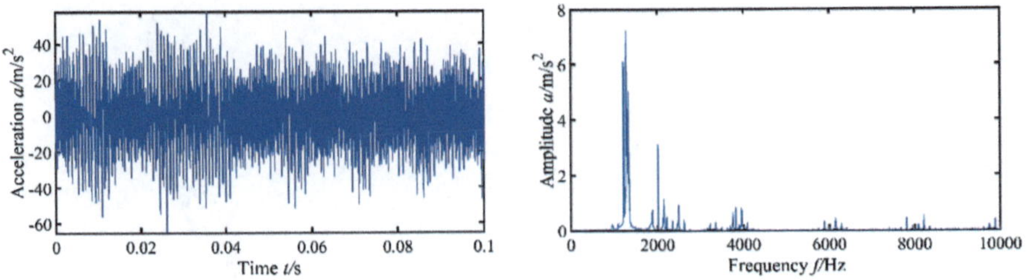

Fig. 3 Response curve of vertical vibration for the intermediate reducer

governing system dynamics. This method is an extension of the linear acceleration approach. The parameters α and β are chosen based on the integration's accuracy and stability criteria. Specifically, when $\alpha = 1/6$ and $\beta = 1/2$, the method corresponds to the linear acceleration technique; whereas $\alpha = 0.25$ and $\beta = 0.5$ indicate the average acceleration method. The algorithm ensures unconditional stability provided that $\beta \geq 0.5$ and $\alpha \leq 0.25(0.5 + \beta)^2$.

The whole system is discretized into 84 nodes by programming in MATLAB, the solution time is 0.1 s, the step length is 1 × 10–6 s, the set working condition is the input speed 4000 r/min, and the input torque is 5 Nm. The vibrational response of each node within the system is acquired. The dynamic response of meshing node is displayed in Figs. 3 and 4.

The vibrational response spectrum of the meshing node in the intermediate reducer exhibits a maximum amplitude at 2000 Hz, corresponding to its meshing frequency, followed by 1200 Hz associated with the meshing frequency of the tail reducer. Conversely, the mesh joint of the tail reducer shows a peak amplitude at 1200 Hz, aligning with its mesh frequency, with the second highest amplitude at 2000 Hz, matching the mesh frequency of the intermediate reducer. These findings are in agreement with the results presented in Reference [3], verifying the correctness of the dynamic model put forward in this study.

5 Conclusion

1. Utilizing the shaft section element methodology, a comprehensive kinematic portrayal of the helicopter's tail propulsion apparatus is constructed, encompassing essential elements like the shaft segment component, aimed at providing an enhanced depiction of the system's dynamic performance.
2. Utilizing the Newmark-β approach, the vibrational response for each of the 84 discrete nodes within the helicopter tail drive system is addressed. The correctness of the dynamic model presented in this paper is substantiated through a comparison with findings from established literature, wherein the primary frequency components depicted in the frequency domain diagram align consistently.

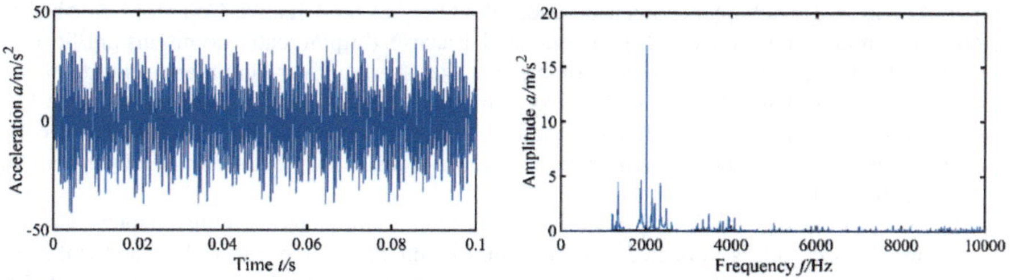

Fig. 4 Response curve of vertical vibration for the tail reducer

References

1. Wu J, Liu X, Song S et al (2022) Dynamic model of helicopter rotor/airframe coupling based on matrix operation. J Naval Aviat Univ 37(04):327–332. https://doi.org/10.7682/j.issn.2097-1427.2022.04.005
2. Li D, Hu Y, Yang G et al (2022) Research on vibration characteristics of helicopter tail drive system based on fine finite element method. Mech Res Appl 35(02):1–5. https://doi.org/10.16576/j.cnki.1007-4414.2022.001
3. Chen G, Tang Q, Li H et al (2024) Prediction modeling and test research on helicopter tail drive system response. J Aerodyn:1–15. https://doi.org/10.13224/j.cnki.jasp.20240364
4. Hu J, Hu N, Shen J et al (2022) Fault dynamics modeling and simulation of typical component of helicopter Main reducer. J Northeast Univ (Nat Sci Edition) 43(12):1732-1739+1747. https://doi.org/10.12068/j.issn.1005-3026.2022.12.009
5. Liu J, Li X, Pang R et al (2023) Dynamic modeling and vibration analysis of a flexible gear transmission system. Mech Syst Signal Process 197:110367. https://doi.org/10.1016/j.ymssp.2023.110367
6. Wei J, Zhang A, Qin D et al (2017) A coupling dynamics analysis method for a multistage planetary gear system. Mech Mach Theory 110:27–49. https://doi.org/10.1016/j.ymssp.2023.110367
7. Zhu H, Chen W, Zhu R et al (2021) Modeling and dynamic analysis of spiral bevel gear coupled system of intermediate and tail gearboxes in a helicopter. Proc Inst Mech Eng C J Mech Eng Sci 235(22):5975–5993. https://doi.org/10.1177/0954406221992798
8. Ma S, Jiang S, Chen Z (2017) Research on structural system and excitation identification based on IMPSCO and improved Newmark—β algorithm. Vib Shock 36(15):22–28. https://doi.org/10.13465/j.cnki.jvs.2017.15.004
9. Chang L, He C, Liu G (2016) Finite element method for dynamic modeling of parallel shaft gear drive system. Vib Shock 35:47–53. https://doi.org/10.13465/j.cnki.jvs.2016.20.008
10. Wu C, Dong Y (2021) Simulation study on longitudinal force of coupler of heavy-haul train based on improved Newmark-β method. J Railw Sci Eng:211–219. https://doi.org/10.19713/j.cnki.43-1423/u.T20200253

Open Access This chapter is licensed under the terms of the Creative Commons Attribution-NonCommercial-NoDerivatives 4.0 International License (http://creativecommons.org/licenses/by-nc-nd/4.0/), which permits any noncommercial use, sharing, distribution and reproduction in any medium or format, as long as you give appropriate credit to the original author(s) and the source, provide a link to the Creative Commons license and indicate if you modified the licensed material. You do not have permission under this license to share adapted material derived from this chapter or parts of it.

The images or other third party material in this chapter are included in the chapter's Creative Commons license, unless indicated otherwise in a credit line to the material. If material is not included in the chapter's Creative Commons license and your intended use is not permitted by statutory regulation or exceeds the permitted use, you will need to obtain permission directly from the copyright holder.

Design and Simulation Optimization of Mechanical Grippers

Anming Pan, Wang Li, Qiang Chen, and Kang Wu

Abstract To address the difficulty of grasping objects of various shapes and the low operational efficiency in traditional equipment, a vacuum-based robotic gripper utilizing negative pressure adsorption was designed, and its jaw mechanism was optimized. By constructing a 3D finite element model and performing dynamic simulations of the internal fluid flow within the gripper using CFD software under given boundary constraints, the flow field distribution of the internal fluid was obtained. The results indicated that the speed and negative pressure at the suction holes decreased progressively from the inside to the outside. Compared to H-type and U-type structures, the Y-type structure exhibited the highest flow velocity and negative pressure at the suction holes, making it more advantageous for workpiece adsorption. This vacuum-based gripper features a simple structure and low cost, providing a design foundation for solving grasping issues in mechanical equipment.

Keywords Mechanical gripper · Vacuum suction · Simulation analysis · Optimization design

1 Introduction

With the rapid development of smart manufacturing and industrial automation technologies, mechanical actuators are playing an increasingly important role in fields such as precision assembly, logistics sorting, and hazardous environment operations

A. Pan · Q. Chen
Zhejiang Guanghe Intelligent Technology Co., Ltd, Zhejiang, China
e-mail: wang.li@cwb.com.cn

W. Li
Cwb Automotive Electronics Co., Ltd, Zhejiang, China

K. Wu (✉)
Huzhou University, Huzhou, China
e-mail: 03313@zjhu.edu.cn

© The Author(s) 2026

Z. Zhang, P. Mercorelli (eds.), *Mechanical Design and Simulation: Insights and Innovations*, Lecture Notes in Mechanical Engineering,
https://doi.org/10.1007/978-981-95-2405-1_124

[1]. Among them, mechanical grippers are core components of end-effectors. In industrial production, mechanical grippers are driven to move stably to a specified location to quickly grasp workpieces. Additionally, using the gripper assembly, workpieces can be transferred to other locations, ensuring stable and orderly operation of industrial production. Therefore, the gripping performance of the mechanical gripper directly impacts the efficiency and reliability of the entire system. Traditional mechanical grippers often use rigid clamping methods, where the distance between two plates is reduced to squeeze and contact the material, completing the grasping task. However, this method cannot grasp irregularly shaped materials, leading to material slippage due to insufficient contact [2, 3]. Furthermore, rigid mechanical grippers exhibit poor adaptability and contact stress concentration when handling delicate parts or porous materials, reducing their application range.

In this context, vacuum-assisted clamping technology, by combining the advantages of contact clamping and non-contact adsorption, demonstrates unique application value. It allows for stable gripping through flexible contact and compensates for insufficient contact force with negative pressure adsorption. For example, Paul Glick and others at Stanford University [4] developed a soft robotic gripper that combines elastomeric sealing and a gecko-like adhesive surface. This adhesive surface enhances the gripper's performance by enabling stronger gripping, lower fluid pressure, and faster operation. Based on this, this paper designs a vacuum-based gripping mechanism that utilizes adsorption-assisted clamping to fix objects, overcoming the problem of object shape irregularities, and significantly improving accuracy and operational efficiency.

2 Numerical Simulation

2.1 Mechanical Gripper Structure and Working Principle

The mechanical gripper mainly consists of a clamping block, soft rubber, suction holes, and a support bracket, as shown in Fig. 1. During operation, the negative pressure generated by suction first adsorbs the workpiece, and then the clamping block action fixes the workpiece. The clamping block is equipped with soft rubber, which prevents damage to the workpiece while holding it. Since the soft rubber deforms when clamping and fixing, it allows for the fixation of irregularly shaped workpieces.

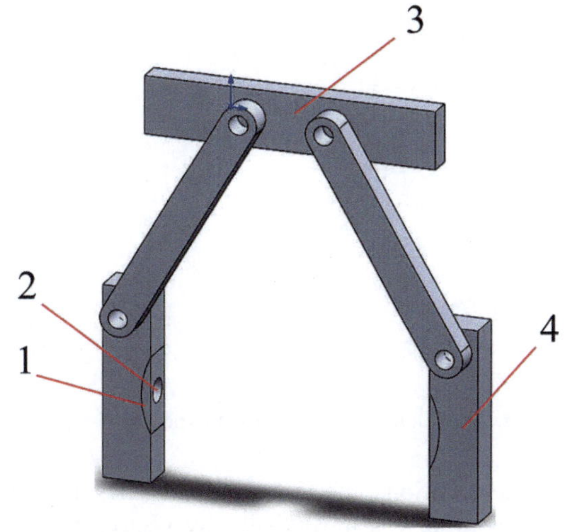

Fig. 1 The mechanical gripper. 1. Soft rubber 2. Suction hole 3.Support bracket 4.Clamping block

2.2 *Model Establishment*

Since the suction hole structure affects the adsorption performance, it is necessary to perform simulation optimization on the internal structure of the suction holes to improve the initial adsorption capacity of the workpiece. During numerical simulation, the size of the computational domain plays a crucial role. Many simulation cases have shown that an inappropriate computational domain can severely impact the accuracy of the simulation results [5, 6]. The computational domain was modeled in SolidWorks 2016, resulting in a 3D model of the suction hole. In this study, three different types of suction hole shapes were designed: Y-type, H-type, and U-type, as shown in Fig. 2.

2.3 *Governing Equation*

It is assumed that the fluid flow in the computational domain is steady, incompressible, three-dimensional turbulence, and the internal flow field must satisfy the mass conservation equation and the momentum conservation equation. Since the fluid is considered incompressible, the energy equation does not need to be considered. When the fluid is turbulent, the turbulence transport equation must also be followed. The mass conservation equation is given by [7].

$$\frac{\partial \rho}{\partial t}+\frac{\partial (\rho u)}{\partial t}+\frac{\partial (\rho v)}{\partial t}+\frac{\partial (\rho w)}{\partial t}=0 \tag{1}$$

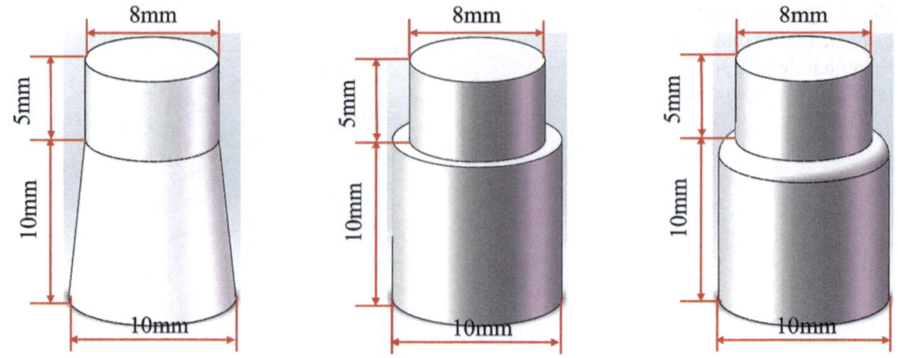

Fig. 2 Suction hole structure. (**a**) Y-type, (**b**) H-type, (**c**) U-type

In the equations, the following variables are defined: ρ means air density, kg/m^3; t means time, s; u means velocity vector, m/s; u, v and w mean components of the velocity vector in the x, y, and z directions, respectively, m/s.

2.4 Grid Generation and Boundary Conditions Setup

Grid generation significantly affects the convergence speed, solution accuracy, and computation time of numerical simulations. A finer mesh helps improve simulation accuracy but also increases computational costs. Therefore, the quality of the grid plays a crucial role in the success of numerical simulations [8, 9]. In this study, the ANSYS pre-processing software ICEM was used to generate the mesh for the airflow field inside the gripper. To enhance the accuracy of the simulation results, local mesh refinement was applied in the region where the flow field's inlet and outlet meet. The specific mesh generation results are shown in Fig. 3.

In turbulence modeling, the standard k-epsilon model is chosen in this study due to its good robustness and computational efficiency. Therefore, the standard k-epsilon model with standard wall functions is employed to calculate the turbulent flow characteristics of the suction system. As for the boundary conditions, the system inlet was set as a pressure inlet with a pressure value of 0 kPa, and the outlet used a pressure outlet boundary condition to ensure the stability and accuracy of the flow field calculation.

Fig. 3 Grid generation

2.5 *Solution Settings*

The simulation of the internal structure of the suction hole is performed on the ANSYS Fluent platform, where the velocity and pressure fields are coupled using the semi-implicit method in the pressure correction method. The interpolation schemes for the diffusion term, source term, and convection term all adopt the first-order upwind difference scheme for convective motion. The calculation used the Reynolds stress model, and the residual convergence accuracy was set to 10^{-6}.

3 Simulation Results Extraction and Discussion

3.1 *Velocity Flow Field Distribution with Different Structural Parameters*

To compare the flow field distribution characteristics under different structural parameters, the longitudinal section at X = 0, located at the midpoint, was set as the flow field monitoring plane, primarily monitoring the average pressure variation of the fluid. The velocity flow field distribution results for different suction hole

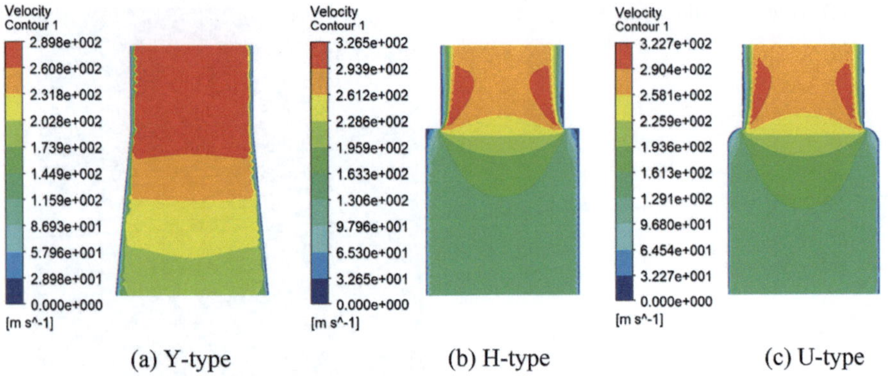

Fig. 4 The velocity flow field distribution results for different suction hole structures. (**a**) Y-type, (**b**) H-type, (**c**) U-type

structures are shown in Fig. 4. From Fig. 4, it can be seen that, compared with the other two structures, the internal flow velocity variation of the Y-type structure is more pronounced. Overall, the flow velocity decreases gradually from the inside to the outside of the suction hole. Among the three suction hole structures, the Y-type structure exhibits the highest flow velocity (180.32 m/s) at the suction hole, while the flow velocities of the H-type (154.18 m/s) and U-type (156.96 m/s) structures are quite similar. The greater the flow velocity, the stronger the adsorption capacity, which is favorable for the suction hole's adsorption of the workpiece.

3.2 Pressure Flow Field Distribution with Different Structural Parameters

The pressure flow field distribution results for different suction hole structures are shown in Fig. 5. From Fig. 5, it can be observed that, overall, as the fluid moves, the negative pressure gradually decreases from the inside to the outside. The H-type structure shows a significant pressure drop at both ends of the diameter change, which is caused by the formation of vortices in this region, increasing energy consumption. Due to the air resistance and vortex formation within the cavity, the U-type structure consumes the most energy, resulting in the most noticeable pressure variation at the suction hole. By calculating the negative pressure at the suction hole, it can be determined that the negative pressure for the H-type structure is 14,560 Pa, and for the U-type structure, it is 14,885 Pa. In comparison to these two structures, the Y-type structure exhibits the highest negative pressure (20,009 Pa) at the suction hole. The greater the negative pressure, the stronger the adsorption capacity, making it easier to achieve firm adsorption of the workpiece. Combining the results from Figs. 4 and 5, it is evident that the Y-type structure has the highest

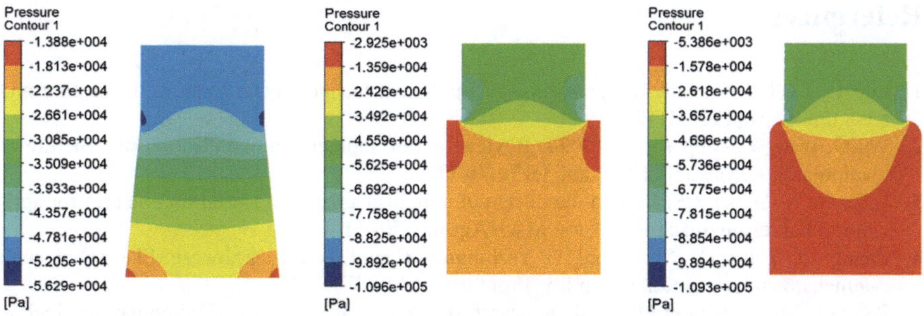

(a) Y-type (b) H-type (c) U-type

Fig. 5 The pressure flow field distribution results for different suction hole structures. (**a**) Y-type, (**b**) H-type, (**c**) U-type

flow velocity and negative pressure at the suction hole, and therefore the best adsorption capacity compared to the other two structures.

4 Conclusions

This study designs an air-suction gripping mechanism, which achieves object fixation through adsorption-assisted clamping. Compared to traditional mechanical clamping methods, air-suction gripping utilizes the controllability and flexibility of airflow, effectively overcoming issues related to irregular object shapes or uneven surfaces, ensuring the stability and safety of the object during the gripping process. During gripping, the air-suction mechanism adjusts the suction force to accommodate objects of different shapes and sizes, avoiding problems of object damage or unstable clamping that can arise with traditional methods.

The study uses the finite element method to simulate the fluid flow field inside the suction hole, analyzing the pressure and velocity distribution characteristics of different suction hole structures. Through simulation analysis, it was found that the flow field inside the suction hole shows significant differences in various regions, especially with the flow velocity and negative pressure gradually decreasing from the inside to the outside of the suction hole. Specifically, compared to the H-type and U-type structures, the Y-type structure exhibits the highest flow velocity and negative pressure at the suction hole, making it more capable of adsorbing workpieces. These results provide a theoretical basis for the optimization design of air-suction gripping mechanisms.

References

1. Luo H, Li R, Li Q (2024) Application of intelligent technology in robot arm grasping and recognition. Appl IC 41(12):8–9
2. Wu Q, Wang Z, Yang Y et al (2024) Review on development and application of structure of manipulator. Aeronaut Sci Technol 35(5):60–73
3. Li L, Gu X (2024) Research on the design of a multi-degree-of-freedom robotic arm for agricultural harvesting machines. Agric Mach Agron 4:50–52
4. Zhang Q, Fang H, Xu J (2021) Yoshimura-origami based earthworm-like robot with 3-dimensional locomotion capability. Front Robot AI 8:738214
5. Peng G, Xu M, Ji G et al (2024) Research on wear characteristics of combination pipeline based on CFD-DEM method. Water Resour Power 42(12):72–76
6. Zhang Y, Kacira M (2022) Analysis of climate uniformity in indoor plant factory system with computational fluid dynamics (CFD). Biosyst Eng 220:73–86
7. Norton T, Sun DW (2006) Computational fluid dynamics (CFD)-an effective and efficient design and analysis tool for the food industry: a review. Trends Food Sci Technol 17:600–620
8. Olatunde G, Atungulu GG, Sadaka S (2016) CFD modeling of air flow distribution in rice bin storage system with different grain mass configurations. Biosyst Eng 151:286–297
9. Zhang C, Kang J, Zhang N et al (2023) Design and experiment of seed separation tray of air-suction roller dibbler for peanut. Trans CSAM 54(06):28–37

Open Access This chapter is licensed under the terms of the Creative Commons Attribution-NonCommercial-NoDerivatives 4.0 International License (http://creativecommons.org/licenses/by-nc-nd/4.0/), which permits any noncommercial use, sharing, distribution and reproduction in any medium or format, as long as you give appropriate credit to the original author(s) and the source, provide a link to the Creative Commons license and indicate if you modified the licensed material. You do not have permission under this license to share adapted material derived from this chapter or parts of it.

The images or other third party material in this chapter are included in the chapter's Creative Commons license, unless indicated otherwise in a credit line to the material. If material is not included in the chapter's Creative Commons license and your intended use is not permitted by statutory regulation or exceeds the permitted use, you will need to obtain permission directly from the copyright holder.

BY NC ND

Development of a High-Pressure Temperature-Controlled Geotechnical Triaxial Testing System

Tengfei Ma, Yong Wang, Pengfei Jia, Guofang Xu, and Jiakai Chen

Abstract The deep-water temperature and pressure environments and their variations significantly influence the mechanical properties of sediments. To enable laboratory experimental studies on deep-sea sediment soils, an independently developed high-pressure temperature-controlled geotechnical triaxial testing system capable of precise volumetric strain measurement has been established. Building upon conventional geotechnical stress-path triaxial testing systems, this innovative apparatus achieves uniform and accurate temperature control through a dual internal and external water bath circulation design. A dual-pressure chamber with internal and external pressure balance enables precise measurement of specimen volumetric strain under high-pressure stress loading and temperature-controlled conditions. The temperature control testing of the experimental system demonstrates that the developed high-pressure temperature-controlled triaxial experimental system can maintain precise and stable temperature control over extended periods. This experimental apparatus provides essential technical support for conducting deep-water soil mechanics research in laboratory settings.

Keywords High-pressure and temperature-controlled system · Geotechnical triaxial system · Dual water bath circulation · Dual pressure chambers · Volumetric strain

T. Ma
Department of Geology, State Key Laboratory of Continental Dynamics, Northwest University, Xi'an, China

State Key Laboratory of Geomechanics and Geotechnical Engineering Safety, Institute of Rock and Soil Mechanics, Chinese Academy of Sciences, Wuhan, China

Y. Wang (✉) · G. Xu · J. Chen
State Key Laboratory of Geomechanics and Geotechnical Engineering Safety, Institute of Rock and Soil Mechanics, Chinese Academy of Sciences, Wuhan, China
e-mail: wangyong@whrsm.ac.cn

P. Jia
Department of Geology, State Key Laboratory of Continental Dynamics, Northwest University, Xi'an, China

© The Author(s) 2026

Z. Zhang, P. Mercorelli (eds.), *Mechanical Design and Simulation: Insights and Innovations*, Lecture Notes in Mechanical Engineering,
https://doi.org/10.1007/978-981-95-2405-1_125

1 Introduction

With the development and utilization of offshore energy and resources in China in recent years, marine soil mechanics has gradually become a research field attracting much attention [1]. Conducting research on the geotechnical mechanical properties of deep-sea seabed sediments is essential for the development of marine soil mechanics, the exploitation of deep-sea resources and energy [2]. Internationally, the seabed with a depth greater than 300 m and less than 1500 m is generally considered the deep-sea area. The deep-sea environment has typical characteristics of high pressure (3 MPa to 15 MPa) and low temperature (−1 to 4 °C). Obtaining the mechanical properties of deep-sea sediments under laboratory conditions is one of the essential research methods [3, 4]. However, currently, conventional indoor geotechnical triaxial test equipment usually has limitations such as a small pressure-bearing range, no temperature control or low temperature-control accuracy, making it difficult to reproduce the temperature and stress conditions of the sediments in the deep-sea environment.

Overseas research on triaxial equipment with controllable temperature and pressure started relatively early. Demars et al. [5] developed a temperature-controlled triaxial apparatus to study the volumetric strain of soil samples under temperature cycling. The temperature control was achieved through an external water-bath temperature-control device and an internal temperature-control metal coil. However, during the testing process of this device, the real-time water temperature was significantly affected by the ambient temperature. Sultan et al. [6] developed a temperature-controlled triaxial device based on GDS (Geotechnical Digital Systems) in the UK to study the influence of temperature on the volumetric strain caused by the pre-consolidation pressure of clay. The temperature control of this device was realized by heating coils installed outside the pressure chamber. The maximum temperature could reach 100 °C, and the maximum pressure could reach 12 MPa. The integrated triaxial experimental equipment for the synthesis and mechanical property testing of hydrate-bearing sediments developed by Heriot-Watt University in the UK [7] controls the temperature by connecting the temperature-control coil installed on the base of the pressure chamber to an external low-temperature thermostat. The temperature-control range of this equipment is −20 to 50 °C, and the maximum confining pressure can reach 40 MPa. In China, professor Chen Zhenghan et al. [8] from the Logistical Engineering University developed the first domestic temperature-controlled triaxial apparatus. This device controls the temperature by placing the triaxial pressure chamber in a large constant-temperature water-bath. The maximum temperature is 100 °C, the temperature fluctuation range of the water-bath is ±1 °C, the maximum pressure is 5 MPa, and the volumetric strain of the specimen is calculated through the change in the volume of the liquid in the confining-pressure chamber. Cai Guoqing et al. [9] developed a temperature-controlled unsaturated soil triaxial apparatus. The temperature is controlled by two circular electric heating plates installed on the inner wall of the pressure chamber, and the temperature in the pressure chamber is feedback-regulated by a temperature controller. The temperature range is 20 to 100 °C, the pressure range is 0 to 1 MPa, and the volumetric strain is measured by measuring the change in the volume of back-pressure water. Yang Dehuan et al. [10] developed a combined testing system for the hydro-mechanical properties of natural gas hydrate-bearing soil. The temperature is controlled by placing the entire pressure chamber in a constant-temperature water-bath. The maximum loading pressure of the system is 30 MPa, with an accuracy of ±0.03 MPa. The temperature-control range of the constant-temperature water-bath is −15 to 30 °C, and the temperature-control accuracy of the water-bath is ±0.1 °C.

The current temperature-control methods of temperature-controlled triaxial devices mainly include four types: (1). Temperature control inside the pressure chamber. The temperature-control accuracy of a single pressure chamber is easily affected by changes in the ambient temperature, leading to poor temperature-control accuracy. (2) Temperature control outside the pressure chamber. Since the heating elements are far from the specimen and the heat-transfer distance is long, the temperature of the specimen cannot be accurately controlled and this method is not suitable for low-temperature conditions. (3) Placing the pressure chamber in a large-scale temperature-control box or temperature-control room for temperature control. Heat needs to be indirectly transferred to the specimen through the wall of the pressure chamber and the liquid inside the pressure chamber. These result in less-than-ideal temperature-control accuracy for the specimen. (4) Combined internal and external temperature control. However, it requires simultaneous heating and temperature control of the pressure chamber, the force-transmitting rod and the piston rod, which is difficult to operate in practice. At the same time, it is not suitable for low-temperature control requirements. In general, the existing temperature-controlled triaxial equipment has the following problems: The temperature-control accuracy is insufficient over a wide temperature range for a long time.

In view of the above-mentioned problems, this paper by introducing a double-water-bath circulation temperature-control system for the inner and outer pressure chambers, a high-pressure temperature-controlled geotechnical triaxial test system is developed. This system can achieve stable temperature control and precise measurement of volumetric strain.

2 Composition and Characteristics of the Experimental System

The test system mainly consists of a high-pressure triaxial pressure chamber, a loading frame, a water-bath circulation temperature-control box, a pressurized plunger pump, a PC host computer, as well as various auxiliary equipment and sensors. The size of the test samples can be set to Φ50 mm × H100 mm and Φ50 mm × H200 mm. The system controls the confining pressure, pore pressure, axial load, and temperature respectively. It can achieve a maximum confining pressure loading of 15 MPa, a maximum axial load of 50 kN with an accuracy of ±0.1 N, a temperature-control range of - 10 to 80 °C, and a temperature-control accuracy of ±0.2 °C.

2.1 Composition of Experimental System

The overall structural design and hardware device composition of the high-pressure temperature-controlled geotechnical triaxial test system is shown in Figs. 1 and 2.

The high-pressure temperature-controlled geotechnical triaxial test system adopts an overall modular and compact structural design [11]. It is mainly composed of a high-pressure triaxial module, a water-bath circulation temperature-control module, a high-pressure loading module, an experiment control and data acquisition module, an auxiliary equipment module, etc.

1. High-pressure triaxial module: This module is mainly composed of inner and outer pressure chambers (D1-D2), a loading frame (B1: KD-50 type), inner and outer water-bath circulation pipelines (P1-P2), an underwater load sensor (C5), inner and outer temperature sensors (T1-T2), inner and outer confining-pressure sensors (C2-C3), a back-pressure sensor (C1), a pore-pressure sensor (C4), and the pipelines of each system. It is the most important component of the developed geotechnical triaxial system.
2. Water-bath circulation temperature-control module: It is mainly composed of a constant-temperature water-bath tank (W), a temperature control and adjustment panel, inner and outer water-bath circulation pipes (P1-P2), and temperature

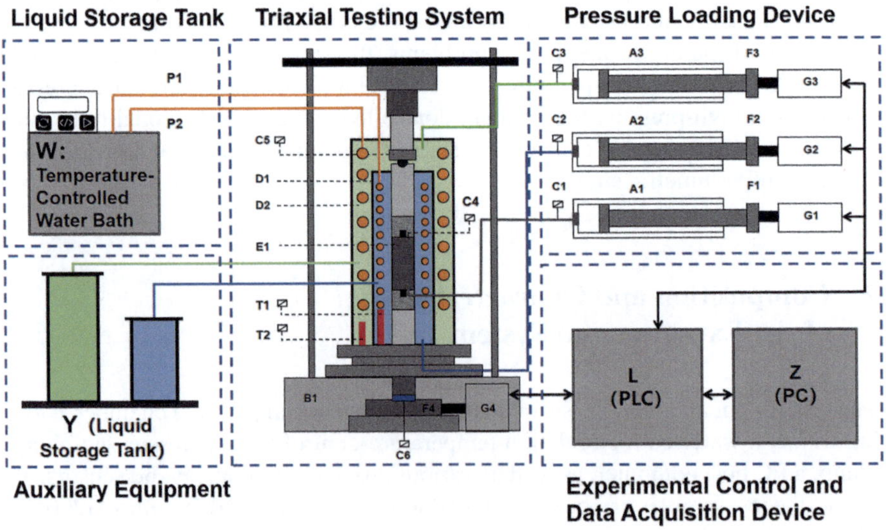

Fig. 1 High-pressure temperature-controlled geotechnical triaxial test system overall structure diagram

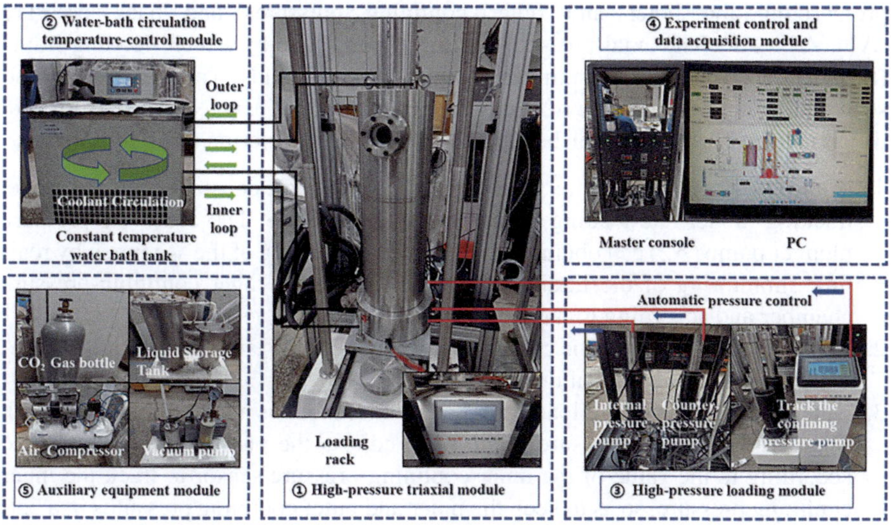

Fig. 2 Composition of the hardware device of the triaxial test system

sensors (T1-T2) for the inner and outer pressure chambers. The water-bath circulation temperature-control system precisely regulates the temperature of the pressure chambers through the water-bath temperature-control tank and the double water-bath circulation pipelines in the inner and outer pressure chambers, thereby controlling the temperature of the specimen.

3. High-pressure loading module: It is mainly composed of an outer confining-pressure tracking pump (A3, with a measuring range of 70 MPa and an accuracy of ±0.01 MPa), an inner confining pressure pump, and a back-pressure pump (A2-A1, with a measuring range of 30 MPa and an accuracy of ±0.01 MPa).
4. Experiment control and data acquisition module: It is mainly composed of a PC host computer (Z) and a general controller PLC (L: Programmable Logic Controller).
5. Auxiliary equipment module: The auxiliary equipment mainly includes liquid-adding and liquid-storing tanks for the inner and outer pressure chambers, CO_2 gas cylinders, air compressors, vacuum pumps, etc.

2.2 *Implementation of High Pressure Loading Module*

The high-pressure loading module mainly includes the loading of inner and outer confining pressures, back pressure, and axial pressure. It can achieve a maximum confining pressure and back-pressure loading of 15 MPa, which can meet the high-pressure loading requirements of the deep-sea environment at a maximum depth of 1500 m. High-precision automatically controlled plunger pumps are used for pressurizing the inner and outer confining pressures and back pressure (A1、A2、A3). The axial-pressure loading is controlled by a force-loading frame (B1) (KD-50 type, with a maximum loading force of 50 kN. The underwater load sensor has a measuring range of 10 kN and an accuracy of ±0.1 N, and sensors with different measuring ranges can be replaced according to experimental requirements).

1. Confining-pressure loading: The control modes include constant-pressure mode, tracking mode, rate-loading mode, and manual mode. The confining-pressure plunger pump (A2) can obtain the volumetric strain data of the specimen by real-time monitoring of the liquid volume change in the inner confining-pressure chamber and according to the corrected calibration curve.
2. Back-pressure loading: The control modes include constant-pressure mode, rate-loading mode, and manual mode.
3. Outer confining-pressure loading: The tracking mode is adopted for the control. The outer confining-pressure pump (A3) adjusts the pump speed in real-time according to the value of the inner confining-pressure sensor to track the inner confining pressure, so as to keep the inner and outer confining pressures in a balanced state.
4. Axial loading: The axial loading is mainly carried out by the loading frame (B1). The control modes include stress-loading mode, strain-loading mode, and manual mode.

2.3 Implementation of Efficient Temperature Control Module

The designed double-pressure-chamber water-bath circulation temperature-control module system is shown in Fig. 3.

The temperature-control system is mainly composed of a water-bath circulation temperature-control box (model: THD-2030; temperature range: −20 to 100 °C, accuracy: ±0.05 °C), water-bath circulation inlet and outlet pipelines, water-bath circulation pipelines for the inner and outer pressure chambers, and temperature sensors for the inner and outer pressure chambers (PT-100 temperature sensors, temperature-measurement range: −50–200 °C, accuracy: ±0.15%). The temperature-control range is −10–80 °C, and the temperature-control accuracy is (±0.2 °C).

As shown in Fig. 3, the water-bath circulation pipes in the inner pressure chamber are evenly distributed around the specimen, enabling the specimen to be heated or cooled evenly. The water-bath circulation pipes in the outer pressure chamber are evenly distributed inside the pressure chamber, allowing for uniform heating or cooling inside the outer pressure chamber. Temperature sensors are installed in both the inner and outer pressure chambers. During the test, the temperature fluctuations in the inner and outer pressure chambers can be monitored simultaneously, and feedback adjustments can be made in a timely manner. When the outer pressure chamber reaches the set temperature and maintains the temperature continuously, it has a good heat-preservation effect on the inner pressure chamber, reducing the

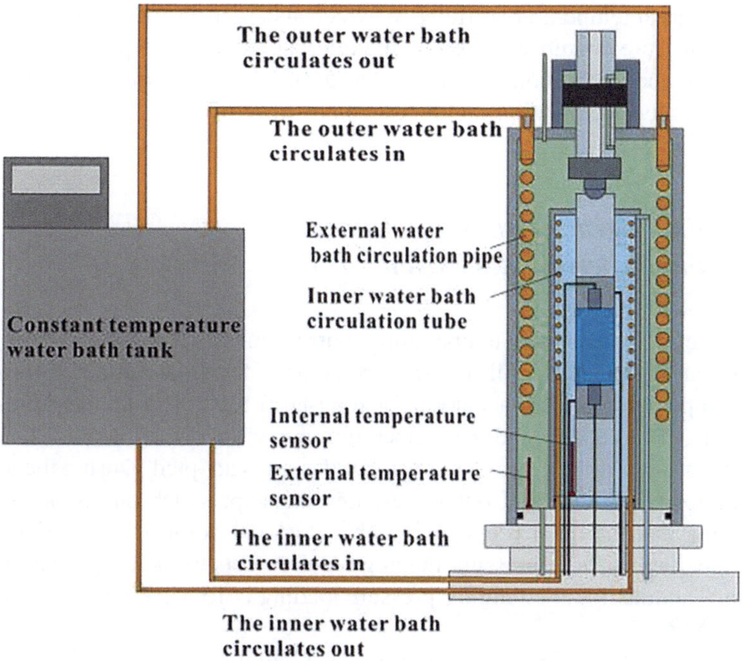

Fig. 3 Dual pressure chamber water bath circulation temperature control system

temperature fluctuations in the inner pressure chamber caused by outdoor temperature changes and maintaining stable temperature control for a long time. Compared with traditional temperature-control methods, the efficiency and temperature-control accuracy of the water-bath circulation temperature-control system are significantly improved.

2.4 Improvement of Volumetric Strain Measurement Module

The volumetric strain measurement module is improved based on the volumetric strain measurement scheme proposed by Wheeler et al. [12]. A design scheme with isolated inner and outer double-pressure chambers is adopted. Each of the inner and outer pressure chambers is externally connected to a plunger pump. The inner confining-pressure pump provides the confining pressure, and the volumetric strain of the specimen is measured through the volume of the inflowing and outflowing liquid. The outer confining-pressure pump is set in the pressure-tracking mode for pressurization to maintain real-time pressure balance with the inner pressure chamber, effectively reducing the self-deformation of the inner pressure chamber caused by the internal-external pressure difference. On the other hand, the inner and outer pressure chambers are temperature-controlled simultaneously to maintain a small internal-external temperature difference, which also reduces the self-deformation of the inner pressure chamber caused by the internal-external temperature difference. Through the above-mentioned methods, the accuracy of the volumetric strain measurement of the specimen during the test can be improved.

2.5 Main Characteristics and Indicators of the Experimental System

1. An external water-bath temperature-control box and an inner-outer double water-bath circulation coil temperature-control design are adopted to improve the temperature-control accuracy for long-term tests. The temperature-control range is - 10 to 80 °C, and the temperature-control accuracy is ±0.2 °C.
2. An inner-outer double-pressure chamber design is adopted. During the test pressurization, the inner and outer pressures are kept synchronized in real-time, ensuring that the inner pressure chamber does not deform under high pressure and meeting the high-pressure loading requirements of the deep-water environment. The maximum confining-pressure loading is 15 MPa, with an accuracy of ±0.01 MPa.
3. The volumetric strain is measured by calculating the volume of the inflowing and outflowing liquid of the inner confining-pressure pump. With the inner-outer double-pressure chamber design, achieving high-precision volumetric strain measurement of the specimen under temperature and pressure control.

3 Performance Testing and Calibration of Experimental Systems

3.1 Testing of Dual Pressure Chamber Water Bath Temperature Control System

A test on the heating, cooling, and heat-preservation performance of the water-bath temperature-control system was carried out. A cooling test (target temperature: −5 °C) is shown in Fig. 4, and a heating test (target temperature: 50 °C) is shown in Fig. 5.

Figure 4 shows the cooling performance test curve of the water-bath circulation temperature-control system. The set temperature is −5 °C, and the laboratory ambient temperature is 10 °C. The antifreeze in the water-bath temperature-control device reaches −5.3 °C after 2 h. At the 60th minute, the water-bath circulation is turned on to start cooling the inner and outer pressure chambers simultaneously. After another 4 hours and 40 minutes, the inner pressure chamber reaches the set temperature, with an average temperature drop of 0.535 °C every 10 minutes. The

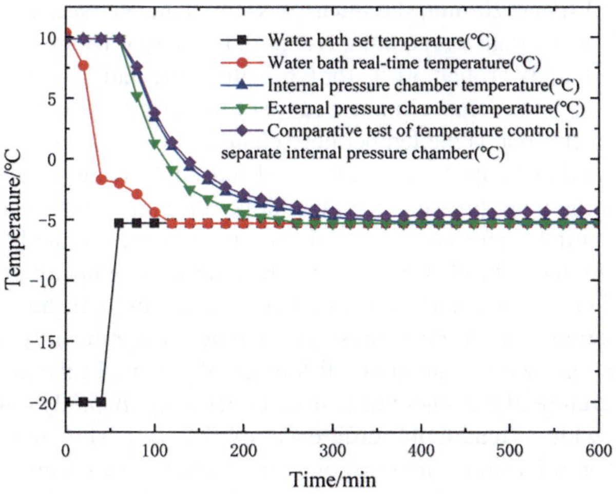

Fig. 4 Cooling test of −5 °C water bath circulation temperature control system

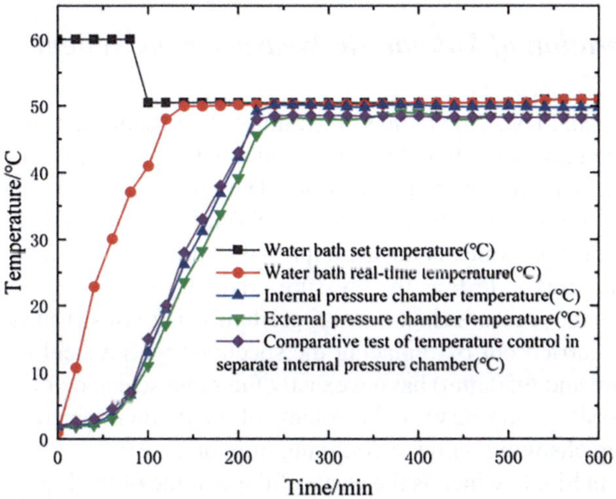

Fig. 5 Temperature rise test of 50 °C water bath circulation temperature control system

temperature is maintained steadily around −5 °C for 4 h and 20 min, with a fluctuation within ±0.2 °C.

Figure 5 shows the heating performance test curve of the water-bath circulation temperature-control system. The set temperature is 50 °C, and the laboratory ambient temperature is 2 °C. The antifreeze in the water-bath temperature-control device reaches 50 °C after 2 h and 20 min. At the 40th minute, the water-bath circulation is turned on to start heating the inner and outer pressure chambers simultaneously.

After another 3 h and 20 min, the inner pressure chamber reaches the preset temperature, with an average temperature rise of 2.4 °C every 10 min. The temperature is maintained steadily around 50 °C for 6 h, with a fluctuation within ±0.2 °C.

In Figs. 4 and 5, the comparison test curve of the temperature control of the inner pressure chamber alone is the temperature change curve when only the inner pressure chamber and its temperature-control coil is used. When using the temperature-control coil of the inner pressure chamber alone for temperature control during the cooling and heating performance tests, due to the influence of the laboratory ambient temperature, it can be clearly seen that the long-term temperature-control accuracy is poor when only the inner pressure chamber coil is used. In the low-temperature test, the temperature of the inner pressure chamber is significantly higher than the preset temperature, with a maximum difference of 0.7 °C. In the high-temperature test, the temperature of the inner pressure chamber is significantly lower than the set temperature, with a maximum difference of 1.7 °C. This indicates that the temperature-control design with inner and outer double water-bath circulation pipelines significantly improves the temperature-control stability and accuracy for long-term tests compared with the single temperature-control pipeline.

3.2 *Calibration of Volumetric Measurement System*

During the triaxial test, the volumetric strain of the unsaturated specimen is calculated by the volume of the liquid flowing in and out of the confining-pressure pump when the system applies confining pressure. The calibration of the volumetric strain measurement system refers to the methods of Yan Rongtao [13], Zhou Jiazuo [14], et al. Assuming that water is an incompressible liquid, when applying confining pressure in the triaxial test, as the pressure gradually increases, the volume of the liquid entering the pressure chamber also gradually increases. The volumetric strain calibration is carried out by replacing the specimen with a steel block (a cylinder with H100 mm and Φ50 mm) having exactly the same size and volume as the specimen. Figure 6 shows the curve of the volume of the liquid injected by the confining-pressure pump changing with the confining pressure.

As shown in Fig. 6, which is the curve of the volume of the liquid injected by the confining-pressure pump changing with the confining pressure, four volumetric strain calibrations were carried out in total, including two calibrations at normal temperature (25 °C), one calibration at a low temperature of 3 °C, and one calibration at a high temperature of 50 °C. At different temperatures, the relative change in the volume of the liquid injected by the confining-pressure pump when the pressure is increased from 3 MPa to 15 MPa is consistent. The change in the volume of the injected liquid is mainly caused by the deformation of the pressure chamber, system pipelines, rubber membrane, etc. Therefore, during the triaxial test, the volumetric strain of the specimen is equal to the change in the volume of the liquid injected by the confining-pressure pump at the same pressure minus the calibrated volume change of the steel block under the corresponding pressure state. Taking 3 MPa as

Fig. 6 Curve of injected liquid volume versus cell pressure variation

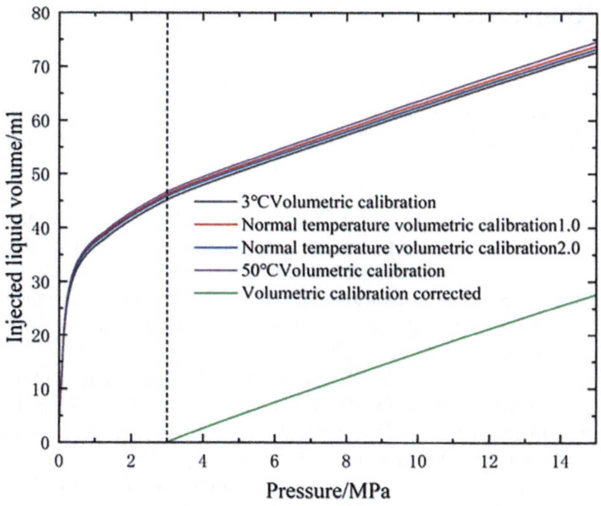

the zero point, the calibration curve is corrected. For the volumetric strain measurement of the specimen, only the relative change in the volume of the liquid injected by the confining-pressure pump is needed, and the corrected calibration curve is used to calculate the volumetric strain of the specimen during the test.

4 Conclusion

1. In this paper, a high-pressure temperature-controlled geotechnical triaxial test system was developed based on the conventional temperature-controlled triaxial apparatus. The system mainly consists of a high-pressure triaxial module, a water-bath circulation temperature-control module, a high-pressure loading module, etc. Through the design of inner and outer double water-bath circulation, long-term uniform and accurate temperature control of the specimen is achieved.
2. The high-pressure temperature-controlled geotechnical triaxial test system can conduct geotechnical triaxial tests under high-pressure and temperature-controlled environments. The temperature-control range is −10 to 80 °C with a temperature-control accuracy of ±0.2 °C. The maximum confining-pressure loading is 15 MPa with an accuracy of ±0.01 MPa. The system features stable performance, complete test functions, and simple operation.

Acknowledgments 1. Project (No.52127815) supported by the Special Fund for Basic Research on Scientific Instruments of the National Natural Science Foundation of China.

2. Project (No.52178372) supported by the National Natural Science Foundation of China.

References

1. Yu H, Qi CZ, Zhu SD et al (2022) Theoretical analysis of shear strength of deep-sea energy soil. J Changjiang River Sci Res Inst 39(5):119-124+131. https://doi.org/10.11988/ckyyb.20210037
2. Zhang JD, Liu XH, Bian PM et al (2022) Numerical simulation on flow behavior in hydrate-bearing sediments. J Cent South Univ (Sci Technol) 53(3):855–863. https://doi.org/10.11817/j.issn.1672-7207.2022.03.008
3. Liu WH, Yang Q, Kong GQ et al (2022) Mechanical properties and elastoplastic constitutive model of undisturbed marine sediment. Chin J Geotech Eng 44(10):1837–1845. https://doi.org/10.11779/CJGE202210009
4. Feng S, Chen P, Zhou JZ et al (2024) The effect of occurrence environments on the mechanical behavior of methane hydrate-bearing sediments. Rock Soil Mech 45(7):1987–1999. https://doi.org/10.16285/j.rsm.2023.1641
5. Demars KR, Charles RD (1982) Soil volume changes induced by temperature cycling. Can Geotech J 19(2):188–194. https://doi.org/10.1016/0376-6349(81)90015-8
6. Sultan N, Delage P, Cui YJ (2002) Temperature effects on the volume change behaviour of boom clay. Eng Geol 64(2):135–145. https://doi.org/10.1016/S0013-7952(01)00143-0
7. Yang J, Hassanpouryouzband A, Tohidi B et al (2019) Gas hydrates in permafrost: distinctive effect of gas hydrates and ice on the geomechanical properties of simulated hydrate-bearing permafrost sediments. J Geophys Res Solid Earth 124(3):2551–2563. https://doi.org/10.1029/2018JB016536
8. Chen ZH, Xie Y, Sun SG et al (2005) Temperature controlled triaxial apparatus for soils and its application. Chin J Geotech Eng 27(8):928–933. https://doi.org/10.3321/j.issn:1000-4548.2005.08.015
9. Cai GQ, Zhao CG, Bai B et al (2012) Development and application of a temperature-controlled triaxial test system for unsaturated soils. Chin J Geotech Eng 34(6):1013–1019. http://www.cgejournal.com/article/id/14600
10. Yang DH, Yan MQ, Lu D et al (2021) Introduction and application of hydro-mechanical united experiment apparatus for hydrate-bearing sediments. J Eng Geol 29(6):1722–1732. https://doi.org/10.13544/j.cnki.jeg.2021-0466
11. Wang LJ, Wang P, Zhu B et al (2024) Development and application of in-flight centrifuge apparatus for modelling hydrate exploitation. Chin J Geotech Eng 46(2):316–324. https://doi.org/10.11779/CJGE20221300
12. Wheeler SJ (1986) The stress-strain behaviour of soils containing gas bubbles. University of Oxford
13. Yan RT, Zhao XY, Yu MB et al (2018) Isotropic compression characteristics of clayey soil saturated by salty solution. Rock Soil Mech 39(1):129–138. https://doi.org/10.16285/j.rsm.2016.2035
14. Zhou JZ, Wei CF, Wei HZ et al (2020) Development and application of multi-functional triaxial test system for hydrate-bearing sediments. Rock Soil Mech 41(1):342–352. https://doi.org/10.16285/j.rsm.2018.1896

Open Access This chapter is licensed under the terms of the Creative Commons Attribution-NonCommercial-NoDerivatives 4.0 International License (http://creativecommons.org/licenses/by-nc-nd/4.0/), which permits any noncommercial use, sharing, distribution and reproduction in any medium or format, as long as you give appropriate credit to the original author(s) and the source, provide a link to the Creative Commons license and indicate if you modified the licensed material. You do not have permission under this license to share adapted material derived from this chapter or parts of it.

The images or other third party material in this chapter are included in the chapter's Creative Commons license, unless indicated otherwise in a credit line to the material. If material is not included in the chapter's Creative Commons license and your intended use is not permitted by statutory regulation or exceeds the permitted use, you will need to obtain permission directly from the copyright holder.

Research on Design Optimization of High-Speed Rail Business Seat from the Semiotic Perspective

Di Zhao, Jian Li, and Hongzhou Dai

Abstract In order to improve the travel experience of high-speed rail business seat design. Combining dynamic symbol analysis and semantic difference method, the symbolic system framework of high-speed rail business seat design is constructed. Through questionnaire survey and SPSS correlation analysis. The findings are as follows: The adjustment effect of dynamic symbols on the user's psychological experience is ignored. In this paper, "dynamic symbol hierarchical model" is proposed to provide theoretical and practical reference for the design of high-speed rail seats.

Keywords High-speed rail business seat · Semiotics · Dynamic symbol analysis · User experience · Semantic difference method

1 Introduction

With the rapid development of high-speed rail in the world, by 2023, the average annual growth rate of business seat users in China's high-speed railways is 15%, but user satisfaction is only 72% (China Railway Corporation, 2023).

This paper proposes to expand the application boundaries of semiotics in dynamic interaction scenarios through the theoretical level. This study introduces dynamic symbol analysis into high-speed rail seat design for the first time, which makes up the gap in the application of semiotics in dynamic interactive scenarios.

D. Zhao · J. Li (✉) · H. Dai
Dalian Jiaotong University, Dalian, Liaoning, China

© The Author(s) 2026

Z. Zhang, P. Mercorelli (eds.), *Mechanical Design and Simulation: Insights and Innovations*, Lecture Notes in Mechanical Engineering,
https://doi.org/10.1007/978-981-95-2405-1_126

2 Data Sources and Methods

2.1 *Semiotic Analysis of High-Speed Rail Business Seats*

Light Changes Because these functions can also achieve similar effects through non-intelligent means, so even if there is no intelligent function, passengers will not feel dissatisfied [1].

Adjustable Seat Layout By adjusting the seat, enrich the comfort of passengers. The adjustment ability of the seat can enhance the sense of surprise and freshness of the passenger mood, and realize a positive interaction between the design developer and the user [2].

Interactive Interface Response The layered interface coupled with detailed control can facilitate passengers to obtain first-hand information, facilitate planning in advance, effectively mitigate the small probability of ensuring delays and delayed information.

2.2 *Research Methods*

Overview of Semiotic Theory Peirce put forward the ternary relation of symbols. As the key point of the symbolic structure, he divided the symbolic structure into three parts: "object, representation and interpretation item", "interpretation item" is the symbolic meaning after the reference interpretation [3].

In this kind of stimulation, sensory stimulation is the primary one, through appealing to the senses [4]. These elements serve as a basis to provide users with convenient services, while the connection of the three parts in the pre-departure stage, the operation stage and the arrival stage requires a certain symbolic logic [5].

Target Audience Analysis Business passengers usually pay attention to the comfort of the ride experience, including the comfort of the seat, the sense of space and quality service facilities. On the other hand, for high-end leisure passengers, they pay more attention to relaxing in the carriage. Passenger behavior decision is produced under the dual promotion of environment and product [6].

Vocabulary Selection and Collection According to the TF-IDF algorithm, high-frequency emotional words were extracted from 100,000 user comments. Through the emotional evaluation of 16 emotional adjectives were collected according to screening, forming 8 pairs of positive and negative semantic adjectives, as shown in Table 1.

Table 1 Word pairs of perceptual adjectives (self-drawn by the author)

Adjective pairs (+)	Adjective pair (−)	Adjective pairs (+)	Adjective pair (−)
Safe	Dangerous	Bright	Dull
Technological	Outdated	Warm	Ice-cold
Laconic	Heavy	Flexible	Bulky
Comfortable	Uncomfortable	Intelligent	Traditional

Experimenter Selection This questionnaire adopts semantic difference method to evaluate the image of 7 samples. The seven-level Likert scale was used, and the scores from weak to strong were −3, −2, −1, 0, 1, 2 and 3 [7].

A total of 180 questionnaires were obtained through random distribution of questionnaires, of which 164 were valid, with an effective rate of more than 91%. The questionnaire quantified the respondents' tendency to word pairs, finally achieving statistics and analysis of the data.

This study adopted the semantic difference method, designed questionnaires and released statistical data, and used SPSS software to conduct correlation analysis on the data of consumers' perceptual evaluation results. The analysis was further based on semiotics theory and explored the necessity of optimization of high-speed rail business seat design from the perspective of dynamic symbols.

3 Correlation Analysis of Data Results

3.1 Data Analysis

The author collected and screened 164 questionnaire data and input them into SPSS software. The extraction is shown in Table 2.

The results of this questionnaire survey indicate that regarding safety. For example, the samples are collectively perceived as safe. All samples have mean scores ranging from 1.53 to 1.96, implying that respondents generally consider these samples to be safe with high confidence. Samples 1 and 7 were slightly cold. The average values are 1.86 and 1.85, respectively, and you can consider adding a warm element to the design.

The score of "sense of science and technology" is high (M = 2.70), but the score of "intelligence" is low (M = 1.77). Combined with semiotic explanation, users equate "sense of science and technology" with visual symbols (such as LED interface), while "intelligence" needs to be reflected by interactive symbols (such as voice control), and there is a symbol fault in the current design.

The seat could automatically adjust its posture based on passenger preferences or enable control of seat functionalities through voice commands. This approach allows passengers not only to visually perceive the presence of advanced technology but also to tangibly experience the convenience brought by intelligent systems.

Table 2 Statistical analysis of mean value, standard deviation and variance obtained from valid questionnaires

	N	Average		Standard deviation	Variance
	Statistics	Statistics	Standard error	Statistics	Statistics
Sample 1—Safe-Dangerous	164	1.59	.084	1.073	1.152
Technological-Outdated	164	2.07	.115	1.478	2.186
Concise-Complex	164	1.95	.109	1.402	1.966
Comfortable-Uncomfortable	164	1.93	.104	1.327	1.761
Bright-Dull	164	2.07	.118	1.515	2.296
Warm-Cold	164	1.86	.095	1.218	1.483
Flexible-Bulky	164	2.36	.146	1.866	3.483
Smart-Traditional	164	2.32	.137	1.758	3.091
Sample 2—Safe-Dangerous	164	1.71	.096	1.227	1.506
Technological-Outdated	164	1.85	.095	1.211	1.467
Concise-Complex	164	1.74	.095	1.212	1.468
Comfortable-Uncomfortable	164	1.85	.094	1.210	1.463
Bright-Dull	164	2.34	.137	1.756	3.085
Warm-Cold	164	2.18	.127	1.633	2.666
Flexible-Bulky	164	2.15	.124	1.584	2.510
Smart-Traditional	164	2.14	.124	1.593	2.538
Sample 3—Safe-Dangerous	164	1.95	.114	1.460	2.133
Technological-Outdated	164	2.58	.145	1.857	3.448
Concise-Complex	164	2.61	.146	1.875	3.515
Comfortable-Uncomfortable	164	2.35	.127	1.630	2.658
Bright-Dull	164	2.61	.141	1.808	3.270
Warm-Cold	164	2.33	.122	1.559	2.431
Flexible-Bulky	164	2.66	.149	1.904	3.625
Smart-Traditional	164	2.70	.158	2.019	4.076
Sample 4—Safe-Dangerous	164	1.53	.078	.993	.987
Technological-Outdated	164	1.71	.098	1.257	1.580
Concise-Complex	164	2.07	.127	1.623	2.633
Comfortable-Uncomfortable	164	1.82	.110	1.405	1.975
Bright-Dull	164	1.96	.113	1.448	2.097
Warm-Cold	164	2.04	.115	1.469	2.158
Flexible-Bulky	164	2.24	.130	1.666	2.775
Smart-Traditional	164	1.77	.098	1.255	1.575
Sample 5—Safe-Dangerous	164	1.60	.088	1.133	1.284
Technological-Outdated	164	1.70	.093	1.195	1.428
Concise-Complex	164	1.93	.112	1.434	2.056
Comfortable-Uncomfortable	164	1.71	.098	1.252	1.568
Bright-Dull	164	1.80	.109	1.393	1.941
Warm-Cold	164	1.89	.106	1.357	1.841

Table 2 (continued)

	N	Average		Standard deviation	Variance
	Statistics	Statistics	Standard error	Statistics	Statistics
Flexible-Bulky	164	2.01	.118	1.517	2.301
Smart-Traditional	164	1.74	.100	1.286	1.655
Sample 6—Safe-Dangerous	164	1.96	.121	1.550	2.404
Technological-Outdated	164	1.64	.090	1.151	1.324
Concise-Complex	164	2.09	.124	1.593	2.538
Comfortable-Uncomfortable	164	1.95	.112	1.439	2.071
Bright-Dull	164	1.85	.105	1.340	1.794
Warm-Cold	164	2.21	.125	1.604	2.574
Flexible-Bulky	164	1.98	.116	1.488	2.214
Smart-Traditional	164	1.73	.099	1.274	1.624
Sample 7—Safe-Dangerous	164	1.69	.097	1.236	1.528
Technological-Outdated	164	1.57	.081	1.034	1.068
Concise-Complex	164	1.83	.103	1.313	1.725
Comfortable-Uncomfortable	164	1.78	.099	1.273	1.620
Bright-Dull	164	1.70	.090	1.152	1.327
Warm-Cold	164	1.85	.107	1.371	1.880
Flexible-Bulky	164	1.96	.117	1.500	2.250
Smart-Traditional	164	1.76	.098	1.254	1.572
Number of valid cases (in columns)	164				

Employing the semantic differential method could further enhance this design strategy. This not only optimizes the user experience but also enriches the symbolic meaning conveyed through the design.

Similarly, the design of facility products in high-speed rail should be aesthetically unified with other designs as far as possible, and the phenomenon of uniformity in design should be avoided as much as possible [8].

3.2 *Dynamic Symbol Analysis*

Regarding the temporal and spatial dimensions of dynamic symbols, the design of lighting in night mode is a direction worth exploring. Warm-colored lighting can create a comfortable and relaxing atmosphere, while cool-colored lighting can enhance concentration. If the lighting color temperature could automatically adjust according to different time periods or passenger needs, it would greatly improve the passenger's travel experience. It might also be possible to introduce a dynamic environment sensing system that adjusts the interior ambiance based on changes in the external environment, providing passengers with a more immersive experience.

For a product, maintaining the unity and continuation of its symbolic meaning and style characteristics can form a continuous information influence on consumers, so that consumers have a sense of affinity and familiarity with the product. Through the discovery of high-speed train design elements, summed up suitable for the local residents of the new travel experience [9].

4 Conclusion

In this study, we construct a hierarchical model of dynamic symbols to reveal the synergistic mechanism of dynamic symbols in the design of high-speed rail business seats. The empirical results show that users' expectations for intelligent and cross-cultural symbols are not fully met, and the experience needs to be optimized through interaction design and cultural integration.

In the future research process, emergency plan should also occupy a part of the proportion of research, it should cover all possible emergency situations, and clarify the responsibilities and action steps of each link [10]. Team members need to receive relevant training, master the disposal skills and coping strategies in emergency situations, and be able to deal with various emergencies quickly and calmly. It is expected that more research will focus on this field in the future to jointly promote the innovation and development of high-speed rail services.

References

1. Ye Z, Jiayue S, Xin Z (2023) Research on the integration of intelligent Technology in Passenger Facilities of high-speed train carriages. Packag Eng 44(12):49–59. https://doi.org/10.19554/j.cnki.1001-3563.2023.12.005
2. Knain E, Fredlund T, Furberg A (2021) Exploring student reasoning and representation construction in school science through the lenses of social semiotics and interaction analysis. Res Sci Educ 51(1):93–111
3. Jian Z (2022) Saussure or Peirce: the choice of product rhetorical symbolic structure. Decoration 10:98–102. https://doi.org/10.16272/j.cnki.cn11-1392/j.2022.10.025
4. Shukla M, Misra R, Singh D (2023) Exploring relationship among semiotic product packaging, brand experience dimensions, brand trust and purchase intentions in an Asian emerging market. Asia Pac J Mark Logist 35(2):249–265
5. Tianyi Z (2024) Research on consistency design strategy of intelligent riding experience in high-speed train business class under scene integration. Design 37(08):106–109. https://doi.org/10.20055/j.cnki.1003-0069.001721
6. Kim H, Kwon S, Wu SK et al (2014) Why do passengers choose a specific car of a metro train during the morning peak hours? Transp Res A Policy Pract 61:249–258
7. Jingyi L, Luyi C, Ye D (2023) Research on perception and image of asparagus Fern pattern based on semantic differential method. J Appl Technol 23(01):90–94
8. Qiang L, Longxiao L (2021) Research on innovative Design of Public Facilities in China's high-speed train stations—taking Beijing Chaoyang Station as an example. Art Panorama 03:132–133

9. Chen Wanpeng X, Jianghua. (2022) Research on safety design of economy class seats for civil aircraft. Design 35(13):150–153
10. Yuan Z, Boyang L, Jiezhen Z et al (2023) Research on rail transit design and improvement based on service quality enhancement. Eng Constr Des 20:51–53. https://doi.org/10.13616/j.cnki.gcjsysj.2023.10.216

Open Access This chapter is licensed under the terms of the Creative Commons Attribution-NonCommercial-NoDerivatives 4.0 International License (http://creativecommons.org/licenses/by-nc-nd/4.0/), which permits any noncommercial use, sharing, distribution and reproduction in any medium or format, as long as you give appropriate credit to the original author(s) and the source, provide a link to the Creative Commons license and indicate if you modified the licensed material. You do not have permission under this license to share adapted material derived from this chapter or parts of it.

The images or other third party material in this chapter are included in the chapter's Creative Commons license, unless indicated otherwise in a credit line to the material. If material is not included in the chapter's Creative Commons license and your intended use is not permitted by statutory regulation or exceeds the permitted use, you will need to obtain permission directly from the copyright holder.

Design of Leek Cutting Machine and Finite Element Analysis of its Chassis

Chen Fang, Changxu Meng, and Jian Wang

Abstract In this article, we designed and conceptualized many parts of a leek-cutting machine. We deconstructed various existing mechanisms, selected the most suitable movement mode for the product, and incorporated a wireless function module for remote control. A cutting mechanism was added for cutting the leeks, and a pulley system was included to transmit the cut leeks. Additionally, we designed a power module and a vision module to support the movement of the mechanism. The motion form of the leek-cutting machine was carefully selected. The related parts were modeled. Furthermore, the chassis was analyzed based on existing conditions.

Keywords Leek cutting machine · Walking mechanism · Vision module · Motion simulation · Three-dimensional modeling

1 Introduction

With the rapid development of modern agriculture, people's living standards are constantly improving. However, the development of China's agricultural technology has not kept pace with these rising standards. This is particularly evident in the case of leeks, which hold a very important position in agriculture. Leeks are leafy vegetables. In today's era of agricultural globalization, they are distinctive Chinese vegetables with strong competitiveness in the global market.

C. Fang · C. Meng
School of Mechanical Engineering, Dalian University of Science and Technology, Dalian, Liaoning, China

J. Wang (✉)
School of Mechanical Engineering, Dalian University of Science and Technology, Dalian, Liaoning, China

Key Laboratory of Digital Design and Intelligent Manufacturing of Lightweight Castings for New Energy Vehicles of Liaoning Province, Dalian University of Science and Technology, Dalian, Liaoning, China

© The Author(s) 2026

Z. Zhang, P. Mercorelli (eds.), *Mechanical Design and Simulation: Insights and Innovations*, Lecture Notes in Mechanical Engineering,
https://doi.org/10.1007/978-981-95-2405-1_127

Currently, leek production and harvesting largely rely on hand-pushed harvesting methods or manual labor, which are labor-intensive and inefficient. Therefore, a leek cutting machine must be designed to address these challenges.

2 Overall Structure and Guiding Transmission Mechanism Design [1]

2.1 Whole Machine Design Method

In terms of the overall structure, we have chosen a modular design. The leek-cutting machine is divided into a cutting mechanism, a transmission mechanism, a control mechanism, a wireless audio and video mechanism, and a guiding transmission mechanism. In future updates, it will only be necessary to update each module to meet different requirements [2].

2.2 Design Method of Guide Rail Transmission Mechanism

The guided transmission mechanism designed in this project refers to various existing transmission mechanisms on the market. We have chosen a synchronous pulley as the main transmission body. We need to consider how to guide the leek into the predetermined position so that the cutting mechanism can complete its work. Additionally, we need to consider how the transmission works and the versatility of the device. With these considerations in mind, we plan to use guides to keep the required crops within range.

Simultaneously, we will add baffles and guide holes on the cutting mechanism to prevent soil splashing. The pilot hole will guide the crop into the range of the cutting mechanism so that it can work in conjunction with the cutting blade to complete the cutting task. After the crop is harvested, it is transferred to the transmission platform via a pulley. If bags are pre-installed on the delivery platform, the crop can be collected for subsequent processing.

A motor drive will be used to stabilize the overall structure's power. The main transmission shaft and passive transmission shaft will be set to ensure the integrity of the transmission mechanism. In the transmission direction, an idler is added between the two gears to make them rotate in the same direction, which can drive the belt drive, as shown in Fig. 1.

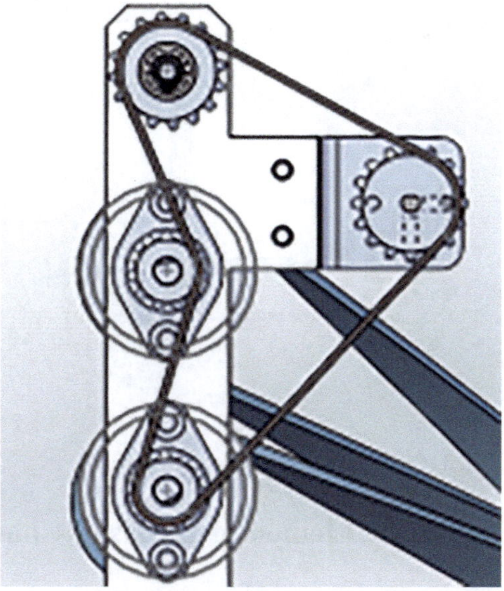

Fig. 1 Transmission mechanism

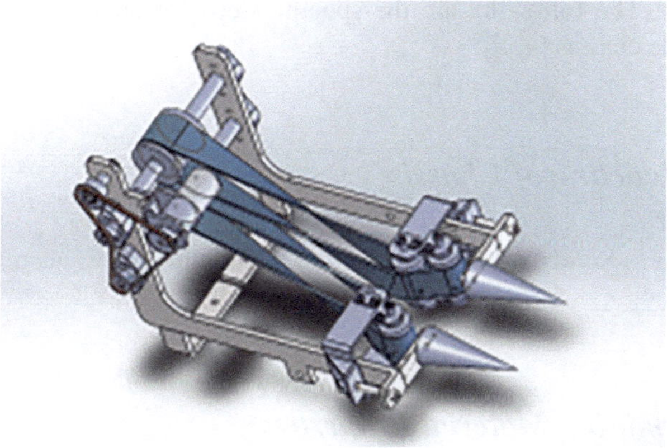

Fig. 2 Transmission mechanism

2.3 *Final Design Results and Discussion*

The final design structure is shown in Figs. 2 and 3. The designed guiding transmission mechanism can gather the leeks to meet the predetermined goal. The structural design of the whole machine reflects the various parts in the movement process, making the working principle clearer. Next, the force on the chassis will be analyzed to verify whether it meets the design requirements. The pulley will be designed. This will make the overall structure design more complete [3].

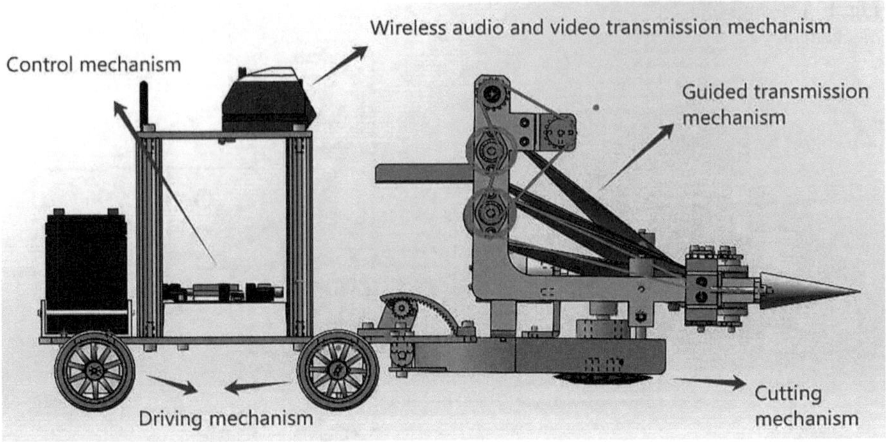

Fig. 3 Overall structure design drawing

3 Finite Element Analysis of Chassis

Finite element analysis of the chassis is an important step in ensuring system stability and safety. The following are the specific steps for conducting finite element analysis on the chassis [4, 5]:

3.1 3D Modeling of Chassis

Software such as SolidWorks can be used to model the chassis accurately in 3D, as shown in Fig. 4.

3.2 Determining Material Properties

No. 45 steel was chosen for its advantages of good heat treatment performance, impact resistance, corrosion resistance, wear resistance, and overall good mechanical properties. These qualities make it suitable for meeting the daily operational needs of the design.

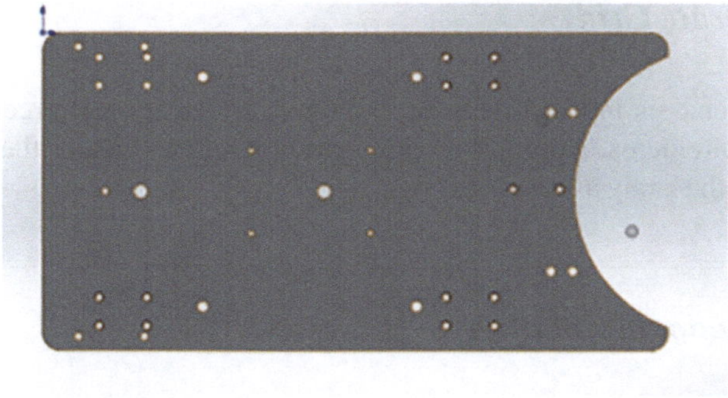

Fig. 4 Chassis modeling

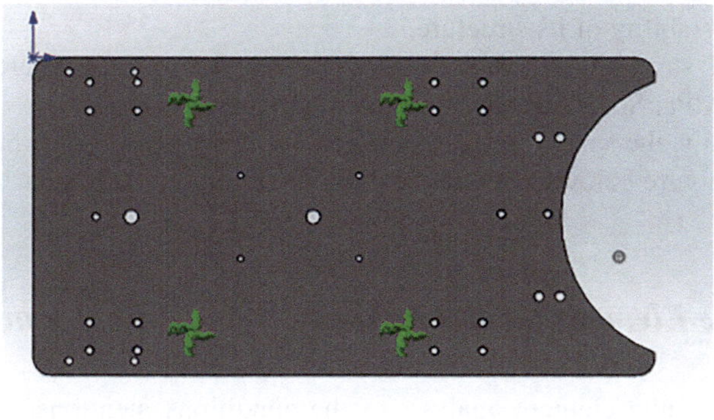

Fig. 5 Adding constraints

3.3 Adding Constraints

In the simulation fixture option, the fixed and free ends of the chassis are determined to accurately simulate the actual fixing and support conditions. We choose to take the contact hole between the chassis and the wheel as the fixed end, restrict the freedom of the chassis to replicate the real force situation, set it as a fixed boundary condition, and ensure the stability of its working state. See Fig. 5.

3.4 Application of Load

We apply appropriate loads to the chassis model according to its design requirements and operating conditions. In daily operation, the chassis is primarily subjected to forces exerted by leeks. The force type is pressure, so we apply pressure loads to the model and conduct static stress analysis. The ground load is set to 50 kg (approximately 500 N), followed by finite element analysis (FEA).

3.5 Generate Grid

We grid the chassis model to ensure that the grid density and precision meet the analysis requirements. Appropriate grid partitioning can ensure the accuracy and reliability of the analysis results.

3.6 Operational Solution

After setting the grid parameters, the finite element analysis solution process is initiated. The software calculates stress, strain, displacement, and other parameters of the chassis based on the applied load and constraint conditions to evaluate the strength and stability of its structure.

The yield strength is 5.3×10^8 MPa. And the maximum pressure is only 2.356×10^7 MPa. So, it fully meets the design requirements.

Additionally, the safety factor is checked, typically set at 1.5 or higher. As illustrated in the figure below, After calculation, the minimum safety factor is 22.

3.7 Finite Element Structure Analysis and Discussion

Based on the finite element analysis of the conditions mentioned above, we can conclude that the chassis design material is No. 45 steel. This material is mainly subjected to pressure during daily operation, so the finite element analysis we performed focused on static stress analysis. Additionally, we determined that its safety factor is well above 1.5, indicating that this design is feasible. Therefore, it can be concluded that the chassis can effectively meet the work requirements in daily operations.

4 Transmission Mechanism Design

Belts come in various types, typically categorized as friction type and mesh type. Friction belt transmission includes flat belt transmission, V-belt transmission, and multi-wedge belt transmission.

After careful consideration, we opt for a flat belt drive. For the motor, we select the ZGA25RP motor, boasting a rated power of 200 W and a rated speed of 4800 r/min.

Firstly, we will proceed with tape selection.

$$P = PX\eta \tag{1}$$

According to Table 1, it can be seen that $\eta = 0.97$. So,

$$P = 0.97X0.97X200 = 0.19KW \tag{2}$$

And Ka choice can be seen in Table 1.

From Table 1, it can be seen that Ka is 1.1. Then we obtained:

$$Pd = KaP = 1.1X0.19 = 0.21KW \tag{3}$$

Then, the belt speed is calculated:

$$V = \frac{\pi d1n1}{60X1000} \tag{4}$$

We set the large belt pulley as the driving pulley, wherein π takes 3.14. d1 is the diameter of the large pulley, and n1 is the speed of the large pulley.

$$d1 = \left(1100 \sim 1350\right) \times 3\sqrt{\frac{P}{n1}} \tag{5}$$

From the above, n1 = 4800 r/min and d1 $\approx$ 54 mm can be obtained. We chose the large pulley diameter of 56 mm.

Then, the belt speed is calculated according to:

$$v = \frac{\pi d1n1}{60x1000} \tag{6}$$

In the equation, π takes 3.14, d1 = 56 mm, and n1 = 4800 r/min.

$$v = \frac{\pi d1n1}{60x1000} = 14.1m/s \tag{7}$$

Because it is a flat belt drive, the large pulley is the driving shaft $i \leq 5$ to meet the design requirements.

So, we designed the gear ratio to be 1:2.

Next, we calculate the diameter of the small pulley:

Table 1 Working condition coefficients

		KA					
		Start with no air and a light load			Heavy load start		
		Number of hours worked per day/h					
Operating conditions		<10	10 ~ 16	>16	<10	10 ~ 16	>16
Load variety Minimal movement	Liquid mixers, ventilators and blowers (≤7.5 kW), centrifugal pumps and compressors, light load Load conveyor	1.0	1.1	1.2	1.1	1.2	1.3

Table 2 Diameter of flat belt pulleys

Basic sized	Limit deviation	Basic sized	Limit deviation	Basic sized	Limit deviation
20	±0.4	125	±1.6	560	±5.0
25		140		630	
32	±0.5	160		710	
40		180	±2.0	800	±6.3

$$d2 = id1(1-\varepsilon) \tag{8}$$

The value ε is 0.01, so the diameter of the small pulley is 27.72 mm. We select the standard value of 32 mm according to Table 2.

And because:

$$d2 = \frac{n1}{n2} d1(1-\varepsilon) \tag{9}$$

So, the speed of the small pulley is 8316 r/min. Next, for the belt length design, we choose the open drive according to:

$$L = 2a + \frac{\pi}{2}(d1 + d2) + \frac{(d1-d2)^2}{4a} \tag{10}$$

where a is the shaft spacing basis:

$$a = (1.5 \sim 2)\cdot(d1 + d2) \tag{11}$$

It can be concluded that a = 116 mm.

Therefore, the belt length is L = 232 + 138.16 + 1.24 = 371 mm. Next, the band thickness calculation can be carried out, which can be based on: $\delta \le \left(\frac{1}{40} \sim \frac{1}{30}\right) d2$.

We can calculate the belt thickness as 1.1 mm and choose 3.6 mm.

To sum up, we designed the diameter of the large pulley to be 56 mm, the diameter of the small pulley to be 32 mm, the belt speed to be 14.1 m/s, the belt length to be 371 mm, the shaft spacing to be 116 mm, and the belt thickness is 3.6 mm.

5 Conclusions

In the first chapter, we introduced the necessity and key aspects of designing a leek-cutting machine. In the second chapter, we designed the overall structure and key components to clarify the working principle. In the third chapter, we conducted a finite element analysis on the chassis, which experiences significant force during operation, to determine its stability. In the fourth chapter, we designed the belt wheel to enhance the transmission mechanism, ensuring the rationality of the overall structural design.

Through the design and analysis presented in these chapters, we have developed a leek-cutting machine capable of operating on varied terrains. We believe that with the advancement of agricultural technology in China, agricultural robots will also make significant progress.

References

1. Lee S (2023) Development and performance test of a simulation-based tractor-attachable wind-blast-type onion stem cutting machine. Machines 11(5):508
2. Feng Y, Yin X, Jin H et al (2023) Design and experiment of a Chinese chive harvester. Int J Agric Biol Eng 16(2):125–131
3. Uğurluay S, Özcan MT, Güzel E et al (2009) Design and manufacture of prototype leek harvesting machine. Tarım Makinaları Bilimi Dergisi 5(1):109–114
4. Rajurkar A, Dangra K, Deshpande A et al (2023) Comparative finite element analysis of a novel robot chassis using structural steel and aluminium alloy materials. Eng Innov 7:61–73
5. Walunj A, Nalawade SM, Kadam PB et al (2023) Design of chassis for self-propelled onion harvester. J Agric Res Technol 48(2):302

Open Access This chapter is licensed under the terms of the Creative Commons Attribution-NonCommercial-NoDerivatives 4.0 International License (http://creativecommons.org/licenses/by-nc-nd/4.0/), which permits any noncommercial use, sharing, distribution and reproduction in any medium or format, as long as you give appropriate credit to the original author(s) and the source, provide a link to the Creative Commons license and indicate if you modified the licensed material. You do not have permission under this license to share adapted material derived from this chapter or parts of it.

The images or other third party material in this chapter are included in the chapter's Creative Commons license, unless indicated otherwise in a credit line to the material. If material is not included in the chapter's Creative Commons license and your intended use is not permitted by statutory regulation or exceeds the permitted use, you will need to obtain permission directly from the copyright holder.

GPSR Compliance

The European Union's (EU) General Product Safety Regulation (GPSR) is a set of rules that requires consumer products to be safe and our obligations to ensure this.

If you have any concerns about our products, you can contact us on ProductSafety@springernature.com

In case Publisher is established outside the EU, the EU authorized representative is:

Springer Nature Customer Service Center GmbH
Europaplatz 3
69115 Heidelberg, Germany

Batch number: 10739616

Printed by Printforce, the Netherlands